Mariano Giaquinta
Giuseppe Modica
Jiří Souček

Cartesian Currents in the Calculus of Variations I

Cartesian Currents

Springer

Mariano Giaquinta
Dipartimento di Matematica
Università di Pisa
Via F. Buonarroti, 2
I-56127 Pisa
Italy

Giuseppe Modica
Dipartimento di Matematica Applicata
Università di Firenze
Via S. Marta, 3
I-50139 Firenze
Italy

Jiří Souček
Faculty of Mathematics and Physics
Charles University
Sokolovská, 83
18600 Praha 8
Czech Republic

Library of Congress Cataloging-in-Publication Data

Giaquinta, Mariano, 1947-
Cartesian currents in the calculus of variations / Mariano Giaquinta, Giuseppe Modica, Jiří Souček.
p. cm. -- (Ergebnisse der Mathematik und ihrer Grenzgebiete ; 3. Folge, v. 37-38)
Includes bibliographical references and index.
ISBN 3-540-64009-6 (v. 1 : hardcover : alk. paper). -- ISBN 3-540-64010-X (v. 2 : hardcover : alk. paper)
1. Calculus of variations. I. Modica, Giuseppe. II. Souček, Jiří. III. Title. IV. Series: Ergebnisse der Mathematik und ihrer Grenzgebiete ; 3. Folge, Bd. 37-38.
QA316.G53 1998
515'.64--dc21 98-18195
CIP

Mathematics Subject Classification (1991): 49Q15, 49Q20, 49Q25, 26B30, 58E20, 73C50, 76A15

ISSN 0071-1136
ISBN 3-540-64009-6 Springer-Verlag Berlin Heidelberg New York

Printed in Germany

Typesetting: Data conversion by Springer-Verlag
SPIN 10639102 44/3143 - 5 4 3 2 1 0 – Printed on acid-free paper

To
Cecilia and Laura,
Giulia, Francesca and Sandra,
Eva and Sonia.

Preface

Non–scalar variational problems appear in different fields. In geometry, for instance, we encounter the basic problems of harmonic maps between Riemannian manifolds and of minimal immersions; related questions appear in physics, for example in the classical theory of σ–models. Non linear elasticity is another example in continuum mechanics, while Oseen–Frank theory of liquid crystals and Ginzburg–Landau theory of superconductivity require to treat variational problems in order to model quite complicated phenomena.

Typically one is interested in finding energy minimizing representatives in homology or homotopy classes of maps, minimizers with prescribed topological singularities, topological charges, stable deformations i.e. minimizers in classes of diffeomorphisms or extremal fields. In the last two or three decades there has been growing interest, knowledge, and understanding of the general theory for this kind of problems, often referred to as *geometric variational problems*.

Due to the lack of a regularity theory in the non scalar case, in contrast to the scalar one – or in other words to the occurrence of singularities in vector valued minimizers, often related with concentration phenomena for the energy density – and because of the particular relevance of those singularities for the problem being considered the question of singling out a *weak formulation*, or completely understanding the significance of various *weak formulations* becames non trivial. Keeping in mind the direct methods of Calculus of Variations, this amounts roughly to the question of identifying *weak maps* or *fields*, and *weak limits* of sequences of weak maps with equibounded energies. As we shall see, the choice of the notions of *weak maps* and *weak convergence* is very relevant, and different choices often lead to different answers concerning equilibrium points.

The aim of this monograph is twofold: discussing a homological theory of weak maps, and in this context treating several typical and relevant variational problems.

The basic idea in defining a weak notion of vector valued maps is to think of them not componentlywise but globally, i.e. as *graphs*. In other words we define weak maps between two oriented and boundaryless Riemannian manifolds $\mathcal{X}$ and $\mathcal{Y}$ similarly to distributions or Sobolev functions, using a standard duality approach, but testing with functions which live in the product space $\mathcal{X} \times \mathcal{Y}$. Thus one is naturally forced to move from the context of Sobolev maps, for instance, to that of *currents* and to allow *vertical parts*, and one is naturally led to the basic notion in this monograph of *Cartesian currents.* One should think of Cartesian currents as weak limits in the sense of currents of graphs of smooth maps, although this is not always true. In particular Cartesian currents satisfy the homological condition of having zero boundary in the cylinder $\mathcal{X} \times \mathcal{Y}$, and they induce a homology map which is continuous for the weak convergence of currents. As the natural context for those notions is *geometric measure theory*, we first develop an elementary introduction to the theory of currents of Federer and Fleming to provide all needed information. In particular we prove the deformation and closure theorems of Federer and Fleming, that, as we shall see, play a relevant role not only to study parametric but also non parametric integrals.

In the first part of our monograph, after a preliminary chapter about measure theory and the phenomenology of weak convergence, we discuss integer rectifiable and normal currents, differentiability properties of the graphs of maps, continuity of Jacobian minors, and how those notions are related in terms of approximate tangent planes, area or mass, and the homological notion of boundary. We also discuss related topics as, for instance, higher integrability of determinants, functions of bounded variations and degree theory, and of course closure, compactness and structure properties of special classes of Cartesian currents. Finally in Vol. I Ch. 5 we deal with the homology theory for currents. There we present classical topics as for instance Hodge theory, Poincaré–Lefschetz and de Rham dualities and intersection numbers, and we conclude by discussing the homology map associated to a Cartesian current in terms of periods and cycles.

In doing this we have tried to keep our treatment elementary, illustrating with simple examples the results, their meaning and their typical use, and we always give detailed proofs. Also, at the cost of some repetition we have tried to make each chapter, and sometimes even sections, readable as far as possible indipendently of the general context, so that parts of this monograph can be easily used separately for example for graduate courses. This we hope justifies the size of our monograph. Open questions are often mentioned and in the final section of each chapter we discuss references to the literature and sometimes supplementary results.

In the second part of our monograph we deal with variational problems. In Vol. II Ch. 1 we discuss general variational problems, their parametric and non parametric formulations, and in connection with them, different notions of ellipticity (parametric ellipticity, polyconvexity, and quasiconvexity). The rest of the second part of this monograph is then dedicated to specific variational problems in the setting of Cartesian currents: in Vol. II Ch. 2 we deal with weak diffeomorphisms and non linear elasticity, in Vol. II Ch. 3, Vol. II Ch. 4, and Vol. II Ch. 5 we discuss some issues of the harmonic mapping problem and of related questions; and, finally, we shortly deal in Vol. II Ch. 6 with the non parametric area problem.

For further information about the content of this monograph we refer the reader to the introductions to each chapter, to the detailed table of contents, and to the index.

In preparing this monograph we have taken advantage from discussions with many friends and colleagues. Among them it is a pleasure for us to thank G. Anzellotti, J. Ball, F. Bethuel, H. Brezis, F. Hélein, S. Hildebrandt, J. Jost, H. Kuwert, F. Lazzeri, D. Mucci, J. Nečás, K. Steffen, M. Struwe, V. Šverák and B. White.

We thank also Mirta Stampella for her invaluable help in the typing and retyping of our manuscript.

We would like to aknowledge supports during the last years from the *Ministero dell'Universitá e della Ricerca Scientifica*, *Consiglio Nazionale delle Ricerche*, EC *European Research Project* GADGET I, II, III, and *Čzech Academy of Sciences.* Parts of this work have been written while the authors were visiting different Universities. We want to extend our thanks to *Alexander von Humboldt Foundation*, *Sonderforschungsbereich 256* of Bonn University, to the *Departments of Mathematics* of the Universities of Bonn, Cachan, Keio, Paris VI, Shizuoka, and to the *Forschungsinstitut ETH*, Zürich, to the *Center for Mathematics and its Applications*, Canberra and to the *Academia Sinica*, Taiwan.

Firenze, October 1997

Mariano Giaquinta
Giuseppe Modica
Jiří Souček

Contents
Cartesian Currents in the Calculus of Variations I and II

Volume I. Cartesian Currents

Volume II. Variational Integrals

Contents
Volume I. Cartesian Currents

Contents
Volume II. Variational Integrals

1. General Measure Theory

This chapter deals with *general measure theory.* In Sec. 1.1 we collect definitions and results from general measure theory that will be freely used later on; in Sec. 1.2, due to its relevance for the sequel, we shall discuss in more details weak convergence of functions and of measures and we shall illustrate some of its main features.

In Ch. 2 we shall then develop some of the basic theory of *n–rectifiable sets* and *integer multiplicity rectifiable currents.*

Of course we do not aim to completeness, for instance Sec. 1.1 of this chapter contains no proof; and in principle, we supply proofs essentially when claims or their proofs are especially relevant for the sequel; sometimes, proofs are postponed to later chapters.

Our goal in these first two chapters is to state precisely results and notations (though we shall usually adopt standard notations) and to illustrate them mainly by examples. In some sense, the first two chapters may be regarded as a simple, and in some regard, rough introduction to the elementary part of *geometric measure theory,* the right context in which the content of the following chapters lives.

At first lecture the reader can start from Ch. 3 and use Ch. 1 and Ch. 2 as reference chapters.

1 General Measure Theory

In this section we collect some basic definitions and results from general measure theory

1.1 Measures and Integrals

Let X be a set and let 2^X denote the collection of all subsets of X.

Definition 1. *A collection $\mathcal{F}$ of subsets of X, $\mathcal{F} \subset 2^X$, is said to be a σ–*algebra *in X if $\mathcal{F}$ has the following three properties*

(i) $X \in \mathcal{F}$,
(ii) *If E, $F \in \mathcal{F}$, then $E \setminus F \in \mathcal{F}$,*

(*iii*) *If* $\{E_k\} \subset \mathcal{F}$, $k = 1, 2, \ldots$, *then* $\cup_{k=1}^{\infty} E_k \in \mathcal{F}$.

Definition 2. *Let* $\mathcal{F}$ *be a* σ*–algebra in* X *and let* μ *be a function defined on* $\mathcal{F}$*, whose range is* $[0, +\infty]$,

$$\mu : \mathcal{F} \to [0, +\infty].$$

We say that μ*, or better* $(\mu, \mathcal{F})$*, is a* σ–measure *on* X *if* μ *is* countably additive *on* $\mathcal{F}$*, that is, if* $\{E_k\}$ *is a disjoint countable collection of members of* $\mathcal{F}$*, then*

$$\mu\left(\bigcup_{k=1}^{\infty} E_k\right) = \sum_{k=1}^{\infty} \mu(E_k) \tag{1}$$

If $(\mu, \mathcal{F})$ is a σ–measure on X, we evidently have

(i) $\mu(\emptyset) = 0$,
(ii) μ is in particular *finitely additive*, i.e., if $E_1, \ldots, E_n$ is a finite collection of disjoint sets in $\mathcal{F}$, then $\mu(E_1 \cup \ldots \cup E_n) = \mu(E_1) + \ldots + \mu(E_n)$,
(iii) if $E_1, E_2 \in \mathcal{F}$, $E_1 \supset E_2$, then $\mu(E_1 \setminus E_2) = \mu(E_1) - \mu(E_2) \geq 0$, hence μ is *monotone* $\mu(E_1) \supset \mu(E_2)$.

Given a σ–measure μ on X, we may define the measure of any subset of X by trying to measure it in the best possible way by means of the elements of $\mathcal{F}$, i.e., by setting

$$\mu^*(E) := \inf\{\sum_{k=1}^{\infty} \mu(F_k) \mid F_k \in \mathcal{F}, \bigcup_{k=1}^{\infty} F_k \supset E\} \tag{2}$$

and, if there is no countable collection $\{F_k\}$ in $\mathcal{F}$ covering E, by setting $\mu^*(E) = +\infty$. This way we define a new set function

$$\mu^* : 2^X \to [0, +\infty]$$

which satisfies, as it is not difficult to see, the following properties

(i) $\mu^*(\emptyset) = 0$,
(ii) μ^* is *monotone*, i.e., if $A \subset B$ then $\mu^*(A) \leq \mu^*(B)$,
(iii) μ^* is *countably sub-additive*, i.e., if $\{E_k\}$ is a countable collection of subsets of X, then

$$\mu\left(\bigcup_{k=1}^{\infty} E_k\right) \leq \sum_{k=1}^{\infty} \mu(E_k) , \tag{3}$$

(iv) μ^* is an extension of μ, i.e., $\mu^*(E) = \mu(E)$ whenever $E \in \mathcal{F}$,
(v) μ^* can also be computed as

$$\mu^*(E) = \inf\{\mu(F) \mid F \in \mathcal{F}, F \supset E\} . \tag{4}$$

Notice that in general μ^* is not countably additive.

Definition 3. *A set function* $\lambda : 2^X \to [0, +\infty]$ *is said to be an* outer measure, *or simply a* measure *on* X *if* $\lambda(\emptyset) = 0$ *and* λ *is monotone and countably subadditive.*

Therefore we can say that every σ–measure $(\mu, \mathcal{F})$ can be extended by (2), or by (4), to a measure μ^* on X. In a sense to be made precise, also the converse is true. To see this, we introduce, following Carathéodory, the important notion of *measurable sets.* Very roughly they are those sets which divide well every subset of X, more precisely

Definition 4 (Carathéodory). *Let* λ *be a measure on* X. *A subset* E *of* X *is called* λ–measurable *if for every subset* F *of* X *we have*

$$\lambda(F) = \lambda(F - E) + \lambda(F \cap E) \tag{5}$$

The collection of all λ*–measurable subsets of* X *is denoted by* $\mathcal{M}_\lambda$,

$$\mathcal{M}_\lambda := \{E \subset X \mid E \text{ is } \lambda\text{-measurable}\}$$

Observe that, for any set E and any set F, we have

$$\lambda(F) \leq \lambda(F - E) + \lambda(E \cap F);$$

also X and $\emptyset$ are measurable and every λ–*null set* E, i.e., every set of zero measure, $\lambda(E) = 0$, is measurable. Actually we have

Theorem 1. *Let* λ *be a measure on* X. *Then* $\mathcal{M}_\lambda$ *is a* σ*–algebra; moreover* $(\lambda_{|\mathcal{M}_\lambda}, \mathcal{M}_\lambda)$ *is a* σ*–measure.*

In particular λ is countably additive on $\mathcal{M}_\lambda$. As an immediate consequence of this we get

Proposition 1. *Let* μ *be a measure on* X *and let* $\{E_k\}$ *be a sequence of measurable sets.*

(i) *If* $E_1 \subset E_2 \subset E_3 \subset \ldots$, *then* $\mu(E_k) \uparrow \mu(\cup_{k=1}^\infty E_k)$

(ii) *If* $E_1 \supset E_2 \supset E_3 \supset \ldots$ *and* $\mu(E_1) < \infty$, *then* $\mu(E_k) \downarrow \mu(\cap_{k=1}^\infty E_k)$.

The process (2) of generating an outer measure from a σ–measure is essentially due to Carathéodory and it is known as *Carathéodory's method*; for this reason outer measures are often called *Carathéodory's measures.* Such a method in fact works starting from *any* set function.

Let $\mathcal{G}$ be any sub-family of 2^X, which contains the empty set, and let $\alpha : \mathcal{G} \to [0, +\infty]$ be *any* set function with $\alpha(\emptyset) = 0$. Define

$$\alpha^*(E) := \inf\{\sum_{k=1}^\infty \alpha(G_k) \mid G_k \in \mathcal{G}, \bigcup_{k=1}^\infty G_k \supset E\}$$

if there exists $\{G_k\} \subset \mathcal{G}$ such that $\cup G_k \supset E$, or $\alpha^*(E) := +\infty$ otherwise. Then it is not difficult to verify that α^* is a measure on X; but, this time, in general we have $\mathcal{G} \not\subset \mathcal{M}_{\alpha^*}$, $\alpha^*(G) \leq \alpha(G)$ and (4) does not hold in general.

Definition 5. *A measure λ on X is said to be $\mathcal{G}$–*regular, *where $\mathcal{G}$ is a subcollection of 2^X, if for every $E \subset X$ we have*

$$\lambda(E) = \inf\{\lambda(G) \mid E \subset G , G \in \mathcal{G}\} .$$

If $\mathcal{G} = \mathcal{M}_\lambda$ we say λ regular *instead of $\mathcal{M}_\lambda$–regular.*

As we have seen, if μ^* is the extension of the σ–measure $(\mu, \mathcal{F})$, then μ^* is $\mathcal{F}$–regular and even regular, but in general a measure on X need not be regular. However to any measure μ we can associate a regular measure $\hat{\mu}$ defined by

$$\hat{\mu}(E) := \inf\{\mu(F)| \ F \supset E , F \in \mathcal{M}_\mu\}$$

In this case one has

(i) *If $A \in \mathcal{M}_\mu$, then $A \in \mathcal{M}_{\hat{\mu}}$; and $\hat{\mu}(A) = \mu(A)$,*
(ii) *if $A \in \mathcal{M}_{\hat{\mu}}$ and $\hat{\mu}(A) < +\infty$, then $A \in \mathcal{M}_\mu$.*

Regular measures are very important; later we shall return on that concept. For the moment we just mention the following:

Proposition 2. *If λ is a regular measure, then* (i) *and* (ii) *of Proposition 1 hold for all sequences $\{E_k\}$ not necessarily in $\mathcal{M}_\lambda$. Moreover the following measurability criterion holds: if $A \cup B$ is measurable and $\lambda(A \cup B) = \lambda(A) + \lambda(B)$, then A and B are λ–measurable*

Finally, we observe that, if $\mathcal{G}$ is a σ–algebra, then $\mathcal{G}$–regularity of λ *is equivalent to the existence, for every $E \subset X$, of an element $F \in \mathcal{G}$ such that* $\lambda(F) = \lambda(E)$. Observe that if λ is regular and $E \notin \mathcal{M}_\lambda$, then there is $F \in \mathcal{M}_\lambda$ with $\lambda(F) = \lambda(E)$, but $\lambda(F - E) > 0$, otherwise $F - E$ would be measurable, so that also $E = F - (F - E)$ is measurable.

We can now conclude our general discussion on the definition of outer measure by saying: *If $(\mu, \mathcal{F})$ is a σ–measure, then μ^* defined by (2) is a measure on X, which is $\mathcal{F}$–regular and regular; moreover the elements of $\mathcal{F}$ are measurable sets and $\mu^*_{|\mathcal{F}} = \mu$. Conversely, if λ is an (outer) measure on X, which is $\mathcal{F}$–regular with respect to a σ–algebra $\mathcal{F}$ contained in $\mathcal{M}_\lambda$, then $(\lambda_{|\mathcal{F}}, \mathcal{F})$ is a σ–measure on X and $(\lambda_{|\mathcal{F}})^* = \lambda$.*

Definition 6. *Let μ be a measure on X and let $A \subset X$. The* restriction of μ *on A is defined as the measure*

$$(\mu \llcorner A)(B) := \mu(A \cap B) \qquad \forall B \subset X.$$

It is easily seen that μ–measurable sets are also $\mu \llcorner A$–measurable; moreover, if μ is regular and A is measurable, then $\mu \llcorner A$ is regular.

We recall that a property $P(x)$ is said to hold *for μ–a.e. $x \in X$, μ–almost every x in X*, if there is a null set $A \subset X$, $\mu(A) = 0$, such that $P(x)$ holds for each $x \in X \setminus A$.

Finally let us recall that *Lebesgue's outer measure* in $\mathbb{R}^n$ is the result of Carathéodory's construction by taking for instance as family $\mathcal{G}$ the family of intervals in $\mathbb{R}^n$ and as set function α the function which associates to each interval its standard measure.

Let us now introduce the *integral* with respect to a measure μ. Here the key point is Lebesgue's point of view which in some sense reverses the classical approach of Cauchy and Riemann. The integral is defined not as limit of sums of terms of the type

$$\mu((x_1, x_2))f(x_1),$$

but instead as limit of sums of terms of the type

$$y_1\ \mu(f^{-1}((y_1, y_2)))$$

Therefore the notion of *measurable functions* plays a fundamental role.

Definition 7. *Let μ be a measure on X, and Y a topological space. A function $f : X \to Y$ is said to be* μ–measurable *if and only if $f^{-1}(\mathcal{U})$ is μ–measurable whenever $\mathcal{U}$ is an open set in Y.*

If $Y = [-\infty, +\infty] := \bar{\mathbb{R}}$, and f_k, $k = 1, 2, 3, \ldots$, are μ–measurable functions from X into $[-\infty, +\infty]$, then $f_1 + f_2$, $f_1 f_2$, $|f_1|$, $\min(f_1, f_2)$, $\max(f_1, f_2)$, $\inf_k f_k$, $\sup_k f_k$, $\liminf_{k\to\infty} f_k$, $\limsup_{k\to\infty} f_k$ are all μ–measurable functions whenever they are defined.

Definition 8. *Let μ be a measure in X. A function $g : X \to [-\infty, +\infty]$ is called a* μ–step function *if g is μ–measurable and its image $g(X) := \{g(x) \,|\, x \in X\}$ is a countable set.*

A μ–step function is said to be μ–integrable, *respectively* μ–summable, *if the sum*

$$I(g; \mu) \ := \ \sum_{y \in g(X)} y\mu(g^{-1}(y))$$

exists in $[-\infty, +\infty]$, respectively in $(-\infty, +\infty)$.

Here we use standard conventions on $\bar{\mathbb{R}}$, in particular $0.\infty = 0$, while $\infty - \infty$ is undefined.

It is not difficult to prove

Proposition 3. *Let f be a non-negative μ–measurable function. Then f is the pointwise limit of a non decreasing sequence of steps functions each of which takes only a finite number of values.*

Let $f : X \to [-\infty, +\infty]$ be any function and let S_μ denote the set of all μ–integrable step functions. We define

$$\int^* f \, d\mu \ := \ \inf\{I(g; \mu) \mid g \in S_\mu \quad f \leq g \quad \mu \text{ -a.e}\}$$

$$\int_* f\,d\mu := \sup\{I(g;\mu) \mid g \in S_\mu \quad g \le f \quad \mu\text{-a.e}\}$$

(as usual we set $\inf \emptyset := +\infty$, $\sup \emptyset := -\infty$).

Definition 9. *The function $f : X \to \bar{\mathbb{R}}$ is said to be μ–*integrable *if and only if*

$$\int f\,d\mu := \int^* f\,d\mu = \int_* f\,d\mu\,;$$

*f is called μ–*summable *if and only if $\int f\,dx \in \mathbb{R}$.*

Evidently, for every μ–integrable step function $g : X \to \bar{\mathbb{R}}$ we have

$$I(g;\mu) = \int g\,d\mu\,.$$

One also verifies that *every non-negative μ–measurable function is μ–integrable* and that *every μ–summable function is μ–measurable.*

It is usual to identify functions which are equal for μ–almost every $x \in X$. The space of all (classes of equivalence of) functions which are μ–summable is denoted by $L^1(X;\mu)$. We have

(i) $L^1(X;\mu)$ is a linear space; moreover $|f| \in L^1(X;\mu)$ if $f \in L^1(X;\mu)$.

(ii) The operator

$$f \in L^1(X;\mu) \to \int f\,d\mu \in \mathbb{R}$$

is linear and continuous in the sense that

$$\left|\int f\,d\mu\right| \le \int |f|\,d\mu\,.$$

(iii) Also, the integral is monotone, i. e., $\int f\,d\mu \ge 0$ whenever $f \ge 0$ a. e. in X.

Next theorem collects some relevant results concerning Lebesgue's integration.

Theorem 2. *Let μ be a measure in X and let $\{f_k\}$, $f_k : X \to [-\infty, +\infty]$, be a sequence of μ–measurable functions. Then we have*

(i) **Egoroff's theorem.** *Suppose that the measurable functions f_k are finite, i.e., $f_k : X \to \mathbb{R}$, and let $\varepsilon > 0$ and $A \subset X$ with $\mu(A) < +\infty$. If*

$$f_k(x) \longrightarrow f(x) \qquad \textit{for a.e. } x \in A\,,$$

then there exists a μ–measurable set $B \subset X$ such that $\mu(A \setminus B) < \varepsilon$ and

$$f_k(x) \longrightarrow f(x) \qquad \textit{uniformly in } B\,.$$

(ii) **Beppo Levi theorem.** *If* $0 \le f_1 \le f_2 \le \dots$ *and* $f(x) := \lim_{k\to\infty} f_k(x)$, *then*

$$\int f_k \, d\mu \uparrow \int f \, d\mu \, .$$

(iii) **Fatou's lemma.** *If* $f_k \ge 0$, $k = 1, 2, \dots$, *then*

$$\int \liminf_{k\to\infty} f_k \, d\mu \le \liminf_{k\to\infty} \int f_k \, d\mu \, .$$

(iv) **Lebesgue's dominated convergence theorem.** *If* $f_k(x) \to f(x)$ *for* μ*–a.e.* $x \in X$ *and if there exists a* μ*–summable function* g *such that*

$$|f_k| \le g \, , \qquad \forall k = 1, 2, \dots \, ,$$

then

$$\int |f_k - f| \, d\mu \to 0 \, ,$$

consequently

$$\int f_k \, d\mu \to \int f \, d\mu \, .$$

(v) **Absolute continuity theorem.** *If* $f : X \to [-\infty, +\infty]$ *is* μ*–summable, then for every* $\varepsilon > 0$ *there exists* $\delta > 0$ *such that*

$$\int_A |f| \, d\mu < \varepsilon$$

for all μ*–measurable sets* A *with* $\mu(A) < \delta$.

Definition 10. *We say that* f_k converge to f in $L^1(X;\mu)$ *if and only if*

$$\int |f_k - f| \, d\mu \longrightarrow 0 \, .$$

We say that f_k converge to f in measure μ *if and only if for every* $\varepsilon > 0$

$$\mu\{x \in X \mid |f_k(x) - f(x)| > \varepsilon\} \longrightarrow 0 \qquad \text{as } k \to \infty$$

Since for every $\varepsilon > 0$ we have

$$\mu\{x \in X \mid |f_k(x) - f(x)| > \varepsilon\} \le \varepsilon^{-1} \int |f_k - f| \, d\mu \, ,$$

we deduce that $f_k \to f$ in measure μ provided $\int |f_k - f| \, d\mu \to 0$, i.e., *$L^1$–convergence implies convergence in measure.*

Also, because of Egoroff's theorem, *convergence a.e. in a set of finite measure implies convergence in measure.*

Proposition 4. *Let $\{f_k\}$ be a sequence of μ–measurable functions, which converges to f in measure μ. Then there exists a subsequence $\{f_{k_i}\}$ such that $f_{k_i}(x) \to f(x)$ for μ–a.e. $x \in X$*

Notice that in general the entire sequence f_k does not converge to f a.e., even if f_k converge to f in L^1.

$\boxed{1}$ For instance let $f(x) = 0$ for $x \in \mathbb{R}$. For each $k \in \mathbb{N}$, write $k = 2^n + m$, where $0 \leq m < 2^n$, $k = 0, 1, \ldots$; then n and m are uniquely determined by k. Let

$$f_k(x) := \begin{cases} 1 & \text{if } \frac{m}{2^n} \leq x < \frac{m+1}{2^n} \\ 0 & \text{otherwise}\ . \end{cases}$$

Then

$$\int |f_k(x) - f(x)|\, dx \;=\; 2^{-n} \to 0 \qquad k \to \infty\ ,$$

but

$$\lim_{k\to\infty} f_k(x)$$

exists for *no* $x \in [0, 1]$. If we instead set

$$f_k(x) := \begin{cases} 2^n & \text{if } \frac{m}{2^n} \leq x < \frac{m+1}{2^n} \\ 0 & \text{otherwise}\ . \end{cases}$$

we see that for every positive ε

$$\text{meas}\,\{\, x \in \mathbb{R} \mid |f_k(x) - f(x)| > \varepsilon \,\} \leq 2^{-n} \longrightarrow 0 \qquad \text{as } k \to \infty$$

but

$$\int |f_k(x) - f(x)| dx \;=\; 1\ ,$$

i.e. the convergence of f_k to f in measure does not imply L^1 convergence. •

Every function $f : X \to Y$ induces a map $f_\#$ which associates to each measure μ on X the *image measure*

$$f_\#\mu$$

on Y by the formula

$$f_\#\mu(B) \;:=\; \mu(f^{-1}(B))\ .$$

One readily verifies that $f^{-1}(B)$ *is μ–measurable if and only if B is $f_\#(\mu \llcorner A)$–measurable for every $A \subset X$*. One also sees that *if $\mu(X) < +\infty$, μ is regular, and $B \subset Y$ is $f_\#\mu$–measurable, then $f^{-1}(B)$ is μ–measurable.*

To state our next results we need the concept of countably μ–measurable sets and functions.

Definition 11. *Let μ be a measure on X.*

(i) *A set $A \subset X$ is called* countably μ–measurable *or* μ–σ–finite *if and only if it is expressible as the union of some countable family of μ–measurable sets A_k, $A = \cup_{k=1}^{\infty} A_k$, with $\mu(A_k) < +\infty$.*

(ii) *A μ–measurable function f with values in a topological vector space is called* countably μ–measurable *or* μ–σ–finite *if and only if the set $\{x \in X \mid f(x) \neq 0\}$ is μ–σ–finite.*

Observe that every μ–measurable function is μ–σ–finite.

Definition 12 (Product measures). *Let μ be a measure on X and let ν be a measure on Y. For any $M \subset X \times Y$ define*

$$(\mu \times \nu)(M) \;:=\; \inf\Big\{\sum_k \mu(A_k) \times \nu(B_k)\Big\}$$

where the infimum is taken over all sequences of μ–measurable sets $A_k \subset X$ and ν–measurable sets $B_k \subset Y$ such that $M \subset \cup_k A_k \times B_k$. The set function $\mu \times \nu$ is called the product measure of μ and ν.

Theorem 3 (Fubini). *Let μ and ν are respectively measures on X and Y.*

(i) *$\mu \times \nu$ is a regular measure in $X \times Y$*

(ii) *If $A \subset X$ is μ–measurable and $B \subset Y$ is ν–measurable, then $A \times B$ is $\mu \times \nu$–measurable and $(\mu \times \nu)(A \times B) = \mu(A)\nu(B)$.*

(iii) *If $M \subset X \times Y$ is countably $\mu \times \nu$–measurable, then the set*

$$M_x \;:=\; \{y \in Y \mid (x,y) \in M\}, \qquad x \in X$$

is ν–measurable for μ–a.e. $x \in X$ and

$$(\mu \times \nu)(M) \;=\; \int_X \nu(M_x)\, d\mu(x)$$

(iv) *If $f : X \times Y \to [-\infty, +\infty]$ is a $\mu \times \nu$–integrable and countably $\mu \times \nu$–measurable function, then $f(x,\cdot) : y \in Y \to f(x,y)$ is ν–integrable for μ–a.e. $x \in X$, the function*

$$h(x) \;:=\; \int_Y f(x,y)\, d\nu(y)$$

is μ–integrable and

$$\int_{X\times Y} f(x,y)\, d\mu \times \nu \;=\; \int_X h\, d\mu \;=\; \int_X \left(\int_Y f(x,y)\, d\nu(y)\right) d\mu(x)$$

However finiteness of iterated integrals does not imply in general that of the multiple integral. But for non-negative f and more generally we have

Theorem 4 (Tonelli). *Let $f : X \times Y \to [0, +\infty]$ be a $(\mu \times \nu)$–measurable and countably $\mu \times \nu$–measurable function. Then the existence of one of the two integrals*

$$\int_{X\times Y} f(x,y)\, d\mu \times \nu \qquad \int_X \left(\int_Y f(x,y)\, d\nu(y) \right) d\mu(x)$$

implies the existence of the other and equality as well.

1.2 Borel Regular and Radon Measures

In this subsection we assume that X is a metric space with distance d and the induced topology.

Definition 1. *Let μ be a measure on the metric space X.*

(i) The smallest σ-algebra containing all open sets in X is called the Borel σ-algebra *and is denoted by $\mathcal{B}(X)$. A function $f : X \to Y$, with values in a topological space Y is called a* Borel function *if, for every open set $\mathcal{U} \subset Y$, $f^{-1}(\mathcal{U}) \in \mathcal{B}(X)$.*
(ii) μ is called a Borel measure *if every Borel set is μ-measurable, i.e., $\mathcal{B}(X) \subset \mathcal{M}_\mu$.*
(iii) μ is called Borel–regular *if μ is Borel, i.e., $\mathcal{B}(X) \subset \mathcal{M}_\mu$, and for every $A \subset X$, there exists $B \in \mathcal{B}(X)$ such that $B \supset A$ and $\mu(B) = \mu(A)$.*

Definition 2. *A measure μ over a locally compact and separable space X is called a* Radon measure *if μ is Borel-regular and for every compact subset K of X we have $\mu(K) < +\infty$.*

An important example of Radon measure in $\mathbb{R}^n$ is given by Lebesgue's n-dimensional outer measure $\mathcal{L}^{n*}$.

One easily verifies that: If μ is Borel-regular and A is μ-measurable, then $\mu \llcorner A$ is Borel-regular.

If μ is Borel-regular and A is a μ-measurable subset of a locally compact and separable space X with $\mu(A) < +\infty$, then $\mu \llcorner A$ is a Radon measure.

Borel measures, in contrast to measures on a generic set, have a quite rich family of measurable sets. The following theorem gives a criterion ensuring that a measure over a metric space is a Borel measure.

Theorem 1 (Carathéodory's criterion). *Let μ be a measure over a metric space X. If*

$$\mu(A \cup B) = \mu(A) + \mu(B)$$

for all A and B with positive distance, i. e.,

$$\operatorname{dist}(A, B) := \inf\{d(x,y) \mid x \in A\,,\, y \in B\} > 0\,,$$

then μ is a Borel measure, i.e., $\mathcal{B}(X) \subset \mathcal{M}_\mu$.

Next theorem states the basic approximation property in terms of Borel measures

Theorem 2. *Let μ be a Borel measure on the metric space X and let $B \subset X$ be a Borel set.*

(i) If $\mu(B) < +\infty$ and $\varepsilon > 0$, then there exists a closed set $C \subset B$ such that $\mu(B \setminus C) < \varepsilon$.
(ii) If $B \subset \cup_{k=1}^{\infty} \mathcal{U}_k$, where $\mathcal{U}_k$ are open sets with $\mu(\mathcal{U}_k) < +\infty$, and $\varepsilon > 0$, then there exists an open set $\mathcal{U} \supset B$ such that $\mu(\mathcal{U} \setminus B) < \varepsilon$.

Moreover, if μ is Borel-regular, then (i) and (ii) hold for all μ-measurable sets $B \subset X$.

The following two propositions follow at once from the previous theorem.

Proposition 1. *Let μ be a Borel-regular measure on the metric space X. Suppose that $X \subset \cup_{k=1}^{\infty} \mathcal{U}_k$ where $\mathcal{U}_k$ are open sets with $\mu(\mathcal{U}_k) < +\infty$. Then*

$$\mu(A) = \inf\{\mu(\mathcal{U}) \mid \mathcal{U} \supset A, \mathcal{U} \text{ open}\}$$

for every subset $A \subset X$, and

$$\mu(A) = \sup\{\mu(C) \mid C \subset A, C \text{ closed}\}$$

for every μ-measurable subset $A \subset X$.

Proposition 2. *Let μ be a Radon measure on a locally compact and separable space X. Then*

$$\mu(A) = \inf\{\mu(\mathcal{U}) \mid \mathcal{U} \supset A, \mathcal{U} \text{ open}\}$$

for every subset $A \subset X$, and

$$\mu(A) = \sup\{\mu(K) \mid K \subset A, K \text{ compact}\}$$

for every μ-measurable A with $\mu(A) < +\infty$.

Let μ be a Borel-regular measure in X. Of course continuous or semicontinuous functions $f : X \to \mathbb{R}$ are μ-measurable and in fact Borel functions; it is also easily seen that the inverse image of a Lebesgue measurable set in $\mathbb{R}$ by a semicontinuous map is measurable. Notice instead that the direct image of a measurable set by a continuous map f in general is not measurable. For instance Cantor–Vitali function in Sec. 1.1.3 below maps the Cantor set, which is a null set, into a set of positive measure, therefore we can obtain non-measurable sets as continuous image of a null set.

Definition 3. *We say that a μ-measurable map $f : X \to \mathbb{R}^m$, $m \geq 1$, satisfies* Lusin's property (N) *if μ–null sets in X are mapped into null sets in $\mathbb{R}^m$.*

The following is then easily proved taking into account Lusin's theorem below: *Let μ be a Borel-regular measure in X and let $f : X \to \mathbb{R}^n$ be a Borel map which has Lusin property* (N). *Then $f(E)$ is Lebesgue measurable provided E is μ-measurable*, compare Step 1 of the proof of Theorem 1 in Sec. 2.1.2. Recall that $f : X \to \mathbb{R}$ is a Borel function iff the inverse image of an open set is a Borel set.

Also recall that not every Lebesgue measurable function $f : \mathbb{R}^n \to \mathbb{R}$ can be obtained as a monotone limit of continuous functions, but we have

Theorem 3 (Vitali-Carathéodory). *Every Lebesgue-measurable function $f : \mathbb{R}^n \to \mathbb{R}$ is the a.e. pointwise limit of a non-decreasing sequence of functions φ_k which are upper semicontinuous and bounded from above. If moreover f is integrable, we have*

$$\lim_{k\to\infty} \int (f - \varphi_k)\, dx = 0 .$$

Finally next theorem shows an important relation between measurable and continuous function.

Theorem 4 (Lusin). *Let μ be a Borel-regular measure on a metric space X and let $f : X \to \mathbb{R}$ be a μ-measurable function.*

(i) If $A \subset X$ is μ-measurable with $\mu(A) < +\infty$ and $\varepsilon > 0$, then there exists a closed set $C \subset A$ such that $\mu(A \setminus C) < \varepsilon$ and $f_{|C}$ is continuous.
(ii) If X is moreover μ–σ-finite, then there exists a Borel function $g : X \to \mathbb{R}$ such that $f = g$ μ-a.e. in X.
(iii) Let $X = \mathbb{R}^n$, let $A \subset \mathbb{R}^n$ be μ-measurable with $\mu(A) < +\infty$, and let $\varepsilon > 0$. Then there exists a continuous function $g : \mathbb{R}^n \to \mathbb{R}$ such that

$$\mu\{x \in A \mid f(x) \neq g(x)\} < \varepsilon .$$

Notice that (iii) follows from (i) by means of *Tietze's extension theorem* for continuous functions on closed sets.

Remark 1. Observing that $\mathbb{R}^n$ is the union of a countable family of closed cubes with disjoint interiors and with side one, one easily sees that Lusin's theorem holds for any measurable set A of $\mathbb{R}^n$, not necessarily of finite measure, if μ is the Lebesgue measure $\mathcal{L}^n$. This remark will be used later.

1.3 Hausdorff Measures

Hausdorff measures are among the most important measures. They allow us to define the dimension of sets in $\mathbb{R}^n$ and provide us with s–dimensional measures in $\mathbb{R}^n$ for any s, $0 \le s \le n$ and actually in any metric space. They will be continuously used in the sequel.

Let s be a nonnegative real number. We denote by ω_s the volume of the unit ball in $\mathbb{R}^s$ for $s = 1, 2, 3, \ldots$, we set $\omega_0 := 1$ and we let ω_s any convenient fixed constant for non-integers s. Since the measure of the unit n–ball is given by

$$\frac{\pi^{n/2}}{(n/2)\Gamma(n/2)} \qquad n = 1, 2, \ldots$$

where $\Gamma(t)$ is *Euler's gamma function*

$$\Gamma(t) := \int_0^\infty x^{t-1} \exp(-x)\, dx ,$$

we may and do take as ω_s

$$\omega_s := \frac{\pi^{s/2}}{(s/2)\Gamma(s/2)} .$$

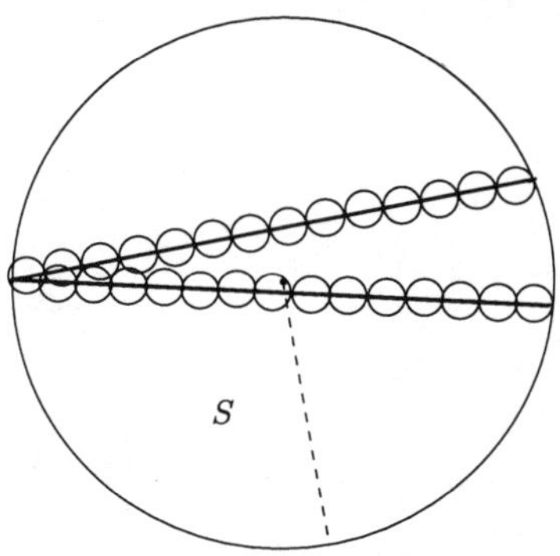

Fig. 1.1. $\widetilde{\mathcal{H}}^1(E) \sim 2$, $\mathcal{H}^1(E) \sim 4$.

Definition 1. *Let X be a metric space. For $\delta > 0$ and $A \subset X$ we set*

(1)
$$\mathcal{H}^s_\delta(A) := \inf\Big\{ \sum_k \omega_s \Big(\frac{\operatorname{diam} C_k}{2}\Big)^s \;\Big|\; \cup_k C_k \supset A\ ,\ \operatorname{diam} C_k < \delta\ ,\ C_k \subset X \Big\}$$

The Hausdorff measure $\mathcal{H}^s$ *is then defined by*

(2)
$$\mathcal{H}^s(A) := \lim_{\delta \to 0} \mathcal{H}^s_\delta(A) , \qquad A \subset X$$

Observe that $\mathcal{H}^s_{\delta_1} \geq \mathcal{H}^s_{\delta_2}$ for $\delta_1 \leq \delta_2$, thus $\mathcal{H}^s$ is well defined and

$$\mathcal{H}^s(A) = \sup_{\delta > 0} \mathcal{H}^s_\delta(A) .$$

Notice that it is essential to use the approximating measures $\mathcal{H}^s_\delta$ in order to define $\mathcal{H}^s$. The naive definition

$$\widetilde{\mathcal{H}}^s(E) := \inf\Big\{ \sum_k \omega_s \Big(\frac{\operatorname{diam} C_k}{2}\Big)^s \;\big|\; E \subset \bigcup_k C_k \Big\}$$

in fact would have very unpleasant features.

In (1) we may also require that the sets C_k be closed or open without changing the value (2).

By the same procedure we can define the so-called *s–dimensional spherical measure* $\mathcal{S}^s$ by requiring in (1) that the sets C_k be balls. Then we obviously have

$$\mathcal{H}^s \leq \mathcal{S}^s \leq 2^s\mathcal{H}^s ,$$

but there exist sets A for which $\mathcal{H}^s(A) < \mathcal{S}^s(A)$.

For $X = \mathbb{R}^n$, one easily verifies

(i) $\mathcal{H}^s(\lambda A) = \lambda^s\mathcal{H}^s(A)$, $\lambda > 0$, *where* $\lambda A := \{\lambda x \mid x \in A\}$

(ii) $\mathcal{H}^0$ *is the so-called* counting measure , *i. e.*,

$$\mathcal{H}^0(A) = \#A := \text{number of elements of } A$$

(iii) $\mathcal{H}^s = 0$ *on* $\mathbb{R}^n$ *for* $s > n$

(iv) $\mathcal{H}^s$ *is not a Radon measure on* $\mathbb{R}^n$ *for* $0 \leq s < n$.

(v) *If* $\mathcal{H}^s_\delta(A) = 0$ *for some* $\delta > 0$, *then* $\mathcal{H}^s(A) = 0$.

Finally, observe that for $0 \leq s < r$

$$\mathcal{H}^r_\delta(A) \leq \left(\frac{\delta}{2}\right)^{r-s} \mathcal{H}^s_\delta(A) ,$$

thus $\mathcal{H}^s(A) = +\infty$ if $\mathcal{H}^r(A) > 0$ and $\mathcal{H}^r(A) = 0$ if $\mathcal{H}^s(A) < \infty$. This in particular shows that $\mathcal{H}^s(A)$ can be positive and finite for at most one (possibly none) $s \geq 0$. This motivates the following definition

Definition 2. *The* Hausdorff dimension *of a set* $A \subset X$ *is defined by*

$$\dim_{\mathcal{H}}(A) := \inf\{s \geq 0 : \mathcal{H}^s(A) = 0\}$$

From the previous remark it follows that if $\dim_{\mathcal{H}} A > 0$, then

$$\dim_{\mathcal{H}}(A) = \sup\{s \geq 0 \mid \mathcal{H}^s(A) = +\infty\}$$

and that, if $0 < \mathcal{H}^s(A) < \infty$, then $\dim_{\mathcal{H}} A = s$.

It is easily seen that $\mathcal{H}^s_\delta$ and $\mathcal{H}^s$ are measures over X. Borel sets are not in general $\mathcal{H}^s_\delta$–measurable for $\delta > 0$: for instance the half-space $\{(x,y) \in \mathbb{R}^2 \mid x > 0\} \subset \mathbb{R}^2$ is not $\mathcal{H}^1_\delta$–measurable. However, observe that $\mathcal{H}^n_\delta = \mathcal{H}^n$ on $\mathbb{R}^n$, and we have

$$\mathcal{H}^s_\delta(A \cup B) = \mathcal{H}^s_\delta(A) + \mathcal{H}^s_\delta(B)$$

if dist $(A, B) > 2\delta$; therefore Borel sets are $\mathcal{H}^s$–measurable by Carathéodory criterion. Actually one proves

Proposition 1. $\mathcal{H}^s_\delta$ *and* $\mathcal{H}^s$ *are measures on* X; *moreover* $\mathcal{H}^s$ *is Borel-regular for all* $s \geq 0$.

If $X = \mathbb{R}^n$ *and* A *is a* H^s*–measurable subset of* X *with* $\mathcal{H}^s(A) < \infty$, *then* $H^s \llcorner A$ *is a Radon measure on* X.

Using the so-called *isodiametric inequality*

$$\mathcal{L}^n(A) \le \omega_n \left(\frac{diam\ A}{2}\right)^n \qquad \forall A \subset \mathbb{R}^n ,$$

which says that among sets $A \subset \mathbb{R}^n$ with a given diameter ρ, the ball with diameter ρ has the largest Lebesgue measure, one also proves the following important theorem

Theorem 1. *The Hausdorff measure $\mathcal{H}^n$ in $\mathbb{R}^n$ coincides with the Lebesgue measure $\mathcal{L}^{n*}$ on $\mathbb{R}^n$. Moreover if M is an s–dimensional smooth submanifold in $\mathbb{R}^n$, $0 < s < n$, then $\mathcal{H}^s \llcorner M$ is the standard volume measure in M, induced by the Euclidean metric in M.*

[1] *Cantor sets.* It is often difficult to determine the Hausdorff dimension of a set and even harder to find its Hausdorff measure. Of course the most difficult part is to give a lower estimate.

The most familiar sets of real numbers of non-integral Hausdorff dimensions are *Cantor's sets.* These are obtained by the following construction. Fix $\delta \in (0, 1/2)$, begin with $E_0 := [0, 1]$ and in the first step remove an open interval of length $1 - 2\delta$ in the middle. Inductively then define E_{k+1} by removing in each interval of E_k a centered open interval of length $\delta^k(1-2\delta)$. Of course $E_k \supset E_{k+1}$ for all k and each E_k is the union of 2^k intervals of length δ. The *Cantor set* associated to δ is then defined as the closed, non denumerable, and dense in itself set

$$C := \bigcap_{k=0}^{\infty} E_k .$$

Classically one chooses $\delta = 1/3$ and the resulting set C is known as Cantor's *middle thirds set.*

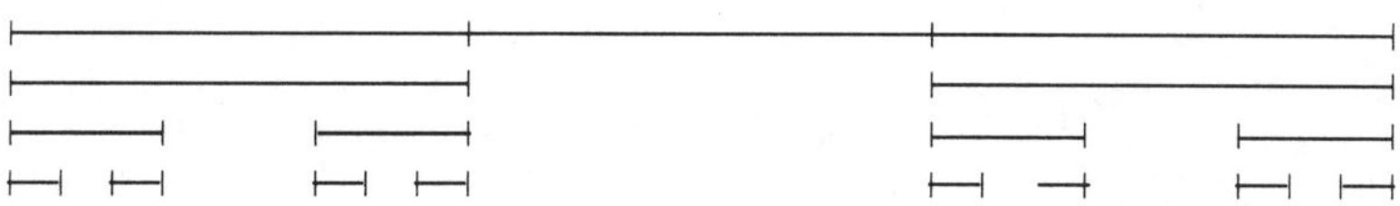

Fig. 1.2. Cantor's middle thirds set.

To be more precise and for future references we can describe Cantor set C as follows. For each $k = 0, 1, \ldots$ and $j = 1, \ldots, 2^k$ define the *base points* $b_{k,j}$ of the Cantor set C inductively by

$$b_{0,1} := 0$$

and

$$b_{k+1,j} := \begin{cases} \delta\, b_{k,j} & \text{if } j = 1, \dots, 2^k \\ (1-\delta) + \delta\, b_{k,j} & \text{if } j = 2^k + 1, \dots, 2^{k+1} \ . \end{cases}$$

Then

$$I_{k-1,j} := b_{k-1,j} + \delta^{k-1}(\delta, 1-\delta) \qquad j = 1, \dots, 2^{k-1}$$

are the intervals removed from E_{k-1} to get E_k at the step k, and

$$J_{k,j} := b_{k,j} + \delta^k\,[0,1] \qquad j = 1, \dots, 2^k$$

are the intervals whose union yields E_k. Consequently

$$C = \bigcap_{k=0}^{\infty} \Big(\bigcup_{j=1}^{2^k} J_{j,k} \Big) \ .$$

From this expression it is easily seen that Cantor set C is *self-similar* in the

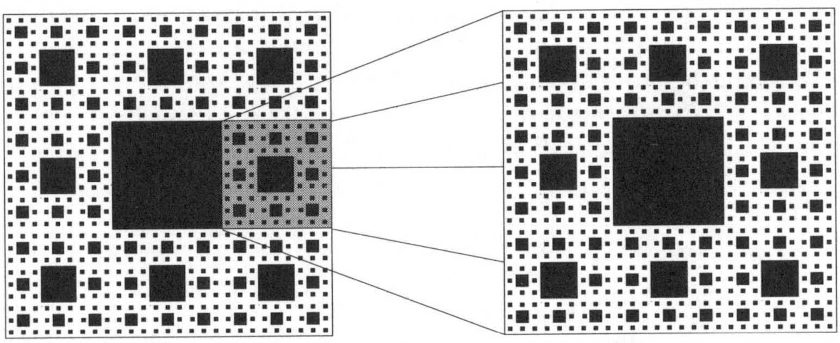

Fig. 1.3. Self-similarity of Cantor sets in two dimensions.

sense that certain homothetic expansions of C are locally identical to the original set; more precisely we have

$$(3) \qquad b_{k,j} + \delta^k C = C \cap J_{k,j} \ .$$

In fact it is readily seen that

$$x \in J_{h,j} \qquad \text{iff} \qquad \delta^k x \in J_{h+k,j} \qquad \text{for all } h, k, j$$

hence

$$\delta^k E_h = E_{h+k} \cap [0, \delta^k] \ ,$$

i.e.,

$$b_{k,j} + \delta^k E_h = E_{h+k} \cap J_{k,j}$$

for all $h, k \geq 0$, $j = 1, \dots, 2^k$, so that (3) follows by taking the intersection in h.

By introducing the map $\tau : \mathcal{P}([0,1]) \to \mathcal{P}([0,1])$ defined by

$$\tau(A) := (\delta A) \cup (1 - \delta + \delta A) ,$$

we also see that $\tau^k([0,1]) = E_k$, $\tau^k(\{0\}) = \{b_{k,1}, \dots, b_{k,2^k}\}$. Therefore the self-similarity of C can be also expressed by

$$C = \bigcap_{k=0}^{\infty} \tau^k([0,1]) , \qquad \text{i.e.,} \qquad \tau(C) = C .$$

Proposition 2. *The Hausdorff dimension of the Cantor set C, actually C_δ, is $s := \log 2 / \log(1/\delta)$. Moreover*

$$\mathcal{H}^s(C) = 2^{-s}\omega_s .$$

Finally for all $x, y \in [0,1]$, $x < y$

$$\mathcal{H}^s(C \cap [x,y]) = \lim_{k\to\infty} \mathcal{L}^1(E_k \cap [x,y]) .$$

Proof. Since $C \subset E_k$ and E_k is the union of 2^k intervals of length δ^k we see that

$$\mathcal{H}^s_{\delta^k}(C) \leq 2^{-s}\omega_s 2^k \delta^{ks} = 2^{-s}\omega_s(2\delta^s)^k = 2^{-s}\omega_s .$$

Letting $k \to \infty$, $\mathcal{H}^s(C) \leq 2^{-s}\omega_s$. To prove the opposite inequality we consider an open covering $\mathcal{F}$ of C. As C is compact we may assume that $\mathcal{F}$ is finite $\mathcal{F} = \{K_1, \dots, K_\ell\}$ and we have $K_1 \cup \dots \cup K_\ell \supset E_{\bar{k}}$ for some $\bar{k} \in \mathbb{N}$. For any $\varepsilon > 0$ choose k large enough so that $4\delta^k \ell < \varepsilon$, and let $K := (a,b)$ denotes one of the intervals in $\mathcal{F}$. For convenience denote by $R(I)$ and $L(I)$ respectively the right and left end points of any interval I. We have

$$[0,1] = \Big(\bigcup_{j=1}^{2^k} J_{k,j}\Big) \bigcup \Big(\bigcup_{m=0}^{k-1} \bigcup_{i=1}^{2^m} I_{m,i}\Big)$$

where the J-type intervals yield E_k and the I-type intervals are the ones which are taken out up to step k. First we slightly change K as follows

$$\begin{aligned} &\text{if } a \in I_{m,i}, \text{ we set } a' = R(I_{m,i});\ a' \geq a \\ &\text{if } a \in J_{k,j}, \text{ we set } a' = L(J_{k,j});\ a' \geq a - \delta^k . \end{aligned}$$

Similarly

$$\begin{aligned} &\text{if } b \in I_{m,i}, \text{ we set } b' = L(I_{m,i});\ b' \leq b \\ &\text{if } b \in J_{k,j}, \text{ we set } b' = R(J_{k,j});\ b' \leq b + \delta^k . \end{aligned}$$

This way for $K' = (a', b')$ we find

$$||K'| - |K|| < 2\delta^k , \qquad \partial K' \subset \partial E_k$$

and of course K' covers $K \cap C$. From now on we can therefore assume that for each K_r in $\mathcal{F}$

$$\partial K_r \subset \partial E_k , \qquad L(K_r) = L(J_{k,j}), \qquad R(K_r) = R(J_{k,\bar{j}})$$

for some $j \leq 2^k$ and $\bar{j} \leq 2^k$.

Next we note that, by the construction of Cantor set, each $J_{k,j}$ is preceded and followed by one of the I-type intervals, one taken out at the k-step and one at least at step $k-1$. More precisely there are (m_1, i_1) and (m_2, i_2) with $m_1, m_2 \leq k-1$, $i_1 \leq 2^{m_1}$, $i_2 \leq 2^{m_2}$ such that

$$R(I_{m_1,i_1}) = L(J_{k,j}) , \qquad R(J_{k,j}) = L(I_{m_2,i_2})$$

and moreover either $m_1 = k-1$, $m_2 \leq k-2$, or $m_1 \leq k-2$, $m_2 = k-1$.

Consider now the largest I-interval $I_{m,i}$ in $\overline{K}_r = [a_r, b_r]$. We claim that

$$a_r \geq L(I_{m,i}) - \delta^{m+1} , \qquad b_r \leq R(I_{m,i}) + \delta^{m+1} . \tag{4}$$

In fact we have the disjoint decomposition

$$b_{m,i} + \delta^m[0,1] = J_{m+1,j} \cup I_{m,i} \cup J_{m+1,j+1}$$

for some j, where

$$\begin{aligned} J_{m+1,j} &= b_{m,i} + [0, \delta^{m+1}] \\ J_{m+1,j+1} &= b_{m,i} + (1-\delta)\delta^m + [0, \delta^{m+1}] . \end{aligned}$$

Suppose that $a_r < b_{m,i} = L(I_{m,i}) - \delta^{m+1}$. Then there exists $I_{m',i'}$ such that

$$R(I_{m',i'}) = L(I_{m+1,j}) ,$$

hence $I_{m',i'} \subset \overline{K}_r$. By the previous remark we then have $m' < m$, i.e., $|I_{m',i'}| > |I_{m,i}|$, that is $I_{m,i}$ is not the largest I-interval in $\overline{K}_r$, a contradiction. Similar we proceed to show that $b_r \geq R(I_{m,i}) + \delta^{m+1}$.

On account of (4) we can define

$$\widetilde{J}_1 := \overline{K}_r \cap [0, L(I_{m,i})] , \qquad \widetilde{J}_2 := \overline{K}_r \cap [R(I_{m,i}), 1]$$

and find the disjoint decomposition

$$\overline{K}_r = \widetilde{J}_1 \cup I_{m,i} \cup \widetilde{J}_2$$

with

$$|\widetilde{J}_1|, |\widetilde{J}_2| \leq \frac{\delta}{1-2\delta} |I_{m,i}| .$$

Therefore using the concavity of t^s and the fact that $2\delta^s = 1$ we deduce

$$\begin{aligned}
|K_r|^s &= (|\widetilde{J}_1| + |I_{m,i}| + |\widetilde{J}_2|)^s \ \geq \\
&= \Big[|\widetilde{J}_1| + \frac{1}{2}\frac{1-2\delta}{\delta}(|\widetilde{J}_1| + |\widetilde{J}_2|) + |\widetilde{J}_2|\Big]^s \\
&= \delta^{-s}(\frac{1}{2}|\widetilde{J}_1| + \frac{1}{2}|\widetilde{J}_2|)^s \ \geq \frac{1}{2}\delta^{-s}(|\widetilde{J}_1|^s + |\widetilde{J}_2|^s) \\
&= |\widetilde{J}_1|^s + |\widetilde{J}_2|^s \ .
\end{aligned}$$

Thus, replacing K_r by the two intervals $\widetilde{J}_1$ and $\widetilde{J}_2$ which still cover $E_k \cap \overline{K}_r$ and satisfy $\partial\widetilde{J}_1, \partial\widetilde{J}_2 \subset \partial E_k$, we do not increase the sum

$$|K_1|^s + \cdots + |K_\ell|^s \ .$$

Proceeding this way we obtain in a finite number of steps

$$|K_r|^s \ \geq \sum_{J_{k,j} \subset K_r} |J_{k,j}|^s$$

and eventually after a finite number of steps

$$|K_1|^s + \cdots + |K_\ell|^s \ \geq \ |E_h|$$

for a suitable h. This yields $\mathcal{H}^s(C) \geq 2^s\omega_s$. □

•

[2] *Cantor–Vitali functions.* A procedure similar to the one which leads to Cantor set with parameter δ allows to construct a non decreasing continuous function V, which we call *Cantor–Vitali function*, whose derivative vanishes almost everywhere. For that, with the notation of [1] we consider for each k the two functions $V_k, g_k \ : \ [0,1] \to \mathbb{R}$ defined by

$$g_k(x) \ := \ 2^{-k}\delta^{-k}\chi_{E_k}(x) \ , \qquad V_k(x) \ := \int_0^x g_k(t)\,dt \ = \ 2^{-k}\delta^{-k}\mathcal{L}^1(E_k \cap [0,x]) \ .$$

Clearly $V_k(0) = 0$, $V_k(1) = 1$, V_k is nondecreasing and

$$\int_{J_{k,j}} g_k(t)\,dt \ = \ 2^{-k} \ .$$

For $x \in [0,1]$ we denote by $j(x,k)$ the largest integer between 0 and 2^k such that $J_{k,j(x,k)}$ is contained in $[0,x]$; if such an interval does not exist we set $j(x,k) = 0$. We then see that for $x \notin E_k$

$$V_k(x) = \sum_{j=1}^{j(x,k)} \int_{J_{k,j}} g_k(t)\,dt = 2^{-k} j(x,k)$$

$$V_{k+1}(x) = \sum_{j=1}^{j(x,k)} \int_{J_{k,j}} g_{k+1}(t)\,dt = 2^{-k} j(x,k)$$

so that

$$\begin{aligned} V_k(x) &= V_{k+1}(x) = 2^{-k} j(x,k) && \text{if } x \notin E_k \\ V_{k+1}(x) &= \tfrac{2j-1}{2^{k+1}} && \text{if } x \in I_{k,j} \end{aligned}$$

while, if $x \in E_k$, so that $x \in J_{k,j}$ for some j, $V_k(x)$ is linear on each $J_{k,j}$, $V_k'(x) = (2\delta)^{-k}$ in $\mathring{E}_k$ and we have

$$|V_{k+1}(x) - V_k(x)| \le \int_{J_{k,j}} |g_{k+1}(t) - g_k(t)|\,dt \le 2^{-k+1} . \tag{5}$$

Equivalently the *approximate Cantor–Vitali functions* V_k can be inductively defined by

$$V_0(x) = x$$

$$V_{k+1}(x) = \begin{cases} \frac{1}{2} V_k(\delta^{-1}x) & \text{if } x \in [0,\delta] \\ \frac{1}{2} & \text{if } x \in [\delta, 1-\delta] \\ \frac{1}{2} + \frac{1}{2} V_k(\delta^{-1}(x-(1-\delta))) & \text{if } x \in [1-\delta, 1] \end{cases}$$

Then again we readily see that $V_k(0) = 0$, $V_k(1) = 1$, $V_k'(x) = (2\delta)^{-k}$ in $\mathring{E}_k$. Moreover, being $J_{k,j} = [b_{j,k}, b_{j,k} + \delta^k]$, we have

$$V_k(b_{j,k}) = \frac{j-1}{2^k} , \qquad V_k(b_{j,k} + \delta^k) = \frac{j}{2^k}$$

and

$$V_k(x) = \frac{2j-1}{2^{m+1}} \quad \text{for } x \in I_{m,j} , \qquad m = 0, \dots, k-1, \ j = 1, \dots, 2^m .$$

By induction we immediately see that for each k we have

$$|V_k(x) - V_k(y)| \le c\,|x-y|^s \tag{6}$$

where

$$c := (1-2\delta)^{-1} \qquad \text{and} \qquad s = \frac{\log 2}{\log(1/\delta)} .$$

From (4) and (6) we infer that V_k converge uniformly to a Hölder-continuous and non decreasing function V, called the *Cantor–Vitali function*.

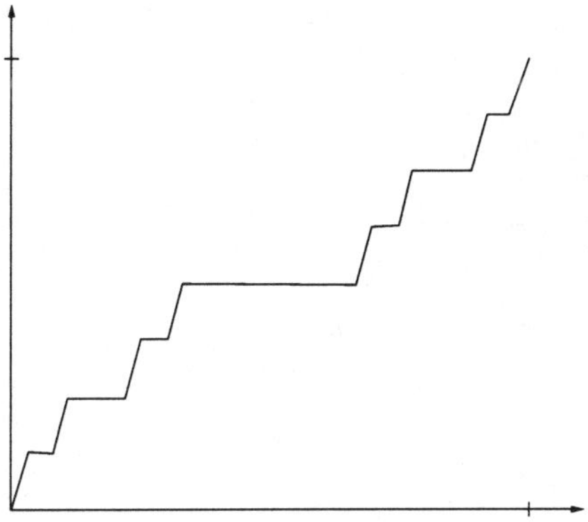

Fig. 1.4. An approximation of Vitali's function.

Notice that

$$V_k(x) = V_{k+1}(x) = \cdots = V(x) \qquad \text{for } x \in [0,1] \setminus E_k$$

and that, by Proposition 2

$$V(x) = \frac{\mathcal{H}^s(C \cap [0,x])}{\mathcal{H}^s(C)} \qquad \forall\, x \in [0,1];$$

self-similarity of V is expressed by

$$V(x) = V(b_{k,j}) + 2^{-k} V(\delta^{-k}(x - b_{k,j})) \qquad \text{for } x \in J_{k,j} := [b_{k,j}, b_{k,j} + \delta^k] .$$

Finally it is easily seen, compare for more details Ch. 4, that

(i) V has bounded variation in $(0,1)$, $V \in BV(0,1)$, since it is non-decreasing, and $V(0) = 0, V(1) = 1$.

(ii) On each interval in $[0,1] \setminus E_k$ we have $V(x) = V_k(x) = constant$. In particular, $V(x)$ is differentiable on each $x \in [0,1] \setminus E$, and (approximately) differentiable almost everywhere in $[0,1]$ since $|E| = 0$, and

$$\text{ap}V'(x) = 0 \qquad \text{for a.e. } x \in [0,1] .$$

(iii) Since V is continuous, V' has no jump part but only a *Cantor part*

$$V' = V'^{(C)};$$

and V' has support in E.

(iv) Obviously $V([0,1]) = [0,1]$, V maps $[0,1] \setminus E$ into the denumerable dyadic set

$$D := \{y \in \mathbb{R} \mid y = \frac{j}{2^k} , \ j \in [0, 2^k], \ j,k \in \mathbb{N}\} .$$

Therefore V maps E onto $[0,1] \setminus D$. In particular V has not Lusin property (N).

•

Let $f : \mathbb{R}^n \to \mathbb{R}^m$, $m \geq 1$, be a map and let A be a subset of $\mathbb{R}^n$. The *graph of f over A* is given by

$$\mathcal{G}_{f,A} := \{(x, f(x)) \mid x \in A\} .$$

Recall also that f is said to be *Lipschitz continuous* if

$$Lip\ f := \sup\{\frac{|f(x) - f(y)|}{|x - y|} \ |x, y \in \mathbb{R}^n , x \neq y\} < \infty .$$

Then we have the following theorem

Theorem 2. *Let $f : \mathbb{R}^n \to \mathbb{R}^m$ and let $A \subset \mathbb{R}^n$*

(i) If $\mathcal{H}^n(A) > 0$, then $\dim_{\mathcal{H}} \mathcal{G}_{f,A} \geq n$
(ii) If f is Lipschitz-continuous, then for $s \geq 0$

$$\mathcal{H}^s(f(A)) \leq (Lip\ f)^s \mathcal{H}^s(A)$$

(iii) If f is Lipschitz-continuous and $\mathcal{H}^n(A) > 0$, then

$$\dim_{\mathcal{H}} \mathcal{G}_{f,A} = n .$$

Remark 1. Hausdorff measure $\mathcal{H}^s$ is the result of a special case of a construction known as *Carathéodory's construction*,conpare Sec. 1.1.1. Given any collection $\mathcal{F}$ of parts of X with $\emptyset \in \mathcal{F}$ and any function $\psi : \mathcal{F} \to [0, +\infty]$ with $\psi(\emptyset) = 0$, we define a measure $\mu = \mu_{\psi,\mathcal{F}}$ on X as follows. First, for each $\delta > 0$, we set

$$\mu_\delta(A) := \inf\{\sum \psi(A_k) \mid \cup_k A_k \supset A , A_k \in \mathcal{F} , diam\ A_k \leq \delta\} , \quad A \subset X ,$$

and then

$$\mu(A) := \sup_{\delta > 0} \mu_\delta(A) , \qquad A \subset X$$

Of course Hausdorff measure $\mathcal{H}^s$ corresponds to the choice $\mathcal{F} = 2^X$ and $\psi(A) = \omega_s(\frac{1}{2} diam\ A)^s$. This way one may construct several interesting geometric measures, but for our purposes it suffices to work with Hausdorff measure.

Having in mind Lebesgue's measure, a natural question is whether $\mathcal{H}^{s+t}(A \times B)$ and $\mathcal{H}^s(A) \times \mathcal{H}^t(B)$ are equal, if $A \subset \mathbb{R}^n$ and $B \subset \mathbb{R}^m$. It turns out that such a result is far from being true and in general we only have

Theorem 3. *Let $A \subset \mathbb{R}^n$ and $B \subset \mathbb{R}^m$. Then*

$$\mathcal{H}^{s+t}(A \times B) \geq b\, \mathcal{H}^s(A) \times \mathcal{H}^t(B)$$

for some constan $b > 0$. In particular $\dim_{\mathcal{H}}(A \times B) \geq \dim_{\mathcal{H}} A + \dim_{\mathcal{H}} B$.

Theorem 3 is a special case of the following inequality

$$(7) \qquad \int^{*} \mathcal{H}^s(A \cap f^{-1}(y))\, d\mathcal{H}^t \;\le\; \frac{\omega_s \omega_t}{\omega_{s+t}} (\operatorname{Lip} f)^s\, \mathcal{H}^{t+s}(A)$$

which holds for any Borel set A, any Lipschitz function f, and non negative real numbers s and t. Its proof is presented in Ch. 2, Theorem 2 in Sec. 2.1.3, step 2, in case s, t are integers, but it works as well for general s, t.

The opposite inequality in (7) does not holds. For instance in Federer [226] one finds for any s general Cantor-type sets C_1 and C_2, $\mathcal{H}^s(C_i) = 1$, $i = 1, 2$, for which $\mathcal{H}^{2s}(C_1 \times C_2) = +\infty$. However, it is easily seen that

$$\mathcal{H}^{2s}(C \times C) \;\le\; c\,\mathcal{H}^s(C) \times \mathcal{H}^s(C)$$

if C is the Cantor set in $\boxed{1}$.

A better result holds if $s = n$

Theorem 4. *For any $B \subset \mathbb{R}^m$ with $\mathcal{H}^t(B) < \infty$ there exists a number c such that*

$$\omega_n^{-1} \;\le\; c\,\omega_t\,\omega_{t+n}^{-1} \;\le\; 2^{-n}\,(n+1)^{(t+n)/2}\,,$$

and

$$\mathcal{H}^{n+t}(A \times B) \;=\; c\,\mathcal{H}^n(A)\mathcal{H}^t(B)$$

for every Lebesgue measurable set A.

In general the constant c in Theorem 4 is different from 1 but it is 1 if B has good tangential properties, i.e. if B is t–rectifiable, compare Theorem 3 in Sec. 2.1.5.

1.4 Lebesgue's, Radon-Nikodym's and Riesz's Theorems

In this subsection we shall assume that X is a locally compact separable metric space. We shall denote by $\mathcal{B}_c(X)$ the collection of all relatively compact Borel sets of X. If X is not compact the collection $\mathcal{B}_c(X)$ is not a σ-algebra, it is only a *local σ-algebra* in the sense that

$$\mathcal{B}_c(X) \llcorner A \;:=\; \{B \in \mathcal{B}_c(X) \mid B \subset A\}$$

is a σ-algebra in A for each $A \in \mathcal{B}_c(X)$.

A set function $\nu : \mathcal{B}_c(X) \to \mathbb{R}$ is called a *σ-additive measure* if

$$\nu\left(\bigcup_{k=1}^{\infty} A_k\right) \;=\; \sum_{k=1}^{\infty} \nu(A_k)$$

for all sequences $\{A_k\}$ of disjoint sets in $\mathcal{B}_c(X)$ such that $\cup_{k=1}^{\infty} A_k \in \mathcal{B}_c(X)$, i.e., if ν restricted over $\mathcal{B}_c(X) \llcorner A$ is σ-additive on $\mathcal{B}_c(X) \llcorner A$ for all $A \in \mathcal{B}_c(X)$.

For any $f : X \to [-\infty, +\infty]$ and $A \subset X$ we set

$$f \llcorner A := \begin{cases} f & \text{in } A \\ 0 & \text{in } X \setminus A \end{cases}$$

and

$$\int_A f \, d\mu := \int_X f \llcorner A \, d\mu \,.$$

Let μ be a Radon measure over X. We denote by $L^1_{\text{loc}}(X;\mu)$ the collection of all locally μ-summable functions $f : X \to [-\infty, +\infty]$, i.e.,

$$L^1_{\text{loc}}(X;\mu) := \{f \mid f \llcorner K \in L^1(X;\mu) \;\; \forall K \subset X \,,\, K \text{ compact}\} \,.$$

For any $f \in L^1_{\text{loc}}(X;\mu)$, the expression

(1) $$(\mu \llcorner f)(A) := \int_A f \, d\mu \qquad A \in \mathcal{B}_c(X)$$

defines a σ-additive measure $\mu \llcorner f$.

Definition 1. *Let μ and ν be two Radon measures over X. We say that ν is* absolutely continuous with respect to μ, *and write $\nu \ll \mu$, if $\nu(A) = 0$ for every A for which $\mu(A) = 0$.*

The measures μ and ν are called mutually singular, *and we write $\nu \perp \mu$, if there exists a Borel set $A \subset X$ such that $\mu \llcorner (X \setminus A) = 0$ and $\nu \llcorner A = 0$.*

One of the most important result of measure theory is contained in the next theorem.

Theorem 1. *Let μ and ν be two Radon measures over X*

(i) **Lebesgue's decomposition theorem.** *There exist unique Radon measures $\nu^{(a)}$ and $\nu^{(s)}$ on X such that*

$$\begin{gathered} \nu = \nu^{(a)} + \nu^{(s)} \\ \nu^{(s)} \perp \nu^{(a)} \,, \quad \nu^{(s)} \perp \mu \,, \quad \nu^{(a)} \ll \mu \,. \end{gathered}$$

$\nu^{(a)}$ is called the absolutely continuous part of ν with respect to μ, and $\nu^{(s)}$ the singular part of ν with respect to μ.

(ii) **Radon-Nikodym's theorem.** *If $\nu \ll \mu$, then there exists a unique Borel function f in $L^1_{\text{loc}}(X;\mu)$ such that $\nu = \mu \llcorner f$, i.e.,*

$$\nu(A) = \int_A f \, d\mu \qquad \forall A \in \mathcal{B}_c(X) \,.$$

The function f is called the Radon-Nikodym derivative of ν with respect to μ and is denoted also by $\frac{d\nu}{d\mu}$.

Obviously, we can write the claim of Theorem 1, as

$$\nu = \mu \llcorner \frac{d\nu^{(a)}}{d\mu} + \nu^{(s)} \tag{2}$$

or, equivalently, *for any Borel function f with compact support*

$$\int f\,d\nu = \int f \frac{d\nu^{(a)}}{d\mu}\,d\mu + \int f\,d\nu^{(s)} . \tag{3}$$

Moreover, *we have for $A \in \mathcal{B}_c(X)$*

$$\nu^{(a)}(A) := \inf\{\nu(B) \mid B \in \mathcal{B}_c(X)\ ,\ \mu(A \setminus B) = 0\} . \tag{4}$$

Definition 2. *We say that the set function $\nu : \mathcal{B}_c(X) \to \mathbb{R}$ is a* signed measure *if there exists a Radon measure μ on X and $f \in L^1_{\text{loc}}(X;\mu)$ such that*

$$\nu = \mu \llcorner f \qquad \textit{on } \mathcal{B}_c(X) .$$

The set function $\nu : \mathcal{B}_c(X) \to \mathbb{R}^m$, $m \geq 1$ is called a vector valued measure *if there exists a Radon measure μ and a vector valued function $f = (f^1, \dots, f^m) \in L^1_{\text{loc}}(X, \mathbb{R}^m;\mu)$, i.e., with $f^k \in L^1(X;\mu)$ for $k = 1, \dots, m$, such that $\nu = \mu \llcorner f$, i.e.,*

$$(\nu^1, \dots, \nu^m) = (\mu \llcorner f^1, \dots, \mu \llcorner f^m)$$

We define the *total variation* $|\nu| : \mathcal{B}_c(X) \to [0, +\infty)$ of the vector valued measure $\nu : \mathcal{B}_c(X) \to \mathbb{R}^m$, $\nu = \mu \llcorner f$, as

$$|\nu| := \mu \llcorner |f|$$

where $|f| = (\sum_k |f^k|^2)^{1/2}$.

We define the *positive* and *negative parts* ν^+ and ν^- of a signed measure $\nu = \mu \llcorner f$, $f : X \to \mathbb{R}$, as

$$\nu^+ = \mu \llcorner f^+ , \qquad \nu^- = \mu \llcorner f^-$$

where $f^+ = \max(f, 0)$, $f^- = \min(f, 0)$.

The collection of Radon, signed, and vector valued measures on X will be denoted respectively by $\mathcal{M}^+(X)$, $\mathcal{M}(X)$, and $\mathcal{M}(X, \mathbb{R}^m)$. A few remarks are now in order.

Trivially we have $\mu(\{x \mid f(x) \neq 0\}) = 0$ if $(\mu \llcorner f)(A) = 0$ for all A. We also remark that $|\nu|$, ν^+ and ν^- *do not depend on the decomposition* $\nu = \mu \llcorner f$. For instance, if $\nu = \mu_1 \llcorner f_1 = \mu_2 \llcorner f_2$, we have $\mu_1, \mu_2 \ll \mu := \mu_1 + \mu_2$. Hence by Radon-Nikodym's theorem we have, for $k = 1, 2$,

$$\mu_k = \mu \llcorner \frac{d\mu_k}{d\mu} , \quad \mu_k \llcorner f_k = (\mu \llcorner \frac{d\mu_k}{d\mu}) \llcorner f_k = \mu \llcorner (f_k \frac{d\mu_k}{d\mu})$$

Thus $f_1 \frac{d\mu_1}{d\mu} = f_2 \frac{d\mu_2}{d\mu} =: f$ μ-a.e. in X and

$$\mu_k \llcorner |f_k| = \mu \llcorner (|f_k| \frac{d\mu_k}{d\mu}) = \mu \llcorner |f| \,.$$

The same argument yields the result for ν^+ and ν^-.

If ν is a signed measured, we obviously have $\nu = \nu^+ - \nu^-$, $|\nu| = \nu^+ + \nu^-$ thus

$$\nu^+ \,,\, \nu^- \ll |\nu.|$$

Therefore if we define

$$\frac{d\nu}{d|\nu|} := \frac{d\nu^+}{d|\nu|} - \frac{d\nu^-}{d|\nu|} \,,$$

then we have

$$\nu = |\nu| \llcorner \frac{d\nu}{d|\nu|}$$

and by definition

$$\left(\frac{d\nu}{d|\nu|}\right)^+ = \frac{d\nu^+}{d|\nu|} \,, \qquad \left(\frac{d\nu}{d|\nu|}\right)^- = \frac{d\nu^-}{d|\nu|}$$

If ν^k, $k = 1, \dots, m$, are signed measures, $\nu^k = \mu_k \llcorner f_k$, defining $\mu := \sum_k \mu_k$ we find

$$\nu^k = \mu \llcorner (f_k \frac{d\mu_k}{d\mu});$$

thus we conclude

Proposition 1. *If $\nu^1, \dots, \nu^m$ are signed measures, then $\nu = (\nu^1, \dots, \nu^m)$ is a vector valued measure in the sense of Definition 2. Therefore $\mu = (\mu^1, \dots, \mu^k)$ is a vector valued measure if and only if each μ^k is a signed measure.*

We also have

Proposition 2. *The total variation $|\nu|$ of a vector valued measure ν can by equivalently defined as*

$$|\nu|(A) = \sup\{\sum_{k=1}^{\infty} |\nu(A_k)| \;\mid\; A_k \in \mathcal{B}_c(X) \,,\, A_k \text{ disjoint} \,,\, \cup_k A_k \subset A\} \,,$$

for $A \in \mathcal{B}_c(X)$.

If ν is a signed measure, we have, for $A \in \mathcal{B}_c(X)$:

$$\nu^{\pm}(A) = \sup\{\pm\nu(B) \mid B \subset A,\; B \in \mathcal{B}_c(X)\}$$

$$|\nu|(A) = \sup\{|\nu(A_1)| + |\nu(A_2)| \mid A_1 \cap A_2 = \emptyset,\, A_1 \cup A_2 \subset A,\, A_1, A_2 \in \mathcal{B}_c(X)\}.$$

Let $\nu = (\nu^1, \dots, \nu^m)$ and μ be respectively a vector valued measure and a Radon measure on X. We say that ν is absolutely continuous with respect to μ, $\nu \ll \mu$, if

$$|\nu| \ll \mu .$$

or equivalently if $|\nu^i| \ll \mu$ for all i. For two vector valued measures ν and λ we define

$$\nu \perp \lambda \quad \text{iff} \quad |\nu| \perp |\lambda| .$$

Then Theorem 1 readily extends to vector valued measures. In particular we see that, if μ is a Radon measure on X, the set of all vector valued measures $\nu \in \mathcal{M}(X;\mathbb{R}^m)$ such that $\nu \ll \mu$ is equal to $\{\mu \llcorner f \mid f \in L^1_{\text{loc}}(X, \mathbb{R}^m; \mu)\}$. Since by definition $\nu \ll |\nu|$, we also get

Proposition 3. *Let ν be a vector valued measure. Then $\nu = |\nu| \llcorner \vec{\nu}$, where*

$$\vec{\nu} := \frac{d\nu}{d|\nu|} \in L^1_{\text{loc}}(X, \mathbb{R}^m; |\nu|)$$

and $|\vec{\nu}| = 1$ $|\nu|$-a.e. in X.

Given a vector valued measure $\nu = (\nu^1, \dots, \nu^m)$ and a vector valued function $g = (g^1, \dots, g^m)$ in $L^1(X, \mathbb{R}^m; |\nu|)$, we define the integral of g with respect to ν by

$$\int <g, d\nu> := \sum_k \int g^k \, d\nu^k$$

where

$$\int g_k \, d\nu^k := \int g^k \, d\nu^{k+} - \int g^k \, d\nu^{k-} .$$

By Proposition 3 we then get

$$\int <f, d\nu> = \int <f, \vec{\nu}> d|\nu| \tag{5}$$

Denoting by $C_c(X, \mathbb{R}^m)$ the space of all compactly supported and continuous functions on X with values in $\mathbb{R}^m$, evidently

$$\lambda_\nu : f \in C_c(X, \mathbb{R}^m) \to \int <f, \vec{\nu}> d|\nu|$$

defines a linear functional on $C_c(X, \mathbb{R}^m)$ for every $\nu \in \mathcal{M}(X, \mathbb{R}^m)$. Next theorem roughly inverts this observation.

Theorem 2 (Riesz). *We have*

(i) *Let $\lambda : C_c(X, \mathbb{R}^m) \to \mathbb{R}$ be a linear functional which is continuous in the sense that it has locally finite norm, i.e.,*

$$\sup\{\lambda(f) \mid f \in C_c(X, \mathbb{R}^m) \, , \, |f| \le 1 \, , \, \text{spt } f \subset K\} < \infty \tag{6}$$

for each compact $K \subset X$. *Then there exists a unique vector valued measure* $\nu = |\nu| \llcorner \vec{\nu} \in \mathcal{M}(X, \mathbb{R}^m)$ *such that*

$$\lambda(f) = \int <f, d\nu> = \int <f, \vec{\nu}> d|\nu| \qquad \forall f \in C_c(X, \mathbb{R}^m)$$

(ii) *Let* $\lambda : C_c(X, \mathbb{R}) \to \mathbb{R}$ *be a linear functional which is positive, i.e.,* $\lambda(f) \geq 0$ *whenever* $f \geq 0$. *Then there exists a Radon measure* μ *on* X, $\mu \in \mathcal{M}^+(X)$, *such that*

$$\lambda(f) = \int f \, d\mu \qquad \forall f \in C_c(X)$$

Notice that in (ii) there is no condition on the norm of λ, the positivity of λ suffices. This expresses the well-known fact that *every positive distribution is a measure.* A consequence of Theorem 2 is that $\mathcal{M}(X, \mathbb{R}^m)$ *is the dual space of* $C_c(X, \mathbb{R}^m)$. We moreover remark that *a distribution* $\lambda : C_c^\infty(X, \mathbb{R}^m) \to \mathbb{R}$ *is a measure (or better extends to a measure) if and only if we have*

$$\sup\{\lambda(f) \mid f \in C_c^\infty(X, \mathbb{R}^m) \, , \, |f| \leq 1 \, , \operatorname{spt} f \subset K\} < \infty$$

for all compact $K \subset X$. In this case it is common to identify λ with μ and write $\lambda(A)$ for $\int_A d\mu$.

We also readily deduce that, for all open sets $\mathcal{U}$ in $\mathcal{B}_c(X)$, we have

$$(7) \qquad |\nu|(\mathcal{U}) = \sup\{\int <f, d\nu> \mid f \in C_c(X, \mathbb{R}^m), \, |f| \leq 1, \, \operatorname{spt} \, f \subset \mathcal{U}\} \, .$$

This can in fact be used as an equivalent definition of $|\nu|$. Given $\nu \in \mathcal{M}(X, \mathbb{R}^m)$ and an open set $\mathcal{U}$ in X, one defines the *mass of* ν *in* $\mathcal{U}$ by

$$(8) \qquad \mathbf{M}_{\mathcal{U}}(\nu) := \sup\{\int <f, d\nu> \mid f \in C_c(X, \mathbb{R}^m), \, |f| \leq 1, \, \operatorname{spt} f \subset \mathcal{U}\}$$

and one simply writes $\mathbf{M}(\nu)$ for $\mathbf{M}_X(\nu)$. Of course, if we extend the set function

$$\mathbf{M}(\nu) : \mathcal{U} \to \mathbf{M}_{\mathcal{U}}(\nu) \qquad \mathcal{U} \text{ open}$$

to the Radon measure on X given by

$$\mathbf{M}_A(\nu) := \inf\{\mathbf{M}_{\mathcal{U}}(\nu) \mid A \subset \mathcal{U} \, , \, \mathcal{U} \text{ open}\} \, ,$$

we find

$$\mathbf{M}_A(\nu) = |\nu|(A) \, .$$

Actually, we should call $\mathbf{M}_{\mathcal{U}}(\nu)$ in (8) the *Euclidean mass*, because we used the Euclidean norm in $|f| \leq 1$. If we use another norm $\| \cdot \|$ in $\mathbb{R}^m$, the resulting mass $\mathbf{M}'_{\mathcal{U}}(\nu)$ can be expressed as

$$\mathbf{M}'_{\mathcal{U}}(\nu) = \| \nu \|'(\mathcal{U}) \, , \qquad \| \nu \|'(A) := \int \| \frac{d\nu}{d|\nu|} \|' \, d|\nu|$$

where $\|\cdot\|'$ is the dual norm of $\|\cdot\|$, i.e.,

$$\|y\|' := \sup\{<y,x> \mid \|x\| \leq 1\,,\ x \in \mathbb{R}^m\}\,.$$

Finally we remark that any linear functional $\lambda : C_c(X, \mathbb{R}^m) \to \mathbb{R}$ satisfying (6), being a vector valued measure $\lambda = |\nu| \llcorner \vec{\nu}$, $\vec{\nu} = 1$ $|\nu|$–a.e., extends to a linear functional on the space $L^1(X, \mathbb{R}^m; |\nu|)$. In particular if $X = \cup_k F_k$, F_k compact and $|\nu|(X) = \mathbf{M}(\nu) \leq \infty$, then constant functions are ν–summable and λ extends to all Borel bounded functions $X \to \mathbb{R}^m$. Such an extension can be computed by Lebesgue's theorem.

1.5 Covering Theorems, Differentiation and Densities

A classical theorem of Vitali, actually in dimension $n = 1$, states the following

Theorem 1 (Vitali). *Let E be a subset of $\mathbb{R}^n$ which is covered by a collection of closed balls $\mathcal{B}$ with the property that for each $x \in E$ and $\varepsilon > 0$ there exists $B \in \mathcal{B}$ such that $x \in B$ and* diam $B < \varepsilon$. *Then we can choose a disjoint sequence $\{B_k\}$ of balls in $\mathcal{B}$ such that*

$$\text{meas}\ (E - \cup_k B_k) = 0\,.$$

It turns out that the property of covering a set in the sense of the previous theorem is the key point for *differentiating* a measure ν with respect to another measure μ. To state precisely such a theorem we need a few preliminaries.

In this subsection X will denote , if not otherwise states, a locally compact separable metric space. We denote, as usual, the ball of radius r and center $x \in X$ by $B(x,r)$.

Given a collection $\mathcal{B}$ of balls in X, the set of the centers of balls in $\mathcal{B}$ is denoted by $C(\mathcal{B})$

$$C(\mathcal{B}) := \{x \in X \mid \exists r > 0 \text{ so that } B(x,r) \in \mathcal{B}\}\,.$$

The collection $\mathcal{B}$ covers a set $A \subset X$ if $A \subset \cup_{B \in \mathcal{B}} B$.

Definition 1. *We say that $\mathcal{B}$* covers $A \subset X$ finely *if for every $x \in A$ and $\varepsilon > 0$ there exists $B(x,r) \in \mathcal{B}$ with $r < \varepsilon$, i.e., if for all $x \in A$*

$$\inf\{r > 0 \mid B(x,r) \in \mathcal{B}\} = 0\,.$$

Definition 2. *Let μ be a Radon measure over X. We say that X has* the symmetric Vitali property relative to μ *if for every collection $\mathcal{B}$ of balls in X satisfying*

(i) $\mathcal{B}$ *covers finely the set of its centers* $C(\mathcal{B})$
(ii) $\mu(C(\mathcal{B})) < +\infty$

there exists a sub-collection $\mathcal{B}'$ of $\mathcal{B}$ satisfying

(i) *the elements of $\mathcal{B}'$ are disjoint*
(ii) *$\mathcal{B}'$ covers μ-almost all of $C(\mathcal{B})$, i.e.,*

$$\mu\,(C(\mathcal{B}) \setminus \cup_{B\in\mathcal{B}'} B) \;=\; 0\;.$$

Next theorem provide us with two important cases in which X has the symmetric Vitali property relative to μ.

Theorem 2. *We have*

(i) *Let μ be a Radon measure over X. If there exists a constant c such that $\mu(B(x,5r)) \le c\mu(B(x,r))$ for all balls $B(x,r)$ in X, then X has the symmetric Vitali property relative to μ.*
(ii) *The Euclidean space $X = \mathbb{R}^n$ has the symmetric Vitali property relative to μ for any Radon measure μ on $\mathbb{R}^n$.*

The proof of Theorem 2 uses the following

Lemma 1 (Vitali's covering theorem). *Let $\mathcal{B}$ be a collection of closed balls in X with equi-bounded radii, i.e., $\sup\{r \mid B(x,r) \in \mathcal{B}\} < \infty$. Then there exists a sub-collection $\mathcal{B}'$ of $\mathcal{B}$ such that*

(i) *the elements of $\mathcal{B}'$ are disjoint, therefore $\mathcal{B}'$ is numerable*
(ii) *for all $B \in \mathcal{B}$ there exists a ball $B' \in \mathcal{B}'$ such that $B \cap B' \neq \emptyset$ and $\hat{B}' \supset B$, where for $B = B(x,r)$ we have set $\hat{B} = B(x,5r)$.*

In particular

$$\bigcup_{B\in\mathcal{B}} B \subset \bigcup_{B\in\mathcal{B}'} \hat{B}$$

From Lemma 1 one deduces that, if A is covered finely by $\mathcal{B}$, then one has

$$A \setminus \bigcup\{B \mid B \in \mathcal{B}''\} \subset \bigcup\{\hat{B} \mid B \in \mathcal{B}' \setminus \mathcal{B}''\}$$

for each finite sub-collection $\mathcal{B}''$ of $\mathcal{B}'$, and this yields Theorem 2 (i).

The proof of Theorem 2 (ii) uses instead the following

Lemma 2 (Besicovitch's covering theorem). *Let E be a subset of $\mathbb{R}^n$ and let $r : E \to \mathbb{R}$ be a positive bounded function defined on E (for instance, let $E = C(\mathcal{B})$ where $\mathcal{B}$ is a collection of closed balls with center in each point of E and radius $r(x)$). Then one can choose an at most countable family of points $\Lambda := \{x_i\}_{i\in\mathbb{N}}$ in E such that*

(i) *$E \subset \cup_i \overline{B(x_i, r(x_i))}$*
(ii) *The balls $\overline{B(x_i, \frac{1}{3}r(x_i))}$, $x_i \in \Lambda$, are mutually disjoint*
(iii) *Moreover, the balls $B(x_i, r(x_i)), x_i \in \Lambda$, can be distributed in $\xi(n)$ families $\mathcal{B}_k$ of disjoint closed balls, where $\xi(n)$ is a constant depending only on n*
(iv) *No point of E belongs to more than $\xi(n)$ balls of the family $\{\overline{B(x_i, r(x_i))} \mid x_i \in \Lambda\}$*

We notice that Lemma 1 with $X = \mathbb{R}^n$ is a consequence of Lemma 2. Later we shall refer to Lemma 1 and Lemma 2 as to Besicovitch's covering theorem.

Let now μ and ν be Radon measures over X. For $x \in X$ we set

$$\bar{D}_\mu \nu(x) := \limsup_{r \to 0} \frac{\nu(B(x,r))}{\mu(B(x,r))}$$

$$\underline{D}_\mu \nu(x) := \liminf_{r \to 0} \frac{\nu(B(x,r))}{\mu(B(x,r))}$$

whenever $\mu(B(x,r)) > 0$ for all $r > 0$, otherwise, i. e., if $\mu(B(x,r)) = 0$ for some $r > 0$, we set

$$\bar{D}_\mu \nu(x) = \underline{D}_\mu \nu(x) = +\infty .$$

If $\bar{D}_\mu \nu(x) = \underline{D}_\mu \nu(x) \in [0, +\infty]$, we simply write $D_\mu \nu(x)$ for $\bar{D}_\mu \nu(x) = \underline{D}_\mu \nu(x)$.

Definition 3. *We say that ν is* differentiable with respect to μ at x *if and only if*

$$\bar{D}_\mu \nu(x) = \underline{D}_\mu \nu(x) \in [0, +\infty] .$$

In this case, $D_\mu \nu$ is called the (symmetric) derivative of ν with respect to μ *or the* density of ν with respect to μ.

The basic facts concerning the differentiation of a Radon measure ν with respect to a Radon measure μ are collected in the next theorem.

Theorem 3 (Radon-Nikodym). *Let μ and ν be two Radon measures over X. Suppose that X has the symmetric Vitali property with respect to μ. Then*

(i) $D_\mu \nu(x)$ exists for μ-a.e. $x \in X$ and the function $D_\mu \nu(x)$ is μ-measurable.

(ii) There exists a Borel set Z with $\mu(Z) = 0$ such that

$$\nu(A) = \int_A D_\mu \nu \, d\mu + (\nu \llcorner Z)(A)$$

holds for all Borel sets $A \subset X$.

(iii) Taking into account Radon-Nikodym's theorem and Lebesgue's decomposition theorem in Sec. 1.1.4 we have

$$\nu^{(a)} = \mu \llcorner D_\mu \nu , \quad \nu^{(s)} = \nu \llcorner Z$$

thus

$$D_\mu \nu = \frac{d\nu^{(a)}}{d\mu} \qquad \mu\text{-a.e. in } X .$$

Moreover, if ν is absolutely continuous with respect to μ, we may take $Z = \emptyset$ and we have

$$D_\mu \nu = \frac{d\nu}{d\mu} \quad \mu\text{ -a.e. in } X .$$

(iv) *If X has also the symmetric Vitali property with respect to ν, then $D_\mu\nu(x)$ exists also for ν-a.e. $x \in X$ and we may take*

$$Z = \{x \in X \mid D_\mu\nu(x) = +\infty\} .$$

In view of Theorem 2 (ii), the claims (i) ... (iv) of Theorem 3 hold if $X = \mathbb{R}^n$.

The proof of Theorem 3 is based on the following lemma whose proof uses in turn Vitali's property.

Lemma 3. *Let A be any subset of X.*

(i) *If $\underline{D}_\mu\nu(x) \leq \alpha$ in A, then $\nu(A) \leq \alpha\mu(A)$*
(ii) *If $\bar{D}_\mu\nu(x) \geq \alpha$ in A, then $\nu(A) \geq \alpha\mu(A)$*

A simple consequence of Theorem 3 is the following theorem to which we shall return in Sec. 3.1.1.

Theorem 4 (Lebesgue-Besicovitch's differentiation theorem). *Let μ be a Radon measure over $\mathbb{R}^n$ and let f be a function in $L^p_{\text{loc}}(\mathbb{R}^n;\mu)$ with $p \geq 1$. Then*

$$\lim_{r\to 0} \frac{1}{\mu(B(x,r))} \int_{B(x,r)} f\, d\mu = f(x) \quad \textit{for } \mu\textit{-a.e. } x$$

and even

$$\lim_{r\to 0} \frac{1}{\mu(B(x,r))} \int_{B(x,r)} |f - f(x)|^p\, d\mu = 0 \quad \textit{for } \mu\textit{-a.e. } x$$

If we apply Theorem 4 to the characteristic function χ_E of a μ–measurable subset E of $\mathbb{R}^n$, we obtain that

$$\lim_{r\to 0} \frac{\mu(B(x,r)\cap E)}{\mu(B(x,r))} = \lim_{r\to 0} \frac{(\mu \llcorner E)(B(x,r))}{\mu(B(x,r))} = \frac{d(\mu \llcorner E)}{d\mu}(x)$$

equals 1 for μ-a.e. $x \in E$ and equals 0 for μ-a.e. $x \in \mathbb{R}^n \setminus E$. The sets

$$Int_\mu(E) := \{x \mid \frac{d(\mu \llcorner E)}{d\mu}(x) = 1\}$$

$$Ext_\mu(E) := \{x \mid \frac{d(\mu \llcorner E)}{d\mu}(x) = 0\}$$

might be referred as to the *measure-theoretic interior* and *exterior of E with respect to μ* and we may refer to

$$\partial_\mu(E) := \mathbb{R}^n \setminus (Int\ (E) \cup Ext\ (E))$$

as to the *measure-theoretic boundary of E*.

Of particular relevance is the case in which μ is the Lebesgue measure in $\mathbb{R}^n$, or equivalently the n-dimensional Hausdorff measure in $\mathbb{R}^n$. We then set

Definition 4. *Let E be a Lebesgue measurable set in $\mathbb{R}^n$. The* upper *and* lower densities *of E at $x \in E$ are defined by*

$$\begin{aligned}\theta^*(E,x) &:= \limsup_{r\to 0} \frac{\mathcal{H}^n(E\cap B(x,r))}{\mathcal{H}^n(B(x,r))} &= \bar{D}_{\mathcal{H}^n}(\mathcal{H}^n \llcorner E)(x)\\ \theta_*(E,x) &:= \liminf_{r\to 0} \frac{\mathcal{H}^n(E\cap B(x,r))}{\mathcal{H}^n(B(x,r))} &= \underline{D}_{\mathcal{H}^n}(\mathcal{H}^n \llcorner E)(x)\ ,\end{aligned}$$

and the density *of E at x by $\theta(E,x) = \theta^*(E,x) = \theta_*(E,x)$ if the upper and lower densities are equal at x.*

Of course $\mathcal{H}^n$-a.e. $x \in E$ is a point of density 1 and $\mathcal{H}^n$-a.e. $x \in \mathbb{R}^n \setminus E$ is a point of density 0 for E.

In terms of densities we can recover interesting relations between measurable and continuous functions

Definition 5. *Let $f : \mathbb{R}^n \to \mathbb{R}$ be a Lebesgue measurable function. The* approximate upper, *respectively* lower, limit of f at x *is defined as*

$$\operatorname*{aplimsup}_{y\to x} f(y) := \inf\{t\in\mathbb{R} \mid \theta^*(\{z\in\mathbb{R}^n \mid f(z)>t\},x)=0\}$$

respectively

$$\operatorname*{apliminf}_{y\to x} f(y) := \sup\{t\in\mathbb{R} \mid \theta^*(\{z\in\mathbb{R}^n \mid f(z)<t\},x)=0\}\ .$$

We speak of approximate limit of f at x *in case*

$$\operatorname*{aplim}_{y\to x} f(y) := \operatorname*{aplimsup}_{y\to x} f(y) = \operatorname*{apliminf}_{y\to x} f(y)\ ,$$

and we say f to be approximately continuous at x *if*

$$\operatorname*{aplim}_{y\to x} f(y) = f(x)$$

We then have, compare Sec. 3.1.4

(i) f is approximately continuous at x if for every open set $\mathcal{U} \subset \mathbb{R}$ with $f(x) \in \mathcal{U}$ we have
$$\theta(f^{-1}(\mathcal{U}),x) = 1$$
(ii) f is approximately continuous at x if there exists a measurable set E with $x \in E$ such that $\theta(E,x) = 1$ and $f_{|E}$ is continuous at x,

(iii) if $f \in L^1_{\text{loc}}(\mathbb{R}^n;\mu)$ and $\operatorname{aplim}_{y\to x} f(y) = f(x)$, then

$$\lim_{r\to 0} \frac{1}{\mathcal{L}^n(B(x,r))} \int_{B(x,r)} |f(y)-f(x)|\,d\mu = 0\ .$$

and finally

Proposition 1. *Let $f : \mathbb{R}^n \to \mathbb{R}$ be a function which is a.e. defined in $\mathbb{R}^n$. Then f is Lebesgue measurable if and only if it is approximately continuous at a.e. $x \in \mathbb{R}^n$.*

The previous results do not apply, at least directly, to lower dimensional Hausdorff measures, as $\mathcal{H}^s$ is not a Radon measure in $\mathbb{R}^n$ for $0 \le s < n$. For this reason one introduces lower-dimensional densities which in some sense extend the notion of Radon-Nikodym derivative to Hausdorff measures.

Definition 6. *Let μ be a Borel measure over a metric space X. For any subset $A \subset X$, and any point $x \in X$, we define the* n-dimensional upper *and* lower densities *by*

$$\begin{aligned} \theta^{n*}(\mu, A, x) &:= \limsup_{r \to 0} \frac{\mu(A \cap \overline{B(x,r)})}{\omega_n r^n} \\ \theta^{n}_{*}(\mu, A, x) &:= \liminf_{r \to 0} \frac{\mu(A \cap \overline{B(x,r)})}{\omega_n r^n} \end{aligned}$$

In case $A = X$ we simply write

$$\theta^{n*}(\mu, x) := \theta^{n*}(\mu, X, x) \,, \qquad \theta^{n}_{*}(\mu, x) := \theta^{n}_{*}(\mu, X, x)$$

In case $\theta^{n*}(\mu, A, x) = \theta^{n}_{*}(\mu, A, x)$ we write $\theta^{n}(\mu, A, x)$ for the common value.

Clearly we have

$$\theta^{n*}(\mu, A, x) = \theta^{n*}(\mu \llcorner A, x) \,, \qquad \theta^{n}_{*}(\mu, A, x) = \theta^{n}_{*}(\mu \llcorner A, x) \,,$$

and, since $\mu(A \cap \overline{B(x,r)})$ is an upper semicontinuous function in x for r fixed, *$\theta^{n*}(\mu, A, x)$ and $\theta^{n}_{*}(\mu, A, x)$ are both Borel functions.* Notice moreover that the same definition of densities can be used for any measure μ.

Next theorem provides useful connections between μ and $\mathcal{H}^n$ based on information about the upper density of μ.

Theorem 5. *Let μ be a Borel regular measure on a metric space X and let $t \ge 0$.*

(i) If $A \subset B \subset X$ and $\theta^{n*}(\mu, B, x) \ge t$ for all $x \in A$, then $\mu(B) \ge t\mathcal{H}^n(A)$. The case $A = B$ is important, we have: if $\theta^{n*}(\mu, A, x) \ge t$ for all $x \in A$, then $\mu(A) \ge t\mathcal{H}^n(A)$.
(ii) If $A \subset X$ and $\theta^{n*}(\mu, A, x) \le t$ for all $x \in A$, then $\mu(A) \le 2^n t\mathcal{H}^n(A)$.

A consequence of Theorem 5 is

Theorem 6. *Let μ be a Borel regular measure on a metric space X, for instance let μ be the n-dimensional Hausdorff measure $\mathcal{H}^n$. If A is a μ-measurable subset of X with $\mu(A) < \infty$, then $\theta^{n*}(\mu, A, x) = 0$ for $\mathcal{H}^n$-a.e. $x \in X \setminus A$.*

Finally we have the following important bounds for densities with respect to Hausdorff measure

Theorem 7. *Let A be any subset of X.*

(i) If $\mathcal{H}^n(A) < \infty$, then $\theta^{n*}(\mathcal{H}^n, A, x) \leq 1$ for $\mathcal{H}^n$-a.e. $x \in A$.
(ii) If $\mathcal{H}^n_\delta(A) < \infty$ for each $\delta > 0$, then $\theta^{n*}(\mathcal{H}^n_\infty, A, x) \geq 2^{-n}$ for $\mathcal{H}^n$-a.e. $x \in A$.

Since $\mathcal{H}^n \geq \mathcal{H}^n_\delta \geq \mathcal{H}^n_\infty$, we deduce in particular

$$2^{-n} \leq \theta^{n*}(\mathcal{H}^n, A, x) \leq 1 \qquad \text{for } \mathcal{H}^n \text{-a.e. } x \in A$$

if $\mathcal{H}^n(A) < \infty$.

Even in the case that X is an Euclidean space we cannot infer in general that the n-dimensional density of A is 1 at $\mathcal{H}^n$ – *a.e.* point of A. This in fact depends upon the rectifiability of A, compare Theorem 5 in Sec. 2.1.4

Remark 1. Consider the measures $\mathcal{H}^k$ and $\mathcal{H}^k_\infty$, $k \leq n$, in $\mathbb{R}^n$. Although they can be very different, it is easily seen that $\mathcal{H}^k(A) = 0$ if and only if $\mathcal{H}^k_\infty(A) = 0$. Moreover, as we have seen,

$$\limsup_{r \to 0} r^{-k} \mathcal{H}^k_\infty(\Sigma \cap B_r) \geq 2^{-k}\omega_k$$

for $\mathcal{H}^k$-a.e. $x \in \Sigma$, and one can show that, if Q, Q_ν $\nu = 1, 2, \ldots$ are compact sets and if every open set $A \supset Q$ contains Q_ν for ν sufficiently large, then

$$\mathcal{H}^k_\infty(Q) \geq \limsup_{\nu \to \infty} \mathcal{H}^k_\infty(Q_\nu) .$$

Notice that the last inequality is in general false for the Hausdorff measure $\mathcal{H}^k$. The properties of $\mathcal{H}^k_\infty$ illustrated above sometimes make the use of $\mathcal{H}^k_\infty$ more convenient than the use of $\mathcal{H}^k$.

A consequence of the differentiation theory of measures is the following theorem

Theorem 8 (Disintegration of measures). *Let μ be a Radon measure on $X \times Y$. Then there exists a measure λ on X and, for λ-a.e. $x \in X$, a measure ν_x on Y such that for any $M \subset X \times Y$ we have*

$$\mu(M) = \int_X \nu_x(M_x)\, d\lambda(x)$$

where $M_x := \{y \in Y \mid (x, y) \in M\}$. For any μ-summable function f on $X \times Y$, the function $f(x, \cdot)$ is ν_x-summable for λ-a.e. $x \in X$ and

$$\int_{X \times Y} f\, d\mu = \int_X \left(\int_Y f(x, y)\, d\nu_x(y) \right) d\lambda(x)$$

Moreover, if $\mu(X \times Y) < \infty$, we can take $\lambda = \pi_{\#}\mu$ where π is the linear projection $\pi(x, y) := x$. In this case ν_x turns out to be a probability measure for λ-a.e. x, i.e., $\nu_x(X) = 1$.

2 Weak Convergence

Due to its relevance for the sequel, in this section we would like to discuss weak convergence of functions and measures, with the aim of illustrating some of its main features.

2.1 Weak Convergence of Vector Valued Measures

Let X be a locally compact separable metric space.

Definition 1. *Let $\nu, \nu_k \in M(X, \mathbb{R}^n)$ be vector valued measures over X, $k = 1, 2, \ldots$*

(i) We say that ν_k converge weakly *to ν, $\nu_k \rightharpoonup \nu$, iff*

$$\int <\varphi, d\nu_k> \longrightarrow \int <\varphi, d\nu> \qquad \forall\, \varphi \in C_c(X, \mathbb{R}^n)$$

(ii) We say that ν_k converge to ν in the mass norm *, $\nu_k \to \nu$, iff $\mathbf{M}(\nu_k - \nu) \to 0$; ν_k converge locally in the mass norm, $\nu_k \overset{\text{loc}}{\to} \nu$, iff*

$$\mathbf{M}_{\mathcal{U}}(\nu_k - \nu) \to 0 \qquad \forall\, \mathcal{U} \subset\subset X \ , \ \mathcal{U} \text{ open.}$$

(iii) We say that ν_k converge to ν weakly with the mass *iff $\nu_k \rightharpoonup \nu$ and $\mathbf{M}(\nu_k) \to \mathbf{M}(\nu) < \infty$.*

We have

Theorem 1. *Let $\{\nu_k\}$ be a sequence of vector valued measures over X satisfying*

$$\sup_k \mathbf{M}_{\mathcal{U}}(\nu_k) < \infty \qquad \forall\, \mathcal{U} \subset\subset X \ , \ \mathcal{U} \text{ open .} \tag{1}$$

where $\mathbf{M}_{\mathcal{U}}(\nu) = |\nu|(\mathcal{U})$ is the mass of ν in $\mathcal{U}$. Then there exist a subsequence $\{\nu_{k_i}\}$ of $\{\nu_k\}$ and a vector valued measure ν such that $\nu_{k_i} \rightharpoonup \nu$.

The converse of Theorem 1 is also valid, i.e., if $\nu_k \rightharpoonup \nu$ then (1) holds. This is consequence of the following

Theorem 2 (Banach-Steinhaus). *Let $\{\nu_k\}$ be a sequence of vector valued measures over X. If*

$$\sup_k \left| \int <\varphi, d\nu_k> \right| < \infty \qquad \forall\, \varphi \in C_c(X, \mathbb{R}^m)$$

then the masses of ν_k are locally equi-bounded, i.e.,

$$\sup_k \mathbf{M}_{\mathcal{U}}(\nu_k) < \infty \qquad \forall\, \mathcal{U} \subset\subset X \ , \ \mathcal{U} \text{ open}$$

We also have

Theorem 3. *Let μ and μ_k, $k = 1, 2, \ldots$, be Radon measures over X. Then the following three statements are equivalent*

(i) $\mu_k \rightharpoonup \mu$
(ii) $\mu_k(B) \to \mu(B)$ *for all relatively compact Borel sets* $B \in \mathcal{B}_c(X)$ *satisfying* $|\mu|(\partial B) = 0$
(iii) *for all open subsets* $\mathcal{U} \subset X$ $\mu(\mathcal{U}) \leq \liminf_{k\to\infty} \mu_k(\mathcal{U})$, *and for all compact subsets* $K \subset X$ $\limsup_{k\to\infty} \mu_k(K) \leq \mu(K)$.

Notice that $\mu_k(B)$ does not converge in general to $\mu(B)$ if $\mu(\partial B) \neq 0$. For example if $\{x_k\}$ is a sequence in $B(0,1) \subset \mathbb{R}^n$ converging to $x_\infty \in \partial B(0,1)$, we have $\delta_{x_k} \rightharpoonup \delta_{x_\infty}$, but $\delta_{x_k}(B(0,1)) = 1 \not\to \delta_{x_\infty}(B(0,1)) = 0$.

Notice also that in Theorem 3 (ii) we can in fact state $\mu_k(B) \to \mu(B)$ for all open set B with $\mu(\partial B) = 0$. In fact every Borel subset can be decomposed as $B = \overset{\circ}{B} \cup (B \cap \partial B)$ with $\mu(B) = \mu(\overset{\circ}{B})$, and $\partial \overset{\circ}{B} \subset \partial B$.

Theorem 3 does not extend immediately to vector valued measures. For example consider the sequence of signed measures

$$\mu_n := \delta_{1/n} - \delta_{-1/n} \in M(\mathbb{R}) .$$

Obviously $\mu_n \rightharpoonup \mu$, $\mu = |\mu| = 0$, thus, for $B = [0,1)$, we have $\mu(\partial B) = |\mu|(\partial B) = 0$, but $\mu_n(B) = \delta_{1/n}(B) = 1$, so $\mu_n(B) \not\to \mu(B)$.

In order to extend Theorem 3 we need a stronger condition.

Definition 2. *We say that a sequence of vector valued measures* $\mu_k \in M(X, \mathbb{R}^n)$ quasi-converges *to* (μ, α), $\mu_k \overset{q}{\rightharpoonup} (\mu, \alpha)$, *where* $\mu \in M(X, \mathbb{R}^n)$ *and* $\alpha \in M_+(X)$ *if and only if* $\mu_k \rightharpoonup \mu$ *and* $|\mu_k| \rightharpoonup \alpha$.

In general $\alpha \neq |\mu|$ and in fact by lower semicontinuity $|\mu| \ll \alpha$. We have

Theorem 4. *Let $\{\mu_k\}$ be a sequence of vector valued measures over X. The following two claims are equivalent*

(i) $\mu_k \overset{q}{\rightharpoonup} (\mu, \alpha)$
(ii) $\mu_k(B) \rightharpoonup \mu(B)$ *and* $|\mu_k|(B) \rightharpoonup \alpha(B)$ *for all relatively compact Borel sets* $B \in \mathcal{B}_c(X)$ *satisfying* $\alpha(\partial B) = 0$.

Notice that, just by definition, *the mass is lower semicontinuous with respect to the weak convergence of measures*, that is,

$$\mathbf{M}_{\mathcal{U}}(\nu) \leq \liminf_{k\to\infty} \mathbf{M}_{\mathcal{U}}(\nu_k) \qquad \text{if } \nu_k \rightharpoonup \nu .$$

Let $\{\mu_k\}$ be a sequence of vector valued measures such that $\mu_k \rightharpoonup \mu$ and $|\mu_k|(X) \to |\mu|(X)$ where $|\mu|(X) < \infty$. Using the lower semicontinuity of the total variation on open sets we have

$$|\mu|(A) \leq \liminf_{k\to\infty} |\mu_k(A)| .$$

On the other hand for any closed set F we have

$$\begin{aligned}\limsup_{k\to\infty}|\mu_k|(F) &= \limsup_{k\to\infty}(|\mu_k|(X)-|\mu_k|(X\setminus F))\\ &= |\mu|(X)-\liminf_{k\to\infty}|\mu_k|(X\setminus F)\geq|\mu|(X)-|\mu|(X\setminus F)=|\mu|(F)\,.\end{aligned}$$

Thus by Theorem 3 we infer

Proposition 1. *Let $\{\mu_k\}$ be a sequence of vector valued measures which converges with the masses to μ, i.e.,*

$$\mu_k \rightharpoonup \mu, \qquad |\mu_k|(X)\to|\mu|(X), \qquad |\mu|(X)<\infty\,.$$

Then μ_k quasi-converges to $(\mu,|\mu|)$.

The next remark will be used later.

Proposition 2. *Suppose that $X=\cup_{k=1}^{\infty}F_h$, F_h compact, $\overset{\circ}{F}_{h+1}\supset\supset F_h$. If $\mu_k\rightharpoonup\mu$ with the mass then $\mu_k(\varphi)\to\mu(\varphi)$, $|\mu_k|(f)\to|\mu|(f)$ for all continuous and bounded functions φ, f on X.*

Proof. In fact, let $\chi_h\in C_c(X,\mathbb{R})$, $0\leq\chi_h\leq1$, $\chi_h=0$ on $X\setminus\overset{\circ}{F}_h$, $\chi_h\uparrow 1$. Since

$$\limsup_{k\to\infty}|\mu|_k(X\setminus\overset{\circ}{F}_h)\leq|\mu|(X\setminus\overset{\circ}{F}_h)\,,$$

we have for any h

$$\limsup_{k\to\infty}|(\mu_k-\mu)(\varphi)|\leq\limsup_{k\to\infty}|(\mu_k-\mu)((1-\chi_h)\varphi)|\leq 2\|\varphi\|\,|\mu|(X\setminus\overset{\circ}{F}_h)$$

and similarly

$$\limsup_{k\to\infty}|(|\mu_k|-|\mu|)(f)|\leq 2\|f\|\,|\mu|(X\setminus\overset{\circ}{F}_h).$$

Letting $h\to\infty$ the result follows. □

Often it is convenient to check the condition

$$\int<\varphi,d\nu_k>\ \longrightarrow\ \int<\varphi,d\nu_k> \tag{2}$$

not for all $\varphi\in C_c(X,\mathbb{R}^m)$, but for φ in suitable subsets Λ of $C_c(X,\mathbb{R}^m)$. In this respect we can state

(i) Suppose that Λ has the following property. For every $\varphi\in C_c(X,\mathbb{R}^m)$ there exists a compact set K such that for every $\varepsilon>0$ there exists $\psi\in\Lambda$ with $\operatorname{spt}\psi\subset K$ and $|\varphi-\psi|<\varepsilon$ on X. Then $\nu_k\rightharpoonup\nu$ if (2) holds for all $\varphi\in\Lambda$ and $\sup_k \mathbf{M}_{\mathcal{U}}(\nu_k)<\infty$ for all open sets $\mathcal{U}\subset\subset X$,
(ii) Suppose that Λ is dense in $C_c(X,\mathbb{R}^m)$ in the C-norm topology. If (2) holds for all $\varphi\in\Lambda$ and $\sup_k \mathbf{M}(\nu_k)<\infty$, then $\nu_k\rightharpoonup\nu$.

The weak convergence in $M(X, \mathbb{R}^m)$ is not induced in general by a metric. But we have the following. Consider for any positive constant c the bounded set in $M(X, \mathbb{R}^m)$

$$\mathcal{K}_c := \{\nu \in M(X, \mathbb{R}^m) \mid \mathbf{M}(\nu) \le c\} .$$

Since $C_c(X, \mathbb{R}^m)$ is separable in the C-norm, then there exists a metric ρ on $\mathcal{K}_c$ such that $\nu_k \rightharpoonup \nu$ (ν_k and ν in $\mathcal{K}_c$) if and only if $\rho(\nu_k, \nu) \to 0$, i.e., *the weak convergence on $\mathcal{K}_c$ is metrizable.* A metric ρ which does it is for instance

$$\rho(\mu_1, \mu_2) := \frac{\sum_{\ell=1}^{\infty} 2^{-\ell} \left|\int <\varphi_\ell, d\mu_1> - \int <\varphi_\ell, d\mu_2>\right|}{1 + \left|\int <\varphi_\ell, d\mu_1> - \int <\varphi_\ell, d\mu_2>\right|}$$

where φ_ℓ, $\ell = 1, 2, \dots$ is a dense set in $C_c(X, \mathbb{R}^m)$.

Let μ be a Radon measure on X and let p be a real number with $1 \le p < \infty$. The space of measurable functions which are p-summable with respect to μ, i.e., for which

$$\int |f|^p \, d\mu < \infty$$

is denoted by $L^p(X; \mu)$. As it is well-known $L^p(X; \mu)$ is a Banach space with norm

$$\| f \|_{L^p(X;\mu)} = \| f \|_p := \left(\int |f|^p \, d\mu \right)^{1/p} , \tag{3}$$

moreover $C_c(X)$ is dense in $L^p(X; \mu)$, and the dual space of $L^p(X; \mu)$ is $L^{p'}(X; \mu)$ where p' is the dual exponent of p, $\frac{1}{p} + \frac{1}{p'} = 1$ if $p > 1$ or $p' = \infty$ if $p = 1$.

Definition 3. *Let μ be a Radon measure on X and let $f, f_1, f_2, \dots$ belong to $L^p(X; \mu)$.*

(i) We say that f_k converge weakly in $L^p(X; \mu)$ *to f, $f_k \rightharpoonup f$ in L^p, iff*

$$\int f_k g \, d\mu \longrightarrow \int f g \, d\mu \qquad \forall\, g \in L^{p'}(X; \mu) \tag{4}$$

We say that f_k converge weakly in the sense of measures, *or* as measures, *to f, $f_k \rightharpoonup f$, iff*

$$\int f_k g \, d\mu \longrightarrow \int f g \, d\mu \qquad \forall\, g \in C_c(X) \tag{5}$$

(ii) We say that the function f_k, $k = 1, 2, \dots$, are equi-summable *or* equi-integrable *if for every $\varepsilon > 0$ there exists a $\delta > 0$ such that*

$$\int_B |f_k| \, d\mu < \varepsilon$$

holds for all k and all Borel sets $B \subset X$ with $\mu(B) < \delta$.

We have

Theorem 5. *Let μ be a Radon measure over X and let $f_1, f_2, \dots$ be functions in $L^p(X;\mu)$.*

(i) Suppose $p > 1$. If $\sup_k \| f_k \|_{L^p(X;\mu)} < \infty$, then there exist $f \in L^p(X;\mu)$ and a subsequence $\{f_{k_i}\}$ such that $f_{k_i} \rightharpoonup f$ weakly in $L^p(X;\mu)$.

(ii) Suppose $p = 1$. Then

a) We have $f_k \rightharpoonup f$ weakly in $L^1(X;\mu)$ iff $f_k \rightharpoonup f$ as measures and the f_k's are equi-integrable in $L^1(X;\mu)$

b) If $\sup_k \| f_k \|_{L^1(X;\mu)} < \infty$ and the f_k's are equi-integrable, then there exist $f \in L^1(X,\mu)$ and a subsequence $\{f_{k_i}\}$ such that $f_{k_i} \rightharpoonup f$ weakly in $L^1(X;\mu)$.

While bounded sequences $\{f_k\}$ in $L^1(X;\mu)$ converge as measures to a measure modulo passing to subsequences, they in general do not converge, even passing to subsequences, weakly in $L^1(X;\mu)$, compare Sec. 1.2.2 and Sec. 1.2.4 below. Finally, concerning the equi-integrability condition we mention the following

Proposition 3. *A sequence $\{f_k\} \subset L^1(X;\mu)$ is equi-integrable if and only if there exists a continuous function $\phi : [0,\infty) \to [0,\infty)$ such that*

$$\sup_k \int \phi(|f_k|)\, d\mu < \infty \quad \textit{and} \quad \phi(t)/t \longrightarrow \infty \ \textit{as}\ t \to \infty.$$

In the remaining part of this section we would like to discuss in more details weak convergence of measures and of functions.

2.2 Typical Behaviours of Weakly Converging Sequences

We begin by illustrating some typical features of the behaviour of sequences which are weakly, but not strongly converging.

Oscillations. Sequences of rapidly oscillating functions provide examples of weakly, but not strongly, converging sequences.

[1] Consider the sequence

$$u_k(x) := \sin kx\,, \qquad x \in (0, 2\pi) \subset \mathbb{R}\,, \quad k \in \mathbb{N}\,.$$

then, by *Riemann-Lebesgue's lemma*, see e.g. Proposition 1 in Sec. 1.2.3 $u_k \rightharpoonup u := 0$ weakly in $L^q(0,2\pi)$ $\forall\, q \geq 1$ but $\| u_k \|_{L^q(0,2\pi)} = c(q) > 0$ so that $\{u_k\}$ does not converge to u in any $L^q(0,2\pi)$, $q \geq 1$, and

$$\| u \|_{L^q(0,2\pi)} < \liminf_{k\to\infty} \| u_k \|_{L^q(0,2\pi)}\,.$$

If $u_k \rightharpoonup u$ weakly in L^q, $q \geq 1$, by the weak lower semicontinuity of the L^q-norm we have

$$(1) \qquad \| u \|_{L^q} \leq \liminf_{k\to\infty} \| u_k \|_{L^q} .$$

If $u_k \to u$ strongly in L^q, we instead trivially have equality in (1), but we may as well have equality under mere weak convergence.

Let $u_k(x) := 1 + \sin kx$, $x \in (0, 2\pi)$, $\kappa \in \mathbb{N}$. Then $u_k \rightharpoonup u := 1$ weakly in L^q for all $q \geq 1$. Clearly

$$\int_0^{2\pi} |u_k|\, dx = \int_0^{2\pi} |u|\, dx = 2\pi \qquad \forall\, k$$

$$\int_0^{2\pi} |u_k - u|\, dx = \int_0^{2\pi} |\sin kx|\, dx = 4$$

so that $u_k \rightharpoonup u$ weakly in L^1, $u_k \not\to u$ strongly in L^1 but

$$\| u_k \|_{L^1(0,2\pi)} \to \| u \|_{L^1(0,2\pi)} .$$

Notice that instead

$$\liminf_{k\to\infty} \int_0^{2\pi} |u_k|^q\, dx > \int_0^{2\pi} |u|^q \qquad \forall\, q > 1 ,$$

because otherwise, passing to a subsequence, we would have $u_k \rightharpoonup u$ weakly in L^q, $\| u_k \|_{L^q} \to \| u \|_{L^q}$, $q > 1$ which implies $u_k \to u$ strongly in L^q, see Sec. 1.2.3.

Notice also that, since $\sqrt{1+t^2}$ is strictly convex, for $k = 1,\ 2,\ \dots$

$$\fint_0^{2\pi} \sqrt{1+u_k^2}\, dx = \fint_0^{2\pi} \sqrt{1+u_1^2}\, dx > \sqrt{1+\left(\fint_0^{2\pi} u_1\right)^2} = \sqrt{2}$$

so that

$$\liminf_{k\to\infty} \int_0^{2\pi} \sqrt{1+u_k^2}\, dx > \int_0^{2\pi} \sqrt{1+u^2}\, dx .$$

•

Concentrations. This is for instance the way Dirac mass appears. But concentration phenomena appear also just in the context of functions.

[2] Let $u_k : (-1, 1) \subset \mathbb{R} \to \mathbb{R}$ be defined by

$$u_k(x) = \begin{cases} k & \text{if } 0 \leq x \leq 1/k \\ 0 & \text{otherwise.} \end{cases}$$

Then

$$(2) \qquad u_k \rightharpoonup \delta_0 \quad \text{weakly as measures in } (-1,1)\ .$$

Of course, we may smooth u_k and still have (2). Notice that $u_k \rightharpoonup 0$ weakly as measures in $(0,1)$ while $u_k \not\rightharpoonup 0$ weakly in $L^1(0,1)$.

By considering instead $v_k : (-1,1) \to \mathbb{R}$ given by

$$v_k := \begin{cases} -k & \text{if } -1/k < x < 0 \\ k & \text{if } 0 < x < 1/k \\ 0 & \text{otherwise} \end{cases}$$

we clearly have $v_k \rightharpoonup v := 0$ weakly as measures in $(-1,1)$ while no subsequence of $\{v_k\}$ does converge weakly in L^1. In fact for $\varphi := \chi_{(0,1)} \in L^\infty(-1,1)$ we have

$$\int_{-1}^{1} v_k \varphi\, dx = 1 \not\to \int_{-1}^{1} v\varphi\, dx = 0\ .$$

This shows that $\{v_k\}$ is not pre-compact in the weak convergence of L^1, though $\int_{-1}^1 |v_k|\, dx = 2 \quad \forall\, k$. Notice also that we have a loss in energy:

$$\| v \|_{L^1} < \liminf_{k\to\infty} \| v_k \|_{L^1}$$

while $\| v_k \|_{L^q} \to \infty$ if $q > 1$. •

[3] A similar in certain respect, and different for others, concentration phenomena shows also in L^q-spaces for $q > 1$.

Let $q > 1$, and let $u_k : (-1,1) \to \mathbb{R}$ be defined by

$$u_k(x) := \begin{cases} k^{1/q} & \text{if } 0 < x < 1/k \\ 0 & \text{otherwise.} \end{cases}$$

We have $\| u_k \|_{L^q} = 1 \quad \forall\, k$, $u_k \rightharpoonup 0$ weakly in L^q but $\{u_k\}$ does not converge strongly in L^q, since the norms $\| u_k \|_{L^q}$ concentrate. But $u_k \to 0$ strongly in $L^p(0,1)$ for any $p < q$. In fact

$$\int_{-1}^{1} |u_k|^p\, dx = k^{\frac{p}{q}-1} \longrightarrow 0\ , \qquad \text{as } \frac{p}{q} - 1 < 0\ .$$

In particular the norms $\| u_k \|_{L^p}$, $p < q$, do not concentrate. •

Distributions. Concentrated measures may distribute producing a distributed measure in the weak limit. But also functions may concentrate and produce in the limit a function.

[4] Consider the Radon measures in $(0,1) \subset \mathbb{R}$ given by

$$\mu_k := \frac{1}{k} \sum_{i=1}^{k} \delta_{i/k}\ .$$

An opposite phenomenon to concentration then appears, as clearly $\mu_k \rightharpoonup \mu = \mathcal{L}^1 \llcorner (0,1)$ weakly as measures, in fact for any smooth ϕ with support in $(0,1)$

$$\frac{1}{k}\sum_{i=1}^{k} \phi(\frac{i}{k}) \longrightarrow \int_0^1 \phi\, dx$$

Replacing each Dirac mass $\delta_{i/k}$ with an "almost Dirac mass" a similar phenomenon may appear in the context of functions. Consider for instance the sequence of functions in $L^1(0,1)$

$$u_k := \frac{1}{k}\sum_{i=1}^{k} k^2 \chi_{(i/k-i/k^2, i/k)} .$$

Then we again have $u_k \rightharpoonup u := \chi_{(0,1)}$ weakly as measures.

This is a typical *concentration-distribution* phenomenon. In fact

$$\mathcal{L}^1(\operatorname{spt} u_k) = \frac{1}{k}$$

so that, outside a small set the u_k's are zero. One could expect then that $\mathcal{L}^1(\operatorname{spt} u) = 0$, but this is not true. The u_k's concentrate in sets of small measure which are almost densely distributed in such a way that the weak limit is distributed.

Note also that $u_k, u \geq 0$, and $\| u_k \|_{L^1(0,1)} = \| u \|_{L^1(0,1)} = 1$. Notice also that the sequence $\{u_k\}$ is not equi-absolutely integrable, therefore by Theorem 5 in Sec. 1.2.1, $\{u_k\}$ (as well as any of its subsequence) does not converge weakly in L^1. •

Nonlinearity destroys the weak convergence. If $\{u_k\}$ converges weakly but not strongly, then for a generic non linear function f the sequence $\{f(u_k)\}$ does not converge weakly to $f(u)$

[5] Let $u_k(x) = \sin kx$, $x \in (0, 2\pi)$ and let $f(t) = t^2$. Then we have $u_k \rightharpoonup u := 0$ in L^q, $q \geq 1$ while

$$f(u_k) = \sin^2 kx = \frac{1}{2}(1 - \cos 2kx) \rightharpoonup \frac{1}{2} \neq f(u) = 0 .$$

Notice that $\{u_k\}$ is equibounded in L^∞.

Similarly, if we let $f(t) := \max(0, t)$, we have

$$f(u_k) \rightharpoonup \frac{1}{\pi} \neq f(0) \qquad \text{weakly in } L^1(0, 2\pi)$$

or, if we let $f(t) = |t|$, $|u_k| \rightharpoonup \frac{2}{\pi} \neq |u|$ weakly in $L^1(0, 2\pi)$. •

We notice instead that, if for instance $f(t)$ is a continuous function satisfying

$$0 \le f(t) \le |t|^p, \qquad p \ge 1$$

and $u_k \to u$ strongly in $L^p(\Omega)$, then

$$f(u_k) \to f(u) \text{ strongly } in L^p(\Omega).$$

This is readily seen taking into account Egoroff theorem and the absolute continuity of Lebesgue integral.

2.3 Weak Convergence in L^q, $q > 1$

Let Ω be an open set in $\mathbb{R}^n$ and let $1 < q < \infty$. For Lebesgue summable functions *weak* convergence just amounts to:

(i) Let $u_k, u \in L^q(\Omega)$. We say that $u_k \to u$ weakly in L^q iff

(1)
$$\int_\Omega u_k \varphi \, dx \longrightarrow \int_\Omega u \varphi \, dx$$

for all $\varphi \in L^{q'}(\Omega)$, $q' = q/(q-1)$.

(ii) Let $u_k, u \in L^\infty(\Omega)$. We say that $u_k \to u$ weakly* in L^∞, $u_k \overset{*}{\rightharpoonup} u$, iff (1) holds for all $\varphi \in L^1(\Omega)$.

(iii) Let $u_k, u \in L^1(\Omega)$. We say that
a) $u_k \to u$ weakly in L^1 iff (1) holds for all $\varphi \in L^\infty(\Omega)$,
b) $u_k \to u$ as measures in Ω iff (1) holds for all $\varphi \in C^0_c(\Omega)$,
c) $u_k \to u$ as distributions in Ω iff (1) holds for all $\varphi \in C^\infty_c(\Omega)$.

Clearly

$$\text{(ii)} \Rightarrow \text{(i)} \Rightarrow \text{(iii)}_a \Rightarrow \text{(iii)}_b \Rightarrow \text{(iii)}_c .$$

As it is known the previous definitions can be subsumed to general definitions in the context of Banach spaces.

Let B be a Banach space and let B^* denotes its dual with duality $<x^*, x> = x^*(x)$ for $x \in B$, $x^* \in B^*$.

(a) Let $x_k, x \in B$. One says that $x_k \rightharpoonup x$ weakly iff $x^*(x_k) \to x^*(x) \quad \forall\, x^* \in B^*$.
(b) Let $x^*_k, x^* \in B^*$. One says that $x^*_k \overset{*}{\rightharpoonup} x^*$ weakly* iff $x^*(x_k) \to x^*(x) \quad \forall\, x \in B$.

Thus the weak convergence is defined in every Banach space, while the weak* convergence is defined only on dual Banach spaces. Of course $B^{**} \supset B$, hence the weak convergence in B^* implies the weak* convergence. In particular, if B is reflexive, $B^{**} = B$, then weak and weak* convergence in B^* (and in B) agree.

Banach-Steinhaus theorem ensures that weakly (weakly*) converging sequences are equibounded in B (in B^*). The norm in B (B^*) is lower semicontinuous with respect to the weak (weak*) convergence. Moreover we recall that bounded sequences in a reflexive Banach space B (in the dual B^* of a separable Banach space B) admit weakly (weakly*) converging subsequences. Finally let us mention *Banach-Saks* or *Mazur lemma* which states that for any $\varepsilon > 0$, if $x_k \rightharpoonup x$ in some Banach space B, then there exist an integer n and real $\alpha_1, \dots, \alpha_n$ with $\alpha_1 + \dots + \alpha_n = 1$ such that

$$\| x - \sum_{i=1}^{n} \alpha_i x_i \| < \varepsilon .$$

A similar result is instead false in general for the weak* convergence.

The definitions in the beginning, as well as many results on weak convergence of functions may then be seen as applications of the previous abstract facts to L^p-spaces, taking into account that

I. If $1 < q < \infty$, then $L^q(\Omega)$ is reflexive and $(L^q(\Omega))^* = L^{q'}(\Omega)$, $q' = q/(q-1)$.
II. $(L^1(\Omega))^* = L^\infty(\Omega)$ and $(L^\infty(\Omega))^* \underset{\neq}{\supset} L^1(\Omega)$.

This explains why one defines only the weak* convergence in L^∞.

In particular we see that we can apply Banach-Steinhaus theorem to all L^p-spaces, we can infer a simple weak compactness theorem in L^q, $1 < q < \infty$, and also in $L^\infty = L^{1*}$; while a compactness theorem in L^1 turns out to be more delicate.

Finally we mention that in order to have $x_k \rightharpoonup x$, ($x_k^* \overset{*}{\rightharpoonup} x^*$) it suffices to check that $\sup_k \| x_k \| < \infty$ ($\sup_k \| x_k^* \| < \infty$) and $z^*(x_k) \to z^*(x)$ ($x_k^*(z) \to x^*(z)$) for a dense set $z^* \in B^*$ (of $z \in B$).

From the previous remark or directly we can now easily deduce

Theorem 1. *Let $1 < q < \infty$, and let $u_k, u \in L^q(\Omega)$. Suppose that*

$$\sup_k \int_\Omega |u_k|^q \, dx < \infty$$

then the following statements are equivalent

(i) $u_k \rightharpoonup u$ *weakly in* $L^q(\Omega)$.
(ii) $u_k \rightharpoonup u$ *as measures in* Ω.
(iii) $u_k \rightharpoonup u$ *as distributions.*
(iv) *For any Borel set* $B \subset \Omega$ *with* $|B| > 0$ $\fint_B u_k \, dx \rightharpoonup \fint_B u \, dx$.
(v) *For any cube* $Q \subset \Omega$ $\fint_Q u_k \, dx \rightharpoonup \fint_Q u \, dx$.

Moreover one sees that a similar equivalence holds also if $q = \infty$, replacing in (i) the weak convergence with the weak* convergence in L^∞.

We can then say that the weak convergence in L^q, $q > 1$, is characterized by the equi-boundedness of the L^q-norms and the convergence in the sense of distributions.

Condition (v) expresses precisely the intuitive idea of weak convergence (in L^q, $q > 1$) as convergence of the mean values.

The meaning of the weak convergence as the process of smoothing oscillations is expressed by the next proposition which extends *Riemann-Lebesgue's lemma* that we have already used in Sec. 1.2.2.

Proposition 1. *Let* $\Omega := \prod_{i=1}^{n}(a_i, b_i) \subset \mathbb{R}^n$ *and let* $u \in L^q(\Omega)$, *where* $1 \leq q \leq \infty$. *Extend* u *by periodicity from* Ω *to* $\mathbb{R}^n$ *and let*

$$u_k(x) \;:=\; u(kx) \qquad x \in \mathbb{R}^n,\; k \in \mathbb{N}\,.$$

Then

$$u_k \;\rightharpoonup\; \overline{u} := \fint_\Omega u\,dx \quad \textit{weakly in } L^q(\Omega),\ \textit{if } 1 \leq p < \infty$$

$$u_k \;\rightharpoonup\; \overline{u} := \fint_\Omega u\,dx \quad \textit{weakly* in } L^\infty(\Omega)\,.$$

Proof. Trivially we have $\| u_k(x) \|_{\infty,\Omega} = \| u \|_{\infty,\Omega}$ and

$$\int_\Omega |u_k(x)|^q\,dx \;=\; \int_\Omega |u(x)|^q\,dx.$$

Next we show that

$$\fint_Q u_k(x)\,dx \;\longrightarrow\; \fint_Q \overline{u}(x)\,dx = \overline{u}\,. \tag{2}$$

This of course concludes the proof in the case $1 < q \leq \infty$. Of course it suffices to show (2) in the case Q has faces parallel to those of Ω. In this case we can find $\alpha, \beta \in \mathbb{R}^n$ such that

$$Q = \alpha + \beta\Omega = \prod_{i=1}^{n}(\alpha_i + \beta_i a_i, \alpha_i + \beta_i b_i)$$

and we have ($[x]$ = integral part of x), using the periodicity,

$$\begin{aligned}
\int_Q (u_k - \overline{u})\,dx &= \int_{\alpha+\beta\Omega} (u(kx) - \overline{u})\,dx = \frac{1}{k^n}\int_{k\alpha+k\beta\Omega} (u(y) - \overline{u})\,dy \\
&= \frac{1}{k^n}\int_{k\alpha+[k\beta]\Omega} (u(y) - \overline{u})\,dy + \frac{1}{k^n}\int_{k\alpha+(k\beta-[k\beta])\Omega} (u(y) - \overline{u})\,dy \\
&= \left(\frac{[k\beta]}{k}\right)^n \int_\Omega (u(y) - \overline{u})\,dy + \frac{1}{k^n}\int_{k\alpha+(k\beta-[k\beta])\Omega} (u(y) - \overline{u})\,dy \\
&= \frac{1}{k^n}\int_{k\alpha+(k\beta-[k\beta])\Omega} (u(y) - \overline{u})\,dy
\end{aligned}$$

therefore

$$\left| \int_Q (u_k - \overline{u})\, dy \right| \le \frac{1}{k^n} \int_\Omega |u(y) - \overline{u}|\, dy$$

which implies (2).

In order to conclude the proof in the case $q = 1$, we need to show that the sequence u_k is equi-absolutely integrable, compare Sec. 1.2.1 and the next subsection. For $\delta > 0$ and $x \in \Omega$ define

$$v_\delta(x) := \max(\min(u, \delta), -\delta), \qquad w_\delta(x) := u(x) - v_\delta(x).$$

Of course for any $\varepsilon > 0$ we can find $\delta = \delta(\varepsilon)$ such that $\int_\Omega |w_\delta|\, dx < \frac{\varepsilon}{2}$. By periodicity we extend v_δ and w_δ to $\mathbb{R}^n$. Then we find for $k = 1, 2, \ldots$

$$\int_\Omega |w_\delta(kx)|\, dx = \int_\Omega |w_\delta(x)|\, dx < \frac{\varepsilon}{2}\,.$$

Therefore

$$\int_E |u_k(x)|\, dx \le \int_E |v_\delta(kx)|\, dx + \int_E |w_\delta(kx)|\, dx \le \delta |E| + \frac{\varepsilon}{2}$$

and we find

$$\int_E |u_k(x)|\, dx \le \varepsilon$$

if we choose meas $E \le \varepsilon / 2\delta$. □

Notice finally that Theorem 1 does not hold for $q = 1$. In fact as we have seen in Sec. 1.2.2 there exist $\{u_k\}$ with

$$\sup_k \| u_k \|_{L^1} < \infty\,, \quad u_k \rightharpoonup 0 \text{ as measures}\,, \quad u_k \not\rightharpoonup 0 \text{ weakly in } L^1\,.$$

The simple remark contained in the next proposition is often quite useful.

Proposition 2. *Let $1 < q < \infty$, and let $u_k, u \in L^q(\Omega)$, $v_k, v \in L^{q'}(\Omega)$, $q' = q/(q-1)$. Suppose that*

$$u_k \rightharpoonup u \text{ weakly in } L^q(\Omega) \text{ and } v_k \rightharpoonup v \text{ strongly in } L^{q'}(\Omega).$$

Then $u_k v_k \rightharpoonup uv$ weakly in $L^1(\Omega)$.

The next proposition shows that if a weakly converging sequence in L^q, $q > 1$, does not converge strongly in L^q, then there must be a loss in the q-norm energy.

Proposition 3 (Radon-Riesz). *Let $u_k, u \in L^q(\Omega)$, $1 < q < \infty$, be such that*

$$u_k \rightharpoonup u \quad \text{weakly in } L^q(\Omega), \qquad \| u_k \|_{L^q(\Omega)} \to \| u \|_{L^q(\Omega)}.$$

Then $u_k \to u$ strongly in $L^q(\Omega)$.

We notice that Proposition 3, is false for $q = 1$, compare [1] in Sec. 1.2.2. It is usually stated as a consequence of the so-called *Clarkson's inequalities*, which express the uniform convexity of the L^q-spaces for $q > 1$. We refer for instance to Riesz and Sz.-Nagy [556, n. 37] for a proof.

The case $q = 2$ is particularly simple. In fact if $u_k \rightharpoonup u$ in $L^2(\Omega)$ we have

$$\lim_{k\to\infty} \left(\int_\Omega |u_k|^2 \, dx - \int_\Omega |u_k - u|^2 \, dx \right) = \int_\Omega |u|^2 \, dx \tag{3}$$

since $|u_k|^2 - |u_k - u|^2 = -|u|^2 + 2u_k u \rightharpoonup |u|^2$ weakly in $L^1(\Omega)$. And clearly (3) implies $\| u_k - u \|_{L^2(\Omega)} \to 0$ under the further assumption $\| u_k \| \to \| u \|$.

As corollary of Proposition 3 or by a direct proof we can also state the following weaker form of Proposition 3

Proposition 4. *Let $u_k, u \in L^q(\Omega)$, $1 < q < \infty$, be such that*

$$u_k \to u \quad \text{a.e.} \quad \textit{and} \quad \| u_k \|_{L^q(\Omega)} \to \| u \|_{L^q(\Omega)}.$$

Then $u_k \to u$ strongly in $L^q(\Omega)$.

Proof. In fact passing to subsequences, u_{k_i} converge weakly to some v, and v must agree a.e. with u. More directly, by Egoroff theorem we infer that $u_k \to u$ uniformly except on an open set of small measure δ. Choosing δ so that by the absolute continuity of the integral

$$\int_\Lambda |u|^q \, dx < \varepsilon$$

for sets Λ with $\operatorname{meas} \Lambda \le \delta$, and observing that also $\int_\Lambda |u_k|^q \, dx \le \varepsilon \quad \forall\, k$, we easily conclude. □

Notice that we have in fact proved that Proposition 4 *holds also for* $q = 1$.

Since the convergence in measure implies a.e. convergence passing to a subsequence, it is also easily seen that Proposition 4 remains true replacing the a.e. convergence by the convergence in measure. However, under this assumption, it does not hold anymore for $q = 1$.

There is an analogous of (3) also for $q \neq 2$.

Proposition 5. *Let $1 \le q < \infty$. Suppose that*

$$u_k \to u \text{ a.e. in } \Omega \textit{ and } u_k \rightharpoonup u \textit{ weakly in } L^q(\Omega). \tag{4}$$

Then

$$\lim_{k\to\infty} \left(\| u_k \|^q_{L^q(\Omega)} - \| u_k - u \|^q_{L^q(\Omega)} \right) = \| u \|^q_{L^q(\Omega)}.$$

Proof. Fix $\varepsilon > 0$ and let $v_k^\varepsilon := \sup(0, | \, |u_k|^q - |u_k - u|^q - |u|^q \, | - \varepsilon |u_k - u|^q)$. Clearly $v_k^\varepsilon \to 0$ a.e.. From the elementary inequality

$$| \, |a+b|^q - |a|^q \, | \; < \; \varepsilon |a|^q + c(\varepsilon, q)|b|^q$$

we also infer

$$v_k^\varepsilon \le \sup(0, | \, |u_k|^q - |u_k - u|^q \, | + |u|^q - \varepsilon |u_k - u|^q) \le (c(\varepsilon, q) + 1) \, |u|^q \; .$$

Thus the dominated convergence theorem yields $\lim_{k\to\infty} \int_\Omega v_k^\varepsilon \, dx = 0$, and we conclude

$$\limsup_{k\to\infty} \int_\Omega | \, |u_k|^q - |u_k - u|^q - |u|^q \, | \, dx$$
$$\le \varepsilon \limsup_{k\to\infty} \int_\Omega |u_k - u|^q) \, dx = O(\varepsilon) \; .$$

□

It is easily seen, compare the proof of Proposition 4, that, if $1 < q < \infty$, we can replace the assumptions (4) in Proposition 5 by

$$(5) \qquad u_k \longrightarrow u \quad \text{a.e.} \, , \qquad \sup_k \| \, u_k \, \|_{L^q(\Omega)} < \infty$$

and the conclusion still holds. However (5) does not suffices, if $q = 1$. For example consider for $\lambda > 0$ and for $k \ge \lambda$

$$v_k(x) := \begin{cases} k & \text{if } x \in (0, 1/k) \\ -k & \text{if } x \in (-1/k, 0) \\ 0 & \text{otherwise} \end{cases}$$

and set $u_k(x) := \lambda + v_k(x)$. We have $u_k(x) \to \lambda$ a.e. and $\| \, u_k \, \|_{L^1} = \| \, u_k - \lambda \, \|_{L^1} = 2$.

Finally we remark that a key ingredient in the previous statement is *convexity*, as we shall see better in Vol. II Ch. 2.

We conclude this subsection by restating the compactness property of L^q-spaces, $1 < q \le \infty$.

Theorem 2. *Let $1 < q \le \infty$. Then*

(i) If $u_k \rightharpoonup u$ weakly in $L^q(\Omega)$ ($u_k \overset{*}{\rightharpoonup} u$ if $q = \infty$), then

$$\sup_k \| \, u \, \|_{L^q(\Omega)} \; < \; \infty \; .$$

(ii) If

$$\sup_k \| \, u \, \|_{L^q(\Omega)} \; < \; \infty$$

then there exist a subsequence $\{u_{k_i}\}$ of $\{u_k\}$ and $u \in L^q(\Omega)$ such that

$$u_{k_i} \rightharpoonup u \qquad \textit{weakly in } L^q(\Omega), \textit{ if } q < \infty$$
$$u_{k_i} \rightharpoonup u \qquad \textit{weakly* in } L^\infty(\Omega), \textit{ if } q = \infty \; .$$

2.4 Weak Convergence in L^1

Most interesting for us will be the notions of weak convergence and convergence as measures in $L^1(\Omega)$ where Ω is a bounded open set in $\mathbb{R}^n$.

As we have seen, compare in particular Sec. 1.2.2, bounded sets in $L^1(\Omega)$ are not compact with respect to the weak convergence of L^1. This is the reason for regarding $L^1(\Omega)$ as embedded in the Banach space of Radon measures with bounded total variation in Ω

$$L^1(\Omega) \hookrightarrow \mathcal{M}(\Omega) .$$

Since $\mathcal{M}(\Omega)$ is the dual space of $C^0_0(\Omega)$, we can then consider in $L^1(\Omega)$ the weak* convergence of $\mathcal{M} = (C^0_0(\Omega))^*$, which is nothing else than the convergence as measures we have already considered. The important outcome is then

Theorem 1. *Let $\{u_k\}$ be a bounded sequence in $L^1(\Omega)$. Then there exist a subsequence $\{u_{k_i}\}$ and a Radon measure $\mu \in \mathcal{M}(\Omega)$ such that $u_k \rightharpoonup \mu$ as measures in Ω or, equivalently, weakly* in $\mathcal{M}(\Omega)$.*

Next theorem, due essentially to Lebesgue and which we have already stated in Sec. 1.2.1, yields a necessary and sufficient condition for the more delicate compactness with respect to the weak-L^1-convergence.

Theorem 2. *Let $\{u_k\}$ be a sequence in $L^1(\Omega)$ such that*

(i) the u_k's are equibounded in $L^1(\Omega)$, $\sup_k \| u_k \|_{L^1(\Omega)} < \infty$
(ii) the u_k's are equi-integrable, or in other words the set functions $E \longrightarrow \int_E |u_k|\, dx$, $E \subset \Omega$ are equi-absolutely continuous, i.e., for any $\varepsilon > 0$ there is $\delta > 0$ such that $\int_E |u_k|\, dx < \varepsilon$ for all k provided $\operatorname{meas} E < \delta$.

Then there exists a subsequence of $\{u_k\}$ which converges weakly in $L^1(\Omega)$. Conversely, if $\{u_k\}$ converges weakly in $L^1(\Omega)$, then (i) and (ii) hold true.

Proof. Since step functions are dense in $L^\infty(\Omega)$, and taking also into account the Radon–Nikodym theorem, it is easy to see that it suffices to show that for a subsequence u_{k_i} the limit

$$\lim_{i\to\infty} \int_B u_{k_i}\, dx$$

exists for all measurable set $B \subset \Omega$.

Suppose (i) holds. From Theorem 1 we infer that there is a subsequence $\{u_{k_i}\}$ and a measure μ such that

$$\int_\Omega u_{k_i} \varphi\, dx \longrightarrow \int_\Omega \varphi\, d\mu =: \mu(\varphi) \qquad \forall\, \varphi \in C^0_c(\Omega) .$$

Using (ii) we shall now show that for such a subsequence and all measurable $B \subset \Omega$ $\{\int_B u_{k_i}\, dx\}$ is a Cauchy sequence.

By Lusin theorem we can find a sequence $\{\varphi_h\} \subset C^0_c(\Omega)$ with

$$\varphi_h \longrightarrow \chi_B \quad \text{a.e. in } \Omega\,, \qquad \|\varphi_h\|_{L^\infty(\Omega)} \le 1\,.$$

In correspondence of $\delta = \delta(\varepsilon)$ in (ii) we find an open set $B_\delta \subset \Omega$ such that $|B_\delta| < \delta$ and $\{\varphi_h\}$ converges uniformly in $\Omega \setminus B_\delta$ to χ_B. Now we have

$$\begin{aligned}
\left|\int_B (u_{k_i} - u_{k_j})\,dx\right| &= \left|\int_\Omega (u_{k_i} - u_{k_j})\chi_B\,dx\right| \\
&\le \left|\int_{B_\delta} (u_{k_i} - u_{k_j})(\chi_B - \varphi_h)\,dx\right| \\
&\quad + \left|\int_{\Omega\setminus B_\delta} (u_{k_i} - u_{k_j})(\chi_B - \varphi_h)\,dx\right| + \left|\int_\Omega (u_{k_i} - u_{k_j})\varphi_h\,dx\right| \\
&\le 2\int_{B_\delta} (|u_{k_i}| + |u_{k_j}|)\,dx + \sup_{\Omega\setminus B_\delta} |\chi_B - \varphi_h| \int_\Omega (|u_{k_i}| + |u_{k_j}|)\,dx \\
&\quad + \left|\int_\Omega (u_{k_i} - u_{k_j})\varphi_h\,dx\right|\,.
\end{aligned}$$

In dependence of ε we now find some h_0 such that

$$\sup_{\Omega\setminus B_\delta} |\chi_B - \varphi_h| \;<\; \varepsilon \qquad \forall\, h \ge h_0\,.$$

Since $\varphi_{h_0} \in C^0_c(\Omega)$ for all i, j larger than some k_0 depending on h_0 and ε we have

$$\left|\int_\Omega (u_{k_i} - u_{k_j})\varphi_{h_0}\,dx\right| \;<\; \varepsilon\,.$$

Therefore we infer

$$\left|\int_B (u_{k_i} - u_{k_j})\,dx\right| \;\le\; 2\cdot 2\varepsilon + 2\varepsilon k + \varepsilon\,.$$

This concludes the proof of the first part of the theorem.

Suppose now that $u_k \rightharpoonup u$ weakly in $L^1(\Omega)$ and assume for simplicity $u = 0$ (otherwise we would consider the sequence $\{u_k - u\}$). Then (i) follows from Banach-Steinhaus's theorem. In order to verify (ii), we first prove the following claim:

If (ii) does not hold, then there exist a positive number z, a sequence of disjoint measurable sets $E_i \subset \Omega$ and an increasing sequence of integers ν_i such that

$$\int_{E_i} |u_{\nu_i}|\,dx \;\ge\; z \qquad \text{for all } i \in \mathbb{N}\,.$$

By assumption, there exists an $\varepsilon > 0$ such that, for all $\delta > 0$, we can find some set $F \subset \Omega$ with $\operatorname{meas} F < \delta$, and some $\nu \in \mathbb{N}$ which may be taken arbitrarily large such that

$$\int_F |u_\nu|\, dx \ \geq\ \varepsilon\ .$$

Since $u_\nu \in L^1(\Omega)$, for any $\sigma > 0$ there is some $\eta > 0$ such that

$$\int_B |u_\nu|\, dx \ <\ \sigma$$

holds true for all measurable sets B with meas $\operatorname{meas} B < \eta$. Choose now $\sigma = \sigma_1 = \varepsilon/2$, $\delta = \delta_1 = \operatorname{meas}\Omega$; then we find $\eta_1 > 0$, $F_1 \subset \Omega$ and $\nu_1 \in \mathbb{N}$ such that $\operatorname{meas} F_1 < \delta_1$, $\int_{F_1} |u_{\nu_1}|\, dx \geq \varepsilon$ and $\int_B |u_{\nu_1}|\, dx < \sigma_1$ for all B with $\operatorname{meas} B < \eta_1$. Next we choose $\sigma_2 = \frac{1}{2}\frac{1}{2}\varepsilon$, $\delta_2 = \delta_1/2$; then we find $\eta_2, F_2 \subset \Omega$ $\nu_2 \in \mathbb{N}$ such that $\operatorname{meas} F_2 < \delta_2$, $\int_{F_2} |u_{\nu_2}|\, dx \geq \varepsilon$ and $\int_B |u_{\nu_2}|\, dx < \sigma_2$ for all B with $\operatorname{meas} B < \eta_2$. Similarly for all $i > 2$, we choose $\sigma_i = \frac{1}{i}\frac{\varepsilon}{2}$, $\delta_i = \min\{\frac{\eta_1}{2^{i-1}}, \dots, \frac{\eta_{i-1}}{2}\}$; then we find $\eta_i > 0$, $F_i \subset \Omega$, $\nu_i > \nu_{i-1}$ such that $\operatorname{meas} F_i < \delta_i$, $\int_{F_i} |u_{\nu_i}|\, dx \geq \varepsilon$, $\int_B |u_{\nu_i}|\, dx < \sigma_i$ for all B with $\operatorname{meas} B < \eta_i$. Set now

$$E_i \ :=\ F_i - \bigcup_{q>i} F_q\ .$$

We have

$$\operatorname{meas}\Big(\bigcup_{q>i} F_q\Big) \leq \sum_{q>i} \operatorname{meas} F_q \leq \sum_{q>i} \frac{\eta_i}{2^{q-i}} = \eta_i\ ,$$

and also

$$\int_{E_i} |u_{\nu_i}|\, dx = \int_{F_i \setminus \cup_{q>i} F_q} |u_{\nu_i}|\, dx \geq \varepsilon - \frac{\varepsilon}{2i} \geq \varepsilon/2 \qquad \text{for all } i \geq 1\ .$$

Since the E_i are disjoint, the claim is proved with $z = \varepsilon/2$.

We now observe that replacing E_i by $E_i \cap \{x : u_{\nu_i}(x) \geq 0\}$ or by $E_i \cap \{x : u_{\nu_i}(x) \leq 0\}$ and z by $z/2$ we can modify the claim as follows: There exist $E_i \subset \Omega$ and ν_i such that

$$\Big|\int_{E_i} u_{\nu_i}\, dx\Big| \ \geq\ z \qquad \text{for all } i \in \mathbb{N}\ .$$

We shall finally construct a function $\varphi \in L^\infty(\Omega)$ for which the sequence

$$\int u_k \varphi\, dx$$

does not converge to zero. This will conclude the proof. The function φ will be defined as 1 on the union of suitable E_i and zero otherwise. We set $E^{(1)} := E_1$ and $\nu^{(1)} := \nu_1$, and we choose $\varepsilon^{(1)}$ so that

$$\int_B |u_{\nu^{(1)}}|\, dx \;<\; z/3 \qquad \text{for all } B \text{ with } \operatorname{meas} B < \varepsilon^{(1)}\,.$$

If $\int_{E^{(1)}} u_k\, dx$ does not converge to zero, the proof is complete as we can take $\varphi = \chi_{E^{(1)}}$. Otherwise we choose $E^{(2)}$ as the first E_i for which the remaining ones, i.e., the E_j with $j > i$, have total measure less than $\varepsilon^{(1)}$ (this is possible since the E_i are disjoint) and the corresponding index ν_i is such that

$$\Big|\int_{E^{(1)}} u_k\, dx\Big| \;<\; z/3 \qquad \text{for all } k \geq \nu_i$$

(this is possible since $\int_{E^{(1)}} u_k\, dx \to 0$). Denoting by $\nu^{(2)}$ the index corresponding to $E^{(2)}$, we choose $\varepsilon^{(2)} > 0$ such that

$$\int_B |u_{\nu^{(2)}}|\, dx \;<\; z/3 \qquad \text{for all } B \text{ with } \operatorname{meas} B < \varepsilon^{(2)}\,.$$

If $\int_{E^{(1)}\cup E^{(2)}} u_k\, dx$ does not converge to zero, the proof is complete. Otherwise we proceed as before and in general we find $E^{(k)}$, $\nu^{(k)}$, $\varepsilon^{(k1)}$ such that

$$\operatorname{meas} \bigcup_{i\geq k} E_i < \varepsilon^{(k-1)}\,, \qquad \Big|\int_{\cup_{i>k} E^{(i)}} u_h\, dx\Big| \;>\; z/3 \quad \text{for all } h \geq \nu^{(k)}$$

and $\int_B |u_{\nu^{(k)}}|\, dx < z/3$ for all B with $\operatorname{meas} B < \varepsilon^{(k)}$. Finally, set $\varphi := \chi_{\cup_{k\geq 1} E^{(k)}}$. Then we find

$$\begin{aligned}\Big|\int_\Omega u_{\nu^{(k)}}\varphi\, dx\Big| &= \Big|\int_{\cup_{i<k} E^{(i)}} u_{\nu^{(k)}}\, dx + \int_{E^{(k)}} u_{\nu^{(k)}}\, dx + \int_{\cup_{i>k} E^{(i)}} u_{\nu^{(k)}}\, dx\Big| \\ &\geq \Big|\int_{E^{(k)}} u_{\nu^{(k)}}\, dx\Big| - \frac{z}{3} - \frac{z}{3} \geq \frac{z}{3} \qquad \text{for all } k \in \mathbb{N}\end{aligned}$$

and this gives a contradiction to the weak convergence of u_k to zero. □

Often Theorem 2 is referred to as the *Dunford-Pettis theorem*, although it was first proved by Lebesgue [425].

It is not difficult to show that paralleling the case $q > 1$, in $L^1(\Omega)$ we have

Theorem 3. *Let $u_k, u \in L^1(\Omega)$. Suppose that $\sup_k \| u_k \|_{L^1(\Omega)} < \infty$ and $\{u_k\}$ is equi-integrable. Then the following statements are equivalent*

(i) $u_k \rightharpoonup u$ *weakly in* $L^1(\Omega)$
(ii) $u_k \rightharpoonup u$ *as measures*
(iii) $u_k \rightharpoonup u$ *as distributions*
(iv) *for any Borel set* $B \subset \Omega$, $⨍_B u_k\, dx \longrightarrow ⨍_B u\, dx$

(v) for any cube $Q \subset \Omega$, $\fint_Q u_k\,dx \longrightarrow \fint_Q u\,dx$.

Next proposition, though simple, is very useful

Proposition 1. *Let* a_k, a, b_k, b *be measurable functions on a measurable set* $\Omega \subset \mathbb{R}^n$.

(i) If $a_k(x) \to a(x)$ *a.e.,* $\| a_k \|_{\infty,\Omega} \leq A \; \forall\, k$, *and* $b_k \rightharpoonup b$ *weakly in* $L^1(\Omega)$, *then* $a_k\, b_k \rightharpoonup ab$ *weakly in* $L^1(\Omega)$.

(ii) If $a_k \to a$ *in* L^1, $\| a_k \|_{\infty,\Omega} \leq A \; \forall\, k$, *and* $b_k \rightharpoonup b$ *weakly in* $L^1(\Omega)$, *then* $a_k\, b_k \rightharpoonup ab$ *weakly in* $L^1(\Omega)$.

Proof. From the weak convergence of b_k to b, we deduce that $\{b_k\}$ is equi-absolutely continuous, i.e., for any $\varepsilon > 0$ there exists $\delta_\varepsilon > 0$ such that if meas $E < \delta_\varepsilon$ then $\int_E |b_k|\,dx < \varepsilon \; \forall\, k$. In correspondence of δ_ε, Egoroff's theorem yields a closed set F such that meas $(\Omega \setminus F) < \delta_\varepsilon$ and $a_k \to a$ uniformly in F. Thus for $\varphi \in L^\infty(\Omega)$ we get

$$\int_{\Omega\setminus F} \varphi(a_k - a)b_k\,dx \;\leq\; 2\,A\|\,\varphi\,\|_{\infty,\Omega} \int_{\Omega\setminus F} |b_k|\,dx \;<\; 2\,A\|\,\varphi\,\|\varepsilon$$

$$\int_F \varphi(a_k - a)b_k\,dx \;\longrightarrow\; 0 \quad \text{as } k \to \infty\,.$$

From this we conclude $(a_k - a)b_k \rightharpoonup 0$ weakly in $L^1(\Omega)$. On the other hand, since $a \in L^\infty$, $a(b_k - b) \rightharpoonup 0$ weakly in $L^1(\Omega)$. Thus (i) is proved.

Part (ii) now follows from (i). In fact from any subsequence of $\{a_k\}$ we can extract a subsequence which converges a.e. to a. Part (i) then yields that any subsequence of $a_k b_k$ has a subsequence which converges weakly in L^1 to ab. Since the limit is independent from the chosen subsequence, (ii) follows. □

Of course (iv) of Theorem 3 does not hold if we merely have $u_k \rightharpoonup u$ as measures in Ω. However one can prove, compare Sec. 1.2.1,

Theorem 4. *Let* $u_k, u \in L^1(\Omega)$ *be such that* $\sup_k \| u_k \|_{L^1(\Omega)} < \infty$ *and* $u_k \rightharpoonup u$ *as measures in* Ω. *Assume moreover that there exists a function* $v \in L^1(\Omega)$ *such that* $u_k \geq v \quad \forall\, k \in \mathbb{N}$. *Then*

(i) for any compact $K \subset \Omega$, $\int_K u\,dx \;\geq\; \limsup_{k\to\infty} \int_K u_k\,dx$,

(ii) for any open set $U \subset \Omega$, $\int_U u\,dx \;\leq\; \liminf_{k\to\infty} \int_U u_k\,dx$,

(iii) for any Borel set $B \subset \Omega$ *with* $|\partial B| = 0$, $\fint_B u_k\,dx \longrightarrow \fint_B u\,dx$.

The second example in [2] in Sec. 1.2.2 shows that the assumption $u_k \geq v$ is essential. The example in [4] in Sec. 1.2.2 shows that strict inequality may occur in (i) and (ii).

As we have seen in the previous subsection

$$u_k \to u \quad \text{a.e.} \quad \text{and} \quad \| u_k \|_{L^1(\Omega)} \to \| u \|_{L^1(\Omega)}$$

imply strong convergence in $L^1(\Omega)$ of u_k to u. Examples in Sec. 1.2.2 show that one cannot replace the a.e. convergence with the weak convergence in L^1. However we shall prove in Proposition 1 in Vol. II Sec. 1.3.4, as consequence of a theorem due to Reshetnyak the following interesting result

Proposition 2. *Let $u_k, u \in L^1(\Omega)$. Suppose that $u_k \rightharpoonup u$ as measures in Ω and*

$$\int_\Omega \sqrt{1+u_k^2}\, dx \longrightarrow \int_\Omega \sqrt{1+u^2}\, dx \,.$$

Then u_k converge strongly in $L^1(\Omega)$ to u.

2.5 Concentration: Weak Convergence of Measures

Here we would like to show that concentration-distribution phenomena associated to weak convergence of measures are in some sense universal, and shortly discuss weak convergence of measures in all of $\mathbb{R}^n$.

[1] Every (concentrated) Radon measure $\mu \in \mathcal{M}(\mathbb{R}^n)$ can be obtained as weak limit (in the sense of measures) of a sequence of smooth functions with equibounded L^1-norm. In fact regularizing μ, i.e., setting

$$u_\varepsilon(x) := \mu(\phi_{x,\varepsilon})$$

where $\phi_{x,\varepsilon}(y) := \frac{1}{\varepsilon^n}\phi\Big(\frac{y-x}{\varepsilon}\Big)$, ϕ being a standard mollifying kernel, we have $u_\varepsilon \in C^\infty(\mathbb{R}^n)$ and $\mathcal{L}^n \llcorner u_\varepsilon \rightharpoonup \mu$ weakly in $\mathcal{M}(\mathbb{R}^n)$. •

[2] Let $u \in C^0([0,1])$. Then the distributed measure $\mathcal{L}^n \llcorner u$ is the weak limit of the concentrated measures

$$\mu_k := \frac{1}{k}\sum_{i=1}^k u\Big(\frac{i}{k}\Big)\,\delta_{i/k} \in \mathcal{M}((0,1))\,.$$

Similarly the sequence of functions

$$u_k := \frac{1}{k}\sum_{i=1}^k k^2 u\Big(\frac{i}{k}\Big)\,\chi_{(\frac{i}{k}-\frac{1}{k^2},\frac{i}{k})}$$

converge in the sense of measure to u. Notice that the measures $\mathcal{L}^1 \llcorner u_k$ are 'almost-concentrated' in fact $\mathcal{L}^1(\operatorname{spt}\mu_k) \le 1/k$. •

If Ω is not bounded, $C^0_c(\Omega)$ is not a Banach space and the space of Radon measures with bounded total variation in Ω turns out to be the dual of the separable space

$$C_0^0(\Omega) := \Big\{u \in C^0(\Omega) \mid \forall\, \varepsilon > 0 \text{ there exists a compact } K \subset \Omega$$
$$\text{such that } |u| < \varepsilon \text{ in } \Omega \setminus K\Big\}\,.$$

Independently of the boundedness or the unboundedness of Ω one says that $\mu_k \rightharpoonup \mu$ *locally in* $\mathcal{M}(\Omega)$ iff

$$\int_\Omega \varphi\, d\mu_k \longrightarrow \int_\Omega \varphi\, d\mu \qquad \forall\, \varphi \in C_c^0(\Omega)\,.$$

The Banach-Steinhaus theorem then says that

$$\sup_k |\mu_k|(V) < \infty \qquad \forall\, V \subset\subset \Omega\,, \tag{1}$$

while, as consequence of the compactness theorem, every sequence $\{\mu_k\}$ satisfying (1) has a subsequence which converges locally to some μ with locally finite total variation.

Consider now a sequence of *probability measures* μ_k in $\mathbb{R}^n$, i.e.,

$$\mu_k \geq 0\,, \qquad \int_{\mathbb{R}^n} d\mu_k = 1\,.$$

Then there is a subsequence μ_{k_i} such that $\mu_{k_i} \rightharpoonup \mu$ locally in $\mathcal{M}(\mathbb{R}^n)$ with $\mu(\mathbb{R}^n) \leq 1$ and, in general, $\mu(\mathbb{R}^n) < 1$. The following result, due to P.L. Lions, describes modulo subsequences what may happen

Theorem 1 (Concentration-compactness lemma). *Let $\{\mu_k\}$ be a sequence of probability measures in $\mathbb{R}^n$. Then there exists a subsequence $\{\mu_{k_i}\}$ such that one of the following three conditions holds*

(i) (Compactness) There exists a sequence $\{x_i\} \subset \mathbb{R}^n$ such that for any $\varepsilon > 0$ there is a radius R with the property that

$$\mu_{k_i}(B(x_i, R)) \geq 1 - \varepsilon \qquad \forall\, i$$

(ii) (Vanishing) For all $R > 0$ we have

$$\sup_{x \in \mathbb{R}^n} \mu_{k_i}(B(x, R)) \longrightarrow 0 \qquad \text{as } i \to \infty$$

(iii) (Dichotomy) There exists $\lambda \in (0,1)$ such that for any $\varepsilon > 0$ there is a radius $R > 0$ and a sequence $\{x_i\}$ satisfying the following property: for any $R' > R$ there exist non-negative measures $\mu^1_{k_i}$ and $\mu^2_{k_i}$ such that

$$0 \leq \mu^1_{k_i} + \mu^2_{k_i}\,, \quad \operatorname{spt} \mu^1_{k_i} \subset B(x_i, R)\,, \quad \operatorname{spt} \mu^2_{k_i} \subset \mathbb{R}^n \setminus B(x_i, R')$$
$$\limsup_{i \to \infty} (|\lambda - \mu^1_{k_i}(\mathbb{R}^n)| + |(1-\lambda) - \mu^2_{k_i}(\mathbb{R}^n)| \leq \varepsilon\,.$$

Proof. The proof is based on the notion of *concentration function*

$$Q(r) := \sup_{x \in \mathbb{R}^n} \mu(B(x,r))$$

of a non-negative measure, introduced by Lévy [431].

Let Q_k denote the concentration function of μ_k. Clearly $\{Q_k\}$ is a sequence of non-decreasing, non-negative, bounded functions on $[0, \infty)$ with

$$\lim_{R \to \infty} Q_k(R) = 1 .$$

In particular $\{Q_k\}$ is a sequence which is locally bounded in $BV(0, \infty)$. Thus there exist a subsequence μ_{k_i} and a bounded, non-negative, non-decreasing function Q with

$$Q_{k_i}(R) \longrightarrow Q(R) \qquad \text{as } i \to \infty \text{ and for a.e. } R .$$

Let

$$\lambda := \lim_{R \to \infty} Q(R) .$$

Clearly $0 \leq \lambda \leq 1$. If $\lambda = 0$, we have vanishing, case (ii). Suppose $\lambda = 1$. Then for some $R_0 > 0$ we have $Q(R_0) > 1/2$. Choose x_i such that

$$Q_{k_i}(R_0) \leq \mu_{k_i}(B(x_i, R_0)) + \frac{1}{i} .$$

Then for $0 < \varepsilon < 1/2$ fix R such that $Q(R) > 1 - \varepsilon > 1/2$ and let y_i satisfy

$$Q_{k_i}(R) \leq \mu_{k_i}(B(y_i, R)) + \frac{1}{i} .$$

For large i we have

$$\mu_{k_i}(B(y_i, R)) + \mu_{k_i}(B(y_i, R_0)) > 1 = \mu_{k_i}(\mathbb{R}^n) .$$

It follows that for such i

$$B(y_i, R) \cap B(x_i, R_0) \neq \emptyset .$$

That is $B(y_i, R) \subset B(x_i, 2R + R_0)$, and hence

$$1 - \varepsilon \leq \mu_{k_i}(B(x_i, 2R + R_0)$$

for large i. This proves (i).

If $0 < \lambda < 1$, given $\varepsilon > 0$ we choose R and a sequence $\{x_i\}$, depending on ε and R, such that

$$\mu_{k_i}(B(x_i, R)) > \lambda - \varepsilon \qquad \text{for all } i \geq i_0(\varepsilon) .$$

Enlarging $i_0(\varepsilon)$, if necessary, we can also find $R_i \to \infty$ such that

$$Q_{k_i}(R) \;\le\; Q_{k_i}(R_i) \;<\; \lambda+\varepsilon \qquad \text{for all } i \ge i_0(\varepsilon)\,.$$

Now let

$$\mu^1_{k_i} := \mu_{k_i} \llcorner B(x_i,R)\,, \qquad \mu^2_{k_i} := \mu_{k_i} \llcorner (\mathbb{R}^n \setminus B(x_i,R_i))\,.$$

Obviously $0 \le \mu^1_{k_i} + \mu^2_{k_i} \le \mu_{k_i}$ and, given $R' > R$, for large i we also have

$$\operatorname{spt}\mu^1_{k_i} \subset B(x_i,R)\,, \quad \operatorname{spt}\mu^2_{k_i} \subset \mathbb{R}^n \setminus B(x_i,R_i) \subset \mathbb{R}^n \setminus B(x_i,R')\,.$$

Finally, for $i \ge i_0(\varepsilon)$ we get

$$\begin{aligned}&|\lambda - \mu^1_{k_i}(\mathbb{R}^n)| + |1-\lambda-\mu^2_{k_i}(\mathbb{R}^n)|\\ &\qquad = |\lambda - \mu_{k_i}(B(x_i,R))| + |\mu_{k_i}(B(x_i,R_i) - \lambda| < 2\varepsilon\end{aligned}$$

which concludes the proof. □

2.6 Oscillations: Young Measures

A very important and fruitful idea, due to L.C. Young in the context of Calculus of Variations and revived more recently among others by L. Tartar, is that one can understand better oscillation phenomena by looking at graphs, or more precisely at the associated *Young measures.* For instance those measures will give information on the weak limit of the square of a weakly converging sequence.

A similar idea is the starting point in the sequel of this monograph for studying currents associated to graphs of maps u. In fact Young measures may be seen as the first component of currents associated to graphs, compare Vol. II Ch. 5. The main difference is that in the context of currents one regards graphs as 1-*graphs*, i.e., as sets which carry at (almost) all of their points tangent planes, while in our present context of measures one regards graphs just as sets.

Definition 1. *Let Ω be a bounded open set in $\mathbb{R}^n$ and let $u \in L^\infty(\Omega,\mathbb{R}^N)$. The measure $\mathcal{Y}_u \in \mathcal{M}^+(\Omega\times\mathbb{R}^N)$ defined for each continuous and bounded function $\varphi : \Omega\times\mathbb{R}^N \to \mathbb{R}$, $\varphi \in C^0_b(\Omega\times\mathbb{R}^N)$, by*

$$\int_{\Omega\times\mathbb{R}^N} \varphi\, d\mathcal{Y}_u \;:=\; \int_\Omega \varphi(x,u(x))\,dx$$

is called the Young measure associated to u.

Clearly

$$\pi_\# \mathcal{Y}_u = \mathcal{L}^n \llcorner \Omega$$

if $\pi : (x,y) \in \Omega\times\mathbb{R}^n \to x \in \Omega$ denotes the standard projection and $\pi_\#\mu$ is given by

$$\pi_\#\mu(E) \;:=\; \mu(E\times\mathbb{R}^N)\,.$$

In fact for $\phi(x) \in C^0_b(\Omega)$ and $\varphi(x,y) := \phi(x)$ we have

$$\int_\Omega \phi(x)\, d\pi_\# \mathcal{Y}_u = \int_{\Omega\times\mathbb{R}^N} \pi^\# \phi\, d\mathcal{Y}_u = \int_{\Omega\times\mathbb{R}^N} \varphi(x,u(x))\, dx = \int_\Omega \varphi(x)\, dx .$$

This motivates the following

Definition 2. *A* Young measure *in* $\Omega \times \mathbb{R}^N$ *is a non-negative Radon measure* $\mu \in \mathcal{M}^+(\Omega\times\mathbb{R}^N)$ *satisfying* $\pi_\#\mu = \mathcal{L}^n \llcorner \Omega$ *i.e.*

$$\mu(E\times\mathbb{R}^N) = \mathcal{L}^n(E) \qquad \textit{for all Borel set } E \subset \Omega .$$

The disintegration theorem for measures then yields at once

Proposition 1. *Let* $\mu \in \mathcal{M}^+(\Omega\times\mathbb{R}^N)$ *be a Young measure. Then for* $\mathcal{L}^n$-a.e. $x \in \Omega$ *there exists a (Radon) probability measure* ν_x *on* $\mathbb{R}^N$*, i.e.,* $\nu_x \geq 0$, $\nu_x(\mathbb{R}^N) = 1$*, such that for all bounded and continuous* $\varphi \in C^0_b(\Omega\times\mathbb{R}^N)$ *we have*

(i) the map

$$x \longrightarrow \int_{\mathbb{R}^N} \varphi(x,y)\, d\nu_x(y)$$

is Lebesgue-measurable

(ii) the following equality holds

$$\int_{\Omega\times\mathbb{R}^N} \varphi(x,y)\, d\mu(x,y) = \int_\Omega dx \int_{\mathbb{R}^N} \varphi(x,y)\, d\nu_x(y) .$$

This way to every Young measure we can associate a collection of probability measures on $\mathbb{R}^N$

(1) $$\{\nu_x\}_{\text{a.e.}x\in\Omega} .$$

Conversely, to every collection (1) we can clearly associate the Young measure defined by (ii), provided (i) holds. Thus one can also define Young measures as a collection (1) for which (i) holds.

Let $\{u_k\}$ be an equibounded sequence in $L^\infty(\Omega\times\mathbb{R}^N)$, and let $\mathcal{Y}_{u_k}$ be the Young measures associated to u_k. We have

$$0 \leq \mathcal{Y}_{u_k}(\Omega\times\mathbb{R}^N) = \pi_\#\mathcal{Y}_{u_k}(\Omega) = \mathcal{L}^n(\Omega) .$$

Therefore for a subsequence we have $\mathcal{Y}_{u_{k_i}} \rightharpoonup \mu$ in $\mathcal{M}^+(\Omega\times\mathbb{R}^N)$. Moreover it is easily seen that μ *is a Young measure* too. It is often called the *Young measure associated to the sequence* $\{u_{k_i}\}$ In fact for any open set V we have

$$\pi_\#\mu(V) = \mu(V\times\mathbb{R}^N) \leq \liminf_{i\to\infty} \mathcal{Y}_{u_{k_i}}(V\times\mathbb{R}^N) = \pi_\#\mathcal{Y}_{u_{k_i}}(V) = \mathcal{L}^n(V) ,$$

and for any compact $K \subset \Omega$ and R such that $\operatorname{spt}\mathcal{Y}_{u_{k_i}} \subset \overline{\Omega}\times\overline{B(0,R)}$ we have

$$\begin{aligned}\pi_{\#}\mu(K) &= \mu(K \times \mathbb{R}^N) = \mu(K \times \overline{B(0,R)}) \\ &\geq \limsup_{i\to\infty} \mathcal{Y}_{u_{k_i}}(K \times \overline{B(0,R)}) = \limsup_{i\to\infty} \mathcal{Y}_{u_{k_i}}(K \times \mathbb{R}^N) \\ &= \pi_{\#}\mathcal{Y}_{u_{k_i}}(K) = \mathcal{L}^n(K).\end{aligned}$$

Let $\{\nu_x\}$ be the collection of probability measures associated to μ, i.e. let $\{\nu_x\}$ be the *disintegration of* μ. For any $\varphi(x,y)$ of the type $\varphi(x,y) = \rho(x)\psi(y)$, $\varphi \in C^0_c(\Omega)$, $\psi \in C^0_c(\mathbb{R}^N)$, we then get from Proposition 1(ii)

$$\begin{aligned}&\lim_{i\to\infty} \int_\Omega \rho(x)\psi(u_{k_i}(x))\,dx \\ &\quad= \lim_{i\to\infty} \int_{\Omega\times\mathbb{R}^N} \varphi(x,y)\,d\mathcal{Y}_{u_{k_i}}(x,y) \\ &\quad= \lim_{i\to\infty} \int_{\Omega\times\mathbb{R}^N} \varphi(x,y)\,d\mu(x,y) = \int_\Omega \rho(x)\Big(\int_{\mathbb{R}^N} \psi(y)\,d\nu_x(y)\Big)\,dx\end{aligned}$$

We can therefore state

Theorem 1. *Let $\{u_k\} \subset L^\infty(\Omega,\mathbb{R}^N)$ be such that $\sup_k \| u_k \|_\infty < \infty$. Suppose that the Young measures $\mathcal{Y}_{u_k}$ associated to u_k converge weakly $\mathcal{Y}_{u_k} \rightharpoonup \mu$ in $\mathcal{M}^+(\Omega \times \mathbb{R}^N)$. Then μ is a Young measure. Moreover for any $\psi \in C^0(\mathbb{R}^N)$ we have*

$$\psi \circ u_k \rightharpoonup v \qquad \text{weakly* in } L^\infty(\Omega)$$

where for almost every $x \in \Omega$

$$v(x) := \int_{\mathbb{R}^N} \psi(y)\,d\nu_x(y)\ ,$$

$\{\nu_x\}$ being the disintegration of μ.

The probability measures $\{\nu_x\}$ can be thought of as giving the limiting probability distribution as $k \to \infty$ of the values of u_k near x.

This shows that all information on the weak* limit of $\psi(u_k)$, $\psi \in C^0$, are just contained in the Young measure $\mu \sim \{\nu_x\}$ associated to the sequence $\{u_k\}$: any information on $\{\nu_x\}$ yields some information about the weak* limit of $\psi(u_k)$.

[1] Suppose for instance that for the limiting Young measure $\{\nu_x\}$ we know that ν_x is a Dirac mass, say $\nu_x = \delta_{u(x)}$ for a.e. $x \in \Omega$. Then we can infer $u_k \to u$ strongly in $L^2(\Omega)$. In fact for $\psi(y) := y^2$ we get

$$\int_\Omega |u_k|^2\,dx \longrightarrow \int_\Omega dx \int_{\mathbb{R}^N} |y|^2\,d\nu_x(y) = \int_\Omega |u(x)|^2\,dx$$

and for $\psi(y) := y$, $\rho \in C^0_c(\Omega)$

$$\int_{\Omega} \rho u_k \, dx \longrightarrow \int_{\Omega} \rho(x) \Big(\int_{\mathbb{R}^N} y \, d\nu_x(y) \Big) \, dx = \int_{\Omega} \rho(x) u(x) \, dx \, .$$

•

Let us discuss now oscillations and associated Young measures in the simple case $n = N = 1$, $\Omega = (0,1)$.

Proposition 2. *Let $u \in L^\infty(0,1)$. Extend u by periodicity, $u(x+m) := u(x)$ for $x \in (0,1)$, $m \in \mathbb{Z}$, and set*

$$u_k(x) := u(kx) \, , \qquad k \in \mathbb{N}, \ x \in (0,1) \, .$$

Then we have $\mathcal{Y}_{u_k} \rightharpoonup \mathcal{L}^1 \llcorner (0,1) \times \lambda \in \mathcal{M}^+((0,1) \times \mathbb{R})$ where λ is the measure defined by

$$\int_{\mathbb{R}} \psi \, d\lambda := \int_0^1 \psi(u(x)) \, dx \qquad \forall \, \psi \in C^0_c(\mathbb{R}) \, . \tag{2}$$

Proof. For $\varphi \in C^0_c(0,1)$ we have

$$\int_0^1 \varphi(x)\psi(u_k(x)) \, dx = \frac{1}{k} \int_0^k \varphi(\xi/k)\psi(u_k(\xi)) \, d\xi$$

$$= \frac{1}{k} \sum_{m=0}^{k-1} \int_m^{m+1} \varphi(\xi/k)\psi(u(\xi)) \, d\xi = \frac{1}{k} \sum_{k=0}^{k-1} \int_0^1 \varphi(\frac{x+m}{k})\psi(u(x)) \, dx.$$

The result then easily follows, since $\frac{1}{k} \sum_{m=0}^{k-1} \varphi\Big(\frac{m+x}{k}\Big) \to \int_0^1 \varphi(\xi) \, d\xi$ uniformly for $x \in (0,1)$. □

The next proposition allows us to compute the measure in (2) for special classes of functions u

Proposition 3. *We have*

(i) Suppose $u \in C^1(0,1)$ and $u' \neq 0$ a.e. in $(0,1)$. Then the measure λ in (1) is given by

$$\lambda = \mathcal{L}^1 \llcorner f \, , \qquad f(y) := \sum_{x \in u^{-1}(y)} \frac{1}{|u'(y)|} \, .$$

(ii) Suppose that $u \in L^\infty(0,1)$ takes only a finite number of values

$$u(x) := \sum_{i=1}^{s} y_i \chi_{A_i}(x) \, , \qquad y_1, \dots, y_s \in \mathbb{R}$$

where $\cup_{i=1}^s A_i$ is a decomposition of $(0,1)$. Then λ in (1) is given by

$$\lambda = \sum_{y \in \mathbb{R}} \mathcal{L}^1(u^{-1}(y))\delta_y = \sum_{i=1}^{s} \mathcal{L}^1(A_i)\delta_{y_i} .$$

Proof. (i) Assume for instance u increasing, $u(0) = 0$, $u(1) = 1$. Changing variable, $x = \hat{u}(y)$, $\hat{u} \circ u = \text{id}$, we get

$$\int \psi \, d\lambda = \int_0^1 \psi(u(x)) \, dx = \int_0^1 \psi(y)\hat{u}'(y) \, dy$$

and $\hat{u}'(y) = \frac{1}{u'(\hat{u}(y))}$. The result then easily follows.

(ii) We have

$$\int_{\mathbb{R}} \psi \, d\lambda = \int_0^1 \psi(u(x)) \, dx = \sum_{i=1}^{s} \int_{A_i} \psi(u(x)) \, dx$$
$$= \sum_{i=1}^{s} \psi(y_i)\mathcal{L}^1(A_i) = \int \psi \, d(\sum_{i=1}^{s} \mathcal{L}^1(A_i)\delta_{y_i}) .$$

□

[2] Let $u(x) := x$, $x \in (0,1)$

$$u_k(x) = kx - [kx] = \text{non integral part of } kx .$$

By Proposition 3 (i) we get $\mathcal{Y}_{u_k} \rightharpoonup \mu := \mathcal{L}^1 \llcorner (0,1)^2$. Notice that the disintegration of $\mathcal{Y}_{u_k}$ is given by $\{\delta_{u_k(x)}\}$ while the disintegration of the limit Young measure μ is given by $\nu_x := \mathcal{L}^1 \llcorner (0,1)$. As consequence of Theorem 1 we obtain $u_k^2 \rightharpoonup 1/3$ weakly* in L^∞; in fact

$$v(x) = \int_{\mathbb{R}} y^2 \, d\mathcal{L}^1 \llcorner (0,1) = \int_0^1 y^2 \, dy = \frac{1}{3} .$$

Notice that $u_k \rightharpoonup 1/2$ as $\int y \, d\mathcal{L}^1 \llcorner (0,1) = \frac{1}{2}$. •

[3] Let $u(x) := \sin 2\pi x$, $x \in (0,1)$, so that $u_k(x) = \sin 2k\pi x$ in $(0,1)$. We have

$$\mathcal{Y}_{u_k} \rightharpoonup \mathcal{L}^1 \llcorner (0,1) \times \lambda \qquad \text{in } \mathcal{M}^+((0,1) \times \mathbb{R})$$

and by Proposition 1 (i) we have

$$\lambda = \mathcal{L}^1 \llcorner f , \qquad f(y) := \frac{1}{\pi} \frac{1}{\sqrt{1-y^2}} \chi_{(-1,1)}(y) .$$

In fact $|u'(x)| = 2\pi\sqrt{1-y^2}$, $y = \sin 2\pi x$ and the multiplicity is 2 if $|y| < 1$ and 0 if $|y| > 1$. Of course

$$\frac{1}{\pi}\int_{-1}^{1}\frac{1}{\sqrt{1-y^2}}\,dy = 1\,.$$

As f is even we have $\int y^m\,d\nu_x = \int y^m f(y)\,dy = 0$, then we find again

$$(\sin 2\pi kx)^m \rightharpoonup 0 \qquad \text{for } m \text{ odd.}$$

For $\psi(y) := y^2$ we instead get as we know

$$\sin^2 2k\pi x \rightharpoonup \int_0^1 \sin^2 2\pi x\,dx$$

in fact

$$\frac{1}{\pi}\int_{-1}^{1}\frac{y^2}{\sqrt{1-y^2}}\,dy = \fint_{(-\pi/2,\pi/2)} \sin^2 x\,dx\,.$$

•

[4] Let $0 < \sigma < 1$,

$$u(x) := \begin{cases} a & \text{if } 0 < x < \sigma \\ b & \text{if } \sigma < x < 1 \end{cases}, \qquad a, b \in \mathbb{R}\,,$$

and, as previously, $u_k(x) := u(kx)$ $x \in (0,1)$. In this case it is easily seen that $\mathcal{Y}_{u_k} \rightharpoonup \mathcal{L}^1 \llcorner (0,1) \times (\sigma\delta_a + (1-\sigma)\delta_b)$ the disintegration of the Young measure associated to u_k is constant in x and given by

$$\nu_x := \sigma\delta_a + (1-\sigma)\delta_b\,.$$

As consequence, for instance, of Theorem 1 we instead have

$$u_k \rightharpoonup \sigma a + (1-\sigma) b \qquad \text{weakly* in } L^\infty(0,1)\,.$$

The weak* convergence of u_k takes into account the mean value distribution of the values, while the convergence of Young measures informs us about values, a and b, and weights, σ and $(\sigma - 1)$, which indicate how much time the graph of u_k spends at a and b. •

We conclude this subsection by stating without proof a quite general theorem on Young measures.

Theorem 2 (Ball). *Let Ω be a bounded measurable set in $\mathbb{R}^n$ and let $K \subset \mathbb{R}^n$ be closed. Let $\{u_k\}$, $u_k : \Omega \to \mathbb{R}^N$, be a sequence of measurable maps satisfying for any open set $U \supset K$*

$$\mathcal{L}^n(\{x \in \Omega \mid u_k(x) \notin U\}) \longrightarrow 0 \qquad \text{as } k \to \infty\,.$$

Then there exists a subsequence $\{u_{k_i}\}$ of $\{u_k\}$ and a family of positive measures $\{\nu_x\}$ on $\mathbb{R}^N$ such that

(i) $\nu_x(\mathbb{R}^N) \le 1$ *for* a.e. $x \in \Omega$
(ii) $\operatorname{spt} \nu_x(\mathbb{R}^N) \subset K$ *for* a.e. $x \in \Omega$
(iii) for any $\psi \in C^0(\mathbb{R}^N)$ *with* $\psi(y) \to 0$ *for* $|y| \to \infty$ *we have*

$$\psi \circ u_{k_i} \rightharpoonup v \qquad \text{weakly* in } L^\infty$$

where

$$v(x) := \int_{\mathbb{R}^N} \psi(y)\, d\nu_x(y) .$$

If moreover we assume that $\sup_k \int_\Omega h(|u_k|)\, dx < \infty$ *for some continuous non decreasing function* $h : [0,\infty) \to \mathbb{R}$ *with* $h(t) \to \infty$ *for* $t \to \infty$ *(for instance, for* $h(t) = t$, $\sup_k \int_\Omega |u_k|\, dx < \infty$*), then* ν_x *is a probability measure, i.e.,* $\nu_x(\mathbb{R}^N) = 1$.

Finally, if furthermore we assume that for some subset $E \subset \Omega$ *and some* $\psi \in C^0(\mathbb{R}^N)$ *the sequence* $\{g \circ u_{k_i} \llcorner E\}$ *is weakly relatively compact in* $L^1(E)$*, then*

$$g \circ u_{k_i} \llcorner E \rightharpoonup v \llcorner E \qquad \text{weakly in } L^1(E) .$$

2.7 More on Weak Convergence in L^1

A bounded sequence $\{u_k\}$ in $L^1(\Omega)$ is not weakly relatively compact in $L^1(\Omega)$, if $\{u_k\}$ is not equi-integrable. To overcome such a difficulty we have discussed two possible ways

(i) regard $L^1(\Omega)$ as embedded in the space of Radon signed measures $\mathcal{M}(\Omega)$ and work with the weak convergence of measures
(ii) associate to u_k its Young measure $\mathcal{Y}_{u_k}$ and work with the weak convergence of the measures $\mathcal{Y}_{u_k}$.

Of course both ways force us to allow limits which do not belong to $L^1(\Omega)$ anymore.

Here we would like to mention two other possibilities which do not require to go outside $L^1(\Omega)$.

(i) the so called *biting convergence*
(ii) the weak convergence in the sense of *absolutely continuous parts*

Definition 1. *Let* $\Omega \subset \mathbb{R}^n$ *be a bounded domain and let* $\{u_k\}$ *be a bounded sequence in* $L^1(\Omega)$

$$\sup_k \| u_k \|_{L^1(\Omega)} < \infty .$$

We say that u_k converge in the biting sense *to* $u \in L^1(\Omega)$, $u_k \overset{b}{\rightharpoonup} u$ *in* Ω *if for every* $\varepsilon > 0$ *there exists a measurable set* $E \subset \Omega$ *such that* $\mathcal{L}^n(E) < \varepsilon$ *and* $u_k \rightharpoonup u$ *weakly in* $L^1(\Omega \setminus E)$.

The following compactness result turns out to hold

Theorem 1 (Biting lemma). *Let $\{u_k\} \subset L^1$ be such that $\sup_k \| u_k \|_{L^1(\Omega)} < \infty$. Then there exist a subsequence $\{u_{k_i}\}$ of $\{u_k\}$ and $u \in L^1(\Omega)$ such that $u_k \overset{b}{\rightharpoonup} u$ in Ω.*

This means that removing (bitten) sets of possible concentrations (of small measure) we have weak convergence.

Theorem 1 can be easily inferred from the following lemma due to Acerbi and Fusco [2]

Lemma 1. *Let $\{\phi_h\} \subset L^1(\Omega)$ be a bounded sequence. Then, for any $\varepsilon > 0$ we can find a measurable set C_ε, a positive number δ and an infinite set $S \subset \mathbb{N}$ such that $\operatorname{meas}(C_\varepsilon) < \varepsilon$ and moreover for any measurable C in $\Omega \setminus C_\varepsilon$ with $\operatorname{meas}(C) < \delta$ we have*

$$\int_C |\phi_h| \, dx \; < \; \varepsilon \qquad \forall \, h \in S$$

Proof. Suppose that the claim is not true. Then there exists $\varepsilon > o$ such that for every $(C_\varepsilon, \delta, S)$ as in the statement we may choose a measurable set B with $B \cap C_\varepsilon = \emptyset$, $\operatorname{meas} B < \delta$, and $k \in S$ such that

$$\int_B |\phi_k| \, dx \; \geq \; \varepsilon \, .$$

This implies that for every set C with $\operatorname{meas} C < \varepsilon$, an every infinite set $S \subset \mathbb{N}$, there exists a set A with $A \cap C = \emptyset$, $\operatorname{meas}(A \cup C) < \varepsilon$, and an infinite subset T of S such that

$$\int_A |\phi_k| \, dx \; \geq \; \varepsilon \qquad \forall k \in T \, .$$

We postpone the proof of the claim, and we prove that this leads to a contradiction. Set $C = \emptyset$, $S = \mathbb{N}$ and let A_1 and T_1 be as above. Starting from $C = A_1$ and $S = T_1$ we pass to A_2 and T_2 where $A_1 \cap A_2 = \emptyset$ and

$$\int_{A_1 \cup A_2} |\phi_k| \, dx = \int_{A_1} |\phi_k| \, dx + \int_{A_2} |\phi_k| \, dx \geq 2 \, \varepsilon \qquad \forall \, k \in T_2 \, .$$

Since $\operatorname{meas}(A_1 \cup A_2) < \varepsilon$, we may set $C = A_1 \cup A_2$, $S = T_2$ and continue with the same argument. If

$$N \; > \; \varepsilon^{-1} \sup_k ||\phi_k||_{L^1} \, ,$$

after N iterations we obtain a contradiction.
Finally, let us prove the claim. Let C and S be as stated, set $S_1 = S$ and take

$$\delta_1 \; < \; \frac{\varepsilon - \operatorname{meas} C}{2} \, .$$

Then there exist B_1 disjoint from C, with meas $B_1 < \delta_1$, and $k_1 \in S_1$ such that $\int_{B_1} |\phi_k|\, dx \geq \varepsilon$. By induction, setting

$$\delta_i := \frac{1}{2}\delta_{i-1}, \qquad S_i := \{k \in S_{i-1} \mid k > k_{i-1}\}\,,$$
$$A = \cup_h B_h, \qquad T = \{k_i \,|\, i \in \mathbb{N}\}$$

it is easily seen that A and T fulfill all requirements. □

The first example in [2] in Sec. 1.2.2 shows that $u_k \rightharpoonup \delta_0$ as measures and $u_k \stackrel{b}{\rightharpoonup} 0$; in the second example we have $u_k \rightharpoonup 0$ as measures and $u_k \stackrel{b}{\rightharpoonup} 0$. But the example in [4] in Sec. 1.2.2 shows that

$$u_k \stackrel{b}{\rightharpoonup} v := 0 \quad \text{and} \quad u_k \rightharpoonup u := 1 \quad \text{as measures,}$$

i.e., in general, there is no connection between the weak limit in the sense of measures and the biting limit. More information can be instead obtained if those limits agree. We state without proof one such a result due to Müller [501]

Proposition 1. *Let Ω be a bounded open set and let $\{u_k\}$ be a non-negative sequence such that*

$$u_k \rightharpoonup u \quad \textit{as measures in } \Omega \quad \textit{and} \quad u_k \stackrel{b}{\rightharpoonup} u.$$

Then for any compact set $K \subset \Omega$ we have $u_k \rightharpoonup u$ weakly in $L^1(K)$.

The second example in [4] in Sec. 1.2.2 shows that the assumption $u_k \geq 0$ is essentially necessary.

Definition 2. *Let Ω be a bounded open set, $\{u_k\} \subset L^1(\Omega)$, $\sup_k \| u_k \|_{L^1(\Omega)} < \infty$, and $u \in L^1(\Omega)$. We say that*

$$u_k \stackrel{ACP}{\rightharpoonup} u$$

if for every subsequence $\{u_{k_i}\}$ of $\{u_k\}$ satisfying $u_{k_i} \rightharpoonup \mu$ as measures, where $\mu \in \mathcal{M}(\Omega)$, the absolutely continuous part of μ (with respect to Lebesgue measure) agrees with u, i.e., $\mu^a = u$.

Remark 1. The rather complicated form of Definition 2 is necessary. In fact we cannot define

$$\widetilde{ACP}\text{-}\lim u_k \;:=\; (\text{weak}^*\text{-}\lim u_k)^a\,.$$

In this case in fact, by considering a sequence $u_k \rightharpoonup \delta_0$ as measure we have

$$u_k \stackrel{ACP}{\rightharpoonup} 0, \qquad u_k \stackrel{\widetilde{ACP}}{\rightharpoonup} 0.$$

But if $v_k \equiv 0$, the sequence

$$\{w_k\} = \{u_1, v_1, v_2, u_3, v_3, \ldots\}$$

would not converge in the $\widetilde{ACP}$ sense. Instead, trivially we have $w_k \overset{ACP}{\rightharpoonup} 0$.

Of course, if $u_k \rightharpoonup u \in \mathcal{M}(\Omega)$ as measures then $u_k \overset{ACP}{\rightharpoonup} \mu^{(a)}$, and, if $u_k \rightharpoonup u \in L^1(\Omega)$ as measures then $u_k \overset{ACP}{\rightharpoonup} u$.

Also a compactness theorem is trivial

Theorem 2. *Let $\{u_k\} \subset L^1(\Omega)$ with $\sup_k \| u_k \|_{L^1(\Omega)} < \infty$. Then there exist a subsequence $\{u_{k_i}\}$ of $\{u_k\}$ and $u \in L^1(\Omega)$ such that $u_{k_i} \overset{ACP}{\rightharpoonup} u$ in Ω. Moreover we can also assume that*

$$u_{k_i} \rightharpoonup \mu \in \mathcal{M}(\Omega) \qquad \textit{as measures}$$

with $\mu^a = u$.

Notice that in the first example of [2] in Sec. 1.2.2 we have

$$\begin{aligned} u_k &\rightharpoonup \delta_0 && \text{as measures in } (-1,1) \\ u_k &\overset{b}{\rightharpoonup} 0 && \text{in } (-1,1) \\ u_k &\overset{ACP}{\rightharpoonup} 0 && \text{in } (-1,1) \end{aligned}$$

in the second example in [2] in Sec. 1.2.2

$$\begin{aligned} u_k &\rightharpoonup 0 && \text{as measures in } (-1,1) \\ u_k &\overset{b}{\rightharpoonup} 0 && \text{in } (-1,1) \\ u_k &\overset{ACP}{\rightharpoonup} 0 && \text{in } (-1,1) \end{aligned}$$

and finally in the example [4] in Sec. 1.2.2

$$\begin{aligned} u_k &\rightharpoonup 1 && \text{as measures in } (0,1) \\ u_k &\overset{b}{\rightharpoonup} 0 && \text{in } (0,1) \\ u_k &\overset{ACP}{\rightharpoonup} 1 && \text{in } (0,1) \end{aligned}$$

3 Notes

1 There are many books which deal with general measure theory, among them we mention e.g. Carathéodory [135], Federer [226], Halmos [338], Hewitt and Stromberg [375], Munroe [505], Rudin [566], Saks [570], Wheeden and Zygmund [662], see also Carathéodory [134]. Concerning Hausdorff measures the reader is referred to the original paper by Hausdorff [362], and to Rogers [564], Falconer [218], and to Federer [226]. Cantor–Vitali functions appears in Vitali [648], compare also Vitali [649]. Concerning the classical differentiability properties of Cantor–Vitali functions the reader is also referred to Darst [174] and its references.

2 The content of Sec. 1.2.1,..., Sec. 1.2.5 is quite classical. We only mention that Proposition 5 in Sec. 1.2.3 is taken from Brezis and Lieb [110], while Theorem 1 in Sec. 1.2.3 is due to P.L. Lions [435]. In connection with weak convergence we would like to mention the lecture notes by Evans [215].

3 Concerning Young measures, a part from the original works by Young, see e.g. Young [678] [682], the reader may consult Tartar [626] where Young measures are developed as a tool for analyzing non linear partial differential equations, Ball [69] (and its references) where Theorem 2 in Sec. 1.2.6 is proved, and Pedregal [530]. The proof of the biting lemma, in Sec. 1.2.7, can be found in Brooks and Chacon [114], compare also Acerbi and Fusco [2] Ball and Murat [72]. Finally, with reference to Proposition 2 in Sec. 1.2.4 the reader may consult e.g. Visintin [647], Amrani, Castaing, and Valadier [40], Balder [59], Kinderleher and Pedregal [412], Brezis [105].

4 Quite interesting results connecting weak convergence in the Hardy space $\mathcal{H}^1(\mathbb{R}^n)$ with a.e.-convergence, biting convergence and weak convergence in L^1 can be found in Coifman, Lions, Meyer, and Semmes [156].

2. Integer Multiplicity Rectifiable Currents

It is the aim of this chapter to develop and illustrate some of the basic theory of *integer multiplicity rectifiable currents.*

In Sec. 2.1 we shall discuss the important *area* and *coarea* formulas and some of the theory of *n-rectifiable* sets. After two preliminary subsections in which we collect relevant definitions and facts on *multivectors, covectors*, and *differential forms*, we shall present, in Sec. 2.2, some of the basic theory of *currents* and especially of *integer multiplicity rectifiable currents.* Because of the relevance of those topics for the sequel we shall give in principle proofs of our claims, though sometimes we refer to later parts or elsewhere; in any case we try to illustrate our claims by simple examples.

1 Area and Coarea. Countably n-Rectifiable Sets

In this section we shall present the important *area* and *coarea formulas* and some of the basic aspects of the theory of *countably n-rectifiable sets.*

1.1 Area and Coarea Formulas for Linear Maps

Let f be a function from $\mathbb{R}^n$ into $\mathbb{R}^N$ and let $A \subset \mathbb{R}^n$. We distinguish two cases

(i) $n \leq N$. In this case we look for an explicit formula, the *area formula*, which expresses the n-dimensional measure of $f(A)$, $\mathcal{H}^n(f(A))$,

(ii) $n \geq N$. In this case we consider the $(n-N)$-dimensional slices $f^{-1}(y) \cap A$ of A for $y \in f(A)$ and we want to relate their $\mathcal{H}^{n-N}$-measure with the total measure of A by a formula, called the *coarea formula.*

Such formulas describe of course transformation properties of volume elements. Therefore we begin by considering first the case in which F is a linear map.

Let us recall a few facts from linear algebra.

A linear map $U : \mathbb{R}^n \to \mathbb{R}^N$ is said to be *orthogonal* if $(Ux) \cdot (Uy) = x \cdot y$ for all $x, y \in \mathbb{R}^n$.

A linear map $S : \mathbb{R}^n \to \mathbb{R}^n$ is said to be *symmetric* if $x \cdot (Sy) = Sx \cdot y$ for all $x, y \in \mathbb{R}^n$.

A linear map $D : \mathbb{R}^n \to \mathbb{R}^n$ is said to be *diagonal* if there exist $d_1, \dots, d_n \in \mathbb{R}$ such that $Dx = (d_1 x^1, \dots, d_n x^n)$ for all $x \in \mathbb{R}^n$.

The adjoint T^* of a linear map $T : \mathbb{R}^n \to \mathbb{R}^N$ is the linear map $T^* : \mathbb{R}^N \to \mathbb{R}^n$ defined by $x \cdot (T^* y) = (Tx) \cdot y$ for all $x \in \mathbb{R}^n$, $y \in \mathbb{R}^N$.

For linear maps R and T one easily verifies that $T^{**} = T$, $(R \circ T)^* = T^* \circ R^*$, $T^* = T^{-1}$ if and only if T is orthogonal, i.e., T is orthogonal if and only if $T^*T = \mathrm{id}$, $T^* = T$ if and only if T is symmetric. Identifying the linear map $T : \mathbb{R}^n \to \mathbb{R}^n$ with an $n \times n$- matrix, we also have

$$\det U = \pm 1$$

if U is orthogonal. Finally, we recall that for every symmetric linear map $S : \mathbb{R}^n \to \mathbb{R}^n$ there exist an orthogonal map $U : \mathbb{R}^n \to \mathbb{R}^n$ and a diagonal map $D : \mathbb{R}^n \to \mathbb{R}^n$ such that

$$S = U \circ D \circ U^{-1} .$$

The following theorem will be useful

Theorem 1 (Polar decomposition). *Let $T : \mathbb{R}^n \to \mathbb{R}^N$ be linear,*

(i) If $n \le N$, there exist a symmetric map $S : \mathbb{R}^n \to \mathbb{R}^n$ and an orthogonal map $U : \mathbb{R}^n \to \mathbb{R}^N$ such that $T = U \circ S$,
(ii) If $n \ge N$, there exist a symmetric map $S : \mathbb{R}^N \to \mathbb{R}^N$ and an orthogonal map $U : \mathbb{R}^N \to \mathbb{R}^n$ such that $T = S \circ U^*$

Proof. The claim (ii) follows applying (i) to T^*. Thus assume $n \le N$. It is easily seen that $C := T^* \circ T : \mathbb{R}^n \to \mathbb{R}^n$ is symmetric, and nonnegative definite, i.e., $(Cx) \cdot x \ge 0$. Hence there exist $\mu_1, \dots, \mu_n \ge 0$ and an orthogonal set $\{x_1, \dots, x_k\}$ in $\mathbb{R}^n$ such that

$$Cx_k = \mu_k x_k , \qquad k = 1, \dots, n .$$

Set $\mu_k = \lambda_k^2$, $\lambda_k \ge 0$. We claim that there exists an orthonormal set $\{z_1, \dots, z_n\}$ in $\mathbb{R}^N$ such that

$$Tx_k = \lambda_k z_k , \qquad k = 1, \dots, n .$$

For that it suffices to define

$$z_k := \frac{1}{\lambda_k} T x_k$$

if $\lambda_k \neq 0$. As for $\lambda_k, \lambda_\ell \neq 0$

$$z_k \cdot z_\ell = \frac{1}{\lambda_k \lambda_\ell} Tx_k \cdot Tx_\ell = \frac{1}{\lambda_k \lambda_\ell} (Cx_k) \cdot x_\ell = \frac{\lambda_k^2}{\lambda_k \lambda_\ell} x_k \cdot x_\ell = \frac{\lambda_k}{\lambda_\ell} \delta_{\ell k}$$

we see that $\{z_k \mid \lambda_k \neq 0\}$ is orthonormal. Then, for $\lambda_k = 0$, we define z_k to be any vector such that $\{z_1, \dots, z_n\}$ is orthonormal. We now define $S : \mathbb{R}^n \to \mathbb{R}^n$ by

$$Sx_k = \lambda_k x_k \qquad k = 1, \dots, n$$

and $U : \mathbb{R}^n \to \mathbb{R}^N$ by

$$Ux_k = z_k .$$

Clearly S is symmetric and U orthogonal, and $U \circ Sx_k = \lambda_k Ux_k = \lambda_k z_k = Tx_k$, so $T = U \circ S$. □

We can now state

Theorem 2 (Area formula for linear maps). *Let $T : \mathbb{R}^n \to \mathbb{R}^N$ be linear, and $n \leq N$. Then for any measurable set $A \subset \mathbb{R}^n$ we have*

$$\mathcal{H}^n(T(A)) = J_T \, \mathcal{L}^n(A) \tag{1}$$

where

$$J_T = \sqrt{\det (T^*T)}$$

Proof. By the polar decomposition theorem we can write

$$T = US$$

and, being S symmetric,

$$S = VS_dV^*$$

where S_d is diagonal, and U and V are orthogonal, that is isometries. Thus we deduce, taking into account *Binet's formula* $\det AB = \det A \det B$,

$$\begin{aligned} \mathcal{H}^n(T(A)) &= \mathcal{H}^n(US(A)) = \mathcal{H}^n(S(A)) = \mathcal{H}^n(S_d(A)) = \mathcal{L}^n(S_d(A)) \\ &= \int_{S_d(A)} dx = \int_A |\det S_d| \, d\mathcal{L}^n = |\det S_d| \mathcal{L}^n(A) = |\det S| \mathcal{L}^n(A) . \end{aligned}$$

It now remain to observe that $T^*T = S^*U^*US = S^*S$, so that

$$(\det S)^2 = \det T^*T .$$

□

It is usual to refer to J_T as to the *Jacobian of the map* T and denote it also by $J(T)$.

Notice that $J_T \neq 0$ if and only if T is injective, i.e., $\dim T(\mathbb{R}^n) = n$, therefore we can also write the area formula (1) as

$$\int_A J_T \, dx = \int_{T(A)} \mathcal{H}^0(A \cap T^{-1}(y)) \, d\mathcal{H}^n(y) . \tag{2}$$

A simple consequence of Theorem 2 is the following

Theorem 3 (Change of variable formula for linear maps). *Suppose that T is a linear map from $\mathbb{R}^n$ into $\mathbb{R}^n$, that for the sake of simplicity we assume invertible and let $u : \mathbb{R}^n \to \mathbb{R}$ be a Lebesgue summable map. Then we have*

$$\int_{\mathbb{R}^n} u(x)\,|\det T|\,dx \;=\; \int_{\mathbb{R}^n} u(T^{-1}y)\,dy\ . \tag{3}$$

Proof. By writing u as the difference of its positive and negative part, we reduce to prove (3) in the case $u \geq 0$. Since in this case u is the limit of a non- decreasing sequence of step functions, the result easily follows applying the theorem of monotone convergence and taking into account the linearity of the integral and the equality

$$\sum_{x\in T^{-1}(y)} \chi_A(x) \;=\; \mathcal{H}^0(A\cap T^{-1}(y))$$

for the characteristic function χ_A of A. □

Suppose now $n \geq N$, and let $T : \mathbb{R}^n \to \mathbb{R}^N$ be linear. In this case the *Jacobian of* T, denoted again J_T or $J(T)$, is defined by

$$J_T \;=\; \sqrt{\det\,(TT^*)} \tag{4}$$

and we have

Theorem 4 (Coarea formula for linear maps). *Let $T : \mathbb{R}^n \to \mathbb{R}^N$ be linear, and $n \geq N$. Then for any Lebesgue measurable set $A \subset \mathbb{R}^n$ the mapping $y \in \mathbb{R}^N \to \mathcal{H}^{n-N}(A \cap T^{-1}(y))$ is Lebesgue measurable and we have*

$$\int_{\mathbb{R}^N} \mathcal{H}^{n-N}(A\cap T^{-1}(y))\,dy \;=\; J_T\,\mathcal{L}^n(A) \tag{5}$$

or equivalently

$$\int_{\mathbb{R}^N} \mathcal{H}^{n-N}(A\cap T^{-1}(y))\,d\mathcal{H}^N(y) \;=\; \int_A J_T\,dx\ . \tag{6}$$

Proof. If $\dim T(\mathbb{R}^n) < N$ the claim is trivial. In fact for $\mathcal{H}^N$-a.e. $y \in \mathbb{R}^N$ we have $A \cap T^{-1}(y) = \emptyset$ and consequently $\mathcal{H}^{n-N}(A \cap T^{-1}(y)) = 0$. Also, as $T = SU^*$ by the polar decomposition theorem, we have $T(\mathbb{R}^n) = S(\mathbb{R}^N)$; thus $\dim S(\mathbb{R}^N) < N$ and hence $J_T = |\det S| = 0$.

Suppose therefore $\dim T(\mathbb{R}^n) = N$. By the polar decomposition theorem we have

$$T \;=\; SU^*$$

where $U : \mathbb{R}^N \to \mathbb{R}^n$ is orthogonal and $S : \mathbb{R}^N \to \mathbb{R}^N$ is symmetric; moreover $\det S = J_T \neq 0$. By considering any orthogonal map $V : \mathbb{R}^n \to \mathbb{R}^n$ such that

$$V^*(x^1,\dots,x^N,0,\dots,0) = U(x^1,\dots,x^N)$$

and denoting by $P^* : \mathbb{R}^N \to \mathbb{R}^n$ the linear inclusion

$$P^*(x^1,\dots,x^N) = (x^1,\dots,x^N,0,\dots,0)$$

we also see that $U^* = PV$, i.e., $T = SPV$.
We now observe that for each $y \in \mathbb{R}^N$ $P^{-1}(y)$ as well as $V^{-1}P^{-1}(y)$ and $T^{-1}(y)$ are the translate of the $(n-N)$-dimensional subspaces of $\mathbb{R}^n$ $P^{-1}(0)$, $V^{-1}P^{-1}(0)$ and $T^{-1}(0)$, respectively. By Fubini's theorem we then infer that the functions of the variable y

$$\mathcal{H}^{n-N}(V(A)\cap P^{-1}(y))\,, \qquad \mathcal{H}^{n-N}(A\cap V^{-1}P^{-1}(y))\,, \qquad \mathcal{H}^{n-N}(A\cap T^{-1}(y))$$

are Lebesgue measurable and

$$\begin{aligned}\mathcal{L}^n(A) = \mathcal{L}^n(V(A)) &= \int_{\mathbb{R}^N} \mathcal{H}^{n-N}(V(A)\cap P^{-1}(y))\,dy \\ &= \int_{\mathbb{R}^N} \mathcal{H}^{n-N}(A\cap V^{-1}P^{-1}(y))\,dy\,.\end{aligned}$$

By changing variable $z = Sy$, the claims follow. □

Notice that for $n = N$ the area and coarea formula are just the same, and the Jacobian of $T : \mathbb{R}^n \to \mathbb{R}^n$ reduces simply to the determinant of the matrix associated to T (with respect to the standard basis of $\mathbb{R}^n$).

We conclude this subsection by stating a formula which expresses the Jacobian of a linear map $T : \mathbb{R}^n \to \mathbb{R}^N$.

First assume $n \le N$, and set

$$I(n,N) := \{\lambda := (\lambda_1,\dots,\lambda_n) \mid \lambda_i \text{ integers, } 1 \le \lambda_1 < \dots < \lambda_n \le N\}$$

For each $\lambda \in I(n,N)$ we denote by S_λ the span of $\{e_{\lambda_1},\dots,e_{\lambda_n}\}$ and by P_λ the projection of $\mathbb{R}^N$ into S_λ defined by

$$P_\lambda(x^1,\dots,x^N) = (x^{\lambda_1},\dots,x^{\lambda_n})\,.$$

Theorem 5 (Cauchy-Binet formula). *Let $T : \mathbb{R}^n \to \mathbb{R}^N$ be a linear map*

(i) If $n \le N$, then

$$J_T^2 = \sum_{\lambda\in I(n,N)} (\det P_\lambda \circ T)^2$$

(ii) If $n \ge N$, then

$$J_T = J_{T^*}\,;$$

thus we can use (i) to compute J_T.

We postpone the proof of Theorem 5 till Sec. 2.2.1, here we only remark that, analytically, it says that J_T is the square root of the sum of the squares of the $n \times n$-subdeterminants of the $(N \times n)$-matrix associated to T, geometrically in view of Theorem 2, it says that the square of the $\mathcal{H}^n$-measure of A equals the sum of the squares of the $\mathcal{H}^n$-measures of the projections of A onto the coordinate planes, thus it may be regarded as a generalization of *Pitagora's theorem* to which it reduces if $n = 1$.

1.2 Area Formula for Lipschitz Maps

In this subsection we shall generalize Theorem 2 in Sec. 2.1.1 to Lipschitz maps.

Let $f : \mathbb{R}^n \to \mathbb{R}^N$ be a Lipschitz map. In Ch. 3 we shall see that

(A) *f is almost everywhere differentiable in the classical sense with differential represented by Jacobi's matrix*

$$Df(x) = \begin{pmatrix} \dfrac{\partial f^1}{\partial x^1} & \cdots & \dfrac{\partial f^1}{\partial f^n} \\ \vdots & & \vdots \\ \dfrac{\partial f^N}{\partial x^1} & \cdots & \dfrac{\partial f^N}{\partial f^n} \end{pmatrix} .$$

Also the partial derivatives $\frac{\partial f^j}{\partial x^i}$, which exist a.e., *agree with the distributional derivatives.*

(B) *For every $\varepsilon > 0$ there exist a closed set F_ε and a C^1-map $g_\varepsilon : \mathbb{R}^n \to \mathbb{R}^N$ such that*

$$\begin{aligned} &f = g_\varepsilon \,, \qquad Df = Dg_\varepsilon \quad \text{in } F_\varepsilon \\ &\mathcal{L}^n(\mathbb{R}^n \setminus F_\varepsilon) < \varepsilon \\ &\| Dg_\varepsilon \|_{\infty,\mathbb{R}^n} \le c\, Lip\, f \end{aligned}$$

for some constant c independent of ε.

Using these results we shall now prove

Theorem 1 (Area formula). *Let $f : \mathbb{R}^n \to \mathbb{R}^N$ be a Lipschitz map, $n \le N$, and let $A \subset \mathbb{R}^n$ be a Lebesgue measurable set. Then*

$$\int_A J_f(x)\, dx = \int_{f(A)} \mathcal{H}^0(A \cap f^{-1}(y))\, d\mathcal{H}^n(y) \tag{1}$$

where $J_f(x)$ is the Jacobian of f

$$J_f(x) := \sqrt{\det\left(Df(x)^* Df(x)\right)} \,. \tag{2}$$

In particular we have

(3) $$\mathcal{H}^n(f(A)) = \int_A J_f(x)\,dx$$

if f is injective on A.

The function

$$y \longrightarrow \mathcal{H}^0(A \cap f^{-1}(y)) = \#\{x \mid f(x) = y,\ x \in A\} =: N(f, A, y)$$

is called the *multiplicity function* or the *Banach indicatrix* or the *counting function* of f over A. Notice that

$$f(A) = \{y \mid N(f, A, y) \neq 0\}$$

Since $\frac{\partial f^j}{\partial x^i}$ are Borel functions, being limits of continuous functions, the Jacobian $J_f(x)$ *is a Borel function.* Below we shall see that the multiplicity function and $f(A)$ are Lebesgue-measurable. Therefore both integrals in (1) exist: Theorem 1 states equality of those integrals; in particular if one of them is finite, the other one is finite, too.

A simple consequence of Theorem 1 will be

Theorem 2 (Change of variable formula). *Let $f : \mathbb{R}^n \to \mathbb{R}^N$ be Lipschitz, $n \le N$, and let $u : \mathbb{R}^n \to [0, +\infty]$ be a Lebesgue measurable map. Then the map*

$$y \longrightarrow \sum_{x \in f^{-1}(y)} u(x)$$

is measurable and

(4) $$\int_{\mathbb{R}^n} u(x)\, J_f(x)\,dx = \int_{\mathbb{R}^N} \Big(\sum_{x \in f^{-1}(y)} u(x) \Big)\, d\mathcal{H}^n(y)\,.$$

In particular, denoting by $N(f, A, y)$ the counting function of f over A,

$$N(f, A, y) := \#\{x \mid f(x) = y,\ x \in A\},$$

for any $\mathcal{H}^n$–measurable function $v : \mathbb{R}^N \to [0, +\infty]$ we have

(5) $$\int_A v(f(x))\, J_f(x)\,dx = \int v(y)\, N(f, A, y)\, d\mathcal{H}^n(y)\,.$$

Proof of Theorem 1. We split it into several steps.

Step 1. $f(A)$ is measurable. We may assume A bounded otherwise we write A as $A = \bigcup_k A_k$, A_k bounded, $A_k \subset A_{k+1}$, and we notice that $f(A) = f(\bigcup_k A_k) = \bigcup_k f(A_k)$.

Let $\{K_h\}$ be a non decreasing sequence of compact sets such that $\mathcal{L}^n(A) = \sup_h \mathcal{L}^n(K_h)$. Then

$$\mathcal{L}^n(\bigcup_h K_h) = \mathcal{L}^n(A) , \qquad \bigcup_h K_h \subset A .$$

Being f continuous, it follows that $f(\bigcup_h K_h) = \bigcup_h f(K_h)$ is a Borel set. Now

$$f(A) = f(\bigcup_h K_h) \cup f(A \setminus \bigcup_h K_h) ,$$

and, since f is Lipschitz, it has Lusin property (N)

$$\mathcal{H}^n(f(A \setminus \bigcup_h K_h)) \le (Lip\ f)^n \mathcal{L}^n(A \setminus \bigcup_h K_h) = 0 ;$$

this shows that $f(A)$ is measurable.

Step 2. $\mathcal{H}^0(A \cap f^{-1}(y))$ *is* $\mathcal{H}^n$*-measurable and*

$$\int_{\mathbb{R}^N} \mathcal{H}^0(A \cap f^{-1}(y))\, d\mathcal{H}^n(y) \le (Lip\ f)^n \mathcal{L}^n(A) .$$

We construct a nondecreasing sequence $\{g_k\}$ of $\mathcal{H}^n$-measurable functions converging a.e. to $\mathcal{H}^0(A \cap f^{-1}(y))$; this shows that $\mathcal{H}^0(A \cap f^{-1}(y))$ is $\mathcal{H}^n$-measurable.

Dividing the unit cube of $\mathbb{R}^n$ into 2^n congruent cubes, iterating the process, and extending by periodicity such a subdivision, we consider a so-called *dyadic decomposition of* $\mathbb{R}^n$ $\{Q_i^k\}$ with side $Q_i^k = 2^{-k}$. Set now

$$g_k(y) := \sum_{i=1}^{\infty} \chi_{f(A \cap Q_i^k)}(y) .$$

As $f(A \cap Q_i^k)$ is $\mathcal{H}^n$-measurable for all i, g_k is measurable for each k. Moreover we clearly have $g_k(y) \le g_{k+1}(y)$. Observe now

1. If $f^{-1}(y) \cap A = \emptyset$, then $g_k(y) = 0 = \mathcal{H}^0(f^{-1}(y) \cap A)$ for all k.
2. If $\mathcal{H}^0(A \cap f^{-1}(y)) < \infty$, then for k sufficiently large every point in $f^{-1}(y) \cap A$ belongs to exactly only one cube Q_i^k, hence

$$g_k(y) = \mathcal{H}^0(A \cap f^{-1}(y)) .$$

3. If $\mathcal{H}^0(A \cap f^{-1}(y)) = +\infty$, we trivially have $\lim_{k\to\infty} g_k(y) = +\infty$.

In conclusion we find

$$g_k(y) \uparrow \mathcal{H}^0(A \cap f^{-1}(y))$$

for each $y \in \mathbb{R}^N$.

Finally by the Monotone Convergence Theorem

$$\int_{\mathbb{R}^N} \mathcal{H}^0(A \cap f^{-1}(y))\, d\mathcal{H}^n(y) = \lim_{k\to\infty} \int_{\mathbb{R}^N} g_k(y)\, d\mathcal{H}^n(y)$$

$$= \lim_{k\to\infty} \sum_i \mathcal{H}^n(f(A \cap Q_i^k)) \le \limsup_{k\to\infty} \sum_i (Lip\ f)^n \mathcal{L}^n(A \cap Q_i^k)$$

$$= (Lip\ f)^n \mathcal{L}^n(A) .$$

Step 3. It suffices to prove (1) for Lipschitz maps which are also of class C^1. In fact assuming (1) valid for C^1-maps and applying (B) above we infer

$$\begin{aligned}\int_A J_f(x)\,dx &= \int_{F_\varepsilon\cap A} J_{g_\varepsilon}(x)\,dx + \int_{A\setminus F_\varepsilon} J_f(x)\,dx \\ &= \int_{f(F_\varepsilon\cap A)} \mathcal{H}^0(F_\varepsilon\cap A\cap f^{-1}(y))\,d\mathcal{H}^n(y) + \int_{A\setminus F_\varepsilon} J_f(x)\,dx\,.\end{aligned}$$

By the Monotone Convergence Theorem letting $\varepsilon \to 0$ we then get (1), as $\mathcal{L}^n(A\setminus F_\varepsilon) \to 0$.

Step 4. In view of Step 3 from now on we may assume f of class C^1. Denote by B the Borel set, actually the open set

$$B := \{x \in \mathbb{R}^n \mid J_f(x) > 0\}\,.$$

We now claim[1]: *Let $t > 1$. There exists a sequence $\{E_j\}$ of disjoint Borel sets such that*

$$B = \bigcup_j E_j\,, \qquad f_{|E_j} \text{ is invertible.}$$

Moreover, there exist linear, symmetric, invertible maps $T_j : \mathbb{R}^n \to \mathbb{R}^n$ and Lipschitz invertible maps $g_j : T_j(E_j) \to f(E_j)$ with Lip $g_j \le t$, Lip $g_j^{-1} \le t$ such that

(6) $$f_{|E_j} = g_j \circ T_j\,.$$

In particular

(7) $$\frac{1}{t^n}\,|\det T_j| \le J_f(x) \le t^n\,|\det T_j| \qquad \forall\, x \in E_j\,.$$

Of course we can split B as union of Borel sets E_j in such a way that $f_{|E_j}$ is invertible. Taking into account the continuity of Df we can also assume that the E_j's be small enough so that, setting

$$T_j := (Df(x_j))^*\,(Df(x_j))$$

where x_j is a point in E_j, the map

$$g_j := f_{|E_j} \circ T_j^{-1}$$

is as close as we like to the identity. This proves the claim.

We are now ready to conclude the proof. For that we distinguish two cases

[1] Such a claim could be proved directly for Lipschitz maps. But its proof is quite more delicate, compare Federer [226, 3.2.2], as in this case one cannot use the continuity of Df.

Step 5. $A \subset B := \{x \in \mathbb{R}^n \mid J_f(x) > 0\}$. Consider the dyadic decomposition $\{Q_i^k\}$ as in Step 2, and let $t > 1$. In view of Step 4 we write $B = \cup_j E_j$ and we define

$$F_{ij}^k := A \cap E_j \cap Q_i^k, \qquad G_k(y) := \sum_{i,j} \chi_{f(F_{ij}^k)}(y).$$

For all k the family $\{F_{ij}^k\}$ defines a partition of A, and the map G_k is $\mathcal{H}^n$-measurable. Moreover $\chi_{f(F_{ij}^k)} = 1$ iff $y \in f(F_{ij}^k)$, that is, iff $f^{-1}(y) \in F_{ij}^k$, and f is invertible on F_{ij}^k. It follows, compare Step 2,

$$G_k(y) \uparrow \mathcal{H}^0(A \cap f^{-1}(y)) \, ,$$

as $k \to \infty$, and the Monotone Convergence Theorem yields

$$\int_{\mathbb{R}^N} G_k(y)\, d\mathcal{H}^n(y) \uparrow \int_{\mathbb{R}^N} \mathcal{H}^0(A \cap f^{-1}(y))\, d\mathcal{H}^n(y) \, .$$

On the other hand

$$\int_{\mathbb{R}^N} G_k(y)\, d\mathcal{H}^n(y) = \sum_{i,j} \mathcal{H}^n(f(F_{ij}^k))$$

and

$$\begin{aligned} \mathcal{H}^n(f(F_{ij}^k)) &= \mathcal{H}^n(g_j(T_j(F_{ij}^k))) \le (Lip\ g_j)^n\, \mathcal{H}^n(T_j(F_{ij}^k)) \\ &\le t^n\, \mathcal{L}^n(T_j(F_{ij}^k)) = t^n\, |\det T_j|\, \mathcal{L}^n(F_{ij}^k) = t^n \int_{F_{ij}^k} |\det T_j|\, dx \\ &\le t^{2n} \int_{F_{ij}^k} J_f(x)\, dx \, , \end{aligned}$$

if we take into account Theorem 2 in Sec. 2.1.1 and (7) above. Also

$$\begin{aligned} \mathcal{H}^n(T_j(F_{ij}^k)) &= \mathcal{H}^n(g_j^{-1}(f(F_{ij}^k))) \le (Lip\ g_j^{-1})^n\, \mathcal{H}^n(f(F_{ij}^k)) \\ &\le t^n\, \mathcal{H}^n(f(F_{ij}^k)) \end{aligned}$$

hence, similarly to the above,

$$\begin{aligned} \mathcal{H}^n(f(F_{ij}^k)) &\ge t^{-n}\, \mathcal{H}^n(T_j(F_{ij}^k)) = t^{-n}\, |\det T_j|\, \mathcal{L}^n(F_{ij}^k) \\ &\ge t^{-2n} \int_{F_{ij}^k} J_f(x)\, dx \, . \end{aligned}$$

Therefore we conclude

$$t^{-2n} \int_A J_f(x)\, dx \;\leq\; \int_{\mathbb{R}^N} G_k(y)\, d\mathcal{H}^n(y) \;\leq\; t^{2n} \int_A J_f(x)\, dx\,,$$

and the result follows first letting $k \to \infty$, and then $t \to 1$.

Step 6. $J_f \equiv 0$ *on* A. In this case it suffices to check that

$$\int_{\mathbb{R}^N} \mathcal{H}^0(A \cap f^{-1}(y))\, d\mathcal{H}^n(y) \;=\; 0$$

or equivalently that $\mathcal{H}^n(f(A)) = 0$. For $\varepsilon > 0$ we define $g_\varepsilon : A \subset \mathbb{R}^n \to \mathbb{R}^N \times \mathbb{R}^n$ by

$$g_\varepsilon(x) \;:=\; (f(x), \varepsilon x)$$

and we factor f as

$$f(x) \;=\; \pi(g_\varepsilon(x))$$

where $\pi(y,z) := y$ for $y \in \mathbb{R}^N$ and $z \in \mathbb{R}^n$. Now we apply Step 5 to g_ε to infer

$$\mathcal{H}^n(f(A)) = \mathcal{H}^n(\pi(g_\varepsilon(A))) \leq (Lip\ \pi)^n\, \mathcal{H}^n(g_\varepsilon(A)) = \int_A J_{g_\varepsilon}(x)\, dx\,.$$

Letting $\varepsilon \to 0$ we then get

$$\mathcal{H}^n(f(A)) \;\leq\; \int_A J_{g_0}(x)\, dx \;=\; 0$$

by the continuity of $J_{g_\varepsilon}(x)$ in x, as $J_{g_0}(x) = 0$.

To conclude the proof of the theorem it suffices now to split A as

$$A \;=\; \{x \in A \mid J_f(x) > 0\} \cup \{x \in A \mid J_f(x) = 0\}$$

and apply Step 5 and Step 6. □

Proof of Theorem 2. If u is the characteristic function of a measurable set (4) simply reduces to (1) as

$$\sum_{x \in f^{-1}(y)} \chi_A(x) \;=\; \mathcal{H}^0(A \cap f^{-1}(y))\,.$$

Consequently (4) holds for all step functions, and therefore for all measurable functions u as they are monotone limits of step functions. In order to prove (5) it suffices to prove that $u(x) := v(f(x))$ is $\mathcal{H}^n$-measurable. This follows from the fact that every Lipschitz map has Lusin property (N), therefore maps measurable sets into measurable sets. □

We shall now state a few important special cases of the area formula, using Cauchy-Binet formula

[1] *Curves in* $\mathbb{R}^N$. Let $\varphi : \mathbb{R} \to \mathbb{R}^N$ be a smooth curve in $\mathbb{R}^N$. Then

$$J_\varphi(t) = |\dot\varphi(t)|$$

and

$$\mathcal{H}^1(\varphi(a,b)) = \int_a^b |\dot\varphi|\, dt \qquad -\infty < a < b < +\infty ,$$

if φ is a one-to-one parametrization. If f is not one-to-one the multiplicity of the covering of $\varphi((a,b))$ comes into play. For example, if $\varphi(t) = (\cos t, \sin t) \in \mathbb{R}^2$, $a = 0$, and $b = 2k\pi + b_1$, $0 \le b_1 < 2\pi$, we have

$$\mathcal{H}^0((a,b) \cap \varphi^{-1}(y)) = \begin{cases} k+1 & \text{if } 0 \le \arg y < b_1 \\ k & \text{if } b_1 < \arg y < 2\pi \end{cases}$$

•

[2] *Graphs of codimension one.* Let $u : \mathbb{R}^n \to \mathbb{R}$ be a Lipschitz function and let Ω be a bounded open set in $\mathbb{R}^n$. The *graph of* u *over* Ω is

$$\mathcal{G}_{u,\Omega} := \{(x,y) \in \mathbb{R}^n \times \mathbb{R} \mid y = u(x)\}$$

and can be regarded as the image of Ω under the *injective* map $\mathcal{U} : \mathbb{R}^n \to \mathbb{R}^n \times \mathbb{R}$ given by

$$\mathcal{U}(x) := (x, u(x)) .$$

Then one easily computes

$$J_{\mathcal{U}}(x) = \sqrt{1 + |Du(x)|^2}$$

hence

$$\mathcal{H}^n(\mathcal{G}_{u,\Omega}) = \mathcal{H}^n(\mathcal{U}(\Omega)) = \int_\Omega \sqrt{1+|Du|^2}\, dx .$$

•

[3] *Parametric hypersurfaces.* Let $u : \mathbb{R}^n \to \mathbb{R}^{n+1}$ be a Lipschitz, and one-to-one into its image map. According to Cauchy-Binet formula

$$J_u^2(x) = \text{sum of the squares of all } n \times n\text{-subdeterminants of } Du(x) .$$

Thus the *area* of the Lipschitz n-dimensional manifold $u(\Omega)$, $\Omega \subset \mathbb{R}^n$, is given by

$$\mathcal{H}^n(u(\Omega)) = \int_\Omega \Big[\sum_{i=1}^{n+1} \Big(\frac{\partial(u^1, \dots, u^{i-1}, u^{i+1}, \dots, u^{n+1})}{\partial(x^1, \dots, x^n)} \Big)^2 \Big]^{1/2} dx .$$

•

[4] *Submanifolds in* $\mathbb{R}^N$. Let $M \subset \mathbb{R}^N$ be an n-dimensional embedded submanifold. Let $\Lambda \subset M$ and let $f : \Omega \subset \mathbb{R}^n \to M$ be a parametrization of Λ over Ω, i.e., a smooth one-to-one map with $f(\Omega) = \Lambda$. Denote by $\{g_{ij}\}$ the *metric tensor* on M

$$g_{ij} := \frac{\partial f}{\partial x^i} \cdot \frac{\partial f}{\partial x^j}$$

and set

$$g := \det(g_{ij}) .$$

Then we have

$$(Df)^* Df = (g_{ij})$$

so that

$$\text{volume of } \Lambda \text{ in } M = \mathcal{H}^n(\Lambda) = \int_\Omega \sqrt{g}\, dx .$$

•

[5] *Graphs of higher codimension.* Let $u : \mathbb{R}^n \to \mathbb{R}^N$ be a Lipschitz map and let Ω a bounded open set in $\mathbb{R}^n$. The *graph of* u *over* Ω is

$$\mathcal{G}_{u,\Omega} := \{(x,y) \in \Omega \times \mathbb{R}^N \mid y = u(x)\}$$

and can be regarded as the image of Ω under the injective map $\mathcal{U} : \mathbb{R}^n \to \mathbb{R}^n \times \mathbb{R}^N$ given by $\mathcal{U}(x) := (x, u(x))$. Set

$$|M_{(0)}(Du)| := 1$$

and for $k = 1, \dots, \underline{n}$, $\underline{n} := \min(n, N)$

$$|M_{(k)}(Du)|^2 := \text{sum of the squares of the determinants of all } k \times k\text{-submatrix of } Du .$$

Applying Cauchy-Binet formula, compare e.g. (17) in Sec. 2.2.1, we then find

$$J_{\mathcal{U}}(x) = \sqrt{\sum_{k=0}^{\underline{n}} |M_{(k)}(Du)|^2} .$$

Consequently,

$$\mathcal{H}^n(\mathcal{G}_{u,\Omega}) = \int_\Omega \sqrt{\sum_{k=0}^{\underline{n}} |M_{(k)}(Du)|^2}\, dx .$$

•

1.3 Coarea Formula for Lipschitz Maps

In this subsection we shall generalize Theorem 2 in Sec. 2.1.1 to Lipschitz maps, proving

Theorem 1 (Coarea formula). *Let $f : \mathbb{R}^n \to \mathbb{R}^N$ be a Lipschitz map, $n \geq N$, and let $A \subset \mathbb{R}^n$ be a Lebesgue measurable set. Then*

$$\int_A J_f(x)\, dx = \int_{\mathbb{R}^N} \mathcal{H}^{n-N}(A \cap f^{-1}(y))\, d\mathcal{H}^N(y) \tag{1}$$

where $J_f(x)$ is the Jacobian of f

$$J_f(x) := \sqrt{\det Df(x)\,(Df(x))^*}\ . \tag{2}$$

A simple consequence of Theorem 1 will be

Theorem 2. *Let $f : \mathbb{R}^n \to \mathbb{R}^N$ be a Lipschitz map, $n \geq N$, and let $u : \mathbb{R}^n \to [0, +\infty]$ be a Lebesgue measurable map. Then*

$$\int_{\mathbb{R}^n} u(x)\, J_f(x)\, dx = \int_{\mathbb{R}^N} \Big[\int_{f^{-1}(y)} u\, d\mathcal{H}^{n-N} \Big]\, d\mathcal{H}^N(y)\ . \tag{3}$$

Before proving those results let us make a few remarks. As we saw in Sec. 2.1.2 $J_f(x)$ is a Borel function, below we shall prove that $y \to \mathcal{H}^{n-N}(A \cap f^{-1}(y))$ is $\mathcal{H}^n$-measurable; therefore both integrals in (1) exist. Theorem 1 states equality of those integrals.

Also, for every $y \in \mathbb{R}^N$ $f^{-1}(y)$ is clearly a Borel set, thus $f^{-1}(y)$ is $\mathcal{H}^{n-N}$-measurable. It is part of the claim in Theorem 2 that

$$y \longrightarrow \int_{f^{-1}(y)} u\, d\mathcal{H}^{n-N}$$

is $\mathcal{H}^N$-measurable, too.

Finally notice that Theorem 1 and Theorem 2 may be regarded as a curvilinear generalization of Fubini's theorem.

Proof of Theorem 1. We split it into several steps.

Step 1. $f(A)$ is $\mathcal{H}^N$-measurable. This can be proved exactly as in Step 1 of the proof of Theorem 1 in Sec. 2.1.2.

Step 2. We have

$$\int^*_{\mathbb{R}^N} \mathcal{H}^{n-N}(A \cap f^{-1}(y))\, dy \leq \frac{\omega_{n-N}\,\omega_N}{\omega_n}\, (Lip\ f)^N\, \mathcal{H}^n(A)\ . \tag{4}$$

For each $j = 1, 2, \dots$ we choose closed balls $\{B_i^j\}$ such that $A \subset \cup_{i=1}^\infty B_i^j$, $\operatorname{diam} B_i^j \le 1/j$ and $\sum_{i=1}^\infty |B_i^j| \le |A| + 1/j$. Then we define

$$g_i^j(y) := \omega_{n-N} \left(\frac{\operatorname{diam} B_i^j}{2} \right)^{n-N} \chi_{f(B_i^j)}(y) .$$

By Step 1 the functions g_i^j are measurable, also for all $y \in \mathbb{R}^N$ we have

$$\mathcal{H}_{1/j}^{n-N}(A \cap f^{-1}(y)) \le \sum_{i=1}^\infty g_i^j(y) .$$

Using Fatou's lemma we therefore get

$$\begin{aligned}
\int^*_{\mathbb{R}^N} \mathcal{H}^{n-N}(A \cap f^{-1}(y))\, dy &= \int^*_{\mathbb{R}^N} \liminf_{j\to\infty} \mathcal{H}_{1/j}^{n-N}(A \cap f^{-1}(y))\, dy \\
&\le \int_{\mathbb{R}^N} \liminf_{j\to\infty} \sum_{i=1}^\infty g_i^j \, dy \le \liminf_{j\to\infty} \sum_{i=1}^\infty \int_{\mathbb{R}^N} g_i^j \, dy \\
&= \liminf_{j\to\infty} \sum_{i=1}^\infty \omega_{n-N} \left(\frac{\operatorname{diam} B_i^j}{2} \right)^{n-N} \mathcal{H}^N(f(B_i^j))
\end{aligned}$$

and, using the isodiametric inequality, compare Sec. 1.1.3, we infer

$$\begin{aligned}
&\int^*_{\mathbb{R}^N} \mathcal{H}^{n-N}(A \cap f^{-1}(y))\, dy \le \\
&\le \liminf_{j\to\infty} \sum_{i=1}^\infty \omega_{n-N} \left(\frac{\operatorname{diam} B_i^j}{2} \right)^{n-N} \omega_N \left(\frac{\operatorname{diam} f(B_i^j)}{2} \right)^N \\
&\le \frac{\omega_{n-N}\, \omega_N}{\omega_n} (Lip\ f)^N \sum_{i=1}^\infty \mathcal{H}^n(B_i^j) \le \frac{\omega_{n-N}\, \omega_N}{\omega_n} (Lip\ f)^N \mathcal{H}^n(A).
\end{aligned}$$

Step 3. The mapping $y \to \mathcal{H}^{n-N}(A \cap f^{-1}(y))$ is $\mathcal{L}^n$-measurable, consequently

$$\int_{\mathbb{R}^N} \mathcal{H}^{n-N}(A \cap f^{-1}(y))\, dy \le \frac{\omega_{n-N}\, \omega_N}{\omega_n} (Lip\ f)^N \mathcal{H}^n(A) . \tag{5}$$

Of course (5) follows from (4) once we establish the measurability of the mapping $y \to \mathcal{H}^{n-N}(A \cap f^{-1}(y))$.

(i) *If A is a compact or an open set, then $y \to \mathcal{H}^{n-N}(A \cap f^{-1}(y))$ is a Borel map.*

First assume A compact. Fix $t \geq 0$, and for each positive integer i, consider the set U_i of all $y \in \mathbb{R}^N$ such that there exist finitely many open sets $S_1, \ldots, S_\ell$ such that

$$A \cap f^{-1}(y) \subset \bigcup_{j=1}^{\ell} S_j\,, \qquad \operatorname{diam} S_j \leq \frac{1}{i}$$

$$\sum_{j=1}^{\ell} \omega_{n-N} \left(\frac{\operatorname{diam} S_j}{2} \right)^{n-N} \leq t + \frac{1}{i}\,.$$

We now claim that the sets U_i are open sets and

$$\{y \mid \mathcal{H}^{n-N}(A \cap f^{-1}(y)) \leq t\} = \bigcap_{i=1}^{\infty} U_i\,. \tag{6}$$

This clearly proves (i) in case A is compact.

To prove that U_i is open, assume $y \in U_i$, $A \cap f^{-1}(y) \subset \bigcup_{j=1}^{\ell} S_j$. Then since f is continuous and A is compact

$$A \cap f^{-1}(z) \subset \bigcup_{j=1}^{\ell} S_j$$

for all z sufficiently close to y.

Let us prove (6). If $\mathcal{H}^{n-N}(A \cap f^{-1}(y)) \leq t$, then for each $\delta > 0$

$$\mathcal{H}_\delta^{n-N}(A \cap f^{-1}(y)) \leq t\,.$$

Given i, choose $\delta \in (0, 1/i)$. Then there exist open sets S_j such that

$$A \cap f^{-1}(y) \subset \bigcup_{j=1}^{\infty} S_j\,, \qquad \operatorname{diam} S_j \leq \delta < \frac{1}{i}$$

$$\sum_{j=1}^{\infty} \omega_{n-N} \left(\frac{\operatorname{diam} S_j}{2} \right)^{n-N} < t + \frac{1}{i}\,.$$

Since $A \cap f^{-1}(y)$ is compact, a finite collection $\{S_1, \ldots, S_\ell\}$ covers $A \cap f^{-1}(y)$, and hence $y \in U_i$. Thus

$$\{y \mid \mathcal{H}^{n-N}(A \cap f^{-1}(y)) \leq t\} \subset \bigcap_{i=1}^{\infty} U_i\,.$$

On the other hand, if $y \in \bigcap_{i=1}^{\infty} U_i$, then for each i

$$\mathcal{H}_{1/i}^{n-N}(A \cap f^{-1}(y)) \leq t + \frac{1}{i}\,,$$

hence $\mathcal{H}^{n-N}(A \cap f^{-1}(y)) \leq t$. This proves (6) and therefore (i) when A is compact.

Finally, suppose A is open. Then there exist compact sets $K_1 \subset K_2 \subset \ldots \subset A$ such that $A = \cup_{i=1}^{\infty} K_i$. Thus, for each $y \in \mathbb{R}^N$,

$$\mathcal{H}^{n-N}(A \cap f^{-1}(y)) = \lim_{i \to \infty} \mathcal{H}^{n-N}(K_i \cap f^{-1}(y)) ,$$

hence the mapping $y \to \mathcal{H}^{n-N}(A \cap f^{-1}(y))$ is again Borel measurable.

(ii) *If A is a Lebesgue measurable set, then $y \to \mathcal{H}^{n-N}(A \cap f^{-1}(y))$ is $\mathcal{L}^n$-measurable.* By writing A as union of an increasing sequence of bounded $\mathcal{L}^n$-measurable sets, we can reduce to the case in which $\mathcal{L}^n(A) < \infty$. Then there exist open sets $V_1 \supset V_2 \supset \ldots \supset A$ such that

$$\lim_{j \to \infty} \mathcal{L}^n(V_j \setminus A) = 0 \qquad \mathcal{L}^n(V_1) < \infty .$$

Now

$$\mathcal{H}^{n-N}(V_i \cap f^{-1}(y)) \leq \mathcal{H}^{n-N}(A \cap f^{-1}(y)) + \mathcal{H}^{n-N}((V_i \setminus A) \cap f^{-1}(y))$$

and by Step 2

$$\limsup_{i \to \infty} \int^*_{\mathbb{R}^N} |\mathcal{H}^{n-N}(V_i \cap f^{-1}(y)) - \mathcal{H}^{n-N}(A \cap f^{-1}(y))| \, d\mathcal{H}^N(y) \leq$$

$$\leq \limsup_{i \to \infty} \frac{\omega_{n-N}\omega_N}{\omega_n} (Lip\ f)^n \, \mathcal{H}^n(V_i \setminus A) = 0 .$$

Consequently

$$\mathcal{H}^{n-N}(V_i \cap f^{-1}(y)) \longrightarrow \mathcal{H}^{n-N}(A \cap f^{-1}(y))$$

$\mathcal{L}^N$-a.e. Therefore (ii) follows from (i).[2]

Step 4. It suffices to prove (1) for Lipschitz maps which are also of class C^1. In fact assuming (1) for C^1 maps and applying (B) in the previous subsection, we infer

$$\int_{F_\varepsilon \cap A} J_f(x)\, dx = \int_{\mathbb{R}^N} \mathcal{H}^{n-N}(F_\varepsilon \cap A \cap f^{-1}(y))\, d\mathcal{H}^n(y) ,$$

and passing to the limit for $\varepsilon \to 0$

$$\int_A J_f(x)\, dx = \int_{\mathbb{R}^N} \mathcal{H}^{n-N}(\bigcup_\varepsilon F_\varepsilon \cap A \cap f^{-1}(y))\, d\mathcal{H}^n(y) .$$

[2] Notice that, since every measurable set can be exhausted apart from a zero-set by an increasing sequence of compact sets, one could slightly shorten the proof by using only Step 2 and the fact that $y \to \mathcal{H}^{n-N}(A \cap f^{-1}(y))$ is a Borel map, if A is compact.

The result then follows since by Step 2

$$\int_{\mathbb{R}^N} |\mathcal{H}^{n-N}(\bigcup_{\varepsilon} F_\varepsilon \cap A \cap f^{-1}(y)) - \mathcal{H}^{n-N}(A \cap f^{-1}(y))| \, d\mathcal{H}^n(y) = 0$$

as $\mathcal{H}^{n-N}(A \setminus (\bigcup_{\varepsilon} F_\varepsilon \cap A)) = 0$.

Step 5. In view of Step 4 from now on we may assume f of class C^1. Denote by B the Borel set, actually the open set,

$$B := \{x \in \mathbb{R}^n \mid J_f(x) > 0\}$$

and assume that $A \subset B$.

The idea is to apply the coarea formula for linear maps. In order to do that it is convenient to introduce for any $\lambda \in I(n-N, n)$ the map $h_\lambda : \mathbb{R}^n \to \mathbb{R}^N \times \mathbb{R}^{n-N}$ given by

$$h_\lambda(x) := (f(x), P_\lambda(x))$$

where $P_\lambda(x)$ is the orthogonal projection

$$P_\lambda(x^1, \ldots, x^n) := (x^{\lambda_1}, \ldots, x^{\lambda_n});$$

in particular $f(x) = \pi \circ h_\lambda(x)$, if $\pi : \mathbb{R}^N \times \mathbb{R}^{n-N} \to \mathbb{R}^N$ is the orthogonal projection $\pi(y, z) = y$. Set also

$$A_\lambda := \{x \in A \mid \det Dh_\lambda \neq 0\}$$

and we observe that

$$A = \bigcup_{\lambda \in I(n-N,n)} A_\lambda ,$$

so we may as well assume that $A = A_\lambda$ for some $\lambda \in I(n-N, n)$. Accordingly, we write h instead of h_λ.

By the local invertibility theorem, for any $t > 1$, we can find disjoint Borel sets $\{D_k\}$ and linear isomorphisms $\{S_k\}$, $S_k : \mathbb{R}^n \to \mathbb{R}^n$, such that

(i) $\mathcal{L}^n(A \setminus \bigcup_{k=1}^{\infty} D_k) = 0$,

(ii) $h_{| D_k}$ is one-to-one,

(iii) the maps $g_k(x) := S_k^{-1} \circ h : D_k \to \mathbb{R}^n$ are one-to-one with

$$Lip\, g_k \leq t \qquad Lip\, g_k^{-1} \leq t ,$$

compare Step 4 of the proof of the area formula in Sec. 2.1.2.

Since

$$Df_{|D_k} = D(\pi \circ h) = \pi \circ S_k \circ Dg_k$$

we have

$$\begin{aligned} J_f^2 &= \det((Df)^* \cdot Df) = \det((Dg_k)^* \cdot (\pi \circ S_k)^* \cdot (\pi \circ S_k) \cdot Dg) \\ &= J^2_{\pi \circ S_k} (\det Dg_k)^2 = J^2_{\pi \circ S_k} J^2_{g_k} . \end{aligned}$$

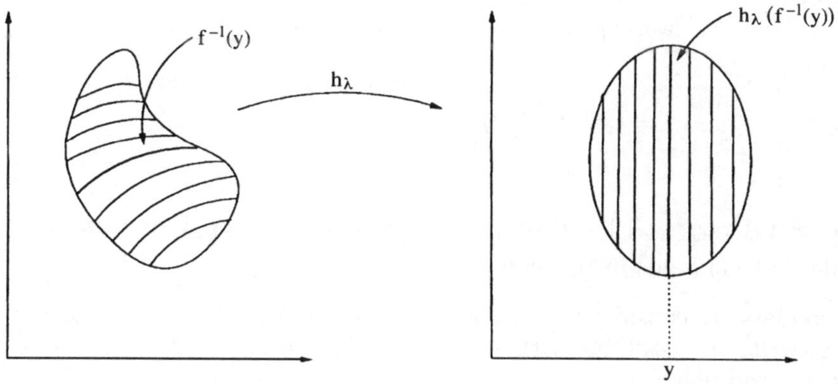

Fig. 2.1. Coarea: the map h_λ.

therefore by the area formula, setting

$$G_k := A \cap D_k$$

we get

$$\int_{G_k} J_f\, dx = J_{\pi\circ S_k}\, \mathcal{H}^n(g_k(G_k)) \,. \tag{7}$$

On the other hand, using the rough estimate for the area we have for any $y \in \mathbb{R}^N$

$$\begin{aligned} t^{N-n}\mathcal{H}^{n-N}(g_k(G_k \cap f^{-1}(y))) &\le \mathcal{H}^{n-N}(G_k \cap f^{-1}(y)) \\ &\le t^{n-N}\mathcal{H}^{n-N}(g_k(G_k \cap f^{-1}(y))), \end{aligned} \tag{8}$$

being *Lip* g_k, *Lip* $g_k^{-1} \le t$. Integrating (8) in y, taking into account the coarea formula for the linear maps $\pi \circ S_k$, compare Theorem 4 in Sec. 2.1.1, using (7) and noting that by construction

$$g_k(G_k \cap f^{-1}(y)) = g_k(G_k) \cap (\pi \circ S_k)^{-1}(y)$$

we then get

$$\begin{aligned} t^{N-n}\int_{G_k} J_f\, dx &= t^{N-n}\, J_{\pi\circ S_k}\, \mathcal{H}^n(g_k(G_k)) \\ &= t^{N-n}\int_{\mathbb{R}^N} \mathcal{H}^{n-N}(g_k(G_k) \cap (\pi\circ S_k)^{-1}(y))\, dy \\ &\le \int_{\mathbb{R}^N} \mathcal{H}^{n-N}(G_k \cap f^{-1}(y))\, dy \end{aligned}$$

$$\begin{aligned} &\leq t^{n-N} \int_{\mathbb{R}^N} \mathcal{H}^{n-N}(g_k(G_k) \cap (\pi \circ S_k)^{-1}(y))\, dy \\ &= t^{n-N} \int_{G_k} J_f \, dx \; . \end{aligned}$$

Now since $\mathcal{H}^n\Big(A \setminus \cup_{k=1}^{\infty} G_k\Big) = 0$ we can sum over k, use Step 3, and let $t \to 1$ to conclude that (1) holds in this case.

Step 6. It remains to consider the case in which $A \subset \{J_f = 0\}$, as in the general case we can write $A = A_1 \cup A_2$ with $A_1 \subset \{J_f > 0\}$ and $A_2 \subset \{J_f = 0\}$.

Fix $\varepsilon > 0$ and define

$$\begin{aligned} &g : \mathbb{R}^n \times \mathbb{R}^N \to \mathbb{R}^N \, , \quad g(x,y) := f(x) + \varepsilon y \, , \quad x \in \mathbb{R}^n, \ y \in \mathbb{R}^N \\ &\widehat{\pi} : \mathbb{R}^n \times \mathbb{R}^N \to \mathbb{R}^N \, , \quad \widehat{\pi}(x,y) := y \; . \end{aligned}$$

Then $Dg = (Df, \varepsilon I)$, therefore $\varepsilon^N \leq J_g \leq C\varepsilon$. Also we observe that for all w

$$\int_{\mathbb{R}^N} \mathcal{H}^{n-N}(A \cap f^{-1}(y))\, dy = \int_{\mathbb{R}^N} \mathcal{H}^{n-N}(A \cap f^{-1}(y - \varepsilon w))\, dy$$

hence

$$\int_{\mathbb{R}^N} \mathcal{H}^{n-N}(A \cap f^{-1}(y))\, dy = \frac{1}{\omega_N} \int_{B(0,1)} dw \int_{\mathbb{R}^N} \mathcal{H}^{n-N}(A \cap f^{-1}(y - \varepsilon w))\, dy \; .$$

We now claim that for $y \in \mathbb{R}^N$, $w \in \mathbb{R}^N$ and $B := A \times B(0,1)$ we have

$$B \cap g^{-1}(y) \cap \widehat{\pi}^{-1}(w) = \begin{cases} \emptyset & \text{if } w \notin B(0,1) \\ (A \cap f^{-1}(y - \varepsilon w)) \times \{w\} & \text{if } w \in B(0,1) \; . \end{cases}$$

In fact $(x,z) \in B \cap g^{-1}(y) \cap \widehat{\pi}^{-1}(w)$ if and only if

$$x \in A \, , \qquad z \in B(0,1) \, , \qquad f(x) + \varepsilon z = y \, , \qquad z = w$$

i.e., if and only if

$$x \in A \, , \qquad z = w \in B(0,1) \, , \qquad f(x) = y - \varepsilon z$$

or, if and only if

$$w \in B(0,1) \, , \qquad (x,z) \in (A \cap f^{-1}(y - \varepsilon w)) \times \{w\} \; .$$

Thus we can continue the computation above using Fubini's theorem and Step 5

$$\begin{aligned}
\int_{\mathbb{R}^N} \mathcal{H}^{n-N}(A \cap f^{-1}(y))\, dy &= \\
&= \frac{1}{\omega_N} \int_{\mathbb{R}^N} dw \int_{\mathbb{R}^N} \mathcal{H}^{n-N}(B \cap g^{-1}(y) \cap \widehat{\pi}^{-1}(w))\, dy \\
&= \frac{1}{\omega_N} \int_{\mathbb{R}^N} \mathcal{H}^{n}(B \cap g^{-1}(y))\, dy = \frac{1}{\omega_N} \int_B J_g \, dx\, dz \\
&\leq \mathcal{H}^n(A) \sup_B J_g \;\leq\; \mathcal{H}^n(A)\,\varepsilon
\end{aligned}$$

and, for $\varepsilon \to 0$, we obtain

$$\int_{\mathbb{R}^N} \mathcal{H}^{n-N}(A \cap f^{-1}(y))\, dy \;=\; 0 \;=\; \int_A J_f \, dx \,.$$

□

Proof of Theorem 2. It is exactly the same as the proof of Theorem 2 in Sec. 2.1.2.
□

[1] *C^1-Sard type theorem.* A consequence of Theorem 1 is that

$$\mathcal{H}^{n-N}(\{J_f = 0\} \cap f^{-1}(y)) \;=\; 0 \qquad \text{for } \mathcal{H}^n\text{-a.e. } y \in \mathbb{R}^N \,.$$

On the other hand if f is of class C^1, $x \in A$, A open, is such that $J_f(x) \neq 0$, then as consequence of the implicit function theorem we can find a neighbourhood V of x such that $V \cap f^{-1}(y)$ is an $(n-N)$-dimensional C^1-submanifold. Therefore we can state

Corollary 1 (C^1–Sard type theorem). *Suppose $f : A \subset \mathbb{R}^n \to \mathbb{R}^N$ is of class C^1, where A is open. Then for $\mathcal{H}^N$-a.e. $y \in f(A)$ we can decompose $f^{-1}(y)$ as*

$$f^{-1}(y) \;=\; (f^{-1}(y) \setminus C) \cup (f^{-1}(y) \cap C)$$

where

$$C \;:=\; \{x \in A \mid J_f(x) = 0\} = \{x \mid \operatorname{rank} Df(x) < N\}$$

$f^{-1}(y) \setminus C$ is an $(n-N)$-dimensional submanifold and

$$\mathcal{H}^{n-N}(C \cap f^{-1}(y)) \;=\; 0 \,.$$

Recall that *Sard theorem* asserts that, if f is of class C^{n-N+1}, then $f^{-1}(y) \cap C = \emptyset$ for $\mathcal{H}^N$-a.e. $y \in \mathbb{R}^N$, so that $f^{-1}(y)$ is itself an $(n-N)$-dimensional C^{n-N+1} submanifold for $\mathcal{H}^N$-a.e. $y \in \mathbb{R}^N$. •

[2] Choosing $f(x) := |x|$ in Theorem 2, we infer at once the following well-known formula for *polar coordinates*

$$\int_{\mathbb{R}^n} u\,dx = \int_{-\infty}^{+\infty} dr \int_{f^{-1}(r)} u\,d\mathcal{H}^{n-1} = \int_0^{\infty} dr \int_{\partial B(0,r)} u\,d\mathcal{H}^{n-1} .$$

In particular we see that for $\mathcal{H}^1$-a.e. r

$$\frac{d}{dr}\Big(\int_{B(0,r)} u\,dx \Big) = \int_{\partial B(0,r)} u\,d\mathcal{H}^{n-1} .$$

In the codimension one case, $f : \mathbb{R}^n \to \mathbb{R}$, as $J_f = |Df|$, Theorem 1 yields at once

$$\int_{\mathbb{R}^n} |Df|\,dx = \int_{-\infty}^{+\infty} \mathcal{H}^{n-1}(\{f = t\})\,dt .$$

More generally, we have

$$\int_{\{f>t\}} |Df|\,dx = \int_t^{\infty} \mathcal{H}^{n-1}(\{f = t\})\,dt$$

so that for $\mathcal{H}^1$-a.e. t

$$\frac{d}{dt} \int_{\{f>t\}} |Df|\,dx = -\mathcal{H}^{n-1}(\{f = t\}) .$$

•

1.4 Rectifiable Sets and the Structure Theorem

The notion of *rectifiable sets* we shall discuss in this section turns out to be the appropriate generalization of n-dimensional C^1-submanifolds. Roughly, n-rectifiable sets are characterized by the equivalent properties of being countable union of measurable pieces of n-dimensional C^1-submanifolds or of possessing *approximate tangent space* $\mathcal{H}^n$-almost everywhere.

Definition 1. *A set $\mathcal{M} \subset \mathbb{R}^{n+N}$ is said to be* countably n-rectifiable *if there exist n-dimensional embedded submanifolds $\mathcal{N}_1, \mathcal{N}_2, \ldots$ and $\mathcal{N}_0 \subset \mathbb{R}^{n+N}$ with $\mathcal{H}^n(\mathcal{N}_0) = 0$ such that*

(1) $$\mathcal{M} \subset \mathcal{N}_0 \cup \bigcup_{k=1}^{\infty} \mathcal{N}_k .$$

Let $\mathcal{M}$ be a countably n-rectifiable set which is also $\mathcal{H}^n$-measurable. Then clearly we can write $\mathcal{M}$ as disjoint union

(2) $$\mathcal{M} = \mathcal{M}_0 \cup \bigcup_{k=1}^{\infty} \mathcal{M}_k$$

where $\mathcal{H}^n(\mathcal{M}_0) = 0$, and each $\mathcal{M}_k$ is a Borel subset of an n-dimensional submanifold. If moreover $\mathcal{H}^n(\mathcal{M}) < \infty$, we then have

$$\mathcal{H}^n(\mathcal{M}) = \sum_{k=1}^{\infty} \mathcal{H}^n(\mathcal{M}_k) . \tag{3}$$

We may also and do assume that at each point $x \in \mathcal{M}_k$ we have (compare Theorem 6 in Sec. 1.1.5)

$$\theta^n(\mathcal{M}_k, x) = 1, \qquad \theta^n(\mathcal{M} \setminus \mathcal{M}_k, x) = 0 .$$

Definition 2. *A countably n-rectifiable set $\mathcal{M}$ is said to be n-*rectifiable *if and only if $\mathcal{M}$ is $\mathcal{H}^n$-measurable and $\mathcal{H}^n(\mathcal{M}) < \infty$, or equivalently, iff it can be written apart for a null $\mathcal{H}^n$-set as disjoint union of Borel subsets of n-dimensional C^1-submanifolds with finite $\mathcal{H}^n$ measure.*

If $\mathcal{M}$ is countably $\mathcal{H}^n$-measurable, then $\mathcal{M}$ is countably n-rectifiable if and only if it is a countable union of n-rectifiable sets. Notice also that an n-rectifiable set apart for sets of small $\mathcal{H}^n$-measure is a *finite* union of Borel pieces of C^1-submanifolds.

Despite this nice structure however n-rectifiable sets can be everywhere dense. Consider for instance a dense set $\{x_k\}$ in $\mathbb{R}^{n+N}$ and denote by $B^n(x_k, r_k)$ the n-dimensional ball in $\mathbb{R}^{n+N}$ of radius r_k and center x_k. Choose r_k so that

$$\sum_{k=1}^{\infty} r_k^n < \infty ,$$

then clearly

$$\mathcal{M} := \bigcup_{k=1}^{\infty} B^n(x_k, r_k)$$

is a dense n-rectifiable set in $\mathbb{R}^{n+N}$.

Taking into account Lusin type property (B) of Sec. 2.1.2 for Lipschitz maps, and also the extension properties of Lipschitz and C^1-maps, compare Ch. 3, it is not difficult to prove also that countably rectifiable sets are characterized by the property of being countable union of Lipschitz images of n-dimensional sets

Proposition 1. *A set $\mathcal{M} \subset \mathbb{R}^{n+N}$ is countably n-rectifiable if and only if $\mathcal{M} \subset A_0 \cup \big(\bigcup_{k=1}^{\infty} f_k(A_k)\big)$ where $\mathcal{H}^n(A_0) = 0$ and $f_k : A_k \to \mathbb{R}^{n+N}$ are Lipschitz maps, $A_k \subset \mathbb{R}^n$.*

We shall now prove that rectifiable sets are characterized by their tangential properties. First let us define the *approximate tangent space* in terms of a blow-up procedure

Definition 3. *Let $\mathcal{M} \subset \mathbb{R}^{n+N}$ be an $\mathcal{H}^n$-measurable set such that $\mathcal{H}^n(\mathcal{M} \cap K) < \infty$ for every compact $K \subset \mathbb{R}^{n+N}$. We say that an n-dimensional subspace P of $\mathbb{R}^{n+N}$ is the* approximate tangent space *for $\mathcal{M}$ at $x \in \mathbb{R}^{n+N}$ if*

$$\lim_{\lambda \to 0+} \int_{\eta_{x,\lambda}(\mathcal{M})} f(y)\, d\mathcal{H}^n(y) = \int_P f(y)\, d\mathcal{H}^n(y) \qquad \forall\, f \in C_c(\mathbb{R}^{n+N}) \tag{4}$$

where $\eta_{x,\lambda}(y) := \frac{y-x}{\lambda}$ for $x, y \in \mathbb{R}^{n+N}$, $\lambda > 0$.

Of course P is unique, if it exists, and will be denoted by $T_x\mathcal{M}$. By definition it is characterized by the property

$$\mathcal{H}^n \llcorner \eta_{x,\lambda}(\mathcal{M}) \rightharpoonup \mathcal{H}^n \llcorner P\,. \tag{5}$$

The condition that $\mathcal{M}$ has locally finite $\mathcal{H}^n$-measure can be removed; and it is convenient to do it. It suffices to suppose the existence of a positive locally $\mathcal{H}^n$-integrable function θ on $\mathcal{M}$. This is equivalent to the requirement that $\mathcal{M}$ can be expressed as the countable union of $\mathcal{H}^n$-measurable sets with locally finite $\mathcal{H}^n$-measure. We then set

Definition 4. *Let $\mathcal{M} \subset \mathbb{R}^{n+N}$ be an $\mathcal{H}^n$-measurable set and let θ be a positive locally $\mathcal{H}^n$-integrable function on $\mathcal{M}$. We say that an n-dimensional subspace P of $\mathbb{R}^{n+N}$ is the* approximate tangent space for $\mathcal{M}$ at $x \in \mathbb{R}^{n+N}$ with respect to *θ if*

$$\lim_{\lambda \to 0} \int_{\eta_{x,\lambda}(\mathcal{M})} f(y)\theta(x+\lambda y) d\mathcal{H}^n(y) = \theta(x) \int_P f(y)\, d\mathcal{H}^n(y) \quad \forall\, f \in C_c(\mathbb{R}^{n+N}). \tag{6}$$

Actually P does not depend on θ in the sense that the approximate tangent spaces of $\mathcal{M}$ with respect to two different positive $\mathcal{H}^n$-integrable functions θ and $\tilde{\theta}$ coincide $\mathcal{H}^n$-a.e. in $\mathcal{M}$. In fact if $\mu := \mathcal{H}^n \llcorner \theta$ and $\mathcal{M}_\varepsilon := \{x \in \mathcal{M} \mid \theta(x) > \varepsilon\}$ we have $\mathcal{H}^n(\mathcal{M}_\varepsilon \cap K) < \infty$ for every compact $K \subset \mathbb{R}^{n+N}$ and, by Theorem 6 in Sec. 1.1.5, $\theta^*(\mu, \mathcal{M} \setminus \mathcal{M}_\varepsilon) = 0$ for $\mathcal{H}^n$-a.e. $x \in \mathcal{M}_\varepsilon$. Hence for $\mathcal{H}^n$-a.e. $x \in \mathcal{M}_\varepsilon$ the approximate space for $\mathcal{M}$ with respect to θ coincide with $T_x\mathcal{M}_\varepsilon$ as defined in Definition 3 if the latter exists.

Let $\mathcal{M} \subset \mathbb{R}^{n+N}$ be an $\mathcal{H}^n$-measurable and countably n-rectifiable set. Then we can write $\mathcal{M}$ as the disjoint union

$$\mathcal{M} = \mathcal{M}_0 \cup \bigcup_{k=1}^{\infty} \mathcal{M}_k$$

where $\mathcal{H}^n(\mathcal{M}_0) = 0$, $\mathcal{M}_k$ is $\mathcal{H}^n$-measurable and $\mathcal{M}_k \subset \mathcal{N}_k$, where $\mathcal{N}_k$ is an embedded n-dimensional C^1-submanifold of $\mathbb{R}^{n+N}$. Let $\mu := \theta\mathcal{H}^n \llcorner \mathcal{M}$ where θ is any positive locally $\mathcal{H}^n$-integrable function on $\mathcal{M}$, and assume without loss of generality that $\mathcal{H}^n(\mathcal{N}_k) < \infty\ \forall\, k$. Since $\mathcal{N}_k$ is C^1 by the differentiation theorem we have

$$\theta^n(\mu, \mathcal{M}_k, x) = \lim_{\rho\to 0} \frac{\mu(\mathcal{M}_k \cap B(x,\rho))}{\mathcal{H}^n(\mathcal{N}_k \cap B(x,\rho))} = \theta(x) \qquad \mathcal{H}^n\text{-a.e. } x \in \mathcal{M}_k$$

while, by Theorem 6 in Sec. 1.1.5,

$$\begin{aligned}\theta^{*n}(\mu, \mathcal{N}_k \setminus \mathcal{M}_k, x) &= 0 \quad \mathcal{H}^n\text{-a.e. } x \in \mathcal{M}_k\\ \theta^{*n}(\mu, \mathcal{M} \setminus \mathcal{M}_k, x) &= 0 \quad \mathcal{H}^n\text{-a.e. } x \in \mathcal{M}_k\,.\end{aligned}$$

From this we easily infer that for a.e. $x \in \mathcal{M}_k$ and for every $f \in C_c(\mathbb{R}^{n+N})$

$$\begin{aligned}\int_{\eta_{x,\lambda}(\mathcal{M}\setminus\mathcal{M}_k)} f(y)\,\theta(x+\lambda y)\,d\mathcal{H}^n(y) &\longrightarrow 0\\ \int_{\eta_{x,\lambda}(\mathcal{N}_k\setminus\mathcal{M}_k)} f(y)\,\theta(x+\lambda y)\,d\mathcal{H}^n(y) &\longrightarrow 0\end{aligned}$$

as $\lambda \to 0$, that is, by Lebesgue theorem,

$$\begin{aligned}\lim_{\lambda\to 0} \int_{\eta_{x,\lambda}(\mathcal{M})} f(y)\,\theta(x+\lambda y)\,dy &= \lim_{\lambda\to 0} \int_{\eta_{x,\lambda}(\mathcal{N}_k)} f(y)\,\theta(x+\lambda y)\,dy\\ &= \theta(x) \int_{T_x\mathcal{N}_k} f(y)\,dy\end{aligned}$$

as $\mathcal{N}_k$ is of class C^1. This shows that *every $\mathcal{H}^n$-measurable and countably n-rectifiable set $\mathcal{M}$ admits $\mathcal{H}^n$-a.e. on $\mathcal{M}$ approximate tangent plane with respect to any positive locally $\mathcal{H}^n$-integrable function θ on $\mathcal{M}$, moreover*

(7) $$T_x\mathcal{M} = T_x\mathcal{N}_k \qquad \mathcal{H}^n\text{-a.e. } x \in \mathcal{M}_k\,.$$

The converse is also true and we can state

Theorem 1. *Suppose $\mathcal{M}$ is $\mathcal{H}^n$-measurable. Then $\mathcal{M}$ is countably n–rectifiable if and only if there is a positive locally $\mathcal{H}^n$-integrable function θ on $\mathcal{M}$ with respect to which the approximate tangent space exists for $\mathcal{H}^n$-a.e. $x \in \mathcal{M}$.*

Theorem 1 characterizes countably n–rectifiable sets as sets with good tangential properties: an approximate tangent plane exists at $\mathcal{H}^n$ a.e. point. We remark that countably n–rectifiable sets can be "fractured".

We need to prove the "if" part. For future purposes we shall derive it from a slightly more general result on the rectifiability of Radon measures, applied to the measure $\mu := \theta\mathcal{H}^n \llcorner \mathcal{M}$.

Definition 4 uses the measure $\mathcal{H}^n \llcorner \theta$, but we can easily extend it to any Radon measure

Definition 5. *Let μ be a Radon measure on $\mathbb{R}^{n+N}$, and for $x \in \mathbb{R}^{n+N}$ and $\lambda > 0$ let $\mu_{x,\lambda}$ be the measure given by*

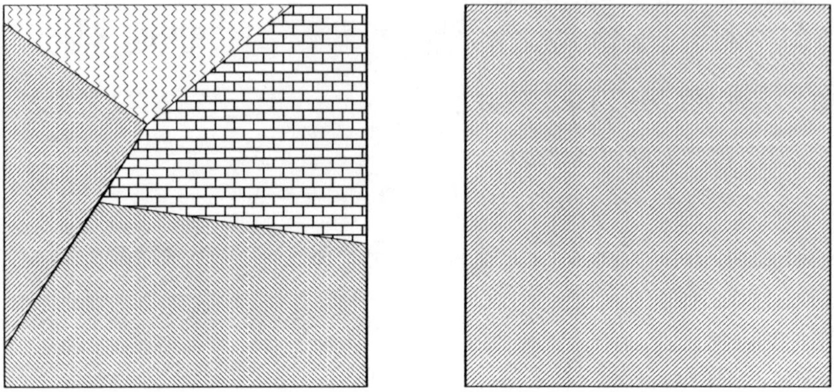

Fig. 2.2. $A = B$ as rectifiable sets.

$$\mu_{x,\lambda}(A) = \lambda^{-n}\mu(x+\lambda A) . \tag{8}$$

We say that the n-dimensional subspace $P \subset \mathbb{R}^{n+N}$ *is the* approximate tangent space for μ at x with multiplicity $\theta(x) \in (0, +\infty)$ *if*

$$\lim_{\lambda\to 0} \int f(y)\, d\mu_{x,\lambda}(y) = \theta(x) \int_P f(y)\, d\mathcal{H}^n(y) . \tag{9}$$

Clearly $0 \in P$, and P and $\theta(x)$ are unique, if they exist. For any Radon measure μ we denote by $\mathcal{M}_\mu$ the set of points in $\mathbb{R}^{n+N}$ for which there exists the approximate tangent space for μ at x with multiplicity $\theta(x) \in (0,\infty)$. Then we have

Theorem 2 (Rectifiability theorem for Radon measures). *Assume that* μ *has an approximate tangent plane for* μ*–a.e.* $x \in \mathbb{R}^{n+N}$ *with multiplicity* $\theta(x) \in (0,\infty)$, *i.e.*

$$\mu(\mathbb{R}^n \setminus \mathcal{M}_\mu) = 0 .$$

Then $\mathcal{M}_\mu$ *is countably n-rectifiable,* θ *is* $\mathcal{H}^n$*-measurable, and* $\mu = \theta\mathcal{H}^n \llcorner \mathcal{M}_\mu$.

Proof. Choose any ball $B(0,R)$. For any $x \in B(0,R)$ and any bounded Borel set A we have

$$(\mu \llcorner (\mathbb{R}^{n+N} \setminus B(0,R)))_{x,\lambda}(A) = 0$$

provided λ is sufficiently small. From this we easily deduce that

$$\mathcal{M}_\mu \cap B(0,R) = \mathcal{M}_{\mu \llcorner B(0,R)} .$$

Therefore replacing μ by $\mu \llcorner B(0,R)$ we may and do assume that $\mu(\mathbb{R}^{n+N}) < \infty$.

Given any N-dimensional subspace $\Pi \subset \mathbb{R}^{n+N}$ and any $\alpha \in (0,1)$ let p_Π denote the orthogonal projection of $\mathbb{R}^{n+N}$ onto Π and $C_\alpha(\Pi, x)$ denote the cone

$$C_\alpha(\Pi,x) = \{y \in \mathbb{R}^{n+N} \mid |p_\Pi(y-x)| > \alpha|y-x|\} .$$

having "axis" Π. Also, recalling that there is a bijection between N-dimensional subspaces and projectors, define the distance between two N-dimensional subspaces Π, Π' as the norm of the difference of their projectors

$$\operatorname{dist}(\Pi, \Pi') = \sup_{|x|=1} |p_\Pi(x) - p_{\Pi'}(x)| .$$

This way we obtain a compact metric space.

For $s \in \mathbb{N}$ choose now $\theta_s > 0$ and a Borel subset $F \subset \mathbb{R}^{n+N}$ such that

$$\mu(\mathbb{R}^{n+N} \setminus F) \leq \frac{1}{s} \mu(\mathbb{R}^{n+N}) \tag{10}$$

and such that for each $x \in F$, μ has an approximate tangent space P_x at x with multiplicity $\theta(x) \geq \theta_s$. In particular, we therefore have

$$\lim_{\rho\to 0} \frac{\mu(B(x,\rho))}{\omega_n \rho^n} \geq \theta_s \qquad \lim_{\rho\to 0} \frac{\mu(C_{1/2}(P_x^\perp, x) \cap B(x,\rho))}{\omega_n \rho^n} = 0. \tag{11}$$

For $i = 1, 2, \dots$ and $x \in F$ define

$$f_i(x) := \inf_{0<\rho<1/i} \frac{\mu(B(x,\rho))}{\omega_n \rho^n}$$

$$q_i(x) := \sup_{0<\rho<1/i} \frac{\mu(C_{1/2}(P_x^\perp, x) \cap B(x,\rho))}{\omega_n \rho^n} .$$

Then

$$\lim_{i\to\infty} f_i(x) \geq \theta_s , \qquad \lim_{i\to\infty} q_i(x) = 0 \qquad \forall\, x \in F \tag{12}$$

and by Egoroff's Theorem we can choose a μ-measurable set $E \subset F$ with

$$\mu(F \setminus E) \leq \frac{1}{s} \mu(\mathbb{R}^{n+N}) \tag{13}$$

and with (12) holding uniformly in E. Therefore for each $\varepsilon > 0$ we find $\delta > 0$ such that

$$\frac{\mu(B(x,\rho))}{\omega_n \rho^n} \geq \theta_s - \varepsilon, \quad \frac{\mu(C_{1/2}(P_x^\perp, x) \cap B(x,\rho))}{\omega_n \rho^n} \leq \varepsilon \text{ if } x \in E,\ 0 < \rho < \delta. \tag{14}$$

According to the compactness property of the metric space of N-planes, we now choose a finite number of N-dimensional planes $\Pi_1, \dots, \Pi_k$ in $\mathbb{R}^{n+N}$ such that for each N-dimensional plane Π there is a $j \in \{1, \dots, k\}$ such that $\operatorname{dist}(\Pi, \Pi_j) < 1/16$, and let $E_1, \dots, E_k$ be the subsets of E defined by

$$E_j := \{x \in E \mid d(\Pi_j, P_x^\perp) < 1/16\} .$$

Clearly $E = \cup_{j=1}^k E_j$. We now claim that for $\varepsilon = \theta_s/16^n$ and δ so that (14) holds we have

$$(15) \qquad C_{3/4}(\Pi_j, x) \cap E_j \cap B_{\delta/2}(x) = \{x\} \qquad \forall\, x \in E_j,\ j = 1, \dots, k\,.$$

In fact otherwise we find $x \in E_j$ and $y \in C_{3/4}(\Pi_j, x) \cap E_j \cap \partial B(x, \rho)$ for some $0 < \rho < \delta/2$. From (14) we then have

$$\mu(C_{1/2}(P_x^\perp, x) \cap B(x, 2\rho)) \leq \varepsilon \omega_n (2\rho)^n$$

and, since

$$|p_{P_x^\perp}(y - x)| \geq |p_{\Pi_j}(y - x)| - \frac{1}{16}|y - x| \geq (\frac{3}{4} - \frac{1}{16})|y - x| = \frac{11}{16}\rho$$

we have $B(y, \rho/8) \subset C_{1/2}(P_x^\perp, x) \cap B(x, 2\rho)$ and, again by (14)

$$\mu(C_{1/2}(P_x^\perp, x) \cap B(x, 2\rho)) \geq \mu(B(y, \rho/8)) \geq (\theta_s - \varepsilon)\omega_n \left(\frac{\rho}{8}\right)^n .$$

This gives $\theta_s \leq (1 + 8^n)\varepsilon$, a contradiction, since $\varepsilon = \theta_s/16^n$.

From (15) we also see, taking into account the extension theorem for Lipschitz maps (compare Theorem 1 in Sec. 3.1.3) that for any fixed $x_0 \in E_j$

$$(16) \qquad E_j \cap B(x_0, \delta/2) \subset q(\operatorname{graph} f)$$

where q is an orthogonal transformation of $\mathbb{R}^{n+N}$ with $q(\Pi_j) = \mathbb{R}^N$ and where $f = (f^1, \dots, f^N)$ is Lipschitz. In fact we may assume $x_0 = (0, 0) \in \mathbb{R}^n \times \mathbb{R}^N$, $\Pi_j = \{0\} \times \mathbb{R}^N$. Let A denote the projection of $E_j \cap B(x_0, \delta/4)$ on $\mathbb{R}^n \times \{0\}$, then for any $(x, 0) \in A$ we claim there is a unique $(x, y) \in E_j \cap B((0, 0), \delta/2)$. In fact suppose that $(x_i, y_i) \in E_j \cap B((0, 0), \delta/2)$, $i = 1, 2$, then from (15) we infer $(x_2, y_2) \notin C_{3/4}(\Pi_j, (x_1, y_1))$, hence

$$|y_2 - y_1| < \frac{3}{4}(|x_2 - x_1|^2 + |y_2 - y_1|^2)^{1/2} \leq \frac{3}{4}(|x_2 - x_1| + |y_2 - y_1|)$$

i.e.

$$|y_2 - y_1| < 3\,|x_1 - x_2|\,.$$

This yields at once uniqueness, and shows that the map

$$(x, 0) \in A \longrightarrow (x, y) \in E_j \cap B((0, 0), \delta/2)$$

is Lipschitz. Extending such a map to a Lipschitz map defined on $\mathbb{R}^n \times \{0\} \supset A$ we then easily get (16).

From the above we conclude that E is an n-rectifiable set and

$$\mu(\mathbb{R}^{n+N} \setminus E) < \frac{2}{s}\mu(\mathbb{R}^{n+N})\,.$$

This shows that

$$\mathcal{M}_\mu = \mathcal{M}_0 \cup \bigcup_{k=1}^{\infty} \mathcal{M}_k$$

with $\mu(\mathcal{M}_0) = 0$ and $\mathcal{M}_k \subset \mathcal{N}_k$, $\mathcal{N}_k$ being n-dimensional C^1-submanifolds of $\mathbb{R}^{n+N}$. Since for any x in $\mathcal{M}_\mu$ the existence of the approximate tangent plane implies that

$$\theta(x) = \lim_{\rho\to 0} \frac{\mu(B(x,\rho))}{\omega_n \rho^n} =: \theta^n(\mu, x) \tag{17}$$

we have

$$\theta^n(\mu, x) > 0 \qquad \text{for } \mu-\text{a.e. } x \in \mathcal{M}_\mu \,.$$

Therefore on account of Theorem 5 in Sec. 1.1.5, we infer $\mathcal{H}^n(\mathcal{M}_0) = 0$, hence $\mathcal{M}_\mu$ is $\mathcal{H}^n$–measurable and countably n–rectifiable. As $\theta^{*n}(\mu, x)$ is a Borel function, we also infer from (17) that $\theta(x)$ is $\mathcal{H}^n$-measurable on $\mathcal{M}_\mu$. Finally (17) yields

$$\lim_{\rho\to 0} \frac{\mu(B(x,\rho))}{\mathcal{H}^n(N_k \cap B(x,\rho))} = \theta(x) \,,$$

which in turn implies $\mu = \theta\mathcal{H}^n \llcorner \mathcal{M}_\mu$ by the differentiation theorem of measures in Sec. 1.1.5. □

We conclude this subsection by stating without proof the important and difficult *Besicovitch-Federer structure theorem* for purely unrectifiable sets.

Let A be an arbitrary set in $\mathbb{R}^{n+N}$ which can be written as countable union of sets of finite $\mathcal{H}^n$-measure. Consider the collection of all countably n-rectifiable subsets of A ordered by inclusion. Since the union of families of countably n–rectifiable sets in A is again a countably n–rectifiable set, by Hausdorff maximal principle there exists a maximal element R, and we have

$$A = R \cup P$$

where R is countably n–rectifiable and P contains no countably n–rectifiable subset of positive $\mathcal{H}^n$-measure.

Definition 6. *A set P which contains no countably rectifiable subset of positive $\mathcal{H}^n$-measure is called* purely n-unrectifiable.

Purely unrectifiable sets Q of $\mathbb{R}^{n+N}$ are characterized by the fact that they have $\mathcal{H}^n$-null projection via almost all orthogonal projection onto n-dimensional subspaces of $\mathbb{R}^{n+N}$. Of course also purely unrectifiable sets Q can be written as countable union of sets of finite $\mathcal{H}^n$-measure.

An example of a purely 1-unrectifiable set in $\mathbb{R}^2$ is given by $C \times C$ where C is the Cantor set in Sec. 1.1.3, of dimension 1/2. In fact, as we have seen, in this case,

$$0 < \mathcal{H}^1(C \times C) < \infty \,.$$

Therefore if $\mathcal{M} \subset C \times C$ is 1-rectifiable we should have either $\mathcal{H}^1(\pi(\mathcal{M})) > 0$ or $\mathcal{H}^1(\widehat{\pi}(\mathcal{M})) > 0$, being π and $\widehat{\pi}$ the projections of $\mathbb{R} \times \mathbb{R}$ into the first and second factor. A contradiction as

$$\mathcal{H}^1(\pi(\mathcal{M})) \leq \mathcal{H}^1(C) = 0\,, \qquad \mathcal{H}^1(\widehat{\pi}(\mathcal{M})) \leq \mathcal{H}^1(C) = 0\,.$$

Theorem 3 (Besicovitch-Federer structure theorem). *Let Q be a purely n–unrectifiable subset of $\mathbb{R}^{n+N}$ with*

$$Q = \bigcup_{k=1}^{\infty} Q_k\,, \qquad \mathcal{H}^n(Q_k) < \infty \qquad \forall\, k\,.$$

Then $\mathcal{H}^n(p(Q)) = 0$ for almost (in the sense of Haar measure for $O(n+N,n)$) all orthogonal projections of $\mathbb{R}^{n+N}$ onto n-dimensional subspaces of $\mathbb{R}^{n+N}$, $p \in O(n+N,n)$.

Notice that only the purely n-unrectifiable subsets could possibly have the null projection property stated in Theorem 3.

An immediate consequence of Theorem 3 is

Theorem 4 (Rectifiability theorem). *Let A be an arbitrary subset of $\mathbb{R}^{n+N}$ which can be written as countable union*

$$A = \bigcup_{k=1}^{\infty} A_k \quad \text{with} \quad \mathcal{H}^n(A_k) < \infty \quad \forall\, k\,.$$

Suppose that every subset $B \subset A$ with positive $\mathcal{H}^n$-measure has the property that $\mathcal{H}^n(p(B)) > 0$ for a set $p \in O(n+N,n)$ of positive Haar measure. Then A is countably n-rectifiable.

Of course the $\mathcal{H}^k$-density at $\mathcal{H}^k$-a.e. point x of a k-rectifiable set is 1 while it is zero if $x \notin \mathcal{M}$

$$\theta^{*k}(\mathcal{H}^k, \mathcal{M}, x) = \begin{cases} 1 & \text{for } \mathcal{H}^k\text{-a.e. } x \in \mathcal{M} \\ 0 & \text{for } \mathcal{H}^k\text{-a.e. } x \notin \mathcal{M}\,. \end{cases}$$

A natural question is to ask for which sets A of finite $\mathcal{H}^k$-measure k non necessarily integer, there exists for $\mathcal{H}^k$-a.e. $x \in A$

$$\theta^k(\mathcal{H}^k, A, x) = \theta^k_*(\mathcal{H}^k, A, x) = \theta^{*k}(\mathcal{H}^k, A, x)\,.$$

The following theorem gives a complete characterization of such sets

Theorem 5 (Preiss). *Let $A \subset \mathbb{R}^n$, with $\mathcal{H}^k(A) < \infty$. Then*

$$\theta^k(\mathcal{H}^k, A, x) \ \textit{exists for}\ \mathcal{H}^k\text{–a.e. } x \in A$$

if and only if k is integer and A is k-rectifiable.

1.5 The General Area and Coarea Formulas

In this last subsection we would like to state the area and coarea formulas for Lipschitz maps $f : \mathcal{M} \to \mathbb{R}^N$, where $\mathcal{M}$ is an $\mathcal{H}^n$-measurable and countably n-rectifiable set.

Recall that an n-dimensional C^r, $r \geq 1$, submanifold of $\mathbb{R}^m$, $m \geq n$, is a set $M \subset \mathbb{R}^m$ such that for any $y \in M$ there are open sets U and V in $\mathbb{R}^m$ with $y \in U$, $0 \in V$ and a C^r-diffeomorphism $\phi : U \to V$ with $\phi(y) = 0$ and

$$\phi(M \cap U) = V \cap \mathbb{R}^n$$

where $\mathbb{R}^n = \{(x^1, \ldots, x^m) \in \mathbb{R}^m \mid x^{n+1} = \ldots = x^m = 0\}$. In this case we have *local representations* of M

$$\psi : V \cap \mathbb{R}^n \to \mathbb{R}^m , \qquad \psi(V \cap \mathbb{R}^n) = M \cap U , \qquad \psi(0) = y ,$$

(indeed $\psi = \phi^{-1}_{|V \cap \mathbb{R}^n}$) such that

$$\frac{\partial \psi}{\partial x^1}(0), \ldots, \frac{\partial \psi}{\partial x^n}(0)$$

are linearly independent vectors in $\mathbb{R}^m$ which form a basis of the tangent space $T_y M$ of M at y. The *tangent space* $T_y M$ of M at y is the subspace of $\mathbb{R}^m$ consisting of those $\tau \in \mathbb{R}^m$ such that $\tau = \dot\gamma(0)$ for some C^1 curve $\gamma : (-1,1) \to M$, $\gamma(0) = y$.

Let $f : M \to \mathbb{R}^N$ be a C^1 function on M. The *directional derivative* $D_\tau f \in \mathbb{R}^N$ in y, $\tau \in T_y M$ is defined by

$$D_\tau f := \frac{d}{dt} f(\gamma(t))_{|t=0}$$

for any C^1 curve $\gamma : (-1,1) \to M$ with $\gamma(0) = y$, $\dot\gamma(0) = \tau$, and it turns out to be independent of the particular curve γ. The induced linear map $df_y : T_y M \to \mathbb{R}^N$ defined by

$$df_y(\tau) := D_\tau f , \qquad \tau \in T_y M$$

yields the *tangent map.*

For scalar functions of class C^1, $f : M \to \mathbb{R}$, the directional derivatives can be represented in terms of the *gradient* $\nabla^M f(y)$ defined as the element in $T_y M$ such that

$$\nabla^M f(y) \cdot \tau = D_\tau f \qquad \forall \, \tau \in T_y M .$$

If $\tau_1, \ldots, \tau_n$ is an orthonormal basis of $T_y M$, the gradient is also given by the formula

$$\nabla^M f(y) = \sum_{k=1}^n (D_{\tau_k} f) \, \tau_k .$$

If $\overline{f}$ is a C^1 function defined in a neighbourhood U of M with $\overline{f} = f$ on M, the gradient $\nabla^M f$ is related to the ordinary gradient of $\overline{f}$ by

$$\nabla^M f(y) = P(\nabla \overline{f}(y))$$

where $P : \mathbb{R}^m \to T_y M$ is the orthogonal projection onto $T_y M$.

The *divergence* of a vector field $X = (X^1, \dots, X^m) : M \to \mathbb{R}^m$ is defined similarly

$$\operatorname{div}_M X := \sum_{k=1}^m e_k \cdot (\nabla^M X^k)$$

where $e_1, \dots, e_m$ is the standard basis of $\mathbb{R}^m$. It is easily seen that

$$\operatorname{div}_M X = \sum_{k=1}^n (D_{\tau_k} X) \cdot \tau_k$$

where $\tau_1, \dots, \tau_n$ is any orthonormal basis for $T_y M$. We have

Theorem 1 (The divergence theorem). *Let M be such that the closure $\overline{M}$ is a smooth compact manifold with boundary $\partial M = \overline{M} \setminus M$. If $X_y \in T_y M$ for each $y \in M$, i.e., X is a tangent vector field, then*

$$\int_M \operatorname{div}_M X \, d\mathcal{H}^n = -\int_{\partial M} X \cdot \eta \, d\mathcal{H}^{n-1} ,$$

where η is the inward pointing unit co-normal of ∂M, i.e., $|\eta| = 1$, η is tangent to M, η is normal to ∂M and points into M. In general, i.e., without any assumption on ∂M, we have

$$\int_M \operatorname{div}_M X = 0$$

for any tangent vector field X such that $\operatorname{spt} X \cap M \subset\subset M$.

The *mean curvature* of a C^2 submanifold $M \subset \mathbb{R}^m$ is expressed in terms of its second fundamental form. For $y \in M$ we denote by $\tau_1, \dots, \tau_n$ an orthonormal basis of $T_y M$ and by $\nu_1, \dots, \nu_{m-n}$ any orthonormal basis of $(T_y M)^\perp$, the orthogonal complement of $T_y M$, that is, $\nu_1, \dots, \nu_{m-n}$ is a basis of the *normal space* of M at y. The *second fundamental form* of M at y is the bilinear form

$$B_y : T_y M \times T_y M \longrightarrow (T_y M)^\perp$$

defined by

$$B_y(\tau, \eta) = -\sum_{\alpha=1}^{m-n} (\eta \cdot D_\tau \nu_\alpha(y)) \, \nu_\alpha(y) .$$

It is not difficult to infer

Proposition 1. *We have*

(i) *Let* $\gamma : (-1,1) \to M$ *be a* C^2 *curve satisfying* $\gamma(0) = y$, $\dot\gamma(0) = \tau$, $|\tau| = 1$. *Then*

$$B_y(\tau,\tau) = (\ddot\gamma(0))^\perp$$

where $v^\perp$ *denotes the orthogonal projection of* v *onto* $(T_yM)^\perp$.

(ii) *Let* U *be a neighbourhood of* $(0,0) \in \mathbb{R}^2$ *and let* $\phi : U \to M$ *is a* C^2 *map with* $\phi(0,0) = y$ *and*

$$\frac{\partial\phi}{\partial t^1}(0,0) = \tau\,, \qquad \frac{\partial\phi}{\partial t^2}(0,0) = \eta\,.$$

Then

$$B_y(\tau,\eta) = \left(\frac{\partial^2\phi}{\partial t^1\partial t^2}(0,0)\right)^\perp.$$

The vector $(\ddot\gamma(0))^\perp$ in (i) is the normal component of the curvature of γ; in other words $B_y(\tau,\tau)$ measures the normal curvature of M in the direction τ. From (ii) it follows that B_y is a symmetric bilinear form with values in $(T_yM)^\perp$.

The *mean curvature* vector field $\mathbf{H}$ of M at y is the trace of B_y

$$\mathbf{H}(y) := \sum_{k=1}^n B_y(\tau_k,\tau_k) \in (T_yM)^\perp$$

or equivalently

$$\mathbf{H}(y) := -\sum_{\alpha=1}^{m-n}\sum_{k=1}^n (\tau_k \cdot D_{\tau_k}\nu_\alpha(y))\,\nu_\alpha(y)$$

so that

$$\mathbf{H} := -\sum_{\alpha=1}^{m-n} (\operatorname{div}_M \nu_\alpha)\,\nu_\alpha\,.$$

In terms of the mean curvature vector $\mathbf{H}$ one can write the divergence theorem for generic vector fields X not necessarily tangent to M.

Theorem 2. *Let M be as in Theorem 1. Let $X : M \to \mathbb{R}^m$ be any smooth vector field. Decompose X into its tangent and normal part*

$$X = X^T + X^\perp\,, \qquad X^\perp = \sum_{\alpha=1}^{m-n}(\nu_\alpha\cdot X)\,\nu_\alpha\,.$$

Then we have

$$\operatorname{div} X^\perp = -X\cdot\mathbf{H}$$

and

$$\int_M \operatorname{div}_M X\,d\mathcal{H}^n = -\int_M X\cdot\mathbf{H}\,d\mathcal{H}^n - \int_{\partial M} X\cdot\eta\,d\mathcal{H}^{n-1}\,.$$

Without any assumption on ∂M *and provided* $\operatorname{spt} X \cap M \subset\subset M$, *we have*

$$\int_M \operatorname{div}_M X \, d\mathcal{H}^n = -\int_M X \cdot \mathbf{H} \, d\mathcal{H}^n .$$

Consider now an $\mathcal{H}^n$-measurable and countably n-rectifiable set $\mathcal{M} \subset \mathbb{R}^m$, $n \leq m$. As we have seen M can be written as

$$\mathcal{M} = \mathcal{M}_0 \cup \bigcup_{k=1}^{\infty} \mathcal{M}_k$$

where $\mathcal{H}^n(\mathcal{M}_0) = 0$, $\mathcal{M}_k$ is $\mathcal{H}^n$-measurable for all k, and each $\mathcal{M}_k$ lies in some n-dimensional C^1 submanifold $\mathcal{N}_k$, $\mathcal{M}_k \subset \mathcal{N}_k$. Suppose $f : U \to \mathbb{R}$ be a locally Lipschitz map from a neighbourhood of $\mathcal{M}$ into $\mathbb{R}$. Of course for any k $f_{|\mathcal{N}_k}$ is also locally Lipschitz, hence $\mathcal{H}^n$-a.e. differentiable (compare (A) of Sec. 2.1.2 and Ch. 3), also it is easy to check that at all points x where $f_{|\mathcal{N}_k}$ is differentiable we have $f_{|L}$ is differentiable on the affine space $L := x + T_x\mathcal{N}_k$ at the point x, and $\operatorname{grad} f_{|L}(x) = \nabla^{\mathcal{N}_k} f(x)$. Thus we can set

Definition 1. *The* gradient *of* f *relative to* $\mathcal{M}$ *is defined for* $\mathcal{H}^n$ *a.e.* $x \in \mathcal{M}_k$ *by*

$$\nabla^{\mathcal{M}} f(x) := \nabla^{\mathcal{N}_k} f(x) .$$

We explicitly note that $\nabla^{\mathcal{M}} f(x)$ exists and belongs to $T_x\mathcal{M}$ for $\mathcal{H}^n$-a.e. $x \in \mathcal{M}$, and, up to a set of $\mathcal{H}^n$-measure zero, is independent of the particular decomposition $\mathcal{M} = \cup_k \mathcal{M}_k$ used in the definition. This is easily seen as $T_x\mathcal{N}_k = T_x\mathcal{M}$ for $\mathcal{H}^n$-a.e. $x \in \mathcal{M}_k$ and $T_x\mathcal{M}$ is independent of the particular decomposition.

For $\mathcal{H}^n$-a.e. point in $\mathcal{M}$ for which $T_x\mathcal{M}$ and $\nabla^{\mathcal{M}} f(x)$ exist we can then define the linear map

$$d^{\mathcal{M}} f_x : T_x\mathcal{M} \to \mathbb{R}$$

by

$$d^{\mathcal{M}} f_x(\tau) = \tau \cdot \nabla^{\mathcal{M}} f(x) , \qquad \tau \in T_x\mathcal{M} .$$

If $f : \mathcal{M} \to \mathbb{R}^N$ is a Lipschitz map we then set

$$d^{\mathcal{M}} f_x(\tau) := \sum_{s=1}^{N} (\tau \cdot \nabla^{\mathcal{M}} f^\ell) \, e_\ell$$

where $e_1, \dots, e_N$ is the standard basis in $\mathbb{R}^N$, and we define the *Jacobian* $J_f^{\mathcal{M}}(x)$ for $\mathcal{H}^n$-a.e. $x \in \mathcal{M}$ as the Jacobian of the linear map $d^{\mathcal{M}} f_x : T_x\mathcal{M} \to \mathbb{R}^N$, i.e.

$$\begin{aligned} J_f^{\mathcal{M}}(x) &:= (\det (d^{\mathcal{M}} f_x)^* (d^{\mathcal{M}} f_x))^{1/2} \qquad \text{if } n \leq N \\ J_f^{\mathcal{M}}(x) &:= (\det (d^{\mathcal{M}} f_x)(d^{\mathcal{M}} f_x)^*)^{1/2} \qquad \text{if } n > N . \end{aligned}$$

Often we shall also use the notation

$$J_n(d^{\mathcal{M}} f(x)) \qquad \text{for} \qquad J_f^{\mathcal{M}}(x) .$$

Taking into account the approximation property of Lipschitz maps (compare (B) of Sec. 2.1.2) and the fact that sets of zero measure are mapped into sets of zero measure by Lipschitz maps it is not difficult to show that the area and coarea formulas hold for Lipschitz maps $f : \mathcal{M} \to \mathbb{R}^N$, i.e.,

Theorem 3. *Let $\mathcal{M}$ be an $\mathcal{H}^n$-measurable and countably n-rectifiable subset of $\mathbb{R}^m$, and let $A \subset \mathcal{M}$ be an $\mathcal{H}^n$-measurable set. Then for any non-negative $\mathcal{H}^n$-measurable function $u : \mathcal{M} \to \mathbb{R}$ we have*

$$\int_A u(x) J_f^{\mathcal{M}}(x)\, d\mathcal{H}^n(x) = \int_{\mathbb{R}^N} \Big(\int_{f^{-1}(y)\cap A} u\, d\mathcal{H}^0 \Big)\, d\mathcal{H}^n(y)$$

if $n \le N$, and

$$\int_A u(x) J_f^{\mathcal{M}}(x)\, d\mathcal{H}^n(x) = \int_{\mathbb{R}^N} \Big(\int_{f^{-1}(y)\cap A} u\, d\mathcal{H}^{n-N} \Big)\, d\mathcal{H}^N(y)$$

if $n > N$.

We finally notice that from the decomposition $\mathcal{M} = \cup_k \mathcal{M}_k$, together with the C^1-Sard type theorem in Sec. 2.1.3 and the approximation property (B) in Sec. 2.1.2 it follows

Proposition 2. *Let $\mathcal{M}$ be an $\mathcal{H}^n$-measurable and countably n-rectifiable subset of $\mathbb{R}^{n+m}$ and let $f : \mathcal{M} \to \mathbb{R}^N$ be a Lipschitz map. Then the slices $\mathcal{M} \cap f^{-1}(y)$ are countably $(n-N)$-rectifiable subsets of $\mathbb{R}^m$ for $\mathcal{H}^n$-a.e. $y \in \mathbb{R}^N$.*

2 Currents

In this section we deal with the general notion of *currents* and in particular of *integer multiplicity rectifiable currents.* We do not aim of course to completeness but only to discuss basic definitions and results which are relevant for the sequel. In particular not all proofs are included, and sometimes they are postponed to later chapters.

After the first two subsections on *multivectors, covectors* and *differential forms*, where we also fix some notation which will be used later on, basic definitions and results, as the important *closure-compactness theorem* for integer multiplicity rectifiable currents, are presented in Sec. 2.2.3 and Sec. 2.2.4. In Sec. 2.2.4 we also present some elementary examples, while further developments are discussed in Sec. 2.2.5, Sec. 2.2.6 and Sec. 2.2.7.

2.1 Multivectors and Covectors

Vectors of a finite dimensional vector space V may be multiplied giving rise to new objects called *multivectors*; for $v_1, \ldots, v_k \in V$ the result

$$\xi = v_1 \wedge \ldots \wedge v_k$$

is called a *k-vector*. But before describing such a procedure, let us first fix some notations.

Usually we shall identify V with $\mathbb{R}^n$ with the standard basis $e_1, \ldots, e_n$; its dual V^* will be again identified with $\mathbb{R}^n$ with dual basis $e^1, \ldots, e^n$, so that

$$< e_i, e^j > = \delta_i^j .$$

We shall use the standard notations for *ordered multi-indices*

(1) $\quad I(k,n) := \{\alpha = (\alpha_1, \ldots, \alpha_k) \mid \alpha_i \text{ integers}, 1 \le \alpha_1 < \ldots < \alpha_k \le n\}$

and for convenience we set

$$I(0,n) = \{0\}$$

and

$$|\alpha| = k \qquad \text{if} \qquad \alpha \in I(k,n) .$$

If $\alpha \in I(k,n)$, $k = 0, 1, \ldots, n$, we denote by $\bar{\alpha}$ the element in $I(n-k,n)$ which complements α in $\{1, 2, \ldots, n\}$ in the natural increasing order; of course we have $\bar{\bar{\alpha}} = \alpha$, $\bar{0} = (1, \ldots, n)$. If $|\alpha| = n-1$ and $|\beta| = 1$, we shall often write i instead of $\bar{\alpha}$, $i = 1, \ldots, n$, and j instead of β, $j = 1, \ldots, n$, and $\bar{i}$ for $(1, \ldots, i-1, i+1, \ldots, n)$.

Often in the sequel we shall have to consider maps from $\mathbb{R}^n$ into $\mathbb{R}^N$. We shall denote the standard basis of $\mathbb{R}^N$ by $\varepsilon_1, \ldots, \varepsilon_N$, so that $(e_1, \ldots, e_n, \varepsilon_1, \ldots, \varepsilon_N)$ is the standard basis of $\mathbb{R}^n \times \mathbb{R}^N$; with respect to those bases the coordinates will be denoted by $(x^1, \ldots, x^n, y^1, \ldots, y^N) = (x,y)$, and in order to stress this we often write $\mathbb{R}^n_x \times \mathbb{R}^N_y$ instead of $\mathbb{R}^n \times \mathbb{R}^N$.

Let $G : \mathbb{R}^n \to \mathbb{R}^N$ be a linear map, identified, with respect to the standard basis with the $N \times n$ matrix

$$G = (G^i_j) , \qquad i = 1, \ldots, N , \ j = 1, \ldots, n$$

and let

(2) $$\underline{n} := \min(n, N) .$$

Given two ordered multi-indices with $|\beta| = k$, $1 \le k \le \underline{n}$, $|\alpha| = n-k$, we shall denote by

(3) $$G^\beta_{\bar{\alpha}}$$

the $k \times k$-submatrix of G with rows $(\beta_1, \ldots, \beta_k)$ and columns $(\bar{\alpha}_1, \ldots, \bar{\alpha}_k)$. Its determinant will be denoted by

(4) $$M^\beta_{\bar{\alpha}}(G) := \det G^\beta_{\bar{\alpha}} ;$$

we shall set

(5) $$M^0_0(G) := 1$$

k-vectors. Let V be a finite dimensional vectors space over $\mathbb{R}$, with basis $e_1, \dots, e_n$. For k, $1 \le k \le n$, generic k vectors can be multiplied by means of the *wedge product*

$$\xi := v_1 \wedge \dots \wedge v_k$$

obtaining a new object. This wedge product is characterized by the following properties

(i) *it is multilinear*

$$\begin{aligned} &v_1 \wedge \dots \wedge (a v_i + b w_i) \wedge \dots \wedge v_k \\ &\quad = a v_1 \wedge \dots \wedge v_i \wedge \dots \wedge v_k + b v_1 \wedge \dots \wedge w_i \wedge \dots \wedge v_k \end{aligned}$$

(ii) *it is alternating*

$$v_1 \wedge \dots \wedge v_i \wedge \dots \wedge v_j \wedge \dots \wedge v_k = - v_1 \wedge \dots \wedge v_j \wedge \dots \wedge v_i \wedge \dots \wedge v_k$$

equivalently

$$v_1 \wedge \dots \wedge v_k = 0 \quad \text{if} \quad v_i = v_j \quad \text{for some } i \neq j$$

or

$$v_1 \wedge \dots \wedge v_k = 0 \quad \text{if} \quad v_1, \dots, v_k \text{ are linearly dependent.}$$

In terms of the basis $e_1, \dots, e_n$, computation of ξ yields an object of the form

$$\xi = \sum_{\alpha \in I(k,n)} \xi^\alpha e_\alpha$$

where $\xi^\alpha \in \mathbb{R}$, and e_α is in

$$\{ e_\alpha = e_{\alpha_1} \wedge \dots \wedge e_{\alpha_k} \mid \alpha \in I(k,n) \} .$$

The set of all linear combinations of the e_α, $\alpha \in I(k,n)$, is the space of *k-vectors*, denoted $\Lambda_k V$. It is a vector space of dimension $\binom{n}{k}$.

Every k-vector $v_1 \wedge \dots \wedge v_k$ defines a *k-linear* and *alternating* map with domain the product of k copies of V^*, the dual space of V

$$v_1 \wedge \dots \wedge v_k : V^* \times \dots \times V^* \longrightarrow \mathbb{R}$$

defined by

$$(v_1 \wedge \dots \wedge v_k)(w^1, \dots, w^k) := \det(<w^i, v_j>) , \qquad w^i \in V^* .$$

In fact it turns out that *$\Lambda_k V$ can be identified with the linear space of all k-linear and alternating maps on V^**. More precisely, by considering the k-linear and alternating forms e_α, $\alpha \in I(k,n)$ defined by

$$e_\alpha(w^1, \dots, w^k) := \det(w^i_{\alpha_j}) , \qquad w^i \in V^* \tag{6}$$

we have for any $\beta = (\beta_1, \ldots, \beta_k) \in I(k,n)$, $e^1, \ldots, e^n$ denoting the dual basis of V^*,

$$e_\alpha(e^{\beta_1}, \ldots, e^{\beta_k}) = \delta_\alpha^\beta , \qquad \delta_\alpha^\beta := \det(\delta_{\alpha_j}^{\beta_i})$$

and the e_α defined this way form a basis of the linear space of all k-linear alternating forms on V^*. This shows that $\Lambda_k V$ is defined intrinsically, i.e., independently of any fixed basis.

k-covectors. They are defined similarly to k-vectors replacing V by its dual V^*. The linear space of k-covectors is denoted by $\Lambda^k V := \Lambda_k V^*$, and is identified with the linear space of k-linear and alternating forms on V, with basis e^α, $\alpha \in I(k,n)$, given by

$$(7) \qquad e^\alpha(v_1, \ldots, v_k) := \det(v_i^{\alpha_j}) , \qquad v_1, \ldots, v_k \in V .$$

Exterior product of multivectors. Given two multivectors $\xi \in \Lambda_k V$, $\eta \in \Lambda_h V$, $h, k \geq 0$, $h + k \leq n$, $\xi = \sum_{\alpha \in I(k,n)} \xi^\alpha e_\alpha$, $\eta = \sum_{\beta \in I(h,n)} \eta^\beta e_\beta$ we define the *exterior product of* ξ *and* η by

$$(8) \qquad \xi \wedge \eta := \sum_{\alpha \in I(k,n)} \sum_{\beta \in I(h,n)} \xi^\alpha \eta^\beta e_\alpha \wedge e_\beta$$

where $e_\alpha \wedge e_\beta$ by (i) and (ii) is given by

$$e_\alpha \wedge e_\beta := \begin{cases} 0 & \text{if } \alpha \cap \beta \neq \emptyset \\ \sigma(\alpha, \beta) e_{\alpha \cup \beta} & \text{if } \alpha \cap \beta = \emptyset \end{cases}$$

where $\alpha \cup \beta \in I(h+k, n)$ is the multi-index obtained reordering the multi-index (α, β) and

$$\sigma(\alpha, \beta)$$

is the sign of the permutation which reorders (α, β) in the natural increasing order, and $\sigma(\bar{0}, 0) = 1$. This way $\xi \wedge \eta \in \Lambda_{h+k} V$. In order to give a meaning to (8) also in the case $h + k > n$ we set

$$e_\alpha \wedge e_\beta = 0 \qquad \text{for } h + k > n .$$

It is not difficult to see that the exterior product defined above has the following properties: *Let* $\alpha, \delta \in \Lambda_k V$, $\beta \in \Lambda_h V$, $\gamma \in \Lambda_s V$, *and* $c \in \mathbb{R}$. *Then*

(i) $(\alpha + \delta) \wedge \beta = (\alpha \wedge \beta) + (\delta \wedge \beta)$, $\qquad \beta \wedge (\alpha + \delta) = (\beta \wedge \alpha) + (\beta \wedge \delta)$
(ii) $c(\alpha \wedge \beta) = (c\alpha) \wedge \beta = \alpha \wedge (c\beta)$
(iii) $\alpha \wedge \beta = (-1)^{hk} \beta \wedge \alpha$
(iv) $(\alpha \wedge \beta) \wedge \gamma = \alpha \wedge (\beta \wedge \gamma)$
(v) *If* $v_1, \ldots, v_k \in V \simeq \Lambda_1 V$, *and* $\xi^1, \ldots, \xi^k \in V^*$, *then*

$$v_1 \wedge \ldots \wedge v_k(\xi^1, \ldots, \xi^k) = \det(<\xi^i, v_j>) .$$

Properties (i),...,(v) in fact characterize the notion of exterior product, so that it is an intrinsic notion independent of any fixed basis.

Of course one can proceed similarly in order to define the exterior product of covectors. In this case (v) reads: *for* $v_1, \dots, v_k \in V$ *we have*

$$\xi^1 \wedge \dots \wedge \xi^k(v_1, \dots, v_k) \;=\; \det\left(<\xi^j, v_i>\right) . \tag{9}$$

Simple k-vectors. A k-vector $\xi \in \Lambda_k V$ is called *simple* or *decomposable* if it can be written as a single wedge product of vectors,

$$\xi \;=\; v_1 \wedge \dots \wedge v_k$$

for some $v_1, \dots, v_k \in V$.

Of course $\Lambda_n V \simeq \mathbb{R}$ and all n-vectors are simple as they are multiple of the base vector $e_1 \wedge \dots \wedge e_n$. $\Lambda_1 V \simeq V$, hence all 1-vectors (vectors) are simple. The two-vector

$$\xi \;:=\; e_1 \wedge e_2 + e_3 \wedge e_4$$

provides an example of a *non decomposable* 2-vector in $\Lambda_2 \mathbb{R}^4$. Assume in fact it is simple $\xi = v_1 \wedge v_2$, then $\xi \wedge \xi = 0$, i.e.

$$e_1 \wedge e_2 \wedge e_3 \wedge e_4 + e_3 \wedge e_4 \wedge e_1 \wedge e_2 = 0$$

and, as

$$e_1 \wedge e_2 \wedge e_3 \wedge e_4 \;=\; (-1)^2\, e_3 \wedge e_4 \wedge e_1 \wedge e_2$$

we infer $e_1 \wedge e_2 \wedge e_3 \wedge e_4 = 0$: a contradiction, since $e_1 \wedge e_2 \wedge e_3 \wedge e_4$ is a basis of $\Lambda_4 \mathbb{R}^4$.

Simple k-vectors play a fundamental role since *they represent oriented k-planes in V*, as we shall see below, and they are one of the main reason to introduce the space of k-vectors. In fact in many respect it is convenient to work with a linear space instead of working with simple k-vectors which do not form a linear space.

Duality between $\Lambda_k V$ and $\Lambda^k V$. Covectors and multivectors are in duality. Given $\xi = \sum_{\alpha \in I(k,n)} \xi^\alpha e_\alpha \in \Lambda_k V$, $w = \sum_{\alpha \in I(k,n)} w_\alpha e^\alpha \in \Lambda^k V$ the duality between $\Lambda_k V$ and $\Lambda^k V$ is defined by

$$<\xi, w> \;:=\; \sum_{\alpha \in I(k,n)} \xi^\alpha w_\alpha \, . \tag{10}$$

It is easily seen that $<\,,\,>$ does not depend on the fixed basis. In fact one can show that on simple n-vectors we have

$$<v_1 \wedge \dots \wedge v_k, w> \;=\; w(v_1, \dots, v_k) \tag{11}$$

and a linear form on $\Lambda_k V$ is uniquely identified by its values on simple vectors.

In particular, from (7) and (11) we infer that the duality product between a simple k-vector and a simple k-covector is given by

$$<v_1\wedge\ldots\wedge v_k, w^1\wedge\ldots\wedge w^k> \;=\; \det(<v_i, w^j>)$$

where $<v_i, w^j>$ is the duality between V and V^*.

The duality above of course allows to identify canonically, i.e., independently from the choice of the bases, $(\Lambda_k V)^*$ and $\Lambda_k V^* =: \Lambda^k V$.

Finally notice that, since the k-vectors e_α are simple, a k-vector ξ is zero if $<\xi, w> = 0$ for all simple k-covectors w.

The inner product of multivectors. Suppose that V is endowed with a inner product $v\cdot w$, and let $e_1,\ldots e_n$ be an orthonormal basis. We can then define an inner product in $\Lambda_k V$ (respectively in $\Lambda^k V$) in such a way that e_α (respectively e^α) be an orthonormal basis of $\Lambda_k V$ ($\Lambda^k V$) by setting

$$(12)\qquad \xi\cdot\eta := \sum_{\alpha\in I(k,n)} \xi^\alpha\eta^\alpha \qquad \omega\cdot\sigma := \sum_{\alpha\in I(k,n)} \omega_\alpha\sigma_\alpha$$

where $\xi = \sum_{\alpha\in I(k,n)} \xi^\alpha e_\alpha$, $\eta = \sum_{\alpha\in I(k,n)} \eta^\alpha e_\alpha$ and $\omega = \sum_{\alpha\in I(k,n)} \omega_\alpha e^\alpha$, $\sigma = \sum_{\alpha\in I(k,n)} \sigma_\alpha e^\alpha$.

The inner product defined this way actually depends only on the scalar product in V and not on the fixed orthonormal basis $e_1,\ldots,e_n$, and it is invariant for orthogonal changes in V. To see that we consider the duality map

$$R: V \longrightarrow V^*\,, \qquad w\to Rw\,, \qquad <v, Rw> = v\cdot w \quad \forall\, v\,,$$

and we define $R_\# : \Lambda_k V \to \Lambda^k V$ by setting

$$R_\#\xi(v_1\wedge\ldots\wedge v_k) \;=\; \xi(Rv_1,\ldots,Rv_k)\,.$$

Since $Re_i = e^i$, $e^1,\ldots,e^n$ being the dual base in V^*, we easily see that

$$R_\# e_\alpha \;=\; e^\alpha\,.$$

As by definition $R_\#$ depends only on the inner product in V,

$$(13)\qquad \xi\cdot\eta \;=\; <\xi, R_\#\eta>\,,$$

and the duality product is independent of the coordinate system, we see that also the inner product is independent of the coordinate system.

Given now two simple k-vectors $v_1\wedge\ldots\wedge v_k$, $w_1\wedge\ldots\wedge w_k$ from (13) we compute

$$\begin{aligned} v_1\wedge\ldots\wedge v_k\cdot w_1\wedge\ldots\wedge w_k &= <v_1\wedge\ldots\wedge v_k, R_\#(w_1\wedge\ldots\wedge w_k)> \\ &= R_\#(w_1\wedge\ldots\wedge w_k)(v_1,\ldots,v_k) = (w_1\wedge\ldots\wedge w_k)(Rv_1,\ldots,Rv_k) \\ &= \det(<v_i, Rw_j>) = \det(v_i\cdot w_j) \end{aligned}$$

hence

$$(14)\qquad (v_1\wedge\ldots\wedge v_k)\cdot(w_1\wedge\ldots\wedge w_k) = \det(v_i\cdot w_j)\,.$$

Consequently the *modulus of a simple vector* is given by

$$|v_1\wedge\ldots\wedge v_k| \;=\; \sqrt{\det\,(v_i\cdot v_j)}\;. \tag{15}$$

Let us interpret geometrically such a formula in the case $V=\mathbb{R}^n$. We associate to the simple k-vector $v_1\wedge\ldots\wedge v_k$ the linear map

$$T:\mathbb{R}^k \longrightarrow \mathbb{R}^n\,, \qquad Tw := \sum w^i v_i\;.$$

Such a map is represented by the matrix with columns the components of the vectors $v_1,\ldots,v_k$

$$T := (v_1 \mid \ldots \mid v_k),$$

i.e., $v_i := Te_i$, and the matrix $(v_i\cdot v_j)$ is exactly T^*T. Therefore we see that

$$|v_1\wedge\ldots\wedge v_k| = |Te_1\wedge\ldots\wedge Te_k| = \sqrt{\det T^*T} \tag{16}$$

One can interpret $|v_1\wedge\ldots\wedge v_k|$ as the k-dimensional measure of the k-parallelepiped generated by the vectors $v_1,\ldots,v_k$, i.e.

$$|v_1\wedge\ldots\wedge v_k| \;=\; \mathcal{H}^k(T[0,1]^k))\;.$$

In fact from the area formula

$$\mathcal{H}^k(T[0,1]^k)) = J_T\mathcal{H}^k([0,1]^k) = \sqrt{\det T^*T} = |v_1\wedge\ldots\wedge v_k|\;.$$

Also, if we write

$$\xi := v_1\wedge\ldots\wedge v_k = \sum_{\alpha\in I(k,n)} \xi^\alpha e_\alpha$$

we have

$$\xi^\alpha = (v_1\wedge\ldots\wedge v_k)(e^{\alpha_1},\ldots,e^{\alpha_k}) = \det\,(<v_i,e^{\alpha_j}>) = \det\,(v_i^{\alpha_j})$$

and $(v_i^{\alpha_j})$ is the matrix obtained from T choosing the rows $\alpha_1,\ldots,\alpha_k$. Hence

$$\begin{aligned}\det TT^* &= \det\,(v_iv_j) = |v_1\wedge\ldots\wedge v_k|^2 = \sum_{\alpha\in I(k,n)} (\xi^\alpha)^2 \\ &= \sum_{\alpha\in I(k,n)} (\det v_i^{\alpha_j})^2\;.\end{aligned} \tag{17}$$

This proves *Cauchy-Binet formula* in Sec. 2.1.1. In fact in terms of the projections

$$P_\alpha:\mathbb{R}^n \longrightarrow \mathbb{R}^k \qquad P_\alpha v = (v^{\alpha_1},\ldots,v^{\alpha_k})$$

the matrix $(v_i^{\alpha_j})$ is nothing else than $P_\alpha\circ T$, hence (17) yields

$$T^*T = \sum_{\alpha\in I(k,n)} (\det\,(P_\alpha\circ T))^2\;.$$

Usually we shall work with the Euclidean norm in $\mathbb{R}^n$ and the induced norm

$$|\xi| = \Big(\sum_{|\alpha|=k} (\xi^\alpha)^2 \Big)^{1/2}$$

in the class of k-vector $\xi \in \Lambda_k\mathbb{R}^n$. However in connection with the calculus of variations it is convenient and natural in many instances to consider the so-called *mass* and *comass* norms on multivectors and covectors. Here we confine ourselves to give just the definitions and mention a property, compare Federer [226] and Vol. II Ch. 1.

The *comass norm* of $\omega \in \Lambda^k V$ is defined by

$$(18) \qquad \|\omega\| := \sup\{<\xi,\omega> \mid \xi \in \Lambda_k V,\ |\xi| \le 1,\ \xi \text{ simple}\}\,.$$

Of course

$$\|\omega\| \le |\omega|\,.$$

The *mass norm* of $\xi \in \Lambda_k V$ is defined by

$$(19) \qquad \|\xi\| := \sup\{<\xi,\omega> \mid \omega \in \Lambda^k V,\ \|\omega\| \le 1\}\,,$$

and of course

$$\|\xi\| \ge |\xi|\,.$$

One can also show that for some $c = c(n,k)$ one has

$$\|\omega\| \ge \frac{1}{c}|\omega|\,, \qquad \|\xi\| \le c|\xi|\,.$$

However in general

$$\|\omega\| < |\omega|\,, \qquad \|\xi\| > |\xi|$$

and we have:

$$\begin{aligned} \|\xi\| &= |\xi| \qquad \textit{if and only if } \xi \textit{ is simple} \\ \|\omega\| &= |\omega| \qquad \textit{if and only if } \omega \textit{ is simple} \end{aligned}$$

For example for the 2–vector $\xi := e_1\wedge e_2 + e_3\wedge e_4$ in $\Lambda_2\mathbb{R}^4$ we have $\|\xi\| = 2$, $|\xi| = \sqrt{2}$.

Simple k-vectors and oriented k-planes. Let us come now to one of the main motivation to introduce k-vectors. We shall show that, similarly to vectors which identify oriented one-dimensional lines, *unit simple k-vectors identify the oriented k-dimensional planes in V*. We shall assume V endowed with an inner product.

First we observe that

k vectors $v_1,\dots,v_k$ are linearly dependent if and only if $v_1\wedge\dots\wedge v_k = 0$.

In fact if $v_1, \dots, v_k$ are linearly dependent we know that $v_1 \wedge \dots \wedge v_k = 0$. Suppose on the contrary that $v_1, \dots, v_k$ are linearly independent. Then we can complete them in order to form a basis of V and consider k elements $\omega^1, \dots, \omega^k$ of the dual basis, so that $< v_i, \omega^j > = \delta_i^j$. We then find

$$< v_1 \wedge \dots \wedge v_k, \omega^1 \wedge \dots \omega^k > \; = \; \det (< v_i, \omega^j >) \; = \; 1$$

hence $v_1 \wedge \dots \wedge v_k \neq 0$.

Let P be an oriented k-dimensional subspace of V and let $\tau_1, \dots, \tau_k$ be a positively oriented orthonormal basis of P. We then set

$$\xi_P := \tau_1 \wedge \dots \wedge \tau_k \; .$$

If $v_1, \dots, v_k$ is another basis of P then we have

$$v_i = \sum_{j=1}^{k} a_{ij} \tau_j \; , \qquad i = 1, \dots, k$$

and it turns out that

$$v_1 \wedge \dots \wedge v_k = \det A \, \tau_1 \wedge \dots \wedge \tau_k$$

as one can easily verify checking on simple covectors. Hence, if $v_1, \dots, v_k$ is another oriented orthonormal basis (so that $\det A = 1$, being $\det A > 0$) we have

$$\xi_P = v_1 \wedge \dots \wedge v_k \; .$$

This shows that ξ_P is uniquely determined by the oriented k-plane P. In general, if $v_1, \dots, v_k$ is any oriented basis of P, then

$$\xi_P = \frac{v_1 \wedge \dots \wedge v_k}{|v_1 \wedge \dots \wedge v_k|} \; .$$

Conversely, let $\xi \in \Lambda_k V$ be a simple non-zero k-vector. Set

$$P_\xi := \{ v \in V \mid \xi \wedge v = 0 \} \; .$$

Of course P_ξ is a linear subspace of V. The dimension of P_ξ is k. In fact, being ξ simple, $\xi = v_1 \wedge \dots \wedge v_k$, and $\xi \neq 0$ implies that $v_1, \dots, v_k$ are linearly independent; moreover, the observation above says that $v \in P_\xi$ if and only if v is a linear combination of $v_1, \dots, v_k$. It follows that

$$P_\xi = \operatorname{span} (v_1, \dots, v_k) \; .$$

Orienting P_ξ in such a way that $v_1, \dots, v_k$ be positively oriented we then see that

$$\xi_{P_\xi} = \frac{\xi}{|\xi|} \; .$$

This proves our claim in the beginning.

We would like now to read the previous characterization of simple vectors in a somewhat more algebraic way. Let $\xi = \xi^\alpha e_\alpha \in \Lambda_k V$ be a simple vector with $\xi \neq 0$. Of course $\xi^\alpha \neq 0$ for some α and by a suitable choice of the basis we may assume that $\alpha = (1, \dots, k)$. This way we may reduce to a Cartesian situation. As such a situation is especially relevant for us in the sequel, we shall discuss it with some details.

Simple n-vectors in the Cartesian product $\mathbb{R}^n \times \mathbb{R}^N$. With the notation above every n-vector in $\Lambda_n \mathbb{R}^{n+N}$ can be written as

$$\xi = \sum_{|\alpha|+|\beta|} \xi^{\alpha\beta}\, e_\alpha \wedge \varepsilon_\beta$$

where α and β are ordered multi-indices. We shall refer to $\xi^{\alpha\beta}$ as to the *components* of ξ and to $\xi^{\bar{0}0}$ as to the $\bar{0}0$-*components* or the *first component* of ξ.

Denoting for any k, $0 \leq k \leq \underline{n} := \min(n, N)$ by $V_{n,k}$ the linear subspace of $\Lambda_n \mathbb{R}^{n+N}$ given by the linear combinations of n-vectors of the type $v_1 \wedge \ldots \wedge v_{n-k} \wedge w_1 \wedge \ldots \wedge w_k$ with $v_i \in \mathbb{R}^n$, $w_i \in \mathbb{R}^N$, so that

$$V_{n,k} := \Lambda_{n-k}\mathbb{R}^n \otimes \Lambda_k \mathbb{R}^N \ , \tag{20}$$

the n-vector ξ can be uniquely decomposed as

$$\xi = \sum_{k=0}^{\underline{n}} \xi_{(k)} \tag{21}$$

where $\xi_{(k)} \in V_{n,k}$ is given by

$$\xi_{(k)} = \sum_{\substack{|\alpha|+|\beta|=n \\ |\beta|=k}} \xi^{\alpha\beta}\, e_\alpha \wedge \varepsilon_\beta \ ;$$

in particular

$$\xi_{(0)} = \xi^{\bar{0}0}\, e_1 \wedge \ldots \wedge e_n \ .$$

Similarly every n-covector splits as

$$\omega = \sum_{|\alpha+|\beta|=n} \omega_{\alpha\beta}\, e^\alpha \wedge \varepsilon^\beta = \sum_{k=0}^{\underline{n}} \omega^{(k)} \ , \qquad \omega^{(k)} \in V^{n,k} = \Lambda^{n-k}\mathbb{R}^n \otimes \Lambda^k \mathbb{R}^N \ ,$$

and the duality $<,>$ reads as

$$<\omega, \xi> = \sum_{|\alpha+|\beta|=n} \omega_{\alpha\beta}\, \xi^{\alpha\beta} = \sum_k <\omega^{(k)}, \xi_{(k)}>$$

$$<\omega^{(k)}, \xi> = <\omega, \xi_{(k)}> = <\omega^{(k)}, \xi_{(k)}> \ .$$

Let $G := (G^j_i)_{i=1,\dots,n,\, j=1,\dots,N}$ be the $N \times n$-matrix associated to the linear transformation that we still denote by G from $\mathbb{R}^n$ into $\mathbb{R}^N$. The vectors

$$e_i + Ge_i \;, \qquad i = 1, \dots, n$$

yields a basis of the tangent plane to the graph of G in $\mathbb{R}^{n+N}$. The unit simple n-vector

$$\xi := \frac{M(G)}{|\,M(G)\,|} \tag{22}$$

where

$$M(G) := (e_1 + Ge_1)\wedge \dots \wedge(e_n + Ge_n) \tag{23}$$

identifies the plane graph of G, and in fact orients such a plane. It will be then called the *tangent n-vector to the graph $\mathcal{G}_G$ of G*. Notice that if we denote by $\mathrm{id} \bowtie G$ the map $\mathrm{id} \bowtie G : \mathbb{R}^n \to \mathbb{R}^{n+N}$ given by

$$(\mathrm{id} \bowtie G)(x) = (x, Gx) \tag{24}$$

then

$$\mathcal{G}_G := (\mathrm{id} \bowtie G)(\mathbb{R}^n) \;.$$

Notice also that ξ depends only on the orientation of $\mathbb{R}^n$ and not on the special fixed basis.

Conversely, if ξ is a unit simple n-vector with positive first component, $\xi^{\bar{0}0} > 0$, then ξ represents an n-plane with no vertical vector, and therefore there exists a linear map $L : \mathbb{R}^n \to \mathbb{R}^N$ such that

$$\xi = \frac{M(L)}{|\,M(L)\,|} \;.$$

More precisely we have

Proposition 1. *Let V be an n-dimensional subspace of $\mathbb{R}^{n+N}$, i.e., an n-plane in $\mathbb{R}^{n+N}$ through the origin oriented by the unitary n-vector $\vec{V}$*

$$\vec{V} = \sum_{|\alpha|+|\beta|=n} V^{\alpha\beta}\, e_\alpha \wedge \varepsilon_\beta \;, \qquad \sum_{|\alpha|+|\beta|=n} (V^{\alpha\beta})^2 = 1$$

and let π_V be the restriction to V of the orthogonal projection π to $\mathbb{R}^n$. We have

(i) V contains vertical vectors, i.e., non zero vectors z such that $\pi(z) = 0$, if and only if $V^{\bar{0}0} = 0$.

(ii) If $V^{\bar{0}0} > 0$, then there exists a linear map $L : \mathbb{R}^n \to \mathbb{R}^N$ such that $V =$ range $(\mathrm{id} \bowtie L)$ and

$$\vec{V} = \frac{M(L)}{|\,M(L)\,|} \;, \qquad M(L) := (e_1 + Le_1)\wedge \dots \wedge(e_n + Le_n) \;.$$

Moreover π_V is the inverse of $\mathrm{id} \bowtie L : \mathbb{R}^n \to V$.

(iii) If J_{π_V} denotes the Jacobian determinant of π_V, then

$$J_{\pi_V} = V^{\bar{0}0} .$$

Proof. Suppose that V contains at least a vertical vector z_1. Let $(z_1, \dots, z_n)$ be a basis for V. We split each z_i as $v_i + w_i$, where $v_i \in \mathbb{R}^n$ and $w_i \in \mathbb{R}^N$. Since z_1 is vertical we have $v_1 = 0$, and, as $z_1 \wedge \dots \wedge z_n$ is a multiple of $\vec{V}$

$$\vec{V} = k\, z_1 \wedge \dots \wedge z_n = k\,(v_1 + w_1) \wedge \dots \wedge (v_n + w_n) .$$

Being $V^{\bar{0}0} = k \det(v_1, \dots, v_n)$, we therefore deduce $V^{\bar{0}0} = 0$. Suppose V does not contain vertical vectors. Choose a basis $z_1, \dots, z_n$ in V and split as previously each z_i as $z_i = v_i + w_i$. Clearly the n vectors v_i are linearly independent, otherwise we could find λ_i such that $\sum \lambda_i v_i = 0$ and therefore $\sum \lambda_i z_i = \sum \lambda_i w_i$ would be vertical. Thus $V^{\bar{0}0} := k \det(v_1, \dots, v_n) \neq 0$. This proves (i).

Define the linear map $L : \mathbb{R}^n \to \mathbb{R}^N$ by $L v_i = w_i$, where $z_i = v_i + w_i$ is a basis for V. Clearly, $v_i + L v_i = z_i$ thus the range of $\mathrm{id} \bowtie L$ is V, moreover π_V is the inverse of $\mathrm{id} \bowtie L$. Finally,

$$\frac{M(L)}{|M(L)|} = \frac{z_1 \wedge \dots \wedge z_n}{|z_1 \wedge \dots \wedge z_n|}$$

$$M^{\bar{0}0}(L) = 1 , \qquad \left(\frac{z_1 \wedge \dots \wedge z_n}{|z_1 \wedge \dots \wedge z_n|}\right)^{\bar{0}0} > 0 ,$$

we therefore conclude that $\vec{V} = M(L)/|M(L)|$. This proves (ii).

If $V^{\bar{0}0} > 0$, being π_V the inverse of $\mathrm{id} \bowtie L$, the Jacobian determinant of π_V is the inverse of the Jacobian determinant of $\mathrm{id} \bowtie L$, thus

$$J_{\pi_V} = \frac{1}{J_{\mathrm{id} \bowtie L}} = \frac{1}{|M(L)|} = V^{\bar{0}0} .$$

If $V^{\bar{0}0} = 0$, then $\pi_V(V)$ is a lower dimensional subspace of $\mathbb{R}^n$, thus $J_\pi^V = 0$. In both cases $V^{\bar{0}0} > 0$ and $V^{\bar{0}0} = 0$, we therefore have $J_{\pi_V} = V^{\bar{0}0}$. This proves (iii) and concludes the proof of the proposition. □

In coordinates it is not difficult to compute, compare (31) (33) below,

(25)
$$M(G) = \sum_{|\alpha| + |\beta| = n} \sigma(\alpha, \bar{\alpha})\, M_{\bar{\alpha}}^{\beta}(G)\, e_\alpha \wedge \varepsilon_\beta$$

where recall $M_{\bar{\alpha}}^{\beta}(G) = \det G_{\bar{\alpha}}^{\beta}$ denotes the determinant of the $(\beta, \bar{\alpha})$-minor of G $G_{\bar{\alpha}}^{\beta} = (G_j^i)_{i \in \beta, j \in \bar{\alpha}}$ and $M_0^0(G) = 1$. In particular we have

(26)
$$\begin{aligned} M_{(0)}(G) &= e_1 \wedge \dots \wedge e_n \\ M_{(1)}(G) &= \sum_{i,j} (-1)^{n-i} G_i^j e_{\bar{i}} \wedge \varepsilon_j \\ M_{(k)}(G) &= \sum_{\substack{|\alpha| + |\beta| = n \\ |\beta| = k}} \sigma(\alpha, \bar{\alpha}) M_{\bar{\alpha}}^{\beta}(G) e_\alpha \wedge \varepsilon_\beta \end{aligned}$$

and in terms of the components $\xi^{\alpha\beta}$ of $\xi := M(G)/|M(G)|$

$$
\begin{aligned}
\xi^{\bar{0}0} &= \frac{1}{|M(G)|} \\
\xi^{\bar{i}j} &= (-1)^{n-i}\xi^{\bar{0}0}G^j_i \qquad (27)\\
\frac{\xi^{\alpha\beta}}{\xi^{\bar{0}0}} &= \sigma(\alpha,\bar{\alpha})M^{\beta}_{\bar{\alpha}}\left(\left((-1)^{n-i}\frac{\xi^{\bar{i}j}}{\xi^{\bar{0}0}}\right)\right).
\end{aligned}
$$

In conclusion we can then state

Proposition 2. *An n-vector $\xi \in \Lambda_n\mathbb{R}^{n+N}$ with non-zero first component is simple if and only if*

$$\frac{\xi^{\alpha\beta}}{\xi^{\bar{0}0}} = \sigma(\alpha,\bar{\alpha})M^{\beta}_{\bar{\alpha}}\left(\left((-1)^{n-i}\frac{\xi^{\bar{i}j}}{\xi^{\bar{0}0}}\right)\right).$$

We also see that the map

$$M : \{N \times n\text{-matrices}\} \longrightarrow \Lambda_n\mathbb{R}^{n+N}$$

is injective, and that the map

$$M_{(1)} : \{N \times n\text{-matrices}\} \longrightarrow V_{n,1} := \Lambda_{n-1}\mathbb{R}^n \otimes \Lambda_1\mathbb{R}^N$$

yields an *isometry* of linear spaces. This allows us to say that

$$\Sigma_1 := \{\xi \in \Lambda_n\mathbb{R}^{n+N} \mid \xi = M(G) \text{ for some } N \times n\text{-matrix } G\}$$

is the *graph of an algebraic function over* $V_{n,1}$.

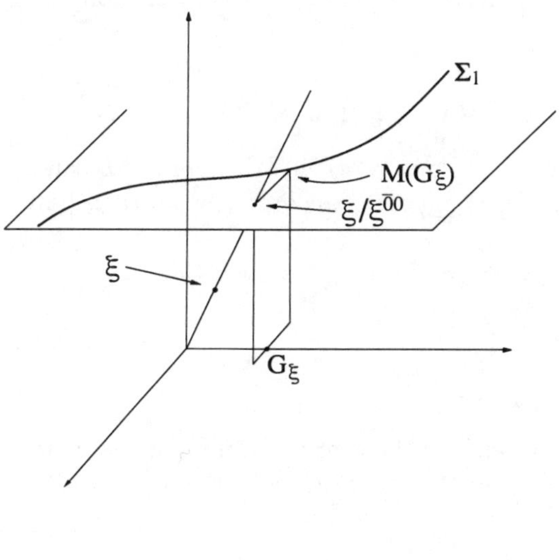

Fig. 2.3. The map M.

For each $\xi \in \Lambda_n\mathbb{R}^{n+N}$ with $\xi^{\bar{0}0} > 0$, we can associate to ξ the $N \times n$-matrix G_ξ defined by

$$G_\xi := M_{(1)}^{-1}\left(\frac{\xi_{(1)}}{\xi^{\bar{0}0}}\right) . \tag{28}$$

Then $\xi \to G_\xi$ is "linear" and $G_{\xi|\Sigma_1} = M^{-1}$, i.e.,

$$G_{M(G)} = G \quad \textit{for all } N \times n\textit{-matrices } G$$

and $\xi = M(G_\xi)$ *if and only if* ξ *is simple and* $\xi \in \Lambda_1$ *where*

$$\Lambda_1 := \{\xi \in \Lambda_n\mathbb{R}^{n+N} \mid \xi^{\bar{0}0} = 1\} .$$

Also

$$G_\xi = 0 \textit{ if and only if } \xi_{(1)} = 0 .$$

In terms of G_ξ Proposition 2 above reads: $\xi \in \Lambda_n\mathbb{R}^{n+N}$ *with* $\xi^{\bar{0}0} > 0$ *is simple if and only if*

$$\xi/\xi^{\bar{0}0} = M(G_\xi) . \tag{29}$$

Finally we notice that, as $\dim \Lambda_n\mathbb{R}^{n+N} = \binom{n+N}{n}$ and simple n-vectors can be parametrized by $\xi_{(1)} \simeq (\xi^{\bar{i}j})$ and by $\xi^{\bar{0}0}$, simple n-vectors form a "small" manifold of dimension $nN+1$; in other words, there are $\binom{n+N}{n} - nN - 1$ conditions for simplicity.

Induced linear transformations. With any linear map $L : V \to W$ between two finite dimensional vector spaces V and W, we can associate the linear maps $\Lambda_k L : \Lambda_k V \to \Lambda_k W$ defined by

$$\Lambda_k L(v_1 \wedge \ldots \wedge v_k) = Lv_1 \wedge \ldots \wedge Lv_k . \tag{30}$$

One can show that the family of linear maps $\Lambda_k L$ is characterized by the following two properties

(i) $\Lambda_1 L = L$
(ii) $\Lambda_{h+k} L(\alpha \wedge \beta) = \Lambda_h L\alpha \wedge \Lambda_k L\beta$, $\quad \alpha \in \Lambda_h V,\ \beta \in \Lambda_k V$

With respect to bases $e_1, \ldots, e_n$ in V and $\varepsilon_1, \ldots, \varepsilon_N$ in W, the map L can be represented by a matrix (a^i_j) so that $Le_j = \sum_i a^i_j \varepsilon_i$. Accordingly we then find

$$\Lambda_k L e_\alpha = \sum_\beta c^\beta_\alpha \, \varepsilon_\beta ,$$

where

$$c^\beta_\alpha = \det\,(a^{\beta_i}_{\alpha_j}) , \tag{31}$$

i.e. the matrix associated to $\Lambda_k L$ has as entries the k-order minors of the matrix A associated to L. In fact, denoting by $\varepsilon^1, \ldots, \varepsilon^N$ the dual basis of $\varepsilon_1, \ldots, \varepsilon_N$ we have

$$\begin{aligned} c_\alpha^\beta &= (\Lambda_k L)(e_\alpha)(\varepsilon^{\beta_1}, \dots, \varepsilon^{\beta_k}) \\ &= (Le_{\alpha_1} \wedge \dots \wedge Le_{\alpha_k})(\varepsilon^{\beta_1}, \dots, \varepsilon^{\beta_k}) \\ &= (\varepsilon^{\beta_1} \wedge \dots \wedge \varepsilon^{\beta_k}) \cdot (Le_{\alpha_1} \wedge \dots \wedge Le_{\alpha_k}) \\ &= \det (< \varepsilon^{\beta_i}, Le_{\alpha_j} >) = \det (< \varepsilon^{\beta_i}, a^i_{\alpha_j} \varepsilon_i >) = \det (a^{\beta_i}_{\alpha_j}) \end{aligned}$$

and, by (16)

(32) $$|\Lambda_k L| = \sqrt{\det T^* T}.$$

It is also readily seen that, if $\xi = v_1 \wedge \dots \wedge v_k$ is a simple vector and P the subspace generated by $v_1, \dots, v_k$, then $\Lambda_k L(\xi)$ is again a simple vector, and, in fact it is the k-vector associated to the plane $L(P)$, if L has maximal rank or 0.

Returning to n-vectors in $\Lambda_n \mathbb{R}^{n+N}$ and linear maps $G : \mathbb{R}^n \to \mathbb{R}^N$, we can then write

$$M(G) = (e_1 + Ge_1) \wedge \dots \wedge (e_n + Ge_n) = \Lambda_n (\mathrm{id} \bowtie G)(e_1 \wedge \dots \wedge e_n) .$$

The computations to prove (31), then yield a proof of the decomposition

(33) $$M(G) = \sum_{|\alpha| + |\beta| = n} \sigma(\alpha, \bar\alpha)\, M^\beta_{\bar\alpha}(G)\, e_\alpha \wedge \varepsilon_\beta .$$

In fact computing $M(G)$ we get terms of the type

$$\dots \wedge e_i \wedge \dots \wedge Ge_j \wedge \dots \wedge e_h \wedge \dots \wedge Ge_k \wedge \dots ;$$

reordering such terms then we get

(34) $$M(G) = \sum_\alpha^n \sigma(\alpha, \bar\alpha) e_\alpha \wedge (\Lambda_{|\alpha|} G)(e_{\bar\alpha})$$

that by (31) is just (33).

Similarly, there is a linear induced map

$$\Lambda^k L : \Lambda^k W \longrightarrow \Lambda^k V$$

characterized by the properties

(i) $\Lambda^1 L = L^*$, $L^* : W \to V$

(ii) $\Lambda^{h+k} L(\omega \wedge \delta) = \Lambda^h L(\omega) \wedge \Lambda^k L(\delta)$, $\omega \in \Lambda^h W$, $\delta \in \Lambda^k V$

and again defined by

$$\Lambda^k L(\omega^1 \wedge \dots \wedge \omega^k) = L^* \omega^1 \wedge \dots \wedge L^* \omega^k .$$

One easily checks that $\Lambda^k L$ is the dual map of $\Lambda_k L$, i.e.,

(35) $$< \Lambda_k L \xi, \omega > = < \xi, \Lambda^k L \omega > \qquad \forall\, \xi \in \Lambda_k V,\ \omega \in \Lambda^k W .$$

and

(36) $$||\Lambda^k L|| = ||\Lambda_k L||.$$

If $G : W \to Z$ is another linear map, we have

(37) $$\Lambda_k(G \circ L) \;=\; \Lambda_k G \circ \Lambda_k L \;.$$

Equalities (35) and (37) express the so–called *Binet formulas* , compare Vol. II Ch. 2, which in the case $V = W = Z$ and $k = \dim V$ amount respectively to

$$\det L = \det L^T \qquad \det G\, L = \det G \det L \;,$$

as one can see using (31).

2.2 Differential Forms

Let $\mathcal{X}$ be an n-dimensional manifold. A *k-vector field* ξ on $\mathcal{X}$ is a map

$$\xi : x \in \mathcal{X} \;\longrightarrow\; \xi(x) \in \Lambda_k T_x M \;.$$

Similarly, a *k-covector field* on $\mathcal{X}$ is a map

$$\omega : x \in \mathcal{X} \;\longrightarrow\; \omega(x) \in \Lambda^k T_x M \simeq \Lambda_k(T_x M)^* \;.$$

A k-covector field is also called a *(differential) k-form.* In fact it can be generated by applying the exterior differential d to functions and using the wedge product.

For the sake of simplicity here we shall only consider k-forms on an open set U of $\mathbb{R}^n$, so that ω is a map

$$\omega : U \subset \mathbb{R}^n \;\longrightarrow\; \Lambda^k \mathbb{R}^n;$$

we shall discuss the general situation in Ch. 5.

Exterior differentiation. As usual, let $e_1, \dots, e_n$ be the standard basis in $\mathbb{R}^n$. We denote by dx^i the 1-form on $\mathbb{R}^n$ defined as the constant map

$$dx^i : x \in \mathbb{R}^n \;\longrightarrow\; e^i \in \Lambda^1 \mathbb{R}^n \simeq (\mathbb{R}^n)^* \;,$$

$e^1, \dots, e^n$ being the dual basis of $\mathbb{R}^n$, so that $dx^i(x) = x^i$. For any smooth function $f : U \to \mathbb{R}$ its *differential* is defined as the 1-form $df : U \to \Lambda^1 \mathbb{R}^n$

$$df : x \in U \;\longrightarrow\; \sum_{i=1}^n f_{x^i}(x) dx^i \;\in\; \Lambda^1 \mathbb{R}^n \;.$$

Of course df does not depend of the fixed basis, in fact for any smooth vector field $\xi : x \in U \to \xi(x) \in \Lambda_1 \mathbb{R}^n \simeq \mathbb{R}^n$ we have

$$\begin{aligned} <\xi(x), df_x> \;&=\; <\sum \xi^i(x) e_i, \sum f_{x^i}(x) e^j> \\ &=\; D_{\xi(x)} f(x) = \frac{d}{dt} f(x + t\xi(x))_{|\,t=0} \;. \end{aligned}$$

Notice also that the definition of dx^i is consistent with the definition of df; in fact if $f(x) := x^k$, then

$$df_x = \sum_{i=1}^{n} f_{x^i}(x)\, dx^i = \delta_{ki} e^i = e_k \, .$$

For any ordered multi-index $\alpha \in I(k,n)$ the k-form

$$dx^\alpha : x \in U \longrightarrow e^\alpha \in \Lambda^k \mathbb{R}^n$$

can be written as

$$dx^\alpha = dx^{\alpha_1} \wedge \ldots \wedge dx^{\alpha_k} \, ,$$

and every k-form $\omega : U \subset \mathbb{R}^n \to \Lambda^k \mathbb{R}^n$ as $\omega(x) = \sum_{|\alpha|=k} \omega_\alpha(x)\, dx^\alpha$; so that

$$\omega = \sum_{|\alpha|=k} \omega_\alpha \, dx^\alpha$$

where the product of a k-form ω by a function is defined by

$$(f\omega)(x) := f(x)\omega(x) \, ,$$

and more generally the exterior product of two forms by

$$(\omega \wedge \delta)(x) := \omega(x) \wedge \delta(x) \, .$$

The *exterior derivative* of a k-form, $k \geq 1$, $\omega = \sum_{|\alpha|=k} \omega_\alpha \, dx^\alpha$ with smooth coefficients $\omega_\alpha(x)$ is defined by

$$d\omega := \sum_{|\alpha|=k} d\omega_\alpha \wedge dx^a = \sum_{\substack{|\alpha|=k \\ i=1,\ldots,n}} \omega_{\alpha x^i} \, dx^i \wedge dx^\alpha \, .$$

This way

(i) *df agrees with the ordinary differential for C^1 function f*
(ii) *$d(\omega + \eta) = d\omega + d\eta$ for all k-forms ω and η*
(iii) *$d(\omega \wedge \eta) = d\omega \wedge \eta + (-1)^k \omega \wedge d\eta$ if ω has degree k*
(iv) *$d(d\omega) = 0$, for any form of class C^2.*

Notice that (iii) generalizes the ordinary differentiation rule for the product of functions, and (iv) expresses in case ω is a function the equality of mixed derivatives

$$D_i D_j f = D_j D_i f \, .$$

Conversely one can easily show that the exterior derivative operator d is characterized by properties (i)...(iv). This in particular implies that $d\omega$ is intrinsically defined, independently of the coordinate system.

The space of infinitely differentiable and compactly supported k-forms in U is defined by

$$\mathcal{D}^k(U) := \{\sum_{|\alpha|=k} \omega_\alpha \, dx^\alpha \mid \omega_\alpha \in C_c^\infty(U)\} .$$

Similarly we define

$$\mathcal{E}^k(U) := \{\sum_{|\alpha|=k} \omega_\alpha \, dx^\alpha \mid \omega_\alpha \in C^\infty(U)\}$$

and we speak of m-differentiable k-forms. We may then regard the exterior derivative as an operator

$$d : \mathcal{E}^k(U) \longrightarrow \mathcal{E}^{k+1}(U) .$$

Of course it increases of one the degree of a form and decreases of one its regularity property.

Pull-back. Let $U \subset \mathbb{R}^n$ and $V \subset \mathbb{R}^N$, and let $f : U \to V$ be a smooth map. To every k-form $\omega = \sum \omega_\beta \, dy^\beta \in \mathcal{D}^k(V)$ we can associate the form $f^\# \omega \in \mathcal{E}^k(U)$, called the *pull-back* of ω and defined by

$$f^\# \omega = \sum_{|\beta|=k} \omega_\beta \circ f \, df^\beta \tag{1}$$

where

$$df^\beta := df^{\beta_1} \wedge \ldots \wedge df^{\beta_k} , \quad df^{\beta_h} = \sum_{i=1}^n f_{x^i}^{\beta_h} \, dx^i .$$

It is not difficult to show that for any $x \in U$ we have

$$f^\# \omega(x) = \Lambda^k df_x(\omega(f(x))) \tag{2}$$

and

$$f^\# (dy^{\beta_1} \wedge \ldots \wedge dy^{\beta_k}) = df^{\beta_1} \wedge \ldots \wedge df^{\beta_k}$$

so that pulling back is an intrinsic operation, and $f^\#$ enjoys the following properties

(i) $f^\#(\omega + \eta) = f^\#(\omega) + f^\#(\eta)$,
(ii) $f^\#(\omega \wedge \eta) = f^\#(\omega) \wedge f^\#(\eta)$,
(iii) $f^\#(dg) = d(g \circ f)$, if $g : V \to \mathbb{R}$,
(iv) $d(f^\# \omega) = f^\#(d\omega)$ if $f \in C^2$.

Differential forms in a Cartesian product. Later we shall be interested in differential r-forms in $U := \Omega \times \mathbb{R}^N \subset \mathbb{R}^n \times \mathbb{R}^N$. As for r-vectors and r-covectors, we can write every r-form in $\mathcal{D}^r(U)$ as

$$\omega = \sum_{|\alpha|+|\beta|=r} \omega_{\alpha\beta}\, dx^\alpha \wedge dy^\beta$$

or

$$\omega = \sum_{k=\max(0,r-n)}^{\min(r,N)} \omega^{(k)}$$

where

$$\omega^{(k)} := \sum_{\substack{|\alpha|+|\beta|=r \\ |\beta|=k}} \omega_{\alpha\beta}\, dx^\alpha \wedge dy^\beta \ .$$

Integration of differential forms. Let M be a smooth, say C^1, k-dimensional manifold that for the sake of simplicity we assume to be embedded in $\mathbb{R}^n$. Let (U_i, ψ_i) define *local charts* in M and $\varphi_i : V \subset \mathbb{R}^k \longrightarrow M \cap U_i$, $\varphi_i = \psi_i^{-1}$, be a *local parametrization* of M with parameter variables u. The *tangent space* of M at $\varphi_i(u)$ is given by

$$T_{\varphi_i(u)}M := \operatorname{span}\{D_1\varphi_i(u), \dots, D_k\varphi_i(u)\}$$

and the unit k-vector

$$\xi(\varphi_i(u)) := \frac{D_1\varphi_i(u)\wedge \dots \wedge D_k\varphi_i(u)}{|\, D_1\varphi_i(u)\wedge \dots \wedge D_k\varphi_i(u)\,|} \qquad u \in U_i$$

defines an *orientation* of $T_{\varphi_i(u)}M$. Recall that two parametrizations yield the same orientation if the Jacobian determinant of the transformation which reduces one chart to the other is positive. In this case we have

$$\frac{D_1\varphi_i(u)\wedge \dots \wedge D_k\varphi_i(u)}{|\, D_1\varphi_i(u)\wedge \dots \wedge D_k\varphi_i(u)\,|} = \frac{D_1\varphi_j(u)\wedge \dots \wedge D_k\varphi_j(u)}{|\, D_1\varphi_j(u)\wedge \dots \wedge D_k\varphi_j(u)\,|} \quad \text{for } \varphi_i(u) = \varphi_j(u).$$

If one can choose charts or parametrizations in such a way that $\xi(\varphi_i(u))$ defines a smooth unitary tangent vector field on M, one says that M is an *oriented manifold with orientation* ξ.

Suppose now ω is a smooth k-form on M (or in a neighbourhood of M in $\mathbb{R}^n$), and suppose $\operatorname{spt}\omega \subset U_i$, then classically the integral of ω on M is defined as the integral of the k-form $\varphi_i^{\#}\omega$ on $\mathbb{R}^k$

(3) $$\int_M \omega := \int_V \varphi_i^{\#}\omega$$

given by

$$\int_M \omega := \int_V <\varphi_i^{\#}\omega(u), e_1\wedge\ldots\wedge e_k> d\mathcal{H}^k(u) .$$

One easily computes

$$\begin{aligned} <\varphi_i^{\#}(\omega(u)), e_1\wedge\ldots\wedge e_k> &= \omega(\varphi_i(u))(D\varphi_i(u)e_1\wedge\ldots\wedge D\varphi_i(u)e_k) \\ &= <D_1\varphi_i(u)\wedge\ldots\wedge D_k\varphi_i(u), \omega(\varphi_i(u))> \\ &= <\xi(\varphi_i(u)), \omega(\varphi_i(u))> |D_1\varphi_i(u)\wedge\ldots\wedge D_k\varphi_i(u)| \\ &= <\xi(\varphi_i(u)), \omega(\varphi_i(u))> J_{\varphi_i}(u) . \end{aligned} \tag{4}$$

In the general case, that is if $\operatorname{spt}\omega$ is not contained in a chart U_i, and M is oriented by $\xi \in \Lambda_n TM$, one shows by means of a decomposition of unity $\{\eta_j\}$ that

$$\int_M \omega = \sum_{i,j} \int_V \varphi_i^{\#}(\eta_j\omega), \qquad \varphi_{i\#}(e_1\wedge\ldots\wedge e_n) = \xi ,$$

is well defined. According to the computation in (4) and taking into account the area formula in Sec. 2.1, $J_\varphi(u)\, d\mathcal{H}^k(u) = d\mathcal{H}^k \llcorner M$, we then find

$$\int_M \omega = \int_M <\xi(x), \omega(x)> d\mathcal{H}^k(x) . \tag{5}$$

This shows that a k-form is a natural object to integrate over an oriented, k-dimensional rectifiable set, because it is sensitive only to both the location $x \in M$ and the tangent plane to M at x.

Given a k-rectifiable set $\mathcal{M} \subset \mathbb{R}^n$, we say that $\mathcal{M}$ is *oriented* by $\xi : \mathcal{M} \to \Lambda_k\mathbb{R}^n$ if

(i) ξ is $\mathcal{H}^k$-measurable,
(ii) for $\mathcal{H}^k$-a.e. x $|\xi(x)| = 1$ and $\xi(x)$ is associated to $T_x\mathcal{M}$, in the sense that ξ is the wedge product of k vectors of a basis of $T_x\mathcal{M}$.

In this case we define the integral of the k-form ω on the oriented rectifiable set $\mathcal{M}$ simply by (5).

Of special relevance for us will be the case in which $\mathcal{M}$ is the graph say of a smooth map $u : \Omega \subset \mathbb{R}^n \to \mathbb{R}^N$. In this case we easily infer from the above

$$\int_{\mathcal{G}_{u,\Omega}} \omega = \int_\Omega (\mathrm{id} \bowtie u)^{\#}\omega = \int_\Omega <\omega(x, u(x)), M(Du(x))> dx . \tag{6}$$

2.3 Currents: Basic Facts

Let $U \subset \mathbb{R}^n$ be an open set, $0 \le k \le n$. We denote by $\mathcal{D}^k(U)$ the space of all infinitely differentiable and compactly supported k-forms in U topologized by the usual topology which is characterized by the assertion that

$$\omega^i := \sum_{\alpha \in I(k,n)} \omega^i_\alpha \, dx^\alpha \longrightarrow \omega := \sum_{\alpha \in I(k,n)} \omega_\alpha \, dx^\alpha \,, \qquad i \to \infty$$

if there is a fixed compact $K \subset U$ such that

(i) $\operatorname{spt} \omega^i_\alpha \subset K \ \forall \, \alpha \in I(k,n)$ and $\forall \, i$
(ii) $\lim_{i\to\infty} D^\beta \omega^i_\alpha = D^\beta \omega_\alpha \ \forall \, \alpha \in I(k,n)$ and every multi-index β.

Definition 1. *A k-dimensional current in U is a continuous linear functional on $\mathcal{D}^k(U)$. The space of k-dimensional currents in U is denoted by $\mathcal{D}_k(U)$.*

Given $T \in \mathcal{D}_k(U)$, we can define for any $\alpha \in I(k,n)$ the *components of T* (with respect to standard basis $e_1, \dots, e_n$ in $\mathbb{R}^n$) by setting for any $\varphi \in C^\infty_c(U)$

$$T^\alpha(\varphi) := T(\varphi \, dx^\alpha) \,.$$

Clearly with a certain abuse of language we have

$$T = \sum_{|\alpha|=k} T^\alpha e_\alpha \,,$$

where we regard e_α as the operator

$$e_\alpha \Big(\sum \omega_\beta \, dx^\beta \Big) := \omega_\alpha$$

i.e.

$$e_\alpha(dx^\beta) = \delta^\beta_\alpha = \delta^{\beta_1}_{\alpha_1} \cdot \delta^{\beta_2}_{\alpha_2} \cdots \delta^{\beta_k}_{\alpha_k} \,,$$

or equivalently

$$T(\omega) = \sum_{\alpha \in I(k,n)} T^\alpha(\omega_\alpha) \qquad \text{if} \qquad \omega = \sum_{\alpha \in I(k,n)} \omega_\alpha \, dx^\alpha \,.$$

From this point of view a current is just a collection of ordinary distributions indexed by $\alpha \in I(k,n)$. Accordingly we have

Definition 2. *Let $T_h, T \in \mathcal{D}_k(U)$. We say that the sequence $\{T_h\}$* converges weakly *to T, $T_h \rightharpoonup T$, if $T_h(\omega) \to T(\omega)$ for all $\omega \in \mathcal{D}^k(U)$.*

The first notion which distinguish currents from ordinary distributions is that of *boundary*. It is simply defined as the dual operation of the exterior derivative.

Definition 3. *The* boundary *of $T \in \mathcal{D}_k(U)$ is defined as the $(k-1)$-current*

(1) $$\partial T(\eta) := T(d\eta) \qquad \forall \, \eta \in \mathcal{D}^{k-1}(U) \,.$$

We set

$$\partial T = 0 \qquad \text{if} \qquad T \in \mathcal{D}_0(U) \,.$$

We shall see that the boundary operator

$$\partial \;:\; \mathcal{D}_k(U) \longrightarrow \mathcal{D}_{k-1}(U)$$

comprises two apparently different concepts

(a) that of derivative of a function
(b) that of boundary of submanifolds.

Before discussing the important subclass of currents with components which are measures in U, let us make a few general remarks.

As for distributions the *support of a current* $T \in \mathcal{D}_k(U)$ is defined by

$$\operatorname{spt} T = \bigcap \{K \subset U \quad | \quad K \text{ relatively closed in } U,\ T(\omega) = 0 \text{ for all } \omega \in \mathcal{D}^k(U) \text{ with } \operatorname{spt}\omega \subset U \setminus K\} \,.$$

One can define the *Cartesian product of currents.* Let $T_i \in \mathcal{D}_{k_i}(U_i)$, $U_i \subset \mathbb{R}^{n_i}$, $i = 1, 2$. The current

$$T \;:=\; T_1 \times T_2 \;\in\; \mathcal{D}_{k_1+k_2}(U_1 \times U_2)$$

is defined as follows. If $\omega \in \mathcal{D}^{k_1+k_2}(U)$ has the form

$$\omega \;=\; \sum_{\substack{|\alpha|=k_1\\ |\beta|=k_2}} \omega_{\alpha\beta}(x,y)\, dx^\alpha \wedge dy^\beta \;,$$

then we set

$$T(\omega) \;=\; T_1\Big(\sum_{|\alpha|=k_1} T_2\Big(\sum_{|\beta|=k_2} \omega_{\alpha\beta}(x,y)\, dy^\beta\Big)\, dx^\alpha\Big) \,. \tag{2}$$

If $|\alpha| \neq k_1$, $|\beta| \neq k_2$, $|\alpha| + |\beta| = k_1 + k_2$, and $\varphi \in C_c^\infty(U)$, we set

$$T(\varphi\, dx^\alpha \wedge dy^\beta) \;=\; 0 \,. \tag{3}$$

One easily shows that

$$\partial(T_1 \times T_2) \;=\; \partial T_1 \times T_2 + (-1)^{k_1} T_1 \times \partial T_2 \,. \tag{4}$$

As a special case consider $T \in \mathcal{D}_k(U)$, $U \subset \mathbb{R}^n$, and the 1-current $[\![(0,1)]\!]$ defined as integration of 1-forms on the oriented segment $(0,1)$ of $\mathbb{R}$. Let $\omega \in \mathcal{D}^{k+1}([0,1] \times U)$. Split ω into its horizontal and vertical parts, $\omega = \omega_H + \omega_V$ where

$$\omega_H(t,x) = \sum_{|\alpha|=k} \omega_{H\alpha}(t,x)\, dt \wedge dx^\alpha \qquad \omega_V(t,x) = \sum_{|\alpha|=k+1} \omega_{V\alpha}(t,x)\, dx^\alpha$$

and denote by $[\omega]_{(0,1)}$ the k-form in U obtained by averaging ω_H with respect to the t variable

$$[\omega]_{(0,1)}(x) := \sum_{|\alpha|=k} \Big(\int_0^1 \omega_{H\alpha}(t,x)\, dt \Big)\, dx^\alpha .$$

From the definition of product of currents we infer, compare (3), that $[\![(0,1)]\!] \times T(\omega_V) = 0$, while from (2)

$$(5) \qquad \| (0,1) \| \times T(\omega) = \int_0^1 T\Big(\sum_{|\alpha|=k} \omega_{H\alpha}(t,x)\, dx^\alpha \Big)\, dt = T([\omega]_{(0,1)}).$$

Also

$$(6) \qquad \begin{aligned} \partial([\![(0,1)]\!] \times T) &= (\delta_1 - \delta_0) \times T - [\![(0,1)]\!] \times \partial T \\ &= \delta_1 \times T - \delta_0 \times T - [\![(0,1)]\!] \times \partial T \end{aligned}$$

where $\delta_0, \delta_1 \in \mathcal{D}_0(\mathbb{R}^1)$ are the Dirac masses at 0 and 1,

$$\delta_0(\varphi) = \varphi(0) \ , \qquad \delta_1(\varphi) = \varphi(1) \qquad \forall\, \varphi \in \mathcal{D}^0(\mathbb{R}^1) \ .$$

Currents which are representable by integration. Those are currents with components which are measures with locally finite total variations instead of mere distributions.

Definition 4. *Let $U \subset \mathbb{R}^n$ and $V \subset U$ be open sets, and let $T \in \mathcal{D}_k(U)$. The mass of T in V is defined by*

$$\underline{\mathbf{M}}_V(T) := \sup\{T(\omega) \mid \omega \in \mathcal{D}^k(U),\ \operatorname{spt}\omega \subset V,\ |\omega(x)| \le 1\ \forall\, x \in U\} \ .$$

Actually there are very good reasons for replacing in the previous definition the Euclidean norm $|\omega(x)|$ by the *comass norm* defined in Sec. 2.2.1, compare Vol. II Ch. 1.

Definition 5. *Let $U \subset \mathbb{R}^n$ and $V \subset U$ be an open set, and let $T \in \mathcal{D}_k(U)$. The* mass *of T in V is defined by*

$$\mathbf{M}_V(T) := \sup\{T(\omega) \mid \omega \in \mathcal{D}^k(U),\ \operatorname{spt}\omega \subset V,\ \|\omega\| \le 1\ \forall\, x \in U\}.$$

The reason for such a definition will be apparent in Vol. II Ch. 1 where dealing with approximation theorems, compare also Sec. 2.2.6. The Euclidean and mass norms are equivalent, in fact

$$\underline{\mathbf{M}}_V(T) \le \mathbf{M}_V(T) \le c\underline{\mathbf{M}}_V(T)$$

where c is an absolute positive constant, see Sec. 2.2.1; but in general we have

$$\underline{\mathbf{M}}_V(T) < \mathbf{M}_V(T).$$

If $V = U$ we shall simply write $\mathbf{M}(T)$ instead of $\mathbf{M}_V(T)$, and we set

$$\begin{aligned} \mathcal{M}_k(U) &:= \{T \in \mathcal{D}_k(U) \mid \mathbf{M}(T) < \infty\} \\ \mathcal{M}_{k,\text{loc}}(U) &:= \{T \in \mathcal{D}_k(U) \mid \mathbf{M}_V(T) < \infty \ \forall\, V \subset\subset U\}\,. \end{aligned}$$

Of course $T \in \mathcal{M}_{k,\text{loc}}(U)$ (respectively $\mathcal{M}_k(U)$) if and only if the T^α's are Radon measures (respectively Radon measures with finite total variation in U).

Introducing the *total variation* of T with respect to the comass and the Euclidean norm

$$\begin{aligned} \|T\|(\varphi) &:= \sup\{T(\omega) \mid \|\omega(x)\| \le \varphi(x)\} \\ |T|(\varphi) &:= \sup\{T(\omega) \mid |\omega(x)| \le \varphi(x)\}, \end{aligned}$$

$\varphi \ge 0$, an immediate consequence of Riesz theorem for measures is

Theorem 1. *Let $T \in \mathcal{M}_{k,\text{loc}}(U)$. We have*

(i) There exists a $|T|$-measurable function f with $1 \le f \le c$ such that

$$\|T\| = |T| \llcorner f.$$

(ii) There exists a unique $\|T\|$-measurable map $\vec{T} : U \to \Lambda_k\mathbb{R}^n$ with $\|\vec{T}\| = 1$ $\|T\|$-a.e. such that for all $\omega \in \mathcal{D}^k(U)$

$$T(\omega) = \int_U <\vec{T}, \omega> d\,\|T\|. \tag{7}$$

Moreover

$$\|T\|(V) = \mathbf{M}_V(T) \qquad \forall\, V \subset\subset U,\ V \text{ open}$$

and $\vec{T}$ is the Radon-Nikodym derivative of T with respect to $\|T\|$

$$\vec{T} = \frac{dT}{d\|T\|}$$

(iii) Also

$$\begin{aligned} T(\omega) &= \int_U <f\vec{T}, \omega> d\,|T| \\ |T|(V) &= \mathbf{M}_V(T) \qquad \forall\, V \subset\subset U,\ V \text{ open} \end{aligned}$$

and $|f\vec{T}| = 1$ $|T|$-a.e.

Notice that conversely, given a non-negative Radon measure μ and a μ-measurable and locally summable map $f : U \to \Lambda_k\mathbb{R}^n$ the current $T = \mu \llcorner f$ defined by

$$T(\omega) := \int <f, \omega> d\mu$$

belongs to $\mathcal{M}_{k,\text{loc}}(U)$ and we have

$$\| T \| = \mu \llcorner \| f \| , \qquad \vec{T} = \frac{f}{\| f \|} \qquad \mu \text{ a.e. } .$$

Let $T \in \mathcal{M}_{k,\text{loc}}(U)$, i.e., $T(\omega) = \int <\vec{T}, \omega> d\| T \|$. Though in general $\underline{\mathbf{M}}_V(T) \neq \mathbf{M}_V(T)$, *if* $\vec{T}$ *is simple*, we infer from Theorem 1

$$1 = f\,|\vec{T}| = f\,\| \vec{T} \| = f \quad |T|\text{–a.e., equivalently } ||T||\text{–a.e..}$$

Therefore *we conclude in this case* $\underline{\mathbf{M}}_V(T) = \mathbf{M}_V(T)$. In the sequel we shall always work with the mass and comass norms, but we understood equalities like

$$\| T \| = |T|, \quad \| \vec{T} \| = |\vec{T}| \quad \| T \|\text{-a.e. or } |T|\text{-a.e.}$$

in the case that $\vec{T}$ is simple.

From the definition of mass we readily infer

Proposition 1 (Lower semicontinuity of the mass). *Let* $T_j, T \in \mathcal{D}_k(U)$. *If* $T_j \rightharpoonup T$ *then for any* $V \subset U$, V *open*, $\mathbf{M}_V(T) \leq \liminf_{j\to\infty} \mathbf{M}_V(T_j)$.

and from the compactness theorem for Radon measures

Proposition 2 (Compactness-closure theorem). *Let* $\{T_j\} \subset \mathcal{M}_{k,\text{loc}}(U)$ *be a sequence satisfying* $\sup_j \mathbf{M}_V(T_j) < \infty\ \forall\ V \subset\subset U$. *Then there exists a subsequence* $\{T_{j'}\}$ *and* $T \in \mathcal{M}_{k,\text{loc}}(U)$ *such that* $T_{j'} \rightharpoonup T$. *Moreover*

$$\mathbf{M}(T) \leq \liminf_{j'\to\infty} \mathbf{M}(T_{j'}) < \infty$$

if the masses of $T_{j'}$ *are equibounded.*

By the monotone convergence theorem any current with finite mass, $T \in \mathcal{M}_k(U)$, can be extended to all k-forms with Borel bounded coefficients, and again we have

$$T(\omega) := \int_U <\vec{T}, \omega> d\| T \| .$$

A similar extension works also for $T \in \mathcal{M}_{k,\text{loc}}(U)$ and for compactly supported k-forms ω in U with bounded Borel coefficients.

In particular, if $T \in \mathcal{M}_k(U)$ and $A \subset U$ is any Borel set we can define the *restriction of* T *on* A, $T \llcorner A$, by

(8) $$T \llcorner A(\omega) := \int_A <\vec{T}, \omega> d\| T \| .$$

More generally, we can define for any bounded Borel function

$$T \llcorner f(\omega) := \int <\vec{T},\omega> f\, d\|T\| . \tag{9}$$

However a certain care is needed when dealing with weak convergence. From $\{T_j\} \subset \mathcal{M}_k(U), \sup_j \mathbf{M}(T_j) < \infty,\ T_j \rightharpoonup T$, i.e.,

$$T_j(\omega) \longrightarrow T(\omega) \qquad \forall\, \omega \in \mathcal{D}^k(U)$$

one can easily infer that $T_j(\omega) \to T(\omega)$ on k–forms with continuous coefficients with compact support, but in general the $T_j(\omega)$ do *not* converge to $T(\omega)$ if ω has only bounded Borel coefficients (even with support contained in U). The reason for that is of course that a sequence of functions in $L^1(\Omega)$ which converges as measures need not converge in L^1, compare Ch. 1.

For future purposes we note that for any $T \in \mathcal{M}_{k,\mathrm{loc}}(U)$, $T := \|T\| \llcorner \vec{T}$, *the current* $[\![(0,1)]\!] \times T$ *has locally finite mass and actually*

$$\| [\![(0,1)]\!] \times T \| = \mathcal{L}^1 \times \|T\| \qquad \overrightarrow{[\![(0,1)]\!] \times T} = e_1 \wedge \vec{T}.$$

For the reader's convenience we include the simple proof. With the same notation as above let $\omega \in \mathcal{D}^{k+1}(U)$, $\omega = \omega_H + \omega_V$. As clearly $\|\omega_H\| \leq \|\omega\|$, we have $\| [\omega]_{(0,1)} \|(x) \leq \int_0^1 \|\omega(t,x)\|\, dt$, hence

$$\begin{aligned} |[\![(0,1)]\!] \times T(\omega)| &= |T([\omega]_{(0,1)})| \leq \|T\|(\| [\omega]_{(0,1)} \|) \\ &\leq \|T\|\Big(\int_0^1 \|\omega(t,x)\|\, dt\Big) = \mathcal{L}^1 \times \|T\|(\|\omega(t,x)\|), \end{aligned}$$

from which we infer $\| [\![(0,1)]\!] \times T \| \leq \mathcal{L}^1 \times \|T\|$. To prove the opposite inequality, for $\varepsilon > 0$ and $f \in C^0([0,1] \times U)$, let $\omega \in \mathcal{D}^k([0,1] \times U)$ be such that for any t

$$T(\omega(t,x)) \geq \|T\|(f(t,\cdot)) - \varepsilon.$$

Integrating with respect to t in $(0,1)$ we get

$$\| [\![(0,1)]\!] \times T \|(f) \geq \int_0^1 T(\omega(t,x))\, dt \geq \mathcal{L}^1 \times \|T\|(f) - \varepsilon,$$

which shows that $\| [\![(0,1)]\!] \times T \| \geq \mathcal{L}^1 \times \|T\|$. To conclude the proof of the claim it suffices to notice the following equality

$$\begin{aligned} [\![(0,1)]\!] \times T(\omega) &= \int_0^1 T(\omega_H)\, dt = \int_0^1 dt \int <\omega_H(t,x), \vec{T}(x)>\, d\|T\| \\ &= \int <\omega(t,x), e_1 \wedge \vec{T}>\, d\mathcal{L}^1 \times \|T\|. \end{aligned}$$

Normal currents. The class of *normal currents* is defined as

$$\begin{aligned}\mathbf{N}_k(U) &:= \{T \in \mathcal{D}_k(U) \mid \mathbf{M}(T) + \mathbf{M}(\partial T) < \infty\} \\ &= \{T \in \mathcal{M}_k(U) \mid \partial T \in \mathcal{M}_{k-1}(U)\}\,,\end{aligned}$$

and the class of *locally normal currents* by

$$\mathbf{N}_{k,\mathrm{loc}}(U) := \{T \in \mathcal{D}_k(U) \mid \mathbf{M}_V(T) + \mathbf{M}_V(\partial T) < \infty \;\forall\, V \subset\subset U\}\,.$$

If $T \in \mathbf{N}_k(U)$ one sets

$$\mathbf{N}_V(T) := \mathbf{M}_V(T) + \mathbf{M}_V(\partial T).$$

Of course also for normal currents a *closure-compactness theorem* holds: If $T_j \in \mathbf{N}_{k,\mathrm{loc}}(U)$ with

$$\sup_j\{\mathbf{M}_V(T_j) + \mathbf{M}_V(\partial T_j)\} < \infty \qquad \forall\, V \subset\subset U\,,$$

then there exists a subsequence $\{T_{j'}\}$ of $\{T_j\}$ and $T \in \mathbf{N}_{k,\mathrm{loc}}(U)$ such that

$$T_{j'} \rightharpoonup T\,.$$

An interesting result for normal currents, which will be proved in Theorem 3 in Sec. 4.3.1, states that they cannot concentrate on small closed sets; more precisely we have

Theorem 2. *Let $T \in \mathbf{N}_{k,\mathrm{loc}}(U)$, $U \subset \mathbb{R}^n$, and let p_α denotes for any $\alpha \in I(k,n)$ the orthogonal projection*

$$p_\alpha : (x^1,\dots,x^n) \longrightarrow (x^{\alpha_1},\dots,x^{\alpha_k})\,.$$

Then, for any closed set $E \subset U$ with $\mathcal{H}^k(p_\alpha(E)) = 0 \;\forall\, \alpha \in I(k,n)$, we have

$$\|T\|(E) = 0$$

i.e. $T \llcorner E = 0$.

n-dimensional currents in $\mathbb{R}^n$. Those currents will be discussed with some details in Ch. 4. Here we shall confine ourselves to state for the reader's convenience a few facts.

Let U be an open set in $\mathbb{R}^n$. Of course

$$\mathcal{D}^n(U) = \{b\, dx^1\wedge\dots\wedge dx^n \mid b \in \mathcal{D}^0(U) = C_c^\infty(U)\}\,.$$

Thus to any n-current T it corresponds the distribution

$$b \longrightarrow T(b\, dx^1\wedge\dots\wedge dx^n)$$

and, conversely, to any distribution $S \in \mathcal{D}_0(U)$ it corresponds the n- current

$$b\,dx^1\wedge\ldots\wedge dx^n \longrightarrow S(b)\ .$$

In particular, to any function $u \in L^1_{\mathrm{loc}}(U)$ we can associate the n-current

$$T_u : \mathcal{D}^n(U) \to \mathbb{R}\ , \qquad T_u(b\,dx^1\wedge\ldots\wedge dx^n) := \int_U bu\,dx\ .$$

Obviously

$$\mathbf{M}_V(T_u) = \int_V |u|\,dx \qquad \forall\, V \subset\subset U\ .$$

As every $\omega \in \mathcal{D}^{n-1}(U)$ can be written as $\omega = \sum_{j=1}^n (-1)^{j-1} a_j\,dx^1\wedge\ldots\wedge\widehat{dx^j}\wedge\ldots\wedge dx^n$, we compute

$$(\partial T_u)(\omega) = \int u\mathrm{div}\, a\,dx\ , \qquad a = (a_1,\ldots,a_n)\ .$$

We therefore see that every function $u \in BV_{\mathrm{loc}}(U)$, i.e. every L^1-function whose distributional derivatives are Radon measures defines the *n-dimensional locally normal current T_u*, moreover for any $V \subset\subset U$, we have

$$\mathbf{M}_V(T_u) = \int_V |u|\,dx\ , \qquad \mathbf{M}_V(\partial T_u) = \int_V d|Du|\ .$$

Conversely we shall see in Ch. 4

Theorem 3. *For any locally normal current $T \in \mathbf{N}_{n,\mathrm{loc}}(\mathbb{R}^n)$ there is a function $u \in BV(U)$ such that $T = T_u$.*

As for any n-current in $U \subset \mathbb{R}^n$ we have $T = T^0 e_1\wedge\ldots\wedge e_n$, and $\partial T = 0$ implies $D_i T^0 = 0$, $i = 1,\ldots,n$, in the sense of distributions, i.e., T^0 =const, we also infer the following characterization of n-currents in U without boundary in U.

Theorem 4 (Constancy theorem). *Suppose U is a connected open set in $\mathbb{R}^n$, $T \in \mathcal{D}_n(U)$ is such that $\partial T = 0$, then T is a real multiple of the current integration on U, i.e.,*

$$T = r[\![U]\!]$$

for some $r \in \mathbb{R}$.

[1] For the reader's convenience we compute here the boundary of k-currents in $\mathbb{R}^n$, $n = 1, 2, 3$.This will show in which sense ∂ means derivative.

1. In $\mathbb{R}^1$ we have 0-dimensional currents S which are simply distributions, and for which $\partial S = 0$, and 1-currents. Clearly 1-currents have the form

$$T = T^1 e_1$$

and according to $d\eta = \eta' \, dx$ for $\eta \in \mathcal{D}^0(\mathbb{R})$, we compute

$$\partial T(\eta) = T(d\eta) = T^1(\eta') = -DT^1(\eta)$$

i.e., $\partial T = -DT^1$.

2. In $\mathbb{R}^2$, apart from 0-currents, we have currents of the form

$$T = T^0 \, e_1 \wedge e_2 \in \mathcal{D}_2(\mathbb{R}^2) \qquad S = S^1 \, e_1 + S^2 e_2 \in \mathcal{D}_1(\mathbb{R}^2).$$

Since for $\eta = \eta_1 \, dx^1 + \eta_2 \, dx^2 \in \mathcal{D}^1(\mathbb{R}^2)$ we have

$$d\eta = \eta_{1,x^2} dx^2 \wedge dx^1 + \eta_{2,x^1} dx^1 \wedge dx^2 = (\eta_{2,x^1} - \eta_{1,x^2}) \, dx^1 \wedge dx^2 \ ,$$

we can compute

$$\begin{aligned} \partial T(\eta) &= T(d\eta) = T^0(\eta_{2,x^1} - \eta_{1,x^2}) = \\ &= -D_{x^1} T^0(\eta_2) + D_{x^2} T^0(\eta_1) = -(D_{x^1} T^0 e_1 - D_{x^2} T^0 e_1)(\eta) \end{aligned}$$

i.e.

$$\partial T = D_{x^2} T^0 e_1 - D_{x^1} T^0 e_2$$

and ∂T may be regarded as minus the gradient of T_0.

For $\eta \in \mathcal{D}^0(\mathbb{R}^2)$ we have $d\eta = \eta_{x^1} dx^1 + \eta_{x^2} dx^2$ hence

$$\partial S(\eta) = S(d\eta) = S^1(\eta_{x^1}) + S^2(\eta_{x^2}) = -D_{x^1} S^1(\eta) - D_{x^2} S^2(\eta)$$

i.e.

$$\partial S = -\operatorname{div}(S^1, S^2) \in \mathcal{D}_0(\mathbb{R}^2) \ .$$

3. In $\mathbb{R}^3$, a part for 0-dimensional currents which are distributions, currents have the form

$$\begin{aligned} T &= T^0 e_1 \wedge e_2 \wedge e_3 \in \mathcal{D}_3(\mathbb{R}^3) \\ S &= S^1 e_2 \wedge e_3 + S^2 e_3 \wedge e_1 + S^3 e_1 \wedge e_2 \in \mathcal{D}_2(\mathbb{R}^3) \\ R &= R^1 e_1 + R^2 e_2 + R^3 e_3 \in \mathcal{D}_1(\mathbb{R}^3) \ . \end{aligned}$$

As in $\mathbb{R}^2$ we easily computes

$$\partial T \sim -\text{ gradient of } T_0 \ , \qquad \partial R \sim -\operatorname{div}(R^1, R^2, R^3) \ .$$

For $\eta = \eta_1 dx^1 + \eta_2 dx^2 + \eta_3 dx^3 \in \mathcal{D}^1(\mathbb{R}^3)$ we have

$$d\eta = (\eta_{2,x^1} - \eta_{1,x^2}) dx^1 \wedge dx^2 + (\eta_{3,x^2} - \eta_{2,x^3}) dx^2 \wedge dx^3 + (\eta_{3,x^1} - \eta_{1,x^3}) dx^1 \wedge dx^3 \ ,$$

we therefore see that

$$\partial S(\eta) = -\operatorname{curl}(S^1, S^2, S^3)(\eta_1, \eta_2, \eta_3) \ .$$

•

Image of a current. For its relevance in the sequel we conclude this subsection by discussing the notion of image or *pushing forward* of a normal current $f_\# T$, by a smooth map f.

Under mild conditions one can define the *pushing forward of a current* $T \in \mathcal{D}_k(U)$, $U \subset \mathbb{R}^n$. Let $f : U \to V$, $V \subset \mathbb{R}^n$, be a C^∞ map such that $f_{|\,\mathrm{spt}\,T}$ is *proper*, i.e., $f^{-1}(K) \cap \mathrm{spt}\,T$ is compact in U for any compact set K. Then the current $f_\# T \in \mathcal{D}_k(V)$ is defined by

$$(10) \qquad f_\# T(\omega) := T(\zeta f^\# \omega) \qquad \omega \in \mathcal{D}^k(V)$$

where $\zeta \in C_0^\infty(U)$ with $\zeta = 1$ in a neighbourhood of $\mathrm{spt}\,T \cap \mathrm{spt}\,f^\#\omega$. One easily checks that such a definition is independent of ζ and

$$(11) \qquad \partial f_\# = f_\# \partial\,, \qquad \mathrm{spt}\,f_\# T \subset f(\mathrm{spt}\,T)\,.$$

Suppose now that T has finite mass, i.e. $T \in \mathcal{M}_k(U)$. As we have seen T extends to all differential forms η with Borel coefficients and $\int |\eta|\, d\|T\| < \infty$. In particular for any map $f : U \to V$ of class C^1 with bounded derivatives in U, $\sup_U |Df| < \infty$, the form $f^\#(\omega)$ has bounded and continuous coefficients, hence T is well defined on $f^\#(\omega)$. Therefore we can conclude that for any f of class C^1 and T in $\mathcal{M}_k(U)$ $f_\# T$ is well defined again by

$$(12) \qquad f_\# T(\omega) := T(f^\# \omega)\,.$$

In particular if $U = V \times \mathbb{R}^N$ and $\pi : U \to V$ is the orthogonal projection onto the first factor, $\pi(x, y) = x$, $\pi_\#$ is well defined for any $T \in \mathcal{M}_k(U)$. However in general $f_\#$ is not continuous with respect to the weak convergence of currents. For example if $U = (0,1) \times \mathbb{R}$, $V = (0,1)$ the sequence of currents T_k defined as integration over the line segment from the point $(0,k)$ to $(1,k)$, we have

$$\mathbf{M}(T_k) = 1, \qquad \mathbf{M}(\partial T_k) = 0, \qquad \pi_\#(T_k) = T_0$$

but $T_k \rightharpoonup 0$ as $k \rightharpoonup \infty$ which says that $\pi_\#$ is not continuous.

We explicitly note that (12) implies that $f_\# T$ has finite mass if T has finite mass and actually

$$\| f_\# T \| = f_\# \| T \|, \qquad \overrightarrow{f_\# T}(f(x)) = \Lambda_k Df(x)(\vec{T}(x)).$$

Similarly, assuming that $\mathbf{M}(T)$, $\mathbf{M}(\partial T) < \infty$ and $f \in C^2(U)$ has bounded second derivatives the pushing forward of ∂T

$$f_\# \partial T(\omega) = \partial T(f^\#(\omega)) = T(d\, f^\#(\omega))$$

is well defined, but in general

$$f_\# \partial T \neq \partial f_\# T\,.$$

Let us consider now mappings $f : U \to V$ which are just Lipschitz continuous. In this case for any ω with smooth coefficients the pull-back $f^\# \omega$ is a form with bounded measurable, but not necessarily Borel coefficients. Consequently we can apply T to $f^{\#}\omega$ only if $\|T\|$ is absolutely continuous with respect to the Lebesgue measure, and in general $T(f^{\#}\omega)$ as in (12) is meaningless.

The following example illustrates the situation. Take $U = V = \mathbb{R}$ and let $f : \mathbb{R} \to \mathbb{R}$ be defined by

$$f(x) := \begin{cases} x+1 & \text{for } x \leq -1 \\ 0 & \text{for } -1 < x < 1 \\ x-1 & \text{for } x \geq 1 \,. \end{cases}$$

If $\omega = \varphi(y)\,dy \in \mathcal{D}^1(\mathbb{R})$, $\varphi \in C_c^\infty(\mathbb{R})$ we have

$$f^{\#}\omega \;=\; \varphi(f(x))\,f'(x)\,dx \;=:\; \psi(x)\,dx$$

where $\psi(x)$ is defined and smooth for $x \neq \pm 1$. But $\psi(1)$ and $\psi(-1)$ are not defined. By taking $T := \delta_1 \in \mathbf{N}_0(\mathbb{R})$ we then see that

$$T(f^{\#}\omega) \;=\; \delta_1(\psi) \;=\; \psi(1)$$

is meaningless.

However it turns out that if T is a locally normal current in $\mathcal{D}_k(U)$ and f is a Lipschitz map which is *proper on* $\operatorname{spt} T$, then $f_\# T$ can be well defined. But before doing that let us show why we need to assume $f_{|\operatorname{spt} T}$ proper.

Let $f : U := (0, +\infty) \hookrightarrow V := \mathbb{R}$ be just the inclusion map, and let $T := [\![(0,b)]\!]$, $b > 0$. We have $\partial T = \delta_b$ so that $T \in \mathbf{N}_1(U)$, consequently $f_\# \partial T = \delta_b$. The push-forward of $[\![(0,b)]\!]$ by f is instead the current $[\![(0,b)]\!] \in \mathcal{D}_1(\mathbb{R})$, $f_\# T = [\![(0,b)]\!]$, hence $\partial f_\# T = \delta_b - \delta_0$. Therefore we have $f_\# \partial T \neq \partial f_\# T$. As one would like to have the equality $f_\# \partial T = \partial f_\# T$, what is wrong is the fact that f is not proper $f^{-1}([-1,1]) = (0,1]$ and $(0,1]$ is not compact in $U = (0, +\infty)$.

Suppose now f Lipschitz and proper on $\operatorname{spt} T$, $T \in \mathbf{N}_{k,\mathrm{loc}}(U)$. Let $\phi \in C_c^\infty(\mathbb{R}^n)$ be a standard mollifier and let

$$\phi_\tau(x) \;:=\; \tau^{-n}\phi(\tau^{-1}x) \qquad \tau > 0$$

and

$$f_\tau \;:=\; f * \phi_\tau$$

be a smoothing of f. We define the *image of* T *by* f as

$$f_\# T(\omega) \;=\; \lim_{\tau \to 0} f_{\tau\#} T(\omega) \qquad \omega \in \mathcal{D}^n(V)\,,$$

i.e. as the weak limit of the well defined currents $f_{\tau\#}T$, compare Sec. 5.1.2. In fact we have

Proposition 3. *If $T \in \mathbf{N}_{k,\mathrm{loc}}(U)$, $f : U \to V$ is Lipschitz and $f_{|\operatorname{spt} T}$ proper, then*

$$f_\# T(\omega) \;:=\; \lim_{\tau \to 0} f_{\tau\#} T(\omega)$$

exists and is independent of the mollifier ϕ *for all* $\omega \in \mathcal{D}^k(V)$. *It agrees with* $f_\# T$ *defined by (12) if* f *is of class* C^1. *Moreover* $\operatorname{spt} f_\# T \subset f(\operatorname{spt} T)$ *and*

$$\mathbf{M}_W(f_\# T) \le \Big(\sup_{f^{-1}(W)} |Df|\Big)^k \mathbf{M}_{f^{-1}(W)}(T)$$

for any open set $W \subset\subset V$. *Finally*

$$f_\# \partial T = \partial f_\# T \,.$$

The proof of Proposition 3 is based on the important *homotopy formula for currents.* Consider two smooth maps $f,\ g : U \to V$ and a smooth homotopy

$$h : [0,1] \times U \to V, \qquad h(0,\cdot) = f, \quad h(1,\cdot) = g.$$

Suppose h is proper in $[0,1] \times \operatorname{spt} T$. Then

$$h_\#([\![(0,1)]\!] \times T)$$

is well defined and we have

$$\begin{aligned} \partial h_\#([\![(0,1)]\!] \times T) &= h_\# \partial([\![(0,1)]\!] \times T) \\ &= h_\#(\delta_1 \times T - \delta_0 \times T - [\![(0,1)]\!] \times \partial T) \\ &= g_\# T - f_\# T - h_\#([\![(0,1)]\!] \times \partial T) \,. \end{aligned}$$

Therefore we can state

Proposition 4 (Homotopy formula). *Let* $U,\ V$ *be open sets in* $\mathbb{R}^n$, $f,\ g : U \to V$ *be smooth maps, and let* $h : [0,1] \times U \to V$ *be a smooth homotopy between* f *and* g*; finally let* $T \in \mathcal{D}_k(U)$. *Assume that* $h^{-1}(K) \cap \operatorname{spt} T$ *is compact in* $[0,1] \times U$ *for any compact set* $K \subset V$, *then we have*

$$g_\# T - f_\# T = \partial h_\#([\![(0,1)]\!] \times T) + h_\#([\![(0,1)]\!] \times \partial T) \tag{13}$$

where $h_\#([\![(0,1)]\!] \times \partial T) := 0$ *if* $k = 0$.

Recalling (5)

$$[\![(0,1)]\!] \times T(\omega) = T([\omega]_{(0,1)})$$

we can write (13) as

$$T(g^\# \omega - f^\# \omega) = T([h^\# d\omega]_{(0,1)} + d[h^\# \omega]_{(0,1)}) \tag{14}$$

where the last summand has to be taken equal to zero if $k = 0$.

It is easy to see that (13) and (14) are also valid even if h is not proper provided T has compact support in U, $T \in \mathcal{E}_k(U)$. By applying (14) to currents of the type

$$\omega \longrightarrow \int \omega \wedge \eta \qquad \eta \in \mathcal{D}^{n-k}(U)$$

we then recover the usual *homotopy formula for forms*

$$g^{\#}\omega - f^{\#}\omega = [h^{\#}d\omega]_{(0,1)} + d[h^{\#}\omega]_{(0,1)}. \tag{15}$$

In the special case of the *affine homotopy*

$$h(t,x) := (1-t)f(x) + tg(x)$$

we have

$$\begin{aligned} h_{\#\,(t,x)}(\overrightarrow{[0,1]\times T}) &= \Lambda_{k+1}Dh(t,x)(e_1\wedge\vec{T}(x)) \\ &= (g(x)-f(x))\wedge\Lambda_k D_x h(t,x)(\vec{T}(x)) \end{aligned}$$

hence

$$h_{\#}([\![(0,1)]\!] \times T)(\omega) = \int <\omega(h(\cdot)), (g-f)\wedge\Lambda_k D_x h(\vec{T})> d(\mathcal{L}^1 \times \|T\|)$$

which implies

$$\mathbf{M}(h_{\#}([\![(0,1)]\!] \times T)) \le c\|T\|(|g-f|(|Df|^k + |Dg|^k))$$

and therefore

$$\mathbf{M}(h_{\#}[\![(0,1)]\!] \times T) \;\le\; \sup_{\operatorname{spt} T}|f-g| \,\sup_{\operatorname{spt} T}(|Df|+|Dg|)^k\,\mathbf{M}(T)\;. \tag{16}$$

Returning to Proposition 3, we consider the affine homotopy

$$h(t,x) \;:=\; tf_\tau(x) + (1-t)f_\sigma(x)$$

corresponding to two parameters $\tau,\sigma > 0$. The homotopy formula yields for any ω with $\operatorname{spt}\omega \subset \overset{\circ}{K}$, $K \subset\subset V$,

$$f_{\tau\#}T(\omega) - f_{\sigma\#}T(\omega) = h_{\#}([\![(0,1)]\!] \times T)(d\omega) + h_{\#}([\![(0,1)]\!] \times \partial T)(\omega)$$

therefore, if moreover $|\omega|, |d\omega| \le 1$, we infer from (16)

$$|f_{\tau\#}T(\omega) - f_{\sigma\#}T(\omega)| \;\le\; c\,\mathbf{M}(T) \sup_{f^{-1}(K)\cap\operatorname{spt} T} |f_\tau - f_\sigma|\,(Lip\ f)^k\;.$$

The proof of Proposition 3 can be now easily concluded.

We conclude this subsection by stating one more simple but useful consequence of the homotopy formula.

Let $T \in \mathcal{D}_k(U)$ with $\operatorname{spt} T \subset U$. Suppose that U is star-shaped relative to the origin, and consider the affine homotopy which joins the null map to the identity map

$$h(t,x) \;=\; tx\;.$$

We define the *cone* over T as the current in $\mathcal{D}_{k+1}(U)$ given by

$$O \mathbin{⋇} T := h_\#([\![(0,1)]\!] \times T) .$$

From the homotopy formula we infer

$$\partial(O \mathbin{⋇} T) = T - (O \mathbin{⋇} \partial T) .$$

In particular, if T is boundaryless, $\partial T = 0$, we then have

$$T = \partial(O \mathbin{⋇} T)$$

which shows that *every boundaryless current T is a boundary*, i.e., if $T \in \mathcal{D}_k(\mathbb{R}^n)$ and $\partial T = 0$, then there exists $R \in \mathcal{D}_{k+1}$ such that $T = \partial R$.

Another immediate consequence of the homotopy formula is the following

Proposition 5 (Poincaré lemma). *Let U be a starshaped domain in $\mathbb{R}^n$ and let ω be a closed form in U. Then ω is exact.*

In fact, assuming U starshaped with respect to the origin and considering the affine homotopy between 0 and the identity, $h(t,x) := tx$, (15) yields

$$\omega = g^\# \omega = d[h^\# \omega]_{(0,1)}.$$

2.4 Integer Multiplicity Rectifiable Currents

In the class of k-dimensional currents in some open set U of $\mathbb{R}^n$ we may identify two typical situations which in some sense are extremal from the point of view of smoothness.

(i) Given a Lebesgue measurable k-vector field ξ in U, $\xi = \sum_{|\alpha|=k} \xi^\alpha e_\alpha$ the k-dimensional current

$$T := \mathcal{L}^n \llcorner \xi = \sum_{|\alpha|=k} \mathcal{L}^n \llcorner \xi^\alpha e_\alpha$$

is a current "distributed" on the n-dimensional set U. Actually by considering a non-negative measure μ and a μ-measurable k-vector field ξ we can realize k-dimensional current

$$T := \mu \llcorner \xi$$

which are "distributed" in sets of any "dimension" less than n.

(ii) Given a smooth k-dimensional oriented submanifold of $\mathbb{R}^n$, M, $M \subset U$, the current integration of infinitely differentiable k-forms with compact support in U, $[\![M]\!]$, defines a k-current which lives on the k-dimensional set M. Moreover, as we have seen in Sec. 2.2.2, the current

$$[\![M]\!](\omega) = \int_M \omega \qquad \omega \in \mathcal{D}^k(U)$$

can be represented as

$$[\![M]\!](\omega) = \int_M <\vec{\xi}(x), \omega(x)> d\mathcal{H}^k$$

where $\vec{\xi}(x)$ orients for $\mathcal{H}^n$-a.e. $x \in M$ the tangent plane $T_x M$ of M at x. One also refers to $[\![M]\!]$ as to the *current carried by* M.

Integer multiplicity rectifiable currents arise as the natural extension of the currents of the type $[\![M]\!]$, in which we allow M to be $\mathcal{H}^k$-measurable and countably rectifiable and we also allow integer multiplicity. As we shall see in Sec. 2.2.6 such a class turns out to characterize the class of the weak limits of the k-dimensional currents carried by smooth submanifolds.

But before turning to precise definitions, let us return to the k-dimensional currents which are integration over smooth graphs, as they shall be specially important for us in the sequel.

Currents carried by smooth graphs. Let Ω be an open set in $\mathbb{R}^n$ and let $u : \Omega \subset \mathbb{R}^n \to \mathbb{R}^N$ be a smooth map. Its graph is defined by

$$\mathcal{G}_{u,\Omega} := \{(x, u(x)) \in \mathbb{R}^{n+N} \mid x \in \Omega\}$$

and the *Cartesian current carried by the graph of* u is given by the n-dimensional current

$$G_u := [\![\mathcal{G}_{u,\Omega}]\!]$$

i.e., by

$$G_u(\omega) = \int_\Omega (\mathrm{id} \bowtie u)^{\#}\omega = \int_\Omega <\omega(x, u(x)), M(Du(x))> dx \qquad \forall\, \omega \in \mathcal{D}_n(U)$$

where $U := \Omega \times \mathbb{R}^N$. By (33) in Sec. 2.2.1 we have

$$G_u(\omega) = \sum_{|\alpha|+|\beta|=n} \sigma(\alpha, \bar{\alpha}) \int_\Omega \omega_{\alpha\beta}(x, u(x))\, M^\beta_{\bar{\alpha}}(Du(x))\, dx$$

for any n-form ω

$$\omega = \sum_{|\alpha+|\beta|=n} \omega_{\alpha\beta}(x, y)\, dx^\alpha \wedge dy^\beta$$

where $\sigma(\alpha, \bar{\alpha})$ is the sign of the permutation which reorders the multi-index $(\alpha, \bar{\alpha})$ in its natural order, and $M^\beta_{\bar{\alpha}}(Du)$ denotes the determinant of the minor of Du with rows $(\beta_1, \dots, \beta_n)$ and columns $(\bar{\alpha}_1, \dots, \bar{\alpha}_n)$. Alternatively this can be seen as follows. We have

$$du^{\beta_1} \wedge \ldots \wedge du^{\beta_k} = \sum u^{\beta_1}_{x^{j_1}} \cdot \ldots \cdot u^{\beta_k}_{x^{j_k}}\, dx^{j_1} \wedge \ldots \wedge dx^{j_k}$$

and those terms will contribute to $\omega_{\alpha\beta}\, dx^\alpha \wedge dy^\beta$ only if the differentials

$$dx^{\alpha_1}, \ldots, dx^{\alpha_{n-k}} \quad \text{and} \quad dx^{j_1}, \ldots, dx^{j_k}$$

are different, so that $(j_1, \ldots, j_k)$ must be a permutation of $\bar{\alpha}$. Reordering $(j_1, \ldots, j_k)$ to $\bar{\alpha}$, and then $(\alpha, \bar{\alpha})$ we then get

$$\begin{aligned} dx^{\alpha_1} \wedge \ldots \wedge dx^{\alpha_{n-k}} \wedge du^{\beta_1} \wedge \ldots \wedge du^{\beta_k} &= M^{\beta}_{\bar{\alpha}}(Du)\, dx^{\alpha} \wedge dx^{\bar{\alpha}} \\ &= \sigma(\alpha, \bar{\alpha})\, M^{\beta}_{\bar{\alpha}}(Du)\, dx^1 \wedge \ldots \wedge dx^n \,. \end{aligned}$$

It is worth noticing that $|M(Du)| = \{\sum M^{\beta}_{\bar{\alpha}}(Du)^2\}^{1/2}$ is the element of area on $\mathcal{G}_{u,\Omega}$ and the n-vector

$$\begin{aligned} M(Du) &= \sum_{|\alpha|+|\beta|=n} \sigma(\alpha, \bar{\alpha})\, M^{\beta}_{\bar{\alpha}}(Du) \\ &= (e_1 + D_1 u^i \varepsilon_i) \wedge \; \ldots \; \wedge (e_n + D_n u^i \varepsilon_i) \end{aligned}$$

or more precisely $M(Du)/|M(Du)|$ is the unitary tangent n-vector orienting $\mathcal{G}_{u,\Omega}$. Thus we can write

$$G_u(\omega) = \int <\omega, \frac{M(Du)}{|M(Du)|}> d\mathcal{H}^n \llcorner \mathcal{G}_{u,\Omega}$$

compare Sec. 3.1.5 and Sec. 3.2.1.

For future purposes we would like to discuss some special cases. This in particular will show in which sense the boundary operator describes the *geometric* boundary.

[1] *Dimension and codimension* 1. Let $n = N = 1$, $\Omega = (0,1)$, $U = (0,1) \times \mathbb{R}$. As a 1-forms in U have the form $\omega = \varphi_1(x,y)\, dx + \varphi_2(x,y)\, dy$, with $\varphi_1, \ \varphi_2 \in C^\infty_c(\Omega \times \mathbb{R})$, we get

$$G_u(\omega) \; = \; \int_0^1 [\varphi_1(x, u(x)) + \varphi_2(x, u(x)) u'(x)]\, dx \,.$$

For any 0-form $\eta = \varphi(x,y)$, we have $d\eta \; = \; \varphi_x\, dx + \varphi_y\, dy$ so that

$$\begin{aligned} \partial G_u(\eta) \; = \; G(d\eta) \; &= \; \int_0^1 [\varphi_x(x,u) + \varphi_y(x,u) u']\, dx \\ &= \; \int_0^1 \frac{d}{dx} \varphi(x, u(x))\, dx \; = \; \varphi(b) - \varphi(a) \end{aligned}$$

where $b = (1, u(1))$, $a = (0, u(0))$. From the previous computations we see that

$$\partial G_u \; = \; 0$$

if we regard G_u as a current in U, as $\eta \in \mathcal{D}_0(U)$, and

$$\partial G_u = \delta_b - \delta_a \in \mathcal{D}_0(\mathbb{R} \times \mathbb{R})$$

if we regard G_u as a current in $\mathbb{R} \times \mathbb{R}$. Notice that $\delta_b - \delta_a \in \mathcal{M}_0(\mathbb{R} \times \mathbb{R})$ and is the current carried by the graph of $u_{|\partial\Omega}$, so that, regarding G_u as a current in $\mathbb{R} \times \mathbb{R}$

$$\partial G_u = [\![\mathcal{G}_{\partial\Omega, u_{|\partial\Omega}}]\!]$$

the graph of $u_{|\partial\Omega}$ being oriented in the induced way. •

[2] *Codimension* 1. Suppose $n > 1$, $N = 1$. Then $\mathcal{G}_{u,\Omega}$ has dimension n and an n-form in $\mathbb{R}^{n+1}$ can be written as

$$\omega = \omega_0(x,y)\, dx^1 \wedge \ldots \wedge dx^n + \sum \omega_i(x,y)\, dx^1 \wedge \ldots \wedge \widehat{dx^i} \wedge \ldots \wedge dx^n \wedge dy\,.$$

Then

$$G_u(\omega) = \int_\Omega \left[\omega_0(x,u(x)) + \sum_{i=1}^n (-1)^{n-i} \omega_i(x,u(x))\, u_{x^i} \right] dx\,.$$

By Gauss-Green theorem we can compute

$$G_u(d\eta) = [\![\partial \mathcal{G}_{u,\Omega}]\!](\eta) = [\![\mathcal{G}_{u_{|\partial\Omega}}]\!](\eta)$$

so that

$$\begin{aligned} &\partial G_u = 0 \ \text{ if we regard } G_u \text{ as an element of } \mathcal{D}_n(\Omega \times \mathbb{R}) \\ &\partial G_u = [\![\mathcal{G}_{u_{|\partial\Omega}}]\!] \ \text{ if we regard } G_u \text{ as an element of } \mathcal{D}_n(\mathbb{R}^n \times \mathbb{R})\,. \end{aligned}$$

This can be easily seen, for instance, if $n = 2$ and $\Omega = \{(x^1, x^2) \mid x^1 < 0\}$. If

$$\eta = \varphi_1\, dx^1 + \varphi_2\, dx^2 + \varphi_3\, dx^3 \in \mathcal{D}_1(\mathbb{R}^2 \times \mathbb{R})$$

one easily computes

$$\begin{aligned} (\mathrm{id} \bowtie u)^\# d\eta &= (-\varphi_{1x^2} + \varphi_{2x^1} - \varphi_{1x^3} u_{x^2} + \varphi_{3x^1} u_{x^2} \\ &\qquad + \varphi_{2x^3} u_{x^1} - \varphi_{3x^2} u_{x^1})\, dx^1 \wedge dx^2 \\ &= \Big\{ - D_2[\varphi_1(x,u(x))] + D_1[\varphi_2(x,u(x))] + D_1[\varphi_3(x,u(x))] u_{x^2} \\ &\qquad - D_2[\varphi_3(x,u(x))] u_{x^1} \Big\}\, dx^1 \wedge dx^2. \end{aligned}$$

Denoting by $\nu = (1,0)$ the exterior normal to $\partial\Omega$, we then get

$$\begin{aligned} \int_\Omega (\mathrm{id} \bowtie u)^\# d\eta &= \int [\varphi_2(0,x^2,u(0,x^2)) + \varphi_3(0,x^2,u(0,x^2)) u_{x^2}]\, dx^2 \\ &= [\![\mathcal{G}_{u_{|\partial\Omega}}]\!](\eta)\,. \end{aligned}$$

Similarly one proceeds in the general case, compare Sec. 3.2.3. •

3 *Mapping from* $\mathbb{R}^2$ *into* $\mathbb{R}^2$. Let $u \in C^1(\overline{\Omega}, \mathbb{R}^2)$, $\Omega \subset \mathbb{R}^2$. Any 2-form in $\mathbb{R}^2 \times \mathbb{R}^2$ can be written as

$$\omega = \omega_0(x,y)\, dx^1 \wedge dx^2 + \sum_{i,j=1}^{2} \omega_{ij}(x,y) dx^i \wedge dy^j + \omega_2(x,y) dy^1 \wedge dy^2 \ .$$

Thus $G_u(\omega)$ is given by

$$\int_\Omega \Big[\omega_0(x,u(x)) + \sum_{i,j=1}^{2} (-1)^{i-1} \omega_{ij}(x,u(x)) u^j_{x^i} + \omega_2(x,u(x)) \det Du(x)\Big]\, dx.$$

•

Rectifiable and integer multiplicity rectifiable currents. Let U be an open set in $\mathbb{R}^n$. Given

(i) an $\mathcal{H}^k$-measurable and countably k-rectifiable set $\mathcal{M} \subset U$
(ii) an $\mathcal{H}^k$-measurable map $\theta : \mathcal{M} \to \mathbb{R}$ which is locally $\mathcal{H}^k \llcorner \mathcal{M}$-summable
(iii) an $\mathcal{H}^k$-measurable map $\xi : \mathcal{M} \to \Lambda_k \mathbb{R}^n$ with $||\xi|| = 1$ $\mathcal{H}^k \llcorner \mathcal{M}$-a.e.

we say that a current $T \in \mathcal{D}_k(U)$ is of the type $\tau(\mathcal{M}, \theta, \xi)$, $T = \tau(\mathcal{M}, \theta, \xi)$ if

$$T(\omega) = \int_{\mathcal{M}} <\xi, \omega> \theta \, d\mathcal{H}^k$$

i.e.

$$T = \theta \xi \mathcal{H}^k \llcorner \mathcal{M} \ .$$

By changing $\xi \to \xi \operatorname{sign} \theta$, $\theta \to \theta \operatorname{sign} \theta$, we can and do assume that $\theta \geq 0$, and we call θ the *multiplicity* of $\tau(\mathcal{M}, \theta, \xi)$. Sometime one refers to

$$\text{set}\,(T) := \{x \in \mathbb{R}^n \mid \theta^k(\| T \|, x) > 0\}$$

as to the *set of* T. Of course T determines set (T), while possible $\mathcal{M}$'s are determined only up to sets of $\mathcal{H}^k$-measure zero.

Definition 1. *We say that a current* $T \in \mathcal{D}_k(U)$ *is* rectifiable *if and only if* $T = \tau(\mathcal{M}, \theta, \xi)$ *for some* $\mathcal{M}, \theta, \xi$ *as previously and moreover* $\xi(x)$ *is a* k*-vector associated to the tangent plane* $T_x\mathcal{M}$ *for* $\mathcal{H}^k \llcorner \mathcal{M}$-a.e. x.
If moreover the density θ *is integer-valued, then we say that* $T = \tau(\mathcal{M}, \theta, \xi)$ *is* integer multiplicity rectifiable, *briefly* i.m. rectifiable.
The class of integer multiplicity rectifiable k*-currents in* U *will be denoted by* $\mathcal{R}_k(U)$.

For currents of type $\tau(\mathcal{M}, \theta, \xi)$ we have

$$\| T \| = \theta \mathcal{H}^k \llcorner \mathcal{M}$$

while, according to our discussion after the definition of the mass and the Euclidean mass in Sec. 2.2.3, for rectifiable currents we have

$$\| \xi \| = |\xi| = 1 \quad \text{and} \quad \| T \| = |T| = \theta \mathcal{H}^k \llcorner \mathcal{M}.$$

Of course the class of rectifiable currents in U form a linear space, also

$$T_1 + T_2 \in \mathcal{R}_k(U) \,, \quad \text{if} \quad T_1, T_2 \in \mathcal{R}_k(U)$$

but *in general* $\lambda T \in \mathcal{R}_k(U)$ only if $T \in \mathcal{R}_k(U)$ and λ is an integer.

The following two theorems of Federer and Fleming make the class of i.m. rectifiable k-currents $\mathcal{R}_k(U)$, very natural and important especially in connection with the calculus of variations.

Theorem 1 (Closure theorem). *Let $\{T_j\} \subset \mathcal{R}_k(U)$ be a sequence of i.m. rectifiable k-currents in some open set $U \subset \mathbb{R}^n$ satisfying*

$$\text{(1)} \qquad \sup_j [\mathbf{M}_V(T_j) + \mathbf{M}_V(\partial T_j)] < \infty \qquad \forall\, V \text{ open, } V \subset\subset U$$

and weakly converging to some current $T \in \mathcal{D}_k(U)$, $T_j \rightharpoonup T$. Then $T \in \mathcal{R}_k(U)$.

This theorem together with the compactness theorem for general currents yields

Theorem 2 (Compactness theorem). *Let $\{T_j\} \subset \mathcal{R}_k(U)$ be a sequence satisfying (1). Then there exists a subsequence $\{T_{j'}\}$ of $\{T_j\}$ and $T \in \mathcal{R}_k(U)$ such that $T_{j'} \rightharpoonup T$.*

Defining the *weak convergence in* $\mathbf{N}_{k,\text{loc}}(U)$ as

$$T_j \rightharpoonup T \qquad \text{weakly in } \mathbf{N}_{k,\text{loc}}(U)$$

iff

$$T_j \rightharpoonup T \qquad \text{in } \mathcal{D}_k(U)$$

$$\sup_j (\mathbf{M}_V(T_j) + \mathbf{M}_V(\partial T_j)) < \infty \qquad \forall\, V \subset\subset U$$

we can state Theorem 1 as: $\mathcal{R}_k(U) \cap \mathbf{N}_{k,\text{loc}}(U)$ *is (sequentially) weakly closed in* $\mathbf{N}_{k,\text{loc}}(U)$.

Notice that not every current in $\mathcal{R}_k(\mathbb{R}^n)$ belongs to $\mathbf{N}_k(\mathbb{R}^n)$. For instance a collection of 2-dimensional disjoint disks in $\mathbb{R}^3$ with radius r_j satisfying

$$\sum_{j=1}^{\infty} r_j^2 < \infty \,, \qquad \sum_{j=1}^{\infty} r_j = \infty$$

clearly yields a current T in $\mathcal{R}_2(\mathbb{R}^3)$ for which $\mathbf{M}(\partial T) = \infty$.

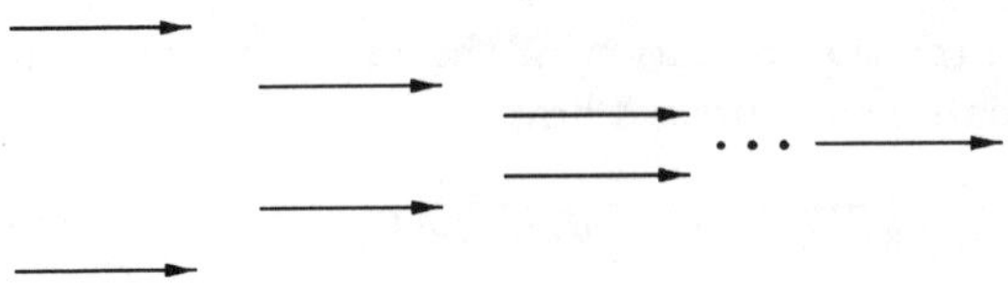

(a)

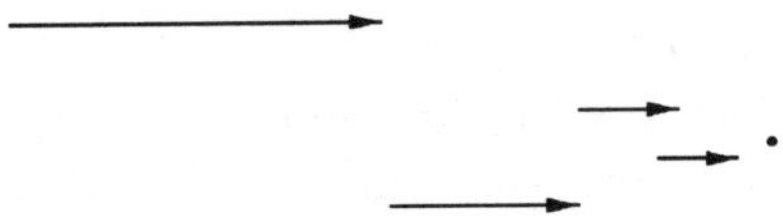

(b)

Fig. 2.4. (a) A sequence of density 1 currents converging to a current of density 2, (b) A rectifiable current with boundary of infinite mass.

We shall return to Theorem 1 later in Sec. 2.2.6 and Sec. 2.2.7. But first we would like to present a few examples with the aim of illustrating Definition 1 and Theorem 1 and Theorem 2.

Of course limits of smooth manifolds M_j, identified with the rectifiable currents $\tau(M_j, 1, [\![\vec{M}_j]\!])$ with density 1, can be in principle rectifiable currents with integer density not necessarily 1, therefore it is natural to allow integer densities in order to identify such limits.

As for distributions we cannot expect any regularity of the limits in the sense of currents of smooth manifolds M_j, if there is no control on the mass of the currents $[\![M_j]\!]$, i.e.,

$$\sup_j \mathbf{M}([\![M_j]\!]) = +\infty .$$

Assuming only $[\![M_j]\!] \rightharpoonup T$ in $\mathcal{D}_k(U)$, the components of T need not even be measures.

4 Consider the sequence of smooth maps

$$u_j : [0,1] \longrightarrow \mathbb{R} , \qquad u_j(x) := \sin 2j\pi x .$$

For the associated i.m. rectifiable currents G_{u_j} carried by the graphs of u_j we trivially have

$$\mathbf{M}(G_{u_j}) \longrightarrow \infty , \qquad \text{as } j \to \infty .$$

Also

$$G_{u_j}(\varphi(x,y)dx) = \int_0^1 \varphi(x, u_j)\, dx$$

$$G_{u_j}(\psi(x,y)dy) = \int_0^1 \psi(x, u_j)\dot{u}_j\, dx$$

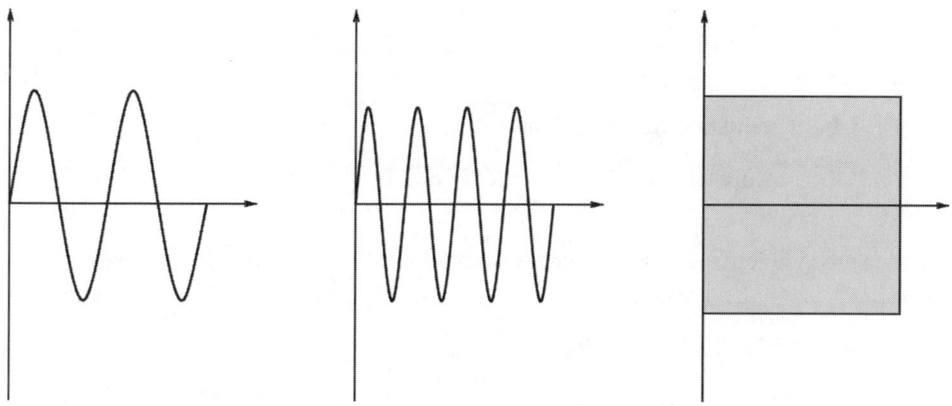

Fig. 2.5. Young measure and distributions as zero and first components of the limit of i.m. currents.

$$= \int_0^1 \left[\frac{d}{dx} \int_0^{u_j} \psi(x,\tau)\, d\tau - \int_0^{u_j} \psi_x(x,\tau)\, d\tau \right] dx$$

$$= -\int_0^1 dx \int_0^{u_j} \psi_x(x,\tau)\, d\tau$$

Therefore, taking into account Proposition 2 in Sec. 1.2.6 we see that the zero component of G_{u_j} converge to the Young measure in $(0,1) \times \mathbb{R}$ given by

$$\mathcal{L}^1 \llcorner (0,1) \times \lambda\,, \qquad \int_{\mathbb{R}} \psi\, d\lambda := \int_0^1 \psi(\sin 2\pi x)\, dx\,,$$

$$G_{u_j}(\varphi(x,y)dx) \longrightarrow \int_0^1 \varphi(x, \sin 2\pi x)\, dx$$

while

$$G_{u_j}(\psi(x,y)dy) \to -\int_0^1 dx \int_0^{\sin 2\pi x} \psi_x(x,\tau)\, d\tau$$

which is not a measure, but just a *distribution* in $\mathbb{R} \times \mathbb{R}$. •

Next we want to show that the closure theorem does not hold for real-multiplicity rectifiable currents.

5 *The closure theorem for i.m. rectifiable* 0*-currents.* Let $k = 0$, $U = \mathbb{R}^n$. It is readily seen that $T \in \mathcal{R}_0(U)$ *if and only if*

$$T = \sum_{i=1}^{s} a_i \delta_{z_i} , \qquad z_i \in \mathbb{R}^n,\ a_i \in \mathbb{Z} .$$

Let $\{T_j\}$ be a sequence in $\mathcal{R}_0(\mathbb{R}^n)$ with

$$\sup_j \mathbf{N}(T_j) < \infty , \qquad \mathbf{N}(T) := M(T) + M(\partial T) .$$

Since the coefficients $a_i^{(j)}$ are integers, we can choose a common s and write each T_j as

$$T_j = \sum_{i=1}^{s} a_i^{(j)} \delta_{z_i^{(j)}}$$

with $\sup_{j,i} |a_i^{(j)}| < \infty$. Therefore passing to a subsequence we have

$$z_i^{(j)} \to z_i \quad \text{or} \quad |z_i^{(j)}| \to \infty , \qquad \text{and} \qquad a_i^{(j)} \to a_i$$

as $j \to \infty$, from which we infer the (elementary) closure theorem

$$T_j \rightharpoonup T = \sum_{i=1}^{s} a_i \delta_{z_i} \in \mathcal{R}_0(U) .$$

Let us allow now real densities. Assume for instance $k = 0$, $n = 1$ and $U := (0,1)$, and consider the sequence of rectifiable currents with real multiplicity

$$T_j := \frac{1}{j} \sum_{i=1}^{j} \delta_{i/j} .$$

We then have, compare Sec. 1.2.5, that T_j converge weakly to $T = \mathcal{L}^1 \llcorner (0,1)$ which belongs to $\mathcal{M}_0(\mathbb{R})$ but is not rectifiable. Of course similar examples occurs in any dimension. For instance the sequence of real-multiplicity rectifiable 1-dimensional currents in $\mathbb{R}^2$

$$T_j := \frac{1}{j} \sum_{i=1}^{j} [\![(0,1) \times \{i/j\}]\!]$$

with $\sup_j \mathbf{N}(T_j) < \infty$ converges to the 1-dimensional normal, but non rectifiable current $T := \mathcal{L}^2 \llcorner (0,1) \times (0,1)\, e_1$. •

[6] *The closure theorem for i.m. rectifiable* 1*-currents in* $\mathbb{R}$. Let $k = n = 1$, and let $T = \tau(\mathcal{M}, \theta, \vec{T}) \in \mathcal{R}_1(\mathbb{R})$. Of course we can assume

$$\vec{T} = \sigma e_1 , \qquad \sigma : \mathcal{M} \longrightarrow \{\pm 1\} .$$

From the decomposition theorem for normal n-currents in $\mathbb{R}^n$, compare Sec. 4.3.1, or by a simpler argument in this case, we see that each $T \in \mathcal{R}_1(\mathbb{R}) \cap \mathbf{N}_1(\mathbb{R})$ has the form

$$T = \sum_{i=1}^{s} q_i \,[\![\, (a_i, b_i) \,]\!]$$

where $\{a_i\} \cap \{b_i\} = \emptyset$, $q_i \in \mathbb{Z}$, $q_i \neq 0$. Since $\partial T = \sum_{i=1}^{s} q_i\,(\delta_{b_i} - \delta_{a_i})$ and $\mathbf{M}(\partial T) = 2\sum_{i=1}^{s} |q_i| \geq 2s$ we get

$$s \leq \mathbf{N}(T) := \mathbf{M}(T) + \mathbf{M}(\partial T) .$$

If $\{T_j\} \subset \mathcal{R}_1(\mathbb{R})$, $\sup_j \mathbf{N}(T_j) < \infty$ we can therefore write

$$T_j = \sum_{i=1}^{s} q_i^{(j)} \,[\![\, (a_i^{(j)}, b_i^{(j)}) \,]\!]$$

for some fixed s independent of j. If moreover we assume that $\operatorname{spt} T_j$ in contained in some fixed compact set we then find as previously

$$a_i^{(j)} \to a_i, \quad b_i^{(j)} \to b_i, \quad q_i^{(j)} \to q_i, \quad T_j \rightharpoonup \sum_{i=1}^{s} q_i \,[\![\, (a_i, b_i) \,]\!].$$

•

We would like to show now that the assumption

$$\sup_j \mathbf{M}(\partial T_j) < \infty$$

is essentially necessary in the closure theorem. We shall see that if the previous bound does not hold, and $T_j \rightharpoonup T$, we can have

(i) *Concentration*: the support of T may be a lower dimensional set
(ii) *Distribution*: $T = \tau(\mathcal{M}, \theta, \vec{T})$ where $\vec{T}$ orients $\mathcal{M}$, but θ is not integer
(iii) $T = \tau(\mathcal{M}, \theta, \xi)$, where θ is integer but ξ is not tangent to $\mathcal{M}$.
(iv) combination of the previous phenomena, compare Ch. 1.

[7] *Concentration.* Consider the sequence of 1-dimensional currents

$$T_j := j \,[\![\, (0, 1/j) \,]\!] \in \mathcal{R}_1(\mathbb{R}) \cap \mathbf{N}_1(\mathbb{R}) .$$

Of course $\mathbf{M}(T_j) = 1$, $\mathbf{M}(\partial T_j) = 2j \to \infty$, and we have

$$T_j \rightharpoonup \delta_0 e_1$$

which belongs to $\mathcal{M}_1(\mathbb{R})$, but is not rectifiable, its support is a zero-dimensional set. •

[8] *Distribution.* Consider the sequence of 1-dimensional currents

$$T_j := \sum_{i=1}^{j} [\![\, (\frac{i-\lambda}{j}, \frac{i}{j}) \,]\!] \in \mathcal{R}_1(\mathbb{R}) \cap \mathbf{N}_1(\mathbb{R})$$

where $0 < \lambda < 1$. Of course $\sup_j \mathbf{M}(T_j) < \infty$, $\mathbf{M}(\partial T_j) = 2j \to \infty$. One easily sees that

$$T_j \rightharpoonup T := \lambda [\![(0,1)]\!]$$

and, trivially, T is not i.m. rectifiable. •

9 Consider the sequence of i.m. rectifiable 1-currents in $U = \mathbb{R} \times \mathbb{R} = \mathbb{R}^2$

$$T_j := \sum_{i=1}^{j} [\![(0, 1/j) \times \{i/j\}]\!] \in \mathcal{R}_1(\mathbb{R}^2) \cap \mathbf{N}_1(\mathbb{R}^2) .$$

Then $T_j = \tau(\mathcal{M}_j, 1, e_1)$, $\mathcal{M}_j = \bigcup_{i=1}^{j} (0, 1/j) \times \{i/j\}$, and $\mathbf{M}(T_j) = 1$, $\mathbf{M}(\partial T_j) =$

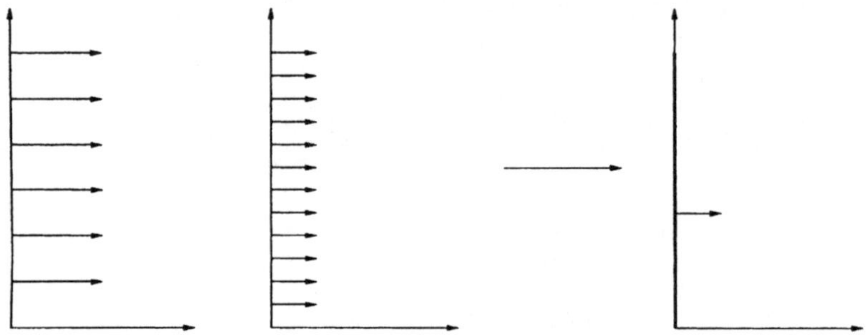

Fig. 2.6. One dimensional i.m. rectifiable currents converging to a non rectifiable current.

$2j \to \infty$. Clearly

$$\mathcal{H}^1 \llcorner \mathcal{M}_j \rightharpoonup \mathcal{H}^1 \llcorner \mathcal{M}, \qquad \mathcal{M} := \{0\} \times (0,1)$$
$$T_j \rightharpoonup T = \tau(\mathcal{M}, 1, e_1) .$$

Since the tangent vector to $\mathcal{M}$ is $\vec{\mathcal{M}} := e_2 \neq e_1$, the limit current T is not rectifiable. •

10 We remark that $\mathbf{M}(\partial T) = +\infty$ in the previous example. More generally, if $T = \tau(\mathcal{M}, \theta, \xi) \in \mathcal{M}_k(\mathbb{R}^n)$ where $\mathcal{M}$ is a smooth k-dimensional submanifold, $\theta > 0$, and ξ is a smooth vector field which is *not* tangent to $\mathcal{M}$, then we have $\mathbf{M}(\partial T) = +\infty$. Let us prove it in the special case

$$T = [\![(0,1) \times \{0\}]\!] e_2 \in \mathcal{M}_1(\mathbb{R}^2) ,$$

as the same argument works in general, compare also Lemma 1 in Sec. 3.3.2. We have

$$T = T^1 e_1 + T^2 e_2 , \qquad T^1 \equiv 0 , \qquad T^2 = \mathcal{H}^1 \llcorner \mathcal{M} , \qquad \mathcal{M} = (0,1) \times \{0\} .$$

consider the 0-form

$$\eta(x_1, \eta_2) \; := \; \zeta(x)\, \varphi(x^2)$$

where

$$\zeta(x) \; := \; \begin{cases} 1 & \text{if } |x| \le 2 \\ 3 - |x| & \text{if } 2 \le |x| \le 3 \\ 0 & \text{if } |x| \ge 3 \end{cases}$$

and, for a fixed $\varepsilon > 0$,

$$\varphi(x^2) \; := \; \begin{cases} -1 & \text{if } x^2 < -\varepsilon \\ (1/\varepsilon)x^2 & \text{if } |x^2| \le \varepsilon \\ 1 & \text{if } x^2 > \varepsilon\,. \end{cases}$$

One computes

$$\partial T(\eta) \; = \; T^2(\eta_{x^2}) \; = \; \int_0^1 \eta_{x^2}(x^1, 0)\, dx \; = \; \frac{1}{\varepsilon}\,.$$

Being $|\eta| \le 1$, we therefore infer $\mathbf{M}(\partial T) \ge \partial T(\eta) = \frac{1}{\varepsilon}$, i.e., $M(\partial T) = \infty$. •

11 *Concentration-distribution.* The different phenomena corresponding to (i) (ii) (iii) may appear in a combined form. Let us see two simple cases.

Set for $j = 2, 3, \ldots$

$$T_j \; := \; \tau(\mathcal{M}_j, j^2, e_1), \qquad \mathcal{M}_j := \bigcup_{i=1}^{j} \left(\frac{i}{j} - \frac{1}{j^3}, \frac{i}{j} \right) \subset (0,1).$$

Clearly $T_j \in \mathcal{R}_1(\mathbb{R}) \cap \mathbf{N}_1(\mathbb{R})$; as $\mathcal{H}^1(\mathcal{M}_j) = 1/j^2$, we have

$$\mathbf{M}(T_j) = 1, \quad \text{while} \quad \mathbf{M}(\partial T_j) = 2j^3 \to \infty.$$

The hypotheses of the closure theorem are not satisfied, but

$$T_j \;\rightharpoonup\; T := \tau((0,1), 1, e_1) = [\![\, (0,1) \,]\!] \in \mathcal{R}_1(\mathbb{R}) \cap \mathbf{N}_1(\mathbb{R})\,.$$

The concentration behaviour is seen from the fact that

$$\mathcal{H}^1(\mathcal{M}_j) \;\longrightarrow\; 0\,, \qquad \sum_j \mathcal{H}^1(\mathcal{M}_j) \;<\; \infty$$

and for any positive ε, we can find a Borel set E with $\mathcal{H}^1(E) < \varepsilon$ such that for large j we have

$$\| T_j \|((0,1) \setminus E) \; = \; 0\,,$$

i.e., $\mathcal{M}_j \subset E$. Such a concentration however produces the i.m. rectifiable current $[\![\, (0,1) \,]\!]$ which is distributed in $(0,1)$.

Consider now the following variant. Set again

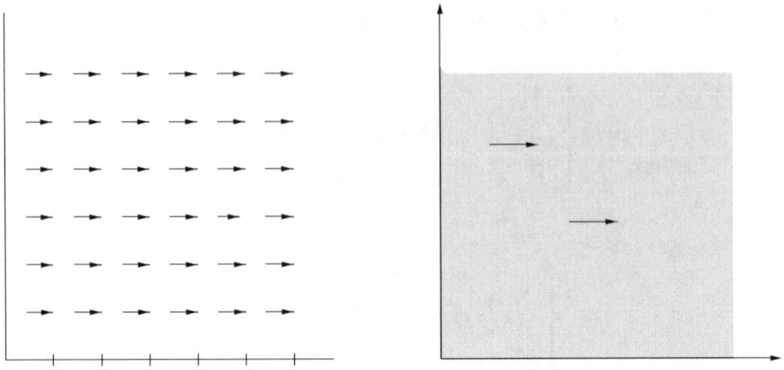

Fig. 2.7. One dimensional i.m. rectifiable currents converging to a 1–dimensional diffused current.

$$T_j \;:=\; \tau(\mathcal{M}_j, j^2, e_1) \;\in\; \mathcal{R}_1(\mathbb{R}^2)$$

where this time

$$\mathcal{M}_j \;:=\; \bigcup_{i,h=1}^{j} \left(\frac{i}{j} - \frac{1}{j^4}, \frac{i}{j}\right) \times \left\{\frac{h}{j}\right\} \;\subset\; \mathbb{R}^2 \,.$$

Then

$$j^2\,\mathcal{H}^1 \llcorner \mathcal{M}_j \;\rightharpoonup\; \mathcal{H}^2 \llcorner Q\,, \qquad Q = (0,1)\times(0,1)$$

and

$$T_j \;\rightharpoonup\; T := \tau(Q, 1, e_1) = \mathcal{H}^2 \llcorner Q\, e_1\,.$$

Again we have concentration, with this time a non rectifiable current distributed on the 2-dimensional set Q. Of course in this case we again have $\mathbf{M}(\partial T_j) = 2j^4 \to \infty$. •

[12] Finally we would like to remark that the vanishing of ∂T in some sense fixes the orientation of T. Let $B(0,1)$ be the unit ball in $\mathbb{R}^2$, trivially $T = \tau(B(0,1), 1, e_1\wedge e_2)$ defines an i.m. rectifiable current in $\mathbb{R}^2$ without boundary in $B(0,1)$. Also $T_1 + T_2$, where

$$\begin{aligned} T_1 \;&:=\; \tau(B(0,1/2), 1, e_1\wedge e_2) \\ T_2 \;&:=\; \tau(B(0,1)\setminus B(0,1/2), 1, -e_1\wedge e_2)\,, \end{aligned}$$

is an i.m. rectifiable current in $\mathbb{R}^2$; however we have

$$\partial(T_1+T_2)\llcorner B(0,1) \;=\; 2[\![\,\partial B(0,1/2)\,]\!] \;\neq\; 0\,.$$

•

Image of a rectifiable current under a Lipschitz map. Let T be an i.m. rectifiable k-current in $U \subset \mathbb{R}^n$, $T = \tau(\mathcal{M}, \theta, \xi) \in \mathcal{R}_k(U)$, $U \subset \mathbb{R}^n$, and let $f : U \subset \mathbb{R}^n \to V \subset \mathbb{R}^N$ be a Lipschitz map such that

$$f_{|\operatorname{spt} T} \text{ is proper.}$$

Then the push-forward or the image of T under f we have already defined in the general context of currents turns out to be an i.m. rectifiable k-current which can be explicitly written as

$$f_\# T(\omega) = \int_{\mathcal{M}} <\omega(f(x)), (\Lambda_k d^{\mathcal{M}} f_x)\xi(x)> \theta(x)\, d\mathcal{H}^k(x)\,. \tag{2}$$

This can be easily seen in the case in which T is the current integration over a smooth submanifold and f of class C^1, and actually in the same way in our present situation, as we can write $\mathcal{M}$ apart from an $\mathcal{H}^k$-zero set as union of $\mathcal{H}^k$-measurable subset of k-dimensional C^1-submanifolds.

Recall in fact that

$$|\Lambda_k d^{\mathcal{M}} f_x(\xi(x))| \;=\; J_f^{\mathcal{M}}(x)$$

and

$$\frac{(\Lambda_k d^{\mathcal{M}} f_x)\xi(x)}{|\,(\Lambda_k d^{\mathcal{M}} f_x)\xi(x)\,|}$$

defines an orientation of the rectifiable set $f(M)$. Therefore from the general area formula we have

$$f_\# T(\omega) = \int_{f(\mathcal{M})} <\omega(y), \sum_{x \in f^{-1}(y) \cap \mathcal{M}_+} \theta(x) \frac{(\Lambda_k d^{\mathcal{M}} f_x)\xi(x)}{|\,(\Lambda_k d^{\mathcal{M}} f_x)\xi(x)\,|} > d\mathcal{H}^k(y) \tag{3}$$

where

$$\mathcal{M}_+ \;=\; \{x \in \mathcal{M} \;\mid\; J_f^{\mathcal{M}}(x) > 0\}\,. \tag{4}$$

In fact choosing any orthonormal basis $(\tau_1(y), \dots, \tau_k(y))$ in $T_y f(\mathcal{M})$ in a $\mathcal{H}^k$-measurable way for $\mathcal{H}^k$-a.e. $y \in f(\mathcal{M})$, setting

$$\eta(y) \;:=\; \tau_1(y) \wedge \ldots \wedge \tau_k(y)$$

and for $\mathcal{H}^k$-a.e. $y \in f(\mathcal{M})$

$$N(y) \;:=\; \sum_{x \in f^{-1}(y) \cap \mathcal{M}_+} \theta(x)\varepsilon(x)$$

where $\varepsilon(x) = \pm 1$ is defined by

$$\frac{(\Lambda_k d^{\mathcal{M}} f_x)\xi(x)}{|\,(\Lambda_k d^{\mathcal{M}} f_x)\xi(x)\,|} \;=\; \varepsilon(x)\,\eta(f(x))$$

we can write

$$f_\# T(\omega) = \int_{f(\mathcal{M})} <\omega, \eta> N \, d\mathcal{H}^k$$

i.e.

$$f_\# T = \tau(f(\mathcal{M}), |N|, \operatorname{sign} N\eta)$$

and clearly

$$|N(y)| = \sum_{x \in f^{-1}(y) \cap \mathcal{M}_+} \theta(x)$$

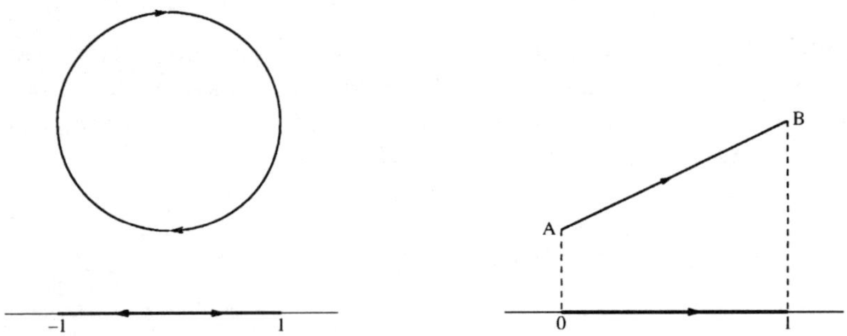

Fig. 2.8. (a) $\pi_\# [\![S^1]\!] = 0$, $\pi_\# ||[\![S^1]\!]|| = 2\pi \mathcal{L}^1 \llcorner (-1,1)$; (b) $\pi_\# [\![\overline{AB}]\!] = [\![(0,1)]\!]$, $\pi_\# ||[\![\overline{AB}]\!]|| = \overline{AB} \mathcal{L}^1 \llcorner (0,1)$.

Later we shall be interested to the case in which f is the projection $\pi : (x,y) \in \mathbb{R}^{n+N} \to x \in \mathbb{R}^n$ and $T = \tau(\mathcal{M}, \theta, \xi) \in \mathcal{R}_n(\mathbb{R}^{n+N})$. As a trivial example consider the case $n = N = 1$

$$\mathcal{M} := \{(x, kx) \mid x \in (0,1)\}, \qquad k > 0$$

and

$$T = \tau\left(\mathcal{M}, 1, \frac{e + k\varepsilon}{\sqrt{1+k^2}}\right) \in \mathcal{R}_1(\mathbb{R}^2).$$

Then we have

$$\pi_\# T = \tau((0,1), 1, e) = [\![(0,1)]\!].$$

Notice that

$$T = \mathcal{H}^1 \llcorner \mathcal{M} \frac{e}{\sqrt{1+k^2}} + \mathcal{H}^1 \llcorner \mathcal{M} \frac{k}{\sqrt{1+k^2}} \varepsilon$$

and, as measures, we have

$$\pi(\mathcal{H}^1 \llcorner \mathcal{M}) = \sqrt{1+k^2}\, \mathcal{H}^1 \llcorner (0,1)$$

while $\pi_{\#}e = e$, $\pi_{\#}\varepsilon = 0$ so that

$$\pi_{\#}T = \sqrt{1+k^2}\,\mathcal{H}^1 \llcorner (0,1) \frac{1}{\sqrt{1+k^2}} e = \mathcal{H}^1 \llcorner (0,1)\; e\ .$$

Remark the difference between projecting a measure and projecting a current!

The Cartesian product of rectifiable currents. It is also not difficult to show that the Cartesian product of two i.m. rectifiable currents we have defined in the general context of currents is also a rectifiable current. More precisely, if

$$T_i = \tau(\mathcal{M}_i, \theta_i, \xi_i) \in \mathcal{R}_{k_i}(U_i)\ , \qquad U_i \in \mathbb{R}^{n_i},\ i = 1,2$$

then

$$T_1 \times T_2 = \tau(\mathcal{M}_1 \times \mathcal{M}_2, \theta_1\theta_2, \xi_1 \wedge \xi_2) \in \mathcal{R}_{k_1+k_2}(U_1 \times U_2)\ .$$

Notice however that T_1 and $T_1 \times T_2$ rectifiable do not imply that T_2 must be rectifiable. For instance take

$$T_1 \;=\; k [\![(0,1)]\!]\ , \qquad T_2 \;=\; \frac{1}{k} [\![(0,1)]\!]$$

so that

$$T_1 \times T_2 \;=\; [\![(0,1) \times (0,1)]\!]\ .$$

2.5 Slicing

Let T be a k-dimensional current in $\mathbb{R}^n$ and let $f : \mathbb{R}^n \to \mathbb{R}^m$ be a Lipschitz map. Corresponding to $y \in \mathbb{R}^m$ one often needs to consider the part of T in $f^{-1}(y)$ as a $(k-m)$-dimensional current $k \geq m$, usually denoted by

$$<T, f, y>\ .$$

There is a measure theoretic method to define the *slice* of a current T by f at least if T has certain regularity property. It has very interesting applications for instance to intersection theory in algebraic geometry and to complex analysis. Later we shall use some elementary properties of the slices. Therefore we shall collect here a few facts concerning slicing.

First let us consider codimension-one-slices, i.e. the case $m = 1$. In order to define the *slice* $<T, f, y>$ *of* T *by* f *at* y we need T to have some regularity property. We shall now see that it can be defined for locally normal or i.m. rectifiable k-currents, and of course the two definitions agree if $T \in \mathcal{R}_k(\mathbb{R}^n) \cap \mathbf{N}_{k,\mathrm{loc}}(\mathbb{R}^n)$.

Definition 1. *Let* $T \in \mathbf{N}_{k,\mathrm{loc}}(U)$ *where* U *is an open set in* $\mathbb{R}^n$*, i.e., let*

$$\mathbf{N}_V(T) \;:=\; \mathbf{M}_V(T) + \mathbf{M}_V(\partial T) \;<\; \infty \qquad \forall\, V \subset\subset U\ ,$$

and let $f : U \to \mathbb{R}$ *be a Lipschitz function. We set*

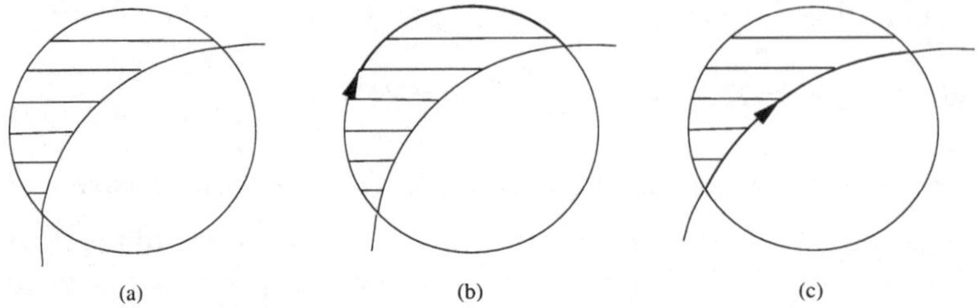

Fig. 2.9. (a) $[\![B(0,1)]\!] \cap \{f < t\}$, (b) $-\partial[\![B(0,1)]\!] \llcorner \{f < t\}$, (c) $< [\![B(0,1)]\!], f, t >$.

$$
(1) \qquad \begin{aligned} <T, f, t^-> &:= \partial(T \llcorner \{f < t\}) - \partial T \llcorner \{f < t\} \\ <T, f, t^+> &:= -\partial(T \llcorner \{f > t\}) + \partial T \llcorner \{f > t\} \end{aligned}
$$

where

$$
\begin{aligned} \{f > t\} &:= \{x \in U \mid f(x) > t\} , \\ \{f < t\} &:= \{x \in U \mid f(x) < t\} . \end{aligned}
$$

Of course we have

$$
<T, f, t> \; := \; <T, f, t^-> \; = \; <T, f, t^+>
$$

if

$$
\mathbf{M}(T \llcorner \{f = t\}) + \mathbf{M}(\partial T \llcorner \{f = t\}) = 0 ,
$$

and this happens for all but countably many values of t. Therefore $<T, f, t>$ *is well defined for* $\mathcal{H}^1$-a.e. *value* t.

Consider now the case of i.m. rectifiable currents. As a consequence of the theory of rectifiable sets and of the general coarea formula, in particular taking into account Sard type lemma in Sec. 2.1.4, we infer

Lemma 1. *Let $\mathcal{M}$ be an $\mathcal{H}^k$-measurable and countably rectifiable k-dimensional set in $\mathbb{R}^n$, $f : \mathbb{R}^n \to \mathbb{R}$ be a Lipschitz function, and*

$$
\mathcal{M}_+ := \{x \in \mathcal{M} \mid J_f^{\mathcal{M}}(x) := |(\Lambda_k d^{\mathcal{M}} f_x)\xi| > 0\}
$$

where $\xi(x)$ is a k-vector orienting $T_x\mathcal{M}$. Then for $\mathcal{H}^1$-a.e. $t \in \mathbb{R}$ we have

(i) *$\mathcal{M}_t := f^{-1}(t) \cap \mathcal{M}_+$ is countably rectifiable and $\mathcal{H}^{k-1}$-measurable.*

(ii) *For $\mathcal{H}^{k-1}$*-a.e. *$x \in \mathcal{M}_t$ $T_x\mathcal{M}_t$ and $T_x\mathcal{M}$ both exist, $T_x\mathcal{M}_t$ is a $(k-1)$-dimensional subspace of $T_x\mathcal{M}$ and*

$$
T_x\mathcal{M} = \{\tau + \lambda \nabla^{\mathcal{M}} f(x) \mid \tau \in T_x\mathcal{M}_t, \ \lambda \in \mathbb{R}\}
$$

where $\nabla^{\mathcal{M}} f(x)$ is the gradient of f projected on $T_x\mathcal{M}$.

Moreover

(2)
$$\int_{\mathbb{R}} dt \int_{\mathcal{M}_t} g\, d\mathcal{H}^{k-1} = \int_{\mathcal{M}} |\nabla^{\mathcal{M}} f| g\, d\mathcal{H}^k$$

for all non-negative $\mathcal{H}^k$-measurable functions g.

Applying (2) to $g \cdot \chi_{\{f<t\}}$, we infer

$$\int_{\mathcal{M}\cap\{f<t\}} |\nabla^{\mathcal{M}} f| g\, d\mathcal{H}^k = \int_{-\infty}^{t} ds \int_{\mathcal{M}_s} g\, d\mathcal{H}^{k-1} ,$$

so that the left hand side is an absolutely continuous function of t and for a.e. $t \in \mathbb{R}$

(3)
$$\frac{d}{dt} \int_{\mathcal{M}\cap\{f<t\}} |\nabla^{\mathcal{M}} f| g\, d\mathcal{H}^k = \int_{\mathcal{M}_t} g\, d\mathcal{H}^{k-1} .$$

Using the scalar product $<,>$ in $\Lambda_k T_x \mathcal{M}$, define now for $\xi \in \Lambda_k T_x \mathcal{M}$ and $w \in T_x \mathcal{M}$ the $(k-1)$-vector $\xi \llcorner w$ by

(4)
$$<\xi \llcorner w, \eta> := <\xi, w \wedge \eta> \qquad \forall\, \eta \in \Lambda_{k-1} T_x \mathcal{M} .$$

Then one sees that for $\mathcal{H}^1$-a.e. $t \in \mathbb{R}$ and $\mathcal{H}^{k-1}$-a.e. $x \in \mathcal{M}_t$

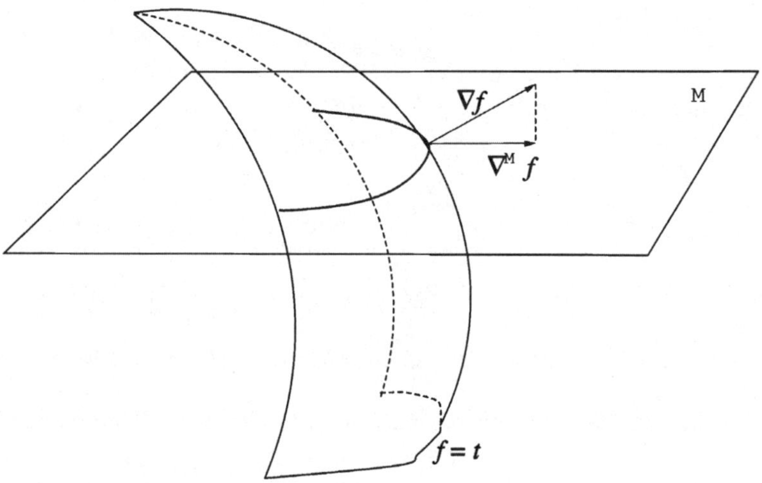

Fig. 2.10. Slice

$$\xi \llcorner \frac{\nabla^{\mathcal{M}} f(x)}{|\nabla^{\mathcal{M}} f(x)|} \in \Lambda_{k-1} T_x \mathcal{M}_t$$

is a unit simple $(k-1)$-vector.

Definition 2. *Let* $T = \tau(\mathcal{M}, \theta, \xi) \in \mathcal{R}_k(U)$. *For* a.e. $t \in \mathbb{R}$ *we define the* slice *of* T *by* f *at* t

$$<T, f, t>$$

as the rectifiable current in $\mathcal{R}_{k-1}(U)$ *given by*

$$<T, f, t> \;:=\; \tau(\mathcal{M}_t, \theta_t, \xi_t)$$

where for $\mathcal{H}^{k-1}$-a.e. $x \in \mathcal{M}_t$

$$\begin{aligned} \xi_t(x) &:= \xi(x) \llcorner \frac{\nabla^{\mathcal{M}} f(x)}{|\nabla^{\mathcal{M}} f(x)|} \\ \theta_t(x) &:= \theta_{+|\mathcal{M}_t}, \qquad \theta_+ := \theta \chi_{\mathcal{M}_+} . \end{aligned}$$

We have

Proposition 1. *Let* $T = \tau(\mathcal{M}, \theta, \xi) \in \mathcal{R}_k(U)$, $U \subset \mathbb{R}^n$. *Then*

(i) *For any open set* $V \subset U$

$$\int_{\mathbb{R}} \mathbf{M}_V(<T, f, t>)\, dt = \int_{\mathcal{M} \cap V} |\nabla^{\mathcal{M}} f| \theta \, d\mathcal{H}^k \le \sup_{\mathcal{M} \cap V} |\nabla^{\mathcal{M}} f| \, \mathbf{M}_V(T) .$$

(ii) *If moreover* $T \in \mathbf{N}_{k,\mathrm{loc}}(U)$, *i.e,* $\mathbf{M}_V(\partial T) < \infty$ *for all* $V \subset\subset U$, *then for* a.e. $t \in \mathbb{R}$

$$<T, f, t> \;=\; \partial(T \llcorner \{f < t\}) - \partial T \llcorner \{f < t\} .$$

(iii) *If also* ∂T *is rectifiable, then for* a.e. $t \in \mathbb{R}$

$$<\partial T, f, t> \;=\; -\partial <T, f, t> .$$

For locally normal currents in $\mathbf{N}_{k,\mathrm{loc}}(U)$ we finally have

Proposition 2. *Let* $T \in \mathbf{N}_{k,\mathrm{loc}}(U)$, $U \subset \mathbb{R}^n$. *Then*

(i) $\operatorname{spt} <T, f, t^{\pm}> \;\subset\; \operatorname{spt} T \cap \{f = t\}$.

(ii) *For all open sets* $V \subset U$

$$\mathbf{M}_V(<T, f, t^+>) \le \sup_V |Df| \, \liminf_{h \to 0} \frac{1}{h} \mathbf{M}_V(T \llcorner \{t < f < t + h\})$$

$$\mathbf{M}_V(<T, f, t^->) \le \sup_V |Df| \, \liminf_{h \to 0} \frac{1}{h} \mathbf{M}_V(T \llcorner \{t - h < f < t\}).$$

(iii) $\mathbf{M}_V(T \llcorner \{f < t\})$ *is trivially non decreasing in* t, *hence* a.e. *differentiable and*

$$\int_a^b \frac{d}{dt} \mathbf{M}_V(T \llcorner \{f < t\})\, dt \;\le\; \mathbf{M}_V(T \llcorner \{a < f < b\}) .$$

Thus

$$\int_a^b \mathbf{M}_V(<T, f, t^{\pm}>)\, dt \le \sup_V |Df| \, \mathbf{M}_V(T \llcorner \{a < f < b\}) .$$

We shall not prove Proposition 1 and Proposition 2, but we would like to make a few remarks on their proofs. Proposition 1 (i) is a consequence of (2). The proof of Proposition 1 (ii) is done by approximating f with its smoothings f_σ. The proof of Proposition 1 (iii) is simple. One applies (ii) of Proposition 1 to ∂T

$$<\partial T, f, t> \;=\; \partial(\partial T \llcorner \{f < t\}) \,,$$

while applying ∂ to the identity in (ii)

$$\partial<T, f, t> \;=\; -\partial(\partial T \llcorner \{f < t\}) \,.$$

In order to prove (2) one first consider the case $f \in C^1$, and then one gets the result by approximation. For $f \in C^1$ and $\gamma : \mathbb{R} \to [0, +\infty)$ smooth and increasing we get

$$\begin{aligned}&\Big[\partial(T \llcorner \gamma \circ f) - (\partial T \llcorner \gamma \circ f)\Big](\omega)\\ &\quad = (T \llcorner \gamma \circ f)(d\omega) - \partial T(\gamma \circ f\, \omega) = T(\gamma \circ f\, d\omega) - T(d(\gamma \circ f\, \omega))\\ &\quad = T(\gamma \circ f\, d\omega - d(\gamma \circ f\, \omega)) \;=\; -T(\gamma' \circ f\, df \wedge \omega) \,.\end{aligned}$$

Choosing γ such that $\gamma = 0$ for $t < a$, $\gamma = 1$ for $t > b$, and

$$0 \;<\; \gamma' \;<\; \frac{1+\varepsilon}{b-a} \qquad \text{for } a < t < b,\ \varepsilon > 0 \,,$$

we infer

$$\partial(T \llcorner \gamma \circ f)(\omega) - (\partial T \llcorner \gamma \circ f)(\omega) \;\longrightarrow\; <T, f, a^+>$$

if $b \downarrow a$ and this implies (i). Also

$$|T(\gamma' \circ f\, df \wedge \omega)| \leq \sup_V |Df| \frac{1+\varepsilon}{b-a} \mathbf{M}_V(T \llcorner \{a < f < b\}) |\omega|$$

which implies (ii) and (iii).

Quite more complicated is to deal with slices of codimension larger that 1. To motivate the procedure let us consider the special case $T = \tau(M, 1, \xi)$ where M is a smooth k-dimensional submanifold in $U \subset \mathbb{R}^n$ oriented by the k-vectorfield ξ, with $\mathcal{H}^k(M) < \infty$, and let

$$f : \mathbb{R}^n \longrightarrow \mathbb{R}^m \,, \qquad k \geq m$$

be smooth. For $x \in M$ we factor

$$\xi(x) \;=\; \eta(x) \wedge \zeta(x)$$

so that $\eta(x)$ is a simple m-vector of $T_x M$, $|\eta(x)| = 1$

$$<\eta(x), f^\# dy^1 \wedge \ldots \wedge dy^m> \;=\; J_f^M(x) \,,$$

$\zeta(x)$ is a simple $(k-m)$-vector of $T_x M$, $|\zeta(x)| = 1$, and

$$\ker d^M f(x) \text{ is associated with } \zeta(x)$$

in case $J_f^M(x) > 0$. Using the coarea formula one finds that if $\omega \in \mathcal{D}^{k-m}(U)$ and φ is any real valued Borel function on $\mathbb{R}^m$, then

$$\begin{aligned} T \llcorner f^\#(\varphi\, dy^1 \wedge \ldots \wedge dy^m)(\omega) &= T(f^\#(\varphi\, dy^1 \wedge \ldots \wedge dy^m) \wedge \omega) \\ &= \int_M <\xi(x), f^\#(\varphi\, dy^1 \wedge \ldots \wedge dy^m) \wedge \omega(x)> d\mathcal{H}^k(x) \\ &= \int_M \varphi(f(x))\, J_f^M(x) <\zeta(x), \omega(x)> d\mathcal{H}^k(x) \\ &= \int_{\mathbb{R}^m} \varphi(y) \int_{M \cap f^{-1}(y)} <\zeta(x), \omega(x)> d\mathcal{H}^{k-m}(x)\, d\mathcal{H}^m(y) \\ &= \int_{\mathbb{R}^m} \varphi(y)\, \tau(M \cap f^{-1}(y), 1, \zeta)(\omega)\, d\mathcal{H}^m(y)\ . \end{aligned}$$

It follows from the differentiation theory that for $\mathcal{H}^m$-a.e. y

$$T \llcorner f^\#(\chi_{B(y,\rho)} dy^1 \wedge \ldots \wedge dy^m)/(\omega_n \rho^n) \longrightarrow \tau(M \cap f^{-1}(y), 1, \zeta)$$

in $\mathcal{D}_{k-m}(U)$ as $\rho \to 0$.

When $T \in \mathbf{N}_k(U)$ is an arbitrary normal current (or even a flat chain), one introduces for $y \in U$ and $\rho > 0$ the $(k-m)$–currents

$$T_\rho := \frac{1}{\omega_m \rho^m} T \llcorner f^\#(\chi_{B(y,\rho)}(t) dt^1 \wedge \ldots \wedge dt^m) \in \mathcal{D}_{k-m}(\mathbb{R}^n)$$

and defines the *slice* $<T, f, y>$ of T at y as the weak limit of T_ρ

$$<T, f, y> := \lim_{\rho \to 0} T_\rho$$

whenever it exists. It turns out that $<T, f, y>$ *exists for* $\mathcal{H}^m$-a.e. y. In fact for $\alpha \in \mathcal{D}^{k-m}(U)$ we have $f_\#(T \llcorner \alpha) \in \mathbf{N}_m(\mathbb{R}^m)$, hence, compare Theorem 2 in Sec. 4.3.1,

$$f_\#(T \llcorner \alpha)(\omega) = \int h_\alpha \omega\, dx, \qquad \omega \in \mathcal{D}^m(\mathbb{R}^m)$$

for some $h_\alpha \in L^1(\mathbb{R}^m)$. Consequently

$$T_\rho(\alpha) = (-1)^{m(k-m)} \fint_{B(y,\rho)} h_\alpha(t)\, d\mathcal{H}^m$$

and $\lim_{\rho \to 0} T_\rho(\alpha)$ exists for each fixed α and $\mathcal{H}^m$–a.e. y by the Lebesgue differentiation theorem. If we choose a denumerable subset $\mathcal{C}$ of $\mathcal{D}^{k-m}(U)$ which is dense with respect to the uniform topology we infer that for $\mathcal{H}^m$–a.e. y $\lim_{\rho\to 0} T_\rho(\alpha)$ exists for all $\alpha \in \mathcal{C}$, and, as

$$|T_\rho(\alpha)| \leq \mathbf{M}(f_\#(T \llcorner \alpha)) \leq c\,\mathbf{M}(T)\,||\alpha||_\infty,$$

for all $\alpha \in \mathcal{D}^{k-m}(U)$.

Also one has

(i) $\operatorname{spt} <T,f,y> \subset f^{-1}(y) \cap \operatorname{spt} T$,
(ii) $\partial<T,f,y> = (-1)^m <\partial T, f, y>\,, \quad k > m.$

Moreover

(iii) $<T,f,y> \in \mathbf{N}_{k-m}(U)$ *for* $\mathcal{H}^m$-a.e. y
(iv) $(T \llcorner f^\# dy^1 \wedge \ldots \wedge dy^m)(\omega) = \int <T,f,y>(\omega)\, d\mathcal{H}^m(y)$, $\omega \in \mathcal{D}^{k-m}(U)$
(v) $\mathbf{M}(T \llcorner f^\# dy^1 \wedge \ldots \wedge dy^m) = \int \mathbf{M}(<T,f,y>)\, d\mathcal{H}^m(y)$,

and finally

(vi) $<T,f,y>$ *is i.m. rectifiable for* $\mathcal{H}^m$ *a.e.* y *if* T *is i.m. rectifiable.*

Finally we remark that there is no problem in extending the theory of slicing to the case in which f takes values in an *oriented* m*-dimensional submanifold of class* C^1.

2.6 The Deformation Theorem and Approximations

In this section we shall state without proofs the important *deformation theorem* and some of its consequences: the *isoperimetric inequality* and the *approximation theorem.*

Consider that ε*-grid* or ε*-skeleton,* $\varepsilon > 0$, of $\mathbb{R}^{n+N}$

$$L_{n,\varepsilon} \;:=\; \{x \in \mathbb{R}^n \;\mid\; \frac{1}{\varepsilon} x \in \mathbb{Z}^{n+N}\}$$

and denote by $\mathcal{L}_{n,\varepsilon}$ the n-dimensional faces of $L_{n,\varepsilon}$, or equivalently the family of all n-dimensional cubes in $\mathbb{R}^{n+N}$ with side ε and vertices in $L_{n,\varepsilon}$.

Theorem 1 (The deformation theorem). *We have*

(i) Let $T \in \mathbf{N}_n(\mathbb{R}^{n+N})$ *and let* ε *be any positive real number. Then there exist a real polyhedral chain* P *made of* n*-faces in* $\mathcal{L}_{n,\varepsilon}$*, i.e.*

$$P \;=\; \sum_{F_j \in \mathcal{L}_{n,\varepsilon}} a_j [\![F_j]\!]\,, \qquad a_j \in \mathbb{R}$$

and

$$R \in \mathbf{N}_{n+1}(\mathbb{R}^{n+N})\,, \qquad S \in \mathcal{M}_n(\mathbb{R}^{n+N})$$

such that

$$T \;=\; P + \partial R + S\,.$$

Moreover P is controlled as follows

$$\mathbf{M}(P) \le c\mathbf{M}(T) , \qquad \mathbf{M}(\partial P) \le c\mathbf{M}(\partial T) ,$$

and the rest $\partial R + S$ is small in the sense that

$$\mathbf{M}(R) \le c\varepsilon\,\mathbf{M}(T) \qquad \mathbf{M}(S) \le c\varepsilon\,\mathbf{M}(\partial T)$$

and

$$\mathbf{M}(\partial R) \le c\{\mathbf{M}(T) + \varepsilon\mathbf{M}(\partial T)\} ,$$

where c is an absolute constant. Finally

$$\operatorname{spt} P, \operatorname{spt} R \subset U_\delta(\operatorname{spt} T) \qquad \operatorname{spt} \partial P, \operatorname{spt} \partial R \subset U_\delta(\operatorname{spt} \partial T)$$

where $\delta = \delta(\varepsilon) = 2\sqrt{n+N}\varepsilon$ and $U_\delta(A) := \{x \mid \operatorname{dist}(x, A) < \delta\}$.

(ii) *If moreover $T \in \mathcal{R}_n(\mathbb{R}^{n+N})$ is i.m. rectifiable, we can choose P and R i.m. rectifiable, i.e. $R \in \mathcal{R}_{n+1}(\mathbb{R}^{n+N})$ and*

$$P = \sum_{F_j \in \mathcal{L}_{n,\varepsilon}} a_j [\![F_j]\!] , \qquad \text{with } a_j \in \mathbb{Z} ;$$

and, if also ∂T is i.m. rectifiable, we can choose S i.m. rectifiable, too.

Such a result is obtained by deforming T into $\mathcal{L}_{n,\varepsilon}$, so that the result is automatically a polyhedral chain. The main error term is ∂R, where R is the surface through which T is deformed; S is a secondary error term due to moving ∂T into the skeleton $L_{n,\varepsilon}$. It is not difficult to see that there is no real problem to perform such a procedure by projecting from points inside the $(n+1)$-cubes defined by $\mathcal{L}_{n,\varepsilon}$, if T does not wind about the center projection and has no auto-intersections. In the general case the key point is choosing the center of projection suitably by observing that in the average the distortion is still controlled. We shall postpone the proof to Sec. 5.1.1.

A first consequence of Theorem 1 is the following

Theorem 2 (Isoperimetric inequality). *Let T be a boundaryless i.m. rectifiable $(n-1)$-dimensional current in $\mathbb{R}^{n+N}$ with compact support. Then there exists an i.m. rectifiable n-current R with compact support such that $\partial R = T$ and*

(1)
$$\mathbf{M}(R) \le c\,\mathbf{M}(T)^{n/(n-1)}$$

where $c = c(n, N)$.

Proof. From the deformation theorem we can write in correspondence of $\varepsilon > 0$

$$T = P + \partial R + S .$$

The result follows by choosing ε, but strangely, ε large. From

$$\mathbf{M}(S) \le c\varepsilon\, \mathbf{M}(\partial T) = 0$$

we infer $S = 0$. Next we observe that if

$$P = \sum_{F_j \in \mathcal{L}_{n,\varepsilon}} a_j [\![F_j]\!] \neq 0,$$

we have

$$\mathbf{M}(P) = N(\varepsilon)\, \varepsilon^{n-1}, \qquad N(\varepsilon) \text{ integer}$$

as $a_j \in \mathbb{Z}$ and $\mathbf{M}([\![F_j]\!]) \ge \varepsilon^{n-1}$. From $\mathbf{M}(P) \le c\mathbf{M}(T)$, and the previous inequality we therefore infer by choosing

$$\varepsilon := (2c\mathbf{M}(T))^{1/(n-1)},$$

that

$$N(\varepsilon)\varepsilon^{n-1} = \mathbf{M}(P) \le c\mathbf{M}(T) = \frac{1}{2}\varepsilon^{n-1},$$

i.e. $N(\varepsilon) = 0$. Then $P = 0$ and $T = \partial R$ with $\operatorname{spt} R$ compact and

$$\mathbf{M}(R) \le c\varepsilon\, M(T) = c'\, (\mathbf{M}(T))^{n/(n-1)}.$$

□

As in the classical case it turns out that the best constant c in (1) is that corresponding to equality in

$$\mathbf{M}(\partial T) \le c\,\mathbf{M}(T)^{n/(n-1)}$$

which arises when T is a k-dimensional ball.

The next consequence of the deformation theorem we want to state is the following

Theorem 3 (Weak polyhedral approximation). *Let $T \in \mathcal{R}_n(U)$, U open in $\mathbb{R}^{n+N}$, be such that $\partial T \in \mathcal{M}_{n-1}(U)$. Then there is a sequence of i.m. polyhedral chains of the form*

$$P_j = \sum_s a_s^{(j)} [\![F_s^{(j)}]\!], \quad a_s^{(j)} \in \mathbb{Z}, \quad F_s^{(j)} \in \mathcal{L}_{n,\varepsilon_j}, \qquad \varepsilon_j \to 0$$

such that

$$\text{(2)} \qquad \begin{aligned} &\sup_j (\mathbf{M}(P_j) + \mathbf{M}(\partial P_j)) < \infty \\ &P_j \rightharpoonup T \qquad \textit{weakly in } \mathcal{D}_n(U) \end{aligned}$$

and consequently

$$\text{(3)} \qquad \partial P_j \rightharpoonup \partial T \qquad \textit{weakly in } \mathcal{D}_{n-1}(U).$$

Proof. First assume $U = \mathbb{R}^n$. We apply the deformation theorem with $\varepsilon = \varepsilon_j$ to obtain

$$T = P_j + \partial R_j + S_j$$

with

$$\sup(\mathbf{M}(P_j) + \mathbf{M}(\partial P_j)) \le c(\mathbf{M}(T) + \mathbf{M}(\partial T)) < \infty$$
$$\mathbf{M}(R_j) \le c\varepsilon_j \, \mathbf{M}(T) , \qquad \mathbf{M}(S_j) \le c\varepsilon_j \, \mathbf{M}(\partial T) .$$

Hence $S_j \rightharpoonup 0$, $R_j \rightharpoonup 0$, consequently $\partial R_j \rightharpoonup 0$, and we get $P_j \rightharpoonup T$.

In the general case we take $\varphi : \mathbb{R}^{n+N} \to \mathbb{R}$ with $\varphi > 0$ in U, $\varphi \equiv 0$ in $\mathbb{R}^{n+N} \setminus U$ such that $\{x \mid \varphi(x) > \lambda\} \subset\subset U \; \forall \, \lambda > 0$. For $\mathcal{H}^1$-a.e. $\lambda > 0$ we infer from Sec. 2.2.5 that $T_\lambda := T \llcorner \{x \mid \varphi(x) > \lambda\}$ is such that $\mathbf{M}(\partial T_\lambda) < \infty$. Since $\operatorname{spt} T_\lambda \subset\subset U$, we can apply the previous argument and conclude the proof by choosing a suitable sequence $\lambda_j \downarrow 0$. □

Notice that in particular Theorem 3 says that every i.m. rectifiable n-dimensional current is the weak limit in the sense of currents of a sequence of smooth n-dimensional submanifolds of $\mathbb{R}^{n+N}$. One can also prove, compare Sec. 5.1.1

Theorem 4 (Strong approximation). *Let T be an i.m. rectifiable current in $\mathbb{R}^{n+N}$, $T \in \mathcal{R}_n(\mathbb{R}^{n+N})$ with compact support and $\partial T \in \mathcal{R}_{n-1}(\mathbb{R}^n)$. Let $\varepsilon > 0$. Then there exist an i.m. rectifiable polyhedral chain P in $\mathbb{R}^{n+N}$ and a C^1-diffeomorphism of $\mathbb{R}^{n+N}$ into $\mathbb{R}^{n+N}$ such that*

$$\mathbf{M}(P - f_\# T) + \mathbf{M}(\partial P - \partial f_\# T) \le \varepsilon . \tag{4}$$

Moreover

$$\begin{aligned} &\operatorname{spt} P \subset \{x \mid \operatorname{dist}(x, \operatorname{spt} T) \le \varepsilon\} =: \overline{U_\varepsilon(\operatorname{spt} T)} \\ &\mathit{Lip}\, f \le 1 + \varepsilon , \qquad \mathit{Lip}\, f^{-1} \le 1 + \varepsilon \\ &|f(x) - x| \le \varepsilon \; \forall \, x \in \mathbb{R}^{n+N} \quad \textit{and} \quad f(x) = x \textit{ for } x \notin \overline{U_\varepsilon(\operatorname{spt} T)} . \end{aligned} \tag{5}$$

We emphasize that inequality (4) means that P and $f_\# T$, as well as ∂P and $f_\# \partial T$ *coincide* up to a piece of small mass.

Usually the *class of i.m. polyhedral chains* of dimension n in $\mathbb{R}^{n+N}$ is denoted by $\mathcal{P}_n(\mathbb{R}^{n+N})$. It is defined as the class of currents of the form

$$\sum a_j \, [\![\Sigma_j]\!]$$

where $a_j \in \mathbb{Z}$ and Σ_j are n-dimensional simplexes in $\mathbb{R}^{n+N}$.

Finally we would like to state a similar *approximation theorem with mass for normal currents.* Here it is essential to use the *mass norm* of n-vectors instead of the Euclidean norm, compare Sec. 2.2.1 and Vol. II Ch. 1.

We denote by $\mathbf{P}_n(\mathbb{R}^{n+N})$ the class of *polyhedral chains with real coefficients.* We have, compare Vol. II Sec. 1.3.4 and Federer [226] for a complete proof

Theorem 5 (Strong approximation theorem for normal currents). *Let T belong to $\mathbf{N}_n(\mathbb{R}^{n+N})$ with $\operatorname{spt} T$ compact. Then there exists a sequence of polyhedral chains P_j with real coefficients, $P_j \in \mathbf{P}_n(\mathbb{R}^{n+N})$, such that*

$$\begin{aligned} P_j &\rightharpoonup T &&\text{weakly in } \mathcal{D}_n(\mathbb{R}^{n+N}) \\ \partial P_j &\rightharpoonup \partial T &&\text{weakly in } \mathcal{D}_{n-1}(\mathbb{R}^{n+N}) \end{aligned}$$

and

$$\mathbf{M}(P_j) \to \mathbf{M}(T) \quad \text{and} \quad \mathbf{M}(\partial P_j) \to \mathbf{M}(\partial T) .$$

2.7 The Closure Theorem

In this final subsection we would like to indicate the main steps in the proof of the closure theorem for i.m. rectifiable currents, Theorem 1 in Sec. 2.2.4.

Let $U \subset \mathbb{R}^{n+N}$ be an open set and let $T \in \mathbf{N}_{n,\text{loc}}(U)$. For $a \in U$ and $r < r_a := \operatorname{dist}(a, \partial U)$ we know from Sec. 2.2.5 that the slice of T over $\partial B(a,r)$ is well defined as

$$<T, a, r> \; := \; \partial(T \llcorner B(a,r)) - (\partial T) \llcorner B(a,r)$$

and is an i.m. rectifiable $(n-1)$-current for $\mathcal{H}^1$-a.e. $r < r_a$ if moreover $T \in \mathcal{R}_n(U)$. We have

Lemma 1 (Slicing lemma). *Let $T_k, T \in \mathbf{N}_{n,\text{loc}}(U)$ be such that*

$$\begin{aligned} &\sup_k \mathbf{N}_V(T_k) < \infty &&\forall\, V \subset\subset U \\ &T_k \rightharpoonup T &&\text{weakly in } \mathcal{D}_n(U) . \end{aligned}$$

Then, given $a \in U$, for $\mathcal{L}^1$-a.e. $r < r_a$ there exists a subsequence, depending on r, $\{T'_k\}$ of $\{T_k\}$ such that

(1) $$<T'_k, a, r> \; \rightharpoonup \; <T, a, r>$$

(2) $$\sup_k \mathbf{N}_V(<T'_k, a, r>) < \infty \qquad \forall\, V \subset\subset U .$$

If moreover for some $V_0 \subset\subset U$ we have $\mathbf{N}_{V_0}(T_k) \to 0$ then $\{T'_k\}$ can be chosen in such a way that also $\mathbf{N}_{V_0}(<T'_k, a, r>) \to 0$.

Proof. Passing to a subsequence we can assume that

$$\| T_k \| + \| \partial T_k \| \rightharpoonup \mu$$

where μ is a positive Radon measure on U. If $\mu(\partial B(a,r)) = 0$, (notice that this holds for all but countably many r at most), we then infer

$$T_k \llcorner B(a,r) \rightharpoonup T \llcorner B(a,r) , \qquad \partial T_k \llcorner B(a,r) \rightharpoonup \partial T \llcorner B(a,r) ,$$

hence

$$\partial(T_k \llcorner B(a,r)) \rightharpoonup \partial(T \llcorner B(a,r)) ,$$

consequently (1) holds. For $0 < r_1 < r_2 < r_a$ and $V \subset\subset U$, we also know from Sec. 2.2.5 that

$$\int_{r_1}^{r_2} \mathbf{M}_V(<T_k, a, r>)\, dr \le c\, \mathbf{M}_V(T_k \llcorner B(a, r_2)) .$$

The same holds for

$$\partial{<}T_k, a, r{>} = -\partial(\partial T_k \llcorner B(a,r)) = -{<}\partial T_k, a, r{>}$$

hence

$$\int_{r_1}^{r_2} f_k(r)\, dr \le c\mathbf{N}_V(T_k \llcorner B(a, r_2)) .$$

where

$$f_k(r) := \mathbf{N}_V(<T_k, a, r>) \ge 0 .$$

Fatou's lemma and the equiboundedness of $\mathbf{N}_V(T_k)$ then yield

$$\int_{r_1}^{r_2} \liminf_{k\to\infty} f_k(r)\, dr \le \liminf_{k\to\infty} \int_{r_1}^{r_2} f_k(r)\, dr < \infty .$$

Therefore for a.e. $r \in (r_1, r_2)$ we have $\liminf_{k\to\infty} f_k(r) < \infty$. Choosing those r's we can finally find a subsequence such that (2) holds. The rest of the claim is trivial. □

A part from the simple remark in Lemma 1, the first key ingredient in the proof of the closure theorem is the following theorem which says that if almost all slices of T by spheres, i.e., by $f(x) := |x - a|$, are rectifiable then T is rectifiable, the second ingredient is the *boundary rectifiability theorem* stated below.

Theorem 1. *Let $T \in \mathcal{M}_{n,\mathrm{loc}}(U)$. Suppose $\partial T = 0$ and that for all $a \in U$ and almost all $r < r_a$ the slices*

$$<T, a, r> := \partial(T \llcorner B(a,r))$$

are i.m. rectifiable. Then T is i.m. rectifiable.

Let us postpone the proof of Theorem 1 and let us give the

Proof of the Closure theorem. We shall prove it by induction on the dimension n of the currents (keeping fixed $n + N$ and U).

Let $T_k \in \mathcal{R}_n(U)$, $\sup_k \mathbf{N}_V(T_k) < \infty$ $\forall$ $V \subset\subset U$, $T_k \rightharpoonup T$ in $\mathcal{D}_n(U)$. By the same argument in Theorem 3 in Sec. 2.2.6 we can assume $U = \mathbb{R}^{n+N}$ and $\sup_k \mathbf{N}(T_k) < \infty$.

The theorem clearly holds for 0-dimensional currents, compare Sec. 2.2.4. Suppose that it holds for $(n-1)$-dimensional currents and let us prove it for n-currents. From the proof of the boundary rectifiability theorem below (where one uses the closure theorem for $(n-1)$- currents) we infer that

$$\partial T_k \in \mathcal{R}_{n-1}(U)$$

and of course $\mathbf{N}(\partial T_k) < \infty$. Using again the closure theorem for $(n-1)$-currents we therefore infer $\partial T \in \mathcal{R}_{n-1}(U)$. By the isoperimetric theorem, for instance, we can then find $S \in \mathcal{R}_{n-1}(U)$ such that $\partial S = \partial T$. By considering

$$\overline{T}_K := T_k - S\,, \qquad \overline{T} := T - S$$

we see that it suffices to prove the theorem in the case $\partial T = 0$.

Fix a. By the slicing lemma, for a.e. $r < r_a$ there is a subsequence $\{T'_k\}$ of $\{T_k\}$ such that

$$<T'_k, a, r> \rightharpoonup <T, a, r>, \qquad \sup_k \mathbf{N}(<T'_k, a, r>) < \infty.$$

Again by the proof of the boundary rectifiability theorem

$$<T'_k, a, r> \in \mathcal{R}_{n-1}(\mathbb{R}^{n+N})$$

and by the inductive assumption we conclude

$$<T, a, r> \in \mathcal{R}_{n-1}(\mathbb{R}^{n+N})\,.$$

The conclusion now follows from Theorem 1. □

Theorem 2 (Boundary rectifiability theorem). *Let $T \in \mathcal{R}_n(U)$, U open in $\mathbb{R}^{n+N}$. Suppose*

$$\mathbf{M}_V(T) < \infty \qquad \forall\, V \subset\subset U\,.$$

Then ∂T is an i.m. rectifiable current in U, $\partial T \in \mathcal{R}_{n-1}(U)$.

Proof. By the polyhedral weak approximation theorem in Sec. 2.2.6 we find a sequence $\{P_k\}$ of polyhedral chains with integer coefficients with

$$\sup_k \mathbf{N}(P_k) < \infty\,, \qquad P_k \rightharpoonup T\,.$$

Trivially

$$\partial P_k \in \mathcal{R}_{n-1}(U)\,, \qquad \sup_k \mathbf{N}(\partial P_k) < \infty\,.$$

Therefore applying the closure theorem (note: for $(n-1)$-currents) we infer that ∂T is i.m. rectifiable, as $\partial P_k \rightharpoonup \partial T$. □

Remark 1. We have seen that the closure theorem for boundaryless i.m. rectifiable $(k-1)$-currents in $\mathbb{R}^n$ implies the boundary rectifiability of i.m. rectifiable k-currents in $\mathbb{R}^n$. Let $\{T_i\}$ be a sequence in $\mathcal{R}_{n-1}(\mathbb{R}^n)$ satisfying

$$\partial T_i = 0\,, \qquad \operatorname{spt} T_i \subset B(0,1)\,, \qquad T_i \rightharpoonup T \text{ in } \mathcal{D}_{n-1}(\mathbb{R}^n)\,.$$

We can write

$$T_i = \partial R_i\,, \qquad R_i \in \mathcal{R}_n(\mathbb{R}^n)\,, \qquad \operatorname{spt} R_i \subset B(0,1)\,, \qquad \sup_k \mathbf{N}(R_i) < \infty\,.$$

Trivially, since the R_i's are n–currents in $\mathbb{R}^n$, then $R_i \rightharpoonup R$, $R \in \mathcal{R}_n(\mathbb{R}^n)$, and

$$T_i \rightharpoonup T = \partial R$$

By the boundary rectifiability theorem we then conclude $T \in \mathcal{R}_{n-1}(\mathbb{R}^n)$. In other words: *The closure theorem for boundaryless i.m. rectifiable currents of codimension* 1 *is equivalent to the boundary rectifiability theorem for i.m. rectifiable currents of maximal dimension.*

Let us discuss now the proof of Theorem 1. It relies on the important and difficult *Besicovitch-Federer structure theorem*, compare Theorem 3 in Sec. 2.1.4.

The first step consists in proving the following lower density lemma.

Theorem 3 (A lower density lemma). *Let* $T \in \mathcal{D}_n(U)$, $U \subset \mathbb{R}^{n+N}$ *with*

$$\mathbf{M}_V(T) + \mathbf{M}_V(\partial T) < \infty \qquad \forall\, V \subset\subset U\,.$$

Then

(i) *For* $\|T\|$-a.e. x

$$\lim_{r\to 0} \frac{\lambda(x,r)}{\|T\|(B(x,r))} = 1$$

where

$$\lambda(x,r) := \inf\{\mathbf{M}(S) \mid \partial S = \partial(T \llcorner B(x,r)),\ S \in \mathcal{D}_n(U)\}\,.$$

(ii) *If moreover* $\partial T = 0$ *and if* $\partial(T \llcorner B(x,r))$ *is i.m. rectifiable for every* x *and almost every* r, *then there is* $\delta > 0$ *such that*

$$\theta^n_*(\|T\|, x) > \delta$$

for $\|T\|$-a.e. $x \in U$.

Proof. Clearly $\lambda(x,r) \le \|T\|(B(x,r))$, so if (i) is false there is an $\varepsilon > 0$ and a set X such that $\|T\|(X) > 0$ and for each $x \in X$ there exist arbitrarily small balls $B(x,r)$ with

$$\lambda(x,r) < (1-\varepsilon)\,\mathbf{M}(T \llcorner B(x,r))\,.$$

We assume

$$X \subset V \subset\subset U .$$

By Besicovitch's covering theorem, for every $\rho > 0$ there exists a disjoint collection $B_1, B_2 \ldots \subset V$ of such balls covering $\| T \|$-almost all of X and having radii $\leq \rho$. Let $S_i \in \mathcal{D}_n(U)$ be a current such that

$$\partial S_i = \partial(T \llcorner B_i) , \qquad \mathbf{M}(S_i) < (1-\varepsilon)\,\mathbf{M}(T \llcorner B_i)$$

and set

$$T_\rho := T - \sum_i T \llcorner B_i + \sum_i S_i .$$

Then, denoting by x_i the center of the ball B_i, we have

$$\begin{aligned} &(T - T_\rho)(\omega) = \sum_i (T \llcorner B_i - S_i)(\omega) \\ = \;& \sum_i \partial(\delta_{x_i} \mathbin{⨳} (T \llcorner B_i - S_i))(\omega) = \sum_i (\delta_{x_i} \mathbin{⨳} (T \llcorner B_i - S_i))(d\omega) \\ \leq \;& \sum_i \mathbf{M}(O \mathbin{⨳} (T \llcorner B_i - S_i)) \sup |d\omega| \leq \sum_i \rho \mathbf{M}(T \llcorner B_i - S_i) \sup |d\omega| \\ \leq \;& \sum_i 2\rho \mathbf{M}(T \llcorner B_i) \sup |d\omega| \leq 2\rho \mathbf{M}_V(T) \sup |d\omega| , \end{aligned}$$

hence $T_\rho \rightharpoonup T$ as $\rho \to 0$ and

(3) $$\mathbf{M}_V(T) \leq \liminf_{\rho \to 0} \mathbf{M}_V(T_\rho) .$$

On the other hand

$$\begin{aligned} \mathbf{M}_V(T_\rho) &\leq \mathbf{M}_V(T - \sum_i T \llcorner B_i) + \sum_i \mathbf{M}_V(S_i) \\ &\leq \mathbf{M}_V(T - \sum_i T \llcorner B_i) + (1-\varepsilon) \sum_i \mathbf{M}_V(T \llcorner B_i) \\ &= \mathbf{M}_V(T) - \varepsilon \sum_i \mathbf{M}_V(T \llcorner B_i) \\ &\leq \mathbf{M}_V(T) - \varepsilon \| T \|(X) \end{aligned}$$

which contradicts (3).

To prove (ii) consider a point x at which (i) holds, and let

$$f(r) := \mathbf{M}(T \llcorner B(x,r)) .$$

Then for r small, $r < R$,

$$f(r) < 2\lambda(x,r) .$$

By the slicing theory in Sec. 2.2.5, as $\partial T = 0$,

$$\mathbf{M}(\partial(T \llcorner B(x,r))) \leq f'(r) \qquad \text{for a.e. } r$$

and by the isoperimetric inequality

$$\lambda(x,r)^{(n-1)/n} \le c f'(r) \qquad \text{for a.e. } r ,$$

since the slice $\partial(T \llcorner B(x,r))$ is rectifiable. Thus

$$f(r)^{(n-1)/n} \le c f'(r)$$

or

$$\frac{d}{dr} f(r)^{1/n} \ge c' .$$

Therefore

$$f(r) \ge (c' r)^n$$

for $r < R$. □

Next step is the following *rectifiability theorem*

Theorem 4 (Rectifiability theorem). *Let $T \in \mathcal{M}_n(U)$, $U \subset \mathbb{R}^{n+N}$, with $\partial T \in \mathcal{M}_{n-1}(U)$. Set*

$$\widetilde{\mathcal{M}} := \{x \in U \mid \theta^{*n}(\| T \|, x) > 0\}$$

and suppose that

$$\| T \|(U \setminus \widetilde{\mathcal{M}}) = 0$$

i.e. that

$$\theta^{*n}(\| T \|, x) > 0 \qquad \textit{for } \| T \| - \text{a.e. } x \in U .$$

Then T is a real-multiplicity rectifiable current.

Proof. First we show that $\widetilde{\mathcal{M}}$ is a countable union of sets of finite $\mathcal{H}^n$-measure. In fact, setting

$$\widetilde{\mathcal{M}}_k := \{x \mid \theta^{*n}(\| T \|, x) > 1/k\} ,$$

by Theorem 5 in Sec. 1.1.5, we obtain

$$\frac{1}{k} \mathcal{H}^n(\widetilde{\mathcal{M}}_k) \le \| T \|(U) < \infty$$

i.e.

$$\widetilde{\mathcal{M}} = \bigcup_k \widetilde{\mathcal{M}}_k \qquad \mathcal{H}^n(\widetilde{\mathcal{M}}_k) < \infty .$$

Let us prove now that $\xi(x)$ orients $T_x\mathcal{M}$, for H^n-a.e. $x \in \mathcal{M}$. Write

$$\mathcal{M} = \mathcal{M}_0 \cup \bigcup_{j=1}^{\infty} \mathcal{M}_j$$

with $H^n(\mathcal{M}_0) = 0$, $\mathcal{M}_j$ pairwise disjoint, and $\mathcal{M}_j \subset \mathcal{N}_j$, $\mathcal{N}_j$ a C^1-submanifold. From

(4) $$\theta^{*n}(\|T\|, \bigcup_{r\neq j} \mathcal{M}_r, x) = 0, \qquad \theta^{*n}(\|T\|, \mathcal{N}_j \setminus \mathcal{M}_j, x) = 0$$

for H^n–a.e. $x \in \mathcal{M}_j$, writing as usual $\eta_{x,\lambda}(y) := x + \lambda^{-1}(y-x)$ we have for any $\omega \in \mathcal{D}^n(\mathbb{R}^{n+N})$, all $x \in \mathcal{M}_j$ for which (4) holds, and all λ sufficiently small

$$\eta_{x,\lambda\#}T(\omega) = T(\eta^{\#}_{x,\lambda}\omega) = \int_{\mathcal{N}_j} <\xi, \eta^{\#}_{x,\lambda}\omega>\theta\, d\mathcal{H}^n + \varepsilon(\lambda)$$

where $\varepsilon(\lambda) \to 0$ as $\lambda \downarrow 0$, i.e.

$$\eta_{x,\lambda\#}T(\omega) = \int_{\eta_{x,\lambda}(\mathcal{N}_j)} <\xi(x+\lambda z), \omega(z)>\theta(x+\lambda z)\, d\mathcal{H}^n(z) + \varepsilon(\lambda)\ .$$

Since $\mathcal{N}_j$ is of class C^1, this gives

$$\lim_{\lambda\downarrow 0} \eta_{x,\lambda\#}T(\omega) = \theta(x) \int_P <\xi(x), \omega(z)>d\mathcal{H}^n(z)$$

for $\mathcal{H}^n$-a.e. $x \in \mathcal{M}_j$, where

$$P \;=\; T_x\mathcal{N}_j \;=\; T_x\mathcal{M}_j\ .$$

On the other hand from $\mathbf{M}(\partial T) < \infty$ we get

$$|\partial\eta_{x,\lambda\#}T(\omega)| = |\eta_{x,\lambda\#}\partial T(\omega)| = |\partial T(\eta^{\#}_{x,\lambda}\omega)| \le \lambda ||\omega|| \omega_n \theta^{*n}(||\partial T||, x)$$

for H^n-a.e. $x \in \mathcal{M}_j$. Thus for such x

$$\lim_{\lambda\to 0} \mathbf{M}(\partial\eta_{x,\lambda\#}T) = 0.$$

Finally we clearly have

$$\lim_{\lambda\to 0} \mathbf{M}(\eta_{x,\lambda\#}T) \;<\; \infty\ .$$

Thus we conclude that for $\mathcal{H}^n$-a.e. $x \in \mathcal{M}$ we can find $\lambda_i \downarrow 0$ such that

$$\eta_{x,\lambda_i\#}T \rightharpoonup S_x\ , \quad S_x \in \mathcal{D}_n(\mathbb{R}^{n+N})\ , \quad \partial S_x = 0$$

where

(5) $$S_x(\omega) \;=\; \theta(x) \int_{T_x\mathcal{M}} <\xi(x), \omega(z), >d\mathcal{H}^n(z)\ .$$

We finally claim that (5) and $\partial S_x = 0$ imply that $\xi(x)$ orients $T_x\mathcal{M}$. Without loss of generality we can assume

$$T_{x_0}\mathcal{M} \;=\; \mathbb{R}^n \times \{0\} \;\subset\; \mathbb{R}^{n+N}\ .$$

Choose

$$\omega := x^j \varphi(x)\, dx^{i_1} \wedge \ldots \wedge dx^{i_{n-1}} \in \mathcal{D}^{n-1}(\mathbb{R}^{n+N})$$

where $j \geq n+1$, $\varphi \in C_c^\infty(\mathbb{R}^{n+N})$. Since $x^j = 0$ on $\mathbb{R}^n \times \{0\}$ we deduce from (5) and $\partial S_{x_0} = 0$

$$\begin{aligned} 0 &= \partial S_{x_0}(\omega) = S_{x_0}(d\omega) = \\ &= \theta(x_0) \int_{T_{x_0}\mathcal{M}} \varphi(x) < \xi(x_0), dx^j \wedge dx^{i_1} \wedge \ldots \wedge dx^{i_{n-1}} > d\mathcal{H}^n(x) \\ &= \theta(x_0) \int_{T_{x_0}\mathcal{M}} \varphi(x) \xi(x_0) \cdot e_j \wedge e_{i_1} \wedge \ldots \wedge e_{i_{n-1}} \, d\mathcal{H}^n(x) \, . \end{aligned}$$

As φ is arbitrary we then get

$$\xi(x_0) \cdot e_j \wedge e_{i_1} \wedge \ldots \wedge e_{i_{n-1}} = 0 \quad \text{whenever } j \geq n+1 \, .$$

This yields at once

$$\xi = \pm e_1 \wedge \ldots \wedge e_n$$

as required, since $|\xi| = 1$. □

Proof of Theorem 1. By the lower density lemma, Theorem 3, there is $\delta > 0$ such that

$$\theta_*^n(\| T \|, x) \geq \delta \qquad \text{for } \| T \|\text{-a.e. } x \, .$$

Thus if we let

$$\mathcal{M} = \{x \in \mathbb{R}^{n+N} \mid \theta_*^n(\| T \|, x) \geq \delta\}$$

we have

(6) $$\| T \|(\mathbb{R}^{n+N} \setminus \mathcal{M}) = 0 \, ,$$

and from Theorem 5 in Sec. 1.1.5

(7) $$\mathcal{H}^n(\mathcal{M}) \leq \delta^{-1} \| T \|(\mathcal{M}) < \infty \, .$$

On the other hand Theorem 3 in Sec. 4.3.1 and (6) imply that $\| T \|$ is absolutely continuous with respect to $\mathcal{H}^n \llcorner \mathcal{M}$, thus we can write

$$\| T \|(f) := \int_{\mathcal{M}} f \theta \, d\mathcal{H}^n$$

and

(8) $$T(\omega) = \int_{\mathcal{M}} < \omega(x), \vec{T}(x) > \theta(x) \, d\mathcal{H}^n(x).$$

One can now apply the rectifiability theorem Theorem 4 and infer that T is a real multiplicity rectifiable current. Therefore it remains to prove that $\theta(x) \in \mathbb{Z}$ $\mathcal{H}^n$ a. e. $x \in \mathcal{M}$.

With the notations in the proof of Theorem 4 we have

$$S_j := \eta_{x,\lambda_j\#}T \rightharpoonup \theta(x)[\![T_x\mathcal{M}]\!] \qquad \text{for } \mathcal{H}^n\text{-a.e } x \in M$$

where $[\![T_x\mathcal{M}]\!]$ is oriented by $\xi(x)$, and $\mathbf{N}(S_j) < \infty$. Let $d(y) := |y|$, $y \in \mathbb{R}^{n+N}$. Applying the slicing lemma we infer that for a.e. r and for a subsequence S_{n_j}, that we again denote by S_j, we have

$$<S_j, d, r> \rightharpoonup \theta(x)[\![T_x\mathcal{M} \cap \partial B(0,r)]\!]$$

with $\sup_j \mathbf{N}(<S_j, d, r>) < \infty$. Choosing now r in such a way that $<T, |\cdot - x|, r>$ is rectifiable, we infer that all the $<S_j, d, r>$ are rectifiable too and the closure theorem for $(n-1)$-currents then yields that

$$\theta(x)[\![T_x\mathcal{M} \cap \partial B(0,r)]\!]$$

is rectifiable, i.e. $\theta(x) \in \mathbb{Z}$. □

A direct proof of the closure theorem.. We close this section giving an alternative proof of the closure theorem, in fact of Theorem 1, which does not use the structure theorem nor the rectifiability theorem, but the rectifiability criterion for measures Theorem 2 in Sec. 2.1.4. A key point is again the lower density lemma Theorem 3

The proof proceeds by induction, and similarly to the previous proof it is enough to prove the following claim:

Theorem 5. *If $T \in \mathcal{D}_n(\mathbb{R}^{n+N})$, $\partial T = 0$, $M(T) < \infty$, and for every Lipschitz function f, the slice $<T, f, r> = \partial(T \llcorner \{f < r\})$ is i.m. rectifiable, then T is i.m. rectifiable.*

In the proof one can and do freely use the closure theorem for $(n-1)$–currents in $\mathbb{R}^{n+N}$ which is available because of the inductive hypotheses.

Proof of Theorem 5. As in the proof of Theorem 4 we infer using the lower density lemma that if we let

$$\mathcal{M} = \{x \in \mathbb{R}^{n+N} \mid \theta^n_*(\|T\|, x) \geq \delta\}$$

then

(9) $$\|T\|(\mathbb{R}^{n+N} \setminus \mathcal{M}) = 0 \,,$$

(10) $$\mathcal{H}^n(\mathcal{M}) \leq \delta^{-1}\|T\|(\mathcal{M}) < \infty \,.$$

Moreover from Theorem 3 in Sec. 4.3.1 and (9), we then infer that $\|T\|$ is absolutely continuous with respect to $\mathcal{H}^n \llcorner \mathcal{M}$, thus we can write

$$T(\omega) = \int_{\mathcal{M}} <\omega(x), \vec{T}(x)> \theta(x)\, d\mathcal{H}^n(x)$$

or

$$T(\omega) = \int_{\mathcal{M}} <\omega(x), \vec{\tau}(x)> d\mathcal{H}^n(x) , \qquad \vec{\tau}(x) := \theta(x)\vec{T}(x)$$

and we need again to show that $\mathcal{M}$ is rectifiable, for $\mathcal{H}^n$-a.e. $x \in \mathcal{M}$, $\vec{T}(x)$ is a simple vector associated to $T_x\mathcal{M}$, and $\theta(x)$ is an integer. We then prove the claim using the rectifiability criterion of measures, Theorem 2 in Sec. 2.1.4 instead of the Besicovitch-Federer structure theorem.

From general measure theory we know that for $\mathcal{H}^n$ almost every $a \in \mathcal{M}$ we have

$$\theta^{n*}(\mathcal{H}^n, \mathcal{M}, a) \leq 1 \tag{11}$$

$$\lim_{r \to 0} \frac{1}{\mathcal{H}^n(\mathcal{M} \cap B(a,r))} \int_{\mathcal{M} \cap B(a,r)} |\vec{\tau}(x) - \vec{\tau}(a)| \, d\mathcal{H}^n(x) = 0 \tag{12}$$

and we fix one such a point $a \in \mathcal{M}$.

By (12) and the definition of $\mathcal{M}$ we have

$$\theta^n_*(\mathcal{H}^n, \mathcal{M}, a) = |\vec{\tau}(a)|^{-1} \theta^n_*(\| T \|, a) > 0 \tag{13}$$

and

$$\begin{aligned} \limsup_{\lambda \to 0} \mathcal{H}^n \llcorner \eta_{a,\lambda}(\mathcal{M})(B(0,R)) &= \limsup_{\lambda \to 0} \lambda^{-n} \mathcal{H}^n(\mathcal{M} \cap B(a, \lambda R)) \\ &\leq \omega_n R^n \theta^{n*}(\mathcal{H}^n, \mathcal{M}, a) < \infty , \end{aligned}$$

where as usual $\eta_{a,\lambda}(x) := \lambda^{-1}(x-a)$. It easily follows passing to subsequences

$$\mathcal{H}^n \llcorner \eta_{a,\lambda_k}(\mathcal{M}) \rightharpoonup \mu \tag{14}$$

where μ is a Radon measure, and

$$\int <\omega, \vec{\tau}(a)> d\mathcal{H}^n \llcorner \eta_{a,\lambda_k}(\mathcal{M}) \longrightarrow \int <\omega, \vec{\tau}(a)> d\mu =: S_a(\omega)$$

for any n-form $\omega \in \mathcal{D}^n(\mathbb{R}^{n+N})$. Setting

$$T_{\lambda_k} = \eta_{a,\lambda_k \#} T ,$$

from

$$\begin{aligned} \mathbf{M}_{B(0,1)}(T_{\lambda_k} - \vec{\tau}(a)\mathcal{H}^n \llcorner \eta_{a,\lambda_k}(\mathcal{M})) &= \lambda_k^{-n} \mathbf{M}_{B(a,\lambda_k R)}(T - \vec{\tau}(a)\mathcal{H}^n \llcorner \mathcal{M}) \\ &= \lambda_k^{-n} \int_{\mathcal{M} \cap B(a,\lambda_k r)} |\vec{\tau}(x) - \vec{\tau}(a)| \, d\mathcal{H}^n(x) \longrightarrow 0 \end{aligned}$$

we also infer

$$\begin{aligned} T_{\lambda_k} &\rightharpoonup S_a \\ \| T_{\lambda_k} \| &\rightharpoonup \mu \llcorner |\vec{\tau}(a)| = \| S_a \| . \end{aligned} \tag{15}$$

Since for every Lipschitz f, almost all slices $<T_\lambda, f, r>$ are rectifiable, from (15) the slicing lemma, and the $(n-1)$-dimensional case of the closure theorem we infer that almost all slices $<\mu \llcorner \vec{\tau}(a), f, r>$ are i.m. rectifiable. Thus by Theorem 3 we have the lower density bound

$$\theta^n_*(\mu, x) \ \geq \ \frac{\delta}{|\vec{\tau}(a)|} \ > \ 0 \tag{16}$$

for μ-a.e. x.

We now claim that μ *is translation invariant in at least* n*-directions*, and, since the set of point where (16) holds has locally finite $\mathcal{H}^n$-measure, in exactly n-directions. Denoting by V the subspace of $\mathbb{R}^{n+N}$ of vectors v such that T is invariant under translation in the direction of v, we shall in fact prove that $\vec{\tau}(a) \in \Lambda_n V$. This implies that $\dim V \geq n$, and by the previous remark, $\dim V = n$ and $\vec{\tau}(a)$ is simple.

By making an orthogonal change of coordinates, we can assume that V has a basis consisting of vectors of the standard basis of $\mathbb{R}^{n+N}$, $e_1, \dots, e_{n+N}$. Of course, in order to prove the claim, it then suffices to show that if $e_1 \notin V$, then $\vec{\tau}(a)$ has no-component of the form $e_1 \wedge e_{i_2} \wedge \dots \wedge e_{i_n}$. By mollification we can also assume that $\| \mu \llcorner \vec{\tau}(a) \| = \mathcal{H}^{n+N} \llcorner \theta$ for a smooth θ. Then for all $f \in C^\infty_c(\mathbb{R}^{n+N})$

$$\begin{aligned} 0 &= (\partial(\mu \llcorner \vec{\tau}(a))(f\, dx^2 \wedge \dots \wedge dx^n) = (\mu \llcorner \vec{\tau}(a))(\sum_i \frac{\partial f}{\partial x^i}\, dx^i \wedge dx^2 \wedge \dots \wedge dx^n) \\ &= \int \sum_i a_i \frac{\partial f}{\partial x^i} \theta(x)\, dx = \int \sum_i a_i \frac{\partial \theta}{\partial x^i} f \end{aligned}$$

where

$$a_i \ := \ \text{the } e_i \wedge e_2 \wedge \dots \wedge e_n \text{ component of } \vec{\tau}(a) \ .$$

Hence

$$\sum_i a_i \frac{\partial \theta}{\partial x^i} \ = \ 0 \ ,$$

i.e. θ, consequently $\mu \llcorner \vec{\tau}(a)$ is invariant under translation in the direction $\sum a_i e_i$. Since $e_1 \notin V$ we conclude that $a_1 = 0$ which means that $\vec{\tau}(a)$ has no component of the type $e_1 \wedge e_2 \wedge \dots \wedge e_n$. The same argument can be used to prove that $\vec{\tau}(a)$ has no component of the type $e_1 \wedge e_{i_2} \wedge \dots \wedge e_{i_n}$.

Returning to the proof of the theorem, we then find a collection $P_1, \dots, P_p$ of n-planes parallel to the n-dimensional subspace determined by $\vec{\tau}(a)$ such that

$$\mu \ = \ \sum_{j=1}^p \alpha_j (\mathcal{H}^n \llcorner P_j) \ .$$

By (11) and (14)

$$\sum_{j=1}^p \alpha_i \ \leq \ \theta^{n*}(\mathcal{H}^n, \mathcal{M}, a) \ \leq \ 1$$

and by the lower bound (16) $\alpha_j \ge \delta/|\vec{\tau}(a)|$, so p must be finite.

We now claim $p = 1$. Suppose not and let

$$3\varepsilon \;<\; \min\{\operatorname{dist}(P_i, P_j) \;\mid\; P_i \neq P_j\}\;.$$

Fix j and let $f(x) := \operatorname{dist}(x, P_j)$. Also assume without loss of generality that P_j is parallel to $\mathbb{R}^n \times \{0\} \subset \mathbb{R}^{n+N}$ and set $U := B(0,R) \subset \mathbb{R}^n$ and $W := U \times \mathbb{R}^N \subset \mathbb{R}^{n+N}$. By the second of (15)

$$\lim_{k\to\infty} \mathbf{M}_W(T_{\lambda_k} \llcorner \{\varepsilon < f < 2\varepsilon\}) \le |\vec{\tau}(a)|\mu(\overline{W} \cap \{\varepsilon \le f \le 2\varepsilon\})$$

so by the slicing lemma we can find a further subsequence and $t \in (\varepsilon, 2\varepsilon)$ such that

(17) $$T_{\lambda_k} \llcorner \{f < t\} \rightharpoonup \vec{\tau}(a)\mu \llcorner \{f < t\} = \alpha_j(\mathcal{H}^n \llcorner P_j)\vec{\tau}(a) = \alpha_j|\vec{\tau}(a)|\,[\![\, P_j \,]\!]$$

and such that

$$\mathbf{M}_W(\partial(T_{\lambda_k} \llcorner \{f < t\})) \;\longrightarrow\; 0\;.$$

In particular

(18) $$\mathbf{M}_U(\partial\pi_\#(T_{\lambda_k} \llcorner \{f < t\})) \;\longrightarrow\; 0\;.$$

Also $\pi_\# T_{\lambda_k} \llcorner \{f < t\}$ is an n-dimensional current in $\mathbb{R}^n$ with bounded mass and boundary mass, hence it is represented by a BV function u_k

$$\pi_\# T_{\lambda_k} \llcorner \{f < t\}(\omega) \;=\; \int_U u_k\omega$$

with

$$\mathbf{M}_U(\partial\pi_\#(T_{\lambda_k} \llcorner \{f < t\})) \;=\; \int_U |Du_k|.$$

By Poincaré inequality for suitable β_k we infer

$$\int_U |u_k - \beta_k|\,dx \;\le\; c\int_U |Du_k|.$$

From (17) (18) it follows that

$$\beta_k \;\to\; \beta \;=\; \alpha_j|\vec{\tau}(a)|, \qquad \int_U |u_k - \beta|\,dx \;\to\; 0\;.$$

But

$$\beta_k\,\mathcal{L}^n(\{\,x \in U \;\mid\; u_k(x) = 0\,\}) \;\le\; \int_U |u_k - \beta_k|\,dx \;\le\; c\int_U |Du_k|$$

so

$$\mathcal{L}^n(\{x \in U \mid u_k(x) = 0\}) \longrightarrow 0$$

Thus using also (14)

$$\begin{aligned}
\mathcal{L}^n(U) &= \lim_{k\to\infty} \mathcal{L}^n(\{x \in U \mid u_k(x) \neq 0\}) \\
&= \lim_{k\to\infty} \mathcal{L}^n(\pi(\eta_{a,\lambda_k}(\mathcal{M}) \cap \{f < t\}) \cap U) \\
&\leq \lim_{k\to\infty} \mathcal{H}^n \llcorner \eta_{a,\lambda_k}(\mathcal{M})(\{f < t\} \cap W) \\
&= \mu(\{f < t\} \cap W) \;=\; \alpha_j \mathcal{H}^n(U)\,.
\end{aligned}$$

It follows that each $\alpha_j \geq 1$, and by (11), (13) and (14)

$$0 < \theta^n_*(\mathcal{H}^n, \mathcal{M}, a) \leq \sum_j \alpha_j \leq \theta^{n*}(\mathcal{H}^n, \mathcal{M}, a) \leq 1\,.$$

Hence $p = 1$ and $\alpha_1 = 1$.

Furthermore P_1 must pass through the origin since for every $r > 0$

$$\begin{aligned}
\mu(B(0,r)) &\geq \liminf_{k\to\infty} \mathcal{H}^n \llcorner \eta_{a,\lambda_k}(\mathcal{M})(B(0,r)) \\
&= \liminf_{\lambda\to 0} \lambda^{-n} \mathcal{H}^n(\mathcal{M} \cap B(a, \lambda r)) \\
&\geq r^{-n} \theta^n_*(\| T \|, a) \;>\; 0
\end{aligned}$$

by (13). Finally note that P_1 is determined by $\vec{\tau}(a)$ and therefore does not depend on particular subsequences.

This shows that $\mathcal{M}$ has an approximate tangent plane for $\mathcal{H}^n$-a.e. $a \in \mathcal{M}$. Hence $\mathcal{M}$ is n-rectifiable by Theorem 2 in Sec. 2.1.4. As in the previous proof, see the proof of Theorem 1, or observing that by the $(n-1)$-dimensional closure theorem $\mu \llcorner \vec{\tau}(a)$ has integer multiplicity slices and $\mu = \mathcal{H}^n \llcorner P_1$, we finally get that $\theta(a) = |\vec{\tau}(a)|$ is an integer.

□

3 Notes

1 The classical references for *geometric measure theory* are Federer's treatise [226], and Simon's book [592]. In our presentation we mostly followed Simon [592] and its terminology and notation, which slightly differ from those of Federer [226]. The reader will surely find also interesting consulting Morgan [485] Hardt and Simon [357]. We also mention Whitney [674], the foundational paper of Federer and Fleming [230], and De Rham [189].

2 The rectifiability theorem for Radon measures Theorem 2 in Sec. 2.1.4 is essentially due to De Giorgi [177]. In our presentation we have follow Simon [592]. The structure theorem is due to Besicovitch [83] for $k = 1$ and to Federer [223]. A presentation of Besicovitch's original proof can be found in Falconer [218], the general case is discussed in Federer [226].

Concerning rectifiable sets, we would like to mention the following interesting criterion, see e.g. Alberti, Ambrosio, and Cannarsa [8] and Simon [596]. Given any set $S \subset \mathbb{R}^n$ and any point $x \in \mathbb{R}^n$, define

$$T(S,x) := \{rp \mid r > 0,\ p = \lim_{h\to\infty} \frac{x - x_h}{|\, x - x_h\,|},\ x_h \in S \setminus \{x\},\ x_h \to x\}$$

and the *tangent space of* S *at* x, $\operatorname{Tan}(S,x)$ as the smallest subspace of $\mathbb{R}^n$ which contains $T(S,x)$. We then have

Theorem 1. *If*

$$\dim \operatorname{Tan}(S,x) \ \leq\ k \qquad \forall\, x \in S\ ,$$

then S *is* $\mathcal{H}^k$*-measurable and* $\mathcal{H}^k$*-countably rectifiable.*

3 The content of Sec. 2.2.1 and Sec. 2.2.2 is quite standard. The reader may consult e.g. Federer [226], and for an elementary presentation Fleming [238]. In Sec. 2.2.3, Sec. 2.2.4 we presented basic definitions and facts, as well as some simple examples. The reader is referred to the books quoted above. There one also finds the proof of the general slicing theorem, of the deformation theorem, and of its consequences. For the isoperimetric inequality the reader is referred to Almgren [22], where the optimal inequality in any codimension is proved, and to Almgren [21]. For further information on isoperimetric inequalities the reader is referred to Burago and Zalgaller [117] and to its bibliography.

4 The closure theorem was first proved for codimension one currents, i.e. $(n-1)$–dimensional currents in $\mathbb{R}^n$ by De Giorgi [177]. We shall present it in Ch. 4 in the context of Caccioppoli sets. The ideas introduced in De Giorgi [177] play an important role in the general case. The first proof of the closure theorem in Sec. 2.2.7 is essentially the classical Federer–Fleming proof in Federer and Fleming [230] as modified by Simon [592], except for one point where we have used the lower density lemma, Theorem 3 in Sec. 2.2.7. Theorem 3 in Sec. 2.2.7 is taken from White [669], compare also Fleming [240] and Solomon [604]. The second proof, which develops in the same spirit as De Giorgi's proof in codimension 1, is due to White [669].

3. Cartesian Maps

In this chapter, we shall be concerned with the notions of *graph*, *tangent plane to a graph*, and *weak convergence of graphs* for non–smooth maps from $\mathbb{R}^n$ into $\mathbb{R}^N$, n, $N \geq 1$.

We begin in Sec. 3.1 by investigating the differentiability properties of Sobolev maps. As it is well known in this context the basic notion, at least for scalar maps, is that of *distributional derivative*, and Sobolev maps do not possess smoothness properties in the usual sense. In particular Sobolev maps are not even almost everywhere differentiable in the classical sense, except in the case that they have gradient summable with a power larger than the dimension n. However a *theorem of Calderón and Zygmund* states that they are almost everywhere differentiable in the so–called L^p-sense, compare Sec. 3.1.2.

Essentially as consequence of this result we shall infer, compare Sec. 3.1.3, that Sobolev maps, $u \in W^{1,p}(\Omega, \mathbb{R}^N)$, $\Omega \subset \mathbb{R}^n$, $p \geq 1$ possess the following *Lusin type properties*

(i) there is a denumerable disjoint family of measurable sets $\{F_k\}$, such that $u_{|F_k}$ is Lipschitz continuous and meas $(\Omega \setminus \cup_k F_k) = 0$.
(ii) except on arbitrarily small open sets, u is of class C^1.

It turns out, compare Sec. 3.1.4, that those Lusin type properties are equivalent to a sort of measure–theoretic pointwise differentiability notion called *approximate differentiability almost everywhere.* Approximate differentiability provides the right way to express local approximability by linear transformations and, just using the equivalence with the Lusin type properties mentioned above, the classical formulas for area and coarea and for the degree can be extended to almost everywhere approximately differentiable maps. All that may be regarded as a special case of the *rectifiability theory* of Besicovitch, De Giorgi, Federer and Fleming, we have partly described in Ch. 2.

For us the important outcoming is that the *graph of a map* which is almost everywhere approximately differentiable is *well defined*, and is a *countably n-rectifiable set.* Our definition of graph, $\mathcal{G}_{u,\Omega}$ in Sec. 3.1.5, more than to the usual notion of graph of a continuous function defined in Ω, is modeled after a kind of 1-*graph* in which each points $z \in \mathcal{G}_{u,\Omega}$ carries the tangent plane to $\mathcal{G}_{u,\Omega}$ at $z = (x, u(x))$. An important consequence of such a definition is *Lusin's property* (N), i.e., that null sets in Ω are mapped into $\mathcal{H}^n$–null sets in $\Omega \times \mathbb{R}^N$, a property which is typical of Lipschitz maps, but which does not hold in general for continuous maps.

This way we see that the graph and its tangent planes are not really related to the distributional derivative, but to the in some cases weaker and in general *different* notion of almost everywhere approximate differentiability of the map, which in fact will be the starting point for us in this chapter. As a simple example the real map $x/|x| : \mathbb{R} \to \mathbb{R}$ has a rectifiable graph with horizontal tangent plane everywhere except at zero, corresponding to the naive, and correct in the sense of approximate differentials, idea that it has zero (approximate) differential everywhere except at zero. A more complicate example which shows a lot of *fractures* in the graph is *Cantor–Vitali function* for which the approximate differential is zero almost everywhere and the graph in the previous sense consists of infinitely many horizontal segments, compare next chapter.

Usually the approximate differential is denoted by apDu. As here it is our main notion of derivative, in the sequel we shall simply write Du for apDu if no confusion with the distributional derivative may arise.

Approximately differentiable maps provide essentially the natural and correct class of maps whose graphs are $\mathcal{H}^n$–measurable and countably n-rectifiable sets in the product $\Omega \times \mathbb{R}^N$. Their tangent planes, at points of approximate differentiability, are described by the *minors* of the Jacobian matrix in terms of the approximate differential exactly as in the classical case. Consequently the n-dimensional area of the graph of an almost everywhere approximately differentiable map is finite if and only if all the minors of the Jacobian matrix are summable.

In Sec. 3.2 we introduce and discuss the class of approximately differentiable maps with summable Jacobian minors, denoted by $\mathcal{A}^1(\Omega, \mathbb{R}^N)$. Each map $u \in \mathcal{A}^1(\Omega, \mathbb{R}^N)$ turns out to be identified completely by the *integer multiplicity rectifiable n-dimensional current* G_u defined as integration of compactly supported differentiable n-forms in $\Omega \times \mathbb{R}^N$ over the graph $\mathcal{G}_{u,\Omega}$ of u in Ω naturally oriented by the independent variables.

Being interested in weak limits of sequences $\{u_k\}$, there are at least three important elements which turn out to be relevant for a satisfactory geometric description of such weak limits, compare Ch. 1 and Ch. 2.

(i) *the tangent plane*, that is the existence almost everywhere of the tangent plane to the graphs of the $\{u_k\}$'s or equivalently the *rectifiability* of the graphs of the $\{u_k\}$'s.
(ii) *the area of the graph* or better the equiboundedness of the areas of the approximating graphs $\{u_k\}$. In fact limits of graphs may be measures in the product $\mathbb{R}^n \times \mathbb{R}^N$ already in the case $n = N = 1$, if they do not have equibounded masses.
(iii) *the boundary in $\Omega \times \mathbb{R}^N$*. The rectifiability of the graphs and the equiboundedness of of their areas are not sufficient to provide a good geometric description of limits of smooth graphs. In fact such limits may again be measures

distributed in the product $\Omega \times \mathbb{R}^N$, if the control of the size of boundaries of the approximating sequence is lost.

With respect to the first two points, the class $\mathcal{A}^1(\Omega, \mathbb{R}^N)$ yields a satisfactory setting. In the sequel of Sec. 3.2 we discuss the *boundary* question.

In the scalar case, $N = 1$, the summability of the distributional derivative turns out to be equivalent to the coincidence of the approximate differential and the distributional derivative and to the non existence of boundaries of the current G_u inside the cylinder $\Omega \times \mathbb{R}$

$$\partial G_u \llcorner \Omega \times \mathbb{R} \;=\; 0\,.$$

This is no longer true in the vector valued case, $N > 1$. It will be true for maps $u \in W^{1,\bar{n}}(\Omega, \mathbb{R}^N)$, with $\bar{n} = \min(n, N)$, that

$$\partial G_u \llcorner \Omega \times \mathbb{R}^N \;=\; 0\,. \tag{1}$$

However (1) does not hold in general for maps $u \in W^{1,p}(\Omega, \mathbb{R}^N)$ if $p < \bar{n}$. For instance the map $x/|x|$ from the unit ball B^3 of $\mathbb{R}^3$ into $\mathbb{R}^3$ belongs to $W^{1,p}$ for all $p < 3$; as $\det D(x/|x|) = 0$, its graph has finite area, but one has

$$\partial G_{\frac{x}{|x|}} \llcorner B^3 \times \mathbb{R}^3 \;=\; -\delta_0 \times \partial[\![B^3]\!]$$

Condition (1) is obviously satisfied by smooth maps, and is also closed with respect to the weak convergence of graphs. In fact (1) can be equivalently stated as

$$G_u(d\omega) \;=\; 0$$

for all $(n-1)$–forms ω supported in $\Omega \times \mathbb{R}^N$. Thus it is natural to require (1) also for reasonable *weak maps*. This way we are naturally led to introduce the class of *Cartesian maps* as

$$\operatorname{cart}^1(\Omega, \mathbb{R}^N) \;=\; \{u \in \mathcal{A}^1(\Omega, \mathbb{R}^N) \,|\, \partial G_u \llcorner \Omega \times \mathbb{R}^N = 0\}$$

In Vol. II Ch. 2 we shall see that (1) expresses in the context of finite elasticity the physical requirement that a perfectly elastic body does not fracture if we deform it. In Sec. 3.2.3 of this chapter we shall see that condition (1) amounts not only to the integration by parts formula for the derivatives (or equivalently to the existence in L^1 of the distributional derivatives) but to all formulas of integration by parts, actually in the product space $\Omega \times \mathbb{R}^N$, for all Jacobian minors, corresponding to the classical differential identities which state that all minors are null Lagrangians or divergence free fields. In Sec. 3.2.5 we shall then briefly compare traces in the sense of Sobolev spaces and boundaries.

A few significant examples will be discussed in Sec. 3.2.2. In Sec. 3.2.4 we shall introduce special subclasses, denoted $\mathcal{A}_{p,q}$, $p \geq n-1$, $q \geq n/(n-1)$ whose elements are Cartesian maps and discuss higher integrability of the Jacobian determinant.

Having always in mind the question of identifying weak limits of smooth maps, we shall discuss in Sec. 3.3 the *weak continuity of the minors*. More precisely we ask under which conditions it happens that for a sequence $\{u_k\}$ in $\mathcal{A}^1(\Omega, \mathbb{R}^N)$ such that

$$\begin{aligned} u_k &\to u \quad \text{strongly in } L^1 \\ Du_k &\rightharpoonup v \quad \text{weakly in } L^1 \\ M_{\bar{\alpha}}^{\beta}(Du_k) &\rightharpoonup v_{\bar{\alpha}}^{\beta} \quad \text{weakly in } L^1 \end{aligned}$$

we have

$$v = Du , \qquad v_{\bar{\alpha}}^{\beta} = M_{\bar{\alpha}}^{\beta}(Du) .$$

After discussing a few examples, we shall show that essentially this happens if $\partial G_{u_k} \llcorner \Omega \times \mathbb{R}^N = 0$, i.e. if the $\{u_k\}$'s are Cartesian maps. For the class $\mathrm{cart}^1(\Omega, \mathbb{R}^N)$ we shall provide in Sec. 3.3.2 a *closure theorem* which essentially relies on Federer-Fleming closure theorem for integer multiplicity rectifiable currents. This way we are able to formulate quite general results concerning the weak continuity of minors which extend results as Reshetnyak's theorem on the weak continuity of determinants in the Sobolev setting.

Of course $\mathrm{cart}^1(\Omega, \mathbb{R}^N)$ misses nice compactness properties, it just inherits the compactness property of the L^1 space. This makes that class not suitable for the calculus of variations, and in fact we postpone to next chapter introducing and discussing the important class of *Cartesian currents*. Better compactness properties are of course enjoyed by the classes

$$\mathrm{cart}^p(\Omega, \mathbb{R}^N) := \{u \in \mathcal{A}^p(\Omega, \mathbb{R}^N) \mid \partial G_u \llcorner \Omega \times \mathbb{R}^N = 0\}$$

when p is strictly larger than 1.This will be discussed in Sec. 3.3.3.

Starting from a sequence of smooth maps $\{u_k\}$, for which we clearly have

$$\partial G_{u_k} \llcorner \Omega \times \mathbb{R}^N = 0 , \tag{2}$$

by Stokes's theorem, and assuming for instance that all Jacobian minors of $\{u_k\}$ are equibounded in some L^p-norm with $p > 1$, we may then infer passing to a subsequence that

$$G_{u_k} \rightharpoonup G_u$$

where $u \in \mathrm{cart}^p(\Omega, \mathbb{R}^N)$. Therefore the non validity of (1) is a *homological obstruction* for the approximability of a map in $\mathcal{A}^p(\Omega, \mathbb{R}^N)$. However it turns out that this is not the only obstruction. There are also homotopic obstructions to the approximability of maps in $\mathrm{cart}^p(\Omega, \mathbb{R}^N)$. This follows from a result concerning the strong and the weak approximability of minors in L^p. In this respect the known results seem to be still at an initial stage and better understanding seems to be needed. We refer the reader to Sec. 3.4 for precise statements and open questions.

1 Differentiability of Non Smooth Functions

In this section we shall investigate pointwise differentiability properties of functions in Sobolev spaces.

Let Ω be a domain in $\mathbb{R}^n$, and let p be a real number in $[1, \infty)$. The *Sobolev space* $W^{1,p}(\Omega, \mathbb{R})$, simply $W^{1,p}(\Omega)$, we recall, can be defined as the class of all L^p-functions in Ω with derivatives in the sense of distributions in $L^p(\Omega)$

$$W^{1,p}(\Omega) := \{u \in L^p(\Omega) \mid Du \in L^p(\Omega, \mathbb{R}^n)\} \tag{1}$$

normed by

$$\| u \|_{W^{1,p}(\Omega)} := \left(\int_\Omega |u|^p \, dx \right)^{1/p} + \left(\int_\Omega |Du|^p \, dx \right)^{1/p} . \tag{2}$$

It turns out that the linear space $W^{1,p}(\Omega)$ is complete, i.e., is a Banach space with respect to the norm in (2), and even a Hilbert space for $p = 2$, with scalar product

$$(u, v)_{W^{1,2}(\Omega)} := \int_\Omega u \, v \, dx + \int_\Omega Du \, Dv \, dx . \tag{3}$$

One also shows that $W^{1,p}(\Omega)$ can be characterized as the closure of smooth functions in $W^{1,p}$, i.e., as the closure in $W^{1,p}(\Omega)$ of

$$\{u \mid u \in C^1(\Omega) \cap W^{1,p}(\Omega)\} .$$

In general one interprets u in $W^{1,p}$ as a class of equivalence of functions in L^p; in this section we shall instead work with a suitable *representative* of u and we are interested in its pointwise behaviour. Of course we cannot expect that Sobolev functions possess smoothness properties in the usual classical sense, but we shall show that a Sobolev function u is almost everywhere differentiable in the *L^p-sense*, which in turn implies a *Lusin type property*, that is Sobolev functions are of class C^1 except an arbitrarily small open sets. Of course the situation is much easier for $p > n$. In this case it turns out that u is "almost everywhere differentiable" in the classical sense, and in fact we can find a representative of u which is continuous in Ω and almost everywhere differentiable: this is the content of the classical *Morrey-Sobolev theorem*. From our point of view the important outcoming of this will be the possibility of defining the *graph* $\mathcal{G}_{u,\Omega}$ of a Sobolev function $u \in W^{1,p}(\Omega, \mathbb{R}^N)$ and of showing that $\mathcal{G}_{u,A}$ is a *countably n-rectifiable set* for all $p \geq 1$. Actually both Lusin type property and the rectifiability of $\mathcal{G}_{u,\Omega}$ turn out to be equivalent to the almost everywhere *approximate differentiability*, Sec. 2.1.4. In fact the notion of approximate differentiability, which is no more related to the distributional derivatives, is the right notion in order to define the graphs of a map and its tangent planes and will be one of the basic notions in this book.

In Sec. 3.1.1 we present a few relevant facts about *Hardy-Littlewood maximal operator* which will be used later. As a consequence we shall then prove *Lebesgue's theorem* about the convergence of averages of a function on balls and an estimate of the Hausdorff dimension of the set of points which are not *Lebesgue points* of Sobolev functions. We conclude Sec. 3.1.1 proving the classical *Calderón-Zygmund decomposition argument* and giving a characterization of *Zygmund class* $L \log L$.

In Sec. 3.1.2 we shall prove the classical *Calderón-Zygmund* L^p*-differentiability theorem*, some of its consequences, and *Morrey-Sobolev theorem.*

In Sec. 3.1.3, after recalling the classical *Whitney extension theorem* and *Rademacher theorem*, we shall present *Lusin type properties* for Sobolev functions.

In Sec. 3.1.4 we shall discuss the notions of *approximate limits, approximate differentials*, and we shall prove equivalence of approximate differentiability and Lusin type properties.

In Sec. 3.1.5, after discussing the validity of the *area formula*, we shall introduce the notions of *graph* and *tangent plane* for vector valued maps which are almost everywhere approximately differentiable and we shall see that the *area formula* holds for such maps. This allows us to define the *degree* for maps with Jacobian determinant in L^1. However we shall see that in such a general case the degree loses homological invariance and it is not stable by weak convergence.

1.1 The Maximal Function and Lebesgue's Differentiation Theorem

In order to study the differentiability properties of Sobolev functions it turns out to be very useful to work with *Hardy-Littlewood maximal operator* which is defined, for every $u \in L^1_{\text{loc}}(\mathbb{R}^n)$ and for all $x \in \mathbb{R}^n$ as

$$M(u)(x) = \sup_{r>0} \fint_{B(x,r)} |u(y)| \, dy \,, \tag{1}$$

where

$$\fint_A f(y)\, dy := \frac{1}{|A|} \int_A f(y)\, dy \,.$$

We shall now prove a few properties of $M(u)$ which will be used in the sequel.

Given a measurable function $u(x)$ in $\mathbb{R}^n$, the function

$$\lambda(t) := \text{meas}\, \{x \in \mathbb{R}^n \mid |u(x)| > t\} \qquad t \in \mathbb{R}_+ \tag{2}$$

is called the *distribution function of* u, and, if $u \in L^p(\mathbb{R}^n), p \geq 1$, one has

$$\int_{\mathbb{R}^n} |u|^p \, dx = p \int_0^\infty t^{p-1} \lambda(t)\, dt \,.$$

Moreover, for any $u \in L^1(\mathbb{R}^n)$ we have

$$\lambda(t) \leq \frac{A}{t} \tag{3}$$

where $A = \| u \|_{L^1}$. Estimate (3) is usually referred as a 1-1-*weak estimate* for u. In fact (3) is weaker than the summability of u, as it is shown for instance by the extension to zero of the function $1/x$ in $(0,1)$ which trivially satisfies (3) but is not in L^1.

One can easily see that the maximal function $M(u)$ of a non zero $u \in L^1(\mathbb{R}^n)$ is not in $L^1(\mathbb{R}^n)$, in other words *the operator* $u \to M(u)$ *is not bounded in* $L^1(\mathbb{R}^n)$. Indeed if $|u| > 0$ on a set of positive measure, then $M(u)$ decreases at infinity no faster than $c\,|x|^{-n}$. However $M(u)$ verifies a *weak* (1-1)-*estimate* which makes it very useful.

Proposition 1 (Maximal theorem). *Let* $u \in L^p(\mathbb{R}^n)$, *with* $p \geq 1$. *Then for any* $t > 0$

$$\operatorname{meas}\{x \in \mathbb{R}^n \mid M(u)(x) > t\} \leq \frac{6^n}{t} \int_{|u|>t/2} |u(y)|\,dy\,. \tag{4}$$

In particular $M(u)(x)$ *is finite almost everywhere.*

Proof. Let $u \in L^1(\mathbb{R}^n)$ and let

$$A_t := \{x \in \mathbb{R}^n \mid M(u)(x) > t\}\,.$$

For every $x \in A_t$ there exists a ball $B(x, r(x))$ such that

$$\fint_{B(x,r(x))} |u(y)|\,dy > t$$

or equivalently

$$\operatorname{meas} B(x, r(x)) < \frac{1}{t} \int_{B(x,r(x))} |u(y)|\,dy\,.$$

Using Besicovitch's covering theorem we can choose a denumerable set of points $\{x_i\}$ in A_t such that $\cup_i B(x_i, r(x_i)) \supset A$ and the balls $B(x_i, r(x_i))$ do not overlap more than 3^n times. Thus

$$\begin{aligned} \operatorname{meas} A_t &\leq \sum_i \operatorname{meas} B(x_i, r(x_i)) \\ &\leq \frac{1}{t} \sum_i \int_{B(x_i,r(x_i))} |u(y)|\,dy \leq \frac{3^n}{t} \int_{\mathbb{R}^n} |u(y)|\,dy\,. \end{aligned} \tag{5}$$

Suppose now that $u \in L^p(\mathbb{R}^n)$, with $p \geq 1$. The function

$$\tilde{u}(x) := \begin{cases} u(x) & \text{if } |u(x)| > t/2 \\ 0 & \text{if } |u(x)| \leq t/2 \end{cases}$$

is easily seen to be in $L^1(\mathbb{R}^n)$. Moreover

$$M(u)(x) \leq M(\tilde{u})(x) + \frac{t}{2} \qquad \forall\, x \in \mathbb{R}^n ,$$

hence

$$\{x \in \mathbb{R}^n \mid M(u)(x) > t\} \subset \{x \in \mathbb{R}^n \mid M(\tilde{u})(x) > t/2\} .$$

Inequality (5) applied to $\tilde{u}$ then yields

$$\begin{aligned}&\operatorname{meas}\{x \in \mathbb{R}^n \mid M(u)(x) > t\}\\ &\quad\leq \frac{2\cdot 3^n}{t}\int_{\mathbb{R}^n} |\tilde{u}(y)|\,dy = \frac{2\cdot 3^n}{t}\int_{|u|>t/2} |u(y)|\,dy .\end{aligned}$$

The second part of the theorem follows easily from (1). □

Remark 1. If $u \in L^p(\mathbb{R}^n)$, $p > 1$, Hölder inequality trivially yields

$$M(u)(x) \leq [M(u^p)(x)]^{1/p} .$$

Therefore we find

$$\begin{aligned}\operatorname{meas}\{x \in \mathbb{R}^n \mid M(u)(x) > t\} &\leq \operatorname{meas}\{x \in \mathbb{R}^n \mid M(u^p)(x) > t^p\}\\ &\leq \frac{6^n}{t^p}\int_{|u|^p>t^p/2} |u|^p\,dy \leq \frac{6^n}{t^p}\int_{|u|>t/2} |u|^p\,dy ,\end{aligned}$$

i.e., the following $(p$-$p)$–weak estimate

(6) $$\operatorname{meas}\{x \in \mathbb{R}^n \mid M(u)(x) > t\} \leq \frac{6^n}{t^p}\int_{|u|>t/2} |u|^p\,dy .$$

Actually we have

Proposition 2. *Suppose* $1 < p \leq +\infty$ *and* $u \in L^p$. *Then* $M(u) \in L^p$ *and*

(7) $$||M(u)||_{L^p(\mathbb{R}^n)} \leq A(p)\, ||u||_{L^p(\mathbb{R}^n)}$$

where $A(p)$ *is a constant which depends only on* p *and on* n, $A(p) \to \infty$ *as* $p \to 1$.

Proof. Using Proposition 1 we infer

$$\begin{aligned}\int_{\mathbb{R}^n} |M(u)|^p\,dx &= p\int_0^\infty t^{p-1}\operatorname{meas}\{\,x \mid M(u)(x) > t\,\}\,dt\\ &= 6^n p\int_0^\infty t^{p-2}\Big(\int_{|u|>t/2} |u(x)|\,dx\Big)\,dt = 6^n p\int_{\mathbb{R}^n}\Big(\int_0^{2|u(x)|} t^{p-2}dt\Big)|u(x)|\,dx\\ &= 6^n 2^{p-1}\,\frac{p}{p-1}\int_{\mathbb{R}^n} |u(x)|^p\,dx\end{aligned}$$

□

Remark 2. In fact we also have $u \in L^p$ if $M(u) \in L^p$, $p \geq 1$, as consequence of Lebesgue's differentiation theorem below.

Proposition 1 is a useful tool in many questions which we are not going to touch. For our purposes, we only point out a trivial consequence of (4) contained in the following proposition, and we derive the fundamental *differentiation theorem of Lebesgue.*

Proposition 3. *Let $u_k, u \in L^p(\mathbb{R}^n)$, with $p \geq 1$.*

(i) If $u_k \to u$ in L^p, then $M(|u_k - u|^p) \to 0$ in measure, i.e., for any $\varepsilon > 0$

$$\text{meas}\,\{x \in \mathbb{R}^n \mid M(|u_k - u|^p)(x) > \varepsilon\} \longrightarrow 0 \quad \textit{as} \quad k \to \infty \,.$$

(ii) We have

$$t^p \text{meas}\,\{x \in \mathbb{R}^n \mid M(u)(x) > t\} \longrightarrow 0 \quad \textit{as} \quad t \to \infty \,.$$

Proof. The weak estimate (6) yields

$$\text{meas}\,\{x \in \mathbb{R}^n \mid M(|u_k - u|^p)(x) > \varepsilon\} \leq \frac{6^n}{\varepsilon^p} \| u_k - u \|^p_{L^p} \,,$$

thus (i) follows; while (6) and the absolute continuity of the integral yield (ii). □

Theorem 1 (Lebesgue's differentiation theorem). *Let $u \in L^p(\mathbb{R}^n)$, $p \geq 1$. Then for almost every x we have*

$$|u(x)| \leq M(u)(x) \tag{8}$$

and

$$\lim_{r \to 0} \fint_{B(x,r)} |u(y) - u(x)|^p \, dy = 0 \,; \tag{9}$$

in particular

$$\lim_{r \to 0} \fint_{B(x,r)} u(y) \, dy = u(x) \tag{10}$$

holds for almost every x.

Proof. Let $\{u_k\}$ be a sequence in $C^\infty_c(\mathbb{R}^n)$ which converges to u in $L^p(\mathbb{R}^n)$. Passing to a subsequence we can assume that

$$u_k(x) \longrightarrow u(x) \qquad \text{for a.e. } x \tag{11}$$

and, since $M(|u_k - u|^p)(x) \to 0$ in measure, that

(12) $$M(|u_k - u|^p)(x) \longrightarrow 0 \qquad \text{for a.e. } x\ .$$

Denote by E the set where (11) and (12) hold

(13) $$E := \{x \in \mathbb{R}^n \mid (11) \text{ and } (12) \text{ hold}\}\ .$$

Clearly meas $(\mathbb{R}^n \setminus E) = 0$. Let us prove that (8) holds for all $x \in E$.

Since u_k is smooth, we have

$$u_k(x) = \lim_{\rho \to 0} \fint_{B(x,\rho)} u_k(y)\, dy$$

thus

$$|u_k(x)| \le \lim_{\rho \to 0} \fint_{B(x,\rho)} |u_k(y)|\, dy \le M(u_k)(x)\ ;$$

from this inequality (8) follows taking into account (11) and (12).

In order to prove (9) define now

$$V_r^p(u)(x) := \sup_{\rho \le r} \fint_{B(x,\rho)} |u(y) - u(x)|^p\, dy\ .$$

For $\rho \le r$ we have

$$\begin{aligned}
&\left(\fint_{B(x,\rho)} |u(y) - u(x)|^p\, dy \right)^{1/p} \\
&\le \operatorname{osc}_{B(x,\rho)} u_k + \left\{ \fint_{B(x,\rho)} [(u(y) - u_k(y)) - (u(x) - u_k(x))]^p\, dy \right\}^{1/p} \\
&\le \operatorname{osc}_{B(x,\rho)} u_k + [M(|u - u_k|^p)(x)]^{1/p} + |u(x) - u_k(x)|\ .
\end{aligned}$$

Thus, taking into account (8) for $x \in E$, we find that for all k and $x \in E$

(14) $$(V_r^p(u)(x))^{1/p} \le \operatorname{osc}_{B(x,r)} u_k + 2[M(|u - u_k|^p)(x)]^{1/p}\ .$$

Now for every k we choose δ_k so that $\operatorname{osc}_{B(x,\delta_k)} u_k < 2^{-k}$ uniformly in x, we then deduce from (14) that for any $x \in E$

$$\lim_{k \to \infty} \left(V_{\delta_k}^p(u)(x) \right)^{1/p} \le \lim_{k \to \infty} 2^{-k} + 2 \lim_{k \to \infty} [M(|u - u_k|^p)(x)]^{1/p} = 0$$

and this implies (9).

Finally (10) is a consequence of (9). □

Remark 3. For future uses, we explicitly state that (8), (9) and (10) of Theorem 1 hold for all $x \in E$, where E is the set defined in (13).

Actually the previous considerations apply to any representative u of a function in $L^1(\mathbb{R}^n)$ since we have used, compare Theorem 1, pointwise values of u.

Definition 1. *Let $u \in L^1_{\mathrm{loc}}(\mathbb{R}^n)$. The* Lebesgue set $\mathcal{L}_u$ of u *is the set of points $x \in \mathbb{R}^n$ for which there exists $\lambda \in \mathbb{R}$ such that*

$$(15) \qquad \fint_{B(x,r)} |u(y) - \lambda|\, dy \longrightarrow 0 \qquad \text{as } r \to 0\,.$$

For $x \in \mathcal{L}_u$, the number $\lambda = \lambda_u(x)$ is uniquely defined by (15) and it is named the Lebesgue value of u at x.

It is clear from the definition that $\mathcal{L}_u$ and $\lambda_u(x)$ are uniquely defined by the class of equivalence of $u \in L^1(\mathbb{R}^n)$ and by Theorem 1

$$(16) \qquad \operatorname{meas}(\mathbb{R}^n \setminus \mathcal{L}_u) = 0\,.$$

Any function $\tilde{u}(x)$ such that

$$\tilde{u}(x) = \lambda_u(x) \qquad \forall x \in \mathcal{L}_u$$

is referred to as a *Lebesgue representative of* u. For instance we can set

$$\tilde{u}(x) = \begin{cases} 0 & \text{if } x \notin \mathcal{L}_u \\ \lambda_u(x) & \text{if } x \in \mathcal{L}_u\,, \end{cases}$$

It is easy to check that $\forall x \in \mathcal{L}_u$

$$|\tilde{u}(x)| \le M(u)(x)\,.$$

If moreover $u \in L^p_{\mathrm{loc}}(\mathbb{R}^n)$ we also have

$$(17) \qquad \fint_{B(x,r)} |\tilde{u}(y) - \tilde{u}(x)|^p\, dy \longrightarrow 0 \qquad \text{as } r \to 0\,.$$

for almost every $x \in \mathbb{R}^n$. We shall denote by $\mathcal{L}_{u,p}$ the set of points where (17) holds.

Points in $\Omega \setminus \mathcal{L}_u$ are often called *exceptional points*, and $\Omega \setminus \mathcal{L}_u$ the *exceptional set of* u. We shall now prove that the Hausdorff dimension of the exceptional set of a function $u \in W^{1,p}_{\mathrm{loc}}(\Omega)$, Ω open in $\mathbb{R}^n$, is not greater then $n-p$. In doing that the next lemma is the key point.

Lemma 1. *Let Ω be an open set of $\mathbb{R}^n$, v be a function in $L^1_{\mathrm{loc}}(\Omega)$ and $0 \le \alpha < n$. Set*

$$E_\alpha := \{\, x \in \Omega \mid \limsup_{\rho \to 0} \rho^{-\alpha} \int_{B(x,\rho)} |v(y)|\, dy > 0 \,\}.$$

Then we have

$$\mathcal{H}^\alpha(E_\alpha) = 0\,.$$

Proof. It suffices to show that for each compact subset $K \subset \Omega$ we have $\mathcal{H}^\alpha(F) = 0$ where $F = E_\alpha \cap K$. Set

$$F^{(s)} := \{ x \in F \mid \limsup_{\rho \to 0} \rho^{-\alpha} \int_{B(x,\rho)} |v(y)|\, dy > s^{-1} \}.$$

Obviously $F = \cup_{s=1}^\infty F^{(s)}$, hence it suffices to show that $\mathcal{H}^\alpha(F^{(s)}) = 0$ for all s. Let Q be a bounded open set with $K \subset\subset Q \subset\subset \Omega$ and let $d := \min(1, \operatorname{dist}(K, \partial Q))$. For all $\varepsilon > 0$, $0 < \varepsilon < d$, and for all $x \in F^{(s)}$ there exists $r(x)$, $0 < r(x) < \varepsilon$, such that

$$r(x)^{-\alpha} \int_{B(x,r(x))} |v(y)|\, dy > \frac{1}{2s}\, .$$

Applying Besicovitch's covering theorem we can then choose a denumerable set of points $\{x_i\}$ in $F^{(s)}$ such that $\cup_i B(x_i, r(x_i)) \supset F^{(s)}$ and the balls $B(x_i, r_i)$, $r_i = r(x_i)$, do not overlap more than 3^n times. Therefore we infer

$$\sum_i r_i^\alpha \le 2s \sum_i \int_{B(x_i,r_i)} |v(y)|\, dy = 2\, 3^n s \int_{\cup_i B(x_i,r_i)} |v(y)|\, dy \tag{18}$$

Since $\alpha < n$, this inequality implies

$$\begin{aligned} \operatorname{meas}\big(\cup_i B(x_i, r_i)\big) &\le \omega_n \sum_i r_i^n \le \omega_n \varepsilon^{n-\alpha} \sum_i r_i^\alpha \\ &\le 2 3^n s \varepsilon^{n-\alpha} \int_Q |v(y)|\, dy\,, \qquad \omega_n = \operatorname{meas}(B(0,1)) \end{aligned} \tag{19}$$

From (19) and the absolute continuity theorem, applied to (18), it follows that the right hand side of (18) goes to zero as $\varepsilon \to 0$, and therefore that $\mathcal{H}^\alpha(F^{(s)}) = 0$. □

We can now state

Theorem 2. *Let Ω be an open set in $\mathbb{R}^n$ and let $u \in W^{1,p}_{\text{loc}}(\Omega)$, $1 \le p < n$. Set*

$$\Sigma_1 := \{x \in \Omega \mid \limsup_{\rho \to 0} \rho^{p-n} \int_{B(x,\rho)} |Du|^p\, dy > 0\}$$

$$\Sigma_2 := \{x \in \Omega \mid \lim_{\rho \to 0} \fint_{B(x,\rho)} u(y)\, dy \text{ does not exist finite}\}$$

Then we have

$$\mathcal{H}^{n-p}(\Sigma_1) = 0 \tag{20}$$

$$\mathcal{H}^{n-p+\varepsilon}(\Sigma_2) = 0 \qquad \forall\, \varepsilon > 0 \,. \tag{21}$$

Moreover the exceptional set of u has Hausdorff dimension not greater than $n-p$, $\dim_{\mathcal{H}}(\Omega \setminus \mathcal{L}_u) \le n-p$, i.e.,

$$\mathcal{H}^{n-p+\varepsilon}(\Omega \setminus \mathcal{L}_u) = 0 \qquad \forall \varepsilon > 0. \tag{22}$$

Proof. Lemma 1 yields at once (20). To prove (21), for a fixed x_0 we consider the function

$$\phi(\rho) := \fint_{B(x_0,\rho)} u(x)\, dx = \fint_{B(0,1)} u(x_0 + \rho y)\, dy \,.$$

The function $\phi(\rho)$ is differentiable in $(0,\infty)$ and

$$\phi'(\rho) = \fint_{B(0,1)} y^i D_i u(x_0 + \rho y)\, dy \,,$$

hence

$$|\phi'(\rho)| \le \Big(\fint_{B(x_0,\rho)} |Du(x)|^p \, dx \Big)^{1/p} \,.$$

Suppose now that $x_0 \notin E_{n-p+\varepsilon}$, i.e.,

$$\rho^{-n+p-\varepsilon} \int_{B(x_0,\rho)} |Du(x)|^p \, dx \longrightarrow 0$$

then

$$|\phi'(\rho)| \le c\, \rho^{-1+\frac{\varepsilon}{p}} \,,$$

and therefore

$$|\phi(\rho) - \phi(s)| \le \Big| \int_s^\rho |\phi'(t)|\, dt \Big| \le \frac{c}{\varepsilon} \, |\rho^{\varepsilon/p} - s^{\varepsilon/p}| \,,$$

which implies that $x_0 \notin \Sigma_2$. Therefore we conclude that

$$\Sigma_2 \subset E_{n-p+\varepsilon} \qquad \forall \varepsilon \,.$$

This implies (21), because of Lemma 1. Finally Poincaré inequality

$$\int_{B(x,\rho)} |u(y) - u_{x_0,\rho}|^p \, dy \le c\, \rho^p \int_{B(x,\rho)} |Du(y)|^p \, dy$$

yields at once that every point in $\Omega \setminus (\Sigma_1 \cup \Sigma_2)$ is a Lebesgue point for u and consequently (22) holds, if we take into account (20) and (21). □

The function

$$u(x) = \log\log|x|^{-1}$$

which belongs to $W^{1,2}(B(0,1))$, $B(0,1) \subset \mathbb{R}^2$, shows that Theorem 2 is optimal for $p > 1$. Later we shall see that the case $p = 1$ is special. We shall in fact see in Sec. 4.1.4, Remark 3 in Sec. 4.1.4, that *the exceptional set of a function* $u \in W^{1,1}(\Omega)$ *has zero* $(n-1)$*–dimensional Hausdorff measure.*

We conclude this subsection by proving a fundamental decomposition theorem which is very useful in many instances from real and complex analysis to partial differential equations. Here we shall use it in order to give a characterization of the class of functions with maximal function in L^1.

Let Q be a cube in $\mathbb{R}^n$. Dividing Q into 2^n congruent cubes and iterating the process one obtain the so–called *dyadic decomposition of* Q.

Lemma 2 (Calderón–Zygmund decomposition argument). *Assume that* u *is a non negative function in* $L^1(Q)$ *and let* t *be a positive number such that*

$$\fint_Q u(x)\,dx \le t\,.$$

Then there exists a countably family $\{Q_i\}$ *of cubes in the dyadic decomposition of* Q *such that*

(i) $$t < \fint_{Q_i} u(x)\,dx \le 2^n\,t$$

(ii) $$u(x) \le t \qquad \text{for a.e. } x \in Q \setminus \cup_i Q_i$$

Proof. By bisection of the sides of Q, we subdivide Q into 2^n congruent and equal subcubes. Those subcubes P which satisfy

$$\fint_P u(x)\,dx \le t$$

are similarly subdivided and the process is repeated indefinitely (or finitely if there is no such cube). Let $\mathcal{Q} = \{Q_i\}$ denote the family of subcubes so obtained which satisfy

$$\fint_{Q_i} u(x)\,dx > t$$

and for each Q_i denote by $\widetilde{Q}_i$ the subcube whose subdivision gives Q_i. Since

$$|\widetilde{Q}_i| = 2^n\,|Q_i|$$

we get immediately (i) as

$$\fint_{\widetilde{Q}_i} u(x)\,dx \le t\,.$$

If $x \in Q \setminus \cup_i Q_i$ and is not on the boundary of some Q_i then clearly it belongs to infinitely many cubes P of the successive subdivision in 2^n subcubes with $|P| \to 0$. Since, on the other hand, at every Lebesgue point x we have

$$\lim_{\substack{|P|\to 0\\ x\in P}} \fint_P |u(y) - u(x)|\, dy \le c(n) \lim_{\delta(P)\to 0} \fint_{B(x,\delta(P))} |u(y) - u(x)|\, dy = 0 ,$$

$\delta(P)$ being the diameter of P, in particular

$$u(x) = \lim_{\substack{|P|\to 0\\ x\in P}} \fint_P u(y)\, dy$$

we infer (ii) at once. □

Remark 4. Actually Calderón–Zygmund decomposition $\mathcal{Q} = \{Q_i\}$ depends on the parameter t. Let $\{Q_i^{(t)}\}$ and $\{Q_i^{(s)}\}$ be the families corresponding to the parameters t and s, $t < s$. Then *each* $Q_i^{(s)}$ *is contained in some* $Q_j^{(t)}$. This is clear from the construction. In fact we have

$$\fint_{Q_i^{(s)}} u(x)\, dx > s > t ,$$

therefore in the chain of cubes $\{P\}$ which are successively divided into 2^n equal cubes and produce $Q_i^{(s)}$ the first cube P for which

$$\fint_P u(x)\, dx > t$$

belongs to $Q_i^{(t)}$.

We shall now prove that the distribution function of the maximal function $M(u)(x)$, the function

$$t \longrightarrow \frac{1}{t} \int_{|u|>t} |u(x)|\, dx ,$$

and the measure of the union of the cubes $Q_i^{(t)}$ of the Calderón–Zygmund decomposition relative to $|u|$ and t are equivalent for large values of t. In fact we have

Proposition 4. *Let Q be a cube in $\mathbb{R}^n$, let $u \in L^1(Q)$ and let t be such that*

$$t > \fint_Q |u(x)|\, dx .$$

Denote by $Q_i^{(t)}$ *the family of subcubes of* Q *in the Calderón–Zygmund decomposition relative to* $|u|$ *and* t. *We have*

$$2^{-n}\, t^{-1} \int\limits_{|u|>t} |u(x)|\, dx \le \sum_i |Q_i^{(t)}| \le 2\, t^{-1} \int\limits_{|u|>t/2} |u(x)|\, dx \tag{23}$$

and for a constant $\gamma = \gamma(n)$

$$\begin{aligned} \frac{1}{t} \int\limits_{|u|>t} |u(x)|\, dx \ &\le\ c_1(n)\, \mathrm{meas}\,\{\, x \in Q \mid M(u)(x) > \gamma(n) t \,\} \\ &\le\ \frac{c_2(n)}{t} \int\limits_{|u|>\gamma(n)t/2} |u(x)|\, dx \end{aligned} \tag{24}$$

Proof. Calderón–Zygmund lemma yields

$$\int\limits_{|u|>t} |u(x)|\, dx \le \sum_i \int\limits_{Q_i^{(t)}} |u(x)|\, dx \le 2^n\, t \sum_i |Q_i^{(t)}|$$

and

$$\begin{aligned} t\, |Q_i^{(t)}| \le \int\limits_{Q_i^{(t)}} |u(x)|\, dx \ &= \int\limits_{Q_i^{(t)} \cap \{|u| \le t/2\}} |u(x)|\, dx + \int\limits_{Q_i^{(t)} \cap \{|u| > t/2\}} |u(x)|\, dx \\ &\le \frac{t}{2}\, |Q_i^{(t)}| + \int\limits_{Q_i^{(t)} \cap \{|u| > t/2\}} |u(x)|\, dx \,. \end{aligned}$$

from which (23) follows at once.

If $x \in Q_i^{(t)}$, then

$$\fint\limits_{Q_i^{(t)}} |u(x)|\, dx \ > \ t \,.$$

By considering the smallest ball B with center x containing $Q_i^{(t)}$, we then get

$$M(u)(x) \ge \fint\limits_{B} |u(y)|\, dy \ge c(n) \fint\limits_{Q_i^{(t)}} |u(y)|\, dy \ge \gamma(n)\, t \,.$$

This proves (24) taking also into account (23) and (4). □

As first consequence we shall now give a characterization of the class of functions with maximal function in $L^1(\mathbb{R}^n)$.

Definition 2. *We say that u belongs to $L\log L$ in Q if u is measurable and*

$$\int_Q |u|\log(2+|u|)\,dx \;<\; \infty\,.$$

Notice that

$$L\log L(\mathbb{R}^n)\subset L^1(\mathbb{R}^n)\,.$$

Proposition 5. *We have*

(i) If $u\in L\log L(\mathbb{R}^n)$ then $M(u)\in L^1_{\mathrm{loc}}(\mathbb{R}^n)$.
(ii) Let Q a cube in $\mathbb{R}^n$, u a measurable function with $M(u)\in L^1(Q)$. Then $u\in L\log L(Q)$.

Usually Proposition 5(ii) is proved under the additional hypotheses that $u=0$ outside of Q. This is not necessary and actually the two claims are equivalent.

Proof. Let K be a compact in $\mathbb{R}^n$. Using Proposition 1 we get

$$\begin{aligned}
\int_K M(u)(x)\,dx &\le 2\,|K| + \int_{M(u)>2} M(u)(x)\,dx\\
&= 2\,|K| + \int_2^\infty |\{M(u)>t\}|\,dt \;\le\; 2\,|K| + c\int_1^\infty \frac{1}{t}\Big(\int_{|u|>t}|u(x)|\,dx\Big)\,dt\\
&= 2\,|K| + c\int_{|u|>1} |u(x)|\Big(\int_1^{|u(x)|}\frac{1}{t}\,dt\Big)\,dx\\
&= 2\,|K| + c\int_{|u|>1} |u|\log|u|\,dx\,.
\end{aligned}$$

This proves (i).

Because of Lebesgue's differentiation theorem,

$$|u(x)| \;\le\; M(u)(x) \tag{25}$$

for a.e. x in Q, hence $u\in L^1(Q)$. Since

$$\int_Q |u|\log(2+|u|)dx \;\le\; \log 3\int_Q |u|\,dx + \int_{|u|>1}|u|\log|u|\,dx \tag{26}$$

it suffices to prove that the last integral is finite. Set

$$t_Q \;:=\; ⨍_Q |u(x)|\,dx\,.$$

We have

$$(27)\qquad \int\limits_{|u(x)|>1} |u|\log|u|\,dx \le \log t_Q \int\limits_{|u|\le t_Q} |u(x)|\,dx + \int\limits_{|u|>t_Q} |u|\log|u|\,dx\ ,$$

Using Proposition 4 we infer

$$(28)\qquad \begin{aligned} \int\limits_{|u|>t_Q} |u|\log|u|\,dx &= \int\limits_{|u|>t_Q} |u(x)|\Big(\int\limits_1^{|u(x)|} \frac{dt}{t}\Big)\,dx \\ &= \int\limits_{t_Q}^{\infty} \frac{dt}{t} \int\limits_{|u|>t} |u(x)|\,dx \le c_1(n)\|\, M(u)\,\|_{L^1(Q)} \end{aligned}$$

We therefore conclude from (25) (26) (27) (28)

$$\int\limits_Q |u|\log|u|\,dx \ \le\ c(n)(\log(3t_Q)+1)\|\, M(u)\,\|_{L^1(Q)}$$

which proves the claim (ii). □

1.2 Differentiability Properties of $W^{1,p}$ Functions

Let $u \in W^{1,p}_{\text{loc}}(\mathbb{R}^n)$, with $p \ge 1$. In this subsection we shall be concerned with pointwise differentiability properties of u, thus it is not restrictive to assume that in fact $u \in W^{1,1}(\mathbb{R}^n)$, or even $u \in W^{1,p}(\mathbb{R}^n)$ and we shall think of u as of a *representative* of the equivalence class of u.

Let $u \in W^{1,p}_{\text{loc}}(\mathbb{R}^n)$. We denote by

$$R_u \ :=\ \mathcal{L}_u \cap \mathcal{L}_{Du} \cap \{\, x \mid u(x) = \lambda_u(x)\,\}$$

the set of Lebesgue points for both u and the *weak derivatives* Du where $u(x)$ is the Lebesgue value of u at x. By Lebesgue differentiation theorem,

$$\operatorname{meas}(\mathbb{R}^n \setminus R_u) \ =\ 0\ .$$

For $u \in W^{1,p}_{\text{loc}}(\mathbb{R}^n)$, $x \in \mathbb{R}^n$ and $\delta > 0$ we define

$$V^p_\delta(Du)(x) \ :=\ \sup_{0<r\le\delta} \ \fint\limits_{B(x,r)} |Du(y) - Du(x)|^p\,dy\ .$$

Notice that, for almost all $x \in R_u$, we have

$$V^p_\delta(Du)(x) \ \longrightarrow\ 0 \qquad \text{as}\ \ \delta \to 0\ .$$

The following proposition contains the basic estimates in order to study the differentiability properties of functions in $W^{1,p}_{\text{loc}}(\mathbb{R}^n)$.

Proposition 1. *Let $u \in W^{1,p}_{\text{loc}}(\mathbb{R}^n)$, $p \geq 1$, $x \in \mathcal{L}_u$ and $u(x)$ be the Lebesgue value of u at x. Then for all $r > 0$ we have*

$$\fint_{B(x,r)} \frac{|u(y)-u(x)|^p}{|y-x|^p}\,dy \leq \int_0^1 dt \fint_{B(x,tr)} |Du(y)|^p\,dy\,. \tag{1}$$

In particular Morrey's inequality

$$\fint_{B(x,r)} |u(y)-u(x)|^p\,dy \leq r^p \int_0^1 dt \fint_{B(x,tr)} |Du(y)|^p\,dy\,. \tag{2}$$

holds. If moreover $p > n$, then

$$\fint_{B(x,r)} |u(y)-u(x)|\,dy \leq \frac{p}{p-n} r \left(\fint_{B(x,r)} |Du(y)|^p\,dy \right)^{1/p}\,. \tag{3}$$

Proof. Of course we may assume $x = 0$ and $u(x) = 0$, thus we have to prove

$$\fint_{B(0,r)} \frac{|u(y)|^p}{|y|^p}\,dy \leq \int_0^1 dt \fint_{B(0,tr)} |Du(y)|^p\,dy \tag{4}$$

provided $u \in W^{1,p}_{\text{loc}}(\mathbb{R}^n)$ and $\fint_{B(0,r)} |u(y)|\,dy \to 0$ as $r \to 0$.

Step 1. (4) holds for $u \in C^1(\mathbb{R}^n)$ such that $u(0) = 0$. In this case in fact

$$u(y) = \int_0^1 Du(ty) \cdot y\,dt$$

so that

$$\fint_{B(0,r)} \frac{|u(y)|^p}{|y|^p}\,dy \quad \fint_{B(0,r)} dy \left[\int_0^1 |Du(ty)|\,dt \right]^p$$

$$\leq \fint_{B(0,r)} dy \int_0^1 |Du(ty)|^p\,dt = \int_0^1 dt \fint_{B(0,tr)} |Du(y)|^p\,dy\,.$$

Step 2. We now prove (4) when $u \in C^1(\mathbb{R}^n \setminus \{0\})$ and $\fint_{B(0,r)} |u(y)|\,dy \to 0$ as $r \to 0$. Assume

$$\int_0^1 dt \fint_{B(0,tr)} |Du(y)|^p \, dy < +\infty$$

otherwise there is nothing to prove. By changing the order of integration we get

$$\fint_{B(0,r)} dz \int_0^1 |Du(tz)|^p \, dt = \int_0^1 dt \fint_{B(0,tr)} |Du(y)|^p \, dy < \infty$$

from which we deduce that for a.e. $z \in B(0,r)$ the function $u_z(t) := u(tz)$, $t \in (0,1)$ is absolutely continuous. In particular there exists

$$\lambda(z) := \lim_{t \to 0} u(tz) .$$

Since $\fint_{B(0,r)} |u(y)| \, dy \to 0$ as $r \to 0$ we infer that $|\lambda(z)| = 0$ for a.e. z. Therefore we conclude

$$u(z) = u(z) - 0 = \int_0^1 Du(tz) \cdot z \, dt .$$

The claim then follows as in the Step 1.

Step 3. Let now $u \in W^{1,p}_{\text{loc}}(\mathbb{R}^n)$ and $\fint_{B(0,r)} |u| \, dy \to 0$ as $r \to 0$. Again we may assume

$$\int_0^1 dt \fint_{B(0,tr)} |Du(y)|^p \, dy < +\infty$$

First we approximate u by a sequence of maps $\{u_k\}$ for which step 2 holds, as stated in the following

Lemma 1. *Let $u \in W^{1,p}_{\text{loc}}(\mathbb{R}^n)$, $p \geq 1$, and suppose that*

$$\fint_{B(0,r)} |u| \, dy \to 0$$

as $r \to 0$. Then there exists a sequence $\{u_k\} \subset W^{1,p}_{\text{loc}}(\mathbb{R}^n) \cap C^1(\mathbb{R}^n \setminus \{0\})$ such that $u_k \to u$ in $W^{1,p}_{\text{loc}}(\mathbb{R}^n)$ and for any $k = 1, 2, \ldots$

$$\fint_{B(0,r)} |u_k| \, dy \longrightarrow 0 \qquad \text{as } r \to 0$$

(5)

$$\fint_{B(0,r)} |Du_k|^p \, dy \leq c \fint_{B(0,2r)} |Du|^p \, dy \qquad \forall r > 0.$$

We postpone the proof of the lemma and conclude the proof of the proposition. By construction all the u_k satisfy the hypotheses in Step 2, hence for all $k = 1, 2, \dots$ we have

$$\fint_{B(0,r)} \frac{|u_k(y)|^p}{|y|}\, dy \le \int_0^1 dt \fint_{B(0,tr)} |Du_k|^p\, dy\,.$$

Passing to subsequences, we can suppose that $u_k(y) \to u(y)$ for a.e. y. Therefore by Fatou's lemma and Lebesgue's dominated convergence theorem, using also (5), we get (4) and finally (1). Morrey's inequality (2) then trivially follows. Similarly we have

$$\begin{aligned}\fint_{B(x,r)} |u(y)-u(x)|\, dy &\le r\int_0^1 dt \fint_{B(x,tr)} |Du|\, dy \\ &\le r\int_0^1 dt \Big(\fint_{B(x,tr)} |Du|^p\, dy\Big)^{1/p}\,;\end{aligned}$$

if $p > n$, we can estimate the last integral by

$$r\int_0^1 \frac{1}{t^{n/p}} \Big(\fint_{B(x,r)} |Du|^p\, dy\Big)^{1/p} dt = \frac{r}{1-n/p}\Big(\fint_{B(x,r)} |Du|^p\, dy\Big)^{1/p}$$

which yields (3). □

Proof of Lemma 1. We construct the sequence u_k by mollifying u so that its "value" at zero does not change. Consider the sequence of balls $B_k = B(0, \alpha^{-k})$, where for reasons to be seen later we choose $\alpha = \sqrt[3]{2}$, and define $A_0 = \mathbb{R}^n \setminus \overline{B_2}$, $A_k := B_k \setminus \overline{B_{k+2}}$, $k = 1, 2, \dots$. The open sets A_k form a locally finite covering of $\mathbb{R}^n \setminus \{0\}$ so that there exists a decomposition of unity $\{\varphi_k\}$ associated with $\{A_k\}$.

Choose a sequence ε_k in such a way that

$$\operatorname{spt}(\varphi_k * \rho_{\varepsilon_k}) \subset B_{k-1} \setminus B_{k+3}$$

where ρ is a standard mollifier and $\rho_\varepsilon(x) := \varepsilon^{-n}\rho(\frac{x}{\varepsilon})$. Then the function

$$\tilde u := \sum_k (u\varphi_k) * \rho_{\varepsilon_k} \tag{6}$$

is well defined and of class C^∞ in $\mathbb{R}^n \setminus \{0\}$.

Fix $r > 0$ and let k be the largest integer such that $\alpha^{-k} > r$. By the choice of α, only the values of u for x with $|x| < \alpha^{2-k} < \alpha^3 r$ enter in the definition of $\tilde u(y)$ with $|y| < r$, thus

$$\fint_{B(0,r)} |\tilde{u}|^p \, dy \;\le\; c \fint_{B(0,2r)} |u|^p \, dy \tag{7}$$

where c is a constant depending only on n. Similarly we get

$$\fint_{B(0,r)} |D\tilde{u}|^p \, dy \;\le\; c \fint_{B(0,2r)} |Du|^p \, dy \tag{8}$$

We shall now prove that for every $\delta > 0$ we can choose $\{\varepsilon_k\}$ in such a way that

$$\| u - \tilde{u} \|_{W^{1,p}} \;\le\; 3\,\delta$$

This together with (7) and (8) clearly concludes the proof of the lemma.

We have

$$\tilde{u} - u \;=\; \sum_k ((u\varphi_k) * \rho_{\varepsilon_k} - u\varphi_k)$$

so that

$$\| \tilde{u} - u \|_{L^p} \;\le\; \sum_k \| (u\varphi_k) * \rho_{\varepsilon_k} - u\varphi_k \|_{L^p}$$

therefore, choosing ε_k in such a way that

$$\| (u\varphi_k) * \rho_{\varepsilon_k} - u\varphi_k \|_{L^p} \;\le\; \delta/2^{k+1}$$

we get

$$\| \tilde{u} - u \|_{L^p} \;\le\; \delta \,.$$

On the other hand

$$\begin{aligned} D\tilde{u} \;&=\; \sum_k D((u\varphi_k) * \rho_{\varepsilon_k}) \\ &=\; \sum_k (Du \cdot \varphi_k) * \rho_{\varepsilon_k} + (u\, D\varphi_k) * \rho_{\varepsilon_k} \\ &=\; Du + \sum_k ((Du\varphi_k) * \rho_{\varepsilon_k} - Du\varphi_k) + \sum_k ((u\, D\varphi_k) * \rho_{\varepsilon_k} - u\, D\varphi_k) \end{aligned}$$

thus, choosing ε_k in such a way that

$$\begin{aligned} \| (u\, D\varphi_k) * \rho_{\varepsilon_k} - (u\, D\varphi_k) \|_{L^p} \;&\le\; \delta/2^{k+1} \\ \| (Du\, \varphi_k) * \rho_{\varepsilon_k} - (Du\, \varphi_k) \|_{L^p} \;&\le\; \delta/2^{k+1} \end{aligned}$$

we conclude

$$\| Du - D\tilde{u} \|_{L^p} \;\le\; 2\delta$$

□

Next theorem is a simple consequence of Proposition 1.

Theorem 1. *Let $u \in W^{1,p}_{\text{loc}}(\mathbb{R}^n)$, with $p \geq 1$. Let x be a Lebesgue point for u and let $u(x)$ be the Lebesgue value of u at x. Then we have*

$$\text{(9)} \qquad \fint_{B(x,r)} \frac{|u(y) - u(x) - Du(x)(y-x)|^p}{|y-x|^p}\, dy \;\leq\; V_r^p(Du)(x) \,.$$

If moreover x and $Du(x)$ satisfy the additional condition

$$V_r^p(Du)(x) \longrightarrow 0 \qquad \text{as } r \to 0$$

then

$$\text{(10)} \qquad \fint_{B(x,r)} \frac{|u(y) - u(x) - Du(x)(y-x)|^p}{|y-x|^p}\, dy \longrightarrow 0 \qquad \text{as } r \to 0 \,.$$

In particular, for all $x \in R_u$, we have

$$\fint_{B(x,r)} \frac{|u(y) - u(x) - Du(x)(y-x)|}{|y-x|}\, dy \longrightarrow 0 \qquad \text{as } r \to 0 \,.$$

provided also $Du(x)$ is the Lebesgue value of Du at x.

Proof. The claims easily follow by applying Proposition 1 to the function $y \to u(y) - Du(x)(y-x)$. □

Theorem 1 yields at once

Theorem 2 (Calderón-Zygmund). *Let $u \in W^{1,p}_{\text{loc}}(\mathbb{R}^n)$, with $p \geq 1$. Let $x \in \mathcal{L}_u$ and $u(x)$ be the Lebesgue value of u at x. Then*

$$\text{(11)} \qquad \frac{1}{r^p} \fint_{B(x,r)} |u(y) - u(x) - Du(x)(y-x)|^p\, dy \;\leq\; V_r^p(Du)(x) \,.$$

Consequently

$$\text{(12)} \qquad \lim_{r \to 0^+} \frac{1}{r^p} \fint_{B(x,r)} |u(y) - u(x) - Du(x)(y-x)|^p\, dy \;=\; 0$$

for a.e. $x \in \mathbb{R}^n$, $Du(x)$ being the Lebesgue value of Du at x.

If (12) holds one says that u *is differentiable in the L^p-sense at x*. Therefore Theorem 2 says that *every function $u \in W^{1,p}_{\text{loc}}(\mathbb{R}^n)$ is almost everywhere differentiable in the L^p sense.*

Remark 1. Let $p^* := \frac{np}{n-p}$ be the Sobolev exponent of p, $1 \leq p < n$. Sobolev's inequality yields

$$
\begin{aligned}
&\Big(\fint_{B(x,r)} |v(y)-v(x)|^{p^*}\,dy \Big)^{1/p^*} \\
&\le c\,r \Big(\fint_{B(x,r)} |Dv(y)|^p\,dy \Big)^{1/p} + c \Big(\fint_{B(x,r)} |v(y)-v(x)|^p\,dy \Big)^{1/p} \\
&\le c\,r \Big(\fint_{B(x,r)} |Dv(y)|^p\,dy \Big)^{1/p} + c\,r \Big(\int_0^1 dt \fint_{B(x,tr)} |Dv(y)|^p\,dy \Big)^{1/p} .
\end{aligned}
$$

Applying the previous inequality to

$$v(y) := u(y) - u(x) - Du(x)(y-x)$$

we infer

$$
\begin{aligned}
&\Big(\fint_{B(x,r)} |u(y)-u(x)-Du(x)(y-x)|^{p^*}\,dy \Big)^{1/p^*} \\
&\le c\,r \Big(\fint_{B(x,r)} |Du(y)-Du(x)|^p\,dy \Big)^{1/p} \\
&\quad + c\,r \Big(\int_0^1 dt \fint_{B(x,tr)} |Du(y)-Du(x)|^p\,dy \Big)^{1/p} \\
&\le 2c\,r \big(V_r^p(Du)(x) \big)^{1/p}
\end{aligned}
$$

Thus, we conclude that *every $u \in W^{1,p}_{\mathrm{loc}}(\mathbb{R}^n)$, with $1 \le p < n$, is almost everywhere differentiable in the L^{p^*}-sense.*

[1] A function $u \in W^{1,p}(\mathbb{R}^n)$, with $1 \le p \le n$, is not in general almost everywhere differentiable as it is shown for instance by the following example

Consider the function $\varphi : \mathbb{R}^2 \to \mathbb{R}$

$$\varphi(x) := \begin{cases} \log\log \frac{1}{|x|} & \text{if } |x| < \frac{1}{e} \\ 0 & \text{otherwise} , \end{cases}$$

u belongs to $W^{1,2}(\mathbb{R}^2)$. For a denumerable set of dense points $\{x_k\}$ in $\mathbb{R}^2$ we now define

$$u(x) := \sum_{k=1}^{\infty} 2^{-k} \varphi(x - x_k) .$$

Then it is easily seen that $u(x) \in W^{1,2}(\mathbb{R}^2)$, and that u is not almost everywhere differentiable in the classical sense since it is unbounded in a neighborhood of every point. •

Denote now for $y \neq x$

$$R(y\,;x) \;:=\; \frac{|u(y) - u(x) - Du(x)(y-x)|}{|y-x|}$$

and by

$$\phi(x,r) \;:=\; \fint_{B(x,r)} R(y\,;x)\,dy$$

Even if u is not differentiable in the classical sense, we have the following

Theorem 3. *Let $u \in W^{1,1}_{\mathrm{loc}}(\mathbb{R}^n)$, and let C be a closed set for which*

$$\lim_{r\to 0} \phi(x,r) \;=\; 0$$

holds uniformly on compact subsets of C. Then $u_{|C}$ is differentiable in the classical sense, i.e.,

$$\lim_{\substack{y\to x\\ y\in C}} R(y;x) \;=\; 0\;,$$

moreover the limit is uniform on compact subsets of C.

Proof. Let $r := |x-y|$. We have

$$\begin{aligned}
&|u(y) - u(x) - Du(x)(y-x)| \\
&\qquad \le \Big|\fint_{B(y,r)} u(z)\,dz - u(x) - Du(x)(y-x)\Big| \\
&\qquad + \Big|\fint_{B(y,r)} u(z)\,dz - u(y)\Big|
\end{aligned}$$

Using (2) of Proposition 1, we deduce

$$\begin{aligned}
&\Big|\fint_{B(y,r)} u(z)\,dz - u(x) - Du(x)(y-x)\Big| \\
&\qquad = \Big|\fint_{B(y,r)} [u(z) - u(x) - Du(x)(z-x)]\,dz\Big| \\
&\qquad \le 2^n \fint_{B(x,2r)} |u(z) - u(x) - Du(x)(z-x)|\,dz \\
&\qquad \le 2^{n+1} r\,\phi(x,2r)
\end{aligned}$$

and, similarly,

$$\Big|\fint_{B(y,r)} u(z)\,dz - u(y)\Big| \;=\; \Big|\fint_{B(y,r)} (u(z) - u(y) - Du(y)(z-y))\,dz\Big| \;\le\; r\,\phi(y,r)\;.$$

Therefore

$$R(y\,;x) \;\le\; c(n)\,\{\phi(x,2r)+\phi(y,r)\}\;, \qquad r=|x-y|\;.$$

The claim easily follows from this inequality. In fact, for any $\varepsilon>0$, $\delta>0$ such that $\phi(y,s)<\varepsilon$ holds for $s<\delta$ and $|y-x_0|\le 1$, we deduce

$$R(y;x) \;\le\; 2c(n)\varepsilon$$

for any x,y with $|x-x_0|\le\frac{1}{2}$, $|y-x|\le\frac{1}{2}\min(1,\delta)$. □

Also, as consequence of Theorem 2, we can deduce that the functions in $W^{1,1}_{\mathrm{loc}}(\mathbb{R}^n)$ possess almost everywhere classical directional derivatives in almost all directions.

Theorem 4. *Let $u\in W^{1,1}_{\mathrm{loc}}(\mathbb{R}^n)$. Then for almost every x we have*

$$\text{(13)} \qquad k\,[u(x+\frac{1}{k}z)-u(x)] \;\longrightarrow\; Du(x)\,z, \qquad k\to\infty\;, \quad z\in\mathbb{R}^n$$

in $L^1_{\mathrm{loc}}(\mathbb{R}^n)$.

Proof. From (12) we deduce that for a.e. x and all R

$$\lim_{k\to\infty}\frac{k}{R}\;\fint_{B(0,R/k)} |u(x+w)-u(x)-Du(x)\;w|\,dw \;=\; 0\;;$$

changing variable, $w=z/k$, we then get

$$\lim_{k\to\infty}\frac{1}{R}\;\fint_{B(0,R)} |k[u(x+\frac{z}{k})-u(x)]-Du(x)\;z|\,dz \;=\; 0\;.$$

□

We shall now prove that functions $u\in W^{1,p}_{\mathrm{loc}}(\mathbb{R}^n)$, with $p>n$, are "almost everywhere differentiable". More precisely we have

Theorem 5 (Morrey-Sobolev). *Let $u\in W^{1,p}_{\mathrm{loc}}(\mathbb{R}^n)$ with $p>n$. Then in the equivalence class of u there is a function $\tilde u$ which is locally Hölder-continuous with exponent $\alpha:=1-n/p$, $\tilde u\in C^{0,\alpha}_{\mathrm{loc}}(\mathbb{R}^n)\cap W^{1,p}_{\mathrm{loc}}(\mathbb{R}^n)$, in particular $\mathcal{L}_u=\mathbb{R}^n$, $\tilde u(x)$ is the Lebesgue value of u at x, and for $x,\ y\in B(0,r)$, $r>0$ we have*

$$|\tilde u(x)-\tilde u(y)|\le c(n,p)\,|x-y|^{1-n/p}\Big(\int_{B(0,r)}|Du(x)|^p\,dx\Big)^{1/p}\;.$$

Moreover $\tilde u$ is almost everywhere differentiable in the classical sense, i.e.

$$\text{(14)} \qquad \lim_{y\to x}\frac{|\tilde u(y)-\tilde u(x)-D\tilde u(x)(y-x)|}{|x-y|} \;=\; 0\;.$$

for a.e. $x\in\mathbb{R}^n$, $D\tilde u(x)$ being the Lebesgue value of Du at x.

Proof. Fix some $R_0 > 0$, and for $x, y \in B(0, R_0)$, set $\delta := |x - y|$ and

$$S = B(x,\delta) \cap B(y,\delta) .$$

Let v any smooth function in $\mathbb{R}^n$. For all $z \in S$ we have

$$|v(x) - v(y)| \leq |v(x) - v(z)| + |v(z) - v(y)| ,$$

hence, integrating over S with respect to z, we deduce

$$|v(x) - v(y)| \leq \fint_S |v(x) - v(z)|\, dz + \fint_S |v(z) - v(y)|\, dz .$$

As in the proof of (3) of Proposition 1, we easily deduce

$$\begin{aligned} \fint_S |v(x) - v(z)|\, dz &\leq c(n) \fint_{B(x,\delta)} |v(x) - v(z)|\, dz \\ &\leq c(n,p)\, \delta \Big(\fint_{B(x,\delta)} |Dv|^p\, dy \Big)^{1/p} \end{aligned}$$

Similarly we estimate $\fint_S |v(z) - v(y)|\, dz$, and we conclude that

$$(15) \qquad |v(x) - v(y)| \leq c(n,p)\, \delta \Big(\fint_{B(x,2\delta)} |Dv|^p\, dy \Big)^{1/p}$$

Let now $\{u_k\}$ be a sequence of smooth mappings which converges to u in $W^{1,p}_{\text{loc}}(\mathbb{R}^n)$. Applying (15) to the approximating sequence $\{u_k\}$,

$$(16) \qquad |u_k(x) - u_k(y)| \leq c(n,p)\, \delta \Big(\fint_{B(x,2\delta)} |Du_k|^p\, dy \Big)^{1/p}$$

we get

$$|u_k(x) - u_k(y)| \leq c(n,p)\, |x - y|^{1-n/p} \Big(\int_{B(0,3R_0)} |Du_k|^p\, dx \Big)^{1/p}$$

Thus we deduce, possibly passing to a subsequence, that $\{u_k\}$ converges uniformly on bounded sets (by Ascoli-Arzelà theorem, and of course in $W^{1,p}_{\text{loc}}(\mathbb{R}^n)$) to a continuous function $\tilde{u}$ (which is in the same class of equivalence of u), which is Hölder continuous with exponent $1 - n/p$; moreover (15) holds for $\tilde{u}$. This proves the first part of the theorem. Since (15) holds also for

$$v(y) := \tilde{u}(y) - \tilde{u}(x) - Du(x)(y - x)$$

we get

$$\begin{aligned} |v(y) - v(x)| &\le c(n,p)\,\delta \left[\fint_{B(x,2\delta)} |Du(y) - Du(x)|^p\, dy \right]^{1/p} \\ &\le c(n,p)\, [V^p_{2\delta}(Du)(x)]^{1/p}\, |x-y| \,. \end{aligned}$$

Thus

$$R(y\,;x) \le c(n,p)\, [V^p_{2\delta}(Du)(x)]^{1/p} \,,$$

and the claim follows. □

1.3 Lusin Type Properties of $W^{1,p}$ Functions

In this subsection we shall prove, as consequence of Theorem 1 in Sec. 3.1.2 the following Lusin type properties for functions in $W^{1,p}_{\text{loc}}(\mathbb{R}^n)$.

(i) the set of regular points $R_u := \mathcal{L}_u \cap \mathcal{L}_{Du} \cap \{\, x \,|\, u(x) = \lambda_u(x) \,\}$ of $u \in W^{1,p}_{\text{loc}}(\mathbb{R}^n)$ can be decomposed in a denumerable union of measurable sets $\{F_k\}$ such that $u_{|F_k}$ is a Lipschitz function,
(ii) except on arbitrarily small open sets u is of class C^1.

Let us begin by discussing property (i). Recall

Definition 1. *Let A be a subset of $\mathbb{R}^n$. A function $u : A \to \mathbb{R}$ is said to be* Lipschitz *in A if there exists a constant L such that*

$$|u(x) - u(y)| \le L|x-y| \qquad \forall\, x, y \in A \,. \tag{1}$$

The infimum of all constants L satisfying (1) is called the Lipschitz constant *of u in A and denoted by $Lip(u, A)$.*

By considering the function $\hat{u} : \mathbb{R}^n \to \mathbb{R}$ defined by

$$\hat{u}(x) := \inf_{y\in A} (u(y) + L|x-y|) \qquad L := Lip(u, A)$$

it is not difficult to prove the following extension theorem

Theorem 1 (Kirszbraun). *Let $A \subset \mathbb{R}^n$ and let $u : A \to \mathbb{R}$ be a Lipschitz map. Then there exists a Lipschitz function $\hat{u} : \mathbb{R}^n \to \mathbb{R}$ such that $\hat{u} = u$ in A and $Lip(\hat{u}, \mathbb{R}^n) = Lip(u, A)$.*

In fact Kirszbraun theorem holds for Lipschitz maps $u : \mathbb{R}^n \to \mathbb{R}^N, N \ge 1$, see e.g. Federer [226].

We also have

Theorem 2. *A function u is Lipschitz in some open set if and only if it has bounded distributional derivatives.*

Proof. Let u be Lipschitz. By mollifying u we find a sequence of smooth maps with equibounded gradients in L^∞ and converging strongly to u in any $W^{1,p}_{\text{loc}}(\mathbb{R}^n)$. Conversely if u has bounded distributional derivatives then u belongs to any $W^{1,p}_{\text{loc}}(\mathbb{R}^n)$, $p > n$, thus the Hölder-continuous representative $\tilde{u}$ of u satisfies (15) in Sec. 3.1.2, hence $\tilde{u}$ is actually Lipschitz. □

Next theorem, which contains the basic differentiability result for Lipschitz maps, is an immediate consequence of Theorem 2 and of Morrey-Sobolev theorem in Sec. 3.1.2.

Theorem 3 (Rademacher). *Let $u : \mathbb{R}^n \to \mathbb{R}$ be a locally Lipschitz function. Then u is almost everywhere differentiable, i.e. for a.e. x*

$$\lim_{y\to x} \frac{|u(y) - u(x) - Du(x)(y-x)|}{|y-x|} = 0 ,$$

where $Du(x)$ is the Lebesgue value at x of the distributional gradient of u.

We are now ready to prove Lusin property (i) for functions $u \in W^{1,1}_{\text{loc}}(\mathbb{R}^n)$. As the content of (i) is local we shall in fact consider functions $u \in W^{1,1}(\mathbb{R}^n)$.

Consider the maximal function of $|Du|$, $M(|Du|)$, and for $\lambda > 0$ define the closed sets F_λ by

$$F_\lambda := \{x \in \mathbb{R}^n \mid M(|Du|)(x) \le \lambda\}$$

Then we have

Theorem 4. *Let $u \in W^{1,p}(\mathbb{R}^n)$. Then the restriction $u_{|F_\lambda \cap R_u}$ is a Lipschitz function with Lipschitz constant not greater than $c(n)\lambda$. Thus for any $k = 1, 2, \ldots$, one can find a closed set F_k and a Lipschitz function $u_k : \mathbb{R}^n \to \mathbb{R}$ with $Lip(u_k) \le k$ such that*

(i) $u_{|R_u \cap F_k} = u_k$
(ii) $\bigcup_k F_k \supset R_u$
(iii) $Du(x) = Du_k(x)$ *for a.e.* $x \in R_u \cap F_k$.

Moreover

$$k^p \, meas(\mathbb{R}^n \setminus F_k) \le c(n,p) \int_{\mathbb{R}^n} |Du|^p \, dx \tag{2}$$

and in fact

$$k^p \, meas(\mathbb{R}^n \setminus F_k) \le c(n,p) \int_{|Du|>k/2} |Du|^p \, dx \to 0 \qquad \text{as } k \to \infty \tag{3}$$

whenever $u \in W^{1,p}(\mathbb{R}^n)$, $p \ge 1$.

Proof. Let $x, y \in R_u \cap F_\lambda$, $\delta := |x-y|$, and let

$$S = B(x,\delta) \cap B(y,\delta)$$

For any $z \in S$ we have

$$|u(x) - u(y)| \le |u(x) - u(z)| + |u(z) - u(y)|$$

thus integrating over S we obtain

$$|u(x) - u(y)| \le \fint_S |u(x) - u(z)|\, dz + \fint_S |u(y) - u(z)|\, dz \,.$$

Using (2) in Sec. 3.1.2 of Proposition 1 in Sec. 3.1.2 we then estimate

$$\begin{aligned}\fint_S |u(x) - u(z)|\, dz &\le c(n) \fint_{B(x,\delta)} |u(x) - u(z)|\, dz \\ &\le c(n)\, \delta \int_0^1 dt \fint_{B(x,t\delta)} |Du(y)|\, dy \le c(n)\, \delta\, M(|Du|)(x) \\ &\le c(n)\, \lambda\, |x - y| \,.\end{aligned}$$

Similarly we estimate $\fint_S |u(y) - u(z)|\, dz$. Hence we infer

$$|u(x) - u(y)| \le c(n)\, 2\lambda\, |x - y|$$

and the first part of the claim is proved. Renaming F_k the set $F_{k/c(n)}$ and using Kirszbraun theorem we extend $u_{|F_k \cap R_u}$ to a Lipschitz function $u_k : \mathbb{R}^n \to \mathbb{R}$ with $Lip(u_k) \le k$. As

$$R_u \subset \{x \mid M(|Du|)(x) < \infty\} = \cup_k F_k$$

(ii) follows trivially, and (iii) follows from Theorem 2 and Theorem 1 in Sec. 3.1.2. In fact both functions u and u_k belong to $W^{1,1}_{\text{loc}}$ and for the difference

$$v_k := u - u_k$$

we have

$$\fint_{B(x,r)} \frac{|v_k(y) - v_k(x) - Dv_k(x)(y - x)|}{|y - x|}\, dy \longrightarrow 0 \qquad \text{as } r \to 0 \,,$$

for every $x \in R_u$, hence for $x \in R_u \cap F_k$

$$\frac{1}{\text{meas}(B(x,r))} \int_{B(x,r) \cap F_k} \frac{|Dv_k(x)(y - x)|}{|y - x|}\, dy \longrightarrow 0 \qquad \text{as } r \to 0 \,.$$

This implies that $Dv_k(x) = 0$, if moreover the density of F_k at x is 1. In fact suppose $Dv_k(x) \neq 0$ and denote by C the cone

$$C := \{y \mid |Dv_k(x)(y-x)| \geq \frac{1}{2}|Dv_k(x)|\,|y-x|\} ,$$

then

$$\begin{aligned}\theta(F_k \cap C, x) &:= \lim_{r\to 0} \frac{\text{meas}\,(F_k \cap C \cap B(x,r))}{\text{meas}\,(B(x,r))} \\ &\leq 2 \lim_{r\to 0} \frac{1}{\text{meas}\,(B(x,r))} \int_{F_k\cap C\cap B(x,r)} \frac{|Dv_k(x)(y-x)|}{|y-x|}\,dy = 0\end{aligned}$$

which contradicts $\theta(F_k, x) = 1$. As almost every point of F_k is of density 1 for F_k, the proof of (iii) is concluded.

Finally estimates (2) and (3) easily follow from (6) in Sec. 3.1.1. □

Remark 1. A similar result can be proved for function $u \in W^{1,1}(\Omega)$ where Ω is a domain in $\mathbb{R}^n$. We leave the statement and its proof to the reader.

In order to discuss Lusin property (ii), let us recall some simple facts. Let A be a subset in $\mathbb{R}^n$ and let $u : A \to \mathbb{R}$ be a function defined on A.

Definition 2. *We say that $u : A \to \mathbb{R}$ is* differentiable *at $x \in A$ if there exists a linear map $L_x : \mathbb{R}^n \to \mathbb{R}$ such that*

$$\lim_{\substack{y\to x\\ y\in A}} R(y\,;x) = 0 \tag{4}$$

where we have set for $x \neq y$, $x, y \in A$

$$R(y\,;x) := \frac{|u(y) - u(x) - L_x(y-x)|}{|y-x|} \tag{5}$$

We say that u is differentiable *in A if u is differentiable at every point $x \in A$.*

Of course the previous definition is meaningless if x is an isolated point of A. We agree upon the fact that on isolated points any function is continuous and every choice of L_x give rise to a differentiable function. Notice that instead L_x is unique, if it exists, on points of A of positive density.

Definition 3. *Let E be a closed set in $\mathbb{R}^n$. We say that a function $u : E \to \mathbb{R}$ belongs to $C^1(E)$, or is of class C^1 in E, if u is differentiable in E, the gradient Du is continuous in E, and $R(y\,;x) \to 0$ as $y \to x$, uniformly on compact subsets of E.*

Notice that if E is also open, for instance $E = \mathbb{R}^n$, the previous definition agrees with the standard one, as it is easily seen from the inequality

$$\sup_{y\in B(x,r)} R(y\,;x) \leq \text{osc}_{\,B(x,r)} Du$$

Next proposition states that any differentiable function is of class C^1 except on suitable small sets.

Proposition 1. *Let A be a measurable subset of $\mathbb{R}^n$ and let $u : A \to \mathbb{R}$ be a differentiable function with measurable differential Du. Then for every $\varepsilon > 0$ there exists a closed set $F_\varepsilon \subset A$ such that $u \in C^1(F_\varepsilon)$ and $\operatorname{meas}(A \setminus F_\varepsilon) < \varepsilon$.*

Proof. Fix $\varepsilon > 0$. Since $Du(x)$ is measurable, we find by Lusin's theorem a closed set B_ε such that $Du(x)$ is continuous in B_ε and $\operatorname{meas}(A \setminus B_\varepsilon) < \varepsilon/2$. Define now for $k = 1, 2, \dots$

$$\eta_k(x) := \begin{cases} 0 & \text{if } B(x, \frac{1}{k}) \cap A \setminus \{x\} = \emptyset \\ \sup\{|R(y\,; x)| \mid y \in B(x, \frac{1}{k}) \cap A \setminus \{x\}\} & \text{otherwise} \end{cases}$$

R being defined in (5). Clearly $\eta_k(x) \to 0$ as $k \to \infty$ for all $x \in A$. Therefore, by Egoroff's theorem we can find a closed set $F_\varepsilon \subset B_\varepsilon$ such that $\operatorname{meas}(B_\varepsilon \setminus F_\varepsilon) < \varepsilon/2$ and $\eta_k \to 0$ as $k \to \infty$, uniformly on compact subsets of F_ε, i.e., $u \in C^1(F_\varepsilon)$; finally

$$\operatorname{meas}(A \setminus F_\varepsilon) < \operatorname{meas}(A \setminus B_\varepsilon) + \operatorname{meas}(B_\varepsilon \setminus F_\varepsilon) < \varepsilon .$$

□

The relevance of C^1-functions on closed sets of $\mathbb{R}^n$ is due to the following extension theorem due to Whitney that we state without proof.

Theorem 5 (Whitney's theorem). *Let E be a closed subset in $\mathbb{R}^n$ and let $u : E \to \mathbb{R}$ be a function in $C^1(E)$. Then there exists a continuously differentiable function $g : \mathbb{R}^n \to \mathbb{R}$ such that*

$$g = u , \qquad Dg = Du \qquad \text{in } E .$$

Moreover one has the following estimates

$$\| g \|_{\infty, \mathbb{R}^n} \le c \, \| u \|_{\infty, E}$$

$$\| Dg \|_{\infty, \mathbb{R}^n} \le c \, \max\{\| Du \|_{\infty, E}, \max_{x,y \in E} \frac{|u(x) - u(y)|}{|x - y|}\}$$

where c is a constant depending only on the dimension n.

Combining Proposition 1 and Whitney's extension theorem we get at once

Proposition 2. *Let $u : A \to \mathbb{R}$ be a differentiable function on a measurable subset A of $\mathbb{R}^n$. Then for every $\varepsilon > 0$ there exists a closed set $F_\varepsilon \subset A$ and a function $g_\varepsilon \in C^1(\mathbb{R}^n)$ such that*

$$\operatorname{meas}(A \setminus F_\varepsilon) < \varepsilon, \quad u = g_\varepsilon \text{ in } F_\varepsilon, \quad Du = Dg_\varepsilon \text{ in } F_\varepsilon .$$

Combining Kirszbraun's theorem, Rademacher's theorem and Proposition 2 we immediately deduce

Proposition 3. *Let $A \subset \mathbb{R}^n$ be a measurable set and let $u : A \to \mathbb{R}$ be a Lipschitz function. Then, for every $\varepsilon > 0$, there is a closed set $F_\varepsilon \subset A$ and a C^1-function $g_\varepsilon : \mathbb{R}^n \to \mathbb{R}$ such that*

$$\text{meas}\,(A \setminus F_\varepsilon) \le \varepsilon, \quad u = g_\varepsilon \textit{ in } F_\varepsilon, \quad Du = Dg_\varepsilon \textit{ in } F_\varepsilon$$

and

$$\| g_\varepsilon \|_{\infty,\mathbb{R}^n} \le c \, \| u \|_{\infty,A}, \qquad \| Dg_\varepsilon \|_{\infty,\mathbb{R}^n} \le c \, Lip(u, A)$$

where c is a constant depending only on the dimension n.

Combining Theorem 4 and Proposition 3 we derive at once the following theorem of which we give also a different proof in the C^1-context avoiding Lipschitz maps.

Theorem 6. *Let $u \in W^{1,1}_{\text{loc}}(\mathbb{R}^n)$. Then for every $\varepsilon > 0$ there exists a closed set F_ε and a function $g_\varepsilon \in C^1(\mathbb{R}^n)$ such that*

$$\text{meas}\,(\mathbb{R}^n \setminus F_\varepsilon) < \varepsilon, \quad u = g_\varepsilon \textit{ in } F_\varepsilon, \quad Du = Dg_\varepsilon \textit{ in } F_\varepsilon \; .$$

Proof. Let $\varepsilon > 0$. By Lusin theorem one can find a closed set C_ε such that u and Du are continuous on C_ε. Since $⨍_{B(x,r)} R(y\,;x)\,dy \to 0$ for $x \in R_u$, compare Theorem 1 in Sec. 3.1.2, by Egoroff theorem one can find a closed set $F_\varepsilon \subset C_\varepsilon \cap R_u$ such that $\text{meas}\,(\mathbb{R}^n \setminus F_\varepsilon) < \varepsilon$ and

$$⨍_{B(x,r)} R(y\,;x)\,dy \longrightarrow 0 \qquad \text{as } r \to 0$$

uniformly on compact subsets of F_ε, therefore by Theorem 3 in Sec. 3.1.2 $u \in C^1(F_\varepsilon)$. Then Whitney extension theorem, Theorem 5, yields the C^1 extension g_ε on the whole of $\mathbb{R}^n$ of $u_{|F_\varepsilon}$. □

Although we shall not need it, we conclude this subsection by proving a quantitative form of Theorem 6, that is, by proving that not only u agrees with $g_\varepsilon \in C^1(\mathbb{R}^n)$ except on some open set of small measure, but also that g_ε is not far from u in the $W^{1,p}$-norm.

Theorem 7 (Liu). *Let Ω be a domain of $\mathbb{R}^n$ and let $u \in W^{1,p}(\Omega)$ with $p \ge 1$. Then for every $\varepsilon > 0$ there exist a closed set $C_\varepsilon \subset \Omega$ and a function $g_\varepsilon \in C^1(\Omega)$ such that*

$$\text{meas}\,(\Omega \setminus C_\varepsilon) < \varepsilon, \quad g_\varepsilon = u \textit{ in } C_\varepsilon, \quad Dg_\varepsilon = Du \textit{ in } C_\varepsilon$$

and

$$\| u - g_\varepsilon \|_{W^{1,p}(\Omega)} \; < \; \varepsilon$$

First we consider the case $\Omega = \mathbb{R}^n$, and we prove

Theorem 8. *Let $u \in W^{1,p}(\mathbb{R}^n)$ with $p \geq 1$. Then for any $\lambda > 0$ there exist a closed set B_λ and a function $g_\lambda \in C^1(\mathbb{R}^n)$ such that*

$$g_\lambda = u\,, \quad Dg_\lambda = Du \quad \text{on } B_\lambda$$

$$\text{meas}\,(\mathbb{R}^n \setminus B_\lambda) \leq c(n)\lambda^{-p} \| u \|_{W^{1,p}(\mathbb{R}^n)}$$

$$\| g_\lambda \|_{1,\infty} := \sup_{\mathbb{R}^n} |g_\lambda| + \sup_{\mathbb{R}^n} |Dg_\lambda| \leq \lambda\,.$$

Moreover

(6) $$\| g_\lambda \|^p_{1,\infty} \text{ meas}\,(\mathbb{R}^n \setminus B_\lambda) \longrightarrow 0 \qquad \text{as } \lambda \to \infty\,,$$

in particular

(7) $$\| u - g_\lambda \|_{W^{1,p}} \longrightarrow 0 \qquad \text{as } \lambda \to \infty\,.$$

Remark 2. Of course relabeling B_λ and g_λ we can arrange so that

$$\text{meas}\,(\mathbb{R}^n \setminus B_\varepsilon) < \varepsilon, \quad g_\varepsilon = u \text{ on } B_\varepsilon \quad Dg_\varepsilon = Du \text{ on } B_\varepsilon$$

and $\| u - g_\varepsilon \|_{W^{1,p}} < \varepsilon$.

In order to prove Theorem 8 we define for $\lambda > 0$

$$H_\lambda := \{x \in \mathbb{R}^n \mid M(|u|)(x) + M(|Du|)(x) \leq \lambda\}\,.$$

As in the proof of Theorem 4 we easily deduce

Lemma 1. *Let $u \in W^{1,p}(\mathbb{R}^n)$, $p \geq 1$. Then the function $u_{|H_\lambda \cap R_u}$ is Lipschitz continuous. Moreover for all $x, y \in H_\lambda \cap R_u$ we have*

$$|u(x)| \leq \lambda, \quad |Du(x)| \leq \lambda, \quad \frac{|u(x) - u(y)|}{|x - y|} \leq c(n)\,\lambda\,.$$

Proof of Theorem 8. From Lemma 1 we have: $u_{|H_\lambda \cap R_u}$ is Lipschitz-continuous and

$$\| u \|_{\infty, H_\lambda \cap R_u} + \max\{\, \| Du \|_{\infty, H_\lambda \cap R_u}, \sup_{x,y \in H_\lambda \cap R_u} \frac{|u(x) - u(y)|}{|x - y|} \,\} \leq c(n)\lambda.$$

If $\text{meas}\,(\mathbb{R}^n \setminus H_\lambda) = 0$, then the result is an immediate consequence of Proposition 1. Suppose now that $\text{meas}\,(\mathbb{R}^n \setminus H_\lambda) > 0$. Applying Theorem 6 we find a closed set $C_\lambda \subset R_u$ such that $u \in C^1(C_\lambda)$ and $\text{meas}\,(\mathbb{R}^n \setminus C_\lambda) \leq \text{meas}\,(\mathbb{R}^n \setminus H_\lambda)$. On $B_\lambda := C_\lambda \cap H_\lambda$ we have $u \in C^1(B_\lambda)$ and

$$\| u \|_{\infty, B_\lambda} + \max\{\, \| Du \|_{\infty, B_\lambda}, \sup_{x,y \in B_\lambda} \frac{|u(x) - u(y)|}{|x - y|} \,\} \leq c(n)\lambda\,.$$

Thus applying Whitney's extension theorem, we find a function g_λ of class $C^1(\mathbb{R}^n)$ such that

$$g_\lambda = u , \quad Dg_\lambda = Du \quad \text{on } B_\lambda$$

and

(8) $$\| g_\lambda \|_{1,\infty} \le c(n)\lambda .$$

On the other hand

$$\begin{aligned}\mathbb{R}^n \setminus H_\lambda &\subset \{x \in \mathbb{R}^n \mid M(u)(x) + M(|Du|)(x) > \lambda\} \\ &\subset \{x \in \mathbb{R}^n \mid M(u)(x) > \frac{\lambda}{2}\} \cup \{x \in \mathbb{R}^n \mid M(|Du|)(x) > \frac{\lambda}{2}\} ,\end{aligned}$$

hence the weak estimate in Proposition 1 in Sec. 3.1.1 yields

$$\text{meas}\,(\mathbb{R}^n \setminus B_\lambda) \le 2\text{meas}\,(\mathbb{R}^n \setminus H_\lambda) \le c(n)\lambda^{-p}\| u \|_{W^{1,p}(\mathbb{R}^n)}$$

while Proposition 3 in Sec. 3.1.1 (ii) yields

$$\| g_\lambda \|^p_{1,\infty} \text{meas}\,(\mathbb{R}^n \setminus B_\lambda) \le c(n)\lambda^{-p}\text{meas}\,(\mathbb{R}^n \setminus H_\lambda) \to 0 \quad \text{as } \lambda \to \infty ,$$

taking also into account (8). Finally, we have

$$\begin{aligned}\| u - g_\lambda \|_{W^{1,p}(\mathbb{R}^n)} \; \| u - g_\lambda \|_{W^{1,p}(\mathbb{R}^n \setminus B_\lambda)} &\le \\ \le \| u \|_{W^{1,p}(\mathbb{R}^n \setminus B_\lambda)} + \| g_\lambda \|_{W^{1,p}(\mathbb{R}^n \setminus B_\lambda)} .\end{aligned}$$

The first integral tends to zero by the absolute continuity of integral, while the second integral tends to zero because of

$$\| g_\lambda \|^p_{W^{1,p}(\mathbb{R}^n \setminus B_\lambda)} \le \| g_\lambda \|^p_{1,\infty} \text{ meas}\,(\mathbb{R}^n \setminus B_\lambda) \to 0 \quad \text{as } \lambda \to \infty .$$

thus also (7) holds. □

Now Theorem 7 follows from Theorem 8 by means of a suitable partition of unity.

Proof of Theorem 7. Let $\{C_i\}$ be a sequence of non empty compact sets such that

$$C_i \subset \overset{\circ}{C}_{i+1} , \qquad \cup_i C_i = \Omega .$$

Set $C_{-1} := \emptyset$. For each i, let φ_i be a C^∞ function on $\mathbb{R}^n$ such that $0 \le \varphi_i \le 1$, $\text{spt}\,\varphi_i \subset \overset{\circ}{C}_{i+1}$ and C_i is a subset of $\{x \mid \varphi_i(x) = 1\}$. Define

$$\psi_0 := \varphi_0 , \quad \psi_i := \varphi_i - \varphi_{i-1} \quad \text{for } i \ge 1 .$$

Then $\psi_i \in C^\infty_c(\mathbb{R}^n)$ and $\text{spt}\,\psi_i \subset \overset{\circ}{C}_{i+1} - C_{i-1}$, moreover, for $x \in \Omega$, $\psi_i(x) \neq 0$ for at most two values of i, and

$$\sum_{i=0}^{\infty} \psi_i(x) = 1 \qquad \forall\, x \in \Omega .$$

For each $i = 0, 1, 2, \dots$ we now define

$$u_i(x) := \begin{cases} u(x)\psi_i(x) & x \in \Omega \\ 0 & x \notin \Omega \end{cases}$$

By Theorem 8, we find function $g_{\varepsilon,i} \in C^1(\mathbb{R}^n)$, which we may assume with support in $\overset{\circ}{C}_{i+1} - C_i$, such that

$$\| g_{\varepsilon,i} - u_i \|_{W^{1,p}(\mathbb{R}^n)} \leq \frac{\varepsilon}{2^{i+1}}$$
$$\text{meas}\, \{x \mid g_{\varepsilon,i}(x) \neq u_i(x)\} < \frac{\varepsilon}{2^{i+1}} .$$

It is now easily seen that the functions

$$g_\varepsilon = \sum_{i=0}^{\infty} g_{\varepsilon,i}, \quad \varepsilon > 0 ,$$

satisfy the claim. □

1.4 Approximate Differential and Lusin Type Properties

We shall now prove that Lusin type properties (i) and (ii) of Sec. 3.1.3 are in fact equivalent to a kind of pointwise differentiability called approximate differentiability.

We recall, compare Sec. 1.2.1, that for any measurable set $A \subset \mathbb{R}^n$, the *upper density* $\theta^*(A, x)$ *of* A *at* x is defined as

$$\limsup_{r \to 0} \frac{\text{meas}\,(B(x,r) \cap A)}{\text{meas}\,(B(x,r))} .$$

Similarly the *lower density* $\theta_*(A, x)$ *of* A *at* x is given by

$$\liminf_{r \to 0} \frac{\text{meas}\,(B(x,r) \cap A)}{\text{meas}\,(B(x,r))} .$$

The *density of* A *at* x is defined whenever $\theta^*(A,x) = \theta_*(A,x)$ as such a common value

$$\theta(A,x) = \theta^*(A,x) = \theta_*(A,x)$$

By Lebesgue's differentiation theorem or Radon-Nikodym theorem, $\theta(A,x) = 1$ for a.e. $x \in A$ and $\theta(A,x) = 0$ for a.e. $x \in \mathbb{R}^n \setminus A$, compare Ch. 1.

Let us introduce the concept of *approximate limit* and *approximate continuity* for a measurable function $u : A \to \mathbb{R}$.

Definition 1. *Let* A *be a measurable subset of* $\mathbb{R}^n$*, and let* $x \in \mathbb{R}^n$ *be such that* $\theta^*(A,x) > 0$*. Let* $u : A \to \mathbb{R}$ *be a measurable function. We say that* ℓ *is the* approximate limit of u at x *when* y *tends to* x *in* A*, and we write*

$$\ell \ := \ \operatorname*{aplim}_{\substack{y \to x \\ y \in A}} u(y) \ ,$$

if for all $\varepsilon > 0$, *the set*

$$A_\varepsilon \ = \ \{y \in A \mid |u(y) - \ell| \geq \varepsilon\}$$

has density 0 *at* x.

We say that $u : A \to \mathbb{R}$ *is approximately continuous at* x, if

(1) $$u(x) \ := \ \operatorname*{aplim}_{\substack{y \to x \\ y \in A}} u(y) \ .$$

From the definition, one easy sees that

(i) The condition $\theta^*(A, x) > 0$ is essential. In fact in the case $\theta^*(A, x) = 0$ every real number would be the approximate limit of u at x for y tending to x.

(ii) $\operatorname{aplim}_{\substack{y \to x \\ y \in A}} u(y) = \ell$ iff for any open set $W \subset \mathbb{R}$ containing ℓ, the set $A \setminus u^{-1}(W)$ has density 0 at x.

(iii) The existence and the value of the approximate limit do not change if we redefine u on a set N of density zero at x. In particular if $\theta(A, x) = 1$, one may think of u as defined on the whole of $\mathbb{R}^n$, since $\theta(\mathbb{R}^n \setminus A, x) = 0$. Thus we may simply write

$$\ell \ := \ \operatorname*{aplim}_{y \to x} u(y)$$

when $\theta(A, x) = 1$.

Proposition 1. *Approximate limits are unique.*

Proof. Let A be a measurable set, $x \in A$ with $\theta^*(A, x) > 0$, and let $u : A \to \mathbb{R}$ be a measurable function. Suppose that for ℓ, ℓ' with $\ell \neq \ell'$ we have

$$\theta(A_\varepsilon, x) \ = \ \theta(A'_\varepsilon, x) \ = \ 0 \qquad \forall\, \varepsilon$$

where

$$A_\varepsilon = \{y \in A \mid |u(y) - \ell| \geq \varepsilon\}, \qquad A'_\varepsilon = \{y \in A \mid |u(y) - \ell'| \geq \varepsilon\}$$

Choosing $\varepsilon = \frac{|\ell - \ell'|}{3}$, we have

$$3\,\varepsilon \ = \ |\ell - \ell'| \ \leq \ |u(y) - \ell| + |u(y) - \ell'|$$

for any $y \in B(x, r) \cap A$, hence

$$B(x, r) \cap A \ \subset \ (A_\varepsilon \cap B(x, r)) \cup (A'_\varepsilon \cap B(x, r)) \ ,$$

consequently

$$\theta(A, x) \ \leq \ \theta(A_\varepsilon, x) + \theta(A'_\varepsilon, x) \ = \ 0$$

a contradiction. $\square$

The next proposition relates standard limits and approximate limits

Proposition 2. *Let A be a measurable subset of $\mathbb{R}^n$, and let $x \in \mathbb{R}^n$ with $\theta^*(A,x) > 0$. Let $u : A \to \mathbb{R}$ be a measurable function. The following two claims are equivalent*

(i) $\ell := \operatorname{aplim}_{\substack{y\to x\\ y\in A}} u(y)$

(ii) *There exists a measurable set $E \subset \mathbb{R}^n$ with $\theta(A \setminus E, x) = 0$ such that*

$$\ell = \lim_{\substack{y\to x\\ y\in E}} u(y) \tag{2}$$

In particular u is approximately continuous at x if and only if there exists a measurable set E with $\theta(A\setminus E, x) = 0$ such that $u_{|E}$ is continuous at x.

Proof. Suppose that (2) holds. Then clearly for any ε and r sufficiently small we have

$$E \cap A_\varepsilon \cap B(x,r) = \{y \in E \mid |u(y) - \ell| \geq \varepsilon\} \cap B(x,r) = \{x\}\,.$$

Thus $\theta(A_\varepsilon, x) = \theta(A_\varepsilon \cap E, x) = 0$.

On the contrary, if $\operatorname{aplim}_{\substack{y\to x\\ y\in A}} u(y) = \ell$ then one can find a decreasing sequence $\{r_k\}$ of radii, such that

$$\operatorname{meas}(\{y \in A \mid |u(y) - \ell| > 1/k\} \cap B(x,r_k)) \leq \frac{\operatorname{meas}(B(x,r_k))}{2^k} \tag{3}$$

Define now

$$E := A \setminus \bigcup_{k=1}^{\infty} (B(x,r_k) \setminus B(x,r_{k+1})) \cap \{y \,|\, |u(y)-\ell| \geq 1/k\}\,.$$

Clearly E is measurable and $\lim_{y\to x} u_{|E} = \ell$. We now prove that $\theta(A\setminus E, x) = 0$. We have

$$A \setminus E = \bigcup_{k=1}^{\infty} (B(x,r_k) \setminus B(x,r_{k+1})) \cap \{y \,|\, |u(y)-\ell| \geq 1/k\}$$

For any $r > 0$ denote by $\underline{k}$ the integer for which $r_{\underline{k}+1} < r < r_{\underline{k}}$. We have

$$\begin{aligned}
&\operatorname{meas}(B(x,r) \cap (A\setminus E)) \leq \\
&\quad \sum_{k=\underline{k}}^{\infty} \operatorname{meas}((B(x,r_k) \setminus B(x,r_{k+1})) \cap \{y \,|\, |u(y)-\ell| \geq 1/k\}) \leq \\
&\quad \sum_{k=\underline{k}}^{\infty} \operatorname{meas}((B(x,r_k) \cap \{y \,|\, |u(y)-\ell| \geq 1/k\}) \leq \\
&\quad \sum_{k=\underline{k}}^{\infty} \frac{\operatorname{meas}(B(x,r_k)}{2^k} \leq \operatorname{meas}(B(x,r)) \sum_{k=\underline{k}(r)}^{\infty} \frac{1}{2^k}
\end{aligned}$$

from (3). Since as $r \to 0$, $\underline{k} = \underline{k}(r) \to \infty$ we get the result. □

Approximate continuity yields also a kind a local characterization of measurability as shown by next proposition and remark.

Proposition 3. *If the function* $u : \mathbb{R}^n \to \mathbb{R}$ *is measurable then* u *is approximately continuous at a.e.* $x \in \mathbb{R}^n$.

Proof. Suppose that u is measurable. Then for each $k \in \mathbb{N}$ by Lusin theorem, one can find a closed set F_k with $\text{meas}\,(\mathbb{R}^n \setminus F_k) < 1/k$ such that $u_{|F_k}$ is continuous. Denote by $E_k := \{x \in F_k \,|\, \theta(F_k, x) = 1\}$ and by $E := \cup_k E_k$.

Let $x \in E_k$, and $\varepsilon > 0$. Since $u_{|F_k}$ is continuous

$$B(x,r) \cup \{y \in F_k | \; |u(y) - u(x)| \geq \varepsilon\} = \emptyset$$

for small r. Thus

$$B(x,r) \cap \{y \in \mathbb{R}^n | \; |u(y) - u(x)| \geq \varepsilon\} \;\subset\; B(x,r) \setminus F_k$$

and the claim follows letting $r \to 0$, since $\theta(F_k, x) = 1$. □

Remark 1. One can define densities and approximate limits even for non measurable sets replacing in the definition Lebesgue's measure with Lebesgue's outer measure $\mathcal{L}^{n*}$. Then one can reverse the previous proposition, proving, compare Federer [226]: *u is measurable if and only if u is approximately continuous almost everywhere.*

We finally observe that the previous proposition is an easy consequence of the Lebesgue differentiation theorem in the case of functions $u \in L^1_{\text{loc}}(\mathbb{R}^n)$. We have in fact

Proposition 4. *Let* $u \in L^1_{\text{loc}}(\mathbb{R}^n)$. *If*

$$\fint_{B(x,r)} |u(y) - \ell| \, dy \to 0$$

as $r \to 0$ *then*

$$\operatorname*{aplim}_{y \to x} u(y) = \ell \; .$$

In particular the approximate limit exists at each Lebesgue point x *and coincides with the Lebesgue value of* u *at* x.

Proof. Simply note that for $\varepsilon > 0$

$$\text{meas}\,(\{y \mid |u(y) - \ell| \geq \varepsilon\} \cap B(x,r)) \;\leq\; \frac{1}{\varepsilon} \int_{B(x,r)} |u(y) - \ell| \, dy \; .$$

□

[1] The converse of Proposition 4 clearly holds, if moreover u is bounded near x. However it is false in general, as the following example shows.

Let $Q(0,1)$ be the cube in $\mathbb{R}^2$ $\{(x,y) \mid |x| \leq 1, |y| \leq 1\}$ and let $u_\varepsilon(x,y)$ be the function given by

$$u_\varepsilon(x,y) := \begin{cases} x^{-2\varepsilon} & \text{if } 0 \leq x \leq 1,\ 0 \leq y \leq x^{1+\varepsilon} \\ 0 & \text{otherwise .} \end{cases}$$

It is easy to check that $u_\varepsilon \in L^p(Q(0,1))$ for $p < \frac{2+\varepsilon}{2\varepsilon}$ and

$$\fint_{Q(x,r)} |u_\varepsilon(x,y)|\,dx\,dy = \frac{1}{4(2-\varepsilon)} r^{-\varepsilon} \longrightarrow \infty \qquad \text{as } r \to 0$$

while

$$\operatorname*{aplim}_{(x,y)\to(0,0)} u_\varepsilon = 0.$$

•

Let $A \subset \mathbb{R}^n$ be a measurable set and let $x \in \mathbb{R}^n$ be such that $\theta^*(A,x) > 0$. For any measurable function $u : A \to \mathbb{R}$ and any $t \in \mathbb{R}$ we consider the measurable sets

$$\begin{aligned} E_t^+ &= \{x \in A \mid u(x) > t\} \\ E_t^- &= \{x \in A \mid u(x) < t\} \end{aligned}$$

Definition 2. *The* approximate upper limit of u at x *is defined as the number (eventually $+\infty$ or $-\infty$) given by*

$$\operatorname*{aplimsup}_{\substack{y\to x\\ y\in A}} u(y) := \inf\{t \mid \theta^*(E_t^+, x) = 0\}$$

and similarly the approximate lower limit *is given by*

$$\operatorname*{apliminf}_{\substack{y\to x\\ y\in A}} u(y) := \sup\{t \mid \theta^*(E_t^-, x) = 0\}$$

Of course

$$\operatorname*{apliminf}_{\substack{y\to x\\ y\in A}} u(y) \leq \operatorname*{aplimsup}_{\substack{y\to x\\ y\in A}} u(y)$$

and whenever

$$\operatorname*{apliminf}_{\substack{y\to x\\ y\in A}} u(y) = \operatorname*{aplimsup}_{\substack{y\to x\\ y\in A}} u(y) = \ell \in \mathbb{R}$$

the approximate limit exists and we have

$$\operatorname*{aplim}_{\substack{y\to x\\ y\in A}} u(y) = \ell$$

In the same spirit of approximate limits and approximate continuity, we may now introduce the notion of *approximate differential.*

Let A be a measurable subset of $\mathbb{R}^n$ and $u : A \to \mathbb{R}$ be a measurable function. Suppose that $x \in \mathbb{R}^n$ be such that $\theta^*(A,x) > 0$.

Definition 3. *We say that* u *is* approximately differentiable at x *if there exists a linear mapping* $L : \mathbb{R}^n \to \mathbb{R}$ *such that*

$$\operatorname*{aplimsup}_{\substack{y\to x\\ y\in A}} \frac{|u(y)-u(x)-L(y-x)|}{|y-x|} = 0 \,. \tag{4}$$

From the definition it is not difficult to show

(i) If u is approximately differentiable at x then u is approximately continuous at x.

(ii) The *approximate differential*, denoted by $\mathrm{ap}Du(x)$ or simply by $Du(x)$ if no confusion may arise, is unique whenever it exists.

(iii) If $u : A \to \mathbb{R}$ is differentiable in the classical sense at x and $\theta^*(A, x) > 0$ then u is approximately differentiable at x and $\mathrm{ap}Du(x)$ is the ordinary differential $Du(x)$ of u at x.

(iv) u is approximately differentiable at x if and only if there exists a measurable set E with $\theta^*(A\setminus E, x) = 0$ such that $u_{|E}$ is differentiable at x in the classical sense. Moreover $D(u_{|E})(x) = \mathrm{ap}Du(x)$.

(v) If u is approximately differentiable at x and ϕ is differentiable at $u(x)$, then $\phi \circ u$ is approximately differentiable at x and

$$\mathrm{ap}D\phi\circ u(x) \;=\; D\phi(u(x))\,\mathrm{ap}Du(x)\,.$$

(vi) If $u : A \to \mathbb{R}$ is approximately differentiable in A, A measurable, then $\mathrm{ap}Du : A \to \mathbb{R}$ is measurable.

(vii) If $u : A \to \mathbb{R}$ is a.e. approximately differentiable in A and $v = u$ a.e. in A then v is almost everywhere approximately differentiable in A and $\mathrm{ap}Dv = \mathrm{ap}Du$ a.e. in A. That is, the notion of almost everywhere differentiability does not depend on the representative chosen in the equivalence class, compare Lemma 1 in Sec. 3.1.5.

For $u : A \to \mathbb{R}$ measurable, A measurable, one can also define for almost every $x \in A$ the *approximate partial derivatives* of u as the numbers $\mathrm{ap}D_i u(x) = g_i(x)$ for which

$$\operatorname*{aplim}_{\substack{t\to 0\\ x+te_i\in A}} \frac{|u(x+te_i)-u(x)-g_i(x)t|}{|t|} = 0$$

Denote by A_i the domain of existence of $\mathrm{ap}D_i u(x)$. Then one can prove

Theorem 1. *Each* A_i *is measurable, the functions* $\mathrm{ap}D_i u : A_i \to \mathbb{R}$ *are measurable,* u *is approximately differentiable at a.e.* x *in* $\cap_i A_i$ *with*

$$\mathrm{ap}Du(x)(\nu) \;=\; \sum_{i=1}^{n} \mathrm{ap}D_i u(x)\,\nu^i$$

whenever $\nu \in \mathbb{R}^n$.

Since in the sequel we shall not use Theorem 1, we omit its proof for which we refer to Federer [226, 3.1.4].

In terms of approximate differentiability Theorem 2 in Sec. 3.1.2 yields now

Theorem 2. *Let $u \in W^{1,1}_{\text{loc}}(\mathbb{R}^n)$. Then any representative $\tilde{u}$ of u is approximately differentiable at each point $x \in R_{\tilde{u}}$ with approximate differential given by the Lebesgue value $Du(x)$ of the distributional gradient Du at x.*

Let us point out that the existence of the approximate differential $\text{ap}Du(x)$ almost everywhere does not imply that $u \in W^{1,1}_{\text{loc}}$. In fact $\text{ap}Du(x)$ does not agree in general with the distributional derivative of u.

[2] The function $\operatorname{sign} x$ has zero approximate differential at all points $x \neq 0$, but $u \notin W^{1,1}_{\text{loc}}$: its distributional derivative is twice the Dirac mass at zero.

An example of a continuous function with zero approximate differential a.e. is given by Cantor–Vitali function, which is not in $W^{1,1}$, compare [2] in Sec. 1.1.3. •

In Ch. 4 we shall see that if u has derivatives Du which are measures of locally finite total variation then u is approximately differentiable and its approximate differential agrees with the absolutely continuous part of the measure Du with respect to Lebesgue measure.

We also observe that Theorem 4 in Sec. 3.1.2 extends as

Proposition 5. *Let u be approximately differentiable at x_0 and let*

$$f_k(z) := \frac{u(x_0 + \frac{z}{k}) - u(x_0) - Du(x_0)\,\frac{z}{k}}{\frac{|z|}{k}}$$

then $f_k \to 0$ in measure. In particular there exists a sequence k_i such that $f_{k_i}(z) \to 0$ for almost every z.

Proof. The claim follows at once since from the approximate differentiability

$$\text{meas}\,\{\, z \in B(0,R) \mid |f_k(z)| \geq \varepsilon \,\} \longrightarrow 0 \quad \text{as } k \to \infty$$

for all $\varepsilon > 0$. □

We conclude this subsection proving the equivalence of Lusin type properties in Sec. 3.1.3 with the almost everywhere approximate differentiability.

Let A be a measurable set in $\mathbb{R}^n$ and let $u : A \to \mathbb{R}$ be a measurable function. Consider the set of points of approximate differentiability of u

$$A_D(u) := \{x \in A \mid u \text{ is approximately differentiable at } x\}$$

and the set

$$A_L(u) := \{x \in A \mid \operatorname*{aplimsup}_{\substack{y \to x \\ y \in A}} \frac{|u(y) - u(x)|}{|y - x|} < \infty\}$$

Trivially $A_D(u) \subset A_L(u)$ and by definition

$$A_D(u) \subset A_L(u) \subset \{x \in A \mid \theta^*(A, x) > 0\}$$

Theorem 3 (Federer). *The following claims are equivalent*

(i) $\text{meas}\,(A \setminus A_D(u)) = 0$.

(ii) $\text{meas}\,(A \setminus A_L(u)) = 0$.

(iii) *There exist a non decreasing sequence* $\{C_j\}$ *of measurable sets and a sequence* $\{u_j\}$ *of Lipschitz functions in* $\mathbb{R}^n$ *such that*

$$\text{meas}\,(A \setminus \cup_j C_j) = 0, \qquad u = u_j \ \textit{on}\ C_j\ .$$

(iv) *There exist a non decreasing sequence* $\{F_j\}$ *of closed sets and a sequence* $\{v_j\}$ *of functions of class* $C^1(\mathbb{R}^n)$ *such that*

$$\text{meas}\,(A \setminus \cup_j F_j) = 0, \qquad u = v_j \ \textit{on}\ F_j\ .$$

Moreover

(5)
$$A_D(u) \ \subset \ A_L(u) \ = \ \cup_j C_j$$

and in (iii) *and* (iv) *we respectively have*

$$\begin{aligned} \text{ap}Du(x) &= Du_j(x) \ \textit{a.e.}\ x \in C_j \\ \text{ap}Du(x) &= Dv_j(x) \ \textit{for all}\ x \in F_j \end{aligned}$$

Proof. As $A_D(u) \subset A_L(u)$, clearly (i) implies (ii). Let us prove that (ii) implies (iii). Let $x,\ y \in \mathbb{R}^n$ and $\delta := |x-y|$. Let $\beta = \beta(n)$ denote the absolute constant given by

$$\beta \ := \ \frac{\text{meas}\,(B(x,\delta) \cap B(y,\delta))}{\text{meas}\,(B(x,\delta))}\ .$$

Define for $x \in A$ and $j = 1, 2, \ldots$ the sets

$$E_j(x) \ := \ \{y \in A \mid |u(y) - u(x)| \geq j\,|y-x|\}$$

and set

$$B_j \ := \ \{x \in A \mid \frac{\text{meas}\,(B(x,r) \cap E_j(x))}{\text{meas}\,(B(x,r))} \ \leq \ \frac{\beta}{4} \quad \forall\, r\ 0 < r \leq j^{-1}\}$$

Clearly B_j is measurable. Let us prove that

(6)
$$A_L(u) \ \subset \ \cup_j B_j$$

Suppose that $x \notin \cup_j B_j$. Then one can find a sequence r_j, $r_j \to 0$ such that

$$\frac{\text{meas}\,(B(x,r_j) \cap E_j(x))}{\text{meas}\,(B(x,r_j))} \ > \ \frac{\beta}{4}$$

and, since for $k < j$ $E_j(x) \subset E_k(x)$,

$$\frac{\text{meas}\,(B(x,r_j) \cap E_k(x))}{\text{meas}\,(B(x,r_j))} \ \geq \ \frac{\beta}{4} \qquad \forall\, j \geq k$$

thus $\theta^*(E_k(x), x) > 0 \; \forall \; k$, therefore $x \notin A_L(u)$.

Next we observe that

$$|u(y) - u(x)| \;\leq\; 2j\,|y - x| \tag{7}$$

whenever $x, y \in B_j$, $|y - x| \leq 1/j$. In fact if $\delta = |y - x|$, from the definition of B_j we get

$$\begin{aligned} &\operatorname{meas}\left((B(x,\delta) \cap E_j(x)) \cup (B(y,\delta) \cap E_j(y))\right) \\ &\qquad \leq \frac{\beta}{2}\operatorname{meas}\left(B(x,\delta)\right) = \frac{1}{2}\operatorname{meas}\left(B(x,\delta) \cap B(y,\delta)\right) \end{aligned}$$

so there exists at least one z for which

$$\begin{array}{cc} |x - z| \leq \delta & |y - z| \leq \delta \\ |u(x) - u(z)| \leq j|x - z|, & |u(y) - u(z)| \leq j|y - z| \end{array}$$

and by the triangle inequality we get (7).

Finally we express B_j as a union of measurable sets $B_{j,1}, \ldots, B_{j,m}, \ldots$ with diameters less than $1/j$. Note that $u_{|B_{j,k}}$ is Lipschitz. We then glue together the $B_{j,k}$ to form the non decreasing family $\{C_j\}$ of measurable sets and $\{u_{|C_j}\}$ is Lipschitz. As u and u_j are approximately differentiable almost everywhere and coincide on C_j we conclude, compare the argument in Theorem 4 in Sec. 3.1.3, that $Du_j(x) = Du(x)$ if $\theta^*(C_j, x) > 0$. Kirszbraun's theorem yields then the extension u_j. Clearly

$$\cup_j C_j \;=\; \cup_j B_j \;\supset A_L(u)$$

Let us prove that (iii) implies (i). By Rademacher's theorem u_j is differentiable on a set N_j with $\operatorname{meas}(\mathbb{R}^n \setminus N_j) = 0$, hence $u_{|N_j}$ is approximately differentiable at every point in

$$N_j \;\cap\; \{x \in C_j \mid \theta(C_j, x) = 1\}\;, \tag{8}$$

As $\theta(C_j, x) = 1$ also u turns out to be approximately differentiable in the set in (8). The conclusion then follows as the set in (8) differs from C_j by a zero set.

The same argument with C_j replaced by F_j proves that (iv) implies (i). Finally Proposition 2 in Sec. 3.1.3 yields at once that (iii) implies (iv). □

Remark 2. Since every Lipschitz function u has partial derivatives almost everywhere which coincide with ap $Du(e_i)$, from Theorem 3 we infer that every almost everywhere approximately differentiable function has approximate partial derivatives almost everywhere. The converse is also true, compare Theorem 1.

1.5 Area Formulas, Degree, and Graphs of Non Smooth Maps

Let $u : \mathbb{R}^n \to \mathbb{R}^N$, $n \leq N$, be a Lipschitz map, and let A be a measurable subset of $\mathbb{R}^n$. The *counting function* or *Banach's indicatrix of u in A* is defined for all $y \in \mathbb{R}^N$ by

$$N(u, A, y) = \#\{x \mid x \in A\,,\ u(x) = y\} \tag{1}$$

Then the classical *area formula* states that $N(u, A, y)$ is $\mathcal{H}^n$–measurable in y and

$$\int_A J_u(x)\,dx = \int_{\mathbb{R}^N} N(u, A, y)\,d\mathcal{H}^n(y)\,, \quad J_u := \sqrt{\det{(Du)^* Du}} \tag{2}$$

compare Sec. 2.1.2. In particular, if u is one-to-one, we deduce

$$\mathcal{H}^n(u(A)) = \int_A J_u(x)\,dx \tag{3}$$

and more generally the following *change of variables formula* holds

$$\int_A f(u(x)) J_u(x)\,dx = \int_{\mathbb{R}^N} f(y) N(u, A, y)\,d\mathcal{H}^n(y) \tag{4}$$

for all measurable $f : u(A) \to \mathbb{R}$, whenever one of the two sides is meaningful.

We shall now show that (2) (3) (4) can be extended to any L^1 map which is almost everywhere approximately differentiable, in particular to $W^{1,1}$–maps, provided $u(A)$ and $N(u, A, y)$ be suitably defined or provided a suitable representative of u is chosen. In doing that the key point is the rectifiability theorem in the form given in Theorem 3 in Sec. 3.1.4 and a relevant role is played by the set $A_D(u)$ or $\mathcal{R}_u$ of points where the approximate differentiability gives a differential control. Of course the set

$$A_D(u) := \{x \mid u \text{ is approximately differentiable at } x\}$$

depends on the pointwise values of u. However we have

Lemma 1. *Let $u, v : \Omega \to \mathbb{R}^n$ be two measurable maps which agree almost everywhere in Ω. Suppose that, for a given x, u is approximately differentiable at x and v is approximately continuous at x. Then v is approximately differentiable at x, too; moreover* $\mathrm{ap}Dv(x) = \mathrm{ap}Du(x)$.

Proof. In fact for any $\varepsilon > 0$ the sets

$$\{y \mid \frac{|u(y) - u(x) - L(y-x)|}{|y-x|} \geq \varepsilon\} \text{ and } \{y \mid \frac{|v(y) - v(x) - L(y-x)|}{|y-x|} \geq \varepsilon\}$$

with $L := Du(x)$ differ by a null set. □

Lemma 1 shows also that the notion of almost everywhere approximate differentiability depends on the equivalence class of u and not on the chosen representative u.

Introducing the set of approximate continuity of a representative of u

$$A_C(u) := \{x \in \Omega \mid u \text{ is approximately continuous at } x\}$$

we can state Lemma 1 in formula as

$$A_D(u)\Delta A_D(v) \subset A_C(u)\Delta A_C(v)$$

where Δ denotes the symmetric difference.

Definition 1. *Let $u : \mathbb{R}^n \to \mathbb{R}^N$ be a map which is almost everywhere approximately differentiable. We define the* Banach indicatrix *of u by*

$$N(u, A, y) := \#\{x \mid x \in A \cap A_D(u),\ u(x) = y\} \tag{5}$$

Then we have

Theorem 1. *Let Ω be an open set on $\mathbb{R}^n$ and u be an almost everywhere approximately differentiable map, in particular let $u \in W^{1,1}(\Omega, \mathbb{R}^N)$. Then for any measurable subset A of Ω we have*

$$\int_A J_u(x)\, dx = \int_{\mathbb{R}^N} N(u, A, y)\, d\mathcal{H}^n(y) , \quad J_u := \sqrt{\det (Du)^* Du} ; \tag{6}$$

more precisely, $N(u, A, \cdot)$ is $\mathcal{H}^n$–measurable and (6) holds. Also $J_u \in L^1(A)$ if and only if $N(u, A, \cdot)$ is $\mathcal{H}^n$–summable in $\mathbb{R}^N$. Moreover, for any non–negative measurable function $f : \mathbb{R}^N \to \mathbb{R}$ or any $f : \mathbb{R}^N \to \mathbb{R}$ such that either

$$f(\cdot)\ N(u, A, \cdot) \quad \textit{is } \mathcal{H}^n\textit{–summable in } \mathbb{R}^N$$

or

$$f(u(\cdot))J_u(\cdot) \in L^1(A)$$

we have

$$\int_A f(u(x))J_u(x)\, dx = \int_{\mathbb{R}^N} f(y)N(u, A, y)\, d\mathcal{H}^n(y) \tag{7}$$

Proof. It suffices to prove (6), (7) follows in a similar way. Using Theorem 3 in Sec. 3.1.4, we can cover $A_D(u)$ by a non decreasing sequence of disjoint measurable sets $\{F_k\}$ and find Lipschitz functions $u_k \in \mathrm{Lip}\,(\mathbb{R}^n)$ such that

$$\cup_k F_k \supset A_D(u), \qquad u_k = u \text{ on } F_k, \qquad Du_k = Du \text{ a.e on } F_k .$$

From the standard area formula for Lipschitz maps, compare Sec. 2.1.2, we infer

$$\int_{A\cap A_D(u)\cap F_k} J_u(x)\, dx = \int_{A\cap A_D(u)\cap F_k} J_{u_k}(x)\, dx$$

$$\int_{\mathbb{R}^N} N(u_k, A \cap A_D(u) \cap F_k, y)\, d\mathcal{H}^n = \int_{\mathbb{R}^N} N(u, A \cap A_D(u) \cap F_k, y)\, d\mathcal{H}^n$$

observing that $A_D(u) \cap F_k \uparrow A_D(u)$, $\mathrm{meas}\,(\Omega \setminus A_D(u)) = 0$, $N(u_k, A \cap A_D(u) \cap F_k, y) \uparrow N(u, A \cap A_D(u), y)$ and, (6) follows at once, taking into account the definition of $N(u, A, y)$. □

If we define the *rectifiable image of A by u* as

$$u(A)^r := u(A \cap A_D(u)) = \{y \mid N(u, A, y) \neq 0\}$$

we can write (7) as

$$\text{(8)} \qquad \int_A f(u(x)) J_u(x)\, dx = \int_{u(A)^r} f(y) N(u, A, y)\, d\mathcal{H}^n(y)$$

Notice that $u(A)^r$ is rectifiable and just the definition in (5) of $N(u, A, y)$ allows us to identify it.

In the definition of the Banach indicatrix and of the rectifiable image of u we have taken out the singular points, i.e., $\Omega \setminus A_D(u)$. This way $u_{|A_D(u)}$ maps null sets into null sets. This is one of the key properties which makes Theorem 1 valid, as clearly appears also in the next proposition. Recall that a map $u : \Omega \subset \mathbb{R}^n \to \mathbb{R}^N$ has *Lusin's property (N) in Ω* if $\mathcal{H}^n(u(E)) = 0$ whenever $\mathcal{H}^n(E) = 0$, $E \subset \Omega$.

Proposition 1. *Let $u : \Omega \subset \mathbb{R}^n \to \mathbb{R}^N$ be an almost everywhere approximately differentiable map. Then*

(i) $u_{|A_D(u)}$ *has Lusin property (N),*
(ii) $u_{|A_D(u)}$ *maps $\mathcal{L}^n$–measurable sets into $\mathcal{H}^n$–measurable and countably n–rectifiable sets,*
(iii) if $u = v$ a.e., then v is a.e. approximately differentiable, and for any $A \subset \Omega$

$$N(v, A, y) = N(u, A, y) \qquad \textit{for a.e. } y \in \mathbb{R}^N \;;$$

in particular $u(A)^r$ and $v(A)^r$ differ by a null set.

Proof. By Theorem 3 in Sec. 3.1.4 we cover $A_D(u)$ with measurable sets $\{B_j\}$ such that $u_{|B_j}$ is Lipschitz. Then (i) and (ii) clearly follow. To prove (iii) denote by S the symmetric difference between $\{x \,|\, x \in A \cap A_D(u), u(x) = y\}$ and $\{x \,|\, x \in A \cap A_D(v), v(x) = y\}$. If $y \in S$, then either $u(x) = y$ or $v(x) = y$ and $u(x) \neq v(x)$, hence

$$S \subset u(E \cap A_D(u)) \Delta v(E \cap A_D(v))$$

where $E := \{x \mid u(x) \neq v(x)\}$. Since $\operatorname{meas}(E) = 0$ we then infer $\mathcal{H}^n(S) = 0$. □

Let $u \in L^1(\Omega, \mathbb{R}^N)$, $\Omega \subset \mathbb{R}^n$. Recall that the *Lebesgue set of u*, denoted by $\mathcal{L}_u$ is the set of points $x \in \Omega$ for which there exists $\lambda = \lambda_u(x) \in \mathbb{R}^N$ such that

$$⨍_{B(x,r)} |u(y) - \lambda|\, dy \longrightarrow 0 \qquad \text{as } r \to 0,$$

$\lambda_u(x)$ is then called the *Lebesgue value of u at $x \in \mathcal{L}_u$*. As a result of the differentiation theorem for integrals we have seen in the previous subsections that $u(x) = \lambda_u(x)$ for almost every $x \in \Omega$, hence $\lambda_u(x)$ is a measurable function on $\mathcal{L}_u$ and meas $(\Omega \setminus \mathcal{L}_u) = 0$. A *Lebesgue representative* of $u \in L^1$ is then just an extension of $\lambda_u(x)$ to the whole of Ω. We finally set

$$\mathcal{R}_u := \mathcal{L}_u \cap A_D(u)$$

Notice that $u(x) = \lambda_u(x)$ for $x \in \mathcal{R}_u$, since u is approximately continuous for every $x \in A_D(u)$.

Definition 2. *Let $u \in L^1(\Omega, \mathbb{R}^N)$. Any representative of u which maps $\mathcal{H}^n$ null sets in Ω into $\mathcal{H}^n$–null sets is called a* Lusin representative *of u.*

As consequence of Proposition 1 we readily get

Corollary 1. *Let $u \in L^1(\Omega, \mathbb{R}^N)$ be almost everywhere approximately differentiable. Then we have*

(i) Any representative $\tilde{u}$ of u restricted on $\mathcal{R}_{\tilde{u}}$ has Lusin property (N) in $\mathcal{R}_{\tilde{u}}$. In particular a Lusin representative of u in Ω is given simply by

$$\tilde{u}(x) := \begin{cases} u(x) & \text{if } x \in \mathcal{R}_u \\ y_0 & \text{if } x \notin \mathcal{R}_u \end{cases}$$

where $y_0 \in \mathbb{R}^N$ is arbitrary.

(ii) Any Lusin representative of u maps measurable sets into $\mathcal{H}^n$–measurable and countably n–rectifiable sets.

(iii) For any two representatives $\tilde{u}$ and $\tilde{v}$ of u we have

$$N(\tilde{u}, A, y) = N(\tilde{v}, A, y) \qquad \text{for } \mathcal{H}^n\text{–a.e. } y$$

and the sets $\tilde{u}(A)^r$ and $\tilde{v}(A)^r$ differ by a $\mathcal{H}^n$–null set.

From Proposition 1 and Theorem 1 we also infer

Corollary 2. *The change of variables formula (4) holds for any map $u \in L^1$ which is almost everywhere approximately differentiable provided in (4) we take as u a Lusin representative of u, and of course one of the two sides is meaningful.*

Remark 1. In many respects it would have been more convenient to define as Lebesgue representative of u the map $\lambda_u(x)$ with domain $\mathcal{L}_u$ and as Lusin representative the restriction of $\lambda_u(x)$ on $\mathcal{R}_u$. This way one could simply state that the area formula hold for the Lusin representative of an a.e. approximately differentiable map. Though sometimes confusing, we preferred however to follow the tradition in introducing the Definition 1 and Definition 2.

Remark 2. We notice that even if every point in $\Omega \setminus \mathcal{R}_u$ is a Lebesgue point, for instance even if u is continuous, the null set $\Omega \setminus \mathcal{R}_u$ may be mapped by u into a set of positive measure, making (4) not true, while of course (8) holds. This is for instance the case for the Cantor–Vitali map V, compare [2] in Sec. 1.1.3. V

is continuous in $[0,1]$ with zero a.e. differential. It maps $[0,1]\setminus E$, E being the Cantor set, into the dyadic set

$$D := \{x = \frac{j}{2^k} \mid k = 1,2,\dots\,,\ j \in \mathbb{N},\ j \in [0,2^k]\}$$

and E into $[0,1]\setminus D$. Consequently

$$N(u,A,y) = 0 \qquad \text{for a.e. } y$$

while

$$\#\{x \mid V(x) = y\} \;=\; 1\ .$$

However as a simple consequence of Theorem 1, or Corollary 2, we have

Theorem 2 (Radó–Reichelderfer). *Let $u : \Omega \to \mathbb{R}^N$ be a continuous map. Assume that u has Lusin property and is differentiable almost everywhere in Ω. Then (4) holds whenever one of the two sides is meaningful.*

Remark 3. Formula (7) does not hold in general for maps $u : \Omega \to \mathbb{R}^n$ which are in $W^{1,n}(\Omega,\mathbb{R}^n)$, if $N(u,A,y)$ is the classical Banach indicatrix in (1). In fact, given a domain Ω in the plane with boundary a simple curve for which the Lebesgue measure of any arc is non zero, one can find a conformal map u of the half disk $\{z \in \mathbb{C} \,|\, |z| \le 1,\ Im\, z \ge 0\}$ onto Ω. The map u is continuous and belongs to $W^{1,2}$. We extend u to the disk $\{\,|z| \le 1\,\}$ by setting $f(z) = f(\bar z)$ if $Im\, z \le 0$. One then sees that u is continuous and of class $W^{1,2}$ in the disk $\{\,|z| \le 1\,\}$. However u has not Lusin property, the segment $Im\, z = 0$, $-1 < Re\, z < 1$, is mapped into some arc of non–zero measure. But we have

Theorem 3. *Let u belong to $W^{1,p}(\Omega,\mathbb{R}^N)$, $\Omega \subset \mathbb{R}^n$, with $p > n$. Then the continuous representative of u still denoted by u has Lusin property (N) and (4) holds for it.*

Proof. Let $A \subset \Omega$ be such that $|A| = 0$. We need to show that $\mathcal{H}^n(u(A)) = 0$. Fix $\varepsilon > 0$ and choose an open set $V \subset \Omega$ with $|V| < \varepsilon$, $V \supset A$. Let $V = \cup_\nu Q_\nu$ be a cube subdivision of V and r_ν the length of an edge of Q_ν. We have

$$u(A) \;\subset\; \bigcup_{\nu=1}^{\infty} u(Q_\nu)$$

and in view of Morrey–Sobolev estimate in Sec. 3.1.2, the essential diameter of the set $u(Q_\nu)$ does not exceed

$$c\, r_\nu^{1-n/p} \Big(\int_{Q_\nu} |Du|^p\, dx \Big)^{1/p}\ .$$

Consequently

$$\mathcal{H}^n(u(A)) \le \sum_1^\infty \mathcal{H}^n(u(Q_\nu)) \le c \sum_{\nu=1}^\infty r_\nu^{n(1-n/p)} \Big(\int_{Q_\nu} |Du|^p \, dx \Big)^{n/p} .$$

Applying Hölder inequality we infer

$$\begin{aligned} \mathcal{H}^n(u(A)) &\le c \Big(\textstyle\sum_{\nu=1}^\infty r_\nu^n \Big)^{1-n/p} \textstyle\sum_{\nu=1}^\infty \Big(\int_{Q_\nu} |Du|^p \, dx \Big)^{n/p} \\ &\le c |V|^{1-n/p} \Big(\int_V |Du|^p \, dx \Big)^{n/p} \le c \varepsilon^{1-n/p} \Big(\int_\Omega |Du|^p \, dx \Big)^{n/p} \end{aligned}$$

Since $\varepsilon > 0$ is arbitrary, this gives us $\mathcal{H}^n(u(A)) = 0$. □

Let us consider now the special case of maps from a domain $\Omega \subset \mathbb{R}^n$ into $\mathbb{R}^{n+N}$ of the type

$$\mathrm{id} \bowtie u : \Omega \longrightarrow \mathbb{R}^{n+N}$$

given by

$$(\mathrm{id} \bowtie u)(x) := (x, u(x)) .$$

Let $u \in L^1(\Omega, \mathbb{R}^N)$ be a map from $\Omega \subset \mathbb{R}^n$ into $\mathbb{R}^N$ which is almost everywhere approximately differentiable in Ω.

Definition 3. *We define the* graph of u *in* $\Omega \times \mathbb{R}^N$ *as*

$$\mathcal{G}_{u,\Omega} := \{(x, u(x)) \mid x \in \mathcal{R}_u \cap \Omega\}$$

Obviously

$$\mathcal{G}_{u,\Omega} = (\mathrm{id} \bowtie u)(\mathcal{R}_u \cap \Omega) = (\mathrm{id} \bowtie u)(\Omega)^r.$$

Notice that again we are taking out the singular points of u, that is we are considering the *rectifiable image*, which also turns out to be the image of a Lusin representative of u, so that the area formula holds. In particular notice that we have

$$\mathcal{H}^n(\mathcal{G}_{u,\Omega} \cap (A \times \mathbb{R}^N)) = 0$$

whenever A is a null set in Ω. Also $\mathcal{G}_{u,\Omega}$ can be computed in the $\mathcal{H}^n$ a.e. sense using any other representative v of u. In fact, if $v = u$ a.e., then v is a.e. approximately differentiable and $\mathcal{R}_v$ and $\mathcal{R}_u$ differ by a null set. By the area formula we then conclude that

$$\mathcal{H}^n(\mathcal{G}_{u,\Omega} \Delta \mathcal{G}_{v,\Omega}) = 0 .$$

With respect to the standard notion of graph of a map

$$\mathrm{graph}\, u := \{ (x, u(x)) \,|\, x \in \Omega \}$$

the graph $\mathcal{G}_{u,\Omega}$ of an approximately differentiable map is clearly a subset of $\mathrm{graph}\, u$, and in general we may have

$$\mathcal{H}^n(\operatorname{graph} u \setminus \mathcal{G}_{u,\Omega}) > 0$$

From this point of view it would be more reasonable to call $\mathcal{G}_{u,\Omega}$ the 1–*graph of* u, or in Federer's terminology the *countably rectifiable graph*, to distinguish it from the usual graph. In fact, as we have seen, $\mathcal{G}_{u,\Omega}$ is $\mathcal{H}^n$–measurable and countably n–rectifiable and therefore for almost every point of $\mathcal{G}_{u,\Omega}$ the (approximate) tangent plane to $\mathcal{G}_{u,\Omega}$ exists and does not contain vertical vectors, compare Sec. 2.2.1 and Sec. 2.2.4. In the sequel we shall in fact use also the terminology 1–graph, but if no confusion may arise we just speak of graphs.

The discussion above can be formally stated as

Theorem 4. *Let $u \in L^1(\Omega, \mathbb{R}^N)$ be an almost everywhere approximately differentiable map in $\Omega \subset \mathbb{R}^n$. Then $\mathcal{G}_{u,\Omega}$ is $\mathcal{H}^n$–measurable and countably n-rectifiable,*

$$\mathcal{H}^n(\mathcal{G}_{u,\Omega}) = \int_\Omega \sqrt{1 + \sum_{|\alpha|+|\beta|=n} |M_{\bar{\alpha}}^{\beta}(Du(x))|^2}\, dx$$

where $\{M_{\bar{\alpha}}^{\beta}(Du(x))\}$ denotes the set of minors of the approximate differential $Du(x)$. In particular $\mathcal{G}_{u,\Omega}$ has finite area, i.e. is n–rectifiable, if and only if all Jacobian minors of u are in $L^1(\Omega)$.

We emphasize that in particular at $\mathcal{H}^n$-a.e. point $z_0 \in (x_0, y_0)$ in $\mathcal{G}_{u,\Omega}$ there exists an approximate tangent plane L_{x_0} which is obtained as limit

$$\mathcal{H}^n \llcorner \eta_{z_0,\lambda}(\mathcal{G}_{u,\Omega}) \rightharpoonup \mathcal{H}^n \llcorner L_{x_0}$$

as $\lambda \to 0$, compare Sec. 2.1.4, where

$$\eta_{z_0,\lambda}(z) := \frac{z - z_0}{\lambda} .$$

For $x_0 \in \Omega$ and $\lambda > 0$, set

$$u_\lambda(x) := \frac{1}{\lambda}(u(x_0 + \lambda x) - u(x_0)) .$$

Then u_λ converge in measure to the *approximate tangent map* $x \to Du(x_0)x$ *at* x_0

$$u_\lambda(x) \longrightarrow Du(x_0)x \qquad \text{in measure} ,$$

if u is approximately differentiable at x_0, compare Proposition 5 in Sec. 3.1.4, and

$$u_\lambda(x) \longrightarrow Du(x_0)x \qquad \text{in } L^1$$

if u is differentiable in the L^1 sense, compare Theorem 4 in Sec. 3.1.2.

We have

Theorem 5. *Let $u \in L^1(\Omega, \mathbb{R}^N)$ be an almost everywhere approximately differentiable map with $\mathcal{H}^n(\mathcal{G}_{u,\Omega}) < \infty$. Then for $\mathcal{H}^n$-a.e. x_0 in Ω or $\mathcal{H}^n$-a.e. $(x_0, u(x_0))$ in $\mathcal{G}_{u,\Omega}$ both the approximate tangent map $Du(x_0)x$ and the approximate tangent plane L_{x_0} to $\mathcal{G}_{u,\Omega}$ at $(x_0, u(x_0))$ exist and agree*

$$L_{x_0} = \{(x,y) \in \mathbb{R}^n \times \mathbb{R}^N \mid y = Du(x_0)x\}.$$

Proof. Being u a.e. approximately differentiable, by (iv) of Theorem 3 in Sec. 3.1.4 we can decompose Ω as $\Omega = \cup \Omega_j$ and find functions $u_j \in C^1(\mathbb{R}^n, \mathbb{R}^N)$ such that

$$\mathcal{L}^n(\Omega_0) = 0, \quad u_j = u_{|\Omega_j}, \quad Du_j = \text{ap}Du \text{ a.e. in } \Omega_j.$$

Denote by $|M(Du(x))|$ the area density which by assumption is in L^1 and for $j \geq 1$ let $x_0 \in \Omega_j$ be such that

$$\text{(9)} \qquad \begin{aligned} &Du_j(x_0) = \text{ap}Du(x_0) \\ &x_0 \text{ is a Lebesgue point for } |M(Du)| \\ &\theta^*(\mu, \mathbb{R}^n \setminus \Omega_j, x_0) = 0 \end{aligned}$$

where μ is the measure

$$\mu(A) = \int_A |M(Du)|\, dx.$$

Of course a.e. point x_0 in Ω_j meets the requirements in (9).

Set

$$\eta_\lambda(x,y) := \left(\frac{x - x_0}{\lambda}, \frac{y - u(x_0)}{\lambda}\right)$$

and let f be any smooth function with support in $B(0,1) \subset \mathbb{R}^n \times \mathbb{R}^N$

$$\begin{aligned} \mathcal{H}^n \llcorner \eta_\lambda(\mathcal{G}_u)(f) &= \int_{\eta_\lambda(\mathcal{G}_u)} f(x,y)\, d\mathcal{H}^n \\ &= \frac{1}{\lambda^n} \int_{\mathcal{G}_u} f\left(\frac{x - x_0}{\lambda}, \frac{y - u(x_0)}{\lambda}\right) d\mathcal{H}^n \\ &= \frac{1}{\lambda^n} \int_{\mathbb{R}^n} f\left(\frac{x - x_0}{\lambda}, \frac{u(x) - u(x_0)}{\lambda}\right) |M(Du(x))|\, dx \\ &= \int_{B(0,1)} f\left(\xi, \frac{u(x_0 + \lambda\xi) - u_0(x_0)}{\lambda}\right) |M(Du(x_0 + \lambda\xi)|\, d\xi. \end{aligned}$$

Set now $F_{j,\lambda} := \{\xi \mid x_0 + \lambda\xi \in \Omega_j\}$ and compute the last integral as

$$\text{(10)} \qquad \int_{B(0,1)} \cdots = \int_{B(0,1) \cap F_{j,\lambda}} \cdots + \int_{B(0,1) \setminus F_{j,\lambda}} \cdots.$$

For $\lambda \to 0$ we have

$$M(Du(x_0+\lambda\xi)) \longrightarrow M(Du(x_0)) \quad \text{in } L^1$$

$$\chi_{F_{j,\lambda}\cap B(0,1)} \longrightarrow \chi_{B(0,1)} \quad \text{pointwisely}$$

$$\frac{u(x_0+\lambda\xi)-u(x_0)}{\lambda} = \frac{u_j(x_0+\lambda\xi)-u_j(x_0)}{\lambda} \longrightarrow Du_j(x_0)\xi \text{ uniformly,}$$

therefore the first integral on the right hand side of (10) converges to

$$\int f(\xi, Du(x_0)\xi)|M(Du(x_0))|\,dx$$

which of course is just $\mathcal{H}^n \llcorner L_{x_0}(f)$, while the second integral can be estimated by

$$\| f \|_\infty \int_{B(0,1)\setminus F_{j,\lambda}} |M(Du(x_0+\lambda\xi))|\,d\xi = \| f \|_\infty \frac{1}{\lambda^n} \int_{B(0,\lambda)\setminus \Omega_j} |M(Du)|\,d\xi$$

which tends to zero by the third condition in (9). □

Consider now a map $u : \Omega \subset \mathbb{R}^n \to \mathbb{R}^n$ which is approximately differentiable almost everywhere. We can define the *degree of u at y with respect to a measurable set $A \subset \Omega$* by

$$\begin{aligned}
\deg(u, A, y) \quad &:= \quad \#\{x \in A \cap \mathcal{R}_u \mid u(x) = y,\ \det Du(x) > 0\} \\
&\qquad - \#\{x \in A \cap \mathcal{R}_u \mid u(x) = y,\ \det Du(x) < 0\} \\
&= \quad N(u, A \cap \{x \,|\, \det Du(x) > 0\}, y) \\
&\qquad - N(u, A \cap \{x \,|\, \det Du(x) < 0\}, y)
\end{aligned}$$

i.e. we set

Definition 4. *The* degree of u at y with respect to a measurable set $A \subset \Omega$ *is the integer defined by*

$$\deg(u, A, y) = \sum_{x \in u^{-1}(y) \cap A \cap \mathcal{R}_u} \operatorname{sign}(\det Du(x))\ . \tag{11}$$

whenever the sum on the right exists.

We then infer

Theorem 6. *Let u be an almost everywhere approximately differentiable map, in particular let $u \in W^{1,1}(\Omega, \mathbb{R}^n)$ and let A be a measurable subset of Ω. Then* $\deg(u, A, y)$ *is measurable and* $\deg(u, A, y) \in L^1(\mathbb{R}^n)$ *if and only if* $\det Du(x) \in L^1(A)$ *and*

$$\int_A \det Du(x)\,dx = \int_{\mathbb{R}^n} \deg(u, A, y)\,dy\ . \tag{12}$$

Moreover $\deg(u, A, \cdot) = \deg(v, A, \cdot)$ *a.e. in $\mathbb{R}^n$ if $u = v$ a.e. in Ω.*

Proof. It suffices to apply the area formula (7) to the subsets

$$\begin{aligned} A_+ &:= A \cap \{x \in \mathcal{R}_u \mid \det Du(x) > 0\} \\ A_- &:= A \cap \{x \in \mathcal{R}_u \mid \det Du(x) < 0\} \end{aligned}$$

□

Remark 4. We observe that, although for every $u \in W^{1,1}$ with $\det Du(x)$ in L^1 the degree $\deg(u, A, y)$ is well defined, it is not very useful in such a generality. In fact an important property of the degree for smooth maps is its *homological invariance*, i.e., if $u, v \in C^1(\Omega, \mathbb{R}^n)$ and $u = v$ on $\partial\Omega$, then

$$\deg(u, \Omega, y) = \deg(v, \Omega, y) \quad \text{for a.e. } y .$$

But this property is not anymore valid for $W^{1,1}$-functions with $\det Du \in L^1$. For instance, consider the functions $u, v : B(0,1) \subset \mathbb{R}^n \to \mathbb{R}^n$

$$u(x) := x , \qquad v(x) := \frac{x}{|x|} .$$

We have $v(x) \in W^{1,p}(B(0,1), \mathbb{R}^n)$ for all $p < n$, $u = v$ on $\partial B(0,1)$, but

$$\deg(u, B(0,1), y) = 1, \quad \deg(v, B(0,1), y) = 0$$

for a.e. $y \in B(0,1)$. We shall return to this point in Sec. 4.3.

2 Maps with Jacobian Minors in L^1

Motivated by the results of Sec. 3.1, and in particular of Sec. 3.1.5, we shall be concerned in this section with maps u in $L^1(\Omega, \mathbb{R}^N)$ which are approximately differentiable almost everywhere in Ω, for instance Sobolev maps, and with Jacobian minors in L^1.

We shall see that the geometric-differential properties of such maps are described by the integer multiplicity rectifiable current G_u defined as integration of compactly supported n-forms in $\Omega \times \mathbb{R}^N$ over the graph $\mathcal{G}_{u,\Omega}$, compare Sec. 3.2.1, i.e. as the current carried by $\mathcal{G}_{u,\Omega}$.

For smooth maps, $u \in C^1(\Omega, \mathbb{R}^N)$ Stokes theorem yields that the current G_u is boundaryless in $\Omega \times \mathbb{R}^N$,

$$\partial G_u \llcorner \Omega \times \mathbb{R}^N = 0 . \tag{1}$$

This is still true for maps $u \in W^{1,\underline{n}}(\Omega, \mathbb{R}^N)$, $\underline{n} := \min(n, N)$. However it turns out that in general (1) does not hold for $u \in W^{1,p}(\Omega, \mathbb{R}^N)$, $p < \underline{n}$, compare Sec. 3.2.2.

The geometric condition (1) is related to suitable analytical formulas of integration by parts. For instance in the scalar case, $N = 1$, it is equivalent to the classical formula of integration by parts for the derivatives, which expresses the coincidence of the approximate differential and the distributional derivatives.

This is no more true in the vector valued case, $N > 1$. For smooth functions several formulas of integration by parts for the minors are known, e.g., the so called *Piola identities* corresponding to the differential identities

$$D_j(\operatorname{adj}(Du)^{\beta}_{\bar\alpha})^i_j = 0 .$$

We shall see in Sec. 3.2.3 that condition (1) is equivalent to the stronger integration by parts formulas

$$\sum_{j\in\bar\alpha} \int D_j[\phi(x,u(x))](\operatorname{adj}(Du)^{\beta}_{\bar\alpha})^i_j \, dx = 0 \qquad \forall\, \phi \in C^\infty_c(\Omega \times \mathbb{R}^N) .$$

In Sec. 3.2.4 we shall also discuss in terms of boundaries the notion of *distributional determinant* and certain *classes* of maps denoted $\mathcal{A}_{p,q}(\Omega, \mathbb{R}^n) \subset W^{1,n-1}(\Omega, \mathbb{R}^n)$ where $p \geq n-1$, $q \geq n/(n-1)$, whose elements turn out to have graphs with no boundary in $\Omega \times \mathbb{R}^N$. For maps in these classes we shall also prove an *isoperimetric inequality for the determinant* from which one can easily deduce higher integrability of the determinant. We conclude Sec. 3.2.4 with a brief discussion of a different approach to higher integrability in terms of Hardy spaces.

Finally relations between traces and boundaries in $\partial\Omega \times \mathbb{R}^N$ will be discussed in Sec. 3.2.5.

2.1 The Class $\mathcal{A}^1(\Omega, \mathbb{R}^N)$, Graphs and Boundaries

Let Ω be a bounded domain in $\mathbb{R}^n$.

Definition 1. *We denote by $\mathcal{A}^1(\Omega, \mathbb{R}^N)$ the class of functions in $L^1(\Omega, \mathbb{R}^N)$ which are approximately differentiable almost everywhere in Ω, with approximate differential Du, and with the property that all minors of the Jacobian matrix Du are in $L^1(\Omega)$, i.e.,*

$$\begin{aligned} \mathcal{A}^1(\Omega, \mathbb{R}^N) := \; & \{u \in L^1(\Omega, \mathbb{R}^N) \mid \\ & u \text{ is approximately differentiable a.e. },\\ & M^{\beta}_{\bar\alpha}(Du) \in L^1(\Omega)\ \forall\, \alpha, \beta \text{ with } |\alpha| + |\beta| = n\} . \end{aligned} \tag{1}$$

Here we have denoted by $M^{\beta}_{\bar\alpha}(Du)$ the $(\beta, \bar\alpha)$-minor of Du, compare Sec. 2.2.1. We set

$$\| u \|_{\mathcal{A}^1} := \int_\Omega (|u| + |M(Du)|)\, dx \tag{2}$$

where we recall that $M(Du)$ is the n–vector in $\Lambda_n \mathbb{R}^{n+N}$ given by

$$\begin{aligned} M(Du) &= (e_1 + D_1 u) \wedge \ldots \wedge (e_n + D_n u) \\ &= (e_1 + \sum_{i=1}^{N} D_1 u^i \varepsilon_i) \wedge \ldots \wedge (e_n + \sum_{i=1}^{N} D_n u^i \varepsilon_i) \\ &= \sum_{|\alpha|+|\beta|=n} \sigma(\alpha, \bar{\alpha})\, M_{\bar{\alpha}}^{\beta}(Du)\, e_\alpha \wedge \varepsilon_\beta . \end{aligned}$$

and

$$|M(G)| := \{1 + \sum_{\substack{|\alpha|+|\beta|=n \\ |\beta|>0}} |M_{\bar{\alpha}}^{\beta}(G)|^2\}^{1/2}.$$

We remark that $\mathcal{A}^1(\Omega, \mathbb{R}^N)$ is not a linear space and $\| \cdot \|_{\mathcal{A}^1}$ is not a norm. Also observe that every function $W^{1,p}(\Omega, \mathbb{R}^N)$, with $p \geq \underline{n}$, where

(3) $$\underline{n} := \min(n, N) ,$$

belongs to $\mathcal{A}^1(\Omega, \mathbb{R}^N)$. In fact any k-minor of Du, being a linear combination of products of k derivatives of u, belongs to $L^{\underline{n}/k}(\Omega)$.

Let $u \in \mathcal{A}^1(\Omega, \mathbb{R}^N)$ and let $\omega \in \mathcal{D}^n(\Omega \times \mathbb{R}^N)$ be an infinitely differentiable n–form with compact support in $\Omega \times \mathbb{R}^N$. We can define the pull-back of ω onto Ω by the map $\mathrm{id} \bowtie u : \Omega \to \Omega \times \mathbb{R}^n$, $(\mathrm{id} \bowtie u)(x) := (x, u(x))$, as in the smooth case, just replacing ordinary derivatives by approximate derivatives. In coordinates, if

$$\omega(x, y) = \sum_{|\alpha|+|\beta|=n} \omega_{\alpha\beta}(x, y)\, dx^\alpha \wedge dy^\beta$$

then for a.e. $x \in \Omega$

$$\begin{aligned} (\mathrm{id} \bowtie u)^{\#}(\omega)(x) &:= \sum_{|\alpha|+|\beta|=n} \omega_{\alpha\beta}(x, u(x))\, dx^\alpha \wedge du^\beta(x) \\ &= \sum_{|\alpha|+|\beta|=n} \sigma(\alpha, \bar{\alpha})\, \omega_{\alpha\beta}(x, u(x))\, M_{\bar{\alpha}}^{\beta}(Du(x))\, dx \\ &= < \omega(x, u(x)), M(Du(x)) > dx \end{aligned}$$

where $< \, , >$ denotes the duality between covectors and vectors in $\mathbb{R}^n \times \mathbb{R}^N$. In particular we find

(4) $$|(\mathrm{id} \bowtie u)^{\#}(\omega)(x)| \leq ||\omega||_\infty \, |M(Du(x))| \quad \text{a.e. in } \Omega .$$

Therefore to any $u \in \mathcal{A}^1(\Omega, \mathbb{R}^N)$ we can associate the n–dimensional current $G_u \in \mathcal{D}_n(\Omega \times \mathbb{R}^N)$ defined for $\omega \in \mathcal{D}^n(\Omega \times \mathbb{R}^N)$ by

(5) $$\begin{aligned} G_u(\omega) &= \int_\Omega (\mathrm{id} \bowtie u)^{\#}\omega = [\![\Omega]\!]((\mathrm{id} \bowtie u)^{\#}\omega) \\ &= \int_\Omega < \omega(x, u(x)), M(Du(x)) > dx . \end{aligned}$$

From (4) we also see that G_u has finite mass given by

$$\text{(6)} \qquad \mathbf{M}(G_u) = \int_\Omega |M(Du)|\, dx \,.$$

As for smooth graphs, and actually for any current in $\mathcal{D}_n(\Omega \times \mathbb{R}^N)$, the *$k$-component of G_u*, $0 \le k \le \min(n, N)$, is defined by testing G_u with the part of ω with k-differentials in $y \in \mathbb{R}^N$,

$$G_{u(k)}(\omega) := G_u(\omega^{(k)})$$

and, obviously

$$\text{(7)} \qquad G_u(\omega^{(k)}) = \int_\Omega <\omega^{(k)}(x, u(x)), M_{(k)}(Du(x))> dx \,.$$

The *(α, β)-component of G_u*, $(G_u)^{\alpha\beta}$, is the distribution, actually the measure, in $\Omega \times \mathbb{R}^N$ given for $\phi \in C_c^\infty(\Omega \times \mathbb{R}^N)$ by

$$\text{(8)} \qquad G_u^{\alpha\beta}(\phi(x,y)) := \sigma(\alpha, \bar{\alpha}) \int_\Omega \phi(x, u(x))\, M_{\bar{\alpha}}^{\beta}(Du(x))\, dx \,.$$

As $\mathbf{M}(G_u) < +\infty$, we can think of G_u as being defined on the space of all n-forms ω with Borel bounded coefficients in $\mathbb{R}^n \times \mathbb{R}^N$. In fact if $\{\omega_k\} \subset \mathcal{D}^n(\Omega \times \mathbb{R}^N)$ converges pointwisely to ω in $\Omega \times \mathbb{R}^N$, and $\sup_k \| \omega_k \|_{\infty, \Omega \times \mathbb{R}^N} < \infty$, by Lebesgue dominated convergence theorem, we can define $G_u(\omega)$ as

$$G_u(\omega) := \lim_{k \to \infty} G_u(\omega_k) \,.$$

compare Sec. 1.1.4.

Let $u \in \mathcal{A}^1(\Omega, \mathbb{R}^N)$, and let $\mathcal{G}_{u,\Omega}$ be the graph of u defined in Sec. 3.1.5. From Theorem 4 in Sec. 3.1.5 we known that $\mathcal{G}_{u,\Omega}$ is n-rectifiable and

$$\mathcal{H}^n(\mathcal{G}_{u,\Omega}) = \int_\Omega |M(Du(x))|\, dx = \mathbf{M}(G_u) \,.$$

and by Theorem 5 in Sec. 3.1.5

$$\vec{\mathcal{G}}_{u,\Omega}(x, u(x)) = \frac{M(Du(x))}{|M(Du(x))|}$$

is an n–vector which orients $\mathcal{G}_{u,\Omega}$. Again the area formula, Theorem 4 in Sec. 3.1.5, then yields that

$$G_u(\omega) = \int < \omega, \vec{\mathcal{G}}_{u,\Omega} > d\mathcal{H}^n \llcorner \mathcal{G}_{u,\Omega} \,,$$

and

$$\mathcal{H}^n(\mathcal{G}_{u,\Omega} \cap (A \times \mathbb{R}^N)) = 0 \quad \text{if and only if } \mathcal{H}^n(A) = 0 \,,$$

therefore we can state

Proposition 1. *Let $u \in \mathcal{A}^1(\Omega, \mathbb{R}^N)$. Then G_u is an i.m. rectifiable current with multiplicity one, more precisely*

$$G_u = \tau(\mathcal{G}_{u,\Omega}, 1, \vec{\mathcal{G}}_{u,\Omega}) .$$

Definition 2. *We say that a sequence $\{u_k\} \subset \mathcal{A}^1(\Omega, \mathbb{R}^N)$ converges weakly in $\mathcal{A}^1$ to $u \in \mathcal{A}^1(\Omega, \mathbb{R}^N)$, $u_k \stackrel{\mathcal{A}^1}{\rightharpoonup} u$, if and only if*

$$\begin{array}{lllr} u_k & \to & u & \textit{strongly in } L^1 \\ M(Du_k) & \rightharpoonup & M(Du) & \textit{weakly in} L^1 . \end{array} \tag{9}$$

Notice that in Definition 2 we require strong convergence of u_k to u in L^1. This is due to the fact that the Du_k are not in general the distributional derivatives of u_k. If this is the case, we can as well replace strong convergence of u_k to u by weak convergence. We also notice that the a.e. convergence of u_k to u would do the same.

The $\mathcal{A}^1$-weak convergence reads as convergence of currents.

Proposition 2. *Let $u_k, u \in \mathcal{A}^1(\Omega, \mathbb{R}^N)$ and suppose that $u_k \stackrel{\mathcal{A}^1}{\rightharpoonup} u$. Then $G_{u_k}(\omega) \to G_u(\omega)$ for any n-form ω with continuous and bounded coefficients in $\mathbb{R}^n \times \mathbb{R}^N$, i.e.*

$$G_{u_k} \rightharpoonup G_u \qquad \textit{in } \mathcal{D}_n(\mathbb{R}^n \times \mathbb{R}^N) ;$$

in particular

$$G_{u_k} \rightharpoonup G_u \qquad \textit{in } \mathcal{D}_n(\Omega \times \mathbb{R}^N) .$$

Proof. Possibly passing to a subsequence from $u_k \stackrel{\mathcal{A}^1}{\rightharpoonup} u$ we deduce

$$\begin{array}{ll} M^{\beta}_{\bar{\alpha}}(Du_k(x)) \rightharpoonup M^{\beta}_{\bar{\alpha}}(Du(x)) & \text{weakly in } L^1 \\ u_k(x) \to u(x) & \text{a.e. in } \Omega \\ \phi(x, u_k(x)) \to \phi(x, u(x)) & \text{a.e. in } \Omega \\ \sup_k |\phi(x, u_k(x))| < +\infty & \end{array}$$

for any α, β with $|\alpha| + |\beta| = n$ and for any $\phi : \mathbb{R}^n \times \mathbb{R}^N \to \mathbb{R}$ which is continuous and bounded. Thus (i) of Proposition 1 in Sec. 1.2.4 implies

$$\begin{aligned} (G_{u_k})^{\alpha\beta}(\phi(x,y)) &= \sigma(\alpha, \bar{\alpha}) \int_\Omega \phi(x, u_k(x)) M^{\beta}_{\bar{\alpha}}(Du_k(x))\, dx \\ \longrightarrow \sigma(\alpha, \bar{\alpha}) \int_\Omega \phi(x, u(x)) M^{\beta}_{\bar{\alpha}}(Du(x))\, dx &= G_u^{\alpha\beta}(\phi(x,y)) . \end{aligned}$$

We then conclude that any subsequence of $\{u_k\}$ has a subsequence $\{u_{h_k}\}$ such that

$$G^{\alpha\beta}_{u_{h_k}}(\phi) \longrightarrow G^{\alpha\beta}_u(\phi) \qquad \forall\, \phi .$$

Since the limit is independent of the chosen subsequence, the claim follows at once. □

Let $u : \Omega \to \mathbb{R}^N$ be a smooth map, for instance $u \in C^2(\bar{\Omega}, \mathbb{R}^N)$. Then $u \in \mathcal{A}^1(\Omega, \mathbb{R}^N)$, and, due to *Stokes's theorem* G_u is boundaryless in $\Omega \times \mathbb{R}^N$, i.e.,

$$\partial G_u \llcorner \Omega \times \mathbb{R}^N = 0 . \tag{10}$$

In fact $(\mathrm{id} \bowtie u)^{\#}\omega$ is an $(n-1)$–form in Ω with C^1_c–coefficients for any $\omega \in \mathcal{D}^{n-1}(\Omega \times \mathbb{R}^N)$, hence

$$(\mathrm{id} \bowtie u)^{\#}\omega = \sum_{i=1}^{n} (-1)^{i-1} f_i(x)\, \widehat{dx^i}, \qquad f_i \in C^1_c(\Omega) ,$$

therefore

$$\partial G_u(\omega) = G_u(d\omega) = [\![\, \Omega \,]\!](d(\mathrm{id} \bowtie u)^{\#}\omega) = \int_{\Omega} \mathrm{div}\, f\, dx = \int_{\partial\Omega} f \cdot \nu_{\partial\Omega}\, d\mathcal{H}^{n-1} = 0 .$$

As we have seen in Ch. 1, and as we shall see later on, property (10) is crucial in a number of situations and especially if we want conserve homological properties of regular maps to their weak limits. But examples in Sec. 3.2.2 below show that, for a generic $u \in \mathcal{A}^1(\Omega, \mathbb{R}^N)$, (10) does not hold. In general, if $u \in \mathcal{A}^1(\Omega, \mathbb{R}^N) \cap W^{1,p}(\Omega, \mathbb{R}^N)$, G_u does have boundary in $\Omega \times \mathbb{R}^N$ whenever $p < \underline{n}$, $\underline{n} := \min(n, N)$. We then set

Definition 3. *Let $\Omega \subset \mathbb{R}^n$ be an open domain. The class of* Cartesian maps *is defined by*

$$\mathrm{cart}^1(\Omega, \mathbb{R}^N) := \{u \in \mathcal{A}^1(\Omega, \mathbb{R}^N) \mid \partial G_u \llcorner \Omega \times \mathbb{R}^N = 0\}$$

We have

Proposition 3. *Let u be a map in $W^{1,\underline{n}}(\Omega, \mathbb{R}^N)$, $\underline{n} := \min(n, N)$. Then $u \in \mathrm{cart}^1(\Omega, \mathbb{R}^N)$.*

Proof. Since any minor of Du is a linear combination of products of k derivatives of u, $1 \le k \le \underline{n}$, it belongs to $L^{\underline{n}/k}$, thus $u \in \mathcal{A}^1(\Omega, \mathbb{R}^N)$. Let $\{u_k\} \subset C^1(\Omega, \mathbb{R}^N) \cap W^{1,\underline{n}}(\Omega, \mathbb{R}^N)$ be a sequence of smooth maps converging to u in $W^{1,\underline{n}}(\Omega, \mathbb{R}^N)$. Then $u_k \rightharpoonup u$ weakly in $\mathcal{A}^1(\Omega, \mathbb{R}^N)$, and Proposition 2 yields $G_{u_k} \rightharpoonup G_u$ weakly in the sense of the currents in $\Omega \times \mathbb{R}^N$. As $\partial G_{u_k} \rightharpoonup \partial G_u$ in $\Omega \times \mathbb{R}^N$ and $\partial G_{u_k} \llcorner \Omega \times \mathbb{R}^N = 0$, the claim is proved. □

Notice that in particular $\partial G_u \llcorner \Omega \times \mathbb{R}^N = 0$, i.e. $u \in \mathrm{cart}^1(\Omega, \mathbb{R}^N)$ if u is a Lipschitz map.

2.2 Examples

We shall present in this section a few examples of maps in $\mathcal{A}^1(\Omega, \mathbb{R}^N)$ whose graphs have *holes* in $\Omega \times \mathbb{R}^N$ and the corresponding currents G_u's have non zero boundaries in $\Omega \times \mathbb{R}^N$.

1 Let us begin by showing that there exist maps in $\mathcal{A}^1(\Omega, \mathbb{R}^N) \cap W^{1,p}(\Omega, \mathbb{R}^N)$, $p < n$, whose graphs have *homological holes* in $\Omega \times \mathbb{R}^N$.

Let $B(0,1)$ be the unit ball in $\mathbb{R}^n$ and let $u : B(0,1) \to B(0,1)$ the map

$$u := \frac{x}{|x|}\ .$$

Clearly $u \in W^{1,p}(B(0,1), \mathbb{R}^n)$ for all $p < n$, hence all minors of order less or

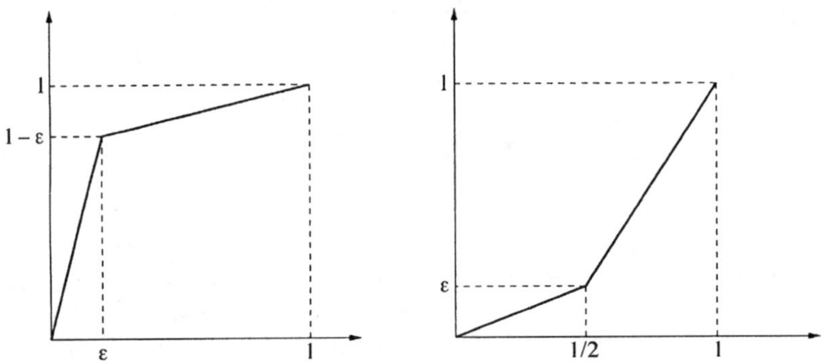

Fig. 3.1. The maps $r_\varepsilon(t)$ and $\rho_\varepsilon(t)$.

equal to $n-1$ belong to $L^{p/(n-1)}(\Omega) \subset L^1(\Omega)$, while for the last minor which is of order n we have

$$\det Du(x) = 0 \qquad \text{for } x \neq 0$$

since u maps $B(0,1) \setminus \{0\}$ into $S^{n-1} \subset \mathbb{R}^n$. Thus $u \in \mathcal{A}^1(\Omega, \mathbb{R}^n)$. We now claim that

$$\text{(1)} \qquad \partial G_u \llcorner \Omega \times \mathbb{R}^n = -\delta_0 \times \partial [\![\{y \in \mathbb{R}^n \mid |y| \leq 1\}]\!]\ ,$$

or in other words, that $\mathcal{G}_{\frac{x}{|x|}, B(0,1)}$ has the set $\{0\} \times B(0,1)$ in $\mathbb{R}^n \times \mathbb{R}^n$ as a hole, which lies above the origin of $\mathbb{R}^n$.

In order to prove (1) we approximate $u(x) := \frac{x}{|x|}$ in $W^{1,p}(B(0,1), \mathbb{R}^n)$, $p < n$, by the Lipschitz-continuous maps

$$u_\varepsilon(x) := r_\varepsilon(|x|)\,\frac{x}{|x|}$$

where $r_\varepsilon(t)$ is the function from $[0,1]$ to $[0,1]$ that is linear in $(0,\varepsilon)$ and in $(\varepsilon,1)$ and satisfies $r_\varepsilon(0) = 0$, $r_\varepsilon(\varepsilon) = 1-\varepsilon$, $r_\varepsilon(1) = 1$, see Figure 3.1.

We observe that an easy computation yields that in the sense of measures

$$\det Du_\varepsilon \rightharpoonup |B_1|\delta_0$$

where δ_0 is the Dirac mass at zero. We shall now prove that in fact

$$\text{(2)} \qquad G_{u_\varepsilon} \rightharpoonup G_u + \delta_0 \times [\![B(0,1)]\!]\ ,$$

therefore, since $\partial G_{u_\varepsilon} \llcorner B(0,1) \times \mathbb{R}^n = 0$, $\forall\, \varepsilon$, we deduce

$$\partial(G_u + \delta_0 \times [\![B(0,1)]\!]) \llcorner B(0,1) \times \mathbb{R}^n = 0$$

i.e.

$$(\partial G_u) \llcorner B(0,1) \times \mathbb{R}^n = -\delta_0 \times \partial[\![B(0,1)]\!]$$

which is (1).

To show (2) we reparametrize $\mathcal{G}_{u_\varepsilon, B(0,1)}$ as follows. Let $\gamma_\varepsilon : B(0,1) \to B(0,1)$ be the map

$$\gamma_\varepsilon(z) := \rho_\varepsilon(|z|)\frac{z}{|z|} ,$$

ρ_ε being the map from $[0,1]$ in $[0,1]$ which is linear in $(0,1/2)$ and in $(1/2,1)$, and satisfies $\rho_\varepsilon(0) = 0$, $\rho_\varepsilon(1/2) = \varepsilon$, $\rho(1) = 1$, see Figure 3.1.

Then defining

$$\phi_\varepsilon : B(0,1) \to B(0,1) \times B(0,1) , \qquad \phi_\varepsilon(z) := (\gamma_\varepsilon(z), u_\varepsilon(\gamma_\varepsilon(z))) ,$$

we have

$$(x, u_\varepsilon(x)) = \phi_\varepsilon(z) , \qquad x = \gamma_\varepsilon(z) ,$$

thus

$$G_{u_\varepsilon} = \phi_{\varepsilon\#}[\![B(0,1)]\!]$$

On the other hand ϕ_ε is a family of equi-Lipschitz maps which converge uniformly to the Lipschitz map ϕ given by

$$\phi(z) := \begin{cases} (0, 2z) & \text{if } |z| \le \frac{1}{2} \\ ((2|z|-1)\frac{z}{|z|}, \frac{z}{|z|}) & \text{if } \frac{1}{2} \le |z| \le 1 \end{cases}$$

$$= \begin{cases} (0, 2z) & \text{if } |z| \le \frac{1}{2} \\ (\gamma(z), u(\gamma(z))) & \text{if } \frac{1}{2} \le z \le 1 \end{cases}$$

where $\gamma(z) = (2|z|-1)\dfrac{z}{|z|}$. Thus we deduce

$$\begin{aligned} G_{u_\varepsilon} \rightharpoonup \phi_\#[\![B(0,1)]\!] &= \phi_\#[\![B(0,1) \setminus B(0,1/2)]\!] + \phi_\#[\![B(0,1/2)]\!] \\ &= G_u + [\![\{0\} \times B(0,1)]\!] \end{aligned}$$

which proves (2).

The previous example shows also that it is not possible to approximate weakly $G_{\frac{x}{|x|}}$ by a sequence of smooth graphs G_{u_k}, since in this case we would have

$$\partial G_{\frac{x}{|x|}} \llcorner B(0,1) \times \mathbb{R}^n = 0$$

while as we have seen

$$\partial G_{\frac{x}{|x|}} \llcorner B(0,1) \times \mathbb{R}^n = -\delta_0 \times [\![B(0,1)]\!] \neq 0 .$$

In a more analytic form, this shows that, while we can approximate $\frac{x}{|x|}$ in $W^{1,p}$, for any $p < n$, by a sequence of smooth functions, no such a sequence $\{u_k\}$ may satisfy

$$\det Du_k \longrightarrow \det Du \qquad \text{in } L^1(B(0,1))$$

or even

$$\det Du_k \rightharpoonup \det Du$$

in the sense of measures. •

[2] Similarly to [1] we can discuss the boundary in $B(0,1) \times \mathbb{R}^N$ of graphs of functions which are homogeneous of degree zero.

Let φ a smooth map from $S^{n-1} \subset \mathbb{R}^n$ into $\mathbb{R}^N$ and let us denote by u the homogeneous extension of φ to $B(0,1)$, i.e.,

$$u : B(0,1) \subset \mathbb{R}^n \to \mathbb{R}^N \,, \qquad u(x) \;:=\; \varphi(\frac{x}{|x|})$$

Clearly u is singular at the origin (except in the case $\varphi =$ constant), moreover $u \in W^{1,p}(B(0,1), \mathbb{R}^N)$ for any $p < n$, and also $u \in \mathcal{A}^1(B(0,1), \mathbb{R}^N)$.

Approximating u by

$$u_\varepsilon(x) \;:=\; r_\varepsilon(|x|)\, \varphi(\frac{x}{|x|})$$

r_ε being the same function in [1], setting as in [1] $\phi_\varepsilon(z) := (\gamma_\varepsilon(z), u_\varepsilon(\gamma_\varepsilon(z)))$, we get

$$\phi_{\varepsilon\#}[\![B(0,1)]\!] \;=\; G_{u_\varepsilon}$$

and

$$G_{u_\varepsilon} \rightharpoonup \phi_\#[\![B(0,1)]\!] \;=\; \phi_\#[\![B(0,1) \setminus B(0,1/2)]\!] + \phi_\#[\![B(0,1/2)]\!]$$

where

$$\phi(z) \;:=\; \begin{cases} (0, 2|z|\varphi(\frac{z}{|z|})) & |z| \le \frac{1}{2} \\ (\gamma(z), \varphi(\frac{\gamma(z)}{|\gamma(z)|})) & \frac{1}{2} \le |z| < 1 \end{cases} \qquad \gamma(z) \;:=\; (2|z| - 1)\frac{z}{|z|}$$

Now, clearly

$$\phi_\#[\![B(0,1) \setminus B(0,1/2)]\!] \;=\; G_u \,,$$

and

$$\partial\phi_\# \,[\![B(0,1)]\!] \llcorner B(0,1) \times \mathbb{R}^N \;=\; 0$$

while

$$\begin{aligned} \partial\phi_\# \,[\![B(0,1/2)]\!] \llcorner B(0,1) \times \mathbb{R}^N \;&=\; \partial\phi_\# \,[\![B(0,1/2)]\!] \;= \\ =\; \phi_\# \,\partial[\![B(0,1/2)]\!] \;&=\; \delta_0 \times \varphi_\#[\![S^{n-1}]\!] \,, \end{aligned}$$

$[\![S^{n-1}]\!]$ denoting the current integration over S^{n-1} oriented so that $[\![S^{n-1}]\!] = \partial[\![B(0,1)]\!]$. Therefore we conclude that

$$\partial G_u \llcorner B(0,1) \times \mathbb{R}^N = -\delta_0 \times \varphi_\# [\![S^{n-1}]\!]$$

and that *there exists a sequence* $\{u_k\}$ *of smooth maps* u_k *such that* $G_{u_k} \rightharpoonup G_u$ *if and only if* $\partial G_u = 0$ *in* $B(0,1)\times\mathbb{R}^N$ *or equivalently if and only if* $\varphi_\#[\![S^{n-1}]\!] = 0$.

Very roughly we can say that even if the graph G_u of the map $u(x)$ has a hole above zero, one can approximate G_u by graphs G_{u_k} of smooth maps u_k if and only if the integration of forms on $\varphi(S^{n-1})$ is identically zero.

In the special case $\varphi : S^{n-1} \to S^{n-1}$, by the constancy theorem, compare Sec. 4.3.1 and Sec. 5.5.4, we have

$$\varphi_\#[\![S^{n-1}]\!] = \deg\varphi\, [\![S^{n-1}]\!]$$

where $\deg\varphi$ is the degree of the map $\varphi : S^{n-1} \to S^{n-1}$, thus

$$\partial G_u \llcorner B(0,1) \times \mathbb{R}^N = -\deg\varphi\ \delta_0 \times [\![S^{n-1}]\!]$$

and we can conclude that *there exists a sequence of smooth maps* $u_k : B(0,1) \subset \mathbb{R}^n \to \mathbb{R}^n$, $u_k(x) = \varphi(x)$ *for* $x \in \partial B(0,1)$ *such that*

$$G_{u_k} \rightharpoonup G_u\ , \qquad u := \varphi(\frac{x}{|x|})$$

if and only if $\deg\varphi = 0$. •

[3] In $\mathbb{R}^2$ which we identify with the complex plan $\mathbb{C}$ we consider

$$\varphi(z) := \frac{z^k}{|z|^k}\ , \quad z\in\mathbb{C}\ , \quad k\in\mathbb{Z}\ .$$

We have $\deg\varphi = k$, thus by [2]

$$\partial G_u \llcorner B(0,1) \times \mathbb{R}^2 = -k\,\delta_0 \times [\![S^1]\!]$$

where

$$u(x) := \varphi(\frac{x}{|x|})$$

In particular we have

$$\mathbf{M}(\partial G_u \llcorner B(0,1) \times \mathbb{R}^2) = 2\pi\,|k|\ ,$$

that is, one can find in $\mathcal{A}^1(\Omega,\mathbb{R}^N)$ mappings with non zero boundary in $\Omega\times\mathbb{R}^N$ of arbitrarily large mass. •

[4] Actually there exists maps u in $\mathcal{A}^1(\Omega,\mathbb{R}^N)$ with $\mathbf{M}(\partial G_u \llcorner \Omega \times \mathbb{R}^N) = \infty$.

Let $\psi : \mathbb{R} \to \mathbb{R}$ be a Lipschitz function which is zero outside the interval $(-1,1)$ and satisfies $\psi(0) = 1$. For $k = 1,2,\dots$ we denote by x_k the points in $\mathbb{R}^2$ with coordinates

$$x_k = (\frac{1}{k}, 0)$$

and we set

$$\lambda_k := \min(|x_k - x_{k-1}|, |x_k - x_{k+1}|) = \frac{1}{k(k+1)} .$$

Then we define $u : B(0,2) \subset \mathbb{R}^2 \to \mathbb{R}^2$ as

$$u(x) := \sum_{k=1}^{\infty} \sqrt{\lambda_k} \frac{x - x_k}{|x - x_k|} \psi(\lambda_k^{-1}|x - x_k|) .$$

An easy computation shows that $u \in W^{1,p}(B(0,2), \mathbb{R}^2)$ for all $p < 2$, $\det Du \in L^1(B(0,2))$, i.e., $u \in \mathcal{A}^1(B(0,2), \mathbb{R}^2)$. According to the computations in Example 3 one also sees that

$$\partial G_u \llcorner B(0,2) \times \mathbb{R}^2 = -\sum_k \delta_{x_k} \times [\![\partial B(0, \sqrt{\lambda_k})]\!]$$

thus

$$\mathbf{M}(\partial G_u \llcorner B(0,2) \times \mathbb{R}^2) = 2\pi \sum_k \sqrt{\lambda_k} = +\infty .$$

•

2.3 Boundaries and Integration by Parts

In this section we discuss in a more analytic way the nature of the condition

$$\partial G_u \llcorner \Omega \times \mathbb{R}^N = 0 \tag{1}$$

which defines the class $\mathrm{cart}^1(\Omega, \mathbb{R}^N)$, for functions $u \in \mathcal{A}^1(\Omega, \mathbb{R}^N)$.

If $\alpha = (\alpha_1, \dots, \alpha_p)$ is a multi–index in $I(p, n)$, $0 \le p \le n$, we say that the positive integer j belongs to α if j is one of the indexes $\alpha_1, \dots, \alpha_p$. If $j \in \alpha$ we denote by $\alpha - j$ the multi–index of length $p - 1$, obtained by removing the index j from α. Similarly, if $j \notin \bar{\alpha}$, we denote by $\alpha + j$ the multi–index of length $p + 1$ obtaining reordering naturally the multi–index $(\alpha_1, \dots, \alpha_p, j)$.

For any submatrix G_α^β of an $N \times n$-matrix G, $1 \le |\alpha| = |\beta| \le \underline{n} := \min(n, N)$, we define the *matrix of adjoints of* G_α^β by the formulas

$$(\mathrm{adj}\, G_\alpha^\beta)_j^i := \sigma(i, \beta - i)\, \sigma(j, \alpha - j) \det G_{\alpha - j}^{\beta - i} \tag{2}$$

where $i \in \beta$ and $j \in \alpha$. Of course, if G is an $n \times n$-matrix and if $|\alpha| = |\beta| = n$, we simply write $\mathrm{adj}\, G$ for $\mathrm{adj}\, G_\alpha^\beta = \mathrm{adj}\, G_{\bar{0}}^{\bar{0}}$. Also we define $\mathrm{adj}\, G_\alpha^\beta := 1$ if $|\alpha| = |\beta| = 1$. Notice that we also use the notation

$$\det G_{\alpha - j}^{\beta - i} = M_{\alpha - i}^{\beta - i}(G) .$$

The previous definition is set up in such a way that the usual *Laplace's formulas* for the determinant hold, i.e, for $i, h \in \beta$

(3) $$\sum_{j\in\alpha} G_j^h (\operatorname{adj} G_\alpha^\beta)_j^i = \delta^{ih} \det G_\alpha^\beta = \delta^{ih} M_\alpha^\beta(G) ,$$

and, for $i, j \in \alpha$

(4) $$\sum_{h\in\beta} G_j^h (\operatorname{adj} G_\alpha^\beta)_i^h = \delta_{ij} M_a^\beta(G)$$

Let u be a function in $\mathcal{A}^1(\Omega, \mathbb{R}^N)$, where Ω is a domain of $\mathbb{R}^n$. Since G_u has finite mass, we may regard G_u as a linear functional on the space of all n-forms with compact support in $\mathbb{R}^n \times \mathbb{R}^N$, and not only in $\Omega \times \mathbb{R}^N$, defined by the same formula

$$G_u(\omega) := \int_\Omega < \omega(x, u(x)), M(Du(x)) > dx ,$$

that is, we may regard G_u as an element of $\mathcal{D}_n(\mathbb{R}^n \times \mathbb{R}^N)$. The measure theoretic boundary of G_u is then defined by

$$\partial G_u(\omega) := G_u(d\omega)$$

for any $(n-1)$-form ω with compact support in $\mathbb{R}^n \times \mathbb{R}^N$. The product structure in $\mathbb{R}^n_x \times \mathbb{R}^N_y$ induces a natural splitting of the exterior differential operator d in $\mathbb{R}^n \times \mathbb{R}^N$ as

$$d = d_x + d_y$$

and ∂G_u splits into its components $(\partial G_u)_{(k)}$, $0 \le k \le \underline{n}$, defined by

$$(\partial G_u)_{(k)}(\omega) := \partial G_u(\omega^{(k)}) ,$$

i.e., by testing ∂G_u on the $(n-1)$-forms with exactly k differentials with respect to y. Since for any form ω with exactly k differentials in y, obviously $d_x\omega$ has again k differentials in y while $d_y\omega$ has $(k+1)$ differentials in y, we can write

(5) $$(\partial G_u)_{(k)}(\omega) = G_{u(k)}(d_x\omega) + G_{u(k+1)}(d_y\omega) .$$

First we would like to discuss the boundary of G_u in $\Omega \times \mathbb{R}^N$. If u is smooth we know that

(6) $$\partial G_u \llcorner \Omega \times \mathbb{R}^N = 0 .$$

Equivalently we can write (6) as: for all $(n-1)$-forms ω with compact support in $\Omega \times \mathbb{R}^N$ and for all k, $0 \le k \le \min(n-1, N)$, we have

(7) $$G_{u(k)}(d_x\omega) + G_{u(k+1)}(d_y\omega) = 0 ,$$

i.e., (6) provides a linking between the k and $k+1$ components of G_u, which looks like, and in fact is, a formula of integration by parts. We shall now restate (5) in a more explicit and analytic way for functions $u \in \mathcal{A}^1(\Omega, \mathbb{R}^N)$ thus interpreting (6) and (7) which, as we have seen in Sec. 3.2.1, do not hold for all $u \in \mathcal{A}^1(\Omega, \mathbb{R}^N)$.

We begin by considering the 0-component $(\partial G_u)_{(0)}$ of ∂G_u.

Proposition 1. *Let $u \in \mathcal{A}^1(\Omega, \mathbb{R}^N)$. The following claims are equivalent*

(i) $(\partial G_u)_{(0)} \llcorner \Omega \times \mathbb{R}^N = 0$,

(ii) $\int_\Omega D_j[\phi(x, u(x))]\, dx \;=\; 0 \qquad \forall\, \phi \in C_c^\infty(\Omega \times \mathbb{R}^N)$,

(iii) *the approximate differential of u is the distributional gradient,*

(iv) $u \in W^{1,1}(\Omega, \mathbb{R}^N)$.

Proof. An $(n-1)$-form with no differentials in the vertical directions can be written as

$$\omega(x,y) \;=\; \sum_{i=1}^{n} (-1)^{i-1}\, \phi_i(x,y)\, dx^{\bar{i}} \;, \qquad \phi_i \in C_c^\infty(\Omega \times \mathbb{R}^N) \;.$$

Thus

$$\begin{aligned} d_x\omega &= \sum_{i=1}^{n} \frac{\partial \phi_i}{\partial x^i}\, dx^{\bar{0}} \\ d_y\omega &= \sum_{i=1}^{n}\sum_{j=1}^{N} (-1)^{n+i-2}\, \frac{\partial \phi_i}{\partial y^j}\, dx^{\bar{i}} \wedge dy^j \\ &= \sum_{i=1}^{n}\sum_{j=1}^{N} (-1)^{n-i}\, \frac{\partial \phi_i}{\partial y^j}\, dx^{\bar{i}} \wedge dy^j \\ &= \sum_{i=1}^{n}\sum_{j=1}^{N} \sigma(\bar{i}, i)\, \frac{\partial \phi_i}{\partial y^j}\, dx^{\bar{i}} \wedge dy^j \;. \end{aligned}$$

Hence

$$\begin{aligned} (\partial G_u)_{(0)}(\omega) &= G_{u(0)}\,(d_x\omega) + G_{u(1)}(d_y\omega) \\ &= \sum_{i=1}^{n} \int_\Omega \frac{\partial \phi_i}{\partial x_i}(x, u(x))\, dx + \sum_{i=1}^{n}\sum_{j=1}^{N} \int_\Omega \frac{\partial \phi_i}{\partial y_j}(x, u(x))\, D_i u^j(x)\, dx \\ &= \sum_{i=1}^{n} \int_\Omega D_i\, \phi_i(x, u(x))\, dx \;. \end{aligned}$$

If $u \in W^{1,1}(\Omega, \mathbb{R}^N)$ and $\phi_i \in C_c^\infty(\Omega \times \mathbb{R}^N)$, then $\phi_i(x, u(x)) \in W_0^{1,1}(\Omega)$, so that $(\partial G_u)_{(0)}(\omega) = 0$.

Denote by $\{v_j^i\}$ the approximate differential of u, and suppose conversely that $(\partial G_u)_{(0)} = 0$. For all $\varphi \in C_c^\infty(\Omega \times \mathbb{R}^N)$ we have

$$\int_\Omega \phi_{x^i}(x, u(x))\, dx + \sum_{j=1}^{N} \int_\Omega \phi_{y^j}(x, u(x))\, v_i^j(x)\, dx \;=\; 0 \tag{8}$$

Apply now (8) to

$$\phi(x,y) := \varphi(x)\, y^j\, \chi_R(y)$$

where $\varphi \in C_c^\infty(\Omega)$, and $\chi_R(y)$ is a cut off function in $\mathbb{R}^N$, i.e., $\chi \in C_c^\infty(\mathbb{R}^N)$

$$\begin{aligned} &0 \le \chi_R \le 1\,,\ \chi_R = 1 \ \text{ on } \ B(0,R)\,, \\ &\chi_R = 0 \quad \text{in } \mathbb{R}^N \setminus B(0,2R) \\ &\|\, D\chi_R \,\|_\infty \ \le \ \frac{2}{R} \quad \text{on } \ B(0,2R)\setminus B(0,R)\,. \end{aligned}$$

We get

$$\begin{aligned} &\int_\Omega D_i\varphi\, u^j\, \chi_R \circ u\, dx + \int_\Omega \varphi\, v_i^j\, \chi_R \circ u\, dx \\ &\quad + \sum_{h=1}^N \int_\Omega \varphi(x)\, u^j(x)\, \frac{\partial \chi_R}{\partial y^h}(u(x))\, v_i^h(x)\, dx \ = \ 0 \end{aligned}$$

and, as $R \to \infty$, we deduce

$$\int_\Omega D_i\varphi\, u^j\, dx + \int_\Omega \varphi\, v_i^j\, dx \ = \ 0 \tag{9}$$

since

$$\begin{aligned} \left| \int_\Omega \varphi(x)\, u^j(x)\, \frac{\partial \chi_R}{\partial y^h}(u(x))\, v_i^h(x)\, dx \right| \ &\le \ \frac{c}{R} \int_{R\le |u| \le 2R} |u|\, |v_i^h|\, dx \\ &\le \ c \int_{R \le |u| \le 2R} |v_i^h|\, dx \end{aligned}$$

and

$$\int_{R\le|u|\le 2R} |v_i^j|\, dx \ \longrightarrow \ 0\,.$$

Finally, (9) implies that v_i^j are the distributional derivatives of u. □

Let us consider the other components $(\partial G_u)_{(k)}$, $1 \le k \le \min(n-1, N)$ of ∂G_u.

Proposition 2. *Let $u \in \mathcal{A}^1(\Omega, \mathbb{R}^N)$ and let α, β are multi–indices with $|\alpha| + |\beta| = n$, $|\beta| = k+1$ where $0 \le k \le \min(n-1, N)$. Then, for any $i \in \beta$ and any function $\phi \in C^1(\mathbb{R}^n \times \mathbb{R}^N)$ with bounded derivatives, the function*

$$x \ \longrightarrow \ \sum_{j \in \bar\alpha} D_j\, [\phi(x, u(x))] (\operatorname{adj}(Du(x))_{\bar\alpha}^{\beta})_j^i$$

belongs to $L^1(\Omega)$ and we have

$$\partial G_u(\phi(x,y)\,dx^\alpha \wedge dy^{\beta-i}) = \tag{10}$$
$$(-1)^{|\alpha|}\,\sigma(\alpha,\bar\alpha)\,\sigma(i,\beta-i)\sum_{j\in\bar\alpha}\int_\Omega D_j\,[\phi(x,u(x))]\,(\mathrm{adj}\,(Du(x))^{\beta}_{\bar\alpha})^i_j\,dx$$

Since every $(n-1)$-form with k differential in y is a linear combination of forms of the type $\phi(x,y)\,dx^\alpha \wedge dy^{\beta-i}$, $|\alpha|+|\beta|=n$, $i\in\beta$, and $|\beta|=k+1$, we immediately deduce from Proposition 2

Corollary 1. *Let $u\in\mathcal{A}^1(\Omega,\mathbb{R}^N)$ and let $1\le k\le \min(n-1,N)$. Then*

$$(\partial G_u)_{(k)} \llcorner \Omega\times\mathbb{R}^N = 0$$

if and only if

$$\int_\Omega \sum_{j\in\bar\alpha} D_j[\phi(x,u(x))](\mathrm{adj}\,(Du(x))^{\beta}_{\bar\alpha})^i_j\,dx = 0 \tag{11}$$

holds for all α,β, with $|\alpha|+|\beta|=n$, $|\beta|=k+1$, for all $i\in\beta$ and for all $\phi\in C^\infty_c(\Omega\times\mathbb{R}^N)$.

Proof of Proposition 2. Fix α,β,i,ϕ as in the assumptions and consider the $(n-1)$-form

$$\omega(x,y) := \phi(x,y)\,dx^\alpha\wedge dy^{\beta-i}\,.$$

We have

$$d_x\omega(x,y) = \sum_{j\in\bar\alpha}\sigma(j,\alpha)\,\phi_{x^j}(x,y)\;dx^{\alpha+j}\wedge dy^{\beta-i}$$
$$d_y\omega(x,y) = (-1)^{|\alpha|}\sum_{h=1}^N\sigma(h,\beta-i)\,\phi_{y^h}(x,y)\;dx^\alpha\wedge dy^{\beta-i+h}\;,$$

thus

$$\begin{aligned}
&G_{u(k)}(d_x\omega)\\
&= \sum_{j\in\bar\alpha}\sigma(\alpha+j,\bar\alpha-j)\,\sigma(j,\alpha)\int_\Omega \phi_{x^j}(x,u(x))\,M^{\beta-i}_{\bar\alpha-j}(Du(x))\,dx\;,
\end{aligned}$$

$$\begin{aligned}
&G_{u(k+1)}(d_y\omega) \qquad (12)\\
&= (-1)^{|\alpha|}\,\sigma(\alpha,\bar\alpha)\sum_{h=1}^N\sigma(h,\beta-i)\int_\Omega\phi_{y^h}(x,u(x))M^{\beta-i+h}_{\bar\alpha}(Du(x))\,dx\\
&= (-1)^{|\alpha|}\,\sigma(\alpha,\bar\alpha)\sum_{h\in\bar\beta+i}\sigma(h,\beta-i)\int_\Omega\phi_{y^h}(x,u(x))M^{\beta-i+h}_{\bar\alpha}(Du(x))\,dx.
\end{aligned}$$

Now we claim that

$$\sigma(\alpha+j,\bar\alpha-j)\,\sigma(j,\alpha) \;=\; (-1)^{|\alpha|}\,\sigma(\alpha,\bar\alpha)\,\sigma(j,\bar\alpha-j)\,. \tag{13}$$

This can be checked using the following strategy to reorder $(\alpha+j,\bar\alpha-j)$: first shift to the left the index j with $\sigma(j,\alpha)$ permutations, then shift j after the α's with $(-1)^{|\alpha|}$ permutations, then insert j in $\bar\alpha-j$ at the right place with $\sigma(j,\bar\alpha-j)$ permutations; now α and $\bar\alpha$ are in increasing order, so with $\sigma(\alpha,\bar\alpha)$ permutations we complete the reordering of $(\alpha+j,\bar\alpha-j)$.

From (12), taking into account (13), we then deduce

$$\begin{aligned}(\partial G_u)_{(k)}(\omega) &= G_{u(k)}(d_x\omega)+G_{u(k+1)}(d_y\omega)\\ &= (-1)^{|\alpha|}\,\sigma(\alpha,\bar\alpha)\,\sigma(i,\beta-i)\\ &\cdot\Big\{\sigma(i,\beta-i)\sum_{j\in\bar\alpha}\sigma(j,\bar\alpha-j)\int_\Omega \phi_{x^j}(x,u(x))\,M^{\beta-i}_{\bar\alpha-j}(Du(x))\,dx\\ &+\sigma(i,\beta-i)\sum_{h=1}^{N}\sigma(h,\beta-i)\int_\Omega \phi_{y^h}(x,u(x))\,M^{\beta-i+h}_{\bar\alpha}(Du(x))\,dx\Big\}\\ &= (-1)^{|\alpha|}\sigma(\alpha,\bar\alpha)\sigma(i,\beta-i)\int_\Omega A(x)\,dx\end{aligned} \tag{14}$$

where

$$\begin{aligned}A(x) \;:=\; &\sum_{j\in\bar\alpha}\phi_{x^j}(x,u(x))\,(\operatorname{adj}(Du(x))^{\beta}_{\bar\alpha})^i_j\\ &+\sigma(i,\beta-i)\sum_{h=1}^{N}\sigma(h,\beta-i)\phi_{y^h}(x,u(x))\,M^{\beta-i+h}_{\bar\alpha}(Du(x))\,.\end{aligned}$$

Clearly $A(x)\in L^1(\Omega)$; we shall now prove that for a.e. x

$$A(x) \;=\; \sum_{j\in\bar\alpha} D_j\,[\phi(x,u(x))]\,(\operatorname{adj}(Du(x))^{\beta}_{\bar\alpha})^i_j \tag{15}$$

which clearly proves (10), taking into account (14).

From Laplace's formula

$$\begin{aligned}M^{\beta-i+h}_{\bar\alpha}(Du) \;&=\; \sum_{j\in\bar\alpha} D_ju^h(\operatorname{adj}(Du)^{\beta-i+h}_{\bar\alpha})^h_j\\ &= \sum_{j\in\bar\alpha} D_ju^h\,\sigma(j,\bar\alpha-j)\,\sigma(h,\beta-i)\,M^{\beta-i}_{\bar\alpha-j}(Du)\end{aligned}$$

we deduce

$$\sigma(h,\beta-i)\,\sigma(i,\beta-i)\,M^{\beta-i+h}_{\bar\alpha}(Du) \;=\; \sum_{j\in\bar\alpha} D_ju^h(\operatorname{adj}(Du)^{\beta}_{\bar\alpha})^i_j$$

thus

$$A(x) = \sum_{j\in\bar\alpha}[\phi_{x^j}(x,u(x)) + \sum_{h=1}^{N}\phi_{y^h}(x,u(x))\,D_j u^h(x)]\,(\mathrm{adj}\,(Du(x))_{\bar\alpha}^{\beta})_j^i$$

which yields at once (15). □

Remark 1. Taking into account Laplace's formulas, it is easily seen that (11) and therefore the condition

$$(\partial G_u)_{(k)} \llcorner \Omega\times\mathbb{R}^N = 0$$

is equivalent, in terms of determinants, to the following family of formulas of integrations by parts: for all $i\in\beta$

$$\text{(16)}\qquad \sum_{j\in\bar\alpha}\sigma(j,\bar\alpha-j)\int_\Omega \phi_{x^j}(x,u(x))\,M_{\bar\alpha-j}^{\beta-i}(Du(x))\,dx + \sigma(i,\beta-i)\int_\Omega \phi_{y^i}(x,u(x))\,M_{\bar\alpha}^{\beta}(Du(x))\,dx = 0\,.$$

Remark 2. It is easily see that (11) and (16) hold also for all $\phi\in C^1(\Omega\times\mathbb{R}^N)$ which have bounded derivatives in $\Omega\times\mathbb{R}^N$ and $\mathrm{spt}\,\phi\subset\tilde\Omega\times\mathbb{R}^N$, with $\tilde\Omega\subset\subset\Omega$, compare the proof of Proposition 1. In fact it suffices in (11) (16) to test with ϕ of the type $\varphi(x)g(y)$, as linear combinations of such products are dense. Moreover (11) (16) hold for any $\phi(x,y)=\varphi(x)g(y)$ where $\varphi(x)$ is a Lipschitz function, $\mathrm{spt}\,\phi\subset\subset\Omega$, and $g\in C^1(\mathbb{R}^N)$ with bounded derivatives. Finally we observe that (10) holds for all $\phi\in C^1(\Omega\times\mathbb{R}^N)$ with bounded derivatives.

A consequence of (11) and of Remark 2 are the classical identities

$$\sum_{j=1}^{N}\frac{\partial}{\partial x^j}\,(\mathrm{adj}\,(Du)_{\bar\alpha}^{\beta})_j^i = 0$$

to be understood now in the sense of distributions.

Corollary 2 (Piola identities). *Let $u\in\mathcal{A}^1(\Omega,\mathbb{R}^N)$ and let α,β be multi-indices with $|\alpha|+|\beta|=n$, $|\beta|=k+1$, $1\le k\le\min(n-1,N)$. Suppose that*

$$(\partial G_u)_{(k)} \llcorner \Omega\times\mathbb{R}^N = 0\,.$$

Then for all $i\in\beta$ and all Lipschitz functions φ in Ω with $\mathrm{spt}\,\varphi\subset\subset\Omega$ we have

$$\text{(17)}\qquad \int_\Omega\sum_{j\in\bar\alpha}D_j\varphi(x)\,(\mathrm{adj}\,(Du(x))_{\bar\alpha}^{\beta})_j^i\,dx = 0\,.$$

Proof. It suffices to apply (11) with $\phi(x,y) = \varphi(x)$, taking into account Remark 2. □

However conditions (17) for $k = 1, 2, \dots, \min(n-1, N)$ are weaker than (6) i.e., they do not imply in general that G_u is boundaryless in $\Omega \times \mathbb{R}^N$. In fact we have

Proposition 3. *Let $u \in W^{1,\underline{n}-1}(\Omega, \mathbb{R}^N)$. Then (17) holds for all k.*

Proof. Consider a sequence of smooth functions converging $W^{1,\underline{n}-1}$ to u. Then all the adjoints of u_k converge in L^1 to the adjoints of u. Since (17) holds for smooth functions, we also get it for u. □

In particular we see that (17) holds for the function in $\mathcal{A}^1$ $u(x) := x/|x|$ for which, however, we have $\partial G_u \llcorner B(0,1) \times \mathbb{R}^n \neq 0$.

Remark 3. Let $u \in W^{1,p}(\Omega, \mathbb{R}^N)$, where p is an integer with $1 \le p \le \underline{n}$. The argument in the proof of Proposition 3 yields also at once that

$$(\partial G_u)_{(k)} \llcorner \Omega \times \mathbb{R}^N = 0$$

for all $k \le p - 1$.
For $p = \underline{n}$, this is nothing else than Proposition 3 in Sec. 3.2.1; for $p = 1$, this is just Proposition 1. For $p = \underline{n} - 1$ we find that

$$(\partial G_u)_{(k)} \llcorner \Omega \times \mathbb{R}^N = 0$$

for all $k \le \underline{n} - 2$, and the only non zero component of the boundary ∂G_u in $B(0,1) \times \mathbb{R}^N$ can be

$$(\partial G_u)_{(\underline{n}-1)} \llcorner \Omega \times \mathbb{R}^N .$$

[1] In (11) it could sound very natural to test only with functions ϕ which depend just on x and on the dependent variables which appear in the submatrix $(Du)^{\beta}_{\bar{\alpha}}$, i.e. with ϕ of the type

$$\phi(x,y) = \varphi(x, y^{\beta_1}, \dots, y^{\beta_{k+1}}) .$$

Of course, as in Corollary 2, we deduce that, if (11) holds, then

$$\int_{\Omega} \sum_{j \in \bar{\alpha}} D_j[\varphi(x, u^{\beta_1}(x), \dots, u^{\beta_{k+1}}(x))] \, (\operatorname{adj}(Du(x))^{\beta}_{\bar{\alpha}})^i_j \, dx = 0 . \tag{18}$$

However (18) is again strictly weaker than (11) as it is shown by the following example.

Consider the map $\varphi : S^1 \subset \mathbb{R}^2 \to \mathbb{R}^3$, defined by

$$\varphi(\theta) := \begin{cases} (\cos 4\theta, \sin 4\theta, 0) & \text{if } 0 \le \theta \le \frac{\pi}{2} \\ (1, 0, \theta - \frac{\pi}{2}) & \text{if } \frac{\pi}{2} \le \theta \le \pi \\ (\cos 4(\theta - \pi), \sin 4(\theta - \pi), \frac{\pi}{2}) & \text{if } \pi \le \theta \le \frac{3}{2}\pi \\ (1, 0, 2\pi - \theta) & \text{if } \frac{3}{2} \le \theta \le 2\pi , \end{cases}$$

and its homogeneous extensions of degree zero $u : B(0,1) \subset \mathbb{R}^2 \to \mathbb{R}^3$, $u(x) := \varphi(x/|x|)$. As we have seen in [2] in Sec. 3.2.2

$$\partial G_u \llcorner B(0,1) \times \mathbb{R}^3 = -\delta_0 \times \varphi_{\#} [\![S^1]\!] .$$

Hence, in our case,

$$\begin{aligned} \varphi_{\#}[\![S^1]\!] &= [\![\{y \in \mathbb{R}^3 \mid (y^1)^2 + (y^2)^2 = 1\,,\ y^3 = 0\}]\!] \\ &- [\![\{y \in \mathbb{R}^3 \mid (y^1)^2 + (y^2)^2 = 1\,,\ y^3 = \frac{\pi}{2}\}]\!] \neq 0 \end{aligned}$$

i.e.,

$$\partial G_u \llcorner B(0,1) \times \mathbb{R}^3 \neq 0 .$$

For any $\beta \in I(2,3)$ set

$$v_\beta(x) := (u^{\beta_1}(x), u^{\beta_2}(x)) .$$

The maps $v_\beta(x)$ are homogeneous of degree zero, and again

$$\partial G_{v_\beta} \llcorner B(0,1) \times \mathbb{R}^2 = -\delta_0 \times v_{\beta\#}[\![S^1]\!]$$

But $v_{\beta\#}[\![S^1]\!]$ is the projection of $\varphi_{\#}[\![S^1]\!]$ over the coordinate plane $(y^{\beta_1}, y^{\beta_2})$, hence $v_{\beta\#}[\![S^1]\!] = 0$ and therefore

$$\partial G_{v_\beta} \llcorner B(0,1) \times \mathbb{R}^2 = 0 .$$

This implies that

$$\int_\Omega \sum_{j=1,2} D_j \left[\varphi(x, u^{\beta_1}(x), u^{\beta_2}(x))\right] (\operatorname{adj} Dv_\beta(x))^i_j \, dx = 0 ,$$

equivalently

$$\int_\Omega \sum_{j=1,2} D_j \left[\varphi(x, u^{\beta_1}(x), u^{\beta_2}(x))\right] (\operatorname{adj} (Du(x))^{\beta}_{\bar 0})^i_j \, dx = 0 ,$$

for any β, with $|\beta| = 2$, i.e. $i \in \beta$, and $\varphi \in C^\infty_c(\Omega \times \mathbb{R}^2)$, while, since $\partial G_u \llcorner B(0,1) \times \mathbb{R}^3 \neq 0$, for any β with $|\beta| = 2$, there is some $i \in \beta$ and $\phi \in C^\infty_c(\Omega \times \mathbb{R}^3)$ such that

$$\int_\Omega \sum_{j=1,2} D_j \left[\varphi(x, u^1(x), u^2(x), u^3(x))\right] (\operatorname{adj} (Du(x))^{\beta}_{\bar 0})^i_j \, dx \neq 0 .$$

•

We shall now discuss an equivalent formulation of the zero boundary condition $\partial G_u \llcorner \Omega \times \mathbb{R}^N = 0$ in terms of the commutator of the exterior derivative and pullback. Recall that for any $\omega \in \mathcal{D}^h(\Omega)$ and $\eta \in \mathcal{D}^k(\Omega)$, $h+k \leq n-1$ we have

$$d(\omega \wedge \eta) = d\omega \wedge \eta + (-1)^h \omega \wedge d\eta \,.$$

This allows us to define the differential $d\gamma$ for any k–form γ with coefficients in $L^1(\Omega)$ by

$$\int_\Omega \omega \wedge d\gamma = (-1)^{n-k} \int_\Omega d\omega \wedge \gamma \qquad \forall \omega \in \mathcal{D}^{n-k-1}(\Omega)$$

Proposition 4. *Let $u \in \mathcal{A}^1(\Omega, \mathbb{R}^N)$, $\omega \in \mathcal{D}^{n-k-1}(\Omega)$ and $\eta \in \mathcal{D}^k(\mathbb{R}^N)$, $0 \leq k \leq n$. Then*

$$[\![\Omega]\!](\omega \wedge (du^\# \eta - u^\# d\eta)) = (-1)^{n-k} \partial G_u(\omega \wedge \eta)$$

Proof. In fact

$$\begin{aligned} &[\![\Omega]\!](\omega \wedge du^\# \eta) = (-1)^{n-k} [\![\Omega]\!](d\omega \wedge u^\# \eta) = (-1)^{n-k} G_u(d\omega \wedge \eta) \\ &= (-1)^{n-k} \partial G_u(\omega \wedge \eta) + G_u(\omega \wedge d\eta) = (-1)^{n-k} \partial G_u(\omega \wedge \eta) + [\![\Omega]\!](\omega \wedge u^\# d\eta) \end{aligned}$$

□

From the density of linear combinations of forms of type $\omega \wedge \eta$ in $\mathcal{D}^{n-1}(\Omega \times \mathbb{R}^N)$ we readily infer

Corollary 3. *Let $u \in \mathcal{A}^1(\Omega, \mathbb{R}^N)$. Then*

$$\partial G_u \llcorner \Omega \times \mathbb{R}^N = 0$$

if and only if

$$u^\# d = d u^\# \,.$$

2.4 More on the Jacobian Determinant

In this subsection we discuss in term of boundaries some subclasses of the Sobolev space $W^{1,n-1}(\Omega, \mathbb{R}^n)$, $\Omega \subset \mathbb{R}^n$, and we collect a number of recent results concerning the Jacobian determinant.

For the reader convenience we first state some of the results of the previous subsection in the special case of maps in $W^{1,n-1}(\Omega, \mathbb{R}^n)$.

Proposition 1. *Let $u \in W^{1,n-1}(\Omega, \mathbb{R}^n)$. Then we have*

(i) u satisfies the Piola identities

$$\int_\Omega \sum_{j=1}^n D_j \phi(x) (\text{adj}\, Du(x))^i_j \, dx = 0$$

for any $i = 1, \ldots, n$ and any $\phi \in C^1_c(\Omega)$.

(ii) $u \in \mathcal{A}^1(\Omega, \mathbb{R}^n)$ *if and only if* $\det Du(x) \in L^1(\Omega)$.
(iii) $\partial G_u \llcorner \Omega \times \mathbb{R}^n = 0$, *i.e.* $u \in \mathrm{cart}^1(\Omega, \mathbb{R}^n)$ *if and only if* $\det Du \in L^1(\Omega)$ *and*

$$\int_\Omega \sum_{i,j=1}^n D_j\phi(x) g^i(u(x))(\mathrm{adj}\, Du(x))^i_j \, dx + \int_\Omega \phi(x) \mathrm{div}\, g(u(x)) \det Du(x)\, dx = 0 \tag{1}$$

for any $\phi \in C^1_c(\Omega)$ *and any* $g \in C^1_c(\mathbb{R}^n, \mathbb{R}^n)$.

Proof. (i) is precisely Proposition 3 in Sec. 3.2.3, and (ii) is obvious. By Remark 3 in Sec. 3.2.3, $\partial_{(k)} G_u \llcorner \Omega \times \mathbb{R}^n = 0$ for $k = 1, \dots, n-2$. Thus (iii) follows since (1) is equivalent to $\partial_{(n-1)} G_u \llcorner \Omega \times \mathbb{R}^n = 0$, i.e.

$$\int_\Omega \sum_{j=1}^n [D_j \psi(x, u(x))](\mathrm{adj}\, Du(x))^i_j \, dx = 0$$

since linear combinations of the functions $\phi(x)g(y)$, $\phi \in C^1_c(\Omega)$, $g \in C^1_c(\mathbb{R}^n)$ are dense in $C^1_c(\Omega \times \mathbb{R}^n)$ (in the C^1 topology). □

Let $u \in \mathcal{A}^1(\Omega \times \mathbb{R}^n) \cap W^{1,n-1}(\Omega, \mathbb{R}^n)$. Considering the n-vector fields $\sigma^i \in L^1(\Omega, \mathbb{R}^n)$, $i = 1, 2, \dots, n$

$$\sigma^i(x) \;:=\; (\mathrm{adj}\, Du(x))^i_j$$

and denoting by Div the divergence operator in the sense of distributions

$$< \mathrm{Div}\, \sigma, \varphi >:= -\int_\Omega \sigma \cdot D\varphi \, dx \,, \qquad \varphi \in C^1_c(\Omega)\,.$$

to distinguish it from the div operator in the sense of approximate differentiability, (1) is equivalent to

$$\mathrm{Div}\,(g^i(u)\sigma^i) \;=\; \mathrm{div}\,(g^i(u)\sigma^i) \tag{2}$$

since, taking into account the Piola identities and Laplace formulas we have

$$\begin{aligned} &\mathrm{div}\,(g^i(u(x))\sigma^i(x)) \\ &= \sum_{j,h} g^i_{u^h}(u(x)) D_j u^h(x) \sigma^i_j(x) + g^i(u(x)) \mathrm{div}\, \sigma^i(x) \\ &= g^i_{u^h}(u(x)) \delta^{ih} \det Du(x) = \mathrm{div}\, g(u(x)) \det Du(x)\,. \end{aligned}$$

A sufficient condition ensuring (2) is given in the next proposition. Let v be a function in $W^{1,p}$ and let σ be a vectorfield in $L^q(\Omega, \mathbb{R}^n)$ where $\frac{1}{p} + \frac{1}{q} - \frac{1}{n} \le 1$. Since $v\sigma \in L^1(\Omega, \mathbb{R}^n)$, we can consider the distribution $\mathrm{Div}\,(v\sigma)$ Then we have

Proposition 2. *Let* $u \in W^{1,p}(\Omega)$ *and* $\sigma \in L^q(\Omega, \mathbb{R}^n)$ *where* $\frac{1}{p} + \frac{1}{q} - \frac{1}{n} \leq 1$. *Suppose that* $\operatorname{Div} \sigma$ *and* $\operatorname{Div}(v\sigma)$ *are locally absolutely continuous with respect to the Lebesgue measure,*

$$\begin{aligned} \operatorname{Div} \sigma &= \operatorname{div} \sigma(x)\, dx\,, \quad \operatorname{div} \sigma(x) \in L^1_{\text{loc}}(\Omega) \\ \operatorname{Div}(v\sigma) &= d(x)\, dx\,, \quad d(x) \in L^1_{\text{loc}}(\Omega)\,. \end{aligned}$$

Then for almost every $x \in \Omega$ *we have*

$$d(x) = Dv(x) \cdot \sigma(x) + v(x) \operatorname{div} \sigma(x) \tag{3}$$

Actually if $\operatorname{Div}(v\sigma)$ *is merely a Radon measure and*

$$\operatorname{Div}(v\sigma) = d(x)\, dx + (\operatorname{Div}(v\sigma))^{(s)}$$

then (3) still holds.

Proof. Suppose first that $\operatorname{div} \sigma = 0$. From Lebesgue's differentiation theorem and Calderón-Zygmund theorem on the L^{p^*}-differentiability of v, $p^* := \frac{np}{n-p}$, of Sec. 3.1.2 we have, for a.e. $x_0 \in \Omega$,

$$\lim_{r \to 0} \fint_{B(x_0,r)} |d(x) - d(x_0)|\, dx = 0\,, \tag{4}$$

$$\lim_{r \to 0} \fint_{B(x_0,r)} |\sigma(x) - \sigma(x_0)|^q\, dx = 0\,, \tag{5}$$

and

$$\lim_{r \to 0} \frac{1}{r} \fint_{B(x_0,r)} |v(x) - v(x_0) - Dv(x_0)(x - x_0)|^{p^*}\, dx = 0 \tag{6}$$

Fix such a point x_0, choose $\psi \in C^\infty_c(B_1)$, $\psi \geq 0$, $\fint_{B_1} \psi(x)\, dx = 1$, and let

$$\psi_r(x) = r^{-n} \psi(\frac{x - x_0}{r})\,.$$

Then we have, by the differentiation theorem for measures, compare Ch. 1,

$$\lim_{r \to 0} < \operatorname{Div}(v\sigma), \psi_r > = \lim_{r \to 0} \fint_{B(x_0,r)} d(x) \psi_r(x)\, dx = d(x_0)$$

On the other hand letting

$$P(x) := v(x_0) + Dv(x_0)\,(x - x_0)$$

we have

$$<\mathrm{Div}\,(v\sigma), \psi_r> = -\int_\Omega D_j\psi_r(x)\, v(x)\, \sigma_j(x)\, dx$$

$$= - \int_{B(x_0,r)} D_j\psi_r\,(v-P)\,\sigma_j\,dx - \int_{B(x_0,r)} D_j\psi_r\, P\,\sigma_j\,dx$$

$$(7) \qquad = - \int_{B(x_0,r)} D_j\psi_r\,(v-P)\,\sigma_j\,dx + \int_{B(x_0,r)} \psi_r(x)\, D_jv(x_0)\,\sigma_j(x)\,dx$$

$$= - \int_{B(x_0,r)} D_j\psi_r\,(v-P)\,\sigma_j\,dx + \int_{B(x_0,r)} \psi_r(x)\, D_jv(x_0)\,(\sigma_j(x)-\sigma_j(x_0))\,dx$$

$$+ \int_{B(x_0,r)} \psi_r(x)\, D_jv(x_0)\,\sigma_j(x_0)\,dx\,.$$

The last term is of course $D_jv(x_0)\sigma_j(x_0) = Dv(x_0)\cdot\sigma(x_0)$. The first term in the last expression is bounded by

$$r^{-n-1}\sup|D\psi| \int_{B(x_0,r)} |v-P|\,|\sigma|\,dx$$

$$\leq \frac{c}{r}\Big(\fint_{B(x_0,r)} |v(x)-P(x)|^{p^*}\,dx\Big)^{1/p^*} \Big(\fint_{B(x_0,r)} |\sigma|^q\,dx\Big)^{1/q}$$

and thus converges to zero as $r \to 0$ by (5) and (6). Finally the second term in the last identity in (7) is bounded by

$$c|Dv(x_0)| \fint_{B(x_0,r)} |\sigma(x)-\sigma(x_0)|\,dx$$

and therefore it converges to zero as $r \to 0$. From this it follows that

$$d(x_0) = \sum_{j=1}^n D_jv(x_0)\sigma_j(x_0)\,.$$

For the general case, $\mathrm{div}\,\sigma \neq 0$, we choose x_0 so that we have

$$(8) \qquad \lim_{r\to 0} \fint_{B(x_0,r)} |\mathrm{div}\,\sigma(x) - \mathrm{div}\,\sigma(x_0)|\,dx = 0$$

in addition to (4) (5) (6). Now in (7) the additional term

$$\int_{B(x_0,r)} \psi_r(x)P(x)\mathrm{div}\,\sigma(x)\,dx$$

appears. We write it as

$$\int_{B(x_0,r)} \psi_r(x)(P(x) - P(x_0))\operatorname{div}\sigma(x)\,dx +$$
$$\int_{B(x_0,r)} \psi_r(x)P(x_0)(\operatorname{div}\sigma(x) - \operatorname{div}\sigma(x_0))\,dx +$$
$$\int_{B(x_0,r)} \psi_r(x)P(x_0)\operatorname{div}\sigma(x_0)\,dx\ .$$

Since the last term is clearly $v(x_0)\operatorname{div}\sigma(x_0)$, while the first two terms can be estimated by

$$c\,|Dv(x_0)|\,r \quad -\!\!\!\!\!\!\int_{B(x_0,r)} |\operatorname{div}\sigma(x)|\,dx$$

and

$$c\,|v(x_0)| \quad -\!\!\!\!\!\!\int_{B(x_0,r)} |\operatorname{div}\sigma(x) - \operatorname{div}\sigma(x_0)|\,dx$$

which go to zero taking into account (8), the proof is concluded. □

[1] *The distributional determinant.* Let u be a smooth map from $\Omega \subset \mathbb{R}^n$ into $\mathbb{R}^n$. From Laplace's formulas we deduce

$$\sum_{j=1}^n \frac{\partial}{\partial x^j}(u^1(\operatorname{adj} Du)^1_j)$$
$$= \sum_{j=1}^n \frac{\partial u^1}{\partial x^j}(\operatorname{adj} Du)^1_j + u^1 \sum_{j=1}^n \frac{\partial}{\partial x^j}(\operatorname{adj} Du)^1_j \;=\; \det Du\ ,$$

i.e., that $\det Du$ can be expressed as a divergence.

Suppose now that $u \in W^{1,\frac{n^2}{n+1}}(\Omega, \mathbb{R}^n)$. From Sobolev's embedding theorem we have $u \in L^{n^2}(\Omega, \mathbb{R}^n)$, in fact n^2 is the Sobolev exponent of $\frac{n^2}{n+1}$; since the dual exponent of $\frac{n^2}{n^2-1}$ is just n^2 and $(\operatorname{adj} Du)^1_j \in L^{\frac{n^2}{n^2-1}}(\Omega)$, we then get

$$u^1\,(\operatorname{adj} Du)^1_j \;\in\; L^1(\Omega)$$

Therefore, for any $u \in W^{1,\frac{n^2}{n+1}}(\Omega, \mathbb{R}^n)$, we can define the *distributional determinant of Du* as the distribution

$$\operatorname{Det} Du \;:=\; \sum_j D_j\,(u^1(\operatorname{adj} Du)^1_j)$$

i.e., as the distribution

$$\varphi \in C^\infty_c(\Omega) \;\longrightarrow\; <\operatorname{Det} Du, \varphi> := -\int_\Omega \sum_{j=1}^n D_j\varphi(x)u^1(x)(\operatorname{adj} Du(x))^1_j\,dx$$

In general Det Du is not a function and

$$\operatorname{Det} Du \neq \det Du ,$$

for instance, if $u(x) := \frac{x}{|x|}$ one easily verifies that $\det Du = 0$ while $\operatorname{Det} Du = |B_1|\delta_0$. Similarly for the map

$$u(x) := (|x|^2 + a^2)^{1/2} \frac{x}{|x|}$$

we have

$$\det Du(x) = 1 \quad \text{and} \quad \operatorname{Det} Du = 1 \cdot dx + |B(0,a)|\delta_0 .$$

But we have

Proposition 3. *Let* $u \in \operatorname{cart}^1(\Omega, \mathbb{R}^n) \cap W^{1,\frac{n^2}{n+1}}(\Omega, \mathbb{R}^n)$. *Then* Det Du *is a function, and actually*

(9) $$\operatorname{Det} Du = \det Du .$$

Proof. Since $\partial G_u \llcorner \Omega \times \mathbb{R}^n = 0$ we have, compare Remark 3 in Sec. 3.2.3 and Corollary 1 in Sec. 3.2.3,

(10) $$\int_\Omega \Big[\sum_j \phi_{x^j}(x, u(x)) (\operatorname{adj} Du(x))^i_j + \sum_{j,\ell} \phi_{y^\ell}(x, u(x)) D_j u^\ell(x) (\operatorname{adj} Du(x))^i_j \Big] \, dx = 0$$

for all $\phi \in C^1(\Omega \times \mathbb{R}^n)$, $\operatorname{spt} \varphi \subset\subset \Omega \times \mathbb{R}^n$ and $|D\varphi| \in L^\infty$. Since $u \in W^{1,\frac{n^2}{n+1}}$ and $D_j u^\ell (\operatorname{adj} Du)^i_j \in L^1$, we now deduce as in Corollary 2 in Sec. 3.2.3 that (10) holds also for $\phi(x,y) = \varphi(x) y^1$, $\varphi(x) \in C^1_c(\Omega)$, therefore

$$\int_\Omega D_j(\varphi\, u^1) \, (\operatorname{adj} Du)^1_j \, dx = 0 \quad \forall\, \varphi \in C^1_c(\Omega) .$$

On the other hand, since $\det Du \in L^1$ we have

$$-\langle \operatorname{Det} Du, \varphi \rangle + \int_\Omega \varphi \det Du \, dx = \int_\Omega D_j(\varphi\, u^1) \, (\operatorname{adj} Du)^1_j \, dx$$

thus (9) holds. □

We also have

Proposition 4 (Müller). *Let u belong to* $W^{1,\frac{n^2}{n+1}}(\Omega, \mathbb{R}^n)$. *If the distributional determinant* Det Du *is a Radon measure then*

(11) $$\operatorname{Det} Du = \det Du \, dx + \mu , \qquad \mu \perp dx .$$

In particular, if Det Du *is absolutely continuous with respect to Lebesgue measure, then*

(12) $$\operatorname{Det} Du = \det Du \, dx .$$

Proof. Apply Proposition 2 with

$$v \;:=\; u^1 \in W^{1,\frac{n^2}{n+1}}$$

and for $j = 1, \dots, n$,

$$\sigma_j \;:=\; (\operatorname{adj} Du)^1_j \in L^{\frac{n^2}{n-1}} \,.$$

Since $\operatorname{Div}\sigma \;=\; 0$, we get by Proposition 2 that

$$\operatorname{Det} Du = \operatorname{Div}(v\sigma) = \sum_{j=1}^{n} D_j v(x)\sigma_j(x)\,dx + \mu = \det Du(x)\,dx + \mu \,.$$

□

Finally we observe that a natural question is whether the converse of Proposition 3 holds, i.e., whether, for functions $u \in \mathcal{A}^1 \cap W^{1,\frac{n^2}{n+1}}$, (9) is equivalent to $\partial G_u \llcorner \Omega \times \mathbb{R}^n = 0$. But we do not know the answer. •

[2] *Isoperimetric inequality for the determinant.* Next we prove an isoperimetric inequality for the determinant of an almost everywhere approximately differentiable map $u : \Omega \subset \mathbb{R}^n \to \mathbb{R}^n$ in which the absence of boundaries play an important role.

Proposition 5. *Let u be an almost everywhere approximately differentiable map for which* $\operatorname{adj} Du$, $\det Du \in L^1(\Omega, \mathbb{R}^n)$ *and for which* $(\partial G_u)_{(n-1)} \llcorner \Omega \times \mathbb{R}^n = 0$, *i.e.,*

$$(13) \qquad \int_\Omega \sum_{i,j=1}^{n} D_j\phi(x) g^i(u(x))(\operatorname{adj} Du(x))^i_j\,dx + \int_\Omega \phi(x)\operatorname{div} g(u(x))\det Du(x)\,dx \;=\; 0$$

holds for any $\phi \in C^1_c(\Omega)$ and any $g \in C^1_c(\mathbb{R}^n, \mathbb{R}^n)$. Then for all $x_0 \in \Omega$ and a.e. r, $0 < r < r_0 := \operatorname{dist}(x_0, \partial\Omega)$ we have

$$(14) \qquad \int_{B(x_0,r)} \operatorname{div} g(u(x))\det Du(x)\,dx = \int_{\partial B(x_0,r)} \sum_{i,j=1}^{n} g^i(u)\,(\operatorname{adj} Du)^i_j\,\nu_j\,d\mathcal{H}^{n-1}$$

for any $g \in C^1_c(\mathbb{R}^n, \mathbb{R}^n)$. Here $\nu = (\nu_1, \dots, \nu_n) := \frac{x - x_0}{|x - x_0|}$.

Moreover for any $\psi \in C^1_c(\mathbb{R}^n)$ and a.e r we have

$$(15)\qquad \Big|\int_{B(x_0,r)} \psi(u(x))\det Du(x)\,dx\Big| \le \|\psi\|_\infty\Big(\int_{\partial B(x_0,r)} |\operatorname{adj} Du|\,d\mathcal{H}^{n-1}\Big)^{\frac{n}{n-1}}$$

Proof. We first prove (14). Fix $g \in C^1_c(\mathbb{R}^n)$. Using Fubini theorem or the coarea formula we find that for a.e. $r \in (0, r_0)$, $\sum_{i,j} g^i(u(\cdot))(\operatorname{adj} Du(\cdot))^i_j \nu_j(\cdot) \in L^1(\partial B(x_0, r))$ and the map

$$r \longrightarrow h(r) := \int_{\partial B(x_0,r)} \sum_{i,j} g^i(u)\,(\operatorname{adj} Du)^i_j \nu_j\, d\mathcal{H}^{n-1}(x)$$

is in $L^1(0, r_0)$. Moreover using (13)

$$\frac{1}{2\varepsilon}\int_{r-\varepsilon}^{r+\varepsilon} h(s)\,ds = \int_{\Omega} \phi_{r,\varepsilon}(x)\operatorname{div} g(u(x))\det Du(x)\,dx$$

where

$$\phi_{r,\varepsilon} = \begin{cases} 1 & \text{if } \quad |x-x_0| < r-\varepsilon \\ \frac{1}{2} - \frac{1}{2\varepsilon}(|x|-r) & \text{if } r-\varepsilon < |x-x_0| < r+\varepsilon \\ 0 & \text{if } \quad |x-x_0| > r+\varepsilon \end{cases}$$

Thus for $\varepsilon \to 0$ we get (14) for a.e. r and fixed g. Now (14) follows taking into account that $C^1_c(\mathbb{R}^n)$ has a denumerable dense subset. We notice that in the proof one uses just the general slicing argument.

We now use (14) to prove (15). By the area formula, compare Sec. 3.1.5, setting $d_r(y) := \deg(u, B(x_0, r), y)$, we have

$$\int_{B(x_0,r)} \operatorname{div} g(u(x))\det Du(x)\,dx = \int_{\mathbb{R}^n} \operatorname{div} g(y)\, d_r(y)\,dy$$

which, together with (14) says that

$$\Big|\int_{\mathbb{R}^n} \operatorname{div} g(y)\, d_r(y)\,dy\Big| \le \|g\|_\infty \int_{\partial B(x_0,r)} |\operatorname{adj} Du|\,d\mathcal{H}^{n-1}\,.$$

In other words for a.e r, d_r is a function of bounded variation in $\mathbb{R}^n$, compare Ch. 4, and

$$|Dd_r|(\mathbb{R}^n) \le \int_{\partial B(x_0,r)} |\operatorname{adj} Du|\,d\mathcal{H}^{n-1}\,.$$

By Sobolev embedding theorem, taking also into account that $d_r(y) \in \mathbb{Z}$, we then get

$$\Big(\int_{\mathbb{R}^n} |d_r(y)|\, dy \Big)^{\frac{n-1}{n}}$$

$$\leq \Big(\int_{\mathbb{R}^n} |d_r(y)|^{\frac{n}{n-1}}\, dy \Big)^{\frac{n-1}{n}} \leq c(n) \int_{\partial B(x_0,r)} |\mathrm{adj}\, Du|\, d\mathcal{H}^{n-1}$$

and therefore

$$(16)\qquad \Big| \int_{B(x_0,r)} \psi(u(x)) \det Du(x)\, dx \Big| = \Big| \int_{\mathbb{R}^n} \psi(y)\, d_r(y)\, dy \Big|$$

$$\leq \|\psi\|_\infty \int_{\mathbb{R}^n} |d_r(y)|\, dy \leq c(n) \|\psi\|_\infty \Big(\int_{\partial B(x_0,r)} |\mathrm{adj}\, Du|\, d\mathcal{H}^{n-1} \Big)^{\frac{n}{n-1}} .$$

□

Remark 1. Notice that (14) and consequently (15) hold for any $r \in (0, r_0)$ if u is a smooth map. Going trough the proof of Proposition 5 it is easily seen that in fact we have

$$(17)\qquad \Big| \int_{B(x_0,r)} \psi(u(x)) \det Du(x)\, dx \Big|$$

$$\leq c(n)\, \|\psi\|_\infty \Bigg(\int_{\partial B(x_0,r)} \Big\{ \sum_i \Big[\sum_j (\mathrm{adj}\, Du)^i_j \nu_j \Big]^2 \Big\}^{1/2} d\mathcal{H}^{n-1} \Bigg)^{\frac{n}{n-1}}$$

$c(n)$ being the isoperimetric constant in $\mathbb{R}^n$. If u is smooth and 1-to-1, (17) amounts just to the isoperimetric inequality

$$|u(B(x_0,r))| \leq c(n)\, \big[\mathcal{H}^{n-1}(u(\partial B(x_0,r)))\big]^{n/(n-1)}$$

since the integral on the right hand–side of (17) is just the $\mathcal{H}^{n-1}$–measure of $u(\partial B(x_0,r))$, compare Sec. 2.1.2.

•

3 *The class* $\mathcal{A}_{p,q}(\Omega, \mathbb{R}^n)$. Consider the class of functions $u \in W^{1,p}(\Omega, \mathbb{R}^n)$ where $p \geq n-1$, such that $\mathrm{adj}\, Du \in L^q(\Omega, \mathbb{R}^n)$ for some $q \geq n/(n-1)$

$$\mathcal{A}_{p,q}(\Omega, \mathbb{R}^n) := \{ u \in W^{1,p}(\Omega, \mathbb{R}^n) \mid \mathrm{adj}\, Du \in L^q(\Omega, \mathbb{R}^{n^2}) \}$$

We claim that

$$\mathcal{A}_{p,q}(\Omega, \mathbb{R}^n) \subset \mathrm{cart}^1(\Omega, \mathbb{R}^n)$$

if $q \geq p/(p-1)$. In fact from (i) of Proposition 1, we have

$$(18)\qquad \int_\Omega \sum_{j=1}^n D_j \varphi\, (\mathrm{adj}\, Du)^i_j\, dx = 0$$

for any $\varphi \in C^1_c(\Omega)$. Since $(\operatorname{adj} Du)^i_j \in L^q(\Omega)$, (18) holds also for all φ in $W^{1,q'}(\Omega)$, where $\frac{1}{q'} + \frac{1}{q} = 1$. On the other hand for any $\phi \in C^\infty_c(\Omega \times \mathbb{R}^n)$ the map $\varphi(x) := \phi(x, u(x))$ belongs to $W^{1,p}(\Omega)$, thus, since $\frac{1}{p} + \frac{1}{q} \leq 1$ one deduces that $u \in \mathcal{A}^1$ and that

$$\int_\Omega \sum_j D_j[\phi(x, u(x))]\,(\operatorname{adj} Du(x))^i_j\,dx = 0\,.$$

By (iii) of Proposition 1 we then conclude that $u \in \operatorname{cart}^1(\Omega, \mathbb{R}^n)$.

Actually we have

Theorem 1 (Müller–Tang–Yan). $\mathcal{A}_{n-1,n/(n-1)} \subset \operatorname{cart}^1(\Omega, \mathbb{R}^n)$.

In order to prove Theorem 1, we need the following lemma

Lemma 1. *Let $u \in W^{1,n-1}(\Omega, \mathbb{R}^n)$, $x_0 \in \Omega$ and $r_0 := \operatorname{dist}(x_0, \partial\Omega)$. Then for a.e. $r \in (0, r_0)$ we have*

$$\Big| \int_{\partial B(x_0,r)} \sum_{i,j=1}^n g^i(u)\,(\operatorname{adj} Du)^i_j \nu_j\,d\mathcal{H}^{n-1} \Big| \leq c(n)\,||\operatorname{div} g||_\infty \Big(\int_{\partial B(x_0,r)} |\operatorname{adj} Du|\,d\mathcal{H}^{n-1} \Big)^{\frac{n}{n-1}} \tag{19}$$

where $\nu = (\nu_1, \ldots, \nu_n) := \frac{x - x_0}{|x - x_0|}$.

Proof. Suppose first that u is a smooth function. Then we may apply Proposition 5 to u and (19) follows from (14) and (15). For the general case, let $\{u_k\}$ be a sequence of smooth maps which converges in $W^{1,n-1}$ to u. Since $u_k \to u$ and $\operatorname{adj} Du_k \to \operatorname{adj} Du$ strongly in L^1, we can select a subsequence u_{h_k} such that

$$\| u_{h_k} - u \|_{L^1} \leq 4^{-k}, \quad \| \operatorname{adj} Du_{h_k} - \operatorname{adj} Du \|_{L^1} \leq 4^{-k}.$$

By Fubini theorem we then infer that for a.e. r

$$\int_{\partial B_r} \Big(\sum_{k=0}^\infty 2^k |u_{h_k} - u| + \sum_{k=0}^\infty 2^k |\operatorname{adj} Du_k - \operatorname{adj} Du \Big)\,dx < \infty,$$

hence

$$u_{h_k} \longrightarrow u, \quad \operatorname{adj} Du_{h_k} \longrightarrow \operatorname{adj} Du \quad \text{strongly in } L^1(\partial B_r).$$

Writing (19) for u_{h_k} and passing to the limit the result then follows at once. □

Proof of Theorem 1. If F denotes any $n \times n$–matrix, by (3) in Sec. 3.2.3 we have

$$F \cdot (\operatorname{adj} F)^T = \det F \cdot \mathrm{id}$$

hence

(20) $$|\det F|^{n-1} \le |\det(\operatorname{adj} F)| \le c(n)\,|\operatorname{adj} F|^n .$$

This proves that if $u \in \mathcal{A}_{n-1,n/(n-1)}$ then $\det Du \in L^1$and $u \in \mathcal{A}^1(\Omega, \mathbb{R}^N)$. We show now that $\partial G_u \llcorner \Omega \times \mathbb{R}^N = 0$. In view of Proposition 1 (iii) it suffices to prove (1).

For any $i = 1, \dots, n$, denote by σ^i the vector field whose components are given by

$$\sigma^i_j(x) := (\operatorname{adj} Du)^i_j .$$

We have $\sigma^i \in L^q(\Omega)$, $q = \frac{n}{n-1}$ and $\operatorname{Div}\sigma^i = 0$. Let $g \in C^1_c(\mathbb{R}^n, \mathbb{R}^n)$. Then

$$v(x) := g(u(x))$$

belongs to $W^{1,p}$, $p = n-1$, and

$$\frac{1}{p} + \frac{1}{q} - \frac{1}{n} = \frac{2-n}{n(n-1)} + 1 \le 1 .$$

If

(21) $$\operatorname{Div}(g^i(u)\,\sigma^i) \in L^1_{\mathrm{loc}}(\Omega)$$

for any $i = 1, \dots, n$, we may apply Proposition 2 and conclude that

$$\begin{aligned}\operatorname{Div}(g^i \circ u\,\sigma^i) &= \textstyle\sum_{i,j,h} g^i_{,y_h}(u(x)) D_j u^h(x) (\operatorname{adj} Du)^i_j\, dx \\ &= \textstyle\sum_{i,j,h} g^i_{y_h}(u(x))\delta^{ih} \det Du(x)\, dx = \operatorname{div} g(u(x)) \det Du(x)\, dx\end{aligned}$$

which is precisely (1).

Therefore it remains to prove (21). Consider a radial mollifier $\rho(x) = \psi(|x|)$ and let $\rho_\varepsilon := \varepsilon^{-n}\rho(\frac{x}{\varepsilon})$. Denote by T the distribution

$$T := \operatorname{Div}(g^i(u)\,\sigma^i)$$

Fix $\Omega_0 \subset\subset \Omega$. Clearly the distributions $T * \rho_\varepsilon$ belongs to $L^1(\Omega_0)$ and converges as distributions to T. Thus (21) holds if we show that in fact $T * \rho_\varepsilon$ converges weakly in $L^1(\Omega_0)$ to T.

Let $\varepsilon < \frac{1}{2}\operatorname{dist}(\Omega_0, \partial\Omega)$ and $z \in \Omega_0$. Then we have

$$\begin{aligned}T * \rho_\varepsilon(z) &= -\int_{\mathbb{R}^n} (g^i(u(x)))(\operatorname{adj} Du(x))^i_j D_j \rho_\varepsilon(x)\, dx \\ &= -\int_{\mathbb{R}^n} (g^i(u(x)))(\operatorname{adj} Du(x))^i_j \psi'(\frac{|x-z|}{\varepsilon})\nu_j\, dx \\ &= \varepsilon^{-(n+1)} \int_0^\varepsilon \Big\{ \int_{\partial B(z,r)} \sum_{i,j} g^i(u(x))(\operatorname{adj} Du(x))^i_j \nu_j\, d\mathcal{H}^{n-1} \Big\} \psi'(\frac{r}{\varepsilon})\, dr .\end{aligned}$$

Thus by Lemma 1 we get

$$|T * \rho_\varepsilon(z)| \le c(n)\,\varepsilon^{-(n+1)} \int_0^\varepsilon \Big\{ \int_{\partial B(z,r)} |\mathrm{adj}\, Du|\, d\mathcal{H}^{n-1} \Big\}^{\frac{n}{n-1}} dr$$

$$\le c(n)\,\varepsilon^{-(n+1)} \int_0^\varepsilon \Big\{ \int_{\partial B(z,r)} |\mathrm{adj}\, Du|^{\frac{n}{n-1}}\, d\mathcal{H}^{n-1} \Big\} r\, dr$$

$$\le c(n)\,\varepsilon^{-n} \int_{B(z,\varepsilon)} |\mathrm{adj}\, Du|^{\frac{n}{n-1}}\, dx\ .$$

and, assuming without loss of generality that

$$\rho_{|B(0,1/2)} \ge |B(0,1)|^{-1}\ ,$$

we deduce that

$$|T * \rho_\varepsilon(z)| \ \le\ c\, |\mathrm{adj}\, Du|^{\frac{n}{n-1}} * \rho_{2\varepsilon}(z) \qquad \text{in } \Omega_0\ .$$

The claim now follows since the functions $|\mathrm{adj}\, Du|^{\frac{n}{n-1}} * \rho_{2\varepsilon}(z)$ are equi-absolutely continuous. □

•

[4] *Higher integrability of the determinant.* As a further consequence of the isoperimetric inequality (15) we now prove

Theorem 2 (Müller). *Let* $u \in W^{1,1}(\Omega, \mathbb{R}^n)$ *be such that*

(i) $\mathrm{adj}\, Du \in L^{n/(n-1)}(\Omega, \mathbb{R}^{n^2})$
(ii) $\partial_{(n-1)} G_u \llcorner \Omega \times \mathbb{R}^N = 0$, *i.e., (13) holds*
(iii) $\det Du \ge 0$.

Then for any $\widetilde{\Omega} \subset\subset \Omega$ *we have* $\det Du \log(2 + \det Du) \in L^1(\widetilde{\Omega})$ *and*

$$\| \det Du \log(2 + \det Du) \|_{L^1(\widetilde{\Omega})} \ \le\ c(\Omega, \widetilde{\Omega}, \| \mathrm{adj}\, Du \|_{L^{n/(n-1)}(\Omega, \mathbb{R}^{n^2})})\ .$$

Proof. The proof of Theorem 2 relies on the characterization of the class $L \log L$, Proposition 5 in Sec. 3.1.1.

It suffices to show that $M(f) \in L^1(\widetilde{\Omega})$, $\widetilde{\Omega} \subset\subset \widehat{\Omega} \subset\subset \Omega$ where

$$f(x) \ :=\ \begin{cases} \det Du(x) & x \in \widehat{\Omega} \\ 0 & x \notin \widehat{\Omega}\ . \end{cases}$$

In fact for $d := \mathrm{dist}\,(\widehat{\Omega}, \partial\Omega)$ it suffices to estimate

$$⨍_{B(x,R)} |f(y)|\, dy$$

for all $x \in \widehat{\Omega}$ and all $R < d/2$, as for $R \geq d/2$ we have

$$\fint_{B(x,R)} |f(y)|\, dy \leq c(d) \int_{\Omega} \det Du\, dy \leq c(d) \int_{\Omega} |\operatorname{adj} Du|^{n/(n-1)}\, dy$$

because of (20).

From (15) we infer that for a.e. r, $R < r < 2R$ $(R < d/2)$ we have

$$\Big| \int_{B(x,r)} \det Du\, dy \Big|^{\frac{n-1}{n}} \leq c(n) \int_{\partial B(x,r)} |\operatorname{adj} Du|\, d\mathcal{H}^{n-1} .$$

Integrating with respect to r from R to $2R$ and using (iii)

$$\begin{aligned} R\Big(\int_{B(x,R)} f(y)\, dy \Big)^{\frac{n-1}{n}} &\leq \int_R^{2R} dr \Big(\int_{B(x,R)} \det Du\, dy \Big)^{\frac{n-1}{n}} \\ &\leq c(n) \int_{B(x,2R)} |\operatorname{adj} Du|\, dy \end{aligned}$$

hence

$$\fint_{B(x,R)} f(y)\, dy \leq c(n) \Big(\fint_{B(x,2R)} |\operatorname{adj} Du|\, dy \Big)^{\frac{n}{n-1}} .$$

Setting

$$g(x) := \begin{cases} |\operatorname{adj} Du(x)| & x \in \Omega \\ 0 & x \notin \Omega , \end{cases}$$

we can then state

$$M(f)(x) \leq c(n)\, [M(g)(x)]^{n/(n-1)} + c(d)\, \| g \|^{n/(n-1)}_{L^{n/(n-1)}} .$$

Integrating on $\widehat{\Omega}$, and taking into account Proposition 2 in Sec. 3.1.1, the result follows at once from Proposition 5 in Sec. 3.1.1. □

One can show, compare Müller [501], that the result of Theorem 2 is optimal in the sense that the sign condition on $\det Du$ cannot be dropped, nor can $\widetilde{\Omega}$ be replaced by Ω. Moreover $\det Du \log(2+\det Du)$ cannot be replaced by $\gamma(\det Du)$ with $\gamma(t)/(t\log(2+t)) \to \infty$ as $t \to \infty$.

An immediate consequence of Theorem 2 is the following

Corollary 1. *Let $u \in W^{1,n}(\Omega, \mathbb{R}^n)$. Suppose that $\det Du \geq 0$. Then*

$$\det Du \log(2 + \det Du) \in L^1_{\text{loc}}(\Omega) .$$

•

[5] *BMO and Hardy space* $\mathcal{H}^1(\mathbb{R}^n)$. There is a more general approach to results of the type of Theorem 2 in terms of the *Hardy one* space.

We recall that a locally summable function $f : \mathbb{R}^n \to \mathbb{R}$ is said to have *bounded mean oscillation* if

$$\| f \|_* := \sup_{x,r} \fint_{B(x,r)} |f(y) - f_{x,r}| \, dy < \infty$$

where

$$f_{x,r} := \fint_{B(x,r)} f(y) \, dy = \frac{1}{|B(x,r)|} \int_{B(x,r)} f(y) \, dy \, .$$

The class of functions of bounded mean oscillation is denoted by BMO and often is referred as *John-Nirenberg space.*

Note that $\| f \|_* = 0$ if and only if f is constant a.e. One then can show that BMO modulo constants is a Banach space with respect to the norm $\| f \|_*$.

Let $g \in L^1(\mathbb{R}^n)$ and let ϕ be any smooth function with support in the unit ball and satisfying

$$\int_{\mathbb{R}^n} \phi(x) \, dx = 1 \, .$$

Set

$$g^*(x) := \sup_{r>0} \left| \frac{1}{r^n} \int g(y) \, \phi(\frac{x-y}{r}) \, dy \right| = \sup_{r>0} |\phi_r * g(x)|$$

where $\phi_r(x) = r^{-n}\phi(x/r)$. Notice that at every Lebesgue point x for g we have

$$|g(x)| \leq |g^*(x)|$$

hence

$$\| g \|_{L^1(\mathbb{R}^n)} \leq \| g^* \|_{L^1(\mathbb{R}^n)} \, .$$

The *Hardy space* $\mathcal{H}^1(\mathbb{R}^n)$ is defined as

$$\mathcal{H}^1(\mathbb{R}^n) := \{ g \in L^1(\mathbb{R}^n) \mid g^* \in L^1(\mathbb{R}^n) \} \, .$$

It turns out that $\mathcal{H}^1(\mathbb{R}^n)$ is a Banach space with norm

$$\| g \|_{\mathcal{H}^1} := \| g^* \|_{L^1(\mathbb{R}^n)},$$

which does not depend on the chosen function ϕ. In fact one can show that $f \in \mathcal{H}^1(\mathbb{R}^n)$ if and only if the function

$$\sup_{r>0} \sup \{ |\phi_r * g| \mid \phi \in C_c^\infty(B(0,1)), \ \| D\phi \|_\infty \leq 1 \}$$

belongs to L^1.[1] Trivially

[1] Notice that in the sequel working with a specific ϕ or with all ϕ with $\| D\phi \|_\infty \leq 1$ is just equivalent.

$$\mathcal{H}^1(\mathbb{R}^n) \subset L^1(\mathbb{R}^n),$$

and the behaviour at infinity together with the summability of f^* yield that every $f \in \mathcal{H}^1(\mathbb{R}^n)$ necessarily has zero average. In particular one easily sees that no function of $L^1(\mathbb{R}^n)$ with compact support and non vanishing integral belongs to $\mathcal{H}^1(\mathbb{R}^n)$. On the contrary one can show that every function $f \in L^p(\mathbb{R}^n)$, $p \in (1, +\infty]$, spt f compact, and $\int f = 0$ belongs to $\mathcal{H}^1(\mathbb{R}^n)$.

Finally we observe that *every bounded sequence $\{f_j\}$ in $\mathcal{H}^1(\mathbb{R}^n)$ has a subsequence which converges in the sense of distributions to a function $f \in \mathcal{H}^1(\mathbb{R}^n)$ with*

$$\| f \|_{\mathcal{H}^1} \le \liminf_{j\to\infty} \| f_j \|_{\mathcal{H}^1} ,$$

in fact passing to a subsequence $\{f_j\}$ converges in the sense of distributions to some f and it is not difficult to see that

$$f^*(x) \le \liminf_{j\to\infty} f_j^*(x) \qquad \forall\, x \in \mathbb{R}^n.$$

A fundamental theorem in the theory of Hardy space $\mathcal{H}^1$ developed by C. Fefferman and E. Stein asserts

Theorem 3 (Fefferman). *The dual space of $\mathcal{H}^1(\mathbb{R}^n)$ is BMO. More precisely, L is a continuous linear functional on $\mathcal{H}^1(\mathbb{R}^n)$ if and only if it can be represented as*

$$L(g) = \int_{\mathbb{R}^n} fg$$

for some function f in BMO, moreover for any $f \in BMO$ and any $g \in \mathcal{H}^1(\mathbb{R}^n)$ we have

$$\left| \int_{\mathbb{R}^n} fg\, dx \right| \le c(n) \| f \|_* \| g \|_{\mathcal{H}^1} . \tag{22}$$

Theorem 3 plays an important role in many fields, among them harmonic analysis, complex analysis, interpolation theory and partial differential equations. By means of it we shall now deduce a few interesting results in our context. In order to do that let us first state a simple lemma.

Lemma 2. *Let $u \in W^{1,p}(\mathbb{R}^n)$ and $v \in L^{p'}(\mathbb{R}^n)$, $p > 1$, $p' = p/(p-1)$. Then we have*

$$\int_{\mathbb{R}^n} \sup_r \frac{1}{r^{n+1}} \Big(\int_{B(x,r)} |u(y) - u_{x,r}|\, |v(y)|\, dy \Big)\, dx \le c(n) \| Du \|_{L^p(\mathbb{R}^n)} \| v \|_{L^{p'}(\mathbb{R}^n)} .$$

Proof. Chose s and q so that

$$\max(1, s_*) < q < p < s < p^* ,$$
$$\frac{1}{p^*} := \frac{1}{p} - \frac{1}{n} , \qquad \frac{1}{s_*} = \frac{1}{s} + \frac{1}{n} .$$

Using Hölder and Sobolev-Poincaré inequalities we get

$$\int_{\mathbb{R}^n} \sup_r \frac{1}{r^{n+1}} \Big(\int_{B(x,r)} |u(y) - u_{x,r}|\, |v(y)|\, dy \Big)\, dx$$

$$\leq \int_{\mathbb{R}^n} \sup_r \frac{1}{r} \Big(\fint_{B(x,r)} |u(y) - u_{x,r}|^s\, dy \Big)^{1/s} \Big(\fint_{B(x,r)} |v(y)|^{s'}\, dy \Big)^{1/s'} dx$$

$$\leq c \int_{\mathbb{R}^n} \sup_r \Big(\fint_{B(x,r)} |Du(y)|^q\, dy \Big)^{1/q} \Big(\fint_{B(x,r)} |v(y)|^{s'}\, dy \Big)^{1/s'} dx$$

$$\leq \int_{\mathbb{R}^n} \Big[M(|Du|^q)(x) \Big]^{\frac{1}{q}} \Big[M(|v|^{s'})(x) \Big]^{\frac{1}{s'}} dx$$

$$\leq c \Big(\int_{\mathbb{R}^n} M(|Du|^q)^{\frac{p}{q}}\, dx \Big)^{1/p} \Big(\int_{\mathbb{R}^n} M(|v|^{s'})^{\frac{p'}{s'}}\, dx \Big)^{1/p'}$$

$$\leq c \,\| Du \|_{L^p}\, \| v \|_{L^{p'}}$$

because of Proposition 2 in Sec. 3.1.1. □

We can now prove

Proposition 6. *Let $u \in W^{1,p}(\mathbb{R}^n)$ and let $v \in L^{p'}(\mathbb{R}^n, \mathbb{R}^n)$ be a vector field with zero divergence in the sense of distributions, $1 < p < \infty$. Then $Du \cdot v$ belongs to $\mathcal{H}^1(\mathbb{R}^n)$ and*

$$\| Du \cdot v \|_{\mathcal{H}^1(\mathbb{R}^n)} \leq c \,\| Du \|_{L^p}\, \| v \|_{L^{p'}} . \tag{23}$$

Proof. Clearly $Du \cdot v \in L^1(\mathbb{R}^n)$ and

$$\int_{\mathbb{R}^n} Du \cdot v\, dx = \int_{\mathbb{R}^n} \operatorname{div}(uv)\, dx = 0 .$$

It remains to estimate the L^1-norm of $(Du \cdot v)^*$. Using the fact that v has zero divergence we have

$$\Big| \int_{\mathbb{R}^n} Du(y) \cdot v(y)\, \phi_r(x-y)\, dy \Big| = \Big| \int_{B(x,r)} \operatorname{div}(uv)(y)\, \phi_r(x-y)\, dy \Big|$$

$$= \Big| \int_{B(x,r)} (u(y) - u_{x,r}) v(y) \cdot D\phi_r(x-y)\, dy \Big| \tag{24}$$

$$\leq \frac{c}{r^{n+1}} \int_{B(x,r)} |u(y) - u_{x,r}|\, |v(y)|\, dy ,$$

whence (23) follows at once from Lemma 2. □

Late we shall need the following localized version of Proposition 6

Proposition 7. *Let* $u \in W^{1,p}(\mathbb{R}^n)$ *with* $\operatorname{spt} u \subset\subset \mathbb{R}^n$*, say* $\operatorname{spt} u \subset B(x_0, R)$*, and let* $v \in L^{p'}(\mathbb{R}^n, \mathbb{R}^n)$ *be a vector field with zero divergence in the sense of distributions,* $1 < p \leq \infty$*. then* $Du \cdot v$ *belongs to* $\mathcal{H}^1(\mathbb{R}^n)$ *and*

$$\| Du \cdot v \|_{\mathcal{H}^1(\mathbb{R}^n)} \leq c \| Du \|_{L^p(B(x_0,R))} \| v \|_{L^{p'}(B(x_0,2R))}$$

Proof. Choosing a standard cut–off function η, $\eta \in C_c^\infty(B(x_0, 2R))$, $\eta = 1$ on $B(x_0, R)$, $|D\eta| \leq 2/R$, we get similarly to (24)

$$\begin{aligned} (Du \cdot v) * \phi_r(x) &= \int_{B(x,r)} u(y)\, \eta^2(y)\, v(y) \cdot D\phi_r(x-y)\, dy \\ &= \int_{B(x,r)} (u - u_{x,r})\eta^2\, v \cdot D\phi_r\, dy + \int_{B(x,r)} u_{x,r}\, v\, \eta^2\, D\phi_r\, dy \\ &= \int_{B(x,r)} (u - u_{x,r}) v\, \eta^2\, D\phi_r\, dy - 2 \int_{B(x,r)} u_{x,r}\, D\eta \cdot v(y)\, \phi_r\, dy \end{aligned}$$

Therefore

$$(25) \qquad |(Du \cdot v) * \phi_r(x)| \leq \int_{\mathbb{R}^n} \sup_r \frac{1}{r^{n+1}} \int_{B(x,r)} |u(y) - u_{x,r}|\, |v(y)\, \eta^2(y)|\, dy + \frac{4}{R} \frac{1}{r^n} |u_{x,r}| \fint_{B(x,r)} |v\eta|\, dy$$

Using Lemma 2 we infer

$$(26) \qquad \int_{\mathbb{R}^n} \sup_r \frac{1}{r^{n+1}} \int_{B(x,r)} |u(y) - u_{x,r}|\, |v(y)\, \eta^2(y)|\, dy\, dx \leq c \, \| Du \|_{L^p(\mathbb{R}^n)} \| v\eta^2 \|_{L^{p'}(\mathbb{R}^n)}$$

while as in the proof of Lemma 2, we also have

$$\begin{aligned} &\frac{4}{R} \int_{\mathbb{R}^n} \sup_r \frac{1}{r^n} |u_{x,r}| \fint_{B(x,r)} |v\eta|\, dy\, dx \\ &\leq \frac{c}{R} \| u \|_{L^p(\mathbb{R}^n)} \| v\eta \|_{L^{p'}(\mathbb{R}^n)} \leq c \, \| Du \|_{L^p(\mathbb{R}^n)} \| v\eta \|_{L^{p'}(\mathbb{R}^n)} . \end{aligned}$$

The claim then follows easily from (25) (26) and the last inequality. □

A similar conclusion to that of Proposition 6 can be inferred for the *commutator* or the *Jacobian minor* of two functions $u \in W^{1,p}(\mathbb{R}^n)$ and $v \in W^{1,p'}(\mathbb{R}^n)$. Set

$$[u,v]_{ij} = u_{x^i}\,v_{x^j} - u_{x^j}\,v_{x^i}$$

which of course belongs to $L^1(\mathbb{R}^n)$. Due to its divergence structure, namely

$$[u,v]_{ij} = (u\,v_{x^j})_{x^i} - (u\,v_{x^i})_{x^j}$$

in the sense of distributions, we have

Proposition 8. *Let $u \in W^{1,p}(\mathbb{R}^n)$, $v \in W^{1,p'}(\mathbb{R}^n)$, $1 < p < \infty$. Then $[u,v]_{ij}$ belongs $\mathcal{H}^1(\mathbb{R}^n)$ and*

$$\|\,[u,v]_{ij}\,\|_{\mathcal{H}^1(\mathbb{R}^n)} \le c\,\|\,Du\,\|_{L^p(\mathbb{R}^n)}\,\|\,Dv\,\|_{L^{p'}(\mathbb{R}^n)}\,. \tag{27}$$

Proof. Clearly $[u,v]_{ij} \in L^1(\mathbb{R}^n)$ and

$$\int_{\mathbb{R}^n} [u,v]_{ij}\,dx = 0\,.$$

As in the proof of Proposition 6 we now have

$$\begin{aligned}\frac{1}{r^n}\int_{\mathbb{R}^n}[u,v]_{ij}\,dy &= -\int_{B(x,r)} (u(y)-u_{x,r})\,v_{y^j}(y)\,\phi_{r,y^i}(x-y)\\ &\quad+\int_{B(x,r)} (u(y)-u_{x,r})\,v_{y^i}(y)\,\phi_{r,y^j}(x-y)\,.\end{aligned}$$

Thus

$$\frac{1}{r^n}\left|\int_{\mathbb{R}^n}[u,v]_{ij}\,dy\right| \le \frac{c}{r^{n+1}}\int_{B(x,r)} |u(y)-u_{x,r}|\,|Dv|\,dy\,,$$

and the claim follows again from Lemma 2. □

Let $u \in W^{1,n}(\mathbb{R}^n,\mathbb{R}^n)$. As

$$\det Du = D_j u^1\, v^1_j$$

where v^1 is the vector field

$$v^1 := (\operatorname{adj} Du)^1_j$$

which is divergence free and

$$v^1 \in L^{n/(n-1)}(\mathbb{R}^n)\,,$$

we can apply Proposition 6 and infer

Corollary 2. *Let $u \in W^{1,n}(\mathbb{R}^n,\mathbb{R}^n)$. Then $\det Du$ belong to the Hardy space $\mathcal{H}^1(\mathbb{R}^n)$.*

Actually in a similar way one can get

Corollary 3. *Let* $u \in \mathcal{A}_{p,q}(\mathbb{R}^n, \mathbb{R}^n)$ *where* $1/p + 1/q \leq 1$. *Then* $\det Du \in \mathcal{H}^1(\mathbb{R}^n)$.

We can localize the previous result recovering this way Corollary 1. If $u \in W^{1,n}(\Omega, \mathbb{R}^n)$ and $\eta \in C_c^\infty(\Omega)$, $\eta \leq 0$ write

$$\eta \det Du \;=\; D_j(u^1\eta)\, v_j^1 - u^1\, D_j\eta\, v_j^1$$

and observe that

$$D_j(u^1\eta)\cdot v_j^1 \in \mathcal{H}^1(\mathbb{R}^n), \qquad u^1 D_j\eta\, v_j^1 \in L^p, \quad 1 < p < \frac{n}{n-1}.$$

Assume now that $\det Du \geq 0$, then

$$M(\eta\det Du)(x) \leq c_1(\eta\det Du)^* \leq c_1(D_j(u^1\eta)v_j^1)^*(x) + c_1(u^1 D_j\eta\, v_j^1)^*$$

$$\leq c_1(D_j(u^1\eta)v_j^1)^*(x) + c_1 M(u^1 D_j\eta\, v_j^1).$$

It follows that $M(\eta\det Du) \in L^1(\mathbb{R}^n)$ and consequently $\eta\det Du\, \log(2+\eta\det Du)$ is summable, recovering this way Corollary 1.

We conclude this subsection by stating one more consequence of Proposition 6. Since every vector field E in $L^p(\mathbb{R}^n)$, $1 < p < n$, with $\operatorname{curl} E = 0$ can be written as

$$E \;=\; Du$$

for some scalar function $u \in W^{1,p}(\mathbb{R}^n)$, we readily get

Corollary 4. *Let* $E \in L^p(\mathbb{R}^n, \mathbb{R}^n)$, $B \in L^{p'}(\mathbb{R}^n, \mathbb{R}^n)$, $1 < p < \infty$, *be two vector fields in* $\mathbb{R}^n$ *satisfying in the sense of distributions*

$$\operatorname{curl} E \;=\; 0 \qquad \operatorname{div} B \;=\; 0\,.$$

Then $E \cdot B \in \mathcal{H}^1(\mathbb{R}^n)$.

•

2.5 Boundaries and Traces

Let Ω be a bounded domain in $\mathbb{R}^n$ with smooth boundary and let u be a map in $\mathcal{A}^1(\Omega, \mathbb{R}^N) \cap W^{1,1}(\Omega, \mathbb{R}^N)$. The trace of u on $\partial\Omega$, $u_{|\partial\Omega}$ is well defined in the sense of Sobolev spaces. Also the current G_u is well defined in $\mathbb{R}^n \times \mathbb{R}^N$ and, when $\partial G_u \llcorner \Omega \times \mathbb{R}^N = 0$, the boundary ∂G_u of G_u has support in $\partial\Omega \times \mathbb{R}^N$. A first question which naturally arises is whether the trace of u on $\partial\Omega$ determines ∂G_u , that is, whether we have

$$\partial G_u \;=\; \partial G_v$$

for $u, v \in \operatorname{cart}^1(\Omega, \mathbb{R}^N)$ satisfying $u_{|\partial\Omega} = v_{|\partial\Omega}$.

The answer to this question is in general negative as shown in [1] below. In fact the trace in the $W^{1,1}$ sense just fixes the 0-component of the boundary.

Proposition 1. *Let $u, v \in \mathcal{A}^1(\Omega, \mathbb{R}^N) \cap W^{1,1}(\Omega, \mathbb{R}^N)$. Then $u = v$ on $\partial\Omega$ if and only if $(\partial G_u)_{(0)} = (\partial G_v)_{(0)}$.*

Proof. First we note that $(\partial G_u)_{(0)} \llcorner \Omega \times \mathbb{R}^N = (\partial G_v)_{(0)} \llcorner \Omega \times \mathbb{R}^N = 0$ by Proposition 1 in Sec. 3.2.3. Now we observe that any $(n-1)$-form with no component in the vertical direction is a sum of forms of type

$$\omega = (-1)^{j-1} \phi(x, y)\, dx^{\bar{j}}$$

when $\phi \in C_c^\infty(\mathbb{R}^n \times \mathbb{R}^N)$ and, by Proposition 2 in Sec. 3.2.3, compare also the proof of Proposition 1 in Sec. 3.2.3,

$$(\partial G_u)_{(0)}(\omega) = \int_\Omega D_j[\phi(x, u(x))]\, dx$$

$$(\partial G_v)_{(0)}(\omega) = \int_\Omega D_j[\phi(x, v(x))]\, dx\,.$$

As $\phi(x, u(x)) \in W^{1,1}(\Omega)$ and $\operatorname{trace} \phi(x, u(x)) = \phi(x, \operatorname{trace} u(x))$, we then infer

$$(\partial G_u)_{(0)}(\omega) - (\partial G_v)_{(0)}(\omega) = \int_{\partial\Omega} [\phi(x, u(x)) - \phi(x, v(x))] \nu_j \, d\mathcal{H}^{n-1} = 0\,.$$

Conversely, if $(\partial G_u)_{(0)} = (\partial G_v)_{(0)}$, then for any $\phi \in C_c^1(\mathbb{R}^n \times \mathbb{R}^N)$ we have

$$\int_\Omega D_j[\phi(x, u(x)) - \phi(x, v(x))]\, dx = 0$$

and by approximation, as in the proof of Proposition 1 in Sec. 3.2.3, we deduce, for $\phi(x, y) = \varphi(x) y^i$, $\varphi(x) \in C_c^\infty(\mathbb{R}^n)$

$$0 = \int_\Omega D_j[\varphi(x)(u^i - v^i)]\, dx = \int_{\partial\Omega} \varphi(x)\, \nu_j (u^i - v^i)\, d\mathcal{H}^{n-1}\,,$$

that is $u = v$ on $\partial\Omega$. □

As consequence of Proposition 3 in Sec. 3.2.1 we instead have

Proposition 2. *Let Ω be a bounded Lipschitz domain in $\mathbb{R}^n$. Let u, v be maps in $W^{1,\underline{n}}(\Omega, \mathbb{R}^N)$, $\underline{n} := \min(n, N)$. Then $u = v$ on $\partial\Omega$ implies $\partial G_u = \partial G_v$.*

Proof. Clearly the claim is local. Fix a point $x_0 \in \partial\Omega$ and choose a neighborhood $\mathcal{U}$ of x_0 for which there exists a bi-Lipschitz map $\varphi : B(0,1) \to \mathcal{U}$ such that

$$\varphi(B^+(0,1)) = \Omega \cap \mathcal{U}, \qquad B^+(0,1) = \{x \in B(0,1) \,|\, x_1 > 0\}$$

and

$$\varphi(\Gamma) = \partial\Omega \cap \mathcal{U}, \qquad \Gamma = B(0,1) \cap \{x_1 = 0\}$$

Denote by $\phi(x,y)$ the map $\phi(x,y) := (\varphi(x), y)$. As

$$\phi_\# G_{u\circ\varphi} = G_u, \qquad \phi_\# G_{v\circ\varphi} = G_v \quad \text{in } \mathcal{U} \times \mathbb{R}^n$$

and

$$u \circ \varphi = v \circ \varphi \qquad \text{on } \Gamma$$

in the sense of traces in $W^{1,\underline{n}}$, it is not restrictive to assume that u, v belong to $W^{1,\underline{n}}(B^+(0,1))$ with $u = v$ on Γ.

Consider now the function w defined on $B(0,1)$ as

$$w(x) = \begin{cases} u(x_1, \overline{x}) & \text{if } x_1 > 0 \\ v(-x_1, \overline{x}) & \text{if } x_1 < 0 \end{cases} \tag{1}$$

Clearly w belongs to $W^{1,\underline{n}}(B(0,1), \mathbb{R}^N)$. Thus by Proposition 3 in Sec. 3.2.1

$$\partial G_w \llcorner B(0,1) \times \mathbb{R}^N = 0 .$$

From this equality one easily infers the claim. For that fix α, β, $|\alpha| + |\beta| = n$, $|\beta| = k+1$, $0 \le k \le \underline{n} - 1$, $g \in C^1_c(\mathbb{R}^N)$ and $\varphi \in C^1_c(B(0,1))$ and set

$$\psi(x_1, \overline{x}) := \begin{cases} \varphi(x_1, \overline{x}) & \text{if } x_1 > 0 \\ \varphi(-x_1, \overline{x}) & \text{if } x_1 < 0 . \end{cases}$$

By Corollary 1 in Sec. 3.2.3 and Remark 2 in Sec. 3.2.3 we have for $i \in \beta$

$$\begin{aligned} 0 &= \int_{B(0,1)} \sum_{j\in\bar{\alpha}} D_j[\psi(x) g(w(x))](\text{adj}\,(Dw(x))^{\beta}_{\bar{\alpha}})^i_j \, dx = \int_{B^+(0,1)} \cdots + \int_{B^-(0,1)} \cdots \\ &= \int_{B^+(0,1)} \sum_{j\in\bar{\alpha}} D_j[\varphi(x) g(u(x))](\text{adj}\,(Du(x))^{\beta}_{\bar{\alpha}})^i_j \, dx \\ &\quad - \int_{B^+(0,1)} \sum_{j\in\bar{\alpha}} D_j[\psi(x) g(v(x))](\text{adj}\,(Dv(x))^{\beta}_{\bar{\alpha}})^i_j \, dx \\ &= (-1)^{|\alpha|}\sigma(\alpha,\bar{\alpha})\sigma(i,\beta-i)(\partial G_u - \partial G_v)(\varphi(x)g(y)dx^\alpha \wedge dy^{\beta-i}). \end{aligned}$$

□

Actually the proof of the previous proposition shows also

Proposition 3. *Let Ω be a bounded Lipschitz domain in $\mathbb{R}^n$. Let u, v be maps in $W^{1,p}(\Omega, \mathbb{R}^N)$ with $1 \le p \le \underline{n}$. If $u = v$ on $\partial\Omega$ then*

$$(\partial G_u)_{(k)} = (\partial G_v)_{(k)}$$

for all k with $0 \le k \le p-1$, $0 \le k \le \min(n-1, N)$.

We also have

Proposition 4. *Let Ω be a bounded Lipschitz domain in $\mathbb{R}^n$. Let u, v be maps in $\mathcal{A}_{n-1,n/(n-1)}$. Then $u = v$ on $\partial\Omega$ in the sense of traces implies $\partial G_u = \partial G_v$.*

Proof. We proceed as in the proof of Proposition 2. By Binet's formula, see e.g. Lemma 3 in Vol. II Sec. 2.5.2, we have

$$\operatorname{adj} D(u \circ \varphi) \;=\; \operatorname{adj} D\varphi \;\cdot \operatorname{adj}(Du) \circ \varphi,$$

which yields at once

$$w \in \mathcal{A}_{n-1,n/(n-1)}(B(0,1), \mathbb{R}^n)\ ,$$

w being the map in (1). By Theorem 1 in Sec. 3.2.4 we again infer

$$\partial G_w \;=\; 0 \qquad \text{on } B(0,1) \times \mathbb{R}^n$$

thus we can conclude as in the proof of Proposition 2. □

[1] Let $u, v \in \operatorname{cart}^1(\Omega, \mathbb{R}^N)$. We can have $u = v$ on $\partial\Omega$ while $\partial G_u \neq \partial G_v$.

Let $\varphi : S^1 \subset \mathbb{R}^2 \to \mathbb{R}^2$ be a smooth map and let $v(x) := \varphi(\frac{x}{|x|})$ be its homogeneous extension to $B(0,1) \subset \mathbb{R}^2$. We choose a cut off function $\eta(|x|)$,

$$\begin{array}{lll} \eta(t) = 1 & \text{for } \ 0 \le t \le \dfrac{1}{2} & 0 \le \eta \le 1\ , \\ \eta(t) = 0 & \text{for } \ \dfrac{3}{4} \le t \le 1\ , & |\eta'(t)| \le 8\ ,\ \eta'(t) \le 0\ , \end{array}$$

and we consider the function $u(x)$, restriction of $v(x)\eta(|x|)$ to the upper half ball $B^+(0,1) = B(0,1) \cap \{(x^1, x^2) \mid x^2 > 0\}$. Evidently $u(x) \in W^{1,p}(B^+(0,1), \mathbb{R}^2)$ for any $p < 2$; moreover, since $\det Du = 0$ on $B^+(0,1/2)$ and u is smooth outside $B^+(0,1/2)$, u belongs to $\mathcal{A}^1(B^+(0,1/2), \mathbb{R}^2) \cap W^{1,1}(B^+(0,1/2), \mathbb{R}^2)$. The trace of $u(x)$ on $\partial B^+(0,1)$, in the sense of $W^{1,p}$, is given by

$$u(x) \;:=\; \begin{cases} \varphi(1,0)\,\eta(|x|) & \text{if } x^2 = 0\ ,\ x^1 > 0 \\ \varphi(-1,0)\,\eta(|x|) & \text{if } x^2 = 0\ ,\ x^1 < 0 \\ 0 & \text{if } (x^1)^2 + (x^2)^2 = 1\ ,\ x^2 > 0 \end{cases}$$

In particular, if we choose φ so that $\varphi(1,0) = \varphi(-1,0) = 0$, we have $u_{|\partial B^+(0,1)} = 0$. For instance, this is the case if we multiply by $\eta(|x|)$ the function

$$\text{(2)} \qquad \bar{u}(x) \;=\; \left(\frac{x^1 x^2}{(x^1)^2 + (x^2)^2}, \frac{(x^2)^2}{(x^1)^2 + (x^2)^2} \right), \quad (x^1, x^2) \in B^+(0,1)\ .$$

We shall now show that, though $u_{|\partial B^+(0,1)} = 0$, the boundary of G_u is different from the boundary of the current carried by the graph of the constant map into $0 \in \mathbb{R}^2$.

$$\partial G_u \neq \partial G_0 = \partial B^+(0,1) \times \delta_0, \qquad \delta_0 = \text{Dirac mass at } 0 \in \mathbb{R}^2 .$$

Observe that $\partial G_u \llcorner B^+(0,1) \times \mathbb{R}^2 = 0$.

As in [2] in Sec. 3.2.2, we set

$$u_\varepsilon(x) = r_\varepsilon(|x|)\,\varphi(\frac{x}{|x|})\,\eta(|x|)$$

where $r_\varepsilon(t)$ is linear in $(0,\varepsilon)$ and $r_\varepsilon(0) = 0$, $r_\varepsilon(\varepsilon) = 1 - \varepsilon$, and $r_\varepsilon(1) = 1$. The functions $u_\varepsilon(x)$ are Lipschitz-continuous, and have trace equal zero on $\partial B^+(0,1)$. Hence

(3) $$\partial G_{u_\varepsilon} = \partial G_0 = \partial B^+(0,1) \times \delta_0$$

for every $\varepsilon > 0$. Reparametrizing $\mathcal{G}_{u_\varepsilon, B^+(0,1)}$ as

$$\phi_\varepsilon(z) := (\gamma_\varepsilon(z), u_\varepsilon(\gamma_\varepsilon(z)))$$

where

$$\gamma_\varepsilon(z) := \rho_\varepsilon(z)\,\frac{z}{|z|}$$

and $\rho_\varepsilon(t)$ is linear in $(0,1/2)$, $(1/2,1)$, and $\rho(0) = 0$, $\rho(1/2) = \varepsilon$, $\rho(1) = 1$, we get

$$G_{u_\varepsilon} = \phi_{\varepsilon\#}[\![B^+(0,1)]\!] .$$

Since ϕ_ε is a family of equi-Lipschitz functions converging uniformly to the Lipschitz function

$$\phi(z) := \begin{cases} (0, 2|z|\varphi(\frac{z}{|z|}) & |z| \leq \frac{1}{2} \\ (\gamma(z), \varphi(\frac{\gamma(z)}{|\gamma(z)|})\,\eta(\gamma(|z|)) & \frac{1}{2} \leq |z| \leq 1 \end{cases}$$

we deduce that

(4) $$G_{u_\varepsilon} \rightharpoonup \phi_\#[\![B^+(0,1) \setminus B^+(0,1/2)]\!] + \phi_\#[\![B^+(0,1/2)]\!]$$

As

$$\phi_\#[\![B^+(0,1) \setminus B^+(0,1/2)]\!] = G_u$$

we deduce from (4), taking into account (3)

$$\partial G_0 = \partial G_u + \partial\phi_\#[\![B^+(0,1/2)]\!]$$

i.e., denoting by $[\![S^{1+}]\!]$ the current integration on the positive half circle

$$\partial G_u = \partial G_0 - \phi_\#\partial[\![B^+(0,1/2)]\!] = \partial G_0 - \varphi_\# [\![S_1^+]\!]$$

if we also take into account the expression of $\phi(z)$ for $|z| = \frac{1}{2}$.

For the map in (2) we therefore get

$$\bar{u}_{|\partial B^+(0,1)} = 0$$

and

$$\partial G_{\bar{u}} = \partial G_0 - \delta_0 \times \partial[\![B((0,1/2),1/2)]\!] \neq \partial G_0 ,$$

since obviously

$$\varphi_{\#} [\![S^{1+}]\!] = \partial[\![B((0,1/2),1/2)]\!] .$$

Essentially a similar phenomenon occurs for the mapping

$$w(z) = (\frac{z}{|z|})^{2k}$$

in complex coordinates. The associated current has a non zero boundary in $\{0\} \times \mathbb{R}^2$ given by $\delta_0 \times 2k[\![S^1]\!]$, while the boundary of the restriction u of $w(z)$ to $Im z > 0$ is given by

$$\partial G_u = G_0 - k\,\delta_0 \times [\![S^1]\!] .$$

Notice finally that $2\bar{u}$ can be obtained from $w(z) = \frac{z^2}{|z|^2}$ by a rotation R of $\frac{\pi}{4}$ degree in the z-plane plus a translation in the target plane

$$2\,\bar{u}(x^1,x^2) = w(Rz) + (0,1) , \qquad z = x^1 + ix^2 .$$

•

The previous considerations show that in terms of graphs we have a large choice in imposing to G_u *Dirichlet's type conditions* in vector valued problems: we may fix the 0-component of ∂G_u

$$(\partial G_u)_{(0)} ,$$

to which sometime we shall refer as to the *weak anchorage condition*, or

$$(\partial G_u)_{(0)}, (\partial G_u)_{(1)}, \ldots, (\partial G_u)_{(k_0)} \qquad 0 < k_0 < n-1$$

or we may fix the entire boundary ∂G_u, i.e.,

$$(\partial G_u)_{(0)}, \ldots, (\partial G_u)_{(n-1)} ,$$

to which we shall refer also as to the *strong anchorage condition.*

Notice that, since

$$\partial G_{u_k} \rightharpoonup \partial G_u$$

and even

$$(\partial G_{u_k})_{(i)} \rightharpoonup (\partial G_u)_{(i)} \qquad \text{for } i = 0, 1, \ldots, n-1$$

whenever

$$u_k \overset{\mathcal{A}^1}{\rightharpoonup} u ,$$

the corresponding subclasses of the u's in $\mathcal{A}^1(\Omega, \mathbb{R}^N)$, for which

$$(\partial G_u)_{(0)}, \ldots, (\partial G_u)_{(k)} \qquad 0 \le k \le n-1$$

are prescribed, are closed with respect to the convergence in $\mathcal{A}^1$.

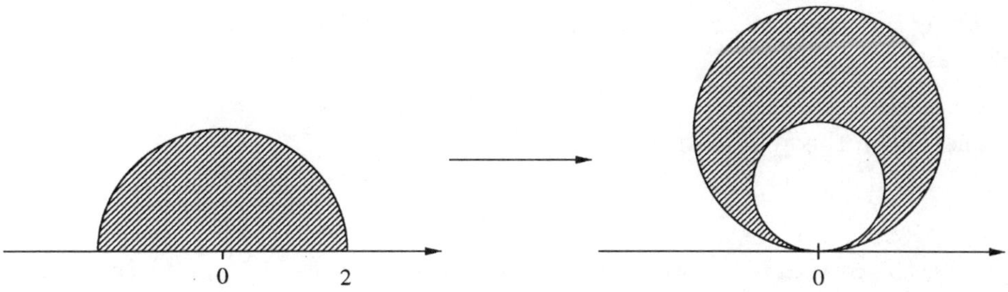

Fig. 3.2. The map v.

[2] It is worth remarking that the weak anchorage condition, i.e. prescribing the trace in the $W^{1,1}$–sense, is far from the intuitive idea of prescribing the boundary. This is better shown by the following modification of the example in [1]

Set

$$\eta(|x|) \;:=\; \begin{cases} 1 & \text{if } 0 \le |x| \le 1 \\ |x|^2 & \text{if } 1 \le |x| \le 2 \end{cases}$$

and consider

$$v(x) \;:= \eta(|x|)\overline{u}(x) \qquad x \in B^+(0,2) \subset \mathbb{R}^2$$

where $\overline{u}$ is the map in (2). Clearly $v \in W^{1,p}(B^+(0,2), \mathbb{R}^2) \cap \mathcal{A}^1(B^+(0,2), \mathbb{R}^2)$, $p < 2$, and one easily sees that

$$v(B^+(0,2)) \;=\; B((0,1),1) \setminus B((0,1/2),1/2)$$

and

$$v(\partial B^+(0,2)) \;=\; \partial B((0,1),1)$$

in the sense of traces. •

Let Ω be a Lipschitz domain in $\mathbb{R}^n$. For $\mathcal{H}^{n-1}$-a.e. $x \in \partial\Omega$ the exterior unit normal vector $\nu(x)$ is well-defined. As it is well-known flux integrals can be transformed into integration of differential forms, or in terms of vectors, to every vector ν we can associate the $(n-1)$-vector $*\nu$ orienting the orthogonal $(n-1)$-plane to ν so that

$$\nu \wedge *\nu \;=\; e_1 \wedge \ldots \wedge e_n$$

defined by

$$*\nu = \sum_{i=1}^{n} (-1)^{i-1} \nu^i e_{\bar{i}} \quad \text{if} \quad \nu = \sum_{i=1}^{n} \nu^i e_i \,.$$

Defining now the i.m. rectifiable $(n-1)$-current $[\![\partial\Omega]\!]$ carried by the boundary of Ω as

$$[\![\partial\Omega]\!] = \tau(\partial\Omega, 1, *\nu)$$

Gauss-Green theorem reads as

$$\partial[\![\Omega]\!] = [\![\partial\Omega]\!] .$$

In fact, as every $(n-1)$-form ω can be written as $\omega = (-1)^{i-1}\varphi_i(x)\,\widehat{dx^i}$ so that $d\omega = \operatorname{div}\varphi\, dx$, we have

$$\begin{aligned}\partial[\![\Omega]\!](\omega) &= \int_\Omega \operatorname{div}\varphi\, dx = \int_{\partial\Omega} \varphi\cdot\nu\, d\mathcal{H}^{n-1} \\ &= \int_{\partial\Omega} <\omega, *\nu> d\mathcal{H}^{n-1} = [\![\partial\Omega]\!](\omega) .\end{aligned}$$

Let u be a map of class C^2 from a neighbourhood of $\overline{\Omega}$ into $\mathbb{R}^N$. The i.m. rectifiable $(n-1)$-current

$$G_{u,\partial\Omega} := (\mathrm{id}\bowtie u)_\# [\![\partial\Omega]\!] \tag{5}$$

is well defined and Stokes theorem says that

$$\partial(G_u \llcorner \Omega\times\mathbb{R}^N) = G_{u,\partial\Omega} . \tag{6}$$

In fact for any differential $(n-1)$-form in $\mathbb{R}^n\times\mathbb{R}^N$ we have

$$\begin{aligned}G_u \llcorner (\Omega\times\mathbb{R}^N)(d\omega) &= [\![\Omega]\!]((\mathrm{id}\bowtie u)^\# d\omega) \\ = [\![\Omega]\!](d(\mathrm{id}\bowtie u)^\#\omega) &= \partial[\![\Omega]\!]((\mathrm{id}\bowtie u)^\#\omega) \\ = [\![\partial\Omega]\!]((\mathrm{id}\bowtie u)^\#\omega) &= G_{u,\partial\Omega}(\omega) .\end{aligned}$$

In coordinates, if $|\alpha|+|\beta| = n$, $|\beta|\ge 1$, $i\in\beta$ and $\omega = \phi(x,y)\,dx^\alpha\wedge dy^{\beta-i}$, with $\phi\in C^\infty(\mathbb{R}^n\times\mathbb{R}^N)$ we have taking into account Proposition 2 in Sec. 3.2.3 and Corollary 2 in Sec. 3.2.3

$$\begin{aligned}\partial G_u(\omega) &= (-1)^{|\alpha|}\sigma(\alpha,\bar\alpha)\,\sigma(i,\beta-i)\sum_{j\in\bar\alpha}\int_\Omega D_j\,[\phi(x,u(x))]\,(\operatorname{adj}(Du)^\beta_{\bar\alpha})^i_j\,dx \\ &= (-1)^{|\alpha|}\sigma(\alpha,\bar\alpha)\,\sigma(i,\beta-i)\int_\Omega\sum_{j\in\bar\alpha} D_j\left[\phi(x,u(x))\,(\operatorname{adj}(Du(x))^\beta_{\bar\alpha})^i_j\right]dx \\ &= (-1)^{|\alpha|}\sigma(\alpha,\bar\alpha)\,\sigma(i,\beta-i)\int_{\partial\Omega}\sum_{j\in\bar\alpha}\phi(x,u(x))\,(\operatorname{adj}(Du(x))^\beta_{\bar\alpha})^i_j\nu_j\,dx .\end{aligned}$$

Recalling the definition of the adjoint minors in Sec. 3.2.3, we therefore conclude for any $\omega\in\mathcal{D}^n(\Omega\times\mathbb{R}^N)$ that

$$G_{u,\partial\Omega}(\omega) = \partial G_u(\omega) = \int_{\partial\Omega} <\omega(x,u(x)), A(Du(x))> d\mathcal{H}^{n-1} \tag{7}$$

where $A(Du(x))$ denotes the $(n-1)$-vector in $\mathbb{R}^{n+N}$ given by

$$\begin{aligned} A(Du(x)) &= \sum_{|\alpha|+|\beta|=n-1} A^{\alpha\beta}(Du(x))\, e_\alpha \wedge \varepsilon_\beta \\ A^{\alpha\beta}(Du(x)) &:= (-1)^{|\alpha|}\sigma(\alpha,\bar{\alpha})\sum_{j=1}^{n} \sigma(j,\bar{\alpha}-j)\, M^{\beta}_{\bar{\alpha}-j}(Du(x))\, \nu_j(x)\ . \end{aligned} \tag{8}$$

Of course $A(Du(x))$ depends only on u and on the tangential derivatives of u on $\partial\Omega$,

$$\vec{G}_{u,\partial\Omega} = \frac{A(Du)}{|\,A(Du)\,|}$$

and

$$\mathbf{M}(G_{u,\partial\Omega}) = \int_{\partial\Omega} |A(Du)|\, d\mathcal{H}^{n-1}\ .$$

Moreover we notice that the same holds, by approximation, under the weaker assumption that u be of class C^1 in a neighbourhood of Ω. Because of that we can now set

Definition 1. *A map $u: \partial\Omega \to \mathbb{R}^N$ is said to be in $\mathcal{A}^1(\partial\Omega,\mathbb{R}^N)$ if u belongs to $L^1(\partial\Omega,\mathbb{R}^N)$ and is $\mathcal{H}^{n-1}$-a.e. approximately differentiable in $\partial\Omega$ with $A(Du) \in L^1(\partial\Omega)$, where, for instance, $\mathcal{H}^{n-1}$-a.e. $x \in \partial\Omega$ $A(Du(x))$ can be computed by extending constantly u in the normal direction to $\partial\Omega$.*

Definition 2. *Let $u \in \mathcal{A}^1(\partial\Omega,\mathbb{R}^N)$. We define $G_{u,\partial\Omega}$ by*

$$G_{u,\partial\Omega}(\omega) := \int_{\partial\Omega} <\omega(x,u(x)), A(Du)> d\mathcal{H}^{n-1}\ .$$

Of course $G_{u,\partial\Omega}$ is a well defined i.m. rectifiable $(n-1)$-current in $\mathbb{R}^n \times \mathbb{R}^N$.

To be more explicit, we can assume that in a suitable neighbourhood of a point of $\partial\Omega$ the boundary of $\partial\Omega$ is flat, for instance, $\Omega = B^+(0,1)$, where

$$B^+(0,1) := \{x \in B(0,1) \mid x_1 > 0\}\ .$$

Let $u \in \mathcal{A}^1(\partial\Omega)$ and let

$$i : \mathbb{R}^{n-1} \longrightarrow \mathbb{R}^n = \mathbb{R} \times \mathbb{R}^{n-1}$$

denote the immersion $\overline{x} \to (0,\overline{x})$, and

$$\tilde{u}(\overline{x}) := u(i(\overline{x})) \qquad |\overline{x}| < 1\ .$$

Then $\tilde{u} \in \mathcal{A}^1(\{\overline{x} \in \mathbb{R}^{n-1} \mid |\overline{x}| < 1\}, \mathbb{R}^N)$, and, extending $\tilde{u}$ constantly in the normal direction to $\overline{x} = 0$, we easily see that

$$(9) \qquad G_{u,\partial\Omega} = i_\# G_{\tilde{u}} \qquad \text{in } \mathcal{D}^{n-1}(B(0,1) \times \mathbb{R}^N) ,$$

i.e., that $G_{u,\partial\Omega}$ is the current carried by the graph of the restriction of u on $\partial\Omega$. More generally, if

$$\varphi : \overline{\Delta} \longrightarrow \overline{\Omega}$$

is a bi-Lipschitz transformation which preserves the orientation,

$$\varphi_\# [\![\Delta]\!] = [\![\Omega]\!] ,$$

by Gauss-Green formula we have

$$(10) \qquad \varphi_\# [\![\partial\Delta]\!] = \varphi_\# \, \partial [\![\Delta]\!] = \partial \varphi_\# [\![\Delta]\!] = \partial [\![\Omega]\!] = [\![\partial\Omega]\!] ,$$

therefore we have

$$(11) \qquad \phi_\# G_{u\circ\varphi,\partial\Delta} = G_{u,\partial\Omega}$$

where

$$\phi(x,y) := (\varphi(x), y) .$$

In fact

$$\begin{aligned} \phi_\# G_{u\circ\varphi,\partial\Delta}(\omega) &= [\![\partial\Delta]\!] ((\mathrm{id} \bowtie u \circ \varphi)^\# \phi^\# \omega) \\ &= [\![\partial\Delta]\!] (\varphi^\# (\mathrm{id} \bowtie u)^\# \omega) \\ &= [\![\partial\Omega]\!] ((\mathrm{id} \bowtie u)^\# \omega) = G_{u,\partial\Omega}(\omega) . \end{aligned}$$

Let $u \in \mathrm{cart}^1(\Omega, \mathbb{R}^N)$ with trace $u_{|\,\partial\Omega}$ in the sense of $W^{1,1}$ in $\mathcal{A}^1(\partial\Omega, \mathbb{R}^N)$. As we have seen in [1], in general $\partial G_u \neq G_{u,\partial\Omega}$. However we have

Theorem 1. *Let Ω be a Lipschitz bounded domain in $\mathbb{R}^n$ and let u belong to $W^{1,\underline{n}}(\Omega, \mathbb{R}^N)$, respectively $u \in \mathcal{A}_{n-1,n/(n-1)}(\Omega, \mathbb{R}^n)$. If the trace of u on $\partial\Omega$ belongs to $W^{1,\underline{n}}(\partial\Omega, \mathbb{R}^N)$, respectively to $\mathcal{A}_{n-1,n/(n-1)}(\partial\Omega, \mathbb{R}^n)$, then*

$$\partial G_u = G_{u,\partial\Omega} .$$

Proof. We proceed as in the proof of Proposition 2 defining w as

$$w(x) := \begin{cases} u(x_1, \overline{x}) & \text{if } x_1 > 0 \\ u(0, \overline{x}) & \text{if } x_1 < 0 \end{cases}$$

instead of as in (1). By Proposition 3 in Sec. 3.2.1 respectively by Theorem 1 in Sec. 3.2.4, we have

$$\partial G_w \llcorner B(0,1) \times \mathbb{R}^N = 0 .$$

For $\varepsilon > 0$ denote by $\rho_\varepsilon : \mathbb{R} \to \mathbb{R}$ the piecewise linear function with $\rho_\varepsilon(t) = 0$ for $t \leq -\varepsilon$, $\rho_\varepsilon(t) = 1$ for $t \geq 0$, and $\rho_\varepsilon(t) = 1 + t/\varepsilon$ for $-\varepsilon \leq t \leq 0$, let $\varphi \in C_c^\infty(B(0,1))$, and let $g \in C_c^1(\mathbb{R}^N)$.

By Corollary 1 in Sec. 3.2.3 and Remark 2 in Sec. 3.2.3 we have for any $\alpha, \beta,\ |\alpha| + |\beta| = n,\ i \in \beta$

$$\begin{aligned}
0 &= \int_{B(0,1)} \sum_{j\in\bar\alpha} D_j[\varphi(x)\,\rho_\varepsilon(x_1)\,g(w)]\,(\text{adj}\,(Dw)_{\bar\alpha}^{\beta})_j^i\,dx \\
&= \int_{B^+(0,1)} \ldots + \int_{B^-(0,1)} \ldots \\
&= \int_{B^+(0,1)} \sum_{j\in\bar\alpha} D_j[\varphi(x)\,g(u)]\,(\text{adj}\,(Du)_{\bar\alpha}^{\beta})_j^i\,dx \\
&\quad + \frac{1}{\varepsilon}\int_{-\varepsilon}^{0} dx^1 \int_{|\bar x|<1} \varphi(x)\,g(u(0,\bar x))\,(\text{adj}\,(Du(0,\bar x))_{\bar\alpha}^{\beta})_1^i\,d\mathcal{H}^{n-1} + o(\varepsilon) \\
&= (-1)^{|\alpha|}\,\sigma(\alpha,\bar\alpha)\,\sigma(i,\beta-1)\,\partial G_u(\varphi(x)\,g(y)\,dx^\alpha \wedge dy^{\beta-i}) \\
&\quad + \int_{|\bar x|<1} \varphi(0,\bar x)\,g(u(0,\bar x))\,(\text{adj}\,(Du(0,\bar x))_{\bar\alpha}^{\beta})_1^i\,d\mathcal{H}^{n-1} + o(\varepsilon)\,.
\end{aligned}$$

This concludes the proof as, we recall,

$$(\text{adj}\,(Du))_{\bar\alpha}^{\beta})_j^i \;=\; \sigma(i,\beta-i)\,\sigma(j,\bar\alpha-j)\,M_{\bar\alpha-j}^{\beta-i}(Du)\,.$$

□

Though

$$\partial G_u \;\neq\; G_{u,\partial B(0,1)}$$

for $u \in \text{cart}^1(B(0,1),\mathbb{R}^N)$, even if $\mathbf{M}(\partial G_u) < \infty$ and u smooth, we have

$$\partial(G_u \llcorner B(0,r)) \;=\; G_{u,\partial B(0,r)}$$

for almost every $r \in (0,1)$. More precisely we have

Theorem 2. *Let $u \in \text{cart}^1(B(0,1),\mathbb{R}^N)$. Then for a.e. r, $0<r<1$, u is H^{n-1} a.e. approximately differentiable in $\partial B(0,r)$, $u(x) = u_{|\partial B(0,r)}(x)$ in the sense of traces, $A(Du) \in L^1(\partial B(0,r))$, and*

$$\partial(G_u \llcorner B(0,r)) \;=\; G_{u,\partial B(0,r)} \tag{12}$$

Proof. Observing that by Fubini theorem u, $A(Du) \in L^1(\partial B_r,\mathbb{R}^n)$ and the restriction of u on ∂B_r is approximately differentiable with approximate differential equal to the tangential derivative of u on ∂B_r for a.e. r, we see that $A(Du)$

depends only on the values of u on ∂B_r; hence (7) holds with Ω replaced by B_r. Equality (12) then follows as in the proof of Proposition 5 in Sec. 3.2.4 or by the slicing theory as (7) reads as

$$G_{u,\partial B_r} = \langle G_u, d, r \rangle$$

where $d(x,y) := |x|$. □

3 Cartesian Maps

In this section we shall deal with the *weak continuity* of minors. More precisely, let $\{u_k\}$ be a sequence of maps in $\mathcal{A}^1(\Omega, \mathbb{R}^N)$ such that

$$\begin{array}{rcll} u_k & \to & u & \text{strongly in } L^1(\Omega, \mathbb{R}^N) \\ \text{ap}Du_k & \rightharpoonup & v & \text{weakly in } L^1 \\ M^{\beta}_{\bar{\alpha}}(\text{ap}Du_k) & \rightharpoonup & v^{\beta}_{\bar{\alpha}} & \text{weakly in } L^1 \end{array} \tag{1}$$

for all multi–indices α and β with $|\alpha| + |\beta| = n$, $|\beta| \geq 2$, and functions u, v and $v^{\beta}_{\bar{\alpha}}$ in $L^1(\Omega)$ [2]. We ask whether almost everywhere

$$v(x) = \text{ap}Du(x) \quad \text{and} \quad v^{\beta}_{\bar{\alpha}}(x) = M^{\beta}_{\bar{\alpha}}(\text{ap}Du(x)) \,.$$

The answer to this question is negative already in the scalar case, as the following example shows.

[1] Consider the sequence of functions $u_k : (0,1) \to \mathbb{R}$ given by

$$u_k(t) := t - \frac{[2^k t]}{2^k} \,,$$

$[x]$ denoting the integer part of x. Clearly $u_k \to 0$ strongly in $L^\infty(0,1)$ while

$$\text{ap}Du_k = 1 \qquad \text{a.e. in } (0,1) \,.$$

•

Recall however that if

$$Du_k \rightharpoonup v \qquad \text{weakly in } L^1(\Omega, \mathbb{R}^n) \,,$$

Du_k being the *distributional* derivative of u_k, then v is the *distributional* derivative of u.

In view of Theorem 2 in Sec. 3.1.4 we therefore conclude that our question has a positive answer provided

---

[2] compare the remark following Definition 2 in Sec. 3.2.1

$$\text{(2)} \qquad \mathrm{ap}Du_k(x) \;=\; Du_k(x)$$

in the L^1 sense, i.e. $\{u_k\}$ is a sequence in $W^{1,1}(\Omega, \mathbb{R})$.

By Proposition 1 in Sec. 3.2.3 condition (2) amounts to

$$\partial G_{u_k} \llcorner \Omega \times \mathbb{R} \;=\; 0 \, .$$

In the vector valued case the situation is totally similar. Even assuming that the approximate derivatives $\mathrm{ap}Du_k$ agree with the distributional derivatives, Du_k, i.e.,

$$\begin{array}{rcll} u_k & \longrightarrow & u & \text{in } L^1(\Omega, \mathbb{R}^N) \\ Du_k & \rightharpoonup & Du & \text{weakly in } L^1 \\ M_{\bar{\alpha}}^{\beta}(\mathrm{ap}Du_k) & \rightharpoonup & v_{\bar{\alpha}}^{\beta} & \text{weakly in } L^1 \end{array}$$

it is *not true in general that* $v_{\bar{\alpha}}^{\beta} = M_{\bar{\alpha}}^{\beta}(Du)$, compare Sec. 3.3.1. Again the key condition is that all G_{u_k} be boundaryless in $\Omega \times \mathbb{R}^N$,

$$\partial G_{u_k} \llcorner \Omega \times \mathbb{R}^N \;=\; 0 \, .$$

Actually, we shall see that $v_{\bar{\alpha}}^{\beta} = M_{\bar{\alpha}}^{\beta}(\mathrm{ap}Du)$, if (1) holds and the boundaries of G_{u_k} in $\Omega \times \mathbb{R}^N$ are controlled in the sense that

$$\sup_k \mathbf{M}(\partial G_{u_k} \llcorner \Omega \times \mathbb{R}^N) < \infty$$

This is just a special case of the closure theorem of Federer and Fleming.

This way we shall generalize a well known theorem of Reshetnyak concerning the continuity of the Jacobian determinant with respect to the weak convergence in $W^{1,n+\varepsilon}$, $\varepsilon > 0$.

For this reason, after presenting a few examples and Reshetnyak's theorems in Sec. 3.3.1, we shall study in Sec. 3.3.2 the space of *Cartesian maps*

$$\mathrm{cart}^1(\Omega, \mathbb{R}^N) \;:=\; \{u \in \mathcal{A}^1(\Omega, \mathbb{R}^N) \;\mid\; \partial G_u \llcorner \Omega \times \mathbb{R}^N = 0\}$$

for which we prove a *closure* and *compactness theorem*, with respect to the weak L^1-convergence. The subclasses $\mathrm{cart}^p(\Omega, \mathbb{R}^N)$ of $\mathrm{cart}^1(\Omega, \mathbb{R}^N)$ of maps with Jacobian minors in L^p will finally be discussed in Sec. 3.3.3.

3.1 Weak Continuity of Minors

Let $\{u_k\}$ be a sequence of maps in $\mathcal{A}^1(\Omega, \mathbb{R}^N) \cap W^{1,1}(\Omega, \mathbb{R}^N)$ such that

$$\text{(1)} \qquad \begin{array}{c} u_k \to u \text{ in } \; L^1 \\ Du_k \rightharpoonup Du \text{ in } \; L^1 \end{array}$$

for some $u \in \mathcal{A}^1(\Omega, \mathbb{R}^N) \cap W^{1,1}(\Omega, \mathbb{R}^N)$. Suppose that for all multi–indices α and β with $|\alpha| + |\beta| = n$, $|\beta| \geq 2$, there exist functions $v_{\bar{\alpha}}^{\beta} \in L^1(\Omega, \mathbb{R})$ such that

$$(2) \qquad M^{\beta}_{\bar{\alpha}}(Du_k) \rightharpoonup v^{\beta}_{\bar{\alpha}} \qquad \text{weakly in } L^1 .$$

We shall see in [3] and in [4] below that in general $v^{\beta}_{\bar{\alpha}}(x) \neq M^{\beta}_{\bar{\alpha}}(Du(x))$ and we shall discuss some circumstances in which equality holds. A more thoroughly discussion is delayed to next section.

In many respects it is also reasonable to replace the convergence in (2) by the convergence in the sense of measures

$$(3) \qquad M^{\beta}_{\bar{\alpha}}(Du_k) \rightharpoonup v^{\beta}_{\bar{\alpha}} \qquad \text{as measures}$$

or by the related convergence of graphs $\{G_{u_k}\} \subset \mathcal{D}_n(\Omega \times \mathbb{R}^N)$

$$(4) \qquad G_{u_k} \rightharpoonup S$$

where S is the current in $\mathcal{D}_n(\Omega \times \mathbb{R}^N)$ whose components are given by

$$(5) \qquad \begin{aligned} S^{\bar{0}0}(\phi) &:= \int_\Omega \phi(x, u(x))\, dx \\ S^{\bar{i}j}(\phi) &:= (-1)^{n-j} \int_\Omega \phi(x, u(x))\, D_j u^i(x)\, dx \\ S^{\alpha\beta}(\phi) &:= \sigma(\alpha, \bar{\alpha}) \int_\Omega \phi(x, u(x))\, v^{\beta}_{\bar{\alpha}}(x)\, dx \end{aligned}$$

where $\phi \in C^\infty_c(\Omega \times \mathbb{R}^N)$ and $|\alpha| + |\beta| = n$, $|\beta| \geq 2$, compare (5) in Sec. 3.2.1. Thus, before going into our main question, let us first compare the convergences in (2), (3), and (4).

Clearly the weak convergence in L^1 implies the convergence in the sense of measures, and the proof of Proposition 2 in Sec. 3.2.1 easily yields that the weak convergence in L^1 implies also the convergence of graphs, i.e.,

$$\begin{array}{ccc} \text{convergence in (2)} & \Rightarrow & \text{convergence in (3)} \\ \Downarrow & & \\ \text{convergence in (4)} & & \end{array}$$

If the minors $M(Du_k)$ are also equi-integrable, then the convergence as measures implies, and therefore is equivalent, to the weak convergence in L^1, compare Theorem 2 in Sec. 1.2.4.

The following two examples show that in general the convergence in (3) or in (4) does not imply the convergence in (2), i.e.,

$$\begin{array}{ccc} \text{convergence in (2)} & \not\Leftarrow & \text{convergence in (3)} \\ \not\Uparrow & & \\ \text{convergence in (4)} & & \end{array}$$

Finally a few words concerning the comparison between the convergences in (3) and (4). If the sequence $\{u_k\}$ is equibounded in L^∞, it is easily seen that the convergence of graphs implies the convergence in the sense of measures of the minors, compare also Proposition 1 in Sec. 4.2.3, i.e.

$$\left.\begin{array}{l}\text{convergence in (4)}\\ \text{equiboundedness in } L^\infty \text{ of } \{u_k\}\end{array}\right. \Longrightarrow \text{convergence in (3)}$$

But it seems to be not clear whether the convergence in (4) imply the convergence in (3), compare Theorem 3 and [6] in Sec. 4.2.5.

[1] Consider the sequence of radial mappings

$$\begin{aligned} u_k &: B(0,1) \subset \mathbb{R}^n \to \mathbb{R}^n , \qquad k = 1, 2, \dots \\ u_k(x) &:= r_k(|x|) \frac{x}{|x|} \end{aligned}$$

where $r_k(t)$ is the continuous and piecewise linear function

$$r_k(t) := \begin{cases} k\,t & \text{if } 0 \le t \le \frac{1}{k} \\ 2 - k\,t & \text{if } \frac{1}{k} \le t \le \frac{2}{k} \\ 0 & \text{if } \frac{2}{k} \le t \le 1 \end{cases}$$

It is easily seen that the sequence $\{u_k\}$ is equibounded in $W^{1,n}(B(0,1), \mathbb{R}^n)$, and

$$\| u_k \|^n_{W^{1,n}(B(0,1),\mathbb{R}^n)} \le c(n) \int_0^1 t^{n-1} \left(|r_k'(t)|^n + |\frac{r_k(t)}{t}|^n \right) dt \le c(n) ,$$

thus, since $u_k \to 0$ a.e. and $|u_k| \le 1$, $u_k \rightharpoonup 0$ weakly in $W^{1,n}(B(0,1), \mathbb{R}^n)$ as $k \to \infty$. Also one easily computes

$$\det Du_k(x) = \left(\frac{r_k(|x|)}{|x|} \right)^{n-1} r_k'(|x|) =: \varphi_k(|x|)$$

where

$$\varphi_k(t) := \begin{cases} k^n & \text{if } 0 < t < \frac{1}{k} \\ -(\frac{2}{t} - k)^{n-1} k & \text{if } \frac{1}{k} < t < \frac{2}{k} \\ 0 & \text{if } \frac{2}{k} < t < 1 , \end{cases}$$

thus $\{\det Du_k\}$ is equibounded in L^1, $\int_{B(0,1)} |\det Du_k| \, dx \le c(n)$. Observing now that

$$\int_{B(0,1)} \det Du_k \, dx = 0$$

for all k, and writing for any $\varphi \in C_c^\infty(B(0,1))$,

$$\begin{aligned} \int_{B(0,1)} \varphi \det Du_k \, dx &= \varphi(0) \int_{B(0,1)} \det Du_k \, dx \\ &\quad + \int_{B(0,1)} [\varphi(x) - \varphi(0)] \det Du_k(x) \, dx , \end{aligned}$$

we easily deduce that

$$\det Du_k \rightharpoonup 0 \quad \text{in the sense of measures.}$$

On the other hand, if $\delta > 0$ and $\frac{2}{k} < \delta$, we have

$$\begin{aligned}\int_{B(0,\delta)} |\det Du_k|\, dx &= c(n) \int_0^\delta t^{n-1} \Big(\frac{r_k(t)}{t}\Big)^{n-1} |r_k'(t)|\, dt \\ &= c(n) \int_0^{2/k} |r_k(t)|^{n-1} |\, |r_k'(t)|\, dt = \frac{2c(n)}{n}\,,\end{aligned}$$

This shows that $\{\det Du_k\}$ is not equi-integrable in $B(0,1)$, and therefore $\{\det Du_k\}$ does not converge weakly in L^1.

One can also show, compare the examples in Sec. 3.2.2, that $G_{u_k} \rightharpoonup G_0$ where $G_0 = [\![B(0,1)]\!] \times \delta_0$ is the graph of the null map from $B(0,1)$ into $\mathbb{R}^n$.

Notice that this example shows also that the map $u \to \det Du$ is *not* sequentially weakly continuous from $W^{1,n}(B(0,1), \mathbb{R}^n)$ into $L^1(B(0,1))$. •

A similar example to [1] is the following

[2] Let $n = 2$, $Q = (-1,1) \times (-1,1)$. Consider the sequence of maps u_k from Q into $\mathbb{R}^2$

$$u_k(x,y) := k^{-1/2}\,(1-|y|)^k\,(\sin kx, \cos kx)\ .$$

We have

$$\begin{aligned}\| u_k \|_\infty &\leq k^{-1/2} \\ \| Du_k \|^2_{L^2} &= 2k \int_{-1}^1 dx \int_0^1 [(1-y)^{2k} + (1-y)^{2(k-1)}]\, dy = \\ &= 4k\,[\frac{1}{2k+1} + \frac{1}{2k-1}] < 8\end{aligned}$$

hence u_k converge weakly in $W^{1,2}$ to zero. Moreover

$$\det Du_k(x,y) = k\,(1-|y|)^{2k-1}\ \mathrm{sign}\, y\ ,$$

Thus one easily sees that in the sense of measures $\det Du_k \rightharpoonup 0$. However for $k \to \infty$

$$\begin{aligned}\int_{-a}^a \int_{-a}^a |\det Du_k|\, dx\, dy &= 2k \int_{-a}^a dx \int_{-a}^a (1-|y|)^{2k-1}\, dy \\ &= 2a\,[1-(1-a)^{2k}] \to 2a\end{aligned}$$

thus $\{\det Du_k\}$ does not converge in L^1 to zero.

Again one could show that $G_{u_k} \rightharpoonup G_0$. •

Let us return to our main question. A by now classical result concerning the weak convergence of determinants is the following theorem

Theorem 1 (Reshetnyak). *Let Ω be a bounded domain of $\mathbb{R}^n$ and let $u_k : \Omega \to \mathbb{R}^n$ be a sequence of continuous maps in $W^{1,n}(\Omega, \mathbb{R}^n)$. Suppose that the maps u_k converge uniformly in Ω to u and that the sequence of Jacobian determinants $\{\det Du_k\}$ is equi-integrable in $L^1(\Omega)$. Then*

$$\det Du_k \rightharpoonup \det Du$$

weakly in $L^1(\Omega)$.

Proof. Recall, compare (3) in Sec. 3.2.3, that Laplace's formulas yield

$$\begin{aligned} M_\alpha^\beta(Du) &= \sum_{j\in\alpha} D_j u^i\, (\operatorname{adj} Du)_\alpha^\beta{}_j^i \\ &= \sum_{j\in\alpha} \sigma(i,\beta-i)\sigma(j,\alpha-j)\, M_{\alpha-j}^{\beta-i}(Du) D_j u^i \end{aligned}$$

where $|\alpha| = |\beta| = \ell$, $1 \le \ell \le n$, and that, in the sense of distribution we have for every $u \in W^{1,n}$, and $i = 1, 2, \dots, n$

$$D_j((\operatorname{adj} Du)_\alpha^\beta)_j^i \;=\; 0 \;,$$

compare Proposition 3 in Sec. 3.2.1 and Corollary 2 in Sec. 3.2.3. Thus for all α, β, $|\alpha| + |\beta| = \ell$, $1 \le \ell \le n$, and $1 \le i \le n$,

$$(6) \qquad \int_\Omega M_\alpha^\beta(Du)\,\varphi\, dx \;=\; -\int_\Omega u^i \sum_{j\in\alpha} D_j\varphi\, (\operatorname{adj}(Du)_\alpha^\beta)_j^i\, dx$$

holds for all $\varphi \in C_c^\infty(\Omega)$.

Since $\{\det Du_k\}$ is equi-integrable, in order to prove the weak convergence of $\{\det Du_k\}$ in L^1 it suffices to show that

$$\int \varphi \det Du_k\, dx \;\longrightarrow\; \int \varphi \det Du\, dx$$

for all $\varphi \in C_c^\infty(\Omega)$.

We shall now prove by induction on the order of the minors that

$$(7) \qquad \int \varphi\, M_\alpha^\beta(Du_k)\, dx \;\longrightarrow\; \int \varphi\, M_\alpha^\beta(Du)\, dx \qquad \forall\, \varphi \in C_c^\infty$$

for all α, β with $|\alpha| = |\beta| = \ell$, $1 \le \ell \le n$.

Obviously (7) holds for $\ell = 1$, $u_k \rightharpoonup u$ in $W^{1,n}$. Suppose that (7) holds for $\ell - 1$. From (6) we deduce that for $|\alpha| = |\beta| = \ell$ we have

$$\int_\Omega M_\alpha^\beta(Du_k)\,\varphi\,dx = -\int_\Omega \sum_{i,j} u_k^i\,(\mathrm{adj}\,(Du_k)_\alpha^\beta)_j^i\,D_j\varphi\,dx$$
$$= -\int_\Omega \sum_{i,j} u^i\,(\mathrm{adj}\,(Du_k)_\alpha^\beta)_j^i\,D_j\varphi\,dx + \int_\Omega \sum_{i,j}(u^i - u_k^i)\,(\mathrm{adj}\,(Du)_\alpha^\beta)_j^i\,D_j\varphi\,dx.$$

By the inductive assumption the first integral on the right tends to

$$-\int \sum_{i,j} u^i\,(\mathrm{adj}\,(Du)_\alpha^\beta)_j^i\,D_j\varphi\,dx$$

which is equal by (6) to

$$\int M_\alpha^\beta(Du)\,\varphi\,dx$$

while the second integral on the right tends to zero. This proves (7) for ℓ and therefore the theorem. □

An immediate corollary of Theorem 1 is the sequential weak continuity of the operator $u \to \det Du$ from $W^{1,p}(\Omega,\mathbb{R}^n)$, $p > n$, into $L^1(\Omega)$. Actually, going through the proof of Theorem 1 one immediately gets the following

Theorem 2. *Let $\{u_k\} \subset W^{1,p}(\Omega,\mathbb{R}^n)$, where $p > n$. Suppose that*

$$u_k \rightharpoonup u \qquad \textit{weakly in } W^{1,p}$$

then

$$\det Du_k \rightharpoonup \det Du \qquad \textit{weakly in } L^{p/n}$$

and actually

$$M_{(\ell)}(Du_k) \rightharpoonup M_{(\ell)}(Du) \qquad \textit{weakly in } L^{p/\ell}$$

Notice that in fact Theorem 2 is a closure theorem. Given an equibounded sequence $\{u_k\} \subset W^{1,p}(\Omega,\mathbb{R}^n)$, $p > n$, passing to subsequences, we may assume that $\{u_{k_h}\}$ converges uniformly to u by the Sobolev embedding theorem, and that $M_{(\ell)}(Du_k) \rightharpoonup v_\ell$ weakly in L^1 and actually in $L^{p/\ell}$. By Theorem 1 $v_\ell = M_{(\ell)}(Du)$. Therefore, if u is also independent of the chosen subsequence, we conclude that the entire sequence converges together with all minors in L^1-weak to the map u and the corresponding minors of Du.

Theorem 1 and Theorem 2 are optimal in the sense that, as we have seen in [1] and [2] det is *not* sequentially weakly continuous from $W^{1,n}$ into L^1. However we have

Theorem 3 (Reshetnyak). *Let $\{u_k\} \subset W^{1,n}(\Omega,\mathbb{R}^n)$. Suppose that*

$$u_k \rightharpoonup u \qquad \textit{weakly in } W^{1,n}\,.$$

Then

$$\det Du_k \rightharpoonup \det Du \qquad \textit{in the sense of measures.}$$

Proof. From Theorem 2 we infer that

$$\operatorname{adj} Du_k \rightharpoonup \operatorname{adj} Du \qquad \text{weakly in } L^{\frac{n}{n-1}}$$

and by Laplace's formula and Piola's identity, compare Sec. 3.2.3, that

$$\int_\Omega \det Du_k \,\varphi\, dx = -\sum_{j=1}^n \int_\Omega u_k^1 (\operatorname{adj} Du_k)_j^1 \, D_j\varphi\, dx$$

holds for any $\varphi \in C_c^1(\Omega)$. As $u_k^1\, D_j\varphi \to u^1\, D_j\varphi$ in $L^n(\Omega)$, we conclude that

$$\int_\Omega \det Du_k \,\varphi\, dx \longrightarrow \int_\Omega \det Du \,\varphi\, dx \tag{8}$$

for any $\varphi \in C_c^1(\Omega)$, and by approximation that (8) hold for any $\varphi \in C_c^0(\Omega)$. □

Taking into account the result on the higher integrability of the Jacobian determinant in Theorem 2 in Sec. 3.2.4 we can also state

Theorem 4. *Let $\{u_k\} \subset W^{1,n}(\Omega, \mathbb{R}^n)$. Suppose that $\det Du_k(x) \geq 0$ a.e. and that*

$$u_k \rightharpoonup u \qquad \textit{weakly in } W^{1,n}\ .$$

Then

$$\det Du_k \rightharpoonup \det Du \qquad \textit{weakly in } L^1\ .$$

The same argument in the proof of Theorem 3 actually yields the following two results, compare [3] in Sec. 3.2.4 and [1] in Sec. 3.2.4.

Proposition 1. *Let $\{u_k\}$ be a sequence such that*

$$u_k \rightharpoonup u \qquad \textit{weakly in } W^{1,p}(\Omega, \mathbb{R}^n)$$
$$\operatorname{adj} Du_k \rightharpoonup \operatorname{adj} Du \qquad \textit{weakly in } L^q$$

with $p \geq n-1$ and $q \geq \frac{p}{p-1}$. Then

$$\det Du_k \rightharpoonup \det Du \quad \textit{in the sense of measures.}$$

Proposition 2. *Let $\{u_k\} \subset W^{1,p}(\Omega, \mathbb{R}^n)$, $p > \frac{n^2}{n+1}$, be a sequence such that $u_k \rightharpoonup u$ weakly in $W^{1,p}$. Suppose that $u_k,\ u \in W^{1,n}(\Omega, \mathbb{R}^n)$. Then the distributional determinants* Det Du_k *converge to the distributional determinant of u in the sense of distributions.*

We finally observe that the determinant is not weakly lower semicontinuous in $W^{1,p}$ for $p < n$, under no extra assumptions, as shown by the next two examples.

3 Let $Q := [-1,1]^n \subset \mathbb{R}^n$, $n \geq 1$. For $\xi \in Q$ we set

$$u(x) := \frac{x}{\| x \|} - x, \qquad x \in Q .$$

where $\| x \| := \max_{1 \leq k \leq n} |x^k|$. The function $u : Q \to \mathbb{R}^n$ is clearly continuous except at zero and vanishes on ∂Q; for $x \in Q \setminus \{0\}$ we have

$$\frac{\partial u^i}{\partial x^k} = \frac{\delta^i_k}{\| x \|} - \frac{x^i}{\| x \|^2} \frac{\partial \| x \|}{\partial x^k} - \delta^i_k$$

$$\frac{\partial \| x \|}{\partial x^k} = \begin{cases} \operatorname{sign} x^k & \text{if } \| x \| = |x^k| \\ 0 & \text{if } \| x \| = |x^i| , \; i \neq k , \end{cases}$$

moreover

$$\left| \frac{\partial u^i}{\partial x^k} \right| \leq \frac{2}{\| x \|} + 1$$

From this one easily deduces that $u \in W^{1,p}(Q, \mathbb{R}^n)$ for all p with $p < n$.

Since $u = 0$ on ∂Q, we may extend u as a Q-periodic function to the whole of $\mathbb{R}^n$, and denoting again by u this extension, evidently we have $u \in W^{1,p}_{\text{loc}}(\mathbb{R}^n, \mathbb{R}^n)$ for all $p < n$.

Let Ω be now a bounded open subset of $\mathbb{R}^n$. We define

$$u_k(x) := x + k^{-1} u(kx) , \qquad x \in \Omega$$

Then, compare Proposition 1 in Sec. 1.2.3, $u_k \rightharpoonup x$ in $W^{1,p}(\Omega, \mathbb{R}^n)$ for $1 \leq p < n$; furthermore $u_k \to x$ strongly in $L^\infty(\Omega, \mathbb{R}^n)$. But for almost all $x \in \Omega$ we have

$$Du_k = \text{id} + Du(kx) = D\frac{y}{\| y \|}\Big|_{y=kx}$$

hence

$$\det Du_k = 0 \qquad \forall\, k .$$

Therefore we have found a sequence $\{u_k\}$ such that

$$\begin{array}{lll} u_k \rightharpoonup x & \text{in } W^{1,p}(\Omega, \mathbb{R}^n) & \forall\, p < n \\ u_k \to x & \text{in } L^\infty(\Omega, \mathbb{R}^n) & \\ \det Du_k = 0 & \forall\, k & \end{array}$$

but

$$\det Du = \det \text{id} = 1 .$$

Passing to the graphs, it is easily seen that

$$\sup_k \mathbf{M}(G_{u_k}) < \infty$$

while

$$\mathbf{M}(\partial G_{u_k}) \longrightarrow \infty$$

As it will clearly appear in the sequel of this section, the unboundedness of the masses of the boundaries ∂G_{u_k} is responsible for the loss of the sequential weak continuity of the determinant. •

[4] Let $Q = [-1,1]^n \subset \mathbb{R}^n$, $n \geq 3$ and $1 < p < n-1$ and let $r := \max_{i=1,n-1} |x_i|$. For $k = 1, 2, \ldots$, let $Q_k := [-1,1]^{n-1} \times [1-2k, 2k-1]$ and $\alpha_k := k^{-p/(n-p-1)}$. Set

$$v_{k,i}(x) = \begin{cases} 0 & r \leq \alpha_k \\ \frac{x_i}{\alpha_k} - \frac{x_i}{r} & \alpha_k < r \leq 2\alpha_k \\ \frac{x_i}{r} & r > 2\alpha_k \end{cases}$$

for $i = 1, \ldots, n-1$, and

$$v_{k,n}(x) = \begin{cases} 0 & r \leq 2\alpha_k \\ x_n(\frac{r}{\alpha_k} - 2) & 2\alpha_k < r \leq 3\alpha_k \\ x_n & r > 3\alpha_k \end{cases}$$

We extend v_k to $[1-2k, 2k-1]^n$ in such a way that $v_{k,i}(x) - x$ is 2–periodic in the variables $x_1, \ldots, x_n$ and set $u_k(x) := \frac{1}{k} v_k(x)$. Then, compare Proposition 1 in Sec. 1.2.3, $\{u_k\}$ converges to the identity map in $L^p(Q)$ and $\det Du_k = 0$ a.e. for every k. Actually

$$u_k \rightharpoonup x \quad \text{weakly in } W^{1,p}, \qquad \det Du_k = 0$$

since one can easily prove that the $W^{1,p}$ norms of the u_k's are equibounded. In fact

$$\int_Q |Du_k|^p \, dx = \frac{1}{k} \int_{Q_k} |Dv_k|^p \, dx$$

and

$$|Dv_k(x)| \leq c_1 \frac{1}{r} \qquad \text{for } r > 3\alpha_k,$$

$$|Dv_k(x)| \leq c_2 \left(\frac{1}{r} + \frac{k}{\alpha_k}\right) \qquad \text{for } r \leq 3\alpha_k$$

hence

$$\int_{Q_k} |Dv_k|^p \, dx \leq c_3 \, k \left(\int_0^1 t^{n-2-p} \, dt + k^p \alpha_k^{n-p-1} \right) \leq c_4 k \, .$$

which implies that $\int_Q |Du_k|^p \, dx < \infty$.

Passing to graphs, we obviously have $\partial G_{u_k} \llcorner Q \times \mathbb{R}^n = 0$, while one can check that

$$\mathbf{M}(G_{u_k}) \to \infty \, .$$

•

3.2 The Class $\mathrm{cart}^1(\Omega, \mathbb{R}^N)$: Closure and Compactness

We shall now discuss in more details the class of Cartesian currents we already introduced in Sec. 3.2.1.

Definition 1. *The class of* Cartesian maps *is defined by*

$$\text{cart}^1(\Omega,\mathbb{R}^N) := \{u \in \mathcal{A}^1(\Omega,\mathbb{R}^N) \mid \partial G_u \llcorner \Omega \times \mathbb{R}^N = 0\} \tag{1}$$

By Proposition 1 in Sec. 3.2.3, we obviously have

$$\text{cart}^1(\Omega,\mathbb{R}^N) \subset W^{1,1}(\Omega,\mathbb{R}^N)$$

In particular every $u \in \text{cart}^1(\Omega,\mathbb{R}^N)$ is $n/(n-1)$ summable by Sobolev embedding theorem, but we cannot expect any extra summability of u as consequence of the summability of Jacobian minors. For instance the map $(u(x),u(x))$ from a domain of $\mathbb{R}^2$ into $\mathbb{R}^2$ belongs to $\text{cart}^1(\Omega,\mathbb{R}^2)$ for any $u \in W^{1,1}(\Omega,\mathbb{R}^2)$. Notice also that in general the i.m. rectifiable current G_u associated to $u \in \text{cart}^1(\Omega,\mathbb{R}^N)$ is *not* a normal current in $\mathbb{R}^n \times \mathbb{R}^N$ since a priori we may have in $\mathbb{R}^n \times \mathbb{R}^N$ $\mathbf{M}(\partial G_u) = +\infty$.

We have

Theorem 1. *Let $\{u_k\}$ be a sequence of maps in* $\text{cart}^1(\Omega,\mathbb{R}^N)$. *Suppose that there is a map $u \in W^{1,1}(\Omega,\mathbb{R}^N)$ and that there exist functions $v^\beta_{\bar\alpha} \in L^1(\Omega)$ for $|\alpha|+|\beta| = n$, $|\beta| \geq 2$ such that*

$$\begin{aligned} u_k &\rightharpoonup u && \textit{weakly in } L^1 \\ Du_k &\rightharpoonup Du && \textit{weakly in } L^1 \\ M^\beta_{\bar\alpha}(Du_k) &\rightharpoonup v^\beta_{\bar\alpha} && \textit{weakly in } L^1 \,. \end{aligned} \tag{2}$$

Then

$$v^\beta_{\bar\alpha}(x) = M^\beta_{\bar\alpha}(Du(x)) \,. \tag{3}$$

Before proving Theorem 1, let us state a proposition which explains the geometric meaning of the conclusion in (3).

Proposition 1. *Let $u \in L^1(\Omega,\mathbb{R}^N)$ be an a.e. approximately differentiable map, and let $v^\beta_{\bar\alpha} \in L^1(\Omega)$, $|\alpha|+|\beta| = n$, $|\beta| \geq 1$. Define the current $S = S_{u,v} \in \mathcal{D}_n(\Omega \times \mathbb{R}^N)$ by*

$$S^{\bar 0 0}(\phi) := \int_\Omega \phi(x,u(x))\,dx \,,$$

$$S^{\alpha\beta}(\phi) := \sigma(\alpha,\bar\alpha)\int_\Omega \phi(x,u(x))\,v^\beta_{\bar\alpha}(x)\,dx \,, \quad |\alpha|+|\beta| = n \,,\ |\beta| \geq 1$$

Then the following claims are equivalent

(i) $v^\beta_{\bar\alpha}(x) = M^\beta_{\bar\alpha}(Du(x))$ *for a.e. $x \in \Omega$ and for all α,β with $|\alpha|+|\beta| = n$, $|\beta| \geq 1$*

(ii) $u \in \mathcal{A}^1(\Omega,\mathbb{R}^N)$ *and* $S = G_u$

(iii) *S is an i.m. rectifiable current in* $\Omega \times \mathbb{R}^N$.

Proof. If $v^\beta_{\bar\alpha}(x) = M^\beta_{\bar\alpha}(Du(x))$ for a.e. x, then $u \in \mathcal{A}^1(\Omega, \mathbb{R}^N)$, $S = G_u$ and, by Proposition 1 in Sec. 3.2.1, S is an i.m. rectifiable current.

Suppose now that S is an i.m. rectifiable current, $S = \tau(\mathcal{S}, \theta, \vec{S})$. Denote by $\vec{v}$ the n-vector given by

$$\vec{v} := \sum_{|\alpha|+|\beta|=n} \sigma(\alpha, \bar\alpha) v^\beta_{\bar\alpha}\, e_\alpha \wedge \varepsilon_\beta,$$

and let $M := \{x \in \Omega \,|\, |MDu(x)| < \infty\}$ As $\mathcal{G}_{u,\Omega}$ is $\mathcal{H}^n$ measurable and countably n–rectifiable from the area formula we deduce

$$\begin{aligned} S(\omega) &:= \int_\Omega < \omega(x, u(x)), \vec{v}(x) > dx \\ &= \int_M < \omega(x, u(x)), \frac{\vec{v}(x)}{|M(Du(x))|} > |M(Du(x))|\, dx \\ &= \int < \omega, \vec{\xi} > \lambda\, d\mathcal{H}^n \llcorner \mathcal{G}_{u,\Omega\cap M} = \tau(\mathcal{G}_{u,\Omega\cap M}, \lambda, \vec{\xi})(\omega) \end{aligned}$$

for any $\omega \in \mathcal{D}^n(\Omega \times \mathbb{R}^N)$. Here we have set

$$\vec{\xi}(x) := \frac{\vec{v}(x)}{|\vec{v}(x)|}, \qquad \lambda(x) := \frac{|\vec{v}(x)|}{|M(Du(x))|}.$$

Since S is i.m. rectifiable $S = \tau(\mathcal{S}, \theta, \vec{\mathcal{S}})$, we have $\mathcal{S} = \mathcal{G}_{u,\Omega\cap M}$, $\vec{\mathcal{S}} = \vec{\mathcal{G}}_{u,\Omega}$, $\lambda = \theta$ $\mathcal{H}^n$-a.e. and,

$$\frac{\vec{v}(x)}{|\vec{v}(x)|} = \vec{\mathcal{S}}$$

thus

$$\frac{\vec{v}(x)}{|\vec{v}(x)|} = \vec{\mathcal{G}}_{u,\Omega}(x, u(x)) = \frac{M(Du(x))}{|M(Du(x))|} \qquad \text{on } M \tag{4}$$

On the other hand for any $\phi \in C^\infty_c(\Omega \times \mathbb{R}^N)$ we have

$$\int_M \phi(x, u(x))\, dx = S(\phi(x, y)\, dx) = \int_\Omega \phi(x, u(x))\, \lambda(x)\, dx$$

thus $M = \Omega$ a.e., $\lambda = 1$ a.e. and (4) yields

$$v^\beta_{\bar\alpha}(x) = M^\beta_{\bar\alpha}(Du(x)) \qquad \text{for a.e. } x \in \Omega.$$

□

Proof of Theorem 1. As in the proof of Proposition 2 in Sec. 3.2.1 we deduce that

$$G_{u_k} \rightharpoonup S \qquad \text{in } \mathcal{D}_n(\Omega \times \mathbb{R}^N) .$$

On the other hand, by Proposition 1 in Sec. 3.2.1, all currents G_{u_k} are i.m. rectifiable, and

$$\sup_k \mathbf{M}(G_{u_k}) = \sup_k \int_\Omega |M(Du_k)|\, dx < \infty \quad \mathbf{M}(\partial G_{u_k} \llcorner \Omega \times \mathbb{R}^N) = 0 .$$

Therefore Federer-Fleming's closure theorem yields that S is an i.m. rectifiable current, and the result follows from Proposition 1. □

As Federer-Fleming's closure theorem holds for sequences of i.m. rectifiable currents G_{u_k} with equibounded masses satisfying also

$$\sup_k \mathbf{M}(\partial G_{u_k} \llcorner \Omega \times \mathbb{R}^N) < +\infty ,$$

the same proof actually yields

Theorem 2. *Let $\{u_k\}$ be a sequence in $\mathcal{A}^1(\Omega, \mathbb{R}^N)$. Suppose that*

$$u_k \to u \qquad \textit{strongly in } L^1$$

and for each α, β, $|\alpha| + |\beta| = n$, $|\beta| \geq 1$

(5) $$M^\beta_{\bar{\alpha}}(Du_k(x)) \to v^\beta_{\bar{\alpha}}(x) \qquad \textit{weakly in } L^1 .$$

If

(6) $$\sup_k \mathbf{M}(\partial G_{u_k} \llcorner \Omega \times \mathbb{R}^N) < +\infty ,$$

then we have

$$v^\beta_{\bar{\alpha}}(x) = M^\beta_{\bar{\alpha}}(Du(x)) \qquad \mathcal{H}^n\textit{-a.e. in } \Omega .$$

Remark 1. Actually it is easily seen from the proof that both Theorem 1 and Theorem 2 hold under the possibly weaker convergence

$$G_{u_k} \rightharpoonup S \qquad \text{in } \mathcal{D}_n(\Omega \times \mathbb{R}^N) .$$

The proof of Theorem 1 and Theorem 2 we have presented strongly relies on Federer-Fleming closure theorem; actually in our situation the proof is much simpler, as by Calderón–Zygmund theorem we already know that the limit current $\tau(\mathcal{G}_{u,\Omega}, 1, \xi)$ has as set of integration the rectifiable set $\mathcal{G}_{u,\Omega}$ and therefore it suffices only to show that ξ is the orienting n–vector of $\mathcal{G}_{u,\Omega}$; see Theorem 4 in Sec. 2.2.7.

Therefore let us give a direct

Proof of Theorem 1 and Theorem 2. From the assumptions the currents G_{u_k} carried by the graphs of the u_k's converge to the n- current in $\Omega \times \mathbb{R}^N$ given by

$$S(\omega) := \int_\Omega <\omega(x, u(x)), v(x)> dx$$

where

$$v(x) := \sum \sigma(\alpha, \bar{\alpha})\, v_{\bar{\alpha}}^{\beta}\, e_\alpha \wedge \varepsilon_\beta, \quad v_{\bar{0}}^{0} := 1.$$

We now consider the blow up of S in $\mathbb{R}^n \times \mathbb{R}^N$ with center $(x_0, u(x_0))$ to be chosen suitably. Set for $\lambda \to 0$

$$\eta_\lambda(x, y) := \left(\frac{x - x_0}{\lambda}, \frac{y - u(x_0)}{\lambda}\right)$$

and let $\omega \in \mathcal{D}^n(\Omega \times \mathbb{R}^N)$. We have

$$\begin{aligned} \eta_{\lambda\#} S(\omega) &= S(\eta_\lambda^{\#}\omega) \\ &= \frac{1}{\lambda^n} \int_\Omega <\omega\left(\frac{x - x_0}{\lambda}, \frac{u(x) - u(x_0)}{\lambda}\right), v(x)> dx \\ &= \int_{\mathbb{R}^n} <\omega\left(z, \frac{u(x_0 + \lambda z) - u(x_0)}{\lambda}\right), v(x_0 + \lambda z)> dz. \end{aligned}$$

If x_0 is such that

$$v(x_0 + \lambda z) \longrightarrow v(x_0) \qquad \text{strongly in } L^1_{\mathrm{loc}} \text{ as } \lambda \to 0$$

$$\frac{u(x_0 + \lambda z) - u(x_0)}{\lambda} \longrightarrow Du(x_0)z \qquad \text{strongly in } L^1_{\mathrm{loc}} \text{ as } \lambda \to 0,$$

(notice that those x_0 are almost all points of Ω, by Lebesgue differentiation theorem and Calderón-Zygmund theorem) we can conclude that

$$S_{x_0,\lambda} := \eta_{\lambda\#} S \rightharpoonup S_{x_0,\infty} \text{ for a.e. } x_0 \in \Omega$$

where $S_{x_0,\infty}$ is the current integration over the tangent plane to u at x_0 but with "orienting" vector the constant n-vector $v(x_0)$, i.e. $S_{x_0,\infty} = (\mathcal{H}^n \llcorner T_{x_0}\mathcal{G}_{u,\Omega}) \llcorner v(x_0)$ or

$$S_{x_0,\infty}(\omega) = \int_{\mathbb{R}^n} <\omega(z, Du(x_0)z), v(x_0)> dz.$$

In the case of Theorem 1 in Sec. 3.3.1 we clearly infer

(7) $$\partial S_{x_0,\infty} = 0 \text{ for a.e. } x_0 \in \Omega;$$

we shall now prove that (7) holds also under the assumptions of Theorem 2. By lower semicontinuity $\mathbf{M}((\partial S \llcorner \Omega \times \mathbb{R}^N) < \infty$, consequently the set $E := \{z \in \mathbb{R}^{n+N} \mid \theta^{n*}(\| \partial S \|, z) = +\infty\}$ has zero $\mathcal{H}^n$-measure, and for a.e. $x_0 \in \Omega$

$$\limsup_{\rho\to 0} \frac{\| \partial S \|(B((x_0, u(x_0)), \rho)}{\omega_n \rho^n} = c(x_0) < \infty.$$

Therefore, if for instance $\omega \in \mathcal{D}^{n-1}(B(0,r))$, we infer

$$|\partial S(\eta_\lambda^\# \omega)| \le \lambda \frac{\| \partial S \|(B((x_0, u(x_0)), \lambda r)}{\lambda^n} \| \omega \|_\infty,$$

hence

$$\partial S_\lambda(\omega) = \eta_{\lambda\#} S(d\omega) = \partial S(\eta_\lambda^\# \omega) = O(\lambda) \text{ as } \lambda \to 0$$

i.e. (7).

The claims in Theorem 1 and Theorem 2 follow at once from (7) and the next lemma. □

Lemma 1. *Let P be an n-plane in $\mathbb{R}^{n+N}$, v a constant n-vector and L the current defined by*

$$L(\omega) := \int <\omega, v> d\mathcal{H}^n \llcorner P.$$

If $\partial L = 0$, then $v = c\vec{P}$ where c is a real constant.

Proof. Since the claim is rotationally invariant we can assume $P = \mathbb{R}^n \times \{0\}$ so that

$$L(\omega) = \int <\omega(x,y), v> dx.$$

Assuming $|\beta| \ge 1$, choose ω as

$$\omega := \phi(x) y^i \, dx^\alpha \wedge dy^{\beta - i}, \qquad |\alpha| + |\beta| = n, \; i \in \beta.$$

Then

$$d\omega_{|P} = \sigma(i, \beta - i) \phi(x) \, dx^\alpha \wedge dx^\beta$$

hence

$$0 = \partial L(\omega) = L(d\omega) = \sigma(i, \beta - i) \int \phi(x) v_{\bar{\alpha}}^{\beta}(x) \, d\mathcal{H}^n$$

consequently $v_{\bar{\alpha}}^{\beta} = 0$ a.e. Being β arbitrary, $|\beta| \ge 1$ we conclude that $v(x) = v^{\bar{0}0} e_1 \wedge \ldots \wedge e_n$. □

A more precise analysis then shows that we may replace condition (6) by the weaker condition (9) in the next proposition. Denote by $\mathcal{F}_p$ the class of all $(n-1)$-forms which are linear combinations of forms of the type

$$\phi(x, y^i) \, dx^\alpha \wedge dy^{\beta - i}$$

where $|\alpha| + |\beta| = n$, $1 \le |\beta| \le p$ and $i \in \beta$, and $\phi(x,t) \in C_c^\infty(\Omega \times \mathbb{R})$.

Proposition 2. *Let $\{u_k\}$ be a sequence of maps in $\mathcal{A}^1(\Omega, \mathbb{R}^N)$ such that $u_k \to u$ in L^1 and*

$$M^{\beta}_{\bar{\alpha}}(Du_k(x)) \rightharpoonup v^{\beta}_{\bar{\alpha}}(x) \qquad \text{weakly in } L^1 \tag{8}$$

where $|\alpha| + |\beta| = n,\ 1 \leq |\beta| \leq p$. If

$$\sup_k \sup\{G_{u_k}(d\omega) \mid \omega \in \mathcal{F}_p,\ |\omega| \leq 1\} < \infty \tag{9}$$

then

$$v^{\beta}_{\bar{\alpha}}(x) = M^{\beta}_{\bar{\alpha}}(Du(x)) \quad \mathcal{H}^n\text{-a.e. } \forall \beta,\ |\beta| \leq p. \tag{10}$$

Proof. As in the proof above we find for a.e. $x_0 \in \Omega$ $\partial S_{x_0,\infty}(\omega) = 0\ \forall\, \omega \in \mathcal{F}_p$. Let $R : \mathbb{R}^{n+N} \to \mathbb{R}^{n+N}$ be the map $R(x, y) := (x, y - Du(x_0)x)$. It is readily seen that $R^{\#}\omega \in \mathcal{F}_p$ if $\omega \in \mathcal{F}_p$. Hence

$$\partial R_{\#} S_{x_0,\infty}(\omega) = R_{\#} \partial S_{x_0,\infty}(\omega) = 0 \quad \forall\, \omega \in \mathcal{F}_p.$$

This means that

$$\partial\Big(\int_{\mathbb{R}^n_x \times \{0\}} <\omega(z), R_{\#} v(x_0)> d\mathcal{H}^n \Big) = 0 \quad \forall\, \omega \in \mathcal{F}_p.$$

As in the proof of Lemma 1 we then find

$$R_{\#} v(x_0) = c\Big(e_1 \wedge \ldots \wedge e_n + \sum_{|\beta|>p} (R_{\#} v)^{\beta}_{\bar{\alpha}} e_\alpha \wedge \varepsilon_\beta\Big)$$

equivalently

$$R_{\#}(v(x_0) - M(Du(x_0))) = \sum_{|\beta|>p} (R_{\#} v)^{\beta}_{\bar{\alpha}} e_\alpha \wedge \varepsilon_\beta,$$

i.e.,

$$v^{\beta}_{\bar{\alpha}}(x_0) = M^{\beta}_{\bar{\alpha}}(Du(x_0)) \qquad \text{for } |\beta| \leq p.$$

□

As a corollary we have

Corollary 1. *Let $\{u_k\}$ be a sequence of maps in $L^1(\Omega, \mathbb{R}^N)$ such that $u_k \to u$ in L^1 and $M^{\beta}_{\bar{\alpha}}(Du_k(x)) \rightharpoonup v^{\beta}_{\bar{\alpha}}(x) \in L^1(\Omega)$ weakly in L^1 where $|\alpha| + |\beta| = n,\ 1 \leq |\beta| \leq p \leq \min(n.N)$. If*

$$\partial_{(j)} G_{u_k} \llcorner \Omega \times \mathbb{R}^N = 0$$

for all j, $0 \leq j \leq p - 1$ and all k, then $v^{\beta}_{\bar{\alpha}}(x) = M^{\beta}_{\bar{\alpha}}(Du(x))$ $\mathcal{H}^n$-a.e. $\forall \beta,\ 1 \leq |\beta| \leq p$.

Remark 2. Notice that, under assumption (6), we may replace the hypothesis (8) by the possibly weaker condition

$$G_{u_k} \rightharpoonup S_{u,v} \qquad \text{in } \mathcal{D}_n(\Omega \times \mathbb{R}^N) \tag{11}$$

and the same proof yields the result (10). The proof instead fails if we replace (8) by (11) and we insist in requiring only (9) instead of (6).

We may consider $\mathcal{A}^1(\Omega, \mathbb{R}^N)$ as embedded in $L^1(\Omega, \mathbb{R}^N) \times L^1(\Omega, \Lambda_n \mathbb{R}^{n+N})$ by the map $u \to (u, M(Du))$. The previous theorems then read as *weak closure theorems* in $L^1(\Omega, \mathbb{R}^N) \times L^1(\Omega, \Lambda_n \mathbb{R}^{n+N})$. In particular we have

Proposition 3. *The class of Cartesian maps*

$$\operatorname{cart}^1(\Omega, \mathbb{R}^N) \ := \ \{u \in \mathcal{A}^1(\Omega, \mathbb{R}^N) \mid \partial G_u \llcorner \Omega \times \mathbb{R}^N = 0\}$$

is sequentially weakly closed in $L^1(\Omega, \mathbb{R}^n) \times L^1(\Omega, \Lambda_n \mathbb{R}^{n+N})$.

We conclude this subsection stating a compactness theorem.

Recall that a well known result that we shall discuss in Vol. II Ch. 1 states that, given a non negative and convex function $F(\xi)$ in $\mathbb{R}^m$ and a sequence $\{u_k\}$in $L^1(\Omega, \mathbb{R}^m)$ with $u_k \rightharpoonup u$ weakly in $L^1(\Omega, \mathbb{R}^m)$, then

$$\int_\Omega F(u)\, dx \ \le \ \liminf_{k \to +\infty} \int_\Omega F(u_k)\, dx\ . \tag{12}$$

Taking into account the L^1-weak compactness criterion in Theorem 2 in Sec. 1.2.4 we can now easily prove

Proposition 4. *Let $\phi : \mathbb{R}_+ \to \mathbb{R}_+$ be an increasing and convex function satisfying*

$$\frac{\phi(t)}{t} \longrightarrow \infty \qquad \text{as } t \to \infty\ .$$

For all $K \in \mathbb{R}_+$ the sets

$$\operatorname{cart}^1_K \ := \ \{u \in \operatorname{cart}^1(\Omega, \mathbb{R}^N) \mid \|u\|_{L^1(\Omega, \mathbb{R}^N)} + \int_\Omega \phi(|M(Du)|)\, dx \ \le \ K\}$$

are sequentially compact with respect to the weak convergence in $L^1(\Omega, \mathbb{R}^N) \times L^1(\Omega, \Lambda_n \mathbb{R}^{n+N})$.

Proof. Let $\{u_k\} \subset \operatorname{cart}^1_K$. Passing to a subsequence we may assume that

$$\begin{aligned} u_k \ \rightharpoonup \ u\ , Du_k \ &\rightharpoonup \ Du \quad \text{weakly in } L^1 \\ M^\beta_{\bar\alpha}(Du_k) \ &\rightharpoonup \ v^\beta_{\bar\alpha} \quad \text{weakly in } L^1\ , \end{aligned}$$

by the sequential compactness criterion in L^1-weak Theorem 2 in Sec. 1.2.4. By the closure theorem $v^\beta_{\bar\alpha} = M^\beta_{\bar\alpha}(Du)$, thus $u_k \rightharpoonup u$ weakly in $\mathcal{A}^1$, and $u \in \operatorname{cart}^1(\Omega, \mathbb{R}^N)$. Finally (12) yields at once

$$\int_\Omega |u|\, dx + \int_\Omega \phi(|M(Du)|)\, dx \;\le\; \liminf_{k\to\infty} \left[\int_\Omega |u_k| + \int_\Omega \phi(|M(Du_k)|)\, dx \right]$$

hence $u \in \mathrm{cart}^1_K$. □

3.3 The Classes $\mathrm{cart}^p(\Omega, \mathbb{R}^N)$, $p > 1$

Analogously to Sec. 3.2.1, we define the class $\mathcal{A}^p(\Omega, \mathbb{R}^N)$, for all $p \ge 1$, as the collection of all mappings from $\Omega \subset \mathbb{R}^n$ into $\mathbb{R}^N$ which have approximate differential $Du(x)$ at a.e. $x \in \Omega$ and which are p-summable together with all minors $M^\beta_{\bar\alpha}(Du(x))$, $|\alpha| + |\beta| = n$,

$$\text{(1)} \qquad \mathcal{A}^p(\Omega, \mathbb{R}^N) := \{u \in \mathcal{A}^1(\Omega, \mathbb{R}^N) \mid |u|,\ |M(Du)| \in L^p(\Omega)\}$$

In $\mathcal{A}^p$ we introduce the "norm"

$$\text{(2)} \qquad \| u \|_{\mathcal{A}^p} := \| u \|_{L^p(\Omega,\mathbb{R}^N)} + \| \, |M(Du)| \, \|_{L^p(\Omega)} ,$$

and we say that a sequence $\{u_k\}$ in $\mathcal{A}^p(\Omega, \mathbb{R}^N)$, *converges weakly in* $\mathcal{A}^p$ *to* $u \in \mathcal{A}^p(\Omega, \mathbb{R}^N)$

$$u_k \rightharpoonup u \quad \text{in } \mathcal{A}^p$$

if and only if

$$u_k \to u \qquad \text{strongly in } L^p(\Omega, \mathbb{R}^N)$$

$$M^\beta_{\bar\alpha}(Du_k) \rightharpoonup M^\beta_{\bar\alpha}(Du) \qquad \text{weakly in } L^p(\Omega)$$

for all α, β with $|\alpha| + |\beta| = n$ and $|\beta| \ge 1$.

Evidently the map $u \to (u, M(Du))$ from $\mathcal{A}^p(\Omega, \mathbb{R}^N)$ into $L^p(\Omega, \mathbb{R}^N) \times L^p(\Omega, \Lambda_n\mathbb{R}^{n+N})$ is injective and $\| u \|_{\mathcal{A}^p}$ is just the norm of $(u, M(Du))$, in $L^p(\Omega, \mathbb{R}^N) \times L^p(\Omega, \Lambda_n\mathbb{R}^{n+N})$. From now on this immersion will be understood.

If p is larger than one, we expect that $\mathcal{A}^p(\Omega, \mathbb{R}^N)$ enjoys better properties than $\mathcal{A}^1(\Omega, \mathbb{R}^N)$. This is in fact true, but only partially. For instance [3] in Sec. 3.3.1 shows that in general $\mathcal{A}^p(\Omega, \mathbb{R}^N)$ *is not weakly closed in* $L^p(\Omega, \mathbb{R}^N) \times L^p(\Omega, \Lambda_n\mathbb{R}^{n+N})$. Therefore, compare Sec. 3.3.2, we are led to introduce the following subclass of $\mathcal{A}^p(\Omega, \mathbb{R}^N)$.

Definition 1. *The class of p-*Cartesian maps *is defined by*

$$\text{(3)} \qquad \begin{aligned} \mathrm{cart}^p(\Omega, \mathbb{R}^N) &:= \mathcal{A}^p(\Omega, \mathbb{R}^N) \cap \mathrm{cart}^1(\Omega, \mathbb{R}^N) \\ &= \{u \in \mathcal{A}^p(\Omega, \mathbb{R}^N) \mid \partial G_u \llcorner \Omega \times \mathbb{R}^N = 0\} \end{aligned}$$

As in Sec. 3.3.2 we have

$$\mathrm{cart}^p(\Omega, \mathbb{R}^N) \subset W^{1,p}(\Omega, \mathbb{R}^N) ,$$

moreover for $u \in \mathrm{cart}^p(\Omega, \mathbb{R}^N)$ the approximate differential agrees with the distributional derivative. Often we shall write

$$||u||_{\mathrm{cart}^p} \quad \text{instead of } ||u||_{\mathcal{A}^p} .$$

From the closure theorem of Sec. 3.3.2 we then deduce

Theorem 1 (Closure theorem). *The class* $\mathrm{cart}^p(\Omega, \mathbb{R}^N)$ *is closed under the weak convergence of* $L^p(\Omega, \mathbb{R}^N) \times L^p(\Omega, \Lambda_n \mathbb{R}^{n+N})$. *That is, if* u_k *and* u *belong to* $\mathrm{cart}^p(\Omega, \mathbb{R}^N)$ *and*

$$\begin{array}{lll} u_k & \rightharpoonup u & \text{weakly in } L^p(\Omega, \mathbb{R}^N) \\ M^{\beta}_{\bar{\alpha}}(Du_k(x)) & \rightharpoonup v^{\beta}_{\bar{\alpha}}(x) & \text{weakly in } L^p(\Omega), \end{array} \tag{4}$$

for all α, β *with* $|\alpha| + |\beta| = n$, *then* $v^{\beta}_{\bar{\alpha}}(x) = M^{\beta}_{\bar{\alpha}}(Du(x))$ *for a.e.* $x \in \Omega$.

The sequential weak compactness of bounded sets in L^p, $p > 1$, together with the previous theorem, readily yields the following

Theorem 2 (Compactness theorem). *Let* $\{u_k\}$ *be a sequence of maps in* $\mathrm{cart}^p(\Omega, \mathbb{R}^N)$, $p > 1$. *Suppose that*

$$\sup_k \| u_k \|_{\mathcal{A}^p} < \infty$$

then there exists a subsequence $\{u_{k_i}\}$ *of* $\{u_k\}$ *and a map* $u \in \mathrm{cart}^p(\Omega, \mathbb{R}^N)$ *such that*

$$u_{k_i} \rightharpoonup u \qquad \text{in } \mathcal{A}^p$$

In contrast with the case $p = 1$, for $p > 1$ the weak convergence of sequence $\{u_k\}$ in $\mathrm{cart}^p(\Omega, \mathbb{R}^N)$ is equivalent to the convergence of the graphs G_{u_k} in $\mathcal{D}_n(\Omega \times \mathbb{R}^N)$ together with the equiboundedness of the norms $\| u_k \|_{\mathcal{A}^p}$.

Theorem 3. *Let* $\{u_k\}$ *be a sequence in* $\mathrm{cart}^p(\Omega, \mathbb{R}^N)$, $p > 1$. *Suppose that*

$$\sup_k \| u_k \|_{\mathcal{A}^p} < \infty$$

and that there exists a current $T \in \mathcal{D}_n(\Omega \times \mathbb{R}^N)$ *such that*

$$G_{u_k} \rightharpoonup T \qquad \text{in } \mathcal{D}_n(\Omega \times \mathbb{R}^N) .$$

Then there exists $u \in \mathrm{cart}^p(\Omega, \mathbb{R}^N)$ *such that*

$$u_k \rightharpoonup u \qquad \text{in } \mathcal{A}^p$$

and $T = G_u$.

In particular $\{u_k\} \subset \mathrm{cart}^p(\Omega, \mathbb{R}^N)$ *converges weakly in* $\mathcal{A}^p$ *to* u *if and only if*

$$\begin{array}{ll} G_{u_k} \rightharpoonup G_u & \text{in } \mathcal{D}_n(\Omega \times \mathbb{R}^N) \\ \sup_k \| u_k \|_{\mathcal{A}^p} < \infty & \end{array} \tag{5}$$

Proof. Since $\sup_k \| u_k \|_{\mathcal{A}^p} < \infty$, from any subsequence of $\{u_k\}$ we can extract a subsequence $\{u_{k_i}\}$ such that $u_{k_i} \rightharpoonup u$ and $u \in \mathrm{cart}^p(\Omega, \mathbb{R}^N)$. Thus $G_{u_{k_i}} \rightharpoonup G_u$ and $T = G_u$. The first part of the claim then follows since the limit u is independent from the chosen subsequence. In fact suppose that for two subsequences $\{v_k^{(1)}\}$ and $\{v_k^{(2)}\}$ of $\{u_k\}$ we had $v_k^{(1)} \overset{\mathcal{A}^p}{\rightharpoonup} v^{(1)}$ and $v_k^{(2)} \overset{\mathcal{A}^p}{\rightharpoonup} v^{(2)}$, $v^{(1)} \neq v^{(2)}$; from $T = G_{v^{(1)}} = G_{v^{(2)}}$ we infer

$$\int \phi(x, v^{(1)})\, dx = \int \phi(x, v^{(2)})\, dx$$

for all $\phi \in C_c^\infty(\Omega \times \mathbb{R}^N)$, which contrasts $v^{(1)} \neq v^{(2)}$.

In particular this shows that (5) implies $u_k \overset{\mathcal{A}^p}{\rightharpoonup} u$. The converse is consequence of Banach-Steinhaus theorem in L^p, in fact, if $u_k \overset{\mathcal{A}^p}{\rightharpoonup} u$, then we have $\sup_k \| u_k \|_{\mathcal{A}^p} < \infty$, and as $\{u_k\} \subset \mathrm{cart}^p(\Omega, \mathbb{R}^N)$, $G_{u_k} \rightharpoonup G_u$. □

Remark 1. Theorem 3 is false for $p = 1$. In fact we have seen in [1] in Sec. 3.2.2 and [2] in Sec. 3.2.2 that

$$G_{u_k} \rightharpoonup T, \qquad \sup \Big(\mathbf{M}(G_{u_k}) + \int_\Omega |u_k|\, dx \Big) < +\infty$$

do not necessary imply weak convergence of u_k in $\mathcal{A}^1$. We shall see in the next section that, even if

$$u_k \rightharpoonup u \text{ in } L^1, \quad \sup_k \| u_k \|_{\mathcal{A}^1} < \infty, \text{ and } G_{u_k} \rightharpoonup T \text{ in } \mathcal{D}_n(\Omega \times \mathbb{R}^N)$$

in general T is only partially related to u.

Finally let us state explicitly the following simple corollary of Theorem 3

Corollary 1. *Let $\{u_k\} \subset \mathrm{cart}^p(\Omega, \mathbb{R}^N)$, $p > 1$. Suppose that*

$$u_k \rightharpoonup u \quad \textit{weakly in } L^p, \qquad \sup_k \int_\Omega |M(Du_k)|^p\, dx < \infty$$

then

$$M_{\bar\alpha}^{\beta}(Du_k) \rightharpoonup M_{\bar\alpha}^{\beta}(Du) \qquad \textit{weakly in } L^p$$

for all α and β with $|\alpha| + |\beta| = n$.

Remark 2. In the definition of the classes $\mathcal{A}^p(\Omega, \mathbb{R}^N)$ we may require that each component of $M(Du)$, $M_{(k)}(Du)$, or, for a fixed system of coordinates, even each $M_{\bar\alpha}^{\beta}(Du)$ be summable with exponent depending on the component itself

$$\int_\Omega |u|^{p_0}\, dx < +\infty\,, \qquad \sum_{k=1}^{\min(n,N)} \int_\Omega |M_{(k)}(Du)|^{p_k}\, dx < +\infty\,.$$

This way we can define the classes

$$\mathcal{A}^p(\Omega,\mathbb{R}^N)\ ,\quad \mathrm{cart}^p(\Omega,\mathbb{R}^N)$$

and so on with p a multi–index

$$p \ :=\ (p_0,p_1,\dots,p_{\min(n,N)})\ .$$

With the obvious changes, the whole theory evidently extends to these classes. As we shall see later this is not just a formal extension, as those kind of spaces arise quite naturally.

4 Approximability of Cartesian Maps

One of our main questions is to characterize weak limits of smooth maps from Ω into $\mathbb{R}^N$ with equibounded energies. It is therefore natural to introduce the class

$$\text{sw-lim}_{\mathcal{A}^p}\left(C^1(\Omega,\mathbb{R}^N)\cap\mathcal{A}^p(\Omega,\mathbb{R}^N)\right)\ ,\quad p\geq 1 \tag{1}$$

consisting of all $u\in\mathcal{A}^p(\Omega,\mathbb{R}^N)$ which can be obtained as *sequential weak limits in* $\mathcal{A}^p$ *of maps* $u\in C^1(\Omega,\mathbb{R}^N)$ *with equibounded* $\mathcal{A}^p$*-norms.*

Of course, by the closure theorem in $\mathrm{cart}^p(\Omega,\mathbb{R}^N)$

$$\text{sw-lim}_{\mathcal{A}^p}\left(C^1(\Omega,\mathbb{R}^N)\cap\mathcal{A}^p(\Omega,\mathbb{R}^N)\right)\ \subset\ \mathrm{cart}^p(\Omega,\mathbb{R}^N)\ ,$$

since for every map $u\in C^1(\Omega,\mathbb{R}^N)\cap\mathcal{A}^p(\Omega,\mathbb{R}^N)$ we have

$$\partial G_u\llcorner\Omega\times\mathbb{R}^N\ =\ 0\ .$$

In particular no function in $\mathcal{A}^p\setminus\mathrm{cart}^p$ can be weakly approximated in $\mathcal{A}^p$ by smooth maps, and the class in (1) agrees with the class of sequential weak limits in cart^p of maps $u\in C^1(\Omega,\mathbb{R}^N)$ with equibounded cart^p-norms

$$\text{sw-lim}_{\mathcal{A}^p}\left(C^1(\Omega,\mathbb{R}^N)\cap\mathrm{cart}^p(\Omega,\mathbb{R}^N)\right)\qquad p\geq 1\ .$$

Remark that, while for $p>1$ every bounded sequence in $\mathcal{A}^p$ has a subsequence weakly converging in $\mathcal{A}^p$, this is not true for $p=1$.

Consider now a map u in

$$\text{sw-lim}_{\mathcal{A}^p}\left(\text{sw-lim}_{\mathcal{A}^p}\left(C^1(\Omega,\mathbb{R}^N)\cap\mathcal{A}^p(\Omega,\mathbb{R}^N)\right)\right)$$

By definition there are $u_{h,k}\in C^1(\Omega,\mathbb{R}^N)\cap\mathcal{A}^p(\Omega,\mathbb{R}^N)$ and u_h belonging to $\text{sw-lim}_{\mathcal{A}^p}\left(C^1(\Omega,\mathbb{R}^N)\cap\mathcal{A}^p(\Omega,\mathbb{R}^N)\right)$ such that

$$u_h\rightharpoonup u\qquad\text{weakly in }\ \mathcal{A}^p$$

and for fixed h

$$u_{h,k} \rightharpoonup u_h \qquad \text{weakly in } \mathcal{A}^p .$$

As a consequence of Banach-Steinhaus theorem, we obtain

$$\| u_h \|_{\mathcal{A}^p} \le c , \qquad \| u_{h,k} \|_{\mathcal{A}^p} \le c(h) .$$

Nevertheless, even in the case $p > 1$, it is not clear whether the whole family $\{\| u_{h,k} \|_{\mathcal{A}^p}\}$ is bounded. Hence it is not clear whether

$$\begin{aligned} &\text{sw-lim}_{\mathcal{A}^p}\big(\text{sw-lim}_{\mathcal{A}^p}\big(C^1(\Omega,\mathbb{R}^N)\cap \mathcal{A}^p(\Omega,\mathbb{R}^N)\big)\big) \\ &\qquad = \text{sw-lim}_{\mathcal{A}^p}\big(C^1(\Omega,\mathbb{R}^N)\cap \mathcal{A}^p(\Omega,\mathbb{R}^N)\big), \end{aligned}$$

equivalently whether $\text{sw-lim}_{\mathcal{A}^p}\big(C^1(\Omega,\mathbb{R}^N)\cap \mathcal{A}^p(\Omega,\mathbb{R}^N)\big)$ is sequentially weakly closed in $\text{cart}^p(\Omega,\mathbb{R}^N)$, and we in fact do not know the answer for $p > 1$. Therefore we set

Definition 1. *We denote by* $\text{Cart}^p(\Omega,\mathbb{R}^N)$ *the smallest subset of* $\mathcal{A}^p(\Omega,\mathbb{R}^N)$ *which contains* $C^1(\Omega,\mathbb{R}^N)\cap \mathcal{A}^p(\Omega,\mathbb{R}^N)$ *and is closed under the sequential weak convergence in* L^p *of the maps and of all their minors.*

A more precise definition of $\text{Cart}^p(\Omega,\mathbb{R}^N)$ is the following. Denote by j the immersion

$$j : \mathcal{A}^p(\Omega,\mathbb{R}^N) \to L^p(\Omega,\mathbb{R}^N) \times L^p(\Omega,\Lambda_n\mathbb{R}^{n+N}) , \quad j(u) = (u, M(Du)) .$$

Let $C := j\big(C^1(\Omega,\mathbb{R}^N)\cap \mathcal{A}^p(\Omega,\mathbb{R}^N)\big)$ and let $\tilde{C}$ be the smallest subset of $L^p(\Omega,\mathbb{R}^N)\times L^p(\Omega,\Lambda_n\mathbb{R}^{n+N})$ which contains C and which is (sequentially) closed under the weak convergence in $L^p(\Omega,\mathbb{R}^N) \times L^p(\Omega,\Lambda_n\mathbb{R}^{n+N})$, then by definition

$$\text{Cart}^p(\Omega,\mathbb{R}^N) := j^{-1}(\tilde{C}) .$$

Since $j(\text{cart}^p(\Omega,\mathbb{R}^N))$ is sequentially weakly closed in $L^p(\Omega,\mathbb{R}^N) \times L^p(\Omega,\Lambda_n\mathbb{R}^{n+N})$, we have $\tilde{C} \subset j(\text{cart}^p(\Omega,\mathbb{R}^N))$, thus

$$\text{Cart}^p(\Omega,\mathbb{R}^N) \subset \text{cart}^p(\Omega,\mathbb{R}^N)$$

Evidently the compactness theorem is valid in $\text{Cart}^p(\Omega,\mathbb{R}^N)$, i.e. Theorem 2 in Sec. 3.3.3 for $p > 1$ and Proposition 4 in Sec. 3.3.2 for $p = 1$ still hold if we replace cart^p by Cart^p both in the hypotheses and in the theses.

As it is well-known, the weak sequential closure of a subset of Banach space is a quite complicated object, so it is in principle $\text{Cart}^p(\Omega,\mathbb{R}^N)$[3]. It is obtained in general by a transfinite inductive process (up to the first uncountable ordinal) of limits. For its relevance for us, we shall now explain with some details this procedure in Sec. 3.4.1.

We have

$$\text{sw-lim}_{\mathcal{A}^p}(C^1(\Omega,\mathbb{R}^N)\cap \mathcal{A}^p(\Omega,\mathbb{R}^N)) \subset \text{Cart}^p(\Omega,\mathbb{R}^N) \subset \text{cart}^p(\Omega,\mathbb{R}^N)$$

and the following two questions arise naturally

[3] Remark that $C^1(\Omega,\mathbb{R}^N)\cap \mathcal{A}^p(\Omega,\mathbb{R}^N)$ is not convex.

I. *Does*

$$\text{sw-lim}_{\mathcal{A}^p}(C^1(\Omega,\mathbb{R}^N)\cap \mathcal{A}^p(\Omega,\mathbb{R}^N)) \;=\; \mathrm{Cart}^p(\Omega,\mathbb{R}^N)$$

hold or not ?

II. *Does*

$$\mathrm{Cart}^p(\Omega,\mathbb{R}^N) \;=\; \mathrm{cart}^p(\Omega,\mathbb{R}^N)$$

hold or not ?

If not in general, in which circumstances the previous equalities hold? Recall that functions in $\mathcal{A}^p \setminus \mathrm{cart}^p$ cannot be weakly approximated in $\mathcal{A}^p$ by smooth maps. Are there reasonable conditions guaranteeing that a given map u in $\mathrm{cart}^p(\Omega,\mathbb{R}^N)$ or in $\mathrm{Cart}^p(\Omega,\mathbb{R}^N)$ could be weakly approximated in $\mathcal{A}^p$ by a sequence of smooth maps?

We might also consider the strong closure of $C^1(\Omega,\mathbb{R}^N)\cap \mathcal{A}^p(\Omega,\mathbb{R}^N)$ in $L^p(\Omega,\mathbb{R}^N)\times L^p(\Omega,\Lambda_n\mathbb{R}^{n+N})$, denoted

$$\text{strong-cl}_{\mathcal{A}^p}(C^1(\Omega,\mathbb{R}^N)\cap \mathcal{A}^p(\Omega,\mathbb{R}^N))$$

or

$$\mathrm{CART}^p(\Omega,\mathbb{R}^N)\,.$$

The strong convergence of $\{u_k\}$ to u in the sense

$$\int_\Omega |u-u_k|^p\,dx + \int_\Omega |M(Du_k)-M(Du)|^p\,dx \;\longrightarrow\; 0 \quad \text{as } k\to\infty$$

implies, and in fact for $p>1$ is equivalent by Radon-Riesz's theorem [4], to the weak approximability of u in $\mathcal{A}^p$ and in p-energy, i.e.,

$$u_k \rightharpoonup u \quad \text{weakly in } \mathcal{A}^p(\Omega,\mathbb{R}^N)$$
$$\int_\Omega |M(Du_k)|^p\,dx \;\to\; \int_\Omega |M(Du)|^p\,dx$$

We may then ask

III. *Does*

$$\mathrm{CART}^p(\Omega,\mathbb{R}^N) := \text{strong}-\mathrm{cl}_{\mathcal{A}^p}(C^1(\Omega,\mathbb{R}^N)\cap \mathcal{A}^p(\Omega,\mathbb{R}^N))$$

agree with $\mathrm{Cart}^p(\Omega,\mathbb{R}^N)$ *or with* $\text{sw-lim}_{\mathcal{A}^p}(C^1(\Omega,\mathbb{R}^N)\cap \mathcal{A}^p(\Omega,\mathbb{R}^N))$?

We do not know the answers to questions I and III, except in the case $p=1$. This will be discussed in Sec. 3.4.2 where we also show that question II has in general a negative answer, i.e.,

$$\mathrm{Cart}^p(\Omega,\mathbb{R}^N) \underset{\neq}{\subset} \mathrm{cart}^p(\Omega,\mathbb{R}^N)\,.$$

---

[4] Recall that this is not true for $p=1$, compare [1] in Sec. 1.2.2.

4.1 The Transfinite Inductive Process

To each set we can associate its *cardinal number* which roughly tells us the number of its elements. To totally well ordered sets we can associate an *ordinal number*. Recall that a *totally ordered set* is a pair $(X, \leq)$ where $\leq$ is a partial order (i.e., a reflexive, antisymmetric, and transitive relation on $X \times X$) satisfying the *trichotomy condition*: for all $x, y \in X$ either $x \leq y$ or $y \leq x$. A totally ordered set $(X, \leq)$ is called *well-ordered* iff every its subset has a smallest element. Every set can be well ordered and this claim is equivalent to the axiom of choice, or to Hausdorff maximality principle, or to Zorn's lemma. Finally the following principle holds

Principle of transfinite induction. *Let $(X, \leq)$ be a well ordered set and let $A \subset X$ be such that $a \in A$ whenever*

$$\{x \in X \mid x \leq a \quad x \neq a\} \subset A .$$

Then $A = X$.

There are many essentially different ways of well order a given set. Each of these ways has its own ordinal number despite the fact that the set has its own cardinal number. For example we can well order $\mathbb{N}$ in its natural way

$$1 < 2 < 3 < 4 < \dots$$

or by

$$2 < 3 < 4 < \dots < 1 ;$$

the first order has no maximum element, while the second does.

Let A and B be totally ordered sets. An *order isomorphism* from A onto B is a one-to-one function f from A onto B such that $x \leq y$ in A implies $f(x) \leq f(y)$ in B. We write $A \approx B$, to say that there is an order isomorphism between A and B. One easily sees that $\approx$ is an equivalence relation: the equivalence classes are called *order type*. This way every set A has an order type noted $\operatorname{ord} A$. If A is well ordered, its order type is called an *ordinal number*.

A natural order is defined among the ordinals. If α and β are ordinals and $\operatorname{ord} A = \alpha$, $\operatorname{ord} B = \beta$, one says $\alpha < \beta$ if there is $x \in B$ such that $A \approx \{y \in B \mid y < x\}$, and we write $\alpha \leq \beta$ to mean that either $\alpha < \beta$ or $\alpha = \beta$. One easily sees that the order defined this way does not depend on the choice of well ordered sets A and B. With the order defined previously any set of ordinals is totally ordered and even well ordered. Moreover, for any ordinal $\alpha > 0$, the order of the set

$$P_\alpha := \{\text{ordinals } \beta \mid \beta < \alpha\}$$

is α,

$$ord\, P_\alpha = \alpha.$$

Also, for any cardinal number a, there exists an ordinal number α such that

$$\text{cardinality of } P_\alpha = a .$$

We say that α *is countable iff* P_α *is countable.*

We set

$$\begin{aligned} \operatorname{ord} \phi &= 0 \\ \operatorname{ord} \{1, 2, \ldots, n\} &= n \\ &\vdots \\ \operatorname{ord} \{1, 2, 3, \ldots\} &= \omega_0 \\ \operatorname{ord} \{2, 3, \ldots, 1\} &= \omega_0 + 1 \\ &\vdots \\ \operatorname{ord} \{n+1, n+2, \ldots; 1, 2, \ldots, n\} &= \omega_0 + n \end{aligned}$$

The first ordinal following these numbers is $\omega_0 + \omega_0$ denoted by $\omega_0.2$

$$\operatorname{ord} \{1, 3, 5, 7, \ldots; 2, 4, 6, \ldots\} = \omega_0 + \omega_0 .$$

The numbers following $\omega \cdot 2$ are

$$\omega_0 \cdot 2 + 1, \omega_0 \cdot 2 + 2, \ldots, \omega_0 \cdot 2 + n, \ldots$$

Continuing this process we find all numbers of the form

$$\omega_0 \cdot n + m$$

The number following those numbers is denoted by ω^2. Continuing this process we define numbers of the form

$$\omega_0^k \cdot n_0 + \omega_0^{k-1} \cdot n_1 + \ldots + \omega_0 \cdot n_{k-1} + n_k$$

The number $\omega_0^{\omega_0}$ follows these numbers, and the process restarts with $\omega_0^{\omega_0} + 1$ till the number $\omega_0^{\omega_0^{\omega_0}}$. The number following numbers of the form $\omega_0^{\omega_0^{\cdot^{\cdot^{\cdot^{\omega_0}}}}}$ is denoted by ε and so on. Notice that all these numbers are countable since only a denumerable set of numbers are included in the process.

One then proves

Proposition 1. *We have*

(i) *There exists a smallest uncountable ordinal number* ω_1*, i.e., such that* P_{ω_1} *is uncountable. In particular* P_α *is countable for all* $\alpha \in P_{\omega_1}$.

(ii) *For every countable subset* C *of* P_{ω_1} *one can find a* $\beta \in P_{\omega_1}$ *such that* $\alpha \leq \beta$ $\forall \alpha \in C$.

(iii) Transfinite induction up to ω_1. *Let* A *be a subset of* P_{ω_1} *such that* $\alpha_0 \in A$ *and* α *belongs to* A *whenever all* β *with* $\alpha_0 \leq \beta < \alpha$ *belong to* A. *Then* $A = P_{\omega_1}$.

We notice that the continuum hypothesis is the assertion that ω_1, or more precisely, P_{ω_1} has the cardinality of $\mathbb{R}$.

Finally we remark that every countable ordinal α, i.e., every $\alpha \in P_{\omega_1}$ either is the successive of another ordinal, that is there exists β such that

$$\alpha = \beta + 1 ,$$

and in this case we say that α is a *non-limit ordinal*, or is a *limit ordinal*, i.e., for all $\beta < \alpha$, there exists γ with $\beta < \gamma < \alpha$.

We are now ready to turn back to the sequential weak closure of a set in a Banach space X.

Let X be a Banach space and let C be a subset of X. We define

$$\text{sw-lim}_X\, C \;:=\; \{x \in X \mid \exists \{x_k\} \subset C\,,\; x_k \rightharpoonup x\}$$

The following example shows that in general sw-lim$_X\, C$ is not sequentially closed in X.

[1] Let H be a separable Hilbert space and let $\{e_n\}$ be an orthonormal basis. Define

$$w_{n,k} := e_n + n\, e_k\,, \qquad C := \{w_{n,k}\}_{n,k=1,2,\dots}$$

Obviously

$$\| e_n + n\, e_k \| \;\geq\; \| n\, e_k \| - \| e_n \| \;=\; n - 1\,.$$

Therefore the only weakly converging sequences in C are the sequences $\{w_{n,k}\}_{k\in\mathbb{N}}$, for $n \in \mathbb{N}$, and we have $w_{n,k} \rightharpoonup e_n$ as $k \to \infty$. Thus

$$\text{sw-lim}_X\, C \;=\; C \cup \{e_n\}_{n=1,2,\dots}$$

But $\{e_n\}_{n=1,2,\dots} \subset$ sw-lim$_X\, C$, $e_n \rightharpoonup 0$, $0 \notin$ sw-lim$_X\, C$. •

Set $C^{(0)} := C$, and for any countable ordinal number α define

$$C^{(\alpha)} \;:=\; \text{sw-lim}_X\, C^{(\beta)}$$

if $\alpha = \beta + 1$ is a non limit ordinal, and

$$C^{(\alpha)} \;:=\; \bigcup_{\beta<\alpha} C^{(\beta)}$$

if α is a limit ordinal. Notice that

$$\begin{aligned} C^{(\alpha)} &\subset C^{(\beta)} \qquad \text{if } \alpha \leq \beta \\ C^{(1)} &= \text{sw-lim}_X\, C\,. \end{aligned}$$

Finally, define the class of *successive limits up to* ω_1 as

$$\text{succw-lim}_X\, C \;:=\; \bigcup_{\alpha<\omega_1} C^{(\alpha)}$$

and denote by

$$\text{sw-cl}_X\, C$$

the *sequential weak closure of* C, i.e., the smallest sequentially weakly closed set in X containing C.

Proposition 2. *We have*

$$\text{sw-cl}_X\, C \;=\; \text{succw-lim}_X\, C\ .$$

Proof. Let $\{x_k\} \subset \text{succw-lim}_X\, C$ with $x_k \rightharpoonup x$. Then for each $k \in \mathbb{N}$ there exists $\alpha_k < \omega_1$ such that $x_k \in C^{(\alpha_k)}$ and by Proposition 1 (ii) there exists $\beta < \omega_1$ such that $\alpha_k < \beta$ for all $k \in \mathbb{N}$. Therefore $\{x_k\} \subset C^{(\beta)}$, hence $x \in C^{(\beta+1)} \subset \cup_{\alpha<\omega_1} C^{(\alpha)}$ as $x_k \rightharpoonup x$. This shows that $\text{succw-lim}_X\, C$ is sequentially weakly closed. Hence

$$\text{sw-cl}_X\, C \;\subset\; \text{succw-lim}_X\, C\ .$$

Let us prove the opposite inclusion by means of the principle of transfinite induction up to ω_1. For simplicity write $\tilde{C}$ for $\text{sw-cl}_X\, C$. Obviously $C^{(0)} = C \subset \tilde{C}$. If $\alpha = \beta+1$ and $C^{(\beta)} \subset \tilde{C}$, then $C^{(\alpha)} \subset \tilde{C}$ as $\tilde{C}$ is sequentially weakly closed. If α is a limit ordinal and if $C^{(\beta)} \subset \tilde{C}\ \forall\, \beta < \alpha$, then obviously $C^{(\alpha)} = \cup_{\beta<\alpha} C^{(\beta)} \subset \tilde{C}$. Therefore by Proposition 1 (iii) we conclude

$$\bigcup_{\alpha<\omega_1} C^{(\beta)} \;\subset\; \tilde{C}\ .$$

□

Remark 1. Suppose that X is the dual of a Banach space. Then we may consider on X also the weak* convergence. Evidently we may repeat the same construction and proof above replacing weak by weak*, in particular we then conclude, with the obvious meaning of the notation

$$\text{sw}^*\text{-cl}_X\, C \;=\; \text{succw}^*\text{-cl}_X\, C$$

[2] Returning to the example in [1], $C := \{w_{n,k}\}_{n,k=1,2,\dots}$ we readily see that

$$C^{(0)} \;=\; C\ , \quad C^{(1)} \;=\; C \cup \{e_n\}_{n\in\mathbb{N}}\ , \quad C^{(2)} \;=\; C \cup \{e_n\} \cup \{0\}$$

Since $C^{(2)}$ is trivially sequentially weakly closed, we conclude in this case

$$\text{sw-cl}_X\, C \;=\; C^{(2)} \;=\; C \cup \{e_n\} \cup \{0\}$$

•

We emphasize the fact that $\text{sw-cl}_X\, C$ is the *sequentially* weak closure of C. The closure of C in the *weak topology* of C is in general strictly larger than $\text{sw-cl}_X\, C$ as shown by a classical example of von Neumann, see [3] below. Recall that the *weak topology* in X is defined as the weakest topology in X for which all linear functionals $x \to <f,x>$ $f \in X'$ are continuous. A base of neighborhoods of a point $x_0 \in X$ is obtained by taking the sets

$$V := \{x \in X \mid |<f_i, x - x_0>| < \varepsilon \,\forall\, i \in I\}$$

where I is a finite family of indexes, $f_i \in X'$, and $\varepsilon > 0$. We shall never work with the weak topology.

[3] *von Neumann's example.* In the Hilbert space ℓ^2 we consider the subset E of vectors e_{mn}, $1 \le m < n < +\infty$, the m-th component of e_{mn} is one, the n-th component of e_{mn} is m, and all other components of e_{mn} are zero. Trivially E is weakly sequentially closed as E contains no weakly converging sequence except for the ones which definitively are constant. However 0, which is not in E, is in the weak closure of E. In fact neighborhoods of 0 in the weak topology of ℓ^2 have the form

$$V := \{x \in \ell^2 \mid |\sum_{n=1}^{\infty} b_n^{(i)} x_n| < \varepsilon \,,\; i = 1, 2, \dots, k \sum_{n=1}^{\infty} (b_n^i)^2 < +\infty\}$$

As for all $b_n^{(i)}$ as above we can choose m, n so that

$$|b_m^i + m b_n^{(i)}| < \varepsilon$$

it follows that 0 belongs to the weak closure of E. •

4.2 Weak and Strong Approximation of Minors

We begin by giving a characterization of the strong approximability property by smooth maps.

The trivial observation contained in the following proposition allows us to replace C^1-maps by Lipschitz-continuous maps.

Proposition 1. *Let $u \in \operatorname{cart}^p(\Omega, \mathbb{R}^N)$. Suppose that there exists a sequence of Lipschitz-continuous maps u_k such that*

$$\begin{array}{rcll} u_k & \to & u & \text{strongly in } L^p(\Omega, \mathbb{R}^N) \\ M(Du_k) & \to & M(Du) & \text{strongly in } L^p(\Omega, \Lambda_n \mathbb{R}^{n+N}) \,. \end{array}$$

Then there exists a sequence of C^1-maps $\tilde{u}_k$ which again are such that

$$\begin{array}{rcll} \tilde{u}_k & \to & u & \text{strongly in } L^p(\Omega, \mathbb{R}^N) \\ M(D\tilde{u}_k) & \to & M(Du) & \text{strongly in } L^p(\Omega, \Lambda_n \mathbb{R}^{n+N}) \,. \end{array}$$

Proof. It suffices to approximate each u_k or each Lipschitz-continuous extension u_k to $\mathbb{R}^n$ by smooth maps in $W^{1,n}(\Omega, \mathbb{R}^N)$ or in $W^{1,n}(\tilde{\Omega}, \mathbb{R}^N)$, $\tilde{\Omega} \supset\supset \Omega$. □

The next theorem contains a key result concerning the approximation property

Theorem 1. *Let $u \in \mathrm{cart}^p(\Omega, \mathbb{R}^N)$, $p \geq 1$. The the following two claims are equivalent*

(i) There exists a sequence of locally Lipschitz maps $\{u_k\} \subset \mathrm{cart}^p(\Omega, \mathbb{R}^N)$ such that

$$u_k \rightharpoonup u \qquad \text{weakly in } \mathcal{A}^p \tag{1}$$

and the family

$$\{|M(Du_k)|^p\} \quad \text{is equi-integrable in } \Omega \tag{2}$$ [5]

(ii) There exists a sequence of locally Lipschitz maps $\{w_k\} \subset \mathrm{cart}^p(\Omega, \mathbb{R}^N)$ such that

$$\begin{array}{rcll} w_k & \to & u & \text{strongly in } L^p(\Omega, \mathbb{R}^N) \\ M(Dw_k) & \to & M(Du) & \text{strongly in } L^p(\Omega, \Lambda_n\mathbb{R}^{n+N}) \ . \end{array} \tag{3}$$

(iii) There exists a sequence of locally Lipschitz maps $\{w_k\} \subset \mathrm{cart}^p(\Omega, \mathbb{R}^N)$ such that

$$\mathrm{meas}\,\{x \,|\, u \neq w_k\} \ < \ 1/k$$

and

$$\int_{w_k \neq u} |M(Dw_k)|^p \, dx \ \longrightarrow \ 0$$

Proof. Trivially (iii) implies (ii) and (ii) implies (i). Let us prove that (i) implies (iii). Let $\{u_k\}$ be a sequence in (i). We have

$$\sup_k \| u_k \|_{\mathcal{A}^p} \ < \ c_1 \ , \quad \| u \|_{\mathcal{A}^p} \ \leq \ c_1$$

for some constant c_1; moreover, possibly passing to a subsequence, we can assume that

$$\begin{array}{lll} u_k & \longrightarrow u & \text{strongly in } L^p(\Omega, \mathbb{R}^N) \\ u_k & \longrightarrow u & \text{a.e. in } \Omega \ . \end{array}$$

First, for any $\delta > 0$, we construct a Lipschitz map w_δ in $\mathbb{R}^n$ and a closed set F_δ such that, if

$$G_\delta \ := \ \Omega \setminus F_\delta \ ,$$

we have

$$\begin{array}{ll} \mathrm{meas}\, G_\delta \ \leq \ \delta & \\ u \ = \ w_\delta & \text{in } F_\delta \\ u_k \ \longrightarrow \ u & \text{uniformly in } F_\delta \\ u^j \ \neq \ w^j_\delta & \text{a.e. in } G_\delta \ , \ \text{for all } j = 1, \dots, N \ . \end{array} \tag{4}$$

By Lusin type theorem for functions in $W^{1,p}$, compare Sec. 3.1.3, we find a closed set $F_{1,\delta}$ and a Lipschitz map f_δ such that

[5] Note that, when $p = 1$, the condition of equi-integrability of $\{|M(Du_k)|\}$ is already contained in (1).

$$\text{meas}\,(\Omega \setminus F_{1,\delta}) < \delta/2 \quad \text{and} \quad u = f_\delta \text{ in } F_{1,\delta}\,,$$

applying Egoroff's theorem we find a closed set $F_\delta \subset F_{1,\delta}$ such that

$$\text{meas}\,(\Omega \setminus F_\delta) < \delta, \quad u = f_\delta \text{ on } F_\delta \text{ and } u_k \to u \text{ uniformly in } F_\delta\,.$$

Then we finally modify f_δ on $G_\delta := \Omega \setminus F_\delta$ by setting, for $j = 1, \dots, N$,

$$w_\delta^j \;=\; f_\delta^j + \lambda_j \,\text{dist}\,(x, F_\delta)$$

where λ_j's are chosen by considering the family H_λ of disjoint sets

$$H_\lambda \;:=\; \{x \in G_\delta \;\mid\; f_\delta^j(x) + \lambda\,\text{dist}\,(x, F_\delta) = u^j(x)\}$$

and choosing λ_j so that $\text{meas}\, H_{\lambda_j} = 0$. Clearly the map w_δ we constructed this way is Lipschitz continuous

$$(5) \qquad |Dw_\delta(x)| \;\le\; c_2(\delta)$$

and satisfies all requirements in (4). Finally, since $\|\, u_k - u \,\|_{\infty, F_\delta} \to 0$ and $\text{meas}\,\{x \mid |u_k(x) - u(x)| > t\}$ tends to zero as $k \to \infty$, passing possibly to a subsequence, we can also assume that

$$(6) \qquad \begin{aligned} \|\, u_k - u \,\|_{\infty, F_\delta} \;&<\; 2^{-k} \\ \text{meas}\,\{x \mid |u_k(x) - u(x)| > 2^{-k}\} \;&\le\; 2^{-k}\,. \end{aligned}$$

We now define the sequence of locally Lipschitz maps $w_k = (w_k^1, \dots, w_k^N)$ as

$$(7) \qquad w_k^j \;:=\; \max\{\min\{w_\delta^j, u_k^j + 2^{-k}\}\,,\; u_k^j - 2^{-k}\}$$

and we shall show that for $k \ge k(\delta)$

$$(8) \qquad \text{meas}\,\{x \,|\, w_k(x) \ne u(x)\} \;=\; o(1)$$

$$(9) \qquad \int_\Omega |M(Dw_k) - M(Du)|^p \, dx \;=\; o(1)$$

as $\delta \to 0^+$. Evidently this implies (iii).

From the expression of w_k we have

$$w_k^j(x) \;:=\; \begin{cases} u_k^j + 2^{-k} & \text{if } w_\delta^j \ge u_k^j + 2^{-k} \\ w_\delta^j & \text{if } |w_\delta^j - u_k^j| < 2^{-k} \\ u_k^j - 2^{-k} & \text{if } w_\delta^j \le u_k^j - 2^{-k}\,, \end{cases}$$

in particular, from (6) and $w_\delta = u$ on F_δ, we deduce that $w_k = u$ on F_δ. In order to compute the minors of Dw_k let us define, for any multi–index $\xi = (\xi_1, \xi_2, \dots, \xi_N) \in \{-1, 0, 1\}^N$, the map $z_k^\xi := (z_k^{\xi_1}, \dots, z_k^{\xi_N}) : \Omega \to \mathbb{R}^N$ as

$$z_k^{\xi_j} := \begin{cases} u_k^j + 2^{-k} & \text{if } \xi_j = +1 \\ w_\delta^j & \text{if } \xi_j = 0 \\ u_k^j - 2^{-k} & \text{if } \xi_j = -1 \,, \end{cases}$$

and the subsets of G_δ

$$E_k^\xi := \{x \in G_\delta \mid w_k(x) = z_k^\xi(x)\} \,.$$

Since the values of $w_k^j(x)$ are just the values of $z_k^{\xi,j}(x)$ for some $\xi \in \{-1,0,1\}^N$, we have

$$G_\delta = \bigcup_\xi E_k^\xi$$

We now divide G_δ into a $(\text{good})_k$ part A_k given by

$$A_k := \bigcup_{\xi \in \{-1,1\}^N} E_k^\xi$$

and a $(\text{bad})_k$ part given by

$$B_k := G_\delta \setminus A_k \,.$$

On A_k we have $Dw_k = Du_k$, hence $M(Dw_k) = M(Du_k)$. On B_k Dw_k is mixed in the sense that some its rows are rows of Dw_δ and others from Du_k. If $x \in E_k^\xi$, $\xi \notin \{-1,1\}^N$, α, β are such that $|\alpha| + |\beta| = n$, $\beta = (\beta_1, \dots, \beta_n)$ and $\beta' = \beta'(\xi)$ is the submulti–index obtained by choosing the β_i's in such a way that $\xi_i \neq 0$, by Laplace's formula we deduce

$$|M_{\bar\alpha}^\beta(Dw_k)| \le \| Dw_\delta \|_{\infty,F_\delta}^{|\beta|-|\beta'|} \sum_{\substack{\bar\alpha' \le \bar\alpha \\ |\alpha'|+|\beta'|=n}} |M_{\bar\alpha'}^{\beta'}(Du_k)| \,.$$

Therefore, taking also into account (5), we deduce

$$\int_{B_k} |M(Dw_k)|^p \, dx \le c_3(\delta) \int_{B_k} |M(Du_k)|^p \, dx \,. \tag{10}$$

We now prove that

$$\lim_{k\to\infty} \text{meas}\, B_k = 0 \,. \tag{11}$$

In fact, if $x \in B_k$, then there exists $j \in \{1, \dots, N\}$ such that $w_k^j(x) = w_\delta^j(x)$, i.e., $|u_k^j(x) - w_\delta^j(x)| < 2^{-k}$. This yields either $|u_k^j(x) - u^j(x)| > 2^{-k}$ or $|w_\delta^j(x) - u^j(x)| \le |w_\delta^j(x) - u_k^j(x)| + |u_k^j(x) - u^j(x)| \le 2^{-k} + 2^{-k} = 2^{-k+1}$. Therefore

$$B_k \subset \{x \in G_\delta \mid |u_k(x) - u(x)| \ge 2^{-k}\} \bigcup \bigcup_{j=1}^N \{x \in G_\delta \mid |u^j(x) - w_\delta^j(x)| \le 2^{-k+1}\} \,.$$

Since, by (4),

$$\bigcap_{k=1}^{\infty} \{x \in G_\delta \mid |u^j - w_\delta^j| \le 2^{-k}\} = \{x \in G_\delta \mid u^j = w_\delta^j\}$$

has measure zero, taking into account (6) we get (11).

Finally, let us prove (8) and (9). We obviously have (8) as

$$\{x \mid w_k(x) \neq u(x)\} \subset G_\delta ,$$

while, taking into account (10),

$$\begin{aligned} \int_{G_\delta} |M(Dw_k)|^p &= \int_{A_k} |M(Du_k)|^p\,dx + \int_{B_k} |M(Dw_k)|^p\,dx \le \\ &\le \int_{G_\delta} |M(Du_k)|^p\,dx + c_3(\delta) \int_{B_k} |M(Du_k)|^p\,dx \end{aligned}$$

The first integral is $o(1)$ in virtue of the equi-integrability of $|M(Du_k)|^p$, while, again by the equi-integrability of $|M(Du_k)|^p$ and (11), also the second integral is $o(1)$ as $\delta \to 0$. □

An immediate consequence of Theorem 1 is the following corollary

Corollary 1. *We have*

$$\begin{aligned} \mathrm{Cart}^1(\Omega, \mathbb{R}^N) &= \text{sw-lim}_{\mathcal{A}^1}(C^1(\Omega, \mathbb{R}^N) \cap \mathcal{A}^1(\Omega, \mathbb{R}^N)) \\ &= \text{strong-cl}_{\mathcal{A}^1}(C^1(\Omega, \mathbb{R}^N) \cap \mathcal{A}^1(\Omega, \mathbb{R}^N)) \\ &= \mathrm{CART}^1(\Omega, \mathbb{R}^N) . \end{aligned}$$

Proof. By Theorem 1 the class

$$\text{strong-cl}_{\mathcal{A}^1}(C^1(\Omega, \mathbb{R}^N) \cap \mathcal{A}^1(\Omega, \mathbb{R}^N))$$

is sequentially weakly closed. In fact if $u_k \rightharpoonup u$ weakly in $\mathcal{A}^1$, and $\{u_k\} \in$ strong-cl$_{\mathcal{A}^1}(C^1(\Omega, \mathbb{R}^N) \cap \mathcal{A}^1(\Omega, \mathbb{R}^N))$, then for each k we can find $\{u_{k,h}\} \subset C^1(\Omega, \mathbb{R}^N)$ such that

$$u_{k,h} \longrightarrow u_k \qquad \text{strongly in } \mathcal{A}^1$$

Thus we can choose $h = h(k)$ so that

$$u_{k,h(k)} \rightharpoonup u \qquad \text{weakly in } \mathcal{A}^1 .$$

Applying Theorem 1, we then see that u belongs to the strong closure in $\mathcal{A}^1$ of smooth maps. Consequently

$$\mathrm{Cart}^1(\Omega, \mathbb{R}^N) \subset \text{strong-cl}_{\mathcal{A}^1}(C^1(\Omega, \mathbb{R}^N) \cap \mathcal{A}^1(\Omega, \mathbb{R}^N))$$

and equality holds as the opposite inclusion is trivial. □

Later in this subsection we shall see that in general

$$\mathrm{Cart}^1(\Omega, \mathbb{R}^N) = \mathrm{CART}^1(\Omega, \mathbb{R}^N) \subsetneq \mathrm{cart}^1(\Omega, \mathbb{R}^N) \,.$$

Remark 1. Though for every $u \in \mathrm{Cart}^1(\Omega, \mathbb{R}^N)$ we can find a sequence of Lipschitz maps u_k such that

$$u_k \to u, \qquad M(Du_k) \to M(Du) \text{ strongly in } L^1 \,,$$

it is not clear whether a *Liu type theorem* holds. In particular it would be of some relevance to decide under which conditions on u we can find for every $\lambda > 0$ a Lipschitz map u_λ *with Lipschitz constant less than* λ such that

$$\{x \,|\, u_\lambda \neq u\} \to 0, \qquad \int_{u_\lambda \neq u} |M(Du_\lambda)|\, dx \to 0$$

as $\lambda \to \infty$.

For $p > 1$ we introduce the following definition

Definition 1. *Let* $u \in \mathrm{cart}^p(\Omega, \mathbb{R}^N)$.

(i) *We say that* u *has the* q-strong approximation property, $q \leq p$, *if there exists a sequence of smooth maps, of class* C^1 *or Lipschitz,* $\{u_k\}$ *such that*

$$u_k \longrightarrow u \qquad \text{strongly in } L^q$$

and

$$M(Du_k) \longrightarrow M(Du) \qquad \text{strongly in } L^q \,.$$

(ii) *We say that* u *has the* p^--strong approximation property *if there exists a sequence of smooth maps* $\{u_k\}$ *such that*

$$\begin{array}{rcll} u_k & \longrightarrow & u & \text{strongly in } L^q(\Omega) \\ M(Du_k) & \longrightarrow & M(Du) & \text{strongly in } L^q(\Omega) \end{array}$$

for all $q < p$.

Of course the p^--strong approximation property is just the q-strong approximation property for all $q < p$.

We also define

Definition 2. *We denote by* $\mathrm{Cart}^{p^-}(\Omega, \mathbb{R}^N)$ *the subclass of elements* u *in* $\mathrm{cart}^p(\Omega, \mathbb{R}^N)$ *which have the* p^-*-strong approximation property.*

As a consequence of Theorem 1 we infer

Corollary 2. $\mathrm{Cart}^{p^-}(\Omega, \mathbb{R}^N)$ *is sequentially weakly closed in* $\mathrm{cart}^p(\Omega, \mathbb{R}^N)$. *In particular*

$$\mathrm{Cart}^p(\Omega, \mathbb{R}^N) \subset \mathrm{Cart}^{p^-}(\Omega, \mathbb{R}^N) \,.$$

Proof. Let $\{u_k\}$ be a sequence of map in $\text{cart}^p(\Omega,\mathbb{R}^N)$ which have the p^--strong approximation property and such that $u_k \overset{\mathcal{A}^p}{\rightharpoonup} u$. Fix $q < r < p$. We can find a sequence of smooth maps $\{v_k\}$ such that

$$\int |M(Dv_k) - M(Du_k)|^r \, dx \; < \; 2^{-k} \, , \quad v_k \longrightarrow u \text{ in } L^r \, .$$

In particular $v_k \overset{\mathcal{A}^r}{\rightharpoonup} u$ and the family $\{|M(Dv_k)|^q\}$ is equi-integrable. Hence, by Theorem 1, there exists a sequence of smooth maps $\{w_k\}$ such that

$$\int |M(Dw_k) - M(Du)|^q \longrightarrow 0 \, .$$

Choosing an increasing sequence $\{q_k\}$ converging to p, we now find by a diagonal process a sequence $\{w_k\}$ such that

$$\int |M(Dw_k) - M(Du)|^q \, dx \longrightarrow 0 \qquad \forall \, q < p \, .$$

□

For $p > 1$ we therefore have

$$\begin{aligned} \text{CART}^p(\Omega,\mathbb{R}^N) \;&:=\; \text{strong} - \text{cl}_{\mathcal{A}^p}(C^1(\Omega,\mathbb{R}^N) \cap \mathcal{A}^p(\Omega,\mathbb{R}^N)) \\ &\subset\; \text{sw} - \lim{}_{\mathcal{A}^p}(C^1(\Omega,\mathbb{R}^N) \cap \mathcal{A}^p(\Omega,\mathbb{R}^N)) \\ &\subset\; \text{Cart}^p(\Omega,\mathbb{R}^N) \;\subset\; \text{Cart}^{p^-}(\Omega,\mathbb{R}^N) \end{aligned}$$

It is an open question to decide whether some of the previous inclusions are actually equalities or strict inclusions.

We shall now prove that in general

$$\text{Cart}^{p^-}(\Omega,\mathbb{R}^N) \underset{\neq}{\subset} \text{cart}^p(\Omega,\mathbb{R}^N)$$

consequently

$$\text{Cart}^{p}(\Omega,\mathbb{R}^N) \underset{\neq}{\subset} \text{cart}^p(\Omega,\mathbb{R}^N)$$

for $p > 1$, and actually also for $p = 1$.

Denote by $B_r = B(0,r)$ the open ball in $\mathbb{R}^2$ of radius r centered at the origin. We shall consider maps $u : B_1 \to \mathbb{R}^2$ satisfying the following condition

(A) *u takes values on a fixed domain*

$$\widehat{\Omega} := C_0 \setminus \cup_{k=1}^h \bar{C}_k$$

where $C_1, \ldots, C_h$ are open balls with disjoint closures contained in a ball C_0 centered at zero.

Let $u \in \mathrm{cart}^1(B_1, \mathbb{R}^2)$ satisfy condition (A). Set

$$R_u := \{ r \in (0,1) \mid \int_{\partial B_r} |Du| \, d\mathcal{H}^1 < \infty , \; u(\partial B_r) \subset \widehat{\Omega} \}$$

and

$$u_{(r)} := u \text{ restricted to } \partial B_r .$$

Clearly $\mathcal{H}^1((0,1) \setminus R_u) = 0$ and, by Sobolev theorem, $u_{(r)} \in C^0(\partial B_r, \widehat{\Omega})$ for all $r \in R_u$. Then we have

Theorem 2. *Let $u \in \mathrm{cart}^1(B_1, \mathbb{R}^2)$ satisfy condition (A). Suppose that there exists a sequence of locally Lipschitz maps $w_k \in \mathrm{cart}^1(B_1, \mathbb{R}^2)$ such that*

$$w_k \to u \qquad \textit{strongly in } L^1$$
$$M(Dw_k) \to M(Du) \quad \textit{strongly in } L^1 .$$

Then $u_{(r)}$ is homotopic in $\widehat{\Omega}$ to a constant map for almost every $r \in R_u$.

Proof. Choose $\varepsilon > 0$ such that

$$\varepsilon < \mathcal{H}^2(C_i) , \qquad i = 1, \dots, h . \tag{12}$$

where C_i, $i = 1, \dots, h$ are the balls in the definition of $\widehat{\Omega}$. From $\det Dw_k \to \det Du$ strongly in L^1 it follows that the sequence of functions $\det Dw_k$ is equi-integrable, hence there exists $\delta > 0$ such that for all k

$$\int_E |\det Dw_k| \, dx < \varepsilon \qquad \text{if } \operatorname{meas} E < \delta . \tag{13}$$

Also, there exists a subsequence of $\{w_k\}$ such that

$$\sum_{k=1}^{\infty} \int_{B_1} (|w_k - u| + |Dw_k - Du|) \, dx < \infty ,$$

thus we can find $R_u^0 \subset R_u$ such that

$$\mathcal{H}^1(R_u \setminus R_u^0) = 0$$

$$\sum_{k=1}^{\infty} \int_{\partial B_r} (|w_k - u| + |Dw_k - Du|) \, dx < \infty \qquad \text{for } r \in R_u^0 . \tag{14}$$

Consequently, for each $r \in R_u^0$ we deduce

$$w_k \longrightarrow u \qquad \text{strongly in } W^{1,1}(\partial B_r, \mathbb{R}^2) ,$$

and by the Sobolev embedding theorem

$$(15) \qquad w_k \longrightarrow u \qquad \text{uniformly on } \partial B_r \text{ for } r \in R_u^0 .$$

We shall now show that for $r_1, r_2 \in R_u^0$, $0 < r_1 < r_2 \leq 1$ satisfying

$$(16) \qquad |r_1 - r_2| < \frac{\delta}{2\pi}$$

we have

$$(17) \qquad u_{(r_1)} \text{ and } u_{(r_2)} \text{ are homotopic in } \widehat{\Omega} .$$

Let $\sigma > 0$ be such that

$$(18) \qquad \widehat{\Omega}_\sigma := \{x \in \mathbb{R}^2 \mid \operatorname{dist}(x, u(\partial B_{r_1}) \cup u(\partial B_{r_2})) < \sigma\} \subset \widehat{\Omega} .$$

For k_0 sufficiently large we deduce from (15)

$$(19) \qquad w_{k_0}(\partial B_{r_1} \cup \partial B_{r_2}) \subset \widehat{\Omega}_\sigma .$$

From (16) it follows $\mathcal{H}^1(B_{r_2} \setminus B_{r_1}) < \delta$, thus by (13)

$$(20) \qquad \int |\det Dw_{k_0(r_1,r_2)}| \, dx < \varepsilon$$

where we have set

$$w_{k_0(r_1,r_2)} := w_{k_0 \mid B_{r_2} \setminus B_{r_1}}$$

From (12) and (20) we then see that the set $w_{k_0(r_1,r_2)}(B_{r_2} \setminus B_{r_1})$ does not cover any of the C_i, $i = 1, 2, \dots, h$. Therefore we can find $\gamma_0, \gamma_1, \dots, \gamma_h > 0$ and $y_1 \in C_1, \dots, y_h \in C_h$ such that

$$(21) \qquad \widetilde{\Omega} := B_{\gamma_0}(0) \setminus \bigcup_{i=1}^{h} B_{\gamma_i}(y_i) \supset \widehat{\Omega}$$

and

$$(22) \qquad \widetilde{\Omega} \supset w_{k_0(r_1,r_2)}(B_{r_2} \setminus B_{r_1}) .$$

Since $w_{k_0(r_1,r_2)}$ is a homotopy between $w_{k_0(r_1)}$ and $w_{k_0(r_2)}$ we deduce that $w_{k_0(r_1)}$ and $w_{k_0(r_2)}$ are homotopic in $\widetilde{\Omega}$ and, as it can be easily seen, in $\widehat{\Omega}$. On the other hand from (18) and (19) we also see that for $i = 1, 2$ $w_{k_0(r_i)}$ and $u_{(r_i)}$ are homotopic, a homotopy being given by

$$h_t(x) := t\, w_{k_0}(x) + (1 - t)\, u(x) \qquad x \in \partial B_{r_i} ,$$

thus (17) is proved.

In a similar way we can prove that $u_{(r)}$ is homotopic to a constant map for $r < \frac{1}{2\pi}\delta$. This together with (17) proves that $u_{(r)}$ is homotopic to a constant map for each $r \in R_u^0$. □

Let $u : B_1 \to \mathbb{R}^2$ be a map which is homogeneous of degree zero and, more precisely, is of the type

$$u(x) = \varphi(\frac{x}{|x|}) \tag{23}$$

where φ is a smooth map from $S^1 = \partial B^1$ into $\mathbb{R}^2$. Clearly, compare [1] in Sec. 3.2.2, $u \in W^{1,p}(B^2, \mathbb{R}^2)$ for all $p < 2$, $\det Du = 0$ in B_1, thus $u \in \mathcal{A}^p(\Omega, \mathbb{R}^2)$ for any $p < 2$ and

$$\partial G_u \llcorner B_1 \times \mathbb{R}^2 = -\delta_0 \times [\![\varphi(S^1)]\!] .$$

In particular, if for instance $\varphi(S^1)$ is covered twice with opposite orientation, it follows that

$$\partial G_u \llcorner B_1 \times \mathbb{R}^2 = 0 ,$$

hence $u \in \text{cart}^p(B_1, \mathbb{R}^2)$ for $p < 2$.

It follows from Theorem 1 that any map with the following properties

(i) u is of the type (23) and $u \in \text{cart}^1(B_1, \mathbb{R}^2)$
(ii) $\mathbb{R}^2 \setminus \varphi(S^1)$ has a finite number of connected components
(iii) φ is not homotopic to a constant map in $\mathbb{R}^2 - \{x_1, \dots, x_k\}$ where $x_1, \dots, x_k$ are points respectively in each bounded connected component of $\mathbb{R}^2 \setminus \varphi(S^1)$.

does not belong to $\text{Cart}^p(B_1, \mathbb{R}^2)$, $p < 2$. An explicit example is given for instance by the map

$$u(x) = \varphi(\frac{x}{|x|}) \tag{24}$$

where φ is defined by

$$\varphi(\theta) = \begin{cases} (-1 + \cos 4\theta, \sin 4\theta) & \text{for } 0 \le \theta \le \pi/2 \\ (\ 1 - \cos 4\theta, \sin 4\theta) & \text{for } \pi/2 \le \theta \le \pi \\ (-1 + \cos 4\pi, -\sin 4\pi) & \text{for } \pi \le \theta \le 3/2\pi \\ (\ 1 - \cos 4\theta, -\sin 4\theta) & \text{for } 3/2\pi \le \theta \le 2\pi \end{cases} \tag{25}$$

Clearly the image of $S^1 = \partial B_1$ through φ in (25) is the boundary of the union of the two unit discs centered at $(-1, 0)$ and $(1, 0)$ of $\mathbb{R}^2$, $\varphi(S^1)$ is covered twice with opposite orientation so that $u \in \text{cart}^1(B_1, \mathbb{R}^2)$, and one easily sees that φ is not homotopic to a constant map in $\mathbb{R}^2 \setminus \{(-1, 0), (1, 0)\}$.

We finally observe that, though u in (24) is not in $\text{Cart}^1(B(0,1), \mathbb{R}^2)$ we can always find by [2] in Sec. 3.2.2 a sequence of *smooth maps* $\{u_k\}$ such that

$$G_{u_k} \rightharpoonup G_u$$

in $B(0, 1) \times \mathbb{R}^2$.

4.3 The Join of Cartesian Maps

Given two maps u, v from a bounded domain Ω of $\mathbb{R}^n$ respectively in $\mathbb{R}^{N_1}$ and $\mathbb{R}^{N_2}$, the *join of* u, v, denoted by $u \bowtie v$, is defined as the map from Ω into $\mathbb{R}^N$, $N = N_1 + N_2$,

$$u \bowtie v(x) \;=\; (u(x), v(x)) \;.$$

In general the join of two Cartesian maps is not a Cartesian map as it is shown by the map $\frac{x}{|x|}$ from $B(0,1) \subset \mathbb{R}^2$ in $\mathbb{R}^2$. In this subsection we shall give sufficient conditions in order that the join of two Cartesian maps be a Cartesian map. These kind of results may be regarded as a generalization of Proposition 3 in Sec. 3.2.1 which, on account of

$$W^{1,p}(\Omega, \mathbb{R}) = \operatorname{cart}^p(\Omega, \mathbb{R}) \;,$$

may be restated as

Proposition 1. *Let* $u : \Omega \to \mathbb{R}^N$ *be a map with components* $u^1, \dots, u^N$ *in* $\operatorname{cart}^p(\Omega, \mathbb{R})$. *Suppose that* $p \geq \underline{n} := \min(n, N)$. *Then* $u \in \operatorname{Cart}^{p/\underline{n}}(\Omega, \mathbb{R}^N)$.

Showing that $u \bowtie v$ is a Cartesian map of course splits into the two questions of guaranteeing that

(i) $u \bowtie v$ belongs to some $\mathcal{A}^r(\Omega, \mathbb{R}^N)$.
(ii) $\partial G_{u \bowtie v} \llcorner \Omega \times \mathbb{R}^N = 0$.

Answers to question (i) are easily given in terms of standard formulas on the minors of the sum of two matrices and Hölder inequality. More delicate is the question of the vanishing of the boundary on $\Omega \times \mathbb{R}^N$ of the join. In fact essentially the only operative way we known to ensure the vanishing of $\partial G_{u \bowtie v} \llcorner \Omega \times \mathbb{R}^N$ is by approximating $G_{u \bowtie v}$ by currents which are boundaryless in $\Omega \times \mathbb{R}^N$. This was actually the the argument in the proof of Proposition 1.

Let us begin by recalling the standard formula for the minors of the sum of two matrices

Lemma 1. *Let* A *and* B *be two* $N \times n$*-matrices. Then for all multi-indices* α, β *with* $|\alpha| = |\beta|$, $0 \leq |\beta| \leq \underline{n} := \min(n, N)$ *we have*

$$M_\alpha^\beta(A + B) \;=\; \sum_{\substack{\alpha' + \alpha'' = \alpha \\ \beta' + \beta'' = \beta \\ |\alpha'| = |\beta'|}} \sigma(\alpha', \alpha'')\, \sigma(\beta', \beta'')\, M_{\alpha'}^{\beta'}(A)\, M_{\alpha''}^{\beta''}(B) \;.$$

Proof. Denote by $e_1, \dots, e_n$ and $\varepsilon_1, \dots, \varepsilon_N$ the standard bases in $\mathbb{R}^n$ and $\mathbb{R}^N$, respectively, and denote still by A and B the linear operators from $\mathbb{R}^n$ into $\mathbb{R}^N$ associated to A and B in those bases. Setting $|\alpha| = p$, we have

$$
\begin{aligned}
&(A+B)\wedge\ldots\wedge(A+B)(e_\alpha)\\
(1)\qquad &= (A+B)(e_{\alpha_1})\wedge\ldots\wedge(A+B)(e_{\alpha_p})\\
&= \sum_{\beta\in I(p,N)} M_\alpha^\beta(A+B)\,\varepsilon_\beta
\end{aligned}
$$

and analogously

$$
(2)\qquad
\begin{aligned}
(A\wedge\ldots\wedge A)(e_\alpha) &= \sum_{\beta\in I(p,N)} M_\alpha^\beta(A)\,\varepsilon_\beta\\
(B\wedge\ldots\wedge B)(e_\alpha) &= \sum_{\beta\in I(p,N)} M_\alpha^\beta(B)\,\varepsilon_\beta
\end{aligned}
$$

Thus from the k-linearity of the wedge product and (2) we infer

$$
\begin{aligned}
&(A+B)\wedge\ldots\wedge(A+B)(e_\alpha) =\\
&\quad = A(e_{\alpha_1})\wedge B(a_{\alpha_2})\wedge\ldots\wedge B(e_{\alpha_p}) + \ldots\\
&\quad = \sum_{\alpha'+\alpha''=\alpha} \sigma(\alpha',\alpha'')\,(A\wedge\ldots\wedge A)(e_{\alpha'})\wedge(B\wedge\ldots\wedge B)(e_{\alpha''})\\
&\quad = \sum_{\alpha'+\alpha''=\alpha}\ \sum_{\beta'\in I(p',N)}\ \sum_{\beta''\in I(p'',N)} \sigma(\alpha',\alpha'')\,M_{\alpha'}^{\alpha''}(A)\,M_{\alpha''}^{\beta''}(B)\,\varepsilon_{\beta'}\wedge\varepsilon_{\beta''}\\
&\quad = \sum_{\beta\in I(p,N)} \Big\{ \sum_{\substack{\alpha'+\alpha''=\alpha\\ \beta'+\beta''=\beta}} \sigma(\alpha',\alpha'')\,\sigma(\beta',\beta'')\,M_{\alpha'}^{\alpha''}(A)\,M_{\alpha''}^{\beta''}(B)\Big\}\varepsilon_\beta
\end{aligned}
$$

where $p' = |\beta'|, p'' = |\beta''|$. The result then follows comparing with (1). □

As

$$
D\,u \bowtie v = \begin{pmatrix} Du \\ 0 \end{pmatrix} + \begin{pmatrix} 0 \\ Dv \end{pmatrix}
$$

an immediate consequence of Lemma 1 is

Lemma 2. *Let u, v be maps in $\mathcal{A}^1(\Omega,\mathbb{R}^{N_1})$ and $\mathcal{A}^1(\Omega,\mathbb{R}^{N_2})$, respectively. Then*

$$
|M(D\,u \bowtie v)| \le c\,|M(Du)|\,|M(Dv)|
$$

where c is a constant depending only on n, N_1 and N_2.

Taking into account Hölder inequality, we can therefore state

Proposition 2. *We have*

(i) Let $u \in \mathcal{A}^r(\Omega,\mathbb{R}^{N_1})$, $r \ge 1$ and let $v : \Omega \to \mathbb{R}^{N_2}$ be Lipschitz. Then $u \bowtie v \in \mathcal{A}^r(\Omega,\mathbb{R}^N)$, $N = N_1 + N_2$.

(ii) Let $u \in \mathcal{A}^p(\Omega,\mathbb{R}^{N_1})$ and $v \in \mathcal{A}^q(\Omega,\mathbb{R}^{N_2})$. If $1/r := 1/p + 1/p \le 1$ then $u \bowtie v$ belongs to $\mathcal{A}^r(\Omega,\mathbb{R}^N)$.

Next lemma in conjunction with the closure theorems for Cartesian maps will allow us to give sufficient conditions in order that $\partial G_{u\bowtie v} \llcorner \Omega \times \mathbb{R}^N = 0$.

Lemma 3. *Let $v, v_k \in \text{cart}^p(\Omega, \mathbb{R}^N)$, $v_k \rightharpoonup v$ weakly in $\mathcal{A}^p$ and let w belong to $\text{cart}^q(\Omega, \mathbb{R}^N)$ where $1/r := 1/p + 1/q \leq 1$. Then*

$$v_k \bowtie w \rightharpoonup v \bowtie w \qquad \textit{weakly in } \mathcal{A}^r .$$

In particular $v \bowtie w \in \text{cart}^r(\Omega, \mathbb{R}^N)$ provided $v_k \bowtie w \in \text{cart}^r(\Omega, \mathbb{R}^N)$.

Proof. Recall that

$$f_k\, g \rightharpoonup fg \qquad \text{weakly in } L^r ,$$

if $f_k \rightharpoonup f$ weakly in L^p, g belongs to L^q and $1/r := 1/p + 1/q \leq 1$. Taking into account Lemma 1 and applying the previous remark to $f_k := M^{\beta'}_{\alpha'}(Dv_k)$ and $g := M^{\beta''}_{\alpha''}(Dw)$ we then infer

$$M^\beta_\alpha(D\, v_k \bowtie w) \rightharpoonup M^\beta_\alpha(D\, v \bowtie w) \qquad \text{weakly in } L^r .$$

This concludes the proof as the second part of the theorem is a trivial consequence of the weak closure of cart^r. □

Proposition 3. *Let $v : \Omega \to \mathbb{R}^{N_2}$ be a Lipschitz map. Then we have*

(i) $u \bowtie v \in \text{cart}^r(\Omega, \mathbb{R}^N)$, *if* $u \in \text{cart}^r(\Omega, \mathbb{R}^{N_1})$
(ii) $u \bowtie v \in \text{Cart}^r(\Omega, \mathbb{R}^N)$, *if* $u \in \text{Cart}^r(\Omega, \mathbb{R}^{N_1})$

Proof. The claim (i) follows at once as the current $G_{u\bowtie v}$ is the Lipschitz image of the boundaryless current G_u by

$$\begin{aligned} V \ :\ \mathbb{R}^{n+N_1} &\longrightarrow \mathbb{R}^{n+N} \\ (x,y) &\mapsto (x,y,v(x)) \end{aligned}$$

$G_{u\bowtie v} = V_\# G_u$, and taking into account Proposition 2(i).

In order to prove (ii) we consider the set

$$C_1 := \{u \in \text{cart}^r(\Omega, \mathbb{R}^{N_1}) \mid u \bowtie v \in \text{Cart}^r(\Omega, \mathbb{R}^N)\}$$

It suffices then to show that C_1 is closed under the weak convergence in $\mathcal{A}^r$, so that

$$\text{Cart}^r(\Omega, \mathbb{R}^{N_1}) \subset C_1 .$$

To show this consider a sequence $\{u_k\}$ in C_1 such that $u_k \to u$ in $\mathcal{A}^r$. From Lemma 3 we infer

$$u_k \bowtie v \rightharpoonup u \bowtie v \qquad \text{in } \mathcal{A}^r$$

and $u \bowtie v \in \text{cart}^r(\Omega, \mathbb{R}^N)$. Since each $u_k \bowtie v$ belongs to $\text{Cart}^r(\Omega, \mathbb{R}^N)$, it follows that also $u \bowtie v$ belongs to $\text{Cart}^r(\Omega, \mathbb{R}^N)$, hence $u \bowtie v \in C_1$. □

We are now ready to state the main result of this subsection.

Theorem 1. *Let p, q be such that $1/r := 1/p + 1/q \leq 1$.*

(i) If $u \in \mathrm{cart}^p(\Omega, \mathbb{R}^{N_1})$ and $v \in \mathrm{Cart}^q(\Omega, \mathbb{R}^{N_2})$ then $u \bowtie v \in \mathrm{cart}^r(\Omega, \mathbb{R}^N)$.

(ii) If $u \in \mathrm{Cart}^p(\Omega, \mathbb{R}^{N_1})$ and $v \in \mathrm{Cart}^q(\Omega, \mathbb{R}^{N_2})$ then $u \bowtie v \in \mathrm{Cart}^r(\Omega, \mathbb{R}^N)$.

Let moreover $\Omega_1 \subset \mathbb{R}^{N_1}$ be a bounded open set

(iii) If $u \in \mathrm{cart}^p(\Omega, \mathbb{R}^{N_1})$, $v \in \mathrm{Cart}^q(\Omega, \mathbb{R}^{N_2})$ and

$$\partial G_u \llcorner \mathbb{R}^n \times \Omega_1 = 0 \tag{3}$$

then

$$\partial G_{u \bowtie v} \llcorner \mathbb{R}^n \times \Omega_1 \times \mathbb{R}^{N_2} = 0 \tag{4}$$

Proof. (i) For any $w \in \mathrm{cart}^q(\Omega, \mathbb{R}^{N_2})$ we set

$$C_2 := \{v \in \mathrm{cart}^p(\Omega, \mathbb{R}^{N_1}) \mid v \bowtie w \in \mathrm{cart}^r(\Omega, \mathbb{R}^N)\} .$$

From Proposition 3 we know that

$$\mathrm{Lip}\,(\Omega, \mathbb{R}^{N_1}) \subset C_2 .$$

We shall now prove that C_2 is sequentially closed with respect to the weak-$\mathcal{A}^p$ convergence, hence

$$\mathrm{Cart}^p(\Omega, \mathbb{R}^{N_1}) \subset C_2 .$$

This clearly proves (i). Let us prove that C_2 is sequentially weakly closed in $\mathrm{cart}^p(\Omega, \mathbb{R}^{N_1})$.

Consider a sequence $\{v_k\} \subset C_2$ weakly converging in $\mathcal{A}^p$ to some v which of course belongs to $\mathrm{cart}^p(\Omega, \mathbb{R}^{N_1})$. By Lemma 3

$$v_k \bowtie w \rightharpoonup v \bowtie w \qquad \text{in } \mathcal{A}^r$$

and $v \bowtie w \in \mathrm{cart}^r(\Omega, \mathbb{R}^N)$.

(ii) For any $w \in \mathrm{Cart}^q(\Omega, \mathbb{R}^{N_2})$ we set

$$C_3 := \{v \in \mathrm{cart}^p(\Omega, \mathbb{R}^{N_1}) \mid v \bowtie w \in \mathrm{Cart}^r(\Omega, \mathbb{R}^N)\}$$

Again from Proposition 3

$$\mathrm{Lip}\,(\Omega, \mathbb{R}^{N_2}) \subset C_3$$

and C_3 is sequentially closed in $\mathrm{cart}^p(\Omega, \mathbb{R}^{N_1})$.

In fact if $v_k \rightharpoonup v$ in $\mathcal{A}^p$, we have $v \in \mathrm{cart}^p(\Omega, \mathbb{R}^{N_1})$ and

$$v_k \bowtie w \rightharpoonup v \bowtie w \qquad \text{in } \mathcal{A}^r, \qquad v \bowtie w \in \mathrm{cart}^r(\Omega, \mathbb{R}^N),$$

by Lemma 3. On the other hand $v_k \bowtie w \in \mathrm{Cart}^r(\Omega, \mathbb{R}^N)$, hence $v \bowtie w \in \mathrm{Cart}^r(\Omega, \mathbb{R}^N)$ by closure theorem.

As $\mathrm{Cart}^r(\Omega, \mathbb{R}^N)$ is the smallest sequentially closed set which contains the class $\mathrm{Lip}\,(\Omega, \mathbb{R}^N)$, we conclude

$$C_3 \supset \mathrm{Cart}^p(\Omega, \mathbb{R}^{N_1})$$

which proves our claim.
(iii) In order to prove (iii), for any $u \in \mathrm{cart}^p(\Omega, \mathbb{R}^{N_1})$ satisfying (3) we consider the class

$$C_4 := \{v \in \mathrm{Cart}^q(\Omega, \mathbb{R}^{N_2}) \mid u \bowtie v \text{ satisfies } (4)\} .$$

It suffices then to show that
(c) $C_4 \supset C^1(\overline{\Omega}, \mathbb{R}^{N_2})$
(d) C_4 is weakly closed in $\mathrm{cart}^q(\Omega, \mathbb{R}^{N_2})$.
To prove (c) we consider associated to $v \in C^1(\Omega, \mathbb{R}^{N_2})$ the map

$$V : (x, y) \in \mathbb{R}^n \times \mathbb{R}^{N_1} \to (x, (y, v(x))) \in \mathbb{R}^n \times \mathbb{R}^{N_1} \times \mathbb{R}^{N_2} .$$

As in the proof of Proposition 3 we see that $G_{u\bowtie v} = V_\# G_u$. For any $(n-1)$-form ω with $\mathrm{spt}\,\omega \subset \mathbb{R}^n \times \Omega_1 \times \mathbb{R}^{N_2}$ we now compute

$$G_{u\bowtie v}(d\omega) = V_\# G_u(d\omega) = G_u(V^\# d\omega) = G_u(d(V^\# \omega))$$

and

$$\mathrm{spt}\, V^\# \omega = V^{-1}(\mathrm{spt}\,\omega) \subset V^{-1}(\mathbb{R}^n \times \Omega_1 \times \mathbb{R}^{N_2}) = \mathbb{R}^n \times \Omega_1 ,$$

consequently $G_u(d(V^\#\omega)) = 0$, i.e., $G_{u\bowtie v}(d\omega) = 0$. This proves (c). Observing that condition (4) is weakly closed, the claim (d) follows immediately from Lemma 3. □

Remark 1. It is an open question to establish whether $u \bowtie v$ belongs to $\mathrm{cart}^r(\Omega, \mathbb{R}^N)$ or not under the weaker assumptions that $u \in \mathrm{cart}^p(\Omega, \mathbb{R}^{N_1})$, $v \in \mathrm{cart}^q(\Omega, \mathbb{R}^{N_2})$ and $1/r := 1/p + 1/q \leq 1$.

We conclude this subsection with a simple but useful remark concerning the continuity of the join of Cartesian maps with respect to the weak convergence in L^1 with equibounded cart-norms.

Theorem 2. *Let* $\{u_k\} \subset \mathrm{cart}^p(\Omega, \mathbb{R}^{N_1})$, $\{v_k\} \subset \mathrm{cart}^q(\Omega, \mathbb{R}^{N_2})$ *where* $1/r := 1/p + 1/q \leq 1$, *be sequences such that*

$$u_k \rightharpoonup u \text{ weakly in } L^1, \quad v_k \rightharpoonup v \text{ weakly in } L^1$$

and

$$\sup_k \| u_k \bowtie v_k \|_{\mathcal{A}^r} < \infty .$$

Then

$$u_k \bowtie v_k \rightharpoonup u \bowtie v \qquad \text{weakly in } \mathcal{A}^r$$

and $u \bowtie v \in \mathrm{cart}^r(\Omega, \mathbb{R}^N)$, *if at least one of the two sequences* $\{u_k\}$ *and* $\{v_k\}$ *is contained in* $\mathrm{Cart}^p(\Omega, \mathbb{R}^{N_1})$ *or in* $\mathrm{Cart}^q(\Omega, \mathbb{R}^{N_2})$ *respectively.*

Finally, the claim remains true also for $r = 1$, *if moreover we assume that the functions*

$$|M(Du_k)|\,|M(Dv_k)|$$

are equi-integrable.

Proof. It is an immediate consequence of Proposition 2, Theorem 1, and of the closure theorem in cart, as

$$u_k \bowtie v_k \rightharpoonup u \bowtie v \qquad \text{weakly in } L^1 .$$

□

Remark 2. Let u and v be two maps from $\Omega \subset \mathbb{R}^n$ into $\mathbb{R}^N$. We can consider the map

$$u + v \; : \Omega \longrightarrow \mathbb{R}^N$$

It is not difficult to see that all results in this subsection remain true with unchanged formulations and proofs if we replace $\bowtie$ by $+$. In fact we could have proved all theorems for the sum, and then deduce them for the join as

$$u \bowtie v \;=\; (u, 0) + (0, v) .$$

5 Notes

1 For the introductory content of Sec. 3.1.1 the reader is referred for instance to Stein [611, Chapter 1], Stein [612], Stein and Weiss [613], Torchinsky [634] and their bibliographical notes. For the theory of Sobolev spaces the reader may consult e.g. Adams [4] Maz'ja [463] Kufner, A. and John, O. and Fučik, S. [416] Ziemer [686].

Theorem 2 in Sec. 3.1.1 appeared in Giusti [304]. A better estimate of the exceptional set in terms of *Riesz* or *Bessel capacities* can be found in Federer and Ziemer [231] Ziemer [686]. From such an estimate for instance it follows that $\mathcal{H}^{n-1}(\Omega \setminus \mathcal{L}_u) = 0$ if $u \in W^{1,1}(\Omega)$, compare Remark 3 in Sec. 4.1.4.

2 The theory of pointwise differentiability of non-smooth functions, as well as the notion of approximate differential, goes back to the beginning of this century in particular to the work of Lebesgue, Khintchine, Denjoy, Rademacher, and Stepanoff among others, compare Saks [570]. Relevant contributions are due to Whitney [673] and Federer [226, chap. 3.1]. Examples of absolutely continuous functions in dimension $n \leq 2$ which are not differentiable on a set of positive measure can be found e.g. in Caccioppoli [128] Saks [569] Cesari [141] and Serrin [585]. In particular in Serrin [585] it is shown a function in $W^{1,n}(B^n) \cap C^0(B^n)$ which is nowhere differentiable in a classical sense.

Pointwise differentiability properties of Sobolev maps from a modern point of view are discussed for example in Ziemer [686], compare also Stein [611, Chapter 8]. The basic L^p-differentiability theorem was proved in Calderón and Zygmund [131], see also Calderón and Zygmund [132]. The Hölder continuity of $W^{1,p}$-functions, $p > n$, is a classical well-known result due to Morrey [486] [487], while the a.e. differentiability seems to be proved by Cesari [141]. From these results the a.e. approximate differentiability of Sobolev maps follows easily.

3 Given an a.e. approximately differentiable function u, Federer in 1945 proved in Federer [220] [221] that the set of points of approximate differentiability of u can be decomposed as a countable family of measurable sets such that the restriction of u to each member of the family is Lipschitz. More precisely he essentially proved the equivalence of the claims (i) (ii) and (iii) in Theorem 3 in Sec. 3.1.4. In conjunction with Whitney extension theorem this yields that every a.e. approximately differentiable function is of class C^1 except on small open sets, Whitney [673].

This way the theory developed in Sec. 3.1.3, apart from the estimates, could be regarded as a consequence of Theorem 3 in Sec. 3.1.4, that is, as a special case of the rectifiability theory of Federer, via Calderón-Zygmund result. However we preferred to present Lusin type properties for Sobolev functions in the context of Sobolev spaces theory as this provides us also nice estimates.

4 The proofs of Lusin type properties in Sec. 3.1.3 are inspired by Liu [440], in particular Theorem 7 in Sec. 3.1.3 and Theorem 8 in Sec. 3.1.3 are taken from Liu [440], in connection with Theorem 4 in Sec. 3.1.3 the reader may also compare Acerbi and Fusco [3]. Related papers are Michael [470] Goffman [312] Chen and Liu [148].

Lusin type properties for Sobolev functions with higher order derivatives are discussed in Ziemer [686] and in Liu and Tai [442] [443].

The proof of Rademacher theorem we have given in Theorem 3 in Sec. 3.1.3 can be found in Morrey [490], compare also Simon [592]. For the proof of Whitney Theorem, Theorem 5 in Sec. 3.1.3, we refer e.g. to Whitney [672], Stein [611] Ziemer [686].

Sec. 3.1.4 closely follows Federer [226, chap. 3.1], compare in particular Federer [226, 3.1.8, 3.1.16, 3.1.17].

Some extensions of the theory to rectifiable sets of higher order of differentiability can be found in Anzellotti and Serapioni [50].

5 Since long time and even nowadays the concept of area, as well as representation formulas, area formulas or the change of variables formula, have been of central importance in many questions. A recent historical analysis of the different contributions seems to be missing, though it would surely be of interest. To quote only a few books we refer the interested reader to Saks [570] Radó [545] Radó and Reichelderfer [546] Cesari [142]. Eventually the most reasonable approach to *area formulas* was formulated by Federer [220], in terms of rectifiable sets. Our presentation in Sec. 3.1.5 follows in fact Federer's ideas. The reader may find interesting consulting also e.g. Besicovitch [84] Besicovitch [85], Goffman and Liu [315] [316], Goffmann and Ziemer [321], Nishiura and Breckenridge [514], Nöbeling [521], Liu [441], Malý and Martio [450].

Motivated by Federer's approach, the notion of graph, or 1–graph, of an a.e. approximately differentiable map in Sec. 3.1.5 is modeled after a similar notion and first appeared in Giaquinta, Modica, and Souček [279], see also Liu [439]. Theorem 2 in Sec. 3.1.5 is taken from Radó and Reichelderfer [546], while we could not trace back Theorem 4 in Sec. 3.1.5. It appears in Marcus and Mizel [458] Rickman [555].

For Remark 3 in Sec. 3.1.5 compare Reshetnyak [554] and also Morrey [490, p. 352] and Besicovitch [85].

6 The classes $\mathcal{A}^p(\Omega, \mathbb{R}^N)$ were essentially introduced in Giaquinta, Modica, and Souček [279], where however it was assumed that their elements were Sobolev functions. The relationship between the old and new definition given here is formulated in Proposition 1 in Sec. 3.2.3. The present definition in terms of a.e. approximately differentiable maps makes clear the independence of the geometric property of rectifiability of graphs from the vanishing of their boundary in $\Omega \times \mathbb{R}^N$.

Apart from the considerations of Sec. 3.2.4, the results of Sec. 3.2 are a reworking of similar results appeared in Giaquinta, Modica, and Souček [280] [279].

The classes $\mathcal{A}_{p,q}(\Omega, \mathbb{R}^n)$ in Sec. 3.2.4 were introduced by Sverák [618] in connection with nonlinear elasticity. Still in connection with nonlinear elasticity, the notion of distributional determinant was introduced by Ball [63]. The div–lemma, Proposition 2 in Sec. 3.2.4 is due to Müller [500]. It is proved in Müller [503] that the singular part of the distributional determinant may concentrate on a set of prescribed Hausdorff dimension α, $0 < \alpha < n$, compare [9] in Sec. 4.2.5. One could also introduce a notion of distributional minors, in particular of distributional adjoint, Adj Du, but we have not pursue this point. For further information about this subject we refer to Ball [63] and Dacorogna [166].

The results concerning the isoperimetric inequality and higher integrability of the determinant are taken from Müller [501]. There one can also find examples showing the optimality of that theorem. The class $\mathcal{A}_{n-1,n/(n-1)}$ has been introduced in Müller, Tang, and Yan [504], where Theorem 1 in Sec. 3.2.4 is proved.

Results in [5] in Sec. 3.2.4 are taken from Coifman, Lions, Meyer, and Semmes [156] [155] where general results connected to the theory of compensated compactness, compare Murat [506] [508] [507] Tartar [626] [627], are developed. A direct proof of Proposition 6 in Sec. 3.2.4, which does not use Fefferman-Stein's duality theorem, can be found in Chanillo [146] Chanillo and Li [147], compare also Wente [658]. Finally, for the proof of Fefferman-Stein duality theorem we refer the reader to Fefferman [232] Fefferman and Stein [233], compare also Torchinsky [634]. For a primer on Hardy spaces the reader may consult Semmes [583].

7 The results in Sec. 3.2.5 are partly new, and suggested by previous similar results in Sverák [618] and Müller, Tang, and Yan [504]. We have only slightly touched the problem of characterizing currents which are boundaries of i.m. rectifiable currents in $\text{cart}^1(\Omega, \mathbb{R}^N)$. Two problems might deserve further studying. First characterize maps in cart^1 such that $\partial G_u = G_{u,\partial\Omega}$, and, secondly, characterize the boundary space of $\text{cart}^1(\Omega, \mathbb{R}^N)$.

8 After the results of Reshetnyak [550] [551], Theorem 1 in Sec. 3.3.1, Theorem 2 in Sec. 3.3.1 and Theorem 3 in Sec. 3.3.1 on the weak continuity of determinants, a new flowishing interest in the weak convergence of minors was motivated by the work of Ball [63] in nonlinear elasticity and Tartar [626] on the so–called compensated-compactness methods. Other motivations were provided by the interest in studying variational problems for functionals with integrands depending on minors, compare e.g. Dacorogna [166] [164].

The relevance of the vanishing of the boundaries $\partial G_{u_k} \llcorner \Omega \times \mathbb{R}^N$ in studying the weak continuity of minors was pointed out in Giaquinta, Modica, and Souček [280] [279] where the classes of Cartesian maps were introduced. In Giaquinta, Modica, and Souček [280], see also [279] we also proved the closure theorem for Cartesian maps relying on Federer-Fleming closure theorem. Later Müller [499] pointed out that a simpler proof, which in fact reminds of part of the proof of the rectifiability theorem, compare with Proposition 2 in Sec. 3.3.2 and Theorem 4 in Sec. 2.2.7, could be given directly.

9 The examples in [1] in Sec. 3.3.1 [3] in Sec. 3.3.1, due to Ball and Murat, and the example in [2] in Sec. 3.3.1, due to Tartar, are taken from Ball and Murat [71]. The example in [4] in Sec. 3.3.1 is taken from Malý [448], compare also Besicovitch [84] Nöbeling [521] Goffman [312]. Modifications of this example provide sequences of smooth diffeomorphisms $\{u_k\}$ which converge weakly in $W^{1,p}(Q)$ for $p \le n-1$ to the identity but

$$\lim_{k\to\infty} \int_Q \det Du_k \, dx \; < \; \int_Q dx \, .$$

More generally, it is proved in Hajlasz [337] that for each u in $\mathcal{A}_{p,q}$ with $n-1 \leq p < n$ but $q < n/(n-1)$ there exists a sequence $\{u_\nu\} \subset \mathcal{A}_{p,q}$ such that each u_ν has range in a $(n-1)$–dimensional simplicial set and u_ν converge weakly in $\mathcal{A}_{p,q}$ to u; in particular $\det Du_\nu = 0$. Proposition 1 in Sec. 3.3.1 and Proposition 2 in Sec. 3.3.1 were proved by Ball [63], see also Ball, Currie, and Oliver [70]. Dacorogna and Murat showed in [170] that Proposition 2 in Sec. 3.3.1 is optimal.

Finally we mention that *biting* type convergence of minors is studied in Zhang [684].

10 The classes $\mathrm{Cart}^p(\Omega, \mathbb{R}^N)$ were introduced in Giaquinta, Modica, and Souček [280], see also [279]. The content of Sec. 3.4.1 is classical, the reader may consult e.g. Banach [73] Natanson [512]. The results of Sec. 3.4.2 are slight improvements of results due to Malý [447]. Finally the content of Sec. 3.4.3 is new, it will be relevant in studying the composition of weak diffeomorphisms in Vol. II Ch. 2.

4. Cartesian Currents in Euclidean Spaces

Let $\{u_k\} \subset C^1(\Omega, \mathbb{R}^N)$ be a sequence of regular maps from a bounded domain Ω into $\mathbb{R}^N$ satisfying

$$(1) \qquad \sup_k \{\| u_k \|_{L^p(\Omega,\mathbb{R}^N)} + \| M(Du_k) \|_{L^p(\Omega,\Lambda_n\mathbb{R}^{n+N})}\} < \infty$$

for some $p > 1$. We saw in Ch. 3 that there exists a subsequence, that for the sake of simplicity we call again $\{u_k\}$, and a map $u \in \mathrm{cart}^p(\Omega, \mathbb{R}^N)$ such that

$$(2) \qquad \begin{array}{lll} u_k \rightharpoonup u & \text{weakly in} & W^{1,p}(\Omega, \mathbb{R}^N) \\ M^\beta_{\bar{\alpha}}(Du_k) \rightharpoonup M^\beta_{\bar{\alpha}}(Du) & \text{weakly in} & L^p(\Omega) \end{array}$$

for all α, β with $|\alpha| + |\beta| = n$, $\beta \geq 2$. Furthermore, we saw that (2) implies the weak convergence in $\mathcal{D}_n(\Omega \times \mathbb{R}^N)$ of the currents G_{u_k} to the n-current G_u and also the equiboundedness condition (1), i.e.

$$(3) \qquad \begin{array}{l} G_{u_k} \rightharpoonup G_u \qquad \text{in } \mathcal{D}_n(\Omega \times \mathbb{R}^N) \\ \sup_k \{\| u_k \|_{L^p(\Omega,\mathbb{R}^N)} + \| M(Du_k) \|_{L^p(\Omega,\Lambda_n\mathbb{R}^{n+N})}\} < \infty\,. \end{array}$$

Actually, we saw that (2) and (3) are equivalent.

All those nice properties are lost in general if we replace the bound (1) by the weaker bound in L^1

$$(4) \qquad \sup_k \{\| u_k \|_{L^1(\Omega,\mathbb{R}^N)} + \| M(Du_k) \|_{L^1(\Omega,\Lambda_n\mathbb{R}^{n+N})}\} < \infty\,.$$

The lack of weak compactness of bounded sets in L^1 forces us to embed L^1 into the space of measures with bounded total variation and to replace, for instance, $W^{1,1}(\Omega, \mathbb{R}^N)$ with the space of functions with bounded total variation $BV(\Omega, \mathbb{R}^N)$. Therefore from (4) we can only infer the existence of a subsequence of $\{u_k\}$, that we again call $\{u_k\}$, of a function $u \in BV(\Omega, \mathbb{R}^N)$ and of measures $\mu^{\alpha\beta}$, $|\alpha| + |\beta| = n$, $\beta \geq 2$ such that

$$(5) \qquad \begin{array}{lll} u_k \rightharpoonup u & \text{weakly in} & BV(\Omega, \mathbb{R}^N) \\ M^\beta_{\bar{\alpha}}(Du_k) \rightharpoonup \mu^{\alpha\beta} & \text{as measures.} & \end{array}$$

Since

$$\mathbf{M}(G_{u_k}) = \int_\Omega |M(Du_k)|\,dx\ ,$$

the n-currents G_{u_k} are equibounded in mass; again passing to subsequences, there exists a current T in $\mathcal{D}_n(\Omega \times \mathbb{R}^N)$ with finite mass such that

$$G_{u_k} \rightharpoonup T \qquad \text{in } \mathcal{D}_n(\Omega \times \mathbb{R}^N)\ . \tag{6}$$

But in general $u, \mu^{\alpha\beta}$, and T are unrelated in the sense that we cannot recover $\mu^{\alpha\beta}$ from u and T from $\mu^{\alpha\beta}$ and u, as shown by the following examples, compare also Sec. 4.2.5.

[1] *Bubbling of circles.* Consider the sequence of C^1-maps u_k from the interval $(-1,1)$ into $S^1 \subset \mathbb{R}^2$ defined by

$$u_k(\theta) := \begin{cases} (\cos k\theta, \sin k\theta) & \theta \in [0, 2\pi/k] \\ (1,0) & \theta \in (-1,1) \setminus [0, 2\pi/k]\ . \end{cases}$$

Clearly,

$$\sup_k \int_{-1}^{1} (|u_k| + |\dot{u}_k|)\,d\theta = \sup_k 2\,(1 + \frac{\pi}{k}) < \infty$$

$$u_k \rightharpoonup u_\infty \equiv (1,0) \qquad \text{weakly in } BV((-1,1), \mathbb{R}^2)\ ,$$

that is

$$u_k \to (1,0) \qquad \text{strongly in } L^1((-1,1),\mathbb{R}^2)$$
$$\dot{u}_k \rightharpoonup (0,0) \qquad \text{as measures.}$$

But the 1-dimensional currents G_{u_k}, which are just integration of 1-forms over the curves $\theta \to (\theta, u_k(\theta))$ in $\mathbb{R}^3$, clearly converge to the current integration over the graph of the constant map u_∞ plus integration over the circle S^1 which lives above $\theta = 0$

$$G_{u_k} \rightharpoonup G_{u_\infty} + \delta_0 \times [\![S^1]\!]\ .$$

In this case, the graph of the limit map u_∞ is only a part of the whole limiting current. Note also that, if we regard the u_k as maps from S^1 into S^1, the limit map u_∞ has degree zero, though all the u_k have degree one, while the limit graph T has "degree one". Also the length of the curves $(\theta, u_k(\theta))$ tends to $2+2\pi$ which is exactly the mass of T, while the length of the curve $(\theta, u_\infty(\theta))$ is 2. •

The previous example shows clearly that the weak limit of G_{u_k} is not the graph of the map u_∞, nor of any other map, and cannot be recovered from the map u_∞. While in codimension one the vertical part of the "graph" of a BV-function u is uniquely determined by its pointwise values, compare Sec. 4.1, in the vector valued case we are forced to regard weak limits of G_{u_k} as currents in $\Omega \times \mathbb{R}^N$.

In Sec. 4.2.1 we shall see that limit currents T of sequences of smooth maps satisfying (4) belong to the class of *Cartesian currents* defined as

$$\begin{aligned}\operatorname{cart}(\Omega \times \mathbb{R}^N) := \{ & T \in \mathcal{D}_n(\Omega \times \mathbb{R}^N) \mid T \text{ is an i.m. rectifiable current} \\ & \text{in } \Omega \times \mathbb{R}^N,\ \mathbf{M}(T) < \infty,\ \|T\|_1 < \infty, \\ & \partial T \llcorner \Omega \times \mathbb{R}^N = 0,\ \pi_\# T = [\![\Omega]\!],\ T^{\bar{0}0} \geq 0 \}\end{aligned}$$

Such a class plays a relevant role in the study of variational problems for vector valued maps as we shall see later. Its properties will be discussed in Sec. 4.2 where, in particular, we prove a *closure* and *compactness theorem.* There we shall also prove a *structure theorem* for Cartesian currents T, which states that each T has the form

$$T = G_{u_T} + S$$

where G_{u_T} is the current associated to the 1-graph of a function u_T in $BV(\Omega, \mathbb{R}^N)$ and S is a *vertical current*, i.e., $\pi_\# S = 0$.

Rightly the class $\operatorname{cart}(\Omega \times \mathbb{R}^N)$ may be regarded as the counterpart of the classes $\operatorname{cart}^p(\Omega \times \mathbb{R}^N)$, $p > 1$, similarly to as $BV(\Omega)$ is the counterpart of Sobolev spaces $W^{1,p}(\Omega)$, $p > 1$, in the scalar case. However in the vector valued case $\operatorname{cart}(\Omega \times \mathbb{R}^N)$ turns out to be the *natural* substitute even of $W^{1,p}(\Omega, \mathbb{R}^N)$, at least for integer values of $p \leq \min(n, N)$, as we shall see in the sequel of this monograph and as the following example already shows.

[2] *Bubbling of spheres.* Let $\sigma : S^2 \subset \mathbb{R}^3 \to \mathbb{R}^2$ denote the stereographic projection from the south pole $P_S := (0, 0, -1)$ into $\mathbb{R}^2$. Its inverse $\sigma^{-1} : \mathbb{R}^2 \to S^2$ being conformal satisfies

$$\frac{1}{2} |D\sigma^{-1}|^2 = |M_{(2)}(D\sigma^{-1})|$$

hence we have

$$\frac{1}{2} \int_{\mathbb{R}^2} |D\sigma^{-1}|^2 \, dx = \int_{\mathbb{R}^2} |M_{(2)}(D\sigma^{-1})| = \mathcal{H}^2(S^2) = 4\pi .$$

Consider now the sequence of maps

$$u_k(x) := \sigma^{-1}(kx) \qquad x \in \mathbb{R}^2, \quad k = 1, 2, \ldots .$$

Then clearly

$$\frac{1}{2} \int_{\mathbb{R}^2} |Du_k|^2 \, dx = 4\pi \qquad \forall\, k$$

and

$$\begin{aligned} u_k(x) &\to u_\infty(x) := \text{south pole} && \text{pointwisely in } \mathbb{R}^2 \\ u_k &\rightharpoonup u_\infty && \text{weakly in } W^{1,2} . \end{aligned}$$

On the other hand, from the isoperimetric inequality for parallelograms we have

$$|M_{(2)}(Du_k)| \leq \frac{1}{2}|Du_k|^2 ,$$

whence we infer

$$\sup_k\{ \| u_k \|_{L^1(\Omega,\mathbb{R}^2)} + \| M(Du_k) \|_{L^1(\Omega,\mathbb{R}^2)} \} < \infty , \quad \Omega \subset\subset \mathbb{R}^2 .$$

Notice that we have no better control on the two by two minors. Consequently the currents G_{u_k} have locally equibounded masses, and it is not difficult to show that

$$G_{u_k} \rightharpoonup G_{u_\infty} + \delta_0 \times [\![S^2]\!] \qquad \text{in } \mathcal{D}_2(\mathbb{R}^2 \times \mathbb{R}^3) ,$$

compare Sec. 4.2.5.

Regarding the maps u_k as maps from S^2 into S^2, $u_k(x) := \sigma^{-1}(k\sigma(x))$, similarly to $\boxed{1}$, all maps u_k have Brower degree one while u_∞ has degree zero. Though all u_k have Dirichlet energy equal to 4π, the Dirichlet energy of u_∞ is zero. On the contrary the current $G_{u_\infty} + \delta_0 \times [\![S^2]\!]$ has "degree one".

We therefore see that maps with equibounded *Dirichlet integrals* provide examples of the same nature as in $\boxed{1}$, and their limits are better described in the context of Cartesian currents than in the Sobolev spaces $W^{1,2}(\mathbb{R}^2, \mathbb{R}^3)$ or $W^{1,2}(S^2, S^2)$. •

Due to the relevance of functions of bounded variations in dealing with Cartesian currents, we shall begin presenting in Sec. 4.1 the theory of BV-space. After a few simple but important functional results, as for instance the *approximation theorem*, we shall mostly deal with structure property of BV-functions. In particular we shall present *De Giorgi's rectifiability theorem* for *Caccioppoli sets*, i.e., sets whose characteristic function are in BV, in Sec. 4.1.3, and, based on it, a *structure theorem* for BV-functions in Sec. 4.1.4, describing the pointwise behaviour of BV-functions. Sec. 4.1.5 provides a link with Federer's approach to BV-functions in terms of boundaries of subgraphs; while in Sec. 4.2.4 we shall see that in codimension one BV-functions can be identified to Cartesian currents. Typical pointwise properties are best visualized by means of the classical Cantor-Vitali function that we shall discuss in Sec. 4.2.4.

Sec. 4.2 is dedicated to the study of *Cartesian currents.* There we prove *closure, compactness*, and *structure theorems*, and we present a few examples in Sec. 4.2.5. Cartesian currents in codimension one are discussed in Sec. 4.2.4, and, finally, in Sec. 4.2.6 we discuss *radial functions* and *radial currents.*

In Sec. 4.3, after discussing n-dimensional currents in $\mathbb{R}^n$ and in particular proving the so-called *constancy theorem*, we shall develop, on the basis of those results, the *degree theory* for currents, and particularly for Cartesian currents. In Sec. 4.3.2 we shall in fact see that currents and BV-functions provide a natural setting for degree theory. We shall then see that the degree is continuous with respect to the weak convergence of currents. Contrary to weak convergence in

Sobolev spaces, in particular we shall see that the weak convergence of Cartesian currents preserves the degree. Further results on degree will be presented in Sec. 4.3.3 and Sec. 4.3.4.

1 Functions of Bounded Variation

In this section we shall present some of the main results of the theory of *functions of bounded variation* in dimension larger than one. The class of such functions, denoted $BV(\Omega)$, is defined as the family of functions $u : \Omega \subset \mathbb{R}^n \to \mathbb{R}$ with distributional derivatives which are measures of finite total variation in Ω.

We shall see that BV-functions arise naturally as weak limits of smooth functions with equibounded L^1-norms and L^1-norms of the gradients. Moreover we shall see that every function $u \in BV(\Omega)$ can be approximated "strongly" by smooth functions in the sense that for every $u \in BV(\Omega)$ there is a sequence of smooth functions u_k in Ω such that

$$u_k \longrightarrow u \qquad \text{in } L^1(\Omega)$$

$$\int_\Omega |Du_k(x)|\, dx \longrightarrow \int_\Omega |Du|$$

where

$$\int_\Omega |Du|$$

denotes the total variation of the vector valued measure Du in Ω. An immediate consequence of this result is the extension of *Sobolev* and *Poincaré type inequalities* , as well as of compactness properties, to BV-functions. In particular one sees that bounded sets in $BV(\Omega)$ with respect to the natural energy or norm

$$\int_\Omega |u|\, dx + \int_\Omega |Du|$$

are relatively compact in $BV(\Omega)$. This makes $BV(\Omega)$ the natural space to work with, instead of $W^{1,1}$, when dealing with energies with *linear growth at infinity* as for instance the area functional

$$\int_\Omega \sqrt{1 + |Du|^2}\, dx$$

for nonparametric surfaces $u : \Omega \to \mathbb{R}$

Those as well as other basic properties of BV-functions will be discussed in Sec. 4.1.1. There we shall also prove the *coarea formula* for BV-functions.

An important subclass of $BV(\Omega)$ is that of characteristic functions χ_E of sets E for which $D\chi_E$ is a measure in Ω. Sets E for which $\chi_E \in BV_{\text{loc}}(\mathbb{R}^n)$ are called

Caccioppoli sets and the total variation of $D\chi_E$ in Ω, $|D\chi_E|(\Omega)$, is called the *perimeter of* E *in* Ω and denoted by $P(E,\Omega)$. Caccioppoli sets were introduced in the fifties by Caccioppoli and named after him by De Giorgi, who essentially developed a full theory, which we shall present in Sec. 4.1.2 and Sec. 4.1.3. Their relevance is mainly, though not only, in connection with the problem of defining a notion of generalized $(n-1)$-surface and of area suited for the purposes of calculus of variation, i.e., in order to find surfaces of minimal area, or, in other words, in order to identify limits of sequences of smooth $(n-1)$-parametric surfaces in Ω with equibounded areas.

The idea is to regard oriented $(n-1)$-surfaces in $\mathbb{R}^n$ as boundaries of n-dimensional sets E. The naive idea would then be to take as area of ∂E, or perimeter of E, just the $\mathcal{H}^{n-1}$-measure of ∂E. However one immediately realizes that one can make the topological boundary ∂E of E as large as one wants, by just changing E by a set of measure zero. In other words the topological boundary is not the right object in order to identify surfaces of finite area. Caccioppoli sets, and the perimeter, instead do it, as they are defined by duality and therefore depend only on the equivalence class of E. Clearly, identifying the $(n-1)$-surface "boundary of E" together with its "differential" properties is related to the possibility of extending the classical Gauss-Green formula

$$(1) \qquad -\int_E \operatorname{div} g\, dx = \int_{\partial E} (g,\nu)\, d\mathcal{H}^{n-1} \quad \forall\, g \in C^1_c(\Omega,\mathbb{R}^n)$$

to Caccioppoli sets, and more precisely to the question of identifying an $(n-1)$-rectifiable set "∂E" together with an "inward normal" ν to E so that (1) holds.

From the definition of Caccioppoli set we have

$$-\int_E D_i g\, dx = \int g\, d\, D_i\chi_E \qquad \forall\, g \in C^1_c(\Omega)\,.$$

Defining the *reduced boundary of* E, $\partial^- E$, in the sense of De Giorgi, as the set of points x at which the Radon-Nikodym derivative of $D\chi_E$ with respect to $|D\chi_E|$ exists

$$\frac{d\, D\chi_E}{d\,|D\chi_E|} := n(x,E) \quad \text{and} \quad |n(x,E)| = 1$$

we shall see that $\partial^- E$ is locally *an* $(n-1)$*-rectifiable set with tangent plane normal to* $n(x,E)$ *for* $\mathcal{H}^{n-1}$-a.e. $x \in \partial^- E$ *and*

$$D\chi_E = n(x,E)\, \mathcal{H}^{n-1} \llcorner \partial^- E\,.$$

This is the celebrated *De Giorgi's rectifiability theorem for Caccioppoli sets.* And now, generalized $(n-1)$-surfaces are identified as the reduced boundaries of Caccioppoli sets, and their perimeters as the $\mathcal{H}^{n-1}$-measure of the reduced boundaries. Existence of minimal surface of dimension $n-1$ in $\mathbb{R}^n$, under for instance prescribed boundary value, clearly becomes a trivial fact, taking into account the semicontinuity of the total variation, i.e., of the perimeter, and the compactness property of the class of Caccioppoli sets.

We shall see in Sec. 4.1.3 that the reduced boundary $\partial^- E$ agrees apart from a set of zero $(n-1)$-dimensional measure with the so called *measure-theoretic boundary* of E, $\partial_\mu E$, which consists of the points which are neither of density zero for E (i.e., of *rarefaction* for E), nor of density one for E (i.e., of *rarefaction for the complement* of E). Later in Sec. 4.1.4 we shall then show that in fact

$$\partial^- E = \partial_\mu E = \partial^* E \qquad \mathcal{H}^{n-1}\text{-a.e.}$$

where $\partial^* E$ is the *set of jump points* , defined again in a measure-theoretic way, for χ_E with jump equal to one.

In Sec. 4.1.4 we turn back to general BV-functions proving a fine structure theorem which states that, accordingly to our simple one-dimensional idea, we can decompose every BV-function in a part which is almost everywhere approximately differentiable, in a jump part which lives on a locally $(n-1)$-rectifiable set, and in a Cantor part.

Finally, in Sec. 4.1.5 we shall prove that a measurable function is in $BV(\Omega)$ if and only if its subgraph is a set of finite perimeter in $\Omega \times \mathbb{R}$. This way we shall link De Giorgi's approach with Federer's approach to BV-functions. Such a result will be used in Sec. 4.2.4 when proving that *the class of Cartesian currents in codimension one agrees with the class of BV-functions* .

It has to be mentioned that, contemporary to De Giorgi's theory of Caccioppoli sets, related measure-theoretic ideas were being developed by Federer and by Reifenberg in order to deal with general k-dimensional surfaces in $\mathbb{R}^n$. Those ideas eventually led to the theory of integer multiplicity rectifiable currents and varifolds.

In this section we address our attention mainly to structure properties of BV-functions. We shall report on other results, without proof, in the notes of Sec. 4.4.

1.1 The Space $BV(\Omega, \mathbb{R})$

Let Ω be an open subset of $\mathbb{R}^n$, $n \geq 1$.

Definition 1. *A function $u \in L^1_{\text{loc}}(\Omega)$ is said to be a* function of locally bounded variation, *$u \in BV_{\text{loc}}(\Omega)$, if its distributional derivatives, $D_i u$, $i = 1, \dots, n$, are signed Radon measures.*

In other words there exists a vector valued measure Du in $\mathcal{M}(\Omega, \mathbb{R}^n)$, $Du = (D_1 u, \dots, D_n u)$, such that

$$\int_\Omega < g, d\,Du > \;=\; -\int_\Omega u\, div g\, dx \tag{1}$$

for any vector field $g \in C^1_c(\Omega, \mathbb{R}^n)$. Here $\operatorname{div} g = \sum_{i=1}^n D_i g^i$, *as usual.*

We say that a function u has *bounded variation in Ω*, if its total variation $|Du|$ is finite in Ω, $|Du|(\Omega) < +\infty$.

We shall also write instead of $|Du|(\Omega)$,

$$\int_\Omega d\,|Du|$$

and even with some abuse of notation

$$\int_\Omega |Du| \,.$$

Definition 2. *The space $BV(\Omega)$ is defined as the space of all functions in $L^1(\Omega)$ with bounded variation.*

Obviously every function $u \in C^1(\Omega)$ belongs to $BV_{\mathrm{loc}}(\Omega)$. In fact, integrating by parts, we find

$$-\int_\Omega u \operatorname{div} g\, dx \;=\; \int_\Omega \sum_{i=1}^n D_i u\, g^i\, dx \,,$$

i.e.,

$$Du \;=\; d\mathcal{L}^n \llcorner Du(x) \,.$$

Similarly we have $W^{1,1}_{\mathrm{loc}}(\Omega) \subset BV_{\mathrm{loc}}(\Omega)$. Notice that actually

$$W^{1,1}(\Omega) \underset{\neq}{\subset} BV(\Omega)$$

as shown by Heaviside function in $(-1,1)$ or Cantor-Vitali function, compare Sec. 4.2.4.

As we saw in Sec. 1.1.4, by Riesz's theorem, the total variation of Du, $|Du|$ can be computed as the mass of the vector valued measure Du. If $\mathcal{U} \subset \Omega$ is an open set, then, compare (7) in Sec. 1.1.4,

$$(2) \qquad |Du|(\mathcal{U}) \;:=\; \sup\Big\{ \int_\Omega u \operatorname{div} g\, dx \;\mid\; g \in C^1_c(\mathcal{U}, \mathbb{R}^n),\; \|\, g\,\|_\infty \le 1 \Big\} \,.$$

Therefore we can state equivalently

$$\begin{aligned} BV_{\mathrm{loc}}(\Omega) &= \{u \in L^1_{\mathrm{loc}}(\Omega) \mid \mathbf{M}_{\mathcal{U}}(Du) < \infty \ \forall\, \mathcal{U} \text{ open } \mathcal{U} \subset\subset \Omega\} \\ BV(\Omega) &= \{u \in L^1(\Omega) \mid \mathbf{M}_{\Omega}(Du) < \infty\} \,. \end{aligned}$$

From (2) one easily infer

Proposition 1 (Semicontinuity). *Let $\{u_k\}$ be a sequence of L^1 functions in Ω which converges weakly in L^1 to u, $u_k \rightharpoonup u$ in L^1. Then*

$$|Du|(\Omega) \;\le\; \liminf_{k\to\infty} |Du_k|(\Omega) \,.$$

In particular $u \in BV(\Omega)$, if all u_k belong to $BV(\Omega)$ and have equibounded total variations.

The next proposition is just a rewriting of general properties of Radon measures, compare Sec. 1.2.1.

Proposition 2. *Let u_k, $k = 1, 2, \dots$, u be functions in $BV(\Omega)$.*

(i) If $u_k \rightharpoonup u$ as distributions, then $Du_k \rightharpoonup Du$ as measures if and only if

$$\sup_k |Du_k|(\mathcal{U}) \le c(\mathcal{U}) < \infty, \qquad \text{for all } \mathcal{U} \subset\subset \Omega$$

(ii) If $Du_k \rightharpoonup Du$ as measures and $\lim_{k\to\infty} |Du_k|(\Omega) = |Du|(\Omega)$, the following equivalent three claims hold
  - *a) $|Du_k| \rightharpoonup |Du|$ as measures,*
  - *b) $|Du_k|(B) \rightharpoonup |Du|(B)$ for any Borel bounded set B in Ω such that $|Du|(\partial B) = 0$,*
  - *c) for any open set $A \subset \Omega$ we have $|Du|(A) \le \liminf_{k\to\infty} |Du_k|(A)$ and for any compact set $F \subset \Omega$ we have $|Du|(F) \ge \limsup_{k\to\infty} |Du_k|(F)$.*

In particular it follows from (b) that $|Du_k|(B(x,r)) \to |Du|(B(x,r))$ for all $x \in \Omega$ and almost every r for which $B(x,r) \subset \Omega$.

It is easy to show that $BV(\Omega)$ equipped with the norm

$$\|u\|_{BV(\Omega)} := \|u\|_{L^1(\Omega)} + |Du|(\Omega) \tag{3}$$

is a Banach space. However such a structure is of no real use since $C^1(\Omega) \cap BV(\Omega)$ *is not dense in $BV(\Omega)$ and $BV(\Omega)$ is not separable.*

[1] By considering the family $\mathcal{F}$ of characteristic functions χ_α of the intervals $(\alpha, 1)$, $\alpha \in (0,1)$, we see that $\mathcal{F} \subset BV(0,1)$ and for $\alpha \neq \beta$ we have

$$\|\chi_\alpha - \chi_\beta\|_{BV(\Omega)} = 2 + |\alpha - \beta| .$$

This of course shows that $BV(\Omega)$ is not separable. Since the closure of $C^1(\Omega) \cap BV(\Omega)$ with respect to the norm (3) is just the Sobolev space $W^{1,1}(\Omega)$ and $W^{1,1}(\Omega) \underset{\neq}{\subset} BV(\Omega)$, we see that $C^1(\Omega) \cap BV$ is not dense in BV. •

It is instead more convenient to regard $BV(\Omega)$ as subspace of L^1, as we did in the semicontinuity theorem, or endowed with the weak*-convergence or the convergence of measures which is induced from the immersion of $BV(\Omega)$ into the space of vector valued Radon measures $\mathcal{M}(\Omega, \mathbb{R}^{n+1})$ by the mapping $u \to (u, Du)$.

Quite a number of useful properties of BV-functions are consequences of the following approximation theorem.

Theorem 1 (Approximation by smooth functions). *We have*

(i) Let $u \in BV_{\mathrm{loc}}(\Omega)$. Then there exists a sequence $\{u_k\} \subset C_c^\infty(\Omega)$ such that

$$\begin{aligned} u_k &\to u \qquad \text{in } L^1_{\mathrm{loc}}(\Omega) \\ d\mathcal{L}^n \llcorner |Du_k| &\rightharpoonup |Du| \quad \text{as measures in } \Omega . \end{aligned}$$

(ii) Let $u \in BV(\Omega)$. Then there exists a sequence $\{v_k\} \subset C^\infty(\Omega)$ such that

$$v_k \to u \qquad \text{in } L^1(\Omega) \qquad \text{and} \qquad \int_\Omega |Dv_k|\, dx \to |Du|(\Omega) .$$

Moreover, if Ω is unbounded, we can choose the v_k's to be zero near infinity.

Proof. (i) Let ρ be any symmetric mollifier. For $\sigma > 0$ and $u \in L^1_{\text{loc}}$ set $\rho_\sigma(x) := \sigma^{-n}\rho(x/\sigma)$ and $u_\sigma := \varphi_\sigma * \tilde{u}$, where $\tilde{u} := u$ in Ω_σ, $\tilde{u} := 0$ outside Ω_σ and

$$\Omega_\sigma := \{x \in \Omega \mid \operatorname{dist}(x, \partial\Omega) > \sigma\} \cap B(0, 1/\sigma) .$$

Then obviously $u_\sigma \to u$ in $L^1_{\text{loc}}(\Omega)$ as $\sigma \to 0$ and it suffices to show that

$$\int_\Omega \varphi(x)\, |Du_\sigma(x)|\, dx \longrightarrow \int_\Omega \varphi(x)\, d|Du|$$

for any $\varphi \in C^\infty_c(\Omega)$, $\varphi \geq 0$.

As

$$\int \varphi(x)\, d|Du| = \sup\Big\{ \int u \operatorname{div} g \mid g \in C^1_c(\Omega),\ |g(x)| \leq \varphi(x)\Big\}$$

we infer at once

$$\int \varphi(x)\, d|Du| \leq \liminf_{\sigma \to 0} \int \varphi(x)\, |Du_\sigma(x)|\, dx .$$

Thus we only need to show that

$$\limsup_{\sigma \to 0} \int \varphi(x)\, |Du_\sigma(x)|\, dx \leq \int \varphi(x)\, d|Du| . \tag{4}$$

For $g \in C^1_c(\Omega)$ with $|g(x)| \leq \varphi(x)$ and $\sigma < \operatorname{dist}(\operatorname{spt} g, \partial\Omega)$ we have

$$\begin{aligned} \int < g(x), Du_\sigma(x) > dx &= -\int u_\sigma(x) \operatorname{div} g(x)\, dv = -\int (\rho_\sigma * u) \operatorname{div} g\, dx \\ &= -\int u\, (\rho_\sigma * \operatorname{div} g)\, dx = -\int u \operatorname{div}(\rho_\sigma * g)\, dx . \end{aligned} \tag{5}$$

For any $\varepsilon > 0$, since

$$|\rho_\sigma * g(x)| \leq \rho_\sigma * |g(x)| \leq \varphi(x) + \varepsilon$$

if we take σ close enough to zero, we can estimate the right hand-side of (5) by

$$\int_{W_\sigma} (\varphi + \varepsilon)\, |Du|$$

where

$$W_\sigma := \{x \in \Omega \mid \operatorname{dist}(x, \operatorname{spt}\varphi) < \sigma\} ,$$

getting this way

$$\int < g(x), Du_\sigma(x) > dx \le \int_{W_\sigma} (\varphi + \varepsilon)\, |Du| .$$

Taking now the supremum in g and letting $\varepsilon \to 0$ we get (4).

(ii) We still use a mollification procedure but stepping down the size of the mollification when going to the boundary. More precisely, let $\{\Omega_k\}_{k\in\mathbb{N}}$ be an increasing sequence of open sets for which $\Omega_0 \subset\subset \Omega_k \subset\subset \Omega_{k+1}$, $\cup_k \Omega_k = \Omega$ and define $A_1 := \Omega_1$, $A_k := \Omega_k \setminus \overline{\Omega}_{k-2}$ for $k \ge 2$. The open sets A_k form a locally finite covering of Ω so that there is a partition of unity $\{\varphi_k\}$ associated to the A_k's. Choose now a sequence $\{\varepsilon_k\}$ in such a way that

$$\operatorname{spt}(\varphi_k * \rho_{\varepsilon_k}) \subset \Omega_{k+1} \setminus \overline{\Omega}_{k-3} , \qquad \Omega_{-1} = \emptyset$$

where ρ is a standard symmetric mollifier and $\rho_\varepsilon(x) = \varepsilon^{-n}\rho(x/\varepsilon)$. Then the function

$$\tilde{u} := \sum_k (u\varphi_k) * \rho_{\varepsilon_k}$$

is well defined and belongs to $C^\infty(\Omega)$. We shall now prove that for every $\delta > 0$ we can choose $\{\varepsilon_k\}$ in such a way that

$$\text{(6)} \qquad \| u - \tilde{u} \|_{L^1(\Omega)} \le \delta \quad \text{and} \quad \int_\Omega |D\tilde{u}(x)|\, dx \le |Du|(\Omega) + \delta .$$

This of course implies that one can construct a sequence $\{v_k\}$ for which

$$v_k \longrightarrow u \quad \text{in } L^1(\Omega) \quad \text{and} \quad \limsup_{k\to\infty} \int_\Omega |Dv_k(x)|\, dx \le |Du|(\Omega) .$$

The semicontinuity of the total variation with respect to the L^1-convergence then yields

$$|Du|(\Omega) \le \liminf_{k\to\infty} \int_\Omega |Dv_k(x)|\, dx$$

and this completes the proof of (ii).

Let us prove (6). We have

$$\tilde{u} - u = \sum_k ((u\varphi_k) * \rho_{\varepsilon_k} - u\varphi_k)$$

so that

$$\| \tilde{u} - u \|_{L^1(\Omega)} \leq \sum_k \| (u\varphi_k) * \rho_{\varepsilon_k} - u\varphi_k \|_{L^1(\Omega)} \cdot$$

Therefore we get the first claim in (6), if we choose $\{\varepsilon_k\}$ in such a way that

(7) $$\| (u\varphi_k) * \rho_{\varepsilon_k} - (u\varphi_k) \|_{L^1(\Omega)} \leq \delta/2^{k+1} .$$

On the other hand in the sense of distributions we have

$$D\tilde{u} = \sum_k (Du\,\varphi_k) * \rho_{\varepsilon_k} + \sum_k ((u\,D\varphi_k) * \rho_{\varepsilon_k} - u\,D\varphi_k)$$

as $\sum_k D\varphi_k = 0$. Therefore, choosing ε_k in such a way that (7) holds and

$$\| (u\,D\varphi_k) * \rho_{\varepsilon_k} - u\,D\varphi_k \|_{L^1(\Omega)} \leq \delta/2^{k+1} ,$$

we also get

$$\| D\tilde{u} \|_{L^1(\Omega)} \leq \sum_k \| (Du\,\varphi_k) * \rho_{\varepsilon_k} \|_{L^1(\Omega)} + \delta .$$

To conclude the proof of (6) we now observe that for any $g \in C^1_c(\Omega, \mathbb{R}^n)$, $\| g \|_\infty \leq 1$, we have

$$\left| \int_\Omega < (Du\,\varphi_k) * \rho_{\varepsilon_k}, g > dx \right| = \left| \int_\Omega \varphi_k (g * \rho_{\varepsilon_k})\, dDu \right| \leq \int \varphi_k\, d|Du| ,$$

thus

$$\sum_k \| (Du\,\varphi_k) * \rho_{\varepsilon_k} \|_{L^1} \leq \sum_k \int_\Omega \varphi_k\, d|Du| = |Du|(\Omega) .$$

□

Remark 1. Notice that the sequences $\{u_k\}$ and $\{v_k\}$ in Theorem 1 also satisfy

$$d\mathcal{L}^n \llcorner |D_i u_k| \rightharpoonup |D_i u| \qquad i = 1, \dots, n$$

as measures in Ω, and

$$\int_\Omega |D_i v_k|\, dx \longrightarrow |D_i u|(\Omega) .$$

Therefore

$$\begin{aligned} d\mathcal{L}^n \llcorner Du_k &\rightharpoonup Du \\ d\mathcal{L}^n \llcorner Dv_k &\rightharpoonup Du \\ d\mathcal{L}^n \llcorner |D_i v_k| &\rightharpoonup |D_i u| \qquad i = 1 \dots, n \\ d\mathcal{L}^n \llcorner |Dv_k| &\rightharpoonup |Du| \end{aligned}$$

in the sense of measures in Ω, because of Proposition 2.

We shall now discuss a number of interesting and useful results which can be easily obtained by the use of Theorem 1.

[2] *A variational characterization of BV-functions.* An immediate consequence of the semicontinuity and approximation theorems is the following characterization of BV-functions

Proposition 3. *A function u belongs to $BV(\Omega)$ if and only if there exists a sequence $\{u_k\}$ of smooth functions such that*

$$u_k \rightharpoonup u \quad \text{in } L^1(\Omega) \quad \text{with} \quad \sup_k \int_\Omega |Du_k|\,dx < \infty .$$

This claim may be formulated as follows. Consider the total variation functional

$$V(u, \Omega) := \int_\Omega |Du(x)|\,dx$$

with domain the functions $u \in C^1(\Omega)$ with $V(u, \Omega) < \infty$. As we have seen $V(u, \Omega)$ is lower semicontinuous in $C^1(\Omega) \cap \{u \mid V(u, \Omega) < \infty\}$ with respect to the L^1-weak convergence. Thus, following Lebesgue, we can extend it to L^1 as the *highest lower semicontinuous extension*, by setting for any $u \in L^1(\Omega)$

$$V(u, \Omega) := \inf\{\liminf_{k\to\infty} \int_\Omega |Du_k|\,dx \mid u_k \in C^1(\Omega), \\ u_k \to u \text{ weakly in } L^1(\Omega)\}. \tag{8}$$

Clearly for $u \in BV(\Omega)$ we have $|Du|(\Omega) \leq V(u, \Omega)$.

Proposition 3 actually says

Proposition 4. *In $L^1(\Omega)$ we have*

$$V(u, \Omega) = \begin{cases} |Du|(\Omega) & \text{if } u \in BV(\Omega) \\ +\infty & \text{otherwise.} \end{cases} \tag{9}$$

This way $BV(\Omega)$ is characterized as the *proper domain* of $V(u, \Omega)$,

$$BV(\Omega) = \{u \in L^1(\Omega) \mid V(u, \Omega) < \infty\} .$$

•

[3] *Sobolev and Poincaré inequalities.* The approximation theorem is a useful tool to prove in a very short way Sobolev-Poincaré type inequalities and embedding theorems for BV-functions.

For instance, for each $u \in BV(\mathbb{R}^n)$ we can find a sequence $\{u_k\} \subset C_c^\infty(\Omega)$ such that

$$u_k \longrightarrow u \text{ in } L^1 , \qquad u_k \longrightarrow u \text{ a.e.}$$
$$\int_{\mathbb{R}^n} |Du_k|\, dx \longrightarrow |Du|(\mathbb{R}^n) .$$

From Sobolev embedding theorem for smooth maps, we then infer

$$\| u_k \|_{L^{n/(n-1)}} \le c(n) \| Du_k \|_{L^1(\mathbb{R}^n)}$$

and passing to the limit: *For any $u \in BV(\mathbb{R}^n)$*

$$\| u \|_{L^{n/(n-1)}} \le c(n) |Du|(\mathbb{R}^n) . \tag{10}$$

In particular

$$BV(\mathbb{R}^n) \hookrightarrow L^{n/(n-1)}(\mathbb{R}^n)$$

with continuous immersion.

Similarly, for a smooth (Lipschitz) domain Ω in $\mathbb{R}^n$ we get

$$\| u \|_{L^{n/(n-1)}(\Omega)} \le c(n,\Omega) (\| u \|_{L^1(\Omega)} + |Du|(\Omega)) . \tag{11}$$

The same procedure yields

Sobolev-Poincaré inequality . Let $u \in BV(\Omega)$. Set

$$u_{x,\rho} := \fint_{B(x,\rho)} u(y)\, dy , \qquad B(x,\rho) \subset \Omega .$$

Then

$$\Big(\fint_{B(x,\rho)} |u(y) - u_{x,\rho}|^{\frac{n}{n-1}}\, dy \Big)^{1-1/n} \le c(n)\, \rho^{1-n}\, |Du|(B(x,\rho))$$

or

Poincaré inequality. Let $u \in BV(B(x,\rho))$. Then

$$\int_{B(x,\rho)} |u(y) - u_{x,\rho}|\, dy \le c(n)\, \rho\, |Du|(B(x,\rho)) .$$

•

[4] *Compactness in BV*. The approximation theorem in conjunction with Rellich's theorem for Sobolev spaces $W^{1,1}$ or $W^{1,1}_{\text{loc}}$ yields immediately

Proposition 5. *Let $\{u_k\} \subset BV_{\text{loc}}(\Omega)$. Suppose that*

$$\sup_k \| u_k \|_{BV(\mathcal{U})} \le K(\mathcal{U}) < \infty$$

for any open set $\mathcal{U} \subset\subset \Omega$. *Then there exist a function* $u \in BV_{\rm loc}(\Omega)$ *and a subsequence* $\{u_{k_i}\}$ *of* $\{u_k\}$ *such that*

$$\begin{array}{lcll} u_{k_i} & \to & u & \textit{in } L^1_{\rm loc}(\Omega) \\ Du_{k_i} & \rightharpoonup & Du & \textit{as measures in } \Omega \,. \end{array}$$

If moreover $\{u_k\} \subset BV(\Omega)$, Ω *is Lipschitz and* $\sup_k \| u_k \|_{BV(\Omega)} < \infty$, *then there exist* $u \in BV(\Omega)$ *and a subsequence* $\{u_{k_i}\}$ *of* $\{u_k\}$ *such that*

$$\begin{array}{lcll} u_{k_i} & \to & u & \textit{in } L^1(\Omega) \\ Du_{k_i} & \rightharpoonup & Du & \textit{as measures in } \Omega \,. \end{array}$$

In other words the immersion $BV(\Omega) \hookrightarrow L^1(\Omega)$ is compact, provided $\partial\Omega$ is sufficiently smooth. Actually one can easily see that the immersion $BV(\Omega) \to L^p(\Omega)$ is continuous for $p \le n/(n-1)$, and compact for $p < n/(n-1)$. •

5 *The coarea formula for BV-functions.* We conclude this subsection proving an extension of the classical coarea formula for smooth functions. In order to do that we first anticipate a remark which actually belongs to next subsection.

Let E be a smooth subset of $\mathbb{R}^n$, say with boundary of class C^2 and let $g \in C^1_c(\Omega, \mathbb{R}^n)$. From the classical Gauss-Green formula we infer

$$\int \chi_E \operatorname{div} g \, dx = \int_E \operatorname{div} g = - \int_{\partial E \cap \Omega} \nu \cdot g \, d\mathcal{H}^{n-1} \tag{12}$$

where χ_E is the characteristic function of E and ν denotes the inward unit normal to ∂E. From (12) we deduce at once

$$|D\chi_E|(\Omega) \le \mathcal{H}^{n-1}(\partial E \cap \Omega) \,.$$

On the other hand, extending ν to a smooth, say C^1, function $\tilde{\nu}$ with $|\tilde{\nu}| \le 1$ in $\mathbb{R}^n$ we find, using (12) with $g = \tilde{\nu}\eta$,

$$\int_{\partial E \cap \Omega} \eta \, d\mathcal{H}^{n-1} = - \int \chi_E \operatorname{div}(\tilde{\nu}\eta) \le |D\chi_E|(\Omega)$$

for any $\eta \in C^\infty_c(\Omega)$ with $|\eta| \le 1$, thus

$$\mathcal{H}^{n-1}(\partial E \cap \Omega) \le |D\chi_E|(\Omega)$$

and we conclude

$$\mathcal{H}^{n-1}(\partial E \cap \Omega) = |D\chi_E|(\Omega) \,. \tag{13}$$

Recall now the classical coarea formula for smooth function u, say C^∞

$$\int_\Omega |Du(x)| \, dx = \int_{-\infty}^{+\infty} \mathcal{H}^{n-1}(\Omega \cap \{x \in \Omega \mid u(x) = t\}) \, dt \,. \tag{14}$$

By Sard's theorem for a.e. t the set $\{x \in \Omega \mid u(x) = t\}$ is a smooth hypersurface which agrees with the boundary of the upper–level set

$$E_t(u) := \{x \in \Omega \mid u(x) > t\} .$$

therefore by (13) we can write (14) as

$$\int_\Omega |Du(x)|\, dx = \int_{-\infty}^{+\infty} |D\chi_{E_t(u)}|(\Omega)\, dt . \tag{15}$$

We shall now prove that (15) actually holds with obvious variants for any $u \in BV(\Omega)$.

First notice that for any function $u \in L^1(\Omega)$ we have

$$u(x) = u^+(x) - u^-(x) := \int_0^{+\infty} \chi_{E_t(u)}(x)\, dt - \int_{-\infty}^0 (1 - \chi_{E_t(u)}(x))\, dt . \tag{16}$$

Let now $\zeta \in C^1_c(\Omega, \mathbb{R}^n)$. Taking into account that $\operatorname{div}\zeta$ has zero mean value in Ω we infer from (16)

$$\begin{aligned}\int_\Omega u(x) \operatorname{div}\zeta(x)\, dx &= \int_0^\infty dt \int_\Omega \chi_{E_t(u)}(x) \operatorname{div}\zeta(x)\, dx \\ &\quad - \int_{-\infty}^0 dt \int_\Omega (1 - \chi_{E_t(u)}(x)) \operatorname{div}\zeta(x)\, dx \\ &= \int_{-\infty}^{+\infty} dt \int_\Omega \chi_{E_t(u)}(x) \operatorname{div}\zeta(x)\, dx ,\end{aligned}$$

which for $u \in BV(\Omega)$ can be written as

$$\int \zeta\, d\, Du = \int_{-\infty}^{+\infty} dt \int \zeta\, d\, D\chi_{E_t(u)} . \tag{17}$$

Theorem 2 (Fleming-Rishel). *Let $u \in L^1_{\mathrm{loc}}(\Omega)$. Then*

$$|Du|(\Omega) = \int_{-\infty}^{+\infty} |D\chi_{E_t(u)}|(\Omega)\, dt . \tag{18}$$

In particular $u \in BV(\Omega)$ if and only if the function

$$t \longrightarrow |D\chi_{E_t(u)}|(\Omega)$$

is summable on $\mathbb{R}$ *and for almost every* t

$$|D\chi_{E_t(u)}|(\Omega) < \infty .$$

Proof. Taking the supremum for all $\zeta \in C_c^1(\Omega, \mathbb{R}^n)$, $|\zeta| \le 1$ in (17), we get

$$|Du|(\Omega) \le \int_{-\infty}^{+\infty} |D\chi_{E_t(u)}|(\Omega)\, dt .$$

To prove the opposite inequality we apply the standard coarea formula for smooth functions to u_k, where $u_k \to u$ in L^1 and

(19) $$\int_\Omega |Du_k|\, dx \longrightarrow |Du|(\Omega) ;$$

thus

(20) $$\int_\Omega |Du_k|\, dx = \int_{-\infty}^{+\infty} |D\chi_{E_t(u_k)}|(\Omega)\, dt .$$

Now we observe that by (16)

$$u_k(x) - u(x) = \int_{-\infty}^{+\infty} (\chi_{E_t(u_k)}(x) - \chi_{E_t(u)}(x))\, dt$$

and actually

$$|u_k(x) - u(x)| = \int_{-\infty}^{+\infty} |\chi_{E_t(u_k)(x)} - \chi_{E_t(u)}(x)|\, dt$$

since $\operatorname{sign}(u_k(x) - u(x)) = \operatorname{sign}(\chi_{E_t(u_k)(x)} - \chi_{E_t(u)}(x))$ for all t. The convergence in L^1 of u_k to u then yields

$$\chi_{E_t(u_k)} \longrightarrow \chi_{E_t(u)} \qquad \text{in } L^1(\Omega)$$

for almost every t.

Passing to the limit in (20), using (19) and the semicontinuity of the total variation we then get

$$\begin{aligned} |Du|(\Omega) &= \lim_{k\to\infty} |Du_k|(\Omega) = \lim_{k\to\infty} \int_{-\infty}^{+\infty} |D\chi_{E_t(u_k)}|(\Omega)\, dt \\ &\ge \int_{-\infty}^{+\infty} |D\chi_{E_t(u)}|(\Omega)\, dt , \end{aligned}$$

which concludes the proof of the theorem. □

An immediate consequence of the coarea formula is that $\max(u,t) \in BV(\Omega)$, if $u \in BV(\Omega)$, moreover

$$|D\max(u,t)|(\Omega) = |Du|(E_t(u)) .$$

Consider in particular the *truncation of* $u \in BV(\Omega)$

$$u_t(x) := \begin{cases} t & \text{if } u(x) > t \\ u(x) & \text{if } -t \leq u(x) \leq t \\ -t & \text{if } u(x) < -t \end{cases}$$

and set

$$h_t(x) := u(x) - u_t(x) .$$

As $u_t \in BV(\Omega)$, we find

$$|Dh_t|(\Omega) = \int_{-\infty}^{+\infty} |D\chi_{E_s(h_t)}|(\Omega) = \int_{|s|>t} |D\chi_{E_t(u)}|(\Omega)\, ds$$

and consequently, taking also into account that h_t converges strongly in L^1 to zero we can state

Proposition 6. *We have*

(i) $|Dh_t|(\Omega) \longrightarrow 0$ *as* $t \to \infty$, *i.e., the total variation cannot concentrate near infinity, i.e. for* $|u| \to \infty$.

(ii) $BV(\Omega) \cap L^\infty(\Omega)$ *is dense in* $BV(\Omega)$ *with respect to the norm topology.*

●

1.2 Caccioppoli Sets

Let E be a measurable set in $\mathbb{R}^n$ and let Ω be an open set in $\mathbb{R}^n$.

Definition 1. *We say that* E *has* finite perimeter, *or* locally finite perimeter, *in* Ω *if its characteristic function* χ_E *belongs to* $BV(\Omega)$, *respectively to* $BV_{\text{loc}}(\Omega)$. *The* perimeter of E in Ω *is defined as*

$$P(E,\Omega) := |D\chi_E|(\Omega) .$$

Definition 2. *A set* E *with locally finite perimeter in* $\mathbb{R}^n$ *is called a* Caccioppoli set.

Clearly $P(E,\cdot)$ defines a measure, therefore for any Borel set B we can compute $P(E,B)$, i.e. the perimeter of E in B. From the definition of BV-functions it is immediately seen that Caccioppoli sets are exactly the sets for which the following *Gauss-Green formula* holds

$$(1) \qquad -\int \chi_E \, D_i g(x)\, dx \;=\; \int g\, dD_i\chi_E \qquad \forall\, g \in C_c^1(\mathbb{R}^n),\ i = 1, 2, \dots, n\,.$$

From the differentiation theory for measures we infer that

$$D\chi_E \;=\; |D\chi_E| \llcorner n(x, E)$$

where

$$(2) \qquad n(x, E) \;:=\; \lim_{\rho \to 0^+} \frac{D\chi_E(B(x, \rho))}{|D\chi_E|(B(x, \rho))}$$

is well defined and $|n(x, E)| = 1$ for $|D\chi_E|$-a.e. point x. Thus (1) can be written as

$$(3) \qquad -\int_E \operatorname{div} g\, dx \;=\; \int g^i(x)\, n_i(x, E)\, d|D\chi_E| \qquad \forall\, g \in C_c^1(\mathbb{R}^n, \mathbb{R}^n)\,.$$

This justifies the terminology of *inward normal* for $n(x, E)$.

The set of points x where the inward normal $n(x, E)$ exists plays an important role. Thus we set

Definition 3. *The* reduced boundary *(in the sense of De Giorgi) $\partial^- E$ of a Caccioppoli set E is defined as*

$$(4) \qquad \partial^- E \;:=\; \Big\{x \in \mathbb{R}^n \;\Big|\; \exists \lim_{\rho \to 0^+} \frac{D\chi_E(B(x, \rho))}{|D\chi_E|(B(x, \rho))} =: n(x, E) \ \text{and}\ |n(x, E)| = 1\Big\}$$

Notice that

$$|D\chi_E|(\mathbb{R}^n \setminus \partial^- E) \;=\; 0\,.$$

Consequently, $\partial^- E$ is $|D\chi_E|$-measurable and

$$(5) \qquad |D\chi_E| \;=\; |D\chi_E| \llcorner \partial^- E\,.$$

For convenience we set $n(x, E) = 0$ for any $x \notin \partial^- E$.

We already saw that

$$P(E, \Omega) \;=\; \mathcal{H}^{n-1}(\partial E \cap \Omega)$$

if E has smooth boundary. It is immediately seen from the classical Gauss-Green formula that in this case $n(x, E)$ agrees with the classical inward normal to ∂E, $\partial^- E = \partial E$, and

$$|D\chi_E| \;=\; \mathcal{H}^{n-1} \llcorner \partial E\,.$$

However, if E is not smooth this may not be true and we may even have $P(E, \Omega) < \infty$ but $\mathcal{H}^{n-1}(\partial E \cap \Omega) = \infty$, see [1] below. But first, let us restate in terms of Caccioppoli sets the compactness theorem for BV-functions and the semicontinuity result for the total variation.

Proposition 1. *Let $\{E_k\}$ be a sequence of Caccioppoli sets such that*

$$\sup_k P(E_k, \Omega) \leq c(\Omega)$$

for any bounded open set Ω. Then there exist a Caccioppoli set E and a subsequence $\{E_{k_i}\}$ of $\{E_k\}$ such that

$$\chi_{E_{k_i}} \to \chi_E \quad \text{in } L^1_{\text{loc}}(\mathbb{R}^n), \qquad D\chi_{E_{k_i}} \rightharpoonup D\chi_E \quad \text{as measures}$$

and

$$P(E, \Omega) \leq \liminf_{i \to \infty} P(E_{k_i}, \Omega) .$$

[1] Let Ω be the open unit cube in $\mathbb{R}^n$, $n \geq 2$, and let $\{q_k\}$ be a denumeration of the points in Ω with rational coordinates. We select a sequence $\{\rho_k\}$ of positive real numbers such that

$$B(q_k, \rho_k) \subset \Omega, \quad \sum_{k=1}^{\infty} \rho_k^{n-1} < \infty, \quad 0 < |B(0,1)| \sum_{k=1}^{\infty} \rho_k^n < 1.$$

For any h the sets

$$A_h := \bigcup_{i=1}^{h} B(q_i, \rho_i)$$

have piecewise C^∞ boundary and obviously

$$P(A_h, \Omega) \leq \mathcal{H}^{n-1}(\partial B(0,1)) \sum_{i=1}^{h} \rho_i^{n-1} .$$

From Proposition 1 we therefore deduce that the open set

$$A := \cup_{i=1}^{\infty} B(q_i, \rho_i)$$

is a Caccioppoli set. On the other hand, since A is dense in Ω and $\mathcal{L}^n(A) < 1$, we have

$$\mathcal{L}^n(\partial A \cap \Omega) = 1 - \mathcal{L}^n(A \cap \Omega) > 0 .$$

Hence $\mathcal{H}^{n-1}(\partial A \cap \Omega) = +\infty$. •

The following result yields in particular an approximation theorem for bounded Caccioppoli sets.

Theorem 1. *Let E be a set of finite perimeter in Ω. Then there exists a sequence of sets E_h such that $\partial E_h \cap \Omega$ is of class C^∞,*

$$\chi_{E_h} \to \chi_E \quad \text{in } L^1 \quad \text{and} \quad P(E_h, \Omega) \to P(E, \Omega) .$$

Proof. By the approximation theorem for BV-functions, Theorem 1 in Sec. 4.1.1, we find a sequence of functions u_h of class C^∞ such that

$$\int_\Omega |u_h - \chi_E|\, dx \longrightarrow 0 \qquad \int_\Omega |Du_h|\, dx \longrightarrow P(E,\Omega) .$$

Replacing u_h by $\varphi_h(u_h)$ where $\varphi_h \in C^\infty(\mathbb{R})$ is such that

$$\varphi_h(t) = t \quad \forall\, t \in (1/h, 1-1/h) \qquad 0 \le \varphi_h \le 1,\ 0 < \dot\varphi_h < 1 ,$$

we may assume that $0 \le u_h \le 1$, and also that u_h converge a.e. to χ_E. From Fatou lemma and the coarea formula we then get

$$\begin{aligned}\int_0^1 \liminf_{h\to\infty} P(E_t(u_h),\Omega)\, dt &\le \liminf_{h\to\infty} \int_0^1 P(E_t(u_h),\Omega)\, dt \\ &= \liminf_{h\to\infty} \int_\Omega |Du_h(x)|\, dx = P(E,\Omega)\end{aligned}$$

where

$$E_t(u_h) := \{x \in \Omega \mid u_h(x) > t\} .$$

As $\chi_{E_h} \to \chi_E$ in $L^1(\Omega)$, by semicontinuity we have for any $t \in (0,1)$

$$P(E,\Omega) \le \liminf_{h\to\infty} P(E_t(u_h),\Omega)$$

and we conclude that

$$P(E,\Omega) = \liminf_{h\to\infty} P(E_t(u_h),\Omega)$$

for a.e. t. On the other hand, by Sard lemma, for almost every $t \in (0,1)$, $\partial E_t(u_h) \cap \Omega$ is of class C^∞. Therefore we can find $t \in (0,1)$ such that the sets $E_t(u_h)$, $h = 1,2,\dots$ have boundaries in Ω of class C^∞ and

$$\liminf_{h\to\infty} P(E_t(u_h),\Omega) = P(E,\Omega) .$$

A suitable subsequence of $\{E_t(u_h)\}$ satisfies the claim in the theorem. □

Remark 1. In general it is not possible to approximate a Caccioppoli set E by C^∞ sets contained inside E, nor it is possible from the outside. This is easily seen from the example in [1].

Let E be a bounded Caccioppoli set in $\mathbb{R}^n$. Applying Sobolev inequality

$$\Big(\int |u|^{n/(n-1)}\, dx\Big)^{(n-1)/n} \le S(n) \int |Du| , \tag{6}$$

valid for all $u \in BV(\mathbb{R}^n)$, to the characteristic function χ_E of E, we find the following

Isoperimetric inequality : Let E be a bounded Caccioppoli set in $\mathbb{R}^n$. Then

$$|E|^{(n-1)/n} \leq I(n)\, P(E, \mathbb{R}^n) \tag{7}$$

with $I(n) = S(n)$.

Actually we can see that (7) implies also (6) with $S(n) = I(n)$. In particular the best Sobolev constant $S(n)$ in (6) and the best isoperimetric constant in (7) agree.

In order to prove that (7) implies (6) consider the set

$$A_t(u) := \{x \mid |u(x)| > t\}$$

and the function u_t obtained from u by truncation at heights t and $-t$ and let

$$f(t) := \Big(\int |u_t|^{n/(n-1)}\, dx \Big)^{(n-1)/n} .$$

For $h > 0$ we have

$$|u_{t+h}| \leq |u_t| + h\, \chi_{A_t(u)}$$

hence

$$f(t+h) \leq f(t) + h\, |A_t(u)|^{(n-1)/n}$$

i.e.

$$f'(t) \leq |A_t(u)|^{(n-1)/n} .$$

From (7) we then deduce

$$\Big(\int |u|^{n/(n-1)}\, dx \Big)^{(n-1)/n} = f(\infty) - f(0) = \int_0^\infty f'(t)\, dt$$

$$\leq \int_0^\infty |A_t(u)|^{(n-1)/n}\, dt \leq I(n) \int_0^\infty P(A_t(u), \mathbb{R}^n)\, dt = I(n) \int |Du| ,$$

taking into account the coarea formula.

From the previous remark, if we also take into account the approximation theorems for BV-functions and Caccioppoli sets, and Sard lemma, it is not difficult to show the following. The isoperimetric inequality for smooth sets implies Sobolev inequality for smooth functions, via the coarea formula for smooth functions. By approximation, Sobolev inequalities for smooth functions implies Sobolev inequality for BV-functions, which in turn implies the isoperimetric inequality for bounded Caccioppoli sets, and in particular for smooth sets. Moreover the best constants in both inequalities are the same. As a symmetrization procedure shows for instance in the context of smooth sets that the best constant in the isoperimetric inequality is realized by the unit ball, we then conclude that

$$(8) \qquad S(n) = I(n) = |B(0,1)|^{(n-1)/n} |\partial B(0,1)|^{-1} = n^{-1} \omega_n^{-1/n}$$

where ω_n is the volume of the unit ball in $\mathbb{R}^n$, $\omega_n = \frac{\Gamma(\frac{1}{2})^n}{\Gamma(n+\frac{1}{2})}$.

We conclude this subsection by proving two more isoperimetric inequalities which will be used later.

Proposition 2. *Let E be a Caccioppoli set in $\mathbb{R}^n$ and let B_r be any ball of radius r in $\mathbb{R}^n$. Then*

$$(9) \qquad \frac{\mathcal{L}^n(E \cap B_r)}{r^n} \cdot \frac{\mathcal{L}^n(B_r \setminus E)}{r^n} \le c_1(n) \frac{P(E, B_r)}{r^{n-1}} .$$

Moreover

$$(10) \qquad \min\{\mathcal{L}^n(B_r \cap E), \mathcal{L}^n(B_r \setminus E)\}^{\frac{n-1}{n}} \le c_2(n) P(E, B_r) .$$

Proof. As the mean value of χ_E on B_r is given by

$$(\chi_E)_{B_r} = \frac{\mathcal{L}^n(E \cap B_r)}{\mathcal{L}^n(B_r)}$$

and

$$\int_{B_r} |\chi_E - (\chi_E)_{B_r}| \, dx = 2 \frac{\mathcal{L}^n(E \cap B_r)\mathcal{L}^n(B_r \setminus E)}{\mathcal{L}^n(B_r)} ,$$

(9) follows at once applying Poincaré inequality of Sec. 4.1.1 to χ_E on B_r.

Similarly we have

$$\int_{B_r} |\chi_E - (\chi_E)_{B_r}|^{n/(n-1)} \, dx = \Big[\frac{\mathcal{L}^n(B_r \setminus E)}{\mathcal{L}^n(B_r)}\Big]^{n/(n-1)} \mathcal{L}^n(B_r \cap E) + \Big[\frac{\mathcal{L}^n(B_r \cap E)}{\mathcal{L}^n(B_r)}\Big]^{n/(n-1)} \mathcal{L}^n(B_r \setminus E)$$

If $\mathcal{L}^n(B_r \setminus E) \ge \mathcal{L}^n(B_r \cap E)$ then $\mathcal{L}^n(B_r \setminus E) \ge \frac{1}{2}\mathcal{L}^n(B_r)$, otherwise we find $\mathcal{L}^n(B_r \cap E) \ge \frac{1}{2}\mathcal{L}^n(B_r)$. Thus in both cases

$$\int_{B_r} |\chi_E - (\chi_E)_{B_r}|^{n/(n-1)} \, dx \ge \Big(\frac{1}{2}\Big)^{\frac{n}{n-1}} \min\{\mathcal{L}^n(B_r \cap E), \mathcal{L}^n(B_r \setminus E)\} .$$

Applying Sobolev-Poincaré inequality of Sec. 4.1.1 to χ_E in B_r we infer at once (10). □

1.3 De Giorgi's Rectifiability Theorem

As we saw in Sec. 4.1.2, the most important object in dealing with Caccioppoli sets E is the reduced boundary

$$\partial^- E := \{x \in \mathbb{R}^n \mid \exists \lim_{\rho \to 0^+} \frac{D\chi_E(B(x,\rho))}{|D\chi_E|(B(x,\rho))} =: n(x,E) \text{ and } |n(x,E)| = 1\}$$

which carries the measure $D\chi_E$, $|D\chi_E| = |D\chi_E| \llcorner \partial^- E$. This subsection is devoted mainly to prove the following important result of De Giorgi

Theorem 1 (De Giorgi). *Let E be a Caccioppoli set in $\mathbb{R}^n$. Then*

(i) *The reduced boundary $\partial^- E$ is locally $(n-1)$-rectifiable and its approximate $(n-1)$-plane at x is normal to $n(x,E)$ for $\mathcal{H}^{n-1}$-a.e. $x \in \partial^- E$.*
(ii) $|D\chi_E| = \mathcal{H}^{n-1} \llcorner \partial^- E$ *and* $D\chi_E = \mathcal{H}^{n-1} \llcorner n(x,E)$.
(iii) *The following Gauss-Green formula*

$$-\int_E \operatorname{div} g \, dx = \int_{\partial^- E} (g, n(x,E)) \, d\mathcal{H}^{n-1} \qquad \forall\, g \in C_c^1(\mathbb{R}^n, \mathbb{R}^n)$$

holds.

The key point in the previous theorem is the $(n-1)$-rectifiability of $\partial^- E$ which will be obtained as consequence of the rectifiability theorem for measures, Theorem 2 in Sec. 2.1.4.

We begin by collecting a few remarks.

Let E be a measurable set in $\mathbb{R}^n$ and let $x \in \mathbb{R}^n$. The set $E \cap \partial B(x,\rho)$ is $\mathcal{H}^{n-1}$-measurable for a.e. ρ and we have

$$\mathcal{L}^n(E \cap B(x,\rho)) = \int_0^\rho \mathcal{H}^{n-1}(E \cap \partial B(x,t)) \, dt \, . \tag{1}$$

As $\rho \to \mathcal{L}^n(E \cap B(x,\rho))$ is nondecreasing, hence a.e. differentiable, for almost every ρ we also have

$$\frac{d}{dt} \mathcal{L}^n(E \cap B(x,t))_{|\,t=\rho} = \mathcal{H}^{n-1}(E \cap \partial B(x,\rho)) \, . \tag{2}$$

Proposition 1. *Let E a Caccioppoli set in $\mathbb{R}^n$. Then*

(i) *For all $u \in C_c^1(\mathbb{R}^n)$ we have $u\chi_E \in BV(\mathbb{R}^n)$ and*

$$\int |D(u\chi_E)| \le \int |u| \, d|D\chi_E| + \int_E |Du| \, dx \tag{3}$$

(ii) *For all $x \in \mathbb{R}^n$ and almost every $\rho > 0$*

$$P(E \cap B(x,\rho), \mathbb{R}^n) \le P(E, B(x,\rho)) + \mathcal{H}^{n-1}(E \cap \partial B(x,\rho)) \tag{4}$$

(iii) For all $\nu \in \mathbb{R}^n$ with $|\nu| = 1$, for all $x \in \mathbb{R}^n$, and for almost every $\rho > 0$

$$|D_\nu \chi_E|(B(x,\rho)) \leq \mathcal{H}^{n-1}(E \cap \partial B(x,\rho)) \tag{5}$$

Here $D_\nu \chi_E$ denotes the distributional derivative of χ_E in the direction $\nu = (\nu_1, \ldots, \nu_n)$

$$D_\nu \chi_E := \sum_{i=1}^n \nu_i D_i \chi_E .$$

Proof. (i) Since $u \in C_c^1(\mathbb{R}^n)$ we can assume that E is bounded, hence $\chi_E \in BV(\mathbb{R}^n)$. We choose $\{\zeta_k\} \subset C_c^\infty(\mathbb{R}^n)$ such that $\zeta_k \geq 0$, $\zeta_k(x) \to \chi_E$ in L^1 and almost everywhere, and $\mathcal{L}^n \llcorner |D\zeta_k(x)| \rightharpoonup |D\chi_E|$ as measures, compare Theorem 1 in Sec. 4.1.1. Obviously

$$\int |D(u\zeta_k)|\, dx \leq \int |u|\, |D\zeta_k|\, dx + \int \zeta_k\, |Du|\, dx .$$

Thus (i) follows passing to the limit.

(ii) We choose ρ in such a way that (2) holds and we approximate the characteristic function of $\overline{B(x,\rho)}$ by $\xi_k(y)$ where $\xi_k(y) := \varphi_k(|x-y|)$ and

$$\varphi_k(t) := \begin{cases} 1 & \text{if } 0 \leq t \leq \rho \\ -k(t-\rho-1/k) & \text{if } \rho \leq t \leq \rho + 1/k \\ 0 & \text{if } t > \rho + 1/k \end{cases}$$

We have

$$\int_E |D\xi_k|\, dx = k\,[\mathcal{L}^n(E \cap B(x,\rho+1/k)) - \mathcal{L}^n(E \cap B(x,\rho))] .$$

Thus applying (i) with $u = \xi_k$ and passing to the limit we infer

$$P(E \cap B(x,\rho), \mathbb{R}^n) \leq P(E, \overline{B(x,\rho)}) + \frac{d}{dt}\mathcal{L}^n(E \cap B(x,t))_{|\,t=\rho} .$$

The claim then follows since $|D\chi_E|(\partial B(x,\rho)) = 0$ for a.e.ρ.

(iii) Of course it is not restrictive to assume that ν is the direction e_1. From

$$\int \zeta\, d D_1\chi_E = -\int_E D_1\zeta\, dx \qquad \forall\, \zeta \in C_c^\infty(\mathbb{R}^n)$$

we deduce

$$\Big|\int \zeta\, d D_1 \chi_E\Big| \leq \int_E |D_1\zeta(x)|\, dx .$$

Using this inequality with $\zeta = \xi_k$, ξ_k as in the proof of (ii), we infer

$$|D_1\chi_E\,(\overline{B(x,\rho)})| \;\le\; \frac{d}{dt}\mathcal{L}^n(E\cap B(x,t))_{|\,t=\rho}$$

and the claim follows since for almost every ρ, $|D\chi_E|(\partial B(x,\rho)) = 0$ and (2) holds. □

A consequence of Proposition 1 (i) is that the derivatives of the characteristic function of a Caccioppoli set are absolutely continuous with respect to the $(n-1)$-dimensional Hausdorff measure.

Proposition 2. *Let E be a Caccioppoli set in $\mathbb{R}^n$. If $\mathcal{H}^{n-1}(C) = 0$ then $|D\chi_E|(C) = 0$, i.e., $|D\chi_E| \ll \mathcal{H}^{n-1}$.*

Proof. It suffices to prove the proposition in the case that C is compact. As $\mathcal{H}^{n-1}(C) = 0$, for any $\varepsilon > 0$ we can find a finite number of balls $B(x_i,\rho_i)$ such that

$$C \subset \bigcup_{i=1}^{p} B(x_i,\rho_i) \qquad \text{and} \qquad \sum_{i=1}^{p} \rho_i^{n-1} < \varepsilon\,.$$

For some $\theta \in (0,1)$ we then have

$$C \subset \bigcup_{i=1}^{p} B(x_i,\theta\rho_i)\,.$$

We now choose a function $\varphi \in C^1_c(B(0,1))$ such that $\varphi \equiv 1$ on $B(0,\theta)$, $0 \le \varphi \le 1$, and $|D\varphi|(\mathbb{R}^n) \le c(n)$, and we set

$$\varphi_\varepsilon(x) \;=\; \prod_{i=1}^{p}\big(1-\varphi(\frac{x-x_i}{\rho_i})\big)\,\chi_E(x)\,.$$

Clearly $\varphi_\varepsilon(x) \equiv 0$ in a neighborhood of C, thus by Proposition 1 (i)

$$\begin{aligned}|D\varphi_\varepsilon|(\Omega) \;&\le\; |D\chi_E|(\Omega\setminus C) + |D\varphi|(\mathbb{R}^n)\sum_{i=1}^{p}\rho_i^{n-1}\\ &\le\; |D\chi_E|(\Omega\setminus C) + c(n)\,\varepsilon\,.\end{aligned}$$

where Ω denotes a bounded open set containing C. Passing to the limit as $\varepsilon \to 0$ we then get

$$|D\chi_E|(\Omega) \;\le\; |D\chi_E|(\Omega\setminus C)\,,$$

since $\varphi_\varepsilon \to \chi_E$ in L^1, and this trivially implies $|D\chi_E|(C) = 0$. □

We begin the proof of the rectifiability of the reduced boundary. The next lemma yields rough upper and lower estimates for densities at points $x \in \partial^- E$.

Lemma 1. *Let E be a Caccioppoli set in $\mathbb{R}^n$ and let $x \in \partial^- E$. Then there exists a positive constant c depending only on the dimension n such that for sufficiently small ρ, $0 < \rho < \rho_0(E,x)$, we have*

$$\frac{\mathcal{L}^n(E \cap B(x,\rho))}{\rho^n} \geq c \tag{6}$$

$$\frac{\mathcal{L}^n(B(x,\rho) \setminus E)}{\rho^n} \geq c \tag{7}$$

and

$$\frac{1}{c} \geq \frac{P(E, B(x,\rho))}{\rho^{n-1}} \geq c\,. \tag{8}$$

Proof. To simplify the notation, we may assume that $x = 0 \in \partial^- E$ and $n(0,E) = (0,0,\dots,1)$; also we write B_ρ for $B(0,\rho)$. First we observe that for a.e. ρ small we have

$$|D\chi_E|(B_\rho) \leq 2\,|D_n\chi_E(B_\rho)|$$

as consequence of

$$\frac{D\chi_E(B_\rho)}{|D\chi_E|(B_\rho)} \longrightarrow (0,0,\dots,1)\,.$$

Using (5), we then infer

$$P(E,B_\rho) \leq 2\mathcal{H}^{n-1}(E \cap \partial B_\rho)\,. \tag{9}$$

In particular we find

$$P(E,B_\rho) \leq c(n)\,\rho^{n-1}\,.$$

Moreover, since $c(n)\rho^{n-1}$ is continuous in ρ and $\rho \to P(E,B_\rho)$ is nondecreasing the previous inequality holds for any ρ, ρ small. This proves the left inequality in (8).

Let us prove (6). Set

$$\alpha(\rho) := \mathcal{L}^n(E \cap B_\rho)\,.$$

From the isoperimetric inequality (10) in Sec. 4.1.2 we infer

$$\alpha(\rho)^{(n-1)/n} \leq c(n)\,P(E \cap B_\rho, \mathbb{R}^n)$$

and by (4), (2) and (9) that

$$\alpha(\rho)^{(n-1)/n} \leq c(n)\,\{P(E,B_\rho) + \alpha'(\rho)\} \leq 2c(n)\,\alpha'(\rho)\,.$$

Integrating this inequality we find (6).

Inequality (7) can be subsumed to (6) as $\mathbb{R}^n \setminus E$ is a set of locally finite perimeter and $D\chi_{\mathbb{R}^n\setminus E} = -D\chi_E$.

Finally, from the isoperimetric inequality (10) in Sec. 4.1.2 we find

$$P(E,B_\rho) \geq c\min\{\mathcal{L}^n(E \cap B_\rho, \mathcal{L}^n(B_\rho \setminus E)\}^{\frac{n-1}{n}} \geq c\,\rho^{n-1}\,,$$

taking into account (6) and (7). □

The next lemma contains the main facts for the proof of Theorem 1. In order to state it we find convenient to introduce the following notation.

Given a point $x \in \mathbb{R}^n$ and a set E, we denote by $E_{x,\lambda}$ the blow-up of E with factor $1/\lambda$ from the point x,

$$E_{x,\lambda} := \{z \in \mathbb{R}^n \mid x + \lambda(z - x) \in E\} .$$

Given an unitary vector field $\nu(x)$, and we have in mind the case $\nu(x) = n(x, E)$, $x \in \partial^- E$, we denote by $E^0_{x,\nu}$ the $(n-1)$-plane through x which is orthogonal to $\nu(x)$

$$E^0_{x,\nu} := \{z \in \mathbb{R}^n \mid (z - x, \nu(x)) = 0\} .$$

The positive and negative halfspaces generated by $\nu(x)$ are denoted by

$$\begin{aligned} E^+_{x,\nu} &:= \{z \in \mathbb{R}^n \mid (z - x, \nu(x)) > 0\} \\ E^-_{x,\nu} &:= \{z \in \mathbb{R}^n \mid (z - x, \nu(x)) < 0\} . \end{aligned}$$

Lemma 2. *Let E be a Caccioppoli set and let $x \in \partial^- E$. Then the Caccioppoli sets $E_{x,\rho}$ converge in L^1_{loc} as $\rho \to 0^+$ to $E^+_{x,\nu(x)}$, $\nu(x) := n(x, E)$. Moreover*

$$\begin{aligned} |D\chi_{E_{x,\rho}}| &\rightharpoonup \mathcal{H}^{n-1} \llcorner E^0_{x,\nu(x)} \\ D\chi_{E_{x,\rho}} &\rightharpoonup \nu(x)\, \mathcal{H}^{n-1} \llcorner E^0_{x,\nu(x)} \end{aligned}$$

in the sense of measures.

Proof. To simplify the notation we may assume $x = 0$ and $\nu(0) = n(0, E) = (0, 0, \ldots, 1)$. We write E_λ for $E_{0,\lambda}$,

$$E_\lambda := \{z \in \mathbb{R}^n \mid \lambda z \in E\} ,$$

and B_r for $B(0, r)$. We divide the proof into four steps

Step 1. Convergence of the blow-up sequence. Observe that

$$\begin{aligned} \mathcal{L}^n(B_r \cap E_\lambda) &= \lambda^{-n}\, \mathcal{L}^n(E \cap B_{\lambda r}) \\ \mathcal{L}^n(B_r \setminus E_\lambda) &= \lambda^{-n}\, \mathcal{L}^n(B_{\lambda r} \setminus E) \\ |D\chi_{E_\lambda}|(B_r) &= \lambda^{1-n}\, |D\chi_E|(B_{\lambda r}) . \end{aligned}$$

From Lemma 1 we therefore deduce

$$\begin{aligned} &\mathcal{L}^n(B_r \cap E_\lambda) \geq c r^n \\ &\mathcal{L}^n(B_r \setminus E_\lambda) \geq c r^n \qquad (10) \\ &\frac{1}{c} r^{n-1} \geq P(E_\lambda, B_r) = |D\chi_{E_\lambda}|(B_r) \geq c r^{n-1} \end{aligned}$$

provided $r\lambda$ is small enough. Using the compactness result in Proposition 1 in Sec. 4.1.2 we can then find a sequence of positive numbers λ_k, $\lambda_k \to 0^+$, and a Caccioppoli set F such that

$$(11) \qquad \chi_{E_{\lambda_k}} \to \chi_F \quad \text{in } L^1_{\text{loc}}(\mathbb{R}^n) \quad \text{and} \quad D\chi_{E_{\lambda_k}} \rightharpoonup D\chi_F \quad \text{as measures.}$$

In particular, for almost every r we have

$$(12) \qquad D\chi_{E_{\lambda_k}}(B_r) \to D\chi_F(B_r)\ .$$

We also have

$$(13) \qquad \frac{D_n\chi_{E_{\lambda_k}}(B_r)}{|D\chi_{E_{\lambda_k}}|(B_r)} = \frac{D_n\chi_E(B_{\lambda_k r})}{|D\chi_E|(B_{\lambda_k r})} \longrightarrow 1 \quad \text{as } k \to \infty$$

being $0 \in \partial^- E$ and $n(0,E) = (0,\dots,0,1)$. Therefore from (12)) and (13), and the semicontinuity of the total variation we infer for a.e. r

$$D_n\chi_F(B_r) = \lim_{k\to\infty} |D\chi_{E_{\lambda_k}}|(B_r) \geq |D\chi_F|(B_r)\ ,$$

and, as the opposite inequality is always true,

$$(14) \qquad D_n\chi_F(B_r) = |D\chi_F|(B_r) \qquad \text{for a.e. } r\ .$$

From (12), (13) and (14) we then finally get

$$(15) \qquad |D\chi_{E_{\lambda_k}}| \rightharpoonup |D\chi_F| \quad \text{as measures.}$$

Step 2. The blow-up limit is a half-space. From (14) we in particular infer that

$$D_i\chi_F = 0 \quad \text{for } i = 1,\dots,n-1 \quad \text{and} \quad D_n\chi_F \geq 0 \quad |D\chi_F|\text{-a.e.}$$

Hence χ_F is independent of $x_1,\dots,x_{n-1}$ and is a nondecreasing function of x_n; but, since it takes only values 0 and 1, we conclude that F must have the form

$$F = \{x \mid x^n > \beta\}$$

for some $\beta \in \mathbb{R}$.

Step 3. The blow-up limit F is $E^0_{0,\nu}$, $\nu = (0,0,\dots,1)$. It suffices to show that $\beta = 0$. Suppose $\beta > 0$. Then for a small ball B centered at 0 we get on account of (10)

$$0 = \mathcal{L}^n(F\cap B) = \lim_{\lambda\to 0+} \lambda^{-n}\mathcal{L}^n(E_\lambda \cap B) \geq c > 0$$

a contradiction. Similarly, if $\beta < 0$

$$0 = \mathcal{L}^n(B\setminus F) = \lim_{\lambda\to 0+} \lambda^{-n}\mathcal{L}^n(B\setminus E_\lambda) \geq c > 0\ .$$

Therefore $\beta = 0$.

Step 4. Observing that in step 1, step 2, step 3 we have in fact proved that from any sequence $\{\lambda_k\}$ converging to zero, we can extract a subsequence $\{\lambda_{k_i}\}$ so that (11) and (15) hold for $E_{\lambda_{k_i}}$, the claim in the lemma follows easily, since $F = E^+_{0,\nu}$ does not depend of the chosen subsequence. □

Lemma 2 and a simple scaling argument easily yield the following precise estimates of densities at a point x in the reduced boundary

Theorem 2. *Let E be a Caccioppoli set in $\mathbb{R}^n$ and let $x \in \partial^- E$. Then*

$$\theta(E, x) = \frac{1}{2}$$

and, setting $\nu(x) := n(x, E)$,

$$\theta(E \cap E^+_{x,\nu(x)}, x) = \frac{1}{2} .$$

Moreover

$$\lim_{\rho \to 0} \frac{P(E, B(x, \rho))}{\omega_{n-1} \rho^{n-1}} = 1 .$$

Proof of Theorem 1. We consider the measure $\mu := |D\chi_E|$, and its blow-up, in the sense of the Definition 5 in Sec. 2.1.4, $\mu_{x,\lambda}$. Clearly

$$\mu_{x,\lambda} = |D\chi_{E_{x,\lambda}}|$$

and Lemma 2 yields

$$\mu_{x,\lambda} \rightharpoonup \mathcal{H}^{n-1} \llcorner E^0_{x,\nu(x)} , \qquad \nu := n(x, E) .$$

Therefore $\partial^- E$ is contained in the set

$$\mathcal{M} := \{x \in \mathbb{R}^n \mid \exists\, \theta(x) \in (0, +\infty) \text{ and an hyperplane } P_x \text{ such that } \mu_{x,\lambda} \rightharpoonup \theta(x)\mathcal{H}^{n-1} \llcorner P_x\} .$$

Since $|D\chi_E| = |D\chi_E| \llcorner \partial^- E$ and $\partial^- E \subset \mathcal{M}$, we have

$$|D\chi_E| \llcorner (\mathbb{R}^n \setminus \mathcal{M}) = 0 .$$

From Theorem 2 in Sec. 2.1.4 we infer that $\mathcal{M}$, and consequently $\partial^- E$ is an $(n-1)$-rectifiable set, moreover

$$|D\chi_E| = \theta \mathcal{H}^{n-1} \llcorner \mathcal{M}$$

hence, using again that $|D\chi_E| = |D\chi_E| \llcorner \partial^- E$ and $\partial^- E \subset \mathcal{M}$, we conclude that $|D\chi_E| = \mathcal{H}^{n-1} \llcorner \partial^- E$.

This complete the proof of Theorem 1, since the other claims follow easily. □

Besides the topological boundary and the reduced boundary we may also introduce the notion of *measure theoretic boundary*

Definition 1. *The* measure theoretic boundary *of a set $E \subset \mathbb{R}^n$ is defined as the complement of the set of points of density* 0 *or* 1 *of E*

$$\partial_\mu E := \{x \in \mathbb{R}^n \mid \theta^*(E, x) > 0 \text{ and } \theta_*(E, x) < 1\} .$$

We have

Proposition 3. *Let E be a Caccioppoli set in $\mathbb{R}^n$. Then*

$$\partial^- E \subset \partial_\mu E$$

and

$$\mathcal{H}^{n-1}(\partial_\mu E \setminus \partial^- E) = 0 .$$

In order to prove Proposition 3 we first state the following

Lemma 3. *Let E be a Caccioppoli set in $\mathbb{R}^n$. If for some $x \in \mathbb{R}^n$*

$$\lim_{\rho\to 0} \frac{P(E, B(x,\rho))}{\rho^{n-1}} = 0 \tag{16}$$

then either $\theta(E,x) = 0$ or $\theta(E,x) = 1$.

Proof. On account of Proposition 2 in Sec. 4.1.2, (16) implies

$$\lim_{\rho\to 0} \frac{\mathcal{L}^n(E\cap B_\rho)}{\rho^n}\,\frac{\mathcal{L}^n(B_\rho \setminus E)}{\rho^n} = 0 . \tag{17}$$

Set now

$$\alpha(\rho) := \frac{\mathcal{L}^n(E\cap B_\rho)}{\rho^n} .$$

From (17) we infer that for any $\varepsilon > 0$, $0 < \varepsilon < \frac{1}{4}$, there exists $\delta > 0$ such that for any $\rho \in (0,\delta)$

$$\alpha(\rho)\,(1-\alpha(\rho)) < \varepsilon$$

i.e.

$$\alpha(\rho) \in [0, 4\varepsilon) \cup (1-4\varepsilon, 1] .$$

Since α is continuous we infer that either

$$0 \le \alpha(\rho) < 4\varepsilon \qquad \text{for all } \rho < \delta$$

or

$$1 - 4\varepsilon \le \alpha(\rho) \le 1 \qquad \text{for all } \rho < \delta .$$

Letting $\varepsilon \to 0^+$ we conclude that either $\alpha(\rho) \to 0$ or $\alpha(\rho) \to 1$ as $\rho \to 0^+$ and this clearly proves the claim. □

Proof of Proposition 3. By Lemma 3 we have $\partial^- E \subset \partial_\mu E$, and

$$\partial_\mu E \subset \cup_{\delta>0} A_\delta , \qquad A_\delta := \{x \mid \theta^{*n-1}(|D\chi_E|, x) > \delta\} .$$

By Theorem 5 in Sec. 1.1.5 we know that

$$\delta \mathcal{H}^{n-1}(A_\delta \setminus \partial^- E) \le |D\chi_E|(A_\delta \setminus \partial^- E) = 0 ,$$

hence $\mathcal{H}^{n-1}(A_\delta \setminus \partial^- E) = 0$. This clearly yields $\mathcal{H}^{n-1}(\partial_\mu E \setminus \partial^- E) = 0$. □

One can also define

$$\bar{\partial} E := \{x \,|\, 0 < |E \cap B(x,\rho)| < |B(x,\rho)| \;\forall \rho > 0\}.$$

Clearly $\bar{\partial} E \subset \partial_\mu E$ and it is easily seen that $\bar{\partial} E$ is closed. In fact it turns out that

$$\bar{\partial} E = \partial E^*$$

where

$$\begin{aligned} E^* &= (E \cup E_1^*) \setminus E_0^* \\ E_1^* &:= \{x \,|\, \exists \rho > 0 \text{ such that } |E \cap B(x,\rho)| = |B(x,\rho)|\} \\ E_0^* &:= \{x \,|\, \exists \rho > 0 \text{ such that } |E \cap B(x,\rho)| = 0\}. \end{aligned}$$

In particular we have

$$\partial^- E \subset \overline{\partial^- E} \subset \bar{\partial} E.$$

Also, if $x \in \mathbb{R}^n \setminus \overline{\partial^- E}$, we find $B(x,\rho)$ with $B(x,\rho) \cap \partial^- E = \emptyset$, and by Gauss-Green formula,

$$< D\varphi_E, g> = \int_{\partial^- E} g D\varphi_E = 0 \qquad \forall\, g \in C_c^\infty(B(x,\rho)),$$

i.e., φ_E is constant in $B(x,\rho)$. It follows that $x \notin \bar{\partial} E$, that is, $\bar{\partial} E \subset \overline{\partial^- E}$. Consequently

$$\bar{\partial} E = \overline{\partial^- E}.$$

From now on we shall in principle think of a Caccioppoli set E as identified with E^*, so that $\partial E = \bar{\partial} E$. Notice that in principle we can have $\mathcal{H}^{n-1}(\bar{\partial} E \setminus \partial^- E) > 0$, but we have $\operatorname{spt} |D\varphi_E| = \bar{\partial} E$.

We conclude this subsection by stating without proof the following characterization of Caccioppoli sets

Theorem 3 (Federer). *Let E be a measurable set in $\mathbb{R}^n$. Then E is a Caccioppoli set if and only if*

$$\mathcal{H}^{n-1}(\partial_\mu E \cap K) \; < \; \infty$$

for any compact set $K \subset \mathbb{R}^n$.

1.4 The Structure Theorem for BV Functions

In this subsection we shall deal with pointwise properties of functions of bounded variation and we shall prove a *structure theorem* for them. The way this is done is by recovering information about u from information on its level subsets

(1) $$E_t(u) \;:=\; \{x \in \Omega \;\mid\; u(x) > t\}\;.$$

Therefore the main tool will be the coarea formula we proved in Sec. 4.1.1 which states that

(2) $$\int_\Omega |Du| = \int_{-\infty}^{+\infty} dt \int_\Omega |D\chi_{E_t(u)}|$$

or, taking into account the rectifiability theorem of Sec. 4.1.3,

(3) $$\int_\Omega |Du| = \int_{-\infty}^{+\infty} \mathcal{H}^{n-1}(\partial^- E_t(u) \cap \Omega)\, dt\,.$$

Given now a function $u \in BV(\Omega)$ we define

(4) $$\begin{aligned} u_-(x) &:= \operatorname*{apliminf}_{y\to x} u(y) = \sup\{t \mid \theta^n(E_t(u), x) = 1\} \\ u_+(x) &:= \operatorname*{aplimsup}_{y\to x} u(y) = \inf\{t \mid \theta^n(E_t(u), x) = 0\} \\ J_u &:= \{x \in \Omega \mid u_-(x) < u_+(x)\} \\ I_u &:= \{x \in \Omega \mid |u_+(x)| = \infty\} \cup \{x \in \Omega \mid |u_-(x)| = \infty\}\,. \end{aligned}$$

Notice that $u_-(x)$ and $u_+(x)$ are Borel functions in Ω. This is easily seen by reminding that for any measurable set E the upper and lower density functions $\theta^{*n}(E,x)$ and $\theta^n_*(E,x)$ are Borel functions (compare Sec. 1.1.5), and by observing that, for instance a point x satisfies $u_-(x) \geq t$ if and only if

$$\theta^n\Big(\{z \mid u(z) < t - 1/i\}, x\Big) = 0$$

for every positive integer i.

Theorem 1. *Let $u \in BV(\Omega)$. Then*

(i) J_u *is* $\mathcal{H}^{n-1}$*–measurable and countably* $(n-1)$*-rectifiable,*
(ii) $|Du| \llcorner (B \setminus J_u) = 0$ *for any* $\mathcal{H}^{n-1}$*-measurable set* B *with* $\mathcal{H}^{n-1}(B) < \infty$,
(iii) We have

$$|Du| \llcorner J_u = (u_+(x) - u_-(x))\mathcal{H}^{n-1} \llcorner J_u\,.$$

(iv) $\mathcal{H}^{n-1}$*-a.e.* J_u *agrees with the set of points where* $|Du|$ *has positive density with respect to* $\mathcal{H}^{n-1}$,

$$J_u = \{x \mid \theta^{n-1}(|Du|, x) > 0\} \qquad \mathcal{H}^{n-1}\text{-a.e.}\,.$$

and in fact

$$\theta^{n-1}(|Du|, x) = u_+(x) - u_-(x) \qquad \mathcal{H}^{n-1}\text{-a.e. in } J_u.$$

Proof. As all claims are local we can assume that $u \in BV(\mathbb{R}^n)$.
(i) From the coarea formula we known

$$|Du|(\mathbb{R}^n) = \int_{-\infty}^{+\infty} |D\chi_{E_t(u)}|(\mathbb{R}^n)\, dt < \infty .$$

We can therefore select a denumerable and dense set Q in $\mathbb{R}$ such that for each $t \in Q$ $E_t(u)$ has finite perimeter, hence, by the results in Sec. 4.1.3, $\partial^- E_t(u)$ is $(n-1)$-rectifiable and

$$\mathcal{H}^{n-1}(\partial_\mu E_t(u) \setminus \partial^- E_t(u)) = 0 .$$

From the definition of $u_-(x)$ and $u_+(x)$ we immediately see that

$$\begin{aligned} \theta^{*n}(E_t(u), x) &> 0 \quad \text{if } t < u_+(x) \\ \theta^n_*(E_t(u), x) &< 1 \quad \text{if } t > u_-(x) . \end{aligned}$$

Thus for each t we have

(5) $$J_t := \{x \in \mathbb{R}^n \mid u_-(x) < t < u_+(x)\} \subset \partial_\mu E_t(u) ,$$

and

$$J_u = \bigcup_{t \in Q} J_t \subset \bigcup_{t \in Q} \partial_\mu E_t(u) .$$

Consequently,

$$\mathcal{H}^{n-1}\Big(J_u \setminus \bigcup_{t\in Q} \partial^- E_t(u)\Big) = 0 .$$

by Proposition 3 in Sec. 4.1.3. This shows that J_u is $\mathcal{H}^{n-1}$–measurable and countably $(n-1)$-rectifiable.

(ii) Let B be a Borel set. We observe that the sets $(B \setminus J_u) \cap \partial^- E_t(u)$ are disjoint for different t. In fact if $x \in \partial^- E_t(u) \cap \partial^- E_s(u)$ and $t \neq s$, $t, s \in Q$, then $\theta(E_t(u), x) = \theta(E_s(u), x) = 1/2$, hence $u_-(x) < u_+(x)$, i.e. $x \in J_u$. From this we infer in particular that

$$\mathcal{H}^{n-1}((B \setminus J_u) \cap \partial^- E_t(u)) = 0 \qquad \text{for a.e. } t ,$$

if $\mathcal{H}^{n-1}(B) < \infty$. Finally, $|Du|(B \setminus J_u) = 0$ for all Borel sets with $\mathcal{H}^{n-1}(B) < \infty$ follows from the coarea formula

$$|Du|(B \setminus J_u) = \int_{-\infty}^{+\infty} \mathcal{H}^{n-1}((B \setminus J_u) \cap \partial^- E_t(u))\, dt .$$

(iii) Using the coarea formula we get for every Borel set B

$$\begin{aligned} |Du|(J_u \cap B) &= \int_{-\infty}^{+\infty} \mathcal{H}^{n-1}(\partial^- E_t(u) \cap J_u \cap B)\, dt \\ &= \int_{-\infty}^{+\infty} dt \int_B \chi_{\partial^- E_t(u) \cap J_u}\, d\mathcal{H}^{n-1} = \int_B (u_+(x) - u_-(x))\, d\mathcal{H}^{n-1} . \end{aligned}$$

(iv) From (iii) and the countably rectifiability of J_u we see that

$$\theta^{n-1}(|Du|, x) \geq u_+(x) - u_-(x) \qquad \mathcal{H}^{n-1}\text{-a.e. in } J_u$$

hence

$$J_u \subset \{x \mid \theta^{n-1}(|Du|, x) > 0\} .$$

On the other hand for any Borel set $B \subset J_u$ with $\mathcal{H}^{n-1}(B) < \infty$, by (iii) we have

$$\text{(6)} \quad \limsup_{\rho \to 0^+} \frac{|Du|(B(x,\rho))}{\mathcal{H}^{n-1}(B(x,\rho))} \leq \limsup_{\rho \to 0^+} \frac{|Du|(B(x,\rho))}{\mathcal{H}^{n-1}(B \cap B(x,\rho))} = u_+(x) - u_-(x)$$

for $\mathcal{H}^{n-1}$-a.e. $x \in B$, consequently for $\mathcal{H}^{n-1}$-a.e. $x \in J_u$ recalling that J_u is σ-finite with respect to $\mathcal{H}^{n-1}$. This concludes the proof. □

Remark 1. Let u be a BV-function with integer values. From Theorem 1 (iii) we infer that J_u is $(n-1)$–rectifiable, as $u_+(x) - u_-(x) \geq 1$. On the other hand for any t

$$E_t(u) = E_s(u)$$

for some $s \neq t$, hence

$$\partial^- E_t(u) \subset J_u$$

and by the coarea formula $|Du|$ is absolutely continuous with respect to $\mathcal{H}^{n-1} \llcorner J_u$. Consequently

$$|Du| = |Du| \llcorner J_u .$$

Let $u \in BV(\Omega)$. By Lebesgue's decomposition theorem for measures, compare Theorem 1 in Sec. 1.1.4, we can decompose the Radon measure Du into its absolutely continuous part $(Du)^a$ with respect to Lebesgue measure and its singular part $(Du)^s$

$$\text{(7)} \qquad Du = (Du)^a + (Du)^s$$

and

$$(Du)^s \perp \mathcal{L}^n , \qquad (Du)^a = \mathcal{L}^n \llcorner F(x), \qquad F \in L^1(\Omega, \mathbb{R}^n) .$$

We now define the *jump part* $(Du)^{(j)}$ and the *Cantor part* $(Du)^{(C)}$ of Du by

$$\text{(8)} \qquad (Du)^{(j)} := Du \llcorner J_u = (Du)^s \llcorner J_u$$

and

$$\text{(9)} \qquad (Du)^{(C)} := (Du)^s \llcorner (\Omega \setminus J_u) ,$$

so that

$$Du = (Du)^a + (Du)^{(j)} + (Du)^{(C)} .$$

Notice that $(Du)^a$, $(Du)^{(j)}$, and $(Du)^{(C)}$ are *mutually orthogonal.* Moreover note that the measure $(Du)^{(j)}$ lives on the countably $(n-1)$-rectifiable set J_u, while $(Du)^{(C)}$ lives on a Cantor type set C, that is,

$$(Du)^{(C)} = (Du)^{(C)} \llcorner C \quad \text{if} \quad \mathcal{H}^n(C) = 0 ,$$

and

$$(Du)^{(C)}(A) = 0 \quad \text{if} \quad \mathcal{H}^{n-1}(A) < \infty .$$

We shall now discuss fine properties of the measures $(Du)^a$ and $(Du)^{(j)}$. Let us begin with $(Du)^{(j)}$. First, we fix some terminology.

Let $u \in L^1_{\text{loc}}(\mathbb{R}^n)$. We say that a point $x \in \mathbb{R}^n$ is a *regular point for* u if the approximate limit of u at x exists and is finite. Non regular points are said to be *singular.* Among the singular points we now identify the *jump points* as follows

Definition 1. *We say that a point $x \in \mathbb{R}^n$ is a* jump point of u with parameters (α, β, ν), *if there exist a unit vector ν and real numbers α and β such that $\alpha > \beta$ and*

$$\operatorname*{aplim}_{\substack{y\to x\\ y\in E^+_{x,\nu}}} u(y) = \alpha \qquad \operatorname*{aplim}_{\substack{y\to x\\ y\in E^-_{x,\nu}}} u(y) = \beta$$

where we set

$$\begin{aligned} E^+_{x,\nu} &:= \{y \in \mathbb{R}^n \mid (y-x,\nu) > 0\} \\ E^-_{x,\nu} &:= \{y \in \mathbb{R}^n \mid (y-x,\nu) < 0\} \end{aligned}$$

and

$$E^0_{x,\nu} := \{y \in \mathbb{R}^n \mid (y-x,\nu) = 0\} .$$

It is easy to see that *the parameters α, β, ν are uniquely defined at each jump point.* Suppose in fact that at the point x there are two different sets of parameters (a, b, η) and (α, β, ν). If $\eta = \nu$, then $a = \alpha$ and $b = \beta$, as consequence of the uniqueness of the approximate limit, compare Proposition 1 in Sec. 3.1.4. Suppose now that η and ν are independent. Then the four sets

$$E^+_{x,\nu} \cap E^+_{x,\eta} , \quad E^+_{x,\nu} \cap E^-_{x,\nu} , \quad E^-_{x,\nu} \cap E^+_{x,\eta} , \quad E^-_{x,\nu} \cap E^-_{x,\eta}$$

are non empty cones which have non zero density at x. From this it follows at once $a = \alpha = b = \beta$, a contradiction. If finally $\eta = -\nu$, then $a = \beta$ and $b = \alpha$, again a contradiction since $a < b$ and $\alpha < \beta$.

Let x be a jump point for u with parameters (α, β, ν). We shall refer to ν as to the *jump direction of* u *at* x and to $\alpha - \beta > 0$ as to the *jump value*, or, simply, to the *jump of* u *at* x. Note that the jump value is always finite and positive.

Let us consider the special, but important case of characteristic functions of measurable sets. We can then characterize jump points in terms of densities

Proposition 1. *Let E be a measurable set in $\mathbb{R}^n$. Then $x \in \mathbb{R}^n$ is a jump point for χ_E with parameters $(1, 0, \nu)$ if and only if*

$$\theta^n(E \cap E^+_{x,\nu}, x) = \frac{1}{2} \qquad \theta^n(E \cap E^-_{x,\nu}, x) = 0 \tag{10}$$

or equivalently

$$(11) \qquad \theta^n(E \cap E^-_{x,\nu}, x) = 0 \qquad \theta^n((\mathbb{R}^n \setminus E) \cap E^+_{x,\nu}, x) = 0$$

or

$$(12) \qquad \theta^n(E \cap E^+_{x,\nu}, x) = \frac{1}{2} \qquad \theta^n(E, x) = \frac{1}{2} .$$

Proof. First, observe that (10), (11) and (12) are equivalent also to

$$(13) \qquad \theta^n(E \cap E^+_{x,\nu}, x) = \frac{1}{2} \qquad \theta^n((\mathbb{R}^n \setminus E) \cap E^-_{x,\nu}, x) = \frac{1}{2} .$$

Now, since χ_E takes only values 0 and 1, the set E agrees with the set

$$A_\varepsilon := \{x \in \mathbb{R}^n \mid |\chi_E(x) - 1| < \varepsilon\}$$

for any ε, $0 < \varepsilon < 1$. Therefore the first condition in (13) is equivalent to

$$\theta^n(A_\varepsilon \cap E^+_{x,\nu}, x) = \frac{1}{2}$$

for any ε, $0 < \varepsilon < 1$. This in turn is equivalent to

$$\operatorname*{aplim}_{\substack{y \to x \\ y \in E^+_{x,\nu}}} \chi_E(y) = 1 ,$$

compare Sec. 3.1.4.

Similarly we see that the second condition in (13) is equivalent to

$$\operatorname*{aplim}_{\substack{y \to x \\ y \in E^-_{x,\nu}}} \chi_E(y) = 0 ,$$

and this clearly concludes the proof. □

Suppose now that E is a Caccioppoli set in $\mathbb{R}^n$. From Theorem 2 in Sec. 4.1.3 we know that (12) with $\nu = n(x, E)$ holds for all point x in the reduced boundary $\partial^- E$ of E. Thus we immediately infer

Corollary 1. *Let E be a Caccioppoli set. Then every $x \in \partial^- E$ is a jump point for χ_E with parameters $(1, 0, n(x, E))$.*

Actually, we have even more. Denote by $\partial^* E$ the set of jump points of χ_E with jump value one,

$$\partial^* E := \{x \in \mathbb{R}^n \mid x \text{ is a jump point of } \chi_E \text{ with jump value } 1\}$$

and recall that $\partial_\mu E$ denotes the measure theoretic boundary of E, compare Definition 1 in Sec. 4.1.3. Then taking into account Proposition 3 in Sec. 4.1.3 we readily deduce

Proposition 2. *Let E be a Caccioppoli set in $\mathbb{R}^n$. Then*

$$\partial^- E \subset \partial^* E \subset \partial_\mu E$$

and

$$\mathcal{H}^{n-1}(\partial^* E \setminus \partial^- E) = 0 , \qquad \mathcal{H}^{n-1}(\partial_\mu E \setminus \partial^* E) = 0 .$$

Of course every jump point of u is in J_u. The next theorem gives a complete description of the set of jump points of a BV-function.

Theorem 2. *Let $u \in BV(\Omega)$. Then*

(i) *For $\mathcal{H}^{n-1}$-a.e. point $x \in J_u$ and all t with $u_-(x) < t < u_+(x)$, x is a jump point for $\chi_{E_t(u)}$ with parameters independent of t and equal to $(1, 0, n(x, J_u))$.*

(ii) *We have*

$$Du \llcorner J_u = n(x, J_u)\,|Du| \llcorner J_u = (u_+(x) - u_-(x))\, n(x, J_u)\, \mathcal{H}^{n-1} \llcorner J_u$$

(iii) *$\mathcal{H}^{n-1}$-a.e. point x in J_u is a jump point of u with parameters $(u_+(x), u_-(x), n(x, J_u))$.*

Proof. (i) We shall prove that (i) holds for all points x in $J_u \setminus N_1$,

$$N_1 = \bigcup_{t \in Q} (\partial_\mu E_t(u) \setminus \partial^- E_t(u))$$

where $Q \subset \mathbb{R}$ is the set defined in the proof of Theorem 1.

First we claim that

(a) If for $t \neq s$ we have

$$\theta^n(E_t(u), x) = \theta^n(E_s(u), x) = \frac{1}{2}$$

and x is a jump point for $\chi_{E_t(u)}$ with parameters $(1, 0, \nu)$, then x is also a jump point for $\chi_{E_s(u)}$ with the same parameters.

In fact, we have $\theta^n(E_t(u) \setminus E_s(u), x) = 0,\ t < s$, as $E_s(u) \subset E_t(u)$. Setting now $\nu = n(x, E_t(u))$ we deduce

$$\begin{aligned} \theta^n(E_s(u) \cap E^+_{x,\nu}, x) &= \theta^n(E_t(u) \cap E^+_{x,\nu}, x) - \theta^n(E_t(u) \setminus E_s(u), x) \\ &= \frac{1}{2} - 0 = \frac{1}{2} \end{aligned}$$

taking into account that $\theta^n(E_t(u) \cap E^+_{x,\nu}) = 1/2$. Thus the claim (a) follows from Proposition 1.

We now observe that

$$x \in \partial_\mu E_t(u) \cap \partial_\mu E_s(u)$$

whenever $x \in J_u$ and $u_-(x) < t < s < u_+(x)$, while

$$x \in \partial^- E_t(u) \cap \partial^- E_s(u)$$

if $x \in J_u \setminus N_1$ and $t, s \in Q$, $u_-(x) < t < s < u_+(x)$. In this last case, in particular, from Theorem 2 in Sec. 4.1.3 and Proposition 1, x is a jump point for both $\chi_{E_t(u)}$ and $\chi_{E_s(u)}$,

$$\theta^n(E_t(u), x) = \theta^n(E_s(u), x) = \frac{1}{2} .$$

and

$$n(x, E_s(u)) = n(x, E_t(u))$$

by the claim (a).

Since Q is dense in $(u_-(x), u_+(x))$ we also have

$$\theta^n(E_r(u), x) = \frac{1}{2}$$

for all $r \in (u_-(x), u_+(x))$. Therefore, again from the claim (a) we deduce that x is a jump point for $\chi_{E_r(u)}$ for any $r \in (u_-(x), u_+(x))$ with jump direction n given by

$$n = n(x, E_s(u)) \qquad \forall\, s \in Q .$$

In particular $n(x, J_u)$ defined as

$$n(x, J_u) = n$$

is orthogonal to the approximate tangent plane $\mathrm{Tan}_x J_u$ to J_u at x.

(ii) Using the coarea formula in its vectorial formulation, (17) in Sec. 4.1.1,

$$\int_B d\,Du = \int_{-\infty}^{+\infty} dt\Big(\int_B d\,D\chi_{E_t(u)}\Big)$$

one derives

$$\begin{aligned} Du(J_u \cap B) &= \int_{-\infty}^{+\infty} dt \int_B \chi_{\partial^- E_t(u) \cap J_u}\, n(x, E_t(u))\, d|D\chi_{E_t(u)}| \\ &= \int_B n(x, J_u)\, d|Du| \llcorner J_u = \int_B (u_+(x) - u_-(x))\, n(x, J_u)\, d\mathcal{H}^{n-1} \end{aligned}$$

taking into account (iii) of Theorem 1.

(iii) First let us show that

$$\mathcal{H}^{n-1}(J_u \cap I_u) = 0 . \tag{14}$$

By Fubini-Tonelli Theorem and (5) we infer

$$\int_{J_u} (u_+(x) - u_-(x))\, d\mathcal{H}^{n-1} = \int_0^\infty \mathcal{H}^{n-1}(J_u \cap \{x : u_-(x) < t < u_+(x)\})\, dt$$

$$\leq \int_0^\infty \mathcal{H}^{n-1}(\partial_\mu E_t(u))\, dt = \int_0^\infty \mathcal{H}^{n-1}(\partial^- E_t(u))\, dt \leq |Du|(\mathbb{R}^n) < \infty .$$

Consequently $u_+(x) - u_-(x)$ is $\mathcal{H}^{n-1}$-a.e. finite in J_u. This yields (13) as $x \in J_u \cap I_u$ if and only if $u_+(x) - u_-(x) = +\infty$.

Fix now a point $x \in J_u \setminus I_u$ for which $\mathrm{Tan}_x J_u$ and $n(x, J_u) = n(x, E_t(u))$ for a.e. t with $u_-(x) < t < u_+(x)$ are well defined. Notice that this happens $\mathcal{H}^{n-1}$-a.e. in J_u. Define

$$J_x^+ := \{y \in \mathbb{R}^n \mid (y - x, n(x, J_u)) > 0\} .$$

For almost all $\varepsilon > 0$ with

$$u_-(x) < u_+(x) - \varepsilon < u_+(x)$$

we then infer

$$\theta^n(E_{u_+(x)+\varepsilon}(u), x) = 0$$

from the definition of $u_+(x)$ and

$$\theta^n(E_{u_+(x)-\varepsilon}(u) \cap J_x^+, x) = \frac{1}{2}$$

since $n(x, J_u) = n(x, \partial^- E_{u_+(x)-\varepsilon}(u))$. Therefore, for

$$A_\varepsilon := \{y \mid |u(y) - u_+(x)| < \varepsilon\} ,$$

we find

$$\theta^n(A_\varepsilon \cap J_x^+, x) = \theta^n(E_{u_+(x)-\varepsilon}(u) \cap J_x^+, x) - \theta^n(E_{u_+(x)+\varepsilon} \cap J_x^+, x) = \frac{1}{2} .$$

This yields at once

$$\operatorname*{aplim}_{\substack{y \to x \\ y \in J_x^+}} u(y) = u_+(x) ,$$

compare the proof of Proposition 1. Similarly one proves

$$\operatorname*{aplim}_{\substack{y \to x \\ y \in \mathbb{R}^n \setminus \overline{J}_x^+}} u(y) = u_-(x) .$$

□

Corollary 2. *Let $u \in BV(\Omega)$. We have*

(i) *For $|Du|$-a.e. x in J_u, equivalently for $\mathcal{H}^{n-1}$-a.e. $x \in J_u$,*

$$n(x, J_u) = \frac{d(Du)^{(j)}}{d|(Du)^{(j)}|}(x) = \frac{d\,Du}{d\,|Du|}(x)$$

(ii) *For $|Du|$-a.e. $x \in \Omega \setminus J_u$*

$$n(x, \partial^- E_{u_+(x)}(u)) = \frac{d\,Du}{d\,|Du|}(x) .$$

Proof. The first equality in (i) follows at once from Theorem 2 (ii), the second is trivial because of the definition of $(Du)^{(j)}$.

Let us prove (ii). First we observe that if $x \in \Omega \setminus J_u \cap \partial^- E_t(u)$, then $u_-(x) = u_+(x)$ and $\theta(E_t(u), x) = 1/2$; hence $u_-(x) = t = u_+(x)$. Being

$$|D\chi_{E_t(u)}| = \mathcal{H}^{n-1} \llcorner \partial^- E_t(u) ,$$

using the coarea formula we infer for any $\varphi \in C_c^\infty(\Omega)$

$$\begin{aligned}
&\int \varphi(x)\, d\,Du \llcorner \Omega \setminus J_u \\
&= \int_{-\infty}^{+\infty} dt \int \varphi(x) n(x, \partial^- E_t(u))\, d\mathcal{H}^{n-1} \llcorner \partial^- E_t(u) \setminus J_u \\
&= \int_{-\infty}^{+\infty} dt \int \varphi(x) n(x, \partial^- E_{u_+(x)}(u))\, d\mathcal{H}^{n-1} \llcorner \partial^- E_t(u) \setminus J_u \\
&= \int \varphi(x) n(x, \partial^- E_{u_+(x)}(u))\, d|Du| \llcorner \Omega \setminus J_u ,
\end{aligned}$$

since $n(x, \partial^- E_t(u))$ is a Borel function in x and t. This of course proves (ii). □

We saw in the proof of the previous theorem that

$$\mathcal{H}^{n-1}(I_u \cap J_u) = 0 .$$

Now we want to show that in fact

$$\mathcal{H}^{n-1}(I_u) = 0 .$$

It is immediately seen that

$$\begin{aligned}
I_u &= (J_u \cap I_u) \cup \widetilde{I}_u \\
\widetilde{I}_u &:= \{x \mid u_+(x) = -\infty\} \cup \{x \mid u_-(x) = +\infty\} .
\end{aligned}$$

Therefore it suffices to show that the set

$$\widetilde{I} := \{x \mid u_-(x) = +\infty\}$$

has zero $\mathcal{H}^{n-1}$-measure. We shall do that by means of the estimate on the approximating measures $\mathcal{H}^{n-1}_\delta$ contained in the next lemma.

Lemma 1. *Let E be a Caccioppoli set with finite measure and finite perimeter. Denote by E_1 the set*

$$E_1 := \{x \in \mathbb{R}^n \mid \theta^n(E,x) = 1\} .$$

Then there exist two constants c_1 and c_2 depending only on the dimension n such that

$$H_\delta^{n-1}(E_1) \leq c_2 \, |D\chi_E|(\mathbb{R}^n)$$

where

$$\delta := \Big[c_1 \, |D\chi_E|(\mathbb{R}^n)\Big]^{\frac{1}{n-1}} .$$

Proof. For any $x \in E_1$ the map $g : \mathbb{R}_+ \to \mathbb{R}$ given by

$$g(\rho) := \frac{\mathcal{L}^n(E \cap B(x,\rho))}{\mathcal{L}^n(B(x,\rho))}$$

is continuous in $(0,+\infty)$, $g(\rho) \to 1$ as $\rho \to 0$, and $g(\rho) \to 0$ as $\rho \to \infty$, since $\mathcal{L}^n(E) < \infty$. Then we can find for every $x \in E_1$ a radius r_x such that

$$\frac{\mathcal{L}^n(E \cap B(x,r_x))}{\mathcal{L}^n(B(x,r_x))} = \frac{1}{2}$$

and from the isoperimetric inequality (10) in Sec. 4.1.3

$$r_x^{n-1} \leq c_1(n) \, |D\chi_E|(\mathbb{R}^n) = \delta^{n-1} .$$

Using Besicovitch covering theorem, Lemma 2 in Sec. 1.1.5, we find a covering of E_1 by balls $B(x_i, r_{x_i})$ such that the balls $B(x_i, r_{x_i}/3)$ are disjoint. Now we have

$$\begin{aligned} |D\chi_E|(\mathbb{R}^n) &\geq \sum_i |D\chi_E|(B(x_i, \frac{1}{3} r_{x_i}) \geq 3^{1-n} \, c_1(n)^{-1} \sum_i r_{x_i}^{n-1} \\ &\geq c_2(n)^{-1} \, H_\delta^{n-1}(E_1) \end{aligned}$$

where the last inequality follows since all radii r_{x_i} are less than δ. □

Theorem 3. *Let $u \in BV(\Omega)$. Then $\mathcal{H}^{n-1}(I_u) = 0$.*

Proof. Without loss of generality we can assume that u belongs to $BV(\mathbb{R}^n)$ and that u has compact support in $\mathbb{R}^n$. Because of

$$|Du|(\mathbb{R}^n) = \int_{-\infty}^{+\infty} |D\chi_{E_t(u)}|(\mathbb{R}^n) \, dt$$

we can choose a sequence $\{t_k\}$, $t_k \to \infty$, such that

$$|D\chi_{E_{t_k}(u)}|(\mathbb{R}^n) \longrightarrow 0 \qquad \text{as } k \to \infty .$$

Since for any t

$$\{x \mid u_-(x) > t\} \subset \{x \mid \theta(E_t(u), x) = 1\} \subset\subset \mathbb{R}^n$$

we can apply Lemma 1, deducing

$$H^{n-1}_{\delta_k}(\{x \mid u_-(x) > t_k\}) \le c_1 |D\chi_{E_{t_k}(u)}|(\mathbb{R}^n)$$

with

$$\delta_k := [c_2 |D\chi_{E_{t_k}(u)}|(\mathbb{R}^n)]^{\frac{1}{n-1}} .$$

Letting $t_k \to \infty$, we then obtain for each $\delta > 0$

$$H^{n-1}_\delta(\{x \mid u_-(x) = +\infty\}) \le c_1 \liminf_{k\to\infty} |D\chi_{E_{t_k}(u)}|(\mathbb{R}^n) = 0 ,$$

hence

$$\mathcal{H}^{n-1}(\{x \mid u_-(x) = +\infty\}) = 0 .$$

□

Concerning the pointwise values of a BV-function u, Theorem 2 (iii) says in particular that $\mathcal{H}^{n-1}$-a.e. *point* $x \in J_u$ is a *jump point with*

$$u_+(x) = \operatorname*{aplim}_{\substack{y\to x\\ y\in E^+_{x,\nu}}} u(y) , \qquad u_-(x) = \operatorname*{aplim}_{\substack{y\to x\\ y\in E^-_{x,\nu}}} u(y) , \qquad \nu = n(x, J_u)$$

while Theorem 3 says that $\mathcal{H}^{n-1}$-a.e. *point* $x \in \Omega \setminus J_u$ *is a regular point*

$$\operatorname*{aplim}_{y\to x} u(y) = u_+(x) = u_-(x) \in \mathbb{R} .$$

These claims can be in fact improved as follows.

Theorem 4. *Let* $u \in BV(\Omega)$. *Set* $1^* = n/(n-1)$

(i) *For* $\mathcal{H}^{n-1}$-a.e $x \in \Omega \setminus J_u$ *we have*

$$u_-(x) = u_+(x) =: \overline{u}(x) \in \mathbb{R}$$

and

$$\lim_{\rho\to 0^+} \fint_{B(x,\rho)} |u(y) - \overline{u}(x)|^{1^*}\, dy = 0 . \tag{15}$$

(ii) *Set*

$$B^+_\nu(x,\rho) := B(x,\rho) \cap E^+_{x,\nu} , \qquad B^-_\nu(x,\rho) := B(x,\rho) \cap E^-_{x,\nu} .$$

Then for $\mathcal{H}^{n-1}$-a.e. $x \in J_u$ *and* $\nu = n(x, J_u)$, *we have*

$$\lim_{\rho\to 0^+} \fint_{B^+_\nu(x,\rho)} |u(y) - u_+(x)|^{1^*}\, dy = 0 \tag{16}$$

$$\lim_{\rho\to 0^+} \fint_{B^-_\nu(x,\rho)} |u(y) - u_-(x)|^{1^*}\, dy = 0 \tag{17}$$

In order to prove Theorem 4 we need the following variant of Sobolev-Poincaré inequality

Proposition 3 (Sobolev-Poincaré inequality). *Let $u \in BV(B(x_0,\rho))$. Set $1^* := n/(n-1)$ and*

$$\alpha_\rho := \frac{\mathcal{L}^n(\{x \in B(x_0,\rho) \mid u(x) = 0\})}{\mathcal{L}^n(B(x_0,\rho))} .$$

Then we have

$$\text{(18)} \qquad \Big(\int_{B(x_0,\rho)} |u(x)|^{1^*} \, dx \Big)^{1/1^*} \le \frac{c(n)}{1-(1-\alpha_\rho)^{1/n}} \frac{1}{\rho^{n-1}} \int_{B(x_0,\rho)} |Du| .$$

Proof. From Sobolev-Poincaré inequality in [3] in Sec. 4.1.1

$$\begin{aligned} \Big(\fint_{B(x_0,\rho)} |u|^{1^*} \, dx \Big)^{1/1^*} &\le \Big(\fint_{B(x_0,\rho)} |u(x) - u_{x_0,\rho}|^{1^*} \, dx \Big)^{1/1^*} + |u_{x_0,\rho}| \\ &\le \frac{c(n)}{\rho^{n-1}} \int_{B(x_0,\rho)} |Du| + |u_{x_0,\rho}| . \end{aligned}$$

The result then follows since by Hölder inequality

$$\begin{aligned} |u_{x_0,\rho}| &\le \frac{1}{|B(x_0,\rho)|} \int_{\{x \in B(x_0,\rho) \mid u(x) \neq 0\}} |u(x)| \, dx \\ &\le (1-\alpha_\rho)^{1/n} \Big(\fint_{B(x_0,\rho)} |u|^{1^*} \, dx \Big)^{1/1^*} . \end{aligned}$$

□

We also need

Lemma 2. *Let $u \in BV(\Omega)$. Define*

$$N_u := \{ x \in \Omega \mid \theta^{* \, n-1}(|Du|, x) > u_+(x) - u_-(x) \} .$$

Then $\mathcal{H}^{n-1}(N_u) = 0$.

Proof. We split N_u as $N_u = N_1 \cup N_2$ where $N_1 := N_u \cap (\Omega \setminus J_u)$ and $N_2 := N_u \cap J_u$, and we prove that N_1 and N_2 have zero $(n-1)$-dimensional Hausdorff measure. By Theorem 1 clearly $\mathcal{H}^{n-1}(N_2) = 0$. To prove that $\mathcal{H}^{n-1}(N_1) = 0$ it suffices to show that for $k = 1, 2, \ldots$ we have $\mathcal{H}^{n-1}(N_k) = 0$ where

$$N_k := \{ x \in \Omega \setminus J_u \mid \theta^{* \, n-1}(|Du|, x) > \frac{1}{k} \} .$$

From Theorem 5 in Sec. 1.1.5 we know that

$$(19) \qquad \mathcal{H}^{n-1}(N_k) \ \leq \ k\,|Du|(N_k)\ ,$$

hence $\mathcal{H}^{n-1}(N_k) < \infty$. Since $N_k \subset \Omega \setminus J_u$, from Theorem 1 (ii) we then infer $|Du|(N_k) = 0$, and therefore $\mathcal{H}^{n-1}(N_k) = 0$ again by (19). □

Proof of Theorem 4. The proofs of (15) (16) (17) are similar. We shall therefore prove (16) and leave the proof of (15) and (17) to the reader. Consider first the case in which $u \in BV(\Omega) \cap L^\infty(\Omega)$. Then we have

$$(20) \qquad u_+(x) \ = \ \underset{\substack{y \to x \\ y \in E^+_{x,\nu}}}{\mathrm{aplim}}\, u(y) \ \in \ \mathbb{R}$$

for $\mathcal{H}^{n-1}$-a.e. $x \in J_u$ by Theorem 2 (iii). As $u \in L^\infty(\Omega)$, it is then easy to show that (20) is equivalent to (16), compare [1] in Sec. 3.1.4.

In the case in which $u \notin L^\infty(\Omega)$, we consider of truncated of u, $\{u_h\}$, defined by

$$u_h(x) \ = \ \max(\min(u(x), h), -h)\ .$$

Of course

$$u_{h\pm}(x) \ = \ \max(\min(u_\pm(x), h), -h)\ .$$

hence

$$J_{u_h} \ \subset \ J_u\ .$$

For each $h = 1, 2, \ldots$ denote by $\Sigma_h \subset J_u$ the set on which (16) fails for u_h. As $u_h \in L^\infty(\Omega)$ we have

$$\mathcal{H}^{n-1}(\Sigma_h) \ = \ 0\ .$$

We also consider the set

$$N_h \ := \ \{x \in \Omega \ \mid \ \theta^{n-1}(|D(u-u_h)|, x) > (u-u_h)_+(x) - (u-u_h)_-(x)\}$$

for which we have

$$\mathcal{H}^{n-1}(N_h) \ = \ 0$$

by Lemma 2. Then we set

$$\Sigma \ = \ \bigcup_h \Sigma_h \qquad N \ = \ \bigcup_h N_h$$

and

$$I \ = \ I_u \ = \ \{x \ \mid \ |u_+(x)| = \infty \ \text{ or } \ |u_-(x)| = \infty\}\ .$$

Of course $\Sigma \subset J_u$, $\mathcal{H}^{n-1}(\Sigma) = \mathcal{H}^{n-1}(N) = 0$, while from Theorem 3 $\mathcal{H}^{n-1}(I) = 0$. Thus it suffices to prove (16) for all x in $\Omega \setminus (\Sigma \cup N \cup I)$. Fix one such an x. We have

$$\Big(\fint_{B_\nu^+(x,\rho)} |u(y) - u_+(x)|^{1^*}\, dy \Big)^{1/1^*} \le \Big(\fint_{B_\nu^+(x,\rho)} |u(y) - u_h(y)|^{1^*}\, dy \Big)^{1/1^*}$$

$$(21) \qquad + \Big(\fint_{B_\nu^+(x,\rho)} |u_h(y) - u_{h+}(x)|^{1^*}\, dx \Big)^{1/1^*} + |u_{h+}(x) - u_+(x)| \,.$$

As $x \notin \Sigma$, for each h we have

$$(22) \qquad \lim_{\rho\to 0} \fint_{B_\nu^+(x,\rho)} |u_h(y) - u_{h+}(x)|^{1^*}\, dx \;=\; 0\,.$$

As $x \notin I$, $\max(|u_+(x)|, |u_-(x)|) \in \mathbb{R}$; hence, if we choose

$$h > \max(|u_+(x)|, |u_-(x)|)\,,$$

we have

$$(23) \qquad u_{h+}(x) \;=\; u_+(x)\,, \qquad u_{h-}(x) \;=\; u_-(x)\,.$$

Therefore

$$(24) \qquad \limsup_{\rho\to 0} \Big(\fint_{B_\nu^+(x,\rho)} |u(y) - u_+(x)|^{1^*}\, dx \Big)^{1/1^*} \le\; 2 \limsup_{\rho\to 0} \Big(\fint_{B(x,\rho)} |u(y) - u_h(y)|^{1^*}\, dy \Big)^{1/1^*}$$

We now observe that the set $\{y \in \Omega \mid |u(y)| < h\}$ has density 1 at x for the chosen h, and $u - u_h$ is zero on that set. We then apply the Sobolev inequality in Proposition 3 and find

$$\limsup_{\rho\to 0} \Big(\fint_{B(x,\rho)} |u(y) - u_h(y)|^{1^*}\, dy \Big)^{1/1^*}$$

$$(25) \qquad \le\; c(n) \limsup_{\rho\to 0} \frac{|D(u-u_h)|\, B(x,\rho))}{\rho^{n-1}}$$

$$=\; c'(n)\, \theta^{*\,n-1}(|D(u-u_h)|, x)$$

$$\le\; c'(n)\, [u_+(x) - u_{h+}(x) - u_-(x) + u_{h-}(x)] \;=\; 0$$

as $x \notin N$. The claim then follows at once from (21)-(25). □

Remark 2. If $u \in L^1(\Omega)$, we know that the convolutions $u * \varphi_\sigma(x)$ converge to $u(x)$ at every Lebesgue point x of u when $\sigma \to 0^+$. From Theorem 4 we therefore see that convolutions with a symmetric kernel φ converge pointwisely to

$$\frac{u_+(x) + u_-(x)}{2}$$

$\mathcal{H}^{n-1}$-a.e. in J_u, provided $u \in BV(\Omega)$.

Remark 3. Let $u \in W^{1,1}(\Omega)$. Of course then $u \in BV(\Omega)$ and $(Du)^{(j)} = 0$, i.e., $\mathcal{H}^{n-1}(J_u) = 0$. Therefore from Theorem 4 we deduce that *for* $\mathcal{H}^{n-1}$*-a.e.* x *in* Ω

$$\fint_{B(x,r)} |u(y) - u(x)|^{n/(n-1)}\, dy \longrightarrow 0 \qquad \text{as } r \to 0\ ,$$

in particular

$$\mathcal{H}^{n-1}(\Omega \setminus \mathcal{L}_u) = 0\ .$$

Next theorem states that BV-functions are almost everywhere approximately differentiable and, indeed, the approximate differential is the density of the absolutely continuous part $(Du)^a$ of Du.

Theorem 5. *Any function* u *in* $BV(\Omega)$ *is approximately differentiable at almost every point* $x \in \Omega$ *and*

$$(Du)^a = \mathcal{L}^n \llcorner \operatorname{ap}Du\ .$$

In particular we have

$$BV(\Omega) \subset \mathcal{A}^1(\Omega, \mathbb{R}) \tag{26}$$

Actually a stronger statement holds, which in fact implies Theorem 5.

Theorem 6. *Let* $u \in BV(\Omega)$ *and let* x *be a Lebesgue point for* u, $x \in \mathcal{L}_u$, $u(x)$ *the Lebesgue value of* u *at* x*. Suppose moreover that the Radon-Nikodym derivative* $\frac{dDu}{d\mathcal{L}^n}(x)$ *exists at* x*, and* x *is a Lebesgue point for* $\dfrac{dDu}{d\mathcal{L}^n}$, $x \in \mathcal{L}_{\frac{dDu}{d\mathcal{L}^n}}$*. Then*

$$\lim_{\rho \to 0} \fint_{B(x,\rho)} \frac{|u(y) - u(x) - L(y-x)|}{|y-x|}\, dy = 0$$

where L *is the Lebesgue value of* $\frac{dDu}{d\mathcal{L}^n}(x)$ *at* x*. In particular*

$$\operatorname{ap}Du(x) = \frac{dDu}{d\mathcal{L}^n}(x)\ .$$

The proof of Theorem 5 proceeds along the same lines of the proof of Calderón-Zygmund L^p-differentiability theorem in $W^{1,p}$, Theorem 2 in Sec. 3.1.2. The key step is the following proposition which replaces Proposition 1 in Sec. 3.1.2.

Proposition 4. *Let* $u \in BV_{\text{loc}}(\mathbb{R}^n)$*, let* $x \in \mathcal{L}_u$ *and let* $u(x)$ *be the Lebesgue value of* u *at* x*. Then for all* $r > 0$ *we have*

$$\fint_{B(x,r)} \frac{|u(y) - u(x)|}{|y-x|}\, dy \le \int_0^1 \frac{|Du|(B(x,tr))}{|B(x,tr)|}\, dt \tag{27}$$

Proof. Of course we may assume that the right hand side of (27) is finite. From Proposition 1 in Sec. 3.1.2 step (i) and (ii) we know that (27) holds for $u \in BV_{\mathrm{loc}}(\mathbb{R}^n) \cap C^1(\mathbb{R}^n \setminus \{x\})$. The same construction in Lemma 1 in Sec. 3.1.2 yields now a sequence $\{u_k\} \subset C^1(\mathbb{R}^n \setminus \{x\})$ such that

$$\begin{aligned} u_k &\to u \qquad \text{in } L^1 \text{ and a.e.} \\ \mathcal{L}^n \llcorner |Du_k(x)| &\rightharpoonup |Du| \quad \text{as measures} \end{aligned}$$

and for $r \to 0$

$$\fint_{B(x,r)} |u_k(y) - u(x)|\, dy \longrightarrow 0$$

$$\int_{B(x,r)} |Du_k|\, dy \le c|Du|(B(x,2r))\,.$$

Applying (27) to u_k, passing to the limit, and taking into account Fatou's lemma and Lebesgue's dominated convergence theorem we then get

$$\begin{aligned} \fint_{B(x,\rho)} \frac{|u(y)-u(x)|}{|y-x|}\, dy &\le \liminf_{k\to\infty} \fint_{B(x,\rho)} \frac{|u_k(y)-u_k(x)|}{|y-x|}\, dy \\ &\le \liminf_{k\to\infty} \int_0^1 dt \fint_{B(x,t\rho)} |Du_k(y)|\, dy = \int_0^1 \frac{|Du|(B(x,t\rho)}{\mathcal{L}^n(B(x,t\rho))}\, dt\,. \end{aligned}$$

□

Proof of Theorem 6. Denote by $F(x)$ the Radon-Nikodym derivative of Du with respect to $\mathcal{L}^n$. The assumptions yield $x \in \mathcal{L}_F$ and

$$\fint_{B(x,\rho)} |F(y)-F(x)|\, dy \longrightarrow 0\,, \qquad \rho^{-n}|(Du)^s|(B(x,\rho)) \longrightarrow 0\,.$$

Thus applying (27) to the function

$$v(y) := u(y) - u(x) - F(x)\,(y-x)$$

we get

$$\begin{aligned} &\fint_{B(x,\rho)} \frac{|u(y)-u(x)-F(x)(y-x)|}{|y-x|}\, dy \\ &\quad \le \int_0^1 \frac{|Du - F(x)\, d\mathcal{L}^n(y)|(B(x,t\rho))}{\mathcal{L}^n(B(x,t\rho))}\, dt \\ &\quad \le \sup_{0\le r\le\rho} \left(\fint_{B(x,r)} |F(y)-F(x)|\, dy + \frac{|(Du)^s|(B(x,r))}{\mathcal{L}^n(B(x,r))} \right) \end{aligned}$$

which concludes the proof, as the right hand-side tends to zero as $\rho \to 0^+$. □

In conclusion we can state

Theorem 7 (Structure theorem). *Let $u \in BV(\Omega)$. Then*

(i) *The set J_u is $\mathcal{H}^{n-1}$ measurable and countably $(n-1)$-rectifiable and agrees $\mathcal{H}^{n-1}$-a.e. with the set of jump points of u.*
(ii) *$\mathcal{H}^{n-1}$-a.e. point x in $\Omega \setminus J_u$ is a Lebesgue point for u.*
(iii) *u is almost everywhere approximately differentiable in Ω.*
(iv) *The gradient measure Du splits as*

$$Du = \mathrm{ap}Du\,dx + (u_+(x) - u_-(x))\,n(x, J_u)\,d\mathcal{H}^{n-1} \llcorner J_u + (Du)^{(C)} .$$

Moreover

$$(Du)^{(C)} = (Du)^{(C)} \llcorner C \quad \text{if} \quad \mathcal{H}^n(C) = 0$$

and

$$(Du)^{(C)}(A) = 0 \quad \text{if} \quad \mathcal{H}^{n-1}(A) < \infty .$$

1.5 Subgraphs of BV Functions

The aim of this subsection is to show that properties of BV-functions can be read equivalently in terms of their subgraphs.

As usual, Ω will denote a bounded domain of $\mathbb{R}^n$, $n \geq 2$, $\pi : \Omega \times \mathbb{R} \to \Omega$ the orthogonal projection of $\Omega \times \mathbb{R}$ in Ω, and we denote by $(x^1, \dots, x^n, y) = (x, y)$, $x \in \Omega$, the coordinates in $\Omega \times \mathbb{R}$.

For any measurable function $u : \Omega \to \mathbb{R}$ the *subgraph of* u is defined as the measurable subset of $\Omega \times \mathbb{R}$ given by

(1) $$\mathcal{SG}_{u,\Omega} := \{(x, y) \in \Omega \times \mathbb{R} \mid y < u(x)\} .$$

When no confusion may arise we simplify the notation writing

$$\mathcal{S}_u \quad \text{for} \quad \mathcal{SG}_{u,\Omega} .$$

Correspondingly, we shall denote by $\chi_{\mathcal{S}_u}$ the characteristic function of $\mathcal{SG}_{u,\Omega}$ in $\Omega \times \mathbb{R}$, and, for the sake of simplicity, we write $D_i\chi_{\mathcal{S}_u}$ instead of $\frac{\partial}{\partial x^i}\chi_{\mathcal{S}_u}$, $i = 1, \dots, n$, and $D_{n+1}\chi_{\mathcal{S}_u}$ instead of $\frac{\partial}{\partial y}\chi_{\mathcal{S}_u}$. Finally we introduce the $\mathbb{R}^{n+1}$-valued distribution in Ω defined by

(2) $$\mu(Du) := (D_1 u, \dots, D_n u, -\mathcal{L}^n) .$$

Our first theorem characterizes BV-functions in terms of their subgraphs.

Theorem 1. *Let $u \in L^1(\Omega)$. Then u belongs to $BV(\Omega)$ if and only if the subgraph of u, $\mathcal{S}_u$, has finite perimeter in $\Omega \times \mathbb{R}$,*

$$P(\mathcal{S}_u, \Omega \times \mathbb{R}) < \infty .$$

Moreover we have

$$\begin{aligned}
&(3) \qquad \pi_{\#} D_i \chi_{S_u} = \mu_i(Du) \qquad i = 1, \dots, n+1\\
&(4) \qquad \pi_{\#} |D_i \chi_{S_u}| = |\mu_i(Du)| \qquad i = 1, \dots, n+1\\
&(5) \qquad \pi_{\#} |D \chi_{S_u}| = |\mu(Du)| \, .
\end{aligned}$$

Proof. Suppose first that $P(\mathcal{S}_u, \Omega \times \mathbb{R}) < \infty$. In this case the measures $D_i \chi_{S_u}$ can be extended as linear functionals acting on continuous and bounded functions in $\Omega \times \mathbb{R}$ by means of Lebesgue theorem. In particular setting

$$\rho_N(y) := \begin{cases} 1 & |y| \le N \\ N + 1 - |y| & N \le |y| \le N+1 \\ 0 & |y| \ge N+1 \end{cases}$$

we get for any $\varphi \in C_c^\infty(\Omega)$ and $i = 1, \dots, n$

$$\begin{aligned}
\int_{\Omega \times \mathbb{R}} \varphi(x) \, D_i \chi_{S_u} &= - \lim_{N \to \infty} \int_{\Omega \times \mathbb{R}} D_i \varphi(x) \, \rho_N(y) \, \chi_{S_u}(x, y) \, dx \, dy \\
(6) \qquad &= - \lim_{N \to \infty} \int_\Omega D_i \varphi(x) \Big(\int_{-\infty}^{u(x)} \rho_N(y) \, dy \Big) \, dx \, .
\end{aligned}$$

As

$$\int_{-\infty}^{z} \rho_N(y) \, dy \le z + N + 1 \qquad \text{if } z > -N - 1,$$

$$\int_{-\infty}^{z} \rho_N(y) \, dy - N - 1 \longrightarrow z \qquad \text{as } N \to \infty \, ,$$

by definition of $\rho_N(y)$, and as $u \in L^1(\Omega)$, we then get

$$(7) \qquad \int_{\Omega \times \mathbb{R}} \varphi(x) \, D_i \chi_{S_u} = - \int_\Omega D_i \varphi(x) \, u(x) \, dx$$

letting $N \to \infty$ in (6). Equivalently we can write

$$(8) \qquad \pi_{\#} D_i \chi_{S_u} = D_i u \qquad i = 1, \dots, n$$

and consequently

$$(9) \qquad |D_i u|(A) \le |D_i \chi_{S_u}|(A \times \mathbb{R})$$

for any open set A and $i = 1, \dots, n$. This proves that $u \in BV(\Omega)$ and that (3) holds for $i = 1, \dots, n$.

On the other hand we have

$$\int_{\Omega\times\mathbb{R}} \varphi(x,y)\, D_{n+1}\chi_{S_u} = -\int_{\Omega\times\mathbb{R}} \varphi_y(x,y)\,\chi_{S_u}(x,y)\,dx\,dy = -\int_{\Omega} \varphi(x,u(x))\,dx\ ;$$

in particular

$$D_{n+1}\chi_{S_u} \le 0\ , \tag{10}$$

while, proceeding as previously, we infer

$$\int_{\Omega\times\mathbb{R}} \varphi(x)\, D_{n+1}\chi_{S_u} = -\int_{\Omega} \varphi(x)\,dx\ . \tag{11}$$

This concludes the proof of (3) and also gives

$$\pi_{\#}|D_{n+1}\chi_{S_u}|(A) = \mathcal{L}^n(A) \tag{12}$$

taking into account (10), and

$$|\mu(Du)|(A) \le |D\chi_{S_u}|(A\times\mathbb{R}) \tag{13}$$

for all open sets $A\subset\Omega$, taking into account (9).

Suppose now that $u\in BV(\Omega)$. By the approximation theorem, Theorem 1 in Sec. 4.1.1 (ii) we choose a sequence of smooth maps u_k such that $u_k\to u$ in L^1 and

$$\int_{\Omega} |D_i u_k|\,dx \to |D_i u|(\Omega), \qquad \int_{\Omega} |Du_k|\,dx \to |Du|(\Omega) \tag{14}$$

as $k\to\infty$, and we observe that for any $\varphi\in C_c^\infty(\Omega\times\mathbb{R})$ we have

$$\begin{aligned}\int_{\Omega\times\mathbb{R}} \varphi(x,y)\, D_i\chi_{S_{u_k}} &= -\int_{\Omega\times\mathbb{R}} \varphi_{x^i}(x,y)\,\chi_{S_{u_k}}(x,y)\,dx\,dy\\ &= -\int_{\Omega} dx \int_{-\infty}^{u_k(x)} \varphi_{x^i}(x,y)\,dy\\ &= -\int_{\Omega} \Big[\frac{d}{dx^i}\int_{-\infty}^{u_k(x)} \varphi(x,y)dy - \varphi(x,u_k(x))D_iu_k(x)\Big]dx\\ &= \int_{\Omega} \varphi(x,u_k(x))\,D_iu_k(x)\,dx\ ,\end{aligned}$$

$$\int_{\Omega\times\mathbb{R}} \varphi(x,y)\, D_{n+1}\chi_{S_{u_k}} = -\int_{\Omega\times\mathbb{R}} \varphi_y(x,y)\,\chi_{S_{u_k}}\,dy\,dx = -\int_{\Omega} \varphi(x,u_k(x))\,dx\ .$$

Taking the supremum for all $\varphi \in C_c^\infty(A \times \mathbb{R})$, $|\varphi| \leq 1$, $A \subset \Omega$, A open, we then get

$$\begin{aligned} |D_i \chi_{\mathcal{S}_{u_k}}|(A \times \mathbb{R}) &\leq |D_i u_k|(A) \\ |D_{n+1} \chi_{\mathcal{S}_{u_k}}|(A \times \mathbb{R}) &\leq \mathcal{L}^n(A) , \end{aligned}$$

and similarly

$$|D \chi_{\mathcal{S}_{u_k}}|(A \times \mathbb{R}) \leq \mathcal{L}^n \llcorner \sqrt{1 + |Du_k|^2}(A) .$$

Letting $k \to \infty$ we then conclude

$$\begin{aligned} |D_i \chi_{\mathcal{S}_u}|(A \times \mathbb{R}) &\leq |D_i u|(A) \\ |D_{n+1} \chi_{\mathcal{S}_u}|(A \times \mathbb{R}) &\leq \mathcal{L}^n(A) \\ |D \chi_{\mathcal{S}_u}|(A \times \mathbb{R}) &\leq |\mu|(A) , \end{aligned} \tag{15}$$

taking into account (14) and the lower semicontinuity of the total variation with respect to the L^1-convergence. This proves that $\mathcal{S}_u$ has finite perimeter in $\Omega \times \mathbb{R}$. Also (15), in conjunction with (9), (10), (12) and (13), clearly concludes the proof of (4) and (5). □

Remark 1. It is worthwhile noticing that the assumption $u \in L^1(\Omega)$ is not necessary: the measurability of u would suffice. The summability of u follows from the isoperimetric inequality. Suppose in fact that $P(\mathcal{S}_u, \Omega \times \mathbb{R}) < \infty$, then

$$\begin{aligned} \int_\Omega |u(x)|\, dx &= \mathcal{L}^{n+1}(\mathcal{S}_{|u|} \cap \{(x,y) \mid y > 0\} \\ &\leq c(n)\, P(\mathcal{S}_{|u|} \cap \{y > 0\})^{1/(n-1)} . \end{aligned}$$

On the other hand, arguing as in Proposition 1 in Sec. 4.1.3,

$$P(\mathcal{S}_{|u|} \cap \{y > 0\}) \leq P(\mathcal{S}_{|u|}, \Omega \times \mathbb{R}) + |\Omega|$$

and the summability of u is proved if one shows that

$$P(\mathcal{S}_{|u|}, \Omega \times \mathbb{R}) \leq P(\mathcal{S}_u, \Omega \times \mathbb{R}) .$$

This actually follows from a slicing formula for the perimeter of a set or more generally for the total variation of a BV-function, compare Miranda [475].

Let u be a function in $BV(\Omega)$. By the previous theorem then, and actually equivalently, the subgraph $\mathcal{S}_u$ has finite perimeter in $\Omega \times \mathbb{R}$. By De Giorgi's rectifiability theorem in Sec. 4.1.3, we have

$$D\chi_{\mathcal{S}_u} = n(z, \mathcal{S}_u)\, \mathcal{H}^n \llcorner \partial^- \mathcal{S}_u , \qquad z = (x,y) . \tag{16}$$

Now we would like to write (16) in terms of u and Du, by recovering the set $\partial^- \mathcal{S}_u$ and the inward normal $n(z, \partial^- \mathcal{S}_u)$ to the subgraph $\mathcal{S}_u$ in terms of u and Du for which we have proved a structure theorem in Sec. 4.1.4.

Recall that

$$u_+(x) := \operatorname*{aplimsup}_{y\to x} u(y) \qquad u_-(x) := \operatorname*{apliminf}_{y\to x} u(y)$$

and that the jump set of u is defined by

$$J_u := \{x \in \mathbb{R}^n \mid u_-(x) < u_+(x)\} .$$

We have seen in Theorem 1 in Sec. 4.1.4 that J_u is a Borel and countably $(n-1)$-rectifiable set so we can single out the part of the "graph" of u related to the jumps of u by setting

$$D\chi^{(j)}_{\mathcal{S}_u} := D\chi_{\mathcal{S}_u} \llcorner J_u \times \mathbb{R} = n(z, \mathcal{S}_u)\,\mathcal{H}^n \llcorner (\partial^- \mathcal{S}_u \cap J_u \times \mathbb{R}) \tag{17}$$

and the *approximately continuous* part of the "graph" of u by

$$D\chi^{(\mathrm{cont})}_{\mathcal{S}_u} := D\chi_{\mathcal{S}_u} - D\chi^{(j)}_{\mathcal{S}_u} = D\chi_{\mathcal{S}_u} \llcorner (\Omega \setminus J_u) \times \mathbb{R} . \tag{18}$$

Also set

$$\begin{aligned} \mathcal{D}_u &:= \{(x,y) \in \Omega \times \mathbb{R} \mid u_-(x) \le y \le u_+(x)\} \\ \mathcal{D}^{(0)}_u &:= \{(x,y) \in \Omega \times \mathbb{R} \mid u_-(x) < y < u_+(x)\} . \end{aligned}$$

Obviously $\mathcal{D}^{(0)}_u \subset J_u \times \mathbb{R}$.

Concerning the approximately continuous part of the "graph" of u we have

Theorem 2. *For any* $\varphi \in C^\infty_c(\Omega \times \mathbb{R})$ *we have*

$$D_i\chi^{(\mathrm{cont})}_{\mathcal{S}_u}(\varphi(x,y)) = \int_{\Omega\setminus J_u} \varphi(x, u_+(x))\, D_i u \qquad i = 1, \dots, n \tag{19}$$

and

$$D_{n+1}\chi^{(\mathrm{cont})}_{\mathcal{S}_u}(\varphi(x,y)) = -\int_{\Omega} \varphi(x, u_+(x))\, dx \tag{20}$$

Concerning the "jump part" of the "graph" of u we have

Theorem 3. *We have*

(i) $\mathcal{H}^n \llcorner (\partial^- \mathcal{S}_u \cap J_u \times \mathbb{R}) = \mathcal{H}^n \llcorner \mathcal{D}^{(0)}_u$

(ii) *For* $\mathcal{H}^n$-a.e. $(x,s) \in \mathcal{D}^{(0)}_u$ *the inward normal to* $\mathcal{S}_u$ *exists,* x *is a jump point for* u *and*

$$n((x,s), \mathcal{S}_u) = (n(x, J_u), 0) \tag{21}$$

(iii) Consequently, for any $\varphi \in C_c^\infty(\Omega \times \mathbb{R})$ we have

$$(22) \qquad D_i \chi^{(j)}_{\mathcal{S}_u}(\varphi(x,y)) = \int_{J_u} n_i(x, J_u)\Big(\int_{u_-(x)}^{u_+(x)} \varphi(x,s)\,ds\Big)\,d\mathcal{H}^{n-1}(x)$$

for $i = 1, \ldots, n$ and

$$(23) \qquad D_{n+1}\chi^{(j)}_{\mathcal{S}_u}(\varphi(x,y)) = 0$$

In order to prove Theorem 2 we first state

Lemma 1. *We have*

$$\partial_\mu \mathcal{S}_u \subset \mathcal{D}_u$$

where $\partial_\mu \mathcal{S}_u$ denotes the measure theoretic boundary of $\mathcal{S}_u$.

Proof. Suppose that $(x,t) \notin \mathcal{D}_u$. Then either $t > u_+(x)$ or $t < u_-(x)$. In the first case we observe that for s with $t > s > u_+(x)$ we have $\theta^n(E_s(u), x) = 0$ by the definition of $u_+(x)$. On the other hand if $(y,\tau) \in \mathcal{S}_u \cap B((x,t),\rho)$ then $y \in E_\tau(u)$, $|y - x| < \rho$ and $|t - \tau| < \rho$. Thus for $0 < \rho < t - s$ we have

$$\mathcal{S}_u \cap B((x,t),\rho) \subset (E_s(u) \cap B(x,\rho)) \times [t-\rho, t+\rho]\ ,$$

Hence

$$\theta^{n+1}(\mathcal{S}_u, (x,t)) = 0\ ,$$

consequently $(x,t) \notin \partial_\mu \mathcal{S}_u$.

In the case $t < u_-(x)$ we similarly prove that

$$\theta^{n+1}(\mathbb{R}^{n+1} \setminus \mathcal{S}_u, (x,t)) = 0\ ,$$

i.e., that again $(x,t) \notin \partial_\mu \mathcal{S}_u$. □

Proof of Theorem 2. Since

$$\partial^- \mathcal{S}_u \subset \partial_\mu \mathcal{S}_u \subset \mathcal{D}_u$$

by Lemma 1, the function $\varphi(x,y)$ and $\varphi(x, u_+(x))$ agree $|D\chi_{\mathcal{S}_u}| \llcorner ((\Omega \setminus J_u) \times \mathbb{R})$ a.e.. Therefore

$$\begin{aligned} D_i \chi^{(\text{cont})}_{\mathcal{S}_u}(\varphi) &= D_i \chi_{\mathcal{S}_u} \llcorner (\mathcal{D}_u \cap (\Omega \setminus J_u) \times \mathbb{R})(\varphi(x,y)) \\ &= D_i \chi_{\mathcal{S}_u} \llcorner ((\Omega \setminus J_u) \times \mathbb{R})(\varphi(x, u_+(x))) \\ &= \pi_\# D_i \chi_{\mathcal{S}_u} \llcorner (\Omega \setminus J_u)(\varphi(x, u_+(x))) \\ &= \int \varphi(x, u_+(x))\, D_i u \llcorner (\Omega \setminus J_u)\ , \end{aligned}$$

taking into account (3) of Theorem 1. Similarly one proves (20). □

In order to prove Theorem 3 we first state

Lemma 2. *Let $x \in J_u$ and $t \in (u_-(x), u_+(x))$ be such that x is a jump point for $\chi_{E_t(u)}$ with parameters $(1, 0, n)$. Then (x,t) is a jump point for $\chi_{\mathcal{S}_u}$ with parameters $(1, 0, (n, 0))$.*

Proof. Recall that $z \in \mathbb{R}^N$ is a jump point for χ_A, $A \subset \mathbb{R}^N$, with parameters $(1, 0, \nu)$, $\nu \in \mathbb{R}^N$, if and only if

$$\theta^N(A \cap E^-_{z,\nu}, z) = 0 \quad \text{and} \quad \theta^N((\mathbb{R}^N \setminus A) \cap E^+_{z,\nu}, z) = 0 ,$$

compare Proposition 1 in Sec. 4.1.4.

Now let $x \in J_u$ and let s be such that $u_-(x) < s < t$. If $(y, \tau) \in \mathcal{S}_u \cap E^-_{(x,t),(n,0)} \cap B((x,t), \rho)$ then $y \in E_\tau(u) \cap E^-_{x,n}$ and $|t - \tau| < \rho$. Thus for $0 < \rho < t - s$ we have

$$\mathcal{S}_u \cap E^-_{(x,t),(n,0)} \cap B((x,t),\rho) \;\subset\; (E_s(u) \cap E^-_{x,n} \cap B(x,\rho)) \times [t-\rho, t+\rho] .$$

Consequently

$$\theta^{n+1}(\mathcal{S}_u \cap E^-_{(x,t),(n,0)}, (x,t)) \;=\; 0 .$$

Similarly using s with $t < s < u_+(x)$ one shows that

$$\theta^{n+1}((\mathbb{R}^{n+1} \setminus \mathcal{S}_u) \cap E^+_{(x,t),(n,0)}, (x,t)) \;=\; 0$$

which concludes the proof. □

Proof of Theorem 3. With the notation in the proof of Theorem 1 in Sec. 4.1.4 set

$$N_1 \;:=\; \bigcup_{r \in Q} (\partial_\mu E_r(u) \setminus \partial^- E_r(u)) ,$$

and recall that $\mathcal{H}^{n-1}(N_1) = 0$, so that for $N_2 := N_1 \times \mathbb{R}$ we have

$$\mathcal{H}^n(N_2) \;=\; 0 .$$

Let now $x \in J_u \setminus N_1$. By Theorem 2 in Sec. 4.1.4 (i), for every t with $u_-(x) < t < u_+(x)$, x is a jump point for $\chi_{E_t(u)}$, hence, by Lemma 2, (x,t) is a jump point for $\chi_{\mathcal{S}_u}$ with parameters $(1, 0, (n(x, J_u), 0))$. Thus

$$\mathcal{D}^{(0)}_u \;\subset\; \partial^* \mathcal{S}_u \cap (J_u \times \mathbb{R}) \cup N_2 \tag{24}$$

On the other hand for the set

$$\begin{aligned} N_3 \;&:=\; (\mathcal{D}_u \setminus \mathcal{D}^{(0)}_u) \cap (J_u \times \mathbb{R}) \;= \\ &=\; \{(x,y) \mid y = u_-(x) < u_+(x)\} \cup \{(x,y) \mid u_-(x) < u_+(x) = y\} \end{aligned}$$

we have

$$\mathcal{H}^n(N_3) \;=\; 0 \tag{25}$$

by the area formula, since N_3 projects in J_u and for any x in such a projection there are at most two real y_1, y_2 such that (x, y_1) and (x, y_2) belong to N_3.

Using now (24), Lemma 1, and the inclusion relations among reduced boundary jump points, and measure theoretic boundary, we get

$$\begin{aligned} \partial^- \mathcal{S}_u \cap J_u \times \mathbb{R} \quad &\subset \quad \partial_\mu \mathcal{S}_u \cap J_u \times \mathbb{R} \subset \mathcal{D}_u \cap J_u \times \mathbb{R} \\ &\subset \quad \mathcal{D}_u^{(0)} \cup N_3 \subset \partial^* \mathcal{S}_u \cap J_u \times \mathbb{R} \cup N_2 \cup N_3 \end{aligned}$$

hence

$$\mathcal{H}^n \llcorner (\partial^- \mathcal{S}_u \cap J_u \times \mathbb{R}) \; = \; \mathcal{H}^n \llcorner \mathcal{D}_u^{(0)} \; = \; \mathcal{H}^n \llcorner (\mathcal{D}_u \cap J_u \times \mathbb{R}) \, .$$

This proves (i) and (ii). (iii) follows easily from the previous two steps taking into account the (17) and the fact that

$$\mathcal{H}^n \llcorner (J_u \times \mathbb{R}) \; = \; \mathcal{H}^{n-1} \llcorner J_u \times \mathcal{L}^1 \, ,$$

which follows from the rectifiability of J_u. □

As a corollary of Theorem 2 and Theorem 3 we can now state

Theorem 4. *Let $u \in BV(\Omega)$. Then*

(i) for $|\mu(Du)|$-a.e. x in $\Omega \setminus J_u$ we have

$$\frac{d\,\mu(Du)}{d\,|\mu(Du)|}(x) \; = \; n((x, u_+(x)), \mathcal{S}_u)$$

(ii) for $|\mu(Du)|$-a.e. x in J_u

$$\frac{d\,\mu(Du)}{d\,|\mu(Du)|}(x) \; = \; (n(x, J_u), 0) \, .$$

Finally, let us note that from Radón-Nikodym theorem we know that

$$(Du)^{(\text{cont})} \; = \; (Du)^a \, dx + (Du)^{(C)}$$

and

$$(Du)^{(C)} \; = \; (Du)^{(\text{cont})} \llcorner Z$$

where

$$Z \; = \; \{x \in \Omega \mid \frac{d(Du)^{(\text{cont})}}{d\mathcal{L}^n}(x) = +\infty\} \, .$$

Then we infer

$$(Du)^{(C)} \; = \; (Du)^{(\text{cont})} \llcorner W$$

where

$$W \; = \; \{x \in \Omega \mid \frac{d|\mu|}{d\mathcal{L}^n} = +\infty\} \; = \; \{x \in \Omega \mid \frac{d\mathcal{L}^n}{d|\mu|} = 0\}$$

and therefore

$$(Du)^{(C)} \; = \; Du \llcorner \big((\Omega \setminus J_u) \cap \{x \in \Omega \mid n_{n+1}((x, u_+(x)), \mathcal{S}_u) = 0\}\big)$$

by Theorem 4.

In conclusion we can then state

Theorem 5. *Let $u \in BV(\Omega)$. Then*

$$\begin{aligned}
(Du)^a &= Du \llcorner \{x \in \Omega \mid u_+(x) = u_-(x),\ n_{n+1}((x, u_+(x)), \mathcal{S}_u) > 0\} \\
(Du)^{(C)} &= Du \llcorner \{x \in \Omega \mid u_+(x) = u_-(x),\ n_{n+1}((x, u_+(x)), \mathcal{S}_u) = 0\} \\
(Du)^{(j)} &= Du \llcorner \{x \in \Omega \mid u_-(x) < u_+(x)\} \\
&= Du \llcorner \{x \in \Omega \mid u_-(x) < u_+(x),\ n_{n+1}((x, u_+(x)), \mathcal{S}_u) = 0\}\ .
\end{aligned}$$

2 Cartesian Currents in Euclidean Spaces

This section in dedicated to the study of *Cartesian currents*. After a few preliminaries in Sec. 4.2.1, we introduce in Sec. 4.2.2 the class of *Cartesian currents*

$$\begin{aligned}
\text{cart}(\Omega \times \mathbb{R}^N) := \{ & T \in \mathcal{D}_n(\Omega \times \mathbb{R}^N) \mid T \text{ is an i.m. rectifiable current in} \\
& \Omega \times \mathbb{R}^N,\ \mathbf{M}(T) < \infty,\ \| T \|_1 < \infty,\ \partial T \llcorner \Omega \times \mathbb{R}^N = 0, \\
& T^{\bar{0}0} \geq 0,\ \pi_\# T = [\![\Omega]\!] \ \}
\end{aligned}$$

and the class of *graphs*

$$\begin{aligned}
\text{graph}(\Omega \times \mathbb{R}^N) := \{ & T \in \mathcal{D}_n(\Omega \times \mathbb{R}^N) \mid T \text{ is an i.m. rectifiable current} \\
& \text{in } \Omega \times \mathbb{R}^N,\ \mathbf{M}(T) < \infty,\ \| T \|_1 < \infty,\ \mathbf{M}(\partial T) < \infty, \\
& T^{\bar{0}0} \geq 0,\ \pi_\# T = [\![\Omega]\!] \ \}
\end{aligned}$$

and we prove *closure* and *compactness properties* for those classes, with respect to the *weak convergence in cart*, i.e., with respect to the weak convergence of currents with equibounded masses and L^1-norms.

In Sec. 4.2.5 we shall see that not every $T \in \text{cart}(\Omega \times \mathbb{R}^N)$ can be weakly approximated in cart by graphs G_{u_k} of smooth maps u_k. For this reason we also introduce the sequential weak closure of the graphs of smooth maps in $\text{cart}(\Omega \times \mathbb{R}^N)$

$$\text{Cart}(\Omega \times \mathbb{R}^N) := \text{sw}^*\text{-cl}\,\{G_u \mid u \in C^1(\Omega, \mathbb{R}^N) \cap \mathcal{A}^1(\Omega, \mathbb{R}^N)\}\ .$$

Of course we have

$$\text{Cart}(\Omega \times \mathbb{R}^N) \subset \text{cart}(\Omega \times \mathbb{R}^N) \subsetneq \text{graph}(\Omega \times \mathbb{R}^N)\ .$$

In Sec. 4.2.3 we prove a *structure theorem* for graphs, showing that for every $T \in \text{graph}(\Omega \times \mathbb{R}^N)$, in particular for every Cartesian current T, we have

$$T = G_{u_T} + S$$

where G_{u_T} is the current associated to the 1-graph of a BV-map u_T and S is a vertical current, i.e., $\pi_\# S = 0$. We shall also discuss relations between the components of T and the minors of $(Du_T)^a$.

In Sec. 4.2.4 we shall consider Cartesian currents of codimension one and we prove that they can be identified with BV-functions in Ω. In particular we then deduce that Cartesian currents of codimension one are just weak limits of smooth graphs. We conclude Sec. 4.2.4 by brief discussion of the Cantor and jump part decomposition of Cartesian currents.

After presenting a few examples of Cartesian currents in Sec. 4.2.5, and in particular discussing with further details the examples in the introduction to this chapter, in Sec. 4.2.6 we shall discuss *radial currents*, proving in particular that they are weakly approximable by smooth graphs.

2.1 Limit Currents of Smooth Graphs

Let $\{u_k\} \subset C^1(\Omega, \mathbb{R}^N)$ be a sequence of smooth maps from a bounded domain Ω in $\mathbb{R}^n$ into $\mathbb{R}^N$ satisfying

$$(1) \qquad \sup_k \left(\| u_k \|_{L^1(\Omega,\mathbb{R}^N)} + \| M(Du_k) \|_{L^1(\Omega,\Lambda_n \mathbb{R}^{n+N})}\right) < \infty .$$

Passing to a subsequence, we may suppose that

$$(2) \qquad u_k \rightharpoonup u \quad \text{weakly in } L^1(\Omega, \mathbb{R}^N), \quad M^{\beta}_{\bar{\alpha}}(Du_k) \rightharpoonup \mu^{\beta}_{\bar{\alpha}} \quad \text{as measures}$$

for $|\alpha| + |\beta| = n$, $|\beta| \geq 1$, where $\mu^{\beta}_{\bar{\alpha}}$ are Radon measures of bounded total variation in Ω. Also, since

$$\mathbf{M}(G_{u_k}) = \| M(Du_k) \|_{L^1(\Omega,\Lambda_n \mathbb{R}^{n+N})} ,$$

and therefore

$$\sup_k \mathbf{M}(G_{u_k}) < \infty$$

we can suppose that

$$(3) \qquad G_{u_k} \rightharpoonup T$$

where T is an n-dimensional current in $\mathcal{D}_n(\Omega \times \mathbb{R}^N)$.

One of our main questions is to characterize such currents T, i.e., all currents T which are limits of regular graphs in the sense (1) (2) (3). Unfortunately, in this general context, this is beyond our possibilities, and it will remain an open question. Therefore, we shall instead try to describe the class of such limits tentatively by successive approximations, by detecting their properties.

The lower semicontinuity of the mass with respect to the weak convergence

$$\mathbf{M}(T) \leq \liminf_{k\to\infty} \mathbf{M}(G_{u_k})$$

yields evidently

$$(4) \qquad \mathbf{M}(T) < \infty .$$

As by Stokes's theorem $\partial G_{u_k} \llcorner \Omega \times \mathbb{R}^N = 0$, we also have

$$(5) \qquad \partial T \llcorner \Omega \times \mathbb{R}^N = 0 .$$

Since $\{G_{u_k}\}$ is a sequence of i.m. rectifiable currents with equibounded masses and no boundary in $\Omega \times \mathbb{R}^N$, Federer-Fleming's closure theorem yields that

$$T \text{ is an i.m. rectifiable current, } T \in \mathcal{R}_n(\Omega \times \mathbb{R}^N) .$$

Consider now the 0-component of T, or equivalently the $\bar{0}0$-component in $\Omega \times \mathbb{R}^N$

$$T^{\bar{0}0}(\varphi(x,y)) := T(\varphi(x,y)\,dx) =: T_{(0)}(\varphi(x,y)\,dx) \quad \forall\, \varphi \in C_c^\infty(\Omega \times \mathbb{R}^N) .$$

Since

$$G_{u_k}^{\bar{0}0} = \int_\Omega \varphi(x, u_k(x))\,dx ,$$

we see that $G_{u_k}^{\bar{0}0}$ is a positive measure in $\Omega \times \mathbb{R}^N$, thus passing to the limit as $k \to \infty$, also $T^{\bar{0}0}$ is a positive measure,

$$(6) \qquad T^{\bar{0}0} \geq 0 .$$

Actually the previous remarks can be stated evidently in the following slightly more general form.

Proposition 1. *Let $\{T_k\}$ be a sequence of i.m. rectifiable currents satisfying*

$$\sup_k \mathbf{M}(T_k) < \infty, \quad \partial T_k \llcorner \Omega \times \mathbb{R}^N = 0, \quad T_k^{\bar{0}0} \geq 0 \quad \forall k.$$

Suppose that

$$T_k \rightharpoonup T \qquad \text{in } \mathcal{D}_n(\Omega \times \mathbb{R}^N) .$$

Then T is an i.m. rectifiable current with finite mass and no boundary in $\Omega\times\mathbb{R}^N$, moreover $T^{\bar{0}0} \geq 0$.

Of course under the mere assumptions of Proposition 1 the sequence of currents G_{u_k} may just disappear at infinity in the y direction. What prevents this to happen is the equiboundedness of the L^1-norms of u_k. For this reason we now introduce the L^1-norm of a current in $\Omega \times \mathbb{R}^N$.

Let T be a current in $\mathcal{D}_n(\Omega \times \mathbb{R}^N)$ whose $\bar{0}0$-component $T^{\bar{0}0}$ is a Radón measure with bounded variation in $\Omega \times \mathbb{R}^N$, i.e.,

$$(7) \quad \|T\|_0 := \sup\{T(\varphi(x,y)\,dx) \mid \varphi \in C_c^\infty(\Omega \times \mathbb{R}^N),\ |\varphi| \leq 1\} < \infty .$$

We define the L^1*-norm of* T by

$$(8) \qquad \|T\|_1 := \sup\{T(\varphi(x,y)|y|\,dx) \mid \varphi \in C_c^0(\Omega \times \mathbb{R}^N),\ |\varphi| \leq 1\} .$$

Notice that, being $T^{\bar{0}0}$ a measure with bounded variation in $\Omega\times\mathbb{R}^N$, $T^{\bar{0}0}$ extends, by Lebesgue's dominated convergence theorem, to all continuous functions with compact support in $\Omega \times \mathbb{R}^N$, and even to all bounded and continuous or bounded and Borel functions in $\Omega \times \mathbb{R}^N$. Also, taking into account that

$$|T^{\bar{0}0}|(f(x,y)) = \sup_{\substack{\varphi \in C^0_c(\Omega\times\mathbb{R}^N) \\ |\varphi|\le 1}} T^{\bar{0}0}(f(x,y)\,\varphi(x,y))$$

for all $f \geq 0$, we easily deduce

$$\begin{aligned} \| T \|_1 &= \sup\{T^{\bar{0}0}(\varphi(x,y)\,|y|) \mid \varphi \text{ Borel in } \Omega \times \mathbb{R}^N,\ |\varphi| \le 1\} = \\ &= \sup\{|T^{\bar{0}0}|(\varphi(x,y)\,|y|) \mid \varphi \text{ Borel in } \Omega \times \mathbb{R}^N,\ 0 \le \varphi \le 1\}\,. \end{aligned} \tag{9}$$

Moreover, if $\| T \|_1 < \infty$, then we can extend $T^{\bar{0}0}$ to all Borel functions $\varphi(x,y)$ in $\Omega \times \mathbb{R}^N$ satisfying $|\varphi(x,y)| \le c(1+|y|)$, compare the end of Sec. 1.1.4 and we easily deduce

$$\| T \|_1 = |T^{\bar{0}0}|(|y|) = \int_{\Omega\times\mathbb{R}^N} |y|\, d|T^{\bar{0}0}|\,. \tag{10}$$

Finally, we have

Proposition 2. *Let $T_k, T \in \mathcal{D}_n(\Omega \times \mathbb{R}^N)$. Suppose that*

$$\sup_k(\| T_k \|_0 + \| T_k \|_1) < \infty, \qquad T_k \rightharpoonup T \quad \text{in } \mathcal{D}_n(\Omega \times \mathbb{R}^N). \tag{11}$$

Then, obviously

$$\| T \|_0 \le \liminf_{k\to\infty} \| T_k \|_0, \qquad \| T \|_1 \le \liminf_{k\to\infty} \| T_k \|_1.$$

In particular $\| T \|_1 < \infty$.

Proof. From (11) we deduce that $T^{\bar{0}0}$ is a measure with bounded variation and

$$T^{\bar{0}0}(\varphi(x,y)|y|) = \lim_{k\to\infty} T_k^{\bar{0}0}(\varphi(x,y)|y|)$$

for all $\varphi \in C^0_c(\Omega \times \mathbb{R}^N)$[1], thus for $|\varphi| \le 1$

$$T(\varphi(x,y)|y|\,dx) \le \liminf_{k\to\infty} \| T_k \|_1\,.$$

□

For $T = G_u$ where $u \in \mathcal{A}^1(\Omega, \mathbb{R}^N)$ we trivially have

$$\begin{aligned} \| G_u \|_1 &:= \sup\Big\{ \int_\Omega \varphi(x,u)\,|u|\,dx \mid \varphi \in C^0_c(\Omega \times \mathbb{R}^N),\ |\varphi| \le 1\Big\} \\ &= \int_\Omega |u|\,dx\,. \end{aligned}$$

[1] Notice that the previous equality is not true in general for $\varphi \in C^0(\Omega \times \mathbb{R}^N)$.

Thus Proposition 2, applied to the sequence $\{G_{u_k}\}$ in (3), yields

$$(12) \qquad \| T \|_1 < \infty .$$

Let T be such that $\| T \|_0 < \infty$, and let π denote the orthogonal projection

$$\pi : \Omega \times \mathbb{R}^N \longrightarrow \Omega , \qquad \pi(x,y) = x .$$

In these circumstances the push forward of T to the n-dimensional current $\pi_\# T$ in $\mathcal{D}_n(\Omega)$ is well defined. In fact, as π is linear and $||T||_0 < \infty$, T extends to all n–forms with continuous and bounded coefficients in $\Omega \times \mathbb{R}^N$, in particular on forms of the type $\varphi(x)\,dx$, $\varphi \in C^0_c(\Omega)$. By the dominated convergence theorem,

$$(13) \qquad \pi_\# T(\varphi(x)\,dx) := \lim_{R\to\infty} T(\varphi(x)\,\chi_R(y)\,dx) .$$

where for $R > 0$, χ_R is a cutoff function in the vertical directions, $\chi_R \in C^\infty_c(\mathbb{R}^N)$ and,

$$0 \le \chi_R \le 1 , \quad \chi_R = 1 \text{ in } B_R(0) , \quad \chi_R = 0 \text{ in } \mathbb{R}^N \setminus B_{2R}(0) , \quad |D\chi_R| \le 2/R .$$

However $\pi_\# : \mathcal{D}_n(\Omega \times \mathbb{R}^N) \to \mathcal{D}_n(\Omega)$ is not continuous, compare Sec. 2.2.3.

Turning back to our initial sequence G_{u_k} we readily see that

$$\pi_\# G_{u_k}(\varphi(x)\,dx) = G^{\bar{0}0}_{u_k}(\varphi(x)) = \int_\Omega \varphi(x)\,dx = [\![\, \Omega \,]\!]\,(\varphi(x)\,dx)$$

i.e. $\pi_\# G_{u_k} = [\![\, \Omega \,]\!]$ for all k. We now claim that, because $G_{u_k} \rightharpoonup T$ and,

$$\sup_k \| G_{u_k} \|_1 := \sup_k \| u_k \|_{L^1} < \infty$$

we have $\pi_\# T = [\![\, \Omega \,]\!]$.

In fact, more generally, we have

Proposition 3. *Let $\{T_k\}$ be a sequence of currents in $\mathcal{D}_n(\Omega \times \mathbb{R}^N)$ such that*

$$T_k \rightharpoonup T \text{ in } \mathcal{D}_n(\Omega \times \mathbb{R}^N) \quad \text{and} \quad \sup_k(\| T_k \|_0 + \| T_k \|_1) < \infty .$$

Then

$$\pi_\# T_k \rightharpoonup \pi_\# T \qquad \text{in } \mathcal{D}_n(\Omega) .$$

In particular $\pi_\# T = [\![\, \Omega \,]\!]$ if $\pi_\# T_k = [\![\, \Omega \,]\!]$ for all k.

Proof. As previously, if $\chi_R(y)$ is a cut off function and $\varphi \in C^0_c(\Omega)$ we have

$$\begin{aligned} &|\pi_\# T_k(\varphi(x)\,dx) - \pi_\# T(\varphi(x)\,dx)| \\ &\le |T^{\bar{0}0}_k(\varphi(x)\chi_R(y)) - T^{\bar{0}0}(\varphi(x)\chi_R(y))| \\ &+ |T^{\bar{0}0}_k(\varphi(x)(1-\chi_R(y))\,dx) - T^{\bar{0}0}(\varphi(x)(1-\chi_R(y))\,dx)| . \end{aligned}$$

The last term can be estimated by

$$(\| T_k^{\bar{0}0} \| + \| T^{\bar{0}0} \|)(|\varphi(x)|\frac{|y|}{R}) \leq \| \varphi \|_{\infty,\Omega}(\| T_k \|_1 + \| T \|_1)\frac{1}{R},$$

thus, for R sufficiently large, it is small uniformly in k, while the first term on the right hand-side tends to zero as $k \to \infty$ for R fixed. From that the claims follows at once. □

In conclusion we can state

Proposition 4. *Let $\{u_k\} \subset C^1(\Omega, \mathbb{R}^N)$ be a sequence for which (1) holds and $G_{u_k} \rightharpoonup T$ in $\mathcal{D}_n(\Omega, \mathbb{R}^N)$. Then T is an i.m. rectifiable current with finite mass, finite norms $\| T \|_0$ and $\| T \|_1$, without boundary in $\Omega \times \mathbb{R}^N$, and satisfying $T^{\bar{0}0} \geq 0$ and $\pi_{\#}T = [\![\Omega]\!]$.*

The previous considerations can be easily extended to sequences of maps $\{u_k\}$ in $\mathcal{A}^1(\Omega, \mathbb{R}^N)$ for which (1) holds, $G_{u_k} \rightharpoonup T$ in $\mathcal{D}_n(\Omega, \mathbb{R}^N)$ and moreover

$$\sup_k \mathbf{M}(\partial G_{u_k} \llcorner \Omega \times \mathbb{R}^N) < \infty . \tag{14}$$

In this case we have

Proposition 5. *Let $\{u_k\} \subset \mathcal{A}^1(\Omega, \mathbb{R}^N)$ satisfy (1). Suppose that*

$$\sup_k \mathbf{M}(\partial G_{u_k} \llcorner \Omega \times \mathbb{R}^N) < \infty, \qquad G_{u_k} \rightharpoonup T \quad \text{in } \mathcal{D}_n(\Omega \times \mathbb{R}^N).$$

Then the current T is an i.m. rectifiable current with finite mass and finite boundary mass; moreover $\| T \|_0$ and $\| T \|_1$ are finite, and $T^{\bar{0}0} \geq 0$, $\pi_{\#}T = [\![\Omega]\!]$.

Of course (14) is essential in order to get in general the rectifiability of T, compare the examples in Sec. 2.2.4.

Finally we observe that in the previous claims we could replace the norm $\| T \|_1$ by

$$\| T \|_h := \sup\{T^{\bar{0}0}(\varphi(x,y)h(|y|)) \mid \varphi \in C_c^\infty, |\varphi| \leq 1\} \tag{15}$$

where h is a non decreasing function $h : [0, \infty) \to \mathbb{R}$, $h(t) \to \infty$ as $t \to \infty$, for instance $h(t) = t^\beta$, $\beta > 0$, without altering the conclusions (as well as the proofs). However we shall not insist on this point.

2.2 The Classes cart($\Omega \times \mathbb{R}^N$) and graph($\Omega \times \mathbb{R}^N$): Closure and Compactness Theorems

Motivated by the considerations in Sec. 4.2.1 we now introduce the class of *Cartesian currents* as follows

Definition 1. *The class of* Cartesian currents *is defined as*

$$\begin{aligned}\mathrm{cart}(\Omega \times \mathbb{R}^N) := \{ \; & T \in \mathcal{D}_n(\Omega \times \mathbb{R}^N) \mid T \textit{ is an i.m. rectifiable current} \\ & \textit{in } \Omega \times \mathbb{R}^N, \; \mathbf{M}(T) < \infty, \; \| T \|_1 < \infty, \\ & \partial T \llcorner \Omega \times \mathbb{R}^N = 0, \; T^{\bar{0}0} \geq 0, \; \pi_\# T = [\![\Omega]\!] \; \}\end{aligned}$$

For future purposes we also introduce a larger class, the element of which are allowed to have non zero boundary in $\Omega \times \mathbb{R}^N$.

Definition 2. *The class of* graphs *is defined as*

$$\begin{aligned}\mathrm{graph}(\Omega \times \mathbb{R}^N) := \{ \; & T \in \mathcal{D}_n(\Omega \times \mathbb{R}^N) \mid T \textit{ is an i.m. rectifiable current} \\ & \textit{in } \Omega \times \mathbb{R}^N, \; \mathbf{M}(T) < \infty, \; \| T \|_1 < \infty, \\ & \mathbf{M}(\partial T) < \infty, \; T^{\bar{0}0} \geq 0, \; \pi_\# T = [\![\Omega]\!] \; \} \, .\end{aligned}$$

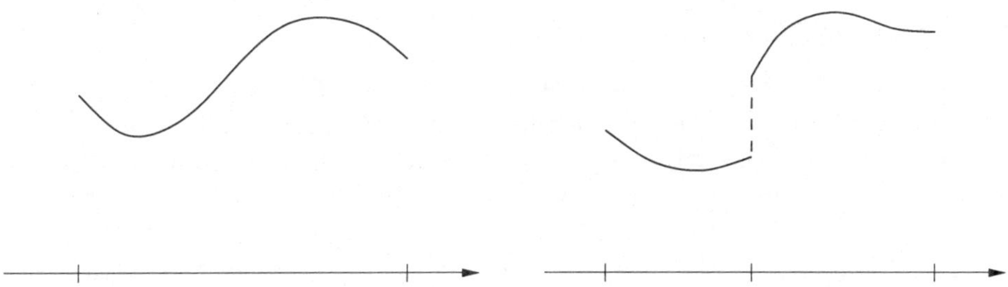

Fig. 4.1. (a) A Cartesian current in $(0,1) \times \mathbb{R}$, (b) A graph in $(0,1) \times \mathbb{R}$.

Of course elements of $\mathrm{graph}(\Omega \times \mathbb{R}^N)$ are in general not approximable by graphs of smooth maps, and

$$\mathrm{cart}(\Omega \times \mathbb{R}^N) \subsetneq \mathrm{graph}(\Omega \times \mathbb{R}^N) \, .$$

We could also introduce intermediate classes, as for instance the class of T in $\mathrm{graph}(\Omega \times \mathbb{R}^N)$ for which we have

$$(\partial T)_{(0)} \llcorner \Omega \times \mathbb{R}^N = \ldots = (\partial T)_{(k)} \llcorner \Omega \times \mathbb{R}^N = 0$$

up to a fixed $k < n - 1$. Particularly important is the case in which

$$(\partial T)_{(0)} \llcorner \Omega \times \mathbb{R}^N = 0$$

as this condition expresses the standard formula of integration by parts for derivatives. But for the moment we do not dwell on that.

An immediate consequence of Proposition 1 in Sec. 4.2.1, Proposition 2 in Sec. 4.2.1 and (3) in Sec. 4.2.1 is the following theorem which makes useful Definition 1

Theorem 1. *Let $\{T_k\}$ be a sequence of currents in* $\mathrm{cart}(\Omega \times \mathbb{R}^N)$. *Suppose that*

$$\sup_k \{\| T_k \|_1 + \mathbf{M}(T_k)\} < \infty, \qquad T_k \rightharpoonup T \quad \text{in } \mathcal{D}_n(\Omega \times \mathbb{R}^N).$$

Then T belongs to $\mathrm{cart}(\Omega \times \mathbb{R}^N)$.

Actually, taking into account Federer-Fleming closure theorem, we readily see that the same theorem holds for sequences of graphs provided the masses of their boundaries in $\Omega \times \mathbb{R}^N$ are equibounded, that is

Theorem 2. *Let $\{T_k\}$ be a sequence of currents in* $\mathrm{graph}(\Omega \times \mathbb{R}^N)$. *Suppose that*

$$\sup_k \{\| T_k \|_1 + \mathbf{M}(T_k)\} \; < \; \infty \, , \qquad \sup_k \mathbf{M}(\partial T_k \llcorner \Omega \times \mathbb{R}^N) \; < \; \infty$$

and

$$T_k \; \rightharpoonup \; T \qquad \text{in } \; \mathcal{D}_n(\Omega \times \mathbb{R}^N) \; .$$

Then $T \in \mathrm{graph}(\Omega \times \mathbb{R}^N)$.

Remark 1. Let T belong to $\mathrm{graph}(\Omega \times \mathbb{R}^N)$. Since $\mathbf{M}(T) < \infty$, T naturally extends, by means of Lebesgue's dominated convergence, to all n-forms $\omega = \sum \omega_{\alpha\beta}\, dx^\alpha \wedge dy^\beta$ with bounded and continuous or even bounded and Borel coefficients in $\Omega \times \mathbb{R}^N$, not necessarily with compact supports in $\Omega \times \mathbb{R}^N$. Notice however that, if

$$T_k \; \rightharpoonup \; T \qquad \text{in } \mathcal{D}_n(\Omega \times \mathbb{R}^N) \; ,$$

then $T_k(\omega)$ does not converge in general to $T(\omega)$ if ω has not compact support in $\Omega \times \mathbb{R}^N$, even if we assume that $\sup_k \{\mathbf{M}(T_k) + \| T_k \|_1\} < \infty$.

Remark 2. From Remark 1 it follows in particular that every Cartesian current, and in fact every graph T in $\Omega \times \mathbb{R}^N$ may be regarded as an i.m. rectifiable current in $\mathbb{R}^n \times \mathbb{R}^N$. In this case however we have no information on ∂T in $\mathbb{R}^n \times \mathbb{R}^N$. In particular Cartesian currents in $\Omega \times \mathbb{R}^N$ are not necessarely *normal* currents in $\mathbb{R}^n \times \mathbb{R}^N$.

Finally we would like to observe that in fact Theorem 1 and Theorem 2 are closure theorems. In order to see this, we introduce the subclass of all n-dimensional currents in $\Omega \times \mathbb{R}^N$

$$\mathcal{C} \; := \; \{T \in \mathcal{D}_n(\Omega \times \mathbb{R}^N) \; | \; \mathbf{M}(T) + \| T \|_1 < \infty\}$$

and we set

$$\| T \|_{\mathcal{C}} \; := \; \| T \|_1 + \mathbf{M}(T) \; .$$

Evidently $\mathcal{C}$ is a linear space and $\| \;\; \|_{\mathcal{C}}$ is a norm on $\mathcal{C}$. For any n-form

$$\omega \; = \; \sum_{|\alpha|+|\beta|=n} \omega_{\alpha\beta}(x,y)\, dx^\alpha \wedge dy^\beta$$

in $\Omega \times \mathbb{R}^N$ we set

$$\begin{aligned} \|\omega\|_{\mathcal{B}} &:= \sup_{\Omega\times\mathbb{R}^N} \Big(\frac{\omega_{\bar{0}0}(x,y)}{1+|y|} + \sum_{\substack{|\alpha|+|\beta|=n \\ |\beta|\geq 1}} |\omega_{\alpha\beta}(x,y)|\Big) \\ \|\omega\|_{\infty} &:= \sup_{\Omega\times\mathbb{R}^N} \sum_{|\alpha|+|\beta|=n} |\omega_{\alpha\beta}(x,y)| \end{aligned}$$

and we denote by $C_0(\Omega\times\mathbb{R}^N)$ the space of continuous functions in $\Omega\times\mathbb{R}^N$ with zero limits as points (x,y) go to infinity or to $\partial\Omega \times \mathbb{R}^N$. We then consider the subclass of all n-forms in $\Omega \times \mathbb{R}^N$

$$\mathcal{B} := \{\omega \mid \frac{\omega_{\bar{0}0}}{1+|y|}\ , \ \omega_{\alpha\beta} \in C_0(\Omega\times\mathbb{R}^N),\ |\alpha|+|\beta| = n\ ,\ |\beta|\geq 1\}\ .$$

Also $\mathcal{B}$ with $\|\cdot\|_{\mathcal{B}}$ is a normed linear space.

Proposition 1. *We have*

(i) $\mathcal{B}$ is a Banach space with the norm $\|\cdot\|_{\mathcal{B}}$. Moreover $\mathcal{B}$ is separable and $\mathcal{D}^n(\Omega\times\mathbb{R}^N)$ is dense in $\mathcal{B}$
(ii) $\mathcal{C}$ is a Banach space with the norm $\|\cdot\|_{\mathcal{C}}$.
(iii) $\mathcal{C}$ is the dual space of $\mathcal{B}$.
(iv) A sequence $\{T_k\}$ converges weakly in $\mathcal{C}$ if and only if*

$$\sup_k(\|T_k\|_1 + \mathbf{M}(T_k)) < \infty, \qquad T_k \rightharpoonup T \quad \text{in } \mathcal{D}_n(\Omega\times\mathbb{R}^N)$$

In virtue of Proposition 1, Theorem 1 now obviously reads

Theorem 3 (Closure theorem). *The class* $\mathrm{cart}(\Omega\times\mathbb{R}^N)$, *and for any* $K>0$ *the class*

$$\mathrm{graph}(\Omega\times\mathbb{R}^N)\cap\{T \mid \mathbf{M}((\partial T)\llcorner\Omega\times\mathbb{R}^N)\leq K\}\ ,$$

which are contained in $\mathcal{C}$, are closed in $\mathcal{C}$ with respect to the weak convergence in $\mathcal{C}$.*

Definition 3. *From now on we shall say that* $\{T_k\}\subset \mathrm{cart}(\Omega\times\mathbb{R}^N)$ *converges* weakly in cart

$$T_k \rightharpoonup T \qquad \text{in } \mathrm{cart}(\Omega\times\mathbb{R}^N)$$

iff

$$T_k \rightharpoonup T \qquad \text{weakly* in } \mathcal{C}$$

or, equivalently, iff

$$T_k \rightharpoonup T \ \text{in } \mathcal{D}_n(\Omega\times\mathbb{R}^N) \quad \text{with} \quad \sup_k(\|T_k\|_1 + \mathbf{M}(T_k)) < \infty\ .$$

Moreover we set for $T\in \mathrm{cart}(\Omega\times\mathbb{R}^N)$

$$\|T\|_{\mathrm{cart}} := \|T\|_{\mathcal{C}} = \|T\|_1 + \mathbf{M}(T)\ .$$

Proof of Proposition 1. Since $C_0(\Omega \times \mathbb{R}^N)$ is a separable Banach space with respect to the sup norm and $C_c^0(\Omega \times \mathbb{R}^N)$ is dense in $C_0(\Omega \times \mathbb{R}^N)$, (i) follows immediately. The claim (ii) is trivial.

Let $\mathcal{B}'$ be the dual of $\mathcal{B}$ and let $T \in \mathcal{B}'$. Then for any $\omega \in \mathcal{B}$ we have

$$|T(\omega)| \leq \| T \|_{\mathcal{B}'} \| \omega \|_{\mathcal{B}} \tag{1}$$

$\| \ \|_{\mathcal{B}'}$ being the norm of $\mathcal{B}'$. For $\omega \in D^n(\Omega \times \mathbb{R}^N)$ we trivially deduce

$$|T(\omega)| \leq \| T \|_{\mathcal{B}'} \| \omega \|_\infty$$

thus $\mathbf{M}(T) < +\infty$ and $\mathbf{M}(T) \leq \| T \|_{\mathcal{B}'}$. Applying (1) to $\omega = f(x,y)|y|\,dx$ with $f \in C_c^0(\Omega \times \mathbb{R}^N)$ and $|f(x,y)| \leq 1$, we get

$$|T(f(x,y)|y|\,dx) \leq \| T \|_{\mathcal{B}'} \sup_{(x,y)} \frac{|f(x,y)|\,|y|}{1+|y|} \leq \| T \|_{\mathcal{B}'}$$

hence we also have $\| T \|_1 \leq \| T \|_{\mathcal{B}'}$. We conclude

$$\mathcal{B}' \subset \mathcal{C}\,, \qquad \| T \|_{\mathcal{C}} = \| T \|_1 + \mathbf{M}(T) \leq 2 \| T \|_{\mathcal{B}'}\,.$$

Conversely suppose that $T \in \mathcal{C}$ and let $\omega \in \mathcal{B}$. We split ω as $\omega = \omega_1 + \omega_2$ where

$$\omega_1 = \frac{\omega_{\bar{0}0}(x,y)}{1+|y|} + \sum_{\substack{|\alpha|+|\beta|=n \\ |\beta|\geq 1}} \omega_{\alpha\beta}\, dx^\alpha \wedge dy^\beta$$

$$\omega_2 = \frac{\omega_{\bar{0}0}(x,y)\,|y|}{1+|y|}\,.$$

From $\mathbf{M}(T) < +\infty$, we get

$$|T(\omega_1)| \leq \mathbf{M}(T) \| \omega_1 \|_{\mathcal{B}}\,,$$

while from $\| T \|_1 < \infty$ we get

$$T(\omega_2) \leq \| T \|_1 \sup_{(x,y)} \frac{|\omega_{\bar{0}0}(x,y)|}{1+|y|} \leq \| T \|_1 \| \omega_1 \|_{\mathcal{B}}\,.$$

Therefore

$$|T(\omega)| \leq (\mathbf{M}(T) + \| T \|_1) \| \omega \|_{\mathcal{B}}\,,$$

that is, T extends to a linear and continuous functional on $\mathcal{B}$. This proves (iii).

Let us finally prove (iv). Suppose that

$$T_k \rightharpoonup T \text{ in } \mathcal{D}_n(\Omega \times \mathbb{R}^N) \quad \text{and} \quad \sup_k \| T_k \|_{\mathcal{C}} < \infty\,.$$

By Proposition 1 in Sec. 4.2.1 and Proposition 2 in Sec. 4.2.1 we get $\| T \|_{\mathcal{C}} < \infty$, hence, using the density of $\mathcal{D}^n(\Omega \times \mathbb{R}^N)$ in $\mathcal{B}$, we deduce

$$T_k(\omega) \longrightarrow T(\omega) \qquad \forall\, \omega \in \mathcal{B}\,,$$

i.e. $\{T_k\}$ converges weakly* to T in $\mathcal{C}$. Conversely, suppose that $T_k \rightharpoonup T$ weakly* in $\mathcal{C}$. Clearly $T_k(\omega) \to T(\omega)$ for all $\omega \in \mathcal{D}^n(\Omega\times\mathbb{R}^N)$, moreover, Banach-Steinhaus theorem yields

$$\sup_k \| T_k \|_{\mathcal{B}'} < \infty$$

hence

$$\sup_k \| T_k \|_{\mathcal{C}} < \infty\,.$$

□

Remark 3. We observe that $\mathcal{C} = \mathcal{B}'$ is not metrizable, while, being $\mathcal{B}$ separable, the weak* topology on bounded sets of $\mathcal{C}$ *is* metrizable, thus defined by the weak* convergence.

It is an open question to decide whether $\operatorname{cart}(\Omega \times \mathbb{R}^N)$ is the smallest sequentially weakly* closed set in $\mathcal{C}$ which contains the graphs of regular maps, i.e., with the terminology of Sec. 3.4 whether

$$\operatorname{cart}(\Omega \times \mathbb{R}^N) = \text{sw}^*\text{-cl}\,(\{G_u \mid u \in C^1(\Omega,\mathbb{R}^N) \cap \mathcal{A}^1(\Omega,\mathbb{R}^N)\})\,.$$

We therefore set

Definition 4. *The smallest sequentially weakly* closed set in $\mathcal{C}$, equivalently, the smallest sequentially weakly closed set in* $\operatorname{cart}(\Omega \times \mathbb{R}^N)$, *which contains the graphs of regular maps with bounded $\mathcal{C}$-norms is denoted by* $\operatorname{Cart}(\Omega \times \mathbb{R}^N)$.

Of course

$$\operatorname{Cart}(\Omega \times \mathbb{R}^N) \subset \operatorname{cart}(\Omega \times \mathbb{R}^N)$$

and $\operatorname{Cart}(\Omega \times \mathbb{R}^N)$ is closed with respect to the weak convergence in cart. The previous question amounts then to ask whether the inclusion above is strict or not.

The class $\operatorname{Cart}(\Omega \times \mathbb{R}^N)$ is in principle obtained by transfinite induction up to the first uncountable ordinal, i.e., in $\mathcal{C}$ with the weak* convergence

$$\operatorname{Cart}(\Omega \times \mathbb{R}^N) = \text{sw}^*\text{-cl}(\{G_u \mid u \in C^1(\Omega,\mathbb{R}^N) \cap \mathcal{A}^1(\Omega \times \mathbb{R}^N)\})\,.$$

We may again ask, compare Sec. 3.4, whether

$$\operatorname{Cart}(\Omega \times \mathbb{R}^N) = \text{sw}^*\text{-}\lim(\{G_u \mid u \in C^1(\Omega,\mathbb{R}^N) \cap \mathcal{A}^1(\Omega,\mathbb{R}^N)\})\,,$$

but we do not know the answer. In Sec. 4.2.5 we shall see that $\operatorname{cart}(\Omega \times \mathbb{R}^N) \neq \text{sw}^*\text{-}\lim(\{G_u \mid u \in C^1(\Omega,\mathbb{R}^N) \cap \mathcal{A}^1(\Omega,\mathbb{R}^N)\})$,

Finally it would be natural try to characterize, compare Sec. 4.1, the currents $T \in \operatorname{cart}(\Omega \times \mathbb{R}^N)$ for which there is a sequence of smooth maps $u_k : \Omega \to \mathbb{R}^N$ such that

$$G_{u_k} \rightharpoonup T \quad \text{and} \quad \mathbf{M}(G_{u_k}) \to \mathbf{M}(T) .$$

Such a question is connected with the problem of *relaxation* of the area integral for non parametric graphs which will be discussed in Vol. II Ch. 6. There we shall see some partial answers.

We also remark that $\operatorname{cart}(\Omega \times \mathbb{R}^N)$ and $\operatorname{Cart}(\Omega \times \mathbb{R}^N)$ are neither linear subspaces nor convex subsets of $\mathcal{C}$ or of $\mathcal{D}_n(\Omega \times \mathbb{R}^N)$. This is trivial as a linear combination of i.m. rectifiable currents is not in general an i.m. rectifiable current.

One could think of defining the sum of two Cartesian currents at least in $\operatorname{Cart}(\Omega \times \mathbb{R}^N)$ by a transfinite approximation process, i.e., defining for instance

$$T \oplus S \;=\; \text{weak limit of } G_{u_k+v_k} \text{ in cart}$$

if

$$G_{u_k} \underset{\mathcal{C}}{\rightharpoonup} T \qquad \text{and} \qquad G_{v_k} \underset{\mathcal{C}}{\rightharpoonup} T ,$$

but in general

$$\sup_k \| G_{u_k+v_k} \|_{\mathcal{C}} \;=\; +\infty .$$

In fact, for the equiboundedness of the masses of $G_{u_k+v_k}$, it is essentially necessary that the L^n-norms of Du_k and Dv_k be equibounded. This simple observation has an important consequence for variational problems with integrand depending on minors of Du. In fact, it roughly means that if one insists on working in a linear space, then one is forced to work essentially with energies which control the L^n-norm of the gradient of the map u.

For future purposes, it is convenient to introduce a more general class of Cartesian currents. Set $\underline{n} = \min(n, N)$ and denote now by p a multi-index $p = (p_0, p_1, \dots, p_{\underline{n}})$ where $p_i \in \mathbb{R}$, $p_i \geq 1$, $i = 0, 1, \dots, \underline{n}$. For $T \in \mathcal{D}_n(U)$ we set

$$\|T\|_{L^{p_0}} := \sup\{\, T(|y|^{p_0}\phi(x,y)\, dx) \mid \phi \in \mathcal{D}(U),\ \sup_U |\phi| \leq 1\, \};$$

for a n-form ω in U, $\omega = \sum_{|\alpha|+|\beta|=n} \omega_{\alpha\beta}(x,y)\, dx^\alpha \wedge dy^\beta$, we set

$$\|\omega\|_{M^{p_k}} := \left\{ \int_\Omega \sup_y \Bigl\{ \sum_{\substack{|\alpha|+|\beta|=n \\ |\beta|=k}} |\omega_{\alpha\beta}(x,y)|^2 \Bigr\}^{p'_k/2} dx \right\}^{1/p'_k} ,$$

p'_k being the dual exponent of p_k, $p'_k = \frac{p_k}{p_k - 1}$, and

$$\|\omega\|_{M^p} := \max_{k=0,\dots,\bar{n}} \|\omega\|_{M^{p_k}} .$$

Finally, for $T \in \mathcal{D}_n(U)$ we define

$$\|T\|_{M^{p_k}} := \sup \Bigl\{ \sum_{\substack{|\alpha|+|\beta|=n \\ |\beta|=k}} T^{\alpha\beta}(\omega_{\alpha\beta}) \mid \omega \in \mathcal{D}^n(U),\ \|\omega\|_{M^{p_k}} \leq 1 \Bigr\}$$

and

$$\|T\|_{M^p} := \max_{k=0,\dots,\bar{n}} \|T\|_{M^{p_k}}, \qquad \|T\|_{\mathrm{cart}^p} := \|T\|_{L^{p_0}} + \|T\|_{M^p}$$

and we consider the Banach space

$$\mathcal{C}^p = \{ T \in \mathcal{D}^n(U) \mid \|T\|_{\mathrm{cart}^p} < \infty \}$$

Observe that $\mathcal{C}^p$ is the dual space of the separable Banach space $\{ \omega \mid \|\omega\|_{M^p} < \infty \}$ and that the weak* convergence in $\mathcal{C}^p$ is equivalent to the weak convergence of currents with equibounded cart^p norms. Notice that if $u \in C^1(\Omega, \mathbb{R}^N), T = [\![G_u]\!]$ and $M_k(Du)$ denote all minors of order k, then

$$\|T\|_{L^{p_0}} = \|u\|_{L^{p_0}(\Omega,\mathbb{R}^N)}, \qquad \|T\|_{M^p} = \max_{k=0,\dots,\bar{n}} \Big\{ \int_\Omega |M_k(Du)|^{p_k}\, dx \Big\}^{1/p_k},$$

and in general $\|T\|_{M^{p_k}}$ is the total variation of the vector valued measure $(T_{\alpha\beta})_{\substack{|\alpha|+|\beta|=n \\ |\beta|=k}}$ if $p_k = 1$, while for $p_k > 1$ $\|T\|_{M^{p_k}}$ is the L^{p_k}-norm of the vector valued Radon-Nykodym derivative of $(\pi_\# |T_{\alpha\beta}|)_{\substack{|\alpha|+|\beta|=n \\ |\beta|=k}}$ with respect to Lebesgue's n-dimensional measure.

Definition 5. *For $p = (p_0, \dots, p_{\underline{n}})$, $p_i \in \mathbb{R}$, $p_i \geq 1$, we set*

$$\mathrm{cart}^p(\Omega \times \mathbb{R}^N) := \mathcal{C}^p \cap \mathrm{cart}(\Omega \times \mathbb{R}^N)$$

and define $\mathrm{Cart}^p(\Omega \times \mathbb{R}^N)$ as the smallest set in $\mathcal{C}^p$ containing the currents in $\mathcal{C}^p$ integration over smooth graphs and sequentially closed with respect to the weak convergence in $\mathcal{C}^p$.*

Of course $\mathrm{cart}^p(\Omega \times \mathbb{R}^N)$ is sequentially weak* closed,

$$\mathrm{Cart}^p(\Omega \times \mathbb{R}^N) \subset \mathrm{cart}^p(\Omega \times \mathbb{R}^N) \subset \mathrm{cart}(\Omega \times \mathbb{R}^N).$$

We shall return to this class in Vol. II Ch. 4 and Vol. II Ch. 5 in connection with the study of harmonic maps with values in S^2.

2.3 The Structure Theorem

In Sec. 4.2.2 we introduced the class of Cartesian currents. Our goal in this subsection is to prove a *structure theorem* which shows that every $T \in \mathrm{cart}(\Omega \times \mathbb{R}^N)$ is in a suitable sense integration over graphs of maps with *vertical parts.* For the sake of clarity we shall split such a structure theorem into several statements.

Let T be a current in graph$(\Omega \times \mathbb{R}^N)$. Since

$$\| T \|_1 + \mathbf{M}(T) \; < \; +\infty$$

we may define an $\mathbb{R}^N$-valued measure in Ω

$$u(T) = (u^1(T), \ldots, u^N(T))$$

by

(1) $$u^j(T)(\varphi(x)) := T(\varphi(x)\, y^j\, dx) = T^{\bar{0}0}(\varphi(x)\, y^j) \qquad \forall\, \varphi \in C^0_c(\Omega),$$

$j = 1, \ldots, N$, and the measures $M^\beta_{\bar{\alpha}}(T)$ in Ω, $|\alpha| + |\beta| = n$, $|\beta| \geq 1$, by

(2) $$M^\beta_{\bar{\alpha}}(T)(\varphi(x)) := \sigma(\alpha, \bar{\alpha})\, T(\varphi(x)\, dx^\alpha \wedge dy^\beta) = \sigma(\alpha, \bar{\alpha})\, T^{\alpha\beta}(\varphi(x))$$

for all $\varphi \in C^0_c(\Omega)$. Since T is an i.m. rectifiable current $T = \tau(\mathcal{M}, \theta, \vec{T}\,)$ where $\mathcal{M}$ is an n-rectifiable set in $\mathbb{R}^{n+N}$, θ an $\mathcal{H}^n$-measurable and integer-valued function, and $\vec{T}$ the n-vector orienting $\mathcal{M}$, we may also assume that the tangent space exists for all $z \in \mathcal{M}$, and that $\theta(z)$ exists and is larger or equal to 1 for any $z \in \mathcal{M}$.

We denote now by $\mathcal{M}_+$ the set of points z in $\mathcal{M}$ for which the tangent space to $\mathcal{M}$ at z does not contain vertical vectors. Proposition 1 in Sec. 2.2.1 yields

(3) $$\mathcal{M}_+ = \{z \in \mathcal{M} \mid \vec{T}^{\bar{0}0}(z) > 0\}\,.$$

Our first theorem describes some properties of T restricted to $\mathcal{M}_+$ and in particular shows that the Young measure $T^{\bar{0}0}$ is a function.

Theorem 1. *Let $T \in \operatorname{graph}(\Omega \times \mathbb{R}^N)$, $T = \tau(\mathcal{M}, \theta, \vec{T}\,)$. Then the measures $u^j(T)$, $j = 1, \ldots, N$, are absolutely continuous with respect to Lebesgue measure $\mathcal{L}^n$, i.e.,*

$$u^j(T) = u^j_T(x)\, dx\,.$$

If we set

$$u_T(x) := (u^1_T(x), \ldots, u^N_T(x)) \in L^1(\Omega, \mathbb{R}^N)\,,$$

then $u_T(x)$ is a.e. approximately differentiable, $u_T \in \mathcal{A}^1(\Omega, \mathbb{R}^N)$, and

$$T \llcorner \mathcal{M}_+ = G_{u_T}$$

Actually we shall prove

Theorem 2. *Let $T \in \operatorname{graph}(\Omega \times \mathbb{R}^N)$, $T = \tau(\mathcal{M}, \theta, \vec{T}\,)$. Then we have*

(i) $\mathcal{L}^n(\Omega \setminus \pi(\mathcal{M}_+)) = 0$. There exists a set $\Omega_1 \subset \pi(\mathcal{M}_+)$ such that $\mathcal{L}^n(\Omega \setminus \Omega_1) = 0$ and for every $x \in \Omega_1$ there exists a unique $\tilde{u}(x) \in \mathbb{R}^N$ such that $(x, \tilde{u}(x)) \in \mathcal{M}_+$ and

$$\sum_{(x,y) \in \pi^{-1}(x) \cap \mathcal{M}_+} \theta(x, y) = 1$$

(ii) *The measures $u^j(T)$, $j = 1, \dots, N$, are absolutely continuous with respect to the Lebesgue measure $\mathcal{L}^n$, i.e.,*

$$u^j(T) = u_T^j(x)\,dx$$

where

$$u_T(x) = (u_T^1(x), \dots, u_T^N(x)) \in L^1(\Omega, \mathbb{R}^N) .$$

Moreover,

$$u_T(x) = \tilde{u}(x) \qquad \text{a.e. } in\ \Omega$$

(iii) *$\mathcal{H}^n(\mathcal{M}_+ \cap A \times \mathbb{R}^N) = 0$, whenever $\mathcal{H}^n(A) = 0$. In particular $\theta(z) = 1$ $\mathcal{H}^n$-a.e. in $\mathcal{M}_+$, the map $\tilde{u}_{|\Omega_1}$ has the Lusin property (N) in Ω_1 and,*

$$\mathcal{M}_+ = \{(x,y) \in \Omega \times \mathbb{R}^N \mid x \in \pi(\mathcal{M}_+),\ y = \tilde{u}(x)\} \quad \mathcal{H}^n \text{ a.e.},$$

(iv) *u_T is a.e. approximately differentiable and $\mathcal{M}_+ = \mathcal{G}_{u_T,\Omega}$ $\mathcal{H}^n$ a.e., $\mathcal{G}_{u_T,\Omega}$ being the 1-graph of u_T.*

(v) *$u_T \in \mathcal{A}^1(\Omega, \mathbb{R}^N)$ and $T \llcorner \mathcal{M}_+ = G_{u_T}$.*

Proof. First observe that

$$\begin{aligned} \|T\|_0 &= \sup\{T(\varphi(x,y)\,dx) \mid \varphi \in C_0^0(\Omega \times \mathbb{R}^N),\ |\varphi| \le 1\} \\ &= \sup\Big\{ \int \varphi(x,y)\,\theta(x,y)\,\vec{T}^{\bar{0}0}(x,y)\,d\mathcal{H}^n \llcorner \mathcal{M} \Big\} \\ &= \int \theta(x,y)\,\vec{T}^{\bar{0}0}(x,y)\,d\mathcal{H}^n \llcorner \mathcal{M}_+ \,, \qquad (4) \\ \|T\|_1 &= \sup\{T(\varphi(x,y)|y|\,dx) \mid \varphi \in C_c^0(\Omega \times \mathbb{R}^N),\ |\varphi| \le 1\} \\ &= \int |y|\,\theta(x,y)\,\vec{T}^{\bar{0}0}(x,y)\,d\mathcal{H}^n \llcorner \mathcal{M}_+ \,. \end{aligned}$$

Since $\|T\|_0 < \infty$, the function $\varphi(x)\,\theta(x,y)\,\vec{T}^{\bar{0}0}(x,y)$ is summable with respect to $\mathcal{H}^n \llcorner \mathcal{M}_+$ for any $\varphi \in C_c^0(\Omega)$. Thus, since by Proposition 1 in Sec. 2.2.1 (iii)

$$J_\pi^{\mathcal{M}} = \vec{T}^{\bar{0}0} ,$$

the area formula, compare Sec. 2.1.5, yields

$$(5) \qquad \int \varphi(x)\,\theta(x,y)\,\vec{T}^{\bar{0}0}(x,y)\,d\mathcal{H}^n \llcorner \mathcal{M}_+ = \int_{\pi(\mathcal{M}_+)} \varphi(x)\,N(x)\,dx$$

where $N(x)$ is the Lebesgue summable function

$$N(x) := \sum_{y \in \pi^{-1}(x) \cap \mathcal{M}_+} \theta(x,y) .$$

Comparing (5) with

$$\int_{\Omega} \varphi(x)\,dx = \int_{\pi(\mathcal{M}_+)} \varphi(x)\,N(x)\,dx ,$$

which follows at once from $\pi_\# T = [\![\Omega]\!]$, we readily deduce that $\mathcal{L}^n(\Omega\backslash\pi(\mathcal{M}_+)) = 0$ and $N(x) = 1$ $\mathcal{H}^n$-a.e. in Ω. Since $\theta(x,y) \geq 1$ for all $(x,y) \in \mathcal{M}$ and $N(x) = 1$ a.e. in $\pi(\mathcal{M}_+)$, we also deduce the existence of a uniquely $\mathcal{H}^n$-a.e. defined function $\tilde{u}(x) = (\tilde{u}^1(x),\dots,\tilde{u}^N(x))$ such that $(x,\tilde{u}(x)) \in \mathcal{M}_+$. This proves (i).

Since $\| T \|_1 < \infty$, also the functions $\varphi(x)\,y^j\,\theta(x,y)\,\vec{T}^{\bar{0}0}(x,y)$ $j = 1,\dots,N$ are $\mathcal{H}^n \llcorner \mathcal{M}_+$-summable for all $\varphi \in C^0_c(\Omega)$. Applying again the area formula we therefore infer that the function

$$u^j_T(x) := \sum_{y\in\pi^{-1}(x)\cap\mathcal{M}_+} y^j\,\theta(x,y) \tag{6}$$

is $\mathcal{H}^n$-summable in Ω and for all $\varphi \in C^0_c(\Omega)$

$$u^j(T)(\varphi) = \int \varphi(x)\,y^j\,\theta(x,y)\,\vec{T}^{\bar{0}0}(x,y)\,d\mathcal{H}^n \llcorner \mathcal{M}_+ = \int_{\pi(\mathcal{M}_+)} \varphi(x)\,u^j_T(x)\,dx .$$

This yields that the measures $u^j(T)$ are absolutely continuous with respect to Lebesgue measure and in conjunction with (6) we get $u^j_T(x) = \tilde{u}(x)$ a.e. in Ω, $u_T \in L^1(\Omega,\mathbb{R}^N)$. This proves (ii).

For any measurable subset A of Ω we have

$$\begin{aligned}\mathbf{M}(T) &\geq \mathcal{H}^n(\mathcal{M}) \geq \mathcal{H}^n(\mathcal{M}_+ \cap A \times \mathbb{R}^N)\\ &= \int \frac{1}{\vec{T}^{\bar{0}0}(x,y)}\,\vec{T}^{\bar{0}0}(x,y)\,d\mathcal{H}^n \llcorner (\mathcal{M}_+ \cap A \times \mathbb{R}^N) .\end{aligned}$$

Since $\mathbf{M}(T) < \infty$, the area formula yields that the function

$$K(x) := \sum_{y\in\pi^{-1}(x)\cap\mathcal{M}_+} \frac{1}{\vec{T}^{\bar{0}0}(x,y)} = \frac{1}{\vec{T}^{\bar{0}0}(x,u_T(x))}$$

is summable in Ω and

$$\mathcal{H}^n(\mathcal{M}_+ \cap A \times \mathbb{R}^N) = \int_A K(x)\,dx .$$

Thus $\mathcal{H}^n(\mathcal{M}_+ \cap A \times \mathbb{R}^N) = 0$ if $\mathcal{H}^n(A) = 0$. In conjunction with step (i) we infer that

$$\theta(z) = 1 \qquad \mathcal{H}^n \text{ a.e. in } \mathcal{M}_+ . \tag{7}$$

and (iii) is proved.

Since $\mathcal{M}$ is rectifiable, taking into account (i), one can decompose $\mathcal{M}_+$ as $\mathcal{M}_+ = \cup_{j=0}^\infty \mathcal{M}_j$, with $\mathcal{H}^n(\mathcal{M}_j) < \infty$, $\mathcal{H}^n(\mathcal{M}_0) = 0$ and for each $j =$

$1, 2, \ldots,\ \mathcal{M}_j = \{(x,y)\,|\, y = \tilde{u}(x),\ x \in \pi(\mathcal{M}_j)\}$, $\pi(\mathcal{M}_j) \subset \Omega_1$, $\mathcal{M}_j \subset \mathcal{N}_j$ where $\mathcal{N}_j$ is a C^1-submanifold with $\mathrm{Tan}_{\mathcal{N}_j}(z) = \vec{T}(z)\ \forall\, z \in \mathcal{M}_j$, and $\tilde{u}$ is the function in Step (i).

Since for every $z = (x, \tilde{u}(x)) \in \mathcal{M}_j$ we have

$$\mathrm{Tan}_{\mathcal{N}_j}(z)^{\bar{0}0} = \vec{T}^{\bar{0}0}(z) > 0 \tag{8}$$

the tangent plane to $\mathcal{N}_j$ at z does not contain vertical vectors. Therefore there exist a neighborhood $\mathcal{U}$ of y and a map $v : U \to \mathbb{R}^N$, $v \in C^1(U)$ such that $\mathcal{N}_j \cap (U \times \mathbb{R}^N)$ is the (smooth) graph of v,

$$\mathcal{N}_j \cap (U \times \mathbb{R}^N) = (\mathrm{id} \bowtie v)(U)$$

and by Proposition 1 in Sec. 2.2.1,

$$\mathrm{Tan}_{\mathcal{N}_j}(x, v(x)) = \frac{M(Dv(x))}{|\,M(Dv(x))\,|}\,. \tag{9}$$

Since $\mathcal{M}_j \subset \mathcal{N}_j, \tilde{u}$ and v agree on $\pi(\mathcal{M}_j) \cap U$.

This way we can decompose $\cup_{j=1}^{\infty} \pi(\mathcal{M}_j) \cap \Omega_1$ as union of measurable pieces in which $\tilde{u}$ coincides with a C^1-function. This implies that $\tilde{u}$ is approximately differentiable at a.e. $x \in \Omega_1$ and therefore at a.e. $x \in \Omega$, by Theorem 3 in Sec. 3.1.4. Moreover, taking into account the Lusin property (N) of $\tilde{u}$, we have

$$\mathcal{G}_{\tilde{u},\Omega_1} = \{(x,y) \mid x \in \Omega_1,\ y = \tilde{u}(x)\} \quad \mathcal{H}^n \text{ a.e.}$$

Therefore u_T is a.e. approximately differentiable, too, $Du_T = D\tilde{u}$ a.e. and,

$$\mathcal{G}_{u_T,\Omega} = \mathcal{G}_{\tilde{u},\Omega_1} = \mathcal{M}_+ \qquad \mathcal{H}^n \text{ a.e.} \tag{10}$$

This proves (iv).

Finally we also have for a.e. $x \in \pi(\mathcal{M}_j) \cap U$ $\mathrm{ap}D\tilde{u}(x) = Dv(x)$. Taking into account (8) (9) we get

$$\vec{T}(x, \tilde{u}(x)) = \frac{M(\mathrm{ap}D\tilde{u}(x))}{|\,M(\mathrm{ap}D\tilde{u}(x))\,|} \qquad \text{a.e. } x \in \Omega. \tag{11}$$

and collecting (7), (10) and (11) we then conclude that

$$T \llcorner \mathcal{M}_+ := \tau(\mathcal{M}_+, \theta, \vec{T}) = \tau(\mathcal{G}_{u_T,\Omega}, 1, \frac{M(Du_T)}{|\,M(Du_T)\,|}) =: G_{u_T}$$

□

Next theorem yields a formula of representation for the zero component $T^{\bar{0}0}$ of T.

Theorem 3. *Let $T \in \mathrm{graph}(\Omega \times \mathbb{R}^N)$, $T = \tau(\mathcal{M}, \theta, \vec{T})$, and let $u_T \in \mathcal{A}^1$ be the map associated to T in Theorem 1. Then we have*

(i) *For all continuous functions* $\phi(x,y)$ *in* $\Omega \times \mathbb{R}^N$ *satisfying* $|\phi(x,y)| \leq c(1+|y|)$ *we have*

$$T^{\bar{0}0}(\phi(x,y)) = \int_\Omega \phi(x, u_T(x))\, dx \,.$$

In particular

$$\| T \|_1 = \int_\Omega |u_T|\, dx \,,$$

(ii) *The function* $u_T \in BV(\Omega, \mathbb{R}^N)$ *if and only if the* L^1*-norm of* ∂T*, defined as*

$$\| \partial T \|_1 := \sup\{\partial T(|y|\varphi_i(x)\, \widehat{dx^i}) \mid \varphi \in C^\infty_c(\Omega, \mathbb{R}^n),\ |\varphi| \leq 1\} \,,$$

is finite. In this case we have

$$\| D_i u^j_T - M^j_{\bar{i}}(T) \| \leq \| \partial T \|_1$$

and

$$\| Du_T \| \leq \mathbf{M}(T) + \| \partial T \|_1 \,.$$

In particular $u_T \in BV(\Omega, \mathbb{R}^N)$ *and* $D_i u^j_T = M^j_{\bar{i}}(T)$ *if* $T \in \operatorname{cart}(\Omega \times \mathbb{R}^N)$.

(iii) *Finally for any integer* k *the truncations at levels* k *of each component of* u_T

$$u^j_{T,k} := \max(\min(u^j_T, k), -k)$$

belong to $BV(\Omega)$*, and*

$$\| D_i u^j_{T,k} \| \leq \mathbf{M}(T) + c(k)\, \mathbf{M}(\partial T) \,.$$

In particular $u_T \in BV(\Omega, \mathbb{R}^N)$ *if* u_T *is bounded.*

Proof. Note that by (4) $\varphi(x,y)\, \theta(x,y)\, \vec{T}^{\bar{0}0}(x,y)$ is $\mathcal{H}^n \llcorner \mathcal{M}_+$-summable for all $\phi \in C^0$ satisfying $|\phi(x,y)| \leq c(1+|y|)$. Moreover we have

$$\begin{aligned} T^{\bar{0}0}(\phi(x,y)\, dx) &= \int \phi(x,y)\, \theta(x,y)\, \vec{T}^{\bar{0}0}\, d\mathcal{H}^n \llcorner \mathcal{M}_+ \\ &= (T \llcorner \mathcal{M}_+)(\phi(x,y)\, dx) = \int_\Omega \phi(x, u_T(x))\, dx \end{aligned}$$

by (iv) of Theorem 2. This proves (i).

Let us prove (ii). Consider the $(n-1)$-forms on $\Omega \times \mathbb{R}^N$ $\omega := y^j\, \varphi(x)\, \widehat{dx^i}$ where $\varphi \in C^1(\Omega)$ and $|D\varphi| \in L^\infty$. We have

$$d\omega = (-1)^{i-1}\, D_i\varphi\, y^j\, dx + \varphi(x)\, dy^j \wedge \widehat{dx^i} \,.$$

Since the coefficient of $(d\omega)^{(0)}$ grows linearly in y and the coefficient of $(d\omega)^{(1)}$ is bounded, we can compute $T(d\omega)$, by approximation, as limit of $T(d\eta)$, the η's being smooth $(n-1)$-forms in $\Omega \times \mathbb{R}^N$ with compact support and $\eta = \eta^{(0)}$. We then deduce

$$\begin{aligned} T(d\omega) &= (-1)^{i-1} T(D_i\varphi\, y^j\, dx) + T(\varphi(x)\, dy^j \wedge \widehat{dx^i}) \\ &= (-1)^{i-1} \Big\{ \int D_i\varphi\, u_T^j(x)\, dx + M_i^j(T)(\varphi(x)) \Big\} . \end{aligned}$$

From this the conclusion follows at once.

To prove (iii) it suffices to choose as ω the form

$$\omega := \max(\min(y^i, k), -k)\, \varphi(x) \widehat{dx^i}$$

and proceed as in the proof of (ii). □

1 It is easily seen that Theorem 3 (vi) is optimal in the sense that in general u_T does not belong to $BV(\Omega, \mathbb{R}^N)$ if T is merely in $\mathrm{graph}(\Omega \times \mathbb{R}^N)$. Consider for instance the function $u : B(0,1) \subset \mathbb{R}^2 \to \mathbb{R}$ given by

$$u(x) = k^2 \quad \text{on } B_{r_k} \setminus B_{r_{k+1}} , \qquad r_k := k^{-2} , \qquad k = 1, 2, \ldots$$

one easily sees that $u \in L^1(B(0,1))$, $G_u \in \mathrm{graph}(B(0,1) \times \mathbb{R})$, but $u \notin BV(B(0,1))$. •

We denote by

$$M_{\bar\alpha}^{\beta}(T) = (M_{\bar\alpha}^{\beta}(T))^a\, dx + (M_{\bar\alpha}^{\beta}(T))^s$$

and, when $u_T \in BV(\Omega, \mathbb{R}^N)$, by

$$Du_T = (Du_T)^a\, dx + (Du_T)^s$$

the Lebesgue decompositions of the measures $M_{\bar\alpha}^{\beta}(T)$ and Du_T with respect to the Lebesgue measure in Ω. Clearly

$$(Du_T)^a(x),\ M_{\bar\alpha}^{\beta}(T)^a(x) \in L^1(\Omega)$$

Theorem 4. *Let $T \in \mathrm{graph}(\Omega \times \mathbb{R}^N)$, $T = \tau(\mathcal{M}, \theta, \vec{T})$. Then we have*

(i) For any α, β with $|\alpha| + |\beta| = n$, $|\beta| \geq 1$,

$$M_{\bar\alpha}^{\beta}(T)^a(x) = M_{\bar\alpha}^{\beta}(\mathrm{ap}Du_T(x)), \qquad \textit{for a.e. } x \in \Omega . \tag{12}$$

Consequently, for any $\phi \in C_c^0(\Omega \times \mathbb{R}^N)$ we have

$$(T^{\alpha\beta} \llcorner \mathcal{M}_+)(\phi(x,y)) = \sigma(\alpha, \bar\alpha) \int_{\Omega} \phi(x, u_T(x))\, M_{\bar\alpha}^{\beta}(T)^a(x)\, dx . \tag{13}$$

(ii) If $(\pi_{\#}\| T^{\alpha\beta} \|)^s = 0$, *then* $\| T^{\alpha\beta} \| \llcorner (\mathcal{M} \setminus \mathcal{M}_+) = 0$, *consequently, in this case,*

$$T^{\alpha\beta}(\phi(x,y)) = \sigma(\alpha,\bar{\alpha}) \int_\Omega \phi(x, u_T(x))\, M^\beta_{\bar{\alpha}}(\text{ap}Du_T(x))\, dx \; .$$

In particular, if $(\pi_{\#}\| T^{\alpha\beta} \|)^s = 0$ *for all* α, β *with* $|\alpha| + |\beta| = n$, *then*

$$\| T \|(\mathcal{M} \setminus \mathcal{M}_+) = 0 \, , \quad i.e. \quad \mathcal{M} = \mathcal{M}_+ \quad \mathcal{H}^n\text{–a.e.},$$

and this implies that $T = G_{u_T}$.

Proof. Taking into account that the total variation of the measure $T^{\alpha\beta} \llcorner \mathcal{M}_+$ is finite, we have

$$(14) \qquad (T^{\alpha\beta} \llcorner \mathcal{M}_+)(\varphi(x)\, dx) = \sigma(\alpha,\bar{\alpha}) \int_\Omega \varphi(x)\, M^\beta_{\bar{\alpha}}(\text{ap}Du_T(x))\, dx \; .$$

Let N be a Borel set such that $\mathcal{H}^n(\Omega \setminus N) = 0$ and $\mathcal{H}^n(\mathcal{M}_+ \setminus (N \times \mathbb{R}^N)) = 0$, which exists because of Theorem 2. Observing that $\pi_{\#}(T^{\alpha\beta} \llcorner (\mathcal{M} \setminus \mathcal{M}_+))$ is concentrated over $\Omega \setminus N$, the identities

$$(15) \qquad \begin{aligned} \pi_{\#} T^{\alpha\beta} &= \pi_{\#}(T^{\alpha\beta} \llcorner \mathcal{M}_+) + \pi_{\#}(T^{\alpha\beta} \llcorner \mathcal{M} \setminus \mathcal{M}_+) \\ \pi_{\#}\| T^{\alpha\beta} \| &= \pi_{\#}(\| T^{\alpha\beta} \| \llcorner \mathcal{M}_+) + \pi_{\#}(\| T^{\alpha\beta} \| \llcorner \mathcal{M} \setminus \mathcal{M}_+) \end{aligned}$$

yield the Lebesgue decompositions of the measures $\pi_{\#} T^{\alpha\beta}$ and $\pi_{\#}\| T^{\alpha\beta} \|$.

From the definition of $M^\beta_{\bar{\alpha}}(T)$ and the uniqueness of Lebesgue's decomposition we therefore deduce that

$$\pi_{\#}(T^{\alpha\beta} \llcorner \mathcal{M}_+) = \sigma(\alpha,\bar{\alpha})\, M^\beta_{\bar{\alpha}}(T)^a(x)\, dx$$

or, equivalently, from (14)

$$(16) \qquad M^\beta_{\bar{\alpha}}(\text{ap}Du_T(x)) = M^\beta_{\bar{\alpha}}(T)^a(x) \qquad \text{for a.e. } x \in \Omega \; .$$

This proves (i).

In order to prove (ii) it suffices to observe that from the second identity in (15) we deduce

$$(\pi_{\#}\| T^{\alpha\beta} \|)^s = \pi_{\#}(\| T^{\alpha\beta} \| \llcorner \mathcal{M} \setminus \mathcal{M}_+) \, ,$$

therefore $(\pi_{\#}\| T^{\alpha\beta} \|)^s = 0$, which yields at once $T^{\alpha\beta} \llcorner \mathcal{M} \setminus \mathcal{M}_+ = 0$. □

Next theorem concerns the convergence of u_{T_k} and $M^\beta_{\bar{\alpha}}(T_k)$ when the Cartesian current T_k converge in cart to T.

Theorem 5. *Let* $\{T_k\} \subset \text{graph}(\Omega \times \mathbb{R}^N)$ *be such that*

$$\sup_k (\mathbf{M}((\partial T_k) \llcorner \Omega \times \mathbb{R}^N) + \| (\partial T_k) \llcorner \Omega \times \mathbb{R}^N \|_1) < \infty;$$

in particular, let $\{T_k\} \subset \text{cart}(\Omega \times \mathbb{R}^N)$*. Suppose that*

$$T_k \rightharpoonup T \text{ in } \mathcal{D}_n(\Omega \times \mathbb{R}^N) \quad \text{and} \quad \sup_k (\| T_k \|_1 + \mathbf{M}(T_k)) < \infty ,$$

then

$$(17) u_{T_k} \longrightarrow u_T \text{ strongly in } L^1 \quad \text{and} \quad Du_{T_k} \rightharpoonup Du_T \text{ as measures in } \Omega .$$

Moreover, for all α, β *with* $|\alpha| + |\beta| = n$ *and* $|\beta| > 1$

$$(18) \qquad M^{\beta}_{\bar{\alpha}}(T_k) \rightharpoonup M^{\beta}_{\bar{\alpha}}(T) \qquad \text{as measures in } \Omega$$

provided the u_{T_k}*'s are equibounded in* $L^\infty(\Omega, \mathbb{R}^N)$.

Proof. Since

$$\| u_{T_k} \|_{L^1} + \| Du_{T_k} \| \le \| T_k \|_1 + \mathbf{M}(T_k) ,$$

the sequence $\{u_{T_k}\}$ is equibounded in $BV(\Omega, \mathbb{R}^N)$. Thus there exists a subsequence $u_{T_{k_h}}$ which converges strongly in $L^1(\Omega, \mathbb{R}^N)$ to some $v \in BV(\Omega, \mathbb{R}^N)$. For $\phi \in C^\infty_c(\Omega \times \mathbb{R}^N)$

$$T^{\bar{0}0}_{k_h}(\phi(x,y)) = T_{k_h}(\phi(x,y)\, dx) \longrightarrow T^{\bar{0}0}(\phi(x,y)) ;$$

on the other hand

$$T^{\bar{0}0}_{k_h}(\phi(x,y)) = \int_\Omega \phi(x, u_{T_{k_h}}(x))\, dx \longrightarrow \int_\Omega \phi(x, u_T)\, dx$$

thus

$$\int_\Omega \phi(x, v(x))\, dx = \int_\Omega \phi(x, u_T(x))\, dx .$$

Since the last equality holds for all $\phi \in C^\infty_c(\Omega \times \mathbb{R}^N)$, we conclude that $v = u_T$, and, as the limit is independent of the subsequence, that the entire sequence $\{u_{T_k}\}$ converges and (17) holds.

If $\{u_k\}$ is equibounded in L^∞, choosing test functions $\phi(x,y)$ of the type $\phi(x,y) = \varphi(x)\chi_R(y)$, $\varphi \in C^\infty_c(\Omega)$ and χ_R a cut off function in $\mathbb{R}^N$, for large R we readily get (18). □

We can in fact prove a slightly more general result on the convergence of minors at least for sequences of maps in $\text{cart}^1(\Omega, \mathbb{R}^N)$. In order to do that we need the notion of *ACP weak convergence* we introduced in Sec. 1.2.7.

Proposition 1. *Let $\{u_k\} \subset \mathrm{cart}^1(\Omega, \mathbb{R}^N)$ be such that*

$$u_k \to u \text{ strongly in } L^1(\Omega, \mathbb{R}^N), \quad \text{and} \quad \sup_k \| u_k \|_{\mathrm{cart}^1} < \infty. \tag{19}$$

Suppose that

$$\sigma_R := \limsup_{k\to\infty} \mathbf{M}(G_{u_k} \llcorner \Omega \times (\mathbb{R}^N \setminus B(0,R))) \longrightarrow 0 \tag{20}$$

as $R \to \infty$. Then $u \in \mathcal{A}^1(\Omega, \mathbb{R}^N)$, $(Du)^a = \mathrm{ap}Du$ and for each α, β, $|\alpha| + |\beta| = n$, $|\beta| \geq 1$

$$M^\beta_{\bar\alpha}(Du_k) \overset{ACP}{\rightharpoonup} M^\beta_{\bar\alpha}(\mathrm{ap}Du) .$$

Proof. First let us assume that moreover

$$G_{u_k} \rightharpoonup T \tag{21}$$

weakly in $\mathrm{cart}(\Omega \times \mathbb{R}^N)$. Then we claim that

$$M^\beta_{\bar\alpha}(Du_k) = M^\beta_{\bar\alpha}(G_{u_k}) \rightharpoonup M^\beta_{\bar\alpha}(T) \qquad \text{as measures.} \tag{22}$$

To prove this for $R > 0$ we consider the function

$$\psi_R(y) := \max(\min(1, R+1-|y|), 0) , \tag{23}$$

which belongs to $C^0_c(\mathbb{R}^N)$ and satisfies $|\psi_k| \leq 1$, and any function $\varphi \in C^0_c(\Omega)$, $|\varphi| \leq 1$. We have

$$\begin{aligned}\limsup_{k\to\infty} |G^{\alpha\beta}_{u_k}(\varphi) - T^{\alpha\beta}(\varphi)| \leq{}& \limsup_{k\to\infty}\big[|G^{\alpha\beta}_{u_k}(\varphi\psi_R) - T^{\alpha\beta}(\varphi\psi_R)| \\ &+ |G^{\alpha\beta}_{u_k}(\varphi(1-\psi_R))| + |T^{\alpha\beta}(\varphi(1-\psi_R))|\big] \leq 2\sigma_R\end{aligned}$$

because from (20) and (21) we have

$$\mathbf{M}(T \llcorner \Omega \times (\mathbb{R}^N \setminus B(0,R)) \leq \sigma_R .$$

Letting $R \to \infty$, we obtain (22).

By Theorem 5 (17) we have $u = u_T$ and by Theorem 4 (i) we have

$$M^\beta_{\bar\alpha}(T)^a = M^\beta_{\bar\alpha}((Du_T)^a) = M^\beta_{\bar\alpha}(\mathrm{ap}Du) . \tag{24}$$

Fix α, β; for every subsequence $\{u_{k_i}\}$, $u_{k_i} \rightharpoonup u$, we can extract a further subsequence $\{u'_{k_i}\}$ such that (21) holds. Applying the above reasoning we deduce from (22) and (24)

$$\mu^a = M^\beta_{\bar\alpha}(T)^a = M^\beta_{\bar\alpha}(\mathrm{ap}Du)$$

and this proves the proposition. □

Remark 1. Proposition 1 extends Theorem 2 in Sec. 3.3.2, showing that equi-integrability of minors *near infinity* is sufficient to ensure convergence. We do not know whether assumption (20) is really necessary. In fact we are not able to prove or disprove the implication (21)⇒(22) without assuming (20). This is related to the following question. Let $u, u_k \in \text{cart}^1(\Omega, \mathbb{R}^N)$, suppose that

$$u_k \to u \text{ in } L^1(\Omega, \mathbb{R}^N) \quad \text{and} \quad \sup_k \| u_k \|_{\text{cart}^1} < \infty .$$

We have two kinds of convergence

(a) $G_{u_k} \rightharpoonup G_u$ as currents
(b) $M(Du_k) \rightharpoonup M(Du)$ as measures

Without assuming equi-integrability of the sequence $\{M(Du_k)\}$ we are not able to prove or disprove complete relations between (a) and (b).

Remark 2. We collect here a few observations which aim to illustrate the meaning of the previous theorems.

(i) Theorem 2, Theorem 4 and Theorem 5 hold for graphs T. In particular they hold for all Cartesian currents. Notice however that we have not used the full information

$$\partial T \llcorner \Omega \times \mathbb{R}^N = 0 . \tag{25}$$

(ii) Examples in [1] in Sec. 4.0 and [2] in Sec. 4.0 of the introduction to this chapter clearly show that a Cartesian current T, even though in sw*-$\lim(C^1(\Omega, \mathbb{R}^N) \cap \mathcal{A}^1(\Omega, \mathbb{R}^N))$, is not determined by its u_T. There may be infinitely many different T's to which it is associated the same map u_T in Theorem 2 and Theorem 4, compare also Sec. 4.2.5 and next chapter.

(iii) The structure theorem yields a complete description of currents T in $\text{cart}(\Omega \times \mathbb{R}^N)$, $T = \tau(\mathcal{M}, \theta, \vec{T})$, when restricted to $\mathcal{M}_+$. Examples in Sec. 4.2.5 show that in general we may have

(i) $\mathcal{H}^n(\mathcal{M} \setminus \mathcal{M}_+) > 0$
(ii) On $\mathcal{M} \setminus \mathcal{M}_+$ the density θ can be any positive integer.

In fact the situation can be even more complicated. Let B^2 be the unit ball in $\mathbb{R}^2$, let y_0, y_1 be the point in $\widehat{\mathbb{R}}^2$, $y_0 := (0,1)$, $y_1 := (0, 1/2)$, and let G_0 be the current in $B^2 \times \widehat{\mathbb{R}}^2$ integration over the graph of the zero constant map. As in [1] in Sec. 4.0 and [2] in Sec. 4.0 of the introduction of this chapter it is easily seen that the currents

$$\begin{aligned} R &:= G_0 + [\![\{0\}]\!] \times \partial [\![B(y_0, 1)]\!] \\ S &:= G_0 + 2[\![\{0\}]\!] \times \partial [\![B(y_1, 1/2)]\!] \end{aligned}$$

are approximable in $\text{cart}(B^2 \times \widehat{\mathbb{R}}^2)$ by smooth graphs, therefore they are Cartesian currents, and obviously different, but $u_R = u_S$, and for all α, β

$$\pi_{\#} R^{\alpha\beta} = \pi_{\#} S^{\alpha\beta}$$

and even

$$\pi_{\#} \| R^{\alpha\beta} \| = \pi_{\#} \| S^{\alpha\beta} \| .$$

Also

$$\pi_{\#} R^{\alpha\beta} = 0 , \qquad \pi_{\#} \| R^{\bar{0}0} \| = \frac{\pi}{2} \delta_0 \neq 0 .$$

This shows that no information on $\pi_{\#} T$ can allow to recover T on $\mathcal{M} \setminus \mathcal{M}_+$, and that Theorem 4 (ii) is optimal.

The *vertical part* $T \llcorner \mathcal{M} \setminus \mathcal{M}_+$ of a Cartesian current is essentially uncontrollable in terms of u_T.

(iv) In terms of functions the previous remarks say the following. Consider a sequence $\{u_k\}$ of smooth maps such that

$$\sup_k \Big(\int_\Omega |u_k| \, dx + \int_\Omega |M(Du_k)| \Big) < +\infty$$

and suppose for simplicity that the u_k's are equibounded in L^∞. Passing to a subsequence, we may assume that there exists a Cartesian current T in $\Omega \times \mathbb{R}^N$ such that

$$u_k \to u_T \qquad \text{strongly in } L^1(\Omega, \mathbb{R}^N)$$

$$\frac{\sigma(\alpha, \bar{\alpha}) \, M_{\bar{\alpha}}^\beta(Du_k)}{|M(Du_k)|} \mathcal{H}^n \llcorner \mathcal{G}_{u_k,\Omega} \rightharpoonup T^{\alpha\beta} \qquad \text{as measures in } \Omega \times \mathbb{R}^N$$

$$\sigma(\alpha, \bar{\alpha}) \, M_{\bar{\alpha}}^\beta(Du_k) \rightharpoonup \sigma(\alpha, \bar{\alpha}) \, M_{\bar{\alpha}}^\beta(T) \qquad \text{as measures in } \Omega$$

$$M_{\bar{\alpha}}^\beta(T) = M_{\bar{\alpha}}^\beta(Du_T(x)) \, dx + M_{\bar{\alpha}}^\beta(T)^s .$$

In general $M_{\bar{\alpha}}^\beta(T)^s \neq 0$; even if $M_{\bar{\alpha}}^\beta(T)^s = 0$ for all α and β, $T^{\alpha\beta}$ need not be equal to

$$\frac{\sigma(\alpha, \bar{\alpha}) \, M_{\bar{\alpha}}^\beta(Du_T(x))}{|M(Du_T(x))|} \mathcal{H}^n \llcorner \mathcal{M}_+ .$$

Moreover, while all formulas of integration by parts are available for u_k and $M_{\bar{\alpha}}^\beta(Du_k)$ as well as for T (compare Sec. 3.2.2), they do not hold for u_T and $M_{\bar{\alpha}}^\beta(Du_T(x))$.

Remark 3. For future purposes we observe that the following claim is an immediate consequence of the area formula: *Let $T \in \mathcal{D}_n(\mathbb{R}^n \times \mathbb{R}^N)$ be an i.m. rectifiable n-current satisfying*

$$\mathbf{M}(T) < \infty , \qquad T^{\bar{0}0} \geq 0 , \qquad \pi_{\#} T = [\![\Omega]\!]$$

for some open set Ω in $\mathbb{R}^n$. Then $\operatorname{spt} T \subset \overline{\Omega} \times \mathbb{R}^N$.

Finally we explicitly state the following easy consequence of the previous results

Proposition 2. *Let* $T \in \text{graph}(\Omega \times \mathbb{R}^N)$, $T = \tau(\mathcal{M}, \theta, \vec{T})$. *Then we have*

(i) There is $u \in \text{cart}^1(\Omega \times \mathbb{R}^N)$ *such that* $T = G_u$ *if and only if*

$$(\pi_\# \| T \|)^s = 0 \qquad \text{and} \qquad \partial T \llcorner \Omega \times \mathbb{R}^N = 0 \, .$$

(ii) There is $u \in \text{cart}^p(\Omega, \mathbb{R}^N)$, $p > 1$, *such that* $T = G_u$ *if and only if*

$$\int (\vec{T}^{\bar{0}0}(z))^{1-p} \, d\mathcal{H}^n \llcorner \mathcal{M} \; < \; \infty \, .$$

and $\partial T \llcorner \Omega \times \mathbb{R}^N = 0$.

2.4 Cartesian Currents in Codimension One

In this subsection we consider more specifically *Cartesian currents in codimension* 1, i.e., elements of $\text{cart}(\Omega \times \mathbb{R})$. We shall prove that they are boundaries of subgraphs of BV-functions, and indeed they can be identified as BV functions in Ω. The theory of Sec. 4.1 then applies and provides us with explicit structure formulas for those currents.

In doing that the simple relevant remark is that, as it is well known, flux integrals can be transformed into integrations of differential forms, or in terms of vectors, to every unit vector $\nu \in \mathbb{R}^n$ we can associate the $(n-1)$-vector $*\nu$ orienting the orthogonal $(n-1)$-plane to ν, so that

$$\nu \wedge * \nu \;=\; e_1 \wedge \ldots \wedge e_n \; ,$$

and defined in coordinates by

$$(1) \qquad * \nu \;=\; \sum_{i=1}^{n} (-1)^{i-1} \nu^i \, e_{\bar{i}}, \qquad \text{if } \nu \;=\; \sum_{i=1}^{n} \nu^i \, e_i \; .$$

Given a measurable set $E \subset \mathbb{R}^n$, we easily see that the distributional derivative $D_i \chi_E$ of the characteristic function χ_E of E are related to the components of the current $\partial[\![E]\!]$ as follows

Proposition 1. *For any* $\varphi \in C_c^\infty(\mathbb{R}^n)$ *we have*

$$(2) \qquad \partial[\![E]\!] \, (\varphi(x) \, \widehat{dx^i}) \;=\; (-1)^i \int \varphi(x) \, d \, D_i \chi_E, \qquad i = 1, \ldots, n \; .$$

Consequently E *is a Caccioppoli set if and only if*

$$\mathbf{M}_{\mathcal{U}}(\partial[\![E]\!]) \; < \; \infty$$

for any open bounded subset $\mathcal{U}$ *of* $\mathbb{R}^n$. *In this case we also have*

$$(3) \qquad \partial[\![E]\!](\omega) \;=\; - \int <\omega, *n(x, E)> d|D\chi_E| \; .$$

Proof. In fact, if $\omega = \varphi(x)\,\widehat{dx^i}$, then $d\omega = (-1)^{i-1}\,\varphi_{x^i}(x)\,dx^1\wedge\ldots\wedge dx^n$ and

$$\begin{aligned}\partial[\![E]\!]\,(\varphi(x)\,\widehat{dx^i}) &= (-1)^{i-1}\int_E D_i\varphi(x)\,dx \\ &= (-1)^i\int \varphi(x)\,d\,D_i\chi_E\,.\end{aligned}$$

If

$$\omega = \sum_{i=1}^n \omega_{\bar{i}}(x)\,dx^{\bar{i}}\,, \qquad d\omega = \sum_{i=1}^n (-1)^{i-1}\,\omega_{\bar{i},x^i}(x)\,dx^1\wedge\ldots\wedge dx^n$$

and E is a Caccioppoli set we find

$$\begin{aligned}\partial[\![E]\!](\omega) &= \int \omega_{\bar{i}}(x)\,(-1)^i\,n^i(x,E)\,d|D\chi_E| \\ &= -\int <\omega, *n(x,E)>\,d|D\chi_E|\,.\end{aligned}$$

□

We note that the sign minus in (3) is due to the fact that $n(x,E)$ is the *inward* normal to $\partial^- E$.

We can now restate De Giorgi's theorem in Sec. 4.1.3 as

Theorem 1. *Let $E\subset\mathbb{R}^n$ be a measurable set for which*

$$\mathbf{M}_{\mathcal{U}}(\partial[\![E]\!]) < +\infty$$

for any open bounded set $\mathcal{U}\subset\mathbb{R}^n$. Then $\partial[\![E]\!]$ is a locally i.m. rectifiable current with density 1, and

$$\partial[\![E]\!](\omega) = -\int <\omega, *n(x,E)>\,d\mathcal{H}^{n-1}\llcorner\partial^- E\,.$$

In particular $\overrightarrow{\partial[\![E]\!]}\,(x) = -*n(x,E)$, and $\partial[\![E]\!] = \tau(\partial^- E, 1, -*n(x,E))$.

Let now $u:\Omega\subset\mathbb{R}^n\to\mathbb{R}$ be a measurable function. We denote by SG_u the $(n+1)$-current in $\mathbb{R}^n\times\mathbb{R}$ defined as integration over the subgraph of u

$$SG_u := [\![\mathcal{SG}_{u,\Omega}]\!]\,.$$

For the sake of simplicity we also write

$$\mathcal{S}_u \quad \text{for } \mathcal{SG}_{u,\Omega}\,.$$

Applying Proposition 1 to SG_u we then get

$$\begin{aligned}\partial SG_u(\varphi(x,y)\,\widehat{dx^i}\wedge dy) &= (-1)^i\textstyle\int\varphi(x,y)\,dD_i\chi_{\mathcal{SG}_{u,\Omega}} \\ \partial SG_u(\varphi(x,y)\,dx) &= (-1)^{n+1}\textstyle\int\varphi(x,y)\,dD_{n+1}\chi_{\mathcal{SG}_{u,\Omega}}\end{aligned} \tag{4}$$

for all $\varphi\in C_c^\infty(\Omega\times\mathbb{R})$ and $i=1,\ldots,n$.

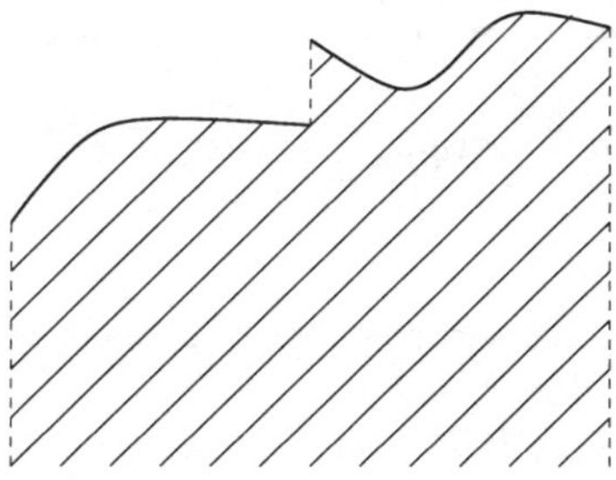

Fig. 4.2. Subgraph of a BV function.

Theorem 2. *Let $u \in L^1(\Omega)$. Then $u \in BV(\Omega)$ if and only if $\mathbf{M}(\partial SG_u) < \infty$. Moreover the current T defined by*

$$T := (-1)^n \, \partial SG_u \llcorner \Omega \times \mathbb{R} ,$$

is a Cartesian current, $T \in \text{cart}(\Omega \times \mathbb{R})$, and

$$T(\omega) = - \int <\omega, *n(x, \mathcal{S}_u)> d\mathcal{H}^n \llcorner \partial^- \mathcal{S}_u .$$

Finally,

$$\pi_\# |T^{\bar{0}0}| = \mathcal{L}^n , \quad \pi_\# |T^{\bar{i}\bar{0}}| = |D_i u| , \quad \pi_\# \| T \| = |\mu(Du)|$$

where $\mu(Du)$ is the measure in Ω $(D_1 u, \dots, D_n u, -\mathcal{L}^n)$; in particular as $\mathbf{M}(T) = \| T \|(\Omega \times \mathbb{R}) = \pi_\# \| T \|(\Omega)$, we have

(5) $$\mathbf{M}(T) = \sup\{T(\omega) \mid \omega = \pi^\#(\eta), \ \eta \in \mathcal{D}^n(\Omega), \ |\eta| \le 1\} .$$

Proof. By Proposition 1 $\mathbf{M}(\partial SG_u \llcorner \Omega \times \mathbb{R}) = P(\mathcal{S}_u, \Omega \times \mathbb{R})$. Therefore all claims, except for $T \in \text{cart}(\Omega \times \mathbb{R})$, follow at once from Theorem 1 in Sec. 4.1.5. We now observe that

$$T^{\bar{0}0}(\varphi(x,y)) = \int_\Omega \varphi(x, u(x)) \, dx$$

by (4), thus $T^{\bar{0}0} \ge 0$, $\| T \|_1 < \infty$ and $\pi_\# T = [\![\Omega]\!]$. Obviously we have in $\Omega \times \mathbb{R}$ $\partial T = (-1)^n \partial SG_u = 0$. In order to conclude that $T \in \text{cart}(\Omega \times \mathbb{R})$ it remains to show that T is an i.m. rectifiable current. This is exactly the content of De Giorgi's theorem as restated in Theorem 2 above. □

We emphasize the fact that (5) says that the mass of T can be computed in terms of the projections of its components. This is typical of codimension one, and in fact not true in codimension larger that 1, compare Remark 2 in Sec. 4.2.3 (ii).

We shall now prove that actually every Cartesian current, $T \in \text{cart}(\Omega \times \mathbb{R})$, can be written as

$$T = (-1)^n \, \partial SG_u \llcorner \Omega \times \mathbb{R}$$

for some u in $BV(\Omega)$. Therefore $\text{cart}(\Omega \times \mathbb{R})$ is characterized as the class of currents which are boundaries of subgraphs of BV-functions.

From Theorem 4 in Sec. 4.2.3 we know that every $T \in \text{cart}(\Omega \times \mathbb{R})$ can be written as $T = G_{u_T} + S_T$ where $u_T \in BV(\Omega)$ is given by

$$u_T = \frac{d\pi_{\#}(T \llcorner y\, dx)}{d\mathcal{L}^n} .$$

We have

Theorem 3. *Let $T \in \text{cart}(\Omega \times \mathbb{R})$. Then $T = (-1)^n \partial SG_{u_T}$.*

Proof. From Theorem 3 in Sec. 4.2.3 (i) we know that

$$T(\phi(x,y)\, dx) = \int_\Omega \phi(x, u_T(x))\, dx ,$$

while a direct computation shows that

$$(-1)^n \, \partial SG_{u_T}(\phi(x,y)\, dx) = \int_\Omega \phi(x, u_T(x))\, dx .$$

The claim then follows at once from Proposition 2 below. □

Proposition 2. *Let $T \in \mathcal{D}_n(\Omega \times \mathbb{R})$ be a boundaryless current, $\partial T \llcorner \Omega \times \mathbb{R} = 0$. Then $T = 0$ provided $T(\varphi(x,y)\, dx) = 0$ for all $\varphi \in C_c^\infty(\Omega \times \mathbb{R})$.*

Proof. It suffices to show that $T(\omega) = 0$ for all form of the type

$$\omega = \varphi(x,y)\, \widehat{dx^i} \wedge dy , \qquad \varphi \in C_c^\infty(\Omega \times \mathbb{R}) , \; i = 1, \dots, n .$$

Set

$$\phi(x,y) := \int_{-\infty}^{y} \varphi(x,s)\, ds , \qquad \alpha(x,y) := (-1)^{i-1} \phi_{x^i}(x,y)$$

and

$$\xi := \phi(x,y)\, \widehat{dx^i}$$

for a fixed i, $i = 1, \dots, n$. Clearly $d\xi = \alpha(x,y)\, dx + (-1)^{n-1}\, \omega$. Being ϕ bounded in $\Omega \times \mathbb{R}$, we then conclude

$$T(\omega) = (-1)^{n-1} \left[T(d\xi) - T(\alpha(x,y)\, dx) \right] = 0 ,$$

since $\partial T \llcorner \Omega \times \mathbb{R} = 0$ and $T(\alpha\, dx) = 0$ by assumption. □

It is worthwhile noticing the homological character of the argument in the proof of Proposition 2.

On account of the structure theorem for BV-functions, compare Sec. 4.1.4 and Sec. 4.1.5, we can now give an explicit representation formula for currents in $\mathrm{cart}(\Omega \times \mathbb{R})$.

Let T be in $\mathrm{cart}(\Omega \times \mathbb{R})$, $T = \tau(\mathcal{M}, 1, \vec{T})$. Denote by $\mathcal{M}_+$ the set of points $z \in \mathcal{M}$ at which the tangent plane $\mathrm{Tan}_z\mathcal{M}$ is not vertical, or equivalently the projection map π restricted to $\mathrm{Tan}_z\mathcal{M}$ has maximal rank n. From Theorem 3 we have

$$T = (-1)^n \, \partial SG_u$$

for some $u \in BV(\Omega)$, actually for $u = u_T$, and from Theorem 2

$$\theta^{n-1}(\pi_\# \| T \|, x) = \theta^{n-1}(|\mu(Du)|, x) = \theta^{n-1}(|Du|, x) \, ,$$

hence the jump set of u can be written, because of Theorem 1 in Sec. 4.1.4, as

$$\begin{aligned} J_u &= \{x \in \Omega \mid u_-(x) < u_+(x)\} \\ &= \{x \in \Omega \mid \theta^{n-1}(\pi_\# \| T \|, x) > 0\} \, . \end{aligned}$$

We now set

$$\begin{aligned} T^{(a)} &:= T \llcorner \mathcal{M}_+ \\ T^{(j)} &:= T \llcorner (J_u \times \mathbb{R}) \\ T^{(C)} &:= T \llcorner (\mathcal{M} \setminus (\mathcal{M}_+ \cup (J_u \times \mathbb{R}))) \, . \end{aligned}$$

Obviously, T decomposes as

$$T = T^{(a)} + T^{(j)} + T^{(C)}$$

and the three measures $\| T^{(a)} \|$, $\| T^{(j)} \|$ and $\| T^{(C)} \|$ are mutually singular.

Taking into account Theorem 2 in Sec. 4.2.2 and the structure theorem for BV-functions, equivalently for subgraphs of BV-functions, we can then easily infer

Theorem 4. *Let $T \in \mathrm{cart}(\Omega, \mathbb{R})$. Then there exists a unique function $u \in BV(\Omega)$, given by $u = u_T$ such that, decomposing T as*

$$T = T^{(a)} + T^{(j)} + T^{(C)} \, ,$$

we have

$$T^{(a)} = G_{u_T}$$

and

$$T^{(j)}(\phi(x,y)\, dx) = T^{(C)}(\phi(x,y)\, dx) = 0 \qquad \forall\, \phi \in C_c^\infty(\Omega \times \mathbb{R}) \, .$$

Moreover, for any n-form $\omega = \phi(x,y)\, \widehat{dx^i} \wedge dy$, $i = 1, \ldots, n$, we have

$$T^{(a)}(\omega) = (-1)^{n-i} \int \phi(x, u_+(x))\,(Du)^a\,dx$$

$$T^{(j)}(\omega) = (-1)^{n-i} \int \Big(\int_{u_-(x)}^{u_+(x)} \phi(x,s)\,ds \Big)\, n^i(x, J_u)\, d\mathcal{H}^{n-1} \llcorner J_u$$

$$T^{(C)}(\omega) = (-1)^{n-i} \int \phi(x, u_+(x))\,(D_i u)^{(C)}$$

An immediate consequence of Theorem 4 is

(6) $$\mathbf{M}(T) = \int_\Omega \sqrt{1 + |(Du)^a|^2}\,dx + \int_\Omega \Big[|(Du)^{(C)}| + |(Du)^{(j)}| \Big]$$

Finally we notice that as consequence of the approximation theorem for BV-functions in Sec. 4.1.1 we infer at once

Proposition 3. *Let $T \in \mathrm{cart}(\Omega \times \mathbb{R})$. Then there exists a sequence of smooth maps $u_k : \Omega \to \mathbb{R}$ such that*

$$G_{u_k} \rightharpoonup T \qquad \textit{and} \qquad \mathbf{M}(G_{u_k}) \to \mathbf{M}(T) \, ,$$

in particular

$$\mathrm{Cart}(\Omega \times \mathbb{R}) = \mathrm{cart}(\Omega \times \mathbb{R}) \, .$$

[1] *Cantor–Vitali functions.* Let V be the Cantor–Vitali function in Sec. 1.1.3. It is easily seen that

(i) $V \in BV(0,1)$, since it is non-decreasing, and $V(0) = 0, V(1) = 1$.

(ii) On each interval in $[0,1] \setminus E_k$ we have $V(x) = f_k(x) =$ *constant*. In particular, $V(x)$ is differentiable on each $x \in [0,1] \setminus C$, and (approximately) differentiable almost everywhere in $[0,1]$ since $|C| = 0$, and

$$\mathrm{ap}V'(x) = 0 \qquad \text{for a.e. } x \in [0,1] \, .$$

(iii) Since V is continuous, V' has no jump part, and therefore

$$V' = V'^{(C)}$$

and V' is concentrated in C, $V' \llcorner C = V'$.

(iv) Obviously $V([0,1]) = [0,1]$, V maps $[0,1] \setminus C$ into the denumerable dyadic set

$$D := \{ y \in \mathbb{R} \mid y = \frac{j}{2^k} \, , \ j \in [0, 2^k], \ j,k \in \mathbb{N} \} \, .$$

Therefore V maps C onto $[0,1] \setminus D$. In particular V has not Lusin property (N).

From the structure theorem we infer that $-\partial SG_V$ is an i.m. rectifiable current $\tau(\mathcal{M}, 1, \xi)$ and, setting

$$\mathcal{M}_+ := \{(x,y) \mid \xi = (\xi_x, \xi_y),\ \xi_x > 0\}$$

we have

$$-\partial SG_V \llcorner \mathcal{M}_+ = G_V = G_V \llcorner ([0,1] \setminus E) \times \mathbb{R} ,$$

whence

$$\mathcal{M}_+ = \sum_{j,k} [\![I_{j,k}]\!] \times \delta_{y_{k,j}} , \qquad y_{k,j} = j/2^k .$$

In particular

$$\mathcal{M}_+ = \{(x,y) \in [0,1] \times [0,1] \mid \xi = (1,0)\} = \{(x,y) \mid \xi_y = 0\}$$

and

$$\mathcal{M} = \{(x,y) \mid \xi_y = 0\} \cup \{(x,y) \mid \xi_x = 0\} .$$

Notice the difference between the 1-graph of V and the graph of V in the sense of continuous functions.

In terms of the points

$$\begin{aligned} A_{k,j} &:= (b_{k,j} + \delta^{k+1}, V(b_{k,j} + \frac{1}{2}\delta^k)) \\ B_{k,j} &:= (b_{k,j} + \delta^k(1-\delta), V(b_{k,j} + \frac{1}{2}\delta^k)) \end{aligned}$$

$k = 0, 1, \dots,\ j = 1, \dots, 2^k$, we can write the 1-dimensional current graph of V as

$$G_V := \sum_{k=0}^{\infty} \sum_{j=1}^{2^k} [\![< A_{k,j}, B_{k,j} >]\!]$$

where $< A, B >$ denotes the oriented line segment from A to B. Its boundary

$$\partial G_V = \sum_{k=0}^{\infty} \sum_{j=1}^{2^k} ([\![B_{k,j}]\!] - [\![A_{k,j}]\!])$$

is a 0-dimensional current, in fact a flat chain. In principle it can act on Lipschitz functions, indeed for Lipschitz functions $\varphi(x,y)$ we have

$$\begin{aligned} |\partial G_V(\varphi)| &\leq \operatorname{Lip} \varphi \sum_{k=0}^{\infty} \sum_{j=1}^{2^k} \operatorname{dist}(B_{k,j}, A_{k,j}) = \sum_{k=0}^{\infty} \sum_{j=1}^{2^k} \delta^k (1 - 2\delta) \\ &\leq c \sum_{k=0}^{\infty} 2^k \delta^k = \frac{c}{1 - 2\delta} < \infty , \end{aligned}$$

but not on Borel functions. In fact we have $\mathbf{M}(\partial G_V) = +\infty$, corresponding to the intuitive idea that the graph of V has infinitely many holes.

In order to prove that $\mathbf{M}(\partial G_V) = +\infty$ we can argue by contradiction. Assume $\mathbf{M}(\partial G_V) < \infty$, then $\partial G_V \llcorner B$ is well defined for any bounded Borel set and obviously we have

$$\mathbf{M}(\partial G_V) \geq \mathbf{M}(\partial G_V \llcorner B) .$$

By taking as B the set

$$\cup\{< A_{k,j}, B_{k,j} > \mid k = 1, \dots, \bar{k},\ j = 1, \dots, 2^k\}$$

we then infer

$$\mathbf{M}(\partial G_V \llcorner B) = 2(2^{\bar{k}+1} - 1)$$

which for $\bar{k} \to \infty$, yields $\mathbf{M}(\partial G_V) = \infty$: a contradiction.

For future reference we would like to present also a direct proof. Let $\eta(x,y) := \varphi(x)\psi(y)$ be a test 0-form with φ and ψ Lipschitz functions, so that $d\eta = \varphi'\psi\, dx + \varphi\psi'\, dy$ and

$$\begin{aligned} \partial G_V(\eta) &= G_V(d\eta) = \int [\varphi'(x)\psi(V(x)) + \varphi(x)\psi'(V(x))V'(x)]\, dx \\ &= \int_0^1 \varphi'(x)\psi(V(x))\, dx \end{aligned}$$

as $V' = 0$ on the 1-graph of V. We then have

$$\mathbf{M}(\partial G_V) = \sup\left\{ \int_0^1 \varphi'(x)\psi(V(x))\, dx \mid |\varphi| \leq 1,\ |\psi| \leq 1 \right\} .$$

We now choose φ and ψ suitably. First we choose φ and ψ as piecewise linear functions satisfying

$$\begin{aligned} &\varphi(0) = \varphi(\delta^2) = \varphi(1-\delta) = \varphi(1) = 0\,, \quad \varphi(\delta(1-\delta)) = \varphi(\delta) = 1 \\ &\psi(0) = \psi(1/4) = \psi(3/4) = \psi(1) = 0\,, \qquad \psi(1/2) = 1\,. \end{aligned}$$

Then $\varphi' = -1$ and $\psi = 1$ on $< A_{0,1}, B_{0,1} >$ and $\varphi' = 1$, $\psi = 0$ on $< A_{1,1}, B_{1,1} >$. Thus

$$\int \varphi'(x)\psi(V(x))\, dx = -1\,.$$

By induction and using the self-similarity of V we now set

$$\varphi_1 := \varphi\,, \qquad \psi_1 = \psi$$

and

$$\varphi_{k+1}(x) := \varphi_k(\delta^{-2}x)\ , \quad \varphi_{k+1} = 0 \quad \text{for } x \geq \delta$$
$$\psi_{k+1}(y) := \psi_k(4y)\ , \quad \psi_{k+1} = 0 \quad \text{for } y \geq 1/4\ .$$

By construction the functions $\varphi_k(x)\psi_k(y)$ have disjoint supports, hence, setting

$$\phi(x,y) \ := \ \varphi_1(x)\psi_1(y) + \cdots + \varphi_{\bar{k}}(x)\psi_{\bar{k}}(y)$$

we have $|\phi| \leq 1$ and

$$\partial G_V(\phi) \ = \ -\bar{k}\ ,$$

from which we infer at once $\mathbf{M}(\partial G_V) = +\infty$. Notice that in fact we have proved that the mass of ∂G_V is infinite in every neighbourhood of the origin. Taking into account the self-similarity of V we actually see that the mass of ∂G_V is infinite in every neighbourhood of $(x, V(x))$ for every $x \in C$.

Consider now the "inverse function of V" denoted by $\widehat{V} : (0,1) \subset \widehat{\mathbb{R}} \to \mathbb{R}$, defined for instance by

$$\partial S\widehat{G}_{\widehat{V}} \llcorner (0,1) \times \mathbb{R} \ = \ \partial S G_V \llcorner (0,1) \times \widehat{\mathbb{R}}$$

where

$$\widehat{\mathcal{G}}_{\widehat{V}} \ := \ \{(x,y) \in (0,1) \times (0,1) \ \mid \ x < \widehat{V}(y)\}\ .$$

Then $\widehat{V}$ is a non-decreasing function which belongs to $BV((0,1))$, and is continuous in $[0,1] \setminus D$, consequently

$$\partial S\widehat{G}_{\widehat{V}} \llcorner (0,1) \times \mathbb{R} \ = \ \tau(\widehat{\mathcal{M}}, 1, \widehat{\xi})\ .$$

It is readily seen that

$$\widehat{\mathcal{M}}_+ \ = \ \{(x,y) \in \mathcal{M} \ \mid \ \xi_y > 0\} \ = \ \mathcal{M} \setminus \mathcal{M}_+$$

and

$$\widehat{\pi}(\widehat{\mathcal{M}}_+) \ = \ (0,1) \setminus D$$

while

$$\widehat{\pi}(\mathcal{M} \setminus \widehat{\mathcal{M}}_+) \ = \ D$$

where $\widehat{\pi} : \mathbb{R} \times \widehat{\mathbb{R}} \to \widehat{\mathbb{R}}$ denotes the orthogonal projection into the second factor. Also, every point in D is a jump point for $\widehat{V}$. Consequently $\widehat{V}$ has no Cantor part, but only a jump part which is concentrate on an everywhere dense set D, which is countably 0–rectifiable but not 0–rectifiable, i.e. finite. Finally, of course $\widehat{V}$ is almost everywhere approximately differentiable with $\widehat{V}' = 0$ a.e. •

2.5 Examples of Cartesian Currents

In this subsection we would like to discuss a few specific examples of Cartesian currents. We shall see that in contrast with the case of codimension one, in general the *vertical part* of a Cartesian current may be quite complicated.

An easy way of producing a Cartesian current with a vertical part is the following. Consider any function u in $\mathcal{A}^1(\Omega, \mathbb{R}^N)$, $\Omega \subset \mathbb{R}^n$, and let S be an n-dimensional i.m. rectifiable current in $\mathbb{R}^{n+N}$ such that

$$\pi_\# S = 0 , \qquad \text{and} \qquad \partial G_u + \partial S = 0 \quad \text{in } \Omega \times \mathbb{R}^N .$$

Then obviously the current

$$T := G_u + S$$

belongs to $\text{cart}(\Omega \times \mathbb{R}^N)$. However what makes Cartesian currents interesting is the fact that they arise as weak limits of smooth graphs: this will be our main point in the following examples.

Let us begin by reconsidering the examples in the introduction to this chapter.

[1] *Bubbling off of circles.* Let $\Omega := (-1, 1)$, $n = 1$ and $N = 2$. We consider the sequence $\{u_k\}$ of Lipschitz maps $u_k : \Omega \to \mathbb{R}^2$ given by

$$u_k(t) := \begin{cases} (\cos kt, \sin kt) & t \in (0, 2\pi/k) \\ (1, 0) & \text{otherwise.} \end{cases}$$

Clearly $u_k \rightharpoonup u_0 := (1, 0)$ in the sense of measures in $(-1, 1)$ and the length of the graph of u_k is given by

$$\mathbf{M}(G_{u_k}) = \int_{-1}^{1} \sqrt{1 + |u'(t)|^2}\, dt$$

which converges to $2 + 2\pi$ as k tends to infinity. As remarked in the introduction to this chapter it is easy to convince ourselves that

$$G_{u_k} \rightharpoonup G_{(1,0)} + \delta_0 \times [\![S^1]\!] \tag{1}$$

where S^1 denotes the current integration over the circle $S^1 := \{y \mid (y^1)^2 + (y^2)^2 = 1\}$ in $\mathbb{R}^2$ with the usual clockcounterwise orientation. Moreover we have $\mathbf{M}(G_{u_k}) \to \mathbf{M}(T) = 2 + 2\pi$. In particular $T \in \text{Cart}(\Omega \times \mathbb{R}^2)$.

Now we would like to supply the proof of (1). We change the "time coordinate" in order to describe the curves $x \to (x, u_k(x))$ in $\mathbb{R} \times \mathbb{R}^2$. Denote by $\gamma_k : [-1, 1] \to [-1, 1]$ the function which is linear on each of the intervals $[-1, 0], [0, 1/2], [1/2, 1]$ and takes the values $\gamma_k(-1) = -1$, $\gamma_k(0) = 0$, $\gamma_k(1/2) = 2\pi/k$, $\gamma_k(1) = 1$, and consider the curves

$$\delta_k(t) := (\gamma_k(t), u_k(\gamma_k(t))) .$$

It is easily seen that the curves δ_k are equi-Lipschitz and converge uniformly to the curve

$$\delta(t) := \begin{cases} (t, 1, 0) & \text{if } -1 \le t \le 0 \\ (0, \cos 4\pi t, \sin 4\pi t) & \text{if } 0 \le t \le 1/2 \\ (2t - 1, 1, 0) & \text{if } 1/2 \le t \le 1 , \end{cases}$$

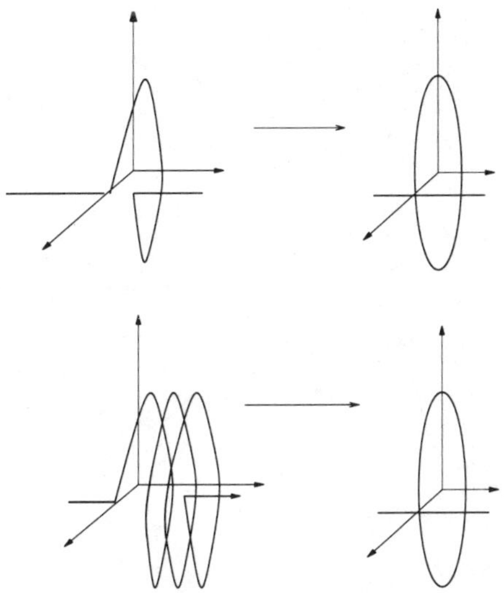

Fig. 4.3. Bubbling off of circles: density of the vertical part equals 1 in (a) and 3 in (b).

moreover

$$G_{u_k} = \delta_{k\#} [\![(-1,1)]\!] .$$

From this it immediately follows that

$$G_{u_k} = \delta_{k\#} [\![(-1,1)]\!] \rightharpoonup \delta_{\#} [\![(-1,1)]\!] = G_{(1,0)} + \delta_0 \times [\![S^1]\!] .$$

Similarly, considering the current in $\operatorname{cart}(\Omega \times \mathbb{R}^2)$ given by

$$T := G_{(1,0)} + p\,\delta_0 \times [\![S^1]\!] , \qquad p \in \mathbb{Z}$$

it is not difficult to construct a sequence of Lipschitz maps u_k such that $G_{u_k} \rightharpoonup T$, and conclude that $T \in \operatorname{Cart}(\Omega \times \mathbb{R}^2)$. It suffices to consider the sequence

$$u_k(t) := \begin{cases} (\cos kpt, \sin kpt) & t \in (0, 2\pi/k) \\ (1,0) & \text{otherwise} \end{cases}$$

and proceed as above. •

[2] More generally, we can choose a sequence $\{\gamma_i\}$ of smooth closed curves in $\mathbb{R}^2$, $\gamma_i : S^1 \to \mathbb{R}^2$, and a sequence of points $\{x_i\} \subset (-1,1)$ and consider the 1-dimensional current T in $(-1,1) \times \mathbb{R}^2$ given by

$$T := G_u + \sum_{i=1}^{\infty} \delta_{x_i} \times \gamma_{i\#} [\![S^1]\!]$$

where u is a smooth map from $(-1,1)$ into $\mathbb{R}^2$. If $\sum_{i=1}^{\infty} \mathbf{M}(\gamma_{i\#} [\![S^1]\!]) < \infty$, then it is easily seen that $T \in \operatorname{cart}(\Omega, \mathbb{R}^2)$. Actually, essentially in a similar way as in the previous example, it is not difficult to construct a sequence of Lipschitz maps u_k such that

$$G_{u_k} \rightharpoonup T\,, \qquad \mathbf{M}(G_{u_k}) \to \mathbf{M}(T)\,,$$

showing therefore that $T \in \mathrm{Cart}(\Omega \times \mathbb{R}^2)$. •

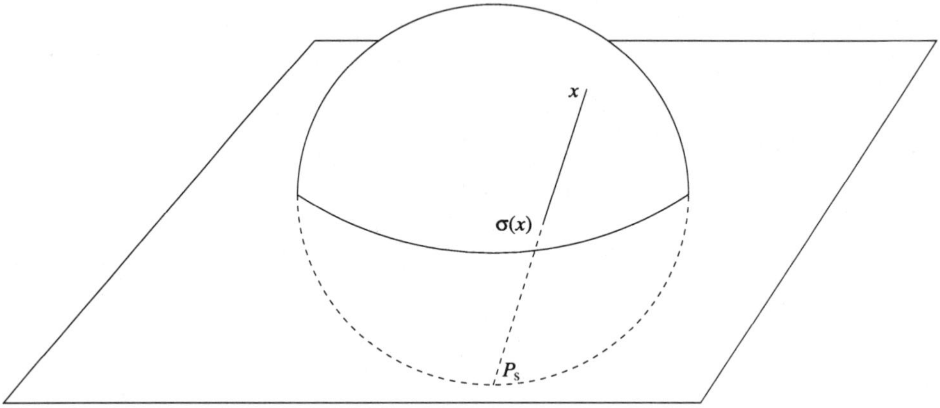

Fig. 4.4. Stereographic projection.

[3] *Bubbling off of spheres.* Consider the stereographic projection $\sigma : S^n \subset \mathbb{R}^{n+1} \to \mathbb{R}^n$ from the south pole P_S. It maps a point $(y, z) \in S^n \subset \mathbb{R}^n \times \mathbb{R}$, $|y|^2 + z^2 = 1$, to $y/(1+z) \in \mathbb{R}^n$, while its inverse $\sigma^{-1} : \mathbb{R}^n \simeq \mathbb{R}^n \times \{0\} \to S^n$ maps $x \in \mathbb{R}^n$ to

$$\sigma^{-1}(x) = \Big(\frac{2}{1+|x|^2}\,x, \frac{1-|x|^2}{1+|x|^2}\Big)\,. \tag{2}$$

A simple computation shows that for $i, j = 1, \dots, n$

$$(D_j\,\sigma^{-1}(x) \mid D_i\,\sigma^{-1}(x))_{\mathbb{R}^{n+1}} = \frac{4}{(1+|x|^2)^2}\,\delta_{ij}\,. \tag{3}$$

In particular

$$|D_i\,\sigma^{-1}(x)| = |D_j\,\sigma^{-1}(x)| \quad \text{and} \quad (D_i\,\sigma^{-1}(x) \mid D_j\,\sigma^{-1}(x))_{\mathbb{R}^{n+1}} = 0 \text{ for } i \neq j\,,$$

i.e., σ^{-1} is *conformal.* Consequently

$$|D\sigma^{-1}|^2 = n\,|D_1\sigma^{-1}|^2 = \frac{4n}{(1+|x|^2)^2}$$

and

$$\begin{aligned}|M_{(n)}(D\sigma^{-1}(x)|^2 &= |D_1\sigma^{-1}\wedge\ldots\wedge D_n\sigma^{-1}|^2\\ &= |D_1\sigma^{-1}|^2\,|D_2\sigma^{-1}|^2\ldots|D_n\sigma^{-1}|^2 = \Big(\frac{1}{n}\,|D\sigma^{-1}|^2\Big)^n .\end{aligned}$$

Hence we have

$$\frac{1}{n^{n/2}}\int_{\mathbb{R}^n}|D\sigma^{-1}|^n\,dx = \int_{\mathbb{R}^n}|M_{(n)}(D\sigma^{-1})|\,dx = \mathcal{H}^n(S^n) \tag{4}$$

as consequence of the area formula, taking into account that σ^{-1} is injective. Finally, a simple checking of the orientation yields that

$$\sigma_{\#}^{-1}[\![\mathbb{R}^n]\!] = (-1)^n\,[\![S^n]\!]$$

where $[\![S^n]\!]$ denotes the current integration over S^n with the usual orientation in which the normal to S^n points outward.

Later, compare Proposition 1 in Vol. II Sec. 4.1.1, we shall see that for any fixed $\varepsilon > 0$ we can modify σ^{-1} outside a large ball $B(0,\delta)$, $\delta = \delta_\varepsilon$, to a map u actually u_ε,

$$u : \mathbb{R}^n \longrightarrow S^n \subset \mathbb{R}^{n+1}$$

in such a way that

$$\begin{gathered}\frac{1}{n^{n/2}}\int_{\mathbb{R}^n}|Du|^n\,dx \le \frac{1}{n^{n/2}}\int_{\mathbb{R}^n}|D\sigma^{-1}|^n\,dx+\varepsilon \\ u = P_s \text{ outside } B(0,2\delta), \qquad u_{\#}[\![\mathbb{R}^n]\!] = (-1)^n\,[\![S^n]\!]\end{gathered} \tag{5}$$

Consider now the sequence of smooth maps $\{u_k\}$ from $B(0,1)$ into S^n given by

$$u_k(x) := u(kx) .$$

Using the conformal invariance of $\int|Du|^n\,dx$ we infer

$$\int_{\mathbb{R}^n}|Du_k(x)|^n\,dx = \int_{\mathbb{R}^n}|Du(x)|^n\,dx$$

and, taking into account the rough isoperimetric inequality

$$|M_{(j)}(Du_k)| \le c_n\,(1+|Du_k|^n) , \qquad j=1,\ldots,n ,$$

that the currents G_{u_k} have equibounded masses. Moreover one has

$$G_{u_k} \rightharpoonup G_{u_0}+(-1)^n\,\delta_0\times[\![S^n]\!] \qquad \text{in } \mathcal{D}_n(\mathbb{R}^n\times\mathbb{R}^{n+1}) \tag{6}$$

where $u_0(x)\equiv(0,\ldots,0,-1)$.

The convergence in (6) can be proved by just changing parametrization in the representation of G_{u_k} as in Sec. 3.2.3. Let in fact $\gamma_k : B(0,1)\to B(0,1)$ be

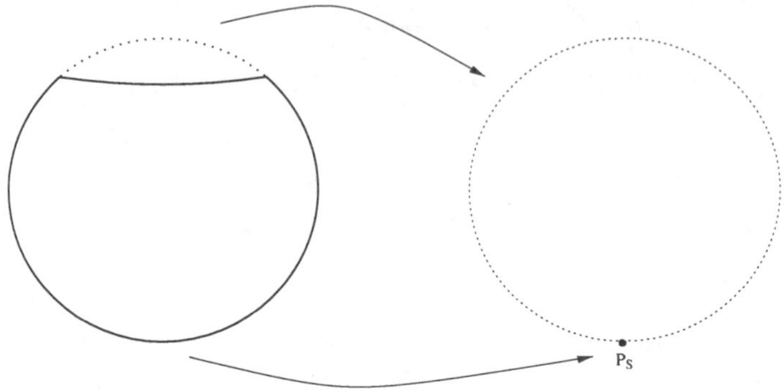

Fig. 4.5. Bubbling off of spheres.

the map $\gamma_k(z) := \Gamma_k(|z|)z/|z|$ where $\Gamma_k(t) : (0,1) \to (0,1)$ is linear in $(0,1/2)$ and in $(1/2,1)$ and takes values $\Gamma(0) = 0$, $\Gamma(1) = 1$, and $\Gamma(1/2) = 1/k$. Set

$$\begin{aligned} v_k \ &: \ B(0,1) \longrightarrow B(0,1) \times \mathbb{R}^{n+1} \\ v_k(z) \ &:= \ (\gamma_k(z), u_k(\gamma_k(z))) \, . \end{aligned}$$

It is easily seen that the maps v_k are equi-Lipschitz and converge uniformly to the map $v : B(0,1) \to B(0,1) \times \mathbb{R}^{n+1}$ given by

$$v(z) \ := \ \begin{cases} (0, \sigma^{-1}(2z)) & 0 \leq |z| \leq \dfrac{1}{2} \\ (2|z|-1)\left(\dfrac{z}{|z|}, P_s\right) & \dfrac{1}{2} \leq |z| \leq 1 \end{cases}$$

Moreover

$$G_{u_k} \ = \ v_{k\#} [\![B(0,1)]\!] \ \rightharpoonup \ v_{\#} [\![B(0,1)]\!]$$

and

$$\begin{aligned} v_{\#} [\![B(0,1/2)]\!] \ &= \ (-1)^n \, \delta_0 \times [\![S^n]\!] \\ v_{\#} [\![B(0,1) \setminus B(0,1/2)]\!] \ &= \ G_{u_0} \, . \end{aligned}$$

This proves that the current

$$G_{u_0} + (-1)^n \, \delta_0 \times [\![S^n]\!]$$

which trivially belongs to $\text{cart}(B(0,1) \times \mathbb{R}^{n+1})$ actually belongs to $\text{Cart}(B(0,1) \times \mathbb{R}^{n+1})$.

Finally, notice that for any $k = 1, 2, \ldots$

$$\frac{1}{n^{n/2}} \int\limits_{B(0,1)} |Du_k|^n \, dx \;\le\; \mathcal{H}^n(S^n) + \varepsilon \;=\; \mathbf{M}(\delta_0 \times [\![S^n]\!]) + \varepsilon \, .$$

Therefore, by a diagonal process, compare Proposition 2 in Vol. II Sec. 4.1.1, we can in fact find a sequence of smooth maps u_k such that

$$G_{u_k} \rightharpoonup G_{u_0} + (-1)^n \, \delta_0 \times [\![S^n]\!]$$

and also satisfy

$$\frac{1}{n^{n/2}} \int\limits_{B(0,1)} |Du_k|^n \, dx \longrightarrow \mathbf{M}((-1)^n \delta_0 \times [\![S^n]\!])$$

$$\mathbf{M}(G_{u_k}) \longrightarrow \mathbf{M}(G_{u_0} + (-1)^n \delta_0 \times [\![S^n]\!]) \, .$$

•

4 *Attaching a cylinder.* In the previous example we have seen that *vertical* spheres can be attached to a graph of a constant map as result of the weak convergence of smooth graphs. We shall now see how a *vertical cylinder* can be attached and in fact produced by weak convergence.

Let Ω be a bounded domain in $\mathbb{R}^{n+p}$. Choose any function $u_0 \in \mathcal{A}^1(\Omega, \mathbb{R}^{n+1})$ and a p-dimensional i.m. rectifiable current L in Ω. Clearly the current

$$T \;:=\; G_{u_0} + L \times [\![S^n]\!]$$

belongs to $\text{cart}(\Omega \times \mathbb{R}^{n+1})$ provided

$$\partial G_{u_0} \;=\; -\partial L \times [\![S^n]\!] \qquad \text{in } \Omega \times \mathbb{R}^{n+1} \, .$$

We now consider the simple case in which $\Omega = B(0,1) \times [0,\ell]^p$, $B(0,1)$ being the unit ball in $\mathbb{R}^n$, $u_0 \equiv (0,0,\dots,0,-1) \in \mathbb{R}^{n+1}$, and L is the p-dimensional current integration over the p-dimensional cube $\{0\} \times [0,\ell]^p \subset \mathbb{R}^n \times \mathbb{R}^p$. Denote the coordinates in $\mathbb{R}^n \times \mathbb{R}^p$ as (x,z), $x \in \mathbb{R}^n$, $z \in \mathbb{R}^p$ and consider the sequence of smooth functions u_k

$$u_k \;:\; B(0,1) \longrightarrow S^n$$

of the previous example. Define now for $x \in B(0,1)$ and $z \in [0,\ell]^p$

$$v_k(x,z) \;:=\; u_k(x) \, .$$

Clearly the graphs of v_k have equibounded masses, moreover

$$G_{v_k} \rightharpoonup G_{u_0} + [\![(0,\ell)^p]\!] \times [\![S^n]\!] \, .$$

This simply follows by rescaling the x-variable as in the previous example and by considering z as a parameter. We therefore see that $G_{u_0} + [\![(0,\ell)^p]\!] \times [\![S^n]\!]$ belongs to $\text{Cart}(\Omega, \times\mathbb{R}^{n+1})$.

Notice that for any $k = 1, 2, \ldots$,

$$\begin{aligned}\frac{1}{n^{n/2}} \int\limits_{B(0,1)\times[0,\ell]^p} |Dv_k|^n \, dx\, dz &= \ell^p \frac{1}{n^{n/2}} \int\limits_{B(0,1)} |Du_k|^n \, dx \le \ell^p \mathcal{H}^n(S^n) + \varepsilon \\ &= \mathbf{M}([\![(0,\ell)^p]\!] \times [\![S^n]\!]) + \varepsilon ,\end{aligned}$$

by construction, hence by a diagonal process, we could in fact find v_k so that

$$G_{v_k} \rightharpoonup [\![(0,\ell)^p]\!] \times [\![S^n]\!]$$

and

$$\frac{1}{n^{n/2}} \int\limits_{B(0,1)\times[0,\ell]^p} |Dv_k|^n \, dx\, dz \longrightarrow \mathbf{M}([\![(0,\ell)^p]\!] \times [\![S^n]\!]) = \ell^p \mathcal{H}^n(S^n) .$$

•

[5] The procedure of the previous two examples can be modified to produce by weak convergence of graphs general surfaces as vertical parts. Let $\{u_k\}$ and u_0 have the same meaning as previously, so that

$$G_{u_k} \rightharpoonup G_{u_0} + \delta_0 \times [\![S^n]\!] .$$

Choose any Lipschitz map $\phi : S^n \longrightarrow \mathbb{R}^{n+1}$ and set

$$v_k(x) := \phi(u_k(x)) .$$

We have $G_{v_k} = (\mathrm{id} \bowtie \phi)_\# G_{u_k}$ and

$$(\mathrm{id} \bowtie \phi)_\# (G_{u_0} + \delta_0 \times [\![S^n]\!]) = G_{\phi \circ u_0} + \delta_0 \times \phi_\# [\![S^n]\!].$$

This way we produce as *vertical part* the n-dimensional i.m. rectifiable current $\phi_\#[\![S^n]\!]$.

For example, if we split $\mathbb{R}^{n+1}$ as $\mathbb{R}^n_y \times \mathbb{R}_z$ and set

$$\phi(y, z) := (y, \varphi(|y|)z)$$

with

$$\varphi(t) := \begin{cases} 0 & 0 \le t \le 1/2 \\ 2t-1 & 1/2 \le t \le 1 \end{cases}$$

then $\phi_\#[\![S^n]\!]$ is just the current integration over standard n-dimensional torus $\mathbf{T}^n$, and conclude that $G_{u_0} + \delta_0 \times \mathbf{T}^n \in \mathrm{Cart}(B(0,1) \times \mathbb{R}^{n+1})$. •

[6] *Concentrations at infinity.* Consider the family of maps u_ε from the unit ball of $\mathbb{R}^2$ in $\mathbb{R}^3$ defined by

$$u_\varepsilon(r, \varphi) := \frac{2\varepsilon}{1+\varepsilon}(\cos\varphi, \sin\varphi, \log r)$$

in the annulus $B(0,1) \setminus B(0,\varepsilon)$, (r,φ) being the polar coordinates in $\mathbb{R}^2$, and by

$$u_\varepsilon(x) = \sigma^{-1}\left(\frac{\varepsilon^2 x}{|x|^2}\right) + \begin{pmatrix} 0 \\ 0 \\ \log\delta - \sqrt{1 - \frac{4\delta^2}{(1+\delta)^2}} \end{pmatrix},$$

in the ball $B(0,\delta)$, σ being the stereographic projection.

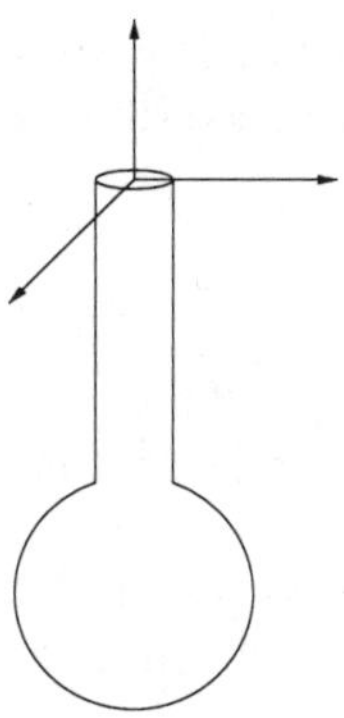

Fig. 4.6. Range of the map u_ε.

It is easily seen that each u_ε is conformal and in particular

$$\frac{1}{2}\int_B |Du_\varepsilon|^2\, dx = \mathcal{H}^2(S^2) + o(\varepsilon).$$

Consequently, by the isoperimetric inequality, we have

$$\sup_{0<\varepsilon<1} \| G_{u_\varepsilon} \|_{\text{cart}} < \infty.$$

Of course $u_\varepsilon \rightharpoonup u_0$, $u_0 \equiv 0$, weakly in $W^{1,2}$, and $G_{u_\varepsilon} \rightharpoonup G_{u_0}$, in particular

$$M_{(2)}(Du_\varepsilon) \rightharpoonup 0 \quad \text{but } |M_{(2)}(Du)| \rightharpoonup \mathcal{H}^2(S^2)\delta_0.$$

•

7 Vertical parts are produced in the weak convergence procedure not only as the result of concentrations, but also as *needed* to close graphs. Consider for

instance the map $x/|x|$ from $B(0,1) \subset \mathbb{R}^n$ into $\mathbb{R}^n$, $n \geq 1$. In [1] in Sec. 3.2.2, we proved that there is a sequence of smooth maps u_k from $B(0,1)$ into $\mathbb{R}^n$ with equibounded L^∞-norms and equibounded masses of their graphs such that

$$G_{u_k} \rightharpoonup G_{x/|x|} + \delta_0 \times [\![B(0,1)]\!] \,.$$

Hence the current

$$T := G_{x/|x|} + \delta_0 \times [\![B(0,1)]\!]$$

belongs to $\mathrm{Cart}(B(0,1) \times \mathbb{R}^n)$.

While of course this is the only way to close the graph of $x/|x|$ by a vertical part with support in $\{0\} \times \mathbb{R}^n$, it is worthwhile noticing that there are other ways of closing the graph of $x/|x|$. For that consider *any* 1-dimensional i.m. rectifiable current L in $B(0,1)$ such that $\partial L = \delta_0$ in $B(0,1)$. Then the current

$$\overline{T} := G_{x/|x|} + L \times [\![S^{n-1}]\!]$$

has no boundary in $B(0,1) \times \mathbb{R}^n$ and $\pi_\#(L \times [\![S^{n-1}]\!]) = 0$ in $B(0,1)$. Therefore $\overline{T} \in \mathrm{cart}(B(0,1) \times \mathbb{R}^n)$. Later, compare Vol. II Sec. 4.2.6, we shall see that actually $\overline{T}$ belongs also to $\mathrm{Cart}(B(0,1) \times \mathbb{R}^n)$.

Of course we can extend the previous considerations to the other examples in Sec. 3.2.2. Consider for instance a smooth map $\varphi : S^{n-1} \to S^{n-1}$ and let $u(x) = \varphi(x/|x|)$. From [2] in Sec. 3.2.2, it follows that

$$\partial G_u = -\deg \varphi \, \delta_0 \times [\![S^{n-1}]\!]$$

and that

$$G_u + \deg \varphi \, \delta_0 \times [\![B(0,1)]\!]$$

belongs to $\mathrm{Cart}(B(0,1) \times \mathbb{R}^n)$. For any 1–dimensional i.m. rectifiable current L in $B(0,1)$ such that

$$\pi_\#(L \times [\![S^{n-1}]\!]) = 0 \,, \qquad \partial(L \times [\![S^{n-1}]\!]) = \deg \varphi \, \delta_0 \times [\![S^{n-1}]\!] \,,$$

(e.g. for a current integration on a line going from the boundary of $B(0,1)$ to the center $\{0\}$ with multiplicity equal to $\deg \varphi$), we have $G_u + L \times [\![S^{n-1}]\!] \in \mathrm{cart}(B(0,1) \times \mathbb{R}^n)$; and in fact one can prove that $G_u + L \times [\![S^{n-1}]\!]$ may be weakly approximated by the graphs of smooth maps, compare Vol. II Sec. 4.2.6.

•

[8] We would like to present two examples of currents which in some sense provide an extension of the notion of vector valued BV-functions. We shall return later on this point.

First we consider a map $u_0 : B(0,1) \to \mathbb{R}^N$ which takes constant values on the upper and lower half balls, i.e.,

$$u_0(x) := \begin{cases} a & \text{if } x \in B_-(0,1) \\ b & \text{if } x \in B_+(0,1) \end{cases}$$

$a, b \in \mathbb{R}^N$, $B_\pm(0,1) := \{x \in B(0,1) \mid x^n \gtrless 0\}$. Of course $u_0 \in \mathcal{A}^1(B(0,1), \mathbb{R}^N)$ and

$$G_{u_0} = [\![B_+(0,1)]\!] \times \delta_b + [\![B_-(0,1)]\!] \times \delta_a \,.$$

Hence

$$\partial G_{u_0} = L \times \delta_b - L \times \delta_a = L \times (\delta_b - \delta_a) \qquad \text{in } B(0,1) \times \mathbb{R}^N$$

where

$$L := \partial [\![B_+(0,1)]\!] \llcorner B(0,1) \,.$$

Choose now a smooth curve $\gamma : [0,1] \to \mathbb{R}^N$ joining b to a and let $\Gamma = \gamma_\# [\![(0,1)]\!]$. Then $L \times \Gamma$ is an n-dimensional i.m. rectifiable current and

$$\partial (L \times \Gamma) = (-1)^{n-1} L \times \partial \Gamma = (-1)^{n-1} L \times [\delta_a - \delta_b]$$

in $B(0,1) \times \mathbb{R}^N$. Thus the current

$$G_{u_0} + (-1)^{n-1} L \times \Gamma$$

has zero boundary and belongs to $\text{cart}(\Omega \times \mathbb{R}^N)$ with $\Omega = B(0,1)$. Actually it is easily seen that $G_{u_0} + (-1)^{n-1} L \times \Gamma$ belongs to $\text{Cart}(\Omega \times \mathbb{R}^N)$; in fact the graphs of the functions

$$u_k(x) := \begin{cases} a & \text{if } x_n < 0 \\ \gamma(k x_n) & \text{if } 0 < x_n < \dfrac{1}{k} \\ b & \text{if } x_n > \dfrac{1}{k} \end{cases}$$

weakly converge to $G_{u_0} + (-1)^{n-1} L \times \Gamma$, and also that

$$\mathbf{M}(G_{u_k}) \longrightarrow \mathbf{M}(G_{u_0} + (-1)^{n-1} L \times \Gamma) \,.$$

We now consider the case in which the function u_0 takes three constant values. For the sake of simplicity we consider the case in which $n = 2$. Denote by (ρ, θ) the polar coordinates in $\mathbb{R}^2$ and set

$$\begin{aligned} \Delta_1 &:= \{(x,y) \in B(0,1) \mid \theta \in (0, 2\pi/3)\} \\ \Delta_2 &:= \{(x,y) \in B(0,1) \mid \theta \in (2\pi/3, 4\pi/3)\} \\ \Delta_3 &:= \{(x,y) \in B(0,1) \mid \theta \in (4\pi/3, 2\pi)\} \end{aligned}$$

and

$$\begin{aligned} E_1 &:= \{(x,y) \in B(0,1) \mid \theta = 4\pi/3\} \\ E_2 &:= \{(x,y) \in B(0,1) \mid \theta = 0\} \\ E_3 &:= \{(x,y) \in B(0,1) \mid \theta = 2\pi/3\} \end{aligned}$$

We shall denote by L_i the currents integration over the segments E_i oriented in the outward direction and we consider the map

$$u_0(x) := \begin{cases} a_1 & \text{if } x \in \Delta_1 \\ a_2 & \text{if } x \in \Delta_2 \\ a_3 & \text{if } x \in \Delta_3 \end{cases}$$

where $a_1, a_2, a_3 \in \mathbb{R}^N$. Clearly $u_0 \in \mathcal{A}^1(B(0,1), \mathbb{R}^N)$ and

$$\begin{aligned} \partial G_{u_0} &= (L_2 - L_3) \times \delta_{a_1} + (L_3 - L_1) \times \delta_{a_2} + (L_1 - L_2) \times \delta_{a_3} \\ &= L_1 \times (\delta_{a_3} - \delta_{a_2}) + L_2 \times (\delta_{a_1} - \delta_{a_3}) + L_3 \times (\delta_{a_2} - \delta_{a_1}) \,. \end{aligned}$$

Choose now three curves $\gamma_1, \gamma_2, \gamma_3$ in $\mathbb{R}^N$, γ_1 from a_2 to a_3, γ_2 from a_3 to a_1, γ_3 from a_1 to a_2 and let $\Gamma_i = \gamma_{i\#}[\![(0,1)]\!]$. Then

$$\begin{aligned} \partial(L_1 \times \Gamma_1) &= \partial L_1 \times \Gamma_1 - L_1 \times \partial \Gamma_1 \\ &= -\delta_0 \times \Gamma_1 - L_1 \times (\delta_{a_3} - \delta_{a_2}) \end{aligned}$$

and similarly

$$\begin{aligned} \partial(L_2 \times \Gamma_2) &= -\delta_0 \times \Gamma_2 - L_2 \times (\delta_{a_1} - \delta_{a_3}) \\ \partial(L_3 \times \Gamma_3) &= -\delta_0 \times \Gamma_3 - L_3 \times (\delta_{a_2} - \delta_{a_1}) \,. \end{aligned}$$

Setting $\Gamma := \Gamma_1 + \Gamma_2 + \Gamma_3$ we therefore have $\partial \Gamma = 0$ and

$$\partial(G_{u_0} + \sum_{i=1}^{3} L_i \times \Gamma_i) = -\delta_0 \times \Gamma \,.$$

Choose now a two-dimensional surface Δ in $\mathbb{R}^N$ so that $\partial \Delta = \Gamma$; then

$$\partial(\delta_0 \times \Delta) = \delta_0 \times \partial \Delta = \delta_0 \times \Gamma \,.$$

In conclusion we then infer that

$$T := G_{u_0} + \sum_{i=1}^{3} L_i \times \Gamma_i + \delta_0 \times \Delta$$

is a Cartesian current, $T \in \text{cart}(B(0,1), \mathbb{R}^N)$. Such a current provides an example in which different vertical parts appear: jump parts $L_i \times \Gamma_i$ over the 1-dimensional sets E_i, and a completely vertical part Δ of dimension two over the origin.

Tough slightly length, one can prove the existence of a sequence of smooth maps u_k such that

$$G_{u_k} \rightharpoonup T := G_{u_0} + \sum_{i=1}^{3} L_i \times \Gamma_i + \delta_0 \times \Delta$$

and also

$$\mathbf{M}(G_{u_k}) \longrightarrow \mathbf{M}(T)$$

so that, in particular, $T \in \text{Cart}(B(0,1), \mathbb{R}^N)$, but we leave it to the reader. •

9 *A Cartesian current with Cantor mass on minors.* We conclude this section by showing a Cartesian current

$$T = G_{u_T} + S_T$$

in $Q \times \mathbb{R}^2$, $Q := (0,1) \times (0,1) \subset \mathbb{R}^2$, which is the weak limit of smooth graphs such that

(i) $u_T \in W^{1,p}(Q)$ for all $p < 2$ and $\det Du_T = 0$ a.e. in Q
(ii) $S_{T(0)} = S_{T(1)} = 0$, $S_{T(2)} \neq 0$, $\operatorname{spt} S_{T(2)} \subset \pi^{-1}(C \times C)$ where C is the Cantor set in Sec. 1.1.3, and actually

$$\pi_{\#} S_{T(2)} = \operatorname{Det} Du_T \, e_1 \wedge e_2 = V' \otimes V' \, e_1 \wedge e_2$$

where $\operatorname{Det} Du_T$ is the distributional determinant of u_T and V is the Cantor–Vitali function.

For that we shall construct a map $u : Q \to \mathbb{R}^2$ which belongs to $W^{1,p} \cap C^{0,s}(Q)$ for all $p < 2$, $s := \log 2 / \log \delta^{-1}$ such that $|Du^1| \, |Du^2| = 0$ a.e. but

$$\operatorname{Det} Du = V' \otimes V' ,$$

i.e. the derivatives of u have no masses, but the distributional determinant has a Cantor-type mass. The current graph of u, G_u, is very similar to the graph of Cantor–Vitali function G_V: the role played by V' in C is here played by $\operatorname{Det} Du$ in $C \times C$. The graph of u has infinitely many holes, and in fact

$$\mathbf{M}(\partial G_u) = +\infty .$$

The current G_u can be "completed" to a boundaryless current, and in fact to a Cartesian current T, so that $u_T = u$, by an approximation argument and this give rise to our example.

To accomplish such a program we need several steps where we use notation and results of Sec. 1.1.3. We begin with

Lemma 1. *Let* $k \geq 0$, $\delta_1 := \frac{1}{2\delta} - 1 \in (0, +\infty)$. *For* $j = 1, \ldots, 2^k$ *consider the open intervals*

$$\widetilde{J}_{k,j} := b_{k,j} + \delta^k(-\delta_1, 1 + \delta_1) .$$

Then $\{\widetilde{J}_{k,j}\}$ *form a disjoint covering of Cantor set* C *and*

$$\operatorname{dist}\Big([0,1] \setminus \bigcup_{j=1}^{2^k} \widetilde{J}_{k,j}, C\Big) \geq \delta_1 \cdot \delta^k . \tag{7}$$

Proof. The $\widetilde{J}_{k,j}$ obviously cover C as $\widetilde{J}_{k,j} \supset J_{k,j}$. Set

$$c_{k,j} := b_{k,j} - \delta_1 \delta^k , \qquad d_{k,j} := b_{k,j} + \delta^k (1 + \delta_1)$$

so that $\widetilde{J}_{k,j} = (c_{k,j}, d_{k,j})$. To show that the cover $\widetilde{J}_{k,j}$ is disjoint it suffices to show that

$$d_{k,j} \leq c_{k,j+1} \qquad \text{for } j = 1, \dots, 2^k - 1 \ . \tag{8}$$

For $k = 0$ there is nothing to prove. Assume (8) holds for k, from

$$c_{k+1,j} = \delta c_{k,j} \ , \quad c_{k+1,j+2^k} = 1 - \delta + \delta c_{k,j}$$
$$d_{k+1,j} = \delta d_{k,j} \ , \quad d_{k+1,j+2^k} = 1 - \delta + \delta d_{k,j}$$

we infer

$$d_{k+1,j} \leq c_{k+1,j+1} \qquad \text{if } j \neq 2^k \ .$$

For $j = 2^k$ we infer from the definition of $b_{k,j}$

$$\begin{aligned} b_{k,2^k} &\leq 1 - \delta^k \\ d_{k,2^k} &= b_{k,2^k} + \delta^k(1 + \delta_1) \leq 1 + \delta_1 \delta^k \end{aligned}$$

hence

$$d_{k+1,2^k} = \delta d_{k,2^k} \leq \delta(1 + \delta_1 \delta^k) \leq \delta(1 + \delta_1) = \delta \frac{1}{2\delta} = \frac{1}{2} \ ,$$

while, as $c_{k,1} = -\delta^k \delta_1$,

$$\begin{aligned} c_{k+1,2^k+1} &= 1 - \delta + \delta c_{k,1} = 1 - \delta - \delta^{k+1}\delta_1 \\ &\geq 1 - \delta(1 + \delta_1) = 1 - \delta \frac{1}{2\delta} = \frac{1}{2} \ . \end{aligned}$$

Finally, since

$$\operatorname{dist}(\mathbb{R} \setminus \widetilde{J}_{k,j}, J_{k,j}) = \delta_1 \delta^k$$

and the $\widetilde{J}_{k,j}$ are disjoint we get (7). □

As next step we shall define an auxiliary function $f : Q = [0,1]^2 \to \mathbb{R}$ which interpolate between Cantor–Vitali function V and the identity. We set

$$\delta_2 := \frac{1 - 2\delta}{1 - \delta}$$

and for $(x^1, x^2) \in [0,1] \times [\delta, 1]$ we set

$$f^{(0)}(x^1, x^2) := \begin{cases} (1 - \delta_2(1 - x^2))^{-1} x^1 & \text{if } x^1 \in [0, \frac{1}{2}(1 - \delta_2(1 - x^2))] \\ \frac{1}{2} & \text{if } x^1 \in [\frac{1}{2}(1 - \delta_2(1 - x^2)), \frac{1}{2}(1 + \delta_2(1 - x^2))] \\ 1 - (1 - \delta_2(1 - x^2))^{-1}(1 - x_1) & \text{if } x^1 \in [\frac{1}{2}(1 + \delta_2(1 - x^2)), 1] \end{cases}$$

$f^{(0)}(x^1, x^2)$ is piecewise linear and $f^{(0)} = \dfrac{1}{2}$ on the segments $[(\delta, \delta), (\frac{1}{2}, 1)]$ and $[(1 - \delta, \delta), (\frac{1}{2}, 1)]$; also

$$f^{(0)}(\cdot, 1) = V_0 := \text{id} \ , \qquad f^{(0)}(\cdot, \delta) = V_1 \ ,$$

$V_k,\ k > 0$, denoting Cantor–Vitali approximate functions. Inductively we now define

$$f^{(k)} \ : \ [0,1] \times [\delta^{k+1}, \delta^k] \ \to \ \mathbb{R}$$

by

$$f^{(k+1)}(x^1,x^2) \ := \ \begin{cases} \frac{1}{2} f^{(k)}(\delta^{-1}x^1, \delta^{-1}x^2) & \text{if } x^1 \in [0,\delta] \\ \frac{1}{2} & \text{if } x^1 \in [\delta, 1-\delta] \\ \frac{1}{2} + \frac{1}{2} f^{(k)}(1-\delta^{-1}(1-x^1), \delta^{-1}x^2) & \text{if } x^1 \in [1-\delta, 1] \ . \end{cases}$$

Taking into account that $f^{(1)}(\cdot,\delta) = f^{(0)}(\cdot,\delta) = V_1$ we infer

$$f^{(k+1)}(\cdot,\delta^k) \ = \ f^{(k)}(\cdot,\delta^k) \ = \ V_k \ ,$$

and we set

$$f \ := \ \begin{cases} f^{(k)} & \text{on } [0,1] \times [\delta^{k+1}, \delta^k] \ , \ k = 0,1,\dots \\ V & \text{on } [0,1] \times \{0\} \ . \end{cases}$$

We also introduce the following approximation of f

$$f_k(x^1,x^2) \ := \ \begin{cases} f(x^1,x^2) & \text{on } [0,1] \times [\delta^k, 1] \\ V_k(x^1) & \text{for } x^2 \in [0,\delta^k] \ . \end{cases}$$

Then

Lemma 2. *We have*

(i) *f, f_k are Hölder-continuous with exponent s and locally Lipschitz in Q,*

(ii) $f(\cdot,1) = f_k(\cdot,1) = \mathrm{id},\ f(0,\cdot) = f_k(0,\cdot) = 0,\ f(1,\cdot) = f_k(1,\cdot) = 1,\ f(\cdot,0) = V,\ f_k(\cdot,0) = V_k,$

(iii) *For $(x^1,x^2) \in J_{k,j} \times [0,\delta^k]$*

$$f(x^1,x^2) = V(b_{k,j}) + 2^{-k} f(\delta^{-k}(x^1 - b_{k,j}), \delta^{-k}x^2),$$

(iv) *For $(x^1,x^2) \in ([0,1] \setminus E_k) \times [0,\delta^k],\ E_k := \bigcup_{j=1}^{2^k} J_{k,j}$, we have*

$$f(x^1,x^2) = V(x^1), \quad \text{and} \quad Df(x^1,x^2) = 0$$

if moreover $x^2 \in (0,\delta^k)$. As usual Df denotes the approximate gradient of f.

(v) *Set $\delta_0 := 1/2 - \delta$. Then*

$$f(x^1,x^2) \ = \ V(x^1) \qquad \text{if} \qquad x^2 \ \le \ \delta_0^{-1} \mathrm{dist}\,(x^1, C) \ .$$

Proof. Assertion (ii) follows at once.

(iii) Inductively we shall prove the following assertion (P_m): If $\delta^{m+1} \le x^2 \le \delta^m$, $k \le m$, $j = 1, \dots, 2^k$, and $0 \le x^1 - b_{k,j} \le \delta^k$, then

$$f(x^1, x^2) = V(b_{k,j}) + 2^{-k} f(\delta^{-k}(x^1 - b_{k,j}), \delta^{-k} x^2) .$$

(P_0) is trivial; suppose (P_{m-1}) holds, and assume moreover that $j \le 2^{k-1}$. Writing

$$x^1 = b_{k,j} + \delta^k \widetilde{x}^1$$

we have

$$b_{k,j} = \delta^{-1} b_{k-1,j} \qquad V(b_{k,j}) = \frac{1}{2} V(b_{k-1,j})$$

and

$$\begin{aligned} f(x^1, x^2) &= \frac{1}{2} f(\delta^{-1} x^1, \delta^{-1} x^2) \\ &= \frac{1}{2} f(b_{k-1,j} + \delta^{k-1} \widetilde{x}^1, \delta^{-1} x^2) \\ &= \frac{1}{2} [V(b_{k-1,j}) + 2^{-(k-1)} f(\widetilde{x}^1, \delta^{-k} x^2)] \\ &= V(b_{k,j}) + 2^{-k} f(\delta^{-k}(x^1 - b_{k,j}), \delta^{-k} x^2) . \end{aligned}$$

Similarly one proceeds in the case $j > 2^{k-1}$.

(iv) Let $0 \le x^2 \le \delta^k$, $x^1 \in [0,1] \setminus E_k$, $E_k = \bigcup_{j=1}^{2^k} J_{k,j}$. As

$$[0,1] = E_k \cup \bigcup_{m=0}^{k-1} \bigcup_{j=1}^{2^m} I_{m,j} ,$$

we find $m \le k-1$ and $i \le 2^m$ such that

$$x^1 \in I_{m,i} = b_{m.i} + \delta^m (\delta, 1 - \delta) .$$

Therefore we can write

$$x^1 = b_{m,i} + \delta^m \widetilde{x}^1 \qquad \text{with} \qquad \delta < \widetilde{x}^1 < 1 - \delta .$$

If $m = 0$, we then have $x^1 = \widetilde{x}^1$, $k \ge 1$, and

$$f(x^1, x^2) = f(\widetilde{x}^1, x^2) = \frac{1}{2} = V(x^1) .$$

If $m > 0$, by (iii) and the self-similarity of V, we get

$$\begin{aligned} f(x^1,x^2) &= V(b_{m,i}) + 2^{-m} f(\widetilde{x}^1, \delta^{-m} x^2) \\ &= V(b_{m,i}) + 2^{-m} V(\widetilde{x}^1) \\ &= V(b_{m,i}) + 2^{-m} V(\delta^{-m}(x^1 - b_{m,i})) = V(x^1) \ . \end{aligned}$$

(v) By (ii) the assertion is trivial if $x^1 \in C$. If $x^1 \notin C$, we find k and j such that

$$x^1 \in I_{k,j} = b_{k,j} + \delta^k (\delta, 1-\delta) \ .$$

Assume first $k = 0$, i.e., $x^1 \in (\delta, 1-\delta)$. Then

$$\text{dist}\,(x^1, C) = \min(|x^1 - \delta|, |x^1 - 1 + \delta|) = \delta_0 - |x^1 - \frac{1}{2}| \ .$$

Therefore $f(x^1,x^2) = \frac{1}{2} = V(x^1)$ if $x^2 \le \delta$. Instead, if $\delta < x^2 \le \delta_0^{-1}\text{dist}\,(x^1, C)$ $= 1 - \delta_0 |x^1 - \frac{1}{2}|$, we have

$$|x_1 - \frac{1}{2}| \ \le \ \delta_0 (1 - x^2) \ \le \ \frac{\delta_2}{2}(1-x^2) \ , \qquad \delta_2 := \frac{1-2\delta}{1-\delta} \ ;$$

hence again $f(x^1,x^2) = \frac{1}{2} = V(x^1)$ by the definition of $f^{(0)}$. Assume now $k > 0$. Then

$$x^2 \ \le \ \delta_0^{-1} \text{dist}\,(x^1, C) \ \le \ \delta^k \ .$$

Setting

$$\widetilde{x}^1 \ := \ \delta^{-k}(x^1 - b_{k,j}) \ , \qquad \widetilde{x}^2 \ := \ \delta^{-k} x^2$$

we then infer

$$\widetilde{x}^2 \ \le \ \delta_0^{-1} \text{dist}\,(\widetilde{x}^1, C) \ , \qquad \widetilde{x}^1 \in (\delta, 1-\delta) \ .$$

Therefore by (iii) and using the case $k = 0$, we get

$$f(x^1,x^2) = V(b_{k,j}) + 2^{-k} f(\widetilde{x}^1, \widetilde{x}^2) = V(b_{k,j}) + 2^{-k} V(\widetilde{x}^1) \ = \ V(x^1) \ .$$

(i) It suffices to show that there is a positive constant c such that for all pairs $(x^1,x^2), (y^1,y^2) \in Q$ we have

$$(i)_1 \qquad |f(y^1,y^2) - f(y^1,x^2)| \ \le \ c\,|y^2 - x^2|^s$$

$$(i)_2 \qquad |f(y^1,x^2) - f(x^1,x^2)| \ \le \ c\,|y^1 - x^1|^s$$

We may assume $y^2 \le x^2$, and $x^2 > 0$ (the claim being trivial for $x^2 = 0$), $x^2 \in (\delta^{k+1}, \delta^k]$ for some k.

Let $k = 0$. Then either $y^2 > \delta^2$, and (i)$_1$ follows from the Lipschitz continuity of f on $[0,1] \times [\delta^2, 1]$, or $y_2 \le \delta^2$; but then $|x^2 - y^2| \ge \delta - \delta^2$ and (i)$_1$ follows as $0 \le f \le 1$.

If $k > 0$ and $y^1 \notin \bigcup_{j=1}^{2^k} J_{k,j}$, (i)$_1$ follows from (iv). If instead $y^1 \in J_{k,j}$ we let $\widetilde{y}^1 := \delta^{-k}(y^1 - b_{k,j})$. Then by (iii) and the case $k = 0$ we infer

$$\begin{aligned}|f(y^1,y^2)-f(y^1,x^2)| &\le 2^{-k}|f(\widetilde{y}_1,\delta^{-k}y^2)-f(\widetilde{y}_1,\delta^{-k}x^2)|\\ &\le 2^{-k}c|\delta^{-k}(y^2-x^2)|^s \le c\,|y^2-x^2|^s\end{aligned}$$

as $2\delta^s = 1$.

This proves (i)$_1$. We shall now prove (i)$_2$ by induction on k assuming $x^2 \in (\delta^{k+1}, \delta^k]$. For $k = 0$ we use the Lipschitz continuity of f on $[0,1]\times[\delta,1]$. Suppose now $k > 0$, i.e., $x^2 \in (\delta^{k+2}, \delta^{k+1}]$.

If $x^1, y^1 \in (\delta, 1-\delta)$, then (i)$_2$ follows from the definition of $f^{(k+1)}$. Assume $x^1 \notin (\delta, 1-\delta)$ and say $x^1 \in [0,\delta]$. If $y^1 \in [0,\delta]$, then by the definition of $f^{(k+1)}$ and the inductive assumption for $x^2 \in (\delta^{k+1}, \delta^k]$ we get

$$\begin{aligned}|f(y^1,x^2)-f(x^1,x^2)| &= \frac{1}{2}|f(\delta^{-1}y^1,\delta^{-1}x^2)-f(\delta^{-1}x^1,\delta^{-1}x^2)|\\ \le\ &\frac{1}{2}c\,|\delta^{-1}(x^1-y^1)|^s = c|x^1-y^1|^s\,.\end{aligned}$$

If $y^1 \in [1-\delta, 1]$, then (i)$_2$ holds as $|x^1 - y^1| \ge 1 - 2\delta > 0$ and $0 \le f \le 1$. Finally, if $y^1 \in (\delta, 1-\delta)$, then $f(y^1, x^2) = \frac{1}{2} = f(\delta, x^2)$ and

$$|f(y^1,x^2)-f(x^1,x^2)| \le c|x^1-\delta|^s \le c|x^1-y^1|^s\,.$$

□

We now abbreviate dist to d and define the map $u = (u^1, u^2) : Q \to \mathbb{R}^2$ by

$$\begin{aligned}u^1(x^1,x^2) &:= f(x^1,\delta_0^{-1}d(x^2,C))\,,\\ u^2(x^1,x^2) &:= f(x^2,\delta_0^{-1}d(x^1,C)) = u^1(x^2,x^1)\,.\end{aligned}$$

With such a definition we can state

Theorem 1. *We have*

(i) $u \in W^{1,p} \cap C^{0,s}(Q,\mathbb{R}^2)$ *for all* $p < 2$
(ii) $|Du^1|\,|Du^2| = 0$ a.e.
(iii) *The distributional determinant of u is a non-negative measure supported on the Cantor-type set $C \times C$, more precisely*

$$\operatorname{Det} Du = V' \otimes V'\ ;$$

moreover there are positive constants c_1 and c_2 such that

$$c_1\,\mathcal{H}^{2s} \llcorner C\times C \le \operatorname{Det} Du \le c_2\,\mathcal{H}^{2s} \llcorner C\times C\,.$$

Proof. (i) As $d(\cdot, C)$ is a Lipschitz function we infer at once that the map u is Hölder-continuous with exponent s and locally Lipschitz in Q, and

$$\begin{aligned}u^1_{x^1}(x^1,x^2) &= f_{y^1}(x^1,\delta_0^{-1}d(x^2,C))\\ u^1_{x^2}(x^1,x^2) &= f_{y^2}(x^1,\delta_0^{-1}d(x^2,C))\,\delta_0^{-1}\frac{\partial}{\partial x^2}d(x^2,C)\,.\end{aligned}$$

As $|Dd(\cdot, C)| \leq 1$ a.e. we only need to show that the function

$$g(x) := |D_y f(x^1, \delta_0^{-1} d(x^2, C))|$$

belongs to $L^p(Q)$ for all $p < 2$.

For $y_2 \in [\delta^{k+1}, \delta^k]$ we have by Lemma 2

$$Df(y^1, y^2) = 0 \qquad if \qquad y^1 \notin E_k := \bigcup_{j=1}^{2^k} [b_{k,j}, b_{k,j} + \delta^k]$$

whereas

$$|Df(y)| \leq c(2\delta)^{-k} = c\delta^{-k(1-s)} \leq c(y^2)^{-(1-s)}$$

if $y^2 \in E_k$. Thus for x^2 with $\delta_0^{-1} d(x^2, C) \in [\delta^{k+1}, \delta^k]$ we infer

$$\int_0^1 |g(x^1, x^2)|^p \, dx^1 \leq c 2^k \delta^k d(x^2, C)^{-(1-s)p} \leq c(d(x^2, C))^{-(1-s)(p-1)} .$$

Taking into account that the set $\{(x^1, x^2) \mid d(x^2, C) = 0 \text{ or } \delta_0\}$ has measure zero, we then conclude

$$\int_Q |g|^p \, dx \leq c \int_0^1 (d(x^2, C))^{-(1-s)(p-1)} \, dx^2 .$$

In order to show that the last integral is finite we consider the distribution function

$$\varphi(\lambda) := \mathcal{H}^1(\{x^2 \in (0,1) \mid (d(x^2, C))^{-(1-s)(p-1)} \geq \lambda\} .$$

We claim that, as consequence of Lemma 1,

$$\varphi(\lambda) \leq c\lambda^{-1/(p-1)} ,$$

and this of course concludes the proof of (i). In fact if

$$\widetilde{J}_{k,j} := b_{k,j} + \delta^k(-\delta_1, 1 + \delta_1) , \qquad \delta_1 = \frac{1 - 2\delta}{2\delta} = \frac{\delta_0}{\delta}$$

is the covering of C in Lemma 2, and

$$\widetilde{E}_k := \bigcup_{j=1}^{2^k} \widetilde{J}_{k,j}$$

we have

$$d((0,1) \setminus \widetilde{E}_k, C) \geq c\delta^k .$$

Since

$$\mathcal{H}^1(\widetilde{E}_k) \le c\,2^k \delta^k = c\,\delta^{k(1-s)}$$

we then deduce

$$\varphi(c\delta^{-k(1-s)(p-1)}) \le c\delta^{(1-s)k}$$

which yields $\varphi(\lambda) \le c\lambda^{-1/(p-1)}$.

(ii) Consider the closed set

$$M := \{(x^1, x^2) \mid d(x^1, C) = d(x^2, C)\}$$

we claim that
 (ii)$_1$ $\mathcal{H}^2(M) = 0$
 (ii)$_2$ for a.e. $(x^1, x^2) \in Q$ either $u^1(x^1, x^2) = V(x^1)$ or $Du^2(x^1, x^2) = 0$
 (ii)$_3$ for a.e. $x \in Q$ either $Du^1(x) = 0$ or $Du^2(x) = 0$.
For that we consider the map

$$h(x^1, x^2) := d(x^1, C) - d(x^2, C)\ .$$

As C is closed with $\mathcal{H}^1(C) = 0$ we have $|h_{x^1}| = |h_{x^2}| = 1$ a.e., hence $|Dh| = \sqrt{2}$ a.e. Setting

$$M_k := h^{-1}\Big(\Big(-\frac{1}{k}, \frac{1}{k}\Big)\Big)$$

the coarea formula yields

$$\sqrt{2}\mathcal{H}^2(M_k) = \int_{M_k} |Dh|\, d\mathcal{H}^2 = \int_{-1/k}^{1/k} \mathcal{H}^1(h^{-1}(t))\, dt\ .$$

As $t \to \mathcal{H}^1(h^{-1}(t))$ is summable, we then infer $\mathcal{H}^2(M_k) \to 0$, which proves (ii)$_1$.

Let us consider $(\bar{x}^1, \bar{x}^2) \in Q \setminus M$. Then $d(\bar{x}^1, C) \ne d(\bar{x}^2, C)$, say $d(\bar{x}^1, C) > d(\bar{x}^2, C)$, and Lemma 2 (v) yields

$$u^1(x) = f(x^1, \delta_0^{-1} d(x^2, C)) = V(x^1)$$

for x close to $\bar{x} = (\bar{x}_1, \bar{x}_2)$. Thus $Du^1 = 0$ a.e. in a neighbourhood of $\bar{x}$. Similarly, we find $Du^2 = 0$ a.e. if $d(\bar{x}^1, C) < d(\bar{x}^2, C)$. Therefore the set of points where (ii)$_2$ or (ii)$_3$ fail has density zero in $Q \setminus M$. As $\mathcal{H}^2(M) = 0$, this proves (ii)$_2$ and (ii)$_3$ and in turn (ii).

(iii) Recall that the distributional determinant is defined by

$$\operatorname{Det} Du(\varphi) = \int_Q (-u^1 u^2_{x^2} \varphi_{x^1} + u^1 u^2_{x^1} \varphi_{x^2})\, dx \qquad \forall\, \varphi \in C^\infty_c(Q)\ .$$

Let

$$\widetilde{I}_{k,j} := I_{k,j} \times [0,1] , \qquad I_{k,j} := b_{k,j} + \delta^k(\delta, 1-\delta) .$$

We observe that the $\widetilde{I}_{k,j}$ are disjoint and

$$\mathcal{H}^2\Big(Q \setminus \bigcup_{k=0}^{\infty} \bigcup_{j=1}^{2^k} \widetilde{I}_{k,j}\Big) = 0 .$$

Consider first $\widetilde{I}_{0,1} = (\delta, 1-\delta) \times [0,1]$. From $V = \frac{1}{2}$ on $(\delta, 1-\delta)$ and $(\mathrm{ii})_2$ we infer

$$u^1(x^1,x^2)u^2_{x^j}(x^1,x^2) = V(x^1)u^2_{x^j}(x^1,x^2) \qquad a.e.$$

thus

$$\int_{\widetilde{I}_{0,1}} (-u^1 u^2_{x^2}\varphi_{x^1} + u^1 u^2_{x^1}\varphi_{x^2})\, dx = \frac{1}{2} \int_{\widetilde{I}_{0,1}} (-u^2_{x^2}\varphi_{x^1} + u^2_{x^1}\varphi_{x^2})\, dx$$

$$= \frac{1}{2} \int_{\widetilde{I}_{0,1}} (u^2\varphi_{x^1x^2} - u^2\varphi_{x^1x^2})\, dx + \frac{1}{2} \int_0^1 u^2\varphi_{x^2}\Big|_{x^1=\delta}^{x^1=1-\delta} dx^2 .$$

Since δ and $1-\delta$ belong to C we have $u^2(\delta, x^2) = V(x^2)$ and $u^2(1-\delta, x^2) = V(x^2)$. Thus we compute the last integral as

$$\frac{1}{2} \int_0^1 V(x^2)\varphi_{x^2}(x^1,x^2)\Big|_{x^1=\delta}^{x^1=1-\delta} dx^2 = \frac{1}{2} \int_{\widetilde{I}_{0,1}} V(x^2)\varphi_{x^1x^2}(x^1,x^2)\, dx$$

$$= \int_{\widetilde{I}_{0,1}} V(x^1)V(x^2)\varphi_{x^1x^2}\, dx .$$

Since the Cantor–Vitali function V is constant on $I_{k,j}$ we obtain similarly

$$\int_{\widetilde{I}_{k,j}} (-u^1 u^2_{x^2}\varphi_{x^1} + u^1 u^2_{x^1}\varphi_{x^2})\, dx = \int_{\widetilde{I}_{k,j}} V(x^1)V(x^2)\varphi_{x^1x^2}\, dx$$

and letting $\varphi(x^1,x^2) = \psi(x^1)\chi(x^2)$

$$\mathrm{Det}\, Du(\varphi) = \int_Q V(x^1)V(x^2)\psi'(x^1)\chi'(x^2)\, dx$$

$$= \int_0^1 V\psi'\, dx^1 \int_0^1 V\chi'\, dx^2 = V'(\psi)V'(\chi) = V' \otimes V'(\varphi) .$$

This proves the first part of (iii). To prove the second part we consider the Radon measures

$$\mu := V' \otimes V' , \qquad \nu := \mathcal{H}^{2s} \llcorner C \times C$$

and an arbitrary rectangle

$$R := [a_1, \bar{a}_1] \times [a_1, \bar{a}_2] \subset Q .$$

Let $\{F_m\}$ be a covering of $R \cap (C \times C)$. Choose a cube $\widetilde{F}_m$ parallel to the coordinate axes with side $\ell_m = 2 \operatorname{diam} F_m$, $\widetilde{F}_m \supset F_m$. By the Hölder continuity of V we have

$$\mu(F_m) \leq \mu(\widetilde{F}_m) \leq \ell_m^{2s} \leq 2^{2s} (\operatorname{diam} F_m)^{2s} .$$

Since $\mu \llcorner (Q \setminus (C \times C)) = 0$ we infer

$$\mu(R) = \mu(R \cap (C \times C)) \leq \sum_m \mu(F_m) \leq 2^{2s} \sum (\operatorname{diam} F_m)^{2s} ,$$

hence

$$\mu(R) \leq c\mathcal{H}^{2s}(R \cap (C \times C)) .$$

To prove the converse bound we take $k \in \mathbb{N}$ and choose integers $m_i, \bar{m}_i, i = 1, 2$ in such a way that

$$b_{k,m_i} \leq a_i \leq b_{k,m_i+1} \qquad b_{k,\bar{m}_i-1} \leq \bar{a}_i \leq b_{k,\bar{m}_i} , \qquad i = 1, 2 .$$

As $V(b_{k,j}) = \frac{j-1}{2}$ and $V' \geq 0$ we have

$$\bar{m}_i - m_i \leq 2^k (V(\bar{a}_i) - V(a_i)) + 2 .$$

Then the collection of cubes

$$\mathcal{F} := \{J_{k,j_1} \times J_{k,j_2} \mid m_1 \leq j_1 \leq \bar{m}_1, \; m_2 \leq j_2 \leq \bar{m}_2\}$$

covers $R \cap (C \times C)$ and each cube in $\mathcal{F}$ has diameter $\sqrt{2}\delta^k$. Thus, being $2\delta^s = 1$,

$$\begin{aligned} \sum_{F \in \mathcal{F}} (\operatorname{diam} F)^{2s} &\leq c\delta^{2sk} [2^k (V(\bar{a}_1) - V(a_1)) + 2] \, [2^k (V(\bar{a}_2) - V(a_2)) + 2] \\ &\leq c(V(\bar{a}_1) - V(a_1))(V(\bar{a}_2) - V(a_2)) + O(2^{-k}) \\ &\leq c\mu(R) + O(2^{-k}) . \end{aligned}$$

Letting $k \to \infty$ it follows

$$\mathcal{H}^{2s}(R \cap (C \times C)) \leq c\mu(R) .$$

Being V continuous, a particular consequence is that $\nu(H) = \mu(H) = 0$ for any straight line H parallel to the coordinate axes. As every open set $U \subset Q$ can be written as countable disjoint union of open rectangles and parts of lines both parallel to the coordinate axes, we then infer

$$\mathcal{H}^{2s}(U \cap (C \times C)) \leq c\mu(U)$$

and this concludes the proof of the second part of (iii). □

Next we shall construct a sequence of Lipschitz maps which approximate the map u in Theorem 1 in terms of the functions f_k introduced before Lemma 2. Notice that

$$f_k(y^1, y^2) = f(y^1, y^2) \qquad \text{if } y^2 \geq \delta^k .$$

We set for $(x^1, x^2) \in [0,1]^2$

$$(9) \qquad u_k^1(x^1, x^2) := f_k(x^1, \delta_0^{-1} d(x^2, C)), \quad u_k^2(x^1, x^2) := u_k^1(x^2, x^1).$$

The maps u_k *are Lipschitz continuous in* Q and $u_k = u$ if

$$d(x^2, C) \geq \delta_0 \delta^k \qquad \text{and} \qquad d(x^1, C) \geq \delta_0 \delta^k .$$

Proposition 1. *The sequence of Lipschitz maps u_k in (9) satisfies*

(i) $\sup_k \| u_k \|_{W^{1,p}(Q,\mathbb{R}^2)} < \infty$ *and* $u_k \to u$ *strongly in* $W^{1,p}(Q, \mathbb{R}^2)$ *for all* $p < 2$, u *being the map in Theorem 1*

(ii) $\sup_k \int_Q |\det Du_k| < \infty.$

Therefore, passing to a subsequence we have $G_{u_k} \rightharpoonup T = G_u + S$, *moreover* $S_{(0)} = S_{(1)} = 0$, $T_{(2)} = S_{(2)} \neq 0$, *and*

$$\operatorname{spt} S \subset (C \times C) \times \mathbb{R}^2, \qquad \pi_{\#} S_{(2)} = \operatorname{Det} Du \, e_1 \wedge e_2 .$$

Proof. Recall that

$$\widetilde{J}_{k,j} := b_{k,j} + \delta^k(-\delta_1, 1 + \delta_1)$$

form a disjoint open cover

$$\widetilde{E}_k := \bigcup_{j=1}^{2^k} \widetilde{J}_{k,j}$$

of C, and

$$\mathcal{H}^2(\widetilde{E}_k) \leq 2^k \delta^k (1 + 2\delta_1) = O(\delta^{(1-s)k}) \to 0, \qquad d((0,1) \setminus \widetilde{E}_k, C) \geq c\delta^k.$$

Set

$$U_k := \{(x^1, x^2) \in Q \mid d(x^1, C) < \delta_0 \delta^k \text{ or } d(x^2, C) < \delta_0 \delta^k\} .$$

We then have $\mathcal{H}^2(U_k) \to 0$, $U_1 \supset U_2 \supset \cdots$, and $u_k = u$ on $Q \setminus U_k$, in fact

$$u_k = u_{k+1} = \ldots = u \quad \text{on } Q \setminus U_k .$$

As in the proof of Theorem 1 (i) one proves that $\{u_k\}$ is equibounded in $W^{1,p}(Q, \mathbb{R}^2)$ for every $p < 2$ and since $\mathcal{H}^2(\{u_k \neq u\}) \to 0$ one easily concludes the proof of (i) by a simple use of Hölder's inequality.

On U_k we have either $u_k^1(\cdot, x^2) = V_k$ or $u_k^2(x_1, \cdot) = V_k$. In the first case we get for instance

$$\det Du_k = V_k'(x^1)(f_k(x^2, \delta_0^{-1} d(x^1, C)))_{x^2} ;$$

and integrating with respect to x^2

$$\int_0^1 \det Du_k \, dx^2 = V_k'(x^1)$$

as $f_k(1,\cdot) = 1$ and $f_k(0,\cdot) = 0$. Therefore, using also that

$$\int_{[0,1]\times[0,\delta^k]} \det Du_k \, dx \le 1$$

we conclude that

$$\sup_k \int_Q |\det Du_k| \, dx < \infty .$$

The second part of the claim now follows easily. In fact (i) yields at once $S_{(0)} = S_{(1)} = 0$. From $\det Du = 0$ we infer $T_{(2)} = S_{(2)}$, from $\operatorname{spt} \det Du_\ell \subset E_k \times E_k$, for $\ell \times k$, we infer $\operatorname{spt} S \subset E_k \times E_k \ \forall\, k$, i.e., $\operatorname{spt} S \subset (C \times C) \times \mathbb{R}^2$. Finally from

$$\int_0^1 \det Du_k \varphi \longrightarrow \pi_\# S_{(2)}(\varphi)$$

we infer $\pi_\# S_{(2)} = \operatorname{Det} Du \, e_1 \wedge e_2$. □

Next proposition concludes the proof of our claims

Proposition 2. *Let u be the map in Theorem 1. Then*

$$\mathbf{M}(\partial G_u \llcorner Q) = +\infty .$$

Proof. We divide the proof in three steps.

Step 1. First we explicitly compute the boundary current ∂G_u. As G_u ha finite mass, we can test ∂G_u on forms with bounded and Lipschitz coefficients, so that

$$\mathbf{M}(\partial G_u) = \sup\{G_u(d\omega) \mid \omega \in Lip(Q, \Lambda^1(\mathbb{R}^2 \times \mathbb{R}^2), \quad \|\omega\|_\infty \le 1\}$$

On forms of the type $\omega := \varphi(x) g(y) dy^1$, where $g \in \operatorname{Lip}(\mathbb{R}^2) \cap L^\infty(\mathbb{R}^2)$ and $\varphi \in C_c^\infty([0,1]^2)$, we have, compare Proposition 2 in Sec. 3.2.3,

$$\partial G_u(\omega) = \int_Q g(u^1, u^2)(\varphi_{x^1} u^1_{x^2} - \varphi_{x^2} u^1_{x^1}) \, dx + \int_Q \varphi(x) g_{y^2}(u^1, u^2) \det Du \, dx .$$

From (ii)$_3$ in the proof of Theorem 1

$$\det Du(x) = 0 \qquad \text{a.e.},$$

while, from $(\text{ii})_2$ and recalling that $u^1(x^1, x^2) = u^2(x^2, x^1)$, we get that $Du^1(x^1, x^2) \neq 0$ implies

$$u^1(x^1, x^2) = u^2(x^2, x^1) = V(x^2) .$$

We then conclude

$$\partial G_u(\omega) = \int_Q g(u^1, V(x^2))(\varphi_{x^1} u^1_{x^2} - \varphi_{x^2} u^1_{x^1})\, dx .$$

Step 2. We now exhibit a form ω of the type $\omega = \varphi(x) g(y)\, dy^1$ with small support for which $\partial G_u(\omega) \neq 0$. Recall that by construction

$$\text{(10)} \qquad u^1(x^1, x^2) = \begin{cases} 1/2 & \text{on } (\delta, 1-\delta) \times (0, \delta) \\ 1/4 & \text{on } (\delta(\delta, 1-\delta)) \times (\delta(1-\delta), 1)) . \end{cases}$$

Denote by L the segment joining the points

$$a_\delta := (\delta(1-\delta), \delta(1 - \frac{\delta}{2})) \qquad b_\delta := (\delta, \delta(1 - \frac{\delta}{2}))$$

and set

$$\begin{aligned} c(\delta) &:= \min(\delta(1-2\delta), \delta^2/2) \\ \varphi(x) &:= \max(0, 1 - \frac{\operatorname{dist}(x, L)}{c(\delta)}) . \end{aligned}$$

Clearly $\varphi \in \operatorname{Lip}(Q)$, $\| \varphi \| \leq 1$, $\operatorname{spt} \varphi \subset (\delta^2, 1-\delta) \times (\delta(1-\delta), \delta)$ and, for each $x^1 \in (\delta(1-\delta), \delta)$, $\varphi(x^1, x^2)$ is constant in x^1, i.e.,

$$\varphi(x^1, x^2) = \rho(x^2)$$

with

$$\rho(t) := \max(0, \frac{|t - \delta(1-\delta/2)|}{c(\delta)}) \qquad t \in \mathbb{R} .$$

Hence for any $(x^1, x^2) \in (\delta(1-\delta), \delta) \times (\delta(1-\delta), \delta)$ we have

$$\text{(11)} \qquad \begin{aligned} \varphi_{x^1}(x^1, x^2) &= 0 \\ \varphi_{x^2}(x^1, x^2) &= \begin{cases} -\dfrac{1}{c(\delta)} & \text{if } x^2 > \delta(1-\delta/2) \\ \dfrac{1}{c(\delta)} & \text{if } x^2 < \delta(1-\delta/2) . \end{cases} \end{aligned}$$

On the other hand the interval $(\delta(1-\delta), \delta)$ is one of the intervals which occurs in the construction of Cantor–Vitali function, and

$$\{x \mid V(x) \in (V(\delta(1-\delta/2)), V(\delta))\}$$

has positive measure. In particular $V(\delta(1-\delta/2)) < V(\delta)$, and for any positive function $g : \mathbb{R} \to \mathbb{R}$ with support in $(V(\delta(1-\delta/2), V(\delta))$, for convenience with $g \leq 1$, we then have

$$\int_{\mathbb{R}} g(V(t))\, dt \ > \ 0 \ . \tag{12}$$

For $\omega = \varphi(x) g(y^2)\, dy^1$, with φ and g defined as above we then compute, taking into account (10), (11) and (12),

$$\begin{aligned}
\partial G_u(\omega) &= \int_Q g(V(x^2))(\varphi_{x^1} u^1_{x^2} - \varphi_{x^2} u^1_{x^1})\, dx^1\, dx^2 \\
&= \frac{1}{c(\delta)} \int_{\delta(1-\delta/2)}^{\delta} dx^2 \int_{\delta(1-\delta)}^{\delta} g(V(x^2) u^1_{x^1}\, dx^1 \\
&= \frac{1}{4c(\delta)} \int_{\delta(1-\delta/2)}^{\delta} g(V(x^2))\, dx^2 \ > \ 0 \ ,
\end{aligned}$$

since for almost all $x^2 \in (\delta(1-\delta/2), \delta)$

$$\int_{\delta(1-\delta)}^{\delta} u^1_{x^1}\, dx^1 = u^1(\delta, x^2) - u^1(\delta(1-\delta), x^2) = \frac{1}{2} - \frac{1}{4} = \frac{1}{4} \ .$$

Step 3. Let φ, g, ω be as in Step 2. Consider the self-similarity transformations

$$\bar{x}^1 := \sigma(1-\delta) + \delta x^1 \ , \qquad \bar{x}^2 = \delta x^2 \ , \qquad \sigma = 0, 1$$

and set

$$\varphi_\sigma(\bar{x}^1, \bar{x}^2) := \varphi(x^1, x^2) \ , \quad g_\sigma(y) = g(y/\delta) \ , \quad \omega_\sigma := \varphi_\sigma(\bar{x}^1, \bar{x}^2) g_\sigma(y)\, dy^1 \ .$$

From

$$f(\bar{x}^1, \bar{x}^2) \ = \ \sigma\frac{1}{2} + \frac{1}{2} f(x^1, x^2)$$

we deduce

$$Df(\bar{x}^1, \bar{x}^2) \ = \ \frac{1}{2\delta} Df(x^1, x^2) \ .$$

Using $\delta C = C \cap [0, \delta]$ and $0 < x^2 < \delta$, we see that

$$d(\bar{x}^2, C) = d(\bar{x}^2, \delta C) = d(\delta x^2, \delta C) = \delta d(x^2, C) \ ,$$

thus

$$\begin{aligned} u^1_{\bar{x}^1}(\bar{x}^1,\bar{x}^2) &= f_{\bar{x}^1}(\bar{x}^1,\delta_0^{-1}d(\bar{x}^2,C)) = \frac{1}{2\delta}f_{x^1}(x^1,\delta^{-1}\delta_0^{-1}d(\bar{x}^2,C)) \\ &= \frac{1}{2\delta}f_{x^1}(x^1,\delta_0^{-1}d(x^2,C)) = \frac{1}{2\delta}u^1_{x^1}(x^1,x^2) \end{aligned}$$

and similarly

$$u^1_{\bar{x}^2}(\bar{x}^1,\bar{x}^2) = \frac{1}{2\delta}u^1_{x^2}(x^1,x^2) .$$

Therefore in both cases $\sigma = 0, 1$, taking into account $V(\bar{x}^2) = \delta V(\bar{x}^2)$, we get

$$\partial G_u(\omega_\sigma) = \int_Q g_\sigma(V(\bar{x}^2))\Big(\varphi_{\sigma,x^1}\, u^1_{x^2} - \varphi_{\sigma,x^2}\, u^1_{x^1}\Big)\, d\bar{x}^1\, d\bar{x}^2 = \frac{1}{2}\partial G_u(\omega) .$$

Moreover, notice that the supports of ω, ω_0, and ω_1 are disjoint.

Iterating the previous construction, i.e., setting $\varphi_{0,1} = \varphi$, $g_0 = g$ and

$$\varphi_{k+1,j}(x^1,x^2) := \begin{cases} \varphi_{k,j}\left(\dfrac{x^1}{\delta},\dfrac{x^2}{\delta}\right) & \text{if } j = 1,\ldots,2^k \\ \varphi_{k,j-2^k}\left(\dfrac{x^1-(1-\delta)}{\delta},\dfrac{x^2}{\delta}\right) & \text{if } j = 2^k+1,\ldots,2^{k+1} \end{cases}$$

$$g_{k+1}(y^2) := \delta g_k\left(\frac{y^2}{\delta}\right)$$

$$\omega_{k,j}(x,y) := \varphi_{k,j}(x) g_k(y^2)\, dy^1$$

we then obtain that the forms $\omega_{k,j}$ have disjoint supports for all k and j and moreover

$$\partial G_u(\omega_{k,j}) = 2^{-k}\partial G_u(\omega) \qquad \forall\, j = 1,\ldots,2^k .$$

Finally, defining for all $\bar{k} = 0, 1, \ldots$

$$\omega^{\bar{k}} := \sum_{k=0}^{\bar{k}}\sum_{j=1}^{2^k}\omega_{k,j}$$

we have $|\omega^{\bar{k}}| \leq 1$ and

$$\partial G_u(\omega^{\bar{k}}) = \sum_{k=0}^{\bar{k}}\sum_{j=1}^{2^k} 2^{-k}\partial G_u(\omega_{k,j}) = (\bar{k}+1)\partial G_u(\omega)$$

i.e. $\partial G_u(\omega^{\bar{k}}) \to \infty$ as $\bar{k} \to \infty$. □

•

[10] cart $\neq$ sw* $-$ lim$(\{G_u \mid u \in C^1(\Omega,\mathbb{R}^N) \cap \mathcal{A}^1(\Omega,\mathbb{R}^N)\})$. Let B^2 be the unit ball in $\mathbb{R}^2$, and let $T \in \text{cart}(B^2 \times \mathbb{R}^2)$ denote the Cartesian current

$$T := G_0 + S_T$$

where G_0 is the current carried by the graph of the null map $u \equiv 0$ and

$$S_T := [\![\partial B(0,r) \times \partial B(y_0,1)]\!]$$

where $0 < r < 1$, $y_0 \in \mathbb{R}^2$, $|y_0| > 2$. We claim that if $u_k : B^2 \to \mathbb{R}^2$ are smooth maps such that

$$G_{u_k} \rightharpoonup T \qquad \text{weakly in } \mathcal{D}_2(B^2 \times \mathbb{R}^2)$$

then

$$\sup_k \mathbf{M}(G_{u_k}) \geq \frac{\pi r}{4} \operatorname{dist}(\partial B(y_0,1),(0,0)_{\mathbb{R}^2_y}). \tag{13}$$

Let us postpone discussing this claim and let us first show how we can then construct a Cartesian current T for which there is no smooth weak approximation, i.e. a current T such that

$$T \in \operatorname{cart}(\Omega \times \mathbb{R}^N) \qquad T \notin \operatorname{sw}^* - \lim(\{G_u \mid u \in C^1(\Omega, \mathbb{R}^N) \cap \mathcal{A}^1(\Omega, \mathbb{R}^N)\}).$$

Consider in fact the current $T := G_0 + S_T$ where $\Omega := B(0,s)$, G_0 is as previously and S_T this time is defined by

$$S_T := \sum_{i=1}^{\infty} T_i, \qquad T_i := [\![\partial B(p_i, r_i) \times \partial B(y_i,1)]\!]$$

where $p_i := (2^{3-i},0)$, $r_i := 2^{-i}$, $y_i := (2^{i+1},0)$.

It is readily seen that T belongs to $\operatorname{cart}(\Omega \times \mathbb{R}^2)$. Actually, one easily sees that T belongs to $\operatorname{Cart}(\Omega \times \mathbb{R}^2)$. Set $U_i := B(p_i, 2^{1-i})$. Clearly $U_i \cap U_j = \emptyset$ and

$$T \llcorner U_i \times \mathbb{R}^2 = G_0 \llcorner U_i \times \mathbb{R}^2 + T_i$$

Suppose now that T be weakly approximable, i.e., assume that there exists a sequence of smooth maps $u_k : \Omega \to \mathbb{R}^2$ such that $G_{u_k} \rightharpoonup T$, then, in particular, we would have

$$\sup_k M(G_{u_k}) < \infty.$$

On the other hand by the claim above

$$\sup_k M(G_{u_k} \llcorner U_i \times \mathbb{R}^2) \geq \frac{\pi}{2}(2^{i+1} - 1)$$

and summing on i

$$\sup_k M(G_{u_k}) = +\infty,$$

a contradiction.

Returning to our initial claim, one easily convinces himself that it holds by considering the simpler example which follows. In $\mathbb{R} \times \mathbb{R}^2$ consider the Cartesian current

$$T := G_0 + [\![\partial B((0,d),1)]\!]$$

Clearly, for any sequence of smooth u_k for which $G_{u_k} \rightharpoonup T$, $\operatorname{spt} G_{u_k}$ must intersect $(-\varepsilon, \varepsilon)$ times a neighborhood of $\partial B((0,d),1)$, therefore

$$\int_{-\varepsilon}^{\varepsilon} |Du_k|\, dx \geq d$$

Consequently $\mathbf{M}(G_{u_k})$ is or the order of d for large d. •

2.6 Radial Currents

A map $u : B(0,1) \subset \mathbb{R}^n \to \mathbb{R}^n$ is said to be *radial* if it has the form

$$u(x) = \mathcal{U}(|x|)\,\frac{x}{|x|}$$

where $\mathcal{U} : (0,1) \to \mathbb{R}$ is a nonnegative function. In this case of course

$$\mathcal{U}(|x|) = |u(x)|\,.$$

The aim of this subsection is to discuss graphs of radial maps and their weak limits in $\operatorname{cart}(B(0,1) \times \mathbb{R}^n)$.

We begin by considering radial maps for which $\mathcal{U} \in C^0([0,1]) \cap C^1(0,1)$, $\mathcal{U} \geq 0$. Our first proposition provides formulas for computing the area and the boundary of their graphs. As usual we denote by ω_n the volume of the unit ball in $\mathbb{R}^n$ so that

$$\mathcal{H}^{n-1}(S^{n-1}) = n\,\omega_n\,.$$

Proposition 1. *Let u be a smooth radial map, possibly singular at zero, i.e.,*

$$u(x) = \mathcal{U}(|x|)\,\frac{x}{|x|}$$

where $\mathcal{U} \in C^0([0,1]) \cap C^1(0,1)$, $\mathcal{U} \geq 0$. Then we have

$$\text{(1)} \qquad \int_{B(0,1)} |u(x)|\, dx = n\,\omega_n \int_0^1 \rho^{n-1}\,\mathcal{U}(\rho)\, d\rho$$

$$\text{(2)} \qquad \mathbf{M}(G_u) = n\,\omega_n \int_0^1 (\rho^2 + \mathcal{U}^2(\rho))^{\frac{n-1}{2}}\,(1 + (\mathcal{U}'(\rho))^2)^{\frac{1}{2}}\, d\rho$$

$$\text{(3)} \qquad \partial G_u \llcorner B(0,1) \times \mathbb{R}^n = -\delta_0 \times \partial [\![\, B(0, \mathcal{U}(0)) \,]\!]$$

Proof. Equality (1) is trivial. To prove (2) we note that

$$D_i u^j = \frac{\mathcal{U}}{\rho}\,\delta_{ij} + (\mathcal{U}' - \frac{\mathcal{U}}{\rho})\,\frac{x_i x_j}{|x|^2}\,, \qquad \rho = |x|$$

so that the Jacobian matrix Du is symmetric with eigenvalues given by

$$\frac{\mathcal{U}(\rho)}{\rho},\dots,\frac{\mathcal{U}(\rho)}{\rho},\mathcal{U}'(\rho)\,.$$

Hence we can rotate coordinates by R in such a way that

$$R\,Du\,R^T \;=\; \operatorname{diag}(\mathcal{U}(\rho)/\rho,\dots,\mathcal{U}(\rho)/\rho,\mathcal{U}'(\rho))$$

and consequently

$$\begin{aligned} M(Du(x)) &= M(R\,Du\,R^T)\\ &= (\tilde{e}_1+\frac{\mathcal{U}(\rho)}{\rho}\,\tilde{\varepsilon}_1)\wedge\dots\wedge(\tilde{e}_{n-1}+\frac{\mathcal{U}(\rho)}{\rho}\,\tilde{\varepsilon}_{n-1})\wedge(\tilde{e}_n+\mathcal{U}'(\rho)\,\tilde{\varepsilon}_n)\end{aligned}$$

where $\tilde{e}_i := Re_i$, $\tilde{\varepsilon}_i = R\varepsilon_i$, $i=1,\dots,n$ and $\rho=|x|$. Since the vectors

$$\tilde{e}_i+\frac{\mathcal{U}(\rho)}{\rho}\,\tilde{\varepsilon}_i\;,\qquad i=1,\dots,n-1\;,\qquad\text{and}\;\;\tilde{e}_n+\mathcal{U}'(\rho)\,\tilde{\varepsilon}_n$$

are mutually orthogonal, we then infer

$$|M(Du)|^2 \;=\; \Big(1+\frac{\mathcal{U}^2(\rho)}{\rho^2}\Big)^{n-1}(1+(\mathcal{U}'(\rho))^2)$$

from which (2) follows taking into account the area formula.

Equality (3) follows as in [1] in Sec. 3.2.2, [2] in Sec. 3.2.2 or in [3] in Sec. 4.2.5 by changing the parametrization of the graph of u. □

[1] Notice that the condition $\mathbf{M}(G_u)<\infty$ does not imply that the length of the graph of $\mathcal{U}$ is finite. For example, if

$$\mathcal{U}(\rho) \;=\; \rho^2\sin\rho^{-\beta}$$

it is easily seen that $\int_0^1(1+(\mathcal{U}')^2)^{1/2}\,d\rho=\infty$ for $\beta>2$ while

$$\mathbf{M}(G_u) \;=\; n\,\omega_n\int_0^1(\rho^2+\mathcal{U}^2)^{\frac{n-1}{2}}(1+(\mathcal{U}')^2)^{1/2}\,d\rho\;<\;\infty\qquad\text{for }\beta<n+1\,.$$

Of course $\mathbf{M}(G_u)<\infty$ implies that the graph of $\mathcal{U}$ over $(\delta,1)$ has finite length for any $\delta>0$. Moreover, if $\mathcal{U}(0)>0$, we have for some positive δ

$$\rho^2+\mathcal{U}^2(\rho)\;>\;\delta^2+\frac{1}{2}\mathcal{U}^2(0)\,,$$

thus, in this case, $\mathbf{M}(G_u)<\infty$ if and only if the graph of $\mathcal{U}$ over $(0,1)$ has finite length. •

In order to characterize the graphs of smooth radial functions we introduce the Lipschitz map $h : \mathbb{R} \times \mathbb{R} \times \mathbb{R}^n \to \mathbb{R}^n \times \mathbb{R}^n$ given by

(4) $$h(\rho, r, z) := (\rho z, rz) .$$

Since h is linear in z, for fixed ρ and r, the tangent map $h_{*,(\rho,r,\theta)}$ at (ρ, r, θ) maps any vector of the type $(0, 0, v)$, $v \in \mathbb{R}^n$ into

$$h_{*,(\rho,r,\theta)}((0,0,v)) = (\rho v, rv)$$

hence

(5) $$|h_{*,(\rho,r,\theta)}((0,0,v))| = \sqrt{\rho^2 + r^2}\, |v|$$

and

(6) $$h_{*,(\rho,r,\theta)}((0,0,v)) \perp h_{*,(\rho,r,\theta)}((0,0,w)) ,$$

if v and w are orthogonal in $\mathbb{R}^n$. Moreover any vector of the type $\xi = (\xi_1, \xi_2, 0)$, $\xi_1, \xi_2 \in \mathbb{R}$, is mapped by $h_{*,(\rho,r,\theta)}$ to

$$h_{*,(\rho,r,\theta)}((\xi_1, \xi_2, 0)) = (\xi_1 \theta, \xi_2 \theta)$$

hence

(7) $$|h_{*,(\rho,r,\theta)}((\xi_1, \xi_2, 0))| = |\xi|\, |\theta|$$

and

(8) $$h_{*,(\rho,r,\theta)}((\xi_1, \xi_2, 0)) \perp h_{*,(\rho,r,\theta)}((0,0,v)) ,$$

if $\xi_1, \xi_2 \in \mathbb{R}$, and v is orthogonal to θ.

In terms of the map h we have

Proposition 2. *Let*

$$u(x) = \mathcal{U}(|x|) \frac{x}{|x|} , \qquad x \in B(0,1)$$

be a radial map with $\mathcal{U} \in C^0([0,1]) \cap C^1(0,1)$, $\mathcal{U} \geq 0$. *Then*

(9) $$h_\#(G_\mathcal{U} \times [\![S^{n-1}]\!]) = G_u \qquad \text{on } \mathcal{D}((\mathbb{R}^n \times \mathbb{R}^n) \setminus \{0,0\}) .$$

Moreover

(10) $$h_\#(\delta_0 \times [\![(0, \mathcal{U}(0))]\!] \times [\![S^{n-1}]\!]) = \delta_0 \times [\![B(0, \mathcal{U}(0))]\!] .$$

Proof. Denote by $e_\mathcal{U} = e_\mathcal{U}(\rho, \mathcal{U}(\rho))$ the tangent vector to the graph of $\mathcal{U}$ in $\mathbb{R}^2$

$$e_\mathcal{U} = \frac{(1, \mathcal{U}'(\rho))}{\sqrt{1 + (\mathcal{U}'(\rho))^2}}$$

and choose a basis $e_{\theta_1}, \ldots, e_{\theta_{n-1}}$ for the tangent plane to $S^{n-1} \subset \mathbb{R}^n$ at $\theta = (\theta_1, \ldots, \theta_n)$ so that the standard orientation of S^{n-1} is given by $e_\theta = e_{\theta_1} \wedge \ldots \wedge e_{\theta_{n-1}}$, i.e., so that

$$\theta \wedge e_{\theta_1} \wedge \ldots \wedge e_{\theta_{n-1}} = e_1 \wedge \ldots \wedge e_n$$

$e_1, \ldots, e_n$ being as usual the standard basis in $\mathbb{R}^n$. Finally we think of $\mathbb{R}^2$ and $\mathbb{R}^n$ as of the factors of $(\mathbb{R} \times \mathbb{R}) \times \mathbb{R}^n$.

The vectors

$$h_{*,(\rho,\mathcal{U}(\rho),\theta)}(e_{\mathcal{U}}) , \qquad h_{*,(\rho,\mathcal{U}(\rho),\theta)}(e_{\theta_i}) , \qquad i = 1, \ldots, n-1$$

generates the tangent plane to G_u at $(x, u(x))$, $x = \rho\theta$, and by (6) and (8) are mutually orthogonal. Also, from (5)) and (7) we infer

$$|h_{*,(\rho,\mathcal{U}(\rho),\theta)}(e_{\mathcal{U}} \wedge e_\theta)| = (\rho^2 + \mathcal{U}^2(\rho))^{\frac{n-1}{2}}$$

Let now ω be an n-form in $\mathcal{D}^n((\mathbb{R}^n \times \mathbb{R}^n) \setminus \{0,0\})$. We have, taking into account (2),

$$G_u(\omega) = \int <\omega, \vec{G}_u> d\mathcal{H}^n \llcorner \mathcal{G}_{u,B(0,1)}$$

$$\int_0^1 (1 + \mathcal{U}'^2)^{1/2} (\rho^2 + \mathcal{U}^2)^{\frac{n-1}{2}} \int_{S^{n-1}} < \omega, \frac{h_{*,(\rho,\mathcal{U}(\rho),\theta)}(e_{\mathcal{U}} \wedge e_\theta)}{|h_{*,(\rho,\mathcal{U}(\rho),\theta)}(e_{\mathcal{U}} \wedge e_\theta)|} > d\mathcal{H}^{n-1}(\theta)\, d\rho$$

$$= \int_0^1 (1 + \mathcal{U}'^2)^{1/2} \int_{S^{n-1}} <\omega, h_{*,(\rho,\mathcal{U}(\rho),\theta)}(e_{\mathcal{U}} \wedge e_\theta)> d\mathcal{H}^{n-1}(\theta)$$

$$= \int_0^1 (1 + \mathcal{U}'^2)^{1/2} \int_{S^{n-1}} <h^\# \omega, e_{\mathcal{U}} \wedge e_\theta> d\mathcal{H}^{n-1}(\theta) = G_{\mathcal{U}} \times [\![S^{n-1}]\!](h^\# \omega)$$

which proves (9). Equality (10) follows easily since for $t \in (0, \mathcal{U}(0))$, $\theta \in S^{n-1}$ we have $h(0, t, \theta) = (0, t\theta) \in \mathbb{R}^n \times \mathbb{R}^n$. □

Motivated by the above we now set

Definition 1. *An n-dimensional current* T *in* $\mathbb{R}^n \times \mathbb{R}^n$ *is said to be* radial *if there exists a* 1*-dimensional current* $L \in \mathcal{D}_1(\mathbb{R} \times \mathbb{R})$ *with support in* $\overline{\mathbb{R}}_+ \times \overline{\mathbb{R}}_+ \subset \mathbb{R} \times \mathbb{R}$, $\overline{\mathbb{R}}_+ := [0, +\infty)$, *such that*

$$T = h_\#(L \times [\![S^{n-1}]\!]) \qquad on \ \mathcal{D}^n((\mathbb{R}^n \times \mathbb{R}^n) \setminus \{0,0\}) .$$

By definition the support of a radial current T is contained in the image of the map h in (4)

$$\text{(11)} \ \operatorname{spt} T \subset \operatorname{Im} h = \{(x,y) \in \mathbb{R}^{2n} \mid x = 0 \text{ or } y = 0 \text{ or } \frac{x}{|x|} = \frac{y}{|y|}\} =: \Sigma .$$

Consider now the map $g : \mathbb{R}^n \times \mathbb{R}^n \setminus \{0,0\} \to ((\overline{\mathbb{R}}_+ \times \overline{\mathbb{R}}_+) \setminus \{0,0\}) \times \mathbb{R}^n$ given by

$$g(x,y) = (|x|, |y|, \frac{x+y}{|x|+|y|}) .$$

Clearly $g \in C^0((\mathbb{R}^n \times \mathbb{R}^n) \setminus \{0,0\})$ and

$$|Dg(x,y)| \le 2 + \frac{1}{|x|+|y|} \qquad \text{on } \mathbb{R}^n \times \mathbb{R}^n \setminus \{0,0\} .$$

When restricted to $\Sigma \setminus \{0,0\}$ the map g turns out to be the inverse of the map h with domain $((\overline{\mathbb{R}}_+ \times \overline{\mathbb{R}}_+) \setminus \{0,0\}) \times S^{n-1}$. Thus for any differential k-form ω in $\mathcal{D}^k(((\overline{\mathbb{R}}_+ \times \overline{\mathbb{R}}_+) \setminus \{0,0\}) \times S^{n-1})$ we have

$$h^\# g^\# \omega = \omega . \tag{12}$$

Finally notice that, if $(x,y) \in \Sigma \setminus \{0,0\}$ and $x \neq 0$, then $g(x,y)$ agrees with the map $g_1(x,y)$ given by

$$g_1(x,y) = (|x|, |y|, \frac{x}{|x|})$$

hence

$$\omega = h^\# g^\# \omega = h^\# g_1^\# \omega \tag{13}$$

for any $\omega \in \mathcal{D}^k((0,1) \times \overline{\mathbb{R}}_+ \times S^{n-1})$.

Given a radial current T, next proposition allows us to recover L from T.

Proposition 3. *Let* $T = h_\#(L \times [\![S^{n-1}]\!]) \in \mathcal{D}^n((\mathbb{R}^n \times \mathbb{R}^n) \setminus \{0,0\})$ *be a radial current with* $\operatorname{spt} L \subset \overline{\mathbb{R}}_+ \times \overline{\mathbb{R}}_+$. *Then*

$$L \times [\![S^{n-1}]\!] = g_\#(T) \qquad \text{on } \mathcal{D}^n(((\overline{\mathbb{R}}_+ \times \overline{\mathbb{R}}_+) \setminus \{0,0\}) \times S^{n-1}) . \tag{14}$$

In particular

$$L(\phi) = n^{-1} \omega_n^{-1} g_\# T(\phi \wedge \omega_{S^{n-1}}) \tag{15}$$

for all $\phi \in \mathcal{D}^1((\overline{\mathbb{R}}_+ \times \overline{\mathbb{R}}_+) \setminus \{0,0\})$, $\omega_{S^{n-1}}$ *being the volume form in* S^{n-1}.

Proof. Let $\omega \in \mathcal{D}^n((\overline{\mathbb{R}}^2_+ \setminus \{0,0\}) \times S^{n-1})$. Then $g^\# \omega \in \mathcal{D}^n(\mathbb{R}^{2n} \setminus \{0,0\})$ and by (12)

$$g_\# T(\omega) = T(g^\# \omega) = L \times [\![S^{n-1}]\!](h^\# g^\# \omega) = L \times [\![S^{n-1}]\!](\omega) .$$

This yield (14) at once. Applying (14) to $\omega = \phi \wedge \omega_{S^{n-1}}$ we infer

$$g_\# T(\phi \wedge \omega_{S^{n-1}}) = L(\phi) [\![S^{n-1}]\!] (\omega_{S^{n-1}}) = n \omega_n L(\phi)$$

which yields (15). □

In general we cannot say very much on the "restriction" of a radial current at zero. A characterization of Cartesian currents, which are radial, is instead given in the next theorem.

Theorem 1. *Let* $T \in \text{cart}(B(0,1) \times \mathbb{R}^n)$ *be a radial current,*

$$T = h_{\#}(L \times [\![S^{n-1}]\!]) \qquad \text{on } \mathcal{D}^n((B(0,1) \times \mathbb{R}^n) \setminus \{0,0\}) .$$

Then we have

(i) *The* BV*-function* u_T *associated to* T *is a radial function,*

$$u_T(x) = \mathcal{U}_T(|x|)\,\frac{x}{|x|}.$$

(ii) $\mathbf{M}(T) = n\,\omega_n \int_{\mathbb{R}_\rho \times \mathbb{R}_r} (\rho^2 + r^2)^{\frac{n-1}{2}}\, d\|L\|$

(iii) *Moreover,* $\mathcal{U}_T \in BV_{\text{loc}}(0,1)$, $\mathcal{U}_T^+(0) := \text{aplim}_{\rho\to 0^+} \mathcal{U}_T(\rho)$ *exists finite and*

$$L = -\partial SG_{\mathcal{U}_T} + \delta_0 \times [\![(0, \mathcal{U}_T^+(0))]\!].$$

Consequently, $T = h_{\#}(-\partial SG_{\mathcal{U}_T} \times [\![S^{n-1}]\!]) + \delta_0 \times [\![B(0, \mathcal{U}_T^+(0))]\!]$.

(iv) *If* $T = G_u$, $u \in \text{cart}^1(B(0,1), \mathbb{R}^N)$, *then* $\mathcal{U}_T^+(0) = 0$.

Proof. We split the proof into several steps.
Step 1. Let Q be a rotation in $\mathbb{R}^n$, $Q \in SO(n)$. Taking into account the form of h and that $\text{spt}\, T \subset \Sigma$ we easily see that in $\mathbb{R}^{2n} \setminus \{0,0\}$

$$\begin{aligned}(Q \bowtie Q)_{\#}T &= (Q \bowtie Q)_{\#}h_{\#}(L \times [\![S^{n-1}]\!]) = h_{\#}(L \times Q_{\#}[\![S^{n-1}]\!]) \\ &= h_{\#}(L \times [\![S^{n-1}]\!]) = T ,\end{aligned}$$

that is, T is $Q \bowtie Q$-invariant. Let u_T be the BV-function associated to $T = \tau(\mathcal{M}, \vartheta, \vec{T})$. We know that $(x, u_T(x)) \in \mathcal{M}$ for a.e. x, hence

$$u_T(x) = |u_T(x)|\,\frac{x}{|x|}$$

as $\mathcal{M} \subset \Sigma$. Finally, by the rotational invariance of T,

$$\begin{aligned}\int |u_T(x)| \varphi(x)\, dx &= T(|y|\varphi(x)\, dx) = (Q \bowtie Q)_{\#}T(|y|\varphi(x)\, dx) \\ &= T(|y|\varphi(Qx)\, dx) = \int |u_T(x)|\, \varphi(Qx)\, dx \\ &= \int |u_T(Q^{-1}x)|\, \varphi(x)\, dx\end{aligned}$$

holds for any $\varphi \in C_c^\infty(B(0,1) \setminus \{0\})$, and actually for any $\varphi \in C_c^\infty(B(0,1))$. Therefore

$$u_T(x) = \mathcal{U}_T(|x|)\,\frac{x}{|x|}$$

for some $\mathcal{U}_T \geq 0$. This proves (i).
Step 2. L is locally i.m. rectifiable in $\overline{\mathbb{R}}^2_+ \setminus \{0,0\}$. In fact, being T of finite mass and i.m. rectifiable, and being g locally Lipschitz on $\mathbb{R}^{2n} \setminus \{0,0\}$, it

follows from Proposition 3 that $L \llcorner (\mathbb{R}^2 \setminus B(0,\varepsilon)) \times [\![S^{n-1}]\!]$, and consequently $L \llcorner (\mathbb{R}^2 \setminus B(0,\varepsilon))$ is i.m. rectifiable for any $\varepsilon > 0$.
Step 3. We now prove (ii). Denote by $(e_L \wedge e_\theta)_{(\rho,r,\theta)}$ the tangent n-vector to $L \times [\![S^{n-1}]\!]$ in $\mathbb{R}^2 \times S^{n-1}$ at (ρ, r, θ), $\rho + r \neq 0$. Since $T = h_\#(L \times [\![S^{n-1}]\!])$, the n-vector

$$h_{*,(\rho,r,\theta)}(e_L \wedge e_\theta)$$

generates the tangent plane to $T = h_\#(L \times [\![S^{n-1}]\!])$ at $(\rho\theta, r\theta)$, compare the proof of Proposition 1. By (5), (7), (6), (8) we have

$$|h_{*,(\rho,r,\theta)}(e_L \wedge e_\theta)| \;=\; (\rho^2 + r^2)^{\frac{n-1}{2}} \, |e_L| \, |e_\theta| \;=\; (\rho^2 + r^2)^{\frac{n-1}{2}} \, .$$

Therefore

$$\begin{aligned} T(\omega) &= L \times [\![S^{n-1}]\!](h^\#(\omega)) \;=\; \int < h^\# \omega, e_L \wedge e_\theta > d\| \, L \times [\![S^{n-1}]\!] \, \| \\ &= \int (\rho^2 + r^2)^{\frac{n-1}{2}} \Big\{ \int_{S^{n-1}} < \omega, \vec{T} >_{| \, h(\rho,r,\theta)} d\mathcal{H}^{n-1}(\theta) \Big\} \, d\| \, L \, \| \, . \end{aligned}$$

taking the supremum for $\omega \in \mathcal{D}^n(\mathbb{R}^{2n} \setminus \{0,0\})$, $|\omega| \leq 1$, we then find

$$\mathbf{M}(T) \;=\; \mathbf{M}(T \llcorner (B(0,1) \setminus \{0\}) \times \mathbb{R}^n) \;\leq\; n\,\omega_n \int (\rho^2 + r^2)^{\frac{n-1}{2}} \, d\| \, L \, \| \, .$$

To prove the opposite inequality we use the characterization in Proposition 3

$$L \times [\![S^{n-1}]\!](\omega) \;=\; T(g^\# \omega) \, .$$

We choose $\omega \in \mathcal{D}^n((\overline{\mathbb{R}}^2_+ \setminus \{0,0\}) \times S^{n-1})$ with $|\omega(\rho, r, \theta)| \leq (\rho^2 + r^2)^{\frac{n-1}{2}}$. Then

$$\begin{aligned} L \times [\![S^{n-1}]\!](\omega) &= \int < g^\# \omega, \vec{T} > d\| \, T \, \| \\ &= \int < \omega, e_L \wedge e_\theta >_{| \, g(x,y)} (|x|^2 + |y|^2)^{\frac{1-n}{2}} \, d\| \, T \, \| \;\leq\; \int d\| \, T \, \| . \end{aligned}$$

Taking the supremum in ω we then infer

$$n\,\omega_n \int (\rho^2 + r^2)^{\frac{n-1}{2}} \, d\| \, L \, \| \;\leq\; \mathbf{M}(T) \, .$$

Step 4. Set

$$L_\varepsilon \;:=\; L \llcorner ((\varepsilon, 1) \times \mathbb{R}) \, , \qquad \varepsilon > 0 \, .$$

We shall prove that $L_\varepsilon \in \operatorname{cart}((\varepsilon, 1) \times \mathbb{R})$ for every $\varepsilon > 0$.

From step 2 we know that L_ε is i.m. rectifiable, $\mathbf{M}(L_\varepsilon) < \infty$. As a consequence of Proposition 3 we also see that $\partial L_\varepsilon = 0$ in $(\varepsilon, 1) \times \mathbb{R}$. Consider now the 1-form $\phi := \varphi(\rho, r) \, d\rho$ in $\mathcal{D}^1((0,1) \times \mathbb{R})$. By (13)

$$\phi\wedge\omega_{S^{n-1}} = h^\# g_1^\#(\phi\wedge\omega_{S^{n-1}}) .$$

As

$$g_1^\#(\phi) = \varphi(|x|,|y|)\, d|x| \quad \text{and} \quad g_1^\#(\omega_{S^{n-1}}) = \sum_{i=1}^n (-1)^i \frac{x^i}{|x|}\, \widehat{dx^i} ,$$

we deduce

$$g_1^\#(\phi\wedge\omega_{S^{n-1}}) = \frac{\varphi(|x|,|y|)}{|x|^{n-1}}\, dx ,$$

hence

$$(16) \qquad n\,\omega_n\, L(\phi) = L \times [\![S^{n-1}]\!](\phi\wedge\omega_{S^{n-1}}) = T\Big(\frac{\varphi(|x|,|y|)}{|x|^{n-1}}\, dx\Big) .$$

This implies that

$$L(\phi) = L^{\bar{0}0}(\varphi) \geq 0 \qquad \text{if } \varphi \geq 0$$

and, choosing $\phi = \varphi(\rho)\, d\rho$

$$L(\phi) = T\Big(\frac{\varphi(|x|)}{|x|^{n-1}}\Big) = \int_0^1 \varphi(\rho)\, d\rho$$

i.e. $\pi_\# L = [\![(\varepsilon,1)]\!]$, π denoting the projection of $(\varepsilon,1)\times\mathbb{R}$ into $(\varepsilon,1)$. As finally (16) implies obviously that the L^1-norm of L is finite in $(\varepsilon,1)\times\mathbb{R}$, we conclude that $L_\varepsilon \in \operatorname{cart}((\varepsilon,1)\times\mathbb{R})$.

Step 5. From step 4 and the results in Sec. 4.2.4 it follows that $L = -\partial SG_{\mathcal{U}}$ for some $\mathcal{U} \in BV_{\text{loc}}((0,1),\mathbb{R})$. We now prove that

$$\mathcal{U} = \mathcal{U}_T \qquad \text{in } (0,1) .$$

This follows from the equalities below which are valid for any $\varphi \in C^1_c(0,1)$

$$\begin{aligned} \int_0^1 \mathcal{U}(\rho)\varphi(\rho)\, d\rho &= L(\varphi(\rho)\, r\, d\rho) = n^{-1}\,\omega_n^{-1}\, T\Big(\frac{\varphi(|x|)}{|x|^{n-1}}|y|\, dx\Big) \\ &= n^{-1}\,\omega_n^{-1} \int_{B(0,1)} |u_T(x)|\, \frac{\varphi(|x|)}{|x|^{n-1}}\, dx \\ &= n^{-1}\,\omega_n^{-1} \int_{B(0,1)} \mathcal{U}_T(|x|)\, \frac{\varphi(|x|)}{|x|^{n-1}}\, dx = \int_0^1 \mathcal{U}_T(\rho)\,\varphi(\rho)\, d\rho . \end{aligned}$$

Step 6. Finally let us prove that

$$L = \delta_0 \times [\![(0,\mathcal{U}_+(0))]\!] - \partial SG_{\mathcal{U}} \qquad \text{in } \overline{\mathbb{R}}^2_+ .$$

Consider the map $\alpha : \overline{\mathbb{R}}^2_+ \to \overline{\mathbb{R}}^2_+$ given by

$$(\rho, r) \longrightarrow (\rho, (\rho + r)^n) .$$

It has range

$$\{(\rho, s) \in \overline{\mathbb{R}}^2_+ \mid s > \rho^n\}$$

and inverse β given by

$$\beta(\rho, s) := (\rho, s^{1/n} - \rho) .$$

Clearly β is Lipschitz continuous away from $(0,0)$. Consider now the map

$$V : (0,1) \longrightarrow \mathbb{R} , \qquad V(\rho) := (\rho + \mathcal{U}(\rho))^n .$$

We claim that $V \in BV(0,1)$. In fact $V \in BV_{\text{loc}}(0,1)$ as $\mathcal{U} \in BV_{\text{loc}}(0,1)$. Moreover, for any 1-form $\omega(\rho, r)$ we have

$$|\alpha^{\#}\omega(\rho, r)| \leq n\,(\rho + r)^{n-1}\,|\omega(\alpha(\rho, r))| ,$$

hence

$$\mathbf{M}(\partial^- SG_V) \leq n \int_0^1 (\rho + r)^{n-1}\, d\mathcal{H}^1 \llcorner \partial^- SG_{\mathcal{U}} < +\infty$$

by step 2.

From the trace theory, the current

$$\delta_0 \times [\![(0, V_+(0))]\!] - \partial SG_V$$

has boundary $-\delta_{(0,0)}$ in $[0,1) \times \mathbb{R}$ and

$$V_+(0) = \operatorname*{aplim}_{\rho \to 0^+} V(\rho) = \operatorname*{aplim}_{\rho \to 0} (\rho + \mathcal{U}(\rho))^n = \mathcal{U}_+^n(0) .$$

Therefore the current

$$\delta_0 \times [\![(0, U_+(0))]\!] - \partial SG_{\mathcal{U}} = \beta_{\#}(\delta_0 \times [\![(0, V_+(0))]\!] - \partial SG_V)$$

has no boundary in $\mathbb{R}^2 \setminus \{0,0\}$. It follows that

$$\widetilde{L} := L - \delta_0 \times [\![(0, \mathcal{U}_+(0))]\!] + \partial SG_{\mathcal{U}}$$

has no boundary in $\mathbb{R}^2 \setminus \{0,0\}$ and by step 5 has support in $\{0\} \times \mathbb{R}$. By the constancy theorem, compare Sec. 4.3.1, we then infer that

$$L = c\,\delta_0 \times \mathbb{R}_+ + \delta_0 \times [\![(0, \mathcal{U}_+(0))]\!] - \partial SG_{\mathcal{U}} \qquad \text{in } \overline{\mathbb{R}}^2_+ \setminus \{0,0\}$$

for some $c \in \mathbb{R}$. Since L has finite mass outside $B((0,0), \varepsilon)$, we finally infer that $c = 0$, which concludes the proof of (iii).

Finally, under the assumptions of (iv), we have

$$[\![\{0\} \times B(0, \mathcal{U}_T^+(0))]\!] = G_u \llcorner \{0\} \times \mathbb{R}^n = 0$$

hence $\mathcal{U}_T^+(0) = 0$. □

An immediate consequence of the definition of radial currents and of Proposition 3 is the following

Theorem 2 (Closure theorem). *Let $\{T_k\}$ be a sequence of radial currents in* $\operatorname{cart}(B(0,1)\times\mathbb{R})$, *suppose that T_k converge weakly in* $\operatorname{cart}(B(0,1)\times\mathbb{R}^n)$ *to T. Then T is a radial Cartesian current.*

Proof. In fact from

$$T_k = h_\#(L_k \times [\![S^{n-1}]\!]) \qquad \text{on } \mathcal{D}^n((\mathbb{R}^n\times\mathbb{R}^n)\setminus\{0,0\})$$

and Proposition 3 we infer that $L_k \rightharpoonup L$ as currents in $\mathbb{R}^2\setminus\{0,0\}$. Then, it follows at once that

$$T = h_\#(L \times [\![S^{n-1}]\!]) \qquad \text{on } \mathcal{D}^n((\mathbb{R}^n\times\mathbb{R}^n)\setminus\{0,0\})\,.$$

□

We conclude this subsection by proving the following *approximation theorem* for radial currents.

Theorem 3. *Let $T\in \operatorname{cart}(B(0,1)\times\mathbb{R}^n)$ be a radial current*

$$T = h_\#(L \times [\![S^{n-1}]\!]) \qquad \textit{on} \qquad \mathcal{D}^n(\mathbb{R}^n\times\mathbb{R}^n\setminus\{0,0\})\,.$$

Then there exists a sequence of smooth radial functions $u_k(x)=\mathcal{U}_k(|x|)\,\frac{x}{|x|}$ with $\mathcal{U}_k\in C^1([0,1])$, $\mathcal{U}_k(0)=0$ such that

$$G_{u_k} \rightharpoonup T \quad \textit{and} \quad \mathbf{M}(G_{u_k}) \to \mathbf{M}(T)$$

Proof. From Theorem 1 we know that

$$L = -\partial SG_{u_T} + \delta_0 \times [\![(0,\mathcal{U}_T^+(0))]\!]\,,$$

$\mathcal{U}_T\in BV_{\text{loc}}(0,1)$ and $\mathcal{U}_T^+(\rho)=\mathcal{U}_T^-(\rho)$, $\lim_{\rho\to 0}\mathcal{U}_T^+(\rho)=\mathcal{U}_T^+(0)$ for ρ in $(0,1)\setminus E$ with $|E|=0$. Therefore we can find a sequence $\varepsilon_k\downarrow 0$ such that $\mathcal{U}_T\in BV((\varepsilon_{k/2},1))$,

$$\mathcal{U}_T^+(\varepsilon_k)=\mathcal{U}_T^-(\varepsilon_k)\,, \qquad \mathcal{U}_T^+(\varepsilon_k) \to \mathcal{U}_T^+(0)$$

and

$$(17) \qquad \|T\|((B(0,\varepsilon_k)\setminus\{0\})\times\mathbb{R}^n) = n\,\omega_n \int_{(0,\varepsilon_k)\times\mathbb{R}} (\rho^2+r^2)^{\frac{n-1}{2}}\, d\|\partial SG_{\mathcal{U}_T}\| \to 0\,,$$

and, by Proposition 3 in Sec. 4.2.4, a sequence of functions $\widetilde{\mathcal{U}}_j\in C^1((\varepsilon_{k/2},1))$ such that

$$-\partial SG_{\widetilde{\mathcal{U}}_j} \rightharpoonup -\partial SG_{\mathcal{U}_T} \quad \text{in } (\varepsilon_{k/2}, 1) \times \mathbb{R}$$
$$\| \partial SG_{\widetilde{\mathcal{U}}_j} \|((\varepsilon_{k/2}, 1) \times \mathbb{R}) \to \| \partial SG_{\mathcal{U}_T} \|(\varepsilon_{k/2}, 1) \times \mathbb{R}) \,.$$

Since $\mathcal{U}_T^+(\varepsilon_k) = \mathcal{U}_T^-(\varepsilon_k)$ we can also assume, compare Note 4 in Sec. 4.4,

(18) $$\widetilde{\mathcal{U}}_j(\varepsilon_k) \longrightarrow \mathcal{U}_T^+(\varepsilon_k) \quad \text{as } j \to \infty$$

(19) $$\int_{(\varepsilon_k,1)} \varphi \, d\| \partial SG_{\widetilde{\mathcal{U}}_j} \| \longrightarrow \int_{(\varepsilon_k,1)} \varphi \, d\| \partial SG_{\mathcal{U}_T} \| \quad \text{as } j \to \infty$$

for any $\varphi \in C_c^0(\mathbb{R}^2)$.

Using (18) and then (18) (19) we see that we can choose a subsequence of ε_k (still denoted by ε_k) and a subsequence $\{j_k\}$ such that, setting

$$\mathcal{U}_k(\rho) := \begin{cases} \widetilde{\mathcal{U}}_{j_k}(\rho) & \rho \geq \varepsilon_k \\ \dfrac{\widetilde{\mathcal{U}}_{j_k}(\varepsilon_k)}{\varepsilon_k}\rho & 0 < \rho < \varepsilon_k \end{cases}$$

and

$$u_k(x) := \mathcal{U}_k(|x|) \frac{x}{|x|} \qquad x \in B(0,1) \,,$$

we have

(20) $$\| G_{u_k} \|((B(0,1) \setminus B(0,\varepsilon_k)) \times \mathbb{R}^n) \to n\,\omega_n \int_{(0,1)} (\rho^2 + r^2)^{\frac{n-1}{2}} \, d\| \partial SG_{u_T} \|$$

and

(21) $$\mathcal{U}_k(\varepsilon_k) \longrightarrow \mathcal{U}_T^+(0) \,.$$

By a direct computation we then see that

$$\begin{aligned} &\| G_{u_k} \| (B(0,\varepsilon_k) \times \mathbb{R}^n) \\ &= n\,\omega_n \int_0^{\varepsilon_k} (\rho^2 + \mathcal{U}_k^2(\rho))^{\frac{n-1}{2}} (1 + \frac{\mathcal{U}_k^2(\varepsilon_k)}{\varepsilon_k^2})^{1/2} \, d\rho \\ &= n\,\omega_n \left(1 + \frac{\mathcal{U}_k^2(\varepsilon_k)}{\varepsilon_k^2}\right)^{\frac{n}{2}} \int_0^{\varepsilon_k} \rho^{n-1} \, d\rho \end{aligned}$$

converge to

$$\omega_n \, (\mathcal{U}_T^+(0))^n = \mathcal{H}^n(B(0, \mathcal{U}_T^+(0)) \,.$$

The result then follows taking into account (20), since by Theorem 1

$$\mathbf{M}(T) = n\,\omega_n \int_{(0,1)\times\mathbb{R}} (\rho^2 + r^2)^{\frac{n-1}{2}} \, d\| \partial SG_{u_T} \| + \mathcal{H}^n(B(0, \mathcal{U}_T^+(0))$$

and by construction trivially $G_{u_k} \rightharpoonup T$. □

3 Degree Theory

In this section we begin by discussing in Sec. 4.3.1 n-dimensional currents in $\mathbb{R}^n$. A key result we shall present is the so-called *constancy theorem* which states that any n-dimensional current T in an open and connected set of $\mathbb{R}^n$ satisfying $\partial T \llcorner \Omega = 0$ is a real, and in fact an integer, in case T is also i.m. rectifiable, multiple of the current integration over Ω, i.e.,

$$T = c [\![\Omega]\!] .$$

Such a result may be regarded as the formulation in terms of currents of the well-known fact that a distribution with zero derivatives is constant.

We shall then see that n-dimensional *normal* currents in $\mathbb{R}^n$, that is, currents T such that

$$\mathbf{M}(T) + \mathbf{M}(\partial T) \ < \ \infty$$

may be *identified* to BV-functions, and that n-dimensional normal currents in $\mathbb{R}^{n+N}$ cannot *concentrate*, i.e. have masses, *on closed sets of dimension smaller than n*. We finally conclude Sec. 4.3.1 discussing some relations between rectifiability of T and of ∂T.

The result of Sec. 4.3.1 are then used in Sec. 4.3.2 where we shall briefly develop a *degree theory for currents.* In fact we shall see that, in some sense, the natural context in which degree theory lives is that of currents, and of Cartesian currents, compare also Ch. 5.

Let $T \in \text{cart}(\Omega \times \widehat{\mathbb{R}}^n)$, $\Omega \subset \mathbb{R}^n$, be a Cartesian current and let $A \subset \Omega$ be a Borel set. Denote by $\widehat{\pi} : \Omega \times \mathbb{R}^n \to \mathbb{R}^n$ the linear projection onto the second factor, and consider the current projection of $T \llcorner \pi^{-1}(A)$, $\widehat{\pi}_\#(T \llcorner \pi^{-1}(A))$ and the open set

$$\Gamma_{T,A} \ := \ \widehat{\mathbb{R}}^n \setminus \text{spt}\, \partial \widehat{\pi}_\#(T \llcorner \pi^{-1}(A)) \ .$$

For any $y \in \Gamma_{T,A}$ consider also its connected component C_y in $\Gamma_{T,A}$. Of course $\widehat{\pi}_\#(T \llcorner \pi^{-1}(A))$ has no boundary in C_y, thus by the constancy theorem

$$\widehat{\pi}_\#(T \llcorner \pi^{-1}(A)) \ = \ m [\![C_y]\!] \ , \qquad m \in \mathbb{Z} \ .$$

Clearly, if $T = G_u$, u being a smooth map from Ω into $\widehat{\mathbb{R}}^n$, then m is the *classical degree of u at y with respect to A*

$$m \ = \ \deg (u, A, y) \ .$$

Such an approach will be pursued in Sec. 4.3.3 where we shall see in which sense the image of an integral current splits into its connected components.

More generally in Sec. 4.3.2 we shall see that the *notion of degree* is build into the area formula and the projection mechanism. For any normal current T of dimension n in $\mathbb{R}^k \times \widehat{\mathbb{R}}^n$, k not necessarily equal n, a *degree mapping of T with respect to the projection $\widehat{\pi}$*, counting with orientation how many times $y \in \widehat{\mathbb{R}}^n$ is covered is well-defined as an integer-valued BV function. It enjoys all properties of the classical degree and agrees with it if $T = G_u$, $k = n$ and u is smooth. Moreover, while as we have seen the degree is not continuous with respect to the weak convergence in Sobolev spaces, one instead sees that the degree mapping is *continuous* with respect to the weak convergence of normal currents provided their supports are equibounded. We shall then see that those results when read for Cartesian currents lead to state that the weak convergence of Cartesian currents preserves the degree.

We shall conclude Sec. 4.3.2 by showing how one can easily recover the classical notion of degree for merely continuous maps from the notion of degree mapping.

3.1 n-Dimensional Currents and BV Functions

It is usual to regard functions, say smooth functions from an open set Ω of $\mathbb{R}^n$ into $\mathbb{R}$, as distributions in $\mathcal{D}'(\Omega)$ by identifying u to the distribution

$$\mu_u(\varphi) := \int \varphi(x)\, u(x)\, dx \,, \qquad \varphi \in C_c^\infty(\Omega) \,.$$

Such an identification turns out to be convenient in many instances; however it depends on the choice of the basis in $\mathbb{R}^n$ as it is clearly shown by the formula for changing variables in integrals. From this point of view it is better to work in the dual context of n-forms and identify functions u as n-dimensional currents acting on n-forms with compact support in Ω. More precisely we identify u to the n-dimensional current in Ω, $T_u := [\![\mathbb{R}^n]\!] \llcorner u$, given by

$$\text{(1)} \qquad T_u(\omega) := [\![\mathbb{R}^n]\!] \llcorner u(\omega) = \int <\omega(x), e_1 \wedge \ldots \wedge e_n> u(x)\, dx$$

or if $\omega(x) = f(x)\, dx^1 \wedge \ldots \wedge dx^n$

$$T_u(\omega) := [\![\mathbb{R}^n]\!] \llcorner u(\omega) = \int f(x)\, u(x)\, dx \,.$$

Since for any change of variables $\phi : \mathbb{R}^n \to \mathbb{R}^n$ we have $\phi^{\#}(\omega)(x) = f(\phi(x)) \det D\phi(x) dx^1 \wedge \ldots \wedge dx^n$, the current T_u is well defined and independent from changes of coordinates which preserve the orientation. Of course, if we fix a basis $\mathbb{R}^n$, we have a one-to-one correspondence between distributions and n-currents given by

$$T_u(\omega) = \mu_u(<\omega(x), e_1 \wedge \ldots \wedge e_n>) \,.$$

From now on this correspondence will be understood.

Next we note that the boundary of T_u, ∂T_u, is clearly related to the distributional derivatives of u. In fact, if ω is an $(n-1)$-form written as

$$\omega(x) = \sum_{i=1}^{n} (-1)^{i-1} f_i\, \widehat{dx_i} \,, \qquad f_i \in C_c^\infty(\Omega) \,,$$

we then have

$$\partial([\![\mathbb{R}^n]\!] \llcorner u)(\omega) = [\![\mathbb{R}^n]\!] \llcorner u(d\omega) = \int_\Omega \sum_{i=1}^n D_i f_i(x)\, u(x)\, dx$$

$$(2) \qquad = -\sum_{i=1}^n \mu_{Du^i}(f_i) = -\sum_{i=1}^n < D_i u, f_i > \ .$$

A basic theorem in the theory of distributions states that any distribution μ on an open set Ω with zero distributional derivatives is constant, i.e., has the form

$$\varphi \longrightarrow c \int \varphi\, dx \ , \qquad c \in \mathbb{R}$$

on each connected component of Ω. In the dual setting of n-dimensional currents this amounts to the following simple but important theorem.

Theorem 1 (Constancy theorem). *Let $T \in \mathcal{D}_n(\Omega)$ be an n-dimensional current in an open set Ω of $\mathbb{R}^n$, and let U be an open set in Ω. Suppose that*

$$\partial T \llcorner U = 0$$

or equivalently $\operatorname{spt} \partial T \subset \overline{\Omega} \setminus U$. *Then for each connected component U_i of U there exists a constant c_i such that*

$$T = \sum_i c_i\, [\![U_i]\!] \qquad \textit{in } U\ .$$

In other words, for any $f \in C_c^\infty(U)$ we have

$$T(f(x)\, dx) = \int f(x)\, c(x)\, dx$$

where $c(x)$ is piecewise constant, $c(x) = c_i$ on U_i.

Proof. We fix an open set $W \subset\subset U$, choose $\sigma < \operatorname{dist}(W, \partial U)$ and a mollifier φ, and we let $\varphi_\sigma := \sigma^{-n}\varphi(x/\sigma)$. For any form $\omega = f(x)dx$ in $\mathcal{D}^n(W)$ the form $\omega_\sigma(x) := f * \varphi_\sigma\, dx$ belongs to $\mathcal{D}^n(U)$, hence we can define the current T_σ in $\mathcal{D}^n(W)$ by

$$T_\sigma(\omega) := T(\omega_\sigma)\ .$$

Since $\omega_\sigma \to \omega$ uniformly together with all derivatives as $\sigma \to 0$, we have

$$(3) \qquad T_\sigma \rightharpoonup T \qquad \text{in } \mathcal{D}_n(W)\ .$$

Also, for every σ we can find a constant c depending on σ such that

$$\| f * \varphi_\sigma \|_{\infty, U} \le c \int_U |f|\, dx\ ,$$

hence

$$|T_\sigma(\omega)| \le c_\sigma \int_U |\omega|\, dx \,. \tag{4}$$

Consequently T_σ extends to a linear continuous functional on forms with $L^p(W)$ coefficients for any $p \ge 1$. By Riesz theorem, compare Ch. 1, then there exists a Radón measure μ_σ on W and a unit n-vector $\nu_\sigma(x)$ which is μ_σ-measurable such that

$$T_\sigma(\omega) = \int <\omega, \nu_\sigma> d\mu_\sigma \,.$$

By (4) μ_σ is absolutely continuous with respect to Lebesgue measure, thus we can write

$$T_\sigma(\omega) = \int f \cdot \theta_\sigma \, d\mathcal{H}^n \,. \tag{5}$$

On the other hand by assumption we have for any $\omega \in \mathcal{D}^{n-1}(U)$ with $\operatorname{spt}\omega \subset W$

$$T_\sigma(d\omega) = \partial T(\omega_\sigma) = 0 \,.$$

Therefore, taking $\omega := f\widehat{dx^j}$ so that $d\omega = (-1)^{j-1} D_j f\, dx$ and using (5), we find

$$\int D_j f \cdot \theta_\sigma \, d\mathcal{H}^n = 0 \qquad j = 1, \dots, n$$

for $f \in C_c^\infty(U)$ with $\operatorname{spt} f \subset W$. This evidently implies that θ_σ =constant depending on σ on each component of W. The result then follows from (5) by letting $\sigma \downarrow 0$ and $W \uparrow U$ if we take into account (3). □

By inspecting the previous proof one realizes that in fact we have the following more general result

Theorem 2. *Let $T \in \mathcal{D}_n(\Omega)$ be an n-dimensional current in a bounded open set $\Omega \subset \mathbb{R}^n$. Suppose that*

$$\mathbf{M}_\Omega(\partial T) < +\infty \,.$$

Then there exists a function $u \in BV_{\text{loc}}(\Omega)$ such that

$$T(f(x)\, dx) = \int_\Omega f(x)\, u(x)\, dx \qquad \forall\, f \in C_c^\infty(\Omega) \,.$$

Moreover

$$\| \partial T \|(A) = |Du|(A) \tag{6}$$

for any open set $A \subset \Omega$.

Proof. We proceed as in the proof of Theorem 1 until we get (5). We now deduce from our assumptions, using $\| f * \varphi_\sigma \|_\infty \le \| f \|_\infty$,

$$\Big| \int f(x)\, \theta_\sigma(x)\, dx \Big| \le c\mathbf{M}(T) \| f \|_{\infty, U}$$

and, for $i = 1, \dots, n$,

$$\Big| \int D_i f(x)\, \theta_\sigma(x)\, dx \Big| \;\le\; c\mathbf{M}(\partial T)\, \| f \|_{\infty,U}$$

where c is independent from σ, $f \in C_c^\infty(U)$, $\operatorname{spt} f \subset W$, $\sigma < \operatorname{dist}(\operatorname{spt} f, \partial U)$. From the compactness theorem for BV_{loc}-functions, we infer the existence of a sequence $\{\sigma_k\}$, $\sigma_k \to 0$, such that

$$\theta_{\sigma_k} \longrightarrow \theta \qquad \text{in } L^1_{\mathrm{loc}}(U)$$

where $\theta \in BV_{\mathrm{loc}}(U)$. From (3) we then get

$$T(f(x)\, dx) \;=\; \int f(x)\, \theta(x)\, dx \qquad \forall\, f \in C_c^\infty(\Omega)\, .$$

Choosing then $\omega = (-1)^{i-1} f_i \widehat{dx^i}$, $f_i \in C_c^\infty(\Omega)$, we finally infer

$$\partial T(\omega) \;=\; \int \operatorname{div} f(x)\, \theta(x)\, dx \tag{7}$$

from which equality (6) follows at once. □

An immediate corollary of Theorem 2 is the following characterization of *normal* n-dimensional currents in Ω

Corollary 1. *An n-dimensional current in $\Omega \subset \mathbb{R}^n$, $T \in \mathcal{D}_n(\Omega)$ satisfies*

$$\mathbf{M}_\Omega(T) + \mathbf{M}_\Omega(\partial T) \;<\; \infty$$

if and only if it is representable as

$$T(f(x)\, dx) \;=\; \int f(x)\, u(x)\, dx$$

for some $u \in BV(\Omega)$. In this case we also have

$$\mathbf{M}_\Omega(T) = \int_\Omega |u(x)|\, dx, \qquad \mathbf{M}_\Omega(\partial T) = \int_\Omega |Du|$$

A consequence of Corollary 1 is that an n-dimensional current in $\Omega \subset \mathbb{R}^n$ with $\mathbf{M}_\Omega(T) + \mathbf{M}_\Omega(\partial T) < \infty$ cannot concentrate on *closed* sets of dimension less than n. More generally we have

Theorem 3. *Let $T \in \mathcal{D}_n(\Omega)$ be a current with $\mathbf{M}_\Omega(T) + \mathbf{M}_\Omega(\partial T) < \infty$, where Ω is an open set in $\mathbb{R}^{n+N}$. Denote by π_α, $\alpha \in I(n, n+N)$, the orthogonal projection of $\mathbb{R}^{n+N}$ into the n-plane coordinate determined by α given by $\pi(x^1, \dots, x^{n+N}) \;=\; (x^{\alpha_1}, \dots, x^{\alpha_n})$, and let E be any closed subset of Ω with*

$$\mathcal{H}^n(\pi_\alpha(E)) \;=\; 0 \qquad \forall\, \alpha \in I(n, n+N)\, .$$

Then $T \llcorner E = 0$.

Proof. It suffices to prove the theorem in the case in which E is a compact set. First we show that

$$(8) \qquad \mathbf{M}_U(T \llcorner E) \le c\,\mathbf{M}_U(T \llcorner (\mathbb{R}^{n+N} \setminus E))$$

for any open set U with $E \subset U \subset \Omega$.

To prove (8) we consider for any form $\omega \in \mathcal{D}^n(U)$, $\omega = \sum_{|\alpha|=n} \omega_\alpha dx^\alpha$ the currents in $\mathcal{D}_n(\mathbb{R}_{\alpha_1} \times \ldots \times \mathbb{R}_{\alpha_n})$ given by

$$\pi_{\alpha\#}(T \llcorner \omega_\alpha(x))(\tau) = T(\omega_\alpha \tau) \qquad \tau \in \mathcal{D}^n(\mathbb{R}_{\alpha_1} \times \ldots \times \mathbb{R}_{\alpha_n}) .$$

Obviously

$$(9) \qquad \mathbf{M}(\pi_{\alpha\#}(T \llcorner \omega_\alpha)) \le \mathbf{M}_U(T) \sup_\Omega |\omega_\alpha| .$$

As for any $\eta \in \mathcal{D}^{n-1}(\Omega)$ we have

$$\partial(T \llcorner \omega_\alpha)(\eta) = T(\omega_\alpha\, d\eta) = T(d(\omega_\alpha \eta)) - T(d\omega_\alpha \wedge \eta)$$

we also infer

$$(10) \qquad \begin{aligned} \mathbf{M}(\partial \pi_{\alpha\#}(T \llcorner \omega_\alpha)) &\le \mathbf{M}(\partial(T \llcorner \omega_\alpha(x))) \\ &\le \mathbf{M}(\partial T) \sup |\omega_k| + \mathbf{M}(T) \sup |d\omega_\alpha| < \infty . \end{aligned}$$

From (9) and (10), taking into account Theorem 2, we deduce the existence of functions $\theta_\alpha \in BV(\mathbb{R}_{\alpha_1} \times \ldots \times \mathbb{R}_{\alpha_n})$ such that

$$\pi_{\alpha\#}(T \llcorner \omega_\alpha)(\tau) = \int < \tau, e_{\alpha_1} \wedge \ldots \wedge e_{\alpha_n} > \theta_\alpha(y)\, dy$$

for every $\tau \in \mathcal{D}^n(\mathbb{R}_{\alpha_1} \times \ldots \times \mathbb{R}_{\alpha_n})$. Hence

$$\pi_{\alpha\#}(T \llcorner \omega_\alpha) \llcorner \pi_\alpha(E) = 0 ,$$

being $\mathcal{H}^n(\pi_\alpha(E)) = 0$.

Since now

$$\begin{aligned} \mathbf{M}(\pi_{\alpha\#}(T \llcorner \omega_\alpha)) &\le \mathbf{M}(\pi_{\alpha\#}(T \llcorner \omega_\alpha) \llcorner (\mathbb{R}^n \setminus \pi_\alpha(E))) \\ &= \mathbf{M}(\pi_{\alpha\#}(T \llcorner \omega_\alpha) \llcorner \pi_\alpha(\mathbb{R}^{n+N} \setminus \pi_\alpha^{-1}\pi_\alpha(E))) \\ &\le \mathbf{M}((T \llcorner \omega_\alpha) \llcorner (\mathbb{R}^{n+N} \setminus \pi_\alpha^{-1}\pi_\alpha(E))) \\ &\le \mathbf{M}(T \llcorner (\mathbb{R}^{n+N} \setminus \pi_\alpha^{-1}\pi_\alpha(E))) \sup |\omega_\alpha| \\ &\le \mathbf{M}(T \llcorner (\mathbb{R}^{n+N} \setminus E)) \sup_\alpha |\omega_\alpha| \end{aligned}$$

and

$$T(\omega) = \sum_\alpha \pi_{\alpha\#}(T \llcorner \omega_\alpha)\, (dx^{\alpha_1} \wedge \ldots \wedge dx^{\alpha_n}) ,$$

we get (8) at once.

Consider now a non increasing sequence of open sets U_k, $U_{k+1} \subset U_k$, such that $\cap_k U_k = E$. Then we have

$$\lim_{k\to\infty} \mathbf{M}_{U_k}(T \llcorner (\mathbb{R}^{n+N} \setminus E)) = 0 ,$$

consequently $T \llcorner E = 0$ if we take into account (8). □

Let us discuss now in which way the previous results may be improved under rectifiability assumptions. The following claim readily follows from Theorem 1.

Theorem 4 (Constancy theorem). *Let T be an i.m. rectifiable n–current in an open set Ω of $\mathbb{R}^n$. Suppose that Ω is connected and that*

$$\partial T \llcorner \Omega = 0 .$$

Then

$$T = c[\![\,\Omega\,]\!]$$

where c is an integer.

Clearly, any i.m. rectifiable n-current $T = \tau(\mathcal{M}, \theta, \vec{T})$ in $\mathbb{R}^n$ must have the form

$$T(f(x)\,dx) = \int_{\mathcal{M}} f(x)\,u(x)\,dx$$

where $u(x)$ is an integer-valued summable function on $\mathcal{M}$. In fact $\vec{T} = \varepsilon(x)$ $e_1\wedge\ldots\wedge e_n$ with $\varepsilon(x) = \pm 1$. Therefore it suffices to take $u(x) := \varepsilon(x)\theta(x)$.

Next theorem allows us to draw an interesting information on n-dimensional currents in $\mathbb{R}^n$ under the assumption that ∂T is i.m. rectifiable.

Theorem 5. *Let T be an n-dimensional current in an open connected subset Ω of $\mathbb{R}^n$, $T \in \mathcal{D}_n(\Omega)$. Suppose that $\partial T = \tau(\mathcal{T}, \theta, \xi)$ in Ω is i.m. rectifiable. There there exists a constant $c \in \mathbb{R}$ such that*

$$T - c[\![\,\Omega\,]\!]$$

is an i.m. rectifiable current. More precisely we have

$$T(f(x)\,dx) = c\int_{\Omega} f(x)\,dx + \int_{\Omega} f(x)\,\nu(x)\,dx$$

where $\nu \in BV(\Omega)$, and for almost every x, $\nu(x)$ is an integer.

Proof. From our assumptions and Theorem 2 we deduce that there exists $u \in BV_{\mathrm{loc}}(\Omega)$ such that

$$T(f(x)\,dx) = \int f(x)\,u(x)\,dx \qquad \forall\, f \in C_c^\infty(\Omega) .$$

Moreover

$$(11) \qquad - < D_i u, f > = \int u(x)\, D_i f(x)\, dx \;=\; \partial T((-1)^{i-1} f(x)\, \widehat{dx^i})$$

If $\xi = \sum \xi^{\hat{\imath}}(x)\widehat{e}_i$, we denote by

$$n(x) \;:=\; \sum_{i=1}^{n} (-1)^{i-1} \xi^{\hat{\imath}} e_i$$

the unit normal vector to the plane defined by the n-vector ξ, oriented in such a way that

$$n(x) \wedge \xi(x) \;=\; e_1 \wedge \ldots \wedge e_n \; .$$

Then we have

$$\begin{aligned} \partial T((-1)^{i-1} f(x)\, \widehat{dx^i}) &= \int (-1)^{i-1}\, f(x)\, \xi^{\hat{\imath}}(x)\, \theta(x)\, d\mathcal{H}^{n-1} \llcorner \mathcal{T} \\ (12) \qquad &= \int f(x)\, n_i(x)\, \theta(x)\, d\mathcal{H}^{n-1} \llcorner \mathcal{T} \; . \end{aligned}$$

From (11) and (12) we then infer

$$D_i u \;=\; -n_i(x)\, \theta(x)\, d\mathcal{H}^{n-1} \llcorner \mathcal{T} \; .$$

In particular the distributional derivatives of u have only jump parts with integral jump values. The conclusion then follows from the result that we state as Proposition 1 below. □

Proposition 1. *Let $u \in BV(\Omega)$, where Ω is an open and connected subset of $\mathbb{R}^n$. Suppose that Du has only jump parts, $Du = (Du)^{(j)}$, and that the jump value is integer-valued. Then there exists a constant $c \in \mathbb{R}$ such that*

$$u(x) \;=\; c + v(x)$$

where $v(x)$ is an integer-valued function in $BV(\Omega)$.

Proof. Clearly it suffices to prove that

$$u^+(y) - u^+(x) \;\in\; \mathbb{Z}$$

for almost every $x,\; y \in B(x_0, R)$. In this case in fact we easily infer that

$$u^+(y) - u^+(x) \;\in\; \mathbb{Z} \qquad \text{for a.e. } x, y \in \Omega$$

by covering Ω with balls, and that for some $x_0 \in \Omega$

$$u^+(y) - u^+(x_0) \;\in\; \mathbb{Z} \qquad \text{for a.e. } y \in \Omega \; ,$$

if we take into account Fubini's theorem.

First we extend u as a function of the variables $x \in \mathbb{R}^n, z \in \mathbb{R}^n, t \in \mathbb{R}$ by setting

$$\overline{u}(x,z,t) := u(z) ;$$

we may think of u as defined on $\mathbb{R}^n$. Then clearly $\overline{u} \in BV(\mathbb{R}^{2n+1})$, $D\overline{u} = (0, Du, 0)$, $u^+(x,z,t) = u^+(z)$, $u^-(x,z,t) = u^-(z)$ and, by the assumption, $D\overline{u}$ has only a jump part $D\overline{u} = (D\overline{u})^{(j)}$ with integral jump value at $\mathcal{H}^{2n}$-a.e. $(x,z,t) \in J_{\overline{u}} = \mathbb{R}^n \times J_u \times \mathbb{R}$.

We now consider the map $\phi(x,y,t) := (x, x+t(y-x), t)$ and the function $v := \overline{u} \circ \phi$. Clearly $v \in BV(\mathbb{R}^{2n+1})$ and Dv has only jump part $Dv = (Dv)^{(j)}$ concentrated along $\phi^{-1}(J_{\overline{u}})$ with integral jump values at $\mathcal{H}^{2n}$-a.e. (x,y,t) in $J_v = \phi^{-1}(J_{\overline{u}})$.

Applying Theorem 6 in Sec. 4.4 in Note 4 in Sec. 4.4 with $i = 2n+1$, so that $\Pi = \{(x,y,t) \mid t = 0\}$ we then infer that for $\mathcal{H}^{2n}$-a.e. $(x,y) \in \Pi$ the function

$$t \longrightarrow v_{x,y}(t) := v(x,y,t)$$

belongs to $BV(\mathbb{R})$. Moreover its gradient has only a jump part, i.e.,

$$(Dv_{x,y})^{(a)} = (Dv_{x,y})^{(C)} = 0 ,$$

and finally

$$Dv_{x,y}(t) = \sum_j k_j \, \delta_{t_j}(t) \qquad k_j \in \mathbb{Z}$$

since

$$\begin{aligned} J_{v_{x,y}} &= \{t \mid (x,y,t) \in J_v\} \\ v^+_{x,y}(t) &= \overline{u}^+(x,y,t) = u^+(x+t(y-x)) \\ v^-_{x,y}(t) &= \overline{u}^-(x,y,t) = u^-(x+t(y-x)) . \end{aligned}$$

The claim then follows since for $\mathcal{H}^{2n}$-a.e. x, y we have

$$u^+(y) - u^+(x) = v^+_{x,y}(1) - v^+_{x,y}(0) = \int \sum_j k_j \, \delta_{t_j} \in \mathbb{Z}$$

□

Remark 1. We notice that the proof of Theorem 5 simplifies considerably if we can prove that there exists an i.m. rectifiable current Σ in Ω such that $\partial\Sigma = \partial T$. In fact we then have $\partial(T - \Sigma) = 0$ and by the constancy theorem

$$T - \Sigma = c[\![\Omega]\!] .$$

This is the case for instance if Ω is star-shaped with respect to a point x_0, as for Σ we could take the cone over ∂T from $\{x_0\}$.

We conclude this subsection proving a special case of the *boundary rectifiability theorem* we stated in Ch. 1, as consequence of De Giorgi's rectifiability theorem Theorem 1 in Sec. 4.2.4, and a *decomposition theorem* for n–currents in $\mathbb{R}^n$.

Theorem 6. *Let T be an i.m. rectifiable n–current in $\Omega \subset \mathbb{R}^n$. If $\mathbf{M}_\Omega(\partial T) < \infty$, then ∂T is i.m. rectifiable in Ω.*

Proof. Since T is i.m. rectifiable T has the form

$$T(f(x)\,dx) = \int f(x)\,u(x)\,dx$$

where $u \in BV(\Omega)$ and $u(x) \in \mathbb{Z}$ for a.e. $x \in \Omega$. From Remark 1 in Sec. 4.1.4 and Theorem 2 in Sec. 4.1.4 we infer

$$Du = (u_+(x) - u_-(x))\,n(x, J_u)\,\mathcal{H}^{n-1} \llcorner J_u\,.$$

Using (7) we can then write

$$\partial T = \tau(J_u, u_+ - u_-, *n(\cdot, J_u))$$

□

Theorem 7. *Let T be an i.m. rectifiable n–current in $\Omega \subset \mathbb{R}^n$ such that $\mathbf{M}_\Omega(\partial T) < \infty$. Then one can find a decreasing sequence of Caccioppoli sets U_j, $j \in \mathbb{Z}$, such that*

$$T = \sum_{j=1}^{\infty} [\![U_j]\!] - \sum_{j=0}^{\infty} [\![\Omega \setminus U_{-j}]\!], \qquad \partial T = \sum_{-\infty}^{+\infty} \partial [\![U_j]\!],$$

and

$$\| \partial T \| = \sum_{-\infty}^{+\infty} \| \partial [\![U_j]\!] \|\,.$$

Proof. Since T is i.m. rectifiable, then $T = \mathcal{L}^n \llcorner u$ with $u \in BV(\Omega, \mathbb{Z})$. For any $j \in \mathbb{Z}$, set

$$U_j := \{x \in \Omega \mid u(x) \geq t\}\,.$$

Being u integer-valued, we can easily check that

$$u(x) = \sum_{j=1}^{\infty} \chi_{U_j}(x) - \sum_{j=0}^{\infty} \chi_{\Omega \setminus U_{-j}}(x)$$

from which the decomposition formula for T follows. On the other hand

$$E_t(u) := \{x \in \Omega \mid u(x) > t\} = U_j \quad \text{for} \quad j < t < j+1$$

and by the Fleming-Rishel formula, Theorem 2 in Sec. 4.1.1 we conclude that the sets U_j are Caccioppoli sets and,

$$\| \partial T \|(B) = |Du|(B) = \int_{-\infty}^{+\infty} |D\chi_{E_t(u)}|(B)\,dt = \sum_{-\infty}^{+\infty} \| \partial[\![U_j]\!] \|(B)$$

for any Borel set B in $\mathbb{R}^n$. The decomposition formula for ∂T follows now from the decomposition of T. □

3.2 Degree Mapping and Degree of Cartesian Currents

Let Ω be an open set in $\mathbb{R}^n$ and let $u : \Omega \to \mathbb{R}^n$ be an almost everywhere approximately differentiable map with Jacobian determinant $\det Du$ in $L^1(\Omega)$. Here $Du(x)$, as usual, denotes the approximate gradient of u at x. In Sec. 3.1.5 we have seen that the area formula leads naturally to the notion of *degree of* u. In fact defining

$$\text{(1)} \qquad \deg(u, \Omega, y) \;:=\; \sum_{x \in u^{-1}(y) \cap \mathcal{A}_D(u)} \operatorname{sign}(\det Du(x))$$

the area formula says that

(i) $\deg(u, \Omega, \cdot)$ is a well defined function in $L^1(\mathbb{R}^n)$, and, as L^1-function, depends only on the equivalence class of u.
(ii) $\deg(u, \Omega, y)$ takes integer values for $\mathcal{H}^n$-a.e. $y \in \mathbb{R}^n$
(iii) for any $\varphi \in C^0(\mathbb{R}^n) \cap L^\infty(\mathbb{R}^n)$

$$\text{(2)} \qquad \int_{\mathbb{R}^n} \varphi(y) \deg(u, \Omega, y)\, dy \;=\; \int_{\Omega} \varphi(u(x)) \det Du(x)\, dx\ .$$

Of course the sum in (1) is extended only to points x where u is approximately differentiable.

Suppose now that $u \in \mathcal{A}^1(\Omega, \mathbb{R}^n)$. Then (2) can be written in a more geometric way in terms of G_u. Consider in fact the current graph of u, G_u, defined on n-forms $\omega \in \mathcal{D}^n(\Omega \times \mathbb{R}^n)$ by

$$G_u(\omega) \;:=\; \int_{\Omega} \langle \omega, M(Du) \rangle\, dx\ .$$

Then for any $\varphi \in C^0_c(\mathbb{R}^n)$ we have

$$\text{(3)} \qquad \int_{\Omega} \varphi(u(x)) \det Du(x)\, dx \;=\; G_u(\varphi(y)\, dy^1 \wedge \ldots \wedge dy^n)\ .$$

Denoting as usual by $\hat{\pi}$ the orthogonal projection on the second factor of $\mathbb{R}^n \times \mathbb{R}^n$, $\hat{\pi}(x, y) = y$, from (2) and (3) we then infer

Proposition 1. *Let* $u \in \mathcal{A}^1(\Omega, \mathbb{R}^n)$. *Then*

$$\mathbf{M}(\widehat{\pi}_\# G_u) = \int_\Omega |\det Du(x)|\, dx \tag{4}$$

$$\widehat{\pi}_\# G_u(\varphi(y)\, dy) = \int \varphi(y) \deg(u, \Omega, y)\, dy \tag{5}$$

Proposition 1 says in particular that $\deg(u, \Omega, y)$ is the Radón-Nykodim derivative of the measure $\mu := \widehat{\pi}_\# G_u \llcorner dy^1 \wedge \ldots \wedge dy^n$ with respect to Lebesgue measure $\mathcal{L}^n$. Therefore counting points over a given y with their sign, that is the degree, is an information totally build into the mechanism of projecting the graph of u on the target space $\mathbb{R}^n$, i.e., into the current $\widehat{\pi}_\# G_u$.

In view of Proposition 1 one can easily extend the notion of degree to any Cartesian current in $\text{cart}(\Omega \times \mathbb{R}^n)$, and actually to any current $T \in \mathcal{D}_n(\mathbb{R}^{k+n})$, where k is not necessarily equal to n, provided $\widehat{\pi}_\# T$ exists.

We shall consider the case of n-dimensional i.m. rectifiable currents $T = \tau(\mathcal{M}, \theta, \vec{T})$ in $\mathbb{R}^k \times \mathbb{R}^n$. Since $\mathbf{M}(T) < \infty$, T extends to forms with bounded and continuous coefficients on the whole of $\mathbb{R}^k \times \mathbb{R}^n$. In particular the current $\widehat{\pi}_\# T$, $\widehat{\pi}_\# T(\omega) = T(\pi^\# \omega)$, is well defined. Denoting now, as usual, by $(e_1, \ldots, e_k)$ and $(\varepsilon_1, \ldots, \varepsilon_n)$ the canonical bases in $\mathbb{R}^k$ and in $\mathbb{R}^n$, the tangent n-vector $\vec{T}(x, y)$ of T at (x, y) can be written as

$$\vec{T}(x, y) = \sum_{|\alpha| + |\beta| = n} \vec{T}^{\alpha\beta}(x, y)\, e_\alpha \wedge \varepsilon_\beta .$$

Clearly, compare Proposition 1 in Sec. 2.2.1,

$$\widehat{\pi}_{\#,(x,y)} \vec{T}(x, y) = \vec{T}^{0\bar{0}}(x, y) , \qquad |J_{\hat{\pi}}^{\mathcal{M}}(x, y)| = |\vec{T}^{0\bar{0}}(x, y)|$$

hence

$$\begin{aligned} \widehat{\pi}_\# T(\varphi(y) dy^1 \wedge \ldots \wedge dy^n) &= \int < \varphi(y)\, dy, \vec{T}(x, y) > \theta(x, y)\, d\mathcal{H}^n \llcorner \mathcal{M} \\ &= \int \varphi(y)\, \theta(x, y)\, \vec{T}^{0\bar{0}}(x, y)\, d\mathcal{H}^n(x, y) \llcorner \mathcal{M} . \end{aligned} \tag{6}$$

We now set

Definition 1. *The* degree mapping, *or simply the* degree of T *at* y *with respect to* $\widehat{\pi}$ *is defined by*

$$\deg(T, \widehat{\pi}, y) := \sum_{x \in \hat{\pi}^{-1}(y) \cap \mathcal{M}} \theta(x, y) \operatorname{sign}(\vec{T}^{0\bar{0}}(x, y)) \tag{7}$$

whenever it exists as a real number.

By definition $\deg(T, \widehat{\pi}, y)$ is an integer whenever it exists. From (6) and the area formula we immediately infer

Proposition 2. *Let $T = \tau(\mathcal{M}, \theta, \vec{T})$ be an n-dimensional i.m. rectifiable current in $\mathbb{R}^k \times \mathbb{R}^n$. Then*

(i) $\deg(T, \widehat{\pi}, y) \in L^1(\mathbb{R}^n)$
(ii) $\widehat{\pi}_\# T(\varphi(y)\, dy) = \int \varphi(y) \deg(T, \widehat{\pi}, y)\, dy$.

In particular $\widehat{\pi}_\# T$ is an n-dimensional i.m. rectifiable current in $\mathbb{R}^n$

$$\widehat{\pi}_\# T = \tau(\widehat{\mathcal{M}}, \lambda, \xi)$$

where

$$\widehat{\mathcal{M}} := \{y \mid \deg(T, \widehat{\pi}, y) \neq 0\}\,, \qquad \lambda(y) := |\deg(T, \widehat{\pi}, y)|$$

and

$$\xi(y) := \operatorname{sign}(\deg(T, \widehat{\pi}, y))\, \varepsilon_1 \wedge \ldots \wedge \varepsilon_n\,.$$

From (5) and (6) we obviously have

$$\deg(u, \Omega, y) = \deg(G_u, \widehat{\pi}, y)$$

whenever $u \in \mathcal{A}^1(\Omega, \mathbb{R}^n)$. Later we shall be mostly interested in the degree of Cartesian current. To simplify notations and to be consistent with the classical notations we set

Definition 2. *Let $T \in \operatorname{cart}(\Omega \times \mathbb{R}^n)$, $\Omega \subset \mathbb{R}^n$, and let A be a Borel set, $A \subset \Omega$. The* degree *of T over A at $y \in \mathbb{R}^n$ is defined by*

$$\deg(T, A, y) := \deg(T \llcorner A \times \mathbb{R}^n, \widehat{\pi}, y)\,.$$

Of course if $T = G_u$, we have

$$\deg(G_u, A, y) = \deg(G_u \llcorner A \times \mathbb{R}^n, \widehat{\pi}, y) = \deg(u, A, y)\,.$$

Before discussing properties of the degree, and in particular before proving that the degree defined above enjoys all properties of the classical degree, and moreover is continuous with respect to the weak convergence of currents, let us consider a few examples.

[1] First we consider the map $u_0(x) := x/|x|$ which belongs to $\mathcal{A}^1(B(0,1), \mathbb{R}^n)$, $B(0,1)$ being the unit ball in $\mathbb{R}^n$. We have

$$\det Du_0(x) = 0 \qquad \text{for a.e. } x$$

hence

$$\deg(u_0, B(0,1), y) = \deg(G_{u_0}, \widehat{\pi}, y) = 0 \qquad \text{for a.e. } y\,.$$

In contrast, if we consider the Cartesian current

$$T := G_{u_0} + \delta_0 \times [\![B(0,1)]\!]$$

we have $\widehat{\pi}_\# T = [\![B(0,1)]\!]$ hence

$$\deg(T, \widehat{\pi}, y) = \deg(T, \Omega, y) = \begin{cases} 1 & \text{if } |y| < 1 \\ 0 & \text{if } |y| > 1 . \end{cases}$$

Finally, for the Cartesian current

$$T := G_{u_0} + L \times [\![S^{n-1}]\!] ,$$

L being a one-dimensional i.m. rectifiable current in $B(0,1)$ such that $\partial L \llcorner B(0,1) = \delta_0$, we have $\widehat{\pi}_\# T = 0$. Hence

$$\deg(T, \widehat{\pi}, y) = 0 \qquad \text{for a.e. } y \in \mathbb{R}^n .$$

•

[2] Consider now the Cantor-Vitali function $V : [0,1] \to [0,1]$, see Sec. 4.2.4. Being V continuous and nondecreasing with $V(0) = 0$ and $V(1) = 1$, its *classical* degree at every point $y \in (0,1)$ is 1. In contrast to that, in the sense of Definition 1 and Definition 2 we have

$$\deg(V, (0,1), y) = 0 \qquad \text{for a.e. } y \in (0,1) ,$$

since $V'(x) = 0$ for almost every $x \in (0,1)$. However, considering its subgraph, SG_V, we have

$$\deg(-\partial SG_V, \widehat{\pi}, y) = \begin{cases} 1 & \text{for } y \in (0,1) \\ 0 & \text{otherwise.} \end{cases}$$

In other words, the notion of degree introduced here distinguishes between the graph of $V(x)$ (in the "almost everywhere differentiability" sense) and the boundary of the subgraph of $V(x)$ as G_V and $-\partial SG_V$ are different. This corresponds to the fact that one can find sequences of smooth maps $u_k : (0,1) \to (0,1)$ such that

$$G_{u_k} \rightharpoonup \partial SG_V ,$$

while there is no sequence of smooth maps v_k such that

$$G_{v_k} \rightharpoonup G_V , \tag{8}$$

the fact that $\partial G_V \neq 0$ in $(0,1) \times \mathbb{R}$ being an obstruction to the approximation in (8). •

In order to state the basic properties of the degree let us define the set

$$\Gamma_T := \{ y \in \mathbb{R}^n \mid y \notin \operatorname{spt}(\partial \widehat{\pi}_\# T) \} .$$

We have

Theorem 1. *Let $T = \tau(\mathcal{M}, \theta, \xi)$ be an n-dimensional i.m. rectifiable current in $\mathbb{R}^k \times \mathbb{R}^n$. Then we have*

(i) $\deg(T, \widehat{\pi}, \cdot)$ *is an integer-valued function in $L^1(\mathbb{R}^n)$, which vanishes outside the support of $\widehat{\pi}_\# T$.*

(ii) $\deg(T, \widehat{\pi}, y)$ *is constant on the connected components of Γ_T.*

(iii) $\mathbf{M}(\partial \widehat{\pi}_\# T) < \infty$ *if and only if* $\deg(T, \widehat{\pi}, \cdot) \in BV(\mathbb{R}^n)$.
In this case $\partial \widehat{\pi}_\# T$ is an i.m. rectifiable $(n-1)$-current and the gradient of $\deg(T, \widehat{\pi}, \cdot)$ *has only jump part with integral jump values.*

(iv) (Excision). *If $\{A_i\}$ is a numerable family of pairwise disjoint $\mathcal{H}^n$-measurable sets in $\mathbb{R}^{k+n}$ and $A := \bigcup_i A_i$, then*

$$\deg(T \llcorner A, \widehat{\pi}, y) = \sum_i \deg(T \llcorner A_i, \widehat{\pi}, y) \qquad \text{for a.e. } y \in \mathbb{R}^n .$$

(v) (Homological invariance). *Let T, S be two i.m. rectifiable n-currents in $\mathbb{R}^{k+n}$. Then*

$$\partial \widehat{\pi}_\# T = \partial \widehat{\pi}_\# S$$

if and only if

$$\deg(T, \widehat{\pi}, y) = \deg(S, \widehat{\pi}, y) \qquad \text{for a.e. } y \in \mathbb{R}^n .$$

(vi) (Homotopic invariance). *Let T_t be a 1-parameter family of i.m. rectifiable n-currents, $t \in [0,1]$. Suppose that the map $t \to \widehat{\pi}_\# T_t(\omega)$ is continuous for any $\omega \in \mathcal{D}^n(\mathbb{R}^n)$. Then* $\deg(T_t, \widehat{\pi}, \cdot)$ *is independent from t*

(vii) (Leray product formula). *Consider any Lipschitz continuous map $\varphi : \mathbb{R}^n \to \mathbb{R}^n$ and the map $\phi(x,y) = (x, \varphi(y))$ in $\mathbb{R}^k \times \mathbb{R}^n$. For $k \in \mathbb{Z}$ and*

$$\Sigma_k := \{y \in \mathbb{R}^n \mid \deg(T, \widehat{\pi}, y) = k\}$$

we have

$$\deg(\phi_\# T, \widehat{\pi}, y) = \sum_{k \in \mathbb{Z}} k \deg(\varphi, \Sigma_k, y) \qquad \text{for a.e. } y \in \mathbb{R}^n .$$

(viii) (Continuity). *Suppose that T_h, $h = 1, 2, \dots$, and T are i.m. rectifiable n-currents in $\mathbb{R}^k \times \mathbb{R}^n$ such that*

$$\widehat{\pi}_\# T_h \rightharpoonup \widehat{\pi}_\# T .$$

Then

$$\deg(T_h, \widehat{\pi}, y) \to \deg(T, \widehat{\pi}, y)$$

in the sense of measures. In particular if $\deg(T_h, \widehat{\pi}, y) = constant = \lambda$ *with respect to h on some open set U, then* $\deg(T, \widehat{\pi}, y) = \lambda$ *on U.*

Proof. The area formula and the constancy theorem yield at once (i) and (ii) respectively. The first part of (iii) follows easily by just computing $\partial\widehat{\pi}_{\#}T$ from (6)). The rest of the claim then follows as in the proof of the Theorem 6 in Sec. 4.3.1. The excision formula in (iv) is an immediate consequence of the trivial equality

$$\widehat{\pi}_{\#}(T \llcorner A) = \sum_{i=1}^{\infty} \widehat{\pi}_{\#}(T \llcorner A_i) ,$$

taking into account that $\deg(T, \widehat{\pi}, y) \in L^1(\mathbb{R}^n)$. To prove (v) we observe that $T - S$ is i.m. rectifiable and $\partial\widehat{\pi}_{\#}(T - S) = 0$, hence by the constancy theorem

$$\widehat{\pi}_{\#}T - \widehat{\pi}_{\#}S = c[\![\mathbb{R}^n]\!] \qquad c \in \mathbb{Z} .$$

From $\mathbf{M}(T - S) \le \mathbf{M}(T) + \mathbf{M}(S) < \infty$, we infer $c = 0$, hence $\widehat{\pi}_{\#}T = \widehat{\pi}_{\#}S$, from which the claim in (v) follows at once. The claim in (vi) follows easily from the continuity in t of $\widehat{\pi}_{\#}T_t$ taking into account the fact that the degree mapping is integer-valued. To prove (vii) we notice that

$$\phi^{\#}\widehat{\pi}^{\#}(f(y)\, dy) = f(\varphi(y)) \det D\varphi(y)\, dy$$

hence

$$\begin{aligned}\int f(y) \deg(\phi_{\#}T, \widehat{\pi}, y)\, dy &= \widehat{\pi}_{\#}\phi_{\#}(T)(f(y)\, dy) \\ &= T(\phi^{\#}\widehat{\pi}^{\#}(f(y)\, dy)) = \int f(\varphi(y)) \det D\varphi(y) \deg(T, \widehat{\pi}, y)\, dy \\ &= \sum_{h\in\mathbb{Z}} h \int_{\Sigma_h} f(\varphi(y)) \det D\varphi(y)\, dy = \sum_{h\in\mathbb{Z}} h \int f(y) \deg(\varphi, \Sigma_h, y)\, dy .\end{aligned}$$

Finally (viii) is trivial. □

Remark 1. We explicitly remark that in the case in which $\partial\widehat{\pi}_{\#}T$ has finite mass, Theorem 1 (iii), the i.m. rectifiable current $\partial\widehat{\pi}_{\#}T$ and the BV-function $\deg(T, \widehat{\pi}, y)$ are related by

$$\partial\widehat{\pi}_{\#}T(\omega) = \int \operatorname{div} f(y) \deg(T, \widehat{\pi}, y)\, dy = -\int f_i(y) D_i \deg(T, \widehat{\pi}, y)\, dy$$

if $\omega = \sum_{i=1}^n (-1)^{i-1} f_i(y)\, \widehat{dy^i}$ compare (7) in Sec. 4.3.1 and Theorem 6 in Sec. 4.3.1.

As consequence of Theorem 1 (ii) and (iii), on each connected component of

$$\Gamma_T := \mathbb{R}^n \setminus \operatorname{spt} \partial\widehat{\pi}_{\#}T$$

the degree of T, $\deg(T, \widehat{\pi}, y)$, can be identified with an integer constant. If moreover $\partial\pi_{\#}T$ has finite mass, such an identification holds in the BV sense, that is, $\mathcal{H}^{n-1}$-almost everywhere, in particular, at every point in which the degree is approximately continuous. That identification being understood, it is then useful to state a pointwise version of Theorem 1.

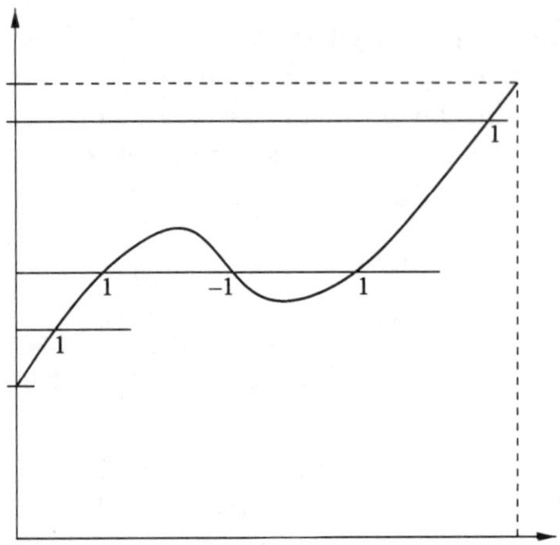

Fig. 4.7. The degree is locally constant on the complement of $\partial u(\Omega)$.

Theorem 2. *Let T be an i.m. rectifiable n-current in $\mathbb{R}^{k+n}$.*

(i) (Excision). *If $\{A_i\}$ is a numerable family of disjoint Borel sets in $\mathbb{R}^n$, and*

$$y \in \cap_{i=1}^{\infty} \Gamma_{T \llcorner A_i} \bigcap \Gamma_{T \llcorner \cup_i A_i} \,, \qquad \mathcal{H}^n\big(\cap_{i=1}^{\infty} \Gamma_{T \llcorner A_i} \bigcap \Gamma_{T \llcorner \cup_i A_i} \big) > 0 \,,$$

then

$$\deg\big(T \llcorner \cup_{i=1}^{\infty} A_i, \widehat{\pi}, y\big) = \sum_{i=1}^{\infty} \deg\left(T \llcorner A_i, \widehat{\pi}, y\right) .$$

(ii) (Homological invariance). *If T, S are i.m. rectifiable n-currents in $\mathbb{R}^{k+n}$ and*

$$\partial \widehat{\pi}_{\#} T = \partial \widehat{\pi}_{\#} S$$

then $\Gamma_S = \Gamma_T$ and

$$\deg\left(T, \widehat{\pi}, y\right) = \deg\left(S, \widehat{\pi}, y\right) \qquad \forall\, y \in \Gamma_T \,.$$

(iii) (Homotopic invariance). *Let T_t, $t \in [0,1]$, be a family of i.m. rectifiable n-currents in $\mathbb{R}^{k+n}$. Suppose that $t \to \widehat{\pi}_{\#} T_t$ is a continuous map from $[0,1]$ into $\mathcal{D}_n(\mathbb{R}^{k+n})$ and that y_0 is an interior point of*

$$\bigcap_{t \in [0,1]} \Gamma_{T_t} \,.$$

Then $\deg\left(T, \widehat{\pi}, y_0\right)$ *is independent of t.*

(iv) (Leray product formula). *Let $\varphi : \mathbb{R}^n \to \mathbb{R}^n$ be a Lipschitz map and let $\phi := (x, \varphi(x))$. Denote by C_i the connected components of Γ_T and by y_i any point in C_i. Then for any y which is not in the closure of $\phi(\mathbb{R}^n \setminus \Gamma_T)$ we have*

$$\deg\left(\phi_{\#} T, \widehat{\pi}, y\right) = \sum_i \deg\left(T, \widehat{\pi}, y_i\right) \deg\left(\phi, C_i, y\right) .$$

(v) (Continuity). *If* T_h, $h = 1, 2, \ldots$, *and* T *are i.m. rectifiable* n*-currents in* $\mathbb{R}^{k+n}$ *such that*

$$\widehat{\pi}_\# T_h \rightharpoonup \widehat{\pi}_\# T \ ,$$

then

$$\deg(T_h, \widehat{\pi}, y) \longrightarrow \deg(T, \widehat{\pi}, y)$$

for any y *which is an interior point of* $\bigcap_h \Gamma_{T_h}$.

Proof. (i) From Theorem 1 we know that for a.e. $y \in \Gamma_{T \llcorner A} \cap \bigcap_i \Gamma_{T \llcorner A_i}$

$$\deg(T \llcorner A, \pi, y) = \sum_i \deg(T \llcorner A_i, \widehat{\pi}, y) \ .$$

The claim then follows since all degrees in the previous formula are constant on the connected components containing y.

(ii) We obviously have $\Gamma_T = \Gamma_S$. For a small ball B around $y \in \Gamma_T$ we have

$$\deg(T, \widehat{\pi}, y) [\![B]\!] = \widehat{\pi}_\# T \llcorner B = \widehat{\pi}_\# S \llcorner B = \deg(S, \widehat{\pi}, y) [\![B]\!]$$

if we take into account the constancy theorem. This proves the claim.

(iii) Considering again a small ball B around y_0 we have

$$\deg(T_t, \widehat{\pi}, y) [\![B]\!] = \widehat{\pi}_\# T_t \llcorner B \ ,$$

from which we see that $\deg(T_t, \widehat{\pi}, y)$ is continuous. The claim then follows since the degree is an integer.

(iv) By the assumption we can find a ball B centered at y and contained in the complement of the closure of $\phi(\mathbb{R}^n \setminus \Gamma_T)$. We can also assume that B is contained in the connected component of $y \in \Gamma_{\phi_\# T}$. Thus $\phi^{-1}(B) \subset \Gamma_T$ and $B \subset \mathbb{R}^n \setminus \phi(\partial C_i)$. We have for $f \in C_c^\infty(B)$

$$\begin{aligned} \deg(\phi_\# T, \widehat{\pi}, y) \int f(y)\, dy &= \int f(y) \deg(\phi_\# T, \widehat{\pi}, y)\, dy \\ &= \sum_h h \int f(y) \deg(\phi, \Sigma_h, y)\, dy \end{aligned}$$

where

$$\Sigma_h := \{ y \mid \deg(T, \widehat{\pi}, y) = h \} \ .$$

Since

$$\Gamma_T \cap \Sigma_h = \bigcup_i \{ C_i \mid \deg(T, C_i, y) = h \}$$

we then find

$$\sum_h h \int f(y) \deg(\phi, \Sigma_h, y)\, dy = \sum_h h \int f(y) \deg(\phi, \Sigma_h \cap \Gamma_T, y)\, dy$$
$$= \sum_i \deg(T, C_i, y) \int f(y) \deg(\phi, C_i, y)\, dy$$

and the claim follows by choosing as y in $\deg(\phi, C_i, y)$ a fixed point $y_i \in C_i$. Since (v) is trivial, this concludes the proof. □

Let us return to the situation we considered at the beginning of this subsection, i.e., let us consider an almost everywhere approximately differentiable map

$$u \,:\, \Omega \subset \mathbb{R}^n \;\to\; \mathbb{R}^n$$

with Jacobian determinant in L^1. We can then consider the n–dimensional current in $\mathbb{R}^n$ defined by

$$\hat{\Pi}_{u,\Omega}(\varphi(y)dy) := \int_\Omega \varphi(u(x)) \det Du(x)\, dx = \int_{\mathbb{R}^n} \varphi(y) \deg(u, \Omega, y)\, dy$$

and it is easily seen that the properties stated in Theorem 1 and Theorem 2 remain true with the following replacements

$$T \to u, \quad \deg(T, \hat{\pi}, y) \longrightarrow \deg(u, \Omega, y), \quad \hat{\pi}_\# T \longrightarrow \Pi_{u,\Omega}.$$

In particular we explicit mention

(i) *Homological invariance* If $\partial \widehat{\Pi}_{u,\Omega} \;=\; \partial \widehat{\Pi}_{v,\Omega}$ then

$$\deg(u, \Omega, y) \;=\; \deg(v, \Omega, y)$$

(ii) We have

$$\partial \widehat{\Pi}_{u,\Omega}(\omega) \;=\; \int_{\mathbb{R}^n} \operatorname{div} g(y) \deg(u, \Omega, y)\, dy \tag{9}$$

if $\omega \;=\; \sum_{i=1}^n (-1)^{i-1} g_i(y) \widehat{dy^i}$, $g \in C^1_c(\mathbb{R}^n, \mathbb{R}^n)$.

Of course if $u \in \mathcal{A}^1(\Omega, \mathbb{R}^n)$, we have $\widehat{\Pi}_{u,\Omega} \;=\; \hat{\pi}_\# G_u$.

Let $u \in \mathcal{A}^1(\Omega, \mathbb{R}^n)$. In dealing with the degree of u a certain precaution is necessary, as the previous general results holds away from $\hat{\pi} \partial G_u$. In particular if $u \in \mathcal{A}^1(\Omega, \mathbb{R}^n)$ we may infer that

$$\deg(u, \Omega, y) \;=\; \deg(v, \Omega, y) \tag{10}$$

if and only if

$$\hat{\pi}_\# \partial G_u \;=\; \hat{\pi}_\# \partial G_v\,. \tag{11}$$

equality of the traces on $\partial\Omega$ in general being not sufficient, compare Remark 4 in Sec. 3.1.5 and [3] below.

Notice that the condition (11) is fulfilled if $\partial G_u = \partial G_v$. This is the case, for instance, if

$$u,\ v \in \operatorname{cart}^1(\tilde{\Omega}, \mathbb{R}^n), \quad \tilde{\Omega} \supset\supset \Omega, \quad u = v \text{ on } \tilde{\Omega} \setminus \bar{\Omega}\ .$$

The results of Sec. 3.2.5 now allow us to read (11) in terms of equality of traces. For instance from Proposition 4 in Sec. 3.2.5, we infer at once

Proposition 3. *Let Ω be a bounded Lipschitz domain in $\mathbb{R}^n$ and let $u,\ v \in \mathcal{A}_{p,q}(\Omega, \mathbb{R}^n)$ with $p \geq n-1$, $q \geq n/(n-1)$. Suppose that $u = v$ on $\partial\Omega$ in the sense of traces. Then*

$$\deg(u, \Omega, y) = deg(v, \Omega, y).$$

[3] Of course in the case of the maps x and $x/|x|$ from $B(0,1)$ into $B(0,1)$ the failure of equality (10) is due to the fact that $G_{x/|x|}$ does not belong to $\operatorname{cart}^1(\Omega, \mathbb{R}^n)$. Example [1] in Sec. 3.2.5 where η is chosen strictly decreasing shows the existence of a one to one map $u \in \operatorname{cart}^1(B^+(0,1), \mathbb{R}^2)$, $B^+(0,1) \subset \mathbb{R}^2$, with

$$\deg(u, B^+(0,1), y) \ = \ 1 \quad \text{on} \quad u(B^+(0,1))\ ,$$

for which

$$u(x) \ = \ (0,0) \qquad \text{on } \partial B^+(0,1)$$

in the sense of $W^{1,p}(B^+(0,1), \mathbb{R}^2)$, $\forall p < 2$. •

Finally, it is worthwhile mentioning that (9) can be written as a boundary integral provided u is sufficiently smooth. In fact we have

Proposition 4. *Let Ω be a bounded Lipschitz domain in $\mathbb{R}^n$ and let u be a map in $\mathcal{A}_{p,q}(\Omega, \mathbb{R}^n)$, $p \geq n-1$, $q \geq n/n-1$. Suppose that also the trace of u on $\partial\Omega$ belongs to $\mathcal{A}_{p,q}(\partial\Omega, \mathbb{R}^n)$, then*

$$\int_{\mathbb{R}^n} \operatorname{div} g(y) \deg(u, \Omega, y)\, dy = \int_{\partial\Omega} \sum_{i,j=1}^{n} g^i(u(x))\, (\operatorname{adj} Du(x))^i_j\, \nu_j(x)\, d\mathcal{H}^{n-1}$$

for all $g \in C^1_c(\mathbb{R}^n, \mathbb{R}^n)$, where ν is the exterior unit normal to $\partial\Omega$.

Proof. In fact from (9) and Theorem 1 in Sec. 3.2.5, we get

$$\begin{aligned}\int_{\mathbb{R}^n} \operatorname{div} g(y) \deg(u, \Omega, y)\, dy \ &= \ \partial\widehat{\Pi}_{u,\Omega}(\omega) \ = \ \partial\hat{\pi}_\# G_u(\omega) \\ &= \ \hat{\pi}_\# \partial G_u(\omega) \ = \ \hat{\pi}_\# G_{u,\partial\Omega}(\omega)\ .\end{aligned}$$

□

Consider now a Cartesian current $T \in \operatorname{cart}(\Omega \times \mathbb{R}^n)$, $\Omega \subset \mathbb{R}^n$. In analogy with Definition 2 we write for any Borel set $A \subset \Omega$

$$\Gamma_{T,A} \ := \ \Gamma_{T \llcorner A \times \mathbb{R}^n} \ = \ A \setminus \operatorname{spt} \hat{\pi}_\# \partial(T \llcorner A \times \mathbb{R}^n)\ .$$

We could then rewrite Theorem 1 and Theorem 2 for Cartesian currents. We leave it to the reader; instead we shall explicitly state two propositions which will be used later on.

Proposition 5. *Let $\Omega \subset\subset \widetilde{\Omega}$ be two bounded domains in $\mathbb{R}^n$ and let $T, T_k \in \operatorname{cart}(\widetilde{\Omega} \times \mathbb{R}^n)$ be such that $T_k \rightharpoonup T$ and $\sup_k \mathbf{M}(T_k) < \infty$. Suppose that*

$$(T_k - T_h) \llcorner \pi^{-1}(\widetilde{\Omega} \setminus \overline{\Omega}) \; = \; 0 \qquad \forall \, h, k \; .$$

Then for all $y \in \Gamma_{T,\overline{\Omega}}$ we have

$$\deg(T_k, \overline{\Omega}, y) \; = \; \deg(T, \overline{\Omega}, y) \; .$$

Proof. For all k we have

$$T_k \llcorner \pi^{-1}(\widetilde{\Omega} \setminus \overline{\Omega}) \; = \; T \llcorner \pi^{-1}(\widetilde{\Omega} \setminus \overline{\Omega}) \; ,$$

hence

$$\begin{aligned} \partial(T_k \llcorner \pi^{-1}(\overline{\Omega})) &= -\partial(T_k \llcorner \pi^{-1}(\widetilde{\Omega} \setminus \overline{\Omega})) \\ -\partial(T \llcorner \pi^{-1}(\widetilde{\Omega} \setminus \overline{\Omega})) &= \partial(T \llcorner \pi^{-1}(\overline{\Omega})) \; . \end{aligned}$$

The result then follows from the homological invariance of the degree. □

Proposition 6. *Let $u, u_k \in \operatorname{cart}^p(\Omega, \mathbb{R}^n)$, $p > 1$, and let A be a Borel set in Ω. Suppose that*

$$\sup_k \| u_k \|_{\operatorname{cart}^p} < \infty \qquad \textit{and} \qquad G_{u_k} \rightharpoonup G_u \; .$$

Then

$$\deg(G_{u_k}, A, y) \; \to \; \deg(G_u, A, y)$$

provided y is an interior point of $\Gamma_{G_u,A} \cap \bigcap_k \Gamma_{G_{u_k},A}$.

Proof. In fact if $\omega = \phi(y)\, dy$ is an n-form with compact support in the connected component of y in $\Gamma_{G_u,A} \cap \bigcap_k \Gamma_{G_{u_k},A}$ with $\omega = 1$, we find

$$\begin{aligned} \deg(G_{u_k}, A, y) \; &= \; \widehat{\pi}_{\#}(G_{u_k} \llcorner \pi^{-1}(A))(\phi(y)dy) \\ &= \; \int_A \phi(u_k(x)) \det Du_k(x)\, dx \end{aligned}$$

thus

$$\lim_{k \to \infty} \deg(G_{u_k}, A, y) = \int_A \phi(u(x)) \det Du(x)\, dx = \deg(G_u, A, y) \; .$$

□

Again a certain care is necessary. Next example may serve to classify the situation.

[4] Consider the sequence of Lipschitz maps $u_k : (0,1) \to (0,1)$ given by

$$u_k(x) = \begin{cases} kx & \text{if } 0 < x < 1/k \\ 1 & \text{if } 1/k \leq x < 1 \,. \end{cases}$$

Clearly

$$\deg(u_k, (0,1), y) = 1 \qquad \forall\, y \in (0,1) \,.$$

The currents G_{u_k} of course belong to $\operatorname{cart}((0,1) \times \mathbb{R})$ and converge weakly in $\operatorname{cart}((0,1) \times \mathbb{R})$ to the Cartesian current in $\operatorname{cart}((0,1) \times \mathbb{R})$ G_{u_0} where $u_0(x) = 1$ for all $x \in (0,1)$, for which

$$\deg(G_{u_0}, (0,1), y) = 0 \qquad \forall\, y \in \mathbb{R} \setminus \{1\} \,.$$

However, if we regard the G_{u_k}'s as i.m. rectifiable currents in $\mathbb{R} \times \mathbb{R}$, we see at once that

$$G_{u_k} \rightharpoonup T := \delta_0 \times [\![(0,1)]\!] + [\![(0,1)]\!] \times \delta_1$$

and

$$\deg(T, \widehat{\pi}, y) = \deg(T, [0,1], y) = 1 \qquad \forall\, y \in (0,1) \,.$$

•

3.3 The Degree of Continuous Maps

In this subsection first we show how one can recover the classical notion of degree for merely continuous maps, and secondly that the degrees in the sense of currents and of continuous maps agree for currents T with $\operatorname{spt} T \subset \{(x, u(x)) \mid x \in \overline{\Omega}\}$, u continuous.

Let Ω be an open bounded set in $\mathbb{R}^n$ and let $u : \overline{\Omega} \to \mathbb{R}^n$ be a continuous map. We extend u outside of $\overline{\Omega}$ to a continuous map that we still denote by u. Chosen a standard mollifier φ and σ, $0 < \sigma < 1$, we let $\varphi_\sigma(x) := \sigma^{-n}\varphi(x/\sigma)$ and

$$u_\sigma = u * \varphi_\sigma \,.$$

Obviously $u_\sigma(x)$ converge uniformly to u in $\overline{\Omega}$. Consider now the continuous map $H : (0,1) \times \overline{\Omega} \to \mathbb{R}^n \times \mathbb{R}^n$ given by

$$H(\sigma, x) := (x, u_\sigma(x)) \,.$$

For any τ and σ, $0 < \tau < \sigma < 1$, H is of class C^∞ in $[\tau, \sigma] \times \overline{\Omega}$ and in $\mathbb{R}^n \times \mathbb{R}^n$ we have

$$\begin{aligned} \partial H_\#([\![(\tau,\sigma)]\!] \times [\![\Omega]\!]) &= H_\#(\partial[\![(\tau,\sigma)]\!] \times [\![\Omega]\!] - [\![(\tau,\sigma)]\!] \times [\![\partial\Omega]\!]) \\ &= H_\#(\delta_\sigma \times [\![\Omega]\!]) - H_\#(\delta_\tau \times [\![\Omega]\!]) - H_\#([\![(\tau,\sigma)]\!] \times [\![\partial\Omega]\!]) \\ &= G_{u_\sigma} - G_{u_\tau} - H_\#([\![(\tau,\sigma)]\!] \times \partial[\![\Omega]\!]) \,. \end{aligned}$$

Consequently the n-current

$$T := G_{u_\sigma} - G_{u_\tau} - H_\#([\![(\tau,\sigma)]\!] \times \partial[\![\Omega]\!])$$

has no boundary and finite mass. Hence

$$\deg(T, \widehat{\pi}, y) = 0 \qquad \text{for a.e. } y .$$

Choose now $y \in \mathbb{R}^n \setminus u(\partial\Omega)$ and let U be a neighborhood of $u(\partial\Omega)$ such that $\operatorname{dist}(U,y) > 0$. Then for $\sigma < \sigma_0 = \sigma_0(y)$ we have

$$u_\sigma(z) \in U \qquad \forall\, z \in \partial\Omega$$

and $\operatorname{spt}(\partial[\![\Omega]\!]) \subset \partial\Omega$. Therefore, since $H([\tau,\sigma] \times \partial\Omega) \subset \overline{U}$

$$\operatorname{spt} H_\#([\![(\tau,\sigma)]\!] \times \partial[\![\Omega]\!]) \subset \overline{U}$$

for $0 < \tau < \sigma < \sigma_0$. This implies that

$$\deg(H_\#([\![(\tau,\sigma)]\!] \times \partial[\![\Omega]\!]), \widehat{\pi}, y) = 0 .$$

Consequently

$$\deg(G_{u_\sigma}, \widehat{\pi}, y) = \deg(G_{u_\tau}, \widehat{\pi}, y)$$

for $0 < \tau < \sigma < \sigma_0$.

More generally, consider any map $v \in C^1(\overline{\Omega}, \mathbb{R}^N)$ sufficiently close to u in the sup norm

$$\| v - u \|_{\infty,\overline{\Omega}} < \varepsilon/2 , \qquad \varepsilon \text{ close to } 0$$

actually such that $v(\partial\Omega)$ is close to $u(\partial\Omega)$ in the sense that

$$\| v - u \|_{\infty,\partial\Omega} < \varepsilon/2$$

and the homotopy

$$H(t,x) := (x, tu_\sigma + (1-t)v)$$

σ close to 0, the argument given above shows that for any point y in $\mathbb{R}^n \setminus \mathcal{U}_\varepsilon$

$$\mathcal{U}_\varepsilon := \{ y \in \mathbb{R}^n \mid \operatorname{dist}(y, u(\partial\Omega)) < \varepsilon \} \tag{1}$$

we have

$$\deg(v, \Omega, y) = \deg(u_\sigma, \Omega, y)$$

for all σ close to zero.

Of course this allows us to define the *degree of* u *at any point* $y \in \mathbb{R}^n \setminus \partial\Omega$ as

$$\deg_c(u, \Omega, y) := \deg(G_{u_\sigma}, \widehat{\pi}, y)$$

for σ sufficiently close to 0 and state that $\deg_c(u, \Omega, y)$ does not depend on the approximating sequence u_σ we chose. Moreover from the previous argument we see that

$$\deg_c(u, \Omega, y) = \deg_c(v, \Omega, y) ,$$

if $u,\ v \in C^0(\overline{\Omega})$ with $u = v$ on $\partial\Omega$.

We then have

Proposition 1. *Let T be a current of finite mass in $(\Omega \times \mathbb{R}^n)$. Suppose that there is a function $u \in C^0(\overline{\Omega}, \mathbb{R}^n)$ such that*

$$\operatorname{spt} \partial T \subset \{(x, u(x)),\ x \in \partial\Omega\}$$

or, in particular, that $T \in \operatorname{cart}(\Omega \times \mathbb{R}^n)$ and $\operatorname{spt} T \subset \{(x, u(x)),\ x \in \overline{\Omega}\}$. Then

$$\deg(T, \Omega, y) = \deg_c(u, \Omega, y)$$

for all $y \in \mathbb{R}^n \setminus u(\partial\Omega)$.

Proof. We shall prove that for any positive ε

$$\deg(T, \Omega, y) = \deg_c(u, \Omega, y) \qquad \forall\, y \in \mathcal{U}_\varepsilon \tag{2}$$

where $\mathcal{U}_\varepsilon$ is the set defined in (1). From this the conclusion follows at once.

In order to prove (2) we consider a map $v \in C^1(\overline{\Omega}, \mathbb{R}^n)$ such that

$$\| v - u \|_{\infty, \overline{\Omega}} < \varepsilon/2 .$$

As we have seen

$$\deg_c(u, \Omega, y) = \deg(v, \Omega, y) = \deg(G_v, \Omega, y) \tag{3}$$

for all $y \in \mathbb{R}^n \setminus \mathcal{U}_\varepsilon$. Next, we consider the homotopy $\phi_t : \overline{\Omega} \times \mathbb{R}^n \to \overline{\Omega} \times \mathbb{R}^n$ given by

$$\phi_t(x, y) := (x, ty + (1-t)v(x)) .$$

It is easily seen that $\{\phi_{t\#}T \mid t \in [0,1]\}$ yields a family of Cartesian currents and in fact a continuous family of currents with $\phi_{0\#}T = \phi_{0\#} \circ \pi_\# T = \phi_{0\#}[\![\, \Omega \,]\!] = G_v$ and $\phi_{1\#}T = T$. Moreover

$$\operatorname{spt} \partial \phi_{t\#} T \subset (\partial\Omega \times \mathbb{R}^n) \cap \{(x, y) \mid x \in \partial\Omega,\ y = tu(x) + (1-t)v(x)\} ,$$

in particular

$$\operatorname{spt} \widehat{\pi}_\# \partial \phi_{t\#} T \subset (tu + (1-t)v)(\partial\Omega) \subset \mathcal{U}_\varepsilon \qquad \forall\, t \in [0,1] .$$

By Theorem 2 in Sec. 4.3.2 (iii) we then infer

$$\deg(T, \Omega, y) = \deg(G_v, \Omega, y) \qquad \forall\, y \in \mathbb{R}^n \setminus \mathcal{U}_\varepsilon ,$$

which together with (3) yields (2). □

Remark 1. Of course from Proposition 1 we infer that if $T \in \operatorname{cart}(\Omega \times \mathbb{R}^n)$ and $T = G_u$ for some $u \in C^0(\overline{\Omega}, \mathbb{R}^n)$, then

$$\deg(G_u, \Omega, y) = \deg_c(u, \Omega, y) \qquad \forall\, y \in \mathbb{R}^n \setminus u(\partial\Omega) .$$

Remark 2. For the Cantor–Vitali function $V(x)$ we then recover the well-known relation

$$\deg_c(V, (0,1), y) = 1 \qquad \forall\, y \in (0,1)$$

and compare [2] in Sec. 4.3.2,

$$\deg_c(V, (0,1), y) = \deg(-\partial S G_V, (0,1), y) \neq \deg(G_V, (0,1), y) .$$

3.4 *h*-Connected Components and the Degree

Let T be an integral current in $\mathbb{R}^{k+n}$, i.e. $T \in \mathcal{R}_n(\mathbb{R}^{k+n})$, $\mathbf{M}(T) + \mathbf{M}(\partial T) < \infty$, so that (by the boundary rectifiability theorem) also $\partial T \in \mathcal{R}_{n-1}(\mathbb{R}^{k+n})$, $\partial T = \tau(\mathcal{M}, \theta, \vec{\partial T})$. We have seen that $\deg(T, \hat{\pi}, y)$ is an integer-valued $BV(\mathbb{R}^n)$-function with $|D \deg(T, \hat{\pi}, y)|(\mathbb{R}^n \setminus \hat{\pi}(\mathcal{M})) = 0$. More precisely, setting

$$\Sigma_k \ := \ \{y \in \mathbb{R}^n \ \mid \ \deg(T, \mathbb{R}^n, y) = k\} \ ,$$

we have seen that Σ_k has finite perimeter and

$$\widehat{\pi}_{\#} T \llcorner \Sigma_k \ = \ k [\![\Sigma_k]\!] \ .$$

We shall now see that we can decompose $\mathbb{R}^n \setminus \hat{\pi}(\mathcal{M})$ into "connected" components on which the degree is constant, paralleling the classical theory. This is achieved by modifying the classical topological notions of closed sets and connected components into corresponding homological ones. A byproduct of those kind of results may be an alternative definition of the degree, but we shall not pursue any further this point.

Definition 1. *Let $F \subset \mathbb{R}^n$ be an $(n-1)$-rectifiable set and let $A \subset \mathbb{R}^n$ be a set of finite perimeter in $\mathbb{R}^n$, $P(A) < \infty$.*

(i) We say that F is homologically closed, *in short h-*closed, *if there exists a current $T \in \mathcal{R}_n(\mathbb{R}^n)$, with $\partial T \in \mathcal{R}_n(\mathbb{R}^n)$, $\partial T = \tau(\mathcal{M}, \theta, \vec{\xi})$, $\theta \geq 1$ on $\mathcal{M}$, such that $F = \mathcal{M}$.*

(ii) We say that A is homologically disconnected *if there exists $B \subset A$ with $P(B) < \infty$, $0 < |B| < \infty$, $|A \setminus B| > 0$ such that $\partial^- B \subset \partial^- A$, and consequently $\partial^-(A \setminus B) \subset \partial^- A$, in the $\mathcal{H}^{n-1}$-sense. Otherwise we say that A is h-*connected.

Equivalently we can state

Proposition 1. *$F \subset \mathbb{R}^n$ is h-closed if and only if there exists $u \in BV(\Omega, \mathbb{Z})$ such that*

$$F \ = \ J_u \qquad \mathcal{H}^{n-1} \ a.e.$$

where J_u is the jump set of u.

Proof. From Corollary 1 in Sec. 4.3.1 we know that T is an n-dimensional i.m. rectifiable current with finite boundary mass if and only if

$$T \ = \ \mathcal{L}^n \llcorner u$$

for some $u \in BV(\Omega)$ with integer values. Theorem 1 in Sec. 4.3.1 then yields

$$\partial T \ = \ \tau(J_u, u^+ - u^-, *n(\cdot, J_u)) \ .$$

□

Proposition 2. *Let $A \subset \mathbb{R}^n$ be a set of finite perimeter, $P(A) < \infty$. Then the following claims are equivalent*

(i) A is h-disconnected

(ii) There is an essentially non-constant function u in $BV(\Omega, \mathbb{Z})$, $\text{ess inf}_A u <$ $\text{ess sup}_A u$, *such that*

$$|Du|((A)_1) = 0 , \qquad (A)_1 := \{x \mid \theta(A,x) = 1\} .$$

(iii) There is $B \subset A$, with $0 < |B| < \infty$ and $|A \setminus B| > 0$ such that

$$P(A) = P(B) + P(A \setminus B) .$$

In particular, if $u \in BV(\Omega, \mathbb{Z})$, A is h-connected and $|Du|((A)_1) = 0$, then u is constant in A.

Proof. For $B \subset A$ we set $B_1 := B,\ B_2 := A \setminus B$. We have

$$\begin{aligned} \partial^- B_i &= (\partial^- B_i \cap \partial^- A) \cup (\partial^- B_i \setminus \partial^- A) \qquad i = 1,2 \\ \partial^- A &= (\partial^- A \cap \partial^- B_1) \cup (\partial^- A \setminus \partial^- B_2) \end{aligned}$$

and

$$\mathcal{H}^{n-1}((\partial^- A \cap \partial^- B_1) \cup (\partial^- A \setminus \partial^- B_2)) = 0$$

as in the $\mathcal{H}^{n-1}$-sense points in $\partial^- A$ cannot belong both to $\partial^- B_1$ and $\partial^- B_2$.

Moreover

$$\partial^- B_1 \setminus \partial^- A = \partial^- B_2 \setminus \partial^- A .$$

This shows that (i) is equivalent to (iii).

Let $B \subset A$ satisfy all conditions in Definition 1 (ii). Then $u := \chi_B$ satisfies $J_u = \partial^- B \subset \partial^- A$, i.e. $|Du|((A)_1) = 0$ and u is not constant in A, thus (i) implies (ii).

Let $u \in BV(\mathbb{R}^n, \mathbb{Z})$ be non-constant in A. Then there is $k \in \mathbb{Z},\ k \neq 0$ such that

$$B := \{x \in A \mid u(x) = k\}$$

satisfies $|B| > 0,\ |A \setminus B| > 0$, and $|B| < \infty$ as $u \in BV$. Then $\partial^- B \subset J_u$ and $|Du|((A)_1) = 0$ imply that $J_u \subset \mathbb{R}^n \setminus (A)_1$ i.e., $\partial^- B \subset \partial^- A$. □

We can now show that, similarly to the topological situation, every set A of finite perimeter can be decomposed into h–connected components.

Theorem 1. *Let $A \subset \mathbb{R}^n$ be a set of finite perimeter, $P(A) < \infty$. Then there exists a (possibly finite) sequence $\{A_k\}$ of disjoint h-connected sets such that $A = \bigcup_k A_k$*

$$P(A) = \sum_k P(A_k) \tag{1}$$

and, consequently,

$$\partial^- A_k \subset \partial^- A \qquad \forall\, k . \tag{2}$$

The proof of Theorem 1 relies on the following two lemmas.

Lemma 1. *Let $A \subset \mathbb{R}^n$, $0 < |A| < \infty$, $P(A) < \infty$. Then there exists $\varepsilon > 0$ such that for each (possible finite) disjoint decomposition $A = \bigcup_k A_k$ satisfying (1) we have $|A_{k_0}| > \varepsilon$ for some k_0.*

Proof. Choose $\varepsilon > 0$ so that

$$|A| \; > \; c_n \, P(A) \, \varepsilon^{\frac{1}{n}}$$

where c_n is the isoperimetric constant

$$c_n \; := \; \sup \frac{|A_k|^{\frac{n-1}{n}}}{P(A)} \; .$$

If $|A_k| < \varepsilon$ for all k, we have

$$|A_k| \; = \; |A_k|^{\frac{n-1}{n}} \, |A_k|^{\frac{1}{n}} \; \leq \; c_n \, P(A_k) \, \varepsilon^{\frac{1}{n}}$$

and by (1)

$$|A| \; \leq \; c_n \sum_k P(A_k) \, \varepsilon^{\frac{1}{n}} \; = c_n \, P(A) \, \varepsilon^{\frac{1}{n}} \; ,$$

a contradiction. □

Lemma 2. *Let $A \subset \mathbb{R}^n$, $0 < A < \infty$, $P(A) < \infty$. Then there exists a subset B of A with $|B| > 0$ which is h-connected.*

Proof. Let ε be the constant in Lemma 1. Assume that each $B \subset A$ with $|B| > 0$, $P(B) < \infty$, $\partial^- B \subset \partial^- A$ is h-disconnected. The collection of all such B's can and will be divided into four families τ_1, τ_2, τ_3 and τ_4

1. $B \in \tau_1$ iff $|B| \geq \varepsilon$ and there is a decomposition $B = B_1 \cup B_2$ satisfying (1) such that $|B_1| \geq \varepsilon$, $|B_2| \geq \varepsilon$
2. $B \subset \tau_2$ iff $B \notin \tau_1$, $|B| \geq \varepsilon$, and there is a decomposition $B = B_1 \cup B_2$ satisfying (1) such that $|B_1| < \varepsilon$ and $|B_2| < \varepsilon$
3. $B \in \tau_3$ iff $B \notin \tau_1 \cup \tau_2$ and $|B| \geq \varepsilon$
4. $B \in \tau_4$ iff $0 < |B| < \varepsilon$.

Clearly for each $B \in \tau_3$ there exists a decomposition such that

$$B = B_1 \cup B_2 \, , \quad P(B) = P(B_1) + P(B_2) \, , \quad |B_1| < \varepsilon \, , \quad |B_2| > 0 \, . \tag{3}$$

First we decompose A by only using decompositions of the type τ_1. After a finite number of steps we arrive at a decomposition satisfying (1)

$$A \; = \; \bigcup_{i=1}^{n_0} B_i \qquad n_0 \in \mathbb{N} \, , \; n_0 \geq 1$$

such that every $B_i \in \tau_2 \cup \tau_3$. Let us call this decomposition $\mathcal{R}_0$. For each ordinal α, $\alpha \geq 1$, we shall define inductively the decomposition $\mathcal{R}_\alpha$:

$$(4) \qquad A = \Big(\bigcup_{i=1}^{n_0} B_i^{(\alpha)}\Big) \cup \Big(\bigcup_{\beta<\alpha} \widetilde{B}_\beta^{(\alpha)}\Big)$$

satisfying (1) and such that

$$(5) \qquad B_i^{(\alpha)} \in \tau_2 \cup \tau_3 \ \ \forall\, i \quad \text{and} \quad \widetilde{B}_\beta^{(\alpha)} \in \tau_4 \ \ \forall\, \beta < \alpha\,,$$

as follows.

For a non limit α: By Lemma 1 some $B_{i_0}^{(\alpha-1)} \in \tau_3$, i.e., there exists a decomposition $B_{i_0}^{(\alpha-1)} = B_1 \cup B_2$ satisfying (3). Then we set

$$B_{i_0}^{(\alpha)} := B_1 \quad B_i^{(\alpha)} := B_i^{(\alpha-1)} \quad \text{for } i \neq i_0$$
$$\widetilde{B}_\beta^{(\alpha)} := \widetilde{B}_\beta^{(\alpha-1)} \quad \text{for } \beta < \alpha - 1\,, \quad \widetilde{B}_{\alpha-1}^{(\alpha)} := B_2$$

For a limit α: We set

$$(6) \qquad B_{i_0}^{(\alpha)} := \bigcap_{\beta<\alpha} B_i^{(\beta)} \qquad \forall\, i = 1, \dots, n_0$$

$$(7) \qquad \widetilde{B}_\beta^{(\alpha)} := \widetilde{B}_\beta^{(\beta+1)} \qquad \forall\, \beta < \alpha\,.$$

By construction we see that (6) is in fact a intersection of a countable decreasing family of sets, so that we have $|B_i^{(\alpha)}| \geq \varepsilon$, because of the inductive hypotheses $|B_{i_0}^{(\beta)}| \geq \varepsilon \ \forall\, \beta < \alpha$. Note also that no $B_{i_0}^{(\alpha)}$ is of type τ_1. This process can be realized for each countable α, but this way we obtain a non-countable family of disjoint sets $\widetilde{B}_\beta^{(\alpha)}$ with $|\widetilde{B}_\beta^{(\alpha)}| > 0$, which is impossible since $|A| < \infty$. □

Proof of Theorem 1. Suppose A be disconnected. Then there exists $B_1 \subset A$ such that $0 < |B_1| < \infty$, $P(B_1) < \infty$, $|A \setminus B_1| > 0$ and $P(A) = P(B_1) + P(A \setminus B_1)$.

Set $A_1 := A \setminus B_1$, and if A_1 is h-disconnected we continue this process (possibly transfinitely). After a countable number of steps we however exhaust A, because A is $\mathcal{H}^n$-σ-finite, compare Ch. 1. This way we obtain a decomposition of A

$$A = \bigcup_k B_k\,, \quad 0 < |B_k| < \infty\,, \quad P(B_k) < \infty \quad \forall\, k$$

and

$$P(A) = \sum_k P(B_k)\,.$$

Using now Lemma 2 we simply decompose each B_k as

$$B_k = \bigcup_j A_j^{(k)}\,, \quad P(B_k) = \sum_j P(A_j^{(k)})\,, \quad A_j^k \ h\text{-connected.}$$

This yields the conclusion since $\sum_{k,j} P(A_j^{(k)}) = P(A)$ which implies that $\partial^- A_j^{(k)} \subset \partial^- A$ for all k. □

Finally, next theorem provides a decomposition of $\mathbb{R}^n \setminus F$ into h-connected components, if F is h-closed.

Theorem 2. *Let F be an $\mathcal{H}^{n-1}$-rectifiable set which is h-closed in the sense of Definition 1. Then the set $G := \mathbb{R}^n \setminus F$ can be decomposed as a countable disjoint union $G := \cup_k G_k$ of h-connected sets G_k, called h-connected components of G, with $\partial^- G_k \subset F$ $\mathcal{H}^{n-1}$-a.e.. Such a decomposition is unique, up to permutations. Moreover for any $A \subset G$ with $\partial^- A \subset F$, we have $A \supset G_k$ $\mathcal{H}^{n-1}$-a.e. if $|A \cap G_k| > 0$.*

Proof. Let $u \in BV(\mathbb{R}^n, \mathbb{Z})$ be the function in Proposition 1, i.e., such that $F = J_u$. Set

$$\mathcal{U}_k := \{x \in \mathbb{R}^n \setminus J_u \mid u^+(x) = k\} \qquad k \in \mathbb{Z} .$$

We have

$$G = \bigcup_{k \in \mathbb{Z}} \mathcal{U}_k , \quad P(\mathcal{U}_k) < \infty \ \forall\, k , \quad J_u = \bigcup_{k \in \mathbb{Z}} \partial^- \mathcal{U}_k$$

consequently $\partial^- \mathcal{U}_k \subset F$. Now we decompose each $\mathcal{U}_k$ with $|\mathcal{U}_k| > 0$, into h-connected components using Theorem 1

$$\mathcal{U}_k = \bigcup_j G_j^{(k)} \qquad \forall\, k .$$

We have

$$\partial^- G_j^{(k)} \subset \partial^- \mathcal{U}_k \subset J_u = F .$$

This proves the first part of the theorem.

If $|A \cap G_k| > 0$, $\partial^- A \subset F$, we set $B_1 := A \cap G_k$, $B_2 := G_k \setminus A$. Then $G_k = B_1 \cup B_2$ is a decomposition of G_k with

$$\partial^- B_1 \subset \partial^- G_k \cup (\partial^- A \cap G_k) = \partial^- G_k ,$$

since $\partial^- A \subset F$, $F \cap G_k = \emptyset$. Similarly we have $\partial^- B_2 \subset \partial^- G_k$. The h-connectivity of G_k then implies $|B_2| = 0$, i.e., $A \supset G_k$ $\mathcal{H}^{n-1}$-a.e.. This proves the last part of the theorem and the uniqueness. □

Theorem 2 can now be easily applied to degree theory.

Let $T \in \mathcal{R}_n(\mathbb{R}^{k+n})$ be a current with $M(T) + M(\partial T) < \infty$. Write

$$\partial T = \tau(\mathcal{M}, \theta, \vec{T}) \in \mathcal{R}_{n-1}(\mathbb{R}^{k+n}) ,$$

set

$$F := \widehat{\pi}(\mathcal{M}) ,$$

and let

$$\mathbb{R}^n \setminus F = \bigcup_k G_k$$

be the decomposition of $\mathbb{R}^n \setminus F$ into h-connected components. The current $\widehat{\pi}_\# T$ belongs to $\mathcal{R}_n(\mathbb{R}^n)$ and can be written as $\hat{\pi}_\# T = \mathcal{L}^n \llcorner u$ for some $u \in BV(\mathbb{R}^n, \mathbb{Z})$, and $J_u \subset F$ $\mathcal{H}^{n-1}$ a.e., i.e.,

$$\partial \widehat{\pi}_\# T = \tau(F, \cdot, \cdot) .$$

Consequently $|Du|(G_k) = 0$ for each k, and u is constant on G_k. In other words

$$\deg(T, \widehat{\pi}, \cdot)$$

is constant on each G_k.

Remark 1. Notice that the rectifiable h-closed set F can be dense say in $B(0,1)$ so that $\operatorname{spt} \widehat{\pi}_\# \partial T \supset B(0,1)$. Nevertheless $|F| = 0$, i.e., $|B(0,1) \setminus \cup G_k| = 0$. Therefore by the above we still have a degree information on $B(0,1)$, providing T and ∂T are both i.m. rectifiable.

4 Notes

1 *Functions of bounded variation in one variable.* The theory of functions of bounded variation in one variable is classical and plays an important role in many fields as for instance in the theory of Fourier series, in the theory of rectifiability of curves, and mainly in the theory of differentiation. It developed with important contributions of Lebesgue, Vitali, Lusin, Denjoy, among others, and in view of a theorem, essentially, due to Jordan, which states that *a function has bounded variation if and only if it is the difference of the two non-decreasing functions*, it can be reduced to the study of monotone functions. For instance the classical Lebesgue's theorem about the almost everywhere differentiability of monotone functions implies then the almost everywhere differentiability of functions of bounded variation. Notice that this is typical of dimension one. In fact in dimension larger than one functions of bounded variations, indeed Sobolev functions, though approximately differentiable almost everywhere, may be not differentiable in the classical sense in any point.

For the one-dimensional theory of BV-functions we refer the interested reader for example to Riesz and Sz.-Nagy [556] Saks [570], see also Rudin [566] Federer [226]. Here we recall some of its special features, with respect to the n-dimensional theory, with also the aim of showing that the classical definition agrees with the one of Sec. 4.1.1, when specialized to $n = 1$.

To each function $f : (a,b) \to \mathbb{R}$, not necessarily continuous, classically one associates its *total variation function* defined for $x \in (a,b)$,by

$$T_f(x) := \sup \sum_{j=1}^{N} |f(x_j) - f(x_{j-1}|$$

where the supremum is taken over all N and over all choices of $\{x_j\}$ such that $a < x_0 < \ldots < x_{N-1} \leq x$. Clearly, $T_f(x)$ is a non-decreasing function, whence

$$V_a^b(f) := \lim_{x \to b^-} T_f(x)$$

exists. If it is finite one says that f is of *bounded variation* and one calls $V_a^b(f)$ the *total variation* of f.

The previous definition has the disadvantage that, if we modify f just at one point, then both $T_f(x)$ and $V_a^b(f)$ change. Hence it is convenient to find "normalized representatives" of functions of bounded variations. This can be done as follows. First one observes that for $x < y$ we have

$$|f(y) - f(x)| \leq T_f(y) - T_f(x) .$$

In particular $\{f(x_i)\}$ is a Cauchy sequence whenever $\{T_f(x_i)\}$ is a Cauchy sequence. Since T_f is a monotone function, and monotone functions have right- and left-limits at all points, and have at most countably many discontinuities, we conclude that the same is true for f. Therefore we can define

$$c := \lim_{t \to a^+} f(t), \qquad g(x) := f(x^-) - c$$

Clearly, $g(x)$ is a left-continuous function, $V_a^b(g) \leq V_a^b(f)$ and we can conclude:

For every function f of bounded variation, $f(x^-)$ exists at every point x of $(a, b]$, $f(x^+)$ exists at every point x of $[a, b)$, the set of points at which f is discontinuous, jump points, is at most countable and there is a unique c and a unique function g of bounded variation, which is left-continuous, satisfies

$$\lim_{x \to a^+} g = 0$$

so that

$$f(x) = c + g(x)$$

at all points of continuity of f.

Using the previous claim we *normalize* every f of bounded variation to $c + g(x)$, and we denote by $NBV(a, b)$ the class of normalized functions of bounded variation. It is easily seen that

$$T_f(x) \in NBV(a, b) \quad \text{whenever} \quad f \in NBV(a, b) .$$

Theorem 1. *We have*

(i) If μ is a Radon measure on (a, b) and if we set

$$f(x) := \int_{(a,x)} d\mu \qquad x \in (a, b) \tag{1}$$

then $f \in NBV(a, b)$

(ii) Conversely, to every $f \in NBV(a, b)$ there corresponds a unique Radon measure μ such that (1) holds; moreover for this μ we have

$$T_f(x) = \int_{(a,x)} d|\mu| .$$

(iii) Finally, if (1) holds, then f is continuous precisely at those points x at which $\mu\{x\} = 0$

Proof. (i) If $x_k \uparrow x$, clearly $\lim_{k \to \infty} \mu((a, x_k)) = \mu(a, x)$, so $f(x_k) \to f(x)$. If $x_k \downarrow a$, then $\mu((a, x_k)) \to 0$, so $f(x_k) \to 0$. While for $a < x_0 < \ldots < x_N = x$

$$\sum_{i=1}^{N} |f(x_i) - f(x_{i-1})| = \sum_{i=1}^{N} |\mu([x_{i-1}, x_i))| \leq |\mu|(a, x)$$

whence

(2) $$T_f(x) \leq |\mu|(a,x) .$$

(ii) We have $f = u - v$ where

$$u := \frac{1}{2}(T_f + f) \qquad v := \frac{1}{2}(T_f - f)$$

are two nondecreasing functions which belongs to $NBV(a,b)$.

We then define two measures μ_u and μ_v as follows. To each $x \in (a,b)$ we associate the set E_x (in dependence of u, respectively of v) defined by

$$E_x := \begin{cases} \{u(x)\} & \text{if } u \text{ is continuous at } x \\ [u(x), u(x^+)] & \text{if } u(x^+) > u(x) \end{cases}$$

and to a set $E \subset (a,b)$ we associate the set $\underset{x\in E}{\cup} E_x$. Then we define

$$\mu_u(E) = \mathcal{L}^1\Big(\bigcup_{x\in E} E_x\Big) .$$

Set now

$$\mu := \mu_u - \mu_v$$

and denote by λ the measure μ_{T_f} associated in the same way to T_f. It is not difficult to show that μ is a Radon measure,

$$\mu([\alpha,\beta)) = f(\beta) - f(\alpha)$$

and

$$\lambda([\alpha,\beta)) = T_f(\beta) - T_f(\alpha) .$$

Hence we get

$$|\mu(E)| \leq \lambda(E)$$

i.e.

$$|\mu|(a,x) \leq \lambda((a,x)) \leq T_f(x)$$

which together with (2) concludes the proof of (ii), and of the theorem, since (iii) follows at once. □

The previous theorem shows of course that functions of bounded variations in the sense of Sec. 4.1.1 just agree with functions of bounded variations in the classical sense.

Denoting by μ the measure associated to the function $f \in NBV$, and using Lebesgue decomposition theorem, we also get

$$\mu(E) = \mu^{(s)}(E) + \int_E \frac{d\mu}{d\mathcal{L}^1}(t)\, dt .$$

Hence, if we set

$$f_s(x) := \mu^{(s)}((a,x)) \qquad x \in (a,b)$$

we conclude that

f_s *is* a.e. *(approximately) differentiable with* $\text{ap}Df_s = 0$.

f *is* a.e. *(approximately) differentiable with* $\text{ap}Df = \dfrac{d\mu}{d\mathcal{L}^1}$.

Finally, $f_s = 0$ if and only if f is *absolutely continuous*.

2 Functions of bounded variation in multidimensional domains were first considered by Cesari [140] in the classical context. Later, after the work of De Giorgi on Caccioppoli sets De Giorgi [176] [177], the modern development of the theory of BV-functions started with the work of Krickeberg [415] and mainly with the work of Miranda [474] [475] [476], and Vol'pert [651]. But, contributions can also be found in the work of many other authors, some of which will be mentioned later.

Starting from the seventies several systematic presentations of the theory of BV-functions and Caccioppoli sets have appeared in books. We mention in particular Anzellotti, Giaquinta, Massari, Modica, and Pepe [49] Giusti [308] Massari and Miranda [461] Maz'ja [463] Vol'pert and Hudjaev [652] Ziemer [686].

Our presentation here focused mostly on structure type theorems. We would like to give now a short account of further relevant results.

3 *Traces of BV-functions.* The theory of traces of BV-functions was essentially stated in Miranda [478], and it is presented for example in Giusti [308] or in Vol'pert [651]. Extension can be found in Burago and Maz'ja [116] Meyers and Ziemer [467] Anzellotti and Giaquinta [45] Souček [605]. Here we shall briefly report on the results of Anzellotti and Giaquinta [45] which however rely on Miranda [478].

We denote by Ω an open set in $\mathbb{R}^n$ such that

$$P(\Omega, \mathbb{R}^n) < \infty\,, \qquad \mathcal{H}^{n-1}(\partial\Omega \setminus \partial^-\Omega) = 0 \tag{3}$$ [2]

Using the coarea formula, as in the proof of Theorem 4 in Sec. 4.1.4 one can then show that the Borel functions

$$\begin{aligned} f_-(x) &:= \sup\{t \in \mathbb{R} \mid \theta(E_t(f), x) = 0\} \\ f_+(x) &:= \inf\{t \in \mathbb{R} \mid \theta(E_t(f), x) = 1\} \\ F(x) &:= \frac{1}{2}(f_-(x) + f_+(x)) \end{aligned}$$

are well defined for $\mathcal{H}^{n-1}$-a.e $x \in \partial\Omega$, and

$$f_-(x) = f_+(x) = F(x) \qquad \mathcal{H}^{n-1}\text{-a.e. in } \partial\Omega$$

whenever $f \in BV(\Omega)$.

Theorem 2. *Let $f \in BV(\Omega) \cap L^\infty(\Omega)$ where Ω is an open set in $\mathbb{R}^n$ satisfying (3). Then*

(i) For $\mathcal{H}^{n-1}$-a.e. $x \in \partial\Omega$

$$F(x) = f_-(x) = f_+(x) = \lim_{\rho\to 0+} \frac{1}{|\Omega \cap B(x,\rho)|} \int_{\Omega\cap B(x,\rho)} f(y)\,dy$$

and

$$\lim_{\rho\to 0+} \frac{1}{|\Omega \cap B(x,\rho)|} \int_{\Omega\cap B(x,\rho)} |f(y) - F(x)|^{\frac{n}{n-1}}\,dy = 0$$

---

[2] Actually the condition $\mathcal{H}^{n-1}(\partial\Omega \setminus \partial^-\Omega) = 0$ is not necessary; the results of Theorem 2, Theorem 3 and Theorem 4 below remain true of course with $\partial\Omega$ replaced by $\partial^-\Omega$.

(ii) Denote by $\widetilde{f}$ the extension of f to zero outside Ω. Then for any Borel set $B \subset \mathbb{R}^n$ we have

$$\int_B D\widetilde{f} = \int_{B\cap\Omega} Df + \int_{B\cap\partial\Omega} F\nu\, d\mathcal{H}^{n-1}$$

or equivalently

$$\int_{\mathbb{R}^n} g\, D_i\widetilde{f} = \int_\Omega g\, D_i f + \int_{\partial\Omega} g\, F\nu\, d\mathcal{H}^{n-1} \quad \forall\, g \in C_c^\infty(\mathbb{R}^n),\ i = 1\dots,n$$

where ν is the inward normal to $\partial\Omega$. Finally,

$$\int_{\mathbb{R}^n} |D\widetilde{f}| = \int_\Omega |Df| + \int_{\partial\Omega} |F|\, d\mathcal{H}^{n-1}$$

and the Gauss-Green formula

$$\int_\Omega Df = -\int_{\partial\Omega} F\nu\, d\mathcal{H}^{n-1}$$

holds.

In order to extend the previous theorem to unbounded functions one needs a *trace estimate*, for which some regularity of $\partial\Omega$ is essential.

For $x \in \partial\Omega$ we set

$$q(x,\Omega) := \limsup_{\rho\to 0+}\left\{ \frac{\int_{\partial\Omega} \chi_A\, d\mathcal{H}^{n-1}}{\int_\Omega |D\chi_A|} \;\middle|\; A \subset \Omega\cap B_\rho(x),\ |A| > 0,\ P(A,\Omega) < \infty \right\}$$

and

$$Q(\Omega) := \sup_{x\in\partial\Omega} q(x)\ .$$

We observe that for open sets with C^1 boundary one has $Q = 1$, while for Lipschitz domains one essentially has $Q = \sqrt{1+L^2}$ where L is the Lipschitz constant of $\partial\Omega$. Finally, $Q < \infty$ for open sets with outward pointing cusps, while the assumption (3) is not sufficient for $Q < \infty$.

Theorem 3. *Let Ω satisfy (3) and $Q(\Omega) < \infty$. Then for any $\varepsilon > 0$ and any $f \in BV(\Omega)$ we have*

$$\int_{\partial\Omega} |F|\, d\mathcal{H}^{n-1} \le (Q(\Omega)+\varepsilon)\int_\Omega |Df| + c(\Omega,\varepsilon)\int_\Omega |f|\, dx \tag{4}$$

where $c(\Omega,\varepsilon)$ does not depend on f. Moreover,

$$\int_\Omega |F|\, d\mathcal{H}^{n-1} \le L\int_\Omega |Df| + c(\Omega)\int_\Omega |f|\, dx \tag{5}$$

holds for a domain Ω satisfying (3) and for all $f \in BV(\Omega)$ if and only if $Q(\Omega) < \infty$, and in this case

$$Q(\Omega) = \inf\{L \mid (5)\ holds\}\ .$$

Rightfully one calls the function $F(x)$ on $\partial\Omega$,

$$F(x) = f_+(x) = f_-(x)$$

the *trace of f on $\partial\Omega$*, denoted also γf, and one proves

Theorem 4. *Let Ω satisfy (3) and $Q(\Omega) < \infty$ and let $f \in BV(\Omega)$. Then*

(i) (Local characterization of the trace). *All conclusions of Theorem 2 hold.*

(ii) (Continuity of the trace operator). *Let $\{f_k\}$ be a sequence in $BV(\Omega)$ such that $f_k \to f$ in $L^1(\Omega)$ and*

$$\int_\Omega |Df_k| \longrightarrow \int_\Omega |Df|$$

then $\gamma f_k \to \gamma f$ in $L^1(\partial\Omega)$.

We observe that the trace operator γ is not continuous with respect to the weak convergence in BV. For instance the functions

$$f_k(x) = \begin{cases} k(x+1) & -1 < x < -1 + 1/k \\ 1 & -1 + 1/k < x < 1 - 1/k \\ k(x-1) & 1 - 1/k < x < 1 \end{cases}$$

have trace zero on $\partial(-1,1)$ and converge weekly in BV to the function identically 1 which has trace on $\partial(-1,1)$ the constant function 1.

Finally we mention that one can show that the trace operator is onto $L^1(\Omega)$, as consequence of the classical *Gagliardo's theorem* on the trace of $H^{1,1}$-functions Gagliardo [256].

4 *Slicing BV functions.* Among the various characterizations of functions in BV [3] one which is of some relevance, also from the historical point of view [4] is the following: *A function u belongs to $BV(\mathbb{R}^n)$ if and only if for almost all lines parallel to the axes its restriction is a function of bounded variation (in one variable)* compare Serrin [585] and Hughs [389]. More precisely, we can state this fact by the following theorem which follows from the approximation Theorem 1 in Sec. 4.1.1.

Denote by Π_i the coordinate $(n-1)$-plane orthogonal to e_i in $\mathbb{R}^n$, $i = 1, \dots, n$. For $u \in L^1(\mathbb{R}^n)$ and $y \in \Pi_i$, $i = 1, \dots, n$, denote by $u_{i,y}(t)$ the function

$$u_{i,y}(t) := u(y + te_i) .$$

By Fubini theorem $u_{i,y}$ is well defined for $\mathcal{H}^{n-1}$-a.e. $y \in \Pi_i$.

Theorem 5. *The function u belongs to $BV(\mathbb{R}^n)$ if and only if for all $i = 1, \dots, n$ and $\mathcal{H}^{n-1}$-a.e. $y \in \Pi_i$ the functions $u_{i,y}(t)$ belong to $BV(\mathbb{R})$. Moreover we have*

$$|D_i u|(\mathbb{R}^n) = \int_{\Pi_i} \Big(\int_{\mathbb{R}} |Du_{i,y}| \Big) \, d\mathcal{H}^{n-1}(y)$$

It is well-known that the previous theorem can be improved under certain circumstances. For instance Calkin [133] showed that the functions $u_{i,y}$ are absolutely continuous for any $i = 1, \dots, n$ and $\mathcal{H}^{n-1}$-a.e. y if and only if $u \in W^{1,1}(\mathbb{R}^n)$ and Vol'pert [651] proved that the $u_{i,y}$ have only jump parts if $u = \sum \lambda_i \chi_{E_i}$, E_i being sets of finite perimeter. In this direction the following general theorem is due to Ambrosio [32]

[3] Results in this number could be state for functions in $BV(\Omega)$, $\Omega \subset \mathbb{R}^n$ open. Only for the sake of simplicity we preferred to state them for $\Omega = \mathbb{R}^n$.

[4] This was in fact the original definition of Cesari.

Theorem 6. *Decompose the measure Du as*

$$Du = \text{ap}Du\,dx + (Du)^{(j)} + (Du)^{(C)} .$$

For any Borel set B define

$$\begin{aligned} B_{i,y} &:= \{t \in \mathbb{R} \mid y + te_i \in B\} \\ \Pi_{i,B_i} &:= \{y \in \Pi \mid B_{i,y} \neq \emptyset\} . \end{aligned}$$

Then we have

$$\begin{aligned} \int_B \text{ap}Du\,dx &= \int_{\Pi_{i,B_i}} d\mathcal{H}^{n-1}(y) \int_{B_{i,y}} \text{ap}Du_{i,y}(t)\,dt \\ (Du)^{(j)}(B) &:= \int_{\Pi_{i,B_i}} (Du_{i,y})^{(j)}(B_{i,y})\,d\mathcal{H}^{n-1}(y) \\ (Du)^{(C)}(B) &:= \int_{\Pi_{i,B_i}} (Du_{i,y})^{(C)}(B_{i,y})\,d\mathcal{H}^{n-1}(y) . \end{aligned}$$

Moreover for almost every $y \in \Pi_i$

$$\text{ap}Du_{i,y}(t) = \text{ap}Du(y + te_i) \qquad \textit{for } \text{a.e. } t \in \mathbb{R} ,$$

the jump set of $u_{i,y}$ is given by

$$J_{u_{i,y}} = \{t \in \mathbb{R} \mid y + te_i \in J_u\}$$

and

$$u^-_{i,y}(t) = u^-(y_i + te_i) , \qquad u^+_{i,y}(t) = u^+(y_i + te_i) .$$

5 *Vector valued BV-functions.* We make only a few remarks. Denote by $BV(\Omega, \mathbb{R}^N)$ the space of maps from $\Omega \subset \mathbb{R}^n$ into $\mathbb{R}^N$, $N \geq 2$, whose components are functions in $BV(\Omega)$. For $u = (u^1, \dots, u^N) \in BV(\mathbb{R}^n, \mathbb{R}^N)$ we define the *jump set of* u as

$$J_u = \bigcup_{i=1}^N J_{u^i} .$$

From the theory of Sec. 4.1.4 one can easily deduce that J_u is a countably $(n-1)$-rectifiable set and the matrix-valued measure Du split as

$$\begin{aligned} Du &= (Du)^{(a)} + (Du)^{(j)} + (Du)^{(C)} \\ (Du)^{(j)} &= Du \llcorner J_u \end{aligned}$$

and

$$|Du| = |(Du)^a| + |(Du)^{(j)}| + |(Du)^{(C)}| .$$

Also one sees that

$$(Du)^a = \mathcal{L}^n \llcorner \text{ap}Du$$

and

$$|Du|(B \setminus J_u) = 0$$

for any Borel set with $\mathcal{H}^{n-1}(B) < \infty$. Finally, one can show that

$$(Du^i)^{(j)} = (u^i_+(x) - u^i_-(x))\, n(x, J_u)\, d\mathcal{H}^{n-1}$$

or in other words that the matrix $(Du)^{(j)}$ is of rank one

$$(Du)^{(j)} = (u_+(x) - u_-(x)) \otimes n(x, J_u)\, d\mathcal{H}^{n-1} .$$

An interesting result of Alberti [7] proves that also $(Du)^{(C)}$ is of rank one, that is

$$\frac{d\,(Du)^{(C)}}{d\,|(Du)^{(C)}|}(x)$$

is a rank-one matrix $|(Du)^{(C)}|$-a.e. In other words, the Cantor part $(Du)^{(C)}$ has a similar geometric structure of the jump part $(Du)^{(j)}$.

6 *Compositions and products.* Let $f : \mathbb{R}^N \to \mathbb{R}$ be a continuously differentiable function and let $u \in BV(\Omega, \mathbb{R}^N)$, $\Omega \subset \mathbb{R}^n$. In Vol'pert [651] Vol'pert and Hudjaev [652] it is proved that the function

$$v := f(u)$$

is in $BV(\Omega)$ and

$$Dv = Df(u)\, Du \qquad \text{on } \Omega \setminus J_u \tag{6}$$

$$Dv = [f(u_+(x) - f(u_-(x))]\, n(x, J_u)\, d\mathcal{H}^{n-1} \qquad \text{on } J_u . \tag{7}$$

More recently in Ambrosio and Dal Maso [37] a general chain rule in the case $f : \mathbb{R}^N \to \mathbb{R}^k$ is a Lipschitz-continuous function is proved. The main difficulty in doing that consists in extending equality in (6), in fact it may happen that the function f is nowhere differentiable in the range of u.

To state the result of Ambrosio and Dal Maso [37], we observe that $|Du|$-a.e. point x in $\Omega \setminus J_u$ is a point of approximate continuity for u and of existence of the Radon-Nikodym derivative of Du with respect to $|Du|$. For such points x we denote by T_x the tangent space

$$T_x := \{y \in \mathbb{R}^N \mid y = u(x) + < \frac{d\,Du}{d\,|Du|}(x), z > \text{ for some } z \in \mathbb{R}^N\} .$$

Then we have

Theorem 7. *Let $u \in BV(\Omega, \mathbb{R}^N)$ and let $f : \mathbb{R}^N \to \mathbb{R}^k$ be a Lipschitz continuous function. Then $v := f(u)$ belongs to $BV(\Omega, \mathbb{R}^k)$ and*

$$(Dv)^{(j)} = [f(u_+(x) - f(u_-(x))] \otimes n(x, J_u)\, d\mathcal{H}^{n-1} .$$

Moreover, $|Du|$-a.e. on $\Omega \setminus J_u$ $f_{|T_x}$ is differentiable at $u(x)$ and

$$Dv = Df_{|T_x}(u(x))\, Du \qquad \text{on } \Omega \setminus J_u .$$

In particular for $N = 1$ we find, similarly to the case of continuously differentiable f, that f is differentiable at $u(x)$ for $|Du|$-a.e. x in $\Omega \setminus J_u$ and (6) holds. Notice also that Theorem 7 extends the classical chain rule for Sobolev functions $W^{1,1}(\Omega)$, essentially due to La Vallée Poussin, compare Serrin and Varberg [587], to a general chain rule for functions in $W^{1,1}(\Omega, \mathbb{R}^N)$, compare Marcus and Mizel [458].

Let us return to the case of continuously differentiable functions f and let us look at the composition $f(u)$, $u \in BV(\Omega, \mathbb{R}^N)$. We can write formulas (6) and (7) in a more suggestive form introducing for each point x (apart for a set of zero $(n-1)$-dimensional measure) the *averaged composition*

$$\overline{f \circ u}(x) = \int_0^1 f(tu_+(x) + (1-t)u_-(x))\, dt .$$

Then it is immediately seen that (6) and (7) can be equivalently written as

$$D_i f(u) = \sum_{k=1}^{N} \overline{f_{u^k} \circ u}(x)\, D_i u^k \tag{8}$$

where the equality has to be understood in the sense of measures. In fact one can show that $f(u) \in BV(\Omega, \mathbb{R}^k)$ *and (8) holds provided the functions* $\overline{f_{u^k} \circ u}(x)$ *be locally summable with respect to the measures* $D_i u^k$. An immediate consequence of this, compare Vol'pert [651] Vol'pert and Hudjaev [652] is: *Let* $u, v \in BV(\Omega)$. *Suppose that* $\overline{u} := \overline{\mathrm{id} \circ u}(x)$ *is locally summable with respect to* Dv *and* $\overline{v} := \overline{\mathrm{id} \circ v}(x)$ *is locally summable with respect to* Du *then* $uv \in BV(\Omega)$ *and*

$$D(uv) = \overline{u}\, Dv + \overline{v}\, Du .$$

As a special case let us consider the case in which $u \in BV(\Omega) \cap L^\infty(\Omega)$ and $v = \chi_E$ where E is a set of finite perimeter: we have $u\chi_E \in BV$ and

$$D(u\chi_E) = \frac{u_+(x) - u_-(x)}{2}\, n(x, E)\, d\mathcal{H}^{n-1} \llcorner \partial^- E + (\chi_{E_1} + \frac{1}{2}\chi_{E_{1/2}})\, Du$$

where

$$E_1 = \{x \in E \mid \theta(E, x) = 1\}, \qquad E_{1/2} = \{x \in E \mid \theta(E, x) = \frac{1}{2}\}.$$

7 *The space* $SBV(\Omega)$. Certainly the most complicated part of the gradient of a BV-function is its Cantor part. In many situations as for instance in problems with free discontinuities or in mathematical problems connected with the study of the image segmentation, see e.g. De Giorgi and Ambrosio [184] Ambrosio [33], one would like to deal with BV-functions which are free of a Cantor part in the gradient.

The space $SBV(\Omega)$ is defined as the space of functions $u \in BV(\Omega)$ such that $(Du)^{(C)} = 0$, or in other words as the subclass of $BV(\Omega)$ of functions $u \in BV(\Omega)$ such that

$$Du = \mathrm{ap}Du(x)\, d\mathcal{L}^n(x) + (u^+(x) - u^-(x))\, n(x, J_u)\, d\mathcal{H}^{n-1} \llcorner J_u .$$

What makes useful this class is the following compactness theorem, conjectured in De Giorgi and Ambrosio [184] and proved in Ambrosio [32], compare also Ambrosio [36] and Alberti and Mantegazza [9] for different proofs.

Theorem 8. *Let* $\{u_k\} \subset SBV(\Omega) \cap L^\infty(\Omega)$ *be a sequence such that*

$$\sup \left\{ \int_\Omega |\mathrm{ap}Du_k|^p\, dx + \mathcal{H}^{n-1}(J_{u_k}) + \| u_k \|_\infty \right\} < \infty$$

for some $p > 1$. *Then there exists a subsequence* $\{u_{k_i}\}$ *converging in* $L^1(\Omega)$ *to* $u \in SBV(\Omega)$ *and*

$$\int_\Omega |\mathrm{ap}Du|^p\, dx \le \liminf_{i \to \infty} \int_\Omega |\mathrm{ap}Du_{k_i}|^p\, dx$$

$$\mathcal{H}^{n-1}(J_u) \le \liminf_{i \to \infty} \mathcal{H}^{n-1}(J_{u_{k_i}})$$

As we know, Cantor or jump parts cannot be produced by weak convergence in $W^{1,1}$. Theorem 8 says in particular that Cantor parts cannot be produced by jumps of controlled length.

For the reader's convenience we give a proof of Theorem 8 emphasizing its geometric aspects. First let us prove the following characterization of subgraphs of SBV functions with jump part of finite size.

Proposition 1. *Let $u \in BV(\Omega, \mathbb{R})$. Then $u \in SBV(\Omega, \mathbb{R})$ with $\mathcal{H}^{n-1}(J_u) < \infty$ if and only if $\mathbf{M}((\partial G_u)_{(0)}) < +\infty$. Moreover $\mathbf{M}((\partial G_u)_{(0)}) = 2\,\mathcal{H}^{n-1}(J_u)$.*

Proof. From Theorem 3 in Sec. 4.2.4 we now that the current associated to the "graph" of a BV function, $T_u := (-1)^n \partial SG_u$, decomposes as

$$T_u = G_u + T^{(C)} + T^{(j)} \tag{9}$$

where G_u is the graph current of u, and $T^{(C)}$ and $T^{(j)}$ are respectively the Cantor and the jump part of T_u. In fact

$$T^{(C)}(\omega) = T^{(j)}(\omega) = 0, \text{ if } \omega = f(x,y)\,dx \tag{10}$$

and

$$T^{(C)}(\omega) = \int_\Omega f(x, u_+(x))\,dD_i u^{(C)} \tag{11}$$

$$T^{(j)}(\omega) = \int_\Omega \Big(\int_{u_-(x)}^{u_+(x)} f(x,y)\,dy \Big) n_i(x, J_u)\,d\mathcal{H}^{n-1} \llcorner J_u \tag{12}$$

if $\omega = f(x,y)\widehat{dx^i}\wedge dy$.

Let $u \in SBV(\Omega, \mathbb{R})$. Clearly (10) and (11) yield $T^{(C)} = 0$. Consequently using (9) and (12) we infer

$$\mathbf{M}((\partial G_u)_{(0)}) = \mathbf{M}((\partial T^{(j)})_{(0)}) = 2\,\mathcal{H}^{n-1}(J_u).$$

Conversely, (9) yields that $\mathbf{M}((\partial T^{(j)} + \partial T^{(j)})_{(0)}) = \mathbf{M}((\partial G_u)_{(0)})$. Assuming that $\mathbf{M}((\partial G_u)_{(0)}) < \infty$, the supremum of $(\partial T^{(j)} + \partial T^{(j)})_{(0)}$ on forms of the type $\omega = \phi(x)\psi(y)\widehat{dx^i}\wedge dy$, $\phi \in C_c^\infty(\Omega)$, $\psi \in C_c^\infty(\mathbb{R})$ is finite and we have

$$\sup_{\substack{|\phi|,|\psi|\le 1 \\ i=1,n}} \Big\{ T^{(C)}(\phi(x)\psi'(y)\widehat{dx^i}\wedge dy) + T^{(j)}(\phi(x)\psi'(y)\widehat{dx^i}\wedge dy) \Big\} \le \mathbf{M}((\partial G_u)_{(0)}) \tag{13}$$

Taking into account that $|Du^C|$ and $\mathcal{H}^{n-1} \llcorner J_u$ are mutually singular, (10), (11), and (12), we finally infer from (13)

$$\sup_{|\psi|\le 1} \int_\Omega |\psi'(u_+(x))|\,d|Du^C| = \mathbf{M}((\partial T^{(C)})_{(0)}) \le \mathbf{M}((\partial G_u)_{(0)}) < \infty. \tag{14}$$

Now, denoting by σ the positive measure on $\mathbb{R}$ given by $\sigma(B) = |Du^C|(u_+^{-1}(B))$ for all Borel sets $B \subset \mathbb{R}$, inequality (14) amounts to

$$\sup_{|\psi|\le 1} \int_{\mathbb{R}} |\psi'|\,d\sigma < \infty.$$

which is clearly impossible unless $\sigma = 0$. Therefore $|Du^C| = 0$, hence $u \in SBV(\Omega, \mathbb{R})$; and, going back to (9) $T_u = G_u + T^{(j)}$ which concludes the proof since trivially

$$\mathbf{M}((\partial T^{(j)})_{(0)}) = 2\mathcal{H}^{n-1}(J_u).$$

□

Proof of Theorem 8. Passing to subsequences, we can assume that $u_k \to u$ in L^1, $Du_k \rightharpoonup Du$ as measures and that $Du^a \to v$ in $L^1(\Omega, \mathbb{R})$, $v \in L^1(\Omega, \mathbb{R}^n)$. Also the currents G_{u_k} converge to a current S in $\Omega \times \mathbb{R}$ whose components are

$$\begin{aligned} S(f(x,y)dx) &= \int_\Omega f(x, u(x))\, dx, \\ S(f(x,y)\widehat{dx^i}\wedge dy) &= (-1)^{n-i} \int_\Omega f(x, u(x)) v^i(x)\, dx. \end{aligned}$$

Moreover, by Proposition 1, $\sup_k \mathbf{M}((\partial G_{u_k})_{(0)}) < \infty$, and this is precisely the condition (9) in Sec. 3.3.2 with $p = 1$. Proposition 2 in Sec. 3.3.2 then yields that $Dv = Du^a$, consequently $S = G_u$ and therefore

$$\mathbf{M}((\partial G_u)_{(0)}) \le \liminf_{k\to\infty} \mathbf{M}((\partial G_{u_k})_{(0)}) < \infty.$$

which yields $u \in SBV(\Omega, \mathbb{R})$ on account of Proposition 1. This concludes the proof as the rest of the claim is trivial. □

We notice that the previous argument extends verbatim to show

Theorem 9. *Consider the class X of Cartesian currents $T = G_{u_T} + S_T$ in $\Omega \times \mathbb{R}^N$ such that $|(Du)^s|(\Omega \setminus J_u) = 0$, i.e. $S^{(C)}_{T(1)} = 0$. Suppose that $\{T_k\}$ be a sequence in X such that*

(i) $\sup_k \mathbf{M}(T_k) < \infty$,
(ii) $M(Du_k)$ *are equi-integrable in* Ω,
(iii) $\sup_k \sup_{|\psi|\le 1} \partial T_k(\psi(x,y)dx^\alpha \wedge dy^\beta) < \infty\ \forall \alpha, \beta\ |\alpha| + |\beta| = n,\ |\beta \ge 0$, *or more generally (9) in Sec. 3.3.2.*

Then passing to a subsequence we have

$$T_k \rightharpoonup T = G_{u_T} + S_T \in X,\ \ i.e.,\ S^{(C)}_{T(1)} = 0$$

and

$$G_{u_{T_k}} \rightharpoonup G_{u_T}, \qquad S_{T_{u_k}(1)} \rightharpoonup S_{T(1)}.$$

Of course in the vector valued case it would be more interesting to find conditions ensuring that in the limit jump parts of the minors do not produce Cantor parts of the minors, but this seems to be a quite complicated question.

8 The approximation theorem, Theorem 1 in Sec. 4.1.1, was proved in Anzellotti and Giaquinta [45], see also Krickeberg [415], and first appeared in Anzellotti, Giaquinta, Massari, Modica, and Pepe [49]. Related Lusin type theorems were proved earlier in Michael [470] Goffman [312], see also Chen and Liu [148]. The coarea formula was proved by Fleming and Rishel [241], and by Federer [224] in the context of Lipschitz maps.

The theory of Caccioppoli sets initiated by Caccioppoli [129] is essentially due to De Giorgi [176] [177] [179], compare also De Giorgi, Colombini, and Piccinini [186].

Originally, sets of finite perimeter in $\mathbb{R}^n$ were defined as sets which can be approximated by polyhedral sets. More precisely, the perimeter of a set was defined as

$$P(E,\mathbb{R}^n) := \inf\{\liminf_{h\to\infty} \mathcal{H}^{n-1}(\partial E_h) \mid E_h \text{ polyhedral set with } (E_h \setminus E) \cup (E \setminus E_h) \to 0\}.$$

More generally, one could define the perimeter of any set, not necessarily measurable, as

$$P(E,\mathbb{R}^n) := \inf\{\liminf \mathcal{H}^{n-1}(\partial E_h) \mid |(E_h \setminus E) \cup (E \setminus E_h)| \to 0\}$$

then show that E is measurable, if $P(E,\mathbb{R}^N) < \infty$, and, in this case, the perimeter coincides with the perimeter defined in Sec. 4.1.2. De Giorgi's rectifiability theorem then amounts to the following. Take as *representative* of a set of finite perimeter E the equivalent set

$$\widetilde{E} := \{x \mid \theta(E,x) = 1\}$$

and set

$$\widetilde{\partial}\widetilde{E} := \{x \mid \theta(E,x) = \frac{1}{2}\} .$$

Then $\widetilde{\partial}\widetilde{E}$ is an $(n-1)$-rectifiable set which $\mathcal{H}^{n-1}$ almost everywhere agrees with $\partial^- E$.

A related approach was developed by Federer [220] [221] [223]. In conjunction with De Giorgi's results this eventually led to regard sets of finite perimeter, and more generally BV-functions, as n-dimensional currents in $\mathbb{R}^n$ and to recover a structure theorem for BV-function as a structure theorem for n-dimensional currents in $\mathbb{R}^n$, compare Federer [226, 4.5]. In particular for Theorem 3 in Sec. 4.1.3 see Federer [226, 4.5.11].

The theory of Caccioppoli sets and of BV-functions was then developed by many contributors including Krickeberg [415] Fleming [238] and especially Miranda [474] [475] and Vol'pert [651].

The results of Sec. 4.1.4 go back to Federer [226, 4.5], see also Vol'pert [651] and Goffman [313]. The results of Sec. 4.1.5 are taken from Miranda [474], and principally from Federer [226, 4.5.9], see also Dal Maso [171]. Actually in Federer [226, 4.5.9(5)] it is also proved that

$$\mathcal{H}^n(\mathcal{D}_u \setminus \partial^- \mathcal{SG}_u) = 0 .$$

Our presentation in Sec. 4.1 is a remaking of previous presentations that can be found in Anzellotti, Giaquinta, Massari, Modica, and Pepe [49] Giusti [308] Vol'pert and Hudjaev [652] Ziemer [686]. In doing that we have taken advantage also of Alberti's thesis [5], respectively for the proof of Theorem 5 in Sec. 4.1.4 and Theorem 6 in Sec. 4.1.4 of Sec. 4.1.4 and of a set of unpublished notes by Ambrosio [34].

9 The notion of *Cartesian current* first appeared in Giaquinta, Modica, and Souček [280]. Basic developments of the theory as structure, closure and compactness theorems, including also examples, appeared in Giaquinta, Modica, and Souček [280], [279] and in the papers Giaquinta, Modica, and Souček [282], [292] [286] which followed. Sec. 4.2, and in particular Sec. 4.2.1, Sec. 4.2.2 and Sec. 4.2.3, provides a reworking and a substantially improved presentation of those basic results.

The characterization of Cartesian currents in codimension one in Sec. 4.2.4 first appeared in Giaquinta, Modica, and Souček [292]. Radial currents were discussed in Giaquinta, Modica, and Souček [280], the approximation theorem is new, however a special version of it was first proved in Marcellini [455].

The construction of Cantor–Vitali type map in [9] in Sec. 4.2.5 due to Müller [503] is related to the one in Ponomarev [534] where it is shown a homeomorphism in $W^{1,p}$ $(p < n)$ which maps a set of measure zero to a set of non zero measure. We mention that recently in Malý and Martio [450] it has been exhibited a map $u \in W^{1,n}$ satisfying

$\det Du = 0$ a.e. (hence $\operatorname{Det} Du = 0$) which also maps null sets into sets of positive measure (compare Morrey [490] Reshetnyak [554]). In our presentation we have followed Müller [503], where in particular Theorem 1 in Sec. 4.2.5 of [9] in Sec. 4.2.5 is proved. The example in [10] in Sec. 4.2.5 was suggested by D. Mucci. Independently it also was pointed out in Coventry [162] where the following result is also proved: *If* $G_u + S_1$ *and* $G_u + S_2$ *are two Cartesian currents in* $\Omega \times \mathbb{R}^N$ *and* $G_u + S_1 \in \operatorname{Cart}(\Omega \times \mathbb{R}^N)$, *then also* $G_u + S_2$ *belongs to* $\operatorname{Cart}(\Omega \times \mathbb{R}^N)$. Actually in Mucci [495] a map $u \in \operatorname{cart}^1(\Omega, \mathbb{R}^2)$, $\Omega \subset \mathbb{R}^2$, is shown such that for any sequence of smooth maps $u_k : \Omega \to \mathbb{R}^2$ with $G_{u_k} \rightharpoonup G_u$ one has $\mathbf{M}(G_{u_k}) = +\infty$.

The sequel of this monograph will show, we hope, the relevance of the notion of Cartesian currents; we believe it should be useful also in other contexts. In this respect we would like to mention the interesting paper Anzellotti, Serapioni, and Tamanini [51] where it is shown in which way Cartesian currents and the set of ideas which brought to such a notion are connected with an approach to generalized curvature.

10 The literature on the classical notion of degree is quite broad. We just mention Alexandroff and Hopf [10], Radó and Reichelderfer [546], Nagumo [509], De Rham [189], Heinz [366], Cronin [163], Fučik, Nečas, Souček, and Souček [255], Nirenberg [513], and the recent papers of Brezis and Nirenberg [112] [113]. A more detailed bibliography of the topic can be found in the above listed papers and books.

The approach in Sec. 4.3.2 is simply based on the area formula, and the definition of degree mapping is modelled after De Rham [189] Federer [226, 4.1.14, 4.1.15, 4.1.16]. Actually one could also define the *degree map* with respect to any Lipschitz map $f : \mathbb{R}^k \times \mathbb{R}^n \to \mathbb{R}^n$

$$\deg(T, f, y)$$

Similarly to $\deg(T, \widehat{\pi}, f)$, compare also Almgren, Browder, and Lieb [24]. The properties of the degree in Sec. 4.3.2 were proved in Giaquinta, Modica, and Souček [279]. Relationships between traces and degree are consequences of Sec. 3.2.5. They are partly new and weaken the assumptions of similar results in Sverák [618] Müller, Tang, and Yan [504].

The results of Sec. 4.3.1, Theorem 1 in Sec. 4.3.1, Theorem 2 in Sec. 4.3.1, Theorem 3 in Sec. 4.3.1 and Theorem 4 in Sec. 4.3.1 are taken from Federer [226], see also Simon [592]; Theorem 5 in Sec. 4.3.1 and Proposition 1 in Sec. 4.3.1 seem to be new, compare Giaquinta, Modica, and Souček [292].

The results of Sec. 4.3.4, in particular Definition 1 in Sec. 4.3.4 (ii) and Theorem 1 in Sec. 4.3.4, are in some sense a special case of a general theorem of Federer [226, 4.2.25] concerning *indecomposable* currents. In Federer's terminology our h-connected sets A regarded as currents $[\![A]\!]$ are called indecomposable.

An i.m. rectifiable current $T \in \mathcal{R}_k(\mathbb{R}^n)$ with $\mathbf{M}(\partial T) < \infty$ is called *indecomposable* if and only if there exists no non zero $R \in \mathcal{R}_k(\mathbb{R}^n)$ with $\mathbf{M}(\partial R) < \infty$ such that

$$T - R \neq 0 \quad \text{and} \quad \mathbf{N}(T) = \mathbf{N}(R) + \mathbf{N}(T - R).$$

The following theorem holds

Theorem 10 (Decomposition theorem). *Let* $T \in \mathcal{R}_k(\mathbb{R}^n)$, $\mathbf{M}(\partial T) < \infty$. *Then there exists a sequence of indecomposable currents* $T_i \in \mathcal{R}_k(\mathbb{R}^n)$, $\mathbf{M}(T_i) < \infty$, *such that*

$$T = \sum_{i=1}^{\infty} T_i, \qquad \mathbf{N}(T_i) = \sum_{i=1}^{\infty} \mathbf{N}(T_i).$$

5. Cartesian Currents in Riemannian Manifolds

In this chapter we deal with non smooth maps and currents between Riemannian manifolds. In Sec. 5.5.1 we discuss Sobolev mappings between manifolds and we focus in particular on topological properties of these maps and on the closely related question of the density of smooth maps. We shall see that a generic Sobolev map $u \in W^{1,p}(X,Y)$ between Riemannian manifolds induces a homotopy map if $p > n$ while for $p \leq n$ they have only a well defined $[p-1]$ homotopy type. In Sec. 5.5.3 and Sec. 5.5.4 we shall see that Cartesian currents and Cartesian maps, which are introduced in Sec. 5.5.2, naturally carry induced homology and cohomology maps. They agree with the classical ones in the case of Cartesian currents associated to smooth maps and depend continuously upon the weak convergence of currents.

In doing that a preliminary question is the description of the homology of manifolds in terms of currents. This is a quite classical topic which started in De Rham [189] Whitney [674] and it is one of the major acheivement of Federer-Fleming's [230] and Federer's [228] [226] theory of currents.

We begin in Sec. 5.1 by proving a number of additional results on currents, in particular the deformation theorem we have already stated in Ch. 2 and used previously, we discuss the mollifying procedure for currents and introduce the notion of *flat chains*.

In Sec. 5.2 we prove the classical *Hodge–Kodaira–Morrey decomposition theorems* for forms in a compact submanifold X of $\mathbb{R}^n$ separately in the cases $\partial X = \emptyset$ and $\partial X \neq \emptyset$. As a consequence we infer the non degeneracy of *Poincaré* and *Poincaré–Lefschetz duality* in real cohomology.

Sec. 5.3 is devoted to discuss the real homology of compact oriented manifolds in terms of currents. As for the cohomology we discuss separately absolute and relative homology focusing on links among *Poincaré-Lefschetz duality* in cohomology, *Poincaré–Lefschetz duality isomorphism* between homology and cohomology, *de Rham theorems*, and the *intersection index*. Similar questions in the case of integral homology are illustrated in Sec. 5.4. In particular there we prove weak closure of integral homology classes defined in terms of integer multiplicity currents.

A particular interesting feature of this approach is that one can easily represent cosets by minimizers of variational integrals, in particular in any homology class one finds a cycle of least mass. Moreover, if X and Y are smooth compact oriented manifolds possibly with boundary and B is a closed Lipschitz retract in Y, we recognize a Cartesian current $T \in \operatorname{cart}(X \times Y)$ with $\operatorname{spt} \partial T \subset \partial X \times B$ as a relative homology cycle,

$$[T]_{\text{rel}} \in H_n(X \times Y, \partial X \times B, \mathbb{Z}), \qquad n := \dim X,$$

and starting from $[T]_{\text{rel}}$ one can easily produce the induced homology and cohomology maps between the two long sequences in homology for $(X,\ \partial X)$ and $(Y,\ B)$ using the Poincaré–Lefschetz duality.

1 More About Currents

In this section we collect a few facts about currents which are relevant or useful for the following. In Sec. 5.1.1 we prove the deformation theorem. In Sec. 5.1.2 we deal with the mollifying procedure for currents and in Sec. 5.1.3 we deal with flat chains.

1.1 The Deformation Theorem

In this subsection we prove *Federer-Fleming deformation theorem* which we already stated in Sec. 2.2.6. We shall see its relevance in studying the homology of manifolds in the sequel of this section. For the sake of completeness we shall also prove the strong approximation theorem, Theorem 4 in Sec. 2.2.6. According to the notation of Sec. 2.2.6 we deal here with n-dimensional currents in $\mathbb{R}^{n+N}$.

First we fix some notation. We decompose $\mathbb{R}^{n+N}$ into cubes with edges parallel to the axes and of unitary length. More precisely we denote by $\mathcal{L}_{n+N}$ the family of cubes

$$\mathcal{L}_{n+N} := \{z \in [0,1]^{n+N} \mid z \in \mathbb{Z}^{n+N}\}$$

so that

$$\mathbb{R}^{n+N} = \bigcup \{F \mid F \in \mathcal{L}_{n+N}\},$$

and for any j, $0 \leq j < n+N$, we denote by $\mathcal{L}_j$ the collection of all j-faces of the cubes in $\mathcal{L}_{n+N}$

$$\mathcal{L}_j := \{F \mid F \text{ is a } j\text{-face of } Q \in \mathcal{L}_{n+N}\}.$$

We instead denote by L_j the j-skeleton of the subdivision

$$L_j := \bigcup \{F \mid F \in \mathcal{L}_j\}$$

and, as we also need the translate of L_j, we also set

$$L_j(a) := a + L_j, \qquad a \in \mathbb{R}^{n+N}.$$

Finally, for any $\alpha \in I(j, n+N)$ we denote by $\mathbb{R}_\alpha$ the coordinate j-plane generated by $e_{\alpha_1} \wedge \ldots \wedge e_{\alpha_j}$ and let $\pi_\alpha : \mathbb{R}^{n+N} \to \mathbb{R}_\alpha$ be the orthogonal projection onto $\mathbb{R}_\alpha$.

Lemma 1. *We have*

(i) For $0 \leq j \leq n+N$

$$L_j(a) = \bigcup_{\alpha \in I(j,n+N)} \bigcup_{z \in \mathbb{Z}^{n+N}} (z + \pi_{\bar{\alpha}}^{-1}(a))$$

(ii) $L_n(a) \cap L_{N-1} = \emptyset$ *if* $a \in (0,1)^{n+N}$.

Proof. The claim (i) is trivial since each j-face $F \in L_j(a)$ is parallel to an α-plane $|\alpha| = j$, hence $F \subset \pi_{\bar{\alpha}}^{-1}\pi_{\bar{\alpha}}(a)$. To prove (ii) we assume by contradiction that $x \in L_n(a) \cap L_{N-1}$. As $x \in L_{N-1}(a)$, at least $n+N-(N-1) = n+1$ coordinates of x are integers, hence at least $n+1$ coordinates of $x-a$ are *not* integer. As $x \in L_n(a)$, the same argument yields that at least $n+N-n = N$ coordinates of $x-a$ are integers: a contradiction. □

As already stated, the geometric idea in the proof the deformation theorem is that of projecting or retracting n-currents into the n-skeleton L_n of the standard subdivision, and doing that by a suitably chosen center of projection. Next lemma contains the construction of a suitable class of retraction maps.

Let $q := (1/2, \dots, 1/2)$ denote the center of the cube $[0,1]^{n+N}$. By reasoning similarly to Lemma 1 (ii) we see that

$$\operatorname{dist}(L_{n-1}(a), L_N) \geq \frac{1}{4} \quad \text{if } a \in B(q, 1/4).$$

For any ρ, $0 < \rho < 1/4$ we denote by

$$L_j(a,\rho) := \{x \in \mathbb{R}^{n+N} \mid \operatorname{dist}(x, L_j(a)) < \rho\}$$

the tubular neighbourhood of L_j of radius ρ.

Lemma 2. *For any* $a \in B(q, 1/4)$ *there is a locally Lipschitz map* $\psi : \mathbb{R}^{n+N} \setminus L_{N-1}(a) \to \mathbb{R}^{n+N} \setminus L_{N-1}(a)$ *such that*

(i) $\psi([0,1]^{n+N} \setminus L_{N-1}(a)) = [0,1]^{n+N} \cap L_n$

(ii) $\psi([0,1]^{n+N} \cap L_n) = \mathrm{id}_{[0,1]^{n+N} \cap L_n}$

(iii) $\psi(z+x) = \psi(x) + z \quad \forall\, x \in \mathbb{R}^{n+N}\ \forall\, z \in \mathbb{Z}^{n+N}$

(iv) $|D\psi(x)| \leq c/\rho$ *for* a.e. $x \in \mathbb{R}^{n+N} \setminus L_{N-1}(a,\rho)$, $0 < \rho < 1/4$.

Proof. We split the proof into three steps.
Step 1. Let F be a j-face in L_j, $j \geq n+1$. Denote by a_F the orthogonal projection of a into F and let $\psi_F : F \setminus \{a_F\} \to \partial F$ be the retraction map which takes $x \in F \setminus \{a_F\}$ to the point $y \in \partial F$ such that for some $\lambda \in (0,1]$

$$x = (1-\lambda)a_F + \lambda y. \tag{1}$$

Of course $\psi_{F|\partial F} = \mathrm{id}_{|\partial F}$. Also for the segment $\overline{aa_F}$ we have

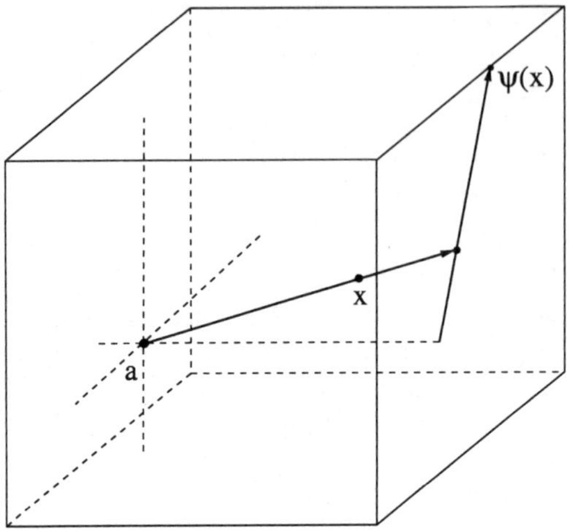

Fig. 5.1. The map ψ.

(2) $$\overline{aa_F} \subset \bigcap_{\alpha \in I(j,n+N)} \pi_{\bar\alpha}^{-1}\pi_{\bar\alpha}(a) \subset \bigcap_{\alpha \in I(n+1,n+N)} \pi_{\bar\alpha}^{-1}\pi_{\bar\alpha}(a),$$

since a_F is the orthogonal projection of a into F. Therefore Lemma 1 (i) yields that $\overline{aa_F}$ lies in $L_{N-1}(a)$. This shows that ψ_F maps $F \setminus L_{N-1}(a)$ into $\partial F \setminus L_{N-1}(a)$ (by (1)),

$$\psi_F : F \setminus L_{N-1}(a) \longrightarrow F \setminus L_{N-1}(a)$$

is onto, ψ_F is the identity on ∂F and is locally Lipschitz.
Step 2. Gluing the ψ_F for each j-face in $[0,1]^{n+N}$, we define a map

$$\psi^{(j)} : [0,1]^{n+N} \cap L_j \setminus L_{N-1}(a) \longrightarrow [0,1]^{n+N} \cap L_{j-1} \setminus L_{N-1}(a),$$

$\psi^{(j)}(x) = \psi_F(x)$ if $x \in F$, which is onto and locally Lipschitz. The composition map

$$\psi_0 := \psi^{(n+1)} \circ \psi^{(n+2)} \circ \cdots \circ \psi^{n+N)}$$

makes sense, maps $[0,1]^{n+N} \setminus L_{N-1}(a)$ onto $[0,1]^{n+N} \cap L_n \setminus L_{N-1}(a)$ which is equal to $[0,1]^{n+N} \cap L_n$ by Lemma 1 (ii), and is the identity on $[0,1]^{n+N} \cap L_n$. Furthermore, if $x, x+z \in [0,1]^{n+N}$, $z \in \mathbb{Z}$, then

(3) $$\psi_0(x+z) = z + \psi_0(x).$$

In fact if $x, x+z \in [0,1]^{n+N}$, either x and $x+z$ belong to L_n where ψ_0 is the identity, or x and $x+z$ belong to the interior of two j-faces F_1, F_2, $j \geq n+1$ which are parallel and $F_2 = z + F_1$. Hence $a_{F_2} = z + a_{F_1}$ and (3) follows from (1). By (2) we can extend by periodicity the map ψ_0 to a map

$$\psi : \mathbb{R}^{n+N} \setminus L_{N-1}(a) \longrightarrow \mathbb{R}^{n+N} \setminus L_{N-1}(a)$$

which satisfies the claims (i), (ii), (iii).

Step 3. Finally, let us prove (iv). We shall prove slightly more: for all $x \in [0,1]^n \setminus L_{N-1}(a)$ we have

$$|\overline{D}\psi(x)| \leq \frac{c}{\operatorname{dist}(x, L_{N-1}(a))} \tag{4}$$

where $c = c(n, N)$ and

$$|\overline{D}\psi(x)| := \limsup_{x' \to x} \frac{|\psi(x') - \psi(x)|}{|x' - x|}.$$

By induction on N, if $N = 1$, (4) is trivial as $y = \psi(x)$ is given by

$$y = a + \lambda(x - a) \quad \text{where } \lambda = \frac{|y - a|}{|x - a|}.$$

Assume (4) holds in case $N - 1$ replaces N and let $\widetilde{\psi} = \psi^{(n+1)} \circ \cdots \circ \psi^{(n+N-1)}$ so that $\psi = \widetilde{\psi} \circ \psi^{(n+N)}$. If $x \in [0,1]^n \setminus L_{N-1}(a)$ and $y = \psi^{n+N}(x)$, as in the case $N = 1$, we immediately infer

$$|D\psi^{(n+N)}(x)| \leq \sqrt{n+N}\frac{|y - a|}{|x - a|}. \tag{5}$$

If $y \in L_n(a)$ then $\widetilde{\psi}(y) = y$ and (4) is trivial. Assume therefore that this is not the case and let F be the $(n + N - 1)$-face of $\mathcal{L}_{n+N-1}$ such that $y \in F$. We can assume that $F \subset \mathbb{R}^{n+N-1} \times \{0\}$. Let π_F be the orthogonal projection of $\mathbb{R}^{n+N}$ into $\mathbb{R}^{n+N-1} \times \{0\}$ and let $a_F \in \operatorname{int} F$ be the projection of a on F. It is easily seen that

$$\widetilde{L}_{N-2}(a_F) = L_{N-1}(a) \cap (\mathbb{R}^{n+N-1} \times \{0\})$$

is the singular set of $\widetilde{\psi}_{|\mathbb{R}^{n+N-1} \times \{0\}}$, therefore by the inductive assumption

$$\left|\overline{D}\widetilde{\psi}(y)\right| \leq \frac{c}{\operatorname{dist}(y, \widetilde{L}_{N-2}(a_F))}. \tag{6}$$

By similarity

$$\frac{|y - a|}{|x - a|} = \frac{\operatorname{dist}(y, \widetilde{L}_{N-2}(a_F))}{\operatorname{dist}(x, \pi_F^{-1}\widetilde{L}_{N-2}(a_F))},$$

while

$$\pi_F^{-1}\widetilde{L}_{N-2}(a_F) \subset L_{N-1}(a),$$

hence

$$\operatorname{dist}(x, \pi_F^{-1}\widetilde{L}_{N-2}(a_F)) \geq \operatorname{dist}(x, L_{N-1}(a)).$$

We then infer

$$\operatorname{dist}(y, \widetilde{L}_{N-2}(a_F)) \geq \frac{|y - a|}{|x - a|}\operatorname{dist}(x, L_{N-1}(a))$$

and

$$|\overline{D}\widetilde{\psi}(y)| \le c\frac{|y-a|}{|x-a|} = \frac{1}{\operatorname{dist}(x, L_{N-1}(a))}, \tag{7}$$

on account of (6). Finally (7) and (5) yield the conclusion. □

Theorem 1 (Deformation theorem, unscaled version). *Let T be a normal current, $T \in \mathbf{N}_n(\mathbb{R}^{n+N})$. Then*

(i) We can decompose T as

$$T = P + \partial R + S$$

where P is a polyhedral chain made of faces of the n-skeleton L_n of the classical subdivision of $\mathbb{R}^{n+N}$

$$P = \sum_i \beta_i [\![F_i]\!] \qquad \beta_i \in \mathbb{R}.$$

Moreover

$$\begin{aligned} &\operatorname{spt} P \subset L_n, \quad \operatorname{spt} \partial P \subset L_{n-1} \\ &\operatorname{spt} R \subset \cup\{\overline{Q} \mid Q \in \mathcal{L}_{n+N}, \ \overline{Q} \cap \operatorname{spt} T \neq \emptyset\} \\ &\operatorname{spt} S \subset \cup\{\overline{Q} \mid Q \in \mathcal{L}_{n+N}, \ \overline{Q} \cap \operatorname{spt} \partial T \neq \emptyset\} \end{aligned}$$

and

$$\begin{aligned} &\mathbf{M}(P) \le c\mathbf{M}(T), \quad \mathbf{M}(\partial P) \le c\mathbf{M}(\partial T) \quad \mathbf{M}(R) \le c\mathbf{M}(T), \\ &\mathbf{M}(S) \le c\mathbf{M}(\partial T) \end{aligned}$$

where $c = c(n, N)$.

Moreover

(ii) If T is i.m. rectifiable, we can choose P, R i.m. rectifiable and $\beta \in \mathbb{Z}$, and, if also ∂T is i.m. rectifiable, we can choose S i.m. rectifiable, too.
(iii) If T is a Lipschitz chain, then P, S, R are Lipschitz chains too.
(iv) If ∂T is a Lipschitz chain, then S is a Lipschitz chain,
(v) If ∂T is supported in L_{n-1} then $S = 0$.

Proof. We would like to project the current T into the n-skeleton L_n according to Lemma 2. But in order to do that efficiently we must choose a in such a way that not much of the mass of T is concentrated close to $L_{N-1}(a)$.
Step 1. Let T be a normal current. We claim that *there is a constant $c_1 = c(n, N)$ and a point $a \in B(q, 1/4)$ such that for any ρ, $0 < \rho < 1/4$*

$$\begin{aligned} &\mathbf{M}(T \llcorner L_{N-1}(a, \rho)) \le c_1 \rho^{n+1} \mathbf{M}(T) \\ &\mathbf{M}(\partial T \llcorner L_{N-1}(a, \rho)) \le c_1 \rho^{n+1} \mathbf{M}(\partial T). \end{aligned} \tag{8}$$

For any $\alpha \in I(n+1, n+N)$ we consider the $(n+1)$-face of $[0,1]^{n+N}$ $F_\alpha := \mathbb{R}_\alpha \cap [0,1]^{n+N}$ and let q_α be its center. In $F_\alpha \cap B(q_\alpha, 1/4)$ we define a good set G_α by $g \in G_\alpha$ iff

$$(9)\quad \mathbf{M}\Big(T \llcorner \bigcup_{z\in\mathbb{Z}^{n+N}\cap\mathbb{R}_\alpha} \pi_\alpha^{-1}(B(g+z,\rho))\Big) \le \beta\rho^{n+1}\mathbf{M}(T) \quad \forall\, \rho\in(0,1/4)$$

β being a constant to be chosen later. We now claim that the bad set $B_\alpha := F_\alpha \cap [0,1]^{n+N} \setminus G_\alpha$ is small, actually

$$\mathcal{H}^{n+1}(B_\alpha) \le c_2\frac{1}{\beta}$$

which of course is small if β is large. In fact for any $b\in B_\alpha$ there is a $\rho_b\in(0,1/4)$ such that

$$\mathbf{M}\Big(T \llcorner \bigcup_{z\in\mathbb{Z}^{n+N}\cap\mathbb{R}_\alpha} \pi_\alpha^{-1}(B(g+z,\rho_b))\Big) \ge \beta\rho_b^{n+1}\mathbf{M}(T);$$

by Besicovitch covering theorem there is a pairwise disjoint and denumerable sub-collection such that

$$B_\alpha \subset \bigcup_k B(b_k, 5\rho_k),$$

but then, setting $b=b_k,\ \rho_b=\rho_k$ in (9) we get $\beta\sum_k \rho_k^{n+1}\mathbf{M}(T)\le\mathbf{M}(T)$ i.e.

$$\sum_k \rho_k^{n+1} \le \frac{1}{\beta},$$

consequently

$$\mathcal{H}^{n+1}(B_\alpha) \le \frac{c_4(n)}{\beta} \quad\text{and}\quad \mathcal{H}^{n+N}(\pi_\alpha^{-1}(B_\alpha)\cap B(q,1/4)) \le \frac{c_5(n,N)}{\beta}.$$

Finally summing on $\alpha\in I(n+1,n+N)$ we get

$$\mathcal{H}^{n+N}\Big(\bigcup_\alpha \pi_\alpha^{-1}(B_\alpha)\cap B(q,1/4)\Big) \le \frac{c_1(n,N)}{\beta},$$

consequently $\cap_\alpha\pi_\alpha^{-1}(G_\alpha)\cap B(q,1/4)$ is most of $B(q,1/4)$ for β large.

Repeating the same argument for ∂T, and denoting by G'_α, B'_α the good and bad sets for ∂T we also find that for β sufficiently large the set

$$\bigcap_\alpha \pi_\alpha^{-1}(G'_\alpha)\cap B(q,1/4)$$

is most of $B(q,1/4)$. Therefore we can find a in the good set $a\in\bigcap_\alpha \pi_\alpha^{-1}(G_\alpha)\cap$ $\cap\pi_\alpha^{-1}(G'_\alpha)\cap B(q,1/4)$, for which (9) holds. Since

$$L_{N-1}(a,\rho) = \bigcup_\alpha \bigcup_{z\in\mathbb{Z}^{n+N}\cap\mathbb{R}_\alpha} \pi_\alpha^{-1}(B(a_\alpha+z,\rho)) \qquad a_\alpha=\pi_\alpha(a)$$

the claim is proved.
Step 2. We now project T into the L_n-skeleton by means of the projection map in Lemma 2 associated to a point $a \in B(q,1/4)$ for which (8) of step 1 hold. For $0<\rho<1/4$ set

$$T_\rho := T \llcorner L_{N-1}(a,\rho), \qquad (\partial T)_\rho := \partial T \llcorner L_{N-1}(a,\rho)$$

so that by (8)

$$\mathbf{M}(T_\rho) \le c\rho^{n+1}\mathbf{M}(T), \qquad \mathbf{M}((\partial T)_\rho) \le c\rho^{n+1}\mathbf{M}(\partial T).$$

Then by Lemma 2 (iv)

$$\text{(10)}\quad \begin{aligned} &\mathbf{M}(\psi_\#(T_\rho - T_{\rho/2})) \le \frac{c}{\rho^n}\rho^{n+1}\mathbf{M}(T) \le c\rho\mathbf{M}(T)\\ &\mathbf{M}(\psi_\#(\partial T)_\rho - \psi_\#(\partial T)_{\rho/2})) \le \frac{\rho}{\rho^{n-1}}\rho^{n+1}\mathbf{M}(T) \le c\rho\mathbf{M}(\partial T). \end{aligned}$$

Similarly, if $h(t,x) := (1-t)x + t\psi(x)$, by the homotopy formula we get

$$\text{(11)}\quad \begin{aligned} &\mathbf{M}(h_\#([\![(0,1)]\!] \times (T_\rho - T_{\rho/2})) \le c\rho\mathbf{M}(T)\\ &\mathbf{M}(h_\#([\![(0,1)]\!] \times ((\partial T)_\rho - (\partial T)_{\rho/2})) \le c\rho\mathbf{M}(\partial T). \end{aligned}$$

Finally, from the slicing theory we find $\rho^* \in (\rho/2,\rho)$ such that

$$\text{(12)}\quad \begin{aligned} &\mathbf{M}(\psi_\# < T, d, \rho^* >) \le \frac{c}{\rho^{n-1}}\rho^{n+1}\mathbf{M}(T_\rho - T_{\rho/2}) \le c\rho\mathbf{M}(T)\\ &\mathbf{M}(h_\#([\![(0,1)]\!] \times < T, d, \rho^* >)) \le c\rho\mathbf{M}(\partial T). \end{aligned}$$

Notice that if T is i.m. rectifiable we can choose ρ^* so that also $< T, d, \rho^* >$ is i.m. rectifiable.

Now we select $\rho_\nu = 2^{-\nu}\rho$, $\rho_\nu^* \in (2^{-\nu-1}\rho, 2^{-\nu}\rho)$ in such a way that (12) holds and we let $T_\nu := T_{\rho_\nu^*}$. From (10), (11), (12) it easily follows that the sequence of currents

$$\begin{aligned} &\psi_\#(T - T_\nu), && h_\#([\![(0,1)]\!] \times (T - T_\nu))\\ &\psi_\#(\partial T - (\partial T)_\nu), && h_\#([\![(0,1)]\!] \times \partial(T - T_\nu)) \end{aligned}$$

are Cauchy sequences with respect to the mass convergence, and moreover

$$\mathbf{M}(< T, d, \rho_\nu^* >) + \mathbf{M}(\psi_\# < T, d, \rho_\nu^* >) \longrightarrow 0.$$

Therefore there are currents $P_1, S_1 \in \mathcal{D}_n(\mathbb{R}^{n+N})$ and $R_1 \in \mathcal{D}_{n+1}(\mathbb{R}^{n+N})$ such that

$$\text{(13)}\quad \begin{aligned} &\mathbf{M}(P_1 - \psi_\#(T - T_\nu)) \longrightarrow 0,\\ &\mathbf{M}(R_1 - h_\#([\![(0,1)]\!] \times (T - T_\nu))) \longrightarrow 0,\\ &\mathbf{M}(S_1 - h_\#([\![(0,1)]\!] \times \partial(T - T_\nu))) \longrightarrow 0, \end{aligned}$$

in particular

$$\text{(14)} \qquad \mathbf{M}(P_1) \le c\mathbf{M}(T), \quad \mathbf{M}(R_1) \le c\mathbf{M}(T), \quad \mathbf{M}(S_1) \le c\,\mathbf{M}(\partial T),$$

and by semicontinuity

$$\text{(15)} \qquad \mathbf{M}(\partial P_1) \le c\,\mathbf{M}(\partial T).$$

Furthermore, by the homotopy formula we have

$$\begin{aligned} T - T_\nu - \psi_\#(T - T_\nu) \\ = \partial h_\#([\![(0,1)]\!] \times (T - T_\nu)) - h_\#([\![(0,1)]\!] \times \partial(T - T_\nu)) \end{aligned}$$

and, as $\partial T_\nu = (\partial T)_\nu - <T, d, \rho_\nu^*>$ compare Sec. 2.2.5, we infer

$$\text{(16)} \qquad T - P_1 = \partial R_1 + S_1.$$

Observe that we also have

$$\text{(17)} \qquad \begin{aligned} &P_1, R_1 \text{ are i.m. rectifiable, if } T \text{ is i.m. rectifiable} \\ &S_1 \text{ is i.m. rectifiable, if } T \text{ is i.m. rectifiable} \\ &P_1, R_1 \text{ are locally Lipschitz chains, if } T \text{ is a Lipschitz chain} \\ &S_1 \text{ is a Lipschitz chain, if } \partial T \text{ is a Lipschitz chain.} \end{aligned}$$

Finally, since ψ retracts $\mathbb{R}^{n+N} \setminus L_{N-1}(a)$ into L_n and ψ is the identity on L_n we have

$$\text{(18)} \qquad \begin{aligned} &\operatorname{spt} P_1 \subset L_n \\ &\operatorname{spt} R_1 \subset \cup\{\overline{Q} \mid Q \in \mathcal{L}_{n+N},\ \overline{Q} \cap \operatorname{spt} T \neq \emptyset\} \\ &\operatorname{spt} S_1 \subset \cup\{\overline{Q} \mid Q \in \mathcal{L}_{n+N},\ \overline{Q} \cap \operatorname{spt} \partial T \neq \emptyset\} \end{aligned}$$

and

$$\text{(19)} \qquad S_1 = 0 \quad \text{if} \quad \operatorname{spt} \partial T \subset L_n.$$

In fact from $L_n \cap L_{N-1}(a, \rho_\nu) = \emptyset$ we infer $(\partial T)_\nu = 0$, hence

$$\partial(T - T_\nu) = \partial T - <T, d, \rho_\nu^*> \ \rightharpoonup \ \partial T$$

consequently

$$R_1 = \lim_{\nu \to \infty} h_\#([\![(0,1)]\!] \times \partial(T - T_\nu)) = h_\#([\![(0,1)]\!] \times \partial T) = 0,$$

being ψ the identity in L_n.

Step 3. Assume now that $\operatorname{spt} \partial T \subset L_{n-1}$ (for instance T is boundaryless). In this case (17), (19) yield

$$T = P_1 + \partial R_1, \qquad \operatorname{spt} P_1 \subset L_n$$

hence $\operatorname{spt}(\partial T - \partial R_1) = \emptyset$. We can then apply the constancy theorem, Theorem 4 in Sec. 2.2.3 and conclude that P_1 is a polyhedral chain

$$P_1 = \sum_i \beta_i [\![F_i]\!] \qquad \beta_i \in \mathbb{R}$$

and easily conclude the proof of the theorem.

In the general case we know that $P_1 \in \mathbf{N}_n(\mathbb{R}^{n+N})$ and is supported in the n-skeleton L_n. By the representation formula for normal currents, Theorem 2 in Sec. 4.3.1, we can represent P_1 over any face $F \in \mathcal{L}_n$ as

$$P_1 \llcorner \overset{\circ}{F}(\omega) = \int <\omega(x), \vec{F}(x)> \theta_F(x)\, d\mathcal{H}^n(x)$$

where $\vec{F}(x)$ orients F and $\theta_F \in BV_{\mathrm{loc}}(\mathbb{R}^n)$. Moreover

$$\mathbf{M}(P_1 \llcorner \overset{\circ}{F}) = \int_{\overset{\circ}{F}} |\theta_F|\, d\mathcal{H}^n, \quad \mathbf{M}(\partial(P_1 \llcorner \overset{\circ}{F}) = \int_{\overset{\circ}{F}} |D\theta_F|;$$

furthermore, if $\beta \in \mathbb{R}$,

$$M(P_1 \llcorner \overset{\circ}{F} - \beta[\![F]\!]) = \int_{\overset{\circ}{F}} |\theta_F - \beta|\, d\mathcal{H}^n$$

$$M(\partial(P_1 \llcorner \overset{\circ}{F} - \beta[\![F]\!])) = \int_{\mathbb{R}^n} |D(\chi_F(\theta_F - \beta))|.$$

As in the proof of Poincaré lemma we select β_F in such a way that

$$\min(\mathcal{L}^n(\theta_F \geq \beta), \mathcal{L}^n(\theta_F < 1/2)) \geq \frac{1}{2}\mathcal{L}^n(F)$$

(notice that θ_F can be chosen integer if θ_F is integer-valued, i.e., if P_1 is i.m. rectifiable) and get

$$|\beta_F| \leq 2 \int |\theta_F|\, d\mathcal{H}^n \llcorner \overset{\circ}{F}$$

$$\int |\theta_F - \beta_F|\, d\mathcal{H}^n \llcorner \overset{\circ}{F} \leq c|D\theta_F| \llcorner \overset{\circ}{F},$$

consequently

$$\mathbf{M}(P_1 \llcorner \overset{\circ}{F} - \beta_F[\![F]\!]) \leq cM(\partial(P_1 \llcorner \overset{\circ}{F}))$$

$$\mathbf{M}(\partial(P_1 \llcorner \overset{\circ}{F} - \beta_F[\![F]\!])) \leq cM(\partial(P_1 \llcorner \overset{\circ}{F}))$$

and $P_1 \llcorner \partial F = 0$, since P_1 is normal. Summing over the faces $F \in L_n$ we set

$$P = \sum_{F \in \mathcal{L}_n} \beta_F [\![F]\!].$$

We have

$$\begin{aligned}&\mathbf{M}(P-P_1)\le c\mathbf{M}(\partial P_1), \qquad \mathbf{M}(\partial(P-P_1))\le c\mathbf{M}(\partial P_1)\\ &\mathbf{M}(P)\le \sum_\beta |\beta_F| \le 2\mathbf{M}(P_1),\end{aligned}$$

therefore, if $S=S_1-P_1-P$, we get

$$\begin{aligned}&\mathbf{M}(S)\le \mathbf{M}(S_1)+\mathbf{M}(P_1-P)\le cM(\partial T)\\ &\mathbf{M}(\partial P)\le cM(\partial P_1)\le cM(T)\end{aligned}$$

and for $T=P+\partial R_1+S$, one easily checks that all claims in the theorem hold true. □

The scaled version of the deformation theorem, Theorem 1 in Sec. 2.2.6 now follows easily by first changing scale $x\to\varepsilon^{-1}x$, then applying Theorem 1 above, and then changing scale back by $x\to\varepsilon x$. This way we actually get

Theorem 2 (Deformation theorem, rescaled version). *Let T be a normal current, $T\in\mathbf{N}_n(\mathbb{R}^{n+N})$. Then we can decompose T as*

$$T=P+\partial R+S$$

where P is a polyhedral chain made of faces of the n-skeleton $L_{n,\varepsilon}$ of the subdivision of $\mathbb{R}^{n+N}$ in cubes of side ε

$$P=\sum_i \beta_i [\![F_i]\!] \qquad \beta_i\in\mathbb{R}.$$

Moreover

$$\begin{aligned}&\operatorname{spt} P\subset L_{n,\varepsilon},\quad \operatorname{spt}\partial P\subset L_{n-1,\varepsilon}\\ &\operatorname{spt} R\subset \cup\{\overline{Q}\mid Q\in\mathcal{L}_{n+N,\varepsilon},\ \overline{Q}\cap\operatorname{spt} T\neq\emptyset\}\\ &\operatorname{spt} S\subset \cup\{\overline{Q}\mid Q\in\mathcal{L}_{n+N,\varepsilon},\ \overline{Q}\cap\operatorname{spt}\partial T\neq\emptyset\}\end{aligned}$$

and

$$\begin{aligned}&\mathbf{M}(P)\le c\mathbf{M}(T),\quad \mathbf{M}(\partial P)\le c\mathbf{M}(\partial T)\quad \mathbf{M}(R)\le c\varepsilon\mathbf{M}(T),\\ &\mathbf{M}(S)\le c\varepsilon\mathbf{M}(\partial T)\end{aligned}$$

where $c=c(n,N)$. Moreover

(i) *If T is i.m. rectifiable, we can choose P,R i.m. rectifiable and $\beta\in\mathbb{Z}$, and, if also ∂T is i.m. rectifiable, we can choose S i.m. rectifiable too*
(ii) *If T is a Lipschitz chain, then P,S,R are Lipschitz chains too*
(iii) *If ∂T is a Lipschitz chain, then S is a Lipschitz chain*
(iv) *If ∂T is supported in $L_{n-1,\varepsilon}$ then $S=0$.*

We conclude this subsection with the proof of the strong approximation theorem Theorem 4 in Sec. 2.2.6. For that we state without proof, and we refer e.g. to Federer [226, 3.1.23], the following

Proposition 1. *Let M be an n-dimensional submanifold of class k, $k \geq 1$, of $\mathbb{R}^{n+N}$, $z \in M$, $0 < t < 1$. Denote by P the n-plane through z which is tangent to M. Then for every sufficiently small positive number r there exists a diffeomorphism $f : \mathbb{R}^{n+N} \to \mathbb{R}^{n+N}$ of class k such that the Lipschitz constants of f and f^{-1} are less than t^{-1}, f is the identity outside $B(z,r)$, and $B(z,tr) \cap M = B(z,tr) \cap f^{-1}(P)$*

Proof of Theorem 4 in Sec. 2.2.6. First we also assume that ∂T *is polyhedral.* We know that $T = \tau(\mathcal{M}, \theta, \vec{T})$ where $\mathcal{M}$ is n-rectifiable and θ integer-valued; also $\mathcal{M}$ is contained in a countable union $\cup M_i$ of C^1 submanifolds of $\mathbb{R}^{n+N}$. Furthermore, as we have seen in Ch. 2, at almost every $x \in \mathcal{M}$ the density of $\mathcal{M}$ and $\cup M_i$ is 1 and there is a single M_i such that $\mathcal{M}$ and M_i coincide at x except for a set of density 0. By a covering argument we can now find open balls $B_i \subset \mathbb{R}^{n+N} \setminus \operatorname{spt} \partial T$ and C^1 submanifolds N_i of B_i such that $\cup N_i$ coincides with $\mathcal{M}$ except for a set of $\| T \|$-measure small. Covering each N_i with small balls centered at N_i of sufficiently small radii in such a way that the conclusion of Proposition 1 holds (with t very close to 1) we obtain a diffeomorphism f of $\mathbb{R}^{n+N}$ which flattens most of $\mathcal{M}$ into n-planes and for which $f_{\#}T$ differs from a polyhedral chain P_1 by a small quantity. However the error $f_{\#}T - P_1$, although small in mass, may have huge boundary. Now we use the deformation theorem to decompose the error

$$f_{\#}T - P_1 = P_2 + \partial R + S.$$

As f leaves $\operatorname{spt} \partial T$ fixed, S is actually zero, and R has small mass if the grid size is small; also, as $f_{\#}T - P_1$ has small mass, so does P_2 and hence so does the remaining term ∂R. If we set

$$P := P_1 + P_2$$

we then see that $f_{\#}T - P$ is equal to ∂R hence has small mass and no boundary.

In the general case in which ∂T is not polyhedral, first we approximate ∂T as previously

$$f_{1\#}\partial T = P_1 + \partial R_1$$

with $\mathbf{M}(R_1)$ and $\mathbf{M}(\partial R_1)$ small. Now $f_{1\#}T - R_1$ has a polyhedral boundary and can be approximated by an integral polyhedral chain P_2

$$f_{2\#}(f_{1\#}T - R_1) = P_2 + \partial R_2$$

with $\mathbf{M}(\partial R_2)$ small. Therefore

$$(f_2 \circ f_1)_{\#}T = P_2 + f_{2\#}R_1 + \partial R_2$$

and the error term $f_{2\#}R_1 + \partial R_2$ and its boundary have small mass as desired. □

1.2 Mollifying Currents

Already in Sec. 4.3.1 we stated a representation formula for n-dimensional currents in $\mathbb{R}^n$ and in doing that we used mollification of currents. Here we discuss these topics for general k-dimensional currents in $\mathbb{R}^n$.

Smoothing kernels. In order to do this we need a standard smoothing kernel $\rho \in C_c^\infty(B(0,1))$, $B(0,1)$ being the unit ball in $\mathbb{R}^n$ with $\int \rho\, dx = 1$, and we set $\rho_\varepsilon := \varepsilon^{-n}\rho(x/\varepsilon)$, $0 < \varepsilon \le 1$. Since we would like to work on open sets U our smoothing process ought to take finer and finer averages when approaching the boundary of U. For that we find convenient (it is in fact necessary for dealing with general currents) to use a *regularized distance* function from the boundary of U, that is a function $d(x) \in C^\infty(U)$ such that

$$0 < d(x) < \operatorname{dist}(x, \partial U), \quad x \in U, \quad \| Dd \|_\infty \le 1$$

and to introduce the following map $H_z : U \longrightarrow \mathbb{R}^n$ defined for all $z \in B(0,1)$

$$H_z(x) := x + d(x)z$$

and the associate homotopy map

$$H^z : [0,1] \times U \longrightarrow \mathbb{R}^n, \qquad H^z(t,x) := H_{tz}(x).$$

It is easily seen that H^z maps $[0,1] \times U$ into U. Actually we have

Proposition 1. *For $z \in B(0,1)$ $H_z : U \to U$ is an orientation preserving diffeomorphism. In particular H_z and its inverse $(H_z)^{-1}$ are proper maps.*

Proof. H_z is injective. In fact, if $H_z(x) = H_z(y)$, then $x - y = z(d(x) - d(y))$. Hence either $x = y$ and the claim is true or $|x-y| < |d(x)-d(y)|$, a contradiction as $\| Dd \|_\infty \le 1$.

H_z is surjective. Let $y \in U$. Set $x_\mu := y - \mu z$, $\mu \in \mathbb{R}$, and $F(\mu) := d(x_\mu)$. Being U an open set F is defined in a segment (a,b) containing 0; being $d > 0$ on U $F(\mu) \to 0$ as $\mu \to a^+, b^-$ and finally $F(0) > 0$, then there is $\overline{\mu} \in (0,b)$ such that $\overline{\mu} = F(\overline{\mu})$. In particular $x_{\overline{\mu}} = y - \overline{\mu}z = y - zd(x_{\overline{\mu}})$. Finally H_z is non singular since

$$DH_z = \mathrm{Id} + z \otimes Dd$$

and

$$\det H_z = 1 + \sum_i D_i d z^i \ge 1 - |z| \| Dd \|_\infty > 0.$$

□

We explicitly note that from the definition of H_z we have

$$|\operatorname{dist}(H_z(x), \partial U) - \operatorname{dist}(x, \partial U)| \le |z| \operatorname{dist}(x, \partial U).$$

Consequently for $V \subset U$ we have

$$\text{dist}\,(H_z(V), \partial U) \geq (1 - |z|)\text{dist}\,(V, \partial U) \tag{1}$$

and

$$\text{dist}\,(H_z^{-1}(V), \partial U) \geq \frac{1}{1+|z|}\text{dist}\,(V, \partial U). \tag{2}$$

Mollification of forms and currents. Let ω be a k-form with coefficients in $L^1_{\text{loc}}(U)$, $\omega \in L^1_{\text{loc}}(U, \Lambda^k\mathbb{R}^n)$. We define the *$\varepsilon$-mollified ω_ε of ω* by

$$\begin{aligned} \omega_\varepsilon(x) &:= \int \rho_\varepsilon(-z) H_z^{\#}(\omega)(x)\, d\mathcal{L}^n(z) \\ &:= d(x)^{-n} \int_U \rho_\varepsilon\Big(\frac{x-y}{d(x)}\Big)\Big(H^{\#}_{\frac{y-x}{d(x)}}(\omega)\Big)(x)\, d\mathcal{L}^n(y) \end{aligned} \tag{3}$$

the second equality following obviously by changing variables in the integral. In the next proposition we state a few simple remarks.

Proposition 2. *Let ω be in $L^1_{\text{loc}}(U, \Lambda^k\mathbb{R}^n)$. Then*

(i) For $0 < \varepsilon < 1$ ω_ε is a well defined form in $L^1_{\text{loc}}(U, \Lambda^k\mathbb{R}^n) \cap \mathcal{E}^k(U)$.
(ii) If $\omega \in L^1(U, \Lambda^k\mathbb{R}^n)$, then $\omega_\varepsilon \to \omega$ in $L^1(U, \Lambda^k\mathbb{R}^n)$.
(iii) If $\omega \in C^0(U, \Lambda^k\mathbb{R}^n)$, then $\omega_\varepsilon \to \omega$ uniformly on compact sets of U and $\|\omega_\varepsilon\|_{\infty,U} \leq c\|\omega\|_{\infty,U}$.
(iv) If $\omega \in \mathcal{D}^k(U)$, then $\omega_\varepsilon \to \omega$ in $\mathcal{D}^k(U)$ and ω_ε has compact support in U, actually

$$\text{dist}\,(\text{spt}\,\omega_\varepsilon, \partial U) \geq \frac{1}{1+\varepsilon}\text{dist}\,(\text{spt}\,\omega, \partial U).$$

Let $T \in \mathcal{D}_k(U)$, we define the *ε-mollified T_ε of T* by duality $T_\varepsilon(\omega) := T(\omega_\varepsilon)$ i.e.

$$T_\varepsilon(\omega) = T\Big(\int \rho_\varepsilon(-z) H_z^{\#}\omega(x)\, d\mathcal{L}^n(z)\Big) = \int \rho_\varepsilon(-z) H_{z\#}(T)(\omega)\, d\mathcal{L}^n(z).$$

Proposition 3. *Let $T \in \mathcal{D}_k(U)$. Then*

(i) For $0 < \varepsilon < 1$, T_ε is a well defined current in $\mathcal{D}_k(U)$.
(ii) $T_\varepsilon \rightharpoonup T$ in $\mathcal{D}_k(U)$
(iii) $\mathbf{M}(T_\varepsilon) \leq c\,\mathbf{M}(T)$
(iv) If T has compact support, then T_ε has compact support in U, actually

$$\text{dist}\,(\text{spt}\,T_\varepsilon, \partial U) \geq (1-\varepsilon)\text{dist}\,(\text{spt}\,T, \partial U).$$

Proof. (i) follows as H_z is proper for any $z \in B(0,1)$ and $0 < \varepsilon < 1$. (ii) and (iii) follow from Proposition 2 (iv) and (iii), respectively. To prove (iv), we observe that, for $|z| < \varepsilon$

$$\text{spt}\,H_{z\#}T \subset H_z(\text{spt}\,T) \subset \{x \mid \text{dist}\,(x, \partial U) \geq (1-\varepsilon)\text{dist}\,(\text{spt}\,T, \partial U)\} =: V_\varepsilon$$

by (1). Consequently $H_{z\#}(T)(\omega) = 0 \;\forall\, \omega \in \mathcal{D}^k(U)$, $\text{spt}\,\omega \subset U \setminus V_\varepsilon$. Multiplying by $\rho_\varepsilon(-z)$ and integrating with respect to z, we get $T_\varepsilon(\omega) = 0$. □

As we have already remarked the map $H^z : [0,1]\times U \to U$ given by $H^z(t,x) = x + tzd(x)$ is an homotopy between $H_0 = \mathrm{id}$ and $H_z = x + zd(x)$. Consequently the homotopy formula for forms, compare (15) in Sec. 2.2.3, yields

$$H_z^{\#}\omega - \omega = d[H^{z\#}\omega]_{(0,1)} + [H^{z\#}(d\omega)]_{(0,1)}$$

for $\omega \in L^1(U, \Lambda^k\mathbb{R}^n) \cap \mathcal{E}^k(U)$. Multiplying by $\rho_\varepsilon(-z)$ and integrating we infer

$$\omega_\varepsilon - \omega = dK_\varepsilon(\omega) - K_\varepsilon(d\omega) \tag{4}$$

where

$$K_\varepsilon(\omega) := \int \rho_\varepsilon(-z)[H^{z\#}(\omega)]_{(0,1)}\, d\mathcal{L}^n(z)$$

is the *smoothing homotopy operator*. Also notice that for $z \in B(0,1)$ $H^z : [0,1] \times U \to U$ is proper, consequently K_ε maps $\mathcal{D}^k(U)$ into $\mathcal{D}^{k-1}(U)$.

Also, the homotopy formula for currents, Sec. 2.2.5, yields

$$H_{z\#}T - T = \partial H_{\#}^{z}([0,1] \times T) - H_{\#}^{z}([0,1] \times \partial T) \quad \text{in } U.$$

Multiplying by $\rho_\varepsilon(-z)$ and integrating with respect to z we then infer

$$T_\varepsilon - T = \partial H_\varepsilon(T) - H_\varepsilon(\partial T) \qquad \text{in } U \tag{5}$$

where $H_\varepsilon : \mathcal{D}_k(U) \to \mathcal{D}_{k+1}(U)$ is the *smoothing homotopy operator* given for all $\omega \in \mathcal{D}^{k+1}(U)$ by

$$H_\varepsilon(T)(\omega) := \int \rho_\varepsilon(-z) H_{\#}^{z}([\![0,1]\!] \times T)(\omega)\, d\mathcal{L}^n(z). \tag{6}$$

Notice that $H_\varepsilon(T)$ has compact support in U if T has compact support in U. Notice also that we have

$$\mathbf{M}(H_\varepsilon(T)) \le c\,\varepsilon\, \mathbf{M}(T), \tag{7}$$

since $\mathbf{M}(H_{\#}^{z}([\![0,1]\!] \times T)) \le c|z|\mathbf{M}(T)$. From (5) and (7) we then infer

$$\begin{aligned} |T_\varepsilon(\omega) - T(\omega)| &\le c\varepsilon\, \{\mathbf{M}(T)\|\, d\omega\,\| + \mathbf{M}(\partial T)\|\,\omega\,\|\} \\ &\le c\varepsilon \sup\{\mathbf{M}(T), \mathbf{M}(\partial T)\}\, (\|\, d\omega\,\| + \|\,\omega\,\|) \end{aligned} \tag{8}$$

for $\omega \in \mathcal{D}^k(U)$.

An immediate consequence of the above is the following

Proposition 4. *We have*

(i) If ω is a closed form in U, then ω_ε and ω are cohomologous, $\omega_\varepsilon - \omega = dK_\varepsilon$. If moreover ω has compact support, ω_ε and ω are cohomologous with compact support.

(ii) *If $T \in \mathcal{D}_k(U)$ is a cycle, $\partial T = 0$, then T and T_ε are homologous, $T_\varepsilon - T = \partial H_\varepsilon(T)$. Moreover, if T has compact support then T_ε and T are homologous with compact support,*

$$\operatorname{spt} T_\varepsilon \cup \operatorname{spt} H_\varepsilon(T) \subset \{x \mid \operatorname{dist}(x, \partial U) \geq (1-\varepsilon)\operatorname{dist}(\operatorname{spt} T, \partial U)\}.$$

Mollifying forms is by definition a weighted mean of pullback of forms of a suitable chosen family of orientation preserving diffeomorphisms of U, $H_z(x) = x + zd(x)$, $x \in U$, $z \in B(0,1)$. As it is clear from the previous considerations, the properties stated in Proposition 2 and the homotopy formula (4) again hold true if we replace the H_z's by another family of orientation preserving diffeomorphisms $H'_z : U \to U$ smoothly varying in z in the sense that

$$\sup_{x\in U} |H'_z(x) - H'_w(x)| + \sup_{x\in U} |\nabla H'_z(x) - \nabla H'_w(x)| \to 0 \tag{9}$$

as $z \to w$. For instance, if we choose $H'_z = (H_z)^{-1}$ the inverse diffeomorphism of H_z, (9) is clearly satisfied. So if we define for $\omega \in \mathcal{E}^k(U) \cap L^1(U, \Lambda^k\mathbb{R}^n)$

$$\omega^\varepsilon(x) := \int \rho_\varepsilon(-z)(H_z^{-1})^\# \omega(x)\, d\mathcal{L}^n(z), \tag{10}$$

then the claims of Proposition 2, (4) and the first part of Proposition 4 hold true replacing ω_ε with ω^ε. In particular if ω is closed then ω^ε and ω are cohomologous, and are cohomologous with compact support if ω has compact support.

It is worthwhile noticing that ω^ε and ω_ε are in general different, but $\omega^\varepsilon = \omega_\varepsilon$ in case $U = \mathbb{R}^n$, $\varphi(x)$ =const and ρ is symmetric. This type of mollified forms appears when we compute the mollified of a current $T = \mathbb{R}^n \llcorner \omega$, $\omega \in L^1(U, \Lambda^{n-k}\mathbb{R}^n)$. In this case we have

Proposition 5. *Let $\omega \in \mathcal{E}^{n-k}(U)$. Then*

$$(\mathbb{R}^n \llcorner \omega)_\varepsilon = \mathbb{R}^n \llcorner \omega^\varepsilon.$$

Proof. Let $\eta \in \mathcal{D}^k(U)$. Then

$$\begin{aligned}(\mathbb{R}^n \llcorner \omega)_\varepsilon(\eta) &= (\mathbb{R}^n \llcorner \omega)(\eta_\varepsilon) = \int \omega \wedge \Big(\int \rho_\varepsilon(-z) H_z^\# \eta(x) \Big) \\ &= \int \rho_\varepsilon(-z) \Big(\int \omega \wedge H_z^\# \eta \Big)\, d\mathcal{L}^n(z).\end{aligned} \tag{11}$$

Since now H_z is an orientation preserving diffeomorphism of U, we have

$$\int_U \omega \wedge H_z^\# \eta = \int_U (H_z^{-1})^\# \omega \wedge \eta. \tag{12}$$

Hence, using (11) and (12) and changing again the order of integration, the claim follows at once. □

We now show that the ε-mollified T_ε are currents that can be simply represented as $\mathcal{L}^n \wedge \xi_{T,\varepsilon}$ where $\xi_{T,\varepsilon}$ is C^∞ k-covector field. To see that first we notice that

$$\begin{aligned} H_z^{\#}\omega(x) &= \Lambda^k(DH_1^z(x))(\omega(x+zd(x))) \\ &= \Lambda^k(\mathrm{id} + z \otimes Dd(x))(\omega(x+zd(x))) \end{aligned}$$

hence, changing variables $z \to y = x + zd(x)$ in (3), we also have

$$\omega_\varepsilon(x) = \int d(x)^{-n} \rho_\varepsilon\left(\frac{x-y}{d(x)}\right) R_{x,y}(\omega(y))\, d\mathcal{L}^n(y) \tag{13}$$

where for $x, y \in U$ $R_{x,y}$ denotes the linear operator on covectors $R_{x,y} : \Lambda^k\mathbb{R}^n \longrightarrow \Lambda^k\mathbb{R}^n$ given by

$$R_{x,y}(\alpha) := \Lambda_k\left(\mathrm{id} + \frac{y-x}{d(x)} \otimes Dd\right)(\alpha).$$

Notice also that the support of the map $\varphi_\varepsilon : x \to \rho_\varepsilon\left(\dfrac{x-y}{d(x)}\right)$ is strictly contained in U since

$$\operatorname{spt} \varphi_\varepsilon \subset \left\{x \in U \mid \frac{d(y)}{1+\varepsilon} \le d(x) \le \frac{d(y)}{1-\varepsilon}\right\} \subset\subset U. \tag{14}$$

Therefore we can find a vector field $\xi_{T,\varepsilon} : U \longrightarrow \Lambda_k\mathbb{R}^n$ such that and in fact defined by

$$<\xi_{T,\varepsilon}(y), \alpha> = T_x\left(d(x)^{-n}\rho_\varepsilon\left(\frac{x-y}{d(x)}\right) R_{x,y}(\alpha)\right) \tag{15}$$

for $y \in U$ and $\alpha \in \Lambda^k\mathbb{R}^n$. It is also seen at once that $\xi_{T,\varepsilon} \in C^\infty(U, \Lambda_k\mathbb{R}^n)$. Corresponding to $\xi_{T,\varepsilon}$ we have the $(n-k)$-form $\omega_{T,\varepsilon} \in \mathcal{E}^{n-k}(U) = C^\infty(U, \Lambda_{n-k}\mathbb{R}^n)$ given by

$$\omega_{T,\varepsilon}(x) := \mathbf{D}_k(\xi_{T,\varepsilon}(x)) \tag{16}$$

where $\mathbf{D}_k(\xi) := \xi \lrcorner\, dx^1 \wedge \ldots \wedge dx^n$ is Hodge operator on $\mathbb{R}^n$, compare Sec. 5.2.2. We can now state

Proposition 6. *Let T be a current in $\mathcal{D}_k(U)$, T_ε be the ε-mollified of T, $\xi_{T,\varepsilon} \in C^\infty(U, \Lambda^k\mathbb{R}^n)$, $\omega_{T,\varepsilon} \in C^\infty(U, \Lambda_{n-k}\mathbb{R}^n)$ be the fields defined respectively in (15) and (16). Then*

$$T_\varepsilon = \mathcal{L}^n \wedge \xi_{T,\varepsilon} = \mathbb{R}^n \llcorner \omega_{T,\varepsilon}. \tag{17}$$

Moreover

$$\mathbf{M}(T_\varepsilon) = \int \| \xi_{T,\varepsilon} \|\, d\mathcal{L}^n = \int \| \omega_{T,\varepsilon} \|\, d\mathcal{L}^n, \tag{18}$$

and T_ε, $\xi_{T,\varepsilon}$, and $\omega_{T,\varepsilon}$ have compact support if T has compact support.

Proof. Let $\omega \in \mathcal{D}^k(U)$. We have

$$\begin{aligned} T_\varepsilon(\omega) &:= T(\omega_\varepsilon) = T_x\Big(\int_U d(x)^{-n} \rho_\varepsilon\Big(\frac{x-y}{d(x)}\Big) R_{x,y}(\omega(y))\, d\mathcal{L}^n(y)\Big) \\ &= \int T_x\Big(d(x)^{-n}\rho_\varepsilon\Big(\frac{x-y}{d(x)}\Big) R_{x,y}(\omega(y))\Big)\, d\mathcal{L}^n(y) \\ &=: \int <\xi_{T,\varepsilon}(y), \omega(y)> d\mathcal{L}^n(y) \end{aligned}$$

which proves the first equality. To prove the second equality we observe, compare Sec. 5.2.2, that for all $\xi \in \Lambda_k\mathbb{R}^n$ and $\eta \in \Lambda^k\mathbb{R}^n$ we have

$$<\mathbf{D}_k(\xi)\wedge\eta, e_1\wedge\ldots\wedge e_n> = <\xi, \eta>,$$

hence, integrating,

$$\int_U \omega_{T,\varepsilon}\wedge\omega = \int <\omega_{T,\varepsilon}\wedge\omega, e_1\wedge\ldots\wedge e_n> d\mathcal{L}^n = \int <\xi_{T,\varepsilon}, \omega> d\mathcal{L}^n.$$

Finally (18) is trivial, and the last claim follows from the definitions and (14). □

Combining Proposition 6 with (5) and Proposition 5 we easily infer

Proposition 7. *Let $T \in \mathcal{D}_k(U)$ be a current with finite mass and $\partial T = 0$. Then T can be decomposed as*

$$T = \mathbb{R}^n \llcorner f + \partial S$$

where $S \in N_{k+1}(U)$ and $f \in \mathcal{E}^{n-k}(U) \cap L^1(U, \Lambda^{n-k}\mathbb{R}^n)$.

A representation formula for normal currents. We also have

Theorem 1. *Let $T \in \mathbf{N}_{k+1}(U)$ and $\delta > 0$. Then there exist two forms $f : U \to \Lambda^{n-k}\mathbb{R}^n$, $g : U \to \Lambda^{n-k-1}\mathbb{R}^n$ with BV coefficients such that*

$$T = \mathbb{R}^n \llcorner f + \partial(\mathbb{R}^n \llcorner g) \qquad \text{in } U. \tag{19}$$

Moreover

$$\mathbf{M}(\mathbb{R}^n \llcorner f) + \mathbf{M}(\partial(\mathbb{R}^n \llcorner g)) \le (1+\delta)\mathbf{M}(T) + \delta\mathbf{M}(\partial T),$$

and f and g have compact support if T has compact support in U.

Proof. We first prove that one can find $\widetilde{S} \in \mathbf{N}_{k+1}(U)$ and $f : U \to \Lambda^{n-k}\mathbb{R}^n$ with BV coefficients such that

$$T = \mathbb{R}^n \llcorner f + \partial\widetilde{S} \tag{20}$$

and

$$(21) \qquad \mathbf{M}(\mathbb{R}^n \llcorner f) + \mathbf{M}(\partial \widetilde{S}) \le (1+\delta)\mathbf{M}(T) + \delta \mathbf{M}(\partial T).$$

To prove this, fix a regularized distance d, a mollified kernel ρ, and let $0 < \varepsilon < 1/2$ to be fixed later. For $i = 0, 1, \dots$ define inductively $\varepsilon_i := \varepsilon^{i+1}$, $T_0 := T$ and

$$(22) \qquad T_{i+1} := H_{\varepsilon_i}(\partial T_i)$$

where H_{ε_i} is the smoothing homotopy operator defined in (6). Then according to (22)

$$(23) \qquad T_{i+1} = T_i - (T_i)_{\varepsilon_i} - \partial H_{\varepsilon_i}(T_i),$$

and, according to (22) and (7)

$$\begin{aligned} &\mathbf{M}(T_{i+1}) \le c\varepsilon^{i+1}\mathbf{M}(\partial T_i) \\ &\mathbf{M}(\partial T_{i+1}) \le \mathbf{M}(\partial T_i - (\partial T_i)_{\varepsilon_i}) \le \varepsilon^{(k-1)(i-1)}\mathbf{M}(\partial T_i). \end{aligned}$$

Therefore

$$\begin{aligned} &\mathbf{M}(\partial T_{i+1}) \le \varepsilon^{(k-1)(i-1)}\mathbf{M}(\partial T) \\ &\sum_{i=0}^{\infty} \mathbf{M}(T_{i+1}) \le c\mathbf{M}(\partial T) \sum \varepsilon^{i(k-1)}\varepsilon^{i+1} \le 2c\varepsilon \mathbf{M}(\partial T). \end{aligned}$$

In particular $T_i \to 0$ in the mass norm. Consequently the currents

$$\widetilde{T} := \sum_{i=0}^{\infty} (T_i)_{\varepsilon_i}, \qquad \widetilde{S} := \sum_{i=0}^{\infty} H_\varepsilon(T_i)$$

belong respectively to $\mathcal{D}_k(U)$ and $\mathcal{D}_{k+1}(U)$, have finite mass and

$$\mathbf{M}(\widetilde{T}) \le (1+\varepsilon^k)\sum_{i=0}^{\infty} M(T_i) \le (1+\varepsilon^k)[\mathbf{M}(T) + 2c\varepsilon\mathbf{M}(\partial T)].$$

Moreover, since $(T_i)_{\varepsilon_i} = \mathbb{R}^n \llcorner f_i$, $f_i = \mathcal{E}^{n-k}(U)$, by Proposition 7, we have $\widetilde{T} = \mathbb{R}^n \llcorner f$, $f = \sum_{i=0}^{\infty} f_i$ and $f \in L^1(U, \Lambda^{n-k}\mathbb{R}^n)$. Adding (23) for $i = 0, 1, \dots$ we then get

$$T = \mathbb{R}^n \llcorner f + \partial \widetilde{S}$$

which also gives $f \in BV(U, \Lambda^{n-k}\mathbb{R}^n)$ as $\partial(\mathbb{R}^n \llcorner f) = \partial T$. Moreover

$$\begin{aligned} \mathbf{M}(\partial \widetilde{S}) &= \mathbf{M}(T - \widetilde{T}) = \mathbf{M}\Big(T - T_{\varepsilon_0} - \sum_{i=1}^{\infty} (T_i)_\varepsilon\Big) \\ &\le \varepsilon^k \mathbf{M}(T) + 2c(1+\varepsilon^k)\mathbf{M}(\partial T). \end{aligned}$$

Choosing ε so that $2c\varepsilon(1+\varepsilon^k) < \delta$ and $1 + 2\varepsilon^k < 1 + \delta$ (20) and (21) follow.

The claim now follows from (20) and (21) by a downward induction on k. It is trivial for $k = n$. Suppose it is true for $k+1$ then

$$\widetilde{S} = \mathbb{R}^n \llcorner g + \partial(\mathbb{R}^n \llcorner h)$$

with $g \in BV(U, \Lambda^{n-k-1}\mathbb{R}^n)$ and $h \in BV(U, \Lambda^{n-k-2}\mathbb{R}^n)$. Consequently $T = \mathbb{R}^n \llcorner f + \partial(\mathbb{R}^n \llcorner g)$. □

1.3 Flat Chains

In this section we discuss the notion of flat chains. Roughly we may think of *flat chain*, respectively *integral flat chains*, as of boundaries of currents of finite mass, respectively of i.m. rectifiable currents. Therefore in general their components are not measures but derivatives of measures. This is a substantial generalization of currents of finite mass we have considered so far. It originated in the work of Whitney and proved its effectiveness in the work of Federer.

Their relevance, especially in connection with homology theory, is due to the following properties

(i) The boundary of a flat chain is again a flat chain
(ii) Flat chains can be pushed-forward by Lipschitz maps.
(iii) The boundary operator commute with push-forward,
(iv) A flat chain with support on a submanifold X is actually a current on X.

This makes flat chains the natural objects to develop a homology theory in Lipschitz category. Such a theory that we shall discuss in the next sections is isomorphic to the classical singular theory but it is preferable for problems involving integration.

In some sense flat chains are the most general currents which carry topological properties of smooth maps. In the Cartesian situation and in codimension 1 a typical flat chain is the "graph" of an L^1-function, i.e., the trace of a BV function. Probably it could be of some interest studying boundaries of Cartesian currents.

Flat rectifiable chains. Let $U \subset \mathbb{R}^n$ be an open set and let $K \subset U$ be a compact set. Following Federer we denote by $\mathcal{R}_{m,K}$ the class of i.m. rectifiable currents with support in K

$$\mathcal{R}_{m,K} := \{T \in \mathcal{R}_m(U) \mid \operatorname{spt} T \subset K\}$$

and by $\mathcal{R}_{m,\mathrm{cpt}}(U)$ the class of all such compactly supported currents

$$\mathcal{R}_{m,\mathrm{cpt}} := \cup\{\mathcal{R}_{m,K}(U) \mid K \subset U,\ K \text{ compact}\}.$$

Definition 1. *The class of* integral flat chains supported in K *and the class of* integral flat chains in U *are respectively defined by*

$$\mathcal{F}_{m,K}(U) := \{R + \partial S \mid R \in \mathcal{R}_{m,K}(U),\ S \in \mathcal{R}_{m+1,K}(U)\}$$

and

$$\mathcal{F}_{m,\mathrm{cpt}}(U) := \cup\{\mathcal{F}_{m,K}(U) \mid K \subset U,\ K \textit{ compact}\}$$

Definition 2. *The* integral flat norm *of an integral flat chain is defined by*

$$\mathcal{F}_K(T) = \inf\{\mathbf{M}(R) + \mathbf{M}(S) \mid T = R + \partial S,\ R \in \mathcal{R}_{m,K}(U),\ S \in \mathcal{R}_{m+1,K}(U)\}.$$

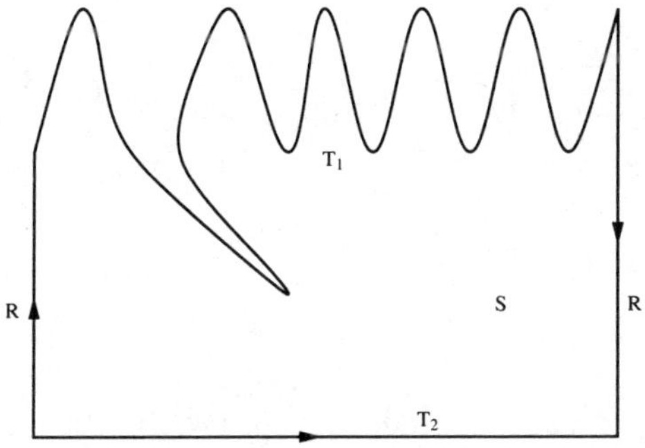

Fig. 5.2. The integral flat distance between T_1 and T_2: $T_2 - T_1 = R + \partial S$.

Clearly $\mathcal{F}_K(T) = +\infty$ if $\operatorname{spt} T \not\subset K$,

$$\mathcal{F}_K(T_1 + T_2) \le \mathcal{F}_K(T_1) + \mathcal{F}_K(T_2),$$

and $\mathcal{F}_K(T_1 - T_2)$ defines a distance, the *flat distance*, in $\mathcal{F}_{m,K}$.

Proposition 1. *We have*

(i) *$\mathcal{F}_{m,K}(U)$ is a complete metric space with respect to the flat metric*
(ii) *The set of integral currents with support in K*

$$\mathcal{I}_{m,K} := \{T \in \mathcal{R}_{m,K}(U) \mid \partial T \in \mathcal{R}_{m-1,K}(U)\}$$

is $\mathcal{F}_K$-dense in $\mathcal{F}_{m,K}(U)$.
Consequently $\mathcal{F}_{m,K}(U)$ can be regarded as the $\mathcal{F}_K$-completion of the space of integral currents with compact supports in K.

Proof. (i) Let $\{T_k\}$ be a Cauchy sequence for $\mathcal{F}_K$. Passing to a subsequence we can assume that

$$\sum_{i=1}^{\infty} \mathcal{F}_K(T_{k_i} - T_{k_{i-1}}) < \infty$$

and consequently that

$$T_{k_i} - T_{k_{i-1}} = R_i + \partial S_i, \quad R_i \in \mathcal{R}_{m,K}(U),\ S_i \in \mathcal{R}_{m+1,K}(U)$$
$$\sum_{i=1}^{\infty} (\mathbf{M}(R_i) + \mathbf{M}(S_i)) < \infty.$$

$\sum_{i=1}^{n} R_i$, $\sum_{i=1}^{\infty} S_i$ converge in mass. Then

$$\sum_{i=1}^{\infty} R_i \in \mathcal{R}_{m,K}(U), \qquad \sum_{i=1}^{\infty} S_i \in \mathcal{R}_{m+1,K}(U),$$

and

$$T := T_{k_1} + \sum_{i=2}^{\infty} R_i + \partial\Big(\sum_{i=2}^{\infty} S_i\Big) \in \mathcal{F}_{m,K}(U).$$

As

$$\mathcal{F}_K(T - T_{k_i}) \le \sum_{j=1+i}^{\infty} (\mathbf{M}(R_j) + \mathbf{M}(S_j)) \longrightarrow 0$$

the conclusion follows at once.

(ii) Let $T = R + \partial S$, $R \in \mathcal{R}_{m,K}(U)$, $S \in \mathcal{R}_{m+1,K}(U)$, and $\varepsilon > 0$. By the slicing theory we know that

$$\mathbf{M}(\partial(R \llcorner B(0,r)) < \infty$$

for a.e. $r > 0$, and, by the boundary rectifiability theorem that $\partial(R \llcorner B(0,r))$ is i.m. rectifiable. As $\mathbf{M}(R - R \llcorner B(0,r)) \to 0$ as $r \to \infty$, we conclude that there is $R_1 \in \mathcal{I}_{m,K}(U)$ with $\mathbf{M}(R - R_1) < \varepsilon$. Since we can do the same for S, the proof is completed. □

Flat chains. We now introduce the notion of (real) flat chain. This is done in a more involved way, as we want to be able to approximate flat chains by normal currents, and the naive decomposition $T = R + \partial S$ with normal currents R and S would not allow this, compare below. Set

$$\mathbf{M}_{m,K}(U) := \{T \in \mathcal{D}_m(U) \mid \mathbf{M}(T) < \infty,\ \operatorname{spt} T \subset K\}$$
$$\mathbf{M}_{m,\text{cpt}}(U) := \cup\{\mathbf{M}_{m,K}(U) \mid K \subset U,\ K \text{ compact}\}$$

and

$$\mathbf{N}_{m,K}(U) := \{T \in \mathbf{M}_{m,K}(U) \mid \mathbf{M}(\partial T) < \infty\}$$
$$\mathbf{N}_{m,\text{cpt}}(U) := \cup\{\mathbf{N}_{m,K}(U) \mid K \subset U,\ K \text{ compact}\}.$$

Recall the notation

$$\mathbf{N}(T) := \mathbf{M}(T) + \mathbf{M}(\partial T).$$

Definition 3. *The* flat seminorm *of an m-form $\varphi \in \mathcal{D}^m(U)$ relatively to K is defined by*

$$\mathbf{F}_K(\varphi) := \max\{\sup_{x\in K} \| \varphi(x) \|, \sup_{x \in K} \| d\varphi(x) \|\}.$$

The flat norm *of $T \in \mathcal{D}_m(U)$ relatively to K is defined by*

$$\mathbf{F}_K(T) := \sup\{T(\varphi) \mid \varphi \in \mathcal{D}^m(U),\ \mathbf{F}_K(\varphi) \le 1\}.$$

Then the space of *flat chains* is defined as the $\mathbf{F}_K$ closure of normal currents. More precisely

Definition 4. *T belongs to $\mathbf{F}_{m,K}(U)$ iff there exists a sequence of normal currents $T_j \in \mathbf{N}_{m,K}(U)$ such that $\mathbf{F}_K(T - T_j) \to 0$. Then we set*

$$\mathbf{F}_{m,\mathrm{cpt}}(U) := \cup\{\mathbf{F}_{m,K}(U) \mid K \subset U,\ K \text{ compact}\}.$$

Often we shall simplify $\mathbf{F}_{m,\mathrm{cpt}}(U)$ in $\mathbf{F}_m(U)$.

We have

Proposition 2. *Let U be an open set in $\mathbb{R}^n$ and K a compact set, $K \subset U$.*

(i) If $T \in \mathcal{D}_m(U)$ and $\mathbf{F}_K(T) < \infty$, then $\operatorname{spt} T \subset K$
(ii) If $\operatorname{spt} T \subset K$, then $\mathbf{F}_K(T) \le \mathbf{M}(T)$.
(iii) If $T \in \mathcal{D}_m(U)$, then $\mathbf{F}_K(\partial T) \le \mathbf{F}_K(T)$.
(iv) If $T \in \mathcal{D}_m(U)$ and $\gamma \in C^\infty_c(U)$, then

$$\mathbf{F}_{K,K\cap \operatorname{spt}\gamma}(T \llcorner \gamma) \le \sup_{x\in K}(|\gamma(x)| + |D\gamma(x)|)\mathbf{F}_K(T).$$

(v) We have

$$\mathbf{F}_K(T) = \min\{\mathbf{M}(R) + \mathbf{M}(S) \mid T = R + \partial S,\ R \in \mathbf{M}_{m,K}(U),\ S \in \mathbf{M}_{m+1,K}(U)\}.$$

Proof. (i) If $x \in \operatorname{spt} T\backslash K$, we find $B(x,r) \subset U\backslash K$, $\varphi \in \mathcal{D}_m(B(x,r))$ with $\operatorname{spt}\varphi \subset B(x,r/2)$ and $T(\varphi) > 0$. Then $\mathbf{F}_K(\lambda\varphi) = 0$ and $T(\lambda\varphi) \to \infty$ a contradiction with $\mathbf{F}_K(T) < \infty$.
(ii) In fact $\mathbf{F}_K(\varphi) \le 1$ for $\varphi \in \mathcal{D}^m(U)$ implies $\|\varphi(x)\| \le 1$ for all $x \in K$. From $\operatorname{spt} T \subset K$ we then infer $T(\varphi) \le \mathbf{M}(T)\sup_K \|\varphi\| \le \mathbf{M}(T)$.
(iii) is trivial and for (iv) it suffices to notice that $\mathbf{F}_K(\gamma\varphi) \le \sup_K|\gamma| + \sup_K|d\gamma|$ if $\mathbf{F}_K(\varphi) \le 1$.
(v) The inequality $\le$ is trivial as whenever $T = R + \partial S$

$$\begin{aligned} T(\varphi) &= R(\varphi) + S(d\varphi) \le \sup_K \|\varphi\|\,\mathbf{M}(R) + \sup_K \|D\varphi\|\,\mathbf{M}(S) \\ &\le \mathbf{F}_K(\varphi)(\mathbf{M}(R) + \mathbf{M}(S)). \end{aligned}$$

To prove the opposite inequality we consider the vector space $P := \mathcal{D}^m(U) \times \mathcal{D}^{m+1}(U)$ with the seminorm

$$\nu(\varphi,\psi) := \max\big(\sup_K \|\varphi\|, \sup_K \|\psi\|\big),$$

the linear map $Q : \mathcal{D}^m(U) \to P$, $Q(\varphi) := (\varphi, d\varphi)$ and the image P_0 of Q

$$P_0 := \{(\varphi,\psi) \in P \mid \psi = d\varphi\}$$

Clearly Q^{-1} is an isomorphism between P_0 and $\mathcal{D}^m(U)$ and the linear functional $T_0 := T \circ Q^{-1} : P_0 \to \mathbb{R}$ satisfies on P_0

$$|T_0(\varphi, \psi)| = |T(\varphi)| \le \mathbf{F}_K(\varphi)\mathbf{F}_K(\varphi) = \nu(\varphi, \psi)\mathbf{F}_K(T), \quad \psi = d\varphi.$$

By Hahn-Banach theorem we can extend T_0 to a functional $L : P \to \mathbb{R}$ satisfying

$$\text{(1)} \quad L(\varphi, \psi) = \nu(\varphi, \psi)\mathbf{F}_K(T) \ \forall\ (\varphi, \psi) \in P, \quad L(\varphi, d\varphi) = T_0(\varphi, d\varphi) = T(\varphi).$$

We now define the currents R and S by

$$R(\varphi) := L(\varphi, 0), \quad S(\psi) := L(0, \psi), \quad \varphi \in \mathcal{D}^m(U)\ \psi \in \mathcal{D}^{m+1}(U).$$

Clearly

$$T(\varphi) = L(\varphi, d\varphi) = R(\varphi) + S(d\varphi) = (R + \partial S)(\varphi)$$

and

$$R(\varphi) + S(\psi) = L(\varphi, \psi) \le \nu(\varphi, \psi)\mathbf{F}_K(T) \le \mathbf{F}_K(T)$$

if $\sup_K \| \varphi \| \le 1$ and $\sup_K \| \psi \| \le 1$. □

Proposition 3. *Let U and K be respectively open and compact in $\mathbb{R}^n$ with $U \supset K$.*

(i) *If $T \in \mathbf{F}_{m,K}(U)$, $m > 0$, then $\partial T \in \mathbf{F}_{m-1,K}(U)$.*
(ii) *If $T \in \mathbf{F}_{m,K}(U)$, $\gamma \in C_c^\infty(U)$, then $T \llcorner \gamma \in \mathbf{F}_{m,K\cap \operatorname{spt}\gamma}(U)$.*

Proof. For any $\varepsilon > 0$ there exists $T_1 \in \mathbf{N}_{m,K}(U)$ such that $\mathbf{F}_K(T - T_1) < \varepsilon$, $\operatorname{spt} T \subset K$. Proposition 2 (ii) yields then $\mathbf{F}_K(\partial T - \partial T_1) < \varepsilon$. Since for $S := \partial T_1$ we have $\operatorname{spt} S \subset K$, $\mathbf{M}(\partial S) = 0$ and $\mathbf{M}(S) = \mathbf{M}(\partial T_1) \le \mathbf{N}(T_1) < \infty$, i.e. $S \in \mathbf{N}_{m-1,K}(U)$. (i) follows. Proposition 2 (iv) yields instead

$$\mathbf{F}_{K\cap \operatorname{spt}\gamma}(T \llcorner \gamma - T_1 \llcorner \gamma) \le \big(\sup_K |\gamma| + \sup_K |d\gamma|\big)\mathbf{F}_K(T - T_1)$$

which clearly implies (ii). □

Weak and flat convergence. For normal currents supported in local Lipschitz neighbourhood retracts the flat convergence amount just to weak convergence. In fact we have

Theorem 1. *Let U be an open set of $\mathbb{R}^n$ and $K \subset U$ be a compact locally Lipschitz neighbourhood retract. Let $\{T_j\} \in \mathbf{N}_{m,K}(U)$ be such that $\sup_j \mathbf{N}(T_j) < +\infty$. Then $\mathbf{F}_K(T - T_j) \to 0$ iff $T_j \rightharpoonup T$.*

Proof. As $|T(\varphi)| \le \mathbf{F}_K(T)\mathbf{F}_{\mathcal{K}}(\varphi)$ we infer that $T_j \rightharpoonup T$ provided $\mathbf{F}_K(T_j - T) \to 0$. The other implication is proved by Theorem 2 below. □

Theorem 2. *Let U be an open set of $\mathbb{R}^n$ and $K \subset U$ be a compact locally Lipschitz neighbourhood retract. Let $T_i, T \in \mathbf{N}_{m,K}(U)$ be normal currents. If $T_i \rightharpoonup T$ with equibounded masses, $\sup_i \mathbf{N}(T_i) = \Lambda < +\infty$. Then one can find sequences $\{R_i\} \subset \mathbf{N}_{m+1,K}(U)$, $S_i \in \mathbf{N}_{m,K}(U)$ such that*

$$\mathbf{M}(R_i) + \mathbf{M}(S_i) \ \to\ 0, \qquad T_i - T = S_i + \partial R_i.$$

Proof. It suffices to consider the case $T = 0$. Moreover it is also obvious that it suffices to prove the claim for a subsequence of $\{T_i\}$. Divide $\mathbb{R}^n$ into congruent cubes of sufficiently small size $\sigma = \sigma_K$ in such a way that the cubes which touch K are contained in U and let us denote by L_K the CW-complex of such cubes. For $i = 1, 2, \dots$, we project T_i into the m-skeleton $L_{m,K}$ of L_K using the deformation theorem and write

$$T_i = P_i + \partial Q_i + P_i'$$

where the P_i are polyhedral chains on $\mathcal{L}_{m,K}$, $Q_i \in \mathbf{N}_{m+1,K}(U)$, $\operatorname{spt} Q_i \subset\subset U$, $P_i' \in \mathbf{N}_k(U)$ and $\mathbf{M}(P_i) \le c\,\mathbf{M}(T_i)$, $\mathbf{M}(Q_i) \le \sigma \mathbf{M}(T_i)$, $\mathbf{M}(P_i') \le \sigma \mathbf{M}(\partial T_i)$.

As chains in $L_{m,K}$ form a finite dimensional space, passing to a subsequence one can and do assume that each P_i is build with the same faces

$$P_i = \sum_h \beta_{ih} [\![F_h]\!], \qquad \beta_{jh} \in \mathbb{R}.$$

Moreover $\forall\ h\ |\beta_{ih}| \le M(P_i) \le \Lambda$; passing again to subsequences, $\beta_{ih} \to \beta_h$, $P_i \to P := \sum \beta_h [\![F_h]\!]$ in the mass norm, therefore for a subsequence of T_i's that we still denote by T_i we have

$$T_i = P + (P_i - P) + P_i' + \partial Q;$$

and $\mathbf{M}(P_i - P + P_i') \le c\sigma\Lambda$, $\mathbf{M}(Q_i) \le c\sigma\Lambda$; consequently,

$$T_i - T_j = (P_i - P_j) + (P_i' - P_j') + \partial(Q_i - Q_j)$$

with $\mathbf{M}(P_i - P_j + P_i' - P_j') \le 4c\sigma\Lambda$, $\mathbf{M}(Q_i - Q_j) \le 2c\sigma\Lambda$. We can now repeat the same construction for $\sigma = \sigma_K, \sigma_K/2, \dots, \sigma_K/2^n, \dots$ and, by a diagonal procedure, we then select a subsequence of $\{T_i\}$ that we name again $\{T_i\}$, and sequences of currents $\{F_i\} \subset \mathbf{N}_m(U)$ and normal currents $G_i \in \mathbf{N}_{m+1}(U)$ such that

$$T_i - T_{i+1} = F_i + \partial G_i$$

with

$$\begin{aligned} M(F_i) &\le c\sigma\Lambda 2^{-i}, \quad \operatorname{spt} F_i \subset \{x \mid \operatorname{dist}(x, K) < \sigma_K \sqrt{n} 2^i\} \\ M(G_i) &\le c\sigma\Lambda 2^{-i}, \quad \operatorname{spt} G_i \subset \{x \mid \operatorname{dist}(x, K) < \sigma_K \sqrt{n} 2^i\}. \end{aligned}$$

Setting

$$S_i = \pi_\#\Big(\sum_{j=i}^{\infty} F_j\Big), \qquad R_i = \pi_\#\Big(\sum_{j=i}^{\infty} G_j\Big)$$

where $\pi : U \to K$ is the retraction onto K we infer

$$M(S_i) + M(R_i) \le c_1 \sigma_A \Lambda \sum_{j=i}^{\infty} 2^{-j} \le c_2 \sigma_A \Lambda 2^{-i}.$$

Therefore we conclude that

$$T_i = \pi_\# T_i = \pi_\# \Big(\sum_{j=1}^{\infty} (T_{j+i} - T_j) \Big) = S_i + \partial R_i$$

which proves the claim. □

Remark 1. Theorem 1 may be regarded as the analogous of the classical fact that L^1 convergence and distributional convergence are equivalent for equibounded sequences of $W^{1,1}$ functions.

Density properties of flat chains. We state

Proposition 4. *Let U be open, K compact, $K \subset U$*

(i) We have

$$\mathbf{F}_{m,K}(U) = \{R + \partial S \mid R \in \mathbf{F}_{m,K}(U),\ S \in \mathbf{F}_{m+1,K}(U) \text{ with } \mathbf{M}(R),\ \mathbf{M}(S) < \infty\}$$

(ii) For any $T \in \mathbf{F}_{m,K}(U)$ there exists $S \in \mathbf{F}_{m+1,K}(U)$ such that

$$\mathbf{F}_K(T) = \mathbf{M}(T - \partial S) + \mathbf{M}(S).$$

(iii) If $T \in \mathbf{F}_{m,K}(U)$ and $\mathbf{M}(T) < +\infty$ then there exists a sequence of normal currents $T_j \in \mathbf{N}_{m,K}(U)$ such that $\mathbf{M}(T_j - T) \to 0$.

Proof. (i) Let $T \in \mathbf{F}_{m,K}(U)$. Clearly we can find a sequence $\{T_k\} \subset \mathbf{N}_{m,K}$ with $\mathbf{F}_K(T - T_k) < 2^{-k}$ so that

$$\sum_{k=1}^{\infty} \mathbf{F}_K(T_k - T_{k-1}) < \infty,$$

and we can write $T_k - T_{k-1} = R_k + \partial S_k$ with

$$\sum_{k=2}^{\infty} (\mathbf{M}(R_k) + \mathbf{M}(S_k)) < \infty. \tag{2}$$

As $\mathbf{M}(T_k) < \infty$, we have $S_k \in \mathbf{N}_{m+1,K}(U)$, hence $S_k \in \mathbf{F}_{m+1,K}(U)$. By Proposition 3 (i) $\partial S_k \in \mathbf{F}_{m,K}(U)$, consequently also R_k belong to $\mathbf{F}_{m,K}(U)$.

By Proposition 2 (ii)

$$\mathbf{F}_K\Big(R - \sum_1^i S_k\Big) = \mathbf{F}_K\Big(\sum_{i+1}^{\infty} R_k\Big) \le \mathbf{M}\Big(\sum_{i+1}^{\infty} R_k\Big) \le \sum_{i+1}^{\infty} \mathbf{M}(R_k) \to 0;$$

similarly

$$\mathbf{F}_K\Big(S - \sum_1^i S_k\Big) \longrightarrow 0.$$

Being $\sum_1^i R_k \in \mathbf{F}_{m,K}(U)$, $\sum_1^i S_k \in \mathbf{F}_{m+1,K}(U)$ we conclude that $R \in \mathbf{F}_{m,K}(U)$ and $S \in \mathbf{F}_{m+1,K}(U)$, and $\mathbf{M}(R), \mathbf{M}(S) < \infty$, because of (2).

On the other hand, if $R \in \mathbf{F}_{m,K}(U)$, $S \in \mathbf{F}_{m+1,K}(U)$ then $\partial S \in \mathbf{F}_{m,K}(U)$ by Proposition 3 (i), consequently $R + \partial S \in \mathbf{F}_{m,K}(U)$.
(ii) Proposition 2 (v) yields $R \in \mathbf{M}_{m,K}(U)$, $S \in \mathbf{M}_{m+1,K}(U)$ such that

$$T = R + \partial S, \qquad \mathbf{F}_K(T) = \mathbf{M}(R) + \mathbf{M}(S),$$

and by the above we get $R_1 \in \mathbf{F}_{m,K}(U)$ and $S_1 \in \mathbf{F}_{m+1,K}(U)$ such that

$$T = R_1 + \partial S_1, \qquad \mathbf{M}(R_1) + \mathbf{M}(S_1) < \infty.$$

Of course $R + \partial S = R_1 + \partial S_1$, and $\mathbf{M}(\partial(S - S_1)) = \mathbf{M}(R - R_1) < \infty$, hence $S - S_1 \in \mathbf{N}_{m+1,K}(U)$, consequently $S \in \mathbf{F}_{m+1,K}(U)$ as $S_1 \in \mathbf{F}_{m+1,K}(U)$.
(iii) Clearly we have $\mathbf{F}_k(T - Q) < \varepsilon$ for some $Q \in \mathbf{N}_{m,K}(U)$, and by (ii) there is $S \in \mathbf{F}_{m+1,K}(U)$ such that

$$\mathbf{F}_K(T - Q) = \mathbf{M}(T - Q - \partial S) + \mathbf{M}(S).$$

From

$$\mathbf{M}(Q + \partial S) \le \mathbf{M}(T) + \mathbf{M}(T - Q - \partial S) < \infty$$
$$\mathbf{M}(\partial(Q + \partial S)) = \mathbf{M}(\partial Q) < \infty$$

we conclude that $T_\varepsilon := Q + \partial S \in \mathbf{N}_{m,K}(U)$ and $\mathbf{M}(T - T_\varepsilon) < \varepsilon$. □

We conclude this subsection with a few remarks. For further information and proofs we refer to Federer [226, 41.12-18].

Remark 2. (i) $\mathcal{F}_{m,K}(U) \subset \mathbf{F}_{m,K}(U)$. This is clear from Proposition 1 (ii) as

$$\mathbf{F}_K(T) \le \mathcal{F}_K(T).$$

More directly one can prove it essentially as in the proof of Proposition 1 (ii) omitting the use of the boundary rectifiability theorem as we are simply interested in approximation with normal currents.

(ii) Flat chains are distributions of order -1. Therefore we notice that claims like $T \llcorner \Omega$ are meaningless for $T \in \mathbf{F}_{m,K}(U)$, and we resort to cut off with C^1 maps, $T \llcorner \gamma$.

(iii) Clearly for every neighbourhood V of $\operatorname{spt} T$ in U we can choose $\gamma \in C_c^\infty(V)$ with $\gamma \equiv 1$ on $\operatorname{spt} T$, hence $T = T \llcorner \gamma$, and, if $T \in \mathbf{F}_{m,K}(U)$, then $T \in \mathbf{F}_{m,K \cap \text{closure}\, V}$. However it can happen that that $T \notin \mathbf{F}_{m,\operatorname{spt} K}$.

(iv) However if $T \in \mathbf{F}_{m,\text{cpt}}(U)$, $\operatorname{spt} T \subset K$, K being a compact Lipschitz neighbourhood retract, then $T \in \mathbf{F}_{m,K}(U)$, see Proposition 7 below. In particular if $T \in \mathbf{F}_{m,\text{cpt}}(\mathbb{R}^n)$, then $T \in \mathbf{F}_{m,C}$, $C = \text{co}(\text{spt T})$ being the convex hull of support of T.

(v) While the Cartesian product of a flat chain times a normal current is a flat chain, the Cartesian product of two flat chains need not be flat.

(vi) Not every current with finite mass is a flat chain.

Push-forward of flat chains. We have already defined the image current of a locally normal current under Lipschitz maps which are proper on the support of T, compare Proposition 3 in Sec. 2.2.3, as a byproduct of the homotopy formula. We shall now extend the push forward to flat chains.

Let $f : U \subset \mathbb{R}^n \to V \subset \mathbb{R}^N$ be a Lipschitz map, U, V open sets in $\mathbb{R}^n$ and $\mathbb{R}^N$ respectively and let K be a compact subset of U. Assuming $T \in \mathbf{N}_{m,K}(U)$, $T = R + \partial S$, $R \in \mathbf{M}_{m,K}(U)$, $S \in \mathbf{M}_{m+1,K}(U)$ such that $\mathbf{F}_K(T) = \mathbf{M}(R) + \mathbf{M}(S)$ we infer that

$$R \in \mathbf{N}_{m,K}(U), \quad S \in \mathbf{N}_{m+1,K}(U), \quad f_\# T = f_\# R + \partial f_\# S$$

and

$$\mathbf{M}(f_\# T) \le c(\operatorname{Lip} f)^m \mathbf{M}(T), \quad \mathbf{M}(f_\# S) \le c(\operatorname{Lip} f)^{m-1}\mathbf{M}(S)$$

consequently

$$\mathbf{F}_{f(K)}(f_\# T) \le c \max\{(\operatorname{Lip} f)^m, (\operatorname{Lip} f)^{m+1}\}\mathbf{F}_K(T). \tag{3}$$

Thus the map $f_\# : \mathbf{N}_{m,K}(U) \to \mathbf{N}_{m,f(K)}(V)$, can be uniquely extended to a linear continuous map still denoted by $f_\#$

$$f_\# : \mathbf{F}_{m,K}(U) \to \mathbf{F}_{m,f(K)}(V)$$

still satisfying the *flat boundedness condition* (3).

Concerning homotopies for flat chains, let us consider $f, g \in C^\infty(U, V)$, U be open in $\mathbb{R}^n$, V be open in $\mathbb{R}^N$ and let $K \subset\subset U$. Set

$$C = \{y \in \mathbb{R}^N \mid y = (1-t)f(x) + tg(x),\ x \in K,\ t \in [0,1]\}.$$

If $\| f - g \|_{\infty,K}$ is sufficiently small, then C is compact and contained in V, consequently for $T \in \mathbf{N}_{m,K}(U)$ we have $h_\#([\![0,1]\!] \times T) \in \mathbf{N}_{m+1,C}(V)$, $h_\#([\![0,1]\!] \times \partial T) \in \mathbf{N}_{m,C}(V)$ and the formula

$$g_\# T - f_\# T = \partial h_\#([\![0,1]\!] \times T) + h_\#([\![0,1]\!] \times \partial T)$$

together with

$$\begin{aligned} \mathbf{M}(h_\#([\![0,1]\!] \times T)) &\le \operatorname{const}\| g - f \|_{\infty,K}\mathbf{M}(T) \\ \mathbf{M}(h_\#([\![0,1]\!] \times \partial T)) &\le \operatorname{const}\| g - f \|_{\infty,K}\mathbf{M}(\partial T) \end{aligned}$$

compare (16) in Sec. 2.2.3, consequently we get

$$\mathbf{F}_C(f_\# T - g_\# T) \le c\| g - f \|_{\infty,K}\mathbf{N}(T) \tag{4}$$

provided $C \subset\subset V$ as for example if $\| g - f \|_{\infty,K}$ is small enough.

Proposition 5. *Let $T \in \mathbf{F}_{m,K}(U)$. The push-forward of T, $f_\# T \in \mathbf{F}_{m,f(K)}(V)$ is characterized by the following property:*

Let $f_j : U \to V$ be Lipschitz maps converging to f uniformly on K with equibounded gradients. Then

$$\mathbf{F}_H(f_{\#h}T - f_\# T) \;\to\; 0, \qquad h \to \infty \tag{5}$$

for any compact set H such that $f(K) \subset \overset{\circ}{H} \subset H \subset\subset V$.

Proof. Let $T \in \mathbf{F}_{m,K}(U)$. Fix $\varepsilon > 0$ one can choose $T_1 \in \mathbf{N}_{m,K}$ such that $\mathbf{F}_K(T - T_1) < \varepsilon$. Since for h large enough the compact set

$$C_h = \{(1-t)f(x) + tg(x),\ t \in [0,1],\ x \in K\}$$

is strictly contained in $\overset{\circ}{H}$, we infer by (4) $\mathbf{F}_H(f_{h\#}T_1 - f_\# T_1) < \varepsilon$ for h large enough. Consequently

$$\begin{aligned}\mathbf{F}_H(f_{h\#}T - f_\# T) \;&\le\; \mathbf{F}_H(f_{h\#}(T - T_1)) + \mathbf{F}_H(f_{h\#}T_1 - f_\# T_1) + \\ &\quad + \mathbf{F}_H(f_\#(T - T_1)) \le (2c+1)\varepsilon.\end{aligned}$$

The convergence then follows from the uniqueness of flat limits. □

We also emphasize that for $f_\#$ we again have whenever $T \in \mathbf{F}_{m,K}(U)$

$$\operatorname{spt} f_\# T \subset f(\operatorname{spt} T), \qquad \partial(f_\# T)) = f_\#(\partial T). \tag{6}$$

It is convenient to define also the space of *locally flat* currents

Definition 5. *We say that $T \in \mathbf{F}^{\mathrm{loc}}_m(U)$, equivalently T is a* locally flat *current if $T \llcorner \gamma \in \mathbf{F}_{m,\mathrm{cpt}}(U)$ for every $\gamma \in C^\infty_c$.*

Then we can also define

$$f_\# T \in \mathbf{F}^{\mathrm{loc}}_m(U) \textit{ whenever } T \in \mathbf{F}^{\mathrm{loc}}_M(U) \textit{ and } f_{|\operatorname{spt} T} \textit{ is proper}$$

in such a way that

$$(f_\# T)(\omega) = [f_\#(T \llcorner \gamma)](\omega)$$

for all $\omega \in \mathcal{D}^m(V)$ and $\gamma \in C^\infty_c(U)$ with $\operatorname{spt} T \cap f^{-1}(\operatorname{spt}\omega) \subset \operatorname{int}\{x \mid \gamma(x) = 1\}$. Again one verifies that (6) holds. From Proposition 4 (iii) we get

Proposition 6. *If $T \in \mathbf{F}_{m,\mathrm{cpt}}(U)$, $\mathbf{M}(T) < \infty$, $\sigma(x)$ is continuous in U, $f : U \to \mathbb{R}^N$ is locally Lipschitz, $|Df(x)| \le \sigma(x)$* a.e., *and f is proper on* $\operatorname{spt} T$, *then*

$$\| f_\# T \| \le f_\#(\| T \| \llcorner \sigma);$$

if also g is locally Lipschitz with $|Dg|(x) \le \sigma(x)$ a.e.,

$$\big\{(1-t)f(x) + tg(x) \;\mid\; x \in \operatorname{spt} T,\ t \in [0,1]\big\} \subset V$$

and h is the affine homotopy from f to g, then

$$\mathbf{M}(h_\#([\![(0,1)]\!] \times T) \le \| T \|(|g - f|\sigma^m).$$

The flatness theorem. The following important support theorem, compare Theorem 1 in Sec. 5.3.1, holds for flat chains.

Proposition 7. *Let $T \in \mathbf{F}_{mK}(U)$ and let $f, g : U \to V$ be Lipschitz maps, U be open in $\mathbb{R}^n$, V be open in $\mathbb{R}^N$. If $f = g$ on $\operatorname{spt} T$, then $f_{\#}T = g_{\#}T$.*

Proof. First suppose that $T \in \mathbf{N}_{m,K}(U)$. Consider the map $\phi_\varepsilon : \mathbb{R}^n \to \mathbb{R}^n$, $\varepsilon > 0$, given by

$$\phi_\varepsilon(y) = \begin{cases} 0 & \text{if } |y| < \varepsilon \\ (|y| - \varepsilon)\dfrac{y}{|y|} & \text{if } |y| \geq \varepsilon. \end{cases}$$

ϕ_ε is Lipschitz, $|D\phi_\varepsilon| \leq 1$ $|\phi_\varepsilon(y) - y| \leq \varepsilon$. Denote by U_ε the open set

$$U_\varepsilon := \{x \in U \mid |f(x) - g(x)| < \varepsilon\}.$$

Clearly $\operatorname{spt} T \cap U \subset U_\varepsilon$ and $h_\varepsilon(x) := \phi_\varepsilon(g(x) - f(x))$ vanishes on U_ε. Therefore h_ε is proper on $\operatorname{spt} T$ and $h_\varepsilon^{\#}(\omega)$ is null in U_ε for all $\omega \in \mathcal{D}_k(\mathbb{R}^n)$, consequently $h_{\varepsilon\#}T = 0$. On the other hand $|f(x) - g(x) - h_\varepsilon(x)| \geq \varepsilon$, $|Dh_\varepsilon| \leq |Df| + |Dg|$, the homotopy formula then yields

$$|(f-g)_{\#}T(\omega) - h_{\varepsilon\#}T(\omega)| \leq c\varepsilon \mathbf{N}(T)(\|\omega\| + \|d\omega\|),$$

and letting $\varepsilon \to 0$ the conclusion follows at once if $T \in \mathbf{N}_{mK}(U)$. For the general case, fix $\varepsilon > 0$ and let $T_1 \in \mathbf{N}_{m,K}(U)$ be such that $\mathbf{F}_K(T - T_1) < \varepsilon$. Being T_1 normal, $f_{\#}T_1 = g_{\#}T_1$ and consequently $\mathbf{F}_K(f_{\#}T - g_{\#}T) < \varepsilon$. The claim follows as ε is arbitrary. □

Theorem 3 (Federer's flatness theorem). *Let $C \subset K \subset U$, C, K be compact and let $r : K \to C$ be a retraction. Then if $T \in \mathbf{F}_{m,K}(U)$ and $\operatorname{spt} T \subset C$ then $T \in \mathbf{F}_{m,C}(U)$.*

Proof. In fact one can extend the retraction to a Lipschitz map $r : U \to C$ which coincides with the identity map on C. Thus

$$T = r_{\#}T \in \mathbf{F}_{m,r(K)}(U) = \mathbf{F}_{m,C}(U).$$

□

A representation formula for flat chains. We then illustrate the mollification procedure for flat chains and extending the representation formula we have proved for normal currents to flat chains.

With the notation of Sec. 5.1.2, if $T \in \mathbf{N}_{m,K}(U)$, then for any $z \in \mathbb{R}^n$, $|z| < \varepsilon$, $H_z(\operatorname{spt} T) \subset K_\varepsilon := \{x \in U \mid \operatorname{dist}(x, K) \leq \varepsilon\}$, hence, using (3), (4) we have

$$\begin{aligned}
&\mathbf{F}_{K_\varepsilon}(H_{z\#}T) \leq c\mathbf{F}_K(T) \\
&\mathbf{F}_{K_\varepsilon}(H^z_{\#}([\![0,1]\!] \times T) \leq c\mathbf{F}_K(T) \\
&\mathbf{F}_{K_\varepsilon}(H_{z\#}T - H_{w\#}T) \leq c|z-w|\mathbf{N}(T) \\
&\mathbf{F}_{K_\varepsilon}(H^z_{\#}([\![0,1]\!] \times T) - H^w_{\#}([\![0,1]\!] \times T)) \leq c|z-w|\mathbf{N}(T)
\end{aligned}$$

for $z, w \in B(0,\varepsilon) \subset \mathbb{R}^n$. Also the mollified T_ε's have support in K_ε, therefore from (7) in Sec. 5.1.2, (8) in Sec. 5.1.2

$$\begin{aligned} &\mathbf{F}_{K_\varepsilon}(H_\varepsilon(T) \leq \mathbf{M}(H_\varepsilon(T)) \leq c\varepsilon\mathbf{M}(T) \ \to\ 0 \\ &\mathbf{F}_{K_\varepsilon}(T - T_\varepsilon) \leq c\varepsilon\mathbf{M}(T) \ \to\ 0. \end{aligned} \tag{7}$$

By density we then have for flat chains $T \in \mathbf{F}_{m,K}(U)$ the following estimates

$$\begin{aligned} &T_\varepsilon \in \mathbf{F}_{m,K_\varepsilon}(U), \quad H_\varepsilon(T) \in \mathbf{F}_{m+1,K_\varepsilon}(U), \\ &\mathbf{F}_{K_\varepsilon}(T_\varepsilon) + \mathbf{F}_{K_\varepsilon}(H_\varepsilon(T)) \leq c\mathbf{F}_K(T) \\ &\mathbf{F}_{K_\varepsilon}(T_\varepsilon - T) + \mathbf{F}_{K_\varepsilon}(H_\varepsilon(T)) \ \to\ 0 \end{aligned} \tag{8}$$

as $\varepsilon \to 0$.

Let $U \subset \mathbb{R}^n$ be an open set. We denote by Ω_m the space of compactly supported L^1-fields with values m-vectors

$$\Omega_m := \{\xi \in L^1(U, \Lambda_m\mathbb{R}^n) \mid \operatorname{spt}\xi \subset\subset U\}.$$

In $\Omega_m \times \Omega_{m+1}$ we introduce the norm

$$\int_U (\|\xi\| + \|\eta\|)\, d\mathcal{L}^n \quad (\xi,\eta) \in \Omega_m \times \Omega_{m+1},$$

and we show that every $T \in \mathbf{F}_{m,\text{cpt}}(U)$ can be represented by an element of $\Omega_m \times \Omega_{m+1}$. More precisely

Theorem 4. *We have*

(i) If $(\xi,\eta) \in \Omega_m \times \Omega_{m+1}$, $\delta > 0$, *and*

$$K_\delta := \{x \in \mathbb{R}^n \mid \operatorname{dist}(x, \operatorname{spt}\xi \cup \operatorname{spt}\eta) \leq \delta\} \subset U,$$

then the current

$$T := \mathcal{L}^n \wedge \xi + \partial(\mathcal{L}^n \wedge \eta)$$

is a flat chain, $T \in \mathbf{F}_{m,K_\delta}(U)$ *and* $\mathbf{F}_{K_\delta}(T) \leq \int_U (\|\xi\| + \|\eta\|)\, d\mathcal{L}^n$.

(ii) If $T \in \mathbf{F}_{m,K}(U)$, $\delta > 0$ *and*

$$K_\delta := \{x \in \mathbb{R}^n \mid \operatorname{dist}(x, K) \leq \delta\} \subset U,$$

then there exists $(\xi,\eta) \in \Omega_m \times \Omega_{m+1}$ *with* $\operatorname{spt}\xi \cup \operatorname{spt}\eta \subset K_\delta$ *such that*

$$T := \mathcal{L}^n \wedge \xi + \partial(\mathcal{L}^n \wedge \eta)$$

and $\int_U (\|\xi\| + \|\eta\|)\, d\mathcal{L}^n \leq \mathbf{F}_K(T) + \delta$.

Proof. (i) The claim is trivial if ξ and η are smooth, as

$$\int_U (\| \rho_\varepsilon * \xi - \xi \| + \| \rho_\varepsilon * \eta - \eta \|) \, d\mathcal{L}^n \longrightarrow 0 \quad \text{as } \varepsilon \to 0$$

it follows in general.
(ii) The proof is close to that of Theorem 1 in Sec. 5.1.2. Set $T_0 := T$, $K_0 := K$ and by induction

$$K_k := \{x \mid \text{dist}\,(x, K_{k-1}) < 2^{-k-1}\delta\} \subset U.$$

Given T_{k-1} for each k choose $R_k \in \mathbf{N}_{m,K_{k-1}}(U)$ and $S_k \in \mathbf{N}_{m+1,K_{k-1}}(U)$ such that

$$\begin{aligned} &\mathbf{F}_{K_{k-1}}(T_{k-1} - R_k - \partial S_k) < 2^{-k-3}\delta \\ &\mathbf{M}(R_k) + \mathbf{M}(S_k) \leq \mathbf{F}_{K_{k-1}}(T_{k-1}) + 2^{-k-2}\delta \end{aligned}$$

and ε_k such that

$$\varepsilon_k \leq 2^{-k-1}\delta, \qquad \varepsilon_k \mathbf{N}(R_k) < 2^{-k-3}\delta.$$

Define $\xi_k \in C_c^\infty(U, \Lambda_m \mathbb{R}^n)$ by

$$\mathcal{L}^n \wedge \xi_k := (R_k)_\varepsilon.$$

Notice that $\operatorname{spt} \xi_k \subset K_k$, and set

$$T_k := T_{k-1} - \mathcal{L}^n \wedge \xi_k - \partial S_k \in \mathbf{F}_{m,K_k}(U).$$

We have $\int_U \| \xi_k \| \, d\mathcal{L}^n \leq \mathbf{M}(R_k)$ and

$$\begin{aligned} \mathbf{F}_{K_k}(T_k) &\leq \mathbf{F}_{K_k}(T_{k-1} - R_k - \partial S_k) + \mathbf{F}_{K_k}(R_k - \mathcal{L}^n \wedge \xi_k) \\ &< 2^{-k-3}\delta + \varepsilon_k \mathbb{N}(R_k) < 2^{-k-2}\delta \end{aligned}$$

consequently

$$\begin{aligned} \sum_{k=1}^\infty \Big[\int_U \| \xi_k \| \, d\mathcal{L}^n + \mathbf{M}(S_k) \Big] &\leq \sum_{k=1}^\infty [\mathbf{M}(R_k) + \mathbf{M}(S_k)] \\ &\leq \sum_{k=1}^\infty \Big[\mathbf{F}_{K_{k-1}}(T_{k-1}) + 2^{-k-2}\delta \Big] \leq \mathbf{F}_K(T) + \delta/2. \end{aligned}$$

This implies that

$$\xi := \sum_{k=1}^\infty \xi_k \in \Omega_m, \quad S := \sum_{k=1}^\infty S_k \in \mathbf{F}_{m+1,K_{\delta/2}}(U)$$

and

$$T = \mathcal{L}^n \wedge \xi + \partial S, \qquad \int_U \| \xi \| \, d\mathcal{L}^n + \mathbf{M}(S) \le \mathbf{F}_K(T) + \delta/2.$$

If $m = n$, we actually have $S_k = 0$ for all k, $S = 0$, hence the proof is completed. If $m < n$, by a downward induction the proof is completed as in Theorem 1 in Sec. 5.1.2. □

As trivial consequences of Theorem 4 we state

Corollary 1. *If $T \in \mathbf{F}_{n,\text{cpt}}(U)$, $U \subset \mathbb{R}^n$, then there exists a Lebesgue summable field $\xi \in L^1(U, \Lambda_n \mathbb{R}^n)$ with compact support such that*

$$T = \mathcal{L}^n \llcorner f, \qquad f \in L^1(U, \mathbb{R}^n).$$

Corollary 2. *If $T \in \mathbf{F}_{n,\text{cpt}}(U)$, $U \subset \mathbb{R}^n$, and $\mathcal{L}^n(\operatorname{spt} T) = 0$, then $T = 0$.*

Actually for flat chains we have the following theorem which extends Theorem 3 in Sec. 4.3.1

Theorem 5 (Federer's support theorem). *Let $T \in \mathbf{F}_{m,\text{cpt}}(U)$, $U \subset \mathbb{R}^n$, $0 < m < n$. Denote by π_α, $\alpha \in I(m,n)$ the orthogonal projection of $\mathbb{R}^n$ into the m-plane coordinate determined by α given by $\pi(x^1, \dots, x^n) = (x^{\alpha_1}, \dots, x^{\alpha_m})$. Suppose that there exists a coordinate system $(x^1, \dots, x^n)$ in $\mathbb{R}^n$ such that*

$$\mathcal{L}^n(\pi_\alpha(\operatorname{spt} T)) = 0$$

for all α. Then $T = 0$.

Proof. Let $\omega \in \mathcal{D}^m(U)$. We have

$$\omega = \sum_{|\alpha|=m} \omega_\alpha \, dx^\alpha = \sum_{|\alpha|=m} \omega_\alpha \wedge \pi_\alpha^{\#}(dx^{\alpha_1} \wedge \dots \wedge dx^{\alpha_m})$$

and

$$\begin{aligned} T(\omega) &= \sum_{|\alpha|=m} T(\omega_\alpha \wedge \pi_\alpha^{\#}(dx^{\alpha_1} \wedge \dots \wedge dx^{\alpha_m})) \\ &= \sum_{|\alpha|=m} \pi_{\alpha\#}(T \llcorner \omega_\alpha)(dx^{\alpha_1} \wedge \dots \wedge dx^{\alpha_m}). \end{aligned}$$

Since $T \llcorner \omega_\alpha$ is a flat chain $\pi_{\alpha\#}(T \llcorner \omega_\alpha)$ is an m-dimensional flat chain in the m-dimensional space $\pi_\alpha \mathbb{R}^n$; moreover

$$\operatorname{spt} \pi_{\alpha\#}(T \llcorner \omega_\alpha) \subset \pi_\alpha(\operatorname{spt} T).$$

Corollary 2 then yields $\pi_{\alpha\#}(T \llcorner \omega_\alpha) = 0$, i.e., $T(\omega) = 0$. □

We state here an approximation theorem a slightly more general form of which will be proved in Vol. II Sec. 1.4.3

Theorem 6. *Let K and C be compact sets and U be an open set such that $K \subset \text{int}C \subset C \subset U \subset \mathbb{R}^n$. If $T \in \mathbf{F}_{m,K}(U)$, $\varepsilon > 0$, then there exists a real polyhedral chain with support in C such that*

$$\mathbf{F}_C(T - P) \leq \varepsilon \quad \text{and} \quad |\mathbf{M}(P) - \mathbf{M}(T)| < \varepsilon.$$

Flat Cochains. We conclude this section introducing the notion of *flat cochain* and proving a representation theorem for flat cochains

Definition 6. *A linear functional $\alpha : \mathbf{F}_{m,\text{cpt}}(U) \to \mathbb{R}$ which is continuous in the sense that there exists a positive constant c for which*

$$\alpha(T) \leq c\mathbf{F}_K(T) \tag{9}$$

*for all compact set $K \subset U$ and all $T \in \mathbf{F}_{m,K}(U)$ is called an m-*dimensional flat cochain, *and the class of m-dimensional cochains in U is denoted by $\mathbf{F}^m(U)$.*

The *flat norm* of a cochain α is defined by

$$\mathbf{F}_\alpha := \inf\{c \in \mathbb{R} \mid (9) \text{ holds } \forall\, K \subset\subset U,\ \forall\, T \in \mathbf{F}_{m,K}(U)\}. \tag{10}$$

The *coboundary* $d\alpha \in \mathbf{F}^{m+1}(U)$ of $\alpha \in \mathbf{F}^m(U)$ is defined by

$$d\alpha(S) := \alpha(\partial S) \qquad \forall\, S \in \mathbf{F}_{m+1}(U). \tag{11}$$

Notice that

$$\mathbf{F}(d\alpha) \leq \mathbf{F}(\alpha) \tag{12}$$

in fact if $S \in \mathbf{F}_{m,K}(U)$

$$d\alpha(S) = \alpha(\partial S) \leq \mathbf{F}(\alpha)\mathbf{F}_K(\partial S) \leq \mathbf{F}(\alpha)\mathbf{F}_K(S)$$

by Proposition 2 (iii).

Flat cochains can be represented by couples of forms with L^∞ coefficients

Theorem 7. *Let $\alpha \in \mathbf{F}^m(U)$ be a flat cochain. The there exist a m-form φ and a $(m+1)$-form ψ on U with bounded and measurable coefficients in U such that*

$$\begin{aligned} \alpha(\mathcal{L}^n \wedge \xi) &= \int_U <\xi, \varphi> d\mathcal{L}^n \qquad && \forall\, \xi \in \Omega_m \\ \alpha(\partial(\mathcal{L}^n \wedge \eta)) &= \int_U <\eta, \psi> d\mathcal{L}^n \qquad && \forall\, \eta \in \Omega_{m+1} \end{aligned} \tag{13}$$

and

$$\int_U (<\text{div}\,\eta, \varphi> + <\eta, \psi>)\, d\mathcal{L}^n = 0 \quad \forall\, \eta \in C_c^\infty(U, \Lambda_{m+1}\mathbb{R}^n), \tag{14}$$

the m-field $\text{div}\,\eta$ *being defined by*

$$\int_U <\text{div}\,\eta, \omega> d\mathcal{L}^n := -\int_U <\eta, d\omega> d\mathcal{L}^n \quad \forall\, \omega \in \mathcal{D}^m(U).$$

Also $\mathbf{F}(\alpha) = \max\left(\sup_U \|\varphi\|, \sup_U \|\psi\|\right)$.

Proof. To the cochain α corresponds a linear functional β on $\Omega_m \times \Omega_{m+1}$ given by

$$\beta(\xi,\eta) = \alpha(\mathcal{L}^n \wedge \xi + \partial(\mathcal{L}^n \wedge \eta)) \quad (\xi,\eta) \in \Omega_m \times \Omega_{m+1}$$

and

$$\mathbf{F}(\alpha) := \sup\left\{\beta(\xi,\eta) \;\mid\; \int_U (\|\xi\| + \|\eta\|)\, d\mathcal{L}^n \le 1\right\}.$$

As the dual of L^1 is L^∞ we get at once L^∞-forms φ and ψ for which (13) holds. For $\eta \in C_c^\infty(U, \Lambda_{m+1}\mathbb{R}^n)$ we then have

$$\int_U <\eta,\psi> = \alpha(\partial(\mathcal{L}^n \wedge \eta)) = \alpha(-\mathcal{L}^n \wedge \operatorname{div}\eta) = -\int_U <\operatorname{div}\eta,\varphi> d\mathcal{L}^n$$

since

$$\begin{aligned}\mathcal{L}^n \wedge \operatorname{div}\eta(\omega) &= \int <\operatorname{div}\eta,\omega> d\mathcal{L}^n = -\int <\eta, d\omega> d\mathcal{L}^n \\ &= -\mathcal{L}^n \wedge \eta(d\omega) = -\partial(\mathcal{L}^n \wedge \eta)(\omega)\end{aligned}$$

□

Finally we remark that in view of the representation formulas for flat chains and flat cochains one also introduce L^p-cochains and L^q-chains by requiring forms to have coefficients in L^q and fields of m-vectors to be in L^q. If q is the dual of p then those are in duality. But we shall not pursue this development.

2 Differential Forms and Cohomology

The goal of this section is mainly to prove *Hodge's* and *Hodge-Morrey* theorems, i.e., of representing cohomology classes of a compact Riemannian manifold by harmonic forms. In Sec. 5.2.5 we look at compact manifolds without boundary where there is a unique harmonic form in each cohomology class. Then in Sec. 5.2.6 we consider manifolds with boundary. In this case we represent cohomology classes of forms tangential at boundary by a unique harmonic form (tangential to the boundary), and show that absolute cohomology classes contain a unique harmonic form normal at the boundary.

We shall prove that as consequence of *the Hodge-Kodaira-Morrey decomposition theorems.* This approach heavily relies on variational methods and strongly uses the regularity theory for elliptic systems.

In the first three subsections we supply some preliminaries. In particular in Sec. 5.2.2 we introduce *Hodge's * operator* and discuss its algebra, and in Sec. 5.2.3 we briefly introduce Sobolev classes of p-forms. Then in Sec. 5.2.4 we discuss *harmonic forms*, and in Sec. 5.2.7 we prove *Weitzenböck formula* expressing the Laplace-Beltrami operator on forms in terms of *covariant derivatives* and of the *curvature tensor.*

Finally in Sec. 5.2.8 we discuss the *Poincaré* and *Poincaré Lefschetz* dualities in cohomology for oriented compact manifolds.

2.1 Forms on Manifolds

Let X be a smooth manifold of dimension m. For $y \in X$ we denote by T_yX the space of tangent vectors to X at y; the space of covectors is denoted by $\Lambda^1 T_yX$, while $\Lambda_p T_yX$ and $\Lambda^p T_yX$ denote the spaces of p-vectors and p-covectors to X at y generated by the exterior algebra construction in Ch. 2. We set also $\Lambda_0 T_yX = \Lambda^0 T_yX = \mathbb{R}$, and $\Lambda_1 T_yX = T_yX$.

A *p-vector field* ξ in X is a map $y \to \xi(y) \in \Lambda_p T_yX$. In the domain of a coordinate chart $U \subset \mathbb{R}^m$, ξ may be represented as a p-vector field ξ_U in U

$$\xi_U(x) = \sum_{\alpha \in I(p,m)} \xi^\alpha(x) e_\alpha$$

and, if two coordinate chart $\varphi : U \to X$, $\psi : V \to X$ overlap in X, then the vector fields ξ_U and ξ_V on U and V are linked by

$$\xi_V(\psi^{-1} \circ \varphi(x)) = \Lambda_p(D\psi^{-1} \circ \varphi)\, \xi_U(x). \tag{1}$$

Similarly, a *p-form* ω on X is a map $y \to \omega(y) \in \Lambda^p T_yX$. In a coordinate chart $\varphi : U \to X$, $U \subset \mathbb{R}^m$, ω can be represented as a *p-form* ω_U on U

$$\omega_U(x) = \sum_{\alpha \in I(p,m)} \omega_\alpha(x)\, dx^\alpha,$$

and the transition formula relating its representations on two overlapping charts U and V is given by

$$\omega_U = (\psi^{-1} \circ \varphi)^{\#} \omega_V \qquad \text{on } U \cap \varphi^{-1} \circ \psi(V) \tag{2}$$

i.e. by

$$\omega_U = \Lambda^p(D\psi^{-1} \circ \varphi)\, \omega_V(\psi^{-1} \circ \varphi) \qquad \text{on } U \cap \varphi^{-1} \circ \psi(V). \tag{3}$$

Smoothness of p-forms and p-vector fields is usually expressed in local coordinates.

Introducing the so-called *p-tangent bundle* we can regard p-vector fields essentially as maps. For that, one defines another manifold $\Lambda_p TX$ as follows. Let $\{U, \varphi_U\}$, $\varphi : U \subset \mathbb{R}^m \to X$, be a system of charts; for any two overlapping charts U, V denote by $\varphi_{VU} : U \cap \varphi^{-1}(\varphi(U) \cap \psi(V)) \to V \cap \psi^{-1}(\varphi(U) \cap \psi(V))$ the change of coordinates from U to V. Consider the new charts $\{U \times \Lambda_p \mathbb{R}^m\}$ together with the change of coordinates given by

$$(x_V, \xi_V) = g_{VU}(x_U, \xi_U) := (\varphi_{VU}(x_U), \Lambda_p D\varphi_{VU}(x_U)\xi_U).$$

Then $\{U \times \Lambda_p \mathbb{R}^m\}$ together with g_{VU} defines a new smooth manifold $\Lambda_p TX$ called the *p-tangent bundle to* X, which also carries a natural projection map

$$\pi : \Lambda_p TX \to X$$

on the first factor of $\Lambda_p\mathbb{R}^m$. A map $s : X \to \Lambda_p TX$, which is compatible with π, i.e., such that

$$\pi \circ s(x) = x \qquad \forall\, x \in X$$

is called a *section of the p-tangent bundle.* Notice that locally $\Lambda_p TX$ is like $X \times \Lambda_p\mathbb{R}^m$ and sections are roughly obtained by gluing together local graphs, $x \to (x, \xi(x))$. Finally, on account of (1) p-vector fields can be identified to sections of $\Lambda_p TX$. Similarly we can define the *p-cotangent bundle* $\Lambda^p TX$ and p-forms can be identified with sections of the p-cotangent bundles.

We shall denote the space of smooth p-forms on X by $\mathcal{E}^p(X)$ and the space of p-forms with compact support in X by $\mathcal{D}^p(X)$, or respectively by $C^\infty(X, \Lambda_p TX)$ and $C^\infty_c(X, \Lambda^p TX)$ if we want to stress their character of sections. The last notation is particular convenient to denote for instance p-forms in a Riemannian manifold with L^2-coefficients, as we can denote them by $L^2(X, \Lambda^p TX)$.

Let ω be a p-form on X, $p \geq 0$, and let $\{\omega_U\}$ be a collection of forms on $\mathbb{R}^m$ which represent locally ω. Since by (2)

$$d\omega_U = d((\psi^{-1} \circ \varphi)^{\#}\omega_V) = (\psi^{-1} \circ \varphi)^{\#} d\omega_V,$$

we readily see that the collection of $(p+1)$-forms on $\mathbb{R}^m$ $\{d\omega_U\}$ represents a $(p+1)$-form on X, denoted $d\omega$. The operator d defined this way is the *exterior differential operator*

$$d = d_X \;:\; \mathcal{E}^p(X) \to \mathcal{E}^{p+1}(X)$$

and it is easy to see that

(i) *If* $f : X \to \mathbb{R}$, *d agrees with the ordinary differential of maps*
(ii) *d is linear*
(iii) $d^2 = d \circ d = 0$
(iv) $d(\omega \wedge \eta) = d\omega \wedge \eta + (-1)^p \omega \wedge d\eta$ *for* $\omega \in \mathcal{E}^p(X)$, $\eta \in \mathcal{E}^q(X)$.

Also, if Y is another smooth manifold and $\phi : X \to Y$ is a smooth map, then we have

(v) $d_X \phi^{\#} = \phi^{\#} d_Y$.

A celebrated theorem by Whitney states that every smooth manifold X of dimension m may be regarded as an embedded submanifold of some $\mathbb{R}^n$ where $m \leq n \leq 2m+1$. In other words there exists a smooth one-to-one map $i : X \to \mathbb{R}^n$ which is a homeomorphism into (i.e., $i : X \to i(X)$, $i(X)$ equipped with the relative topology induced by $\mathbb{R}^n$, is open) and for which the linear tangent map

$$i_{\#y} : T_y X \to T_{i(y)}\mathbb{R}^n \simeq \mathbb{R}^n, \qquad y \in X$$

is of maximal rank m. Consequently, by the implicit function theorem, $i(X)$ is diffeomorphic to X, and, since i is an homeomorphism into, $i(X)$ *is a submanifold of* $\mathbb{R}^n$, i.e, for any $z \in i(X)$ the are two open neighbourhoods U and V respectively of z and 0, and a diffeomorphism $\phi : U \to V$ such that

$$\phi(U \cap i(X)) = V \cap \{x \in \mathbb{R}^n \mid x^{m+1} = \cdots = x^n = 0\};$$

in particular the inverse of ϕ gives a local chart

$$\varphi : W \subset \mathbb{R}^m \to U \cap i(X) \tag{4}$$

by $\varphi(x) := \phi^{-1}(x, 0)$, $W := \{x \in \mathbb{R}^m \mid (x, 0) \in V\}$.

Now we discuss relationships between the tangent and cotangent spaces of $i(X)$ and of $\mathbb{R}^n$. Clearly the tangent space $T_z i(X)$, $z \in i(X)$ is the subspace of $T_z\mathbb{R}^n \simeq \mathbb{R}^n$ generated by the columns of the Jacobian matrix of φ in (4) at 0, hence of dimension m; as $i_{\#y} T_y X \subset T_{i(y)} i(X)$, the image of $i_{\#y} : T_y X \to \mathbb{R}^n$ is exactly $T_{i(y)} i(X)$, and similarly

$$i_{\#y}(\Lambda_p T_y X) = \Lambda_p T_{i(y)} X = \Lambda_p \mathbb{R}^n. \tag{5}$$

The dual map $i_y^\# : \Lambda^p \mathbb{R}^n \to \Lambda^p T_y X$ is surjective, hence $i_y^\#$ is an isomorphism between $\Lambda^p \mathbb{R}^n / \ker i_y^\#$ and $\Lambda^p T_y X$

$$i_y^\# : \Lambda^p \mathbb{R}^n / \ker i_y^\# \longrightarrow \Lambda^p T_y X, \tag{6}$$

and we say that the p-covectors in $\ker i_y^\#$ are *null to* $i(X)$ *at* $i(y)$. Extensions of (5) and (6) to vector fields and forms depend upon relations between smooth maps on $i(X)$ and on $\mathbb{R}^n$. The following proposition is central in this respect

Proposition 1. *Let X be a compact submanifold of $\mathbb{R}^n$. Then X is a smooth neighbourhood retract, i.e., there is a neighbourhood U of X in $\mathbb{R}^n$ and a smooth map $\pi : U \to X$ which is a right inverse of $i : X \hookrightarrow \mathbb{R}^n$, that is, $\pi \circ i = \mathrm{id}_X$.*

U is often referred as a *tubular neighbourhood of* X and π as the *projection map*.

Using the projection map π, every smooth function $f : i(X) \to \mathbb{R}$ extends to a smooth function defined in the tubular neighbourhood U of $i(X)$ by $g(x) := f(\pi(x))$. Consequently p-vector fields on $i(X)$ extend smoothly to p-vector fields on a tubular neighbourhood U of $i(X)$ and the tangent map $i_\#$ turns out to be an isomorphism between smooth p-vectors fields to X and smooth p-vector fields ξ to U for which $\xi(x) \in \Lambda_p T_x i(X)$, $x \in i(X)$. Similarly the pull-back map $i^\# : \mathcal{E}^p(U) \to \mathcal{E}^p(X)$ is surjective, and

$$\mathcal{E}^p(U)/\mathrm{Null}_X^p(U) \simeq \mathcal{E}^p(X)$$

where

$$\begin{aligned} \mathrm{Null}_X^p(U) &:= \ker i^\# = \{\omega \in \mathcal{E}^p(U) \mid i^\# \omega = 0\} \\ &= \{\omega \in \mathcal{E}^p(U) \mid <\omega(i(y)), \xi> = 0 \ \forall\, \xi \in \Lambda_p T_{i(y)} i(y)\}. \end{aligned}$$

denotes the *null forms* to $i(X)$ in $\mathcal{E}^p(U)$.

Finally, the exterior differentiation operators d_X on X and d on U commute, i.e.,

(7) $$i^{\#}d = d_X i^{\#} \text{ on } \mathcal{E}^p(U) \qquad \pi^{\#}d_X = d\pi^{\#} \text{ on } \mathcal{E}^p(X)$$

and

(8) $$d_X = i^{\#}d\pi^{\#} \qquad \text{on } \mathcal{E}^p(X).$$

In particular the first equality in (7) implies that *$d\omega$ is a null form to X in U whenever ω is a null form to X in U.*

On account of the above we can identify X and $i(X)$ and understood the injection map $i : X \hookrightarrow \mathbb{R}^n$. This way tangent vectors to X are seen as a subspace of $\mathbb{R}^n$,

$$\Lambda_p T_y X \simeq \Lambda_p T_{i(y)} i(X) \subset \Lambda_p \mathbb{R}^n,$$

while p-forms in X are seen as p-forms in a tubular neighbourhood U of X modulo null forms, i.e., $\omega \in \mathcal{E}^p(X)$ if $\omega \in \mathcal{E}^p(U)$ and $<\omega(y), \xi> = 0\ \forall\ \xi \in \Lambda_p TX$. Also, the exterior differential operator is nothing but the ordinary exterior differential operator in $\mathbb{R}^n$, because of (7), (8). Finally we can identify the projection map $\pi : U \to X$ with the map $P := i \circ \pi : U \to \mathbb{R}^n$;

$$P^{\#}\omega = \pi^{\#} \circ i^{\#}\omega = 0$$

whenever ω is a null form to X, and for any $\zeta \in C^\infty(U)$ with $\zeta = 1$ on X we have

$$i^{\#}(\omega - \zeta P^{\#}\omega) = i^{\#}(\omega) - i^{\#}\pi^{\#}i^{\#}\omega = 0,$$

that is, *$\omega - P^{\#}\omega$ is a null form to X for any $\omega \in \mathcal{E}^p(U)$*, or, in other words, $\pi^{\#}\omega$ yields a canonical representative of ω modulo null forms.

2.2 Hodge Operator

In this subsection we shall collect a few more fact on exterior algebra. Let V be a vector space of dimension n, and let V^* be its dual

The operators $\mathbf{D}_p$ and $\mathbf{D}^p$. Given an n-covector $\Omega^{(n)} \in \Lambda^n V$ and an integer p, $0 \le p \le n$, we may consider the linear map $\mathbf{D}_p : \Lambda_p V \longrightarrow \Lambda^{n-p} V$ defined for $\xi \in \Lambda_p V$ and $\eta \in \Lambda_{n-p} V$ by

(1) $$<\mathbf{D}_p(\xi), \eta> := <\Omega^{(n)}, \eta \wedge \xi>.$$

If $e_1, \dots, e_n$ is a basis in V and $e^1, \dots, e^n$ is the dual basis in V^* we have

(2) $$\Omega^{(n)} = \lambda e^1 \wedge \dots \wedge e^n, \qquad \lambda := <\Omega^{(n)}, e_1 \wedge \dots \wedge e_n>,$$

hence

$$<\mathbf{D}_p(e_\alpha), e_\beta> = \lambda <e^1 \wedge \dots \wedge e^n, \varepsilon_\beta \wedge e_\alpha> = \begin{cases} 0 & \text{if } \beta \neq \bar{\alpha} \\ \lambda\sigma(\bar{\alpha}, \alpha) & \text{if } \beta = \bar{\alpha} \end{cases}$$

consequently

(3) $$\mathbf{D}_p(e_\alpha) = \lambda\sigma(\bar{\alpha}, \alpha) e^{\bar{\alpha}}, \qquad \alpha \in I(p, n).$$

Using (2), (3) and the linearity of $\mathbf{D}_p$ we easily infer

Proposition 1. *We have*

(i) $\mathbf{D}_p$ *is an isomorphism between* $\Lambda_p V$ *and* $\Lambda^{(n-p)}V$
(ii) $\mathbf{D}_p(\xi)$ *is simple if and only if* $\xi \in \Lambda_p V$ *is simple*
(iii) $\mathbf{D}_p(\xi)\wedge\phi = <\phi, \xi>\Omega^{(n)}$
(iv) $<\mathbf{D}_p(\xi), \eta> = (-1)^{p(n-p)}<\xi, \mathbf{D}_{n-p}(\eta)> \ \forall\ \xi \in \Lambda_p V,\ \eta \in \Lambda_{n-p}V.$

Proof. (i), (ii) and (iv) are trivial. To prove (iii) it suffices to observe that for $\alpha, \beta \in I(p,n)$ we have

$$\mathbf{D}_p(e_\alpha)\wedge e^\beta = \lambda\sigma(\bar{\alpha},\alpha)e^{\bar{\alpha}}\wedge e^\beta = \lambda e^1\wedge\ldots\wedge e^n\delta^\alpha_\beta = <e^\beta, e_\alpha>\delta^\alpha_\beta.$$

□

Similarly, given an n-vector $\Omega_{(n)} \in \Lambda_n V$ and an integer p, $0 \le p \le n$, we may consider the linear map $\mathbf{D}^p : \Lambda^p V \longrightarrow \Lambda_{n-p}V$ given for $\phi \in \Lambda^p V$ and $\psi \in \Lambda^{n-p}V$ by

(4) $$<\mathbf{D}^p(\phi), \psi> = <\Omega_{(n)}, \phi\wedge\psi>.$$

We have

(5) $$\Omega_{(n)} = \mu e_1\wedge\ldots\wedge e_n, \qquad \mu := <e^1\wedge\ldots\wedge e^n, \Omega_{(n)}>$$

and

(6) $$\mathbf{D}^p(e^\alpha) = \mu\sigma(\alpha,\bar{\alpha})e_{\bar{\alpha}}, \qquad \alpha \in I(p,n).$$

Also

Proposition 2. *We have*

(i) $\mathbf{D}^p$ *is an isomorphism between* $\Lambda^p V$ *and* $\Lambda_{n-p}V$
(ii) $\mathbf{D}^p(\phi)$ *is simple if and only if* $\phi \in \Lambda^p V$ *is simple*
(iii) $v\wedge\mathbf{D}^p(\phi) = <\phi, v>\Omega_{(n)} \ \forall\ v \in \Lambda_p V,\ \phi \in \Lambda^p V$
(iv) $<\mathbf{D}^p(\phi), \psi> = (-1)^{p(n-p)}<\phi, \mathbf{D}^{n-p}(\psi)> \ \forall\ \phi \in \Lambda^p V,\ \psi \in \Lambda^{n-p}V.$

[1] *Geometric interpretation.* First observe that, assuming $\Omega^{(n)} \neq 0$, (1) yields at once

$$<\mathbf{D}_p(\xi), \eta> = 0 \qquad \text{iff} \qquad \xi\wedge\eta = 0.$$

In particular, if ξ is simple, $\xi = v^1\wedge\ldots\wedge v^p$ and η is an $(n-p)$-vector of the type $\eta = v^i\wedge\alpha$ with $\alpha \in \Lambda_{n-p-1}V$, then

(7) $$<\mathbf{D}_p(\xi), \eta> = 0.$$

Write now $\mathbf{D}_p(\xi)$ with ξ simple as $\mathbf{D}_p(\xi) = f^1\wedge\ldots\wedge f^{n-p}$, then (7) reads

$$f^i(v^j) = 0 \qquad \forall\ i = 1,\ldots,n-p, \text{ and } j = 1,\ldots,p.$$

Therefore, if we denote by $f : V \to \mathbb{R}^{n-p}$ the map $f := (f^1,\ldots,f^p)$ we conclude that $\ker f$ *is the linear space associated to* ξ

(8) $$\ker f = \{v \in V \mid v\wedge\xi = 0\}.$$

Conversely, let $f : V \to \mathbb{R}^p$ be a non singular linear map, i.e., $\phi := f^1\wedge\ldots\wedge f^p \neq 0$. Choose coordinates $e_1, \ldots, e_n$ in V in such a way that e_1, e_{n-p} generate $\ker f$, $f = (f^1, \ldots, f^p)$, then $<\phi, v> = 0$ for any v of the form $v = e_i \wedge w$, $i = 1, \ldots, n-p$, $w \in \Lambda_{n-p-1}V$. Proposition 2 (iii) then yields

$$e_i\wedge w\wedge \mathbf{D}^p(\phi) = 0 \qquad \forall\, w \in \Lambda_{n-p-1}V,\ i = 1, \ldots, n-p,$$

hence $e_i\wedge\mathbf{D}^p(\phi) = 0 \qquad i = 1, \ldots, n-p$, that is, $\mathbf{D}^p(\phi)$ *is tangent to* $\ker f$. •

Interior multiplications. The homeomorphisms $\mathbf{D}_p$ and $\mathbf{D}^p$ are special cases of the so-called interior multiplications between vectors and covectors.

The *interior multiplication of p-vectors by q-covectors*, $0 \leq q \leq p \leq n$, is defined as the bilinear form $\llcorner : \Lambda_p V \times \Lambda^q V \longrightarrow \Lambda_{p-q}V$ given for $\xi \in \Lambda_p V$ and $\phi \in \Lambda^q V$ by

(9) $$<\xi \llcorner \phi, \omega> = <\xi, \phi\wedge\omega> \qquad \forall\, \omega \in \Lambda^{p-q}V.$$

In terms of a basis $e_1, \ldots, e_n$ of V it is easily seen that

(10) $$e_\alpha \llcorner e^\beta = \begin{cases} 0 & \text{if } \beta \not\subset \alpha \\ \sigma(\beta, \alpha-\beta)e_{\alpha-\beta} & \text{if } \beta \subset \alpha. \end{cases}$$

Often one refers to $\xi \llcorner \phi$ as to the *slice of* ξ *by* ϕ. This is motivated by the following

Proposition 3. *Let $\xi \in \Lambda_p V$ be a non-zero tangent vector to a p-subspace Σ of V, let $\widetilde{\phi} := (\phi^1, \ldots, \phi^q)$ be a linear map from V to $\mathbb{R}^q$, $p \geq q$, and let $\phi := \phi^1\wedge\ldots\wedge\phi^q$. Then $\xi \llcorner \phi$ is non-zero iff the restriction $\widetilde{\phi}_\Sigma$ of $\widetilde{\phi}$ to Σ has maximal rank; moreover, $\xi \llcorner \phi$ is simple and tangent to* $\ker \widetilde{\phi}_\Sigma$.

Proof. Without loss of generality we may assume $\xi = e_1\wedge\ldots\wedge e_p$. If $\phi = \sum_{|\beta|=q} \phi_\beta e^\beta$ then

$$e_1\wedge\ldots\wedge e_p \llcorner \phi = \sum_{\beta\supset\{1,\ldots,p\}} \sigma(\beta, \beta')\phi_\beta e_{\beta'}, \quad \beta' = \beta - \{1, \ldots, p\}$$

therefore $\xi \llcorner \phi \neq 0$ iff $\phi_\beta = <\phi, e_\beta> \neq 0$ for some $\beta \supset \{1, \ldots, p\}$. If we represent $\widetilde{\phi}$ by a $q \times n$-matrix $\overline{\phi}$, this implies that the $q \times p$-submatrix consisting of the first p columns of $\overline{\phi}$ is non singular. Therefore $\widetilde{\phi}_\Sigma$ has maximal rank q.

In order to prove the second part of the claim we observe that $\ker \widetilde{\phi}_\Sigma$ has dimension $p - q$. Choose now a basis of V in such a way that $e_1, \ldots, e_p$ generate Σ and $e_1, \ldots, e_{p-q}$ generate $\Sigma \cap \ker \widetilde{\phi} = \ker \widetilde{\phi}_\Sigma$. In these coordinates ϕ writes as $\phi = \sum_{|\beta|=q} \phi_\beta e^\beta$, $\phi_\beta = 0$ if $\beta \not\supset \{1, \ldots, p-q\}$ and, for $j = 1, \ldots, p-q$, we have

$$e_j\wedge(\xi \llcorner \phi) = \sum_\beta \sigma(\beta, \beta')\phi_\beta e_j\wedge e_{\beta'} = 0,$$

as each factor in the sum is zero since either $j \in \beta'$ or $\phi_\beta = 0$. Consequently

$$\Sigma \cap \ker \phi \subset \{v \mid v \wedge (\xi \llcorner \phi) = 0\} =: T_{\xi \llcorner \phi},$$

which implies that $\dim T_{\xi \llcorner \phi} \geq p - q$. On the other hand $\dim T_{\xi \llcorner \phi} \leq p - q$, hence $\dim T_{\xi \llcorner \phi} = p - q$. We then conclude that $\xi \llcorner \phi$ is simple and

$$\Sigma \cap \ker \widetilde{\phi} = \{v \mid v \wedge (\xi \llcorner \phi) = 0\}$$

which shows that $\xi \llcorner \phi$ is tangent to $\ker \widetilde{\phi}_\Sigma$. □

The *interior multiplication of p-covectors by q-vectors*, $0 \leq q \leq p \leq n$ is defined as the bilinear map $\lrcorner : \Lambda_q V \times \Lambda^p V \longrightarrow \Lambda^{p-q} V$ given for $\xi \in \Lambda_q V$ and $\phi \in \Lambda^p V$ by

(11) $$<\xi \lrcorner \phi, \eta> = <\phi, \eta \wedge \xi> \quad \forall\, \eta \in \Lambda_{p-q} V.$$

In coordinates it is easily seen that for $\alpha \in I(q,n)$, $\beta \in I(p,n)$

$$e_\alpha \lrcorner e^\beta = \begin{cases} 0 & \text{if } \alpha \not\subset \beta \\ \sigma(\alpha, \beta - \alpha) e^{\beta - \alpha} & \text{if } \alpha \subset \beta. \end{cases}$$

In terms of interior multiplications it is readily seen that

(12) $$\mathbf{D}_p(\xi) = \xi \lrcorner \Omega^{(n)}, \qquad \mathbf{D}^p(\phi) = \Omega_{(n)} \llcorner \phi;$$

one also easily checks that

(13) $$\mathbf{D}_{r+s}(\xi \wedge \eta) = \xi \lrcorner \mathbf{D}_s(\eta) \qquad \mathbf{D}^{r+s}(\phi \wedge \psi) = \mathbf{D}^r(\phi) \llcorner \psi.$$

Hodge operator. We assume now that V is a linear space endowed with a scalar product $(\ \mid\ )$. Then, as we have seen in Ch. 2 a scalar product is also induced on each $\Lambda_p V, \Lambda^p V, 0 \leq p \leq n$. If $e_1, \ldots, e_n$ is a basis in V, $g_{ij} := (e_i \mid e_j)$, and G denotes the $n \times n$-matrix (g_{ij}), the scalar product on $\Lambda_p V$ and $\Lambda^p V$ is defined respectively by

$$(e_\alpha \mid e_\beta) = M_\alpha^\beta(G) =: g_{\alpha\beta}$$

and

$$(e^\alpha \mid e^\beta) = M_\alpha^\beta(G^{-1}) =: g^{\alpha\beta}.$$

By classical inversion formulas, compare (5) in Vol. II Sec. 2.3.1 and (6) in Vol. II Sec. 2.3.1, we have

(14) $$g^{\alpha\beta} := M_\alpha^\beta(G^{-1}) = \sigma(\beta, \bar{\beta}) \sigma(\bar{\alpha}, \alpha) \frac{g_{\bar{\alpha}\bar{\beta}}}{g_{\bar{0}\bar{0}}}$$

where we recall $g_{\bar{0}\bar{0}} = \det G$. Also the scalar product induces isomorphisms

$$\beta_p : \Lambda_p V \longrightarrow \Lambda^p V, \qquad \beta^p : \Lambda^p V \longrightarrow \Lambda_p V$$

which are one the inverse of the other and are defined, compare (13) in Sec. 2.2.1, by

$$<\beta_p(\xi), \eta> := (\xi \,|\, \eta)_{\Lambda_p V}, \qquad <\beta^p(\phi), \psi> := (\phi \,|\, \psi)_{\Lambda^p V},$$

i.e., in coordinates

$$\beta_p(e_\alpha) = \sum_{|\beta|=|\alpha|} g_{\alpha\beta} e^\beta \tag{15}$$

$$\beta^p(e^\alpha) = \sum_{|\beta|=|\alpha|} g^{\alpha\beta} e_\beta. \tag{16}$$

Finally, if $e_1, \dots, e_n$ is an orthogonal basis, then $|e_1 \wedge \dots \wedge e_n| = 1$ and every n-vector has the form $\lambda e_1 \wedge \dots \wedge e_n$, $\lambda \in \mathbb{R}$. Therefore in $\Lambda_n V$ there are exactly two n-vectors of length one, and choosing one of the two amounts to fix the *orientation* of V.

Fix now an n-vector on V of length one, denote it by $\Omega_{(n)}$ and its dual by $\Omega^{(n)}$. If $\mathbf{D}_p$ and $\mathbf{D}^p$ are the operators associated to $\Omega_{(n)}$ and $\Omega^{(n)}$ by (1) and (5) the *Hodge* $*$ *operators*

$$* : \Lambda_p V \longrightarrow \Lambda_{n-p} V, \qquad * : \Lambda^p V \longrightarrow \Lambda^{n-p} V$$

are defined for $\xi \in \Lambda_p V$ and $\phi \in \Lambda^p V$ by

$$* \, \xi = \mathbf{D}^p(\beta_p(\xi)), \qquad * \phi = \beta_{n-p}(\mathbf{D}^p(\phi)). \tag{17}$$

If $e_1, \dots, e_n$ is an orthogonal basis and

$$\text{sign}(e) := <e_1 \wedge \dots \wedge e_n, \Omega^{(n)}> = \pm 1$$

denotes the sign of the basis $e_1, \dots, e_n$ with respect to $\Omega_{(n)}$, we get

$$* \, e_\alpha = \text{sign}(e) \sigma(\alpha, \bar{\alpha}) e_{\bar{\alpha}}, \qquad * e^\alpha = \text{sign}(e) \sigma(\alpha, \bar{\alpha}) e^{\bar{\alpha}} \tag{18}$$

so that

$$* * e_\alpha = \sigma(\alpha, \bar{\alpha}) \sigma(\bar{\alpha}, \alpha) e_\alpha, \qquad * * e^\alpha = \sigma(\alpha, \bar{\alpha}) \sigma(\bar{\alpha}, \alpha) e^\alpha \tag{19}$$

which yields

$$** = (-1)^{p(n-p)}$$

since $\sigma(\alpha, \bar{\alpha}) \sigma(\bar{\alpha}, \alpha) = (-1)^{p(n-p)}$. In terms of a generic basis $v_1, \dots, v_n$ of V we have $v_1 \wedge \dots \wedge v_n = \text{sign}(v) (\det G)^{1/2} \Omega_{(n)}$ where $G = (g_{ij})$, $\quad g_{ij} = (v_i \,|\, v_j)$. Therefore

$$<e^1 \wedge \dots \wedge e^n, \Omega_{(n)}> = \frac{\text{sign}(v)}{\sqrt{\det G}}$$

and

$$\begin{aligned} *v_\alpha &= \frac{\text{sign}(v)}{\sqrt{\det G}} \sigma(\beta, \bar{\beta}) g_{\alpha\beta} v_{\bar{\beta}} \\ *v^\alpha &= \frac{\text{sign}(v)}{\sqrt{\det G}} \sigma(\alpha, \bar{\alpha}) \beta_{n-p}(v_{\bar{\alpha}}) = \frac{\text{sign}(v)}{\sqrt{\det G}} \sigma(\alpha, \bar{\alpha}) g_{\bar{\alpha}\bar{\beta}} v^{\bar{\beta}} \end{aligned} \tag{20}$$

or by (14)

$$*v^\alpha = (-1)^{p(n-p)}\mathrm{sign}(v)\sqrt{\det G}g^{\beta\alpha}\sigma(\beta,\bar\beta)v^{\bar\beta}.$$

It is easily seen in orthogonal coordinates that

$$\begin{array}{rcll} \xi\wedge *\eta &=& (\xi\,|\,\eta)\Omega_{(n)} & \forall\xi,\eta\in\Lambda_p V \\ \phi\wedge *\psi &=& (\phi\,|\,\psi)\Omega^{(n)} & \forall\phi,\psi\in\Lambda^p V \\ <*\xi,*\phi> &=& <\xi,\phi> & \forall\xi\in\Lambda_p V,\ \phi\in\Lambda^p V \\ <\xi,*\phi> &=& (-1)^{p(n-p)}<*\xi,\phi> & \forall\xi\in\Lambda_p V,\ \phi\in\Lambda^{n-p} V \end{array} \tag{21}$$

and, as $\Omega_{(n)} = *e_0$, $\Omega^{(n)} = *e^0$,

$$*\,(\xi\wedge *\eta) = *(\eta\wedge *\xi) = (\xi\,|\,\eta) \qquad \forall\xi,\eta\in\Lambda_p V \tag{22}$$

Integration on submanifolds of $\mathbb{R}^n$. Let $X\subset\mathbb{R}^{n+k}$ be an *oriented* submanifold of dimension n. The metric and the orientation of $\mathbb{R}^{n+k}$ induce a Riemannian metric g on X, a metric $(g_{ij}(y))$ on T_yX and a unit tangent n-vector field $\vec T$, $\vec T(y)\in\Lambda_nT_yX$, which now plays the role of $\Omega_{(n)}$ above. The corresponding Hodge operator $*$ yields a $*$ operator $*:\Lambda^pT_yX \longrightarrow \Lambda^{n-p}T_yX$ which lifts to a $*$ isomorphism between forms $*:\mathcal{E}^p(X)\longrightarrow\mathcal{E}^{n-p}(X)$. Denote by $\Omega^{(n)}$ the dual of $\Omega_{(n)}=\vec T$. As

$$\int_X \omega = \int_X <\omega,\vec T>d\mathcal{H}^n$$

we see that $\Omega^{(n)}$ is the volume form on X, i.e.

$$\int_X \Omega^{(n)} = \int_X <\Omega^{(n)},\vec T>d\mathcal{H}^n = \mathcal{H}^n(X) \tag{23}$$

and, since by the second equality in (21)

$$(\alpha(y)\,|\,\beta(y)) = (\alpha(y)\,|\,\beta(y))<\Omega^{(n)}(y),\vec T(y)> = <\alpha\wedge *\beta(y),\vec T(y)>,$$

we get

$$\int_X (\alpha\,|\,\beta)\,d\mathcal{H}^n = \int_X \alpha\wedge *\beta. \tag{24}$$

2.3 Sobolev Spaces of Forms

Let X be an n-dimensional compact submanifold of $\mathbb{R}^{n+N}$. The class of functions $u: X\to\mathbb{R}$ which are $\mathcal{H}^n \llcorner X$-measurable and with

$$\int |u|^s\,d\mathcal{H}^n \llcorner X < \infty, \qquad s\geq 1$$

is denoted by $L^s(X)$. $L^s(X)$ is clearly a Banach space, and for $s = 2$, $L^2(X)$ is an Hilbert space with the obvious scalar product. If (V_i, φ_i), $\varphi_i : V_i \to X$ is a system of charts for X, we define the class $W^{1,s}(X)$, $s \geq 1$ as the class of $u \in L^s(X)$ such that $u \circ \varphi_i \in W^{1,s}(V_i)$, and we set

(1)
$$\| u \|^s_{W^{1,s}(X)} := \| u \|^s_{L^s(X)} + \sum_i \int_{V_i} |D(u \circ \varphi_i)|^s \, dx.$$

It is readily seen that the definition of $W^{1,s}(X)$ is independent of the system of charts, while the norm does depend. However different system of charts give rise to equivalent norms on $W^{1,s}(X)$. An intrinsic norm on $W^{1,s}(X)$ can be defined in terms of approximate tangential derivatives D^X to X: one easily verifies that

$$\| u \|_{L^s(X)} + \left(\int |D^X u|^s \, d\mathcal{H}^n \llcorner X \right)^{1/s}$$

is a norm equivalent to (1).

Properties of the standard Sobolev spaces on open sets of Euclidean spaces are immediately transferred to analogous properties of $W^{1,2}(X)$, we in particular state

(i) *$W^{1,2}(X)$ is a Hilbert space with the scalar product*

$$(u \,|\, v)_{W^{1,2}(X)} := (u \,|\, v)_{L^2(X)} + \int (D^X u \,|\, D^X v)_{TX} \, d\mathcal{H}^n \llcorner X.$$

(ii) *The inclusion $W^{1,s}(X) \hookrightarrow L^s(X)$ is compact*

(iii) *$C^\infty(X)$ is dense in $W^{1,s}(X)$.*

A differential p-form ω on X is said to be measurable iff $<\omega(y), v(y)>$ is $\mathcal{H}^n$-measurable for any smooth vector field v, equivalently iff for any local chart $\varphi_i : V_i \to X$, $\varphi_i^\# \omega$ is $\mathcal{L}^n$-measurable, i.e., all coefficients of $\varphi_i^\# \omega$ are measurable.

Definition 1. *Let $s \geq 1$. The space of measurable p-forms on X such that*

$$\int |\omega|^s \, d\mathcal{H}^n \llcorner X < \infty$$

is denoted by $L^s_p(X)$.

Clearly $L^s_0(X) = L^s(X)$. In terms of local charts (V_i, φ_i), $\omega \in L^s_p(X)$ iff $\varphi_i^\# \omega \in L^s(U, \Lambda^p \mathbb{R}^{n+N})$ from which we infer that $L^s_p(X)$ is a Banach space, and for $s = 2$ an Hilbert space. As

$$<\omega, \eta>_{\Lambda^p T_y X} = <\pi^\# \omega, \pi^\# \eta>_{\Lambda^p \mathbb{R}^{n+N}} \quad \forall \, \omega, \eta \in \Lambda^p T_y X,$$

a scalar product on $L^2_p(X)$ is given by

$$(\omega, \eta)_{L^2_p} := \int_X <\omega(y), \eta(y)>_{\Lambda^p T_y X} d\mathcal{H}^n = \int_X <\pi^\# \omega, \pi^\# \eta>_{\Lambda^p \mathbb{R}^{n+N}} d\mathcal{H}^n.$$

Definition 2. *The class of p-forms ω in $L^s_p(X)$ such that $\varphi^{\#}\omega$ have coefficients in $W^{1,s}(V,\mathbb{R})$ for every coordinate chart (V,φ) are denoted by $W^{1,s}_p(X)$.*

A norm on $W^{1,s}_p(X)$ can be defined by choosing a coordinate system of charts (V_i,φ_i) and setting

$$\|\,\omega\,\|_{W^{1,s}_p(X)} := \|\,\omega\,\|_{L^s_p(X)} + \sum_i |\varphi_i^{\#}\omega|_{W^{1,s}_p(V_i)}.$$

Though the definition of $W^{1,s}_p(X)$ is independent of the coordinate system, the norm does depend on it, but one easily verifies that different coordinate systems produce equivalent norms.

Again properties of $W^{1,s}_p(X)$ are inferred from properties of standard Sobolev spaces; we state

(i) *$W^{1,s}_p(X)$ is a Hilbert space.*
(ii) *The immersion $W^{1,s}_p(X) \hookrightarrow L^s_p(X)$ is compact.*
(iii) *Smooth forms are dense in $W^{1,s}_p(X)$.*

Finally we remark that similarly one can introduce the spaces $W^{2,s}_p(X)$ and more generally $W^{m,r}_p(X)$, but we shall not dwell any further on this point.

2.4 Harmonic Forms

Let X be a smooth compact n-dimensional submanifold of $\mathbb{R}^{n+k}$ possibly with boundary. We assume that X is a C^∞ submanifold even if the claims on this subsection hold for $C^{1,1}$ submanifolds with trivial changes. Let d be the differential operator on forms.

$$d : C^\infty(X, \Lambda^p TX) \longrightarrow C^\infty(X, \Lambda^{p+1}TX).$$

Definition 1. *The* codifferential *operator*

$$\delta : C^\infty(X, \Lambda^p TX) \longrightarrow C^\infty(X, \Lambda^{p-1}TX)$$

is the (formal) adjoint of d, i.e., is the operator defined for all $\alpha \in C^\infty(X,\Lambda^pTX)$ and $\beta \in C^\infty_c(X\setminus\partial X, \Lambda^{p-1}TX)$ by

$$(\delta\alpha\,|\,\beta)_{L^2(X,\Lambda^{p-1}TX)} := -(\alpha\,|\,d\beta)_{L^2(X,\Lambda^{p-1}TX)}.$$

Clearly $\delta^2 = 0$, and δ extends to all $W^{1,2}_p$ forms.

Proposition 1. *$\delta : C^\infty(X,\Lambda^pTX) \to C^\infty(X,\Lambda^{p-1}TX)$ is a differential operator. If $x^1,\dots,x^n$ are local coordinates and $G = (g_{ij})$ the metric tensor of X, then for $\omega = \sum_{|\alpha|=p}\omega_\alpha\, dx^\alpha$ we have $\delta\omega = \sum_{|\alpha|=p}(\delta\omega)_\gamma dx^\gamma$ where*

$$(\delta\omega)_\gamma = \frac{1}{\sqrt{g}} \sum_{|\alpha|=p-1}\sum_{i\notin\alpha}\sum_{|\beta|=p} g_{\gamma\alpha}\sigma(i,\alpha)D_i(\sqrt{g}g^{\beta(\alpha+i)}\omega_\beta). \tag{1}$$

Proof. Let $\eta \in C_c^\infty(X \setminus \partial X, \Lambda^{p-1}TX)$ be a $(p-1)$-form with support in coordinate chart. We have

$$\begin{aligned}
-(\omega, d\eta)_{\Lambda^p T_y X} &= - \sum_{|\beta|=|\gamma|=p} g^{\beta\gamma}\omega_\beta (d\eta)_\gamma \\
&= - \sum_{|\beta|=|\gamma|=p} g^{\beta\gamma}\omega_\beta \Big(\sum_{\alpha+i=\gamma} \sigma(i,\alpha)\eta_{\alpha,i} \Big) \\
&= - \sum_{|\beta|=p} \sum_{|\alpha|=p-1} \sum_{i\notin\alpha} g^{\beta(\alpha+i)}\sigma(i,\alpha)\omega_\beta \eta_{\alpha,i}.
\end{aligned}$$

Integration by parts on X then yields

$$\begin{aligned}
-(\omega \,|\, d\eta)_{L^2_p} &= - \int (\omega, d\eta)\sqrt{g}\, dx \\
&= \int \sum_{|\alpha|=p=1} \Big(\frac{1}{\sqrt{g}} \sum_{|\beta|=p} \sum_{i\notin\alpha} \sigma(i,\alpha) D_i(\sqrt{g} g^{\beta(\alpha+i)}\eta_\alpha) \Big) \sqrt{g}\, dx,
\end{aligned}$$

consequently $g^{\gamma\alpha}(\delta\omega)_\gamma = \frac{1}{\sqrt{g}} \sum_{|\beta|=p} \sum_{i\notin\alpha} \sigma(i,\alpha) D_i(\sqrt{g} g^{\beta(\alpha+i)}\omega_\beta)$, i.e. (1). □

Proposition 2. *If X is oriented and $*$ is the Hodge operator on X then*

$$\delta = (-1)^{n(p+1)} * d *$$

Proof. For $\alpha \in C^\infty(X, \Lambda^p TX)$ and $\beta \in C_c^\infty(X \setminus \partial X, \Lambda^{p-1}TX)$ we have

$$\begin{aligned}
d\beta \wedge * \alpha &= d(\beta \wedge * \alpha) - (-1)^{p-1}\beta \wedge d * \alpha \\
&= d(\beta \wedge * \alpha) - (-1)^{p-1}(-1)^{[n-(n-p+1)](n-p+1)}\beta \wedge * * d * \alpha \\
&= d(\beta \wedge * \alpha) - (-1)^{n(p+1)}\beta \wedge * * d * \alpha.
\end{aligned}$$

Therefore, by Stokes theorem and (24) in Sec. 5.2.2, we get

$$\begin{aligned}
(\delta\alpha \,|\, \beta) &= -(\alpha \,|\, d\beta) = - \int d\beta \wedge * \alpha \\
&= (-1)^{n(p+1)} \int \beta \wedge * * d * \alpha = (-1)^{n(p+1)}(* d * \alpha \,|\, \beta).
\end{aligned}$$

□

Definition 2. *The* Laplace-Beltrami *operator on $C^\infty(X, \Lambda^p TX)$ is the operator*

$$\Delta = d\delta + \delta d \,:\, C^\infty(X, \Lambda^p TX) \longrightarrow C^\infty(X, \Lambda^p TX);$$

a form ω is called harmonic *if $\Delta\omega = 0$.*

The following claims follow at once

Proposition 3. *We have*

(i) Δ *is formally self-adjoint, i.e.,* $(\Delta\alpha \mid \beta) = (\alpha \mid \Delta\beta)$ *for* $\alpha,\ \beta \in C_c^\infty(X \setminus \partial X, \Lambda^p TX)$.

(ii) If $\omega \in C_c^\infty(X \setminus \partial X, \Lambda^p TX)$ *then* $\Delta\omega = 0$ *iff* $d\omega = \delta\omega = 0$, *in fact* $(\omega \mid \Delta\omega) = |\delta\omega|^2 + |d\omega|^2$.

(iii) Δ *is well defined on* $W^{2,2}(X, \Lambda^p TX)$.

Definition 3. *The* Dirichlet integral *associated to a p-form is*

$$\mathcal{D}(\omega) := \int_X (|d\omega|^2 + |\delta\omega|^2)\, d\mathcal{H}^n$$

[1] *The Laplace-Beltrami operator on functions.* First we compute δ on 1-forms. In a chart U, if $\varphi \in C_c^\infty(U)$ and $\omega = \sum \omega_i dx^i$ we have

$$\begin{aligned}(\delta\omega \mid \varphi) &= -\int (\omega, d\varphi)\, dy = -\int_U \sqrt{g} g^{ij} \omega_i \frac{\partial \varphi}{\partial x^j} \\ &= \int_U \frac{1}{\sqrt{g}} D_j(\sqrt{g} g^{ij} \omega_i) \varphi \sqrt{g}\, dx\end{aligned}$$

where (g_{ij}) is the metric tensor on X, $g^{ij} := (g_{ij})^{-1}$, and $g := \det (g_{ij})$. Hence

$$\delta\omega = \frac{1}{\sqrt{g}} D_j(\sqrt{g} g^{ij} \omega_i)$$

and for any function $f \in C^\infty(X)$

$$\Delta f = (\delta d + d\delta) f = \delta df = \frac{1}{\sqrt{g}} D_j\left(\sqrt{g} g^{ij} \frac{\partial f}{\partial x^i}\right).$$

Also in local coordinates, i.e., for $f \in C^\infty(X)$ with compact support in a coordinate chart we have

$$\mathcal{D}(f) = \int_X |df|^2\, d\mathcal{H}^n = \int \sqrt{g} g^{ij} \frac{\partial f}{\partial x^i} \frac{\partial f}{\partial x^j}\, dx.$$

Notice that, by the maximum principle, any harmonic function on a compact and boundaryless manifold must be constant. •

[2] *The Laplace-Beltrami operator on p-forms in* $\mathbb{R}^n$. Given $\omega = \sum_{|\alpha|=p} \omega_\alpha dx^\alpha$ we have

$$d\omega = \sum_{|\alpha|=p} \sum_{i \notin \alpha} \omega_{\alpha,i}\, \sigma(i, \alpha)\, dx^{\alpha+i}$$

and, from (1)

$$\delta\omega = (-1)^{n(p+1)} * d * \omega = \sum_{|\beta|=p-1} \Big(\sum_{i\in\beta} \sigma(i,\beta)\omega_{\beta+i,i} \Big) dx^\beta. \tag{2}$$

We now compute the Laplace-Beltrami operator. Of course

$$d\delta\omega = \sum_\alpha (d\delta\omega)_\alpha \, dx^\alpha,$$

and, according to the above

$$\begin{aligned}(d\delta\omega)_\alpha &= \sum_{\beta+i=\alpha} \sigma(i,\beta)(\delta\omega)_{\beta,i} = \sum_{\beta+i=\alpha} \sum_{j\notin\beta} \sigma(i,\beta)\sigma(j,\beta)\omega_{\beta+j,ji} \\ &= \sum_{i\in\alpha} \sum_{j\notin\alpha-i} \sigma(i,\alpha-i)\sigma(j,\alpha-i)\omega_{\alpha-i+j,ij} \ ;\end{aligned}$$

now we split the sum in j as

$$\sum_{j\notin\alpha-i} = \sum_{j=i} + \sum_{j\notin\alpha}$$

to get

$$\begin{aligned}(d\delta\omega)_\alpha &= \sum_{i\in\alpha} \sigma(i,\alpha-i)\sigma(i,\alpha-i)\omega_{\alpha,ii} \\ &\quad + \sum_{\substack{i\in\alpha\\ j\notin\alpha}} \sigma(i,\alpha-i)\sigma(j,\alpha-i)\omega_{\alpha+j-i,ij}.\end{aligned} \tag{3}$$

Also $\delta d\omega = \sum (\delta d\omega)_\alpha \, dx^\alpha$, and

$$\begin{aligned}(\delta d\omega)_\alpha &= \sum_{j\notin\alpha} \sigma(j,\alpha)(d\omega)_{\alpha+j,j} = \sum_{j\notin\alpha} \sum_{\beta+i=\alpha+j} \sigma(j,\alpha)\sigma(i,\beta)\omega_{\beta,ij} \\ &= \sum_{j\notin\alpha} \sum_{i\in\alpha+j} \sigma(j,\alpha)\sigma(i,\alpha+j-i)\omega_{\alpha+j-i+j,ij}.\end{aligned}$$

Similarly we split the sum in i as the sum of terms in which $i \in \alpha$ and terms in which $i = j$ to get

$$\begin{aligned}(\delta d\omega)_\alpha &= \sum_{j\notin\alpha} \sigma(j,\alpha)\sigma(j,\alpha)\omega_{\alpha,jj} \\ &\quad + \sum_{\substack{i\in\alpha\\ j\notin\alpha}} \sigma(j,\alpha)\sigma(i,\alpha+j-i)\omega_{\alpha+j-i,ij}.\end{aligned} \tag{4}$$

Summing (3) and (4) we then get

$$\begin{aligned}(\Delta\omega)_\alpha &= (d\delta\omega + \delta d\omega)_\alpha = \Delta\omega_\alpha \\ &\quad + \sum_{\substack{i\in\alpha\\ j\notin\alpha}} [\sigma(i,\alpha-i)\sigma(j,\alpha-i) + \sigma(j,\alpha)\sigma(i,\alpha+j-i)]\omega_{\alpha+j-i,ij}.\end{aligned}$$

As for $i < j$

$$\sigma(j, \alpha - i) = -\sigma(j, \alpha), \qquad \sigma(i, \alpha - i) = \sigma(i, \alpha - i + j) \tag{5}$$

while for $i > j$

$$\sigma(j, \alpha - i) = \sigma(j, \alpha), \qquad \sigma(i, \alpha - i) = -\sigma(i, \alpha - i + j) \tag{6}$$

we see that the term in square brackets in (5) vanish, therefore we conclude

$$\Delta\omega = \sum_{|\alpha|=p} (\Delta\omega_\alpha)\, dx^\alpha, \tag{7}$$

i.e., *a p-form in the flat manifold* $\mathbb{R}^n$ *is harmonic if and only if its coefficients are harmonic functions.*

Obviously

$$|d\omega|^2 + |\delta\omega|^2 = \sum_{|\alpha|=p} |D\omega_\alpha|^2 + L(\omega)$$

$L(\omega)$ being a non-zero quadratic form of the derivatives of ω_α. Since the Euler equations of the two integrals

$$\mathcal{D}_0(\omega, U) := \int_U (|d\omega|^2 + |\delta\omega|^2)\, dx \quad \text{and} \quad \int_U \sum_{|\alpha|=p} |D\omega_\alpha|^2,$$

as we have seen agree with $\Delta\omega_\alpha = 0$, we infer that

$$\mathcal{D}_0(\omega, U) = \int_U |D\omega_\alpha|^2\, dx \qquad \forall\, \omega \in W_0^{1,2}(U)\,. \tag{8}$$

Moreover $L(\omega)$ *is a null Lagrangian*; in particular $\int_U L(\omega)\, dx$ depends only on the values of ω (ω_α) at the boundary ∂U of U and vanishes if $\omega \in C_c^1(U)$, compare e.g. Giaquinta and Hildebrandt [275]. •

Of course one can compute $\delta\omega$, $\Delta\omega$, $|\delta\omega|^2 + |d\omega|^2$ explicitly in terms of the metric tensor and of its derivative, but, since we do not need it in full generality, we postpone this topic to Sec. 5.2.7 and here we confine ourselves to the following remark.

From (1) one sees that $(\delta\omega)_\gamma$ has the form

$$(\delta\omega)_\gamma(x) = a_{\beta i}(G)\omega_{\beta,i} + b_\beta(G, DG)\omega_\beta \tag{9}$$

where $a_{\beta i}(G)$ and $b_\beta(G, DG)$ are smooth functions which pointwise depend respectively on the metric tensor and on the metric tensor and its derivatives. Indeed $a_{\beta i}(G(x))$ are Lipschitz functions and $b_\beta(G(x), DG(x))$ are L^∞-functions on X if the metric tensor G is Lipschitz. Moreover we notice that, in case G is the Euclidean metric,

$$(\delta\omega)_\gamma = \sum_{|\beta|=p-1} \sum_{i\notin\beta} \sigma(i,\beta) D_i \omega_{\beta+i}$$

consequently $b_\beta(\mathrm{Id}\,,0) = 0$. On account of the previous considerations it is then easy seen that in local coordinates the integrand of the Dirichlet integral has the form

$$(|d\omega|^2 + |\delta\omega|^2)\sqrt{g}\,dx = A^{ij}_{\alpha\beta}\omega_{\alpha,i}\omega_{\beta,j} + B_{\alpha\beta i}\omega_{\alpha,i}\omega_\beta + C_{\alpha\beta}\omega_\alpha\omega_\beta \tag{10}$$

where $A^{ij}_{\alpha\beta}(x) = A^{ij}_{\alpha\beta}(G(x))$, $B_{\alpha\beta i}(x) = B_{\alpha\beta i}(G(x), DG(x))$, $C_{\alpha\beta}(x) = C_{\alpha\beta}(G(x), DG(x))$ and

$$A^{ij}_{\alpha\beta}(\mathrm{Id}\,) = \delta^{ij}\delta_{\alpha\beta}, \quad B_{\alpha\beta i}(\mathrm{Id}, 0) = C_{\alpha\beta}(\mathrm{Id}, 0) = 0.$$

2.5 Representing Cohomology Classes by Harmonic Forms: Hodge and Hodge-Kodaira-Morrey Theorems

We are now ready to prove the celebrated theorem of Hodge which states that every cohomology class of a compact smooth manifold without boundary contains exactly one harmonic form. More generally we shall prove a decomposition theorem for forms, Hodge-Kodaira-Morrey theorem, from which Hodge result follows. The proof, based on variational methods and more precisely on Dirichlet principle, shows the entire theory of linear elliptic partial differential equations at work.

In this subsection X is a compact, possibly non orientable, boundaryless Riemannian manifold of dimension n that is thought without loss in generality as isometrically embedded in some $\mathbb{R}^N$. We assume that X is a C^∞ manifold even if one can work on $C^{1,1}$ manifolds with trivial changes.

One of the key ingredients of the variational proof we are going to present is the following simple but important lemma

Lemma 1 (Gaffney). *There is a constant C such that for each point $y \in X$ there exist a ball $B(0,\rho) \subset \mathbb{R}^n$ and a chart (U_y, φ), $\varphi : B(0,\rho) \to U_y \subset X$, $\varphi(0) = y$, such that*

$$\mathcal{D}(\omega) \geq \frac{1}{2} \int_{B(0,\rho)} \sum_{i,\alpha} |D_i\omega_\alpha|^2\,dx - c\|\,\omega\,\|^2_{L^2(U_y)}$$

whenever $\varphi^\#\omega = \sum \omega_\alpha\,dx^\alpha$ has support in U_y.

Proof. Fix a point $y \in X$ and choose coordinates so that the metric tensor at y is the identity, $g_{ij}(0) = \delta_{ij}$. Then using the representation (10) in Sec. 5.2.4 for $\mathcal{D}(\omega)$ we infer

$$\begin{aligned}\mathcal{D}(\omega) \;\; \geq \;\; & \int A^{ij}_{\alpha\beta}(0)\omega_{\alpha,i}\omega_{\beta,j}\,dx - \int [A^{ij}_{\alpha\beta}(0) - A^{ij}_{\alpha\beta}(x)]\omega_{\alpha,i}\omega_{\beta,j}\,dx \\ & - \int |B_{\alpha\beta i}(x)|\,|\omega_{\alpha,i}|\,|\omega_\beta|\,dx - \int |C_{\alpha\beta}|\,|\omega_\alpha|\,|\omega_\beta|\,dx.\end{aligned}$$

As

$$\int A^{ij}_{\alpha\beta}(0)\omega_{\alpha,i}\omega_{\beta,j}\,dx = \mathcal{D}_0(\varphi^{\#}\omega, B(0,\rho)) = \int_{B(0,\rho)} \sum_{i,\alpha} |D_i\omega_\alpha|^2\,dx,$$

by (8) in Sec. 5.2.4 the result follows at once choosing ρ sufficiently small so that $|A^{ij}_{\alpha\beta}(x) - A^{ij}_{\alpha\beta}(0)|$ be small. □

Choosing a finite cover of X by charts for which Lemma 1 holds we readily infer that

(i) $\mathcal{D}(\omega) + c\|\omega\|^2_{L^2_p(X)}$ *is an equivalent norm on* $W^{1,2}(X, \Lambda^p TX)$ *for a suitable constant* c *depending on the covering, and in fact there is a positive constant* c_1 *and a non- negative constant* c_2 *such that*

$$\mathcal{D}(\omega) \geq c_1\|\omega\|^2_{W^{1,2}} - c_2\|\omega\|^2_{L^2} \tag{1}$$

and, being $\|\omega\|^2_{L^2(X)}$ continuous with respect to the weak convergence in $W^{1,2}$ by Rellich theorem,

(ii) $\mathcal{D}(\omega)$ *is lower semicontinuous with respect to the weak convergence in* $W^{1,2}(X, \Lambda^p TX)$.

Definition 1. *We say that a p-form* $\omega \in W^{1,2}(X, \Lambda^p TX)$ *is* harmonic *if* $d\omega = \delta\omega = 0$. *The linear space of harmonic p-forms is denoted by* $\mathbf{H}^p$.

Theorem 1. $\mathbf{H}^p$ *is finite dimensional.*

Proof. Otherwise we could find a sequence $\{\omega_k\} \subset \mathbf{H}^p$ which is orthonormal in L^2. As $\|\omega_k\|_{L^2} = 1$ and $\mathcal{D}(\omega_k) = 0$ a subsequence will converge strongly in L^2, which is impossible since ω_k is orthonormal in L^2. □

Theorem 2. *There exists a positive constant c such that*

$$\mathcal{D}(\omega) \geq c\|\omega\|^2_{L^2(X,\Lambda^p TX)} \tag{2}$$

for any $\omega \in W^{1,2}(X, \Lambda^p TX)$ *which is orthogonal to* $\mathbf{H}^p$.

Proof. Otherwise we find a sequence $\omega_k \in W^{1,2}(X, \Lambda^p TX) \cap \mathbf{H}^{p\perp}$ such that

$$\mathcal{D}(\omega_k) \leq k\|\omega_k\|^2_{L^2}.$$

If $\eta_k := \omega_k/\|\omega_k\|_{L^2}$, we have $\|\eta_k\|_{L^2} = 1$, $\mathcal{D}(\eta_k) \leq 1/k$. Passing to a subsequence $\eta_k \rightharpoonup \eta$ weakly in $W^{1,2}$, $\eta_k \to \eta$ strongly in L^2, hence $\|\eta\|_{L^2} = 1$, by semicontinuity $\eta \in H^{p\perp}$ and $\mathcal{D}(\eta) = 0$, i.e., $\eta \in \mathbf{H}^p \cap \mathbf{H}^{p\perp}$ or in other words $\eta = 0$: a contradiction. □

Given a p-form $f \in L^2(X, \Lambda^p TX)$ we are now interested in finding a weak solution of the equation

$$\Delta\omega = f,$$

that is a p-form $\omega \in W^{1,2}(X, \Lambda^p TX)$ such that

$$(3) \qquad (d\omega \,|\, d\eta)_{L^2} + (\delta\omega \,|\, \delta\eta)_{L^2} = (f \,|\, \eta)_{L^2} \quad \forall\, \eta \in W^{1,2}(X, \Lambda^p TX).$$

We have

Theorem 3. *Equation (3) is solvable if and only if f belongs to the orthogonal in $L^2(X, \Lambda^p TX)$ of $\mathbf{H}^p$, denoted by $\mathbf{H}^{p\perp}$. Moreover there exists a unique solution ω_0 in $W^{1,2}(X, \Lambda^p TX) \cap \mathbf{H}^{p\perp}$ and the linear transformation from f to ω_0 is a bounded linear transformation from $L^2(X, \Lambda^p TX)$ into $W^{1,2}(X, \Lambda^p TX)$.*

Proof. Clearly if ω satisfies (3) then f is orthogonal to $\mathbf{H}^p$. Conversely assume $f \in L^2(X, \Lambda^p TX) \cap \mathbf{H}^{p\perp}$ and consider the integral

$$\mathcal{F}(\omega) := \mathcal{D}(\omega) - 2(f \,|\, \omega)_{L^2}$$

defined on $W^{1,2}(X, \Lambda^p TX) \cap \mathbf{H}^{p\perp}$. Trivially $\mathcal{F}$ is lower semicontinuous with respect to the weak convergence in $W^{1,2}$, $W^{1,2} \cap \mathbf{H}^{p\perp}$ is weakly closed and, by Theorem 2,

$$\mathcal{F}(\omega) \geq c\,\|\,\omega\,\|^2_{L^2} - 2\|f\|_{L^2}\,\|\,\omega\,\|_{L^2} \geq \frac{c}{2}\,\|\,\omega\,\|^2_{L^2} - \frac{2}{c}\|\,f\,\|_{L^2}.$$

Therefore, taking also into account (1) we immediately infer that there exists a minimizer $\omega_0 \in W^{1,2}(X, \Lambda^p TX) \cap \mathbf{H}^{p\perp}$.

From the minimality of ω_0 we now infer

$$(4) \qquad 0 = \frac{d}{dt}\mathcal{F}(\omega_0 + t\zeta) = (d\omega_0 \,|\, d\zeta) + (\delta\omega_0 \,|\, \delta\zeta) - (f \,|\, \zeta)$$

for all $\zeta \in W^{1,2}(X, \Lambda^p TX) \cap \mathbf{H}^{p\perp}$ and hence for any $\zeta \in W^{1,2}(X, \Lambda^p TX)$ since any such ζ can be written uniquely in the form $\zeta = h + \zeta_0$ where ζ_0 is orthogonal to $\mathbf{H}^p$ and $h \in \mathbf{H}^p$. From (4) we also get

$$\mathcal{D}(\omega_0) \leq \|\,f\,\|_{L^2}\|\,\omega_0\,\|_{L^2}$$

which together with (1) and (2) proves the boundedness of the map $f \to \omega_0$. Finally, if ω_1 is another solution of (3) we have for $\zeta := \omega_0 - \omega_1$

$$|d\zeta|^2 + |\delta\zeta|^2 = 0$$

i.e. $\omega_0 - \omega_1 \in \mathbf{H}^p$, and if ω_1 also belongs to $\mathbf{H}^{p\perp}$, $\omega_0 - \omega_1 = 0$. □

Equation (3) is the Euler equation of the integral

$$\mathcal{D}(\omega) - 2(f \,|\, \omega)_{L^2}$$

therefore on account of the representation (10) in Sec. 5.2.4, we see that in local coordinates the coefficients of any solution of (3) are in fact solutions of an elliptic system. The regularity theory for elliptic systems then applies, and in particular we can infer

Theorem 4. *Let X be of class $C^{1,1}$ and let ω_0 solves (3). We have*

(i) If $f \in L^2(X, \Lambda^p TX)$, then $\omega_0 \in W^{2,2}(X, \Lambda^p TX)$.
If $f \in L^p(X, \Lambda^p TX)$, then $\omega_0 \in W^{2,p}(X, \Lambda^p TX)$, $p > 1$.
If $f \in C^{0,\mu}(X, \Lambda^p TX)$, then $\omega_0 \in C^{2,\mu}(X, \Lambda^p TX)$.

Higher regularity follows, if we assume higher regularity of X and f. For instance, if X and f are of class C^∞ or analytic, then ω_0 is C^∞ or analytic.

(ii) In particular we see that ω is weakly harmonic iff ω is smooth and harmonic.

For each p set now

$$
\begin{aligned}
\operatorname{Im} d &= \{d\alpha \mid \alpha \in W^{1,2}(X, \Lambda^{p-1}TX)\} \\
\operatorname{Im} \delta &= \{\delta\alpha \mid \alpha \in W^{1,2}(X, \Lambda^{p+1}TX)\} \\
\mathbf{H} &= \{\alpha \in W^{1,2}(X, \Lambda^{p-1}TX) \mid d\alpha = \delta\alpha = 0\}.
\end{aligned}
$$

Then we have

Theorem 5 (Hodge-Kodaira-Morrey). *The linear subspaces* $\operatorname{Im} d$, $\operatorname{Im}\delta$, $\mathbf{H}$ *of $L^2(X, \Lambda^p TX)$ are mutually orthogonal and closed. Moreover*

$$L^2(X, \Lambda^p TX) = \mathbf{H} \oplus \operatorname{Im}\delta \oplus \operatorname{Im} d,$$

i.e., any $\omega \in L^2(X, \Lambda^p TX)$ decomposes uniquely as

$$\omega = h + \delta\alpha + d\beta \tag{5}$$

with $\alpha \in W^{1,2}(X, \Lambda^{p+1}TX)$, $\beta \in W^{1,2}(X, \Lambda^{p-1}TX)$ and $h \in \mathbf{H}$, and

$$\| h \|_{L^2} + \| \alpha \|_{W^{1,2}} + \| \beta \|_{W^{1,2}} \leq c \| \omega \|_{L^2} \tag{6}$$

and in fact, if X is of class C^∞, $h \in C^\infty$, and for all k

$$\| \alpha \|_{W^{1,k}} + \| \beta \|_{W^{1,k}} \leq c_k \| \omega \|_{W^{1,k-1}}. \tag{7}$$

Proof. Let $\omega \in L^2(X, \Lambda^p TX)$. Denote by h its projection on $\mathbf{H}$ which is finite dimensional. Then $\omega - h \in \mathbf{H}^\perp$ and by Theorem 3 and Theorem 4 there is $\omega_0 \in W^{1,2}(X, \Lambda^p TX)$ such that

$$\Delta\omega_0 = \omega - h.$$

Writing $\alpha := d\omega_0$, $\beta = \delta\omega_0$, (5) and (6) follow at once. Let us prove now that the three subspaces $\operatorname{Im}\delta$, $\operatorname{Im} d$ and $\mathbf{H}$ are mutually orthogonal and closed, from which one easily infer that the decomposition (5) is unique. Let $\omega \in \mathbf{H}$, for $\alpha \in W^{1,2}(X, \Lambda^{p-1}TX)$ we have

$$(d\alpha \,|\, \omega) = -(\alpha, \delta\omega) = 0$$

and for $\beta \in W^{1,2}(X, \Lambda^{p+1}TX)$

$$(\delta\alpha \,|\, \omega) = -(\alpha \,|\, d\omega) = 0;$$

while we also have

$$(d\alpha \,|\, \delta\beta) = -(d^2\alpha \,|\, \beta) = 0$$

provided α belongs to $W^{2,2}$. If α belongs to $W^{1,2}$ we choose a sequence $\alpha_k \in W^{1,2}$ such that $\alpha_k \to \alpha$ strongly in $W^{1,2}$ and we get

$$(d\alpha \,|\, \delta\beta) = \lim_{k\to\infty} (d\alpha_k \,|\, \delta\beta) = 0.$$

Finally let us prove that $\operatorname{Im}\delta$ is closed. Let $\omega_k := \delta\eta_k$, $\eta_k \in W^{1,2}(X, \Lambda^{p+1}TX)$, and assume that $\omega_k \to \omega$ in L^2. By (5) we have

$$\omega_k = h_k + \delta\alpha_k + d\beta_k,$$

the uniqueness of the decomposition yields $\delta\alpha_k = \omega_k$, moreover

$$\| \alpha_k \|_{W^{1,2}} \le c \| \omega_k \|_{L^2} \le \text{const.}$$

Therefore we infer passing to a subsequence that

$$\alpha_k \rightharpoonup \alpha \text{ weakly in } W^{1,2}, \qquad \alpha_k \to \alpha \text{ strongly in } L^2,$$

and $\alpha \in W^{1,2}$. Consequently $\delta\alpha = \omega$, which proves that $\operatorname{Im}\delta$ is closed. Similarly one proves that $\operatorname{Im} d$ is closed. □

Before stating Hodge theorem, let us recall the definition of the *De Rham cohomology groups* . A smooth p-form α is called *closed* if $d\alpha = 0$, *exact* if there exists η with $d\eta = \alpha$. Because of $d \circ d = 0$, exact forms are always closed. Two closed forms α, β are called *cohomologous* if $\alpha - \beta$ is exact. This property determines an equivalence relation on the space of closed p-forms and the set of equivalence classes is a vector space over $\mathbb{R}$, called the p-th *de Rham cohomology group* and denoted by

$$H^p_{\mathrm{dR}}(X).$$

Theorem 6 (Hodge). *In every cohomology class in $H^p_{\mathrm{dR}}(X)$ there exists a unique harmonic p-form.*

Proof. Let ω_0 be a closed differential form representing a cohomology class. All forms homologous to ω_0 have the form

$$\omega = \omega_0 + d\beta;$$

as $\omega_0 \in \operatorname{Im}\delta^\perp$, we can write by the decomposition theorem

$$\omega_0 = h_0 + d\beta_0$$

where h_0 is harmonic, therefore every ω cohomologous to ω_0 has the form

$$\omega = h_0 + d(\beta + \beta_0).$$

□

It is worth noticing again that if $\omega = h + d\beta$ is the decomposition of a closed form ω, h being harmonic, we also have, assuming X of class C^∞

$$\| \beta \|_{W^{k+1,r}} \le c_k \| \omega \|_{W^{k,r}} \qquad \| \beta \|_{C^{k+1,\mu}} \le c_k \| \omega \|_{C^{k,\mu}} .$$

Later we shall be interested in closed forms in the product manifold $\Omega \times X$, where Ω is an open set of an oriented Riemannian manifold. The product structure in $\Omega \times X$ induces a canonical splitting of the exterior differential operator d in $\Omega \times X$ as

$$d = d_x + d_y$$

with respect to local coordinates (x, y), $x \in \Omega$ and $y \in X$ and the splitting of $\mathcal{D}^p(\Omega \times X)$ as direct sum

$$\mathcal{D}^p(\Omega \times X) = \bigoplus_{k=\max(0,p-n)}^{\min(p,n)} \bar{\mathcal{D}}^{p,k}(\Omega \times X)$$

where $\bar{\mathcal{D}}^{p,k}(\Omega \times X)$ denotes the class of p-forms in the product $\Omega \times X$ with exactly k-differentials in y. A simple consequence of the decomposition theorem is the following result which will be used later

Proposition 1. *Let ω be a compactly supported p-form in $\Omega \times X$ which has exactly ℓ-differentials in y, $\omega \in \bar{\mathcal{D}}^{p,\ell}(\Omega \times X)$, and is d_y-closed, i.e., $d_y\omega = 0$. Then ω can written as*

$$\omega = \sum_{s=1}^{\bar{s}} \varphi^s(x) \wedge \sigma^s(y) + d_y \eta(x, y)$$

where η is a compactly supported $(p-1)$-form in $\Omega \times X$ with at most $\ell - 1$ differentials in y, $\varphi^s \in \mathcal{D}^{p-\ell}$, and $[\sigma^1], \dots, [\sigma^s]$, σ^i harmonic, form a basis of the cohomology group $H^p_{\mathrm{dR}}(X)$.

Proof. Choose an open covering $\{U_i\}$ of U so that on each U_i there exists a basis of simple tangent $(n-\ell)$-vector fields $e_{\alpha,i}$. Denote by η_i^α the dual form of $e_{\alpha,i}$ and by $\{\psi_i\}$ a decomposition of unity associated to the covering $\{U_i\}$. On each U_i $\psi_i\omega$ can be seen as

$$\psi_i \omega = \sum \eta_i^\alpha \wedge \widetilde{\omega}_i^\alpha$$

where $\widetilde{\omega}_i^\alpha \in C_c^\infty(U_i, \mathcal{D}^\ell(X))$. Thinking of x as a parameter, Theorem 5 applied to $\widetilde{\omega}_i^\alpha$ yields

$$\widetilde{\omega}_i^\alpha(x) = \sum_{s=1}^{\bar{s}} \varphi_i^{\alpha,s}(x) \sigma^s + d_y \eta_i^\alpha$$

with $\varphi_i^{\alpha,s} \in C_c^\infty(U_i)$ and $\eta_i \in C_c^\infty(U_i, \mathcal{D}^{\ell-1}(X))$. The result then follows summing on i and α. □

2.6 Relative Cohomology: Hodge-Morrey Decomposition

Let X be a smooth compact, possibly non orientable, Riemannian manifold of dimension n with smooth boundary. We think that X is isometrically embedded in an Euclidean space $\mathbb{R}^{n+N}$ and we also assume that X is a C^∞ manifold with a C^∞ boundary ∂X even if our presentation extends to $C^{1,1}$ manifolds with Lipschitz boundary with only trivial changes.

For $x \in \partial X$, let $\nu(x)$ be the *inward normal vector to ∂X at X*, i.e., the unit vector $\nu(x) \in T_x X$ orthogonal to $T_x X$ at x and pointing inside X. We infer the existence of special charts for X near each point $x_0 \in \partial X$. For $\rho > 0$ set

$$B^+(0,\rho) := \{x = (x_1, \dots, x_n) \mid |x| < \rho,\ x_n \geq 0\}$$
$$\Gamma(0,\rho) := \{x \in \mathcal{R}^{n-1} \mid (x', 0) \in B(0,\rho),\ x' := (x_1, \dots, x_{n-1})\}$$

Each point $x_0 \in \partial X$ has then a chart $\varphi : B^+(0,\rho_0) \to X$ such that $\varphi(0) = x_0$, $\varphi : \Gamma(0,\rho_0) \to \partial X$ is a local chart for ∂X, and, for $x' \in \Gamma(0,\rho_0)$, $D\varphi(x',0)e_n = \nu(\varphi(x',0))$ is the inward unit normal vector field to ∂X. We refer to these charts as to *admissible charts.* .

Tangential and normal traces of a form. Let us define the tangential and normal parts to $\partial\Omega$ of a form. Denote by $\mathfrak{r}$ the restriction map

$$\mathfrak{r} : C^\infty(X, \Lambda^p TX) \to C^\infty(\partial X, \Lambda^p TX)$$

which associate to each $\omega \in \mathcal{C}^\infty(X, \Lambda^p TX)$ its *trace* $\mathfrak{r}\,\omega \in C^\infty(\partial X, \Lambda^p TX)$, $\mathfrak{r}\,\omega(x) = \omega(x)$, $x \in \partial X$. As ∂X is a submanifold of X, the Riemannian structure on X yields a canonical decomposition of $T_x X$ in two orthogonal components

$$T_x X = T_x \partial X \oplus \mathbb{R}\,\nu(x).$$

Such a decomposition then extends to p-vectors, and we can decompose every p-vector $\xi \in \Lambda_p T_x X$ in a unique way as

$$\xi = \xi_1 + \xi_2 \wedge \nu, \quad \xi_1 \in \Lambda_p T_x \partial X, \quad \xi_2 \in \Lambda_{p-1} T_x \partial X.$$

Thus, give a form $\omega \in C^\infty(X, \Lambda^p TX)$ we can define *the tangential* and *the normal parts of the trace of* ω, respectively $\mathfrak{t}\,\omega$ and $\mathfrak{n}\,\omega$, as the functions in $C^\infty(\partial X, \Lambda^p TX)$ defined by

$$\begin{aligned} &< \mathfrak{t}\,\omega(x), \xi > = < \mathfrak{r}\,\omega(x), \xi_1 > \\ &< \mathfrak{n}\,\omega(x), \xi > = < \mathfrak{r}\,\omega(x), \xi_2 \wedge \nu(x) > \end{aligned} \tag{1}$$

for $\xi = \xi_1 + \xi_2 \wedge \nu(x)$. Obviously we have $\mathfrak{r}\,\omega = \mathfrak{t}\,\omega + \mathfrak{n}\,\omega$ and

$$|\mathfrak{r}\,\omega|^2 = |\mathfrak{t}\,\omega|^2 + |\mathfrak{n}\,\omega|^2. \tag{2}$$

Also, if $i : \partial X \to X$ is the immersion and $\pi : X \cap U_{\varepsilon_0} \to \partial X$ where U_{ε_0} is a neighborhood of ∂X is a smooth right inverse of i, then

(3) $$\mathfrak{t}\omega(x) = \pi^{\#} i^{\#} \omega(x).$$

In local coordinates $(x^1, \dots, x^n, t)$ around $x_0 \in \partial X$ we can write $\omega \in \mathcal{D}^p(X)$ as $\omega = \sum_{|\alpha|=p} \omega_{\alpha 0}(x,t)\, dx^\alpha + \sum_{|\alpha|=p-1} \omega_{\alpha\bar{0}}\, dx^\alpha \wedge dt$, then

$$\mathfrak{t}\omega(x,t) = \sum_{|\alpha|=p} \omega_{\alpha 0}(x,0)\, dx^\alpha, \quad \mathfrak{n}\,\omega(x,t) = \sum_{|\alpha|=p-1} \omega_{\alpha\bar{0}}(x,0)\, dx^\alpha \wedge dt.$$

We recall now that the restriction at the boundary $f \to f_{|\partial\Omega}$ of smooth functions f defined on a Lipschitz domain Ω extends to a *continuous linear operator* from $W^{1,s}(\Omega)$ to $L^s(\partial\Omega)$ for $s \geq 1$. Based on that it is then easy to see using local coordinates that the trace operator $\mathfrak{r} : C^\infty(X, \Lambda^p TX) \to C^\infty(\partial X, \Lambda^p TX)$ extends to a *continuous linear operator*, the *trace operator*

$$\mathfrak{r} : W^{1,s}_p(X) \to L^s_p(\partial X)$$

for any $s \geq 1$.

From (2) we then infer that the *tangential* and *normal* operators extend to linear continuous operators

$$\mathfrak{t}, \mathfrak{n} : W^{1,s}_p(X) \to L^s_p(X)$$

for all $s \geq 1$.

In an *oriented manifold*, the associated Hodge operator $*$ commutes tangential and normal operators. We have

Proposition 1. *Let X be an oriented manifold, let $*$ be the Hodge operator on X, and let δ be the formal adjoint of d. Then we have*

(i) For $\omega \in \mathcal{D}^p(X)$

$$*\mathfrak{n}\,\omega = \mathfrak{t}\,(*\omega), \quad *\mathfrak{t}\,\omega = \mathfrak{n}\,(*\omega), \quad \mathfrak{t}\,(d\omega) = d(\mathfrak{t}\,\omega), \quad \mathfrak{n}\,(\delta\omega) = \delta(\mathfrak{n}\,\omega).$$

(ii) For $\omega \in \mathcal{D}^p(X)$, $\eta \in \mathcal{D}^q(X)$, $q \leq p$, $\mathfrak{t}\,(\omega \wedge *\eta) = \mathfrak{t}\,\omega \wedge *\,\mathfrak{n}\,\eta$.

Proof. Since Hodge's operator commutes with orientation preserving diffeomorphisms, it suffices to prove the claim in an admissible system of charts which is also orientation preserving. In such a system, if $\omega = \sum \omega_{\alpha 0}(x,t)\, dx^\alpha + \sum \omega_{\alpha\bar{0}}\, dx^\alpha \wedge dt$ as above, one easily computes

$$*\mathfrak{t}\,\omega = \sum_{|\alpha|=p} \sigma(\alpha,\bar{\alpha}) \omega_{\alpha 0}(x,0)\, dx^\alpha \wedge dt = \mathfrak{n}\,(*\omega)$$
$$*\mathfrak{n}\,\omega = \sum_{|\alpha|=p-1} \sigma(\alpha,\bar{\alpha})(-1)^{n-p} \omega_{\alpha\bar{0}}(x,0)\, dx^\alpha = \mathfrak{t}\,(*\omega).$$

Moreover, from (3) we infer

$$\mathfrak{t}\,(d\omega) = \pi^{\#} i^{\#} d\omega = d(\pi^{\#} i^{\#}\omega) = d(\mathfrak{t}\,\omega).$$

Equality $\mathfrak{n}\,\delta\omega = \delta\mathfrak{n}\,\omega$ then follows as $\delta = (-1)^{p(n-p)} * d*$. Finally, again from (3) we infer

$$\mathfrak{t}\,(\omega\wedge * \eta) = \pi^{\#}i^{\#}(\omega\wedge * \eta) = \pi^{\#}i^{\#}\omega\wedge\pi^{\#}i^{\#} * \eta = \mathfrak{t}\,\omega\wedge\mathfrak{t}\,(*\eta) = \mathfrak{t}\,\omega\wedge * \mathfrak{n}\,\eta.$$

□

Recall that if X is an oriented manifold, then one fixes an orientation on ∂X by choosing $\vec{\partial X}(x)$ as the unit $(n-1)$-vector field orienting $T_x\partial X$ in such a way that

$$\nu(x)\wedge\overrightarrow{\partial X}(x) = \vec{X}(x) \qquad \text{for } x \in \partial X.$$

We then set for $\gamma \in C^\infty(\partial X, \Lambda^{n-1}TX)$

$$\int_{\partial X} \gamma := \int_{\partial X} <\gamma, \overrightarrow{\partial X}> d\mathcal{H}^{n-1}.$$

Proposition 2. *Suppose that X is oriented and let $*$ be the associated Hodge operator. We have*

(i) If $\omega \in W^{1,1}_{n-1}(X)$, then $\int_X d\omega = \int_{\partial X} i^{\#}\omega = \int_{\partial X} \mathfrak{t}\,\omega$.

(ii) If $\omega \in W^{1,2}_p(X)$ and $\eta \in W^{1,2}_{p+1}(X)$, then

$$(d\omega \mid \eta)_{L^2_{p+1}(X)} + (\omega \mid \delta\eta)_{L^2_p(X)} = \int_{\partial X} \mathfrak{t}\,\omega\wedge * \mathfrak{n}\,\eta.$$

(iii) If $\omega \in W^{2,2}_p(X)$ and $\zeta \in W^{1,2}_p(X)$, then

$$\begin{aligned}(d\omega \mid d\zeta)_{L^2_{p+1}(X)} &+ (\delta\omega \mid \delta\zeta)_{L^2_{p-1}(X)} \\ &= -(\Delta\omega \mid \zeta)_{L^2_p(x)} + \int_{\partial X} \{\mathfrak{t}\,\zeta\wedge * \mathfrak{n}\; d\omega + \mathfrak{t}\,\delta\omega\wedge * \mathfrak{n}\,\zeta\}.\end{aligned}$$

Proof. Since

$$\mathcal{H}^n \llcorner X\text{-a.e } \mathfrak{r}^{\#}\omega(x) = \mathfrak{r}\,\omega(x)|_{\Lambda_p T_x\partial X} \text{ and } i^{\#}\omega(x) = \mathfrak{t}\,\omega|_{\Lambda_p T_x \partial X},$$

(i) follows from a similar claim for smooth ω, on account of the continuity of the trace and of the tangential part operators. To prove (ii) we observe that $\omega\wedge * \eta$ belongs to $W^{1,1}_{n-1}(X)$, and

$$d(\omega\wedge * \eta) = d\omega\wedge * \eta + (-1)^p\omega\wedge d(*\eta) = d\omega\wedge * \eta + \omega\wedge * \delta\eta.$$

Integrating we then get on account of (i)

$$\begin{aligned}(d\omega \mid \eta)_{L^2} + (\omega \mid \delta\eta)_{L^2} &= \int_X d\omega\wedge * \eta + \int_X \omega\wedge * \delta\eta \\ &= \int_{\partial X} d(\omega\wedge * \eta) = \int_{\partial X} i^{\#}(\omega\wedge * \eta) = \int_{\partial X} i^{\#}\omega\wedge i^{\#} * \eta.\end{aligned}$$

On the other hand $i^{\#}\omega = \mathfrak{t}\,\omega|_{\Lambda_p T\partial X}$, hence

$$i^{\#}\omega\wedge i^{\#}(*\eta) = \mathfrak{t}\,\omega\wedge\mathfrak{t}\,(*\eta)|_{\Lambda_p TX} = \mathfrak{t}\,\omega\wedge *\,\delta\mathfrak{n}\,\eta|_{\Lambda_p T\partial X}.$$

Finally (iv) follows at once integrating by parts, $(\Delta\omega \mid \zeta) = (d\delta\omega \mid \zeta)+(\delta d\omega \mid \zeta)$. □

We now observe that Proposition 2 extends to *non orientable manifolds.* In fact as any non orientable manifold X is nevertheless orientable in the small, the scalar function

$$<\alpha,\beta>(x) := <\alpha(x)\wedge *\,\beta(x), \overrightarrow{\partial X}(x)>, \qquad x\in\partial X$$

is well defined if either $\alpha \in C^\infty(X,\Lambda^p TX)$ or $\beta\in C^\infty(X,\Lambda^{p+1}TX)$ are supported in a coordinate patch U, and is independent on the local orientation on U. Therefore $<\alpha,\beta>$ is a well defined function in $C^\infty(\partial X)$ even if X is non orientable.

Proposition 3. *We have*

(i) *If* $\omega\in W^{1,2}_p$ *then* $\mathfrak{t}\,(d\omega) = d(\mathfrak{t}\,\omega)$, $\mathfrak{n}\,(\delta\omega) = \delta(\mathfrak{n}\,\omega)$.

(ii) *If* $\omega\in W^{1,2}_p(X)$ *and* $\eta\in W^{1,2}_{p+1}(X)$, *then*

$$(d\omega \mid \eta)_{L^2_{p+1}(X)} + (\omega \mid \delta\eta)_{L^2_p(X)} = \int_{\partial X} <\mathfrak{t}\,\omega,\mathfrak{n}\,\eta>\,d\mathcal{H}^{n-1}$$

(iii) *If* $\omega\in W^{1,2}_p(X)$ *and* $\eta\in W^{1,2}_{p+2}(X)$, *then*

$$(d\omega \mid \delta\eta)_{L^2_{p+1}(X)} = 0$$

provided either $\mathfrak{t}\,\omega = 0$ *or* $\mathfrak{n}\,\eta = 0$.

(iv) *If* $\omega\in W^{2,2}_p(X)$ *and* $\eta\in W^{1,2}_p(X)$, *then*

$$\begin{aligned}(d\omega \mid d\eta)_{L^2_{p+1}(X)} &+ (\delta\omega \mid \delta\eta)_{L^2_{p-1}(X)}\\ &= -(\Delta\omega \mid \eta)_{L^2_p(x)} + \int_{\partial X}\{<\mathfrak{t}\,\eta,\mathfrak{n}\,d\omega> + <\mathfrak{t}\,\delta\omega,\mathfrak{n}\,\eta>\}\,d\mathcal{H}^{n-1}.\end{aligned}$$

Proof. (i) follows from (i) of Proposition 1 as the claim is local. (ii) and (iv) of Proposition 2 proves respectively (ii) and (iv) if we assume moreover that η is supported in a local chart. The general cases of (ii) and (iv) then follows using a partition of unity. Finally if $\mathfrak{t}\,\omega = 0$ and $\eta_j\to\eta$ in $W^{1,2}$, the η_j being smooth, we have by (ii)

$$(d\omega \mid \delta\eta_j) = (d\omega \mid \delta\eta_j) + (\omega \mid \delta^2\eta_j) = \int_{\partial X} <\mathfrak{t}\,\omega,\mathfrak{n}\,\eta_j>\,d\mathcal{H}^{n-1} = 0.$$

(iii) then follows letting j tend to infinity. Similarly one proceeds if $\mathfrak{n}\,\eta = 0$ by approximating ω. □

The Hodge-Morrey decomposition theorem. A key step toward a decomposition theorem similarly to the boundaryless case is a Gaffney type lemma at boundary points. We have

Lemma 1 (Gaffney). *For each $x_0 \in \partial X$ there are constants $c, \rho > 0$ and an admissible chart $\varphi : B^+(0,\rho) \to U \subset X$, $\varphi(0) = x_0$, such that*

$$\mathcal{D}(\omega, U) \geq \frac{1}{2} \int_{B^+(0,\rho)} \sum_{i,\alpha} |D_i \omega_\alpha|^2 \, dx - c \| \omega \|^2_{L^2_p(U)} \tag{4}$$

whenever $\varphi^{\#}\omega = \sum \omega_\alpha \, dx^\alpha$ is compactly supported in $B(0,\rho)$ and we have either $\mathfrak{t}\omega = 0$ or $\mathfrak{n}\omega = 0$ on ∂X.

Proof. Proceeding as in the proof of Gaffney lemma in the interior, we can bound from below the Dirichlet integral $\mathcal{D}(\omega, U)$ by the flat Dirichlet's integral $\mathcal{D}_0(\varphi^{\#}\omega, B^+(0,\rho))$ getting

$$\mathcal{D}(\omega, U) \geq \mathcal{D}_0(\varphi^{\#}\omega, B^+(0,\rho)) - \frac{1}{2} \int_{B^+(0,\rho)} \sum_{i,\alpha} |D_i \omega_\alpha|^2 \, dx - c \| \omega \|^2_{L^2_p(U)}.$$

The coercivity inequality (4) then follows at once from the following lemma. □

Lemma 2. *We have*

$$\mathcal{D}_0(\omega, B^+(0,\rho)) = \int_{B^+(0,\rho)} \sum_{i,\alpha} |D_i \omega_\alpha|^2 \, dx,$$

if $\omega = \sum \omega_\alpha \, dx^\alpha$ is compactly supported in $B(0,\rho)$ and moreover either $\mathfrak{t}\omega = 0$ or $\mathfrak{n}\omega = 0$ on Γ_ρ.

Proof. From Proposition 2 (iv) we have

$$\mathcal{D}_0(\omega, B^+(0,\rho)) + \int_{B^+(0,\rho)} (\Delta\omega \mid \omega) = \int_{\Gamma_\rho} \{\mathfrak{t}\omega \wedge * \mathfrak{n} \, d\omega + \mathfrak{t}\,\delta\omega \wedge * \mathfrak{n}\,\omega\}. \tag{5}$$

On the other hand in $\mathbb{R}^n$ the Laplacian on forms is just the Laplacian on coefficients, i.e.,

$$\Delta\omega = \sum_\alpha \Delta\omega_\alpha \, dx^\alpha,$$

consequently, integrating by parts and using Gauss-Green formula

$$\int_{B^+(0,\rho)} D_i g_i \, dx = \int_{\partial B^+(0,\rho)} \sum_i (-1)^{i-1} g_i \, \widehat{dx^i} \quad \forall\, g \in C^1(\overline{B_+(0,\rho)}),$$

we infer

$$(6)\qquad \int_{B^+(0,\rho)} (\Delta\omega \mid \omega)\, dx$$

$$= -\int_{B^+(0,\rho)} |D\omega|^2 + (-1)^{n-1} \int_{\Gamma_\rho} \sum_\alpha \omega_\alpha \omega_{\alpha,n}\, dx^1 \wedge \ldots \wedge dx^{n-1}.$$

From (5) and (6) we then get

$$(7)\quad \mathcal{D}_0(\omega, B^+(0,\rho)) - \int_{B^+(0,\rho)} |D\omega|^2\, dx$$

$$= \int_{\Gamma_\rho} \{\mathfrak{t}\omega \wedge *\, \mathfrak{n}\, d\omega + \mathfrak{t}\,\delta\omega \wedge *\, \mathfrak{n}\,\omega + (-1)^n \sum_\alpha \omega_\alpha \omega_{\alpha,n}\, dx^1 \wedge \ldots \wedge dx^{n-1}\}.$$

We now claim that the right hand side of (7) vanishes if either $\mathfrak{t}\,\omega = 0$ or $\mathfrak{n}\,\omega = 0$ on Γ_ρ.

In fact, if $\mathfrak{t}\,\omega = 0$, then, in Γ_ρ we have $\omega = \mathfrak{n}\,\omega = \sum_{\alpha \ni n} \omega_\alpha\, dx^\alpha$, consequently

$$\mathfrak{t}\,\delta\omega = \sum_{\alpha - i \not\ni n} \sum_{\alpha \ni n} \sum_i \sigma(i, \alpha - i) \omega_{\alpha,i}\, dx^{\alpha - i} = (-1)^{p-1} \sum_{\alpha \ni n} \omega_{\alpha,n}\, dx^{\alpha - n}.$$

Hence

$$\mathfrak{t}\,\delta\omega \wedge *\, \mathfrak{n}\,\omega = (-1)^{p-1} \sum_{\alpha \ni n} \sum_{\beta \ni n} \omega_\beta \omega_{\alpha,n} \sigma(\beta, \overline{\beta})\, dx^{\alpha - n} \wedge dx^{\overline{\beta}}.$$

As $\alpha \ni n$ and $\beta \ni n$, we have $dx^{\alpha-n} \wedge dx^{\overline{\beta}} \neq 0$ if and only if $\alpha = \beta$, therefore

$$\begin{aligned} \mathfrak{t}\omega \wedge *\, \mathfrak{n}\,\omega &= (-1)^{p-1} \sum_{\alpha \ni n} \omega_\alpha \omega_{\alpha,n} \sigma(\alpha, \bar{\alpha}) \sigma(\alpha - n, \bar{\alpha})\, dx^1 \wedge \ldots \wedge dx^{n-1} \\ &= (-1)^{p-1+n-p} \sum_{\alpha \ni n} \omega_\alpha \omega_{\alpha,n}\, dx^1 \wedge \ldots \wedge dx^{n-1} \\ &= (-1)^{n-1} \sum_{\alpha \ni n} \omega_\alpha \omega_{\alpha,n}\, dx^1 \wedge \ldots \wedge dx^{n-1} \end{aligned}$$

which proves the claim in this case.

If $\mathfrak{n}\,\omega = 0$, then in Γ_ρ we have $\omega = \mathfrak{t}\,\omega = \sum_{\alpha \not\ni n} \omega_\alpha\, dx^\alpha$,

$$\mathfrak{n}\,\delta\omega = \sum_{\alpha + i \ni n} \sum_{\alpha \not\ni n} \sigma(i, \alpha) \omega_{\alpha,i}\, dx^{\alpha + i} = (-1)^p \sum_{\alpha \not\ni n} \omega_{\alpha,n}\, dx^{\alpha + n}$$

and

$$*\mathfrak{n}\, d\omega = (-1)^p \sum_{\alpha \not\ni n} \omega_{\alpha,n} \sigma(\alpha + n, \bar{\alpha} - n)\, dx^{\bar{\alpha} - n}.$$

Therefore

$$\mathfrak{t}\,\omega\wedge*\,\mathfrak{n}\,d\omega=(-1)^p\sum_{\beta\not\ni n}\sum_{\alpha\not\ni n}\omega_\beta\omega_{\alpha,n}\sigma(\alpha+n,\bar{\alpha}-n)\,dx^\beta\wedge dx^{\bar{\alpha}-n}.$$

As $\beta\not\ni n$ and $\alpha\not\ni n$, we have $dx^\beta\wedge dx^{\bar{\alpha}-n}=0$ unless $\alpha=\beta$, hence

$$\begin{aligned}&\mathfrak{t}\,\omega\wedge*\,\mathfrak{n}\,d\omega\\ &\quad=(-1)^p\sum_{\alpha\not\ni n}\omega_\alpha\omega_{\alpha,n}\sigma(\alpha,\bar{\alpha})(-1)^{n-p-1}\sigma(\alpha,\bar{\alpha}-n)\,dx^1\wedge\ldots\wedge dx^{n-1}\\ &\quad=(-1)^{n-1}\sum_{\alpha\not\ni n}\omega_\alpha\omega_{\alpha,n}\,dx^1\wedge\ldots\wedge dx^{n-1}\end{aligned}$$

which concludes the proof. □

Gaffney's lemmas in the interior and at the boundary now yield at once

Theorem 1. *Let X be a compact Riemannian manifold with boundary. There exist constants $c_1>0,\ c_2>0$ such that*

$$\mathcal{D}(\omega)\geq c_1\|\,\omega\,\|^2_{W^{1,2}_p(X)}-c_2\|\,\omega\,\|^2_{L^2_p(X)}$$

for every $\omega\in W^{1,2}_p(X)$ satisfying one of the boundary conditions

$$\mathfrak{n}\,\omega=0\ \text{ on }\partial X,\qquad \mathfrak{t}\,\omega=0\ \text{ on }\partial X.$$

Set now

$$\begin{aligned}W^{1,2}_{p,T}(X)&:=\{\omega\in W^{1,2}_p(X)\mid \mathfrak{t}\,\omega=0\}\\ W^{1,2}_{p,N}(X)&:=\{\omega\in W^{1,2}_p(X)\mid \mathfrak{n}\,\omega=0\}\\ \mathbf{H}^p_T(X)&:=\{\omega\in W^{1,2}_{p,T}(X)\mid d\omega=0,\ \delta\omega=0\}\\ \mathbf{H}^p_N(X)&:=\{\omega\in W^{1,2}_{p,N}(X)\mid d\omega=0,\ \delta\omega=0\}.\end{aligned}$$

Then, as in the boundaryless case, one proves

Theorem 2. *$\mathbf{H}^p_T$ and $\mathbf{H}^p_N$ are finite dimensional vector spaces.*

Theorem 3. *There exists a positive constant c such that $\mathcal{D}(\omega)\geq c\|\,\omega\,\|_{L^2_p(X)}$ for any $\omega\in W^{1,2}_{p,N}(X)$ which is orthogonal in L^2 to $\mathbf{H}^p_N$ and for any $\omega\in W^{1,2}_{p,T}(X)$ which is orthogonal in L^2 to $\mathbf{H}^p_T$.*

Given a form $f\in L^2_p(X)$, we can now seek a weak solution of the problem

$$\begin{cases}\Delta\omega=f & \text{in } X\\ \mathfrak{n}\,\omega=0 & \text{on } \partial X\end{cases}\tag{8}$$

that is we seek a form $\omega\in W^{1,2}_{p,N}(X)$ such that

(9) $$(d\omega \mid d\zeta)_{L^2} + (\delta\omega \mid \delta\zeta)_{L^2} = (f \mid \zeta) \quad \forall\, \zeta \in W^{1,2}_{p,N}(X)$$

or a weak solution of

(10) $$\begin{cases} \Delta\omega = f & \text{in } X \\ \mathrm{t}\omega = 0 & \text{on } \partial X \end{cases}$$

equivalently $\omega \in W^{1,2}_{p,T}(X)$ such that

(11) $$(d\omega \mid d\zeta)_{L^2} + (\delta\omega \mid \delta\zeta)_{L^2} = (f \mid \zeta)_{L^2} \qquad \forall\, \zeta \in W^{1,2}_{p,T}(X).$$

As in the boundaryless case, an easy consequence of Theorem 1 and Theorem 3 is the following

Theorem 4 (Hodge-Morrey). *We have*

(i) Problem (8), equivalently equation (9), is solvable if and only if f is orthogonal in L^2_p to $\mathbf{H}^p_N$. Moreover there is a unique solution $\omega_0 \in W^{1,2}_{p,N}(X) \cap \mathbf{H}^{p\perp}_N$ of equation (9), and the linear transformation from f to ω_0 is a bounded linear operator from $L^2_p(X) \cap \mathbf{H}^{p\perp}_N$ into $W^{1,2}_{p,N}(X) \cap \mathbf{H}^{p\perp}_N$.

(ii) Problem (10), equivalently equation (11), is solvable if and only if f is orthogonal in L^2_p to $\mathbf{H}^p_T$. Moreover there is a unique solution $\omega_0 \in W^{1,2}_{p,T}(X) \cap \mathbf{H}^{p\perp}_T$ of equation (11), and the linear transformation from f to ω_0 is a bounded linear operator from $L^2_p(X) \cap \mathbf{H}^{p\perp}_T$ into $W^{1,2}_{p,T}(X) \cap \mathbf{H}^{p\perp}_T$.

Problems (8) and (10) are easily seen to be *elliptic* boundary value problems, where one has zero Dirichlet data for suitable coefficients of ω and zero Neumann data for the others. Elliptic regularity theory then allows to complement the existence theorem in Theorem 4 with the following Regularity Theorem

Theorem 5. *Let ω in $W^{1,2}_{p,N}(X)$ or in $W^{1,2}_{p,T}(X)$ be respectively a solution of (9) or of (11). Then we have*

(i) $\omega \in W^{2,2}_p(X)$ if $f \in L^2_p(X)$.

(ii) $\omega \in W^{1,s}_p(X) \cap C^{0,\alpha}(X, \Lambda^p TX)$, $\alpha = 1 - \frac{n}{p}$, if $f \in L^s_p(X)$, $s > n$.

(iii) $\omega \in C^{2,\alpha}(X, \Lambda^p TX)$ if $f \in C^{0,\alpha}(X, \Lambda^p TX)$.

(iv) Higher regularity of ω follows from higher regularity of f, for instance $\omega \in C^\infty$ or ω is analytic if $f \in C^\infty$ or is analytic.

(v) In particular harmonic normal forms, i.e., $\omega \in \mathbf{H}^p_N$, or harmonic tangential forms, $\omega \in \mathbf{H}^p_T$, are C^∞ or analytic up to the boundary if X and ∂X are of class C^∞ or analytic.

Define now

$$\operatorname{Im} d_N := \{d\alpha \mid \alpha \in W^{1,2}_{p-1,N}(X)\} \quad \operatorname{Im} d_T := \{d\alpha \mid \alpha \in W^{1,2}_{p-1,T}(X)\}$$
$$\operatorname{Im} \delta_N := \{\delta\alpha \mid \alpha \in W^{1,2}_{p+1,N}(X)\} \quad \operatorname{Im} \delta_T := \{\delta\alpha \mid \alpha \in W^{1,2}_{p+1,T}(X)\};$$

then we can state

Theorem 6 (Hodge-Morrey decomposition). *We have*

(i) The spaces $\mathbf{H}^p_N, \operatorname{Im} d_N$ and $\operatorname{Im}\delta_N$ are mutually orthogonal and closed in $L^2_p(X)$, and one has the decomposition

$$L^2_p(X) = \mathbf{H}^p_N \oplus \operatorname{Im} d_N \oplus \operatorname{Im}\delta_N,$$

that is, every $f \in L^2_p(X)$ decomposes uniquely as

$$f = h + d\alpha + \delta\beta \tag{12}$$

where $h \in \mathbf{H}^p_N$, $\alpha \in W^{1,2}_{p-1,N}(X)$, $\beta \in W^{1,2}_{p+1,N}(X)$ and

$$\| h \|_{W^{1,2}} + \| \alpha \|_{W^{1,2}} + \| \beta \|_{W^{1,2}} \le c \| f \|_{L^2}.$$

(ii) The spaces $\mathbf{H}^p_T, \operatorname{Im} d_T$ and $\operatorname{Im}\delta_T$ are mutually orthogonal and closed in $L^2_p(X)$, moreover

$$L^2_p(X) = \mathbf{H}^p_T \oplus \operatorname{Im} d_T \oplus \operatorname{Im}\delta_T,$$

i.e., every $f \in L^2_p(X)$ uniquely decomposes as

$$f = h + d\alpha + \delta\beta \tag{13}$$

where $h \in \mathbf{H}^p_T$, $\alpha \in W^{1,2}_{p-1,T}(X)$, $\beta \in W^{1,2}_{p+1,T}(X)$ and

$$\| h \|_{W^{1,2}} + \| \alpha \|_{W^{1,2}} + \| \beta \|_{W^{1,2}} \le c \| f \|_{L^2}.$$

Moreover in both cases $h \in C^\infty(X, \Lambda^p TX)$ and also α and β are of class C^∞ up to the boundary if f is of class C^∞ up to the boundary. Finally, $\delta\beta = 0$ if $df = 0$.

Proof. The proof follows the same path of the analogous proof of the decomposition theorem in the boundaryless case: one has only to take into account the formulas of integration by parts in Proposition 6 in Sec. 5.1.3. For the reader's convenience, we only prove (12). Let h be the projection (in L^2_p) of f into $\mathbf{H}^p_N$ and let $\omega_0 \in W^{1,2}_{p,N}(X) \cap \mathbf{H}^{p\perp}_N$ be the unique solution of

$$(d\omega \mid d\zeta)_{L^2} + (\delta\omega \mid \delta\zeta)_{L^2} = (f \mid \zeta)_{L^2} \quad \forall\, \zeta \in W^{1,2}_{p,N}(X) \tag{14}$$

which exists by Theorem 4. By the regularity theorem ω_0 actually belongs to $W^{2,2}_p$, therefore (iv) of Proposition 3 yields

$$\begin{aligned}(d\omega \mid d\zeta)_{L^2} + (\delta\omega \mid \delta\zeta)_{L^2} &= (\Delta\omega \mid \zeta)_{L^2} \\ &\quad + \int_{\partial X} \{<\mathfrak{t}\zeta, \mathfrak{n}\, d\omega> + <\mathfrak{t}\,\delta\omega, \mathfrak{n}\zeta>\}\, d\mathcal{H}^{n-1}\end{aligned} \tag{15}$$

for all $\zeta \in W^{1,2}_p(X)$. From (14), (15) we infer by taking $\zeta \in C^\infty_c(X, \Lambda^p TX)$ that

$$\Delta\omega_0 = f - h$$

i.e.,

$$f = h + d\alpha + \delta\beta \quad \text{if } \alpha := d\omega_0, \ \beta := \delta\omega_0$$

where $\alpha \in W^{1,2}_{p-1}(X)$, $\beta \in W^{1,2}_{p+1}(X)$. As $\mathfrak{n}\,\alpha = \mathfrak{n}\,\delta\omega_0 = \delta\mathfrak{n}\,\omega_0 = 0$, we see that $\alpha \in W^{1,2}_{p-1,N}(X)$. From (14), (15) we also infer

$$\int_{\partial X} <\mathfrak{t}\,\zeta, \mathfrak{n}\,d\omega_0> d\mathcal{H}^n = 0 \qquad \forall\, \zeta \in W^{1,2}_{p,N}(X),$$

and, being $\mathfrak{t}\,\zeta$ arbitrary on X, that $\mathfrak{n}\;d\omega_0 = 0$, hence $\mathfrak{n}\,\beta = \mathfrak{n}\;d\omega_0 = 0$, i.e. $\beta \in W^{1,2}_{p+1,N}(X)$. □

Remark 1. It is worth remarking that, since Hodge operator transforms p-harmonic forms into $(n-p)$-harmonic forms and

$$*\mathfrak{n}\,\omega = \mathfrak{t}\,(*\omega) \qquad *\,\mathfrak{t}\,\omega = \mathfrak{n}\,(*\omega),$$

actually on *oriented* manifolds Theorem 4 (ii) follows from Theorem 4 (i) and Theorem 6 (ii) follows from Theorem 6 (i).

Remark 2. One might consider other boundary value problems for the Laplace-Beltrami operator and consequently infer other decomposition theorems. However we shall not pursue this topic any further.

For $k = 0, \dots, n$, define

$$Z^k(X, \partial X) := \{\omega \in C^0(X, \Lambda^k TX) \mid i^{\#}\omega = 0, \ d\omega = 0\}$$

to be the vector space of *relatively closed k-forms*, i.e. the space of closed k-forms which vanish on $\partial\Omega$, and

$$B^k(X, \partial X) := \{\omega = d\alpha \mid \alpha \in C^0(X, \Lambda^k TX), \ i^{\#}\alpha = 0, \ \text{or } \omega = 0 \text{ if } k = 0\}.$$

the vector space of *relatively exact k-forms*. The *relative de Rham cohomology groups* are then defined by

$$H^k_{\mathrm{dR}}(X, \partial X) := Z^k(X, \partial X)/B^k(X, \partial X).$$

An immediate consequence of Theorem 6 is

Corollary 1. *We have*

(i) *Each cohomology class $[\omega] \in H^k_{\mathrm{dR}}(X)$ contains exactly one harmonic form h with $\mathfrak{n}\,h = 0$. In particular $H^k_{\mathrm{dR}}(X)$ is finite dimensional.*
(ii) *Each relative cohomology class $[\omega] \in H^k_{\mathrm{dR}}(X, \partial X)$ contains exactly one harmonic form h with $\mathfrak{t}\,h = 0$. In particular $H^k_{\mathrm{dR}}(X, \partial X)$ is finite dimensional.*

2.7 Weitzenböck Formula

In this section we derive a more intrinsic expression for the Laplace Beltrami operator on a Riemannian manifold X in terms of the *covariant* derivatives and of the *curvature of* X.

Denote by $\Gamma(TX)$ the smooth vector fields on X and if $p: E \to X$ is a fiber bundle on X, denote by $\Gamma(E)$ the space of sections over E. A *connection* ∇ on E is an operator

$$\nabla : \Gamma(E) \otimes \Gamma(TX) \to \Gamma(E)$$

denoted by $\nabla_X \xi$ for $X \in \Gamma(TX)$ and $\xi \in \Gamma(E)$ which satisfies the following properties

(i) $\nabla_X \xi$ *is* tensorial *in* X *i.e.*

$$\nabla_{X+Y}\xi = \nabla_X\xi + \nabla_Y\xi, \qquad \nabla_{fX}\xi = f\nabla_X\xi$$

for $x, y \in \Gamma(TX)$, $\xi \in \Gamma(E)$, $f \in C^\infty(X)$.

(ii) $\nabla_X \xi$ *is* linear *in* ξ

(iii) $\nabla_X \xi$ *satisfies the following chains rule*

$$\nabla_X(f\xi) = <df, X>\xi + f\nabla_X\xi \tag{1}$$

for $X \in \Gamma(TX)$, $f \in C^\infty(X)$, $\xi \in \Gamma(E)$.

Notice that (i) implies that $\nabla_X\xi$ at a point $x \in M$ depends only on the tangent field X at point x and moreover that ∇ can be also seen as an operator

$$\nabla : \Gamma(E) \to \Gamma(E) \otimes \Gamma(\Lambda^1 TX).$$

Moreover, (iii) implies that $\nabla_X\xi(x)$ depends only on ξ near x.

Given a connection ∇ on a fiber bundle E one defines a *connection* ∇^* on the dual bundle E^* by the formula

$$<\nabla_X\xi, \omega> + <\xi, \nabla^*_X\omega> = <d<\xi,\omega>, X> \tag{2}$$

for $\xi \in \Gamma(E)$, $\omega \in \Gamma(E^*)$, $X \in \Gamma(TX)$ and also on all exterior algebra of E. In fact sections $\xi \in \Gamma(\Lambda_p E)$ can be represented locally as

$$\xi = \sum_{\alpha \in I(p,n)} \xi^\alpha(x) e_{\alpha_1}(x) \wedge \ldots \wedge e_{\alpha_p}(x)$$

where $\xi^\alpha \in C^\infty(X)$ and $e_{\alpha_1}, \ldots, e_{\alpha_p}$ are smooth sections in E. Consequently, by setting

$$\nabla_X f(x) := <df(x), X>, \qquad f \in C^\infty(X) \tag{3}$$

$$\nabla_X(\xi_1 \wedge \ldots \wedge \xi_p) := \sum_{i=1}^p \xi_1 \wedge \ldots \xi_{i-1} \wedge \nabla_X\xi_i \wedge \xi_{i+1} \wedge \ldots \wedge \xi_p \tag{4}$$

for $\xi \in \Gamma(E)$ we define an operator

$$\nabla : \Gamma(\Lambda_p E) \otimes \Gamma(TX) \to \Gamma(\Lambda_p E)$$

which is easily seen to be a connection. Starting with ∇^* on $\Gamma(E^*)$ a similar construction produces a connection on $\Lambda_p E^* \simeq \Lambda^p E^*$

$$\nabla_X(\omega^1 \wedge \ldots \wedge \omega^p) := \sum_{i=1}^{p} \omega^1 \wedge \ldots \wedge \omega^{i-1} \wedge \nabla_X \omega^i \wedge \omega^{i+1} \wedge \ldots \wedge \omega^p, \tag{5}$$

Proposition 1. *Let* $\xi \in \Gamma(\Lambda_p E)$, $\sigma \in \Gamma(\Lambda_q E)$, $\omega \in \Gamma(\Lambda^p E)$, $\eta \in \Gamma(\Lambda^q E)$. *We have*

$$\nabla_X(\xi \wedge \sigma) = \nabla_X \xi \wedge \sigma + \xi \wedge \nabla_X \sigma \tag{6}$$

$$\nabla_X(\omega \wedge \eta) = \nabla_X \omega \wedge \eta + \omega \wedge \nabla_X \eta \tag{7}$$

$$<\omega, \nabla_X \xi> + <\nabla^*_X \omega, \xi> = <d<\omega, \xi>, X>. \tag{8}$$

Also, if $p \leq q$

$$\nabla^*_X(\xi \lrcorner\, \eta) = \xi \lrcorner\, \nabla^*_X \eta + \nabla_X \xi \lrcorner\, \eta. \tag{9}$$

Proof. Let $e_1, \ldots, e_n$ be a local basis of sections near x_0 and $f_1, \ldots, f_n$ be the dual basis in $\Gamma(E^*)$ so that

$$< f_i(x), e_j(x) > = \delta_{ij} \qquad \text{near } x_0.$$

If $\alpha \in I(p, n)$, $\beta \in I(q, n)$, from (4) we infer

$$\nabla_X(e_\alpha \wedge e_\beta) = \nabla_X e_\alpha \wedge e_\beta + e_\alpha \wedge \nabla_X e_\beta$$

hence (6) follows. Analogously (7) follows from (5). To prove (8) notice that

$$\nabla^* f_\alpha(e_\beta) = f_\alpha(\nabla_X e_\beta) = 0 \quad \text{if } \alpha \neq \beta$$

while for $\alpha = \beta$

$$\begin{aligned} \nabla^*_X f_\alpha(e_\beta) + f_\alpha(\nabla_X e_\alpha) &= \sum_{i=1}^{p} \{\nabla^*_X f_{\alpha_i}(e_{\alpha_i} + f_{\alpha_i}(\nabla_X e_{\alpha_i})\} \\ &= \sum_i < d< f_{\alpha_i}, e_{\alpha_i} >, X> = 0. \end{aligned}$$

Let us prove (9). Using (6) and (8), for $\sigma \in \Gamma(\Lambda_{q-p} E)$

$$\begin{aligned} \nabla^*_X(\xi \lrcorner\, \eta)(\sigma) &= -\xi \lrcorner\, \eta(\nabla_X \sigma) + <d<\eta, \sigma \wedge \xi>, X> \\ &= -\eta(\nabla_X \sigma \wedge \xi) + <d<\eta, \sigma \wedge \xi>, X> \\ &= -\eta(\nabla_X(\sigma \wedge \xi)) + <d<\eta, \sigma \wedge \xi>, X> + <\eta, \sigma \wedge \nabla_X \xi> \\ &= \nabla^*_X \eta(\sigma \wedge \xi) + <\eta, \sigma \wedge \nabla_X \xi> \\ &= \xi \lrcorner\, \nabla^* \eta(\sigma) + \nabla_X \xi \lrcorner\, \eta(\sigma). \end{aligned}$$

□

A basic result on Riemannian geometry is that there is only one connection ∇ on TX, the *Levi-Civita connection* , which satisfies the following additional requirements

(1) ∇ *is* metric, *i.e.*

$$(\nabla_X Y|Z) + (Y|\nabla_X Z) = <d(Y|Z), X> \tag{10}$$

$\forall\, X, Y, Z \in \Gamma(TX)$

(2) ∇ *is* torsion free, *i.e. for* $X, Y \in \Gamma(T, X)$

$$T(x, y) := \nabla_X Y - \nabla_Y X - [X, Y] \equiv 0 \tag{11}$$

being $[X, Y]$ *the* Lie bracket of X and Y.

In this case,

$$[X, Y] := X^j \frac{\partial Y^i}{\partial X^j} e_i - Y^j \frac{\partial X^i}{\partial X^j} e_i$$

where $X := X^i e_i$, $Y := Y^i e_i$.

One can compute ∇ in terms of the metric on $\Gamma(TX)$, and one has

$$\begin{aligned}(\nabla_X X|Z) = \frac{1}{2}\{&X(X|Z) - Z(X|X) + X(X|Z)\\ &-(X|[X, Z]) + (Z|[X, Y]) + (X|[Z, X])\}.\end{aligned} \tag{12}$$

In local coordinates, denoting by G_{ij} the metric tensor of X and by $e_1, \dots, e_n$ a local basis for $\Gamma(TX)$, we have

$$\nabla_{e_i} e_j := \sum_{k=1}^{n} \Gamma_{ij}^k e_k \tag{13}$$

where $\{\Gamma_{ij}^k\}$ are the *Christoffel symbols of the connection* in the given basis; (12) is then equivalent to

$$\Gamma_{ij}^k = \frac{1}{2} G^{kl}\{G_{il,j} + G_{jl,i} - G_{ij,l}\}. \tag{14}$$

The torsion free condition (11) implies

$$\Gamma_{ij}^k = \Gamma_{ji}^k \qquad \forall\, i, j, k.$$

Proposition 2. *Let X be a Riemannian manifold. Denote by ∇ the Levi-Civita connection in X and also its extension to the algebra of p-vector fields and forms defined on an open set $U \subset X$ and let $dx^1, \dots, dx^n$ be the dual frame of p-forms. Then*

$$d\omega = \sum_j dx^j \wedge \nabla_{e_j} \omega \quad \textit{in } U. \tag{15}$$

If moreover e_j are orthonormal at $x_0 \in U$ then at x_0

$$\delta\omega = \sum_j e_j \lrcorner\, \nabla e_j \omega. \tag{16}$$

Proof. The idea is to prove the invariance of both sides in (15) and (16) with respect to changes of coordinates and then proves (15) and (16) in a suitable frame.

Let us prove (15). First we note the the expression

$$\sum_j dx^j \wedge \nabla_{e_j} \omega$$

is independent of the choice of local frame $e_1, \ldots, e_n$. In fact if $f_1, \ldots, f_n$ is another frame field, we have $f_j = A_j^k e_k$ for some $A_j^k \in C^\infty(U, \mathbb{R})$. For the corresponding dual frame $f^1, \ldots, f^n$ we then have $f^j = B_k^j dx^k$ with $B := A^{-1}$. Consequently

$$\sum_j f^j \wedge \nabla_{f_j} \omega = \sum_{j,k,i} B_k^j A_j^i dx^k \wedge \nabla_{e_i} \omega = \sum_{i,k} \delta_k^i dx^k \wedge \nabla_{e_i} \omega = \sum_i dx^i \wedge \nabla_{e_i} \omega.$$

Since d is also independent on the frame field, it suffices to check (15) for a special frame field.

We select around x_0, normal coordinates for which we have $G_{ij}(x_0) = \delta_{ij}$, $G_{ij,l}(x_0) = 0 \; \forall \; i, j, l$. Hence for the coordinate frame fields $e_1, \ldots, e_n$, we have

$$\nabla_{e_i} e_j(x_0) = \Gamma_{ij}^k(x_0) e_k(x_0) = 0$$

from (13) (14). Thus, if $\omega = \varphi dx^\alpha$, $\alpha \in I(p, n)$, $f \in C^\infty(X)$, we obtain at x_0

$$\sum_j dx^j \wedge \nabla_{e_j}(\varphi dx^\alpha) = \sum_j dx^j \wedge \frac{\partial \varphi}{\partial x_j} dx^\alpha = d(\varphi dx^\alpha).$$

To prove (16) we observe that the right hand side is independent of the chosen of the normal frame at x_0. In fact if $f_1, \ldots, f_n$ is another orthonormal frame field, $f_j = A_j^k e_k$, A_j^k being an orthogonal matrix, $A_j^k A_j^i = \delta^{ik}$. Then we have

$$\sum_j f^j \lrcorner \nabla_{f_j} \omega = \sum_{j,k,i} A_j^k A_j^i e_k \lrcorner \nabla_{e_i} \omega = \sum_i e_i \lrcorner \nabla_{e_i} \omega.$$

Again we chose normal coordinates at x_0. The coordinate $(e_1, \ldots, e_n)$ frame is orthogonal at x_0 and $\nabla_{e_i} e_j = 0$. Consequently

$$\sum_j e_j \lrcorner \nabla_{e_j}(\varphi dx^\alpha) = \sum_j \frac{\partial \varphi}{\partial x_j} e_j \lrcorner dx^\alpha. \tag{17}$$

On the other hand from (1) in Sec. 5.2.4, taking into account that $G_{ij}(x_0) = G^{ij}(x_0) = \delta^{ij}$, $G_{ij,l}(x_0) = 0$ we have at x_0

$$\begin{aligned} \delta(\varphi dx^\alpha) &= \sum_\gamma (\delta(\varphi dx^\alpha))_\gamma \, dx^\gamma = \\ &= \sum_{\gamma + i = \alpha} \sigma(i, \gamma) \frac{\partial \varphi}{\partial x_i} \, dx^\alpha = \sum_{i \in \alpha} \frac{\partial \varphi}{\partial x_i} \sigma(i, \alpha) \, dx^{\alpha - i} \end{aligned} \tag{18}$$

and the claim follows comparing (17) (18) since

$$e_i \lrcorner\, dx^\alpha = \begin{cases} 0 & \text{if } i \notin \alpha \\ \sigma(i,\alpha)\, dx^{\alpha - i} & \text{if } i \in \alpha. \end{cases}$$

□

In order to express the Laplace-Beltrami operator in terms of covariant derivatives, we introduce the *second covariant derivative* defined for $\omega \in \Gamma(\Lambda^p X)$ and $X, Y \in TX$ by

(19) $$\nabla^2_{XY}\omega := \nabla_X(\nabla_Y \omega) - \nabla_{\nabla_X Y}\omega$$

and the *curvature tensor associated with a connection*

(20) $$R(X,Y) := \nabla_X \nabla_Y - \nabla_Y \nabla_X : \Lambda^p TX \to \Lambda^p TX.$$

It is easy to check that $\nabla^2_{XY}\omega$ is tensorial in X, Y, i.e. that ∇^2_{XY} is linear in X and Y and for all $x_0 \in Y$

$$\nabla^2_{XY}(\omega)(x_0)$$

depends only on the values $X(x_0), Y(x_0)$ of the vector fields X and Y at x_0. In fact, for $f, g \in C^\infty$, X, Y smooth vector fields,

$$\nabla_{fX}(\nabla_{gY}\omega) = f{<}dg, X{>}\nabla_Y \omega + fg\nabla_X(\nabla_Y \omega)$$

and

$$\nabla_{\nabla_{fX}(gY)}\omega = \nabla_{f<dg,X>X + fg\nabla_X Y} = f{<}dg, X{>}\nabla_Y\omega + fg\nabla_{\nabla_X Y}\omega$$

so

(21) $$\nabla^2_{fX\, gY}\omega = f(x)g(x)\nabla^2_{XY}\omega.$$

We need also the following algebraic lemma.

Lemma 1. *Let* $e_1, \dots, e_n \in \Lambda_1 V$, $dx^1, \dots, dx^n$ *the dual basis in* $\Lambda^1 V$, $\omega \in \Lambda^p V$. *Then*

(22) $$\sum_{i \neq j} \{ e_i \lrcorner\, dx^j \wedge \omega + dx^j \wedge (e_i \lrcorner\, \omega) \} = 0$$

(23) $$e_i \lrcorner\, dx^i \wedge \omega + dx^i \wedge (e_i \lrcorner\, \omega) = \omega \qquad \forall\, i = 1, \dots, n.$$

Proof. Assume first $i \neq j$

$$e_i \lrcorner\, (dx^j \wedge dx^\alpha) = \begin{cases} 0 & \text{if } j \in \alpha, \\ 0 & \text{if } j \notin \alpha,\ i \notin \alpha \\ \sigma(j,\alpha)\sigma(i, \alpha + j - i)\, dx^{\alpha + j - i} & \text{if } j \notin \alpha,\ i \in \alpha. \end{cases}$$

$$dx^j A(e^i \lrcorner\, dx^\alpha) = \begin{cases} 0 & \text{if } i \notin \alpha, \\ 0 & \text{if } i \in \alpha,\ j \in \alpha \\ \sigma(i, \alpha - i)\sigma(j, \alpha - i)\, dx^{\alpha + j - i} & \text{if } j \notin \alpha,\ i \in \alpha. \end{cases}$$

As $\sigma(j,\alpha)\sigma(i,\alpha+j-i) = -\sigma(i,\alpha-i)\sigma(j,\alpha-i)$, compare (5) in Sec. 5.2.4, (6) in Sec. 5.2.4, we see that

$$e_i \lrcorner\, (dx^j \wedge dx^\alpha) + dx^j \wedge (e_i \lrcorner\, dx^\alpha) = 0$$

$\forall\, \alpha \in I(p,n)$, consequently (22) is proved. To prove (23) it suffices to compute

$$e_i \lrcorner\, dx^i \wedge dx^\alpha = \begin{cases} 0 & \text{if } i \in \alpha, \\ dx^\alpha & \text{if } i \notin \alpha, \end{cases} \qquad dx^i \wedge e_i \lrcorner\, dx^\alpha = \begin{cases} dx^\alpha & \text{if } i \in \alpha, \\ 0 & \text{if } i \notin \alpha, \end{cases}$$

thus

$$e_i \lrcorner\, dx^i \wedge \omega + dx^i \wedge e_i \lrcorner\, \omega = \omega.$$

□

Theorem 1 (Wietzenböck formula). *Let X be a Riemannian manifold and let ∇ be the Levi-Civita connection on forms. If $e_1, \dots, e_n$ is a local frame on X which is orthonormal at x_0 and $dx^1, \dots, dx^n$ the dual frame on T^*X, then*

$$\Delta\omega = \nabla^2_{e_i e_i}\omega + dx^i \wedge (e_j \lrcorner\, R(e_i, e_j)\omega). \tag{24}$$

Proof. By definition, $\Delta\omega$ is intrinsic to the Riemannian manifold and depends on the Riemannian structure on X. Also we check that both terms in the right hand side of (24) are intrinsic. In fact, if $f_1, \dots, f_n$ is another orthonormal frame field, then $f_j = A^k_j e_k$, $A^k_j(x_0)$ being orthogonal. Thus (21) implies

$$\sum_i \nabla^2_{f_i f_i}\omega = \sum_{j,h,k} A^k_i A^h_i \nabla^2_{e_k e_h}\omega = \sum_h \nabla^2_{e_h e_h}\omega.$$

Also if f^j is the dual basis of f_j, $f^j = B^j_k dx^k$, $B := A^{-1}$

$$R(f_i, f_j) = \sum_{h,k} A^k_i A^h_j R(e_k, e_h)$$

$$\sum_i f_j \lrcorner\, R(f_i, f_j) = \sum_{hkj,s} A^s_j A^k_i A^h_j e_s \lrcorner\, R(e_k, e_h) = \sum_{k,h} A^k_i e_h \lrcorner\, R(e_k, e_h).$$

Therefore

$$\begin{aligned} \sum_i f^i \wedge (f_j \lrcorner\, R(f_i, f_j)) &= \sum_{h,k,s,i} B^i_s A^k_i dx^s \wedge (e_h \lrcorner\, R(e_k, e_h)) \\ &= \sum_{h,k} dx^k \wedge (e_h \lrcorner\, R(e_k, e_h)). \end{aligned}$$

Consequently it suffices to prove (24) for a special frame. Choose normal coordinates around x_0, and dual coordinate frames $e_1, \dots, e_n$, and $dx^1, \dots, dx^n$. At x_0 we have $\nabla_{e_i} e_j(x_0) = 0$, therefore $\nabla_{e_i}(dx^j)(x_0) = 0$, $\nabla^2_{e_i e_i}\omega(x_0) = \nabla_{e_i}\nabla_{e_i}\omega(x_0)$ and $R(e_i, e_j) = \nabla_{e_i}\nabla_{e_j} - \nabla_{e_j}\nabla_{e_i}$.

Consequently, using Proposition 2 at x_0

$$d\delta\omega = \sum_{i,j} dx^i \wedge \nabla_{e_i}(e_j \lrcorner \nabla_{e_j}\omega) = \sum_{i,j} dx^i \wedge (e_j \lrcorner \nabla_{e_i}\nabla_{e_j}\omega) \tag{25}$$

and also using Lemma 1

$$\begin{aligned} \delta d\omega &= \sum_{i,j} e_j \lrcorner \nabla_{e_j}(dx^i \wedge \nabla_{e_i}\omega) \\ &= \sum_{i,j} e_j \lrcorner (dx^i \wedge \nabla_{e_j} \nabla e_i \omega) - \sum_{i=j} dx^i \wedge (e_j \lrcorner \nabla_{e_j}\nabla_{e_i}\omega) \\ &\quad + \sum_{i=1}^{n} \{e_i \lrcorner (dx^i \wedge \nabla_{e_j}\nabla_{e_i}\omega) + dx^i \wedge (e_i \lrcorner \nabla_{e_i}\nabla_{e_i}\omega)\} \\ &= -\sum_{i,j} dx^i \wedge (e_j \lrcorner \nabla_{e_j}\nabla_{e_i}\omega) + \sum_i \nabla_{e_i}\nabla_{e_i}\omega. \end{aligned} \tag{26}$$

Adding (25) and (26) we get the claim. □

Remark 1. If $p = 0$, i.e. if $\omega = f \in C^\infty(X)$ then $R(e_i, e_j)f = fR(e_i, e_j)1 = 0$ because of the tensorial property of R, consequently for functions

$$\Delta f = \sum_i \nabla^2_{e_i e_i} f.$$

2.8 Poincaré and Poincaré-Lefschetz Dualities in Cohomology

Compact manifolds without boundary. Let X be a compact manifold without boundary. Recalling Hodge's theorem Theorem 6 in Sec. 5.2.5 we first notice that Hodge operator $*$ induces naturally an isomorphism between the homology classes $H^p_{\mathrm{dR}}(X)$ and $H^{n-p}_{\mathrm{dR}}(X)$, in fact since

$$\Delta * \omega = *\Delta\omega$$

we see that $*\omega$ is harmonic if ω is harmonic, hence the map $* : \omega \to *\omega$ defines an isomorphism

$$* : H^p_{\mathrm{dR}}(X) \longrightarrow H^{n-p}_{\mathrm{dR}}(X).$$

However such an isomorphism depends not only on the orientation of X, but on the metric. An algebraic isomorphism independent from the metric between $H^p_{\mathrm{dR}}(X)$ and the dual of $H^{n-p}_{\mathrm{dR}}(X)$ is obtained as follows.

Given two closed forms $\omega \in W^{1,2}(X, \Lambda^p TX)$ and $\eta \in W^{1,2}(X, \Lambda^{n-p}TX)$ it is readily seen that

$$<\omega, \eta> := \int_X \omega \wedge \eta$$

does not change if we replace ω and η by cohomologous forms, hence $<,>$ defines a bilinear map over $H^p_{\mathrm{dR}}(X) \times H^{n-p}_{\mathrm{dR}}(X)$

$$\mathrm{Poinc}< , > : H^p_{\mathrm{dR}}(X) \times H^{n-p}_{\mathrm{dR}}(X) \longrightarrow \mathbb{R}.$$

We have

Theorem 1 (Poincaré duality). *The bilinear form* Poinc< , > *is non degenerate, that is for every non zero cohomology class* $[\omega]$ *there is a non zero class* $[\eta]$ *such that*

$$\mathrm{Poinc}< [\omega], [\eta] > \neq 0,$$

consequently $H^p_{\mathrm{dR}}(X)$ *is isomorphic to the dual* $(H^{n-p}_{\mathrm{dR}}(X))^*$ *of* $H^{n-p}_{\mathrm{dR}}(X)$.

Proof. For each non zero cohomology class that we may assume represented by an harmonic form ω we see that $*\omega$ represents a cohomology class in $H^{n-p}_{\mathrm{dR}}(X)$ for which

$$\int \omega \wedge * \omega = (\omega, \omega) \neq 0.$$

The second part of the claim follows at once from Lemma 1 below as de Rham's cohomology classes are finite dimensional vector spaces. □

Lemma 1. *Let* E, F *be Hilbert spaces and let* $a(\cdot, \cdot) : E \times F \to \mathbb{R}$ *be a bilinear map satisfying*

(i) $|a(u, v)| \leq c \| u \| \| v \|$
(ii) $\exists u_0 \in E,\ u_0 \neq 0,$ *and* $c_0 > 0$ *such that* $a(u_0, v) \geq c_0 \| v \| \ \forall\ v \in F$
(iii) $a(u, v) = 0\ \forall\ v \in F$ *implies* $u = 0$

Then $F \simeq E^*$.

Proof. Consider the linear map $T : F \to E^*$ defined by $T_v(u) := a(u, v)$. By (i) T is continuous and $\| T_v \|_{E^*} \leq k \| v \|_F$; also, by (ii)

$$\frac{c_0}{\| u_0 \|} \| v \|_F \leq \| T_v \|_{E^*}.$$

It follows that T has closed range, is invertible, and T^{-1} is continuous. It remains to show that range $T = E^*$. Assume not, then there is $\phi \in E^*$, $\phi \neq 0$ such that $(\phi | T_v)_{E^*} = 0\ \forall\ v \in F$. Representing ϕ as $\phi(u) = (u | \varphi_0)$, $\varphi_0 \in E$ we then find

$$0 = (\phi | T_v)_{E^*} = T_v(\varphi_0) = a(\varphi_0, v) \qquad \forall\ v \in F,$$

(iii) then implies $\varphi_0 = 0$, i.e., $\phi = 0$: a contradiction. □

Notice that in particular Theorem 1 yields

Proposition 1. *Let* ω *be a closed form in* X. *If* $\int_X \omega \wedge \eta = 0$ *for all* η *with* $d\eta = 0$, *then* ω *is exact.*

Compact manifolds with boundary. Given a smooth compact Riemannian manifold X with boundary, we may define, as we have seen, three sets of cohomology groups depending on $k = 0, \dots, n$:

The *de Rham cohomology groups*

$$H^k_{\mathrm{dR}}(X) := \frac{\text{closed forms of degree } k \text{ in } X}{\text{exact forms of degree } k \text{ in } X},$$

no boundary condition is imposed; for $\omega \in Z^k(X)$, we denote by $[\omega]_X$ the cohomology class of ω,

The *de Rham cohomology groups* of the boundary ∂X

$$H^k_{\mathrm{dR}}(\partial X) := \frac{\text{closed forms of degree } k \text{ in } \partial X}{\text{exact forms of degree } k \text{ in } \partial X};$$

for $\omega \in Z^k(\partial X)$, we denote by $[\omega]_{\partial X}$ the cohomology class of ω in $H^k_{\mathrm{dR}}(\partial X)$,

The *relative cohomology group* of *relative closed forms* modulo *relative exact forms* i.e.

$$H^k_{\mathrm{dR}}(X, \partial X) := \frac{\text{closed forms of degree } k \text{ in } X \text{ which are null on } \partial X}{\text{exact forms of degree } k \text{ in } X \text{ which are null on } \partial X};$$

that we have already defined in Sec. 5.2.6. The relative cohomology class of $\omega \in Z^k(X, \partial X)$ in $H^k_{\mathrm{dR}}(X, \partial X)$ is denoted by $[\omega]_{\mathrm{rel}}$.

The inclusion map $i : \partial X \to X$ clearly induces a map $i^\# : H^k_{\mathrm{dR}}(X) \to H^k_{\mathrm{dR}}(\partial X)$. As forms in $Z^k(X, \partial X)$ and in $B^k(X, \partial X)$ are just special closed and exact forms, it is easily seen that an inclusion map $j^\# : H^k_{\mathrm{dR}}(X, \partial X) \to H^k_{\mathrm{dR}}(X)$ is well defined by

$$j^\#([\omega]_{\mathrm{rel}}) = [\omega]_X.$$

Finally, one can define a *coboundary operator* $\delta : H^k_{\mathrm{dR}}(\partial X) \to H^{k+1}(\partial X)$ by

$$\delta([\omega]_{\partial X}) = [d\widetilde{\omega}]_{\mathrm{rel}}$$

where $\widetilde{\omega}$ is an extension of ω in X. In fact, if $\widetilde{\omega}$ extends ω to X, we have

$$i^\#(d\widetilde{\omega}) = d(i^\#\widetilde{\omega}) = d\omega = 0 \quad \text{on } \partial X,$$

i.e., $d\widetilde{\omega} \in Z^{k+1}(X, \partial X)$; moreover, if $\overline{\omega}$ is another extension of ω, then

$$d\overline{\omega} - d\widetilde{\omega} = d(\overline{\omega} - \widetilde{\omega}) \quad \text{and} \quad i^\#(\overline{\omega} - \widetilde{\omega}) = 0,$$

i.e., $[d\widetilde{\omega}]_{\mathrm{rel}}$ is independent of the choice of the extension $\widetilde{\omega}$ of ω. Also, if $\omega \in B^k(\partial X)$, $\omega = d\alpha$, $\alpha \in \mathcal{D}^{k+1}(\partial X)$, and $\widetilde{\alpha} \in \mathcal{D}^{k+1}(X)$ is an extension of α, we have

$$d\widetilde{\omega} = d(\widetilde{\omega} - d\widetilde{\alpha}) \quad \text{and} \quad i^\#(\widetilde{\omega} - d\widetilde{\alpha}) = \omega - \omega = 0.$$

Recalling now that a sequence of vector spaces

$$\cdots \longrightarrow V_{i-1} \xrightarrow{f_{i-1}} V_i \xrightarrow{f_i} V_{i+1} \longrightarrow \cdots$$

is said to be *exact* if for all i the kernel of f_i is equal to the image of its predecessor f_{i-1}

$$\ker f_i = \operatorname{Im} f_{i-1},$$

we can state

Theorem 2. *The long sequence*

$$\cdots \xrightarrow{j^\#} H^k_{\mathrm{dR}}(X) \xrightarrow{i^\#} H^k_{\mathrm{dR}}(\partial X) \xrightarrow{\delta} H^{k+1}_{\mathrm{dR}}(X, \partial X) \xrightarrow{j^\#} H^{k+1}(X) \xrightarrow{i^\#} \cdots$$

is exact.

Proof. For the reader's convenience we give the simple proof.
(i) First we show that the composition of two successive maps gives zero. Let $[\omega]_{\mathrm{rel}} \in H^k_{\mathrm{dR}}(X, \partial X)$. We have

$$i^\#(j^\#([\omega]_{\mathrm{rel}})) = i^\#([\omega]_X) = [i^\#\omega]_X = 0$$

as $i^\#\omega = 0$. Let $[\omega]_X \in H^k(X)$. We have

$$\delta(i^\#[\omega]_X) = \delta([i^\#\omega]_{\partial X}) = [di^\#\widetilde{\omega}]_{\mathrm{rel}} = [di^\#\omega]_{\mathrm{rel}} = [i^\# d\omega]_{\mathrm{rel}} = 0.$$

Let $[\omega]_{\partial X} \in H^k(\partial X)$. We have

$$j^\#(\delta([\omega]_{\partial X})) = j^\#([d\widetilde{\omega}]_{\mathrm{rel}}) = [d\omega]_X = 0$$

as $d\widetilde{\omega}$ is exact in X (though not relatively exact).
(ii) Suppose that $i^\#([\omega]_X) = 0$, we need to construct $\widetilde{\omega}$ so that $j^\#([\widetilde{\omega}]_{\mathrm{rel}}) = [\omega]_X$. From the assumption $[i^\#\omega]_{\partial X} = 0$, i.e., $i^\#\omega$ is exact in ∂X, $i^\#\omega = d\varphi$ in ∂X. We extend φ to $\widetilde{\varphi}$ in X, define $\widetilde{\omega} = \omega - \widetilde{\varphi}$, and verify

$$i^\#(\omega - d\widetilde{\varphi}) = 0, \qquad j^\#([\widetilde{\omega}]_{\mathrm{rel}}] = [\omega - d\widetilde{\varphi}]_X = [\omega].$$

(iii) Suppose that $[\omega]_{\partial X} \in H^k(\partial X)$ and $\delta([\omega]_{\partial X}) = 0$. Extend ω to $\widetilde{\omega}$ in X, $i^\#\widetilde{\omega} = \omega$; since $[d\widetilde{\omega}]_{\mathrm{rel}} = \delta([\omega]_{\partial X}) = 0$, $d\widetilde{\omega}$ is relatively exact, i.e., $d\widetilde{\omega} = d\varphi$ with $i^\#\varphi = 0$. If we define $\overline{\omega} = \widetilde{\omega} - \varphi$ we have

$$\begin{aligned} &d\overline{\omega} = d\widetilde{\omega} - d\varphi = 0 \\ &i^\#[\overline{\omega}]_X = [i^\#\overline{\omega}]_{\partial X} = [i^\#\widetilde{\omega}]_{\partial X} - [i^\#\varphi]_{\partial X} = [\omega]_{\partial X}. \end{aligned}$$

(iv) Finally, let $[\omega]_{\mathrm{rel}} \in H^k_{\mathrm{dR}}(X, \partial X)$ and $j^\#[\omega]_{\mathrm{rel}} = [\omega]_X = 0$, ω being relatively closed, $d\omega = 0$ and $i^\#\omega = 0$. Then ω is exact in X, $\omega = d\varphi$, and $[i^\#\varphi]_{\partial X} \in H^{k-1}(\partial X)$, as $di^\#\varphi = i^\#\omega = 0$,

$$\delta(i^\#\varphi]_{\partial X}) = [d\varphi]_{\mathrm{rel}} = [\omega]_{\mathrm{rel}}.$$

□

Returning to harmonic forms, we recall Remark 1 in Sec. 5.2.6. i.e., that Hodge $*$ operator yields an isomorphism

$$*: \mathbf{H}^k_N \to \mathbf{H}^{n-k}_T$$

we easily infer that $H^k_{\rm dR}(X)$ and $H^{n-k}_{\rm dR}(X, \partial X)$ are isomorphic. However the duality given by

$$*: H^k_{\rm dR}(X) \to H^{n-k}_{\rm dR}(X, \partial X)$$

depends on the metric (and on the orientation) of X. On an *oriented manifold*, an algebraic isomorphism between $H^k_{\rm dR}(X)$ and the dual of $H^{n-k}_{\rm dR}(X, \partial X)$ is instead given by the map

$$(\omega, \eta) \to \int_X \omega\wedge\eta.$$

It is in fact easily seen that $\int_X (\omega + d\eta)\wedge(\xi + d\sigma) = \int_X \omega\wedge\xi$ if $\omega \in Z^k(X)$, $\eta \in \mathcal{D}^{k-1}(X)$, $\xi \in Z^{n-k}(X, \partial X)$, and $\sigma \in B^{n-k-1}(X, \partial X)$. Therefore such a map induces a bilinear pairing

$$\text{Poinc}{<}[\omega], [\xi]{>} : H^k_{\rm dR}(X) \times H^{n-k}_{\rm dR}(X, \partial X) \to \mathbb{R}$$

between the k-cohomology and the relative $(n-k)$-cohomology. We have

Theorem 3 (Poincaré-Lefschetz duality). *The pairing*

$$\text{Poinc}{<}[\ ], [\ ]{>} : H^k_{\rm dR}(X) \times H^{n-k}_{\rm dR}(X, \partial X) \to \mathbb{R}$$

given by

$$\text{Poinc}{<}[\omega], [\xi]{>} := \int_X \omega\wedge\xi$$

is non degenerate. Moreover the diagram

$$\begin{array}{ccccccccc}
\cdots \stackrel{i^\#}{\leftarrow} & H^k_{\rm dR}(X) & \stackrel{j^\#}{\leftarrow} & H^k_{\rm dR}(X, \partial X) & \stackrel{\delta}{\leftarrow} & H^{k-1}_{\rm dR}(\partial X) & \stackrel{i^\#}{\leftarrow} \cdots \\
 & \times\text{Poinc} & & \times\text{Poinc}^T & & \times\text{Poinc} & \\
\cdots \stackrel{\delta}{\rightarrow} & H^{n-k}_{\rm dR}(X, \partial X) & \stackrel{j^\#}{\rightarrow} & H^{n-k}_{\rm dR}(X) & \stackrel{i^\#}{\rightarrow} & H^{n-k}_{\rm dR}(\partial X) & \underset{\delta}{\rightarrow} \cdots
\end{array}$$

is commutative. Poinc^T *denotes the transposed duality,* $\text{Poinc}^T{<}[\eta]_{\rm rel}, [\omega]{>} := \text{Poinc}{<}[\omega], [\eta]_{\rm rel}{>}$.

The commutativity of

$$\begin{array}{ccc}
A_1 & \stackrel{f}{\longleftarrow} & B_1 \\
\times F_1 & & \times F_2 \\
A_2 & \stackrel{g}{\longrightarrow} & B_2
\end{array}$$

means $F_2(f(a_1), b_2) = F_1(a_1, g(b_2))$, $\forall a_1 \in A_1$, $\forall b_2 \in B_2$.

Proof. If $[\omega]$ is a non trivial cohomology class we can represent it by a non-zero harmonic form h with $\mathfrak{n}\, h = 0$. The form $*h$ is harmonic and $\mathfrak{t}\,(*h) = *\mathfrak{n}\, h = 0$, consequently $*h$ represents a relative cohomology class $[*h] \in H^{n-k}_{\mathrm{dR}}(X, \partial X)$, and we have

$$\mathrm{Poinc}{<}\,[\omega], [*h]\,{>} = \int_X h \wedge * h = \int_X |h|^2 \neq 0.$$

Similarly one proves that, if $[\xi]_{\mathrm{rel}} \neq 0$, we can find $[\omega]$ in $H^k_{\mathrm{dR}}(X)$ such that $\mathrm{Poinc}{<}\,[\omega], [\xi]_{\mathrm{rel}}\,{>} \neq 0$. The commutativity of the diagram is trivial. □

3 Currents and Real Homology of Compact Manifolds

Let X be an oriented compact smooth submanifold of dimension n in $\mathbb{R}^{n+N}$ without boundary. We introduce in Sec. 5.3.2 for $k := 0, 1, \dots$ the k-homology group $H_k(X, \mathbb{R})$ of a X setting

$$\begin{aligned} Z_k(X) &:= \{T \in \mathbf{N}_k(U) \mid \operatorname{spt} T \subset X,\ \partial T = 0\} \\ B_k(X) &:= \{\partial S \mid S \in \mathbf{N}_{k+1}(U),\ \operatorname{spt} S \subset X\} \end{aligned}$$

and

$$H_k(X, \mathbb{R}) := Z_k(X)/B_k(X)$$

where U is a tubular neighbourhood of X, i.e., compare Sec. 5.3.1, as normal cycles in X modulo boundaries of normal currents on X.

Two sets of pairings are naturally associated to homology and cohomology. One is the set of *de Rham pairings* between currents and forms which factors to homology and cohomology

$$<\,,\,> : H_k(X, \mathbb{R}) \times H^k_{\mathrm{dR}}(X) \to \mathbb{R}$$

defined by $<[T], [\omega]> := T(\omega)$. The others are the pairings

$$(\omega, \eta) \in \mathcal{D}^k(X) \times \mathcal{D}^{n-k}(X) \ \to \ \int_X \omega \wedge \eta$$

which, by Stokes theorem, induce *Poincaré pairings* in cohomology

$$\mathrm{Poinc}<\,,\,> : H^k_{\mathrm{dR}}(X) \times H^{n-k}_{\mathrm{dR}}(X) \to \mathbb{R}$$

given by $\mathrm{Poinc}<[\omega], [\eta]> := \int \omega \wedge \eta$. Classical theorems, respectively de Rham's and Poincaré's duality theorems, state in fact that de Rham and Poincaré pairings are *non degenerate*; consequently there are isomorphisms, *Poincaré duality isomorphisms*

$$P : H^{n-k}_{\mathrm{dR}}(X) \ \to \ H_k(X, \mathbb{R}).$$

between $H^{n-k}_{\mathrm{dR}}(X)$ and $H_k(X,\mathbb{R})$. In Sec. 5.3.2 we show that the Poincaré' duality isomorphism is actually induced by the map again denoted by P, $P : Z^{n-k}(X) \to Z_k(X)$ given by $\omega \to \int_X \cdot \wedge \omega$. Moreover, by regularizing a k-cycle $T \in Z_k(X)$ we construct an $(n+N-k)$-form $\omega \in \mathcal{D}^{n+N-k}(U)$, and integrating such a form on the fibers of the projection map π, we construct a so-called *Poincaré dual of* T, $P_T \in \mathcal{D}^{n-k}(X)$. The cohomology class of P_T turns out to depend only on the homology class of T, therefore $T \to P_T$ yields an explicit way to describe surjectivity of P. Non degeneracy of Poincaré duality, which we already proved in Sec. 5.2.8 as consequence of Hodge decomposition theorem, then yields injectivity of the Poincaré' duality isomorphism and also yields non-degeneracy of de Rham pairing, as

$$< [T], [\omega] > = \mathrm{Poinc} < [\omega], [P_T] > = \int_X \omega \wedge P_T,$$

compare Sec. 5.3.2. We then deduce the weak closure of the real homology classes, hence in particular the existence of mass minimizing real cycles in each homology class.

A completely parallel theory holds in the case of compact manifolds with boundary. One has to introduce the *relative homology groups*, compare (1) in Sec. 5.3.3. The *Poincaré–Lefschetz duality isomorphism* this time consists of three sets of maps

$$\begin{array}{rcl} P^\sharp : H^{n-k}_{\mathrm{dR}}(X,\partial X) & \to & H_k(X,\mathbb{R}) \\ P^\flat : H^{n-k}_{\mathrm{dR}}(X) & \to & H_k(X,\partial X,\mathbb{R}) \\ P : H^{n-k-1}_{\mathrm{dR}}(\partial X) & \to & H_k(\partial X,\mathbb{R}) \end{array}$$

which actually give a morphism between the long exact sequences in cohomology and homology, compare (9) in Sec. 5.3.3. The duality forms-currents $<T,\omega> = T(\omega)$ induces three sets of de Rham pairings at the homological level

$$\begin{array}{rcl} <,>^\sharp : H_k(X,\mathbb{R}) \times H^k_{\mathrm{dR}}(X) & \to & \mathbb{R} \\ <,>^\flat : H_k(X,\partial X,\mathbb{R}) \times H^k_{\mathrm{dR}}(X,\partial X) & \to & \mathbb{R} \\ <,> : K_k(\partial X,\mathbb{R}) \times H^k_{\mathrm{dR}}(\partial X) & \to & \mathbb{R} \end{array}$$

which are related each other by the commutative diagram in de Rham's theorem, Theorem 5 in Sec. 5.3.3. As consequence we deduce the closure of absolute and relative homology classes and consequently the existence of mass minimizing cycles in each homology class.

Still another set of pairings is relevant in homology: these are the so-called *intersection indices* of cycles of complementary dimension.

In Sec. 5.3.4 we shall define *Kronecker's index* for two arbitrary currents $S \in \mathbf{N}_k(X)$ and $T \in \mathbf{N}_{n-k}(X)$ for which $\operatorname{spt} \partial S \cap \operatorname{spt} T = \operatorname{spt} \partial T \cap \operatorname{spt} S = 0$ and see that it induces a pairing

$$i_X(\cdot,\cdot) : H_k(X,\mathbb{R}) \times H_{n-k}(X,\partial X,\mathbb{R}) \to \mathbb{R}.$$

It turns out that

$$i_X(S,T) = \int_X P_S^\sharp \wedge P_T^\flat,$$

therefore non degeneracy of Poincaré duality is equivalent to non-degeneracy of the intersection index.

3.1 Currents on Manifolds

Let X be a smooth manifold of dimension m. As already mentioned, by Whitney embedding theorem we may and do assume that X is a smooth (C^∞) submanifold of some $\mathbb{R}^n$, $X \subset \mathbb{R}^n$. Moreover we can assume that X has a tubular neighbourhood $U \supset X$, $U \subset \mathbb{R}^n$, and there is a C^∞ retraction map $\pi : U \to X$, $\pi(U) = X$, $\pi(y) = y \ \forall\, y \in X$.

Definition 1. *A k-dimensional current $T \in \mathcal{D}_k(U)$ is said to be a* current in X *if $T(\omega) = 0$ for any null form $\omega \in \mathcal{D}^k(U)$ to X. The class of k-dimensional currents in X is denoted by $\mathcal{D}_k(X)$.*

From the characterization of null forms to X, compare the end of Sec. 5.2.1, we easily infer

Proposition 1. *Let $T \in \mathcal{D}_k(U)$. Then $T \in \mathcal{D}_k(X)$ if and only if for some $\zeta \in C_c^\infty(U)$ with $\zeta = 1$ on X we have $T = \pi_\#(T \llcorner \zeta)$. Actually, if $T \in \mathcal{D}_k(X)$, we have*

$$T = \pi_\#(T \llcorner \zeta) \qquad \forall\, \zeta \in C_c^\infty(U),\ \zeta = 1 \text{ on } X.$$

Proof. In fact for any $\omega \in \mathcal{D}^k(U)$ the form $\omega - \zeta\pi^\#\omega \in \mathcal{D}^k(U)$ is a null form to X, hence $T(\omega) - T(\zeta\pi^\#\omega) = 0$. Conversely if η is a null form to X, then $\pi^\#\omega = 0$, hence $T(\eta) = \pi_\#(T \llcorner \zeta)(\eta) = T(\zeta\pi^\#\eta)$=0. □

Remark 1. Proposition 1 extends to the case of C^k manifolds, $k \geq 1$, provided T is a vector valued distribution of order $-k$. In this case one in fact can assume that the retraction map is C^k hence $\pi_\#(T \llcorner \zeta)$ is well defined. In particular Proposition 1 holds true for flat chains on C^1 manifolds.

Trivially $\operatorname{spt} T \subset \overline{X}$ if $T \in \mathcal{D}_k(X)$. In fact, if ω is supported outside X and ω is a null form to X, we have $T(\omega) = 0$. But in general it is not true that $\operatorname{spt} T \subset X$ implies $T \in \mathcal{D}_k(X)$. Consider for instance the 1-dimensional current in $\mathbb{R}^2$ integration of 1-forms along the real axis e_1 against the normal vector e_2

$$L(\omega) := \int <\omega, e_2> dx^1.$$

Clearly $\operatorname{spt} L \subset \mathbb{R}_{x^1}$, while for example on the null form $\varphi(x^1)\, dx^2$ we have

$$T(\varphi(x^1)dx^2) = \int \varphi(x^1)\, dx^1.$$

Notice that $\mathbf{M}(T) < \infty$, while $\mathbf{M}(\partial T) = +\infty$, compare [10] in Sec. 2.2.4. However we have

Theorem 1 (Flatness theorem). *Let X be a closed submanifold of $\mathbb{R}^n$. If T is a flat chain in U, $T \in \mathbf{F}_k(U)$, and $\operatorname{spt} T \subset X$, then $T \in \mathcal{D}_k(X)$.*

Proof. Theorem 1 is a consequence of Proposition 7 in Sec. 5.1.3. In fact the retraction map $\pi : U \to X$ and the identity map $\mathrm{id} : U \to U$ trivially agree on X, hence on $\operatorname{spt} T$. Proposition 7 in Sec. 5.1.3 then yields $T = \pi_\#(T \llcorner \zeta)\ \forall\, \zeta \in C_c^\infty(U)$, $\zeta = 1$ on X, and the conclusion follows from Proposition 1. □

By Theorem 1 normal currents on a submanifold X on $\mathbb{R}^n$ may be identified to currents in the ambient space $\mathbb{R}^n$ with support in X. We shall now see that a constancy theorem holds also for currents on submanifolds.

Let X be an m-dimensional C^∞ submanifold of $\mathbb{R}^n$ with locally finite area. Suppose that X is oriented by a smooth m-vector field $\vec{X}$. Then the current

$$[\![X]\!] := \tau(X, 1, \vec{X})$$

is a locally m-rectifiable current in $\mathbb{R}^n$.

Theorem 2 (Constancy theorem). *Let Ω be an open set in $\mathbb{R}^n$ such that $\Omega \cap X$ is connected and let $T \in \mathcal{D}_m(X)$, $m = \dim X$, be such that $\operatorname{spt} \partial T \cap \Omega = \emptyset$. Then there exists a constant $c \in \mathbb{R}$ such that*

$$T = c[\![X]\!] \quad in \quad \mathcal{D}^m(\Omega \cap U)$$

U being a tubular neighbourhood of X. Moreover c is an integer if T is i.m. rectifiable.

Proof. As usual let $\pi : U \to X$ be the retraction map, and let $\zeta \in C_c^\infty(U)$, $\zeta \equiv 1$ on X. We cover $\Omega \cap X$ by a system of coordinates $\varphi_i : V_i \longrightarrow \Omega_i \cap X \quad \Omega_i \subset \Omega$ so that $\varphi_{i\#}\vec{V}_i = \vec{X}$; the maps φ_i have inverses, and more precisely there are smooth maps $\psi_i : U \cap \pi^{-1}(\Omega_i \cap X) \longrightarrow V_i$ which are onto and such that

$$\varphi_i \circ \psi_i = \pi, \qquad \psi_i \circ \varphi_i = \mathrm{id}_{V_i}.$$

Therefore the currents $\psi_{i\#}(T \llcorner \zeta)$ belong to $\mathcal{D}_m(V_i)$ and, as $\operatorname{spt} \partial T \cap \Omega = \emptyset$, we have $\partial \psi_{i\#}(T \llcorner \zeta) = 0$ on V_i. It follows from the constancy theorem in the flat case, Theorem 1 in Sec. 4.3.1 that

$$\psi_{i\#}(T \llcorner \zeta) = c_i[\![V_i]\!] \quad \text{on } V_i,$$

hence

$$T = \pi_\#(T \llcorner \zeta) = \varphi_{i\#}\psi_{i\#}(T \llcorner \zeta) = c_i[\![X]\!] \quad \text{on } \Omega_i \cap U,$$

since $\varphi_{i\#}\vec{V}_i = \vec{X}$. As the Ω_i's overlap on X, we infer that the constant c_i are all equal to a c independent of the chart

$$T = c[\![X]\!] \qquad \text{on } \mathcal{D}^m(\Omega_i \cap U),$$

and the conclusion follows by means of a decomposition of unity associated to the covering $\{\Omega_i\}$. □

Remark 2. Theorem 1 holds also for flat chains on C^1 manifolds, compare Remark 1.

More generally we have

Theorem 3 (Constancy theorem). *Let Σ be an r–dimensional oriented, connected and embedded submanifold of U of class C^1, where U is an open set in $\mathbb{R}^n$, and let $T \in \mathbf{F}_r^{\mathrm{loc}}(U)$. If $(\operatorname{spt} T)\setminus\Sigma$ is closed relative to U and $\operatorname{spt}\partial T \subset U\setminus\Sigma$, then there exists a real number ρ such that*

$$\operatorname{spt}(T - \rho[\![\Sigma]\!]) \subset U \setminus \Sigma.$$

Moreover ρ is an integer if T is locally integer rectifiable.

If $f : X \to Y$ is a Lipschitz map between C^1 manifolds, every $T \in \mathbf{N}_k(X)$ has a Lipschitz image $f_\# T \in N_k(Y)$. Also if $f_0, f_1 : X \to Y$ are two maps and $h : [0,1] \times X \to Y$ is an homotopy map between f_0 and f_1, $h(0,x) = f_0(x)$, $h(1,x) = f_1(x)$, then the homotopy formula holds

$$f_{1\#}T - f_{0\#}T = \partial h_\#([\![0,1]\!] \times T) + h_\#([\![0,1]\!] \times \partial T) \tag{1}$$

for $T \in \mathbf{N}_k(X)$. Also a corresponding homotopy formula for forms holds

$$f^{1\#}\omega - f^{0\#}\omega = d[h^\#(\omega)]_{(0,1)} + \left[h^\#(d\omega)\right]_{(0,1)}, \tag{2}$$

$[\eta]_{(0,1)}$ being the integration along the fibers of the components of η which are not tangential to X. Also the standard estimates for homotopies hold true, compare Sec. 2.2.3.

3.2 Manifolds Without Boundary: Poincaré and de Rham Dualities

By *Poincaré duality isomorphism* one usually refers to the duality between the k-homology group and the $(n-k)$-de Rham cohomology group of an oriented compact smooth manifold without boundary. In this subsection we shall describe such an isomorphism in terms of currents and forms.

Poincaré duality isomorphism on open sets. Let U be an open set in $\mathbb{R}^n$. For $0 \le k \le n$ consider the vector spaces of currents

$$Z_k(U,\mathbb{R}) = \{T \in \mathcal{D}_k(U) \mid \partial T = 0,\ \mathbf{M}(T) < +\infty\}$$
$$B_k(U,\mathbb{R}) = \{\partial S \mid S \in \mathcal{D}_{k+1}(U),\ \mathbf{M}(S) < +\infty\}$$

and set

$$H_k(U,\mathbb{R}) = Z_k(U,\mathbb{R})/B_k(U,\mathbb{R}).$$

Analogously we define *de Rham's cohomology groups* in U by

$$H^k_{\mathrm{dR}}(U,\mathbb{R}) := Z^k(U,\mathbb{R})/B^k(U,\mathbb{R}).$$

where

$$Z^k(U,\mathbb{R}) = \{\omega \in L^1(U,\Lambda^k\mathbb{R}^n) \cap \mathcal{E}^k(U) \mid d\omega = 0\}$$
$$B^k(U,\mathbb{R}) = \{d\eta \mid \eta \in L^1(U,\Lambda^k\mathbb{R}^n) \cap \mathcal{E}^{k-1}(U)\}$$

Then we have

Theorem 1. *$H^{n-k}_{\mathrm{dR}}(U,\mathbb{R})$ and $H_k(U,\mathbb{R})$ are isomorphic.*

To prove this we consider the map $P : L^1(U,\Lambda^{n-k}\mathbb{R}^n) \cap \mathcal{E}^{n-k}(U) \to \{T \in \mathcal{D}_k(U) \mid \mathbf{M}(T) < +\infty\}$ given by

$$P(\omega) := (-1)^{k(n-k)}\mathbb{R}^n \llcorner \omega = \int \cdot \wedge\omega.$$

As $\partial(\mathbb{R}^n \llcorner \omega) = (-1)^{k-1}\mathbb{R}^n \llcorner d\omega$ the map P factorizes to a map still denoted by P

$$P : H^{n-k}_{\mathrm{dR}}(U,\mathbb{R}) \;\to\; H_k(U,\mathbb{R}).$$

We shall prove that P is an isomorphism, indeed the isomorphism of Theorem 1. Of course this amount to prove

Proposition 1. *We have*

(i) Let T be a cycle of finite mass, $\partial T = 0$. Then there is a $(n-k)$-form $P_T \in L^1(U,\Lambda^{n-k}\mathbb{R}^n) \cap \mathcal{E}^{n-k}(U)$ such that for some $S \in \mathcal{D}_{k+1}(U)$, $\mathbf{M}(S) < +\infty$ we have

(1) $$T = \int_U \cdot \wedge P_T + \partial S.$$

(ii) Let $\omega \in L^1(U,\Lambda^{n-k}\mathbb{R}^n) \cap \mathcal{E}^{n-k}(U)$ be such that $\mathbb{R}^n \llcorner \omega = \partial S$ for some $S \in \mathcal{D}_{k+1}(U)$, $\mathbf{M}(S) < +\infty$. Then $\omega = d\eta$, $\eta \in L^1(U,\Lambda^{n-k-1}\mathbb{R}^n) \cap \mathcal{E}^{n-k-1}(U)$.

Proof. We prove both claims by regularizing T. (i) is in fact already proved in Proposition 7 in Sec. 5.1.2, by taking $P_T := (-1)^{k(n-k)} f$.

To prove (ii), we choose a convolution kernel ρ and a regularized distance d. For $0 < \varepsilon < 1$ we have

(2) $$\mathbb{R}^n \llcorner \omega^\varepsilon = (\mathbb{R}^n \llcorner \omega)_\varepsilon = (\partial S)_\varepsilon = \partial S_\varepsilon.$$

On the other hand by Proposition 6 in Sec. 5.1.2

(3) $$S_\varepsilon = \mathbb{R}^n \llcorner \eta_\varepsilon \qquad \eta_\varepsilon \in L^1(U,\Lambda^{n-k-1}\mathbb{R}^n) \cap \mathcal{E}^{n-k-1}(U)$$

From (2) and (3) we conclude that $\omega^\varepsilon = (-1)^{n-k} d\eta_\varepsilon$. As ω^ε and ω are also cohomologous to a form $L^1(U,\Lambda^{n-k-1}\mathbb{R}^n) \cap \mathcal{E}^{n-k-1}(U)$ the claim follows. □

Remark 1. We observe that the same reasoning of Proposition 1 implies the following:

(i) if the current T in (i) of Proposition 1 has compact support in U, then the decomposition (1) holds with $\operatorname{spt} P_T$ and $\operatorname{spt} S$ arbitrarily close to $\operatorname{spt} T$;

(ii) if the form ω and the current S in (ii) of Proposition 1 have compact support in U, then one finds a potential for ω with support arbitrarily to $\operatorname{spt}\omega \cup \operatorname{spt} S$.

(iii) In particular we infer that de Rham's $(n-k)$-cohomology group of smooth forms with compact support is isomorphic to the k-homology group of currents with compact supports

$$\{T \in \mathcal{D}_k(U) \mid \partial T = 0,\ \operatorname{spt} T \subset\subset U\}/\{\partial S \mid S \in \mathcal{D}_{k+1}(U)\}.$$

From now on we shall refer to the map $P : H^{n-k}_{\mathrm{dR}}(U,\mathbb{R}) \to H_k(U,\mathbb{R})$ as to the *Poincaré duality isomorphism* and to $(n-k)$–forms P_T satisfying (1) as to *Poincaré's dual forms of* T. Poincaré's dual forms P_T, $T \in Z_k(U)$ are constructed mollifying T, compare Sec. 5.1.2, in particular there are not unique. However they preserve some of the properties of T, e.g., $P_T \in L^1(U, \Lambda^{n-k}\mathbb{R}^n)$ if $\mathbf{M}(T) < \infty$, $\operatorname{spt} P_T \subset\subset U$ if $\operatorname{spt} T \subset\subset U$. Also (ii) of Proposition 1 states that the cohomology class of any Poincaré dual form $P_T \in Z^{n-k}(U)$ of a given cycle $T \in Z_k(U)$ is uniquely defined by the homology class of T.

Let X be a compact oriented and boundaryless submanifold in U of dimension k. A form $\omega \in Z^{n-k}(U)$ with compact support in U such that

$$[\![X]\!](\eta) = \int_U \eta\wedge\omega$$

for all *closed* forms $\eta \in Z^k(U)$ is classically referred as to a Poincaré's dual form of X. In principle therefore our notion of Poincaré dual form is slightly stronger, as we require equality (1). Actually this is not the case, in fact one can show that every classical Poincaré form is actually a Poincaré form in our sense. This amounts to show that for every $(n-k)$-form ω for which

$$\mathbb{R}^n \llcorner \omega(\alpha) = \int_U \omega\wedge\alpha = 0 \qquad \forall\, \alpha \in \mathcal{D}^k(U),\ d\alpha = 0 \tag{4}$$

(in particular ω is closed) we can find a current $S \in \mathcal{D}_{k+1}(U)$ such that $\mathbb{R}^n \llcorner \omega = \partial S$. This is equivalent to say that the linear functional $S' : B_{k+1}(U) \to \mathbb{R}$

$$S'(\beta) := \int \omega\wedge\alpha \qquad \text{if } \beta = d\alpha \tag{5}$$

extends as a linear *continuous* functional on $\mathcal{D}^{k+1}(U)$. We have in fact

Proposition 2. *Let U be a smooth bounded open set in $\mathbb{R}^n$ and let ω be a form in $L^p(U, \Lambda^{n-k}\mathbb{R}^n) \cap \mathcal{E}^{n-k}(U)$, $p \geq 1$, such that*

$$\int_U \omega \wedge \alpha = 0 \qquad \forall\, \alpha \in \mathcal{D}^k(U) \text{ with } d\alpha = 0,$$

in particular $d\omega = 0$. *Then there exists* $S \in \mathcal{D}_{k+1}(U)$ *such that* $\mathbb{R}^n \llcorner \omega = \partial S$.

Proof. First assume $p > 1$, and let $\omega = \sum \omega_\alpha(x) dx^\alpha$. We solve the problem

$$\begin{cases} \Delta \Omega_\alpha = \omega_\alpha \\ \Omega_\alpha \in W_0^{1,p}(U). \end{cases} \tag{6}$$

By the elliptic regularity we find a unique Ω_α which actually belongs to $C^\infty(U) \cap W^{2,p}(U)$. Consequently the form $\Omega := \sum \Omega_\alpha \, dx^\alpha$ solves

$$d\delta\Omega + \delta d\Omega = \omega,$$

compare (7) in Sec. 5.2.4, δ being the adjoint of d with respect to zero values on ∂U. Since $d\omega = 0$, we infer $\delta d\Omega \in \ker d \cap \ker \delta$ and by the assumption, taking into account Proposition 2 in Sec. 5.2.4, $\delta d\Omega = \omega - d\delta\Omega \in \ker \delta^\perp$, hence $\delta d\Omega = 0$. Therefore we find $\omega = d\eta$, $\eta \in \mathcal{E}^{n-k-1}(U) \cap W^{1,p}(U, \Lambda^{n-k-1}\mathbb{R}^n)$, and consequently

$$\mathbb{R}^n \llcorner \omega = \mathbb{R}^n \llcorner d\eta = (-1)^{n-k} \partial(\mathbb{R}^n \llcorner \eta).$$

If $p = 1$, we can still solve (6), compare Stampacchia [609], and find a weak solution which belongs to $C^\infty(U) \cap W^{1,q}(U)$ for all $q < n/(n-1)$. The proof is then concluded as previously. □

Integration along the fibers and the lift of currents. Let X be a C^2 compact, oriented n-dimensional manifold without boundary in $\mathbb{R}^{n+N}$ and denote by $\vec{X} : X \to \Lambda_n \mathbb{R}^{n+N}$ and by $\mathcal{N} : X \to \Lambda_N \mathbb{R}^{n+N}$ respectively the unit tangent n-vector field to X and a unit *normal* vector field to X so that

$$\vec{X} \wedge \mathcal{N}$$

has the same orientation of $\mathbb{R}^{n+N}$. Recall now that X is a tubular neighbourhood retract, and more precisely we can find an $\varepsilon_0 > 0$ such that each point in

$$U = \{x \in \mathbb{R}^{n+N} \mid \operatorname{dist}(x, X) < \varepsilon_0\}$$

has a unique point of least distance $\pi(x)$ in X, and that $\pi : U \to X$ is a retraction. Moreover, we have a system of charts $\varphi_\alpha : \Omega_\alpha \to V_\alpha \subset X$ of X, and on each chart a system of N independent vectors $v_1^{(\alpha)}, \dots, v_N^{(\alpha)}$ in $\mathbb{R}^{n+N}$ such that

$$v_1^{(\alpha)} \wedge \ldots \wedge v_N^{(\alpha)} = \mathcal{N} \qquad \text{on } V_\alpha;$$

consequently the maps $\phi_\alpha : U_\alpha \times B(0, \varepsilon_0) \subset U_\alpha \times \mathbb{R}^N \to \pi^{-1}(V_\alpha)$ given by

$$\phi_\alpha(x, t) = \varphi_\alpha(x) + \sum_{i=1}^N t_i v_i^{(\alpha)}(x) \tag{7}$$

are diffeomorphisms. Of course, taking into account the special form of ϕ_α and denoting by $\varepsilon_1, \dots, \varepsilon_N$ the standard basis of $\mathbb{R}^N$, we see that

$$\Lambda_N D\phi_\alpha(x,t)(\varepsilon_1\wedge\ldots\wedge\varepsilon_N) = v_1\wedge\ldots\wedge v_N(\varphi(x)) = \mathcal{N}(\varphi(x))$$

for $x \in \Omega_\alpha$, $t \in B(0,\varepsilon_0)$. Also note that

$$\phi_\beta \circ \phi_\alpha^{-1}(x,b) = (\varphi_\beta^{-1} \circ \varphi_\alpha(x), A(x)t),$$

where $A(x) \in M_{N\times N}$ and $\det A > 0$, or in other words $\pi : U \to X$ *is an oriented fiber bundle over* X.

To any $(k+N)$-form $\omega \in \mathcal{D}^{k+N}(U)$ we now associate a k-form $\pi_*\omega$ in $\mathcal{D}^k(X)$ by "integrating ω along the fibers of the fibration $\pi : U \to X$". We define $\pi_*\omega$ by

$$(8) \qquad <\pi_*\omega, v> := \int_{\pi^{-1}(x)} <\omega(z), v\wedge\mathcal{N}(x)> dH^N(z) \quad \forall\, v \in \Lambda^k T_x X.$$

Clearly π_* defines a map from $\mathcal{D}^{k+N}(U)$ into $\mathcal{D}^k(X)$. In the special case in which $X = \Omega$ is an open set of $\mathbb{R}^n$, $U = \Omega \times \mathbb{R}^N$ introducing coordinates $e_1, \dots, e_n$ in Ω and $\varepsilon_1, \dots, \varepsilon_N$ in $\mathbb{R}^N$, we have

$$\vec{X} = e_1\wedge\ldots\wedge e_n \qquad \mathcal{N} = \varepsilon_1\wedge\ldots\wedge\varepsilon_N,$$

and any form $\omega \in \mathcal{D}^{k+N}(\Omega \times \mathbb{R}^N)$ has the form

$$\omega = \sum_{|\alpha|+|\beta|=k+N} \omega_{\alpha\beta}(x,y)\, dx^\alpha \wedge dy^\beta.$$

Then $\pi(x,y) = x$ and

$$(9) \qquad \pi_*(\omega)(x) := \sum_{|\alpha|=k} \Big(\int_{\mathbb{R}^N} \omega_{\alpha\bar{0}}(x,s)\, ds \Big)\, dx^\alpha.$$

Notice that the formula for a product of currents gives

$$(10) \quad T \times \mathbb{R}^N(\omega) = T(\pi_*(\omega)) \qquad \text{for } T \in \mathcal{D}_k(\Omega),\ \omega \in \mathcal{D}^{k+N}(\Omega \times \mathbb{R}^N).$$

Proposition 3. *We have*

(i) $\pi_* d = d\pi_*$.

(ii) $\pi_*(\pi^\# \eta\wedge\omega) = \eta\wedge\pi_*\omega$, $\forall\, \omega \in \mathcal{D}^{k+N}(U)$, $\forall\, \eta \in \mathcal{D}^p(X)$, $p + k \le n$.

(iii) For $\eta \in \mathcal{D}^{n-k}(X)$ *and* $\omega \in \mathcal{D}^{k+N}(U)$

$$\int_X \eta\wedge\pi_*\omega = \int_U \pi^\#\eta\wedge\omega,$$

i.e., $\pi_\#(\mathbb{R}^{n+N} \llcorner \omega) = (-1)^{N(n-k)} [\![X]\!] \llcorner \pi_*\omega$.

(iv) *If P_X is a Poincaré dual form of X in U then $\pi_* P_X = 1$.*

Proof. The first two claims are local, consequently it suffices to prove them for forms η and ω respectively with compact support in a chart V and in $\pi^{-1}(V)$, we omit here the index α, compare (7). Also we first prove the claims in the case that $V =: \Omega$ is an open set of $\mathbb{R}^n$ so that $\pi^{-1}(\Omega) = \Omega \times \mathbb{R}^N$. In this case we can write $\pi(x, y) = x$ and

$$\pi_* \omega := \sum_{|\alpha|=k} \Big(\int \omega_{\alpha\bar{0}}(x, y)\, dy \Big)\, dx^\alpha .$$

Now each component of ω takes the form $\xi = a(x, y)\, dx^\alpha \wedge dy^\beta$. Therefore

$$d\xi = a_{x_i}(x, y) dx^i \wedge dx^\alpha \wedge dy^\beta + \sigma(i, \beta)(-1)^k a_{y_j}(x, y) dx^\alpha \wedge dy^{\beta + j}$$

hence

$$\pi_* d\xi = \begin{cases} 0 & |\beta| < N - 1 \\ \sigma(i, \beta)(-1)^k \Big(\int a_{y_j}(x, y)\, dy \Big)\, dx^\alpha = 0 & |\beta| = N - 1 \\ \sum_{i=1}^{n} \Big(\int a_{x_i}(x, y)\, dy \Big)\, dx^i \wedge dx^\alpha & |\beta| = N \end{cases}$$

and we conclude

$$\pi_* d\omega = \sum_{i=1}^{n} \sum_{|\alpha|=k} \Big(\int a_{x_i}(x, y)\, dy \Big)\, dx^i \wedge dx^\alpha = d\pi_* \omega .$$

Writing $\pi^\# \eta = \sum_{|\gamma|=n} \eta_\gamma(x)\, dx^\gamma$ the second claim follows easily as

$$\begin{aligned} \pi_*(\pi^\#(\gamma) \wedge \omega)(x) &= \sum_{\substack{|\gamma|=h \\ |\alpha|=k}} \Big(\int \eta_\gamma(x) a_{\alpha\bar{0}}(x, y)\, dy \Big)\, dx^\gamma \wedge dx^\alpha \\ &= \eta(x) \wedge \pi_* \omega(x). \end{aligned}$$

The general case follows from the flat case because

$$\pi_*(\phi^\# \omega) = \varphi^\#(\pi_* \omega), \tag{11}$$

ϕ and φ being the diffeomorphisms in (7) (index omitted). In fact

$$\begin{aligned} \varphi^\#[d\pi_* \omega] = d\varphi^\# \pi_* \omega &= \text{(by (11))}\ d\pi_* \phi^\# \omega \\ &= \text{(by (i) in the flat case)}\ \pi_* d\phi^\# \omega \\ &= \pi_* \phi^\#(d\omega) \\ &= \text{(again by (11))}\ \varphi^\#(\pi_* d\omega) \end{aligned}$$

which proves that in any coordinate chart $d\pi_* - \pi_* d = 0$. Similarly one proves (9).

Let us prove (11). We have $\phi^\# \omega(x,t) = \Lambda^{k+N} D\phi(x,t)\omega(\phi(x,t))$, therefore for $v \in \Lambda_k \mathbb{R}^n$,

$$<\phi^\# \omega, v \wedge \varepsilon_1 \wedge \ldots \wedge \varepsilon_N> = <\omega(\phi(x,t)), \Lambda_{k+N} D\phi(x,t)(v \wedge \varepsilon_1 \wedge \ldots \wedge \varepsilon_N)>.$$

But on the account of the particular form of ϕ

$$\Lambda_k D\phi(x,t)(v) = \Lambda_k D\varphi(x)(v)$$
$$\Lambda_N D\phi(x,t)\varepsilon_1 \wedge \ldots \wedge \varepsilon_N = |\Lambda_N D\phi(x,t)| \mathcal{N}(\phi(x,t))$$

$\Lambda_k D\phi(v)$, $\Lambda_N D\phi(\varepsilon_1 \wedge \ldots \varepsilon_N)$ are orthogonal, hence

$$<\phi^\# \omega, v \wedge \varepsilon_1 \wedge \ldots \wedge \varepsilon_N> = <\omega(\phi(x,t)), \Lambda_k D\varphi(x)(v) \wedge \mathcal{N}(\varphi(x,t))>,$$

consequently

$$\begin{aligned} &<\pi_*(\phi^\# \omega), v> \\ &= \int <\omega(\phi(x,t)), \Lambda_k D\varphi(x) v \wedge \mathcal{N}(\varphi(x,t))> |\Lambda_N D\phi(x,t)| dt \\ &= \int_{\pi^{-1}(\varphi(x))} <\omega(z), \Lambda_k D\varphi(x) v \wedge \mathcal{N}(\varphi(x))> d\mathcal{H}^N(z) \\ &= <\pi_* \omega(\varphi(x)), \Lambda_k D\varphi(x) v> = <\varphi^\# \omega(x), v>. \end{aligned}$$

Proof of (iii). We use the general coarea formula

$$\int_X g |\nabla^N f| \, d\mathcal{H}^n = \int_X d\mathcal{H}^n \int_{\pi^{-1}(x)} g(z) \, d\mathcal{H}^N(z)$$

where $f : U \to X$ and $\nabla^N f(x) =$ projection of ∇f onto the level line of f through x. In this case we have for $\eta \in \mathcal{D}^{n-k}(X)$

$$\begin{aligned} \int_X \eta \wedge \pi_*(\omega) &= \int_X \pi_*(\pi^\# \eta \wedge \omega) = \int_X d\mathcal{H}^n \int_{\pi^{-1}(x)} d\mathcal{H}^N <\pi^\# \eta \wedge \omega, \vec{e}_X \wedge \mathcal{N}> \\ &= \int_U <\pi^\# \eta \wedge \omega, e_1 \wedge \ldots \wedge e_{n+N}> |\nabla^N \pi| \, dz = \int_U \pi^\# \eta \wedge \omega \end{aligned}$$

since $|\nabla^N \pi| = 1$. This proves the claim together with

$$\begin{aligned} [\![X]\!] \llcorner \pi_*(\omega) &= (-1)^{(n-k)k} \int_X \eta \wedge \pi_*(\omega) \\ &= (-1)^{(n-k)k} \int_U \pi^\#(\eta) \wedge \omega = (-1)^{(n-k)[k+N-k]} [\![U]\!](\omega \wedge \pi^\#(\eta)) \\ &= (-1)^{(n-k)k} \pi_\#(\mathbb{R}^{n+N} \llcorner \omega). \end{aligned}$$

Finally (iv) follows at once as for $\eta \in \mathcal{D}^n(X)$ by (iii) and (1) we have

$$\int_X \eta\, \pi_*(P_X) = \int_U \pi^{\#}(\eta) \wedge P_X = \int_X \eta,$$

hence $\pi_*(P_X) = 1$. □

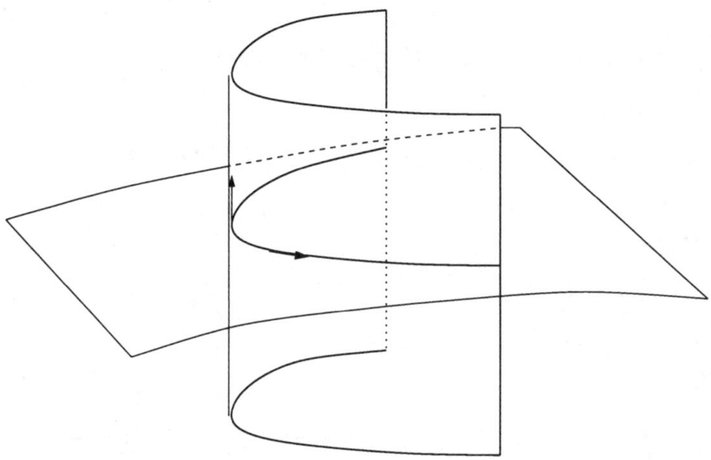

Fig. 5.3. The lift of a current.

Remark 2. Let $\varphi : \Omega \to \Delta$ be a diffeomorphism of two open sets of $\mathbb{R}^n$ and let $\phi : \Omega \times \mathbb{R}^N \to \Delta \times \mathbb{R}^N$ be of the type

$$(12)\, \phi(x,t) = (\varphi(x), h(x,t)), \quad h : \Omega \times \mathbb{R}^N \to \mathbb{R}^N \text{ smooth and } \frac{\partial h}{\partial y}(x,t) > 0.$$

Then it is easily seen that $\pi_* \phi^{\#} = \varphi^{\#} \pi_*$ simply by changing variables. In particular we obtain that if $\pi : E \to X$ is an oriented fiber bundle, then the integration along the fibers is well defined and gives a map

$$\pi_* : \mathcal{E}^{k+N}(E) \to \mathcal{E}^k(X)$$

since the transition functions of E are maps of the type (12).

One then defines *a lift operator* $\pi^* : \mathcal{D}_k(X) \to \mathcal{D}_{k+N}(U)$ on currents by duality

$$\pi^*(S)(\omega) := S(\pi_* \omega).$$

In case $X = \mathbb{R}^n \times \{0\} \subset \mathbb{R}^{n+N}$ and $\pi : \mathbb{R}^{n+N} \to \mathbb{R}^n \times \{0\}$ is the orthogonal projection on the first factor, $\pi(x,y) = x$, then

$$\pi^* S = S \times \mathbb{R}^N \quad \text{for } S \in \mathcal{D}_k(\mathbb{R}^n \times \{0\}).$$

Proposition 4. *We have*

(i) $\partial\pi^* S = \pi^*\partial S$ *in* U,
(ii) $\pi^*([\![X]\!] \llcorner \omega) = [\![U]\!] \llcorner \pi^\# \omega$.
(iii) *If* $\mathbf{M}(S) < +\infty$, *then* $\mathbf{M}(\pi^* S) = vol(B_N(0,1))\varepsilon_0^N \mathbf{M}(S)$.
(iv) *If* $\operatorname{spt} S \subset X$, *and* S *is i.m. rectifiable, then* $\pi^* S$ *is i.m. rectifiable in* U.

Proof. (i) and (ii) follows from (i) and (iii) of Proposition 3.
(iii) For $\zeta \in \Lambda^k T_x X$,

$$|<\pi_*\omega(x), \zeta>| = \Big| \int_{\pi^{-1}(x)} <\omega(x,y), \zeta\wedge\mathcal{N}(x)> d\mathcal{H}^N(y)\Big|.$$

Taking the sup, we get $\| \pi_*\omega(x) \| = \| \omega(x) \| \mathrm{vol}(B_N(0,1))\varepsilon_0^N$ hence the conclusion.
(iv) We claim that if $S = \tau(\mathcal{M}, \theta, \vec{S})$, $\mathcal{M} \subset X$ then $\pi^* S = \tau(\pi^{-1}(\mathcal{M}), \theta \circ \pi, \vec{\pi}^{-1}(\mathcal{M}))$. In fact if $\omega \in \mathcal{D}^{k+N}(U)$ we have

$$\begin{aligned}
\pi^*(S)(\omega) &= \tau(\mathcal{M}, \theta, \vec{s})(\pi_*\omega) = \\
&= \int_{\mathcal{M}} <\pi_*\omega(x), \vec{S}(x)>\theta(x)\, d\mathcal{H}^k(x) = \\
&= \int_{\mathcal{M}} \int_{\pi^{-1}(x)} <\omega(x,y), \vec{S}(x)\wedge\mathcal{N}(x)>\theta(x)\, d\mathcal{H}^N(y)\, d\mathcal{H}^k(x) = \\
&= \int_{\pi^{-1}(\mathcal{M})} <\omega(x,y), \vec{S}(x)\wedge\mathcal{N}(x)>\theta(x)\, d\mathcal{H}^{N+k}(x,y)
\end{aligned}$$

and the claim follows as $\vec{S}\wedge\mathcal{N}(x)$ is a unit vector orienting $\pi^{-1}(\mathcal{M})$ at $(x,y) \in \pi^{-1}(\mathcal{M})$. □

Poincaré duality isomorphism on compact manifolds without boundary. Let us discuss now Poincaré duality between $(n-k)$-cohomology and k-homology groups of oriented, compact and boundaryless submanifolds of dimension n in $\mathbb{R}^{n+N}$ and more precisely prove the following

Theorem 2. *Let* X *be a smooth compact oriented and boundaryless smooth submanifold of dimension* n *in* $\mathbb{R}^{n+N}$. *Then the map*

$$P : \omega \in \mathcal{D}^{n-k}(X) \;\to\; P(\omega) := \int_X \cdot \wedge \omega \in \mathcal{D}_k(X)$$

induces an isomorphism between the $(n-k)$ *de Rham cohomology group of* X *and the* k*-homology group of currents* $H_k(X, \mathbb{R}) := Z_k(X)/B_k(X)$ *where*

$$Z_k(X) = \{T \in \mathbf{N}_k(X) \mid \partial T = 0\}, \quad B_k(X) = \{\partial S \mid S \in \mathbf{N}_{k+1}(X)\}.$$

More precisely

(i) *for any cycle* $T \in Z_k(X)$ *one can find a closed* $(n-k)$*–form* $P_T \in Z^{n-k}(X)$ *such that*

$$T = \int_X \cdot \wedge P_T + \partial S \tag{13}$$

for some $S \in \mathbf{N}_{k+1}(X)$,

(ii) *if for some* $\omega \in Z^{n-k}(X)$ *we have* $\int_X \cdot\wedge\omega = \partial S$, $S \in \mathbf{N}_{k+1}(X)$, *then* ω *is exact in* X.

We refer to P as to the *Poincaré duality isomorphism in* X and to any form P_T satisfying (13) as a *Poincaré dual form* for T.

Proof. Let $T \in Z_k(X)$. Then, if ω_T is a Poincaré dual of T in U, we have

$$T(\alpha) = \int_U \alpha\wedge\omega_T + \partial\Sigma, \quad \Sigma \in N_{k+1}(U),\ \operatorname{spt}\Sigma \subset\subset U.$$

Consequently, taking into account (iii) of Proposition 3, we get

$$\begin{aligned} T(\alpha) &= \pi_{\#}(T)(\alpha) = \int_U \pi^{\#}(\alpha)\wedge\omega_T + \partial\pi_{\#}\Sigma(\alpha) \\ &= \int_X \alpha\wedge\pi_*\omega_T + \partial\pi_{\#}\Sigma(\alpha), \end{aligned}$$

which proves (i).

(ii) From $[\![X]\!] \llcorner \omega = (-1)^{k(n-k)} \int_X \cdot\wedge\omega = \partial S$, $S \in \mathbf{N}_{k+1}(U)$, $\operatorname{spt} S \subset X$ we get

$$[\![U]\!] \llcorner \pi^{\#}\omega = \pi^*([\![X]\!] \llcorner \omega) = \partial\pi^* S \quad \text{in } U$$

by Proposition 4. We then infer from Proposition 1 that $\pi^{\#}\omega = d\beta$, $\beta \in \mathcal{E}^{n-k-1}(U) \cap L^1(U, \Lambda^{n-k-1}\mathbb{R}^{n+N})$ which in turns implies that ω is exact on X. □

Remark 3. It is worthwhile noticing that yet another proof of the non degeneracy of the Poincaré duality in cohomology on a compact manifold X can be recovered from Proposition 2 using the integration along the fibers. In fact, assume that

$$\int_X \omega\wedge\alpha = 0 \qquad \forall\alpha \in Z^{n-k}(X).$$

Then for $\beta \in Z^{n+N-k}(U)$ we have

$$\int_U \pi^{\#}\omega\wedge\beta = \int_X \omega\wedge\pi_*\beta = 0$$

Then $\pi^{\#}\omega$ is exact in U by Proposition 2 and consequently ω is exact in X.

Remark 4. Classically a Poincaré dual of T is a closed $(n-k)$-form in $Z^{n-k}(X)$ such that

$$T = \int_X \cdot \wedge \alpha \qquad \text{on } Z^k(X).$$

If P_T is a Poincaré dual of T we also have

$$T = \int_X \cdot \wedge P_T + \partial S, \qquad S \in \mathbf{N}_{k+1}(X).$$

therefore

$$\int_X \eta \wedge (\alpha - P_T) = 0 \qquad \forall \eta \in Z^k(X).$$

Consequently $\alpha - P_T = d\beta$, and

$$T = \int_X \cdot \wedge \alpha + \partial S - \int_X \cdot \wedge d\beta = \int_X \cdot \wedge \alpha + \partial S_1,$$

which proves that the classical notion of Poincaré dual and the one we have used in this section agree.

de Rham theorem. As we have seen in Sec. 5.2.8, Hodge decomposition theorem yields an isomorphism between $H^k_{\mathrm{dR}}(X)$ and $\mathrm{Hom}\,(H^{n-k}_{\mathrm{dR}}(X), \mathbb{R})$,

$$H^k_{\mathrm{dR}}(X) \;\to\; \mathrm{Hom}\,(H^{n-k}_{\mathrm{dR}}(X), \mathbb{R}).$$

On the other hand Poincaré duality isomorphism gives also an isomorphism between $H^{n-k}_{\mathrm{dR}}(X)$ and the k-homology group $H_k(X, \mathbb{R})$ of X

$$P : H^{n-k}_{\mathrm{dR}}(X) \;\to\; H_k(X, \mathbb{R}).$$

Therefore we can state

Theorem 3 (de Rham). *$H^k_{\mathrm{dR}}(X)$ and $\mathrm{Hom}\,(H_k(X, \mathbb{R}), \mathbb{R})$ are isomorphic.*

Actually we shall now show that the induced non degenerate pairing between homology and cohomology $H^k_{\mathrm{dR}}(X) \times H_k(X, \mathbb{R}) \to \mathbb{R}$ is the integration of closed forms over cycles. In fact the bilinear map

$$<T, \omega> : Z_k(X) \times Z^k(X) \;\to\; \mathbb{R}, \quad <T, \omega> := T(\omega)$$

clearly depends only on the homology class of T and the cohomology class of ω, therefore it induces a bilinear map, *de Rham's duality*

$$<[T], [\omega]> : H_k(X, \mathbb{R}) \times H^k_{\mathrm{dR}}(X) \;\to\; \mathbb{R}$$

and we have

$$<[T],[\omega]> = T(\omega) = \int_X \omega \wedge P_T = \text{Poinc}<[\omega],[P_T]>. \tag{14}$$

where P_T is a Poincaré dual form of T. In particular (14) clearly shows that *de Rham's duality is non degenerate* i. e.

(i) *If $\omega \in Z^k(X)$ and $T(\omega) = 0$ for all $T \in Z_k(X)$, then ω is exact.*
(ii) *If $T \in Z_k(X)$ and $T(\omega) = 0$ for all $\omega \in Z^k(X)$, then $T = \partial S$ for $S \in$* $\mathbf{N}_{k+1}(U)$, $\operatorname{spt} S \subset X$.

De Rham's theorem can be equivalently stated in terms of *periods*. Let $\gamma_1, \ldots, \gamma_s \in Z_k(X)$ be cycles such that their homology classes $[\gamma_1], \ldots, [\gamma_s]$ form a basis of $H_k(X, \mathbb{R})$. If $\omega \in Z^k(X)$, the numbers

$$\gamma_i(\omega)$$

are said to be the *periods of ω with respect to the basis $[\gamma_i]$*. Theorem 3 can be equivalently stated as

Theorem 4 (de Rham). *Let $\{\gamma_i\}$, $i = 1, \ldots, s$ be a basis of k-cycles of X. Then*

(i) *If ω is closed and all its periods of ω are zero, then ω is exact.*
(ii) *Given $\alpha_1, \ldots, \alpha_s \in \mathbb{R}$. Then there exists a closed form $\omega \in Z^k(X)$ with given periods $\alpha_1, \ldots, \alpha_s$, i.e., $\gamma_i(\omega) = \alpha_i$, $i = 1, \ldots, s$.*

In particular homology classes are identified by the *periods* of a basis of harmonic forms. This yields in particular the closure of real homology classes.

Proposition 5. *$B_k(X, \mathbb{R})$ is weakly closed.*

The following corollary is then an immediate consequence of Proposition 5 and of the lower semicontinuity of the mass

Corollary 1. *In each class $\gamma \in H_k(X, \mathbb{R})$ there is a mass minimizing current, that is $T \in Z_k(X, \mathbb{R})$, $[T] = \gamma$ such that*

$$\mathbf{M}(T) = \inf\{\mathbf{M}(S) \mid [S] = \gamma\}.$$

In particular

$$\mathbf{M}(\gamma) := \inf\{\mathbf{M}(T) \mid [T] = \gamma\}$$

defines a norm on $H_k(X, \mathbb{R})$, called the sphere norm, *consequently*

$$\mathbf{M}(\gamma) > 0 \quad \textit{for all } \gamma \in H_k(X, \mathbb{R}),\ \gamma \neq 0$$

and the coset map $Z_k(X, \mathbb{R}) \to H_k(X, \mathbb{R})$ is continuous.

Normal currents and singular real homology. Another path to get the closure of homology classes which also yields a representation of homology classes by Lipschitz chains can be followed by using the *deformation theorem* of Sec. 5.1.1.

Let X be a compact smooth submanifold without boundary of dimension n in $\mathbb{R}^{n+N}$. As it is well known *X is a neighborhood C^1 retract*, i.e. there is an open set $U \supset X$ and a C^1 map $\pi : U \to X$ which retracts U onto X and actually one can think of U as a tubular neibourhood,

$$\{x \in \mathbb{R}^{n+N} \mid \operatorname{dist}(x, X) < \varepsilon_0\} \tag{15}$$

for some positive ε_0 and of π as *homotopic to the identity* in U.

Fix now $\varepsilon < \varepsilon_0/\sqrt{n}$ and denote by $\mathcal{L}_{n,X}$ the collection of all standard cubes of size ε which meet X. Of course if $Q \in \mathcal{L}_{n,X}$, then $\overline{Q} \subset U$ because of (15), and one can modify π by composing π with a projection onto $\mathcal{L}_{n,X}$ to another retraction map again denoted by π, $\pi : \mathcal{L}_{n,X} \to X$ which is homotopic to the identity in $\mathcal{L}_{n,X}$.

Given an integer k, $0 \le k \le n$, we denote by $\mathcal{L}_{k,X}$ the collection of all k-faces of cubes in $\mathcal{L}_{n,X}$, and by $L_{k,X}$ the k-skeleton of $\mathcal{L}_{n,X}$

$$L_{k,X} := \cup\{F \mid F \in \mathcal{L}_{n,X}\}.$$

If $F \in \mathcal{L}_{k,X}$ we orient F in the natural way, i.e., if F is parallel to the plane associated to $e_{\alpha_1} \wedge \ldots \wedge e_{\alpha_k}$, $|\alpha| = k$, then $\vec{F} := e_{\alpha_1} \wedge \ldots \wedge e_{\alpha_k}$. Finally we denote by L_X the CW-complex of cubes in $L_{n,X}$.

Simplicial homology. Given an integer k, $0 \le k \le n$, the simplicial homology of L_X is defined as follows. A *chain* on $L_{k,X}$ is simply a polyhedral chain

$$P := \sum \beta_i [\![F_i]\!], \quad \beta_i \in \mathbb{R}, \ F_i \in \mathcal{L}_{k,X}.$$

Then we set

$$\begin{aligned} Z_{k,\mathrm{simpl}}(L_X, \mathbb{R}) &:= \{P \mid P \text{ is a polyhedral chain in } L_{k,X},\ \partial P = 0\} \\ B_{k,\mathrm{simpl}}(L_X, \mathbb{R}) &:= \{P = \partial Q \mid Q \text{ is a polyhedral chain on } L_{k+1,X}\} \end{aligned}$$

and the *k-simplicial homology of L_X* is defined by

$$H_{k,\mathrm{simpl}}(L_X, \mathbb{R}) := Z_{k,\mathrm{simpl}}(L_X)/B_{k,\mathrm{simpl}}(L_X).$$

Clearly *$H_{k,\mathrm{simpl}}(L_X, \mathbb{R})$ is a vector space over $\mathbb{R}$ of finite dimension*, since the collection $\mathcal{L}_{k,X}$ is finite.

Homology groups of Lipschitz chains. Similarly to the simplicial homology or to the real homology, by considering Lipschitz chains in X, i.e., Lipschitz images of polyhedral chains, we can define the homology groups $H_{k,\mathrm{Lip}}(X, \mathbb{R})$.

We have

Theorem 5. *The spaces $H_k(X, \mathbb{R})$, $H_{k,\mathrm{Lip}}(X, \mathbb{R})$ and $H_{k,\mathrm{simpl}}(L_X, \mathbb{R})$ are isomorphic. In particular $H_k(X, \mathbb{R})$ has finite dimension.*

Proof. Let $\pi : U \to X$ be the retraction map. If P is a cycle in the k-skeleton L_X then $\pi_\#(P)$ is a Lipschitz cycle on X, and, as any Lipschitz cycle is also a normal cycle, we have the following maps

$$H_{k,\mathrm{simpl}}(L_X, \mathbb{R}) \xrightarrow{\pi_*} H_{k,\mathrm{Lip}}(X, \mathbb{R}) \xrightarrow{i_*} H_k(X, \mathbb{R}).$$

Let $T \in Z_k(X, \mathbb{R})$. By the deformation theorem we can write $T = P + \partial R$ where P is a k-cycle on L_X and $R \in \mathbf{N}_{k+1}(U)$. We then have

$$T = \pi_\# T = \pi_\# P + \partial \pi_\# R,$$

i.e. T and $i_* \circ \pi_* P$ are homologous, consequently $i_* \circ \pi_*$ is onto. Suppose now that L is a Lipschitz k-cycle in X which is a boundary regarded as a current in U, $L = \partial R$, $R \in \mathbf{N}_{k+1}(U)$, $\operatorname{spt} R \subset X$. Deforming R into L_X we get

$$R = Q + \partial R_1 + S,$$

with $Q \in L_{k+1,X}$ and S is a Lipschitz chain in U. Therefore $L = \partial R = \partial \pi_\#(Q + S)$ is a Lipschitz boundary which clearly proves injectivity of i_*. Finally let us prove that π_* is injective. Let $L \in L_{k,X}$ such that $\pi_\# L$ is a boundary in X. Then being π homotopic to the identity in $\mathcal{L}_{n,X}$, we have

$$L = \pi_\# L + \partial S, \qquad S \in \mathbf{N}_{k+1}(U),\ \operatorname{spt} S \subset \mathcal{L}_{n,X}$$

hence L is a boundary in $\mathcal{L}_{n,X}$, $L = \partial R$, $R \in \mathbf{N}_{k+1}(U)$, $\operatorname{spt} R \subset \mathcal{L}_{n,X}$. Deforming R into $L_{k+1,X}$ we get

$$R = Q + \partial R_1 + S,$$

with $Q \in L_{k+1,X}$ and $S = 0$, being $\partial R = L \in L_{k,X}$. Therefore $L = \partial R = \partial \pi_\#(Q)$ is a simplicial boundary, which clearly proves injectivity of π_*. □

Remark 5. If A is a local Lipschitz retract, one can do the same constructions and define simplicial homology of the skeleton $L_{n,X}$, and Lipschitz homology. However it is easy to find examples in which no retraction homotopic to the identity exists. The previous proof applies however partially and we get

Theorem 6. *The spaces $H_k(A, \mathbb{R})$, $H_{k,\mathrm{Lip}}(A, \mathbb{R})$ are isomorphic. Moreover the retraction map π yields a surjective linear map $\pi_* : H_{k,\mathrm{simpl}}(L_A, \mathbb{R}) \to H_k(A, \mathbb{R})$. Again $H_k(A, \mathbb{R})$ has finite dimension.*

As we have seen, compare Proposition 5, one of the advantages in dealing with homology groups defined in terms of normal currents is that the cosets are closed with respect to the weak convergence of currents, and this allows of course the use of variational methods. Another proof of Proposition 5 can be done using the following *rough isoperimetric inequality*, which works also for compact neighbourhood Lipschitz retracts.

Proposition 6. *Let A be a compact neighborhood Lipschitz retract. Every current $T \in Z_k(A, \mathbb{R})$ which is the boundary of a normal $(k+1)$-current in U, $T = \partial S_1$ with $S_1 \in \mathbf{N}_{k+1}(U)$, is in fact a boundary in A, i.e., there exists $S \in \mathbf{N}_{k+1}(U)$ with $\operatorname{spt} S \subset A$ and $T = \partial S$, and moreover*

$$\mathbf{M}(S) \leq \tau(k,n)\mathbf{M}(T).$$

Proof. Since simplicial boundaries on L_A form a finite dimensional space, we can easily construct a linear map

$$\Gamma : B_{k,\text{simpl}}(L_A, \mathbb{R}) \longrightarrow \mathbf{N}_k(U)$$

which is continuous with respect to the mass norm. For that, choose a basis $P_1, \dots, P_N \in L_A$ and for each element P_i select a simplicial $(k+1)$-chain $\Gamma(P_i)$ in $L_{k+1,A}$ so that $\partial\Gamma(P_i) = P_i$, then extend Γ by linearity. For any P in $B_{k,\text{simpl}}(L_A, \mathbb{R})$ we of course have

$$\partial\Gamma(P) = P, \qquad \mathbf{M}(\Gamma(P)) \leq \tau_1(k,n)\mathbf{M}(P).$$

Applying the deformation theorem to $T = \partial S_1$ we find $R \in \mathbf{N}_{k+1}(U)$, $\operatorname{spt} R \subset\subset U$, such that

$$T = P + \partial R = \partial(\Gamma(P) + R)$$

and

$$\begin{aligned} \mathbf{M}(\Gamma(P) + R) &\leq \mathbf{M}(\Gamma(P)) + \mathbf{M}(R) \leq \tau_1\mathbf{M}(P) + c(k,n)\varepsilon M(T) \\ &\leq c(k,n)(\varepsilon + \tau_1)\mathbf{M}(T). \end{aligned}$$

It is now readily seen that $S := \pi_\#(\Gamma(P) + R)$ has the desired properties. □

Remark 6. Notice that in fact we have proved the existence of a current $\overline{S}$ in $\mathbf{N}_{k+1}(U)$, $\operatorname{spt} S \subset\subset U$ such that

$$T = \partial\overline{S}, \qquad \mathbf{M}(\overline{S}) \leq \tau(k,n)\mathbf{M}(T)$$

independently from the fact that A is a Lipschitz retract. We have then used the last property to retract $\overline{S}$ to A and find $S := \pi_\#\overline{S}$ with

$$T = \partial S, \qquad \mathbf{M}(S) \leq \tau(k,n)\mathbf{M}(T), \qquad \operatorname{spt} S \subset A.$$

We can now state

Theorem 7. *$B_k(A, \mathbb{R})$ is weakly closed.*

Proof. Let $T_i \in B_k(A, \mathbb{R})$, $T_i \rightharpoonup T$. Of course $T_i = \partial R_i$, $\operatorname{spt} R_i \subset A$. Proposition 6 then yields $S_i \in \mathbf{N}_{k+1}(U)$, $\operatorname{spt} S_i \subset A$ such that $T_i = \partial S_i$ and

$$\mathbf{M}(S_i) \leq \tau(k,n)\mathbf{M}(T_i) \leq \tau(k,n)\sup_i \mathbf{M}(T_i) < \infty.$$

Passing to a subsequence we get $S_i \rightharpoonup S$, $\operatorname{spt} S \subset A$, $T = \partial S$. □

The following corollary is an immediate consequence of Theorem 7 and of the lower semicontinuity of the mass

Corollary 2. *In each class $\gamma \in H_k(A,\mathbb{R})$ there is a mass minimizing current, that is $T \in Z_k(A,\mathbb{R})$, $[T]=\gamma$ such that*

$$\mathbf{M}(T) = \inf\{\mathbf{M}(S) \mid [S]=\gamma\}.$$

In particular

$$\mathbf{M}(\gamma) := \inf\{\mathbf{M}(T) \mid [T]=\gamma\}$$

defines a norm on $H_k(A,\mathbb{R})$, called the sphere norm, *consequently*

$$\mathbf{M}(\gamma) > 0 \quad \textit{for all } \gamma \in H_k(A,\mathbb{R}),\ \gamma \neq 0$$

and the coset map $Z_k(A,\mathbb{R}) \to H_k(A,\mathbb{R})$ is continuous.

3.3 Manifolds with Boundary: Poincaré-Lefschetz and de Rham Dualities

The Poincaré-Lefschetz duality isomorphism. Let X be a compact, oriented smooth n-dimensional submanifold of $\mathbb{R}^{n+N}$ with boundary ∂X. We define *relative cycles* in X as

$$Z_k(X,\partial X) := \{T \in \mathbf{F}_{k,X}(\mathbb{R}^{n+N}) \mid \operatorname{spt}\partial T \subset \partial X \text{ or } \partial T = 0 \text{ if } k=0\},$$

relative boundaries as

$$B_k(X,\partial X) := \{R + \partial S \mid R \in \mathbf{F}_{k,\partial X}(\mathbb{R}^{n+N})\ S \in \mathbf{F}_{k+1,X}(\mathbb{R}^{n+N})\},$$

and the *real relative homology groups* as

$$(1) \qquad H_k(X,\partial X,\mathbb{R}) := Z_k(X,\partial X)/B_k(X,\partial X), \qquad k=0,\dots,n.$$

Since X and ∂X are compact and Lipschitz neighbourhood retracts in $\mathbb{R}^{n+N}$, Federer's flatness theorem yields

$$\mathbf{F}_{k,\partial X}(\mathbb{R}^{n+N}) = \mathbf{F}_{k,X}(\mathbb{R}^{n+N}) \cap \{T \mid \operatorname{spt} T \subset \partial X\};$$

also observe that

$$\mathbf{F}_{k,X}(\mathbb{R}^{n+N}) = \mathbf{F}_{k,X}(U), \quad \mathbf{F}_{k,\partial X}(\mathbb{R}^{n+N}) = \mathbf{F}_{k,\partial X}(V),$$

U and V being neighbourhoods respectively of X and ∂X which retract onto X and ∂X.

A few words are in order to explain why we are working with flat chains while when dealing with the homology of boundaryless manifold we worked with normal currents. First we should remark that in the boundaryless case, though cosets defined in terms of flat chains are larger than cosets constructed with normal currents, *the homology groups defined in terms of flat chains or of normal currents are the same.* This is not difficult to be seen on account of the density of normal currents among flat chains and the closure properties of homology classes. However, while i.m. rectifiable cycles are obviously normal currents, relative cycles, which in general do have boundaries, are not in general normal as in principle we have no control on the masses of their boundaries: this is the reason why we are forced to work with the more general flat relative cycles. In the next section we shall illustrate relative homology in an even more general setting.

Paralleling the similar situation in cohomology, we therefore may define three homology groups depending on $k = 0, 1, \ldots$ on X.
The *homology group* $H_k(X, \mathbb{R})$

$$H_k(X, \mathbb{R}) := \frac{\text{flat cycles in } X}{\text{flat boundaries in } X},$$

if $T \in Z_k(X)$ is a flat cycle, we denote by $[T]_X$ its homology class in $H_k(X, \mathbb{R})$.
The *homology group* $H_k(\partial X, \mathbb{R})$

$$H_k(\partial X, \mathbb{R}) := \frac{\text{flat cycles supported on } \partial X}{\text{flat boundaries in } \partial X};$$

for $T \in Z_k(\partial X)$, we denote by $[T]_{\partial X}$ its homology class in $H_k(\partial X, \mathbb{R})$.
The *relative homology group* $H_k(X, \partial X, \mathbb{R})$

$$H_k(X, \partial X, \mathbb{R}) := \frac{\text{flat relative cycles}}{\text{flat relative boundaries}};$$

for $T \in Z_k(X, \partial X)$ we denote by $[T]_{\text{rel}}$ its relative class in $H_k(X, \partial X, \mathbb{R})$.

The inclusion map $i : \partial X \to X$ induces a map

$$i_\# : H_k(\partial X, \mathbb{R}) \to H_k(X, \mathbb{R}).$$

As cycles are special relative cycles and boundaries are special relative boundaries an inclusion map

$$j_\# : H_k(X, \mathbb{R}) \to H_k(X, \partial X, \mathbb{R})$$

is well defined by

$$j_\#([T]_X) = [T]_{\text{rel}}.$$

Finally, the boundary operator $\partial : \mathbf{F}_{k,X}(\mathbb{R}^{n+N}) \to \mathbf{F}_{k-1,X}(\mathbb{R}^{n+N})$ induces naturally a map

$$\partial : H_k(X, \partial X, \mathbb{R}) \to H_{k-1}(\partial X, \mathbb{R})$$

defined by

$$\partial([T]_{\text{rel}}) = [\partial T]_{\partial X}.$$

Then it is not difficult to prove

Theorem 1. *The long sequence in homology*

$$\cdots \underset{i_\#}{\rightarrow} H_k(X,\mathbb{R}) \underset{j_\#}{\rightarrow} H_k(X,\partial X,\mathbb{R}) \underset{\partial}{\rightarrow} H_{k-1}(\partial X,\mathbb{R}) \underset{i_\#}{\rightarrow} H_{k-1}(X,\mathbb{R}) \underset{j_\#}{\rightarrow} \cdots$$

is exact.

It turns out that the long sequence in homology and the long sequence in cohomology are in duality, such a duality being given by *Poincaré-Lefschetz duality*. In the sequel of this section we shall discuss this topic which will lead us to a *de Rham theorem for manifolds with boundary*.

For $k = 0, 1, \ldots, n$ we first consider the pairing

$$<,>: \mathcal{E}^{k-1}(\partial X) \times \mathcal{E}^{n-k}(X) \to \mathbb{R}$$

given by

(2) $$<\omega,\eta> := \int_{\partial X} \omega\wedge\eta.$$

As we saw in Sec. 5.3.2 (2) induces a map

$$P: Z^{n-k}(\partial X) \to Z_{k-1}(\partial X)$$

given by

$$P\omega := <\cdot,\omega> = \int_{\partial X} \cdot\wedge\omega$$

which factorizes to the *Poincaré duality isomorphism*

$$P: H^{n-k}_{\mathrm{dR}}(\partial X) \to H_{k-1}(\partial X,\mathbb{R})$$

compare Theorem 2 in Sec. 5.3.2.

Of course there are other pairings on X

$$<,>: \mathcal{E}^{k}(X) \times \mathcal{E}^{n-k}(X) \to \mathbb{R}$$

again given by

(3) $$<\omega,\eta> = \int_X \omega\wedge\eta.$$

But, as by Stokes theorem

(4) $$\int_X d\sigma\wedge\eta + (-1)^k \int_X \sigma\wedge d\eta = \int_{\partial X} \sigma\wedge\eta \quad \sigma\in\mathcal{E}^{k-1}(X),\ \eta\in\mathcal{E}^{n-k}(X),$$

we see that there is no way of inducing a map between the absolute cohomology and homology of X unless $\partial X = 0$. However, if $\eta \in Z^{n-k}(X,\partial X)$ the k-dimensional current

$$P^\sharp\eta(\omega) := \int_X \omega\wedge\eta = (-1)^{k(n-k)}[\![X]\!] \llcorner \eta(\omega), \quad \omega \in \mathcal{E}^k(X) \tag{5}$$

is a cycle in X, $P^\sharp\eta \in Z_k(X)$ by (4). Also the homology class $[P^\sharp\eta]_X$ of $P^\sharp\eta$ depends only on $[\eta]_{\text{rel}}$. In fact, if $\eta = d\alpha$ with $i^\# \alpha = 0$, then by (4)

$$P^\sharp\eta(\omega) = \int_X \omega\wedge d\alpha = (-1)^{k+1} \int_X d\omega\wedge\alpha = (-1)^{k+1}\partial P^\sharp\alpha(\omega).$$

In conclusion we can say that the map $P^\sharp : Z^{n-k}(X, \partial X) \to Z_k(X)$ induces a map again denoted by $P^\sharp$

$$P^\sharp : H^{n-k}_{\text{dR}}(X, \partial X) \to H_k(X, \mathbb{R}). \tag{6}$$

Finally, in a similar way, one proves that for $\eta \in Z^{n-k}(X)$ the current

$$P^\flat := (-1)^{k(n-k)}[\![X]\!] \llcorner \eta \tag{7}$$

is a relative cycle, $P^\flat\eta \in Z_k(X, \partial X)$, and moreover the map $P^\flat : Z^{n-k}(X) \to Z_k(X, \partial X)$ induces a map

$$P^\flat : H^k_{\text{dR}}(X) \to H_k(X, \partial X, \mathbb{R}). \tag{8}$$

We have

Theorem 2 (Poincaré-Lefschetz isomorphisms). *The maps P, $P^\sharp$, $P^\flat$ are isomorphisms, and the diagram*

$$\begin{array}{ccccccccc}
\cdots \xrightarrow{\delta} & H^{n-k}_{\text{dR}}(X, \partial X) & \xrightarrow{j^\#} & H^{n-k}_{\text{dR}}(X) & \xrightarrow{i^\#} & H^{n-k}_{\text{dR}}(\partial X) & \xrightarrow{\delta} \cdots \\
& \downarrow P^\sharp & & \downarrow P^\flat & & \downarrow P & \\
\cdots \xrightarrow[i_\#]{} & H_k(X, \mathbb{R}) & \xrightarrow[j_\#]{} & H_k(X, \partial X, \mathbb{R}) & \xrightarrow[\partial]{} & H_{k-1}(\partial X, \mathbb{R}) & \xrightarrow[i_\#]{} \cdots
\end{array} \tag{9}$$

is commutative.

We shall refer to the maps $P^\sharp$, $P^\flat$, P as to the *Poincaré Lefschetz duality isomorphisms.* It is quite simple to prove the commutativity of the diagram, though quite long. We already known, compare Sec. 5.3.2 that P is an isomorphism, Poincaré duality isomorphism. We shall therefore prove only that $P^\sharp$ and $P^\flat$ are isomorphisms. This will be done in the next propositions below.

Let X be a compact, oriented, smooth n–dimensional manifold with smooth boundary ∂X. To be precise we think of X as a C^2 manifold with a C^2 boundary. Observing that ∂X has codimension 1 in X, and X is oriented, the tubular neighbourhood theorem on ∂X yields

Theorem 3 (Collar theorem). *There is an $\varepsilon_0 > 0$ such that the normal geodesic flow to $T_x\partial X$ at $x \in \partial X$, $g(t, x) : [0, \varepsilon_0] \times \partial X \to X$ is a diffeomorphism onto a neighbourhood of ∂X in x with $g(0, x) = x$ for all $x \in \partial X$.*

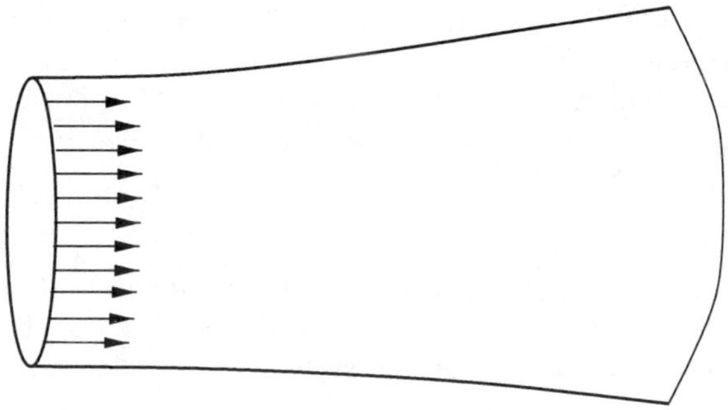

Fig. 5.4. The collar map.

A consequence of the collar theorem is then

Theorem 4. *We have*

$$H^k_{\rm dR}(X, \partial X) = H^k_{\rm cpt}(X \setminus \partial X)$$

where

$$H^k_{\rm cpt}(X \setminus \partial X) := \frac{\{\omega \in C^\infty_c(X \setminus \partial X, \Lambda^k T(X \setminus \partial X)) \mid d\omega = 0\}}{\{\omega = d\alpha \mid \alpha \in C^\infty_c(X \setminus \partial X, \Lambda^k T(X\sigma, \partial X))\}}.$$

In fact

Proposition 1. *We have*

(i) *Each relative cohomology class contains a form $\xi \in Z^k(X, \partial X)$ which is compactly supported in $X \setminus \partial X$.*

(ii) *Each homology class of X contains a cycle $T \in Z_k(X)$ of finite mass such that* $\operatorname{spt} T \cap \partial X = \emptyset$.

Proof. Let ε_0 and g be as in the collar theorem, Theorem 3. Define $r(t)$: $[-\varepsilon_0, 0] \to [-\varepsilon_0, 0]$ to be a non-decreasing function such that $r(-\varepsilon_0) = -\varepsilon_0$, $r(-\varepsilon_0/2) = r(0) = 0$, $r'(-\varepsilon_0) = 1$. Of course $(1-s)t + sr(t)$ yields a homotopy with the identity which is fixed at $-\varepsilon_0$ and 0. Consider then the mapping

$$\phi(x) := \begin{cases} x & \text{if } x \in X_{-\varepsilon_0} \\ g(r(t), y) & \text{if } g(t, y) = x \in X \setminus X_{-\varepsilon_0} \end{cases}$$

which maps $X \setminus X_{-\varepsilon_0}$ onto ∂X and satisfies $\phi \circ i = i$ on ∂X. Then $h(s, x) := (1-s)x + s\phi(x)$ yields a homotopy of ϕ with the identity which keeps the boundary fixed. Consequently, if $\eta \in Z^k(X, \partial X)$, the form $\omega = \phi^\# \eta$ vanish on $X \setminus X_{-\varepsilon_0/2}$, and by the homotopy formula

$$\omega - \eta = d[h^\# \eta]_{(0,1)}.$$

This concludes the proof of the first claim as $i^{\#}[h^{\#}\eta]_{(0,1)} = 0$ being $h(s,x) = x$ $\forall\, x \in \partial X$ and $\forall\, s \in [0,1]$. Analogously, if $T \in Z_k(X)$, we infer from (i) of Proposition 4 in Sec. 5.1.3 that T is homologous to a relative cycle $T' \in Z_k(X)$ of finite mass and consequently $r_{\#}T' \in Z_k(X)$ has finite mass. Moreover $r_{\#}T'$ is homologous to T' and $\operatorname{spt} r_{\#}T' \cap \partial X = \emptyset$. □

Actually without loss of generality we can and do assume that, in case $\partial X \neq \emptyset$, X is actually a compact piece of a larger manifold Y and the geodesic flow on X normal to ∂X extends to both sides of ∂X in Y. Denoting such a flow by $g(t,X)$, the collar theorem Theorem 3 then yields, possibly for a smaller ε_0, a diffeomorphism

$$g(t,x) : [-\varepsilon_0, \varepsilon_0] \times \partial X \to Y$$

onto a neighbourhood of ∂X in Y such that $g(0,x) = x$ for $x \in \partial X$ and $g(t,x) \in X$ iff $t \geq 0$. The equation $g(t,x) = y$ means that x is the point of least geodesic distance in ∂X from y, $|t|$ is the geodesic distance of y from ∂X, and $t \geq 0$ iff $y \in X$.

For $0 < \varepsilon \leq \varepsilon_0$ we now set

$$\begin{aligned} X_{-\varepsilon} &:= \{x \in X \mid x \notin g([0,\varepsilon] \times \partial X\} \\ X_{+\varepsilon} &:= g([-\varepsilon, \varepsilon_0] \times \partial X) \cup X \end{aligned} \tag{10}$$

and $X_0 = X$. Of course any X_ε, $\varepsilon \in [-\varepsilon_0, \varepsilon_0]$, is a compact oriented manifold with boundary. Given any smooth increasing function $r(t) : [-\varepsilon_0, \varepsilon_0] \to [-\varepsilon_0, \varepsilon_0]$ with $r(-\varepsilon_0) = -\varepsilon_0$, $r(\varepsilon_0) = \varepsilon_0$, $r'(-\varepsilon_0) = 1$, $r(-\varepsilon_0/2) = 0$, and $r(0) = \varepsilon_0/2$, it is convenient to introduce the map

$$e : X_{+\varepsilon_0} \to X_{+\varepsilon_0}$$

defined by

$$e(x) := \begin{cases} x & \text{if } x \in X_{-\varepsilon_0} \\ g(r(t), y) & \text{if } g(t,y) = x \in X_{+\varepsilon_0} \setminus X_{-\varepsilon_0}. \end{cases}$$

It is easily seen that e yields an isomorphism which is the identity on $X_{-\varepsilon_0}$, maps $X \setminus X_{-\varepsilon_0/2}$ onto $X_{+\varepsilon_0/2} \setminus X$, $\partial X_{-\varepsilon_0/2}$ onto ∂X and ∂X onto $\partial X_{+\varepsilon_0/2}$. We can also assume that e is homotopic to the identity with a homotopy which keeps the boundary values fixed. The inverse map of e denoted by r

$$r : X_{+\varepsilon_0} \to X_{+\varepsilon_0}$$

which maps the collar $X_{+\varepsilon_0/2} \setminus X$ onto $X \setminus X_{-\varepsilon_0/2}$ and $\partial X_{+\varepsilon_0/2}$ onto ∂X, is then also homotopic to the identity with a homotopy which leaves $\partial X_{\varepsilon_0}$ fixed.

Finally, we shall also use the distance function from X. Setting

$$U_\varepsilon := \{x \in \mathcal{R}^{n+N} \mid \operatorname{dist}(x, X) < \varepsilon\}$$

we have

Proposition 2. *There exists $\varepsilon_0 > 0$ such that any point $x \in U_{\varepsilon_0}$ has a unique point $\pi(x)$ of least distance on X. Moreover the map $\pi : U_{\varepsilon_0} \to X$ is Lipschitz in U_ε and smooth in $U_\varepsilon \setminus X$. Finally, for $x \in X \setminus \partial X$, $\pi^{-1}(x)$ is just a flat ball of center x and radius ε_0 lying in the normal space to $T_x X$.*

For every ε in $(-\varepsilon_0, \varepsilon_0)$ we consider the enlarged manifold X_ε and

$$U_\varepsilon := \{x \in \mathbb{R}^{n+N} \mid \operatorname{dist}(x, X_{\varepsilon_0/2}) < \varepsilon\}.$$

Choosing ε_1 so that $\varepsilon_0 = 4\varepsilon_1$ and taking ε_1 smaller if necessary a least distance projection map

$$\pi : U_{\varepsilon_1} \to X_{2\varepsilon_1}$$

is well defined: it maps each $x \in U_{\varepsilon_1}$ to the least distance point in $X_{2\varepsilon_1}$.

Proposition 3. *$P^\sharp$ is injective, i.e., if $\omega \in Z^{n-k}(X, \partial X)$ and*

$$[\![X]\!] \llcorner \omega = \partial S \tag{11}$$

$S \in \mathbf{F}_{k+1,X}(U_{\varepsilon_1})$, then ω is relatively exact, that is there exists $\alpha \in \mathcal{D}^{n-k-1}(X)$ with $i^\# \alpha = 0$ such that $\omega = d\alpha$.

Proof. On account of (i) of Proposition 1, we can and do assume that $\omega = 0$ on $X \setminus X_{-\varepsilon_1}$. Moreover on account of Proposition 4 in Sec. 5.1.3 we also assume that S has finite mass. Consequently, as in the boundaryless case, we infer

$$[\![\pi^{-1}(X)]\!] \llcorner \pi^\# \omega = \pi^*([\![X]\!] \llcorner \omega) = \partial \pi^* S \quad \text{in } U_{\varepsilon_1},$$

hence $\pi^\# \omega$ is exact, $\pi^\# \omega = d\beta$, $\operatorname{spt} \beta \subset \pi^{-1}(X_{-\varepsilon_1/2})$, compare Proposition 1 in Sec. 5.3.2 and Remark 1 in Sec. 5.3.2. In particular ω is exact on X and zero near ∂X. □

Proposition 4. *$P^\sharp$ is surjective, i.e., if $T \in Z_k(X)$ there exist a relative closed form $\omega \in Z^{n-k}(X, \partial X)$ and a current $S \in \mathbf{F}_{k+1,X}(U_{\varepsilon_1})$ such that*

$$T = [\![X]\!] \llcorner \omega + \partial S.$$

Proof. First notice that one can suppose that $\mathbf{M}(T) < \infty$ and $\operatorname{spt} T \subset X_{-\varepsilon_1}$ on account of Proposition 1. We then regularize T in U_{ε_1} with a parameter $\varepsilon \ll \varepsilon_1/2$. This way we have

$$T = \int_{U_{\varepsilon_1}} \cdot \wedge \xi + \partial S'$$

where $\xi \in \mathcal{D}^{n+N-k}(U_{\varepsilon_1})$, $\operatorname{spt} \xi \subset \pi^{-1}(X_{-\varepsilon_1/2})$, $S' \in \mathbf{F}_{k+1,\mathrm{cpt}}(U_{\varepsilon_1})$, $\operatorname{spt} S' \subset \pi^{-1}(X_{-\varepsilon_1/2})$, $\mathbf{M}(S') < \infty$. Projecting onto X we get

$$T = \pi_\# T = \pi_\#(\int_{U_{\varepsilon_1}} \cdot \wedge \xi) + \partial \pi_\# S' \tag{12}$$

with $\pi_{\#}S' \in \mathbf{F}_{k+1,X}(U_{\varepsilon_1})$; as for $\eta \in \mathcal{E}^k(X)$ we have $\int_X \pi^{\#}\eta \wedge \xi = \int_X \eta \wedge \pi_* \xi$, the conclusion follows since (12) yields also $d\xi = 0$, hence $d(\pi_*\xi) = \pi_* d\xi = 0$, and $\xi = 0$ near $\pi^{-1}(\partial X)$, hence $\pi_*\xi = 0$ near ∂X. □

Proposition 5. *$P^\flat$ is injective, i.e., if $\omega \in Z^{n-k}(X)$ and*

$$[\![X]\!] \llcorner \omega = R + \partial S \tag{13}$$

with $R \in \mathbf{F}_{k,\partial X}(U_{\varepsilon_1})$, $S \in \mathbf{F}_{k+1,X}(U_{\varepsilon_1})$, then ω is exact, that is there is $\alpha \in B^{n-k-1}(X)$ such that $\omega = d\alpha$.

Proof. On account of Proposition 4 in Sec. 5.1.3 we can and do assume that $\mathbf{M}(R) + \mathbf{M}(S) < \infty$. Applying the extension map e we get

$$[\![X_{\varepsilon_1}]\!] \llcorner r^{\#}\omega = R' + \partial S' \tag{14}$$

where $R' = e_{\#}R \in \mathbf{F}_{k,\partial X_{\varepsilon_0/2}}(U)$, $S' = e_{\#}(S) \in \mathbf{F}_{k,X_{\varepsilon_0/2}}(U)$, $\mathbf{M}(R') + \mathbf{M}(S') < \infty$. Since $r_{|X}$ is homotopic to the identity and $d\omega = 0$, the homotopy formula for forms yields that $\omega - r^{\#}\omega$ is exact, hence it suffices to prove that $r^{\#}\omega_{|X}$ is exact too. For that notice that (14) yields

$$[\![X_{\varepsilon_1/2}]\!] \llcorner r^{\#}\omega = \partial S' \quad \text{in } U_{\varepsilon_1} \cap \pi^{-1} X_{\varepsilon_0/2}.$$

We then proceed as in Proposition 3 and conclude that $\pi^{\#}r^{\#}\omega$ is exact in $U_{\varepsilon_1} \cap \pi^{-1}X_{\varepsilon_0/2}$. This implies that $r^{\#}\omega$ is exact on X. □

Proposition 6. *$P^\flat$ is surjective, i.e., if $T \in Z_k(X, \partial X)$ then there are $\omega \in Z^{n-k}(X)$, $R \in \mathbf{F}_{k,\partial X}(U_{\varepsilon_1})$, $S \in \mathbf{F}_{k,X}(U_{\varepsilon_1})$ such that*

$$T = [\![X]\!] \llcorner \omega + R + \partial S.$$

Proof. We may and do assume that $\mathbf{M}(T) < \infty$, compare Proposition 4 in Sec. 5.1.3. Regularizing T with a small parameter, we infer that

$$T = \int_{U_{\varepsilon_1}} \cdot \wedge \xi + \partial S \tag{15}$$

with $\xi \in \mathcal{D}^{n+N-k}(U_{\varepsilon_1})$, $S \in \mathbf{N}_{k+1}(U_{\varepsilon_1})$, $\operatorname{spt} S \subset U_{\varepsilon_1} \cap \pi^{-1}(X_{\varepsilon_0/2})$. Therefore, projecting on X_{ε_0}

$$\begin{aligned} T = \pi_{\#}T &= \int_{U_{\varepsilon_1}} \pi^{\#}(\cdot)\wedge\xi + \partial \pi_{\#} S \\ &= \int_{X_{\varepsilon_0/2}} \cdot \wedge \pi_*\xi + \partial\pi_{\#}S = \int_X \cdot \wedge \pi_*\xi + R' + \partial S' \end{aligned}$$

where $R' := \int_{X_{\varepsilon_0/2} \setminus X} \cdot \wedge \pi_*\xi$, $S' := \pi_{\#}S$. As X is a deformation retract of $X_{\varepsilon_0/2}$, denoting by $p : X_{\varepsilon_0/2} \to X$ the retraction, we infer

$$T = p_\# T = \int_X \cdot \wedge \pi_* \xi + p_\# R' + \partial p_\# S'$$

with $R' \in \mathbf{N}_{k+1,\partial X}(U_{\varepsilon_1})$, $p_\# S' \in \mathbf{N}_{k+1,X}(U_{\varepsilon_1})$. The claim now follows observing that being $\partial T = 0$ in $X \setminus \partial X$, (15) implies that $d\xi = 0$ on $\pi^{-1}(X \setminus \partial X)$ which in turns implies $d\pi_*\xi = 0$ on $X \setminus \partial X$ and by continuity $d\pi_*\xi = 0$ on X. □

Remark 1. We emphasize the fact that surjectivity of $P^\sharp$ amounts to the existence for a given absolute cycle $T \in Z_k(X)$ of a *relative Poincaré dual form* $P_T^\sharp \in Z^{n-k}(X, \partial X)$ such that $P^\sharp[P_T^\sharp]_{\mathrm{rel}} = [T]$, i.e.

$$\begin{cases} T(\omega) = \int_X \omega \wedge P_T^\sharp & \forall \omega \in Z^k(X) \\ i^\# P_T^\sharp = 0 \end{cases}$$

It is reasonable to find for $T \in Z_k(X)$ a Poincaré dual with null value at ∂X since one can deform with an homotopy T into $T' \in Z_k(X)$ in such a way that $\operatorname{spt} T' \cap \partial X = \emptyset$. Also the surjectivity of $P^\flat$ means the existence for a given *relative cycle* $T \in Z_k(X, \partial X)$ of a *Poincaré dual form* $P_T^\flat \in Z^{n-k}(X)$ such that $P^\flat[P_T^\flat] = [T]_{\mathrm{rel}}$, i.e.

$$T(\omega) = \int_X \omega \wedge P_T^\flat + R + \partial S \qquad \forall \omega \in Z^k(X)$$

where $R \in \mathbf{F}_{k,\partial X}(U_{\varepsilon_1})$, $\operatorname{spt} R \subset \partial X$, $S \in \mathbf{F}_{k+1}(U_{\varepsilon_1})$.

Remark 2. The proof of Theorem 2 can follow a different path. First one proves that P and $P^\sharp$ are isomorphisms, as previously, then that the diagram (9) is commutative, and finally one proves the bijectivity of $P^\flat$ using the so–called *five-lemma*

Proposition 7 (Five Lemma). *Suppose we have the following commutative diagram*

$$\begin{array}{ccccccccc} A_1 & \overset{a_1}{\to} & A_2 & \overset{a_2}{\to} & A_3 & \overset{a_3}{\to} & A_4 & \overset{a_4}{\to} & A_5 \\ \downarrow f_1 & & \downarrow f_2 & & \downarrow f_3 & & \downarrow f_4 & & \downarrow f_5 \\ B_1 & \overset{b_1}{\to} & B_2 & \overset{b_2}{\to} & B_3 & \overset{b_3}{\to} & B_4 & \overset{b_4}{\to} & B_5 \end{array} \tag{16}$$

consisting of vector spaces A_i, B_i, $i = 1, \dots, 5$ and of linear maps a_i, b_i, $i = 1, \dots, 4$ and f_i, $i = 1, \dots, 5$. If the rows are exact, i.e. $\operatorname{Im} a_i = \operatorname{Ker} a_{i+1}$, $\operatorname{Im} b_i = \operatorname{Ker} b_{i+1}$ *then*

(i) if f_2, f_4 are surjective and f_5 is injective, then f_3 is surjective,
(ii) if f_2, f_4 are injective and f_1 is surjective, then f_3 is injective
(iii) if f_1, f_2, f_4, f_5 are isomorphisms, then f_3 is an isomorphism.

de Rham duality on manifolds with boundary. As consequence of the non-degeneracy of Poincaré-Lefschetz pairings, compare Theorem 3 in Sec. 5.2.8 and taking into account Poincaré–Lefschetz isomorphisms, we get

Theorem 5 (de Rham). *The pairings*

$$<[T]_X, [\varphi]_X>^\sharp : H_k(X, \mathbb{R}) \times H^k_{\mathrm{dR}}(X) \to \mathbb{R}$$
$$<[T]_{\mathrm{rel}}, [\varphi]_{\mathrm{rel}}>^\flat : H_k(X, \partial X, \mathbb{R}) \times H^k_{\mathrm{dR}}(X, \partial X) \to \mathbb{R}$$

induced by the pairings $<T, \varphi> : \mathcal{E}_k(X) \times \mathcal{E}^k(X) \to \mathbb{R}$, $<T, \varphi> := T(\varphi)$ *are non degenerate. In particular*

$$H^k(X) = H_k(X, \mathbb{R})^*, \quad H^k_{\mathrm{dR}}(X, \partial X) = H_k(X, \partial X, \mathbb{R})^*.$$

Moreover, if we also set

$$<[T]_{\partial X}, [\varphi]_{\partial X}> := T(\varphi) \text{ for } [T]_{\partial X} \in H_k(\partial X, \mathbb{R}) \text{ and } [\varphi]_{\partial X} \in H^k_{\mathrm{dR}}(\partial X),$$

the following diagram is commutative

$$\begin{array}{ccccccc} \cdots \overset{\delta}{\leftarrow} & H^k_{\mathrm{dR}}(\partial X) & \overset{i^\#}{\leftarrow} & H^k_{\mathrm{dR}}(X) & \overset{j^\#}{\leftarrow} & H^k_{\mathrm{dR}}(X, \partial X) & \overset{\delta}{\leftarrow} \cdots \\ & \times <,> & & \times <,>^\sharp & & \times <,>^\flat & \\ \cdots \overset{\partial}{\rightarrow} & H_k(\partial X, \mathbb{R}) & \overset{i_\#}{\rightarrow} & H_k(X, \mathbb{R}) & \overset{j_\#}{\rightarrow} & H^n_k(X, \partial X, \mathbb{R}) & \overset{\partial}{\rightarrow} \cdots \end{array}$$

Let $\gamma_1, \dots, \gamma_s$ be cycles in $Z_k(X)$ such that $[\gamma_1]_X, [\gamma_2]_X, \dots, [\gamma_s]_X$ form a basis of $H_k(X, \mathbb{R})$. Given a form $\omega \in Z^k(X)$, the *periods of* ω *with respect to* $\gamma_1 \dots, \gamma_s$ are the numbers

$$\gamma_i(\omega), \qquad i = 1, \dots, s.$$

Also if $\delta_1, \dots, \delta_r$ be relative cycles in $Z_k(X, \partial X)$ such that $[\delta_1]_{\mathrm{rel}}, \dots, [\delta_r]_{\mathrm{rel}}$ form a basis for $H_k(X, \partial X, \mathbb{R})$, and $\omega \in Z^k(X, \partial X)$, the *relative period of* ω *with respect to* $\delta_1, \dots, \delta_r$ are the numbers

$$\delta_i(\omega), \qquad i = 1, \dots, r.$$

Then it is easily seen that equivalently de Rham theorem can be stated as follows

Theorem 6. *Let* $[\gamma_i]$, $i = 1, \dots, s$ *be cycles in* X *such that* $[\gamma_i]_X$ *forms a basis in homology. Then*

(i) *If* $\omega \in Z^k(X)$ *and all the periods of* ω *with respect to* γ_i *are zero, then* ω *is exact.*

(ii) *Given numbers* $\alpha_i \in \mathbb{R}$, $i = 1, \dots, s$. *Then there is a form* $\omega \in Z^k(X)$ *with given periods* α_i, *i.e.* $\gamma_i(\omega) = \alpha_i$, $i = 1, \dots, s$.

Also, if $\{\delta_i\}$, $i = 1, \ldots, s$ *are relative cycles in* $Z_k(X, \partial X)$ *such that* $[\delta_i]_{\mathrm{rel}}$, $i = 1, \ldots, s$ *form a basis for* $H_k(X, \partial X, \mathbb{R})$, *then*

(iii) *If* $\omega \in Z^k(X, \partial X)$ *and all the periods of* ω *are zero,* $\delta_i(\omega) = 0$, $i = 1, \ldots, s$, *then* ω *is relatively exact, i.e.* $\omega = d\alpha$, $\alpha \in \mathcal{E}^k(X)$ *and* $i^{\#}\omega = 0$.

(iv) *Given numbers* $\alpha_i \in \mathbb{R}$, $i = 1, \ldots, s$. *Then there is a form* $\omega \in Z^k(X, \partial X)$ *with periods* α_i, *i.e.* $\delta_i(\omega) = \alpha_i$, $i = 1, \ldots, s$.

Also, the dual statements holds true, in particular

Theorem 7. *We have*

(i) $T \in B_k(X)$ *iff* $T(\omega) = 0 \ \forall\, \omega \in Z^k(X)$

(ii) $T \in B_k(X, \partial X)$ *iff* $T(\omega) = 0 \ \forall\, \omega \in Z^k(X, \partial X)$

Of course for $T \in Z_k(X)$ we have $T(\omega) = 0 \ \forall\, \omega \in Z^k(X)$ iff $T(\omega_i) = 0 \ \forall\, \omega_i \in Z^k(X)$, $i = 1, \ldots, s$, $[\omega_i]_X$ being a basis of $H^k_{\mathrm{dR}}(X)$, and for $T \in Z_k(X, \partial X)$ $T(\omega) = 0 \ \forall\, \omega \in Z^k(X, \partial X)$ iff $T(\omega_i) = 0 \ \forall\, \omega_i \in Z^k(X, \partial X)$, $i = 1, \ldots, r$, $[\omega_i]_{\mathrm{rel}}$ being a basis of $H^k_{\mathrm{dR}}(X, \partial X)$.

A consequence of Theorem 7 is then

Theorem 8. $B_k(X)$ *and* $B_k(X, \partial X)$ *are closed with respect to the weak convergence of currents respectively in* $Z_k(X)$, $Z_k(X, \partial X)$.

Consequently we can state

Theorem 9. *Homology classes can be represented by mass minimizing currents. That is, given* $\gamma \in H_k(X, \mathbb{R})$ *and* $\delta \in H_k(X, \partial X, \mathbb{R})$ *there are currents* $T \in Z_k(X)$ *and* $R \in Z_k(X, \partial X)$ *such that*

$$\begin{aligned} \mathbf{M}(T) &= \inf\{M(S),\ S \in Z_k(X),\ [S]_X = \gamma\} \\ \mathbf{M}(R) &= \inf\{M(S),\ S \in Z_k(X, \partial X),\ [S]_{\mathrm{rel}} = \gamma\}. \end{aligned}$$

3.4 Intersection of Currents and Kronecker Index

The aim of this subsection is to discuss some extensions to currents of the classical notion of *intersection of oriented surfaces in generic positions* and to introduce the *intersection index* of homology classes.

Intersection of currents in $\mathbb{R}^n$. Let U be an open set of $\mathbb{R}^n$ and let $S \in \mathcal{D}_k(U)$ and $T \in \mathcal{D}_h(U)$, where $h + k \geq n$. Denote by $f : \mathbb{R}^n \times \mathbb{R}^n \to \mathbb{R}^n$ the map

$$f(x, y) := x - y,$$

and let $g : U \to U \times U$ be the diagonal map

$$g(x) := (x, x).$$

Suppose that the slice, compare Sec. 2.2.5, $<S \times T, f, 0>$ exists and defines a current in $\mathbf{N}_{h+k-n}(U \times U)$. Then its support lies in $\{(x, y) \mid x = y\} = \operatorname{Im} g$, consequently, by Federer flatness theorem, see Sec. 5.3.1, there exists a unique current, denoted $S \cap T$ and called the *intersection of S and T*, in $\mathbf{N}_{h+k-n}(U)$ such that

$$g_\#(S \cap T) = (-1)^{(n-k)h} \langle S \times T, f, 0 \rangle. \tag{1}$$

Actually it suffices that $\langle S \times T, f, 0\rangle$ be a flat chain, compare Sec. 5.1.3 below. In the next proposition we collect a few simple properties of the intersection.

Proposition 1. *We have*

(i) If $S \cap T$ exists, then $\operatorname{spt}(S \cap T) \subset \operatorname{spt} S \cap \operatorname{spt} T$.
(ii) $S \cap T$ exists iff $T \cap S$ exists, and $T \cap S = (-1)^{(n-k)(n-h)} S \cap T$.
(iii) Assume that $S \cap T$, and either $S \cap \partial T$ or $\partial S \cap T$ exist. Then

$$\partial(S \cap T) = S \cap \partial T + (-1)^{h-k} \partial S \cap T. \tag{2}$$

Proof. (i) follows as

$$\operatorname{spt} \langle S \times T, f, 0\rangle \subset \operatorname{spt} S \times \operatorname{spt} T) \cap \{x = y\} = g(\operatorname{spt} S \cap \operatorname{spt} T).$$

(ii) Using the map $r(x, y) = (y, x)$, which reverse the role of the factors, we infer $f \circ r = -f$, $r_\#(T \times S) = (-1)^{hk} S \times T$, consequently

$$\begin{aligned} (-1)^{hk} \langle S \times T, f, 0\rangle &= \langle r_\#(T \times S), f, 0\rangle = r_\# \langle T \times S, f \circ r, 0\rangle \\ &= (-1)^n r_\# \langle T \times S, f, 0\rangle = (-1)^n \langle T \times S, f, 0\rangle. \end{aligned}$$

It follows $S \cap T = (-1)^\tau T \cap S$ where $\tau = -hk + n + (n-h)k + (n-k)h = (n-h)(n-k) \pmod 2$.
(iii) From $\partial(S \times T) = \partial S \times T + (-1)^k S \times \partial T$, $S \in \mathcal{D}_k(U)$, we infer that

$$\begin{aligned} \partial \langle S \times T, f, 0\rangle &= (-1)^n \langle \partial(S \times T), f, 0\rangle = \\ &= (-1)^n \langle \partial S \times T, f, 0\rangle + (-1)^{n-k} \langle S \times \partial T, f, 0\rangle \end{aligned}$$

provided $\langle S \times T, f, 0\rangle$ and either $\langle \partial S \times T, f, 0\rangle$ or $\langle S \times \partial T, f, 0\rangle$ exist. Therefore $\partial(S \cap T) = (-1)^\sigma \partial S \cap T + (-1)^\tau S \cap \partial T$ where

$$\begin{aligned} \sigma &= n + (n-h)h + (n-k-1)h = n - h \pmod 2 \\ \tau &= n - k + (n-h)h + (n-k)(h-1) = 0 \pmod 2). \end{aligned}$$

□

Intersection theory of currents is in general quite complicated as the intersection of currents may fail to exist. In fact while the existence of the slice $\langle S \times T, f, z\rangle$ is granted for a.e. z by the Lebesgue differentiation theorem, it may fail to exist at a single point, e.g. $z = 0$. The following examples show some of the difficulties.

1 It is clear by the definition that the intersection defines a bilinear map on $\mathbf{N}_k(U) \times \mathbf{N}_h(U)$ whenever it exists. It is also worth noticing that in general the density of intersection current is the product of the densities of the factors. For example let $\varphi \in L^p(\mathbb{R}^n)$, $\psi \in L^q(\mathbb{R}^n)$ be two functions. Consider the n-dimensional currents $S := \mathbb{R}^n \llcorner \varphi$ and $T = \mathbb{R}^n \llcorner \psi$. Then the current $S \times T \in \mathcal{D}_{2n}(\mathbb{R}^n \times \mathbb{R}^n)$ is given by

$$S \times T(\omega) = \int_{\mathbb{R}^n \times \mathbb{R}^n} \varphi(x)\psi(y)\omega(x,y), \qquad \omega \in \mathcal{D}^{2n}(\mathbb{R}^n \times \mathbb{R}^n).$$

If $f(x,y) = x - y$, $\Omega_\rho(t) = \chi_{B_\rho}(t) dt^1 \wedge \ldots dt^n$, for

$$\omega = \sum_{|\alpha|+|\beta|=n} \omega_{\alpha\beta}(x,y)\, dx^\alpha \wedge dy^\beta$$

we have

$$\frac{1}{\omega_n \rho^n} S \times T(f^\#(\Omega_\rho)\wedge\omega) = \int_{\mathbb{R}^n} dx \fint_{B_\rho(x)} \varphi(x)\psi(y) \sum_{|\gamma|\le n} \sigma(\overline{\gamma},\gamma)\omega_{\bar{\gamma}\gamma}(x,y)\, dy.$$

Since

$$\fint_{B_\rho(x)} \psi(y) \sum_{|\gamma|\le n} \sigma(\overline{\gamma},\gamma)\omega_{\bar{\gamma}\gamma}(x,y) \;\to\; \psi(x) g^\#(\omega)(x)$$

in $L^q(\mathbb{R}^n)$ we infer that, assuming $1/p + 1/q \le 1$, $<S \times T, f, 0>$ exists,

$$<S \times T, f, 0>(\omega) = \lim_{\rho\to 0} \frac{1}{\omega_n \rho^n} S \times T(f^\#(\Omega_\rho)\wedge\omega) = \int_{\mathbb{R}^n} \varphi(x)\psi(x) g^\#(\omega)(x),$$

hence

$$S \cap T(\omega) = \int \varphi(x)\psi(x)\omega(x) \qquad \omega \in \mathcal{D}^n(\mathbb{R}^n).$$

Simple examples of course show that $S \cap T$ may fail to exist. •

2 Let $Q = [\![(0,1) \times (0,1)]\!] \in \mathcal{I}_2(\mathbb{R}^2)$ be the integral current integration over the square in $\mathbb{R}^2$ and let $\pi(x,y) = x$ be the projection on the first factor. Then it is easy to compute

$$<Q, \pi, x> = \begin{cases} \delta_x \times [\![(0,1)]\!] & \text{if } 0 < x < 1 \\ \frac{1}{2}\delta_x \times [\![(0,1)]\!] & \text{if } x = 0, 1 \\ 0 & \text{otherwise} \end{cases}$$

which shows that at specific values the slice need not be rectifiable. Notice that such a phenomena would not appear if we project into a plane which is not perpendicular to any of the edges of Q. •

However, by slightly moving one of the two currents, we can and do ensure the existence of the intersection. In fact denoting by $\tau_a(x) = x + a$, x, $a \in \mathbb{R}^n$ we have

Proposition 2. *Let $S \in \mathbf{N}_k(U)$ be a normal current compactly supported in U, and let $T \in \mathbf{N}_h(U)$, $h + k \geq n$. Then for a.e. $a \in \mathbb{R}^n$ close to zero*

(i) $S \cap \tau_{a\#}T$ exists in $\mathbf{N}_{h+k-n}(U)$,
(ii) $\partial(S \cap \tau_{a\#}T) = S \cap \tau_{a\#}\partial T + (-1)^k(\partial S) \cap \tau_{a\#}T$,
(iii) if moreover $S \in \mathcal{I}_k(U)$, $T \in \mathcal{I}_h(U)$ then $S \cap \tau_{a\#}T \in \mathcal{I}_{h+k-n}(U)$.

Proof. From the slicing theory $<S \times T, f, z>$ exists in $\mathbf{N}_{h+k-n}(U)$ for a.e. $z \in \mathbb{R}^n$ close to zero. On the other hand,

$$<S \times \tau_{a\#}T, f, 0> = <S \times \tau_{a\#}, f \circ (\mathrm{id} \times \tau_a), 0> = \mathrm{id} \times \tau_{a\#}<S \times T, f, -a>$$

whenever $<S \times T, f, -a>$ exists. This proves (i). (ii) then follows from (iii) of Proposition 1 and (iii) follows similarly to (i) on account of $<S \times T, f, z> \in \mathcal{I}_{h+k-n}(U)$ for a.e. $z \in \mathbb{R}^n$. □

Also the intersection is well defined, essentially if we are able to prove continuity of $<S \times T, f, z>$ in z. We have in fact

Proposition 3. *Let $S \in \mathbf{N}_k(U)$ and $\tau \in \mathcal{E}^{n-h}(U)$. Then $S \cap \mathbb{R}^n \llcorner \tau$ exists and*

$$S \cap (\mathbb{R}^n \llcorner \tau) = S \llcorner \tau \qquad \textit{in } U.$$

Proof. Let $\omega \in \mathcal{D}^{k+h-n}(U \times U)$. Consider the k-form $\phi(x,y) = \tau(y) \wedge \omega(x,y) \in \mathcal{D}^k(U \times U)$. As $f^\#(dt^1 \wedge \ldots \wedge dt^n) = \sum (-1)^{|\bar\gamma|}\sigma(\gamma, \bar\gamma)\, dx^\gamma \wedge dy^{\bar\gamma}$ we have

$$\tau(y) \wedge f^\#(dt^1 \wedge \ldots \wedge dt^n) \wedge \omega(x,y) = \sum_{|\gamma| \leq n} (-1)^\delta \sigma(\gamma, \bar\gamma)\, dx^\gamma \wedge \phi(x,y) \wedge dy^{\bar\gamma},$$

where $\delta = (n-j) + j(n-h) + (n-j)(h+k-n)$, $j := |\gamma|$.

Writing $\phi = \sum \phi_{\alpha\beta}\, dx^\alpha \wedge dy^\beta$, we get as $\rho \to 0$

$$⨍_{B(x,\rho)} \phi_{\alpha\beta}(x,y)\, dy^\beta \wedge dy^{\bar\gamma} \;\to\; \begin{cases} 0 & \text{if } \beta \neq \gamma \\ \sigma(\gamma, \bar\gamma)\phi_{\alpha\gamma}(x,x) & \text{if } \beta = \gamma \end{cases}$$

by continuity. Consequently the averages

$$⨍_{B(x,\rho)} \tau(y) \wedge f^\#(dt^1 \wedge \ldots \wedge dt^n) \wedge \omega(x,y)$$

converges, as ρ goes to zero, to

$$\sum_{|\gamma| \leq n} (-1)^\delta (-1)^{|\gamma||\alpha|} \phi_{\alpha\gamma}(x,x)\, dx^\alpha \wedge dx^\gamma = (-1)^{(h+k)n} g^\#(\phi)(x) \tag{3}$$

since $\delta + |\gamma|\,|\alpha| = \delta + j(k-j) = (h+k)n \mod 2$. Since now

$$\begin{aligned} S \times \mathbb{R}^n \llcorner \tau(f^\#(\Omega_\rho)\wedge\omega)/\omega_n\rho^n &= (-1)^{(n-h)k}\frac{1}{\omega_n\rho^n} S\times\mathbb{R}^n(\tau\wedge f^\#(dt^1\wedge\ldots\wedge dt^n)\wedge\omega) \\ &= (-1)^{(n-h)k} S\Big(\fint_{B(x,\rho)} \tau(y)\wedge f^\#(dt^1\wedge\ldots\wedge dt^n)\wedge\omega\Big) \end{aligned}$$

from (3) we infer that $< S\times\mathbb{R}^n \llcorner \tau, f, 0>(\omega)$ exists and agrees with $(-1)^{(n-h)k}S(g^\#\phi) = (-1)^{(n-k)h}S(\tau\wedge g^\#\omega)$. In other words

$$g_\#(S\cap\mathbb{R}^n\llcorner\tau) = g_\#(S\llcorner\tau).$$

□

As immediate corollary we notice

Corollary 1. *Assume* $\sigma \in \mathcal{E}^{n-k}(U)$, $\tau\in\mathcal{E}^{n-h}(U)$, $h+k\geq n$. *Then*

$$(\mathbb{R}^n\llcorner\sigma)\cap(\mathbb{R}^n\llcorner\tau) = \mathbb{R}^n\llcorner(\sigma\wedge\tau) \qquad \text{in } U.$$

Of course one is also interested in intersecting currents when both factors are not integration against form, the case of intersection of rectifiable currents being relevant. We shall not go into these topics and confine ourselves to mention the following result taken from Federer [226] which applies for instance to the intersections of smooth surfaces in general position. We have

Theorem 1. *Let* $S\in\mathcal{I}_k(\mathbb{R}^n)$, $T\in\mathcal{I}_h(\mathbb{R}^n)$ *be two integral currents,* $h+k\geq n$. *Assume that* $\operatorname{spt} S$ *and* $\operatorname{spt} T$ *are contained respectively in two smooth submanifolds* $\mathcal{X}$ *and* $\mathcal{Y}$ *of dimension respectively* k *and* h *so that*

$$S = \tau(\mathcal{M},\theta,\vec{\mathcal{X}}), \qquad T=\tau(\mathcal{N},\varphi,\vec{\mathcal{Y}})$$

where $\mathcal{M}\subset\mathcal{X}$, $\mathcal{N}\subset\mathcal{Y}$. *Assume furthermore that*

(i) θ *is bounded on* $\mathcal{M}$, φ *is bounded on* $\mathcal{N}$.
(ii) *For each* $x\in\mathcal{X}\cap\mathcal{Y}$, $T_x\mathcal{X}\cap T_y\mathcal{Y}$ *has dimension* $h+k-n$.
(iii) *Either* $k=0$ *or* $\operatorname{spt}\partial S\cap\operatorname{spt} T$ *has a neighborhood* $E\subset\mathbb{R}^n$ *such that* $E\cap\operatorname{spt}\partial S$ *is covered by a finite family* Γ *of* $(k-1)$*-dimensional smooth submanifolds* G *of* $\mathcal{X}$ *with* $E\cap\operatorname{clos} G\subset G$

$$T_xG\cap T_x\mathcal{Y} \text{ has dimension } h+k-n-1\ \forall\, x\in G\cap\mathcal{Y}.$$

(iv) *Either* $h=0$ *or* $\operatorname{spt} S\cap\operatorname{spt}\partial T$ *has a neighborhood* $F\subset\mathbb{R}^n$ *such that* $F\cap\operatorname{spt}\partial T$ *is covered by a finite family* Δ *of* $h-1$*-dimensional smooth submanifolds* D *of* $\mathcal{Y}$ *with* $F\cap\operatorname{clos} D\subset D$ *and*

$$T_xD\cap T_x\mathcal{X} \text{ has dimension } h+k-n-1\ \forall\, x\in D\cap\mathcal{X}.$$

Then the $(h+k-n)$-vector $\zeta(x)$ given by

$$g_{\#,x}(\zeta(x)) = \vec{\mathcal{X}}(x)\wedge\vec{\mathcal{Y}}(x) \llcorner (df^1\wedge\ldots\wedge df^n)(x,x)$$

is non zero and simple, whenever $x \in \operatorname{spt} T \cap \operatorname{spt} S \setminus (\operatorname{spt}\partial T \cup \operatorname{spt}\partial S)$. Moreover

$$\mathcal{H}^{h+k-n}(\operatorname{spt} T \cap \operatorname{spt} S \setminus (\operatorname{spt}\partial T \cup \operatorname{spt}\partial S)) = 0$$

and

$$S \cap T = \tau\Big(\mathcal{M}\cap\mathcal{N}, \theta\varphi|\zeta|, \frac{\zeta}{|\zeta|}\Big).$$

We omit the proof of Theorem 1 and we refer to Federer [226]. We just make a few remarks on the role of transversality of $\mathcal{X}$ and $\mathcal{Y}$. If $\vec{\mathcal{X}}(x), \vec{\mathcal{Y}}(x)$ are the two unit tangent vectors to $\mathcal{X}$ and $\mathcal{Y}$ at x, then $\vec{\mathcal{X}}(x)\wedge\vec{\mathcal{Y}}(x)$ generates $T_x(\mathcal{X})\oplus T_x(\mathcal{Y})$ in the product $\mathbb{R}^n \times \mathbb{R}^n$. As $\operatorname{Im} g = \ker f$, we have

$$T_x\mathcal{X} \cap T_x\mathcal{Y} \text{ has dimension } h+k-n$$

iff

$$f : T_x\mathcal{X} \oplus T_x\mathcal{Y} \text{ has a kernel of dimension } h+k-n$$

iff

$$f_{|\,T_x\mathcal{X}\oplus T_x\mathcal{Y}} \text{ has maximal rank } n.$$

From Proposition 3 in Sec. 5.2.2, the last claim is equivalent to say that

$$\vec{\mathcal{X}}(x)\wedge\vec{\mathcal{X}}(x) \llcorner (df^1\wedge\ldots\wedge df^n) \neq 0.$$

Intersection in real homology. In contrast to the many difficulties in defining the intersection of currents, *homologically the intersection of cycles is always well defined.* Let X be a compact, oriented, n–dimensional submanifold of $\mathbb{R}^{n+N}$ possibly with boundary. We first consider the intersection of relative and absolute homology classes. Let $[S]_{\mathrm{rel}} \in H_k(X, \partial X, \mathbb{R})$ and $[T] \in H_h(X, \mathbb{R})$, $h+k \geq n$, and let $P_T^\sharp$ be a Poincaré dual form of T, $P_T^\sharp \in Z^{n-h}(X, \partial X)$. It is easy to see that the absolute homology of the $(h+k-n)$–cycle

$$S(\cdot \wedge P_T^\sharp)$$

depends only on the homology classes $[S]_{\mathrm{rel}}$ of S and the absolute homology class $[T]$ of T. Then we define the *intersection map*

$$\cap_X^\sharp : H_k(X, \partial X, \mathbb{R}) \times H_h(X, \mathbb{R}) \;\to\; H_{h+k-n}(X, \mathbb{R})$$

by

(4) $$[S]_{\mathrm{rel}} \cap_X^\sharp [T] := (-1)^{(n-k)(n-h)}[S(\cdot \wedge P_T^\sharp)].$$

Similarly one defines the *intersection map*

$$\cap_X : H_k(X,\mathbb{R}) \times H_h(X,\mathbb{R}) \;\to\; H_{h+k-n}(X,\mathbb{R})$$

by $[S] \cap_X [T] := (-1)^{(n-k)(n-h)}[S(\cdot \wedge P_T^\sharp)]$ for $[S] \in H_k(X,\mathbb{R})$, $[T] \in H_h(X,\mathbb{R})$, $h+k \geq n$. Finally denoting by $P_T^\flat \in Z^{n-h}(X)$ a Poincaré dual form of a relative cycle $T \in Z_h(X, \partial X)$, the formula

$$[S]_{\mathrm{rel}} \cap_X^\flat [T]_{\mathrm{rel}} := (-1)^{(n-k)(n-h)}[S(\cdot \wedge P_T^\flat)]_{\mathrm{rel}}$$

defines a third set of intersection maps

$$\cap_X^\flat : H_k(X,\partial X,\mathbb{R}) \times H_h(X,\partial X,\mathbb{R}) \;\to\; H_{h+k-n}(X,\partial X,\mathbb{R}).$$

From the definitions it is easily seen that *intersection of cycles in homology is Poincaré dual of the wedge product in cohomology*. For instance for the map $\cap_X^\sharp$ one has the following commutative diagram

$$\begin{array}{ccccc}
H_{\mathrm{dR}}^{n-k}(X) & \times & H_{\mathrm{dR}}^{n-h}(X,\partial X) & \xrightarrow{\wedge} & H^{2n-h-k}(X,\partial X) \\
\downarrow P^\flat & & \downarrow P^\sharp & & \downarrow P^\sharp \\
H_k(X,\partial X,\mathbb{R}) & \times & H_h(X,\mathbb{R}) & \xrightarrow{\cap} & H_{h+k-n}(X,\mathbb{R})
\end{array}$$

where $[\omega]\wedge[\eta] := [\omega\wedge\eta]$.

Kronecker index in $\mathbb{R}^n$. Finally, let us introduce the *Kronecker index* $S \in \mathbf{N}_k(\mathbb{R}^n)$ and $T \in \mathbf{N}_{n-k}(\mathbb{R}^n)$ be two normal currents of complementary dimensions. Assume that

$$\text{either } k=0 \text{ or } \operatorname{spt} S \cap \operatorname{spt} \partial T = \emptyset \quad \text{and} \quad \text{either } h=0 \text{ or } \operatorname{spt} T \cap \operatorname{spt} \partial S = \emptyset,$$

so that $\operatorname{spt}(\partial(S \times T))$ does not meet the diagonal of $\mathbb{R}^{2n}$, i.e.,

$$0 \notin f(\operatorname{spt} \partial(S \times T)).$$

Then the current $(-1)^{n-k} f_\#(S \times T)$ belongs to $\mathbf{N}_n(\mathbb{R}^n)$ and 0 does not belong to $\operatorname{spt} \partial((-1)^{n-k} f_\#(S \times T))$. The constancy theorem then yields a constant $k(S,T) \in \mathbb{R}$ such that in a neighborhood of 0

$$(5) \qquad (-1)^{n-k} f_\#(S \times T) = k(S,T)[\![\mathbb{R}^n]\!].$$

The number $k(S,T)$ is called the *Kronecker index of S and T*, and it is actually an *integer* if S and T are i.m. rectifiable.

For any $\varphi \in C^0(\mathbb{R})$ and ρ sufficiently small, applying (5) to $\varphi(t)\chi_{B_\rho} dt^1 \wedge \ldots \wedge dt^n =: \varphi(t)\Omega_\rho(t)$, we get

$$\frac{(-1)^{n-k}}{|B_\rho|} S \times T \llcorner f^\#(\Omega_\rho)(\varphi \circ f) = k(S,T) \fint_{B(0,\rho)} \varphi(t)\, dt.$$

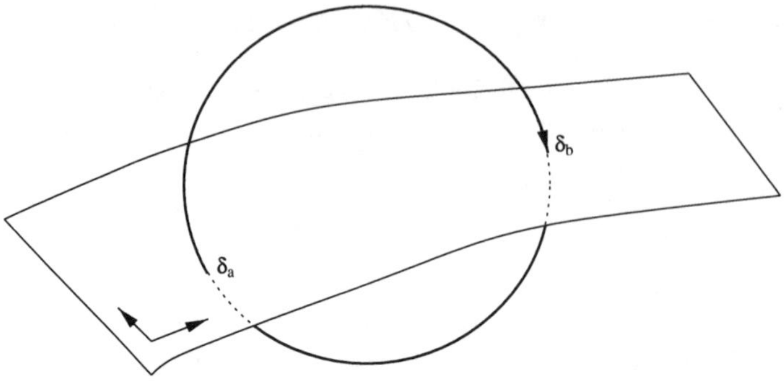

Fig. 5.5. $S \cap T = \delta_a - \delta_b$, $k(S,T) = 0$.

Taking $\varphi = 1$, we then conclude in the case that $S \cap T$ exists that

$$(S \cap T)(1) = g_{\#}(S \cap T)(1) = (-1)^{(n-k)^2+(n-k)} < S \times T, f, 0 > (1) = k(S,T).$$

A special interesting case is that of two oriented submanifolds S and T in generic position. By Theorem 1, in this case $S \cap T$ exists and $S \cap T = \sum_i n_i \delta_{P_i}$ where the n_i are defined by

$$\vec{S}(P_i) \wedge \vec{T}(P_i) = n_i e_1 \wedge \ldots \wedge e_n.$$

Therefore we conclude

$$k(S,T) = \sum n_i,$$

i.e., Kronecker index of S and T is the sum with sign of the intersections.

Another interesting case is that in which

$$S = (-1)^{k(n-k)} \mathbb{R}^n \llcorner \sigma, \qquad \sigma \in \mathcal{D}^{n-k}(\mathbb{R}^n)$$
$$T = (-1)^{k(n-k)} \mathbb{R}^n \llcorner \tau, \qquad \tau \in \mathcal{D}^{k}(\mathbb{R}^n).$$

In this case, as we have seen, $S \cap T$ exists and, by Proposition 4 in Sec. 5.3.2,

$$k(S,T) = S \cap T(1) = \mathbb{R}^n \llcorner (\tau \wedge \sigma)(1) = \int_{\mathbb{R}^n} \tau \wedge \sigma.$$

Kronecker index on manifolds. Of course one can extend the notion of Kronecker index to currents on oriented manifolds. More precisely let $S \in \mathbf{N}_k(X)$, $T \in \mathbf{N}_{n-k}(X)$ where X is a n-dimensional, compact oriented submanifold of $\mathbb{R}^{n+N}$. Assume

$$k = 0 \text{ or } \operatorname{spt} \partial S \cap \operatorname{spt} T = \emptyset \quad \text{and} \quad k = n \text{ or } \operatorname{spt} \partial T \cap \operatorname{spt} S = \emptyset.$$

Recall that the lift $\pi^*(T)$ on a tubular neighborhood U of X of bounded cross-section has finite mass, and that

$$\operatorname{spt} \partial \pi^* T \cap \operatorname{spt} S = \emptyset$$

since $\operatorname{spt} \partial \pi^* T$ lives on $\pi^{-1}(\operatorname{spt} \partial T)$ or outside U. Then we can set

Definition 1. *The* Kronecker index *of S and T in X is defined by*

$$k_X(S,T) := k(S, \pi^* T),$$

i.e. as the number such that

$$(-1)^{n-k} f_\#(S \times \pi^* T) = k_X(S,T) [\![\mathbb{R}^{n+N}]\!] \tag{6}$$

near the origin.

From (6) it is easy seen that

(i) $k_X(S,T)$ *is integer-valued if S and T are i.m. rectifiable*
(ii) $k_X(\partial R, T) = k_X(S, \partial R) = 0$.
(iii) $k_X(S,T) = 0$ if $\operatorname{spt} S \cap \operatorname{spt} T = \emptyset$.

In particular, if $S \in Z_k(X, \partial X)$ is a relative normal cycle and $T \in Z^{n-k}(X)$ is an absolute normal cycle which do not meet ∂X , $k(S,T)$ is well defined and depends only on the relative homology class of S and on the homology class of T, thus it actually defines a map

$$i_X(S,T) : H_k(X, \partial X, \mathbb{R}) \times H_{n-k}(X, \mathbb{R}) \ \rightarrow \ \mathbb{R}$$

which is the *intersection index of S and T* .

As we have essentially seen, if $S \cap \pi^* T$ exists, then

$$S \cap \pi^* T(1) = k_X(S,T) = i_X(S,T).$$

In particular if $S = (-1)^{k(n-k)} [\![X]\!] \llcorner P^\flat_S$, i.e. $P^\flat_S$ is Poincaré dual of S null to the boundary of X, and $T = (-1)^{k(n-k)} [\![X]\!] \llcorner P^\sharp_T$, i.e. $P^\sharp_T$ is a Poincaré dual of T, then

$$\int_X P^\flat_S \wedge P^\sharp_T = S \cap \pi^* T(1) = i_X(S,T). \tag{7}$$

Therefore we can state that *the intersection index is Poincaré-Lefschetz dual of Poincaré Lefschetz duality on forms.*

Of course one has also an intersection index on the compact oriented manifold without boundary ∂X

$$i_{\partial X} : H_k(\partial X, \mathbb{R}) \times H_{n-k}(\partial X, \mathbb{R}) \rightarrow \mathbb{R}$$

defined by $i_{\partial X}(S,T) := K_{\partial X}(S,T)$.

This yields another way to state non degeneracy of Poincaré duality of forms.

Theorem 2. *Let X be a compact oriented manifold. The intersection index*

$$i_X : H_k(X, \partial X, \mathbb{R}) \times H_{n-k}(X, \mathbb{R}) \ \to \ \mathbb{R}$$

is non degenerate. Moreover, denoting by $i_X^T(T, S) := i_X(S, T)$, the following diagram

$$\begin{array}{ccccccccc}
\cdots \overset{i_\#}{\to} & H_k(X,\mathbb{R}) & \overset{j_\#}{\to} & H_k(X,\partial X,\mathbb{R}) & \overset{\partial}{\to} & H_{k-1}(\partial X,\mathbb{R}) & \overset{i_\#}{\to} \cdots \\
& \times\, i_X^T & & \times\, i_X & & \times\, i_{\partial X} & \\
\cdots \underset{\partial}{\leftarrow} & H_{n-k}(X,\partial X,\mathbb{R}) & \underset{j_\#}{\leftarrow} & H_{n-k}(X,\mathbb{R}) & \underset{i_\#}{\leftarrow} & H_{n-k}(\partial X,\mathbb{R}) & \underset{\partial}{\leftarrow} \cdots
\end{array}$$

is commutative.

3.5 Relative Homology and Cohomology Groups

We conclude this section reporting partially and briefly on Federer and Fleming approach to relative homology groups in the setting of local Lipschitz neighbourhood retracts.

Let A and B be local Lipschitz neighbourhood retracts in $\mathbb{R}^n$. Suppose $B \subset A$ and B relatively closed in A. Denote by $\pi : U \to A$ and $p : V \to B$ the retractions of neighbourhoods U of A and V of B, and assume, without loss in generality, that $V \subset U$ and moreover that $(1-t)x + tp(x) \in V$ whenever $x \in V$ and $t \in [0, 1]$. For $k = 0, \dots, n$, we define *relative cycles* by

$$Z_k(A, B, \mathbb{R}) := \{T \in \mathbf{F}_{k,\text{cpt}}(U) \mid \operatorname{spt} T \subset A, \ \operatorname{spt} \partial T \subset B, \ \text{ or } k = 0\},$$

relative boundaries by

$$\begin{aligned} &B_k(A, B, \mathbb{R}) \\ &\quad := \{R + \partial S \mid R \in \mathbf{F}_{k,\text{cpt}}(V), \ \operatorname{spt} R \subset B, \ S \in \mathbf{F}_{k+1,\text{cpt}}(U), \ \operatorname{spt} S \subset A\}, \end{aligned}$$

and the *real relative homology groups* by

$$H_k(A, B, \mathbb{R}) = Z_k(A, B, \mathbb{R}) / B_k(A, B, \mathbb{R}).$$

We point out that all currents involved have compact supports and that the retractions π and p induce continuous maps

$$\pi_\# : \mathbf{F}_{k,\text{cpt}}(U) \to \mathbf{F}_{k,\text{cpt}}(A), \quad p_\# : \mathbf{F}_{k,\text{cpt}}(V) \to \mathbf{F}_{k,\text{cpt}}(B).$$

The relative homology groups defined above satisfy the *Eilenberg-Steenrod axioms* for a homology theory in the *local Lipschitz category*. We recall that the local Lipschitz category consists of couples (A, B) of local Lipschitz neighbourhood retracts, with $A \supset B$, B relatively closed in A, and , morphisms f between couples

$$f : (A, B) \to (A', B')$$

are Lipschitz maps from A to A' satisfying $f(B) \subset B'$. It is not difficult to prove using Federer's flatness theorem that any such morphism induces a map

$$f_{\#} : H_k(A, B, \mathbb{R}) \to H_k(A', B', \mathbb{R})$$

compare Federer [226, 4.4.1]. Also if $C \subset B \subset A$ are local Lipschitz retracts in $\mathbb{R}^n$, the boundary operator acts as

$$\partial : Z_k(A, B) \to Z_{k-1}(B, C), \quad \partial : B_k(A, B) \to B_{k-1}(B, C),$$

thus induces a map again denoted by ∂

$$\partial : H_k(A, B) \to H_{k-1}(B, C).$$

Here we understood the dependence on $\mathbb{R}$, $H_k(A, B) := H_k(A, B, \mathbb{R})$.

Theorem 1. *We have*

(i) *If $f : (A, B) \to (A, B)$ is the identity map, then $f_* : H_k(A, B) \to H_k(A, B)$ is the identity map.*

(ii) *For $f : (A, B) \to (A', B')$ and $g : (A', B') \to (A'', B'')$ we have*

$$(g \circ f)_* = g_* \circ f_*.$$

(iii) *If $C \subset B \subset A$, $C' \subset B' \subset A'$, $f : (A, B) \to (A', B')$, $f_{|B} : (B, C) \to (B', C')$ and $k > 0$, then*

$$f_{|B*}\partial = \partial f_*.$$

(iv) *If $i : (B, C) \to (A, C)$ and $j : (A, C) \to (A, B)$ denote the inclusion maps, then the sequence of homomorphism*

$$\cdots \xrightarrow{\partial} H_k(B, C) \xrightarrow{i_*} H_k(A, C) \xrightarrow{j_*} H_k(A, B) \xrightarrow{\partial} H_{k-1}(B, C) \xrightarrow{i_*} \cdots$$

is exact.

(v) *If $h : ([0, 1] \times A, [0, 1] \times B) \to (A', B')$ and*

$$h_* : (A, B) \to (A', B'), \quad h_*(x) = h(t, x), \ 0 \le t \le 1, \ x \in A,$$

then $h_{0*} = h_{1*}$.

(vi) *If $i : (A, B) \to (A', B')$ is an inclusion map such that $B = A \cap B'$ and $\operatorname{clos}(A' \backslash B') \cap (\operatorname{clos}(A' \backslash A) = \emptyset$, or equivalently $B \subset A \subset A'$, $B' = B \cup (A' \backslash A)$ and $\operatorname{clos}(A \setminus B) \cap \operatorname{clos}(A' \setminus A) = \emptyset$, then i_* is an isomorphism.*

(vii) *If $x \in \mathbb{R}^n$, then $H_0(\{x\}) \simeq \mathbb{Z}$ and $H_k(\{x\}) = \{0\}$ for $k > 0$.*

Suppose now that A and consequently B are *compact*. Then one can construct a CW-complex L_A surrounding A and contained in U which has a sub complex L_B surrounding B and contained in V by selecting cubes not far from A of a standard subdivision of $\mathbb{R}^n$ in cubes with a suitable small mesh size. As in the case $B = \emptyset$ one proves

Theorem 2. *Let A be a compact. Then $H_k(A, B, \mathbb{R})$ is isomorphic to the relative homology of Lipschitz chains of (A, B). Moreover $H_{k,\text{simpl}}(L_A, L_B, \mathbb{R})$ can be mapped linearly onto $H_k(A, B, \mathbb{R})$. In particular the vector space $H_k(A, B, \mathbb{R})$ has finite dimension.*

In this more general context, following the same path of Sec. 5.3.2 one derives an isoperimetric inequality

Theorem 3. *Let $A \supset B$ be compact neighbourhood Lipschitz retracts. There exists a constant $\tau > 0$ such that, if $T \in Z_k(A, B)$ is a relative homology cycle of finite mass, $\mathbf{M}(T) < \infty$, with $T = X + \partial Y$ with $X \in \mathbf{F}_{k,\text{cpt}}(V)$, $Y \in \mathbf{F}_{k+1,\text{cpt}}(U)$, then $T = B_k(A, B)$, i.e.,*

$$T = R + \partial S,$$

with $R \in \mathbf{F}_{k,\text{cpt}}(V)$, $\operatorname{spt} R \subset B$, $S \in \mathbf{F}_{k,\text{cpt}}(U)$, $\operatorname{spt} S \subset A$, and

$$\mathbf{M}(R) + \mathbf{M}(S) \le \tau \mathbf{M}(T);$$

in particular $S \in \mathbf{N}_{k+1}(U)$.

As already remarked it is useful to work with homology group defined as equivalence classes of flat chains as no control on the mass of the boundaries is required. We remark however that as far as homology groups are concerned this is quite irrelevant: replacing flat chains by normal currents leads in fact to isomorphic relative homology groups. This follows from the following proposition, compare Sec. 5.1.3,

Proposition 1. *If $T' \in Z_k(A, B)$, then there is $T \in Z_k(A, B) \cap \mathbf{N}_k(U)$ such that $T - T' \in B_k(A, B)$.*

As in the boundaryless case, a consequence of the isoperimetric inequality is the weak closure of relative homology classes. In fact we have

Theorem 4. *Let $\{T_i\} \subset Z_k(A, B)$ be a sequence which is equibounded in mass and weakly converging to T. Then $T \llcorner (A \setminus B) \in \mathbf{F}_{k,\text{cpt}}(U)$ and $T \llcorner (A \setminus B)$ belongs to $Z_k(A, B)$. Moreover if the $T_i \in B_k(A, B)$ are relative boundaries, then $T \llcorner (A \setminus B)$ is a relative boundary.*

Proof. First notice that by semicontinuity T has finite mass, hence $T \llcorner (A \setminus B)$ is well defined. In order to prove that $T \llcorner (A \setminus B)$ is a flat chain we set $g(x) = \operatorname{dist}(x, B)$, $x \in U$, and for $r \ge 0$

$$H(r) := \{x \in U \mid g(x) > r\}.$$

Being $\operatorname{spt} \partial T \subset B = U \setminus H(0)$, from the slicing theory we have for a.e. r, $0 < r < \varepsilon_0$,

$$<T, g, r> = \partial(T \llcorner H(r)) - \partial T \llcorner H(r) = \partial(T \llcorner H(r))$$

and

$$\int_0^{1/j} \mathbf{M}(\partial(T \llcorner H(r)))\, dr = \int_0^{1/j} \mathbf{M}(<T, g, r>)\, dr \le c\mathbf{M}(T).$$

Consequently we can find a sequence of numbers $\{r_j\}$, $r_j \to 0$, such that $\mathbf{M}(\partial(T \llcorner H(r_j)) < \infty$, i.e., we have

$$T \llcorner H(r_j) \in \mathbf{N}_k(U).$$

Also

$$\begin{aligned} \lim_{j\to\infty} \mathbf{M}(T \llcorner H(r_j) - T \llcorner (A \setminus B)) &= \lim_{j\to\infty} \mathbf{M}(T \llcorner (H(r_j) - H(0))) \\ &= \| T \|(\cap_j (H(r_j) - H(0))) = 0, \end{aligned}$$

hence $\mathbf{F}(T \llcorner H(r_j) - T \llcorner (A \setminus B)) \to 0$, which proves that $T \llcorner (A \setminus B) \in \mathbf{F}_{k,\mathrm{cpt}}(U)$ As trivially $\partial(T \llcorner (A \setminus B)) \subset B$, the first part of the claim is proved too.

If the T_i's are relative boundaries, the isoperimetric inequality yields

$$T_i = R_i + \partial S_i$$

where $R_i \in \mathbf{F}_{k,\mathrm{cpt}}(V)$, $\operatorname{spt} R_i \subset B$, $S_i \in \mathbf{N}_{k+1}(U)$, $\operatorname{spt} S_i \subset A$, and

$$\mathbf{M}(R_i) + \mathbf{M}(S_i) + \mathbf{M}(\partial S_i) \le c\mathbf{M}(T_i).$$

Passing to the limit on a suitable subsequence, we then infer $T = R + \partial S$ with $R \in \mathcal{D}_k(V)$, $\operatorname{spt} R \subset B$, $S \in \mathbf{N}_{k+1}(U)$, $\operatorname{spt} S \subset A$; therefore

$$T \llcorner (A \setminus B) = R - T \llcorner B + \partial S.$$

Being S normal in U with $\operatorname{spt} S \subset A$, S is a flat chain, hence $\partial S \in \mathbf{F}_k(U)$; by the first part of our claim $R - T \llcorner B \in \mathbf{F}_{k,\mathrm{cpt}}(U)$. Finally, since $\operatorname{spt}(R - T \llcorner B) \subset B$, we conclude that $T \llcorner (A \setminus B) \in B_k(A, B)$. □

We notice explicitly that we cannot expect in general that T belongs to $\mathbf{F}_{k,\mathrm{cpt}}(U)$. As corollary of Theorem 4 we can now state

Theorem 5. *We have*

(i) In each relative homology class $\gamma \in H_k(A, B, \mathbb{R})$ there is a homological minimizer, that is a $T \in Z_k(A, B, \mathbb{R})$, $[T]_{\mathrm{rel}} = \gamma$ such that

$$\mathbf{M}(T) = \inf\{\mathbf{M}(R) \mid R \in Z_k(A, B, \mathbb{R}),\ [R]_{\mathrm{rel}} = \gamma\}$$

(ii) In particular the function

$$\mathbf{M}(\gamma) := \inf\{\mathbf{M}(R) \mid R \in Z_k(A, B, \mathbb{R}),\ [R]_{\mathrm{rel}} = \gamma\}$$

defines a norm on $H_k(A, B, \mathbb{R})$.

Proof. First we observe that in every relative homology class there is a current with finite mass, hence

$$\lambda := \inf\{\mathbf{M}(R) \mid R \in Z_k(A, B, \mathbb{R}),\ [R]_{\text{rel}} = \gamma\} < \infty.$$

The claim in (i) then follows at once from the semicontinuity of the mass and from Theorem 5. The claim in (ii) is trivial. □

Actually the previous results extend to the quite more general situation in which one does not assume A compact, but with a number of changes and technical difficulties. For instance homology groups are vector spaces but in general of infinite dimension, in fact they are inductive limits of Banach spaces endowed with a flat norm. Also, in dealing with such a general situation one cannot avoid some compact control on the location of the supports of involved currents. In some sense the homology vector space can be exhausted by a sequence of finite dimensional vector spaces. In fact one has that *the vector subspace* of $H_k(A, B, \mathbb{R})$ consisting of those homology classes which meet $\mathbf{F}_{k,K}(\mathbb{R}^n)$, K fixed compact, has finite dimension. To keep our extension simple we shall not discuss this general approach and we refer the interested readers to Whitney [674], Federer [226] [228].

Relative de Rham cohomology. Let A, B local Lipschitz neighborhood retracts, $B \subset A$, B closed relatively to A, and let $\pi : U \to A$ the retraction. One then defines the *de Rham cohomology groups*

$$H^k_{\text{dR}}(A, B) := Z^k(U, B)\big/\sim$$
$$H^k_{\text{dR}}(A) := H^k(A, \emptyset)$$

where

$$Z^k(U, B) := \Big\{\omega \in Z^k(U) \,|\, \omega = 0 \text{ near } B\Big\}$$

and for $\omega, \eta \in Z^k(U)$ $\omega \sim \eta$ iff there exist an open set V, $A \subset V \subset U$ and a form $\alpha \in Z^{k-1}(V, B)$ such that $\omega - \eta = d\alpha$ in V. One has of course the *immersion map* $j^\# : H^k_{\text{dR}}(A, B) \to H^k_{\text{dR}}(A)$, the *restriction map* $i^\# : H^k_{\text{dR}}(A) \to H^k_{\text{dR}}(B)$ which restrict forms in $H^k_{\text{dR}}(A)$ to their values in B, and a *coboundary operator* $\delta : H^k_{\text{dR}}(B) \to H^{k+1}_{\text{dR}}(A, B)$ defined as follows. Choose an open set U' such that $B \subset\subset U' \subset\subset U \cap V$ and extend $\omega_{|U'}$ to $\tilde{\omega} \in \mathcal{D}^k(U)$. Of course $d\tilde{\omega} = 0$ near B, hence $d\tilde{\omega} \in Z^{k+1}(A, B)$; one defines

$$\delta([\omega]_B) := [d\tilde{\omega}]_{\text{rel}}$$

noticing that the relative homology class of $\tilde{\omega}$ depends only on the cohomology of ω.

It is easy to prove that in case A is a compact oriented submanifold X and $B = \partial X$, that the above definitions agree with the ones in Sec. 5.2.6, compare Proposition 1 in Sec. 5.3.3. Moreover a rewriting of Theorem 2 in Sec. 5.2.8 yields

Theorem 6. *The long sequence*

$$\cdots \xrightarrow{j^\#} H^k_{\mathrm{dR}}(A) \xrightarrow{i^\#} H^k_{\mathrm{dR}}(B) \xrightarrow{\delta} H^{k+1}_{\mathrm{dR}}(A,B) \xrightarrow{j^\#} H^{k+1}_{\mathrm{dR}}(A) \xrightarrow{i^\#} \cdots$$

is exact.

de Rham duality in the Lipschitz category. In the attempt to generalize de Rham duality to the Lipschitz category, Whitney and then Federer introduced the notion of *flat cochains* which play the role of non smooth differential forms. In order to give a simple presentation we assume A compact, and refer to Federer [228] for general results and proofs. Set

$$\mathbf{F}_k(A) := \mathbf{F}_{k,A}(U),$$

we would have to write instead

$$\mathbf{F}_k(A) := \cup\{\mathbf{F}_{k,K}(U) \mid K \subset A,\ K \text{ compact}\}$$

for general local Lipschitz neighbourhood retract A.

Definition 1. *We define*

(i) flat cochains *as linear continuous functionals on* $\mathbf{F}_k(A)$

$$\mathbf{F}^k(A) := \{\alpha : \mathbf{F}_k(A) \to \mathbb{R} \mid \alpha \textit{ linear and continuous}\}$$

(ii) relative cochains, cocycles *and* coboundaries *as*

$$\begin{aligned} \mathbf{F}^k(A,B) &:= \{\alpha \in \mathbf{F}_k(A) \mid \alpha = 0 \textit{ on } \mathbf{F}_k(B)\} \\ Z^k(A,B) &:= \{\alpha \in \mathbf{F}_k(A) \mid \alpha = 0 \textit{ on } B_k(A,B)\} \\ B^k(A,B) &:= \{\alpha \in \mathbf{F}_k(A) \mid \alpha = 0 \textit{ on } Z_k(A,B)\} \end{aligned}$$

(iii) cohomology groups *as*

$$H^k(A,B) = Z^k(A,B)/B^k(A,B).$$

We recall, see Sec. 5.1.3, that $\mathbf{F}_k(A)$ endowed with the flat norm is a Banach space. One sees immediately that $Z_k(A,B)$ *is closed in* $\mathbf{F}_k(A)$, and, using the isoperimetric inequality, that also $B_k(A,B)$ *is closed in* $\mathbf{F}_k(A)$. It follows that

$$\inf\{\mathbf{F}_k(S) \mid S \in \gamma\}, \quad \gamma \in H_k(A,B,\mathbb{R})$$

defines a norm on $H_k(A,B,\mathbb{R})$ with respect to which the coset map

$$p : Z_k(A,B,\mathbb{R}) \to H_k(A,B,\mathbb{R})$$

is continuous, being $H_k(A,B,\mathbb{R})$ finite dimensional. It is worth to notice that instead *relative boundaries with equibounded masses are in general not compact with respect to the flat norm.* For instance take $A := \mathbb{R}^2$ and $B := \mathbb{R}^2 \cap \{x \mid x_2 = 0\}$. Then the sequence of flat relative boundaries

$$T_i := i[\![(p_1, p_2)]\!] + i[\![(p_2, p_3)]\!] + i[\![(p_3, p_4)]\!] \quad i = 1, 2, \dots,$$

where

$$p_1 := (0,0),\ p_2 := (0, i^{-2}),\ p_3 := (i^{-1}, i^{-2}),\ \text{and } p_4 := (i^{-1}, 0),$$

converge to the current $\delta_0 \wedge e_1$ which is not flat.

Clearly every relative cocycle $\alpha \in Z^k(A, B)$ induces by restriction a linear map

$$\alpha : H_k(A, B, \mathbb{R}) \ \to\ \mathbb{R}.$$

Consequently there is a natural morphism

$$H^k(A, B) \ \to\ \operatorname{Hom}(H_k(A, B, \mathbb{R}), \mathbb{R})$$

which is obviously injective. On the other hand if $\ell \in \operatorname{Hom}(H_k(A, B, \mathbb{R}), \mathbb{R})$, $\ell \circ p$ defines a linear continuous map from $Z_k(A, B)$ into $\mathbb{R}$ which extends by Hahn-Banach theorem to a linear continuous map still denoted ℓ

$$\ell : \mathbf{F}_k(A) \ \to\ \mathbb{R}.$$

Therefore we conclude

Proposition 2. $H^k(A, B) \simeq \operatorname{Hom}(H_k(A, B, \mathbb{R}), \mathbb{R})$

Notice that every smooth form ω defines by

$$\widetilde{\omega}(T) := T(\omega)$$

a flat cochain $\widetilde{\omega} \in \mathbf{F}^k(A)$. The main question becomes now that of realizing $H^k(A, B)$. A key point in this respect is the following

Theorem 7. *We have*

$$B^k(A, B) = \{d\beta \mid \beta \in \mathbf{F}^{k-1}(A, B)\}.$$

On the other hand, a regularization procedure, compare Federer [228], then allows to prove

Proposition 3. *We have*

$$H^k(A, B) = H^k_{\mathrm{dR}}(A, B)$$

Consequently Theorem 7 yields the following extension of the de Rham theorem

Theorem 8. *We have*

(i) For every $\nu \in \mathrm{Hom}\,(H_k(A,B,\mathbb{R}),\mathbb{R})$ *there exists a closed differential form* $\omega_\nu \in \mathcal{E}^k(U)$ *which vanish in a neighbourhood of* B *such that*

$$\nu([T]) = T(\omega_\nu).$$

(ii) There exists neighbourhood $U' \subset U$ *and* $V' \subset V$ *such that, if* $\omega \in \mathcal{E}^k(U)$, $d\omega = 0$, $\omega = 0$ *in a neighbourhood of* B*, and*

$$T(\omega) = 0 \qquad \forall\, T \in Z_k(A,B)$$

then ω *is exact in* U'*, i.e.*

$$\omega = d\varphi, \quad \varphi \in \mathcal{E}^{k-1}(U') \text{ and } \operatorname{spt}\varphi \subset U' \setminus V'.$$

In particular the duality between forms and currents yields to a non degenerate bilinear form

$$<,> : H_k(A,B,\mathbb{R}) \times H^k_{\mathrm{dR}}(A,B) \;\to\; \mathbb{R}.$$

Finally the diagram

$$\begin{array}{ccccccccc}
\cdots \overset{\delta}{\leftarrow} & H^k_{\mathrm{dR}}(B) & \overset{i^\#}{\leftarrow} & H^k_{\mathrm{dR}}(A) & \overset{j^\#}{\leftarrow} & H^k(A,B) & \overset{\delta}{\leftarrow} \cdots \\
& \times<,> & & \times<,> & & \times<,> & \\
\cdots \overset{\partial}{\rightarrow} & H_k(B,\mathbb{R}) & \overset{i_\#}{\rightarrow} & H_k(A,\mathbb{R}) & \overset{j_\#}{\rightarrow} & H_k(A,B,\mathbb{R}) & \overset{\partial}{\rightarrow} \cdots
\end{array}$$

is commutative.

4 Integral Homology

In this section we discuss the integral homology of a compact, oriented manifold in terms of currents.

4.1 Integral Homology Groups

In this subsection we discuss the integral homology of a compact, oriented manifold in terms of currents, in particular we shall prove that homology classes are weakly closed. Actually we shall see that the cosets of homology classes are in fact connected components of integral cycles.

Integral absolute homology groups. Let A be a compact neighborhood retract of $\mathbb{R}^n$, let $\pi : U \to A$ be the retraction. We can and do assume that

$$U = \{x \in \mathbb{R}^n \mid \operatorname{dist}(x,A) < \varepsilon_A\}$$

for some $\varepsilon_A > 0$. By decomposing $\mathbb{R}^n$ into cubes of size $\varepsilon_A/2\sqrt{n}$, and denoting by $\mathcal{L}$ the finite family of those cubes which cover A, we can see that

$$\bigcup\{Q \mid Q \in \mathcal{L}\} \subset\subset U.$$

From now on we denote by L_A the CW cubical complex over cubes in $\mathcal{L}$ and by $\mathcal{L}_{k,A}$ the k-skeleton of L_A, $k = 0, \dots, n$.

Similarly to Sec. 5.3.2 an *integral chain* P on $L_{k,A}$ is simply a polyhedral chain with integral coefficients

$$P = \sum \beta_i [\![F_i]\!], \quad \beta_i \in \mathbb{Z}, \ F_i \in \mathcal{L}_{k,A}.$$

Setting

$$Z_{k,\text{simpl}}(L_A, \mathbb{Z}) := \{P \mid P \text{ integral polyhedral chain on } L_{k,A}, \ \partial P = 0\}$$
$$B_{k,\text{simpl}}(L_A, \mathbb{Z}) := \{P = \partial Q \mid Q \text{ integral polyhedral chain on } L_{k,A}\}$$

we define the *simplicial integral k-homology of* L_A by

$$H_{k,\text{simpl}}(L_A, \mathbb{Z}) := Z_{k,\text{simpl}}(L_A, \mathbb{Z}) / B_{k,\text{simpl}}(L_A, \mathbb{Z})$$

which is of course a finitely generated $\mathbb{Z}$–module. We then set for $k = 0, \dots, n$

$$Z_k(A, \mathbb{Z}) := \{T \in \mathcal{I}_k(U), \ \partial T = 0, \ \operatorname{spt} T \subset A\}$$
$$B_k(A, \mathbb{Z}) := \{\partial S, \ S \in \mathcal{I}_{k+1}(U), \ \operatorname{spt} S \subset A\}$$

Then the *integral k-homology group of* A is defined by

$$H_k(A, \mathbb{Z}) := Z_k(A, \mathbb{Z}) / B_k(A, \mathbb{Z}).$$

Proceeding as for Theorem 6 in Sec. 5.3.2, taking into account that the deformation theorem produces i.m. rectifiable decomposition of i.m. rectifiable currents, we get

Theorem 1. *The module $H_k(A, \mathbb{Z})$ is isomorphic to the k-homology of Lipschitz chains in A with integral coefficients. Moreover the retraction map π induces an epimorphism $\pi_* : H_{k,\text{simpl}}(L_A, \mathbb{Z}) \to H_k(A, \mathbb{Z})$. In particular $H_k(A, \mathbb{Z})$ is finitely generated.*

Of course, if A is a C^1 manifold then $H_k(A, \mathbb{Z})$ and $H_{k,\text{simpl}}(L_A, \mathbb{Z})$ are isomorphic, compare Theorem 5 in Sec. 5.3.2.

Again a relevant advantage in representing the singular homology in terms of homology groups of currents is that the cosets in $Z_k(A, \mathbb{Z})$ turn out to be closed with respect to the weak convergence of currents. Actually in the case of integral homology we are dealing now, we shall prove that in fact the cosets are connected components of $Z_k(A, \mathbb{Z})$. The key step in doing that is an isoperimetric inequality which is more stringent than in the real case.

[1] The following example shows that one cannot expect isoperimetric inequality for general currents on manifolds, a size condition is necessary.

Let $S^2 = \{x \in \mathbb{R}^3 \,|\, |x| = 1\}$ be the two sphere and let $\varepsilon > 0$ be small enough. Let

$$\mathcal{M} = S^2 \cap \left\{x \in \mathbb{R}^3 \,|\, x_3 \geq -1 + \varepsilon\right\}$$

and $C = S^2 \cap \{x \in \mathbb{R}^3 \,|\, x_3 = -1 + 2\varepsilon\}$. Clearly C is a 1–cycle in $\mathcal{M}$ and bounds the surface $S := S^2 \cap \{x \in \mathbb{R}^3 \,|\, x > -1 + 2\varepsilon\}$ in $\mathcal{M}$, $C = \partial S$, but $\mathbf{M}(S) \simeq 4\pi$, while $\mathbf{M}(C) \simeq O(\sqrt{\varepsilon})$. If one consider

$$\widetilde{\mathcal{M}} = S^2 \cap \left\{x \in \mathbb{R}^3 \,|\, x \leq 1 - \varepsilon\right\}$$

then the cycle C does not bound any surface lying in $\widetilde{\mathcal{M}}$. •

Theorem 2. *For $k = 0, \dots, n$ there is a constant $\varepsilon = \varepsilon(k, n, \varepsilon_A)$ such that if $T \in Z_k(A, \mathbb{Z})$ with $\mathbf{M}(T) \leq \varepsilon$ then $T \in B_k(A, \mathbb{Z})$; more precisely $T = \partial R$, $R \in \mathcal{I}_{k+1}(U)$, $\operatorname{spt} R \subset A$ and*

$$\mathbf{M}(R) \leq c\,\mathbf{M}(T)^{1+1/k}$$

where $c = c(k, n)$.

Proof. Using the deformation theorem one can deform T over $L_{k,A}$ and write $T = P + \partial R'$ where $R' \in \mathcal{I}_{k+1}(U)$, $\operatorname{spt} R' \subset\subset U$, $P = \sum \beta_i [\![F_i]\!]$, $F_i \in \mathcal{L}_{k,A}$, $\beta_i \in \mathbb{Z}$ and $\mathbf{M}(P) \leq c\,\mathbf{M}(T)$, $\mathbf{M}(R') \leq c\sigma_A \mathbf{M}(T)$. As $\beta_i \geq 1$, $M(P) \geq \sigma_A^k$. Therefore choosing $\varepsilon < \sigma_A^k / c$, we infer that if $\mathbf{M}(T) \leq \varepsilon$ then $P = 0$. Consequently, setting $R = \pi_\#(R')$, we infer

$$T = \pi_\# T = \partial R, \quad R \in \mathcal{I}_{k+1}(U), \quad \operatorname{spt} R \subset A$$

and

$$\mathbf{M}(R) \leq c_1 \mathbf{M}(R') \leq c_1 c\,\mathbf{M}(T) \mathbf{M}(P)^{1/k} \leq c_1 (c\,\mathbf{M}(T))^{1+1/k}.$$

□

The following theorem yields the key decomposition argument, which is relevant also in the context of flat chains, compare Theorem 2 in Sec. 5.1.3.

Proposition 1. *Under the assumption of Theorem 2 in Sec. 5.1.3, if moreover all T_i and T are i.m. rectifiable, we can find integral currents $R_i \in \mathcal{I}_{k+1}(A)$ and $S_i \in Z_k(A, \mathbb{Z})$ such that*

$$\mathbf{M}(R_i) + \mathbf{M}(S_i) \to 0, \qquad T_i - T = S_i + \partial R_i.$$

Proof. We proceed as in the proof Theorem 2 in Sec. 5.1.3 observing that all currents P_i, Q_i, F_i, G_i which are involved in the proof of Theorem 2 in Sec. 5.1.3 this time are i.m. rectifiable and the convergences

$$\sum_{j=1}^{k} F_j \to \sum_{j=1}^{\infty} F_i, \qquad \sum_{j=1}^{k} G_j \to \sum_{j=1}^{\infty} G_i$$

are in the *mass norm*. □

From Theorem 2 and Proposition 1 we readily infer

Theorem 3. *Let $T_i, T \in Z_k(A, \mathbb{Z})$ be integral k-cycles, $T_i \to T$. Then $T_i - T = \partial R_i$, $R_i \in \mathcal{I}_{k+1}(A)$ for large i. Moreover $\mathbf{M}(R_i) \to 0$.*

Proof. In fact using Proposition 1 we can write

$$T_i - T = S_i + \partial R_i'$$

$S_i \in Z_k(A)$, $R_i' \in \mathcal{I}_{k+1}(A)$ and $\mathbf{M}(R_i') + \mathbf{M}(S_i) \to 0$. The isoperimetric inequality of Theorem 2 implies that $S_i = \partial \Sigma_i$, $\Sigma_i \in \mathcal{I}_{k+1}(A)$, $\mathbf{M}(\Sigma_i) \le c\mathbf{M}(S_i)^{1+1/k}$ for large i. Thus

$$T_i - T = \partial(R_i' + \Sigma_i)$$

and the claim follows. □

The following corollary collects some important consequences of Theorem 3.

Corollary 1. *We have*

(i) *$B_k(A, \mathbb{Z})$ is weakly closed.*
(ii) *$Z_k(A, \mathbb{Z}) \setminus B_k(A, \mathbb{Z})$ is weakly closed.*
(iii) *In each integral homology class γ there is a mass minimizing i.m. rectifiable cycle, that is, there is $T \in Z_k(A, \mathbb{Z})$, $[T] = \gamma$, such that*

$$\mathbf{M}(T) := \inf\{M(S) \mid [S] = \gamma\}.$$

(iv) *Given $c > 0$ there are at most a finite number of integral homology classes γ such that*

$$\inf\{\mathbf{M}(S) \mid [S] = \gamma\} \le c.$$

(v) *For a non trivial integral homology class*

$$\inf\{\mathbf{M}(S) \mid [S] = \gamma\} > 0.$$

Proof. (i) and (ii) follows trivially from Theorem 3. (iii) trivially follows from (i). To prove (iv) assume on the contrary that one has a sequence $\{T_i\}$ of cycles of equibounded masses. Passing to a subsequence $T_i \rightharpoonup T$ and by Theorem 3, all the T_i but a finite number belong to the same homology class, a contradiction. (v) is just the isoperimetric inequality, Theorem 2. □

Relative integral homology. A parallel theory for *integral relative homology* can be developed. Let A, B be two local Lipschitz neighbourhood retracts in $\mathbb{R}^n$, $B \subset A$, B relatively closed in A. For $k = 0, \dots, n$, we define the *integral relative cycles* by

$$Z_k(A, B, \mathbb{Z}) := \{T \in \mathcal{F}_k(\mathbb{R}^n) \mid \operatorname{spt} T \subset A, \ \operatorname{spt} \partial T \subset B, \ \text{or } k = 0\},$$

the *integral relative boundaries* by

$$B_k(A, B, \mathbb{Z}) := \{T + \partial S \mid T \in \mathcal{F}_k(\mathbb{R}^n), \ \operatorname{spt} T \subset B, \ S \in \mathcal{F}_{k+1}(\mathbb{R}^n), \ \operatorname{spt} S \subset A\},$$

and the *integral relative homology* by

$$H_k(A,B,\mathbb{Z}) = Z_k(A,B,\mathbb{Z})/B_k(A,B,\mathbb{Z}).$$

One shows that the $\mathbb{Z}$–modules $H_k(A,B,\mathbb{Z})$ satisfy the Eilenberg-Steenrod axioms for an homology theory with integral coefficients in the Lipschitz category, compare Federer [226, 4.4.1], and, being A compact that we have as consequence of the deformation theorem

Theorem 4. *$H_k(A,B,\mathbb{Z})$ is isomorphic to the k-homology of integral Lipschitz chains in (A,B). Moreover one has an epimorphism of the k-simplicial homology $H_k(L_A,L_B,\mathbb{Z})$ of (L_A,L_B) onto $H_k(A,B,\mathbb{Z})$. In particular $H_k(A,B,\mathbb{Z})$ is finitely generated.*

From the deformation theorem we also infer the following *isoperimetric inequality*, compare Federer [226, 4.4.2]

Theorem 5. *Let $T \in \mathcal{R}_k(\mathbb{R}^n) \cap Z_k(A,B) \cap B_k(U,V)$. There exists an integral current S such that* $\operatorname{spt} S \subset A$, $\operatorname{spt}(T-\partial S) \subset B$, *and*

$$\mathbf{M}(S)^{k/(k+1)} + \mathbf{M}(T-\partial S) \le c\mathbf{M}(T),$$

c being an absolute constant.

Proceeding as in the real case and taking into account Federer-Fleming closure theorem one infers

Theorem 6. *Suppose that $T_i \in \mathcal{R}_k(U) \cap Z_k(A,B,\mathbb{Z})$ are relative cycles converging weakly to T. Then T belongs to $\mathcal{R}_k(A\setminus B)$ and $T \llcorner (A\setminus B) \in Z_k(A,B,\mathbb{Z})$. If moreover the T_i's are relative boundaries, then T is also a relative boundary, $T \in B_k(A,B,\mathbb{Z})$*

and consequently, compare Federer [226, 4.4.2]

Theorem 7. *We have*

(i) In each relative homology class $\gamma \in H_k(A,B,\mathbb{Z})$ there is a homological mass minimizer, i.e. a $T \in Z_k(A,B,\mathbb{Z})$ with $[T]_{\mathrm{rel},\mathbb{Z}} = \gamma$ and

$$\mathbf{M}(T) = \inf\{\mathbf{M}(R) \mid R \in Z_k(A,B,\mathbb{Z}),\ [R]_{\mathrm{rel}} = \gamma\}$$

(ii) There are finitely many integral relative homology classes which meet the set $\{T \in Z_k(A,B,\mathbb{Z}),\ \mathbf{M}(T) \le c\}$.

Real and Integral homology. Let us discuss briefly relationships between integral and real homology in terms of cycles.

Let A, B be compact neighbourhood retracts of $\mathbb{R}^n$. According to Theorem 1 the integral homology groups are Abelian groups of finite type. The general *presentation theorem* for such groups says that $H_k(A,B,\mathbb{Z})$ splits as a direct sum of groups with a single generator; more precisely, one can choose exactly q elements

$\gamma_1, \dots, \gamma_q \in H_k(A,B,\mathbb{Z})$, where q is the rank of $H_k(A,B,\mathbb{Z})$, independent over $\mathbb{Z}$ such that

$$H_k(A,B,\mathbb{Z}) = \bigoplus_1^q \mathbb{Z}\gamma_i \oplus T_k, \tag{1}$$

where T_k is the *torsion subgroup of* $H_k(A,B,\mathbb{Z})$, i.e.,

$$T_k := \{\gamma \in H_k(A,B,\mathbb{Z}) \mid \exists\, p \in \mathbb{Z},\ p\gamma = 0\}.$$

In other words each $\gamma \in H_k(A,B,\mathbb{Z})$ decomposes uniquely as

$$\gamma = \sum_i n_i\gamma_i + \omega, \qquad n_i \in \mathbb{Z},\ \omega \in T_k.$$

It is also a very well known fact of homological algebra that *the integral homology groups fix the homology with any group coefficient.* In fact the *universal coefficient theorem*, compare e.g. Spanier [607, p. 222 and p.243] express the homology with coefficients in G as

$$H_k(A,B;G) = H_k(A,B,\mathbb{Z}) \otimes G \oplus \mathrm{Tor}(H_{k-1}(A,B,\mathbb{Z}),G), \tag{2}$$

where the Tor groups are the "torsion" subgroups. In case $G = \mathbb{R}$ (2) reduces to

$$H_k(A,B,\mathbb{R}) = H_k(A,B,\mathbb{Z}) \otimes \mathbb{R}, \tag{3}$$

which in terms of the generators of $H_k(A,B,\mathbb{Z})$ means

Theorem 8. *Let* $q = \operatorname{rank} H_k(A,B,\mathbb{Z})$ *and let* $S_1, \dots, S_q$ *be* q *integral relative cycles in* $Z_k(A,B,\mathbb{Z})$ *such that*

$$[S_1]_{\mathrm{rel},\mathbb{Z}}, \dots, [S_q]_{\mathrm{rel},\mathbb{Z}}$$

are independent over $\mathbb{Z}$ *and*

$$H_k(A,B,\mathbb{Z}) = \sum_1^q \mathbb{Z}[S_i]_{\mathrm{rel},\mathbb{Z}} + T_k(A,B).$$

Then

$$[S_1]_{\mathrm{rel},\mathbb{R}}, \dots, [S_q]_{\mathrm{rel},\mathbb{R}}$$

yield a basis for $H_k(A,B,\mathbb{R})$. *In particular*

$$\dim H_k(A,B,\mathbb{R}) = \operatorname{rank} H_k(A,B,\mathbb{Z})$$

A simple consequence is

Theorem 9. *Let* $\dim H_k(A,B,\mathbb{R}) = q$. *Then we can always choose* q *integral cycles* T_i *in* $Z_k(A,B,\mathbb{Z})$ *such that* *(i)* $[T_1]_{\mathbb{Z}}, \dots, [T_q]_{\mathbb{Z}}$ *are independent over* $\mathbb{Z}$ *and generate the free part of the homology group* $H_k(A, B, \mathbb{Z})$.

(ii) $[T_1]_{\mathbb{R}}, \dots, [T_q]_{\mathbb{R}}$ *form a basis for* $H_k(A, B, \mathbb{R})$.

For the sake of completeness, we give a direct proof of Theorem 8 in terms of currents without any mention of the algebraic theory in the case of the absolute homology. Modulo additional technicalities, the general case of relative homology can be treated in the same way, compare Federer [228, 5.7].

From now on we assume $B = \emptyset$. It is enough to prove Theorem 8 for the simplicial homology $H_{k,\mathrm{simpl}}(L_A, \mathbb{Z})$. For that we need the following lemma

Lemma 1. *Let* $N > 1$ *and* $\Sigma \subset \mathbb{Q}^N$ *be a subset of* $\mathbb{Q}^N$, *and let* $r \in \mathbb{R}^N$ *be such that* $(r \mid \sigma) = 0 \quad \forall\, \sigma \in \Sigma$. *Then for any* $\varepsilon > 0$ *one can find* $q \in \mathbb{Q}^N$ *such that* $|q - r| < \varepsilon$ *and* $(q \mid \sigma) = 0\ \forall \sigma \in \Sigma$.

Proof. We denote by $s_1, \dots, s_k$ a basis for the $\mathbb{Q}$ span of $r_1, \dots, r_N$ and we write

$$r_i = \sum_{j=1}^{k} a_{ij} s_j, \qquad s_j \in \mathbb{R},\ a_{ij} \in \mathbb{Q},\ i = 1, \dots, N.$$

Then $\sum_{i=1}^N r_i \sigma_i = 0$ is equivalent to $\sum_{i=1}^N a_{ij} \sigma_i = 0\ j = 1, \dots, k$. Choosing now $\tau = (\tau_1, \dots, \tau_k)$ in such a way that

$$\sum_{i=1}^{N} \sum_{j=1}^{k} |a_{ij}|\, |\tau_j - s_j| < \varepsilon$$

we see that $q \in \mathbb{Q}^N$ given by $q_i := \sum_{j=1}^k a_{ij} \tau_j$ has the required properties. □

Notice that we can choose $q_i = r_i$ if r_i happen to be rational. Next proposition contains one of the key points in the proof of Theorem 8.

Proposition 2. *Let* $T \in Z_k(L_A, \mathbb{Z}) \cap B_k(L_A, \mathbb{R})$, *i.e.* T *is an integral polyhedral cycle, which is a boundary in the real sense,* $T = \partial R$, $R \in P_{k+1}(L_A, \mathbb{R})$. *Then, for each* $\varepsilon > 0$, *one can find a polyhedral chain* R' *with rational coefficients such that* $T = \partial R'$ *and* $\mathbf{M}(R - R') < \varepsilon$.

Proof. Denote by $\{F_i\}$, $i = 1, \dots, I$, and $\{G_j\}$, $j = 1, \dots, J$ respectively the k-faces and the $(k+1)$ faces of L_A with some orientation. We must have

$$\partial[\![G_j]\!] = \sum_{i=1}^{I} \theta_{ji} [\![F_i]\!], \quad \theta_{ji} \in \mathbb{Z}.$$

Moreover, if we write

$$T = \sum n_i [\![F_i]\!], \quad n_i \in \mathbb{Z}, \qquad R = \sum r_j [\![G_j]\!], \quad r_j \in \mathbb{R}$$

$T = \partial R$ means $n_i = \sum_{j=1}^{J} r_j \theta_{ji}$, $\forall\, i = 1, \dots, I$. Consequently by Lemma 1, given $\varepsilon > 0$, one finds $s_j \in \mathbb{Q}$ such that

$$n_i = \sum_{j-1}^{J} s_j \theta_{ji}.$$

Therefore, if $R' := \sum_j s_j [\![G_j]\!]$, we have $\partial T = R'$ and

$$\mathbf{N}(R - R') \le \sum_j \mathbf{N}([\![G_j]\!]) |r_j - s_j| \le c\varepsilon.$$

□

Remark 1. We observe that T is a polyhedral chain with rational coefficients iff for some $p \in \mathbb{Z}$, pT is an integral polyhedral chain. Accordingly we could restate Proposition 2.

Denote by $j_k : H_k(L_A, \mathbb{Z}) \to H_k(L_A, \mathbb{R})$ the group homomorphism which associate to each integer homology class $[T]_{\mathbb{Z}}$ the corresponding real homology class

$$j_*([T]_{\mathbb{Z}}) = [T]_{\mathbb{R}}.$$

Corollary 2. *The kernel of j_*, $\ker j_*$, is the torsion subgroup of $H_k(L_A, \mathbb{Z})$, i.e., for $T \in Z_k(L_A, \mathbb{Z})$*

$$[T]_R = 0 \quad \textit{iff} \quad [T]_{\mathbb{Z}} \textit{ is a torsion class.}$$

Proof. If $[T]_{\mathbb{Z}}$ belong to the torsion subgroup of $H_k(L_A, \mathbb{Z})$ then for some $p \neq 0$, $p[T]_{\mathbb{Z}} = 0$. Consequently

$$[T]_{\mathbb{R}} = \frac{1}{p} p[T]_{\mathbb{R}} = \frac{1}{p}[pT]_{\mathbb{R}} = \frac{1}{p} j_*([pT]_{\mathbb{Z}}) = 0.$$

Conversely assume that $[T]_{\mathbb{R}} = 0$, that is $T \in Z_k(A, \mathbb{Z})$ and $T = \partial R$, $R \in P_{k+1}(L_A, \mathbb{R})$. By Proposition 2 we find $R' \in P_{k+1}(L_A, \mathbb{Z})$ and an integer $p \neq 0$ such that

$$pT = \partial R'$$

therefore $[pT]_{\mathbb{Z}} = p[T]_{\mathbb{Z}} = 0$. □

Proof of Theorem 8. We split the proof into three steps.
Step 1. $[T_1]_{\mathbb{R}}, \dots, [T_q]_{\mathbb{R}}$ are independent over $\mathbb{R}$.

Assume $\sum_{i=1}^{q} r_i [T_i]_{\mathbb{R}} = 0$. Since $\mathbb{R}$ is a vector space over $\mathbb{Q}$, denoting by $s_1, \dots, s_k$ a basis for the $\mathbb{Q}$-span of $r_1, \dots, r_q$, we write $r_i = \sum_{j=1}^{k} a_{ij} s_j$ $a_{ij} \in \mathbb{Q}$, $s_j \in \mathbb{R}$. Consequently

$$\sum_{i=1}^{q} a_{ij} [T_i]_{\mathbb{R}} = 0 \qquad \forall\, j = 1, \dots, k.$$

Denote by p the l.c.m. of the denominators of the a_{ij}. The current $T := \sum_{i=1}^{k} qa_{ij}T_i$ is an integral cycle and

$$[T]_{\mathbb{R}} = q\sum_{i=1}^{k} a_{ij}[T_i]_{\mathbb{R}} = 0.$$

By Corollary 1 $[T]_{\mathbb{Z}}$ belongs to the torsion subgroup of $H_k(L_A, \mathbb{Z})$ hence

$$0 = [T]_{\mathbb{Z}} = \sum (qa_{ij})[T_i]_{\mathbb{Z}}.$$

Consequently $a_{ij} = 0$, $\forall\, i, j$ and therefore $r_i = 0$.

Step 2. Real homology class of rational polyhedral cycles are linear combinations of $[T_1]_{\mathbb{R}}, \dots, [T_q]_{\mathbb{R}}$ with rational coefficients.

In fact we can write any polyhedral cycle T with rational coefficients as

$$T = \frac{1}{p}S$$

where $p \in \mathbb{Z}$ and $S \in Z_k(L_A, \mathbb{Z})$. Since $[S]_{\mathbb{Z}} = \sum n_i[T_i]_{\mathbb{Z}} + \omega$, $n_i \in \mathbb{Z}$, $\omega \in T_k$ we get

$$[S]_{\mathbb{R}} = j_*([S]_{\mathbb{Z}}) = \sum_{i=1}^{q} n_i[T_i]_{\mathbb{R}}.$$

Consequently

$$[S]_{\mathbb{R}} = \sum_{i=1}^{q} \frac{n_i}{p}[T_i]_{\mathbb{R}}.$$

Step 3. Real homology classes of rational polyhedral cycles are dense in $H_k(A, \mathbb{R})$.

In fact for $T \in Z_k(A, \mathbb{R})$ we find, using Proposition 2, a sequence of rational polyhedral cycles T_j such that $\mathbf{M}(T_j - T) \to 0$. As

$$\| [T]_{\mathbb{R}} - [T_j]_{\mathbb{R}} \| \le \mathbf{M}(T - T_j) \;\to\; 0$$

the proof is concluded. □

Combining Theorem 8 and de Rham's theorems, compare Sec. 5.3.2 and Sec. 5.3.3, we also get

Theorem 10. *Let X be a n dimensional oriented compact manifold. Assume one has q cycles $T_1, \dots, T_q \in Z_k(X, \mathbb{Z})$ such that the corresponding homology classes are independent over $\mathbb{Z}$ and generate the free part of $H_k(X, \mathbb{Z})$. Then one can find q closed k-forms $\omega^1, \dots, \omega^q \in Z^k(X)$ such that*

$$T_i(\omega^j) = \delta_i^j.$$

Analogously, given q' relative cycles $T'_1, \dots, T'_{q'} \in Z_k(X, \partial X, \mathbb{Z})$ such that the corresponding relative homology classes are independent over $\mathbb{Z}$ and generate the free part of $H_k(X, \partial X, \mathbb{Z})$, then one can find q' relatively closed $k-forms$ $\eta^1, \dots, \eta^{q'} \in Z^k(X, \partial X)$ such that

$$T'_i(\eta^j) = \delta_i^j.$$

Proof. By Theorem 8, $[T_i]_{\mathbb{R}}$ generate the real homology $H_k(X,\mathbb{R})$ of X. By de Rham's duality $H^k_{\mathrm{dR}}(X,\mathbb{R})$ is dual to $H_k(X,\mathbb{R})$ therefore choosing dual cohomology class

$$[\omega^1],\dots,[\omega^q]$$

of $[T_1]_{\mathbb{R}},\dots,[T_q]_{\mathbb{R}}$ yields the right forms. The second part of the claim is identical. □

In terms of the forms $\omega^1,\dots,\omega^q$, $\eta^1,\dots,\eta^{q'}$ of Theorem 10 we can state

Corollary 3. *We have*

(i) Let $T\in Z_k(X,\mathbb{R})$. Then T is (real) homologous to an integer cycle iff $T(\omega^j)\in\mathbb{Z}\ \forall\, j=1,\dots,q$.
(ii) Let $T\in Z_k(X,\mathbb{Z})$. Then $T(\omega^j)=0\ \forall\, j=1,\dots,q$ if and only if $[T]_{\mathbb{Z}}$ belongs to the torsion subgroup of $H_k(X,\mathbb{Z})$.
(iii) Let $T\in Z_k(X,\partial X,\mathbb{R})$. Then T is (real) homologous to an integer cycle iff $T(\omega^j)\in\mathbb{Z}\ \forall\, j=1,\dots,q$.
(iv) Let $T\in Z_k(X,\partial X,\mathbb{Z})$. Then $T(\eta^j)=0\ \forall\, j=1,\dots,q$ if and only if $[T]_{\mathbb{Z},\mathrm{rel}}$ belongs to the torsion subgroup of $H_k(X,\partial X,\mathbb{Z})$.

4.2 Intersection in Integral Homology

Let X be a n-dimensional oriented compact submanifold of $\mathbb{R}^{n+N}$. The intersection maps in real homology that we have defined in Sec. 5.3.4 extends to integral homology. The idea is to define $[S]\cap[T]$ by taking the homology class of the intersection of S with almost any translation $\tau_{a\#}T$ of T. We first do that in the case of boundaryless manifolds.

Intersection of cycles on boundaryless manifolds. Assume that $\partial X=\emptyset$ and for convenience set for $T\in\mathcal{D}_k(X)$ and $x,a\in\mathbb{R}^{n+N}$

$$T_a:=\tau_{a\#}\pi^*T.$$

Proposition 1. *Let $S\in Z_k(X,\mathbb{R})$ and $T\subset Z_h(X,\mathbb{R})$, $h+k\geq n$ be two normal cycles. Then we have*

(i) For a.e. a close to zero $S\cap T_a$ exists in $Z_{h+k-n}(X)$ and

$$[S\cap T_a]_{\mathbb{R}}=[S]_{\mathbb{R}}\cap_X[T]_{\mathbb{R}}.$$

(ii) If moreover S and T are integral, then for a.e. a close to zero $S\cap T_a$ is an integral cycle, $S\cap T_a\in Z_{h+k-n}(X,\mathbb{Z})$ and its integral homology class depends only on the integral homology classes $[S]\in H_k(X,\mathbb{Z})$ and $[T]\in H_h(X,\mathbb{Z})$.

Proof. (i) Let $P_T\in Z^{n-h}(X)$ be a Poincaré dual form of T so that $T=\int_X\cdot\wedge P_T^\sharp+\partial\Sigma$, $\Sigma\in\mathbf{N}_{h+1}(X)$, or, setting $\omega:=(-1)^{h(n-h)}P_T$,

$$T=[\![X]\!]\llcorner\omega+\partial\Sigma.$$

Taking into account (iii) of Proposition 3 in Sec. 5.3.2

$$\pi^*([\![X]\!] \llcorner \omega_T) = [\![U]\!] \llcorner \pi^\# \omega_T$$

and therefore

$$\pi^* T = [\![U]\!] \llcorner \pi^\# \omega_T + \partial \pi^* \Sigma \quad \text{in } U.$$

By slightly translating we then infer

$$T_a = [\![U]\!] \llcorner \tau_{-a}^\# \pi^\# \omega_T + \partial \Sigma_a \quad \text{near } X.$$

By Proposition 2 in Sec. 5.3.4 we then infer that $S \cap \Sigma_a$, $S \cap \partial \Sigma_a$, $S \cap T_a$ exists as normal currents on X for a.e. a close to zero. Consequently being $\partial S \cap \Sigma_a = 0$ we infer

$$\partial(S \cap \Sigma_a) = S \cap \partial \Sigma_a$$

and

(1) $$S \cap T_a = S \cap [\![U]\!] \llcorner \tau_{-a}^\# \pi^\# \omega_T + \partial(S \cap \Sigma_a) \quad \text{near } X.$$

for a.e. a close to zero. Since on the other hand $\tau_{-a}^\# \pi^\# \omega_T$ is cohomologous to $\pi^\# \omega_T$ with compact support in $X \setminus \partial X$, we obtain from (1) that $S \cap T_a$ is homologous to

$$S \cap [\![U]\!] \llcorner \pi^\# \omega_T = S \llcorner \pi^\# \omega_T = S \llcorner \omega_T.$$

(i) now follows if we compute

$$S \llcorner \omega_T = (-1)^{(n-k)(n-h)} S(\cdot \wedge \pi^\# P_T^\sharp) = (-1)^{(n-k)(n-h)} S(\cdot \wedge P_T^\sharp).$$

(ii) We already known, compare (iii) of Proposition 2 in Sec. 5.3.4, that $S \cap T_a$ is an integral current for a.e. a close to zero. We now prove that *the integral homology class of $S \cap T_a$ is independent of a for a.e. a close to zero.* In fact, being T_a and T homotopic,

$$T_a = T_0 + \partial \Sigma \quad \text{near } X,$$

$\Sigma \in \mathcal{I}_{h+N+1}(U)$. Therefore for $|b|$ close to zero,

$$T_{a+b} = T_b + \partial \Sigma' \quad \text{near } X,$$

$\Sigma' \in \mathcal{I}_{h+N+1}(U)$. Proposition 2 in Sec. 5.3.4 then yields that for a.e. b the currents $S \cap T_{a+b}$, $S \cap T_b$, $S \cap \Sigma_b$ exist as integral currents in X, and $\partial(S \cap \Sigma_b) = S \cap \partial \Sigma_b$. Consequently

$$S \cap T_{a+b} - S \cap T_b = \partial(S \cap \Sigma') \text{ in } X$$

for each a and a.e. b close to zero. In particular the integral homology of $S \cap T_a$ is independent of a for a.e. a.

On the other hand, if $T = \partial R$, $R \in \mathcal{I}_{h+1}(X)$ or $S = \partial Q$, $Q \in \mathcal{I}_{k+1}(X)$, Proposition 2 in Sec. 5.3.4 again yields that for a.e. a close to zero

$$S \cap (\partial R_a) = \partial(S \cap R_a), \qquad (\partial Q) \cap T_a = (-1)^{k+1} \partial(Q \cap T_a)$$

being $\partial T_a = \partial S = 0$. □

The claim (ii) of Proposition 1 clearly allows us to define the intersection map in integral homology

$$\cap_X : H_k(X,\mathbb{Z}) \times H_h(X,\mathbb{Z}) \to H_{h+k-n}(X,\mathbb{Z})$$

setting

$$[S]_{\mathbb{Z}} \cap_X [T]_{\mathbb{Z}} := [S \cap T_a]_{\mathbb{Z}} \qquad \text{for a.e. } a,$$

while (i) of Proposition 1 show that such a map is compatible with the real version of the intersection map defined in Sec. 5.3.4.

Intersection of cycles on manifolds with boundary. More tricky is the definition of the intersection of cycles for manifolds with boundary. Let us start with the intersection of a relative with an absolute homology class.

Let X be a compact oriented manifold of dimension n in $\mathbb{R}^{n+N}$ possibly with boundary. Let X_1, X_2 be two smooth domains in X and let $A_1 \subset X_1$ and $A_2 \subset \partial X_2$ be compact. Since we deal with the relative homology, we suppose even if this is not important for the sequel that A_1 and A_2 are compact Lipschitz neighborhood retracts.

Proposition 2. *Let $S \in Z_k(X_1, A_1)$, $T \in Z_h(X_2, A_2)$, $h+k \geq n$ be two relative normal cycles, and assume that*

$$X_1 \cap A_2 = X_2 \cap A_1 = \emptyset, \qquad \operatorname{spt} T \cap X_1 \cap \partial X_2 = \emptyset. \tag{2}$$

Then for a.e. a close to zero

(i) $S \cap T_a$ exists as an absolute normal cycle, $S \cap T_a \in Z_{h+k-n}(X_1)$.

(ii) T has a Poincaré dual $P_T^\flat \in Z^{n-h}(X_2)$ such that $P_T = 0$ on $\partial X_2 \cap X_1$ and

$$S \cap T_a = (-1)^{(n-k)(n-k)} S(\cdot \wedge P_T^\flat) + \partial \Sigma, \qquad \text{on } X_1,$$

$\Sigma \in \mathbf{N}_{h+k-n+1}(X_1)$.

(iii) If S and T are i.m. rectifiable, then $S \cap T_a \in Z_{h+k-n}(X_1, \mathbb{Z})$ and its integral homology class depends only on the homology classes $[S]_{\mathbb{Z},\mathrm{rel}} \in H_k(X_1, A_1, \mathbb{Z})$ and $[T]_{\mathbb{Z},\mathrm{rel}} \in H_h(X_2, A_2, \mathbb{Z})$.

Proof. (i) follows from Proposition 2 in Sec. 5.3.4 taking into account that for a close to zero $\operatorname{spt} S \cap \operatorname{spt} \partial T_a$ is still empty being $X_1 \cap A_2 = \emptyset$.

(ii) Since $\operatorname{spt} T \cap X_1 \cap \partial X_2 = \emptyset$ by (2) then clearly one constructs a Poincaré dual form $P_T^\flat$ with $\operatorname{spt} P_T^\flat = 0$ on $\partial X_2 \cap X_1$. Setting $\omega = (-1)^{h(n-h)} P_T^\flat$, and for $\delta := \operatorname{dist}(X_1, A_2)$,

$$X_\delta := \{x \mid \operatorname{dist}(x, X_1) \leq \delta/2\},$$

ω extends as a smooth form on X_δ, $\omega = 0$ on $X_1 \setminus X_2$,

$$T = [\![X_1]\!] \llcorner \omega + R + \partial \Sigma \qquad \text{on } X$$

where $R \in \mathbf{N}_h(X_2)$, $\operatorname{spt} R \cap X_\delta = \emptyset$, $\Sigma \in \mathbf{N}_{h+1}(X_2)$. In particular

$$T = [\![X_1]\!] \llcorner \omega + \partial \Sigma \quad \text{on } X_\delta.$$

Arguing now as in the proof of (ii) of Proposition 1 we then conclude that for a.e. a close to zero

$$S \cap T_a = S \cap [\![\pi^{-1}(X_1)]\!] \llcorner \omega^a + S \cap \Sigma_a = S \llcorner \omega^a + S \cap \Sigma_a \tag{3}$$

$$S \cap \partial \Sigma = \partial (S \cap \Sigma_a), \tag{4}$$

where we have denoted by ω^a the form $\omega^a := \tau_{-a\#}\pi^{\#}\omega$. On the other hand $\omega^a - \omega = d\beta$, $\beta \in \mathcal{D}^{n-h-1}(X_2)$, $\operatorname{spt}\beta \subset X_2 \setminus X_1$. We then have $\partial S \llcorner \beta = 0$ and

$$S \llcorner (\omega^a - \omega) = (-1)^k \partial (S \llcorner \beta) \text{ on } X_1 \tag{5}$$

Putting together (3) (4) (5) we conclude

$$S \cap T_a = [\![X_2]\!] \llcorner \omega + \Sigma \quad \text{on } X_2$$

as required.

(iii) If now S and T are i.m. rectifiable, then $S \cap T_a$ is an i.m. rectifiable current for a close to zero by Proposition 2 in Sec. 5.3.4. Being T_a and T_0 clearly homotopic we have

$$T_a := T_0 + R + \partial \Sigma$$

where $\Sigma \in \mathbf{N}_{h+N+1}(\pi^{-1}(X_2))$, $R \in \mathbf{N}_{h+N}(\pi^{-1}(X_2))$, $\operatorname{spt} R \cap \pi^{-1}(X_\delta) = \emptyset$, X_δ as in step (ii). Consequently for b close to zero $T_{a+b} = T_b + \tau_{b\#}R + \tau_{b\#}\Sigma$ near X_2, $\tau_{b\#}\Sigma \in \mathbf{N}_{h+N+1}(\pi^{-1}(X_2))$, $\tau_{b\#}R \in \mathbf{N}_{h+N}(\pi^{-1}(X_2))$ and $\operatorname{spt} \tau_{b\#}R \cap X_1 = \emptyset$. On the other hand $S \cap \tau_{b\#}\Sigma$, $S \cap \partial\tau_{b\#}\Sigma$, $\partial S \cap \tau_{b\#}\Sigma$, $S \cap T_{a+b}$, $S \cap T_b$ exist for a.e. b as integral currents,

$$S \cap T_{a+b} = S \cap T_b + S \cap \partial\tau_{b\#}\Sigma \text{ on } X_1 \tag{6}$$

and

$$S \cap \partial\tau_{b\#}\Sigma = \partial(S \cap \tau_{b\#}\Sigma) + \partial S \cap \tau_{b\#}\Sigma = \partial(S \cap \tau_{b\#}\Sigma) \tag{7}$$

being $\operatorname{spt} \tau_{b\#}\Sigma \cap \operatorname{spt} \partial S = \emptyset$ for sufficiently small a and b. Putting together (6) and (7) we then conclude that for a.e. a and b $S \cap T_a$ and $S \cap T_b$ are homologous.

To conclude the proof it suffices to note that that for a.e. a close to zero clearly

$$\partial R \cap T_a = (-1)^{k+1} \partial (R \cap T_a), \qquad S \cap \partial R_a = \partial (S \cap R_a). \tag{8}$$

□

The claim (iii) of Proposition 2 shows that by setting

$$[S]_{\text{rel}} \cap [T] := [S \cap T_a]$$

for a.e. a close to zero, $[S]_{\text{rel}} \in H_k(X, \partial X, \mathbb{Z})$, and $[T] \in H_k(X, \mathbb{Z})$ defines actually the *intersection map*

$$\cap_X : H_k(X, \partial X, \mathbb{Z}) \times H_h(X, \mathbb{Z}) \to H_{h+k-n}(X, \mathbb{Z})$$

between the relative and absolute homology classes. Also (ii) of Proposition 2 shows that such a definition is compatible with the intersection map in real homology, compare Sec. 5.3.4.

In order to deal with the intersection of two relative cycles, we need the following variant of Proposition 2. As before we suppose that X_1, X_2 are two smooth domains in a compact oriented manifold X of dimension n that we think to be a submanifold in $\mathbb{R}^{n+N}$. Let $A_1 \subset X_1$ and $A_2 \subset \partial X_2$ be compact Lipschitz neighborhood retracts.

Proposition 3. *Let $S \in Z_k(X_1, A_1)$, $T \in Z_h(X_2, A_2)$, $h+k \geq n$, be two relative normal cycles, and assume that*

(9) $$X_1 \cap A_2 = \emptyset, \qquad \operatorname{spt} T \cap X_1 \cap \partial X_2 = \emptyset.$$

Then for a.e. a close to zero

(i) *$S \cap T_a$ exists as a normal current and $S \cap T_a \in Z_{h+k-n}(X_1, A_1)$.*
(ii) *T has a Poincaré dual $P_T^\flat \in Z^{n-h}(X_2)$ such that $P_T = 0$ on $\partial X_2 \cap X_1$ and*

$$S \cap T_a = (-1)^{(n-k)(n-k)} S(\cdot \wedge P_T^\flat) + R + \partial \Sigma, \qquad \text{in } X_1,$$

$R \in \mathbf{N}_{h+k-n}(X_1)$, $\operatorname{spt} R \subset A_1$, $\Sigma \in \mathbf{N}_{h+k-n+1}(X_1)$.
(iii) *If S and T are i.m. rectifiable, then $S \cap T_a \in Z_{h+k-n}(X_1, A_1, \mathbb{Z})$ and its integral homology class in $H_{h+k-n}(X_1, A_1, \mathbb{Z})$ depends only on the homology classes $[S]_{\mathbb{Z},\mathrm{rel}} \in H_k(X_1, A_1, \mathbb{Z})$ and $[T]_{\mathbb{Z},\mathrm{rel}} \in H_h(X_2, A_2, \mathbb{Z})$.*

Proof. The proof is quite similar to the proof of Proposition 2. Because of the different hypotheses, we may substitute (4), (5) and (7) respectively by

(10) $$S \cap \Sigma_a = \partial(S \cap \Sigma_a) + (-1)^k \partial S \cap \Sigma_a = \partial \Sigma' + R'$$

$R' \in \mathbf{N}_{h+k-n}(X_1)$, $\operatorname{spt} R' \subset \operatorname{spt} S \subset A_1$, $\Sigma' \in \mathbf{N}_{h+k-n+1}(X_1)$, by

(11) $$S \llcorner (\omega^a - \omega) = \partial S \llcorner \beta + (-1)^k \partial(S \llcorner \beta) = R'' + \partial \Sigma''$$

$R'' \in \mathbf{N}_{h+k-n}(X_1)$, $\operatorname{spt} R'' \subset \operatorname{spt} S \subset A_1$, $\Sigma'' \in \mathbf{N}_{h+k-n+1}(X_1)$, and by

(12) $$S \cap \partial \tau_{b\#} \Sigma = \partial(S \cap \tau_{b\#} \Sigma) + \partial S \cap \tau_{b\#} \Sigma = R''' + \partial \Sigma'''$$

$R''' \in \mathbf{N}_{h+k-n}(X_1)$, $\operatorname{spt} R''' \subset \operatorname{spt} S \subset A_1$, $\Sigma''' \in \mathbf{N}_{h+k-n+1}(X_1)$. □

In order to define the intersection of two relative cycles of X, we recall the notation of Sec. 5.1.2, particularly the extension map

$$e : X \to X_{\varepsilon_0}, \qquad \varepsilon_0 > 0.$$

which is homotopic to the identity on X. Let $[S] \in H_k(X, \partial X, \mathbb{Z})$, $[T] \in H_h(X, \partial X, \mathbb{Z})$ be two relative cycles. Then $e_\# T \in Z_h(X_{\varepsilon_0}, \partial X_{\varepsilon_0}, \mathbb{Z})$, and $X \cap$

$\partial X_{\varepsilon_0} = \emptyset$. We then can apply (iii) of Proposition 3 to S and T with $X_1 := X$, $A_1 := \partial X$, $X_2 := X_{\varepsilon_0}$, $A_2 := \partial X_{\varepsilon_0}$ and infer that the homology class of

$$S \cap (e_{\#}T)_a$$

in $H_{h+k-n}(X, \partial X, \mathbb{Z})$ depends only on the relative homology class of S and T for a.e. a close to zero. This then defines the intersection map between relative cycles,

$$\cap_X : H_k(X, \partial X, \mathbb{Z}) \times H_h(X, \partial X, \mathbb{Z}) \to H_{h+k-n}(X, \partial X, \mathbb{Z})$$

by

$$[S]_{\text{rel}} \cap_X [T]_{\text{rel}} := [S \cap (e_{\#}T)_a]_{\text{rel}}.$$

Moreover (ii) of Proposition 3 shows that this definition is compatible with the intersection of relative cycles in real homology that we have defined in Sec. 5.3.4.

Intersection index in integral homology. Let $\{S_i^{(k)}\}$, $\{T_j^{(n-k)}\}$ be two systems of integral cycles which are independent and generate respectively the free part of $H_k(X, \partial X, \mathbb{Z})$ and of $H_{n-k}(X, \mathbb{Z})$. Then the intersection index $I_{ij} := i_X(S_i, T_j)$ takes values in $\mathbb{Z}$, compare Sec. 5.3.4, and non degeneracy is equivalent to the non degeneracy of the real bilinear form

$$\left(\sum r_i S_i, \sum s_j T_j\right) \ \to \ i_X(S, T) = \sum_{i,j} r_i s_j I_{ij}.$$

Thus, being non degeneracy of Poincaré duality and of the intersection index equivalent, we have

Proposition 4. *Non degeneracy of Poincaré duality is equivalent to*

$$\dim H_k(X, \partial X, \mathbb{R}) = \dim H_{n-k}(X, \mathbb{R}) =: q, \qquad \det I_{ij} \neq 0. \tag{13}$$

We can also state

Proposition 5. *Non degeneracy of Poincaré–Lefschetz duality is equivalent to one of the following equivalent statements:*

(i) $\operatorname{rank} H_k(X, \partial X, \mathbb{Z}) = \operatorname{rank} H_{n-k}(X, \mathbb{Z}) =: q$, $\det I_{ij} \neq 0$.
(ii) *given* $S \in Z_k(X, \partial X, \mathbb{Z})$ $i_X(S, T) = 0$ $\forall$ $T \in Z_{n-k}(X, \mathbb{Z})$ *iff* $[S]_{\mathbb{Z},\text{rel}}$ *is a torsion class.*
(iii) *given* $T \in Z_{n-k}(X, \mathbb{Z})$ $i_X(S, T) = 0$ $\forall$ $S \in Z_k(X, \partial X, \mathbb{Z})$ *iff* $[T]_{\mathbb{Z}}$ *is a torsion class.*

In fact in case of (ii) non degeneracy of the intersection index on integral cycles amounts to the non degeneracy of the map

$$\mathbb{Z}^p \times \mathbb{Z}^q \ \to \ \mathbb{Z}, \quad (n, m) \ \to \ \sum n_i m_j I_{ij},$$

that is again to $p = q$ and $\det I_{ij} \neq 0$. Therefore we have $[S]_{\mathbb{R},\text{rel}} = 0$, consequently $[S]_{\mathbb{Z},\text{rel}}$ is a torsion class. The same argument applies for the third claim (iii).

We emphasize the fact that a good basis for $H_k(A,B,\mathbb{R})$ can be obtained only starting from free generators of $H_k(A,B,\mathbb{Z})$. In fact considering a generic basis of $H_k(A,B,\mathbb{R})$, $[T_1]_\mathbb{R},\dots,[T_q]_\mathbb{R}$ of integral cycles we cannot be sure in principle that $[T_1]_\mathbb{Z},\dots,[T_q]_\mathbb{Z}$ generate the free part of $H_k(A,B,\mathbb{Z})$.

An algebraic view of integral homology. Although integral currents are useful in finding representatives of integral homology classes by means of variational integrals, their use in proving the integral version of the Poincaré duality isomorphism between homology and integral cohomology is not completely clear.

In describing the integral homology of a given manifold X, one usually starts with a triangulation of a manifold and then define the integral chains of $k-cells$ of that subdivision, a boundary operator, hence the homology groups

$$H_k(X,\mathbb{Z}), \qquad H_k(X,\partial X,\mathbb{Z}).$$

Then, by duality one defines cochains as morphisms on chains of $k - cells$, a notion of coboundary, and the integral cohomology

$$H^k(X,\mathbb{Z}), \qquad H^k(X,\partial X,\mathbb{Z})$$

of the integral cochains. One then proves

Theorem 1.

$$\begin{aligned} H^k(X,\mathbb{Z}) &= \operatorname{Hom}_\mathbb{Z}(H_k(X,\mathbb{Z}),\mathbb{Z}) \oplus \operatorname{Ext}(H_{k-1}(X,\mathbb{Z}),\mathbb{Z}) \\ H^k(X,\partial X,\mathbb{Z}) &= \operatorname{Hom}_\mathbb{Z}(H_k(X,\partial X,\mathbb{Z}),\mathbb{Z}) \oplus \operatorname{Ext}(H_{k-1}(X,\partial X,\mathbb{Z}),\mathbb{Z}) \end{aligned}$$

In other words $\operatorname{Hom}_\mathbb{Z}(H_k(X,\mathbb{Z}),\mathbb{Z})$ and $\operatorname{Hom}_\mathbb{Z}(H_k(X,\partial X,\mathbb{Z}),\mathbb{Z})$ are the free parts respectively of $H^k(X,\mathbb{Z})$ and $H^k(X,\partial X,\mathbb{Z})$. If moreover $H_{k-1}(X,\mathbb{Z})$ (resp. $H_{k-1}(X,\partial X,\mathbb{Z})$) is free, then $H^k(X,\mathbb{Z})$ (resp. $H^k(X,\partial X,\mathbb{Z})$) is free, too.

Next point consists in defining a pair of morphisms

$$\begin{aligned} \llcorner &: H_k(X,\partial X,\mathbb{Z}) \times H^{n-h}(X,\partial X,\mathbb{Z}) \to H_{h+k-n}(X,\mathbb{Z}) \\ \llcorner &: H_k(X,\partial X,\mathbb{Z}) \times H^{n-h}(X,\mathbb{Z}) \to H_{h+k-n}(X,\partial X,\mathbb{Z}) \end{aligned}$$

known as the *cap product* which mimic the role of the interior product between forms and currents in real homology. Applying such maps with the fundamental cycle $[\![X]\!] \in H_n(X,\partial X)$ as first factor, one constructs the two Poincaré-Lefschetz morphisms

$$\begin{aligned} P^\sharp &: H^{n-h}(X,\partial X,\mathbb{Z}) \to H_h(X,\mathbb{Z}) \\ P^\flat &: H^{n-h}(X,\mathbb{Z}) \to H_h(X,\partial X,\mathbb{Z}) \end{aligned}$$

The celebrated theorems of Poincaré and Poincaré–Lefschetz then state

Theorem 2. *$P^\sharp$ and $P^\flat$ are isomorphisms of $\mathbb{Z}$–modules.*

Finally, combining Theorem 1 and Theorem 2 we get

$$\mathrm{Hom}_{\mathbb{Z}}(H_k(X,\mathbb{Z}),\mathbb{Z}) = H^{k,\mathrm{free}}(X,\mathbb{Z}) = H^{\mathrm{free}}_{n-k}(X,\partial X,\mathbb{Z})$$
$$\mathrm{Hom}_{\mathbb{Z}}(H_k(X,\partial X,\mathbb{Z}),\mathbb{Z}) = H^{k,\mathrm{free}}(X,\partial X,\mathbb{Z}) = H^{\mathrm{free}}_{n-k}(X,\mathbb{Z})$$

In other words defining the *intersection index* on X

$$i_X : H_{n-k}(X,\partial X,\mathbb{Z}) \times H_k(X,\mathbb{Z}) \to \mathbb{Z}$$

by

(14) $$i_X(S,T) := [S]_{\mathrm{rel}} \llcorner (P^\sharp)^{-1}[T],$$

we conclude that

Theorem 3. *i_X is unimodular.*

It turns out that the intersection index that we have defined in terms of currents agrees with the one defined in (14). From the current theory we already known that the intersection matrix $\{I_{ij}\}$ for two given systems of generators $\{S_i^{(k)}\}$ of $H_k^{\mathrm{free}}(X,\partial X,\mathbb{Z})$ and $\{T_j^{(n-k)}\}$ of $H^{\mathrm{free}}_{n-k}(X,,\mathbb{Z})$ have integral non zero determinant, $0 \neq \det I_{ij} \in \mathbb{Z}$. The algebraic integral homology theory adds to this information that $\det I_{ij} = \pm 1$. In particular for $k = 0,\dots,n$ one can choose a basis $\{S_i^{(k)}\}$ of the free part of $H_k(X,\partial X,\mathbb{Z})$ and a basis $\{T_j^{(n-k)}\}$ of the free part of $H_{n-k}(X,\mathbb{Z})$ such that

$$i_X(S_i^{(k)}, T_j^{(n-k)}) = \delta_{ij}.$$

5 Maps Between Manifolds

In this final section we discuss Sobolev and Cartesian maps and currents between Riemannian manifolds focusing on some topological properties of those maps.

In Sec. 5.5.1 we deal with Sobolev maps, reviewing some of Bethuel's results on density of smooth maps and of White's results on the d–homotopy type of Sobolev maps.

After introducing Cartesian maps and currents between Riemannian manifolds and stating a few simple extensions of results we have already proved in the previous chapters for Cartesian currents in Euclidean spaces, we illustrate in Sec. 5.5.3 and Sec. 5.5.4 the homological content of Cartesian maps and currents. In particular we shall prove that, similarly to the case of smooth maps, every Cartesian current induces sets of maps in cohomology and homology.

5.1 Sobolev Classes of Maps Between Riemannian Manifolds

Let Ω be a bounded domain in $\mathbb{R}^n$. If $\partial\Omega$ is regular, say Lipschitz-continuous, and $\tilde{\Omega} \supset\supset \Omega$, then every map $u \in W^{1,p}(\Omega, \mathbb{R}^N)$ can be extended to a map $\bar{u} : \tilde{\Omega} \to \mathbb{R}^N$ in such a way that $\bar{u} \in W^{1,p}(\tilde{\Omega}, \mathbb{R}^N)$, $\bar{u} = u$ in Ω, and even $\bar{u} \in W_0^{1,p}(\tilde{\Omega}, \mathbb{R}^N)$, with $\operatorname{spt} \bar{u} \subset\subset \tilde{\Omega}$. A sequence of mollifiers of $\bar{u}$ is then easily seen to converge in the norm of $W^{1,p}$ to u. Therefore we deduce

Theorem 1. *Suppose that $\partial\Omega$ is Lipschitz-continuous. Then every map $u \in W^{1,p}(\Omega, \mathbb{R}^N)$ is the $W^{1,p}$-limit of a sequence $u_i \in C^\infty(\bar{\Omega}, \mathbb{R}^N)$*

Actually, the following *Meyers-Serrin $H = W$ theorem* holds independently of the smoothness of $\partial\Omega$

Theorem 2. *The class $C^\infty(\Omega, \mathbb{R}^N) \cap W^{1,p}(\Omega, \mathbb{R}^N)$ is dense in $W^{1,p}(\Omega, \mathbb{R}^N)$.*

This is proved by taking *weighted mollifiers* according to the distance of points $x \in \Omega$ from the boundary $\partial\Omega$, compare also the proof of (ii) of Theorem 1 in Sec. 4.1.1.

Introducing the space

(1)
$$H^{1,p}(\Omega, \mathbb{R}^N) := \text{strong closure of } C^1(\Omega, \mathbb{R}^N) \cap W^{1,p}(\Omega, \mathbb{R}^N) \text{ in } W^{1,p}(\Omega, \mathbb{R}^N)$$

we can state Theorem 2 as

(2)
$$W^{1,p}(\Omega, \mathbb{R}^N) = H^{1,p}(\Omega, \mathbb{R}^N).$$

Moreover, recalling that convex sets of a Banach space are sequentially weakly closed if and only if they are strongly closed, one easily deduces that $H^{1,p}(\Omega, \mathbb{R}^N)$, also agrees with the sequential weak closure of $C^\infty(\Omega, \mathbb{R}^N) \cap W^{1,p}(\Omega, \mathbb{R}^N)$, i.e., with the notation of Sec. 3.4, (sw-cl- and strong-cl standing respectively for sequential weak and strong closure in $W^{1,p}$), we have

(3)
$$\begin{aligned} H^{1,p}(\Omega, \mathbb{R}^N) &= \text{sw-cl}_{W^{1,p}}\Big(C^1(\Omega, \mathbb{R}^N) \cap W^{1,p}(\Omega, \mathbb{R}^N)\Big) \\ &= \text{strong-cl}_{W^{1,p}}\Big(C^1(\Omega, \mathbb{R}^N) \cap W^{1,p}(\Omega, \mathbb{R}^N)\Big) \end{aligned}$$

This is what makes the space $H^{1,p}(\Omega, \mathbb{R}^N)$, equivalently $W^{1,p}(\Omega, \mathbb{R}^N)$, very useful in finding minima of energy functionals.

Let now Ω be a bounded domain of an n-dimensional Riemannian manifold $\mathcal{X}$ and let us replace the flat space $\mathbb{R}^N$ by a compact and boundaryless submanifold $\mathcal{Y}$ of $\mathbb{R}^N$ of dimension m. We set

(4)
$$W^{1,p}(\Omega, \mathcal{Y}) := \{u \in W^{1,p}(\Omega, \mathbb{R}^N) \mid u(x) \in \mathcal{Y} \text{ for a.e. } x \in \Omega\}.$$

We shall see that in general

$$(5) \qquad W^{1,p}(\Omega, \mathcal{Y}) \underset{\neq}{\supset} \text{strong-cl}_{W^{1,p}}\Big(C^1(\Omega, \mathcal{Y}) \cap W^{1,p}(\Omega, \mathbb{R}^N)\Big),$$

the obstruction to equality being the topology of $\mathcal{Y}$.

Of course it is to be expected that

$$H^{1,p}_{\text{strong}}(\Omega, \mathcal{Y}) := \text{strong-cl}_{W^{1,p}}\Big(C^1(\Omega, \mathcal{Y}) \cap W^{1,p}(\Omega, \mathbb{R}^N)\Big)$$

has better properties compared to $W^{1,p}(\Omega, \mathcal{Y})$. On the other hand this class in principle lacks nice compactness properties with respect to the weak convergence and therefore it is not suitable for finding minima of energy functionals. For this reason it is convenient to introduce also the class

$$(6) \qquad H^{1,p}_{\text{weak}}(\Omega, \mathcal{Y}) \;:=\; \text{sw-cl}_{W^{1,p}}\Big(C^1(\Omega, \mathcal{Y}) \cap W^{1,p}(\Omega, \mathbb{R}^N)\Big),$$

i.e., the sequential weak closure of $C^1(\Omega, \mathcal{Y}) \cap W^{1,p}(\Omega, \mathbb{R}^N)$ in $W^{1,p}(\Omega, \mathbb{R}^N)$. In general one finds then

$$W^{1,p}(\Omega, \mathcal{Y}) \underset{\neq}{\supset} H^{1,p}_{\text{weak}}(\Omega, \mathcal{Y}) \underset{\neq}{\supset} H^{1,p}_{\text{strong}}(\Omega, \mathcal{Y}).$$

In the sequel of this subsection we shall report on some results about the classes $W^{1,p}(\Omega, \mathcal{Y})$, $H^{1,p}_{\text{weak}}(\Omega, \mathcal{Y})$ and $H^{1,p}_{\text{strong}}(\Omega, \mathcal{Y})$ in principle omitting proofs.

The class $W^{1,p}(\mathcal{X}^n, \mathcal{Y}^m)$ and density of smooth maps. Let $\mathcal{X}$ and $\mathcal{Y}$ be two oriented Riemannian manifolds of dimension respectively n and m. As usual we shall assume that $\mathcal{Y}$ is compact and without boundary, while $\mathcal{X}$ may or may not have boundary. Moreover, without loss of generality, we think of $\mathcal{Y}$ as of a submanifold of $\mathbb{R}^N$, and, if we want to emphasize the dimensions, we write $\mathcal{X}^n$ and $\mathcal{Y}^m$ instead of $\mathcal{X}$ and $\mathcal{Y}$. For any real number $p \geq 1$, but actually we shall only consider the case $p > 1$, we set

$$(7) \qquad W^{1,p}(\mathcal{X}, \mathcal{Y}) \;:=\; \{u \in W^{1,p}(\Omega, \mathbb{R}^N) \,|\, u(x) \in \mathcal{Y} \text{ for a.e. } x \in \mathcal{X}\}.$$

The nonlinear space $W^{1,p}(\mathcal{X}, \mathcal{Y})$ depends in general on the embedding of $\mathcal{Y}$ into $\mathbb{R}^N$. This is not essential in case $\mathcal{X} \cup \partial\mathcal{X}$ and $\mathcal{Y}$ are compact, because different embeddings give rise to homeomorphic spaces $W^{1,p}(\mathcal{X}, \mathcal{Y})$. However, since the definition (7) works also for non compact $\mathcal{X}$ and $\mathcal{Y}$ an intrinsic definition should be more suitable. But we shall not pursue this point.

Our first question is whether $C^\infty(\mathcal{X}, \mathcal{Y}) \cap W^{1,p}(\Omega, \mathbb{R}^N)$ is dense in $W^{1,p}(\mathcal{X}, \mathcal{Y})$.

Suppose for a moment that $\mathcal{X}$ is a domain of $\mathbb{R}^n$ and, for the sake of simplicity, that $\mathcal{X} = \mathbb{R}^n$. Given $u \in W^{1,p}(\mathcal{X}, \mathcal{Y})$, the easiest and natural approach to our question is to consider the mollifiers

$$u_\varepsilon := \int \varphi_\varepsilon(x - y)\, u(y)\, dy \;, \quad \varphi_\varepsilon := \varepsilon^{-n}\, \varphi(\frac{x}{\varepsilon})$$

where φ is a standard $C^\infty_c(\mathbb{R}^n)$ kernel. Then the maps u_ε converge in $W^{1,p}(\mathcal{X}, \mathbb{R}^N)$ to u. But in general the range of u_ε is not $\mathcal{Y}$. Assuming however that $u_\varepsilon(x),\ x \in \mathcal{X}$, lies in a normal neighbourhood V_ε of $\mathcal{Y}$ in $\mathbb{R}^N$, which shrinks to $\mathcal{Y}$ as ε tends to zero, we might consider the maps

$$\Pi \circ u_\varepsilon : \mathcal{X} \longrightarrow \mathcal{Y}$$

where Π is the nearest point projection map from V_ε into $\mathcal{Y}$. In this case $\Pi \circ u_\varepsilon$ is readily seen to converge in $W^{1,p}$ to u.

Unfortunately, in general we cannot expect that $u_\varepsilon(x),\ x \in \Omega$, lies in a small normal neighbourhood of $\mathcal{Y}$; for instance, any mollification with a symmetric kernel of the map $x/|x|$ from B^3 into S^2 has value zero at zero. The previous argument however does work, with minor changes, if p is larger than the dimension n of $\mathcal{X}$ because, by Sobolev-Morrey embedding theorem, we know that every $u \in W^{1,p}(\mathcal{X},\mathcal{Y})$ is actually Hölder-continuous and the mollifiers u_ε of u converge uniformly to u. In fact, essentially the same argument works also for $p = n$, and we have

Theorem 3 (Schoen-Uhlenbeck). *Let $\mathcal{X}^n$ be a Riemannian manifold with smooth boundary and let $\mathcal{Y}^m$ be a compact Riemannian manifold without boundary. Then $C^1(\bar{\mathcal{X}},\mathcal{Y})$ is dense in $W^{1,n}(\mathcal{X},\mathcal{Y})$.*

Proof. By standard extension theorems we may assume that $\mathcal{X}$ is compactly contained in a Riemannian manifold $\tilde{X}$ and that any $u \in W^{1,n}(\mathcal{X},\mathcal{Y})$ is the restriction of a map $\tilde{u} \in W^{1,n}(\tilde{\mathcal{X}},\mathcal{Y})$ to $\mathcal{Y}$. We can also assume that $\tilde{\mathcal{X}}$ is isometrically embedded in some $\mathbb{R}^k,\ k > n$. Let $\mathcal{U}$ be a normal neighbourhood of $\tilde{\mathcal{X}}$ in $\mathbb{R}^k$ and let V be a normal neighbourhood of $\mathcal{Y}$ in $\mathbb{R}^N$.

First we observe that, for $\varepsilon < \operatorname{dist}(\mathcal{X},\partial\tilde{\mathcal{X}})$, the functions

$$G_\varepsilon(x) := \int_{B^n(x,\varepsilon)} |d\tilde{u}|^n \, d\,\mathrm{vol}_{\mathcal{X}} ,$$

where $B^n(x,\varepsilon)$ denotes the geodesic ball in $\mathcal{X}$ around x with radius ε, is a continuous function of $x \in \Omega$. Also, $G_\varepsilon(x)$ clearly decreases when ε decreases, and $\lim_{\varepsilon\to 0} G_\varepsilon(x) = 0$ for all $x \in \mathcal{X}$. Therefore it follows, by the so-called Dini's theorem, that G_ε converges uniformly to zero in $\mathcal{X}$.

Now we extend $\tilde{u}$ to a map $\bar{u} \in W^{1,n}(\mathcal{U},\mathbb{R}^N)$ by setting

$$\bar{u}(x) = \tilde{u}(Px)$$

where $P : \mathcal{U} \to \tilde{\mathcal{X}}$ denotes the nearest point projection onto $\tilde{\mathcal{X}}$. Since the metric on $\mathcal{U}$ is uniformly equivalent to a product metric on $\tilde{\mathcal{X}} \times \mathbb{R}^{k-n}$, we deduce for $x \in P^{-1}(\tilde{\mathcal{X}})$

$$(8) \qquad \int_{B^k(x,\varepsilon)} |D\bar{u}|^n \, dx^1 \ldots dx^k \leq c\varepsilon^{k-n} \int_{B^n(Px,\varepsilon)} |d\tilde{u}|^n \, d\,\mathrm{vol}_{\mathcal{X}} = c\varepsilon^{k-n} G_\varepsilon(x).$$

Note also that $\bar{u}(x) \in \mathcal{Y}$ for a.e. $x \in P^{-1}(\mathcal{X})$.

Let now φ be a standard mollifier on $B^k(0,1)$, $\varphi_\varepsilon(x) := \varepsilon^{-k}\varphi(\frac{x}{\varepsilon})$, and

$$\bar{u}_\varepsilon(x) \ := \ \int \varphi_\varepsilon(x-y)\,\bar{u}(y)\,dy$$

for $x \in \mathcal{X}$. By Poincaré inequality and (8) we deduce

$$\text{(9)} \qquad \begin{aligned} \varepsilon^{-k} \int\limits_{B^k(x,\varepsilon)} & |\bar{u}(y) - \bar{u}_\varepsilon(x)|^n\,dy \\ & \le c\,\varepsilon^{n-k} \int\limits_{B^k(x,\varepsilon)} |D\bar{u}|^n\,dx^1 \ldots dx^k \le c\,G_\varepsilon(x). \end{aligned}$$

Since $\bar{u} : P^{-1}(\mathcal{X}) \to \mathcal{Y}$, this inequality implies that for all $x \in \Omega$ we have

$$\text{(10)} \qquad \operatorname{dist}(\bar{u}_\varepsilon(x), \mathcal{Y}) \ \le \ c\,G_\varepsilon^{1/n}(x)\ .$$

Finally, let $\Pi : V \to \mathcal{Y}$ be the nearest point projection map, and observe that (10) and the fact that G_ε converge uniformly to zero. Then we can define a smooth map $v_\varepsilon : \mathcal{X} \to \mathcal{Y}$ by setting $v_\varepsilon(x) := \Pi \circ \bar{u}_\varepsilon(x)$. It is readily seen that

$$\lim_{\varepsilon\to 0} \|\, v_\varepsilon - u\,\|_{W^{1,n}(\mathcal{X},\mathcal{Y})} \ = \ 0\ .$$

□

Remark 1. Taking into account the proof of Meyers-Serrin's $H = W$ theorem, it is not difficult to adapt the proof Theorem 3 to prove that *$C^\infty(\mathcal{X},\mathcal{Y}) \cap W^{1,n}(\mathcal{X},\mathcal{Y})$ is dense in $W^{1,n}(\mathcal{X},\mathcal{Y})$*, independently of the regularity of $\partial\Omega$. Actually one then shows that *each $u \in W^{1,n}(\mathcal{X},\mathcal{Y})$ can be obtained as $W^{1,n}$-limit of a sequence of functions in $C^\infty(\mathcal{X},\mathcal{Y}) \cap W^{1,n}(\mathcal{X},\mathcal{Y})$ which all have the same trace of u on $\partial\mathcal{X}$.*

Next example shows that Theorem 3 does not hold for $p < n$, in particular we have

[1] The map $u(x) := x/|x|$ from B^3 into S^2, which belongs to $W^{1,2}(B^3,S^2)$ is *not* a $W^{1,2}$-limit of a sequence $u_i \in C^1(B^3,S^2)$.

To see this suppose that such a sequence $\{u_i\}$ exists. Then for almost every $r \in (\frac{1}{2},1)$ we would have $W^{1,2}(\partial B^3(0,r),S^2)$-convergence of u_i to the map $x/|x|$. In particular the sequence of maps from S^2 into S^2 $v_i(x) \ := \ u_i(rx)$ would converge to the identity map of S^2 and each v_i would have degree zero, since each u_i provides an homotopy of v_i to a constant map; in particular for the Jacobian determinant we have

$$\text{(11)} \qquad \int\limits_{S^2} J_2(dv_i) \ = \ 0 \quad \forall\, i$$

On the other hand, by taking a subsequence, we could assume dv_i converge pointwise a.e. to the identity. Thus $J_2(dv_i) \to 1$ a.e. on S^2. Since

$$|J_2(dv_i)| \le \frac{1}{2} |dv_i|^2$$

and $|dv_i|^2$ converge in L^1 to a limit, Lebesgue's dominated convergence theorem would imply that

$$\lim_{i\to\infty} \int_{S^2} J_2(dv_i) = 4\pi$$

a contradiction to (11). •

The previous example rises the question of finding in which circumstances smooth maps between two manifolds are dense in the Sobolev classes $W^{1,p}(\mathcal{X}, \mathcal{Y})$ for $1 < p < n$.

The answer to this question is provided by the following two theorems.

Theorem 4 (Bethuel). *Let $1 < p < n$. Then smooth maps between $\mathcal{X}^n$ and $\mathcal{Y}^m$ are dense in $W^{1,p}(\mathcal{X}^n, \mathcal{Y}^m)$ if and only if the homotopy group of order $[p]$ of $\mathcal{Y}^m$ is trivial, i.e., $\pi_{[p]}(\mathcal{Y}^m) = 0$. Here $[p]$ denotes the largest integer less or equal to p.*

For the proof of Theorem 4 we refer the reader to Bethuel [87]. We shall present the proof of Theorem 4 and of related results in the special case in which $\mathcal{Y}^m$ is the two-dimensional standard sphere S^2 of $\mathbb{R}^3$ in Vol. II Sec. 4.2.2.

A variant of Theorem 4 in the case $\mathcal{X}^n$ has a non zero boundary, in particular when $\mathcal{X}^n$ is replaced by a subdomain of $\mathcal{X}^n$ with smooth boundary is given by

Theorem 5 (Bethuel). *Let $\partial\mathcal{X}^n \neq \emptyset$, $1 < p < n$, and $\pi_{[p]}(\mathcal{Y}^n) = 0$. Suppose that $u \in W^{1,p}(\mathcal{X}^n, \mathcal{Y}^m)$, is such that its trace belongs to $W^{1,p}(\partial\mathcal{X}^n, \mathcal{Y}^m) \cap C^0(\partial\mathcal{X}^n, \mathcal{Y})$, respectively $C^\infty(\partial\mathcal{X}^n, \mathcal{Y}^m)$, and that there exists $v \in C^0(\mathcal{X}^n, \mathcal{Y}^m)$, respectively $v \in C^\infty(\partial\mathcal{X}^n, \mathcal{Y}^m)$, such that $u = v$ on $\partial\mathcal{X}^n$. Then u can be approximated in $W^{1,p}(\mathcal{X}^n, \mathcal{Y}^m)$ by maps in $W^{1,p}(\mathcal{X}^n, \mathcal{Y}^m) \cap C^0(\mathcal{X}^n, \mathcal{Y}^m)$, respectively $C^\infty(\mathcal{X}^n, \mathcal{Y}^m)$, which coincide with u on $\partial\mathcal{X}^n$.*

As an immediate corollary of the previous result we can state in the case $\mathcal{Y}^m = S^2$ in which $\pi_2(S^2) = \mathbb{Z}$

Proposition 1. *We have*

(i) Suppose $n = 2$. Then $C^\infty(\mathcal{X}^2, S^2)$ is dense in $W^{1,p}(\mathcal{X}^2, S^2)$ for all $p > 1$.
(ii) Suppose $n = 3$. Then $C^\infty(\mathcal{X}^3, S^2)$ is dense in $W^{1,p}(\mathcal{X}^3, S^2)$ for all $p < 2$ and $C^\infty(\mathcal{X}^3, S^2)$ is not dense in $W^{1,p}(\mathcal{X}^3, S^2)$ for all p with $2 \le p < 3$.

When $\pi_{[p]}(\mathcal{Y}^m) \neq 0$, smooth maps are not dense in $W^{1,p}(\mathcal{X}^n, \mathcal{Y}^m)$ for $1 < p < n$. However, in this case, each map $u \in W^{1,p}(\mathcal{X}^n, \mathcal{Y}^m)$ can be approximated by maps which are regular except on low dimensional sets.

Definition 1. *We say that $u \in W^{1,p}(\mathcal{X}^n, \mathcal{Y}^m)$ is in $R_p^\infty(\mathcal{X}^n, \mathcal{Y}^m)$, respectively in $R_p^0(\mathcal{X}^n, \mathcal{Y}^m)$ if and only if u is of class C^∞, respectively continuous, except on a singular set*

$$\Sigma(u) = \bigcup_{i=1}^{r} \Sigma_i$$

where, for $i = 1, 2, \dots, r$, Σ_i is a subset of an $(n - [p] - 1)$-manifold and the boundary of Σ_i is smooth; if $p \geq n-1$, Σ_i is a point.

In the case that $\mathcal{X}^n$ is some domain of $\mathbb{R}^n$, we also require that each Σ_i is a subset of an affine subspace of $\mathbb{R}^n$ of dimension $n - [p] - 1$, and $\partial\Sigma_i$ is a subset of an affine subspace of dimension $n - [p] - 2$.

Theorem 6 (Bethuel). *We have*

(i) let $\mathcal{X}^n$ be compact and $\partial\mathcal{X}^n = \emptyset$. For every p with $1 < p < n$, $R_p^0(\mathcal{X}^n, \mathcal{Y}^m)$, respectively $R_p^\infty(\mathcal{X}^n, \mathcal{Y}^m)$ is dense in $W^{1,p}(\mathcal{X}^n, \mathcal{Y}^m)$.

(ii) Suppose $\partial\mathcal{X}^n \neq \emptyset$ and that the assumptions of Theorem 5 hold. Then each $u \in W^{1,p}(\mathcal{X}^n, \mathcal{Y}^m)$ can be approximated in $W^{1,p}(\mathcal{X}^n, \mathcal{Y}^m)$ by maps in $R_p^0(\mathcal{X}^n, \mathcal{Y}^m)$, respectively in $R_p^\infty(\mathcal{X}^n, \mathcal{Y}^m)$, which coincide with u on $\partial\mathcal{X}^n$.

In particular every map in $W^{1,2}(B^3, S^2)$, which is regular on ∂B^3, can be approximated in $W^{1,2}$ by maps from B^3 into S^2 which are regular except at a finite number of points in B^3. Notice that in principle the number of singular points may tend to infinity.

The class $H^{1,p}(\mathcal{X}^n, \mathcal{Y}^m)$. A suited class of Sobolev functions for finding minima of energy functionals is clearly given by the sequential weak closure of smooth functions in $W^{1,p}(\mathcal{X}^n, \mathcal{Y}^m)$ that we denote by

$$H^{1,p}_{\text{weak}}(\mathcal{X}, \mathcal{Y}) := \text{sw-cl}_{W^{1,p}}\Big(C^1(\mathcal{X}, \mathcal{Y}) \cap W^{1,p}(\mathcal{X}, \mathcal{Y})\Big). \tag{12}$$

In general we have

$$W^{1,p}(\mathcal{X}, \mathcal{Y}) \supset H^{1,p}_{\text{weak}}(\mathcal{X}, \mathcal{Y}) \supset H^{1,p}_{\text{strong}}(\mathcal{X}, \mathcal{Y}) \tag{13}$$

As we have seen, if p is larger than or equal to the dimension n of $\mathcal{X}$ or if $1 < p < n$ and $\pi_{[p]}(\mathcal{Y}) = 0$, the inclusions in (13) are actually equality.

In the case $\pi_{[p]}(\mathcal{Y}) \neq 0$, we shall prove in Vol. II Ch. 4 that for integer p and $\mathcal{Y} = S^p$ every element in $W^{1,p}(\mathcal{X}, \mathcal{Y})$ can be *weakly* approximated by a sequence of smooth maps in $W^{1,p}(\mathcal{X}, \mathcal{Y})$. This in particular shows that for $\mathcal{X} = B^3, \mathcal{Y} = S^2, p = 2$

$$W^{1,p}(\mathcal{X}, \mathcal{Y}) = H^{1,p}_{\text{weak}}(\mathcal{X}, \mathcal{Y}) \underset{\neq}{\supset} H^{1,p}_{\text{strong}}(\mathcal{X}, \mathcal{Y}). \tag{14}$$

It is an open question to decide whether (14) holds for all *integer* p, all $\mathcal{X}^n$ and $\mathcal{Y}^m$ with $\pi_{[p]}(\mathcal{Y}) \neq 0$.

For non integral p's we instead have

Theorem 7 (Bethuel). *Let $\partial\mathcal{X} = \emptyset$, $\pi_{[p]}(\mathcal{Y}) \neq 0$ and p be not an integer. Then smooth maps are not sequentially dense for the weak topology in $W^{1,p}(\mathcal{X},\mathcal{Y})$. Moreover every map in $W^{1,p}(\mathcal{X},\mathcal{Y})$ which is a weak limit of smooth maps, i.e., which is in $H^{1,p}_{\text{weak}}(\mathcal{X},\mathcal{Y})$, is in the strong closure of smooth maps in $W^{1,p}(\mathcal{X},\mathcal{Y})$*

Theorem 8 (Bethuel). *Let $\partial\mathcal{X} \neq \emptyset$, $\pi_{[p]}(\mathcal{Y}) \neq 0$ and p be not integer. Suppose that $u \in W^{1,p}(\mathcal{X},\mathcal{Y})$ is the weak limit of a sequence of smooth maps u_k with traces all equals to u, $u_{k|\partial\mathcal{X}} = u$, and $u_{k|\partial\mathcal{X}} \in W^{1,p}(\partial\mathcal{X},\mathcal{Y}) \cap C^0$, and all homotopic to a constant map. Then u can be approximated strongly in $H^{1,p}_{\text{weak}}(\mathcal{X},\mathcal{Y})$ by a sequence of smooth maps which coincide with u on $\partial\mathcal{X}$.*

In particular, for non integral p's and $\pi_{[p]}(\mathcal{Y}) \neq 0$, we deduce

$$W^{1,p}(\mathcal{X},\mathcal{Y}) \underset{\neq}{\supset} H^{1,p}_{\text{weak}}(\mathcal{X},\mathcal{Y}) = H^{1,p}_{\text{strong}}(\mathcal{X},\mathcal{Y}). \tag{15}$$

Sobolev maps and homotopy. A natural question to ask is about the homotopy properties of Sobolev maps. It turns out that the homotopy class of a generic map in $W^{1,p}(\mathcal{X},\mathcal{Y})$ is not well defined. However, as B. White has shown in White [667] White [668], Sobolev maps between manifolds $\mathcal{X}$ and $\mathcal{Y}$, that we shall assume for the sake of simplicity to be compact and without boundary, have a well defined *k–homotopy type* for suitable k.

Definition 2. *Two continuous maps $u, v : \mathcal{X} \to \mathcal{Y}$ are said to be k–*homotopic *if and only if their restrictions to the k-skeleton of a triangulation of X are homotopic.*

If Σ is a polyhedral complex of dimension k and $\varphi : \Sigma \to \mathcal{X}$ is continuous, by an argument similar to the one in the deformation theorem one sees that, whenever X is a k–skeleton of $\mathcal{X}$, φ is homotopic to a map $\bar{\varphi}$ such that $\bar{\varphi}(\Sigma) \subset X$. Consequently the k-homotopy of continuous maps is well defined, and u, v induce the same homomorphism

$$\pi_i(\mathcal{X}) \longrightarrow \pi_i(\mathcal{Y}) \qquad i = 1, \ldots, k$$

whenever they are k–homotopically equivalent.

The two classes $H^{1,p}_{\text{strong}}(\mathcal{X},\mathcal{Y})$ and $H^{1,p}_{\text{weak}}(\mathcal{X},\mathcal{Y})$ play a different role also with respect to homotopy considerations.

Let d be the greatest integer strictly less than p. By Sobolev embedding theorem the restriction of a map in $W^{1,p}$ to a dimensional skeleton of (a triangulation of) $\mathcal{X}$ which is in generic position give rise to a continuous map which has a d–homotopy type. This however depends a priori on the triangulation and on the position of the k–skeleton. B. White shows that for each $K > 0$ there is an $\varepsilon > 0$ such that if $u_1, u_2 \in C^1(\mathcal{X},\mathcal{Y}) \cap W^{1,p}(\mathcal{X},\mathcal{Y})$, $\| u_1 - u_2 \|_{L^p} < \varepsilon$ and $\| Du_i \|_{L^p} \leq K$, then u_1 and u_2 have the same homotopy type, consequently, compare White [668],

Theorem 9 (White). *Let d be the greatest integer strictly less than p. Then each $u \in H^{1,p}_{\text{strong}}(\mathcal{X}, \mathcal{Y})$ has a well defined d-homotopy type, and d-homotopy types are preserved by bounded weak convergence in $H^{1,p}(\mathcal{X}, \mathcal{Y})$.*

He also shows, compare White [669]

Theorem 10 (White). *The infimum of*

$$\int_{\mathcal{X}^n} |du|^p \, dvol_{\mathcal{X}^n}$$

among all smooth maps homotopic to a given smooth map φ depends only on the $[p]$-homotopy type of φ. In particular the infimum is zero if and only if φ has the $[p]$-homotopy type of a constant map.

For the weak classes the following instead holds

Theorem 11 (White). *Let d be the greatest integer less or equal to $p-1$, $d = [p-1]$. Then each $u \in H^{1,p}_{\text{weak}}(\mathcal{X}, \mathcal{Y})$ has a well-defined d-homotopy type, and d-homotopy types are preserved by bounded weak convergence.*

Notice that both theorems say that the $(p-1)$-homotopy type is well defined and preserved by bounded weak convergence if p is integer. While for non integer p's the $[p]$ and $[p-1]$-homotopy classes respectively in $H^{1,p}_{\text{weak}}$ and $H^{1,p}_{\text{strong}}$ are well defined and preserved by the bounded weak convergence.

Next examples show that the previous two theorems are optimal; in particular, if p is an integer, the p-homotopy type is *not* preserved by bounded weak convergence.

[2] Let $1 < p < n$ and let $u_i : \partial B^3 \simeq S^2 \to \partial B^3 \simeq S^2$ be a sequence of conformal diffeomorphisms that converges almost everywhere to a constant map, see [3] in Sec. 4.2.5. Then $\int_{S^2} |du_i|^p$ is uniformly bounded and the u_i's, each of which is 2-homotopic to the identity map, converge to a constant map, which is not homotopic to the identity. •

[3] Consider the map $u : B^2 \to \partial B^2 \simeq S^1$, $u(x) = x/|x| \in H^{1,p}_{\text{weak}}(B^2, \partial B^2)$ for $p < 2$. By Theorem 7 u is not in $H^{1,p}_{\text{strong}}(B^2, \partial B^2)$. Moreover there are curves in B^2, for instance two circles one enclosing the origin and one not enclosing the origin, which are homotopic in B^2 but whose images under u are not homotopic in S^1. •

Remark 2. The previous results imply also that each u in $H^{1,p}_{\text{weak}}(\mathcal{X}, \mathcal{Y})$ or in $H^{1,p}_{\text{strong}}(\mathcal{X}, \mathcal{Y})$ induces homeomorphisms of the k-homotopy groups for $0 \le k \le d$, d being the number defined respectively in Theorem 11 and Theorem 9. In particular if p is an integer u induces homeomorphisms between the (real or integral) k-homology groups of $\mathcal{X}$ and $\mathcal{Y}$ for all k with $0 \le k \le p-1$.

There seems to be deep relationships among the approximability results of Bethuel and the k–homotopy type of maps, in fact as remarked in Baldo and Orlandi [61], from their works it follows

Theorem 12. *Suppose that $p > 1$ is not an integer, and $u \in W^{1,p}(\mathcal{X}, \mathcal{Y})$. Then the following statements are equivalent*

(i) u is the limit in $W^{1,p}$–norm of smooth maps between $\mathcal{X}$ and $\mathcal{Y}$,
(ii) u has a well defined $[p]$–homotopy type
(iii) u induces well defined homeomorphisms $u_* : \pi_i(\mathcal{X}) \to \pi_i(\mathcal{Y})$, $0 \le i \le [p]$.

5.2 Cartesian Currents Between Manifolds

Extending the notions of Cartesian currents and maps to the case in which $\mathbb{R}^n$ and $\mathbb{R}^N$ are replaced by Riemannian manifolds X and Y, as well as generalizing local properties is quite simple. However it is worth to state explicitly a few facts and fix notation in this more general setting.

Approximate differentiability and the area formula. In Sec. 3.1.4 and in Sec. 3.1.5 we have already discussed these notions, clearly they extend to manifolds using local charts. Let X be a n-dimensional submanifold of $\mathbb{R}^{n+k}$ and let $u : X \to \mathbb{R}$ be an $\mathcal{H}^n$ measurable map. we say that u *is approximately differentiable at* $x \in X$ is for some chart $\varphi : U \to X$, $\varphi(0) = x$, $u \circ \varphi$ is approximately differentiable at 0. The approximate differential $D^X u(x)$ is the linear map

$$D^X u(x) : T_x X \longrightarrow \mathbb{R}$$

characterized by

$$\text{ap}D(u \circ \varphi)(0) = D^X u(x) D\varphi(0).$$

One easily checks that the notion of approximate differentiability and of approximate gradient is independent of the chosen local charts. We then also denote by $A_D^X(u)$ the set of points of X where u is approximately differentiable.

Federer's theorem, compare Sec. 3.1.4, extends to manifolds.

Theorem 1. *Let X be a n-dimensional submanifold of $\mathbb{R}^{n+k}$. The following two facts are equivalent*

(i) u is a.e. approximately differentiable on X, $\mathcal{H}^n(X \setminus A_D^X(u)) = 0$.
(ii) There exist a decomposition of X into $\mathcal{H}^n$–measurable pieces

$$X = \bigcup_{i=1}^{\infty} X_i \cup X_0$$

such that $\mathcal{H}^n(X_0) = 0$ and for $i = 1, 2, \dots$, $u_{|X_i}$ is Lipschitz.

Moreover

$$A_D^X(u) \subset \bigcup_{i=1}^{\infty} X_i$$

and $D^X(u_{|X_i})(x) = D^X u(x)$ *for a.e.* $x \in X_i$.

Let X be a n-dimensional submanifold in $\mathbb{R}^{n+k}$ and let $u : X \to \mathbb{R}^N$, $N \geq n$, be an a.e. approximately differentiable map on X. The *approximate Jacobian* is defined at every $x \in A_D^X(u)$ by

$$J_u^X(x) = \sqrt{\det{(D^X u)^*(D^X u)}} = \| \Lambda^n D^X u(x) \|.$$

If A is a measurable subset of X one defines also the *Banach indicatrix of A*

$$N_X(u, A, y) := \#\{x \in A \cap A_D^X(u) \mid u(x) = y\}$$

for every $y \in \mathbb{R}^N$.

The area formula, Theorem 1 in Sec. 3.1.5, then extends to manifolds.

Theorem 2. *We have*

(i) the map $y \to N_X(u, A, y)$ *is* $\mathcal{H}^n$*–measurable in* $\mathbb{R}^N$ *and the equality*

$$\int_A J_u^X(x)\, d\mathcal{H}^n = \int_{\mathbb{R}^N} N_X(u, A, y)\, d\mathcal{H}^n(y)$$

holds.

(ii) The set $\{y \in \mathbb{R}^N \mid N_X(u, A, y) \neq 0\} = u(A \cap A_D^X(u))$ *is countably* n*-rectifiable,*

(iii) $J_u^X \in L^1(A; \mathcal{H}^n)$ *iff* $N_X(u, A, \cdot)$ *is* $\mathcal{H}^n$*-summable in* $\mathbb{R}^N$.

Moreover for any non negative $\mathcal{H}^n$*-measurable function* $f : \mathbb{R}^N \to \mathbb{R}$, *or any* $f : \mathbb{R}^N \to \mathbb{R}$ $\mathcal{H}^n$*-measurable such that either*

(a) $f(\cdot)N_X(u, A, \cdot)$ *is* $\mathcal{H}^n$*-summable or*
(b) $f(u(\cdot))J_u^X(\cdot)$ *is* $\mathcal{H}^n \llcorner A$*-summable*

we have the area formula

$$\int_A f(u(x)) J_u^X(x)\, d\mathcal{H}^n = \int_{\mathbb{R}^N} f(y) N_X(u, A, y)\, d\mathcal{H}^n.$$

Proof. First assume that A is contained in a coordinate chart $\varphi : V \to U$ $A \subset U$ and let $B := \varphi^{-1}(A)$. Then $u \circ \varphi$ is $\mathcal{H}^n$ a.e. approximately differentiable on V and

$$\text{ap}D(u \circ \varphi)(y) = D^X u(\varphi(y)) D\varphi(y) \qquad \mathcal{H}^n \text{ a.e. } y \in V.$$

Therefore $J_{u\circ\varphi}(y) = J_u^X(\varphi(y)) \cdot J_\varphi(y)$. The classical area formula and Theorem 1 in Sec. 3.1.5 then yield

$$\begin{aligned}\int_A J_u^X \, d\mathcal{H}^n &= \int_B J_u^X(\varphi) J_\varphi(y) \, d\mathcal{H}^n = \int_B J_{u\circ\varphi} d\mathcal{H}^n \\ &= \int_{\mathbb{R}^N} N_X(u\circ\varphi, B, z) \, d\mathcal{H}^n = \int_{\mathbb{R}^N} N_X(u, A, z) \, d\mathcal{H}^n(z).\end{aligned}$$

as required. Then we prove by approximation the second part of the claim for functions which are supported in a coordinate chart. For general A's it suffices to use a partition of unity associated with a system of coordinate charts. □

Graphs of approximately differentiable maps. Let X be a n-dimensional *oriented* submanifold of $\mathbb{R}^{n+k}$ and denote by $\vec{X}$ it orienting unit n-vector field. Let Ω be an open set in X, let $u : \Omega \to \mathbb{R}^N$, $N \geq 1$ be an $\mathcal{H}^n \llcorner \Omega$ a.e. approximately differentiable map, and let $A_D^*(u)$ be the set of points of Ω where u is approximately differentiable.

Definition 1. *The* graph $\mathcal{G}_{u,\Omega}$ *of* u *is defined as the subset of* $\Omega \times \mathbb{R}^N$ *given by*

$$\mathcal{G}_{u,\Omega} = \{(x, u(x))), \; x \in \Omega \cap A_D^X(u)\}.$$

The area formula, Theorem 3 in Sec. 5.1.3, then says that $\mathcal{G}_{u,\Omega}$ is countably rectifiable set and

$$\mathcal{H}^n(\mathcal{G}_{u,\Omega}) = \int J^X_{\mathrm{id}\bowtie u} \, d\mathcal{H}^n \llcorner \Omega.$$

Moreover it is also clear that Theorem 5 in Sec. 3.1.5 extends to this more general situation, hence we conclude

Proposition 1. *The following holds*

(i) $\mathcal{H}^n(\mathcal{G}_{u,\Omega}) < +\infty$ *and* $\mathcal{G}_{u,\Omega}$ *is n-rectifiable iff* $\Lambda_n(\mathrm{id} \bowtie D^X u(x))\vec{X}(x) \in L^1(\Omega)$.

(ii) *If* $\mathcal{H}^n(\mathcal{G}_{u,\Omega}) < +\infty$ *then the unit n-vector in* $T_x\Omega \times \mathbb{R}^N$

$$\vec{\mathcal{G}}_{u,\Omega}(x) := \frac{\Lambda_n(\mathrm{id} \bowtie D^X u(x))\vec{X}(x)}{\| \Lambda_n(\mathrm{id} \bowtie D^X u(x))\vec{X}(x) \|}$$

orients the n-tangent plane to $\mathcal{G}_{u,\Omega}$ *at* $(x, u(x))$.

As in the case $\Omega \subset \mathbb{R}^n$ we introduce the class

$$\mathcal{A}^1(\Omega, \mathbb{R}^N) = \{u \in L^1(\Omega, \mathbb{R}^N) \mid u \text{ is a.e. approx. diff.}, \mathcal{H}^n(\mathcal{G}_{u,\Omega} < \infty\}.$$

Of course for the maps in $\mathcal{A}^1$ the conclusions of the previous Proposition 1 hold. Also to each $u \in \mathcal{A}^1(\Omega, \mathbb{R}^N)$ we can associate the rectifiable n-current $G_u \in \mathcal{D}_n(\Omega \times \mathbb{R}^N)$ given by

$$G_u = \tau(\mathcal{G}_{u,\Omega}, 1, \vec{\mathcal{G}}_{u,\Omega}).$$

Clearly

$$(1)\qquad M(G_u) = \mathcal{H}^n(\mathcal{G}_{u,\Omega}) = \int_\Omega \| \Lambda_n(\mathrm{id} \bowtie D^X u(x)) \| \, d\mathcal{H}^n,$$

and, using the area formula, we can express G_u as integration over the base domain. In fact if $\omega \in \mathcal{D}^n(\Omega \times \mathbb{R}^N)$ we compute

$$(2)\qquad \begin{aligned} G_u(\omega) &= \int <\omega, \vec{\mathcal{G}}_{u,\Omega}> d\mathcal{H}^n \llcorner \mathcal{G}_{u,\Omega} = \int <\omega, \vec{\mathcal{G}}_{u,\Omega}> J_u^X \, d\mathcal{H}^n \llcorner \Omega \\ &= \int <\omega, \Lambda_n(\mathrm{id} \bowtie D^X u(x))> d\mathcal{H}^n \llcorner \Omega \\ &= \int <\Lambda^n(\mathrm{id} \bowtie Du(x))\omega(x, u(x)), \vec{X}(x)> d\mathcal{H}^n \llcorner \Omega \\ &=: \int_\Omega (\mathrm{id} \bowtie u)^\# \omega \end{aligned}$$

which expresses, in the case u is Lipschitz, that $G_u = (\mathrm{id} \bowtie u)_\# [\![\Omega]\!]$.

The weak continuity theorems in Sec. 3.3.2 on minors extend trivially to maps defined on manifolds. For example, we have

Theorem 3. *Let $\{u_k\}$ be a sequence of maps in $\mathcal{A}^1(\Omega, \mathbb{R}^N)$ such that $\Lambda_n(\mathrm{id} \bowtie D^X u(x))\vec{X}(x) \in L^1(\Omega)$. Assume that*

$$u_k \rightharpoonup u \qquad \textit{weakly in } L^1$$

and that

$$\Lambda_n(\mathrm{id} \bowtie D^X u_k(x))\vec{X}(x) \rightharpoonup \xi \qquad \textit{weakly in } L^1.$$

If

$$(3)\qquad \sup_k \mathbf{M}(\partial G_{u_k} \llcorner (\Omega \times \mathbb{R}^n)) < +\infty$$

then

$$u \in \mathcal{A}^1(\Omega, \mathbb{R}^N) \quad \textit{and } \Lambda_n(\mathrm{id} \bowtie D^X u(x))\vec{X}(x) = \xi(x).$$

This theorem applies in particular to maps in the class

$$\mathrm{cart}^1(\Omega, \mathbb{R}^N) := \{u \in \mathcal{A}^1(\Omega, \mathbb{R}^N) \mid \partial G_u \llcorner \Omega \times \mathbb{R}^N = 0\}.$$

Actually a stronger result, compare Proposition 2 in Sec. 3.3.2, holds true. Denote by $\mathcal{F}_p$ the class of $(n-1)$-forms which are linear combinations of forms of the type

$$\omega(x) \wedge g(y^i) \, dy^{\beta - i}$$

for $|\beta| = 0 - \min(n, N)$, $i \in \beta$, where $\omega \in \mathcal{D}^k(\Omega)$, $k = n - |\beta|$, and $g \in C_c^\infty(\mathbb{R}, \mathbb{R})$. Then one can substitute the boundary control (3) by

$$(4)\qquad \sup_k \sup\{\partial G_u(\omega) \mid \omega \in \mathcal{F}_p, \ |\omega| \leq 1\} < +\infty$$

and the claim of Theorem 3 holds true.

Cartesian currents in $X \times \mathbb{R}^N$. Let X be an n- dimensional oriented submanifold of $\mathbb{R}^{n+k}$ and let Ω be an open set in X of finite volume. It is a trivial matter to single out the class of Cartesian currents in $X \times \mathbb{R}^N$ and prove for this class closure and structure theorems similar to Theorem 3 in Sec. 4.2.2 and Theorem 1 in Sec. 4.2.3. But in order to fix notation and for completeness let us state a few simple facts. Let us begin with a few remarks on forms in the product $\Omega \times \mathcal{Y}$ where more generally we take as $\mathcal{Y}$ any oriented, compact and boundaryless Riemannian manifold.

The product structure of $\Omega \times \mathcal{Y}$ yields a way to decompose for $x \in \Omega$, $y \in \mathcal{Y}$, the vector space of tangent n-vectors to $T_{(x,y)}(\Omega \times \mathcal{Y})$ which are linear combination of n-vectors of the type $v_1 \wedge \ldots v_{n-k} \wedge w_1 \wedge \ldots \wedge w_k$, with $v_i \in \Lambda_{n-k} T_x \Omega$ and $w_i \in \Lambda_k T_y \mathcal{Y}$. Set

$$V_k := \Lambda_{n-k} T_x \Omega \otimes \Lambda_k T_y \mathcal{Y}.$$

These vector spaces are independent and yield a decomposition of $\Lambda_n T_{(x,y)}(\Omega \times \mathcal{Y})$,

$$\Lambda_n T_{(x,y)}(\Omega \times \mathcal{Y}) = \bigoplus_{k=0}^{\min(n,m)} \Lambda_{n-k} T_x \Omega \otimes \Lambda_k T_y \mathcal{Y}$$

as we can verify using local basis in $T_x \Omega$ and $T_y \mathcal{Y}$. If $v^1, \ldots, v^n, w^1, \ldots, w^n$ are basis respectively for $T_x \Omega$ and $T_y \mathcal{Y}$, one can uniquely express $\xi \in \Lambda_n T_{(x,y)}(\Omega \times \mathcal{Y})$ as

$$\xi = \sum_{|\alpha|+|\beta|=n} \xi_{\alpha\beta} v^\alpha \wedge w^\beta = \sum_{k=0}^{n} \xi_{(k)}$$

where

$$\xi_{(k)} := \sum_{\substack{|\alpha|+|\beta|=n \\ |\beta|=k}} \xi_{\alpha\beta} v^\alpha \wedge w^\beta \in V_k .$$

The decomposition then yields by duality a decomposition of covectors and forms and consequently of currents by setting for $\omega \in \Lambda^n T_{(x,y)}(\Omega \times \mathcal{Y})$

$$\omega = \sum_{k=0}^{\min(n,m)} \omega^{(k)}, \quad <\omega^{(k)}, \xi> := <\omega, \xi_{(k)}> \quad \forall\, \xi \in \Lambda_n T_{(x,y)}(\Omega \times \mathcal{Y}),$$

for $\omega \in \mathcal{D}^n(\Omega \times \mathcal{Y})$

$$\omega = \sum_{k=0}^{\min(n,m)} \omega^k, \quad <\omega^{(k)}(x,y), \xi> := <\omega(x,y), \xi_{(k)}> \quad \forall\, \xi \in \Lambda_n T_{(x,y)}(\Omega \times \mathcal{Y})$$

and for $T \in \mathcal{D}_n(\Omega \times \mathcal{Y})$, $T = \sum_{k=0}^{\min(n,m)} T_{(k)}$ where

$$T_{(k)}(\omega) := T(\omega^{(k)}), \quad \omega \in \mathcal{D}^n(\Omega \times \mathbb{R}^N).$$

Let us denote by $\vec{X}$ the orienting n-vector field of X, by Ω_x the dual n-volume form, and by $\pi : \Omega \times \mathbb{R}^N \to \Omega$, $\widehat{\pi} : \Omega \times \mathbb{R}^N \to \mathbb{R}^N$ the two orthogonal projections on the factors.

Let $T \in \mathcal{D}_n(\Omega \times \mathbb{R}^N)$. We define

$$\| T \|_1 := \sup\{T(f(x,t)|y|\pi^{\#}\Omega_X(x,y)) \mid f \in C_c^\infty(\Omega \times \mathbb{R}^N),\ |f| \le 1\}$$

and the class of *Cartesian currents* in $\Omega \times \mathbb{R}^N$ as

$$\begin{aligned}\operatorname{cart}(\Omega \times \mathbb{R}^N) := \{ \ & T \in \mathcal{D}_n(\Omega \times \mathbb{R}^N),\ T \text{ is i.m. } n\text{-rectifiable}, \\ & M(T) < +\infty,\ \| T \|_1 < +\infty,\ \partial T \llcorner \Omega \times \mathbb{R}^N = 0, \\ & \pi_{\#} T = [\![\Omega]\!],\ T \llcorner \pi^{\#}\Omega_X \ge 0\}.\end{aligned}$$

We have, compare Theorem 1 in Sec. 4.2.2

Theorem 4. *Let $T_k \in \operatorname{cart}(\Omega \times \mathbb{R}^N)$. Assume that $T_k \rightharpoonup T$ in $\mathcal{D}_n(\Omega \times \mathbb{R}^N)$ and $\sup_k(M(T_k) + \| T_k \|_1 < +\infty$. Then $T \in \operatorname{cart}(\Omega \times \mathbb{R}^N)$.*

Let now $T = \tau(\mathcal{M}, \theta, \vec{T}) \in \operatorname{cart}(\Omega \times \mathbb{R}^N)$. Define

$$\mathcal{M}_+ = \{(x,y) \in \mathcal{M} \mid <\pi^{\#}\Omega_X(x,y), \vec{T}(x,y)> > 0\}$$

and for $j = 1, 2, \dots, N$, the measures

$$u^j(T) := T \llcorner y^j \pi^{\#} \Omega_X$$

then we have Theorem 1 in Sec. 4.2.3

Theorem 5. *Let $T \in \operatorname{cart}(\Omega \times \mathbb{R}^N)$. The measures $u^j(T)$ are absolutely continuous with respect to the $\mathcal{H}^n \llcorner \Omega$ measure, $u^j(T) = u^j_T(x)\, d\mathcal{H}^n \llcorner \Omega$. Setting*

$$u_T(x) := (u_T^1(x), \dots, u_T^N(x)) \qquad \textit{for a.e. } x \in \Omega$$

we have

(i) u_T *is a.e. approximately differentiable and* $T \llcorner \mathcal{M}_T = G_{u_T}$, *in particular* $T_{(0)} = G_{u_{T(0)}}$.

(ii) $u_T \in BV(\Omega, \mathbb{R}^N)$.

Thus defining also

$$S_T := T \llcorner (\mathcal{M} \setminus \mathcal{M}_+)$$

we have a canonical decomposition of T as

$$T = G_{u_T} + S_T, \qquad \mathbf{M}(T) = \mathbf{M}(G_{u_T}) + \mathbf{M}(S_T)$$

with $u_T \in BV(\Omega, \mathbb{R}^N)$ and $S_{T(0)} = 0$. Consequently we can state

Proposition 2. $T \in \operatorname{cart}(\Omega \times \mathbb{R}^N)$ *if and only if one can write* $T = G_u + S$ *where*

(i) $u \in BV(\Omega, \mathbb{R}^N) \cap \mathcal{A}^1(\Omega, \mathbb{R}^N)$
(ii) $M(S) < +\infty,\ S_{(0)} = 0$
(iii) $\partial(G_u + S) = 0.$

Moreover the decomposition is unique and

$$\begin{cases} G_u = T \llcorner \mathcal{M}_+ \\ S = T \llcorner (\mathcal{M} \setminus \mathcal{M}_+). \end{cases}$$

Cartesian maps and currents between manifolds. It is now natural to extend the definition of Cartesian maps and currents when the target $\mathbb{R}^N$ is replaced by a m-dimensional oriented submanifold $\mathcal{Y}$ of $\mathbb{R}^N$: it suffices to require that the corresponding graphs be currents in $\Omega \times \mathcal{Y}$. Since our currents are rectifiable, we can impose such a condition just by restricting the support of the graph current, on account of Federer's flatness theorem, Theorem 1 in Sec. 5.3.1. Therefore we set

Definition 2. *We set*

$$\begin{aligned} \operatorname{cart}^1(\Omega, \mathcal{Y}) &:= \{u \in \operatorname{cart}^1(\Omega, \mathbb{R}^N) \mid G_u \in \mathcal{D}_n(\Omega \times \mathcal{Y})\} \\ &= \{u \in \operatorname{cart}^1(\Omega, \mathbb{R}^N) \mid u(x) \in \mathcal{Y}\ \mathcal{H}^n\text{-}a.e.\} \end{aligned}$$

and

$$\begin{aligned} \operatorname{cart}(\Omega \times \mathcal{Y}) &:= \operatorname{cart}(\Omega \times \mathbb{R}^N) \cap \mathcal{D}_n(\Omega \times \mathcal{Y}) \\ &= \{T \in \operatorname{cart}(\Omega \times \mathbb{R}^N) \mid \operatorname{spt} T \subset \overline{\Omega} \times \mathcal{Y}\}. \end{aligned}$$

Let Ω be an open subset of an oriented, n-dimensional submanifold $\mathcal{X}$ in $\mathbb{R}^{n+k}$ and let $\mathcal{Y}$ be a closed submanifold of $\mathbb{R}^N$ of dimension m. In dealing with the Dirichlet integral for mappings between Ω and $\mathcal{Y}$, $u : \Omega \to \mathcal{Y}$, the class

$$\operatorname{cart}^{2,2,1,\dots,1}(\Omega \times \mathcal{Y}) := \operatorname{cart}^{2,2,1,\dots,1}(\Omega \times \mathbb{R}^N) \cap \{T \mid \operatorname{spt} T \subset \overline{\Omega} \times \mathcal{Y}\}$$

briefly

$$\operatorname{cart}^{2,1}(\Omega \times \mathcal{Y})$$

will be relevant

Proposition 3. *We have*

$$\begin{aligned} \operatorname{cart}^{2,1}(\Omega \times \mathcal{Y}) = \{T = G_{u_T} + S_T \in \operatorname{cart}(\Omega \times \mathcal{Y}) \mid u_T \in W^{1,2}(\Omega, \mathcal{Y}), \\ S_{T(0)} = S_{T(1)} = 0\} \\ = \{T = G_u + S \mid u \in W^{1,2}(\Omega, \mathcal{Y}),\ M(S) < +\infty, \\ S_{(0)} = S_{(1)} = 0 \text{ and } \partial(G_u + S) = 0 \text{ on } \partial\Omega \times \mathcal{Y}\} \end{aligned}$$

Proof. Since the $\operatorname{cart}^p(\Omega, \mathbb{R}^N)$-norm of T is finite, being $p = (2, 2, 1)$, we deduce from Sec. 4.2.3 that $M_{\bar{i}}^{j}(T)$ is absolutely continuous with respect to Lebesgue's measure and

$$u_T(x) \in S^2 \text{ a.e.}\,, \quad T^{\bar{0}0}(\phi(x,y)) = \int_\Omega \phi(x, u_T(x))\, dx$$

$$T^{ij}(\phi(x,y)) = (-1)^{i-1} \int_\Omega \phi(x, u_T(x))\, D_i u_T^j\, dx$$

hence $u \in W^{1,1}(\Omega, S^2)$, since $\partial T \llcorner \Omega \times S^2 = 0$. As

$$\| u \|_{W^{1,2}} \leq \| T \|_{\mathrm{cart}^{2,1}}$$

we infer $u_T \in W^{1,2}(\Omega, S^2)$. The isoperimetric inequality

$$|J_2(Du_T)| = |M_{(2)}(Du_T)| \leq \frac{1}{2} |Du_T|^2$$

yields $u_T \in \mathcal{A}^1(\Omega, \mathbb{R}^3)$. □

The class $\mathrm{cart}^{2,1}(\Omega \times \mathcal{Y})$ has the following closure property

Theorem 6. *We have We have Suppose that $\{T_k\}$ is a sequence of currents in* $\mathrm{cart}^{2,1}(\Omega \times \mathcal{Y})$*, $T_k = G_{u_k} + S_k$, for which*

$$\sup_k \mathbf{M}(T_k) = \sup_k (\mathbf{M}(G_{u_k}) + \mathbf{M}(S_k)) < +\infty$$

$$\sup_k \| u_k \|_{W^{1,2}} < +\infty.$$

If $T_k \rightharpoonup T$ as currents and $u_k \rightharpoonup u$ in $W^{1,2}$ then $T \in \mathrm{cart}^{2,1}(\Omega \times \mathcal{Y})$ and $T = G_u + S$, with $S_{(0)} = S_{(1)} = 0$.

Proof. Passing to subsequences

$$G_{u_k} \rightharpoonup P, \qquad S_k \to R \quad \text{in } \mathcal{D}_n(\Omega \times \mathcal{Y})$$

and obviously $R_{(0)} = R_{(1)} = 0$. On the other hand, since $Du_k \rightharpoonup Du$ in L^1,

$$G_{u_k(0)} \rightharpoonup G_{u(0)}, \qquad G_{u_k(1)} \rightharpoonup G_{u(1)}$$

so that

$$G_{u_k} \rightharpoonup P = G_u + Q, \quad \text{and} \quad Q_{(0)} = Q_{(1)} = 0.$$

Hence $T = G_u + S,\ S = Q + R$ and $S_{(0)} = S_{(1)} = 0$ as required. □

Limiting graphs with such vertical pieces S may occur as weak limits of maps with bounded Dirichlet energies. We have already discussed several examples in Sec. 4.2.5. However in the special case of $\dim \Omega = \dim \mathcal{Y} = 2$, such a concentration may occur only if $\mathrm{vol}(\mathcal{Y}) < +\infty$. In fact we have

Theorem 7. *Let $T \in \mathrm{cart}^{2,1}(\Omega \times \mathcal{Y})$ where $\dim \Omega = \dim \mathcal{Y} = 2$ and $\mathcal{H}^2(\mathcal{Y}) = +\infty$. Then $T = G_u$, $u \in W^{1,2}(\Omega, \mathcal{Y})$.*

Proof. We already know that $T = G_u + S$, $u \in W^{1,2}(\Omega, \mathcal{Y})$ and $S_{(0)} = S_{(1)} = 0$. Since $\dim \Omega = 2$, $\partial G_u = 0$ hence ∂S is zero. Being S a 2-current, it suffices to prove only that $S_{(2)} = 0$.

Let α be a 1-form in $\Omega \times \mathcal{Y}$ and $f \in C_c^\infty(\Omega)$. From

$$d(f\alpha) = df \wedge \alpha + f \wedge d\alpha,$$

we infer

$$S \llcorner f(d\alpha) = S(d(f\alpha)) - S(df \wedge \alpha) = \partial S(f\alpha) - S_{(1)}(df \wedge \alpha) = 0$$

i.e. $\forall\, f \in C_c^\infty(\Omega)$, $\partial(S \llcorner f) = 0$. Consequently by the constancy theorem

$$\widehat{\pi}_\#(S \llcorner f) = k[\![\mathcal{Y}]\!].$$

Since $\mathbf{M}(S) < +\infty$ and $\mathbf{M}([\![\mathcal{Y}]\!]) = \mathcal{H}^2(\mathcal{Y}) = +\infty$ we infer $\widehat{\pi}_\#(S \llcorner f) = 0$ $\forall f \in C_c^\infty(\Omega)$, hence

$$S(f(x)\omega(y)) = 0 \qquad \forall f \in C_c^\infty(\Omega),\ \forall \omega \in \mathcal{D}^2(\mathcal{Y}).$$

By density

$$S(\omega) = 0 \qquad \forall\, \omega \in \mathcal{D}^2(\Omega \times \mathcal{Y}),\ \omega = \omega^{(2)}.$$

i.e. $S_{(2)} = 0$. □

5.3 Homology Induced Maps: Manifolds Without Boundary

In this section we briefly discuss different ways of tracking the homological content of a given Cartesian current between oriented, compact and boundaryless Riemannian manifolds. This is actually a classical topic in homology theory, and in fact in this section we shall describe a number of constructions in homology that one can perform starting from a given homology class in the product manifold.

Let X, Y be two compact, oriented and boundaryless Riemannian manifolds of dimension n and m respectively. As usual we denote by $\pi : X \times Y \to X$ and $\widehat{\pi} : X \times Y \to Y$ the orthogonal projections of the product manifold $X \times Y$ onto its factors.

If $u : X \to Y$ is a smooth map between X and Y, a way of tracking its homology is to consider the *homology maps associated to* u, i.e. the maps

$$u_* : H_k(X, \mathbb{Z}) \longrightarrow H_k(Y, \mathbb{Z})$$

defined for $k = 0, \dots, n$ by

$$u_*([R]) := [u_\#(R)]$$

for any k-cycle with finite mass $R \in Z_k(X, \mathbb{Z})$; representing homology classes by k-cycles with finite mass is particularly convenient. If $T \in \text{cart}(X \times Y)$, then T

is an *n-cycle of finite mass*, $T \in Z_n(X \times Y, \mathbb{Z})$. In the sequel we shall see how we can associate to any class $\gamma = [T]_\mathbb{R} \in H_n(X \times Y, \mathbb{R})$ a *homology map*

$$\gamma_* : H_k(X, \mathbb{R}) \longrightarrow H_k(Y, \mathbb{R}) \qquad k = 0, \dots, n$$

and to $\gamma = [T]_\mathbb{Z} \in H_n(X \times Y, \mathbb{Z})$ a *homology map*

$$\gamma_* : H_k(X, \mathbb{Z}) \longrightarrow H_k(Y, \mathbb{Z}) \qquad k = 0, \dots, n$$

which in particular extend the homology maps from the classical case of smooth maps to Cartesian currents. Such a map turns out to be stable with respect to the weak convergence of Cartesian currents just by definition.

As we have proved in Sec. 5.3, on any compact manifold without boundary one has the two non degenerate pairings of de Rham and Poincaré

$$\begin{aligned} <[T],[\omega]> &:= T(\omega), \quad T \in Z_k(X),\ \omega \in Z^k(X), \\ \mathrm{Poinc}_X<[\omega],[\eta]> &:= \int_X \omega \wedge \eta, \qquad \omega \in Z^{n-k}(X),\ \eta \in Z^k(X), \end{aligned}$$

the two being related to the Poincaré isomorphism

$$P_X : H_k(X, \mathbb{R}) \longrightarrow H^{n-k}_{\mathrm{dR}}(X)$$

by

(1) $$<[T],[\omega]> = \mathrm{Poinc}_X<\omega, P_X[T]>.$$

Let now $\gamma \in H_p(X \times Y, \mathbb{R})$, equivalently let $\gamma = [T]_\mathbb{R}$, $T \in Z_p(X \times Y)$. Then T yields for $k = \max(p-n, 0), \dots, \min(p, m)$ bilinear maps

$$\widehat{T} : Z^{p-k}(X) \times Z^k(Y) \longrightarrow \mathbb{R}$$

given by $\widehat{T}(\omega, \eta) = T(\pi^\# \omega \wedge \widehat{\pi}^\# \eta)$. It is easy to check that the value of $\widehat{T}(\omega, \eta)$ depends only on the cohomology classes of ω and η and on the homology class $\gamma = [T]_\mathbb{R}$ of T. This way each $\gamma \in H_p(X \times Y, \mathbb{R})$ fixes the bilinear maps

(2) $$\begin{aligned} &\widehat{\gamma} : H^{p-k}_{\mathrm{dR}}(X) \times H^k_{\mathrm{dR}}(Y) \longrightarrow \mathbb{R}, \quad k \in (\max(p-n, 0), \min(p, m)) \\ &\widehat{\gamma}([\omega],[\eta]) := T(\pi^\# \omega \wedge \widehat{\pi}^\# \eta) \quad \text{if } \omega \in Z^{p-k}(X),\ \eta \in Z^k(Y) \end{aligned}$$

which are the *pairings in cohomology induced by* γ.

Starting from the pairings $\widehat{\gamma} : H^{p-k}_{\mathrm{dR}}(X) \times H^k_{\mathrm{dR}}(Y) \to \mathbb{R}$ one easily obtain equivalent maps by combining $\widehat{\gamma}$ with Poincaré and de Rham dualities.

(i) One defines the *cohomology map* $\gamma^* : H^k_{\mathrm{dR}}(Y) \longrightarrow H^{k+n-p}_{\mathrm{dR}}(X)$ by

(3) $$\mathrm{Poinc}_X<[\omega], \gamma^*[\eta]> := \widehat{\gamma}([\omega],[\eta]) = T(\pi^\# \omega \wedge \widehat{\pi}^\# \eta)$$

for $\eta \in Z^k(Y)$, $\omega \in Z^{p-k}(X)$.

(ii) The so called **D**-*map* $\gamma_\mathbf{D} : H^k_{\mathrm{dR}}(Y) \to H_{p-k}(X, \mathbb{R})$ by

$$<\gamma_{\mathbf{D}}[\eta], \omega> := \widehat{\gamma}([\omega], [\eta]) \tag{4}$$

for $\omega \in Z^{p-k}(X)$, $\eta \in Z^k(Y)$. Of course for $\eta \in Z^k(Y)$, $\gamma_{\mathbf{D}}[\eta]$ is the real homology class of the current $\omega \to T(\pi^{\#}\omega \wedge \widehat{\pi}^{\#}\eta)$, $\omega \in \mathcal{D}^{p-k}(X)$ and

$$P(\gamma_{\mathbf{D}})[\eta] = \gamma^*[\eta].$$

(iii) Also one defines the *homology map* induced by γ, $\gamma_* : H_{k+n-p}(X, \mathbb{R}) \to H_k(Y, \mathbb{R})$ by duality, i.e.

$$<\gamma_*[S], [\eta]> = <[S], \gamma^*[\eta]> \tag{5}$$

for $S \in Z_{k+n-p}(X), \eta \in Z^k(Y)$.

Proposition 1. *Let $T \in Z_p(X \times Y)$ and $\gamma = [T] \in H_p(X \times Y, \mathbb{R})$. Then, for $S \in Z_{k+n-p}(X)$, $\gamma_*[S]$ is the real homology class of the k-current*

$$\eta \longrightarrow (-1)^{(n-p)(p-k)} T(\widehat{\pi}^{\#}\eta \wedge \pi^{\#} P_S),$$

P_S being a Poincaré dual form of S.

Proof. If $\eta \in Z^k(Y)$ and ξ represents the cohomology class of $\gamma^*[\eta]$, $[\xi] = \gamma^*[\eta]$, then we compute

$$\begin{aligned} <[S], \gamma^*[\eta]> &= S(\xi) = \int_X \xi \wedge P_S = \\ &= (-1)^{(k+n-p)(p-k)} \int_X P_S \wedge \xi = (-1)^{(p-k)(k+n-p)} \mathrm{Poin}([P_S], \gamma^*[\eta]) \\ &= (-1)^{(k+n-p)(p-k)} T(\pi^{\#} P_S \wedge \widehat{\pi}^{\#}\eta) = (-1)^{(n-p)(p-k)} T(\widehat{\pi}^{\#}\eta \wedge \pi^{\#} P_S). \end{aligned} \tag{6}$$

□

We now show that γ^* and γ_* are extensions of the classical homology and cohomology maps induced by a smooth map

Proposition 2. *Let $u : X \to Y$ be a smooth map. Then $[G_u]_*$ and $[G_u]^*$ are respectively the ordinary homomorphisms in (real) homology and cohomology*

$$[G_u]_* = u_*, \qquad [G_u]^* = u^*.$$

Proof. Let $\eta \in Z^k(Y)$, $R \in Z_k(X)$ and let P_R be a Poincaré dual form of R. Then

$$G_u(\widehat{\pi}^{\#}\eta \wedge \pi^{\#} P_R) = \int_X u^{\#}\eta \wedge P_R = R(u^{\#}\eta) = u_{\#}(R)(\eta).$$

Thus $<[G_u]_*([R]), [\eta]> = <[u_{\#}R], [\eta]> = <u_*[R], [\eta]>$. Analogously for $\eta \in Z^k(Y)$, $w \in Z^{n-k}(X)$ $G_u(\pi^{\#}\omega \wedge \widehat{\pi}^{\#}\eta) = \int_X \omega \wedge u^{\#}\eta$ and therefore

$$\mathrm{Poinc}_X<[\omega],[G_u]^*[\eta]> = \int_X \omega\wedge u^{\#}\eta \;=\; \mathrm{Poinc}_Y<[\omega],,>[u^{\#}\eta]$$
$$= \mathrm{Poinc}_Y<[\omega],u^*[\eta]>.$$

□

It is worthwhile noticing that if one starts with an integral cycle $T \in Z_p(X \times Y, \mathbb{Z})$ and $\gamma := [T]_{\mathbb{R}}$, the above construction yields actually an homology map $\gamma_* : H_{k+n-p}(X,\mathbb{Z}) \to H_k(Y,\mathbb{Z})$, $\gamma := [T]_{\mathbb{R}}$ provided $H_k(Y,\mathbb{Z})$ is free. In fact from Proposition 1 in Sec. 5.4.2 the (real) intersection class

$$[T]_{\mathbb{R}} \cap_{X\times Y} [S\times Y]_{\mathbb{R}}$$

$S \in Z_{k+n-p}(X,\mathbb{Z})$, contains an integral cycle obtained by intersecting T with a suitable translation of $S \times Y$. Consequently $\gamma_*[S]_{\mathbb{R}} := \widehat{\pi}_*([T]_{\mathbb{R}} \cap_{X\times Y} [S\times Y]_{\mathbb{R}})$ contains an integral cycle, too. Being $H_k(Y,\mathbb{Z})$ free, $\gamma_*[S]_{\mathbb{R}}$ defines a unique integral homology class $\gamma_*[S] \in \mathcal{H}_k(X,\mathbb{Z})$. This way we have defined a map

$$\gamma_* : H_{k+n-p}(X,\mathbb{Z}) \to H_{k+n-p}(X,\mathbb{R}) \to H_k(Y,\mathbb{Z}).$$

In the general case in which the target manifold Y have some torsion, one can construct nevertheless an homology map between the integral homologies of X and Y associated to an *integral* cycle $T \in Z_p(X \times Y, \mathbb{Z})$.

In fact, given $\gamma = [T]_{\mathbb{Z}} \in H_p(X \times Y, \mathbb{Z})$, one defines $\gamma_* : H_{k+n-p}(X,\mathbb{Z}) \to H_k(Y,\mathbb{Z})$ by

$$\gamma_*[T]_{\mathbb{Z}} := \widehat{\pi}_*([T]_{\mathbb{Z}} \cap_{X\times Y} [S\times Y]_{\mathbb{Z}})$$

where $\cap$ is the intersection operator in integral homology for $X \times Y$, compare Sec. 5.4. It is easy to show that *if* $u : X \to Y$ *is a smooth map, then* $[G_u]_* : H_k(X,\mathbb{Z}) \to H_k(Y,\mathbb{Z})$ *agrees with the standard homology map* u_*. In fact, since G_u is smooth the intersection current $G_u \cap_{X\times Y} \tau_{a\#}(S\times Y)$ is well defined for all a close to zero and

$$G_u \cap_{X\times Y} \tau_{a\#}(S\times Y) = (\mathrm{id} \bowtie u)_{\#}(S).$$

Consequently

$$\gamma_*[T]_{\mathbb{Z}} := \pi_*[(\mathrm{id} \bowtie u)_{\#}(S)]_{\mathbb{Z}} = [u_{\#}S]_{\mathbb{Z}} = u_*([S]_{\mathbb{Z}}).$$

Also such a homology map extends the one in real homology we have defined before, on account of (ii) of Proposition 1 in Sec. 5.4.2 and of the relations between intersection in real and integral homology.

Künneth formula. As shown by their definitions (2) (3) (4) and (5) the maps $\widehat{\gamma}$, γ^*, $\gamma_{\mathbf{D}}$ and γ_* can be recovered each other and all depend on γ. Actually γ can be recovered from $\widehat{\gamma}$, hence the previous five ways of describing the (real) homological content of a p-cycle in $X \times Y$ are equivalent. This is a consequence of the *Künneth formula*. This formula yields an explicit relation between the homology of the product $X \times Y$ and the homology of the factors, modulo torsion.

We first state and prove *Künneth formula* for the simplicial integral homology where it reduces to a simple combinatorial argument. So let us assume that X and Y are two finite CW cubical complexes in some $\mathbb{R}^N$ and $\mathbb{R}^{N'}$ and assume that their integral homology groups

$$H_*(X, \mathbb{Z}), \qquad H_*(Y, \mathbb{Z})$$

are torsion free, i.e. free Abelian groups. Then one can choose polyhedral rectifiable cycles $R_i^{(h)} \in Z_h(X, \mathbb{Z})$, $S_j^{(k)} \in Z_k(Y, \mathbb{Z})$, in such a way that the corresponding homology classes yield respectively a basis for $H_h(X, \mathbb{Z})$ and $H_k(Y, \mathbb{Z})$. Künneth formula states that the cycles

$$R_i^{(h)} \times S_j^{(k)} \in Z_p(X \times Y, \mathbb{Z}), \quad h + k = p$$

yield a system of independent generators for $H_p(X \times Y, \mathbb{Z})$. More precisely

Theorem 1 (Künneth formula). *Let $T \in Z_p(X \times Y)$. Then T decomposes uniquely as*

$$T = \sum_{i,j,\ h+k=p} a_{ij}^{hk} R_i^{(h)} \times S_j^{(k)} + \partial \Sigma$$

where $a_{ij}^{hk} \in \mathbb{R}$ and $\Sigma \in B_{p+1}(X \times Y, \mathbb{R})$. If moreover the integral homology groups of X and Y are free and T is rectifiable, $T \in Z_p(X \times Y, \mathbb{Z})$, then the coefficients $a_{ij}^{hk} \in \mathbb{Z}$ and Σ is rectifiable.

In the proof we will use the following simple lemma of commutative algebra whose proof is straightforward, see Bott and Tu [102, 14.13.1, 14.17]

Lemma 1. *Let A, B be free Abelian groups, where $A \subset B$ and the factor group B/A is also free. Then $B \simeq A \oplus B/A$.*

In particular each basis of A (an independent system of generators of A) can be completed by elements from B/A in such a way that the resulting set forms a basis for B.

Proof of Künneth theorem. We first prove the claim for integral cycles, $T \in Z_p(X \times Y, \mathbb{Z})$.

Denote by $B_k(X, \mathbb{Z})$, $Z_k(X, \mathbb{Z})$, $\mathbf{P}_k(X, \mathbb{Z})$ respectively the sets of k-simplicial boundaries with i.m., k-cycles with i.m. and k-chains with i.m. and by $B_k(Y, \mathbb{Z})$, $Z_k(Y, \mathbb{Z})$, $\mathbf{P}_k(Y, \mathbb{Z})$ the corresponding ones for Y. First we construct suitable bases for all $\mathbf{P}_k(X, \mathbb{Z})$ and $\mathbf{P}_k(Y, \mathbb{Z})$.

We start with $k = n$. Note that $B_n(X, \mathbb{Z}) = 0$, choose a basis $\{S_\alpha^{(n)}\}$ for $Z_n(X, \mathbb{Z})$. Complete now $\{S_\alpha^{(n)}\}$ with $\{C_\beta^{(n)}\} \subset \mathbf{P}_n(X, \mathbb{Z})$ so that $\{S_\alpha^{(n)}, C_\beta^{(n)}\}$ be a basis for $\mathbf{P}_n(X, \mathbb{Z})$. To do this we use Lemma 1. Since $Z_n(X, \mathbb{Z}) \simeq H_n(X, \mathbb{Z})$ and $\mathbf{P}_n(X, \mathbb{Z})$ is free we need to show that the factor space $\mathbf{P}_n(X, \mathbb{Z})/Z_n(X, \mathbb{Z})$ is free. But this is trivial since from $C \in \mathbf{P}_n(X,\mathbb{Z}) \setminus Z_n(X,\mathbb{Z})$ and $k \cdot C \in Z_n(X,\mathbb{Z})$ for some $k \in \mathbb{Z}$ it follows

$$0 = \partial(kC) = k\partial C, \qquad \text{i.e. } \partial C = 0.$$

Since $\{S_\alpha^{(n)}, C_\beta^{(n)}\}$ form a basis for $\mathbf{P}_n(X,\mathbb{Z})$ $\{\partial C_\beta^{(n)}\}$ generates $B_{n-1}(X,\mathbb{Z})$. We shall now show that they are independent. Assume that for some $\{b_\beta\} \subset \mathbb{Z}$ we have

$$\sum b_\beta \partial C_\beta^{(n)} = 0, \quad \{b_\beta\} \text{ non-trivial.}$$

Then $\sum b_\beta C_\beta^{(n)}$ is a cycle in $Z_n(X,\mathbb{Z})$ and, using the basis $\{S_\alpha^{(n)}\}$ for $Z_n(X,\mathbb{Z})$, we have

$$\sum b_\beta C_\beta^{(n)} = \sum a_\alpha S_\alpha^{(n)}$$

for some integers a_α. But this contradicts the fact that $\{S_\alpha^{(n)}, C_\beta^{(n)}\}$ form a basis for $Z_n(X,\mathbb{Z})$.

We now proceed by downward induction. We select for $k = n, n-1, \dots$ cycles $S_\beta^{(k)} \in Z_k(X,\mathbb{Z})$ and chains $C_\gamma^{(k)} \in \mathbf{P}_k(X,\mathbb{Z})$ such that

(7) $\{\partial C_\alpha^{(k+1)}\}$ is a basis for $B_k(X,\mathbb{Z})$,

(8) $\{\partial C_\alpha^{(k+1)}, S_\beta^{(k)}\}$ is a basis for $Z_k(X,\mathbb{Z})$,

(9) $\{\partial C_\alpha^{(k+1)}, S_\beta^{(k)}, C_\gamma^{(k)}\}$ is a basis for $\mathbf{P}_k(X,\mathbb{Z})$,

and we prove that

(i) $\{\partial C_\gamma^{(k)}\}$ are independent. If not

$$\sum c_\gamma \partial C_\gamma^{(k)} = 0, \quad c_\gamma \in \mathbb{Z}, \quad \{C_\gamma\} \text{ non trivial}$$

then $\sum c_\gamma C_\gamma^{(k)} \in Z_k(X,\mathbb{Z})$ and by (8) we have $\sum c_\gamma C_\gamma^{(k)} = \sum a_\alpha \partial C_\alpha^{(k+1)} + \sum b_\beta S_\beta^{(k)}$ for some integers a_α, b_β: a contradiction with (9).

(ii) Since $Z_{k-1}(X,\mathbb{Z})/B_{k-1}(X,\mathbb{Z}) \simeq H_{k-1}(X,\mathbb{Z})$ and $B_{k-1}(X,\mathbb{Z})$ free, we can choose $\{S_\beta^{(k-1)}\} \subset Z_{k-1}(X,\mathbb{Z})$ by means of Lemma 1, to obtain a basis

$$\{\partial C_\alpha^{(k)}, S_\beta^{(k-1)}\} \text{ for } Z_{k-1}(X,\mathbb{Z})$$

(iii) We complete the basis for $\mathbf{P}_{k-1}(X,\mathbb{Z})$ by adding suitable polyhedral chains $\{C_\gamma^{(k-1)}\}$, which is possible again by Lemma 1 since $\mathbf{P}_{k-1}(X,\mathbb{Z})/Z_{k-1}(X,\mathbb{Z})$ is trivially free.

Similarly we construct basis $\{\partial \widehat{C}_\alpha^{(h+1)}, \widehat{S}_\beta^{(h)}, \widehat{C}_\gamma^{(h)}\}$ for $B_h(Y,\mathbb{Z}), Z_h(Y,\mathbb{Z})$ and $\mathbf{P}_h(Y,\mathbb{Z})$ respectively. Then each cycle $T \in Z_p(X \times Y,\mathbb{Z})$ can be expressed as a linear combination (with integer coefficients) of products of basis elements in $\mathbf{P}_k(X,\mathbb{Z})$ and $\mathbf{P}_h(Y,\mathbb{Z})$, $k+h=p$. We now observe that the terms involving products of the type

$$\partial C_\alpha^{(k+1)} \times \partial \widehat{C}_{\hat{\alpha}}^{(h+1)}, \ \partial C_\alpha^{(k+1)} \times \widehat{S}_{\hat{\beta}}^{(h)}, \ S_\beta^{(k)} \times \partial \widehat{C}_{\hat{\alpha}}^{(h+1)}$$

are boundaries in $B_p(X \times Y, \mathbb{Z})$, and terms of the type $\partial C_\alpha^{(k)} \times \widehat{C}_\beta^{(h)}$ are homologous to terms like $C_\alpha^{(k)} \times \partial \widehat{C}_\beta^{(h)}$ e.g.

$$\partial(C_\alpha^{(k)} \times \widehat{C}_\beta^{(h)}) = \partial C_\alpha^{(k)} \times \widehat{C}_\beta^{(h)} + (-1)^k C_\alpha^{(k)} \times \partial \widehat{C}_\beta^{(h)}.$$

Therefore, modulo a boundary, we get the decomposition of T,

$$\begin{aligned} T - \partial R \ = \ & \sum a_\alpha^{kh} S_\alpha^{(k)} \times \widehat{S}_\beta^{(h)} + \\ & + \sum b_{\alpha\beta}^{kh} S_\alpha^{(k)} \times \widehat{C}_\beta^{(h)} + \sum c_{\alpha\beta}^{kh} C_\alpha^{(k)} \times \widehat{S}_\beta^{(h)} + \\ & + \sum d_{\alpha\beta}^{kh} C_\alpha^{(k)} \times \widehat{C}_\beta^{(h)} + \sum e_{\alpha\beta}^{kh} C_\alpha^{(k)} \times \partial \widehat{C}_\beta^{(h+1)}, \end{aligned}$$

where coefficients are integers and $R \in \mathbf{P}_{p+1}(X \times Y, \mathbb{Z})$.

Taking the boundary, we obtain

$$\begin{aligned} 0 = \partial T \ = \ & \sum (-1)^k b_{\alpha\beta}^{kh} S_\alpha^{(k)} \times \partial \widehat{C}_\beta^{(h)} + \sum c_{\alpha\beta}^{kh} \partial C_\alpha^{(k)} \times \widehat{S}_\beta^{(h)} + \\ & + \sum d_{\alpha\beta}^{kh} \partial C_\alpha^{(k)} \times \widehat{C}_\beta^{(h)} + \sum (-1)^k d_{\alpha\beta}^{kh} C_\alpha^{(k)} \times \partial \widehat{C}_\beta^{(h)} + \\ & + \sum e_{\alpha\beta}^{kh} \partial C_\alpha^{(k)} \times \partial \widehat{C}_\beta^{(h+1)}. \end{aligned}$$

Since all the products on the right hand side are independent, we infer

$$b_{\alpha\beta}^{kh} = c_{\alpha\beta}^{kh} = d_{\alpha\beta}^{kh} = e_{\alpha\beta}^{kh} = 0$$

hence

$$T - \partial R = \sum a_{\alpha\beta}^{kh} S_\alpha^{(k)} \times \widehat{S}_\beta^{(h)}, \quad a_{\alpha\beta}^{kh} \in \mathbb{Z}.$$

This concludes the proof of the Künneth formula for integral cycles.

If now $T \in Z_p(X \times Y, \mathbb{R})$ is a real cycle we follow the same path and we get

$$T - \partial R = \sum a_{\alpha\beta}^{kh} S_\alpha^{(k)} \times \widehat{S}_\beta^{(h)},$$

for some $a_{\alpha\beta}^{kh} \in \mathbb{R}$ and $R \in \mathbf{P}_{p+1}(X \times Y, \mathbb{R})$.

Finally in order to show that the decomposition is unique, assume that

$$\sum a_{\alpha\beta}^{kh} S_\alpha^{(k)} \times \widehat{S}_\beta^{(h)} = \partial R$$

for some $R \in \mathbf{P}_{p+1}(X \times Y, \mathbb{Z})$ or $R \in \mathbf{P}_{p+1}(X \times Y, \mathbb{R})$. Representing R in the same basis as before and taking the boundary we obtains that ∂R does not contain terms of the type $S_\alpha^{(k)} \times \widehat{S}_\beta^{(h)}$. Using the independence of the elements of the basis (9) we conclude that $a_{\alpha\beta}^{kh} = 0$. □

Since that simplicial homology of a cubical complex inside of a tubular neighborhood of a compact deformation retract X is equivalent to the current homology of X, we infer at once from Theorem 1

Theorem 2. *Let X, Y be two compact deformation retracts in $\mathbb{R}^N$, and assume that the homology groups $H_k(X,\mathbb{Z})$, $H_h(Y,\mathbb{Z})$, $k, h = 0.1\dots$ have no torsion, i.e. are free Abelian groups. Then there exist rectifiable cycles $S_i^{(k)} \in Z_k(X,\mathbb{Z})$, $\widehat{S}_j^{(h)}(Y,\mathbb{Z})$, $k = 0,\dots,n$, $h = 0,\dots,m$ such that the homology classes $\{[S_i^{(k)}]\}$, $\{[\widehat{S}_j^{(h)}]\}$ form bases in $H_k(X,\mathbb{Z})$ and $H_h(Y,\mathbb{Z})$ respectively, and such that each current $T \in Z_p(X \times Y, \mathbb{Z})$ can be uniquely decomposed as*

$$T = \sum a_{ij}^{kh} S_i^{(k)} \times \widehat{S}_j^{(h)} + \partial R$$

with integers a_{ij}^{kh}.

Now we shall assume both X, Y without torsion and fix generators $\{S_i^{(k)}\}$, $\{\widehat{S}_j^{(h)}\}$ as in Theorem 2. Let $\{\omega_\alpha^{(k)}\}$, $\{\widehat{\omega}_\beta^{(h)}\}$ be dual generators in $H^k_{\mathrm{dR}}(Y)$ and $H^h_{\mathrm{dR}}(Y)$, i.e.

$$S_i^{(k)}(\omega_\alpha^{(k)}) = \delta_{i\alpha}, \qquad \widehat{S}_j^{(h)}(\omega_\beta^{(h)}) = \delta_{j\beta}.$$

By duality we obtain the following decomposition of closed forms on $X \times Y$.

Proposition 3. *For each closed p-form $\phi \in \mathcal{Z}^p(X \times Y)$ there exist real numbers $b_{\alpha\beta}^{kh}$ and a form $\psi \in \mathcal{D}^{p-1}(X \times Y)$ such that*

$$\phi = \sum b_{\alpha\beta}^{kh} \pi^{\#}\omega_\alpha^{(k)} \wedge \widehat{\pi}^{\#}\widehat{\omega}_\beta^{(h)} + d\psi.$$

Using the dual basis $\{\omega_i^{(h)}\}$, $\{\eta_j^{(k)}\}$ on forms just defined it is easy to check that the coefficients in Künneth formula are given by

$$a_{ij}^{hk} = T(\pi^{\#}\omega_i^{(h)} \wedge \widehat{\pi}^{\#}\eta_j^{(k)}) = \widehat{\gamma}([\omega_i^{(h)}], [\eta_j^{(k)}]) \tag{10}$$

if $\gamma := [T]$. In particular the coefficients in Künneth formula depends on $\widehat{\gamma}$ and we conclude

Proposition 4. *Let $T \in Z_p(X \times Y, \mathbb{R})$, then the real homology class γ of T, the pairings in cohomology $\widehat{\gamma}$, the real homology maps γ_* and the cohomology maps γ^*, are all equivalent ways of describing the (real) homological content of T.*

Proof. In fact as we have seen $\widehat{\gamma}, \gamma_{\mathbf{D}}, \gamma_*, \gamma^*$ can be recovered one from the other. Now since the coefficients in Künneth formula for T in (10) depend only on $\widehat{\gamma}$, then the homology class γ of T depends only on $\widehat{\gamma}$. □

It is also interesting to compute the homology map γ_* in terms of Künneth's coefficients. We have

Proposition 5. *Let $T \in Z_p(X \times Y, \mathbb{R})$, $\gamma = [T]_{\mathbb{R}}$ and let $R \in Z_{k+n-p}(X, \mathbb{R})$ be a cycle in X. Then we have*

$$\gamma_*[R] = \sum_{i,j} a_{ij}^{p-k,k} i_X(R_i^{(p-k)}, R)[S_j^{(k)}] \tag{11}$$

$i_X(R_i^{(p-k)}, R)$ being the intersection index of $R_i^{(p-k)}$ and R. In particular if T and R are both integral cycles, then $\gamma_*[R]$ has a representative which is integral.

Proof. Let us assume for a moment that R is one of the basis cycles $R_\alpha^{(k+m-p)}$ in $Z_{k+n-p}(X,\mathbb{Z})$, and let us denote by P_R a Poincaré dual form in $Z^{p-k}(X)$ of R. Then we have for $\eta \in Z^k(Y)$

$$\begin{aligned} <\gamma_*[R],[\eta]> &= T(\widehat{\pi}^\# \eta \wedge \pi^\# P_R) = (-1)^{k(p-k)} T(\pi^\# P_R \wedge \widehat{\pi}^\# \eta) \\ &= (-1)^{k(p-k)} \sum_{i,j} a_{ij}^{p-k,k} R_i^{(p-k)}(P_R) S_j^{(k)}(\eta). \end{aligned}$$

Since, compare (7) in Sec. 5.3.4,

$$R_i^{(p-k)}(P_R) = \int_X P_E \wedge P_{R_i^{(p-k)}} = i_X(R, R_i^{(p-k)}),$$

(11) is proved for the cycles of the given basis and by linearity for all cycles. If now T and R are rectifiable, then $a_{ij}^{p-k} \in \mathbb{Z}$, and $i_X(R, R_i^{(p-k)}) \in \mathbb{Z}$, consequently $\gamma_*[R]$ contains a rectifiable cycle. □

[1] *Degree.* Let X and Y be two closed, connected, oriented manifolds of the same dimension n and let $T \in Z_n(X \times Y, \mathbb{Z})$. One can computes the homology map induced by T. Being X connected, $H_n(X,\mathbb{Z})$ has one generator, the fundamental class of X, that we can represent by $[[\![X]\!]]$. Then, taking into account that a Poincaré dual of $[\![X]\!]$ in X is the constant 1, one computes for $\eta \in Z^n(Y)$

$$[T]_*([\![X]\!])(\eta) = T(\widehat{\pi}^\# \eta \wedge 1) = \widehat{\pi}_\#(T)(\eta)$$

and by constancy theorem, $[T]_*([\![X]\!]) = \deg T [\![Y]\!]$. The number $\deg T$ is called the *mapping degree* of T and evidently $\deg T \in \mathbb{Z}$ if $T \in Z_n(X \times Y, \mathbb{Z})$. Notice that $\deg T = 0$ if $\mathrm{vol}(Y) = \infty$.

•

We conclude this section with a few remarks on the *cup product* in the context of the de Rham cohomology.

Let X and Y be two compact manifolds and let $n := \dim X$. As we have already seen in Sec. 5.3.4, the wedge product of closed forms actually induce a product on the de Rham cohomology ring, the *cup product*,

$$[\omega] \wedge [\eta] := [\omega \wedge \eta] \in H_{\mathrm{dR}}^{h+k}(X)$$

for $\omega \in H_{\mathrm{dR}}^k(X)$, $\eta \in H_{\mathrm{dR}}^h(X)$. If $u : X \to \mathcal{Y}$ is a C^1 map, then clearly $u^\# \omega \wedge u^\# \eta = u^\#(\omega \wedge \eta)$, hence

$$u^*[\omega] \wedge u^*[\eta] = u^*([\omega] \wedge [\eta])$$

for $\omega \in Z^k(X)$, $\eta \in Z^h(X)$, $h, k \geq 0$. Since in this case $[G_u]^* = u^*$, we then conclude that

Proposition 6. *If* $\gamma \in H_n(X \times Y, \mathbb{Z})$ *is representable as a graph of a* C^1 *map* $v : X \to Y$, $[G_v] = \gamma$ *then*

$$\gamma^*[\omega]\wedge\gamma^*[\eta] = \gamma^*([\omega]\wedge[\eta]), \qquad \forall\omega \in Z^k(X),\ \forall\eta \in Z^h(X).$$

Definition 1. *Define*

$$\widetilde{\text{cart}}(X \times Y) := \Big\{T \in \text{cart}(X \times Y) \quad | \quad \exists v \in C^1(X, Y) \text{ such that } [G_v] = [T] \in H_n(X \times Y, \mathbb{Z})\Big\}$$

Proposition 7. $\widetilde{\text{cart}}(X \times Y)$ *is closed for the mass bounded weak convergence of currents.*

Proof. In fact, let $T_j \in \widetilde{\text{cart}}(X \times Y)$, $\sup_j \mathbf{M}(T_j) < \infty$, and let $T_j \rightharpoonup T$. From Theorem 4 in Sec. 5.5.2, taking also into account that Y is compact, we infer that $T \in \text{cart}(X \times Y)$. On the other hand, Theorem 1 in Sec. 5.1.3 shows that T_j and T belong to the same homology class for large j hence

$$[T] = [T_j] = [G_{v_j}] \quad \text{for large } j.$$

□

In particular defining

$$\text{Cart}(X \times Y)$$

as the smallest set containing the graphs of $C^1(X, Y)$ maps which is closed under the mass bounded weak convergence of currents, we conclude from Proposition 7 that

Corollary 1. *If* $T \in \text{Cart}(X \times Y)$ *then there is a* C^1 *map* $v : X \to Y$ *such that* $[G_v] = [T]$ *in* $H_n(X \times Y, \mathbb{Z})$.

The same reasoning yields also that

Proposition 8. *Let* $u \in \text{Cart}^1(X, Y)$. *Then there exists* $v \in C^1(X, Y)$ *such that* $[G_u] = [G_v]$ *in* $H_n(X \times Y, \mathbb{Z})$.

[2] Not every homology class represented by a Cartesian current can be represented by a smooth graph.

Let T^2 denotes the standard torus in $\mathbb{R}^3$ with its usual orientation. Consider the product manifold $S^2 \times T^2$ and denote as usual by $\hat{\pi} : S^2 \times T^2 \to T^2$ the orthogonal projection onto T^2. We have

Proposition 9. *If* $T \in \widetilde{\text{cart}}(S^2 \times T^2)$ *then* $\hat{\pi}_\# T = 0$. *In particular if* $u \in \text{Cart}^1(S^2, T^2)$ *then* $\deg u = 0$.

Proof. Recall that any two form ω in T^2 can be decomposed as $\omega = \eta_1 \wedge \eta_2$, $\eta_1, \eta_2 \in \mathcal{D}^1(T^2)$. Since the homology class of T can be represented by a smooth function by Corollary 1, and $H^1_{\mathrm{dR}}(S^2) = 0$, we infer that $T^*[\omega] = T^*[\eta_1] \wedge T^*[\eta_2] = 0$. Consequently

$$\widehat{\pi}_\# T(\omega) = T(\widehat{\pi}^\# \omega) = \mathrm{Poinc}{<}1, T^*[\omega]{>} = 0.$$

If now $u \in \mathrm{Cart}^1(X,Y)$ then for $\eta \in \mathcal{D}^2(Y)$,

$$<(\deg u) T^2, [\eta]> = <[G_u]_* S^2, [\eta]> = G_u(1 \cdot \widehat{\pi}^\# \eta) = 0.$$

□

The current $T := G_c + \delta_q \times T^2$, $c \in T^2$, $q \in S^2$ belongs to $\mathrm{cart}(S^2 \times T^2)$, but $\widehat{\pi}_\# T = T^2$. In particular $T \notin \widetilde{\mathrm{cart}}(S^2 \times T^2)$. •

5.4 Homology Induced Maps: Manifolds with Boundary

Let X, Y be two compact, smooth oriented submanifolds respectively of dimension n, m possibly with boundary and let $B \subset Y$ be a closed Lipschitz neighborhood retract (relevant cases are $B = \emptyset$, $B = \partial Y$, $B = Y$). As usual we think of X and Y as submanifolds of orthogonal Euclidean spaces and we denote by $\pi : X \times Y \to X$, $\widehat{\pi} : X \times Y \to Y$ the orthogonal projections on the factors.

Let $T \in \mathbf{F}_p(X \times Y)$ with $\mathbf{M}(T) < \infty$ and $\mathrm{spt}\, \partial T \subset \partial X \times B$, and let $\gamma := [T]_{\mathrm{rel}}$ be its homology class, $\gamma \in H_p(X \times Y, \partial X \times B)$. It is easy to see that the values of

$$T(\pi^\# \omega \wedge \widehat{\pi}^\# \eta)$$

when either $\omega \in Z^{p-k}(X)$, $\eta \in Z_k(Y,B)$ or $\omega \in Z^{p-k}(X, \partial X)$, $\eta \in Z_k(Y)$ depend only on the cohomology class of ω and η and of the relative homology class γ of T. Therefore one can define three sets of bilinear forms

$$\begin{aligned}
\widehat{\gamma}^\flat &: H^{p-k}_{\mathrm{dR}}(X) \times H^k_{\mathrm{dR}}(Y,B) \longrightarrow \mathbb{R} \\
\widehat{\gamma}^\sharp &: H^{p-k}_{\mathrm{dR}}(X, \partial X) \times H^k_{\mathrm{dR}}(Y) \longrightarrow \mathbb{R} \\
\widehat{\partial\gamma} &: H^{p-k-1}_{\mathrm{dR}}(\partial X) \times H^k_{\mathrm{dR}}(B) \longrightarrow \mathbb{R}
\end{aligned}$$

respectively by

$$(1) \qquad \begin{aligned}
\widehat{\gamma}^\flat([\omega], [\eta]_{\mathrm{rel}}) &:= T(\pi^\# \omega \wedge \widehat{\pi}^\# \eta), \quad \omega \in Z^{p-k}(X),\ \eta \in Z^k(Y,B), \\
\widehat{\gamma}^\sharp([\omega]_{\mathrm{rel}}, [\eta]) &:= T(\pi^\# \omega \wedge \widehat{\pi}^\# \eta), \quad \omega \in Z^{p-k}(X, \partial X),\ \eta \in Z^k(Y), \\
\widehat{\partial\gamma}([\omega], [\eta]) &:= \partial T(\pi^\# \omega \wedge \widehat{\pi}^\# \eta) \quad \omega \in Z^{p-k-1}(\partial X),\ \eta \in Z^k(B).
\end{aligned}$$

The last one is well defined since $\partial T \in \mathbf{F}_{p-1}(\partial X \times B)$ by Federer's support theorem.

The maps $\widehat{\gamma}^\sharp, \widehat{\gamma}^\flat, \widehat{\partial\gamma}$ are compatible with the long exact sequences in cohomology of the pairs $(X, \partial X)$ and (Y, B); in fact we have

Proposition 1. *The diagram*

$$
(2)\qquad
\begin{array}{ccccccccc}
\xrightarrow{i^\#} & H_{\mathrm{dR}}^{p-k-1}(\partial X) & \xrightarrow{\delta} & H_{\mathrm{dR}}^{p-k}(X,\partial X) & \xrightarrow{j^\#} & H_{\mathrm{dR}}^{p-k}(X) & \xrightarrow{i^\#} \\
 & \times\widehat{\partial\gamma} & & \times\widehat{\gamma}^\sharp & & \times\widehat{\gamma}^\flat & \\
\xleftarrow{\delta} & H_{\mathrm{dR}}^{k}(B) & \xleftarrow{i^\#} & H_{\mathrm{dR}}^{k}(Y) & \xleftarrow{j^\#} & H_{\mathrm{dR}}^{k}(Y,B) & \xleftarrow{\delta}
\end{array}
$$

is commutative.

Proof. Recall that the commutativity of

$$
\begin{array}{ccc}
A_1 & \xrightarrow{f} & A_2 \\
\times F_1 & & \times F_2 \\
B_1 & \xleftarrow{g} & B_2
\end{array}
$$

means $F_2(f(a_1),b_2) = F_1(a_1,g(b_2))\ \forall\ a_1 \in A_1,\ \forall\ b_2 \in B_2$.
(i) Let $\omega \in Z^{p-k}(X,\partial X)$, $\eta \in Z^k(Y,B)$. Then $j^\#[\omega]_{\mathrm{rel}} = [\omega]_X$, $j^\#[\eta]_{\mathrm{rel}} = [\eta]_Y$, consequently

$$\widehat{\gamma}^\sharp([\omega]_X,[\eta]_{\mathrm{rel}}) = T(\pi^\#\omega\wedge\widehat{\pi}^\#\eta) = \widehat{\gamma}^\flat([\omega]_{\mathrm{rel}},[\eta]_Y).$$

(ii) Let now $\omega \in Z^{p-k-1}(\partial X)$ and $\eta \in Z^k(Y)$. Then by definition $\delta[\omega] := [d\widetilde{\omega}]_{\mathrm{rel}}$ where $\widetilde{\omega}$ is an extension of ω to X, i.e. $i^\#\widetilde{\omega} = \omega$. Therefore

$$
\begin{aligned}
\widehat{\gamma}^\flat([\delta\omega]_{\mathrm{rel}},[\eta]_Y) &:= T(\pi^\# d\widetilde{\omega}\wedge\widehat{\pi}^\#\eta) = T(d(\pi^\#\widetilde{\omega}\wedge\widehat{\pi}^\#\eta)) \\
&= \partial T(\pi^\#\widetilde{\omega}\wedge\widehat{\pi}^\#\eta) =: \partial\gamma([\omega]_{\partial X}, i^\#[\eta]_Y).
\end{aligned}
$$

(iii) The third case in which $\omega \in Z^{p-k-1}(\partial X)$ and $\eta \in Z^k(B)$ is similar to (ii). □

Representing the bilinear forms $\widehat{\gamma}^\sharp$, $\widehat{\gamma}^\flat$, $\widehat{\partial\gamma}$ by means of the Poincaré-Lefschetz duality in cohomology on X, compare Theorem 3 in Sec. 5.2.8, one obtains the three sets of *induced maps in cohomology*

$$
\begin{aligned}
\gamma^\sharp &: H_{\mathrm{dR}}^k(Y) \longrightarrow H_{\mathrm{dR}}^{k+p-n}(X) \\
\gamma^\flat &: H_{\mathrm{dR}}^k(Y,B) \longrightarrow H_{\mathrm{dR}}^{k+p-n}(X,\partial X) \\
(\partial\gamma)^* &: H_{\mathrm{dR}}^k(B) \longrightarrow H_{\mathrm{dR}}^{k-1+p-n}(\partial X)
\end{aligned}
$$

defined by

$$
(3)\quad
\begin{aligned}
&\mathrm{Poinc}_X\!<[\omega]_{\mathrm{rel}},\gamma^\sharp[\eta]> := T(\pi^\#\omega\wedge\widehat{\pi}^\#\eta), \quad \omega \in Z^{p-k}(X,\partial X),\ \eta \in Z^k(Y) \\
&\mathrm{Poinc}_X\!<[\omega],\gamma^\flat[\eta]_{\mathrm{rel}}> := T(\pi^\#\omega\wedge\widehat{\pi}^\#\eta), \quad \omega \in Z^{p-k}(X),\ \eta \in Z^k(Y,B) \\
&\mathrm{Poinc}_{\partial X}\!<[\omega],(\partial\gamma)^*[\eta]> := \partial T(\pi^\#\omega\wedge\widehat{\pi}^\#\eta), \quad \omega \in Z^{p-k-1}(\partial X), \eta \in Z^k(B)
\end{aligned}
$$

One can show easily that the three sets of maps $\gamma^\sharp$, $\gamma^\flat$, $(\partial\gamma)^*$ yield actually a morphism of the two long exact sequence in cohomology of the pairs $(X,\partial X)$ and (Y,B), i.e.

Proposition 2. *The diagram*

$$
\begin{array}{lccccccc}
 & \cdots \overset{\delta}{\leftarrow} & H_{\mathrm{dR}}^{k-1+p-n}(\partial X) & \overset{i^{\#}}{\leftarrow} & H_{\mathrm{dR}}^{k+p-n}(X) & \overset{j^{\#}}{\leftarrow} & H_{\mathrm{dR}}^{k+p-n}(X,\partial X) & \overset{\delta}{\leftarrow} \cdots \\
(4) & & \uparrow (\partial\gamma)^* & & \uparrow \gamma^{\sharp} & & \uparrow \gamma^{\flat} & \\
 & \cdots \overset{\delta}{\leftarrow} & H_{\mathrm{dR}}^{k}(B) & \overset{i^{\#}}{\leftarrow} & H_{\mathrm{dR}}^{k}(Y) & \overset{j^{\#}}{\leftarrow} & H_{\mathrm{dR}}^{k}(Y,B) & \overset{\delta}{\leftarrow} \cdots
\end{array}
$$

is commutative.

Proof. In order to prove the claim let us prove the following simple fact. Let A_i, B_i, C_i, $i = 1,2$ be finite vector spaces and for $i = 1,2$, let $F_i : B_i \times C_i \to \mathbb{R}$ be a bilinear map, $G_i : A_i \times B_i \to \mathbb{R}$, a duality and $f_i : C_i \to A_i$ be a representation of F_i by means of the duality G_i, i.e.

$$G_i(f_i(a_i), b_i) = F_i(b_i, a_i) \qquad \forall\, a_i \in A_i\ \forall\, b_i \in B_i.$$

Given now three linear maps $g : A_1 \to A_2$, $h : B_1 \to B_2$ and $k : C_1 \to C_2$, form the the following diagram

$$
\begin{array}{ccccc}
 & A_1 & \overset{g}{\longleftarrow} & A_2 & \\
 & \times G_1 & & \times G_2 & \\
f_1 & B_1 & \overset{h}{\longrightarrow} & B_2 & f_2 \\
 & \times F_1 & & \times F_2 & \\
 & C_1 & \overset{k}{\longleftarrow} & C_2 &
\end{array}
$$

We claim that if the small squares are commutative, i.e., if

$$
\begin{cases}
G_1(g(a_2), b_1) = G_2(a_2, h(b_1)) & \forall\, b_1 \in B_1,\ \forall\, a_2 \in A_2 \\
F_1(b_1, k(c_2)) = F_2(h(b_1), c_2) & \forall\, b_1 \in B_1,\ \forall\, c_2 \in C_2
\end{cases}
$$

then the big square

$$
\begin{array}{ccc}
A_1 & \overset{g}{\longleftarrow} & A_2 \\
\uparrow f_1 & & \uparrow f_2 \\
C_1 & \overset{k}{\longleftarrow} & C_2
\end{array}
$$

is commutative too, as in fact for $c_2 \in C_2$ and $b_1 \in B_1$,

$$
\begin{aligned}
G_1(f_1 \circ k(c_2), b_1) &= F_1(b_1, k(c_2)) = F_2(h(b_1), c_2) \\
&= G_2(f_2(c_2), h(b_1)) = G_1(g \circ f_2(c_2), b_1).
\end{aligned}
$$

Putting together the diagram of the Poincaré-Lefschetz duality Theorem 3 in Sec. 5.2.8 and (4) we obtain the diagram

$$\begin{array}{ccccccccc}
\cdots \xrightarrow{\delta} & H^{k+n-p}_{\mathrm{dR}}(\partial X) & \xrightarrow{i^\#} & H^{k+n-p}_{\mathrm{dR}}(X) & \xrightarrow{j^\#} & H^{k+n-p}_{\mathrm{dR}}(X,\partial X) & \xrightarrow{\delta} \cdots \\
 & \times\mathrm{Poinc} & & \times\mathrm{Poinc} & & \times\mathrm{Poinc} & \\
\cdots \xrightarrow{i^\#} & H^{p-k-1}_{\mathrm{dR}}(\partial X) & \xrightarrow{\delta} & H^{p-k}_{\mathrm{dR}}(X,\partial X) & \xrightarrow{j^\#} & H^{p-k}_{\mathrm{dR}}(X) & \xrightarrow{i^\#} \cdots \\
 & \times\widehat{\partial\gamma} & & \times\widehat{\gamma}^\flat & & \times\widehat{\gamma}^\sharp & \\
\cdots \xrightarrow{\delta} & H^{k}_{\mathrm{dR}}(B) & \xrightarrow{i^\#} & H^{k}_{\mathrm{dR}}(Y) & \xrightarrow{j^\#} & H^{k}_{\mathrm{dR}}(Y,B) & \xrightarrow{\delta} \cdots
\end{array}$$

Since all the small squares are commutative the claim follows from the above observation. □

It is also easy to show that in case $\gamma = [G_u]_{\mathrm{rel}}$, u being a smooth function $u : X \to Y$ with $u(\partial X) \subset B$, the maps $[G_u]^\sharp$, $[G_u]^\flat$ and $(\partial[G_u])^*$ agree with the standard induced maps in cohomology by the pullback of forms. Let us prove the claim for $[G_u]^\flat$, the others being similar. Fix $\eta \in Z^k(Y,B)$ and let $\xi \in Z^k(X,\partial X)$ be a form representing the cohomology class of $[G_u]^\flat[\eta]_{\mathrm{rel}}$. Then

$$\int_X \omega\wedge\xi = \mathrm{Poinc}_X < [\omega], [G_u]^\flat[\eta]_{\mathrm{rel}} > = G_u(\pi^\#\omega\wedge\widehat{\pi}^\#\eta) = \int_X \omega\wedge u^\#\eta$$

for $\omega \in Z^{n-k}(X)$. Therefore ξ and $u^\#\eta$ are in the same relative cohomology class, i.e.

$$[G_u]^\flat[\eta] = [\xi]_{\mathrm{rel}} = [u^\#\eta]_{\mathrm{rel}}.$$

Starting again with the bilinear maps in cohomology $\widehat{\gamma}^\sharp$, $\widehat{\gamma}^\flat$, $\widehat{\partial\gamma}$, one then constructs the *induced maps in homology*

$$\begin{aligned}
\gamma_\sharp : H_{k+n-p}(X,\mathbb{R}) &\longrightarrow H_k(Y,\mathbb{R}) \\
\gamma_\flat : H_{k+n-p}(X,\partial X,\mathbb{R}) &\longrightarrow H_k(Y,B,\mathbb{R}) \\
(\partial\gamma)_* : H_k(\partial X,\mathbb{R}) &\longrightarrow H_k(B,\mathbb{R})
\end{aligned}$$

setting respectively

$$\begin{aligned}
&\gamma_\sharp([S]) := T(\cdot\wedge P^\sharp_S), && S \in Z_{k+n-p}(X) \\
(5)\qquad &\gamma_\flat([S]) := T(\cdot\wedge P^\flat_S), && S \in Z_{k+n-p}(X,\partial X) \\
&(\partial\gamma)_*[S] := \partial T(\cdot\wedge P_S), && S \in Z_{k+n-p}(\partial X)
\end{aligned}$$

and $P^\sharp_S$, $P^\flat_S$, P_S denotes respectively a Poincaré dual of S in the three cases, $S \in Z_{k+n-p}(X)$, $S \in Z_{k+n-p}(X,\partial X)$ and $S \in Z_{k+n-p}(\partial X)$.

One can prove, but we leave to the reader, that the couples of maps $\gamma^\sharp$, $\gamma_\sharp$, $\gamma^\flat$, $\gamma_\flat$ and $(\partial\gamma)^*$, $(\partial\gamma)_*$ are dual maps via the de Rham duality,

$$\begin{aligned}
&<\gamma_\sharp[S],[\eta]>_Y = <[S],\gamma^\sharp[\eta]>^\sharp_Y, && S \in Z^{k+n-p}(X),\ \eta \in Z^k(Y) \\
(6)\qquad &<\gamma_\flat[S]_{\mathrm{rel}},[\eta]_{\mathrm{rel}}>_Y = <[S],\gamma^\flat[\eta]>^\flat_X && S \in Z^{k+n-p}(X,\partial X),\ \eta \in Z_k(Y,B) \\
&<(\partial\gamma)_*[S],[\eta]>_B = <[S],(\partial\gamma)^*[\eta]>_{\partial X}, && S \in Z^{k+n-p}(\partial X),\ \eta \in Z_k(B).
\end{aligned}$$

It is also easy to show that in case $\gamma = [G_u]_{\mathrm{rel}}$, u being a smooth function $u : X \to Y$, $u(\partial X) \subset B$, the maps $[G_u]_\sharp$, $[G_u]_\flat$ and $(\partial[G_u])_*$ agree with the standard induced maps in cohomology by the push-forward of currents.

Finally the maps in homology define a morphism between the long exact sequences in homology for the couples $(X, \partial X)$ and (Y, B). In fact one proves that

Proposition 3. *The diagram*

$$
(7)\qquad
\begin{array}{ccccccccc}
\cdots \xrightarrow{i_\#} & H_k(X,\mathbb{R}) & \xrightarrow{j_\#} & H_k(X,\partial X,\mathbb{R}) & \xrightarrow{\partial} & H_{k-1}(\partial X,\mathbb{R}) & \xrightarrow{i_\#} \cdots \\
 & \downarrow \gamma_\sharp & & \downarrow \gamma_\flat & & \downarrow (\partial\gamma)_* & \\
\cdots \xrightarrow{i_\#} & H_k(Y,\mathbb{R}) & \xrightarrow{j_\#} & H_k(Y,B,\mathbb{R}) & \xrightarrow{\partial} & H_{k-1}(B,\mathbb{R}) & \xrightarrow{i_\#} \cdots
\end{array}
$$

is commutative.

Finally starting with an integral cycle $T \in \mathcal{I}_p(X \times Y)$ with $\operatorname{spt} T \subset \partial X \times B$ one sees that the induced real homology maps $T_\sharp$, $T_\flat$, $(\partial T)_*$ extend to maps on the integral homology. Such extensions can be found as in the boundaryless case in terms of intersection of cycles in $X \times Y$. In order to do that, recall the notations of Sec. 5.1.2 and particularly the extension diffeomorphisms

$$e : X \to X_{\varepsilon_0}, \qquad \widehat{e} : Y \to Y_{\varepsilon_O}, \quad \varepsilon_0 > 0$$

which are homotopic respectively to the identity on X and Y, $e(\partial X) = \partial X_{\varepsilon_0}$, $\widehat{e}(\partial Y) = \partial Y_{\varepsilon_0}$, and their inverses

$$r : X_{\varepsilon_0} \to X \qquad \widehat{r} : Y_{\varepsilon_0} \to Y.$$

Let $S \in Z_{k+n-p}(X, \mathbb{Z})$ be a rectifiable cycle. Evidently $r_\# S$ is in the same homology class of S in $H_{k+n-p}(X, \mathbb{Z})$ and $\operatorname{spt} r_\# S \cap \partial X = \emptyset$. Given now an integral relative cycle $T \in Z_p(X, B, \mathbb{Z})$ one can apply (iii) of Proposition 2 in Sec. 5.4.2 to the currents

$$T \in Z_p(X \times Y, \mathbb{Z}), \qquad r_\# S \times \widehat{e}_\# [\![Y]\!] \in Z_{k+n-p+m}(X \times Y_{\varepsilon_0}, X \times \partial Y_{\varepsilon_0}, \mathbb{Z})$$

and find that for a.e. a close to zero, $T \cap (r_\# S \times \widehat{e}_\# [\![Y]\!])_a$ is an integral current with no boundary in $Z_k(X \times Y, \emptyset, \mathbb{Z})$ whose integral homology class depends only on the integral homology classes of S and T. In this way one defines

$$[T]_\sharp : H_{k+n-p}(X, \mathbb{Z}) \to H_k(Y, \mathbb{Z})$$

setting

$$(8)\qquad [T]_\sharp [S]_{\mathbb{Z}} := [\widehat{\pi}_\# (T \cap (r_\# S \times \widehat{e}_\# [\![U]\!])_a)]_{\mathbb{Z}}.$$

Notice that (ii) of Proposition 2 in Sec. 5.4.2 shows that such a definition is compatible with the definition of T_* on real homology.

Similarly one checks that for $[S] \in H_{k+n-p-1}(\partial X, \mathbb{Z})$,

$$(\partial[T])_*[S]_{\mathbb{Z}} := \left[\widehat{\pi}_\#(T \cap (S \times \widehat{e}_\#[\![Y]\!])_a)\right]_{\mathbb{Z}}$$

for a.e. a close to zero yields a boundary homology map

$$(\partial[T])_* : H_{k+n-p}(\partial X, \mathbb{Z}) \to H_k(B, \mathbb{Z})$$

which is compatible with the real one.

Finally, if $S \in Z_{k+n-p}(X, \partial X, \mathbb{Z})$ is a relative normal cycle, then one can apply Proposition 3 in Sec. 5.4.2 to the currents

$$\begin{aligned} &T \in Z_p(X \times Y, \partial X \times B, \mathbb{Z}), \\ &e_\# S \times \widehat{e}_\#[\![Y]\!] \in Z_{k+n+m-p}(X_{\varepsilon_0} \times Y_{\varepsilon_0}, \partial(X_{\varepsilon_0} \times Y_{\varepsilon_0}), \mathbb{Z}) \end{aligned}$$

and obtain for a.e. a close to zero a relative cycle

$$T \cap (e_\# S \times \widehat{e}_\#[\![Y]\!]) \in Z_k(X \times Y, \partial X \times B, \mathbb{Z})$$

whose homology class depends only on the homology classes of S and T. In this way one has a homology induced map $[T]_\flat : H_{k+n-p}(X, \partial X, \mathbb{Z}) \to H_k(Y, B, \mathbb{Z})$ setting

$$[T]_\flat[S]_{\text{rel}} := [\widehat{\pi}_\#(T \cap (e_\# S \times \widehat{e}_\#[\![Y]\!])_a)]_{\text{rel}}$$

for a.e. a. Using (ii) of Proposition 3 in Sec. 5.4.2 one checks that such a definition is compatible with the corresponding map in real homology. Finally it is easy to check that the diagram

$$\begin{array}{ccccccccc} \cdots \xrightarrow{i_\#} & H_k(X, \mathbb{Z}) & \xrightarrow{j_\#} & H_k(X, \partial X, \mathbb{Z}) & \xrightarrow{\partial} & H_{k-1}(\partial X, \mathbb{Z}) & \xrightarrow{i_\#} \cdots \\ & \downarrow [T]_\sharp & & \downarrow [T]_\flat & & \downarrow (\partial[T])_* & \\ \cdots \xrightarrow{i_\#} & H_k(Y, \mathbb{Z}) & \xrightarrow{j_\#} & H_k(Y, B, \mathbb{Z}) & \xrightarrow{\partial} & H_{k-1}(B, \mathbb{Z}) & \xrightarrow{i_\#} \cdots \end{array} \tag{9}$$

is commutative.

6 Notes

1 The proof of the deformation theorem in Sec. 5.1.1 follows quite closely Simon [592], compare also Federer and Fleming [230], Federer [226]; for improved versions with applications to **Q**–valued functions see Almgren [21]. An application of the same ideas to the problem of non parametric area functional can be found in Mucci [498].

Sec. 5.1.2 and Sec. 5.1.3 follow closely Federer [228] and [226].

2 Hodge decomposition theorems are a classical subject and go back to Hodge [382] [381], De Rham [189] [188], Weyl [661], Kodaira [414], Friedrichs [247], Morrey [490] [491]. We followed the variational approach of Morrey [490]. For further information on this topic the reader may consult e.g. Schwarz [582] and its bibliography.

Weizenböck formula in Sec. 5.2.7 gives a link between bounds on curvature and non existence of harmonic forms, this way producing global vanishing theorems in cohomology, see e. g. Bochner [95] [96] and Goldberg [322].

Hodge theorem has been probably the first example of the relations between analysis and topology, another one is describing homology in terms of currents. It is out of question to quote even a few more contributions in the development of the interplay between linear and nonlinear analysis, topology and geometry. This includes among others invariant forms, K-theory, Atiyah-Singer index, Bott periodicity theorem, differential topology, Yang-Mills equations, topology of lower dimensional manifolds, Floer cohomology etc.. Probably these relations are one of the most fruitful research field in the mathematics of this century. The reader may start e.g. with Gilkey [303], Freed and Uhlenbeck [246] and their bibliographies.

3 The approach to homology evolved into a purely algebraic theory based on a set of axioms, the Eilenberg–Steenrood axioms, which characterizes the homology groups for triangulable pairs of spaces. A key point in this approach, even from a computational point of view, is the Meier-Vietoris sequence. Poincaré and Poincaré-Lefschetz theorems on the unimodularity of the intersection index can be proved in this way, compare e.g. Bott and Tu [102] Dold [195] Eilenberg and Steenrood [206] Greenberg [325] Griffits and Harris [326] Spanier [607] Vick [645].

We instead preferred to concentrate on a more analytic point of view to the real homology and cohomology, stressing the duality forms–currents at work, in the spirit of De Rham [189], Whitney [672], Federer [226] and [228]. Thus we define the homology as classes of currents and the Poincaré and Poincaré–Lefschetz duality isomorphism for the real homology are proved by mollifying currents and using Hodge theory. An obvious consequence is then de Rham theorem.

Another analytic-combinatoric proof of de Rham's theorem, at least for compact manifolds without boundary, can be found in Whitney [672], see also Singer and Thorpe [600].

de Rham theorem implies in particular the weak closure of the homology classes. An alternative proof which also gives a representation of homology clases by Lipschitz chains is given in Sec. 5.3.2 following Federer and Fleming [230], Federer [226].

For the integral homology we confined ourselves to represent integral homology classes in terms of currents, and in fact by least mass integral currents, and discuss following Federer [228] the relationships between real and integral cycles. This way we proved that the intersection index is an integer, however the unimodularity of the intersection index apparently seems to require the algebraic theory.

4 There are several results concerning least area representatives of homology classes, see also Vol. II Ch. 6. General partial regularity results has been proved by Almgren [18]. For a brief discussion of properties of optimal representatives of integral homology classes we refer the reader to Almgren [23].

There has been quite some work in finding explicit least area representatives of homology classes. In this context the *calibration theory* turns to be useful. For a short survey the reader may consult e.g. Morgan [484].

5 As general references for Sobolev spaces on $\mathbb{R}^n$ or on Riemannian manifolds the reader may consult e.g. Adams [4], Maz'ja [463] and Hebey [364].

Theorem 3 in Sec. 5.5.1 was proved in Schoen and Uhlenbeck [577]. For the proofs of Theorem 4 in Sec. 5.5.1 to Theorem 8 in Sec. 5.5.1 we refer the reader to Bethuel and Zheng [94], Bethuel [87] Bethuel [86], Hélein [369]. When $C_c^\infty(\mathcal{X}, \mathcal{Y})$ is not dense in $W^{1,p}(X, \mathcal{Y})$, i.e. when $\dim \mathcal{X} > p$ and $\pi_{[p]}(\mathcal{X}) \neq 0$, the strong closure of $C_c^\infty(\mathcal{X}, \mathcal{Y})$ in $W^{1,p}(\mathcal{X}, \mathcal{Y})$ has been studied in Bethuel [86] Bethuel, Coron, Demengel, and Hélein [91] Demengel [191] Giaquinta, Modica, and Souček [295], Isobe [393], Zhou [685], and in particular in Isobe [395].

As already mentioned for the proofs of Theorem 9 in Sec. 5.5.1, Theorem 11 in Sec. 5.5.1 and Theorem 12 in Sec. 5.5.1 we refer respectively to White [667] and White [668], and for the proof of Theorem 12 in Sec. 5.5.1 to Baldo and Orlandi [61].

6 The results of Sec. 5.5.2 are simple extensions of similar results proved in previous chapters. The discussion of the homology maps associated to Cartesian currents in the generality of Sec. 5.5.3 and Sec. 5.5.4 should be considered known to experts though we have no specific references. The proof of Künneth formula is taken from Griffits and Harris [326].

7 *Spaces of currents:*

$$
\begin{aligned}
&\mathcal{D}_k(\Omega) := \{k\text{–dimensional currents in } \Omega\}\\
&\mathcal{E}_k(\Omega) := \{T \in \mathcal{D}_k(\Omega) \,|\, \operatorname{spt} T \subset \Omega\}\\
&\mathbf{P}_k(\Omega) := \{\text{real polyhedral chains}\}\\
&\mathcal{M}_k(\Omega) := \{T \in \mathcal{D}_k(\Omega) \,|\, \mathbf{M}(T) < \infty\}\\
&\mathbf{N}_k(\Omega) := \{T \in \mathcal{D}_k(\Omega) \,|\, \mathbf{M}(T) + \mathbf{M}(\partial T) < \infty\}\\
&\mathbf{F}_k(\Omega) := \{\text{real flat chains}\}\\
&\mathcal{P}_k(\Omega) = \mathbf{P}_k(\Omega, \mathbb{Z}) := \{\text{integral polyhedral chains}\}\\
&\mathcal{R}_k(\Omega) := \{\text{rectifiable currents with integer multiplicity }\}\\
&\mathcal{I}_k(\Omega) := \{T \in \mathcal{R}_k(\Omega) \,|\, \partial T \in R_{k-1}(\Omega)\}\\
&\mathcal{F}_k(\Omega) := \{\text{integral flat chains}\}
\end{aligned}
$$

$$
\begin{array}{ccccccccc}
\mathcal{P}_k & \subset & \mathcal{I}_k & \subset & \mathcal{R}_k & \subset & \mathcal{F}_k & & \\
\cap & & \cap & & & & \cap & & \\
\mathbf{P}_k & \subset & \mathbf{N}_k & \subset & & \subset & \mathbf{F}_k & \subset & \mathcal{D}_k
\end{array}
$$

Bibliography

[1] Acerbi, E. and Dal Maso, G.: New lower semicontinuity results for polyconvex integrals. *Calc. Var.*, **2**, 329–371, 1994.

[2] Acerbi, E. and Fusco, N.: Semicontinuity problems in the calculus of variations. *Arch. Rat. Mech. Anal.*, **86**, 125–145, 1986.

[3] Acerbi, E. and Fusco, N.: An approximate lemma for $W^{1,p}$ functions. In *Material instability in continuum mechanics*, edited by Ball, J. M. Oxford University Press, Oxford, 1988.

[4] Adams, R.: *Sobolev spaces.* Academic Press, New York, 1975.

[5] Alberti, G.: *Proprietà fini delle funzioni a variazione limitata.* Tesi di laurea, 1988, Università di Pisa.

[6] Alberti, G.: A Lusin type theorem for gradients. *J. Funct. Anal.*, **100**, 110–118, 1991.

[7] Alberti, G.: Rank one property for derivatives of functions with bounded variation. *Proc. Roy. Soc. Edinburg*, **123**, 239–274, 1993.

[8] Alberti, G., Ambrosio, L., and Cannarsa, P.: On the singularities of convex functions. *manuscripta math.*, **76**, 421–435, 1992.

[9] Alberti, G. and Mantegazza, C.: A note on the theory of SBV functions. *preprint*, 1996.

[10] Alexandroff, P. and Hopf, H.: *Topologie.* Springer, Berlin, 1935.

[11] Alibert, J. J. and Dacorogna, B.: An example of a quasiconvex function not polyconvex in dimension two. *Arch. Rat. Mech. Anal.*, **117**, 155–166, 1992.

[12] Allard, W. K.: First variation of a varifold. *Ann. of Math.*, **95**, 417–491, 1972.

[13] Allard, W. K.: First variation of a varifold, boundary behaviour. *Ann. of Math.*, **95**, 417–491, 1975.

[14] Almgren, F. J.: *The theory of varifolds.* Mimeographed notes, Princeton, 1965.

[15] Almgren, F. J.: *Plateau's problem. An invitation to varifold geometry.* Benjamin, Inc., New York, 1966.

[16] Almgren, F. J.: Some interior regularity theorems for minimal surfaces and an extension of Bernstein's theorem. *Ann. of Math.*, **84**, 277–292, 1966.

[17] Almgren, F. J.: Existence and regularity almost everywhere of solutions to elliptic variational problems among surfaces of varying topological type and singuarity structure. *Ann. of Math.*, **87**, 321–391, 1968.

[18] Almgren, F. J.: Q valued functions minimizing Dirichlet's integral and the regularity of area minimizing rectifiable currents up to codimension two. *Bull.AMS*, **8**, 327–328, 1983.

[19] Almgren, F. J.: Q valued functions minimizing Dirichlet's integral and the regularity of area minimizing rectifiable currents up to codimension two, 1984. Notes, Princeton, NJ.

[20] Almgren, F. J.: Optimal isoperimetric inequalities. *Bull. Amer. Math. Soc.*, **13**, 123–126, 1985.

[21] ALMGREN, F. J.: Deformations and multiple-valued functions. In *Geometric measure theory and the calculus of variations*, Proc. Sympos. Pure Math., 44,, pp. 29–130. Amer. Math. Soc., Providence, R.I., 1986.

[22] ALMGREN, F. J.: Optimal isoperimetric inequalities. *Indiana Univ. Math. J.*, **35**, 451–547, 1986.

[23] ALMGREN, F. J.: Questions and answers about area minimizing surfaces and geometric measure theory. In *Differential geometry: partial differential equations on manifolds*, Proc. Sympos. Pure Math., 54, Part 1, pp. 29–53. Amer. Math. Soc., Providence, RI, 1993.

[24] ALMGREN, F. J., BROWDER, W., and LIEB, E. H.: Co-area, liquid crystals, and minimal surfaces. In *Partial differential equations*, Lecture Notes in Math., 1306,, pp. 1–22. Springer, Berlin, 1988.

[25] ALMGREN, F. J. and LIEB, E. H.: Singularities of energy-minimizing maps from the ball to the sphere. *Bull. Amer. Math. Soc.*, **17**, 304–306, 1987.

[26] ALMGREN, F. J. and LIEB, E. H.: Singularities of energy minimizing maps from the ball to the sphere: examples, counterexamples, and bounds. *Ann. of Math.*, **128**, 483–530, 1988.

[27] ALMGREN, F. J. and SIMON, L.: Existence of embedded solutions of Plateau's problem. *Ann. Sc. Norm. Sup. Pisa*, **6**, 447–495, 1979.

[28] ALMGREN, F. J. and THURSTON, W. P.: Examples of unknotted curves which bound only surfaces of high genus within their convex hulls. *Ann. Math.*, **105**, 527–538, 1977.

[29] ALT, W.: Verzweigungspunkte von H-Flächen. Part I. *Math. Z.*, **127**, 333–362, 1972.

[30] ALT, W.: Verzweigungspunkte von H-Flächen. Part II. *Math. Ann.*, **201**, 33–55, 1973.

[31] AMBROSIO, L.: New lower semicontinuity results for integral functionals. *Rend. Accad. Naz. XL*, **11**, 1–42, 1987.

[32] AMBROSIO, L.: A compactness theorem for a special class of functions of bounded variation. *Boll. UMI (3-B)*, **7**, 857–881, 1989.

[33] AMBROSIO, L.: Variational problems in SBV. *Acta Applicandae Math.*, **17**, 1–40, 1989.

[34] AMBROSIO, L.: Corso introduttivo alla teoria delle funzioni BV, le correnti ed i varifolds. Unpublished, see also: *Corso introduttivo alla Teoria geometrica della misura e alle superfici minime.* Scuola Normale Superiore, Pisa, 1997.

[35] AMBROSIO, L.: On the lower semicontinuity of quasi-convex integrals in SBV. *Nonlinear Anal.*, **23**, 405–425, 1994.

[36] AMBROSIO, L.: A new proof of the SBV compactness theorem. *Calc. Var.*, **3**, 127–137, 1995.

[37] AMBROSIO, L. and DAL MASO, G.: A general chain rule for distributional derivatives. *Proceedings Am. Math. Soc.*, **108**, 691–702, 1990.

[38] AMBROSIO, L. and DAL MASO, G.: On the representation in $BV(\Omega, \mathbf{R}^m)$ of quasi-convex integrals. *J. Funct. Anal.*, **109**, 76–97, 1992.

[39] AMBROSIO, L., MORTOLA, S., and TORTORELLI, V. M.: Functionals with linear growth defined on vector valued BV functions. *J. Math. Pures Appl.*, **70**, 269–323, 1991.

[40] AMRANI, A., CASTAING, C., and VALADIER, M.: Méthodes de troncature appliquées à des problèmes de convergence faible ou forte dans L^1. *Arch. Rat. Mech. Anal.*, **117**, 167–191, 1992.

[41] ANTMAN, S.: Geometrical and analytical questions in nonlinear elasticity. In *Seminar on Nonlinear Partial Differential Equations*, edited by S.S.Chern. Springer, Berlin, 1984.

[42] ANTMAN, S. S.: *Nonlinear problems of elasticity.* Springer, New York, 1995.
[43] ANZELLOTTI, G.: Dirichlet problem and removable singularities for functionals with linear growth. *Boll. UMI(C)*, **18**, 141–159, 1981.
[44] ANZELLOTTI, G.: Parametric and non-parametric minima. *manuscripta math.*, **48**, 103–115, 1984.
[45] ANZELLOTTI, G. and GIAQUINTA, M.: Funzioni *BV* e tracce. *Rend. Sem. Mat. Univ. Padova*, **60**, 1–21, 1978.
[46] ANZELLOTTI, G. and GIAQUINTA, M.: Existence of the displacements field for an elasto-plastic body subject to Hencky's law and von Mises yield condition. *manuscripta math.*, **32**, 101–136, 1980.
[47] ANZELLOTTI, G. and GIAQUINTA, M.: On the existence of the fields of stresses and displacements for an elasto-perfectly plastic body in static equilibrium. *J. Math. Pures et Appl.*, **61**, 219–244, 1982.
[48] ANZELLOTTI, G. and GIAQUINTA, M.: Convex functionals and partial regularity. *Arch. Rat. Mech. Anal.*, **102**, 243–272, 1988.
[49] ANZELLOTTI, G., GIAQUINTA, M., MASSARI, U., MODICA, G., and PEPE, L.: *Note sul problema di Plateau.* Editrice Tecnico Scientifica, Pisa, 1974.
[50] ANZELLOTTI, G. and SERAPIONI, R.: C^k-rectifiable sets. *J. Reine Angew. Math.*, **453**, 1–20, 1994.
[51] ANZELLOTTI, G., SERAPIONI, R., and TAMANINI, I.: Curvatures, Functionals, Currents. *Indiana Univ. Math. J.*, **39**, 617–669, 1990.
[52] ARONSZAJN, N.: A unique continuation theorem for solutions of elliptic partial differential equations or inequalities of second order. *J. Math. Pures et Appl.*, **36**, 235–249, 1957.
[53] AUBIN, T.: Equations différentielles non linéaires et problème de Yamabe concernent la courbure scalaire. *J. Math. Pures Appl.*, **55**, 269–296, 1976.
[54] AUBIN, T.: *Nonlinear analysis on manifolds. Monge-Ampère equations.* Grundleheren math. Wiss. 252. Springer, Berlin, 1982.
[55] AVILEZ, P. and GIGA, Y.: Variational integrals on mappings of bounded variation and their lower semicontinuity. *preprint*, ????
[56] BAKEL'MAN, I. Y.: Mean curature and quasilinear elliptic equations. *Sibirskii Mat. Ž*, **9**, 752–771, 1968. (English transl. Siberian Math. J. 9 (1968) 752-771).
[57] BAKEL'MAN, I. Y.: Geometric problems in quasilinear elliptic equations. *Uspehi Mat. Nauk*, **25**, 49–112, 1970.
[58] BALDER, E. J.: A general approach to lower semicontinuity and lower closure theorems in optimal control theory. *SIAM J. Cont. Optim.*, **22**, 570–598, 1984.
[59] BALDER, E. J.: On weak convergence implying strong convergence in L^1 spaces. *Bull. Austral. Math. Soc.*, **33**, 363–368, 1986.
[60] BALDES, A.: Stability and uniqueness properties of the equator map from a ball into an ellipsoid. *Math. Z.*, **185**, 505–516, 1984.
[61] BALDO, S. and ORLANDI, G.: Homotopy types for tamely approximable maps between manifolds. *Calc. Var.*, **4**, 369–384, 1996.
[62] BALL, J. M.: Constitutive inequalities and existence theorems in nonlinear elastostatics. In *Nonlinear Analysis and Mechanics: Heriot-Watt Symposium, Vol I.* Pitman Research Notes in Math. 17, London, 1977.
[63] BALL, J. M.: Convexity conditions and existence theorems in nonlinear elasticity. *Arch. Rat. Mech. Anal.*, **63**, 337–403, 1977.
[64] BALL, J. M.: Global invertibility of Sobolev functions and the interpenetration of matter. *Proc. Roy. Soc. Edinburgh*, **88A**, 315–328, 1981.
[65] BALL, J. M.: Discontinuous equilibrium solutions and cavitation in nonlinear elasticity. *Philos. Trans. Roy. Soc.*, **A 306**, 557–611, 1982.
[66] BALL, J. M.: Differentiability properties of symmetric and isotropic functions. *Duke Math. J.*, **50**, 699–727, 1984.

[67] BALL, J. M.: On the paper "Basic calculus of variations". *Pacific J. Math.*, **116**, 7–10, 1985.

[68] BALL, J. M.: Does rank one convexity imply quasiconvexity? In *Metastability and Incompletely Posed Problems*, edited by S. Antman, J.L.Ericksen, D. Kinerleherer and I. Müller, pp. 17–32. Springer, New York, 1987.

[69] BALL, J. M.: A version of the fundamental theorem for Young measures. In *PDEs and continuum models of phase transitions*, Lecture Notes in Phys., 344,, pp. 207–215. Springer, Berlin-New York, 1989.

[70] BALL, J. M., CURRIE, J. C., and OLIVER, P. J.: Null Lagrangians, weak continuity and variational problems of arbitrary order. *J. Funct. Anal.*, pp. 135–174, 1981.

[71] BALL, J. M. and MURAT, F.: $W^{1,p}$-quasiconvexity and variational problems for multiple integrals. *J. Funct. Anal.*, **58**, 225–253, 1988. Erratum,*J. Funct. Anal.* **66** 1986, 439.

[72] BALL, J. M. and MURAT, F.: Remarks on Chacon's biting lemma. *Proc. Amer. Math. Soc.*, **107**, 655–663, 1989.

[73] BANACH, S.: *Théorie des opérations linéaires.* Monografje Matematyczne, Warsawa, 1932. Republished by Chelsea, New York 1978.

[74] BAUMAN, P., OWEN, N., and PHILLIPS, D.: Maximal smoothness of solutions to certain Euler-Lagrange equations from nonlinear elasticity. *Proc. R. Soc. Edinb.*, **119**, 241–263, 1991.

[75] BAUMAN, P., OWEN, N., and PHILLIPS, D.: Maximum principles and a priori estimates for a class of problems from nonlinear elasticity. *Commun. Partial Differential Equations*, **17**, 1185–1212, 1992.

[76] BENCI, V. and CORON, J. M.: The Dirichlet problem for harmonic maps from the disk into the Euclidean n-sphere. *Ann. I.H.P., Analyse Non Linéaire*, **2**, 1985, 1985.

[77] BENNETT, C. and SHARPLEY, R.: Weak-type inequalities for H^p and BMO. *Proc. Symposia Pure Math.*, **35**, 201–229, 1979.

[78] BERKOWITZ, L. D.: Lower semicontinuity of integral functionals. *Trans. Am. Math. Soc.*, **192**, 51–57, 1974.

[79] BERNSTEIN, S.: Sur les surfaces définie au moyen de leur courbure moyenne et totale. *Ann. Ecole Norm. Sup.*, **27**, 233–256, 1910.

[80] BERNSTEIN, S.: Sur les équations du calcul des variations. *Ann. Sci. Ecole Norm. Sup.*, **29**, 431–485, 1912.

[81] BERNSTEIN, S.: Sur un théorème de géométrie et ses applications aux équations aux dérivées partielles du type elliptique. *Comm. Soc. Math. Kharkov*, **15**, 38–45, 1915-1917. (German transl.:Über ein geometrisches Theorem und seine Anwendung auf die partiellen Differentialgleichungen von elliptischen Typus, Math. Z. 26 (1927) 551-558).

[82] BERS, L.: Isolated singularities of minimal surfaces. *Ann. of Math.*, **53**, 364–386, 1951.

[83] BESICOVITCH, A.: On the fundamental geometric properties of linearly measurable plane sets of points III. *Math. Ann.*, **116**, 349–357, 1939.

[84] BESICOVITCH, A.: On the definition and value of the area of a surface. *Quart. J. Math.*, **16**, 86–102, 1945.

[85] BESICOVITCH, A.: Parametric surfaces. *Bull. Amer. Math. Soc.*, **56**, 288–296, 1950.

[86] BETHUEL, F.: A characterization of maps in $H^{1,2}(B^3, S^2)$ which can be approximated by smooth maps. *Ann. Inst. H. Poincaré, Anal. Non Linéaire*, **7**, 269–286, 1990.

[87] BETHUEL, F.: The approximation problem for Sobolev maps between two manifolds. *Acta Math.*, **167**, 153–206, 1992.

[88] BETHUEL, F.: On the singular set of stationary harmonic maps. *manuscripta math.*, **78**, 417–443, 1993.
[89] BETHUEL, F. and BREZIS, H.: Regularity of minimizers of relaxed problems for harmonic maps. *J. Funct. Anal.*, **101**, 145–161, 1991.
[90] BETHUEL, F., BREZIS, H., and CORON, J. M.: Relaxed energies for harmonic maps. In *Variational methods*, edited by Berestycki, H., Coron, J., and Ekeland, J. Birkhäuser, Basel, 1990.
[91] BETHUEL, F., CORON, J. M., DEMENGEL, F., and HÉLEIN, F.: A cohomological criterion for density of smooth maps in Sobolev spaces between two manifolds. In *Nematics, Mathematical and Physical Aspects*, edited by Coron, J. M., Ghidaglia, J. M., and Hélein, F., NATO ASI Series C, 332, pp. 15–23. Kluwer Academic Publishers, Dordrecht, 1991.
[92] BETHUEL, F. and DEMENGEL, F.: Extensions for Sobolev mappings between manifolds. *Calc. Var.*, **3**, 475–491, 1995.
[93] BETHUEL, F., MODICA, G., STEFFEN, K., and WHITE, B.: *Lectures notes on geometric measure theory and geometrical variational problems.* Dipartimento di Matematica, Università di Trento, 1995.
[94] BETHUEL, F. and ZHENG, X. M.: Density of smooth functions between two manifolds in Sobolev spaces. *J. Funct. Anal.*, **80**, 60–75, 1988.
[95] BOCHNER, S.: Curvature and Betti numbers I. *Ann. of Math.*, **49**, 379–390, 1948.
[96] BOCHNER, S.: Curvature and Betti numbers II. *Ann. of Math.*, **50**, 77–93, 1949.
[97] BOMBIERI, E.: Regularity theory for almost minimal currents. *Arch. Rat. Mech. Anal.*, **78**, 99–130, 1982.
[98] BOMBIERI, E., DE GIORGI, E., and GIUSTI, E.: Minimal cones and the Bernstein problem. *Inv. Math.*, **7**, 243–268, 1969.
[99] BOMBIERI, E., DE GIORGI, E., and MIRANDA, M.: Una maggiorazione a priori relativa alle ipersuperfici minimali non parametriche. *Arch. Rat. Mech. Anal.*, **32**, 255–267, 1965.
[100] BOMBIERI (ED.), E.: *Seminar on minimal submanifolds.* Ann. of Math. Studies n. 103. Princeton Univ. Press, Princeton, 1983.
[101] BONNESEN, T. and FENCHEL, W.: *Theorie der konvexen Körper.* Ergebnisse der Math. u. ihrer Grenzgebiete 3. Springer, Berlin, 1934.
[102] BOTT, R. and TU, L. W.: *Differential forms in Algebraic Topology.* Springer-Verlag, New York, 1982.
[103] BREZIS, H.: Large harmonic maps in two dimensions. In *Nonlinear variational problems*, Res. Notes in Math., 127, pp. 33–46. Pitman, Boston, Mass.-London, 1985.
[104] BREZIS, H.: S^k–valued maps with singularities. In *Topics in calculus of Variations*, edited by Giaquinta, M., Lecture notes 1365. Springer-Verlag, Berlin, 1989.
[105] BREZIS, H.: Convergence in $\mathcal{D}'$ and in L^1 under strict convexity. In *Boundary values problems for PDE and applications. Dedicated to E. Magenes on the occasion of his 70th birthday*, edited by Lions, J. L., pp. 43–52. Masson, Paris, 1993.
[106] BREZIS, H. and CORON, J. M.: Large solutions for harmonic maps in two dimensions. *Commun. Math. Phys.*, **92**, 203–215, 1983.
[107] BREZIS, H. and CORON, J. M.: Multiple solutions of H-systems and Rellich's conjecture. *Comm. Pure Appl. Math.*, **37**, 149–187, 1984.
[108] BREZIS, H. and CORON, J. M.: Convergence of solutions of H-sistems or how to blow bubbles. *Arch. Rat. Mech. Anal.*, **89**, 21–56, 1985.
[109] BREZIS, H., CORON, J. M., and LIEB, E. H.: Harmonic maps with defects. *Comm. Math. Phys.*, **107**, 649–705, 1986.

[110] BREZIS, H. and LIEB, E.: A relation between pointwise convergence of functions and convergence of functionals. *Proc. Amer. Math. Soc.*, **88**, 486–490, 1983.

[111] BREZIS, H. and NIRENBERG, L.: Positive solutions of nonlinear elliptic equations involving critical Sobolev exponents. *Comm. Pure Appl. Math.*, **36**, 437–477, 1983.

[112] BREZIS, H. and NIRENBERG, L.: Degree theory and BMO I. Compact manifolds without boundaries. *Selecta Math.*, **1**, 197–263, 1995.

[113] BREZIS, H. and NIRENBERG, L.: Degree theory and BMO Part II. Compact manifolds with boundaries. *preprint*, 1997.

[114] BROOKS, J. and CHACON, R.: Continuity and compactness of measures. *Advances in Math.*, **37**, 16–26, 1980.

[115] BROTHERS (ED.), J.: Some open problems in geometric measure theory and its applications suggested by partecipants of the 1984 AMS Summer Institute. *Proc. Symp. Pure Math.*, **44**, 441–464, 1986.

[116] BURAGO, J. D. and MAZ'JA, V. G.: Some questions of potential theory and function theory for domains with non-regular boundaries. *Zap. Nauchn. Sem. Leningr. Mat. Inst. Steklova*, **3**, 1–152, 1967. In Russian. English transl.: Sem. Math. Steklov Inst., Leningrad 3 (1969) 1-68.

[117] BURAGO, Y. D. and ZALGALLER, V. A.: *Geometric inequalities.* Grundlehren math. Wiss. 285. Springer-Verlag, Berlin, 1988.

[118] BURSTAL, F., LEMAIRE, L., and RAWNSLEY, J.: Harmonic maps bibliography. `http://www.bath.ac.uk/~masfeb/harmonic.html`.

[119] BUSEMANN, H., EWALD, G., and SHEPHARD, G. C.: Convex bodies and Convexity on Grassmann Cones V. *Arch. Math.*, **13**, 512–526, 1962.

[120] BUSEMANN, H., EWALD, G., and SHEPHARD, G. C.: Convex bodies and Convexity on Grassmann Cones I-IV. *Math. Ann.*, **151**, 1–41, 1963.

[121] BUSEMANN, H., EWALD, G., and SHEPHARD, G. C.: Convex bodies and Convexity on Grassmann Cones IX. *Math. Ann.*, **157**, 219–230, 1964.

[122] BUSEMANN, H., EWALD, G., and SHEPHARD, G. C.: Convex bodies and Convexity on Grassmann Cones VI. *J. London Math. Soc.*, **39**, 307–319, 1964.

[123] BUSEMANN, H., EWALD, G., and SHEPHARD, G. C.: Convex bodies and Convexity on Grassmann Cones VII. *Abh. Math. Sem. Univ. Hamburg*, **27**, 167–170, 1964.

[124] BUSEMANN, H., EWALD, G., and SHEPHARD, G. C.: Convex bodies and Convexity on Grassmann Cones VIII. *J. London Math. Soc.*, **39**, 417–423, 1964.

[125] BUSEMANN, H., EWALD, G., and SHEPHARD, G. C.: Convex bodies and Convexity on Grassmann Cones X. *Ann. Math. Pura e Appl.*, **70**, 271–294, 1965.

[126] BUSEMANN, H. and SHEPARD, G.: Convexity on nonconvex sets. In *Proc. Coll. on Convexity*, pp. 20–33. Copenhagen, 1965.

[127] BUTTAZZO, G.: *Semicontinuity, Relaxation and Integral Representation Problems in the Calculus of Variations.* Pitman Res. Notes in Math. 207. Longman, Harlow, 1989.

[128] CACCIOPPOLI, R.: Sul carattere infinitesimale delle superfici quadrabili. *Atti R. Accad. Lincei*, **7**, 901–905, 1928.

[129] CACCIOPPOLI, R.: Misura e integrazione sugli insiemi dimensionalmente orientati. *Rend. Acad. Naz. Lincei*, **12**, 3–11, 137–146, 1952.

[130] CACCIOPPOLI, R. and SCORZA DRAGONI, G.: Necessità della condizione di Weierstrass per la semicontinuità di un integrale doppio sopra una data superficie. *Memorie Acc. d'Italia*, **9**, 251–268, 1938.

[131] CALDERÓN, A. P. and ZYGMUND, A.: On the differentiability of functions which are of bounded variation in Tonelli's sense. *Revista Union Mat. Arg.*, **20**, 102–121, 1960.

[132] CALDERÓN, A. P. and ZYGMUND, A.: Local properties of solutions of elliptic partial differential equations. *Studia Mat.*, **20**, 171–225, 1961.

[133] CALKIN, J. W.: Functions of several variables and absolute continuity I. *Duke math. J.*, **6**, 170–185, 1940.

[134] CARATHÉODORY, C.: Über das lineare Mass von Punktmengen, eine Verallgemeinerung des Längenbegriffs. In *Nach. Ges. Wiss.* , pp. 404–426. Göttingen, 1914.

[135] CARATHÉODORY, C.: *Vorlesungen über reelle Functionen.* Teubner, Leipzig, 1927.

[136] CARBOU, G.: Applications harmoniques à valeurs dans un cercle. *C. R. Acad. Sci. Paris*, **314**, 359–362, 1992.

[137] CASADIO TARABUSI, E.: An algebraic characterization of quasiconvex functions. *Ricerche di Mat.*, **42**, 11–24, 1993.

[138] CASTAING, C. and VALADIER, M.: *Convex analysis and measurable multifunctions.* Lecture Notes in Math. 580. Springer, Berlin, 1977.

[139] CELADA, P. and DAL MASO, G.: Further remarks on the lower semicontinuity of polyconvex integrals. *Ann. Institut H. Poincaré Anal. Non Linéaire*, **11**, 661–691, 1994.

[140] CESARI, L.: Sulle funzioni a variazione limitata. *Ann. Sc. Norm. Sup. Pisa*, **5**, 299–313, 1936.

[141] CESARI, L.: Sulle funzioni assolutamente continue in due variabili. *Ann. Sc. Norm. Pisa*, **10**, 91–101, 1941.

[142] CESARI, L.: *Surface area.* Annals of Math. Studies 35. Princeton Univ. Press, Princeton, 1956.

[143] CESARI, L.: Lower semicontinuity and lower closure theorems without seminormality condition. *Ann. Mat. Pura e Appl.*, **98**, 381–397, 1974.

[144] CHANDRESEKHAR, S.: *Liquid crystals.* Cambridge Univ. Press, Cambridge, 1977.

[145] CHANG, S. X.-D.: Two dimensional area minimizing integral currents are classical minimal surfaces. *J. Amer. Math. Soc.*, **1**, 699–778, 1988.

[146] CHANILLO, A.: Sobolev inequalities involving divergence free maps. *Comm. PDE*, **16**, 1969–1994, 1991.

[147] CHANILLO, A. and LI, A.: Continuity of solutions of uniformly elliptic equations in $\mathbf{R}^2$. *manuscripta math.*, **77**, 415–433, 1992.

[148] CHEN, N. K. and LIU, F. C.: Approximation of nonparametric surfaces of finite area. *Chinese J. Math.*, **1981**, 25–35, 1981.

[149] CHEN, Y. and LIN, F. H.: Remarks on approximate harmonic maps. *Comment. Math. Helvetici*, **70**, 161–169, 1995.

[150] CHERN, S. S., ed.: *Global differential geometry.* The Mathematical Association of America, 1989.

[151] CIARLET, P.: *Mathematical elasticity.* North Holland, Amsterdam, 1988.

[152] CIARLET, P. and NEČAS, J.: Injectivité presque partout, auto contact et non interpénétrabilité en élasticité non-linéaire tridimensionelle. *C. R. Acad. Sci. Paris*, **301**, 621–624, 1985.

[153] CIARLET, P. and NEČAS, J.: Unilateral problems in nonlinear three-dimensional elasticity. *Arch. Rat. Mech. Anal.*, **87**, 319–338, 1985.

[154] CIARLET, P. and NEČAS, J.: Injectivity and selfcontact in nonlinear elasticity. *Arch. Rat. Mech. Anal.*, **97**, 171–188, 1987.

[155] COIFMAN, R., LIONS, P. L., MEYER, Y., and SEMMES, S.: Compacité par compensation et espaces de Hardy. *C. R. Acad. Sc. Paris*, **311**, 519–524, 1990.

[156] COIFMAN, R., LIONS, P. L., MEYER, Y., and SEMMES, S.: Compensated compactness and Hardy spaces. *J. Math. Pures Appl.*, **72**, 247–286, 1993.

[157] CORON, J. and GULLIVER, R.: Minimizing p–harmonic maps into spheres. *J. reine angew. Math.*, **401**, 82–100, 1989.

[158] Coron, J. M., Ghidaglia, J. M., and Hélein (Ed.), F.: *Nematics.* NATO series vol. 332. Kluwer academic Publ., Dodrecht, 1991.

[159] Courant, R.: On the problem of Plateau. *Proc. Nat. Acad. Sci. USA*, **22**, 367–372, 1936.

[160] Courant, R.: Plateau's problem and Dirichlet's Principle. *Ann. Math.*, **38**, 679–724, 1937.

[161] Courant, R.: *Dirichlet's principle, conformal mapping, and minimal surfaces.* Interscience, New York, 1950.

[162] Coventry, A.: Cartesian currents and weak convergence phenomena. *Australian National University, BSc Honours Thesis*, 1994.

[163] Cronin, J.: *Fixed points and topological degree in nonlinear analysis.* Amer. Math. Soc., Providence, 1964.

[164] Dacorogna, B.: *Weak continuity and weak lower semicontinuity of non-linear functionals.* Springer-Verlag, Berlin, 1982.

[165] Dacorogna, B.: Remarques sur les notions de polyconvexité, quasi-convexité et convexité de rang 1. *J. Math. Pures Appl.*, **64**, 403–438, 1985.

[166] Dacorogna, B.: *Direct methods in the calculus of variations.* Springer-Verlag, Berlin, 1989.

[167] Dacorogna, B. and Marcellini, P.: A counterexample in the vectorial calculus of variations. In *Material Instabilities in Continuum Mechanics*, edited by Ball, J. M., pp. 77–83. Clarendon Press, Oxford, 1988.

[168] Dacorogna, B. and Marcellini, P.: Semicontinuité pour des intégrales polyconvexes sans continuité des déterminants. *C.R.A.S. Paris*, **311**, 393–396, 1990.

[169] Dacorogna, B. and Moser, J.: On a partial differential equation involving the Jacobian determinant. *Ann. Institut H. Poincaré Anal. Non Linéaire*, **7**, 1–26, 1990.

[170] Dacorogna, B. and Murat, F.: On the optimality of certain Sobolev exponents for the weak continuity of determinants. *J. Funct. Anal.*, **105**, 42–62, 1992.

[171] Dal Maso, G.: Integral representation on $BV(\Omega)$ of Γ–limits of variational integrals. *manuscripta math.*, **30**, 387–416, 1980.

[172] Dal Maso, G.: *An Introduction to Γ-convergence.* Birkhäuser, Boston, 1993.

[173] Dal Maso, G. and Sbordone, C.: Weak lower semicontinuity of polyconvex integrals: a borderline case. *Math. Z.*, **218**, 603–609, 1995.

[174] Darst, R.: The Hausdorff dimension of the nondifferentiability set of the Cantor function is $(\log 2/\log 3)^2$. *Proc. Am. Math. Soc.*, **119**, 105–108, 1993.

[175] De Gennes, P.: *The physics of liquid crystals.* Oxford Univ. Press, Oxford, 1974.

[176] De Giorgi, E.: Su una teoria generale della misura $(r-1)$-dimensionale in uno spazio ad r dimensioni. *Ann. Mat. Pura Appl.*, **36**, 191–213, 1954.

[177] De Giorgi, E.: Nuovi teoremi relativi alle misure $(r-1)$-dimensionali in uno spazio ad r dimensioni. *Ricerche di Matematica*, **4**, 95–113, 1955.

[178] De Giorgi, E.: Sulla differenziabilità e l'analiticità delle estremali degli integrali multipli regolari. *Mem. Accad. Sci. Torino*, **3**, 25–43, 1957.

[179] De Giorgi, E.: Complementi alla teoria della misura $(n-1)$-dimensionale in uno spazio n dimensionale. *Seminario di Mat. della Scuola Normale Superiore di Pisa*, 1960-61.

[180] De Giorgi, E.: Frontiere orientate di misura minima. *Seminario di Mat. della Scuola Normale Superiore di Pisa*, 1960-61.

[181] De Giorgi, E.: Una estensione del teorema di Bernstein. *Ann. Sc. Norm. Sup. Pisa*, **19**, 79–85, 1965.

[182] De Giorgi, E.: *Teoremi di Semicontinuità nel calcolo delle variazioni.* Istit. Naz. Alta Mat., Roma, 1968-69.

[183] DE GIORGI, E.: Sulla convergenza di alcune successioni di integrali del tipo dell'area. *Rendic. Mat.*, **8**, 277–294, 1975.

[184] DE GIORGI, E. and AMBROSIO, L.: Un nuovo tipo di funzionale del calcolo delle variazioni. *Rend. Accad. Naz. Lincei, Cl. Sci. Fis. Mat. Natur.*, **82**, 199–210, 1988.

[185] DE GIORGI, E., BUTTAZZO, G., and DAL MASO, G.: On the lower semicontinuity of certain integral functionals. *Atti Accad. Lincei*, **74**, 274–282, 1983.

[186] DE GIORGI, E., COLOMBINI, F., and PICCININI, L. C.: *Frontiere orientate di misura minima e questioni collegate.* Scuola Normale Superiore, Pisa, 1972.

[187] DE GIORGI, E. and STAMPACCHIA, G.: Sulle singolarità eliminabili delle ipersuperfici minimali. *Atti Accad. Naz. Lincei*, **38**, 352–357, 1965.

[188] DE RHAM, G.: Sur l'analysis situs des varietés à n–dimensions. *J. Math. Pures et Appl.*, **10**, 115–200, 1951.

[189] DE RHAM, G.: *Variétés différentiables, formes, courants, formes harmoniques.* Act. Sci. et Ind., 1222. Hermann, Paris, 1955.

[190] DELLADIO, S.: Lowersemicontinuity and continuity of measures with respect to the strict convergence. *Proc. Roy. Soc. Edinburgh*, **119**, 265–278, 1991.

[191] DEMENGEL, F.: Une caractérization des applications de $W^{1,p}(S^N, S^1)$ qui peuvent être approchées par des fonctions C^∞. *C. R. Acad. Sci. Paris*, **330**, 553–557, 1990.

[192] DEMENGEL, F. and HADIJI, R.: Relaxed energies for functionals on $W^{1,2}(B^2, S^1)$. *Nonlinear Analysis. Theory, Methods and Applications*, **19**, 625–641, 1992.

[193] DENJOY, A.: Sur les fonctions dérivée sommable. *Bull. Soc. Mat. France*, **43**, 161–248, 1916.

[194] DIERKES, U., HILDEBRANDT, S., KUSTER, A., and WOHLRAB, O.: *Minimal surfaces I, II.* Grundlehren math. Wiss. 295, 296. Springer–Verlag, Berlin, 1992.

[195] DOLD, A.: *Lectures on algebraic topology.* Springer, New York, 1972.

[196] DOUGLAS, J.: The mapping theorem of Koebe and the Plateau problem. *J. Math. Phys.*, **10**, 106–130, 1930-31.

[197] DOUGLAS, J.: Solution of the problem of Plateau. *Trans. Am. Math. Soc.*, **33**, 263–321, 1931.

[198] DUZAAR, F. and KUWERT, E.: Minimization of conformally invariant energies in homotopy classes. *Calc. Var.*, 1998. To appear.

[199] EELLS, J. and LEMAIRE, L.: A report on harmonic maps. *Bull. London Math. Soc.*, **10**, 1–68, 1978.

[200] EELLS, J. and LEMAIRE, L.: *Selected topics in harmonic maps.* Published for the Conference Board of the Mathematical Sciences, Washington, DC; by the American Mathematical Society, Providence, RI, 1983.

[201] EELLS, J. and LEMAIRE, L.: Examples of harmonic maps from disks to hemispheres. *Math. Z.*, **185**, 517–519, 1984.

[202] EELLS, J. and LEMAIRE, L.: Another report on harmonic maps. *Bull. London Math. Soc.*, **20**, 385–524, 1988.

[203] EELLS, J. and RATTO, A.: Harmonic maps between spheres and ellipsoids. *Internat. J. Math.*, **1**, 1–27, 1990.

[204] EELLS, J. and SAMPSON, J.: Harmonic mappings of Riemannian manifolds. *American J. Math.*, **86**, 109–160, 1964.

[205] EELLS, J. and WOOD, J. C.: Restrictions on harmonic maps of surfaces. *Topology*, **15**, 263–266, 1976.

[206] EILENBERG, S. and STEENROOD, N.: *Foundations of algebraic topology.* Princeton Univ. Press, Princeton, NJ, 1952.

[207] EISEN, G.: A counterexample for some lower semicontinuity results. *Math. Z.*, **162**, 241–243, 1978.

[208] EISEN, G.: A selection lemma for sequences of measurable sets, and lower semicontinuity of multiple integrals. *manuscripta math.*, **27**, 73–79, 1979.

[209] EKELAND, I. and TEMAM, R.: *Convex analysis and variational problems.* North-Holland, Amsterdam, 1976.

[210] ERICKSEN, J. L.: Equilibrium theory of liquid crystals. In *Advances in Liquid Crystals, vol. 2*, edited by Brown, G. H., pp. 233–299. Academic Press, New York, 1976.

[211] ERICKSEN, J. L. and KINDERLEHRER (ED.), D.: *Theory and Applications of Liquid Crystals.* Springer, New York, 1987.

[212] ESTEBAN, M. J.: A direct variational approach to Skyrme's model for mesons fields. *Commun. Math. Phys.*, **105**, 571–591, 1986.

[213] ESTEBAN, M. J.: Existence of symmetric solutions for the Skyrme's problem. *Ann. Mat. Pura. Appl.*, **147**, 187–195, 1987.

[214] ESTEBAN, M. J. and MÜLLER, S.: Sobolev maps with integer degree and applications to Skyrme's problem. *Proc. R. Soc. London*, **436**, 197–201, 1992.

[215] EVANS, L. C.: *Weak convergence methods for nonlinear partial differential equations.* CBMS Regional conference ser. in Math. n.74. AMS, Providence, 1990.

[216] EVANS, L. C.: Partial regularity for stationary harmonic maps into spheres. *Arch. Rat. Mech. Anal.*, **116**, 101–, 1991.

[217] EVANS, L. C. and GARIEPY, R. F.: *Measure theory and fine properties of functions.* CRC Press, Boca Raton, FL, 1992.

[218] FALCONER, K.: *The geometry of fractal sets.* Cambridge Univ. Press, Cambridge, 1985.

[219] FALCONER, K.: *Fractal geometry. Mathematical foundations and applications.* John Wiley & Sons, Ltd., Chichester, 1990.

[220] FEDERER, H.: Surface area (I), (II). *Trans. Am. Math. Soc.*, **55**, 420–456, 1944.

[221] FEDERER, H.: The Gauss-Green theorem. *Trans. Am. Math. Soc.*, **58**, 44–76, 1945.

[222] FEDERER, H.: The (ϕ, k) rectifiable subsets of n space. *Trans. Am. Math. Soc.*, **62**, 114–192, 1947.

[223] FEDERER, H.: A note on the Gauss-Green theorem. *Proc. Am. Math. Soc.*, **9**, 447–451, 1958.

[224] FEDERER, H.: Curvature measures. *Trans. Am. Math. Soc.*, **93**, 418–491, 1959.

[225] FEDERER, H.: Currents and area. *Trans. Am. Math. Soc.*, **98**, 204–233, 1961.

[226] FEDERER, H.: *Geometric measure theory.* Grundlehren math. Wissen. 153. Springer-Verlag, Berlin, 1969.

[227] FEDERER, H.: The singular set of area minimizing rectifiable currents with codimension one and area minimizing flat chains modulo two with arbitrary codimension. *Bull. Am. Math. Soc.*, **76**, 767–771, 1970.

[228] FEDERER, H.: Real flat chains, cochains and variational problems. *Indiana Univ. Math. J.*, **24**, 351–407, 1974.

[229] FEDERER, H.: Colloquium lectures on geometric measure theory. *Bull. Amer. Math. Soc.*, **84**, 291–338, 1978.

[230] FEDERER, H. and FLEMING, W. H.: Normal and integral currents. *Ann. of Math.*, **72**, 458–520, 1960.

[231] FEDERER, H. and ZIEMER, W.: The Lebesgue set of a function whose distributional derivatives are p-th power summable. *Indiana Univ. Math. J.*, **22**, 139–158, 1972.

[232] FEFFERMAN, C.: Characterization of bounded mean oscillation. *Bull. AMS*, **77**, 585–587, 1971.

[233] FEFFERMAN, C. and STEIN, E.: H^p spaces of several variables. *Acta Math.*, **129**, 137–193, 1972.

[234] FINN, R.: Remarks relevant to minimal surfaces of prescribed mean curvature. *J. Analyse Math.*, **14**, 139–160, 1965.

[235] FINN, R.: *Equilibrium capillary surfaces.* Springer, New York, 1986.

[236] FISCHER-COLBRIE, D.: Some rigidity theorems for minimal submanifolds of sphere. *Acta Math.*, **145**, 29–46, 1980.

[237] FLEMING, W. H.: An example in the problem of least area. *Proc. Amer. math. Soc.*, **7**, 1063–1074, 1956.

[238] FLEMING, W. H.: Functions whose partial derivatives are measures. *Ill. J. Math.*, **4**, 452–478, 1960.

[239] FLEMING, W. H.: On the oriented Plateau problem. *Rend. Circ. Mat. Palermo*, **11**, 1–22, 1962.

[240] FLEMING, W. H.: Flat chains over a finite coefficient group. *Trans. Am. Math. Soc.*, **121**, 160–186, 1966.

[241] FLEMING, W. H. and RISHEL, R.: An integral formula for total gradient variation. *Arch. Mat.*, **11**, 218–222, 1960.

[242] FONSECA, I. and MARCELLINI, P.: Relaxation of multiple integrals in subcritical Sobolev spaces. *preprint*, 1994.

[243] FONSECA, I. and MÜLLER, S.: Quasiconvex integrands and lower semicontinuity in L^1. *SIAM J. Math. Anal.*, **23**, 1081–1098, 1992.

[244] FONSECA, I. and MÜLLER, S.: Relaxation of quasiconvex functionals in $BV(\Omega, \mathbf{R}^p)$ for integrands $f(x, u, Du)$. *Arch. Rat. Mech. Anal.*, **123**, 1–49, 1993.

[245] FRANK, F. C.: On the theory of liquid crystals. *Discuss. Faraday Soc.*, **28**, 19–28, 1938.

[246] FREED, D. S. and UHLENBECK, K. K.: *Instantons and four-manifolds.* Springer, New York, 1984.

[247] FRIEDRICHS, K.: Differential forms on Riemannian manifolds. *Comm. Pure Appl. Math.*, **8**, 551–590, 1955.

[248] FUCHS, M.: p-harmonic obstacle problems. Part II: Extension of maps and applications. *manuscripta math.*, **63**, 381–419, 1989.

[249] FUCHS, M.: p-harmonic obstacle problems. Part I: Partial regularity theory. *Ann. Mat. Pura Appl.*, **156**, 127–158, 1990.

[250] FUCHS, M.: p-harmonic obstacle problems. Part III: Boundary regularity. *Ann. Mat. Pura Appl.*, **156**, 159–180, 1990.

[251] FUCHS, M. and SEREGIN, G.: Hölder continuity for weak extremals of some two-dimensional variational problems related to nonlinear elasticity. *Adv. Math. Sci. Appl.*, **7**, 413–425, 1997.

[252] FUSCO, N. and HUTCHINSON, J.: Partial regularity in problems motivated by nonlinear elasticity. *SIAM J. Math. Anal.*, **22**, 1516–1551, 1991.

[253] FUSCO, N. and HUTCHINSON, J.: Partial regularity and everywhere continuity for a model problem from nonlinear elasticity. *J. Austral. Math. Soc.*, **57**, 158–169, 1994.

[254] FUSCO, N. and HUTCHISON, J. E.: A direct proof for lower semicontinuity of polyconvex integrals. *manuscripta math*, **87**, 35–50, 1995.

[255] FUČIK, S., NEČAS, J., SOUČEK, J., and SOUČEK, V.: *Spectral analysis of nonlinear operators.* Lecture notes 346. Springer, Berlin, 1973.

[256] GAGLIARDO, E.: Caratterizzazione delle tracce sulla frontiera relative ad alcune classi di funzioni in n variabili. *Rend. Sem. Mat. Univ. Padova*, **27**, 284–305, 1957.

[257] GANGBO, W.: On the weak lower semicontinuity of energies with polyconvex integrands. *J. Math. Pures Appl.*, **73**, 455–469, 1994.

[258] GARDING, L.: Dirichlet problem for linear partial differential equations. *Math. Scand.*, **1**, 55–72, 1953.

[259] Gastel, A.: Regularity theory for minimizing equivariant (p-)harmonic mappings. *Calc. Var.*, 1998. To appear.

[260] Gehring, F.: The L^p-integrability of the partial derivatives of a quasi conformal mapping. *Acta Math.*, **130**, 265–277, 1973.

[261] Gerhardt, C.: On the regularity of solutions to variational problems in $BV(\Omega)$. *Math. Z.*, **149**, 281–286, 1976.

[262] Giaquinta, M.: On the Dirichlet problem for surfaces of prescribed mean curvature. *manuscripta math.*, **12**, 73–86, 1974.

[263] Giaquinta, M.: *Multiple integrals in the calculus of variations and nonlinear elliptic systems.* Ann. of Math. Studies n.105. Princeton Univ. Press, Princeton, 1983.

[264] Giaquinta, M.: The problem of the regularity of minimizers. In *Proceedings of the International Congress of Mathematicians, voll. II*, pp. 1072–1083, 1986.

[265] Giaquinta, M.: Quasiconvexity, growth conditions and partial regularity. In *Partial Differential Equations and Calculus of Variations*, edited by Hildebrandt, S. and Leis, R., Lecture notes 1357. Springer-Verlag, Berlin, 1988.

[266] Giaquinta, M.: Problemi variazionali per applicazioni vettoriali. Aspetti geometrici ed analitici. *Boll. UMI (7)*, **6-A**, 1–34, 1992.

[267] Giaquinta, M.: *Introduction to regularity theory for nonlinear elliptic systems.* Lectures in Math., ETH Zürich. Birkhäuser, Basel, 1993.

[268] Giaquinta, M.: Analytic and geometric aspects of variational problems for vector valued mappings. In *Proceedings of the First European Congress of Mathematicians, Paris, July 1992.* Birkhäuser, Basel, 1994.

[269] Giaquinta, M. and Giusti, E.: Nonlinear elliptic systems with quadratic growth. *manuscripta math.*, **24**, 323–349, 1978.

[270] Giaquinta, M. and Giusti, E.: On the regularity of the minima of variational integrals. *Acta Math.*, **148**, 31–46, 1982.

[271] Giaquinta, M. and Giusti, E.: Differentiability of minima of nondifferentiable functionals. *Invent. Math.*, **72**, 285–298, 1983.

[272] Giaquinta, M. and Giusti, E.: The singular set of the minima of quadratic functionals. *Ann. Sc. Norm. Sup. Pisa*, **11**, 45–55, 1984.

[273] Giaquinta, M. and Hildebrandt, S.: *Direct methods in the calculus of variations.* In preparation.

[274] Giaquinta, M. and Hildebrandt, S.: A priori estimates for harmonic mappings. *J. Reine Angew. Math.*, **336**, 124–164, 1982.

[275] Giaquinta, M. and Hildebrandt, S.: *Calculus of Variations, voll.2.* Grundlehren math. Wiss. 310, 311. Springer, Berlin, 1996.

[276] Giaquinta, M. and Modica, G.: Regularity results for some classes of higher order nonlinear elliptic systems. *J. für reine u. angew. Math.*, **311/312**, 145–169, 1979.

[277] Giaquinta, M. and Modica, G.: Nonlinear systems of the type of the stationary Navier- Stokes system. *J. für reine u. angew. Math.*, **330**, 173–214, 1982.

[278] Giaquinta, M., Modica, G., and Souček, J.: Functionals with linear growth in the calculus of variations. *Comm. Math. Univ. Carolinae*, **20**, 143–171, 1979.

[279] Giaquinta, M., Modica, G., and Souček, J.: Cartesian currents and variational problems for mappings into spheres. *Ann. Sc. Norm. Sup. Pisa*, **16**, 393–485, 1989.

[280] Giaquinta, M., Modica, G., and Souček, J.: Cartesian currents, weak diffeomorphisms and existence theorems in nonlinear elasticity. *Arch. Rat. Mech. Anal.*, **106**, 97–159, 1989. Erratum and addendum, *Arch. Rat. Mech. Anal.* **109** 1990, 385-392.

[281] Giaquinta, M., Modica, G., and Souček, J.: Partial regularity of Cartesian Currents which minimize certain variational integrals. In *P.D.E. and the Calcu-*

lus of Variations, Essays in Honor of Ennio De Giorgi, edited by Colombini, F., Marino, A., Modica, L., and Spagnolo, S., pp. 563–587. Birkauser, Boston, 1989.

[282] GIAQUINTA, M., MODICA, G., and SOUČEK, J.: The Dirichlet energy of mappings with values into the sphere. *manuscripta math.*, **65**, 489–507, 1989.

[283] GIAQUINTA, M., MODICA, G., and SOUČEK, J.: Liquid crystals: relaxed energies, dipoles, singular lines and singular points. *Ann. Sc. Norm. Sup. Pisa*, **17**, 415–437, 1990.

[284] GIAQUINTA, M., MODICA, G., and SOUČEK, J.: Cartesian Currents and Liquid Crystals: dipoles, singular lines and singular points. In *Nematics, Mathematical and Physical Aspects*, edited by Coron, J. M., Ghidaglia, J. M., and Hélein, F., NATO ASI Series C, 332, pp. 113–127. Kluwer Academic Publishers, Dordrecht, 1991.

[285] GIAQUINTA, M., MODICA, G., and SOUČEK, J.: Area and the area formula. *Rend. Sem. Mat. Milano*, **62**, 53–87, 1992.

[286] GIAQUINTA, M., MODICA, G., and SOUČEK, J.: The Dirichlet integral for mappings between manifolds: Cartesian currents and homology. *Math. Ann.*, **294**, 325–386, 1992.

[287] GIAQUINTA, M., MODICA, G., and SOUČEK, J.: The gap phenomenon for variational integrals in Sobolev spaces. *Proc. Roy. Soc. Edinburg*, **120**, 93–98, 1992.

[288] GIAQUINTA, M., MODICA, G., and SOUČEK, J.: Variational problems for the conformally invariant integral $\int |du|^n$. In *Progress in partial differential equations: calculus of variations, applications*, edited by Bandle, C., Bemelmans, J., M., C., Grüter, M., and Paulin, J. S. J., Pittman Research Notes In Math. 267, pp. 27–47. Pittman, 1992.

[289] GIAQUINTA, M., MODICA, G., and SOUČEK, J.: Connectivity properties of the range of weak diffeomorphisms. *Ann. Institut H. Poincaré, Anal. Non Linéaire*, **12**, 61–73, 1993.

[290] GIAQUINTA, M., MODICA, G., and SOUČEK, J.: Graphs of finite mass which cannot be approximated in area by smooth graphs. *manuscripta math.*, **78**, 259–271, 1993.

[291] GIAQUINTA, M., MODICA, G., and SOUČEK, J.: Singular perturbations associated to the Dirichlet energy of maps with values in S^2. *J. Func. Anal.*, **118**, 188–197, 1993.

[292] GIAQUINTA, M., MODICA, G., and SOUČEK, J.: Variational problems for maps of bounded variation with values in S^1. *Calc. Var.*, **1**, 87–121, 1993.

[293] GIAQUINTA, M., MODICA, G., and SOUČEK, J.: A weak approach to finite elasticity. *Calc. Var.*, **2**, 65–100, 1994.

[294] GIAQUINTA, M., MODICA, G., and SOUČEK, J.: Remarks on the degree theory. *J. Func. Analysis*, **125**, 172–200, 1994.

[295] GIAQUINTA, M., MODICA, G., and SOUČEK, J.: Some remarks about the p–Dirichlet integral. *Comm. Mat. Univ. Carolinae*, **35**, 55–62, 1994.

[296] GIAQUINTA, M., MODICA, G., and SOUČEK, J.: Remarks on quasiconvexity and lower semicontinuity. *NoDEA*, **2**, 573–588, 1995.

[297] GIAQUINTA, M., MODICA, G., and SOUČEK, J.: Compositions of weak diffeomorphisms. *Math. Z.*, **224**, 385–402, 1997.

[298] GIAQUINTA, M., MODICA, G., and SOUČEK, J.: Energy minimizing maps from a domain of $\mathbb{R}^3$ into S^2. *preprint*, 1997.

[299] GIAQUINTA, M., MODICA, G., and SOUČEK, J.: Some regularity properties of weakly invertible maps. In *Nonlinear Analysis and Continuum Mechanics*, edited by Galdi, P. and Pucci, P. Springer, New-York, to appear. Preprint 1993.

[300] GIAQUINTA, M. and SOUČEK, J.: Esistenza per il problema dell'area e controesempio di Bernstein. *Boll. UMI*, **9**, 807–817, 1974.

[301] GIAQUINTA, M. and SOUČEK, J.: Harmonic maps into a hemisphere. *Ann. Scuola Norm. Sup. Pisa*, **12**, 81–90, 1985.
[302] GILBARG, D. and TRUDINGER, N.: *Elliptic partial differential equations of second order.* Grundlehren mat. Wiss. Springer, Berlin, 1977. Second edition 1994.
[303] GILKEY, P. B.: *Invariance theory, the heat equation, and the Atiyah–Singer Index theorem.* Publish or Perish, Wilmington, Delaware, 1984.
[304] GIUSTI, E.: Precisazione delle funzioni $H^{1,p}$ e singolarità delle soluzioni deboli di sistemi ellittici non lineari. *Boll. UMI*, **2**, 71–76, 1969.
[305] GIUSTI, E.: Superfici cartesiane di area minima. *Rend. Sem. Mat. Milano*, **40**, 1–21, 1970.
[306] GIUSTI, E.: Boundary behavior of nonparametric minimal surfaces. *Indiana Univ. Math. J.*, **22**, 435–444, 1972.
[307] GIUSTI, E.: Boundary value problems for non-parametric surfaces of prescribed mean curvature. *Ann. Sc. Norm. Sup. Pisa*, **3**, 501–548, 1976.
[308] GIUSTI, E.: *Minimal surfaces and functions of bounded variation.* Birkhäuser, Basel, 1984.
[309] GIUSTI, E.: *Metodi diretti nel Calcolo delle Variazioni.* Unione Matematica Italiana, Bologna, 1994.
[310] GIUSTI, E. and MIRANDA, M.: Sulla regolarità delle soluzioni deboli di una classe di sistemi ellittici quasilineari. *Arch. Rat. Mech. Anal.*, **31**, 173–184, 1968.
[311] GOFFMAN, C.: Lower semicontinuity and area functionals. I. The nonparametric case. *Rend. Circ. Mat. Palermo*, **2**, 203–235, 1953.
[312] GOFFMAN, C.: Approximation of nonparametric surfaces of finite area. *J. Math. Mech.*, **12**, 737–745, 1963.
[313] GOFFMAN, C.: A characterization of linearly continuous functions whose partial derivatives are measures. *Acta Math.*, **117**, 165–190, 1967.
[314] GOFFMAN, C.: An example in surface area. *J. Math. Mech.*, **18**, 321–326, 1969.
[315] GOFFMAN, C. and LIU, F. C.: Discontinuous mappings and surface area. *Proc. London Math. Soc.*, **20**, 237–248, 1970.
[316] GOFFMAN, C. and LIU, F. C.: The area formula for Sobolev mappings. *Indiana Univ. Math. J.*, **25**, 871–876, 1976.
[317] GOFFMAN, C. and LIU, F. C.: Derivatives measures. *Proc. Amer. Math. Soc.*, **78**, 218–220, 1980.
[318] GOFFMAN, C., NEUGEBAUER, C. J., and NISHIURA, T.: Density topology and approximate continuity. *Duke Math. J.*, **28**, 497–505, 1961.
[319] GOFFMAN, C. and WATERMAN, D.: Approximately continuous transformations. *Proc. Am. Mat. Soc.*, **12**, 116–121, 1961.
[320] GOFFMANN, C. and SERRIN, J.: Sublinear functions of measures and variational integrals. *Duke Math. J.*, **31**, 159–178, 1964.
[321] GOFFMANN, C. and ZIEMER, W. P.: Higher dimensional mappings for which the area formula holds. *Ann. of Math.*, **92**, 482–488, 1970.
[322] GOLDBERG, S.: *Curvature and homology.* Dover Publications, 1962.
[323] GREEN, A. and ADKINS, J.: *Large elastic deformations.* Oxford Univ. Press, London, 1970.
[324] GREEN, A. and ZERNA, W.: *Theoretical elasticity.* Oxford Univ. Press, London, 1968.
[325] GREENBERG, M.: *Lectures on algebraic topology.* W. A. Benjamin, New York, 1966.
[326] GRIFFITS, P. and HARRIS, P.: *Principles of Algebraic Geometry.* J. Wiley & Sons, New York, 1978.
[327] GROTOWSKI, J. F., SHEN, Y., and YAN, S.: On various classes of harmonic maps. *Arch. Mat.*, **64**, 353–358, 1995.

[328] GRÜTER, M.: Regularity of weak H-surfaces. *J. reine u. ang. Math.*, **329**, 1–15, 1981.

[329] GULLIVER, R.: Regularity of minimizing surfaces of prescribed mean curvature. *Ann. Math.*, **97**, 275–305, 1973.

[330] GULLIVER, R. and LESLEY, F. D.: On boundary branch points of minimizing surfaces. *Arch. Rat. Mech. Anal.*, **52**, 20–25, 1973.

[331] GULLIVER, R., OSSERMAN, R., and ROYDEN, H. L.: A theory of branched immersions of surfaces. *Amer. J. Math.*, **95**, 750–812, 1973.

[332] GULLIVER, R. and SPRUCK, J.: On embedded minimal surfaces. *Ann. Math.*, **103**, 331–347, 1976. Correction in *Ann. Math.* **109** (1979) 407–412.

[333] GÜNTHER, M.: On the perturbation problem associated to isometric embeddings of Riemannian manifolds. *Ann. Global. Anal. Geom.*, **7**, 69–77, 1989.

[334] GURTIN, M.: *An introduction to continuum mechanics.* Academic Press, New York, 1981.

[335] HAAR, A.: Über das Plateausche Problem. *Math. Ann.*, **97**, 124–258, 1927.

[336] HADAMARD, J.: Résolution d'un question relative aux determinants. *Bull. Sci. Math.*, **17**, 240–248, 1893.

[337] HAJLASZ, P.: A note on weak approximation of minors. *Ann. Inst. H. Poincaré, Analyse Non Linéaire*, **12**, 415–424, 1995.

[338] HALMOS, P.: *Measure theory.* Van Nostrand, New York, 1950.

[339] HAMILTON, R.: *Harmonic maps of manifolds with boundary.* Lecture notes n.471. Springer, Berlin, 1975.

[340] HARDT, R.: Singularities in some geometric variational problems. In *Proceedings of the International Congress of Mathematicians, vol. I, Berkeley*, pp. 540–550, 1986.

[341] HARDT, R.: Singularities of harmonic maps. *Bull. AMS*, **34**, 15–34, 1997.

[342] HARDT, R. and KINDERLEHRER, D.: Mathematical questions of liquid crystal theory. In *Theory and applications of liquid crystals*, edited by Ericksen, J. L. and Kinderlehrer, D., pp. 151–184. Springer, New York, 1987.

[343] HARDT, R., KINDERLEHRER, D., and LIN, F.: Stable defects of minimizers of constrained variational principles. *Ann. Inst. H. Poincaré, Anal. Non Linéaire*, **5**, 297–322, 1988.

[344] HARDT, R., KINDERLEHRER, D., and LIN, F. H.: Existence and partial regularity of static liquid crystals configurations. *Comm. Math. Phys.*, **105**, 547–570, 1986.

[345] HARDT, R., KINDERLEHRER, D., and LIN, F. H.: Existence and partial regularity of static liquid crystal configurations. *Comm. Math. Phys.*, **105**, 547–570, 1986.

[346] HARDT, R., KINDERLEHRER, D., and LIN, F. H.: The variety of configurations of liquid crystals. In *Variational Methods*, edited by Beresticky, H., Coron, J. M., and Ekeland, I., PNDE, pp. 115–131. Birkhäuser, Basel, 1990.

[347] HARDT, R., KINDERLEHRER, D., and LUSKIN, M.: Remarks about the mathematical theory of liquid crystals. In *Calculus of Variations and P.D.E.*, edited by Hildebrandt, S., Kinderlehrer, D., and Miranda, M., Springer Lecture Notes 1340, pp. 123–138. Springer, New York, 1988.

[348] HARDT, R., LAU, C. P., and LIN, F. H.: Non minimality of minimal graphs. *Indiana Math. J.*, **36**, 849–855, 1987.

[349] HARDT, R. and LIN, F. H.: Tangential regularity near C^1-boundary. In *Geometric Measure Theory and Calculus of Variations*, edited by Allard, W. K. and Almgren, F. J. Proc. Symp. Pure Math. 44 AMS, Providence, 1985.

[350] HARDT, R. and LIN, F. H.: A remark on H^1 mappings. *manuscripta math.*, **56**, 1–10, 1986.

[351] Hardt, R. and Lin, F. H.: Mappings minimizing the L^p norm of the gradient. *Comm. Pure Appl. Math.*, **40**, 555–588, 1987.

[352] Hardt, R. and Lin, F. H.: Stability of singularities of minimizing harmonic maps. *J. Differential Geom.*, **29**, 113–123, 1989.

[353] Hardt, R. and Lin, F. H.: The singular set of an energy minimizing map from B^4 to S^2. *Manuscripta Math.*, **69**, 275–289, 1990.

[354] Hardt, R., Lin, F. H., and Poon, C. C.: Axially symmetric harmonic maps minimizing a relaxed energy. *Comm. Pure Appl. Math.*, **45**, 417–459, 1992.

[355] Hardt, R. and Pitts, J.: Solving the Plateau's problem for hypersurfaces without the compactness theorem for integral currents. In *Geometric Measure Theory and the Calculus of Variations*, edited by Allard, W. K. and Almgren, F. J., Proc. Symp. Pure Math., 44, pp. 255–295. Am. Math. Soc., Providence, 1986.

[356] Hardt, R. and Simon, L.: Boundary regularity and embedded solutions for the oriented Plateau problem. *Ann. of Math.*, **110**, 439–486, 1979.

[357] Hardt, R. and Simon, L.: *Seminar on geometric measure theory.* Birkhauser Verlag, Basel-Boston, MA, 1986.

[358] Hartman, P.: On homotopic harmonic maps. *Can. J. Math.*, **19**, 673–687, 1967.

[359] Hartman, P. and Stampacchia, G.: On some nonlinear eliptic differential-functional equations. *Acta Math.*, **115**, 271–310, 1966.

[360] Harvey, R. and Lawson, H. B.: Extending minimal varieties. *Inventiones Math.*, **28**, 209–226, 1975.

[361] Harvey, R. and Lawson, H. B.: Calibrated geometries. *Acta math.*, **148**, 47–157, 1982.

[362] Hausdorff, F.: Dimension und äusseres Mass. *Math. Ann.*, **79**, 157–179, 1918.

[363] Hayman, W. and Kennedy, P.: *Subharmonic functions.* Academic Press, New York, 1976.

[364] Hebey, E.: *Sobolev spaces on Riemannian manifolds.* Springer, Berlin, 1996.

[365] Heinz, E.: Über die Lösungen der Minimalflächengleichung. In *Nachr. Akad. Wiss.*, pp. 51–56. Göttingen, 1952.

[366] Heinz, E.: An elementary analytic theory of the degree of mappings in n-dimensional spaces. *J. Math. Mech.*, **8**, 231–247, 1959.

[367] Heinz, E.: Ein Regularitätssatz für schwache Lösungen nichtlinearer elliptische Systeme. *Nachr. Akad. Wiss. Göttingen, Math.-Phys. Kl.*, **1**, 1–13, 1975.

[368] Hélein, F.: Minima de la fonctionelle énergie libre des crystaux liquides. *Comptes Rend. Acad. Sci. Paris*, **305**, 565–568, 1988.

[369] Hélein, F.: Approximation of Sobolev maps between an open set and an Euclidean sphere, boundary data, and singularites. *Math. Ann.*, **285**, 125–140, 1989.

[370] Hélein, F.: Regularité des applications faiblement harmoniques entre une surface et une sphère. *C. R. Acad. Sci. Paris Ser. I Math.*, **311**, 519–524, 1990.

[371] Hélein, F.: Regularité des applications faiblement harmoniques entre une surface et une variété riemannienne. *C. R. Acad. Sci. Paris Ser. I Math.*, **312**, 591–596, 1991.

[372] Hélein, F.: Regularity of weakly harmonic maps from a surface into a manifold with symmetries. *Manuscripta Math.*, **70**, 203–218, 1991.

[373] Hélein, F.: *Applications harmoniques, lois de conservation et repères mobiles.* Diderot editeur, Paris, 1996.

[374] Herz, C.: The Hardy-Littlewood maximal theorem. In *Symposium on Harmonic Analysis.* Univ. of Warwick, 1968.

[375] Hewitt, E. and Stromberg, K.: *Real and abstract analysis.* Springer, Berlin, 1965.

[376] Hildebrandt, S.: Boundary behaviour of minimal surfaces. *Arch. Rat. Mech. Anal.*, **35**, 47–82, 1969.

[377] HILDEBRANDT, S.: Harmonic mappings of Riemannian manifolds. In *Harmonic mappings and minimal immersions*, Lecture Notes in Math., 1161, pp. 1–117. Springer, Berlin-New York, 1985.

[378] HILDEBRANDT, S., JOST, J., and WIDMAN, K. O.: Harmonic mappings and minimal submanifolds. *Invent. Math.*, **62**, 269–298, 1980.

[379] HILDEBRANDT, S. and KAUL, H.: Two dimensional variational problems with obstructions, and Plateau's problem for H-surfaces in a Riemannian manifold. *Comm. Pure Appl. Math.*, **25**, 187–223, 1972.

[380] HILDEBRANDT, S., KAUL, H., and WIDMAN, K.-O.: An existence theorem for harmonic mappings of Riemannian manifolds. *Acta Math.*, **138**, 1–16, 1977.

[381] HODGE, W.: A Dirichlet problem for harmonic functions with applications to analytic varieties. *Proc. London Math. Soc.*, **36**, 257–303, 1934.

[382] HODGE, W.: *The theory and applications of harmonic integrals.* Cambridge University Press, Cambridge, 1952.

[383] HONG, M. C.: On the Jäger–Kaul theorem concerning harmonic maps. *preprint CMA, Canberra*, 1996.

[384] HOPF, E.: On S. Bernstein's theorem on surfaces $z(x,y)$ of nonpositive curvature. *Proc. Amer. Math. Soc.*, **1**, 80–85, 1950.

[385] HOVE, L. V.: Sur l'extension de la condition de Legendre du calcul des variations aux intégrales multiples à plusieurs fonctions inconnues. *Proc. Koninkl. Ned. Akad. Wetenschap*, **50**, 18–23, 1947.

[386] HOVE, L. V.: Sur le signe de la variation seconde des intégrales multiples à plusieurs fonctions inconnues. *Koninkl. Belg. Acad. Klasse der Wetenschappen*, **24**, 65, 1949.

[387] HU, S. T.: *Homotopy theory.* Academic Press, New York, 1959.

[388] HUGHES, T. and MARSDEN, J.: *Mathematical foundations of elasticity.* Prentice-Hall, Englewood Cliffs, 1983.

[389] HUGHS, R.: Functions of BVC type. *Proc. Amar. Math. Soc.*, **12**, 698–701, 1961.

[390] HUSEMOLLER, D.: *Fiber bundles.* Springer, New York, 1993.

[391] IOFFE, A. D.: On lower semicontinuity of integral functionals II. *SIAM J. Cont. Optim.*, **15**, 991–1000, 1977.

[392] IOFFE, A. D.: On lower sermicontinuity of integral functionals I. *SIAM J. Cont. Optim.*, **15**, 521–538, 1977.

[393] ISOBE, T.: Characterization of the strong closure of $C^\infty(B^4, S^2)$ in $W^{1,p}(B^4, S^2)$ $(16/5 < p < 4)$. *J. Math. Anal. Appl.*, **190**, 361–372, 1995.

[394] ISOBE, T.: Energy gap phenomenon and the existence of infinitely many weakly harmonic maps for the Dirichlet problem. *J. Funct. Anal.*, **129**, 243–267, 1995.

[395] ISOBE, T.: Some new properties of Sobolev mappings: intersection theoretical approach. *Proc. Roy. Soc. Edinburgh*, **127A**, 337–358, 1997.

[396] JÄGER, W. and KAUL, H.: Uniqueness and stability of harmonic maps and their Jacobi fields. *Manuscripta Math.*, **28**, 269–291, 1979.

[397] JÄGER, W. and KAUL, H.: Uniqueness of harmonic mappings and of solutions of elliptic equations on Riemannian manifolds. *Math. Ann.*, **240**, 231–250, 1979.

[398] JÄGER, W. and KAUL, H.: Rotationally symmetric harmonic maps from a ball into a sphere and the regularity problem for weak solutions of elliptic systems. *J. Reine Angew. Math.*, **343**, 146–161, 1983.

[399] JENKINS, H. and SERRIN, J.: The Dirichlet problem for minimal surface equation in higher dimensions. *J. reine u. angew. Math.*, **229**, 170–187, 1968.

[400] JOHN, F. and NIRENBERG, L.: On functions of bounded mean oscillation. *Comm. Pure Appl. Math.*, **14**, 415–426, 1961.

[401] JÖRGENS, K.: Über die Lösungen der Differentialgleichung $rt - s^2 = 1$. *Math. Ann.*, **127**, 130–134, 1954.

[402] JOST, J.: A maximum principle for harmonic mappings which solve a Dirichlet problem. *Manuscripta Math.*, **38**, 129–130, 1982.

[403] JOST, J.: Existence proofs for harmonic mappings with the help of a maximum principle. *Math. Z.*, **184**, 489–496, 1983.

[404] JOST, J.: *Harmonic mappings between Riemannian manifolds.* Australian National University, Centre for Mathematical Analysis, Canberra, 1984.

[405] JOST, J.: *Harmonic maps between surfaces.* Springer-Verlag, Berlin-New York, 1984.

[406] JOST, J.: The Dirichlet problem for harmonic maps from a surface onto a 2-sphere with nonconstant boundary values. *J. Diff. Geom.*, **19**, 393–401, 1984.

[407] JOST, J.: Lectures on harmonic maps (with applications to conformal mappings and minimal surfaces). In *Harmonic mappings and minimal immersions*, Lecture Notes in Math., 1161,, pp. 118–192. Springer, Berlin-New York, 1985.

[408] JOST, J.: *Two dimensional geometric variational problems.* Wiley-Interscience, Chichester, 1991.

[409] JOST, J.: *Riemannian geometry and geometric analysis.* Springer, Berlin, 1995.

[410] JOST, J. and MEIER, M.: Boundary regularity for minima of certain quadratic functionals. *Math. Ann.*, **262**, 549–561, 1983.

[411] KAPITANSKY, L. B. and LADYZHENSKAYA, O. A.: On Coleman's principle concerning the stationary points of invariant integrals. *Zapiski. nauk. sem. LOMI*, **127**, 84–102, 1983.

[412] KINDERLEHER, D. and PEDREGAL, P.: Gradient Young measures generated by sequences in Sobolev spaces. *J. Geometrical Anal.*, **4**, 59–90, 1994.

[413] KINDERLERHER, D.: Recent developments in liquid crystal theory. In *Frontiers in pure and applied mathematics*, pp. 151–178. North–Holland, Amsterdam, 1991.

[414] KODAIRA, K.: Harmonic fields in Riemannian manifolds. *Ann. of Math.*, **50**, 587–665, 1949.

[415] KRICKEBERG, K.: Distributionen, Funktionen beschränkter Variation und Lebesguescher Inhalt nichtparametrischer Flächen. *Ann. Mat. Pura Appl.*, **44**, 105–134, 1957.

[416] KUFNER, A. AND JOHN, O. AND FUČIK, S.: *Function spaces.* Academia, Praha, 1977.

[417] KUWERT, E.: Minimizing the energy of maps from a surface into a 2-sphere with prescribed degree and boundary values. *manuscripta math.*, **83**, 31–38, 1994.

[418] KUWERT, E.: Area-minimizing immersions of the disk type with boundary in a given homotopy class: a general existence theory, 1995. Habilitationsschrift, 1995, University of Bonn.

[419] LADYZHENSKAYA, O. A.: On finding symmetrical solutions of field theory variational problems. In *Proc. of Intern. Congr. of Mathematicians.* Warszawa, 1983.

[420] LADYZHENSKAYA, O. A. and URAL'TSEVA, N. N.: *Linear and quasilinear elliptic equations.* Nauka, Moscow, 1964. (Engl. transl.: Academic Press, new York, 1968; Second Russian edition: Nauka, 1973).

[421] LAWSON, H. B.: *Lectures on minimal surfaces.* INPA, Rio de Janeiro, 1973.

[422] LAWSON, H. B.: Minimal varieties. In *Differential Geometry.* Proc. Symp. Pure Math., AMS, Providence, 1975.

[423] LAWSON, H. B. and OSSERMAN, R.: Non-existence, non-uniqueness and irregularity of solutions to the minimal surface system. *Acta Math.*, **139**, 1–17, 1977.

[424] LEBESGUE, H.: Intégral, longueur, aire. *Ann. Mat. Pura Appl.*, **7**, 231–359, 1902.

[425] LEBESGUE, H.: Sur les intégrales singulières. *Ann. Fac. Sc. Toulouse*, pp. 25–117, 1909.

[426] LEMAIRE, L.: Applications harmoniques de surfaces riemanniennes. *J. Differential Geom.*, **13**, 51–78, 1978.

[427] LEMAIRE, L.: Boundary value problems for harmonic and minimal maps of surfaces into manifolds. *Ann. Scuola Norm. Sup. Pisa*, **9**, 91–103, 1982.

[428] LEONETTI, F.: Maximum principle for functionals depending on minors of the Jacobian matrix of vector-valued mappings. Tech. rep., Centre for Mathematical Analysis, Australian National University, 1990.

[429] LERAY, J.: Discussion d'un problème de Dirichlet. *J. Math. Pure Appl.*, **18**, 249–284, 1939.

[430] LESLIE, F. C.: Theory of flow phenomena in liquid crystals. In *Advances in Liquid Crystals, vol. 2*, edited by Brown, G. H., pp. 1–81. Academic Press, New York, 1976.

[431] LÉVY, P.: *Théorie de l'addition des variables aléatoires.* Gauthier-Villars, Paris, 1954.

[432] LIEB, E. H.: Remarks on Skyrme model, ????

[433] LIN, F. H.: A remark on the map $x/|x|$. *C. R. Acad. Sci. Paris Ser. I Math.*, **305**, 529–531, 1987.

[434] LIN, S. Y.: Numerical analysis for liquid crystal problems. *Thesis*, 1987.

[435] LIONS, P. L.: The concentration-compactness principle in the calculus of variations. The locally compact case. I. *Ann. Inst. H. Poincaré Anal. Non Linéaire*, **1**, 109–145, 1984.

[436] LIONS, P. L.: The concentration-compactness principle in the calculus of variations. The locally compact case. II. *Ann. Inst. H. Poincaré Anal. Non Linéaire*, **1**, 223–283, 1984.

[437] LIONS, P. L.: The concentration-compactness principle in the calculus of variations. The limit case. II. *Rev. Mat. Iberoamericana*, **1**, 45–121, 1985.

[438] LIONS, P. L.: The concentration-compactness principle in the calculus of variations. The limit case. I. *Rev. Mat. Iberoamericana*, **1**, 145–201, 1985.

[439] LIU, F. C.: Essential multiplicity function for a.e. approximately differentiable mapping. *Proc. Acad. Sinica*, 1976.

[440] LIU, F. C.: A Lusin type property of Sobolev functions. *Indiana Univ. Math. J.*, **26**, 645–651, 1977.

[441] LIU, F. C.: Approximately differentiable mappings and surface area. *In "Studies and essays in Commemoration of the Golden Jubilee of Academia Sinica" Taipei*, 1978.

[442] LIU, F. C. and TAI, W. S.: Approximate Taylor polynomials and differentiation of functions. *preprint*, 1991.

[443] LIU, F. C. and TAI, W. S.: Maximal mean steepness and Lusin type properties. *preprint*, 1991.

[444] LUCKHAUS, S.: Partial Hölder continuity for minima of certain energies among maps into a Riemannian manifold. *Indiana Math. J.*, **37**, 349–367, 1988.

[445] LUCKHAUS, S.: Convergence of minimizers for the p-Dirichlet integral. *Math. Z.*, **213**, 449–456, 1993.

[446] LUKEŠ, J., MALÝ, J., and ZAJÍČEK, L.: *Fine Topology Methods in Real Analysis and Potential Theory.* Lecture notes 1189. Springer-Verlag, Berlin, 1986.

[447] MALÝ, J.: L^p-approximation of Jacobians. *Comment. Math. Univ. Carolinae*, **32**, 659–666, 1991.

[448] MALÝ, J.: Weak lower semicontinuity of polyconvex integrals. *Proc. Roy. Soc. Edinburgh*, **123A**, 681–691, 1993.

[449] MALÝ, J.: Lower semicontinuity of quasiconvex integrals. *manuscripta math.*, **85**, 419–428, 1994.

[450] MALÝ, J. and MARTIO, O.: Lusin's condition (N) and mappings of the class $W^{1,n}$. *J. Reine Angew. Math.*, **458**, 19–36, 1995.

[451] MANTON, N. S. and RUBECK, P. J.: Skyrmions in flat space and curved space. *Phys. Lett.*, **B181**, 137–140, 1986.

[452] MARCELLINI, P.: Quasiconvex quadratic forms in two dimensions. *Appl. Math. Optim.*, **11**, 183–189, 1984.

[453] MARCELLINI, P.: Approximation of quasiconvex functions and lower semicontinuity of multiple integrals. *manuscripta math.*, **51**, 1–28, 1985.

[454] MARCELLINI, P.: On the definition and the lower semicontinuity of certain quasiconvex integrals. *Ann. Inst. H. Poincaré, Anal. Non Linéaire*, **3**, 391–409, 1986.

[455] MARCELLINI, P.: The stored-energy for some discontinuous deformations in nonlinear elasticity. In *PDE and the Calculus of Variations*, edited by Colombini, F., Marino, A., Modica, L., and Spagnolo, S., pp. 767–786. Birkhäuser, Boston, 1989.

[456] MARCELLINI, P. and SBORDONE, C.: Semicontinuity problems in the calculus of variations. *Nonlinear Anal.,Theory, Meth. and Appl.*, **4**, 241–257, 1980.

[457] MARCUS, M. and MIZEL, V.: Transformations by functions in Sobolev spaces and lower semicontinuity for parametric variational problems. *Bull. Amer. Math. Soc.*, **79**, 790–795, 1973.

[458] MARCUS, M. and MIZEL, V. J.: Absolute continuity on tracks and mappings of Sobolev spaces. *Arch. Rat. Mech. Anal.*, **45**, 294–320, 1972.

[459] MASSARI, U.: Esistenza e regolarità delle ipersuperfici di curvatura media assegnata in $\mathbb{R}^n$. *Arch. Rat. Mech. Anal.*, **55**, 357–382, 1974.

[460] MASSARI, U.: Frontiere orientate di curvatura media assegnata in L^p. *Rend. Sem. Mat. Padova*, **53**, 37–52, 1975.

[461] MASSARI, U. and MIRANDA, M.: *Minimal surfaces of codimension one.* North-Holland, Amsterdam, 1984.

[462] MATTILA, P.: *Geometry of sets and measures in Euclidean spaces.* Cambridge Univ. Press, 1995.

[463] MAZ'JA, V. G.: *Sobolev spaces.* Springer-Verlag, Berlin, 1985.

[464] MCSHANE, B. J.: On the necessary condition of Weierstrass in the multiple integral problem of the calculus of variations. *Ann. Math.*, **32**, 578–590, 1931.

[465] MEEKS, W. H. and YAU, S.-T.: The classical Plateau problem and the topology of three-dimensional manifolds. *Topology*, **21**, 409–440, 1982.

[466] MEYERS, N. G.: Quasiconvexity and lower semicontinuity of multiple variational integrals of any order. *Trans. A.M.S.*, **119**, 125–149, 1965.

[467] MEYERS, N. G. and ZIEMER, W. P.: Integral inequalities of Poincaré and Wirtinger type for *BV* functions. *Amer. J. Math.*, **99**, 1345–1360, 1977.

[468] MICALLEF, M. and WHITE, B.: The structure of branch points in minimal surfaces and to pseudoholomorphic curves. *Ann. of Math.*, **141**, 35–85, 1995.

[469] MICHAEL, E.: Continuous selections. *Ann. Math.*, **63**, 361–381, 1956.

[470] MICHAEL, J. H.: The equivalence of two areas for nonparametric discontinuous surfaces. *Illinois J. Math.*, **7**, 59–78, 1963.

[471] MICKLE, E. J.: A remark on a theorem of Serge Bernstein. *Proc. Amer. Math. Soc.*, **1**, 86–89, 1950.

[472] MILNOR, J.: On manifolds homeomorphic to the 7-sphere. *Ann. of Math.*, **64**, 399–405, 1956.

[473] MILNOR, J.: *Topology from the differentiable viewpoint.* Univ. Press of Virginia, 1965.

[474] MIRANDA, M.: Distribuzioni aventi derivate misure. Insiemi di perimetro finito. *Ann. Sc. Norm. Sup. Pisa*, **18**, 27–56, 1964.

[475] MIRANDA, M.: Superfici cartesiane generalizzate ed insiemi di perimetro finito sui prodotti cartesiani. *Ann. Sc. Norm. Sup. Pisa*, **18**, 515–542, 1964.

[476] MIRANDA, M.: Sul minimo dell'integrale del gradiente di una funzione. *Ann. Sc. Norm. Sup. Pisa*, **19**, 627–665, 1965.

[477] MIRANDA, M.: Un teorema di esistenza e unicità per il problema dell'area in n variabili. *Ann. Sc. Norm. Sup. Pisa*, **19**, 233–249, 1965.

[478] MIRANDA, M.: Comportamento delle successioni convergenti di frontiere minimali. *Rend. Sem. Mat. Univ. Padova*, **38**, 238–257, 1967.

[479] MIRANDA, M.: Un principio di massimo forte per le frontiere minimali e sua applicazione alle superfici di area minima. *Rend. Sem. Mat. Padova*, **45**, 355–366, 1971.

[480] MIRANDA, M.: Grafici minimi completi. *Ann. Univ. Ferrara*, **23**, 269–272, 1977.

[481] MIRANDA, M.: Sulle singolarità eliminabili delle soluzioni dell'equazione delle superfici minime. *Ann. Sc. Norm. Sup. Pisa*, **4**, 129–132, 1977.

[482] MIRANDA, M.: Superfici minime illimitate. *Ann. Sc. Norm. Sup. Pisa*, **4**, 313–322, 1977.

[483] MORGAN, F.: Area-minimizing currents bounded by higher multiples of curves. *Rend. Circ. Mat. Palermo*, **33**, 37–46, 1984.

[484] MORGAN, F.: Area-minimizing surfaces, faces of Grassmannians, and calibrations. *Am. Math. Monthly*, **95**, 813–822, 1988.

[485] MORGAN, F.: *Geometric measure theory. A beginner's guide.* Academic Press, Inc., Boston, MA, 1988. second edition, 1995.

[486] MORREY, C. B.: On the solutions of quasi-linear elliptic partial differential equations. *Trans. Am. Math. Soc.*, **43**, 126–166, 1938.

[487] MORREY, C. B.: Multiple integral problems in the calculus of variations and related topics. *Univ. of California Publ. in Math.*, **1**, 1–130, 1943.

[488] MORREY, C. B.: The problem of Plateau on a Riemannian manifold. *Ann. of Math.*, **49**, 807–851, 1948.

[489] MORREY, C. B.: Quasi-convexity and the lower semicontinuity of multiple integrals. *Pacific J. Math.*, **2**, 25–53, 1952.

[490] MORREY, C. B.: *Multiple integrals in the calculus of variations.* Grundlehren math. Wissenschaften 130. Springer-Verlag, Berlin, 1966.

[491] MORREY, C. B.: Partial regularity results for nonlinear elliptic systems. *Jour. Math. and Mech.*, **17**, 649–670, 1968.

[492] MOSER, J.: A new proof od De Giorgi's theorem concerning the regularity problem for elliptic differential eqiuations. *Comm. Pure Appl. Math.*, **13**, 457–468, 1960.

[493] MOSER, J.: On Harnack's theorem for elliptic differential equations. *Comm. Pure Appl. Math.*, **14**, 577–591, 1961.

[494] MOU, L.: Harmonic maps with prescribed finite singularities. *Comm. in PDE*, **14**, 1509–1540, 1989.

[495] MUCCI, D.: Graphs of finite mass which cannot be approximated by smooth graphs with equibounded area. *J. Func. Anal., to appear*, **??**, ??, ??

[496] MUCCI, D.: Approximation in area of graphs with isolated singularities. *manuscripta math.*, **88**, 135–146, 1995.

[497] MUCCI, D.: *Approssimazione in area di grafici continui in dimensione e codimensione qualunque.* Ph.D. thesis, Università di Firenze, 1996.

[498] MUCCI, D.: Approximation in area of continuous graphs. *Calc. Var.*, **4**, 525–557, 1996.

[499] MÜLLER, S.: Weak continuity of determinants and nonlinear elasticity. *C. R. Acad. Sci. Paris*, **307**, 501–506, 1988.

[500] MÜLLER, S.: Det=det. A remark on the distributional determinant. *C. R. Acad. Sci Paris*, **311**, 13–17, 1990.

[501] MÜLLER, S.: Higher integrability of determinants and weak convergence in L^1. *J. Reine Angew. Math.*, **412**, 20–34, 1990.

[502] MÜLLER, S.: On quasiconvex functions which are homogeneous of degree 1. *Indiana Univ. Math. J.*, **41**, 295–301, 1992.

[503] MÜLLER, S.: On the Singular Support of the Distributional Determinant. *Ann. Inst. H. Poincaré Anal. Non Linéaire*, **10**, 657–696, 1993.

[504] MÜLLER, S., TANG, Q., and YAN, B.: On a new class of elastic deformations not allowing for cavitation. *Ann. Inst. H. Poincaré Anal. Non Linéaire*, **11**, 217–243, 1994.

[505] MUNROE, M.: *Measure and integration.* Addison-Wesley, Reading, 1971. Second edition.

[506] MURAT, F.: Compacité par compensation. *Ann. Sc. Norm. Sup. Pisa*, **5**, 489–507, 1978.

[507] MURAT, F.: Compacité par compensation, II. In *Proceedings of the International Meeting on Recent Methods in Nonlinear Analysis*, edited by De Giorgi, E., Magenes, E., and Mosco, U. Pitagora, Bologna, 1979.

[508] MURAT, F.: Compacité par compensation: condition nécessaire et suffisante de continuité faible sous une hypothèse de rang constant. *Ann. Sc. Norm. Sup. Pisa*, **8**, 69–102, 1981.

[509] NAGUMO, M.: Degree of mappings in convex linear topological spaces. *Amer. J. Math.*, **73**, 497–511, 1951.

[510] NASH, J.: The imbedding problem for riemannian manifolds. *Ann. of Math.*, **63**, 20–63, 1956.

[511] NASH, J.: Continuity of solutions of parabolic and elliptic equations. *Amer. J. Math.*, **8**, 931–954, 1958.

[512] NATANSON, I. P.: *Theory of functions of real variable I,II.* Frederick ungar Publ., New York, 1964.

[513] NIRENBERG, L.: *Topics in nonlinear functional analysis.* Courant Institute, New York University, New York, 1974.

[514] NISHIURA, T. and BRECKENRIDGE, J. C.: Differentiation, Integration, and Lebesgue Area. *Imdiana Univ. Math. J.*, **26**, 515–536, 1977.

[515] NITSCHE, J. C. C.: Elementary proof of Bernstein's theorem on minimal surfaces. *Ann. of Math.*, **66**, 543–544, 1957.

[516] NITSCHE, J. C. C.: On new results in the theory of minimal surfaces. *Bull. AMS*, **71**, 195–270, 1965.

[517] NITSCHE, J. C. C.: Über ein verallgemeinertes Dirichletsches Problem für die Minimalflächengleichung und hebbare Unstetigkeiten ihrer Lösungen. *Math. Ann.*, **158**, 203–214, 1965.

[518] NITSCHE, J. C. C.: The boundary behaviour of minimal surfaces. Kellog's theorem and branch points on the boundary. *Invent. Math.*, **8**, 313–333, 1969. Addendum, Invent. Math. **9** (1970) 270.

[519] NITSCHE, J. C. C.: *Vorlesungen über Minimaflächen.* Grundlehren 199. Springer-Verlag, Berlin, 1975.

[520] NITSCHE, J. C. C.: *Lectures on minimal surfaces.* Cambridge Univ. Press, 1989.

[521] NÖBELING, G.: Über die Flächenmasse im Euclidischen Raum. *Math. Ann.*, **118**, 687–701, 1943.

[522] OLECH, C.: Existence theory in optimal control problems, the underlying ideas. In *International Conference on Differential Equations* , pp. 612–629. Academic Press, New York, 1975.

[523] OLECH, C.: A characterization of L^1-weak lower semicontinuity of integral functionals. *Bull. Acad. Pol. Sci.*, **25**, 135–142, 1977.

[524] OSEEN, C. W.: The theory of liquid crystals. *Trans. Faraday Soc.*, **29**, 883–889, 1933.

[525] OSSERMAN, R.: *A survey of minimal surfaces.* Van Nostrand, New York, 1969. (Dover Edition, 1986).

[526] OSSERMAN, R.: Minimal varieties. *Bull. AMS*, **75**, 1092–1120, 1969.

[527] OSSERMAN, R.: A proof of the regularity everywhere of the classical solution to Plateau problem. *Ann. Math.*, **91**, 550–569, 1970.

[528] OSSERMAN, R.: Some properties of solutions to the minimal surfaces system for arbitrary codimension. In *Global Analysis.* Proc. Symp. Math. 15, AMS, Providence, 1970.

[529] OSSERMAN, R.: On Bers' Theorem on isolated singularities. *Indiana Univ. Math. J.*, **23**, 337–342, 1973.

[530] PEDREGAL, P.: *Parametrized measures and variational principles.* Birkäuser, 1997.

[531] PERELOMOV, A. M.: Instanton–like solutions in chiral models. *Physica 4D*, pp. 1–25, 1981.

[532] PODIO-GUIDUGLI, P. and VERGARA CAFFARELLI, G.: Discontinuous energy minimizers in nonlinear elastostatics: an example of J. Ball revisited. *J. Elasticity*, **15**, 75–96, 1986.

[533] POHOZAEV, S. I.: Eigenfunctions of the equation $\Delta u + \lambda f(u) = 0$. *Soviet Math. Doklady*, **6**, 1408–1411, 1965. (Transl. from the *Russian Dokl. Akad. Nauk SSSR* 165, **1965**, 33-36).

[534] PONOMAREV, S. P.: Property N of homeomorphisms of class $W^{1,p}$. *Sib. Math. J.*, **28**, 291–298, 1987.

[535] POON, C. C.: Some new harmonic maps from B^3 to S^2. *J. Diff. Geom.*, **34**, 165–168, 1991.

[536] PREISS, D.: Geometry of measures in $\mathbf{R}^n$: distribution, rectifiability, and densities. *Ann. of Math. (2)*, **125**, 537–643, 1987.

[537] QING, J.: Remark on the Dirichlet problem for harmonic maps from the disc into the 2-sphere. *Proc. Roy. Soc. Edinburgh*, **122A**, 63–67, 1992.

[538] RADEMACHER, H.: Über partielle und totale Differentierbarkeit I. *Math. Ann.*, **79**, 340–359, 1919.

[539] RADÓ, T.: Geometrische Betrachtungen über zweidimensionale reguläre Variationsprobeme. *Acta Litt. Sci. Szeged*, **2**, 228–253, 1926.

[540] RADÓ, T.: Zu einem Satze von S. Bernstein über Minimalflächen in Grossen. *Math. Z.*, **26**, 559–565, 1927.

[541] RADÓ, T.: Über zweidimensionale reguläre Variationsprobleme. *Math. Ann.*, **101**, 620–632, 1929.

[542] RADÓ, T.: On Plateau's problem. *Ann. Math.*, **32**, 457–469, 1930.

[543] RADÓ, T.: The problem of least area and the problem of Plateau. *Math. Z.*, **32**, 763–796, 1930.

[544] RADÓ, T.: *On the problem of Plateau.* Springer, Berlin, 1933.

[545] RADÓ, T.: *Length and area.* Colloquium Publications 30. Amer. Math. Soc., New York, 1948.

[546] RADÓ, T. and REICHELDERFER, P.: *Continuous transformations in analysis.* Springer, Berlin, 1955.

[547] REIFENBERG, R. E.: Solution of the Plateau problem for m-dimensional surfaces of varying topological type. *Acta Math.*, **104**, 1–92, 1960.

[548] REIFENBERG, R. E.: An epiperimetric inequality related to the analyticity of minimal surfaces. *Ann. of Math.*, **80**, 1–14, 1964.

[549] REIFENBERG, R. E.: On the analyticity of minimal surfaces. *Ann. of Math.*, **80**, 15–21, 1964.

[550] RESHETNYAK, Y. G.: On the stability of conformal mappings in multidimensional spaces. *Sib. Math. Zhu.*, **8**, 91–114, 1967. In Russian. English transl.: Siberian Math. J. 8 (1967) 69-85.

[551] RESHETNYAK, Y. G.: Stability theorems for mappings with bounded excursion. *Sib. Math. Zhu.*, **9**, 667–684, 1968. In Russian. English transl.: Siberian Math. J. 9 (1968) 499-512.

[552] RESHETNYAK, Y. G.: Weak convergence of completely additive vector functions on a set. *Sib. Math. J.*, **9**, 1039–1045, 1968.

[553] RESHETNYAK, Y. G.: General theorems on semicontinuity and on convergence with a functional. *Sib. Math. J.*, **9**, 801–816, 1969.

[554] RESHETNYAK, Y. G.: *Space mappings with bounded distorsion.* Transl. Math. Monographs. Amer. Math. Soc., Providence, 1989.

[555] RICKMAN, S.: On the number of omitted values of entire quasiregular mappings. *J. Analyse Math.*, **37**, 100–117, 1980.

[556] RIESZ, F. and SZ.-NAGY, B.: *Leçons d'analyse fonctionnelle*, 1952. 4th edition, Gauthier-Villars, Paris, 1965.

[557] RIVIÈRE, T.: Applications harmoniques de B^3 dans S^2 ayant une ligne de singularites. *C. R. Acad. Sci. Paris Ser. I Math.*, **313**, 583–587, 1991.

[558] RIVIÈRE, T.: Applications harmoniques de B^3 dans S^2 partout discontinues. *C. R. Acad. Sci. Paris Ser. I Math.*, **314**, 719–723, 1992.

[559] RIVIÈRE, T.: *Applications harmoniques entre variétés.* Thèse de Doctorat, 1993, Université de Paris VI.

[560] RIVIÈRE, T.: Harmonic maps from B^3 into S^2 having a line of singularities. In *Applications harmoniques entre variétés*, Thèse de Doctorat, 1993, Université de Paris VI.

[561] RIVIÈRE, T.: Everywhere discontinuous maps into spheres. *Acta Math.*, **175**, 197–226, 1995.

[562] RIVIÈRE, T.: Minimizing fibrations and p–harmonic maps in homotopy classes from S^3 into S^2. *Preprint*, 1996.

[563] ROCKAFELLAR, R.: *Convex analysis.* Princeton Univ. Press, Princeton, 1970.

[564] ROGERS, C.: *Hausdorff measure.* Cambridge Univ. Press, Cambridge, 1970.

[565] ROUBIČEK, T.: *Relaxation in optimization theory and variational calculus.* Walter de Gruyter, Berlin, 1997.

[566] RUDIN, W.: *Real and complex analysis.* McGraw-Hill, New York, 1966.

[567] RUH, E. and VILMS, J.: The tensor field of the Gauss map. *Trans. Amer. Math. Soc.*, **149**, 569–573, 1970.

[568] SACKS, J. and UHLENBECK, K.: The existence of minimal immersions of 2-spheres. *Ann. of Math.*, **113**, 1–24, 1981.

[569] SAKS, S.: On the surfaces without tangent planes. *Ann. of Math.*, **34**, 114–124, 1933.

[570] SAKS, S.: *Théorie de l'intégrale.* Monografje Matematyczne, Warsawa, 1933.

[571] SCHOEN, R.: Analytic aspects of the harmonic map problem. In *Seminar on nonlinear partial differential equations*, Math. Sci. Res. Inst. Publ., 2,, pp. 321–358. Springer, New York-Berlin, 1984.

[572] SCHOEN, R.: The effect of curvature on the behaviour of harmonic functions and mappings. In *Nonlinear Partial Differential Equations in Differential Geometry*, edited by Hardt, R. and Wolf, M., pp. 127–183. Amer. Math. Soc., Providence, RI, 1996.

[573] SCHOEN, R. and SIMON, L.: A new proof of the regularity theorem for rectifiable currents which minimize parametric elliptic functionals. *Indiana Univ. Math. J.*, **31**, 415–434, 1982.

[574] SCHOEN, R., SIMON, L., and ALMGREN, F. J.: Regularity and singularity estimates on hypersurfaces minimizing parametric elliptic variational integrals. *Acta Math.*, **139**, 217–265, 1977.

[575] SCHOEN, R., SIMON, L., and YAU, S. T.: Curvature estimates for minimal hypersurfaces. *Acta Math.*, **134**, 276–288, 1975.

[576] SCHOEN, R. and UHLENBECK, K.: A regularity theory for harmonic maps. *J. Differential Geom.*, **17**, 307–335, 1982.

[577] SCHOEN, R. and UHLENBECK, K.: Boundary regularity and the Dirichlet problem for harmonic maps. *J. Differential Geom.*, **18**, 253–268, 1983.

[578] SCHOEN, R. and UHLENBECK, K.: Correction to: "A regularity theory for harmonic maps". *J. Differential Geom.*, **18**, 1983.

[579] SCHOEN, R. and UHLENBECK, K.: Regularity of minimizing harmonic maps into the sphere. *Invent. Math.*, **78**, 89–100, 1984.

[580] SCHOEN, R. and YAU, S.: Lectures on harmonic maps. International Press. *To appear.*

[581] SCHOEN, R. and YAU, S.: Existence of incompressible minimal surfaces and the topology of three dimensional manifolds with non-negative scalar curvature. *Ann. of Math.*, **110**, 127–142, 1979.

[582] SCHWARZ, G.: *Hodge decomposition. A method for solving boundary value problems.* Lecture Notes 1607. Springer, Berlin, 1995.

[583] SEMMES, S.: A primer on Hardy spaces, and some remarks on a theorem of Evans and Müller. *Comm. Part. Diff. Eq.*, **19**, 277–319, 1994.

[584] SERRE, D.: Formes quadratiques et calcul des variations. *J. Math. Pures Appl.*, **62**, 117–196, 1983.

[585] SERRIN, J.: On the differentiability of functions of several variables. *Arch. Rat. Mech. Anal.*, **7**, 359–372, 1961.

[586] SERRIN, J.: The problem of Dirichlet for quasilinear elliptic differential equations with many independent variables. *Phil. Trans. Roy. Soc. London*, **264**, 413–496, 1969.

[587] SERRIN, J. and VARBERG, D. E.: A general chain rule for derivatives and the change of variables formula for the Lebesgue integral. *Amer. Math. Monthly*, **76**, 514–520, 1969.

[588] SIMON, L.: Boundary regularity for solutions of the non-parametric least area problem. *Ann. of Math.*, **103**, 429–455, 1976.

[589] SIMON, L.: Remarks on curvature estimates for minimal hypersurfaces. *Duke Math. J.*, **43**, 545–553, 1976.

[590] SIMON, L.: On a theorem of De Giorgi and Stampacchia. *Math. Z.*, **115**, 199–204, 1977.

[591] SIMON, L.: Asymptotics for a class of non-linear evolution equations, with applications to geometric problems. *Ann. of Math.*, **118**, 525–572, 1983.

[592] SIMON, L.: *Lectures on geometric measure theory.* The Centre for mathematical Analysis, Canberra, 1983.

[593] SIMON, L.: Survey lectures on minimal submanifolds. In *Seminar on minimal submanifolds*, edited by Bombieri, E. Princeton Univ. Press., Princeton, 1983.

[594] SIMON, L.: Isolated singularities of extrema of geometric variational problems. In *Harmonic Mappings and Minimal Immersions*, Lecture notes in Math. n.1161. Springer, Berlin, 1985.

[595] SIMON, L.: Rectifiability of the singular set of energy minimizing maps. *Calc. Var.*, **3**, 1–65, 1995.

[596] SIMON, L.: *Theorems on regularity and singularity of energy minimizing maps.* ETH Lectures on Mathematics. Birkäuser, Basel, 1995.

[597] SIMON, L.: Proof of the basic regularity theorem for harmonic maps. In *Nonlinear Partial Differential Equations in Differential Geometry*, edited by Hardt, R. and Wolf, M., pp. 225–256. Amer. Math. Soc., Providence, RI, 1996.

[598] SIMON, L.: Singularities of geometric variational problems. In *Nonlinear Partial Differential Equations in Differential Geometry*, edited by Hardt, R. and Wolf, M., pp. 185–224. Amer. Math. Soc., Providence, RI, 1996.

[599] SIMONS, J.: Minimal varieties in riemannian manifolds. *Ann. of Math.*, **88**, 62–105, 1968.

[600] SINGER, I. M. and THORPE, J. A.: *Lectures Notes on elementary topology and geometry.* Scott, Foresman and C., Genview, Illinois, 1967.

[601] SIVALOGANATHAN, J.: Uniqueness of regular and singular equilibria for spherically symmetric problems of nonlinear elasticity. *Arch. Rat. Mech. Anal.*, **96**, 97–136, 1986.

[602] SKYRME, T. H. R.: A nonlinear filed theory. *Proc. Roy. Soc.*, **A260**, 127–138, 1961.

[603] SKYRME, T. H. R.: A unified field theory of mesons and baryons. *Nuc. Phys.*, **31**, 556–559, 1962.

[604] SOLOMON, B.: A new proof of the closure theorem for integral currents. *Indiana Univ. Math. J.*, **33**, 393–418, 1984.

[605] SOUČEK, J.: Spaces of functions on the domain Ω whose k-th derivatives are measures defined in $\bar{\Omega}$. *Čas. Pestovani Mat.*, **97**, 10–46, 1972.

[606] SOYEUR, A.: The Dirichlet problem for harmonic maps from the disc into the 2-sphere. *Proc. Roy. Soc. Edinburgh*, **113A**, 229–234, 1989.

[607] SPANIER, E. H.: *Algebraic topology.* McGraw-Hill, New York, 1966.

[608] STAMPACCHIA, G.: On some regular multiple integral problems in the calculus of variations. *Comm. Pure Appl. Math.*, **16**, 383–421, 1963.

[609] STAMPACCHIA, G.: *Equations elliptiques du second ordre à coefficients discontinus.* Les Presses de l'Université de Montréal, Montréal, 1966.

[610] STEFFEN, K. and WENTE, H.: The non–existence of branch points in a solution to certain classes of Plateau type variational problems. *Math. Z.*, **163**, 211–238, 1978.

[611] STEIN, E. M.: *Singular integrals and differentiability properties of functions.* Princeton Univ. Press, Princeton, 1970.

[612] STEIN, E. M.: *Harmonic analysis.* Princeton Univ. Press, Princeton, 1993.

[613] STEIN, E. M. and WEISS, G.: *Introduction to Fourier analysis on Euclidean spaces.* Princeton Univ. Press, Princeton, 1971.

[614] STEPANOFF, W.: Sur les conditions de l'existence de la différentielle totale. *Rec. Math. Soc. Moscou*, **32**, 511–526, 1925.

[615] STRUWE, M.: *Plateau's problem and the Calculus of Variations.* Mathemathical Notes, 35. Princeton Univ. Press, Princeton, 1988.

[616] STRUWE, M.: *Variational methods.* Springer–Verlag, Berlin, 1990. Second edition 1996.

[617] STUART, C. A.: Radially symmetric cavitation for hyperelastic materials. *Ann IHP, Analyse Non Linéaire*, **2**, 33–66, 1985.

[618] SVERÁK, V.: Regularity properties of deformations with finite energy. *Arch. Rat. Mech. Anal.*, **100**, 105–127, 1988.

[619] SVERÁK, V.: Quasiconvex functions with subquadratic growth. *Proc. Roy. Soc. London*, **A 433**, 723–725, 1991.

[620] SVERÁK, V.: New examples of quasiconvex functions. *Arch. Rat. Mech. Anal.*, **119**, 293–300, 1992.

[621] SVERÁK, V.: Rank-one convexity does not imply quasiconvexity. *Proc. Roy. Soc. Edinburgh*, **120A**, 185–189, 1992.

[622] SYCHEV, M. A.: Characterization of weak-strong convergence property of integral functionals in terms of integrands. *preprint*, 1994.

[623] SYCHEV, M. A.: A criterion for continuity of an integral functional on a sequence of functions. *Sib. Math. J.*, **36**, 146–156, 1995.

[624] TAMANINI, I.: Boundaries of Caccioppoli sets with Hölder-continuous normal vector. *J. reine u. angew. Math.*, **334**, 27–39, 1982.

[625] TANG, Q.: Almost everywhere injectivity in non linear elasticity. *Proc. Roy. Soc. Edinburg*, **109 A**, 79–95, 1988.

[626] TARTAR, L.: The compensated compactness method applied to systems of conservation laws. In *Systems of nonlinear partial differential equations. Proc. NATO Advanced Study Inst., Oxford 1982*, edited by Ball, J. M. Reidel, Dordrecht, 1983.

[627] TARTAR, L.: Weak limits of semilinear hyperbolic systems with oscillating data. In *Macroscopic modelling of turbulent flows*, vol. 230 of *Lecture Notes in Physics*. Springer, Berlin, 1985.

[628] TAUBES, C.: The existence of a non-minimal solution to the $SU(2)$ Yang-Mills-Higgs equations on R^3. *Comm. Math. Phys.*, **86**, 257–320, 1982.

[629] TERPSTRA, F. J.: Die Darstellung biquadratischer Formen als Summen von Quadraten mit Anwendung auf die Variationsrechnung. *Math. Ann.*, **116**, 166–180, 1938.

[630] TOMI, F. and TROMBA, A. J.: Extreme curves bound an embedded minimal surface of disk type. *Math. Z.*, **158**, 137–145, 1978.

[631] TONELLI, L.: *Fondamenti di Calcolo delle Variazioni, voll. 2.* Zanichelli, Bologna, 1921-23.

[632] TONELLI, L.: Sul problema di Plateau, I-II. *Rend. R. Accad. Lincei*, **24**, 333–339, 393–398, 1936. (See also Opere scelte, vol.3, 328-341).

[633] TONELLI, L.: *Opere scelte I-IV.* Edizioni Cremonese, Roma, 1960-63.

[634] TORCHINSKY, A.: *Real–Variable Methods in Harmonic Analysis.* Academic Press, New York, 1986.

[635] TRISCARI, D.: Sul'esistenza di cilindri con frontiere di misura minima. *Ann. Sc. Norm. Sup. Pisa*, **17**, 387–399, 1963.

[636] TRISCARI, D.: Sulle singolarità delle frontiere minime orientate di misura minima. *Ann. Sc. Norm. Sup. Pisa*, **17**, 349–371, 1963.

[637] TRISCARI, D.: Sulle singolarità delle frontiere orientate di misura minima nello spazio euclideo a 4 dimensioni. *Le Matematiche*, **18**, 139–163, 1963.

[638] TRUDINGER, N. S.: A new proof of the interior gradient bound for the minimal surface equation in n variables. *Proc. Nat. Acad. Sci. USA*, **69**, 821–823, 1972.

[639] TRUESDELL, C.: *The elements of continuum mechanics.* Springer Verlag, Berlin, 1966.

[640] TRUESDELL, C.: *A first course in rational continuum mechanics.* Academic Press, New York, 1977.

[641] TRUESDELL, C. and NOLL, W.: The non-linear field theories of mechanics. In *Handbuch der Physik, Vol III/3.* Springer Verlag, Berlin, 1965.

[642] UHLENBECK, K.: Harmonic maps: a direct method in the calculus of variations. *Bull. AMS*, **76**, 1082–1087, 1970.

[643] UHLENBECK, K.: Minimal spheres and other conformal variational problems. In *Seminar on minimal submanifolds*, Ann. of Math. Stud., 103,, pp. 169–176. Princeton Univ. Press, Princeton, NJ, 1983.

[644] VALENT, T.: *Boundary value problems of finite elasticity.* Springer, Heidelberg, 1988.

[645] VICK, J. W.: *Homology theory.* Academic Press, New York, 1973.

[646] VIRGA, E.: *Variational theories for liquid crystals.* Chapmans & Hall, London, 1994.

[647] VISINTIN, A.: Strong convergence results related to strict convexity. *Comm. P.D.E.*, **9**, 439–466, 1984.

[648] VITALI, G.: Sulle funzioni integrali. *Atti Acad. Sci. Torino*, **40**, 3–16, 1905.

[649] VITALI, G.: *Opere.* Edizioni Cremonese, Bologna, 1984.

[650] VODOPYANOV, S. K. and GOLDSTEIN, V. M.: Quasiconformal mappings and spaces of functions with generalized first derivatives. *Siberian Math. J.*, **17**, 515–531, 1977.

[651] VOL'PERT, A. I.: The space BV and quasi-linear equations. *Mat. Sb.*, **73**, 255–302, 1967. In Russian. English transl.: Math. USSR-Sb. 2 (1967) 225-267.

[652] VOL'PERT, A. I. and HUDJAEV, S. I.: *Analysis in classes of discontinuous functions and equations of mathematical physics.* Nijhoff, Dordrecht, 1985.

[653] VOLTERRA, V.: *Leçons sur l'intégration des équations différentielles aux dérivées partielles.* Almqvist&Sviksell, Upsalä, 1906. Second edition: Hermann, Paris, 1912.

[654] VOLTERRA, V.: Sur l'équilibre des corps élastiques multiplement connexes. *Ann. Sc. Norm. Paris*, **24**, 401–518, 1907.

[655] VOLTERRA, V.: *Opere Matematiche, voll.5.* Accademia Nazionale dei Lincei, Roma, 1954-62.

[656] VOLTERRA, V. and VOLTERRA, E.: *Sur les distorsion des corps élastiques. Théorie et applications.* Gauthier-Villars, Paris, 1960.

[657] ŠILHAVÝ, M.: *The mechanics and thermodynamics of continuous media.* Springer, Berlin, 1996.

[658] WENTE, H.: An existence theorem for surfaces of constant mean curvature. *J. Math. Anal. Appl.*, **26**, 318–344, 1969.

[659] WENTE, H.: The Dirichlet problem with a volume constaint. *manuscripta math.*, **11**, 141–157, 1974.

[660] WENTE, H.: The differential equation $\Delta x = 2H x_u \wedge x_v$ with vanishing values. *Proc. AMS*, **50**, 59–77, 1975.

[661] WEYL, H.: The method of orthogonal projection in potential theory. *Duke Math. J.*, **7**, 411–440, 1940.

[662] WHEEDEN, R. L. and ZYGMUND, A.: *Measure and integral.* M. Dekker, New York, ????

[663] WHITE, B.: Existence of least area mappings of N–dimensional domains. *Ann. Math.*, **113**, 179–185, 1983.

[664] WHITE, B.: Regularity of area-minimizing hypersurfaces at boundaries with multiplicity. In *Seminar on minimal submanifolds*, edited by Bombieri, E. Princeton Univ. Press, Princeton, 1983.

[665] WHITE, B.: Mappings that minimize area in their homotopy classes. *J. Diff. Geom.*, **20**, 433–446, 1984.

[666] WHITE, B.: The last area bounded by multiples of a curve. *Proc. Am. Math. Soc.*, **90**, 230–232, 1984.

[667] WHITE, B.: Infima of energy functionals in homotopy classes of mappings. *J. Diff. Geom.*, **23**, 127–142, 1986.

[668] WHITE, B.: Homotopy classes in Sobolev spaces and the existence of energy minimizing maps. *Acta Math.*, **160**, 1–17, 1988.

[669] WHITE, B.: A new proof of the compactness theorem for integral currents. *Comment. Math. Helv.*, **64**, 207–220, 1989.

[670] WHITE, B.: , 1993. Private communication.

[671] WHITEHEAD, G. W.: *Homotopy Theory.* MIT Press, Cambridge, 1966.

[672] WHITNEY, H.: Analytic extensions of differentiable functions defined in closed sets. *Trans. Am. Math. Soc.*, **36**, 63–89, 1934.

[673] WHITNEY, H.: On totally differentiable and smooth functions. *Pacific J. Math.*, **1**, 143–159, 1951.

[674] WHITNEY, H.: *Geometric integration theory.* Princeton Univ. Press, Princeton, 1957.

[675] WIENHOLTZ, D.: A method to exclude branch points of minimal surfaces. *Calc. Var., to appear*, 1997.

[676] YAU, S. T.: Survey on partial differential equations in differential geometry. In *Seminar on differential geometry*, edited by Yau, S. T. Princeton Univ. Press, Princeton, 1982.

[677] YAU (ED.), S. T.: *Seminar on differential geometry.* Princeton univ. Press, Princeton, 1982.

[678] Young, L. C.: Generalized curves and the existence of an attained absolute minimum in the calculus of variations. *Comptes Rendus Soc. Sci. et Lettres Varsovie*, **30**, 212–234, 1937.
[679] Young, L. C.: Generalized surfaces in the Calculus of Variations. *Ann. of Math.*, **43**, 84–103, 1942.
[680] Young, L. C.: Generalized surfaces in the Calculus of Variations II. *Ann. of Math.*, **43**, 530–544, 1942.
[681] Young, L. C.: Some extremal questions for simplicial compexes V : The relative area of a Klein bottle. *Rend. Circ. Mat. Palermo*, **12**, 257–274, 1963.
[682] Young, L. C.: *Lectures on the Calculus of Variations and optimal Control Theory.* Saunders, 1969. Reprinted by Chelsea 1980.
[683] Zhang, D.: The existence of nonminimal regular harmonic maps from B^3 to S^2. *Ann. Sc. Norm. Sup. Pisa*, **16**, 355–365, 1989.
[684] Zhang, K.: Biting theorems for Jacobians and their applications. *Ann. Institut H. Poincaré Anal. Non Linéaire*, **7**, 581–621, 1990.
[685] Zhou, Y.: *On the density of smooth maps in Sobolev spaces between two manifolds.* Ph.D. thesis, Columbia University, 1993.
[686] Ziemer, W. P.: *Weakly differentiable functions.* Springer-Verlag, New York, 1989.

Index

Symbols

Printing: Mercedesdruck, Berlin
Binding: Buchbinderei Lüderitz & Bauer, Berlin